Encyclopedia of Plant Physiology

New Series Volume 13 B

Editors

A. Pirson, Göttingen
M.H. Zimmermann, Harvard

Plant Carbohydrates II

Extracellular Carbohydrates

Edited by

W. Tanner and F. A. Loewus

Contributors

G. O. Aspinall C. E. Ballou B. B. Bohlool E. Cabib J. W. Catt
A. E. Clarke R. E. Cleland R. E. Cohen J. R. Colvin
A. D. Elbein M. V. Elorza K. E. Espelie G. B. Fincher
G. H. Fleet G. Franz U. Heiniger T. Higuchi K. Katō
H. Kauss P. E. Kolattukudy T. Kosuge D. T. A. Lamport
G. Larriba L. Lehle R. H. McDowell E. Percival H. J. Phaff
D. G. Robinson M. Rougier U. G. Schlösser E. L. Schmidt
R. Sentandreu J. H. Sietsma C. L. Soliday B. A. Stone
J. G. H. Wessels J. H. M. Willison

With 124 Figures

Springer-Verlag Berlin Heidelberg New York 1981

Professor Dr. WIDMAR TANNER
Institut für Botanik
Universität Regensburg
8400 Regensburg/FRG

Professor Dr. FRANK A. LOEWUS
Institute of Biological Chemistry
Washington State University
Pullman, WA 99164/USA

ISBN-13: 978-3-642-68236-0 e-ISBN-13: 978-3-642-68234-6
DOI: 10.1007/978-3-642-68234-6

Library of Congress Cataloging in Publication Data. Main entry under title: Plant carbohydrates. (Encyclopedia of Plant Physiology; new ser., v. 13) Bibliography: v.2,p. Includes index. Contents: 2. Extracellular carbohydrates. 1. Carbohydrates. 2. Botanical chemistry. I. Thanner, W. (Widmar), 1938–. II. Loewus, Frank Abel, 1919–. III. Aspinall, O. IV. Series. QK711.2.E5 new ser., vol., 13, etc.581s 81-18453 [QK898.C3] [581.19′248] AACR2.

2131/3130-543210

Preface

In 1958, a single volume in the original series of this Encyclopedia adequately summarized the state of knowledge about plant carbohydrates. Expansion into two volumes in the New Series highlights the explosive increase in information and the heightened interest that attended this class of compounds in the intervening years. Even now the search has just begun. Much remains to be accomplished; e.g., a full description of the plant cell wall in chemical terms. Why this growing fascination with plant carbohydrates? Clearly, much credit goes to those who pioneered the complex chemistry of polyhydroxylated compounds and to those who later sorted out the biochemical features of these molecules. But there is a second aspect, the role of carbohydrates in such biological functions as host-parasite and pollen-pistil interactions, the mating reaction in fungi, symbiosis, and secretion to name a few. Here is ample reason for anyone concerned with the plant sciences to turn aside for a moment and consider how carbohydrates, so many years neglected in favor of the study of proteins and nucleic acids, contribute to the physiological processes of growth and development in plants.

In somewhat arbitrary fashion, these two volumes are divided into two subject areas, intracellular and extracellular carbohydrates. The first (Vol. 13 A) addresses such topics as the biochemistry and physiology of monosaccharides, disaccharides, oligosaccharides, and polysaccharides within the plant cell. This volume summarizes current knowledge as it pertains to extracellular carbohydrates of fungi, algae, and higher plants. Obviously, there is much in common between these two areas and the volume editors have attempted to bridge this separation by means of judicious assignments to individual authors, through inter-communication among authors as a group, and by inserting cross-references at appropriate points within the text.

This volume contains five sections which address cell walls of higher plants and of algae and fungi, export of carbohydrate across the cell wall, cell surface interactions and carbohydrate-lectin interactions. Cell surface constituents of eubacteria and archaebacteria are not treated here since this topic has been repeatedly and extensively reviewed in recent years. Although several chapters on the functional aspects of macromolecular carbohydrates as related in a broad sense to the phenomena of cell–cell interactions are included here, this whole area will be covered in much greater detail in a forthcoming volume of this series.

"Extracellular" is defined here as the space outside the plasma membrane (plasmalemma), a terminology generally accepted today in cell biology. All molecules destined for this space, therefore, are considered to be secretory substances. For this reason, three chapters on secretion appear in this volume and where possible, parallels to this process as it occurs in animals are pointed out.

Not all of the current recommendations regarding nomenclature were enforced during the editing process. Authors were permitted to use certain terms (e.g., synthetase instead of synthase) as long as this use was consistent and in keeping with their use of that term in prior publications.

Finally, the editors would like to thank all the authors for their excellent cooperation and the publisher for continuous help and efficient production.

Regensburg and Pullman, October 1981

W. Tanner
F.A. Loewus

Contents

I. Cell Walls of Higher Plants

1 Constitution of Plant Cell Wall Polysaccharides

G.O. ASPINALL

1 The Classification of Plant Polysaccharides 3
2 The Main Structural Features 6
2.1 Cellulose, Other β-D-Glucose and Xyloglucans 6
2.2 Rhamnogalacturonans and Associated Arabinans and Arabinogalactans . 6
2.3 Mannans, Glucomannans and Galactoglucomannans 7
2.4 Xylans 7
2.5 Other Cell Wall Polysaccharides 7
3 Covalent and Non-Covalent Inter-Polymeric Linkages in the Cell Wall . . . 7
References 8

2 Ultrastructure of the Plant Cell Wall: Biophysical Viewpoint

J.R. COLVIN

1 Introduction 9
1.1 What a Plant Cell Wall Is 9
1.2 What a Plant Cell Wall Is Not 10
1.3 Functions and Biological Significance of the Plant Cell Wall 10
2 Idealized Plant Cell Wall: Structure 11
2.1 Intercellular Layer 11
2.2 Primary Cell Wall 12
2.2.1 Matrix Substances 12
2.2.2 Microfibrils of Cellulose 12
2.2.3 Local Structure of the Primary Wall 13
2.3 Secondary Cell Wall 14
2.3.1 Bordered Pits 14
2.4 Tertiary Wall 14
3 Actual Plant Cell Walls: Structure 15
3.1 Intercellular Layers of Various Species 15
3.2 Thickened Primary Wall of Axial Parenchyma Cells of Trembling Aspen . 15
3.3 Secondary Wall Layers of Elm Parenchyma Cells 16
3.4 Ultrastructure of Cell Walls of *Fusarium sulphureum* 16
3.5 Ultrastructure of the Cell Wall of Yeasts 16
3.6 Cell Wall Structure of Algae 17
3.7 Physical Structure of Cell Walls from Protoplasts 18
3.8 Physics of Specialized Structures of Plant Cell Walls 19
4 Macromolecular Problems of Formation of Components of Plant Cell Walls . 19
4.1 Physical Formation of Cellulose Microfibrils 19
4.2 Physical Formation of Chitin Microfibrils 21
4.3 Deposition of Hemicelluloses 21
4.4 Physical Self-Assembly of Plant Cell Walls 21
5 Resumé 22
References 22

3 The Assembly of Polysaccharide Fibrils

D.G. ROBINSON (With 3 Figures)

References 28

4 Ultrastructure of the Plant Cell Wall: Biochemical Viewpoint

K. KATŌ

1 Introduction 29
2 Microfibrillar Component 29
3 Matrix Noncellulosic Polysaccharides 30
3.1 Hemicelluloses 31
3.1.1 Xylans 31
3.1.2 Mixed β-Glucans 31
3.1.3 Xyloglucans 32
3.1.4 Glucomannans 33
3.2 Pectic Polysaccharides 33
3.2.1 Rhamnogalacturonan 33
3.2.2 Arabinogalactans 34
3.2.3 Arabinans and Galactans 34
3.3 Glycoprotein of the Walls 35
3.4 Lignin 36
4 Polysaccharidic Association Within the Primary Cell Wall 36
4.1 Dicot Primary Cell Walls 36
4.2 Monocot Primary Cell Walls 38
5 Chemical Changes in the Cell Wall During Growth and Differentiation . . . 39
6 Conclusion 41
References 42

5 Biosynthesis and Metabolism of Cellulose and Noncellulosic Cell Wall Glucans

G. FRANZ and U. HEINIGER (With 1 Figure)

1 Introduction: Various Aspects of Cellulose Formation in Vivo and in Vitro . 47
2 Possible Substrates for Cellulose Biosynthesis 49
2.1 Occurrence of NDP-Sugars (Sugar Nucleotides) in Tissues Actively Forming Cell Walls 49
2.2 Sucrose Synthetase and Pyrophosphorylase Activities 49
2.3 Different NDP-Glucoses as Hypothetical Substrates for the in Vitro Biosynthesis of Cellulose 50
2.4 In Vivo Studies on Cellulose Biosynthesis 54
3 Primer Requirement for Cellulose Biosynthesis 55
4 Involvement of Lipid-Intermediates in Cellulose Biosynthesis 57
5 Endogenous and Exogenous Factors Which Might Influence the Biosynthesis of Cellulose 59
5.1 Hormonal Control of Cellulose Biosynthesis 59
5.2 Inhibitors in the Process of Cellulose Formation 60
6 Conclusion 61
References 61

6 Metabolism of Noncellulosic Polysaccharides

G.B. FINCHER and B.A. STONE (With 7 Figures)

1 Cell Walls and Cell Wall Metabolism 68
2 Molecular Aspects of Polysaccharide Synthesis 69

2.1 Origins of Monosaccharides and Their Activated Forms 69
2.2 Polysaccharide Assembly 70
2.2.1 Polymerizing Systems 70
2.2.2 Lipid-Saccharide Intermediates 70
2.2.3 Chain Initiation 73
2.2.4 Chain Elongation and Direction of Growth 74
2.2.5 Sequences of Linkages and Monosaccharides, and Insertion of Side Branches 74
2.2.6 Chain Termination 75
2.3 Biosynthesis of Glycosyl Ethers and Esters 75
2.4 Control 76
2.4.1 Supply of Monosaccharide Precursors 76
2.4.2 Activation and Interconversion of Monosaccharides 76
2.4.3 Polymerization 80
3 Cellular Aspects of Polysaccharide Synthesis and Processing 80
3.1 Location of Enzymes Leading to the Establishment of the Nucleotide Sugar Pool 80
3.2 Location of Glycosyl Transferases Involving Lipid Intermediates 81
3.3 Location of Polysaccharide Synthetases 81
3.4 Subcellular Routes of Polysaccharide Assembly 82
3.4.1 Golgi Dictyosome – Golgi Vesicle – Plasma Membrane Route 82
3.4.2 Endoplasmic Reticulum – Plasma Membrane Route 84
3.4.3 Assembly at the Plasma Membrane 85
3.4.4 Deposition of Polysaccharides in the Wall 86
4 Metabolism During Cell Division, Growth, Differentiation, Senescence and Germination 87
4.1 Methodological Considerations and Interpretations 87
4.2 Molecular Strategies in Morphogenesis 89
4.3 Cell Plate Formation 90
4.4 Primary Cell Wall Deposition and Expansion 91
4.5 Secondary Cell Wall Deposition and Growth 93
4.6 Cell Wall Metabolism During Differentiation in Specific Tissues 94
4.6.1 Tracheids, Laticifers, Aerenchyma and Lateral Root Initiation . . . 94
4.6.2 Tyloses 97
4.6.3 Formation of Plasmodesmata and Differentiation of Sieve Plates . . 97
4.6.4 Gametogenic Tissues 99
4.6.5 Mucilage (Slime)- and Gum-secreting Cells 101
4.7 Senescence 102
4.7.1 Abscission 102
4.7.2 Fruit Senescence 103
4.7.3 Flower and Leaf Senescence 105
4.8 Germination 105
5 Metabolism in Response to Environmental Changes and Pathogenesis 108
5.1 Tropisms 108
5.1.1 Geotropism and Phototropism 108
5.1.2 Reaction Wood 108
5.1.3 Hypogravity 109
5.2 Environmental Stress 109
5.3 Wounding Responses 110
5.4 Pathogenesis 111
References 114

7 Glycoproteins and Enzymes of the Cell Wall

D.T.A. LAMPORT and J.W. CATT

1 Introduction 133
2 Structural or Matrix Proteins 134

2.1 Higher and Lower Plants 134
2.1.1 Glycopeptide Linkages 134
2.1.2 Hydroxyproline Glycosides 135
2.1.3 Peptide Sequence and Conformation 135
2.1.4 Role of Glycosylation 136
2.1.5 Assembly 137
2.1.6 Possible Roles for the Matrix Protein Extensin 138
2.2 Algae 144
2.2.1 Occurrence of Matrix Proteins 144
2.2.2 The Chlamydomonas Type Wall 145
2.2.3 Phylogenetic Considerations 146
3 Cell Wall Enzymes 149
3.1 Introduction 149
3.2 Biological Role 149
3.2.1 Involvement in Cell Wall Assembly and Cell Extension 149
3.2.2 Involvement in Transport 151
3.2.3 Involvement in Recognition Phenomena and Disease Resistance . . 151
4 Arabinogalactan Proteins (AGP's) and β-Lectins 152
4.1 General Properties 152
4.2 Chemistry, Physical Properties, and Biosynthesis 153
4.3 Biological Role 154
4.3.1 Water Relations 154
4.3.2 Mechanical Cell-Cell Interactions 154
4.3.3 Cell Recognition 155
5 Concluding Remarks 155
References 156

8 The Role of Lipid-Linked Saccharides in the Biosynthesis of Complex Carbohydrates

A.D. Elbein (With 9 Figures)

1 Introduction 166
2 Nature of the Lipid Carrier 167
3 Glycoproteins or Other Complex Carbohydrates 168
4 Lipid-linked Monosaccharides 170
4.1 Mannose 170
4.2 Glucose 171
4.3 N-Acetylglucosamine 172
4.4 Other Monosaccharide Lipids 173
5 Oligosaccharide Derivatives 174
6 Protein Glycosylation 178
7 Further Reactions – Processing of Proteins 181
8 Effect of Antibiotics and Other Inhibitors 183
9 Subcellular Location of the Enzymes of the Dolichol Pathway 187
10 Conclusions 188
References 188

9 Biosynthesis of Lignin

T. Higuchi (With 22 Figures)

1 Occurrence of Lignin in Plants 194
2 Morphological Distribution of Lignins in Plant Cell Walls 197
3 Biogenesis of Lignin Precursors 201
3.1 Shikimic Acid – Phenylalanine Pathway 201
3.2 Cinnamic Acid Pathway 203

3.2.1 Hydroxylation of Cinnamic Acids . . . 204
3.2.2 Methylation of Hydroxycinnamic Acids . . . 205
3.2.3 Reduction of Hydroxycinnamic Acids . . . 207
4 Dehydrogenative Polymerization of Hydroxycinnamyl Alcohols to Lignins . . 209
4.1 Role of Peroxidase in the Dehydrogenative Polymerization of Hydroxycinnamyl Alcohols (Monolignols) to Lignins . . . 212
4.2 Structural Variation in Dehydrogenation Polymers . . . 215
5 Formation and Distribution of Syringyl Lignin in Angiosperm Woods . . . 217
6 Differences Between Gymnosperms and Angiosperms in Lignin Biosynthesis . 218
7 Regulation of Lignin Biosynthesis . . . 220
References . . . 221

10 Hydrophobic Layers Attached to Cell Walls. Cutin, Suberin and Associated Waxes

P.E. Kolattukudy, K.E. Espelie and C.L. Soliday (With 9 Figures)

1 Introduction . . . 225
2 Location and Ultrastructure of Cutin, Suberin, and Waxes . . . 225
3 Composition and Structure of Cutin and Suberin . . . 227
3.1 Isolation and Depolymerization of Cutin and Suberin . . . 227
3.2 Composition of the Monomers of Cutin and Suberin . . . 228
3.3 Intermolecular Linkages in Cutin and Suberin . . . 231
4 Biosynthesis of Cutin and Suberin . . . 232
4.1 Biosynthesis of the C_{16} Family of Monomers . . . 232
4.2 Biosynthesis of the C_{18} Family of Monomers . . . 234
4.3 Synthesis of the Cutin Polymer from Monomers . . . 234
4.4 Biosynthesis of the Aliphatic Components of Suberin . . . 235
4.5 Regulation of Suberization . . . 236
4.6 Site of Synthesis of the Monomers and the Polymers . . . 236
5 Biodegradation of Cutin and Suberin . . . 237
5.1 Fungal Cutinase . . . 237
5.2 Pollen Cutinase . . . 238
6 Isolation and Analysis of Waxes . . . 239
7 Composition of Waxes . . . 239
8 Biosynthesis of Waxes . . . 243
8.1 Biosynthesis of Very Long Fatty Acids . . . 243
8.2 Biosynthesis of Fatty Alcohols . . . 244
8.3 Biosynthesis of Wax Esters . . . 244
8.4 Biosynthesis of Hydrocarbons and Derivatives . . . 245
8.5 Biosynthesis of β-Diketones . . . 246
9 Function of Cutin, Suberin and Associated Waxes . . . 247
References . . . 248

11 Wall Extensibility: Hormones and Wall Extension

R.E. Cleland (With 2 Figures)

1 Introduction . . . 255
2 Cellular Parameters Which Control Cell Elongation . . . 255
3 Do Hormones Control Wall Extension via Changes in Wall Extensibility? . 256
3.1 WEx and the Mechanical Properties of Cell Walls . . . 256
3.1.1 What is WEx? . . . 256
3.1.2 Mechanical Properties of Cell Walls . . . 257
3.1.3 Changes in WEx in Vivo . . . 258
3.2 Measurement of WEx . . . 258
3.2.1 Measurement of the Mechanical Properties of Isolated Walls . . . 258

3.2.2 Direct Measurement of WEx . . . 260
3.2.3 Other Mechanical Testing Procedures . . . 260
3.3 Cases Where Hormones Affect WEx . . . 261
3.3.1 Auxin . . . 261
3.3.2 Gibberellin . . . 262
3.3.3 Other Hormones . . . 263
3.3.4 Conclusions . . . 263
4 The Mechanism of Auxin-Induced Wall Loosening . . . 263
4.1 Wall Structure, Wall Synthesis and Wall Loosening . . . 263
4.2 Possible Mechanisms for Wall Loosening . . . 264
4.2.1 Breakage of Hydrogen Bonds . . . 264
4.2.2 Calcium Cross-Links . . . 265
4.2.3 Nonenzymatic Wall Loosening . . . 265
4.2.4 Enzymatic Wall Loosening . . . 265
4.2.5 Conclusions . . . 269
References . . . 269

II. Cell Walls of Algae and Fungi

12 Algal Walls – Composition and Biosynthesis

E. Percival and R.H. McDowell (With 6 Figures)

1 Introduction . . . 277
2 Cellulose and Other Glucans . . . 279
2.1 Chlorophyceae . . . 280
2.2 Rhodophyceae . . . 281
2.3 Phaeophyceae . . . 281
2.4 Biosynthesis . . . 281
2.5 $(1 \rightarrow 3)$-β-glucan . . . 281
3 Xylans . . . 282
3.1 Rhodophyceae . . . 282
3.2 Chlorophyceae . . . 282
4 Mannans . . . 283
4.1 Chlorophyceae . . . 283
4.2 Bacillariophyceae . . . 283
5 Alginic Acid . . . 284
5.1 Constitution and Structure . . . 284
5.2 Ion Exchange . . . 285
5.3 Conformation . . . 286
5.4 Gel Formation . . . 286
5.5 Variations in Structure . . . 287
5.6 Biosynthesis . . . 288
6 Galactans . . . 289
6.1 Agar and Related Molecules . . . 291
6.2 Carrageenan and Related Polysaccharides . . . 292
6.3 Conformation . . . 293
6.4 Gametophyte and Sporophyte Carrageenans . . . 294
6.5 Immunochemistry . . . 294
6.6 Cell Wall Localization . . . 294
6.7 Biosynthesis . . . 295
7 Fucans . . . 295
7.1 Structure . . . 297
7.2 Site of Sulfate . . . 297

7.3 Biosynthesis . . . 298
7.4 Location and Function . . . 298
8 Sulfated Polysaccharides of the Chlorophyceae . . . 299
9 Polysaccharides of the Bacillariophyceae . . . 303
10 Extracellular Polysaccharides . . . 303
10.1 From Phaeophyceae . . . 303
10.2 From Rhodophyceae . . . 303
10.3 From the Xanthophyceae . . . 304
10.4 From Bacillariophyceae . . . 305
11 Conclusions . . . 305
11.1 Functions . . . 305
11.2 Taxonomy . . . 306
References . . . 306

13 Algal Walls – Cytology of Formation

D.G. Robinson (With 13 Figures)

1 Matrix Polysaccharide and Slime Production . . . 317
2 Microfibril Synthesis and Orientation . . . 317
2.1 The Formation and Secretion of Scales . . . 317
2.1.1 Scale Structure . . . 317
2.1.2 The Golgi Apparatus and Scale Production . . . 318
2.1.3 Scale Transport and Liberation . . . 320
2.2 Microfibril Deposition in Cellulosic Algae . . . 320
2.2.1 Cladophorales, Siphonocladales . . . 320
2.2.2 Chlorococcales . . . 322
2.2.3 Conjugales (Placoderm Desmids) . . . 325
2.3 The Production of Chitin Microfibrils . . . 326
2.3.1 Poterioochromonas . . . 326
2.3.2 Centric Diatoms . . . 326
3 Glycoprotein Wall Formation . . . 328
References . . . 329

14 Algal Wall-Degrading Enzymes – Autolysines

U.G. Schlösser (With 7 Figures)

1 Introduction . . . 333
2 Autolysines Found in Algae . . . 334
2.1 Chlamydomonas (Chlorophyceae, Volvocales) . . . 334
2.1.1 Evidence . . . 334
2.1.2 Specificity of Action . . . 338
2.1.3 Isolation and Properties . . . 338
2.1.4 Regulation . . . 340
2.1.5 Reproduction Cell Release Mechanism . . . 343
2.2 Volvox (Chlorophyceae, Volvocales) . . . 343
2.3 Chlorella (Chlorophyceae, Chlorococcales) . . . 344
2.4 Geminella (Chlorophyceae, Ulotrichales) . . . 346
3 Indication of Autolysine Action in Other Algae . . . 346
3.1 Cell Wall Changes in the Reproduction Phase . . . 346
3.2 Lysis by Exogenous Enzymes . . . 347
3.3 Dependency on Divalent Cations . . . 348
4 Application of Autolysines . . . 348
References . . . 348

15 Fungal Cell Walls: A Survey

J.G.H. Wessels and J.H. Sietsma (With 1 Figure)

1 Introduction 352
1.1 Methodological Difficulties 352
2 Survey of Wall Polymers 353
2.1 Distribution Among Fungi 353
2.2 Individual Polymers 355
2.2.1 (1 → 4)-β-D-Glycosaminoglycans (Chitin, Chitosan) 355
2.2.2 (1 → 4)-β-D-Glucan (Cellulose) 357
2.2.3 (1 → 3)-β-D/(1 → 6)-β-D-Glucan 358
2.2.4 (1 → 3)-α-D-Glucan with Variable Amounts of (1 → 4)-α-Linkages . 360
2.2.5 Homo- and Hetero-Glucuronans 361
2.2.6 Glycoproteins 362
3 Ultrastructural Localization of Wall Polymers 363
3.1 Methodological Difficulties 363
3.2 Mature Walls 363
3.2.1 Filamentous Fungi 363
3.2.2 Yeasts 367
3.3 Newly Formed Walls 368
3.3.1 Reverted Protoplasts and Germinated Zoospores 368
3.3.2 Growing Areas of the Wall 369
4 Wall Composition and Cellular Morphology 370
4.1 Does Wall Composition Determine Cellular Morphology? 370
4.2 Changes in Wall Composition Accompanying Changes in Cell Morphology 371
5 Wall-Degrading Enzymes 373
5.1 Wall Components as Substrates for Degrading Enzymes 373
5.2 Net Degradation of Wall Components in Relation to Development . . 375
5.3 "Wall-Loosening" Enzymes 377
6 Synthesis of the Wall 379
6.1 Introduction 379
6.2 Biosynthesis of Individual Wall Components 379
6.2.1 Chitin 379
6.2.2 Glucan 381
6.2.3 Glycoprotein 381
6.3 Wall Synthesis and Morphogenesis 381
References 384

16 Chitin: Structure, Metabolism, and Regulation of Biosynthesis

E. Cabib (With 8 Figures)

1 Introduction 395
2 Chemical and Physical Structure 395
3 Distribution and Localization 397
4 The Enzymatic Synthesis of Chitin and its Regulation 397
4.1 Biosynthesis of Precursors 397
4.2 Chitin Synthetase and its Regulation 398
4.2.1 General Properties of Chitin Synthetase 398
4.2.2 Regulation of Chitin Synthetase Activity 401
4.2.3 Subcellular Distribution of Chitin Synthetase 403
5 Chitin Degradation 405
6 Regulation and Localization of Chitin Synthesis in Vivo 405
6.1 Introduction 405
6.2 Primary Septum Formation in Budding Yeasts 406
6.3 Hyphal Growth 408
6.4 Chitin Synthesis During Round Cell Formation in *Blastocladiella emersonii* 410
References 411

17 Fungal Glucans – Structure and Metabolism

G.H. Fleet and H.J. Phaff

1 Introduction . . . 416
2 Glucans from Yeasts . . . 417
2.1 Cell Envelope Glucans . . . 417
2.1.1 Structure . . . 417
2.1.2 Degradation . . . 420
3 Glucans from Filamentous Fungi . . . 426
3.1 Structure . . . 426
3.1.1 β-Linked Glucans . . . 426
3.1.2 α-Linked Glucans . . . 428
3.2 Fungal Glucanases and Glucan Degradation . . . 429
4 Biosynthesis . . . 431
References . . . 432

18 Mannoproteins: Structure

R.E. Cohen and C.E. Ballou (With 3 Figures)

1 Definition and Occurrence . . . 441
2 Isolation and Criteria of Homogeneity . . . 441
3 Methods for Structural Analysis . . . 442
3.1 Selective Chemical Degradations . . . 442
3.2 Selective Enzymic Degradations . . . 443
3.3 Mutant Analysis . . . 443
3.4 Immunochemical Methods . . . 444
3.5 Physical Methods . . . 445
4 *Saccharomyces cerevisiae* Mannoprotein . . . 445
4.1 Bulk Cell Wall Mannoprotein . . . 445
4.2 Characteristics of Mannan Mutants . . . 447
4.3 Invertase, an Example of a "Homogeneous" Extracellular Glycoprotein . 449
4.4 Carboxypeptidase Y, an Example of a "Homogeneous" Intracellular Mannoprotein . . . 451
5 *Hansenula wingei* Sexual Agglutination Factors . . . 453
5.1 Description of the Mating Reaction in Yeasts . . . 453
5.2 Isolation and Structure of the 5-Cell Agglutinin . . . 453
5.3 Isolation and Structure of 21-Factor . . . 454
5.4 Comparison with Agglutination Factors in Other Yeasts . . . 454
References . . . 455

19 Biosynthesis of Mannoproteins in Fungi

L. Lehle (With 2 Figures)

1 Introduction . . . 459
2 Biosynthesis of Mannoproteins in *Saccharomyces* Species . . . 459
2.1 Early in Vitro Studies with Guanosine Diphosphate Mannose . . . 459
2.2 Participation of Lipid-Linked Intermediates . . . 460
2.2.1 Identification of the Glycosylated Lipid Intermediate . . . 460
2.2.2 Role of Dolichyl Monophosphate in the Formation of Mannooligosaccharide Chains Linked Through the Hydroxyl Group of Serine and Threonine . . . 460
2.2.3 Role of Dolichyl Diphosphate-Linked Oligosaccharides in the Formation of the Core Structure Containing the *N*-Acetylglucosaminyl-Asparagine Linkage . . . 461
2.3 Mannosyl Transfer Reactions Involved in the Assembly of the Outer Chain Structure . . . 463

3 Biosynthesis of Mannoproteins in Other Species 465
3.1 *Hansenula* Species 465
3.2 *Kuyveromyces lactis* 465
3.3 *Cryptococcus laurentii* 466
3.4 *Aspergillus* Species 466
3.5 *Penicillium* Species 467
3.6 Other Species 468
4 Subcellular Sites of Glycosylation 468
5 Possible Control of Mannoprotein Biosynthesis 471
6 Attempts to Study Solubilized Glycosyl Transferases 473
7 A Summary of Mannoprotein Biosynthesis; How It Might Occur: a Model . 474
References 477

III. Export of Carbohydrate Material

20 Secretory Processes – General Considerations and Secretion in Fungi

R. Sentandreu, G. Larriba and M.V. Elorza (With 5 Figures)

1 Introduction 487
2 Synthesis and Segregation of Export Polymers 488
2.1 The Signal Hypothesis 488
2.1.1 Mammalian Cells 488
2.1.2 Bacteria 490
2.1.3 Fungi 491
2.2 Mechanisms of Glycosylation 493
2.2.1 Mammalian Cells 493
2.2.2 Fungi 493
3 Transport of Secretion Products 496
3.1 Mammalian Cells 496
3.2 Fungi 498
4 Discharge of Secretion Products 501
5 Synthesis and Secretion of Cell Wall Polysaccharides 504
6 Concluding Remarks 506
References 507

21 Secretion of Cell Wall Material in Higher Plants

J.H.M. Willison (With 9 Figures)

1 Introduction 513
2 Cell Plate Formation 514
2.1 Role of the Phragmoplast 514
2.2 Role of Organelles 516
3 Secretion of Primary Wall Matrix Substances 518
4 Secretion of Microfibrils 521
4.1 Site of Synthesis 521
4.2 Mechanism of Synthesis 522
4.3 Control of Microfibril Orientation 525
4.4 Cellulose Secretion by *Acetobacter xylinum* 526
5 Cell Wall Assembly 528
6 Secretion of Lipidic Wall Materials 528
6.1 The Cuticle 529
6.2 Epicuticular Waxes 530
6.3 Suberinized Layers 531
References 532

22 Secretory Activity of the Root Cap

M. Rougier (With 9 Figures)

1 Introduction 542
2 Root Cap Architecture and Root Cap Secretory Activity 542
2.1 Occurrence and Localization of Secretory Cells in the Root Cap 542
2.2 Life and Differentiation of Secretory Cells 544
2.3 Ultrastructure of Secretory Cells 545
3 Characteristics of the Secretory Products 549
3.1 Collection and Chemical Analysis of Root Cap Slimes 549
3.1.1 Monosaccharide Components 551
3.1.2 Other Components 552
3.2 Structure of Root Cap Slimes 552
3.3 In Situ Identification of Slime Components 553
4 Secretory Pathways 556
4.1 Biosynthesis of Slime Polysaccharides 556
4.2 Transport via Granulocrine Process 558
4.3 Slime Discharge 560
4.4 Model of Secretion 561
5 Physiology of the Secretion Processes 562
5.1 Control of the Polysaccharide Droplet Formation and Size 562
5.2 Control of Vesicle Production and Discharge 563
5.3 Characteristics of the Secretory Cycle 564
5.4 Participation of Enzymes in the Secretion Processes 565
6 Function of the Root Cap Slime 565
6.1 Slime as a Constituent of Mucigel 565
6.2 Function of the Mucigel at the Root Level 565
6.3 Function of Mucigel at the Root-Soil Interface 566
6.3.1 Slime and Sloughed Cells as a Source of Organic Carbon and Nitrogen in the Rhizosphere 566
6.3.2 Action of the Mucigel on Soil Aggregation and Stability 567
6.3.3 Role of Root Cap Slime on Microbial Colonization 567
7 Concluding Remarks 569
References 569

IV. Cell Surface Phenomena

23 Defined Components Involved in Pollination

A.E. Clarke (With 1 Figure)

1 Introduction 577
2 Arabinogalactans as Pistil Components of *Gladiolus* and *Lilium* 577
3 S-Allele Associated Style Components 580
4 Callose as a Response to Self-Incompatible Matings 581
5 Callose as a Pollen Tube Wall Component 581
References 582

24 Carbohydrates in Plant – Pathogen Interactions

T. Kosuge (With 5 Figures)

1 Introduction 584
2 Role of Polysaccharides in the Early Interactions Between Plant and Pathogen 584

2.1 Role of Recognition in Plant – Pathogen Interactions 584
2.2 Recognition in the Infection Process 585
2.3 Recognition in Resistance Reactions 586
2.4 Specificity Through Suppressors and Protectors 588
3 Carbohydrate Metabolism of Pathogens in Host Tissue 588
3.1 Production of Polysaccharide-Degrading Enzymes 588
3.1.1 Nature and Action in Host Tissue 588
3.1.2 Conditions Affecting Production 591
3.1.3 Role of Polysaccharide-Degrading Enzymes in Vivo 592
3.2 Carbohydrates as a Source of Energy of Pathogens in Host Tissue . . . 593
3.2.1 Catabolite Repression in Pathogens in Host Tissue 593
3.2.2 Acquisition of Energy Sources by Pathogens in Host Tissue . . . 594
3.2.3 Pathways of Carbon Catabolism in Host Tissue 595
3.3 Production of Toxic Carbohydrates by Pathogens in Host Tissue 595
3.3.1 Structure of Toxins 595
3.3.2 Mode of Action of Toxins 600
3.3.3 Role of Toxins in Disease 601
3.3.4 Conditions Affecting Production of Toxins 602
4 Carbohydrates in Host Response to Infection 602
4.1 Polysaccharide Elicitors of the Phytoalexin Response 602
4.1.1 Structure and Mode of Action of Elicitors 602
4.1.2 Specificity of Elicitors 603
4.2 Altered Carbohydrate Metabolism in Host in Response to Infection . . 606
4.2.1 Altered Carbon Metabolism in Host 606
4.2.2 Induction of Polysaccharide-Degrading Enzymes by Host Tissue in Response to Infection 608
4.2.3 Production of Polysaccharides by Host in Response to Infection . 608
4.3 Toxic Host Glycosides in Plant – Pathogen Interactions 609
4.3.1 Structure and Metabolism of Toxic Glycosides 609
4.3.2 Role of Toxic Glycosides in Host – Pathogen Interactions 612
4.4 Concluding Remarks 614
References 615

V. Lectin – Carbohydrate Interaction

25 Lectins and Their Physiological Role in Slime Molds and in Higher Plants
H. Kauss

1 What are Lectins? 627
2 Biochemical Properties of Lectins 629
2.1 Assay and Isolation 629
2.2 Carbohydrate Binding Specificity 630
2.3 Structure 634
3 Applications and Biological Properties of Plant Lectins in Animal Systems . 636
4 Role of Lectins in Slime Mold Aggregation 638
5 Lectins in Higher Plants 641
5.1 Recognition Phenomena Possibly Mediated by Lectins 641
5.2 Distribution of Lectins in Various Tissues 644
5.3 Subcellular Localization of Lectins 645
5.3.1 Lectins Associated with Membranes 645
5.3.2 Lectins Associated with Walls 647
5.3.3 Phloem Lectins 648
5.4 Plant Toxins Containing Lectin Subunits 649
5.5 Possible Relations Between Lectins and Enzymes 651
References 652

26 The Role of Lectins in Symbiotic Plant – Microbe Interactions
E.L. SCHMIDT and B.B. BOHLOOL (With 2 Figures)

1 Introduction . . . 658
2 Pre-Recognition Events of Concern . . . 659
2.1 Rhizosphere Competence . . . 659
2.2 Host Rhizosphere Stimulation . . . 661
2.3 Arrival at the Recognition Site . . . 662
3 Lectin-Mediated Recognition . . . 663
3.1 Lectin Recognition Hypothesis . . . 663
3.2 Lectin-Binding Site of Symbiont . . . 664
3.3 Lectin-Binding Site of Host . . . 667
3.4 Lectin-Binding Competence in Rhizobia . . . 668
4 Post-Recognition Events . . . 669
4.1 Attachment . . . 670
4.2 Post-Recognition Function of Lectins . . . 671
5 Concluding Comments . . . 672
References . . . 674

Author Index . . . 679

Subject Index . . . 747

List of Contributors

G.O. Aspinall
Dept. of Chemistry
York University
4700 Keele St.
Downsview, Ontario, M3J 1P3/Canada

C.E. Ballou
Dept. of Biochemistry
University of California
Berkeley, California 94720/USA

B.B. Bohlool
Dept. of Microbiology
University of Hawaii at
Manoa Honolulu, Hawaii 96822/USA

E. Cabib
National Institute
of Arthritis, Diabetes,
Digestive and Kidney Diseases
Bethesda, MD 20205/USA

J.W. Catt
Dept. of Plant Sciences
University of Leeds
Leeds LS2 9JT/United Kingdom

A. Clarke
School of Botany
University of Melbourne
Parkville Vic 3052/Australia

R.E. Cleland
Botany Department
University of Washington
Seattle, WA 98195/USA

R.E. Cohen
Dept. of Molecular Biology
and the Virus Laboratory
University of California
Berkeley, California 94720/USA

J.R. Colvin
Cell Biophysics Group
Division of Biological Sciences
National Research Council of Canada
Ottawa, Ontario, K1A OR6/Canada

A.D. Elbein
Dept. of Biochemistry
University Texas
Health Science Center
San Antonio, Texas 78284/USA

M.V. Elorza
Departamento de Microbiología
Facultad de Farmacia
Universidad de Valencia
Valencia/Spain

K.E. Espelie
Institute of Biological Chemistry
Washington State University
Pullman, Washington 99164/USA

G.B. Fincher
Dept. of Biochemistry
La Trobe University
Bundoora, 3083, Vic/Australia

G.H. Fleet
School of Food Technology
University of New South Wales
P.O. Box 1
Kensington, NSW 2033/Australia

G. Franz
Naturwissenschaftliche Fakultät IV
Chemie und Pharmazie
Universität Regensburg
Universitätsstraße 31
8400 Regensburg/FRG

U. Heiniger
Zollikerstr. 107
Institut für Pflanzenbiologie
Universität Zürich
8008 Zürich/Switzerland

T. Higuchi
Research Section of Lignin Chemistry
Wood Research Institute
Kyoto University
Uji, Kyoto 611/Japan

K. Katō
Central Research Institute
The Japan Tobacco and Salt
Public Corporation
6-2 Umegaoka, Mindori-ku,
Yokohama, Kanagawa 227/Japan

H. Kauss
Fachbereich Biologie der
Universität Kaiserslautern
Postfach 3049
6750 Kaiserslautern/FRG

P.E. Kolattukudy
Institute of Biological Chemistry
Washington State University
Pullman, Washington 99164/USA

T. Kosuge
Dept. of Plant Pathology
University of California
Davis, California 95616/USA

D.T.A. Lamport
MSU-DOE Plant Research Laboratory
Michigan State University
East Lansing, Michigan 48824/USA

G. Larriba
Departamento de Microbiología
Facultad de Ciencias
Universidad de Extremadura
Badajoz/Spain

L. Lehle
Naturwissenschaftliche Fakultät III
Biologie und Vorklinische Medizin
Universität Regensburg
8400 Regensburg/FRG

R.H. McDowell
Consultant Alignate Industries
20 Oak Tree Close, Virginia Water
Surrey GU 25 4JF/UK

E. Percival
Chemistry Department
Royal Holloway College, Egham Hill
Egham, Surrey, TW 20 OEX/UK

H.J. Phaff
Dept. of Food Science and Technology
University of California, Davis
Davis, California 95616/USA

D.G. Robinson
Abteilung Cytologie des
Pflanzenphysiologischen Instituts
der Universität Göttingen
Untere Karspüle 2
3400 Göttingen/FRG

M. Rougier
Laboratoire d'Histophysiologie
de la Sécrétion végétale
Dépt. de Biologie Végétale
Université Lyon I
43 Bd. du 11 Novembre 1918
69622 Villeurbanne Cedex/France

U.G. Schlösser
Sammlung von Algenkulturen
Pflanzenphysiologisches Institut
der Universität
Nikolausberger Weg 18
3400 Göttingen/FRG

E.L. Schmidt
Dept. of Soil Science and Microbiology
University of Minnesota
St. Paul, Minnesota, 55108/USA

R. Sentandreu
Departamento de Microbiología
Facultad de Farmacia
Universidad de Valencia
Valencia/Spain

J.H. Sietsma
Dept. of Developmental Plant Biology
Biological Centre
University of Groningen,
9751 NN Haren/The Netherlands

S.L. Soliday
Institute of Biological Chemistry
Washington State University
Pullman, Washington 99164/USA

B.A. Stone
Dept. of Biochemistry
La Trobe University
Bundoora 3083 Vic/Australia

J.G.H. Wessels
Dept. of Developmental Plant Biology
Biological Centre
University of Groningen
9751 NN Haren/The Netherlands

J.H.M. Willison
Biology Department
Dalhousie University
Halifax, Nova Scotia
Canada B3H 4J1/Canada

I. Cell Walls of Higher Plants

1 Constitution of Plant Cell Wall Polysaccharides

G.O. ASPINALL

1 The Classification of Plant Polysaccharides

Plant polysaccharides associated with cellulose in the cell wall are commonly isolated using selective extraction procedures. The terms *pectic substances*, denoting those materials extracted with water alone or with hot aqueous solutions of chelating agents for calcium, such as disodium ethylenedinitrilotetraacetate or sodium hexametaphosphate, and *hemicelluloses*, denoting polysaccharides extracted with dilute alkali alone or with added borate, are useful in describing broad groups of polysaccharides. However, a more complete and precise classification based on structure is necessary to avoid the difficulties which arise when obviously related polysaccharides, differing in some small and apparently trivial feature such as extent of branching or degree of esterification, show marked differences in physical properties such as solubility.

With the exception of cellulose and other β-D-glucans, most cell wall polysaccharides are heteropolysaccharides which fall into a limited number of structural families. Each family contains species with more or less regularly repeating features, usually in the interior chains, but the individual members may show considerable variations in the nature and proportions of other sugar units in side-chains, and of other modifying features such as ether and ester functions. In addition to purely quantitative variations some, notably the xylan and rhamnogalacturonan families, show discrete variations in the nature of the sugar units in side-chains. Table 1 summarizes the main structural features, both constant and variable, of each polysaccharide family. This classification brings together substances which serve both structural and reserve functions (see MEIER and REID, Chap. 11, Vol. 13A, this Series), and some whose role as cell wall components is unproven. Furthermore some structural families, especially the rhamnogalacturonans and the arabinogalactans of type II, can be extended to include several mucilages and exudate gums which, although yet more complex, have clear structural relationships to known cell wall components (ASPINALL 1969). The gums are often easily isolated in pure and chemically homogeneous form and, in several instances, investigations on the more complex exudate gums preceded those on the cell wall polysaccharides and shed light on the structure of cell wall components when these were examined later.

Table 1. Classification of plant cell wall polysaccharides by structural family[a]

Polysaccharide	Interior chains	Exterior chains and other features
Glucans		
Cellulose	→4)-Glc*p*-(1 $\xrightarrow{\beta}$ 4)-Glc*p*-(1 $\xrightarrow{\beta}$ 4)-Glc*p*-(1 $\xrightarrow{\beta}$ 4)-Glc*p*-(1 $\xrightarrow{\beta}$ 4)-Glc*p*-1 $\overset{\beta}{—}$	
Callose	→3)-Glc*p*-(1 $\xrightarrow{\beta}$ 3)-Glc*p*-(1 $\xrightarrow{\beta}$ 3)-Glc*p*-(1 $\xrightarrow{\beta}$ 3)-Glc*p*-(1 $\xrightarrow{\beta}$ 3)-Glc*p*-1 $\overset{\beta}{—}$	
Cereal *β*-D-glucans	→3)-Glc*p*-(1 $\overset{\beta}{—}$[→4)-Glc*p*-(1 $\overset{\beta}{—}$]$_{2-3}$→3)-Glc*p*-1 $\xrightarrow{\beta}$ 4)-Glc*p*-(1 $\overset{\beta}{—}$	
Xyloglucans	→4)-Glc*p*-(1 → 4)-Glc*p*-(1 → 4)-Glc*p*-(1 → 4)-Glc*p*-(1 → 4)-Glc*p*-1 —	R = Xyl*p*-(1 $\overset{\alpha}{—}$,
	6 6 6 6	Gal*p*-(1 $\xrightarrow{\beta}$ 2)-Xyl*p*-(1 $\overset{\alpha}{—}$, or
	↑ ↑ ↑ ↑	Fuc*p*-(1 $\xrightarrow{\alpha}$ 2)-Gal*p*-(1 $\xrightarrow{\beta}$ 2)-Xyl*p*-(1 $\overset{\alpha}{—}$
	R R R R	
Rhamnogalacturonans and associated arabinans and arabinogalactans		
→5)-Ara*f*-(1 $\xrightarrow{\alpha}$ 5)-Ara*f*-(1 $\overset{\alpha}{—}$ Arabinan [[Ara*f*]$_m$—]		
3 ↑ 1 Ara*f* (branch at O-3)		
→4)-Gal*p*-(1 $\xrightarrow{\beta}$ 4)-Gal*p*-(1 $\xrightarrow{\beta}$ 4)-Gal*p*-(1 $\overset{\beta}{—}$ Arabinogalactan *I* [[Ara*f*]$_x$-[Gal*p*]$_y$—]		
3 ↑ R′ [R′ = Arabinan or Ara*f*-(1—[→5)-Ara*f*-(1—]$_n$]		
Rhamnogalacturonans	→4)-Gal*p*A $\overset{\alpha}{—}$(1 $\overset{\alpha}{—}$[→4)-Gal*p*A-(1 $\overset{\alpha}{—}$]$_n$→4)-Gal*p*A-(1 $\xrightarrow{\alpha}$ 2)-Rha*p*-(1 —	R = Xyl*p*-(1 $\overset{\beta}{—}$,
	3 3 4	Gal*p*-(1 $\xrightarrow{\beta}$ 2)-Xyl*p*-(1 —,
	↑ ↑ ↑	Fuc*p*-(1 $\xrightarrow{\alpha}$ 2)-Xyl*p*-(1 —, or
	R R R′	Api*f*-(1 → 3)-Api*f*-(1 —
		R or R′ = Ara*f*-(1 —,
		[Ara*f*]$_m$—, or
		[Ara*f*]$_x$-[Gal*p*]$_y$—
		Gal*p*A residues as methyl esters
		O-Acetyl groups (location not known)

Arabinogalactan II	$\rightarrow$3)-Gal*p*-(1$\xrightarrow{\beta}$3)-Gal*p*-(1$\xrightarrow{\beta}$3)-Gal*p*-(1$\xrightarrow{\beta}$3)-Gal*p*-(1$\xrightarrow{\beta}$3)-Gal*p*-(1$\xrightarrow{\beta}$ branches (6 ← 1), in chain order: 1st: R$\rightarrow$3)-Gal*p*, itself carrying Gal*p* at 6 (6 ← 1) 2nd: Gal*p*, itself carrying Gal*p* at 6 (6 ← 1) 3rd: R 4th: Gal*p* 5th: Gal*p*, itself carrying Gal*p* at 6 (6 ← 1)	R = Ara*f*-(1—, Ara*p*-(1$\xrightarrow{\beta}$3-Ara*f*-(1—, or Gal*p*-(1$\xrightarrow{\beta}$[$\rightarrow$6(-Gal*p*-)1$\xrightarrow{\beta}$]$_n$
Mannans, glucomannans and galactoglucomannans		
Mannans	$\rightarrow$4)-Man*p*-(1$\xrightarrow{\beta}$4)-Man*p*-(1$\xrightarrow{\beta}$4)-Man*p*-(1$\xrightarrow{\beta}$4)-Man*p*-(1$\xrightarrow{\beta}$4)-Man*p*-(1$\xrightarrow{\beta}$	R = Gal*p*-(1$\xrightarrow{\alpha}$
(Galacto)Glucomannans	$\rightarrow$4)-Man*p*-1$\xrightarrow{\beta}$4)-Glc*p*-(1$\xrightarrow{\beta}$[$\rightarrow$4)-Man*p*-(1$\xrightarrow{\beta}$]$_n$$\rightarrow$4)-Glc*p*-(1$\xrightarrow{\beta}$ 6 ← R on the first Man*p*; 6 ← R on the final Glc*p*	*O*-Acetyl groups
Xylans [arabinoxylans and 4-*O*-methylglucuronoxylans]		
Xylans	$\rightarrow$4)-Xyl*p*-(1$\xrightarrow{\beta}$4)-Xyl*p*-(1$\xrightarrow{\beta}$4)-Xyl*p*-(1$\xrightarrow{\beta}$4)-Xyl*p*-(1$\xrightarrow{\beta}$4)-Xyl*p*-(1$\xrightarrow{\beta}$ branches: 3 ← 1 Ara*f* on the 1st Xyl*p*; 2 ← 1 (4-Me)Glc*p*A on the 3rd Xyl*p*; 3 ← 1 Ara*f* with R$\longrightarrow$Ara*f* on the 5th Xyl*p*	R = Xyl*p*-(1$\xrightarrow{\beta}$2), Xyl*p*-(1$\xrightarrow{\alpha}$3), Gal*p*-(1$\longrightarrow$5) or Gal*p*-(1$\longrightarrow$4)-Xyl*p*-(1$\xrightarrow{\beta}$2)[b] *O*-Acetyl groups
Glucuronomannans	$\rightarrow$4)-Glc*p*A-(1$\xrightarrow{\beta}$2)-Man*p*-(1$\xrightarrow{\alpha}$4)-Glc*p*A-(1$\xrightarrow{\beta}$2)-Man*p*-(1$\xrightarrow{\alpha}$ on each Man*p*: R → 6 and 3 ← R′	R = Xyl*p*-(1$\xrightarrow{\beta}$ R′ = Ara*f*-(1— or [Gal*p*-(1—]$_n$$\rightarrow$3)-Ara*p*-(1—

[a] For simplicity enantiomeric configurations of sugars have been omitted. Unless otherwise indicated the sugars are D-apiose, L-arabinose, L-fucose, D-galactose, D-galacturonic acid, D-glucose, D-glucuronic acid, D-mannose, L-rhamnose and D-xylose

[b] In separate polysaccharides both D- and L-galactose units have been encountered

2 The Main Structural Features

2.1 Cellulose, Other β-D-Glucans and Xyloglucans

For convenience the glucans are grouped together, but those of different linkage types probably arise by different biosynthetic pathways. The xyloglucans form members of a distinct group of polysaccharides, all with a cellulosic main chain, but no direct biosynthetic relationship to cellulose has been established.

2.2 Rhamnogalacturonans and Associated Arabinans and Arabinogalactans

Pectins are rarely simple $(1 \rightarrow 4)$-α-D-galacturonans, but normally contain neutral sugars as integral constituents. Of the neutral sugars L-rhamnose is unique in being present in the interior chain and interrupting sequences of ~10–20 galacturonic acid residues. Other sugars, notably D-galactose and L-arabinose, but also D-xylose, L-fucose and D-apiose, and occasionally rare sugars such as 2-*O*-methyl-D-xylose and 2-*O*-methyl-L-fucose, occur in short side-chains of not more than three units (ASPINALL 1973; DARVILL et al. 1978). However, D-galactose and L-arabinose are also present in polymeric form as galactans and arabinans or as arabinogalactans of type I (4-linked arabinogalactans). The situation is complicated by the fact that polysaccharides of these structural types may be isolated as discrete neutral polymers. It is still an open question as to how far these neutral polysaccharides may arise *either* as biosynthetic sub-units en route to incorporation in a gross macromolecule *or* as adventitious degradation products resulting from depolymerization during the isolation process. Rhamnogalacturonans from different sources may show considerable variations in types of side-chains, but also rhamnogalacturonans with markedly different side-chains have been isolated from the same source (DARVILL et al. 1978). The physical properties and hence the biological functions of pectins, whilst greatly influenced by the nature and proportions of side-chain units, may be further modified by esterification of galacturonic acid residues, where up to 80% may be present as methyl esters, and less frequently by acetylation of as yet undesignated sugar hydroxyl groups (as in sugar beet pectin).

In addition to the above-mentioned arabinogalactans of type I a second polysaccharide of similar composition but quite different linkage type is found in abundance in coniferous woods, especially in larches, where it probably does not function as a cell wall component (TIMELL 1965). However, arabinogalactans of type II (or 3,6-linked arabinogalactans) are now recognized as of more widespread occurrence since they are found (a) in association with pectins in dicot cell walls (KEEGSTRA et al. 1973), (b) apparently covalently linked to protein in cereals (CLARKE et al. 1979), and (c) with further ramifications of outer chains as the major components of many exudate gums (STEPHEN 1980). The highly branched structures of these polysaccharides are too complex to be accounted for in terms of a linear backbone of uniform linkage type with short side-chains (CHURMS et al. 1978).

2.3 Mannans, Glucomannans and Galactoglucomannans

These polysaccharides are conveniently classed together since they all contain (1→4)-β-D-glycan chains, whether of D-mannose alone or of D-glucose and D-mannose. Residues of D-galactose, when present, occur as single-unit side-chains of varying frequency. The mannans are probably glucomannans with very small proportions of galactose residues (see Sect. 2.2.2 in Chap. 11, Vol. 13A, this Series). Glucomannans and galactoglucomannans, which are the main hemicellulose components of coniferous woods, occur as partially acetylated polysaccharides (TIMELL 1965).

2.4 Xylans

Xylans from higher plants are all (1→4)-β-D-glycans with few branch points in the main chains. The xylan family extends from the arabinoxylans, such as those from cereals and grasses (WILKIE 1979), to the 4-*O*-methyl-glucuronoxylans of dicots and by gymnosperms (TIMELL 1964), in which the characteristic single-unit side-chains are attached by (1→3)- and (1→2)-linkage respectively. Several xylans carry both types of single unit side-chains and some arabinoxylans from cereals and grasses carry more extended di- or trisaccharide units. The 4-*O*-methylglucuronoxylans from hardwoods are partially acetylated.

2.5 Other Cell Wall Polysaccharides

The preceding categories include the most abundant and commonly encountered cell wall polysaccharides. Some polysaccharides, however, may be present infrequently and/or only in small quantities and thus escape ready detection. An example is the glucuronomannan recently isolated from suspension-cultured tobacco cells (KATO et al. 1977). This polysaccharide was shown to be structurally similar to leiocarpan A (ASPINALL 1969), one of a small group of exudate polysaccharides whose structural features had not been previously encountered in cell wall materials.

3 Covalent and Non-Covalent Inter-Polymeric Linkages in the Cell Wall

Classification of polysaccharides by structure provides the most satisfactory basis for understanding the molecular architecture of the plant cell wall. Many of the polysaccharides whose structures have been studied in detail have been isolated under very mild conditions with little chance of inadvertent degradation. In other cases the much more drastic procedures used to solubilize polysaccharides have been such that covalent bonds might have been broken. The close association of the various polysaccharides with each other and with glycoprotein (see Elbein, Chap. 8, this Vol.) raises the question as to whether the individual components are held together exclusively by secondary or noncovalent forces or whether covalent inter-polymeric linkages exist (Aspinall 1980). A model for the primary cell wall in dicots has been developed by Albersheim and his collaborators (see McNeil et al. 1979). This model, for which there is good but not conclusive evidence, postulates covalent linkages between most of the polysaccharide and glycoprotein components but association with cellulose by noncovalent forces only.

References

Aspinall GO (1969) Gums and mucilages. Adv Carbohydr Chem Biochem 24:333–379

Aspinall GO (1973) Carbohydrate polymers of plant cell walls. In: Loewus F (ed) Biogenesis of plant cell wall polysaccharides. Academic Press, New York, pp 95–115

Aspinall GO (1980) Chemistry of cell wall polysaccharides. In: Preiss J (ed) The biochemistry of plants. Academic Press, New York, Vol III, pp 473–500

Churms SC, Marrifield EH, Stephan AM (1978) Regularity within the molecular structure of arabinogalactan from Western larch (*Larix occidentalis*). Carbohydr Res 64:C1–C2

Clarke AE, Anderson RL, Stone BA (1979) Form and function of arabinogalactans and arabinogalactan-proteins. Phytochemistry 18:521–540

Darvill A, McNeill M, Albersheim P (1978) Structure of plant cell walls VIII. A new pectic polysaccharide. Plant Physiol 62:418–442

Kato K, Watanabe F, Eda S (1977) Interior chains of glucuronomannan from extracellular polysaccharides of suspension-cultured tobacco cells. Agric Biol Chem 41:539–542

Keegstra K, Talmadge KW, Bauer WD, Albersheim P (1973) The structure of plant cell walls III. A model of the walls of suspension-cultured sycamore cells based on the interconnection of the macromolecular components. Plant Physiol 51:188–196

McNeil M, Darvill AG, Albersheim P (1979) The structural polymers of the primary cell walls of dicots. Fortschr Chem Org Naturst 37:191–249

Stephen AM (1980) Plant carbohydrates. In: Bell EA, Charlwood BV (eds) Encyclopedia of Plant Physiology. Springer, Berlin Heidelberg New York Vol. VIII, pp 555–584

Timell TE (1964) Wood hemicelluloses: Part I. Adv Carbohydr Chem 19:247–302

Timell TE (1965) Wood hemicelluloses: Part II. Adv Carbohydr Chem 20:409–483

Wilkie KCB (1979) The hemicelluloses of grasses and cereals. Adv Carbohydr Chem Biochem 36:215–264

2 Ultrastructure of the Plant Cell Wall: Biophysical Viewpoint

J.R. Colvin

1 Introduction

The purpose of the present chapter is to give the interested reader a summary of existing ideas (1979) on the physical fine structure of the plant cell wall. In contrast to most, if not all, of the subjects of the other chapters, this topic is approaching a state of maturity because of the extensive and intensive work of the past three decades. Although there is still a great deal of tidying-up to do in particular instances, it is unlikely that present overall concepts of the physical ultrastructure of plant cell walls will change much in the future. For an introduction to the earlier literature of a subject which is more than 300 years old, one cannot do better than begin with Chapter 3 of Esau's classic treatise (1965). For comprehensive, detailed, recent consideration of the same topic, the monographs by Preston (1974) and by Frey-Wyssling (1976) are indispensable. In preparation of the present chapter, relevant studies which were published before the middle of 1979 have been reviewed.

The introductory paragraph ought not to leave the impression that work on and interest in the subject of the plant cell wall is in the process of being finished. Far from it; the foci of investigations have only shifted from the determination of the mature, static architecture of these envelopes to the dynamic processes by which the cells bring them into being. As shown by nearly all the other chapters in this volume, the emphasis in plant cell wall research is now on the biochemistry, development and differentiation of these entities, rather than on the establishment of their physical structure or composition (for example, see the recent review by Robinson 1977).

An explanation for some omissions is necessary. A fraction of the many hundreds of investigations of the plant cell wall or related substances has led to postulation of entities or processes for which the evidence is far less than complete. Because some of these suggestions have been markedly heuristic, the concepts behind them have become tightly woven into much recent literature, even though their verity is more than questionable. For this reason, in the following, Ockham's razor has been applied rigorously (multiplicity ought not to be posited without necessity). Therefore, in the present article, a hypothesis or an entity has not been accepted unless the *experimental* evidence for it is irrefutable.

1.1 What a Plant Cell Wall Is

First, we shall try to make clear what a plant cell wall is considered to be, in modern usage. It is the continuous, more or less tough integument of nearly all plant cells, formed mainly by a fibrillar framework within which the interstices are filled with an amorphous material. Both the framework and the amorphous "filler" vary in morphology and in composition between phyla, genera, species and tissues (see Chaps. 4, 6, 15 and 17, this Vol.). The wall is immediately outside the cytoplasmic membrane (plasmalemma) and is therefore non-protoplasmic but certainly cannot be considered inert, i.e., without metabolism. On the outside surface of the wall of the cells of terrestrial plants (including fungi) which are exposed to the atmosphere is a layer of lipidic substances which is called the cuticle. There seems to be a corresponding layer of protein-linked mannans (FREY-WYSSLING 1976) on the outside of yeasts and of mannans on the outside of cells of marine algae (MACKIE and PRESTON 1974). However, because the lipidic cuticle and the corresponding layers in yeasts and algae are exceptional components of the general cell wall, their microstructure and composition will not be dealt with in this chapter (see Chap. 10 by Kolattukudy et al. this Vol.). The microfibrillar component of the wall is prominent in all cells (when examined properly) except in yeasts, where its presence is minimal and the filler portion is correspondingly evident.

It will be easily deduced from the above that all plant cell walls are products of a process of evolution in which the advantages of lightness, strength, flexibility, and the variable permeability of a heterogeneous material were utilized in the biosphere long before they became evident in synthetic composites (see MARK 1967, for a detailed analysis of the similarities). In fact, plant cell walls have often been compared in structure to reinforced concrete. However, except for extreme instances such as fully lignified, rigid walls (FREY-WYSSLING 1976), this comparison is not particularly apt. It overemphasizes the properties of inertness and rigidity while not suggesting the equally important properties of strength, flexibility, and permeability of the wall. Although it too has limitations, a better model may well be a sheet of modern, bonded plywood which is laminated, fibrillar, flexible, light but strong, permeable to some substances but impermeable to others.

1.2 What a Plant Cell Wall Is Not

It is important to stress here that the plant cell wall in modern usage does not include the bacterial cell wall or its capsule (SALTON 1964), pellicles (SCHOLTYSECK 1979), or membranes (LEIVE 1973). In connection with the last word, the reader should be careful about the use of *Membran* (wall) in the German literature, especially in older articles; fortunately, the present tendency is to use the unambiguous word, *Wand.*

1.3 Functions and Biological Significance of the Plant Cell Wall

Incontestably, the primary function of the plant cell wall is protection of the protoplast against injury by abrasion, dessication, osmotic shock, ultraviolet irradiation, microbial attack and probably other forms of insult. However, from this primary function evolved a second use which is equally important, namely skeletal support and rigidity. Probably by the use of lignin, the evolution of reproducible associations of tough, resistant cells into relatively strong, upright, conducting tissues and organs permitted the invasion of the land from the sea. The predominant, characteristic forms and properties of these tissues and organs reflect the fibrillar morphology of major components of the cell walls.

A third significant function of plant cell walls in the biosphere is to act as a form of storage of energy and nutriment. Because of their greater resistance to degradation and the enormous quantities of plant cell walls produced each year (WOODWELL et al. 1978) they tend to smooth out fluctuations in the supply of carbon and nitrogen compounds to the biosphere. This has clear advantages for any biological community regarded as a unit and it is easy to see that this dampening effect may have advantages even for a single species.

The price paid by plants, even unicellular organisms, for these advantages of protection, durability and long-range conduction is that required by immobility and lack of ability to respond. In general, the inert armor with which the plant protoplast has surrounded itself, and with which it may construct large, useful associations, of necessity limits the ability of the same cell or group of cells to adapt to changes about it. That is, the means of protection which is an advantage under stable conditions becomes a disadvantage under changing circumstances. Obviously, the stability of most ecological niches in the biosphere has been sufficient to enable plant cells, organisms, or communities to more than pay the costs of immobility.

2 Idealized Plant Cell Wall: Structure

Three centuries of observation by optical microscope and three decades by electronmicroscope (supplemented by other techniques) have led to a widely accepted, general view of the structure of plant cell walls. This view was developed primarily for rationalization of observations on the cells of terrestrial plants and, as a result, may usually be adapted readily to any particular species within that category. However, its suitability for the description of cell walls of aquatic plants or of microbial plants such as yeasts may be much less. The conditions under which such plants have evolved and grow are so different that it is not surprising that their walls do not correspond closely to those of terrestrial plants. Nonetheless, the structures of the filamentous green, brown, and red algae foreshadow the separation of the wall into two layers (MACKIE and PRESTON 1974). Even for land plants, it is essential to remember that hardly any single class of cells will correspond in all details to the standard, classic, formal scheme outlined below.

2.1 Intercellular Layer

Within many-celled plants, adjacent cell walls are separated from each other by a variable, amorphous layer which is often called the middle lamella but a better, more descriptive name would be the intercellular layer (FREY-WYSSLING 1976). The intercellular layer in land plants is usually a site of heavier-than-average deposition of lignin (a complex, variable, three-dimensional polymer of phenyl-propane or its derivatives; see chapter 9, this Vol.) which serves as a glue between cells, as a stiffening material, and as a protective substance.

2.2 Primary Cell Wall

The amorphous material of the intercellular layer merges gradually with the amorphous matrix of the outermost, "primary" layer of the wall of the cell. This matrix fills the interstices between the microfibrils of cellulose, chitin, mannan, noncellulosic glucan, or xylan (depending on the species) which form a framework or skeleton for the primary wall. As a consequence, the undigested or undisturbed primary wall usually appears compact and not lamellated in cross-section. It is only after part of the matrix is removed by some method that the microfibrillar framework and its laminar character usually become evident.

2.2.1 Matrix Substances

The matrix (filler) of the primary wall in all terrestrial plants seems to be a mixture of complex hemicelluloses and polyuronic acids (see Chaps. 4 and 6, this Vol.). It is probable that each category of polymer is in itself a mixture and it is certain that the proportions of each main category of substance vary profoundly not only between species but within species as a function of age, tissue, treatment, temperature of growth, and certainly other factors. Because the importance of these variations in land plants is more chemical than physical, the reader is referred to the above-cited chapters for a resumé of present knowledge. Much less is known at the present time about the matrix of the cell walls of yeasts (see Chap. 17, this Vol.) or algae (MACKIE and PRESTON 1974) although these are now active areas of study (see also Chaps. 12, 13 and 16, this Vol.).

2.2.2 Microfibrils of Cellulose

From the physical point of view the microfibrillar framework of the primary wall, irrespective of its composition, is the most important component because it largely determines the anisotropic and mechanical properties of the wall. Furthermore, because cellulose is by two orders of magnitude the more abundant of the two commoner skeletal microfibrils (TRACEY 1957), to a first approximation one may describe the physical morphology and functions of all microfibrils in cell walls by describing those of cellulose. Details vary but not significantly for the view of the present chapter.

Native cellulose occurs in plant cell walls as very fine threads (or ribbons) called microfibrils, which vary from 10–25 nm in breadth depending upon species. Each microfibril cross-section contains roughly 10^3 molecular chains of

(1 → 4)-β-D-glucan, which are in the extended form with their axes parallel to the axis of the microfibril. Each glucan chain may be very long (more than 15,000 glucose residues) (MARX-FIGINI 1971) and different parts of each chain may associate with adjacent chains to form either amorphous or crystalline domaines (or something between the two). Within the crystalline domaines (usually called crystallites) the hydroxyl groups of the glucose residues cause the chains to be ordered in sheets (GARDNER and BLACKWELL 1974). A sheet is associated with other sheets by London dispersion forces to build up the crystallite. The portions of the chains in the nonperiodic parts of the microfibril serve to connect these crystallites (of varying size) into a coherent thread which is on the average highly resistant to biological and chemical degradation (GASCOIGNE and GASCOIGNE 1960), and which has a tensile strength approximately that of mild steel (SAKURADA et al. 1962; MARK 1967). Both the biochemical and physical mechanisms of assembly of microfibrils are an active area of study and dispute at the moment (see Chaps. 3, 5, 13 and 16, this Vol.; also COLVIN 1977; BURGESS 1979).

In the primary cell wall these long, thin, stiff, but flexible and strong threads are deposited in the plane of the wall; usually but not always with their axes randomly disposed with respect to each other. In very young and/or thin walls there may be only one layer of microfibrils which is embedded in and supports the amorphous and hydrophilic components of the matrix. Usually, however, there are several distinguishable layers of microfibrils giving the whole wall a lamellated appearance. Since the innermost layer is deposited last and the cell is often being extended during this deposition, the microfibrils in the outer layer may become markedly oriented by this process of mechanical extension (ROELOFSEN 1959; GERTEL and GREEN 1977). There is therefore usually an increasing gradient of orientation of the microfibrils from the inside of the primary wall to the outside. While this extension of the wall is taking place, there is concomitant deposition of the matrix but the mechanism of control of the synthesis is still unknown (see Chap. 6, this vol.).

2.2.3 Local Structure of the Primary Wall

The foregoing may leave the impression that a typical, primary, cell wall has a relatively smooth uniform surface. Such is not the case. Distributed irregularly over the surface are limited areas called primary pit-fields where the density of deposition of the matrix and its supporting microfibrils is less, the wall is thinner and therefore tenuous strands of cytoplasm may pass through small openings in the wall from one cell to a contiguous one (plasmadesmata). These primary pit-fields and the associated plasmadesmata are an important mechanism of transport within the organism (see Vol. 2A, this series).

Probably due to the influence of workers with animal cells, it is often implied that the only structure in a cell which exercises much control over what gets in or out is the periplasmic membrane. This is certainly true for animal cells but not for plant cells. Clearly, the cell wall must screen mechanically the substances which approach the periplasmic membrane. The net effect of this screening is most likely an exclusion of relatively coarse particles. However, equally clear, components of the wall, especially those of the matrix, will modify the flow of soluble substances toward the membrane and into the cell by nonspecific adsorption and ion-exchange. Plant cell walls therefore exercise a part of the function which in animal cells is confined to the plasmalemma (i.e., semipermeability).

2.3 Secondary Cell Wall

Many plant cells have no other sort of integument than that sketched above. However, for others and particularly for those which are related to products with commercial importance, there may be a marked development of an additional layer, the so-called secondary cell wall. About the time cell extension ceases and therefore the primary wall (both framework and matrix) is complete, some cells begin to deposit massive amounts of cellulose (or one of the other fibrillar substances) within the first layer. On completion, the substance of a secondary wall may represent more than 90% of the dry weight of the cell. For terrestrial plants, this new layer is normally composed of a higher proportion of cellulose microfibrils and a lower proportion of matrix substances than the primary wall. In addition, the microfibrils are deposited in layers, often with substantial, common orientation of the microfibrils within a layer and a regular change of orientation between layers. On both optical microscope evidence and observations from the electronmicroscope, the secondary cell wall of terrestrial plants is commonly divided into three subdivisions, S_1 (outermost), S_2, and S_3. Normally, the S_2 layer contains by far the most material. At the present time, the molecular cause of deposition of these layers and of the orientation of the microfibrils within them is a matter of discussion (ROBINSON 1977).

2.3.1 Bordered Pits

In some highly specialized cells with very thick secondary cell walls such as tracheids, communication between cells is almost wholly by a special kind of pore called bordered pits. These are a prominent feature of the physical structure of such cells. By mechanisms which are wholly unknown, the thick, opposed, secondary walls on each side arch over the thin primary wall and intercellular layer to form on each side a pit chamber with a smaller opening to the lumen of each cell. They obviously represent an adpatation whereby the tracheid or a similar cell may contribute strength to the tissue while maintaining means of communication across the dense, thick, supporting wall.

2.4 Tertiary Wall

It is generally assumed that there is within the secondary wall, next to the lumen of all plant cells, a thin, indefinite layer which, by extension of the foregoing terminology, is usually called the tertiary wall (sometimes warty layer). This layer is often thin and easily missed. When observable and tested, it is insoluble in basic solvents and digested only slowly by cellulolytic fungi. In view of its thinness, variability in position, and other properties, there must still be some suspicion that this layer is merely composed of dried residues from the dead cytoplasm and/or unused lignin precursors.

3 Actual Plant Cell Walls: Structure

Within the context of the idealized structure outlined above, the purpose of this section is to describe briefly the static, physical structure of representative, actual, plant cell walls, none of which conforms in all details to the model. Whenever possible, these examples are taken from work published within the last 5 years. Monographs by PRESTON (1974) and FREY-WYSSLING (1976) deal comprehensively with earlier work.

3.1 Intercellular Layers of Various Species

The prevalent view of the amorphous, structureless nature of the intercellular layer (middle lamella) is still generally valid experimentally but has had to be revised in particular instances. As an example, LEPPARD and COLVIN (1971) have shown that in aggregates of suspension-cultured cells of *Daucus carota, Ipomoea* sp. and *Phaseolus vulgaris* there is fine fibrillar or rodlike material between the cells, which has the properties of pectin (COLVIN and LEPPARD 1973). The fibrillar substance is observed between the cells of roots of *Daucus, Ipomea, Pinus resinosa,* and perhaps in *Pseudotsuga taxifolia* (LEPPARD and COLVIN 1972). In addition, massive, but irregular deposition of electron-opaque granules is a common feature of the middle lamellae of many branch and leaf tissues of the same species (LEPPARD and COLVIN 1972). More recently, CARR and CARR (1975) have shown the presence of rows of pectic strands between mesophyll cells of eucalypts and between parenchyma cells of ferns. All these investigations demonstrate that the intercellular layer between cells has a recognizable fine structure when it is sought by proper means and, in addition, they suggest that pectin or its derivatives is a major component of this layer. The function of pectin is as yet a subject for speculation (CARR and CARR 1975) but it cannot fail to influence the permeability of the walls.

3.2 Thickened Primary Wall of Axial Parenchyma Cells of Trembling Aspen

These cells, which represent a transition between typical parenchyma cells and xylem fiber cells, have been described by CHAFE (1974). They are formed at the outer margin of the annual growth ring after summer wood formation has stopped. The wall is not lignified and is somewhat thicker than usual without definite structure. It is, however, still evidently capable of extension. The outer secondary wall is only deposited and lignified following activation of the cambium in the spring. After the outer secondary wall is laid down, an isotropic or protective layer is deposited, which is followed rapidly by another inner secondary wall. The whole process is a superb illustration of how closely the genetic information in the original parenchyma cell with its thin primary wall is controlled and modulated by both the internal and external environment.

3.3 Secondary Wall Layers of Elm Parenchyma Cells

A good example of how the classical model of deposition of the secondary wall layers, S_1-S_2-S_3, may be followed initially and then additional features added to it is given in the study by OUELLETTE (1978) of American elm parenchyma cells. In both healthy and diseased cells the primary wall is apparent with the normal, fibrillar, lignified S_1-S_2-S_3 layers within. However, within the S_3 layer is deposited a fibrillar, slightly opaque, unlignified layer. Within this entity is another, thicker, fibrillar, lignified stratum ($El1_1$). There may be still another thinner, loosely oriented fibrillar layer, the $El1_2$. The lamellae separating the extra lignified layers are similar in texture to the primary wall of the cell. Surprisingly, some of the cells contiguous to cells with such layers lack the normal wall structure.

Why and how this multilayering of the secondary wall occurs is not yet known. Obviously this study is a good illustration of the necessity for prudence in the application of the classical model.

3.4 Ultrastructure of Cell Walls of *Fusarium sulphureum*

For an extensive survey of the physical and chemical structure of fungal walls, the reader is referred to Chapter 15 by WESSELS and SIETSMA, this Volume, and to the study of cell walls of *Fusarium sulphureum* by SCHNEIDER and WARDROP (1979), a good, example of recent work on fungal ultrastructure. In this fungus, the microfibrillar framework of the wall is highly crystalline chitin. The robust wall (matrix and microfibrils) has two layers of approximately equal thickness but the outer one adsorbs heavy metals more intensely. Orientation of the chitin microfibrils is random in the outer, mid-portion, and inner part of the wall, and there are no differences in morphology so that one may infer reasonably that the differential staining characteristics are due to the matrix component(s). Essentially, the microfibrillar component is a single, thick layer (no secondary wall) of superimposed threads bonded together by a matrix which differs from inside to outside. The differences between this experimental result and the idealized structure are evident.

3.5 Ultrastructure of the Cell Wall of Yeasts

Yeasts are fungi, but the physical and chemical nature of their walls is so distinctive that they ought to be considered separately. Because yeast cell wall investigation is an active area of study at the present time, the reader is referred to Chapters 15 and 17, this Volume for a detailed consideration. Here are mentioned only two reports which illustrate differences between factual and idealized structures. KOPECKÁ et al. (1974) demonstrated convincingly by a combination of enzymatic techniques and electronmicroscopy that, contrary to previous conclusions, there is a fibrillar component in the cell walls of *Saccharomyces cerevisiae,* which is composed of separate, densely interwoven microfibrils. As an inference from enzymatic digestion studies, they concluded that these microfibrils were composed of (1 → 3)-β-D-glucan. There was no physical evidence of distinct layers within the wall but partial digestion of the walls by pronase indicated that the outer layers of the wall contained a protein or proteins.

A recent study of the ultrastructure of the wall of blastospores of *Candida albicans* by CASSONE et al. (1978) extends some of the above conclusions. They showed definitively that the mannan polymers which have been known for a long time to be in the wall extend throughout the wall of the blastospore; i.e., they are not present just in one layer as previously surmised. Furthermore, the layering of the wall essentially reflects the distribution of the various alkali-soluble complexes at different levels, both *over* and *in* the rigid, glucan-chitin matrix. This latter conclusion is similar to that noted above for *Fusarium* and again emphasizes the difference between factual observations and deductions from the generally accepted model.

3.6 Cell Wall Structure of Algae

For a detailed consideration of composition, cytology, and formation of algal cell walls, the reader should refer to Chapters 12 and 13, this Volume. The sole purpose of the present section is to give two examples of work from the physical point of view.

The physical description of algal cell wall ultrastructure played a prominent part in pioneer investigations of the plant cell wall with the electronmicroscope (PRESTON 1974). Later, interest in the static architecture of algal walls became muted and was replaced by studies of the mode of differentiation of these integuments. Recently, there has been a tendency to combine the two points of view and this tendency is exemplified by an investigation of the wall of the Valonia-type alga, *Glaucocystis,* by WILLISON and BROWN (1978a). They showed that the wall of this organism is deposited in three phases. First, a thin layer is laid down which would correspond to the primary wall except that it is nonfibrillar. Second, cellulose microfibrils are deposited without interweaving on the thin layer in a helically crossed polylamellate fashion to form what might be called a secondary layer. The microfibrils taper to a point. Finally, the interstices between the microfibrils are filled by matrix substances. Cellular extension apparently results in spreading and in rupture of the microfibrils in the secondary layer. The authors invoke a mechanism for formation of the wall which is speculative (WILLISON and BROWN 1978b), but there is no doubt about the physical structure of the wall itself.

An example of the care, circumspection, and restraint in interpretation of limited data which is presently necessary (but not always practiced) in cell wall structure studies is given by a recent investigation of the alga *Hydrodictyon africanum* by BAILEY and NORTHCOTE (1977). They found that the inner wall microfibrils did not appear to end in aggregates. There did not appear to be any distinct structures on the inner surface of the cytoplasmic membrane which reflected the orientation of the microfibrils in the adjacent wall and there was only a suggestion that perhaps particles shown on the inner face of the outer leaflet of the plasma membrane follow the orientation of microfibrils in the near-by wall. These predominantly negative facts are an accurate reflection of present knowledge of mechanisms of control of wall synthesis.

3.7 Physical Structure of Cell Walls from Protoplasts

A wholly new area of investigation in the last decade has been that of the physical structure of cell walls regenerated by protoplasts (PRAT 1973). Clearly, the motivation of nearly all this work has been not to study the walls per se but to establish methods to find how natural cell walls are deposited or changed. Some workers recognized explicitly that the wall which forms around plant protoplasts in the early stages of regeneration has an aberrant ultrastructure (HORINE and RUESINK 1972; WILLISON and COCKING 1972; HANKE and NORTHCOTE 1974) because of the highly nonphysiological conditions. Nonetheless, this line of work is capable of giving valuable new information which cannot be attained in any other way, and has in fact already begun to do so.

A typical example of this new direction is given by the work of BURGESS and FLEMING (1974) on cell wall regeneration by tobacco protoplasts, which are now known to be among the most useful for this purpose (KINNERSLEY et al. 1978). They showed that a tenuous, fibrillar layer formed over most of the surface of the protoplast after 72 h of culture under the appropriate conditions. This fibrillar material, which behaved like cellulose, was first oriented predominantly at right angles to the plasmalemma surface, but later became parallel to it. A second component, which stained more strongly with heavy metals, had the appearance of a reticulum or of branched fibrils. It was suggested to be a polysaccharide of the pectin type. After 120 h of culture, a normal cell wall was well established and clumps of cells formed from which it proved possible to regenerate whole plants of tobacco. No pattern of structure was detected to explain the orientation or method of synthesis of the microfibrillar part of the cell.

This part of the problem of cell wall synthesis by protoplasts was advanced considerably by GROUT (1975), also working with tobacco cells. Using the deep-etch modification of the freeze-fracture procedure on the early stages of cell wall deposition, he showed that there was a sudden appearance of cellulose microfibrils at the outer protoplast surface after 16 h of culture. These microfibrils were not associated with any structured particles or differentiated regions of the plasmalemma. He interpreted the formation of these cellulose microfibrils as a precipitation or crystallization of cellulose precursors on the outer surface of the plasmalemma, without the participation of enzymatic granules or of enzymes with an assembly function. This study has broken new ground in consideration of the major fibrillar component of plant cell walls and should be extended.

By using tobacco protoplasts once again SCHILDE-RENTSCHLER (1977) has shown that when cell wall regeneration is inhibited by cellulase, nuclear and cell division are also stopped. When this inhibition is relieved by cellobiose, nuclear and cell division proceed. There is thus clear evidence for a link between the three processes, but its nature is not yet known.

In summary, study of the physical structure of the cell walls regenerated by protoplasts is not likely to add much which is new to knowledge of plant cell wall static architecture. However, it promises to help immensely our knowledge of the mechanisms of deposition and differentiation of these organelles.

3.8 Physics of Specialized Structures of Plant Cell Walls

Reference to the monographs of PRESTON (1974) and FREY-WYSSLING (1976) will give many examples of highly specialized structures which are an essential part of plant cell walls. These include ribs (SETTERFIELD and BAYLEY 1958), pits of many different kinds (Plate 15, GUNNING and STEER 1975), sieves (Plate 8, GUNNING and STEER 1975), and especially polylamellate spirals of cellulose microfibrils. These structures have a vital role in the plant but their physical description was essentially complete a decade ago. One point which is common to the physical description of each of them is that they all involve a marked, oriented deposition of cellulose microfibrills. Sometimes the orientation is parallel as in ribs, circular as in bordered pits, reticulate as in sieves, or spiral as in the secondary wall. However, almost nothing is known about the physical mechanism of the mode of deposition of the cellulose microfibrils in each case. It is evident that for each instance there has to be a special means of controlling the deposition of the component microfibrils but, so far, none of these special means is known. This gap in knowledge applies equally well to chitin microfibrils, to mannans, and to other glucans. In each instance a better understanding of the mechanics of formation and deposition of the microfibril is required.

4 Macromolecular Problems of Formation of Components of Plant Cell Walls

One of the purposes of the foregoing was to describe the physical, static structure of plant cell walls or their parts in general. In this final section, the purpose is to outline present ideas on how these entities came into being. None of these ideas are certain and some are simply speculative. However, they will serve to indicate the direction in which investigations are likely to move in the next decade.

4.1 Physical Formation of Cellulose Microfibrils

A definitive statement for the mechanism of the assembly of the cellulose microfibril has yet to be given. PRESTON (1964) advanced a tentative proposal for a mechanism in which an enzyme-complex (island of synthesis) on the outside of the plasmalemmal membrane attached glucose residues in a coordinated way to the ends of (1→4)-β-D-glucans at the tips of microfibrils. This form of end-growth required translation of either the microfibril or the enzyme complex (or perhaps both). This proposal is plausible and has been extremely heuristic but still lacks conclusive support. One major reservation about the general validity of the proposal is concerned with the fact that it postulates an entity which ought to be visible at the end of each growing microfibril but these aggregates are rarely seen (WILLISON 1976; ROBINSON 1977). In those instances in which aggregates have been observed in contact with microfibrils (ROLAND 1967) the possibility of chance superposition cannot be wholly excluded. In addition, this mechanism obviously requires translation of the microfibril or the enzyme-complex (granule?) and it is very difficult to see how such translation is possible in interwoven layers (e.g., Plate 4, GUNNING and STEER 1975). At present, the only valid decision must be that appropriate to Scottish courts; that is to say, unproven. A variation of this proposal, which has been advanced so far only for assembly of bacterial cellulose microfibrils, combines the notion of an active, granular, enzyme-complex with a multiple spinneret mechanism. BROWN et al. (1976) have suggested on the basis of electronmicroscope photographs that bundles or ribbons of microfibrils are produced end-wise from individual, synthesizing sites which are arranged longitudinally along the axis of the bacterium. Each microfibril is supposed to terminate independently in the bacterial envelope at a particle or pore which is responsible for synthesis. ZAAR (1979) has added to this conceptual scheme by interpreting longitudinal rows of particles and craters on the bacterial membrane surfaces of *Acetobacter xylinum* as evidence for pits from which are extruded microfibrils that ultimately are supposed to end in a 10-nm diameter multienzyme complex. Clearly, this scheme has the same difficulties for translation of the microfibrils as the preceding one. A more serious reservation, however, is that there is no direct evidence linking the observed microfibrils with these particles, craters, or pits in the bacterial cell membrane. The inferred, lateral points of attachment of the microfibrils to the bacterial envelope have often been seen before and have quite properly been interpreted as points of adhesion of the outer layer of the bacterial cell wall to adjacent microfibrils (BROWN et al. 1976). In addition, as stated by ZAAR (1979), the observed particles may only be an accumulation of an intermediate which is synthesized generally within the bacterial envelope and without any direct, physical relation to any granular, enzyme complex. In summary, the foregoing scheme too is unproven.

A third proposal, which also has not yet been proved, is distinct from the two foregoing suggestions in that it does not postulate any synthetic, granular, enzyme complexes or their equivalent. Polymeric, hydrated intermediates of glucose, which are certainly involved in cellulose biosynthesis (KJOSBAKKEN and COLVIN 1975), are extruded from a transient, terminal opening in the bacterial cell (COLVIN and LEPPARD 1977). These highly hydrated $(1\rightarrow4)$-β-D-glucans then associate spontaneously and progressively by hydrogen-bond formation between chains to form an entity which has been called the nascent microfibril (LEPPARD et al. 1975). When the water has been withdrawn from the nascent microfibril by time and/or desiccation, the normal mature form is observed. There is evidence for the same intermediate entity in green plant cellulose (LEPPARD and COLVIN 1978). Furthermore, the formation of cellulose microfibrils by tobacco protoplasts without the apparent intervention of any enzymes for assembly or any granular enzyme complexes has been described by GROUT (1975). In that system too, the formation of the microfibrils is thought to be solely due to the association (crystallization) of high-polymeric intermediates on the surface of the plasmalemma, without enzymes.

In summary, one proposal for the physical mechanism of assembly of the cellulose microfibril postulates an elaborate biological machinery; the other proposal attributes microfibril formation simply to the operation of known physicochemical forces. At the moment of writing neither scheme has been proved definitively (BURGESS 1979).

4.2 Physical Formation of Chitin Microfibrils

Our knowledge of chitin microfibril biosynthesis is in some ways more advanced than of cellulose. This process is treated in detail from the biochemical viewpoint in Chapter 16, this Volume. The purpose here is to consider only the physical aspects of chitin microfibril assembly. Working with an enzyme preparation from *Mucor rouxii*, RUIZ-HERRERA and BARTNICKI-GARCÍA (1974) and later RUIZ-HERRERA et al. (1975) were able to polymerize UDP-*N*-acetylglucosamine to a fibrillar product which has the properties of chitin. It appears that the effective catalytic agents are spheroidal granules (diameter 100 nm) which are called chitosomes. When UDP-*N*-acetylglucosamine is added to the chitosomes, coiled microfibrils of chitin or a precursor of chitin are formed within granule, its shell then breaks and extended microfibrils emerge (BRACKER et al. 1976). After incubation the coiled microfibrils are no longer seen and the extended microfibrils become thicker by physical association of smaller threads to form larger ones. No synthetic complexes or enzyme granules are reported. This mechanism of assembly of chitin microfibrils is clearly similar to the second process discussed above for cellulose.

4.3 Deposition of Hemicelluloses

From the physical point of view almost nothing can be said about the deposition of these substances in the plant cell wall except that they are initially in the form of a variable hydrogel, sometimes added by apposition (MULLIS et al. 1976), and sometimes have an oriented structure which changes during extension growth and mechanical extension of the walls (MORIKAWA et al. 1978). The chemical aspects of this component are treated extensively in Chapters 4 and 6, this Volume.

4.4 Physical Self-Assembly of Plant Cell Walls

One new, potentially very important aspect of study of the physics of plant cell wall structure has begun during the last decade. Formerly, it was the practice to take away progressively the components of the wall and deduce its structure from the residue. An excellent example is that of REIS and ROLAND (1974). Now, however, due to increasing understanding and better techniques, it is becoming possible to reconstitute in vitro at least parts of some plant cell walls. This means that, with care and discretion in interpretation, we now have a new general approach to study both the structure and the mode of differentiation of some plant cell walls.

So far the outstanding example of this type of work has been that on the cell wall of the green alga, *Chlamydomonas reinhardi* by the group at the John Innes Institute, England. After establishment of a highly ordered, complex, glycoprotein lattice in the normal structure of the multilayered wall by Roberts et al. (1972) and by Hills et al. (1973), Hills (1973) demonstrated that two components of the cell wall could be dissociated in aqueous 8 M lithium chloride. Dialysis of the cell wall subunits against water caused reassembly of a product with the same morphology and chemical composition as the original cell walls. About the same time, Davies and Lyall (1973) showed that the lattice reassembly will only occur in the presence of a particular kind of physical structure and of medium. They also demonstrated the role of a heritable factor in the process. In an extensive summary of this work Hills et al. (1975) showed that, in addition, successful wall reconstitution requires both the salt-insoluble and salt-soluble (glycoprotein) components of the wall. However, the salt-soluble glycoproteins alone can self-assemble to entities that have the crystalline structure characteristic of the outer layers of the complete cell wall. As the authors indicate, "this is the first case of reconstruction of a plant cell wall in vitro and it suggests that components of very large cellular structures are capable of being built by a simple self-assembly process." One may expect and hope that other instances will follow.

5 Resumé

In summary, the static, mature, physical structure of plant cell walls is now reasonably well understood and no great changes in our understanding are to be expected. Complex, almost infinitely variable, and varying at the macromolecular level, the overall picture of a microfibrillar framework (different composition in different plants) embedded in a polysaccharide matrix will remain.

For the future, the dynamic aspects of the plant cell wall will dominate research; how are their components formed, how is the deposition of the components coordinated, and especially how does the whole entity function dynamically in the organism?

References

Bailey DS, Northcote DH (1977) An ultrastructural study of the relationship between the plasma membrane and the cell wall of the coenocytic alga *Hydrodictyon africanum*. J Cell Sci 23:141–149

Bracker CE, Ruiz-Herrera J, Bartnicki-García S (1976) Structure and transformation of chitin synthetase particles (chitosomes) during microfibril synthesis *in vitro*. Proc Natl Acad Sci USA 73:4570–4574

Brown RM, Willison JHM, Richardson CL (1976) Cellulose biosynthesis in *Acetobacter xylinum*: Visualization of the site of synthesis and direct measurement of the *in vivo* process. Proc Natl Acad Sci USA 73:4565–4569

Burgess J (1979) Cellulose microfibrils – an end in sight? Nature 278:212
Burgess J, Fleming EN (1974) Ultrastructural observations of cell wall regeneration around isolated tobacco protoplasts. J Cell Sci 14:439–449
Carr SGM, Carr DJ (1975) Intercellular pectic strands in parenchyma: Studies of plant cell walls by scanning electron microscopy. Aust J Bot 23:95–105
Cassone A, Mattia E, Boldrini L (1978) Agglutination of blastospores of *Candida albicans* by concanavalin A and its relationship with the distribution of mannan polymers and the ultrastructure of the cell wall. J Gen Microbiol 105:263–273
Chafe SC (1974) Cell wall formation and "protective layer" development in the xylem parenchyma of trembling aspen. Protoplasma 80:335–354
Colvin JR (1977) A new look at cellulose biosynthesis in relation to structure and industrial use. Tappi 60:59–62
Colvin JR, Leppard GG (1973) Fibrillar, modified polygalacturonic acid in, on and between plant cell walls. In: Loewus F (ed) Biogenesis of plant cell wall polysaccharides. Academic Press, New York
Colvin JR, Leppard GG (1977) The biosynthesis of cellulose by *Acetobacter xylinum* and *Acetobacter acetigenus*. Can J Microbiol 23:701–709
Davies DR, Lyall V (1973) The assembly of a highly ordered component of the cell wall: The role of heritable factors and of physical structure. Mol Gen Gent 124:21–34
Esau K (1965) Plant anatomy, 2nd edn. John Wiley and Sons, New York
Frey-Wyssling A (1976) The plant cell wall. Borntraeger, Berlin
Gardner KH, Blackwell J (1974) The structure of native cellulose. Biopolymers 13:1975–2001
Gascoigne JA, Gascoigne MM (1960) Biological degradation of cellulose. Butterworths, London
Gertel ET, Green PB (1977) Cell growth pattern and wall microfibrillar arrangement. Plant Physiol 60:247–254
Grout BWW (1975) Cellulose microfibril deposition at the plasmalemma surface of regenerating tobacco mesophyll protoplasts: A deep-etch study. Planta 123:275–282
Gunning BES, Steer MW (1975) Ultrastructure and the biology of plant cells. Edward Arnold, London
Hanke DE, Northcote DH (1974) Cell wall formation by soybean callus protoplasts. J Cell Sci 14:29–50
Hills GJ (1973) Cell wall assembly *in vitro* from *Chlamydomonas reinhardii*. Planta 115:17–23
Hills GJ, Gurney-Smith M, Roberts K (1973) Structure, composition and morphogenesis of the cell wall of *Chlamydomonas reinhardii*. II. Electron microscopy and optical diffraction analysis. J Ultrastruct Res 43:179–192
Hills GJ, Phillips JM, Gay MR, Roberts K (1975) Self-assembly of a plant cell wall *in vitro*. J Mol Biol 96:431–441
Horine RK, Ruesink AW (1972) Cell wall regeneration around protoplasts isolated from Convolvulus tissue culture. Plant Physiol 50:438–445
Kinnersley AM, Racusen RH, Galston AW (1978) A comparison of regenerated cell walls in tobacco and cereal protoplasts. Planta 139:155–158
Kjosbakken J, Colvin JR (1975) New evidence for an intermediate polymer of glucose in cellulose biosynthesis by *Acetobacter xylinum*. Can J Microbiol 21:111–120
Kopecká M, Phaff JH, Fleet GH (1974) Demonstration of a fibrillar component in the cell wall of the yeast *Saccharomyces cerevisiae* and its chemical nature. J Cell Biol 62:66–76
Leive L (1973) (ed) Bacterial membranes and walls. M Dekker, New York
Leppard GG, Colvin JR (1971) Fine structure of the middle lamella of aggregates of plant cells in suspension culture. J Cell Biol 50:237–246
Leppard GG, Colvin JR (1972) Electron-opaque fibrils and granules in and between the cell walls of higher plants. J Cell Biol 53:695–703
Leppard GG, Colvin JR (1978) Nascent cellulose fibrils in green plants. J Microsc 113:181–184
Leppard GG, Sowden LC, Colvin JR (1975) Nascent stage of cellulose biosynthesis. Science 189:1094–1095

Mackie W, Preston RD (1974) Cell wall and intercellular region polysaccharides In: Stewart WDP (ed) Algal physiology and biochemistry. Bot Monogr 10:40–85
Mark RE (1967) Cell wall mechanics of tracheids. Yale Univ Press, New Haven, London
Marx-Figini M (1971) Investigations on biosynthesis of cellulose: $\overline{DP}_w$ and yield of the alga *Valonia* in the presence of colchicine. Biochim Biophys Acta 237:75–77
Morikawa H, Hayashi R, Senda M (1978) Infra-red analysis of pea stem cell walls and oriented structure of matrix polysaccharides in them. Plant Cell Physiol 19:1151–1159
Mullis RH, Thompson NS, Parham RA (1976) The localization of pentosans within the cell wall of aspen (*Populus tremuloides Michx.*) by high resolution autoradiography. Planta 132:241–248
Ouellette GB (1978) Unusual cell wall layers in elm parenchyma of secondary xylem. Can J Bot 56:2109–2113
Prat R (1973) Contribution à l'étude des protoplastes végétaux II. Ultrastructure du protoplaste isolé et régénération de sa paroi. J Microsc (Paris) 18:56–86
Preston RD (1964) Structural and mechanical aspects of plant cell walls with particular reference to synthesis and growth. In: Zimmermann MH (ed) The formation of wood in forest trees. Academic Press, New York
Preston RD (1974) The physical biology of plant cell walls. Chapman and Hall, London
Reis D, Roland JC (1974) Mise en évidence de l'organisation de parois des cellules végétales en croissance par extractions ménagées des polysaccharides associées a la cytochimie ultrastructurale. J Microsc (Paris) 20:271–284
Roberts K, Gurney-Smith M, Hills GJ (1972) Structure, composition and morphogenesis of the cell wall of *Chlamydomonas reinhardii*. 1 Ultrastructure and preliminary chemical analysis. J Ultrastruct Res 40:599–613
Robinson DG (1977) Plant cell wall synthesis. Adv Bot Res 5:89–150
Roelofsen PA (1959) The plant cell-wall. Borntraeger, Berlin
Roland JC (1967) Sur la sécrétion de la cellulose dans les parois des cellules végétales. CR (Paris) D 264:2757–2760
Ruiz-Herrera J, Bartnicki-García S (1974) Synthesis of cell wall microfibrils *in vitro* by a "soluble" chitin synthetase from *Mucor rouxii*. Science 186:357–359
Ruiz-Herrera J, Sing VO, Van Der Woude WJ, Bartnicki-García S (1975) Microfibril assembly by granules of chitin synthetase. Proc Natl Acad Sci USA 72:2706–2710
Sakurada I, Nukushina Y, Ito T (1962) Experimental determination of the elastic modulus of crystalline regions in oriented polymers. J Polym Sci 57:651–660
Salton MRJ (1964) The bacterial cell wall. Elsevier, Amsterdam
Schilde-Rentschler L (1977) Role of the cell wall in the ability of tobacco protoplasts to form callus. Planta 135:177–181
Schneider EF, Wardrop AB (1979) Ultrastructural studies on the cell walls in *Fusarium sulphureum*. Can J Microbiol 25:75–85
Scholtyseck E (1979) Fine structure of parasitic protozoa. Springer, Berlin Heidelberg New York
Setterfield G, Bayley ST (1958) Arrangement of cellulose microfibrils in walls of elongating parenchyma cells. J Biophys Biochem Cytol 4:377–381
Tracey MV (1957) Chitin. Rev Pure Appl Chem 7:1–14
Willison JHM (1976) An examination of the relationship between freeze-fractured plasmalemma and cell-wall microfibrils. Protoplasma 88:187–200
Willison JHM, Brown RM (1978a) Cell wall structure and deposition in *Glaucocystis*. J Cell Biol 77:103–119
Willison JHM, Brown RM (1978b) A model for the pattern of deposition of microfibrils in the cell wall of *Glaucocystis*. Planta 141:51–58
Willison JHM, Cocking EC (1972) The production of microfibrils at the surface of isolated tomato-fruit protoplasts. Protoplasma 75:397–403
Woodwell GM, Whittaker RH, Reiners WA, Likens GE, Delwiche CC, Botkin DB (1978) The biota and the world carbon budget. Science 199:141–146
Zaar K (1979) Visualization of pores (Export sites) correlated with cellulose production in the envelope of the Gram-negative bacterium *Acetobacter xylinum*. J Cell Biol 80:773–777

3 The Assembly of Polysaccharide Fibrils

D.G. Robinson

Natural cellulose and chitin are microfibrillar in form. Since both are crystalline aggregates of laterally associated homopolymeric chains their synthesis could be regarded as being a two-step process: first the polymerization of the monomer followed by the assembly of the polymer chains to form a microfibril. This possibility, sometimes called the "intermediate high-polymer hypothesis" (Colvin 1964) has recently received some support from work on bacterial cellulose (Colvin 1980 and Colvin, Chap. 2, this Vol.). Although this work might be disputed on technical grounds there are several reasons why a crystallization of preformed polymer chains is untenable for the formation of polysaccharide microfibrils. These are (see also Colvin 1972):

1. when native cellulose is dissolved and allowed to recrystallize from solution, the free polymer chains reassociate as cellulose II rather than cellulose I;
2. twisting of microfibrils around one another, or interweaving of microfibrils from one lamella to another precludes an aggregation of preformed chains;
3. microfibrils have different diameters but in any one cell wall or stage of development they are produced with a constant size and do not show anastomoses;
4. in many cell walls the microfibrils are oriented parallel to one another and also show occasional changes in this orientation during wall development.

It seems inescapable therefore that microfibril formation, in eukaryotic cells at least, is under cellular control and involves more than just the simple production of glycan chains.

The alternative hypothesis ("end synthesis"), whereby polymerization and crystallization are considered to be simultaneous events achieved by a plasma membrane-localized enzyme complex must now, in the light of recent work, be revised. The pertinent experimental observations provide the most significant advance for a number of years in the understanding of this problem of macromolecular synthesis.

Essentially the work has involved the usage of the polysaccharide binding dyes calcofluor white and congo red, and has given rise to more or less the same result for chitin synthesis in *Poterioochromonas* (Herth 1980) and for cellulose synthesis in *Acetobacter* (Haigler et al. 1980, Benziman et al. 1980) and *Oocystis* (Quader 1981, Robinson and Quader 1981). Their addition leads in all cases to an interruption in normal microfibril production. Instead much smaller, fibrillar entities are deposited at the cell surface (cf. Figs. 1a with 1b and c, 2a with 2b, 3a with 3b).

Chemically defined in *Acetobacter* these structures are assumed to be noncrystalline polyglycan chains in the other organisms, too. From the results with

Fig. 2a, b. *Poterioochromonas malhamensis.* **a** Negatively stained chitin microfibrils in the lorica. ×240,000; *Bar*=0.1 µm. **b** Portion of a lorica produced in the presence of 1 $mg\,ml^{-1}$ calcofluor white. ×240,000; *Bar*=0.1 µm. (Micrographs courtesy of W. HERTH)

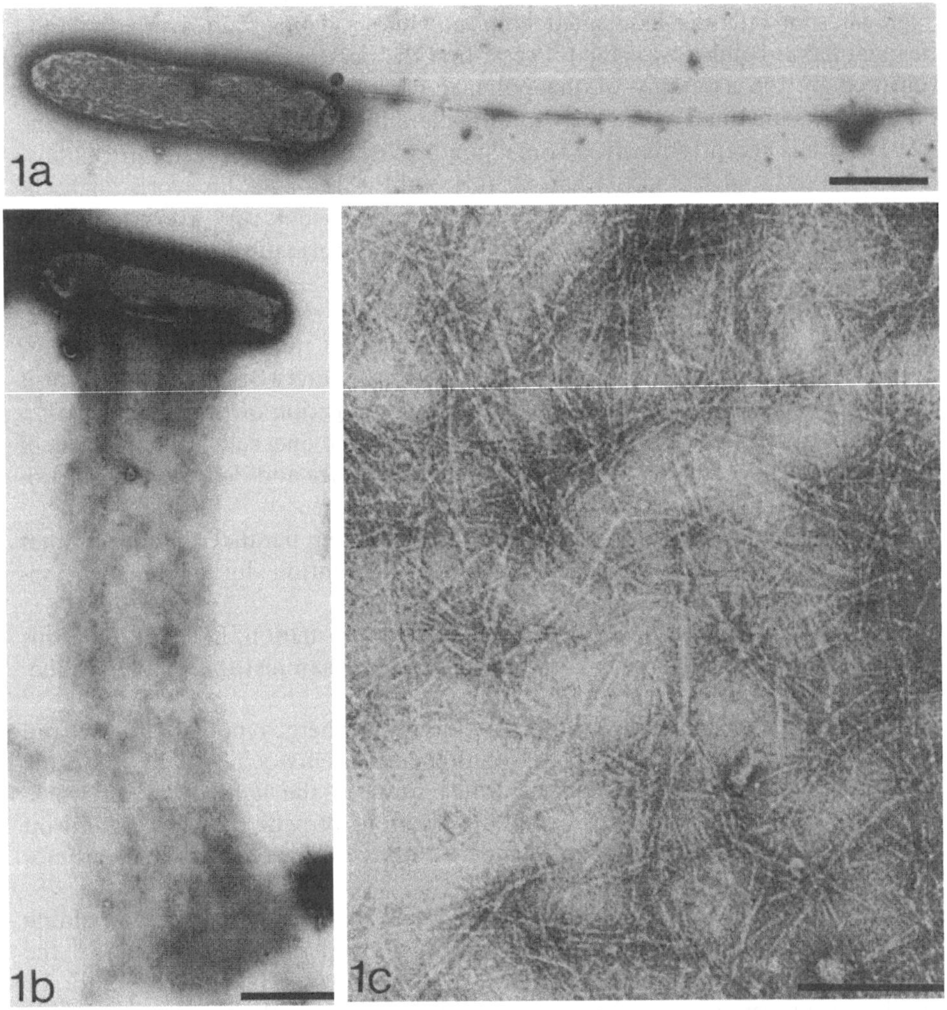

Fig. 1a–c. *Acetobacter xylinum.* **a** Negatively stained bacterium with polar extruded ribbon of twisted cellulose microfibrils. ×12,000; *Bar*=1 µm. **b** Bacterium treated with 0.025% calcofluor white ST for a period of 30 min. In contrast to the normal cell a broad lateral band of fibrous material is produced. ×12,000; *Bar*=1 µm. **c** High resolution micrograph of the fibrillar material produced in the presence of calcofluor white. These structures are on the average less than 2 nm thick. ×178,000; *Bar*=0.1 µm. (Micrographs courtesy of R.M. BROWN, C. HAIGLER and M. BENZIMAN)

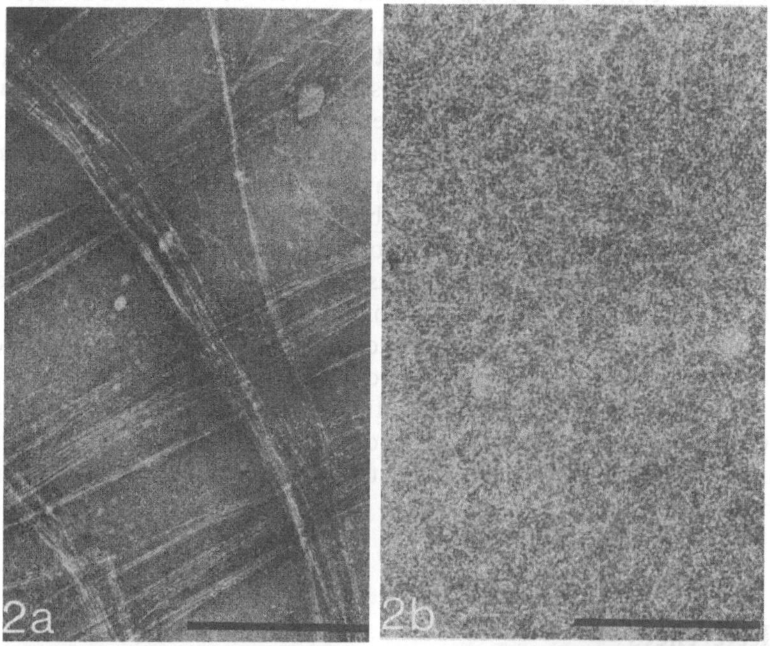

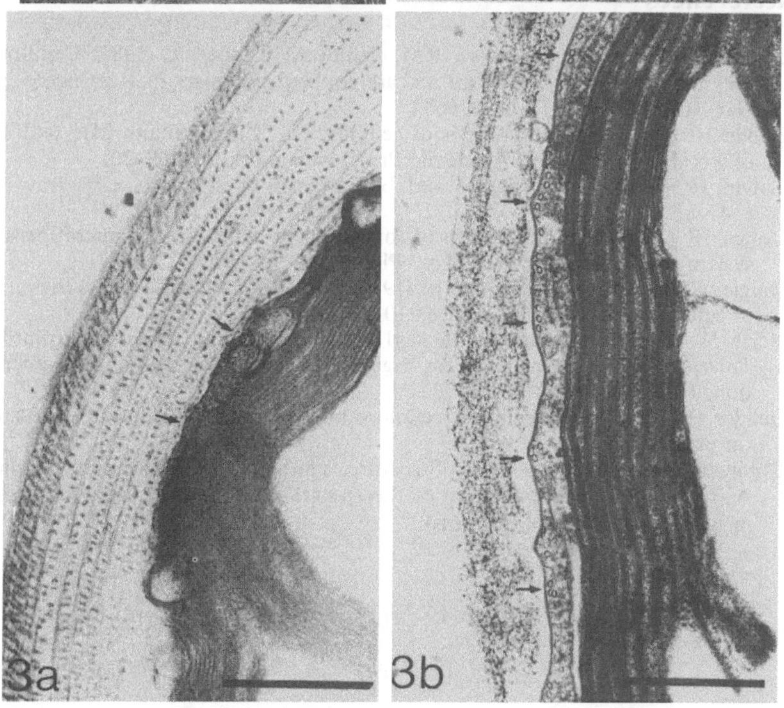

Fig. 3a, b. *Oocystis solitaria*. **a** Section through the cell wall and cell cortex. ×40,000; *Bar*=0.5 μm. **b** Cell wall produced in the presence of 1 mgml^{-1} congo red. Although cortical microtubules are clearly recognizable (*arrows*) the crossed microfibrillar texture of the normal wall is replaced by a different fibrillar network. ×40,000; *Bar*=0.5 μm

Oocystis neither the terminal complexes which are involved in microfibril synthesis (see ROBINSON, Chap. 13, this Vol.), nor the cortical microtubules which control the orientation of this synthesis are removed by these treatments.

One could interpret these observations as being in accordance with the intermediate high polymer hypothesis where the dyes, by binding to polysaccharide chains, prevent their lateral aggregation. The above-mentioned workers, however, all agree that if such a gap between polymerization and crystallization exists it must be extremely short both in temporal and spatial terms, in order to comply with the prerequisites for "end-synthesis". Thus polymerization and crystallization are not simultaneous but rather sequential, coupled events which, under appropriate experimental conditions, can be uncoupled.

It is therefore now possible to postulate that, in addition to the glycan synthetases, the plasma membrane-localized enzyme complexes also possess a "zippering" capability. Whether the creation of interchain hydrogen bonding, as a result of this zippering, is under enzymatic control or is a physical effect, resulting from the specific geometry of the synthetic microregion, remains to be seen.

References

Benziman M, Haigler C, Brown RM, White A, Cooper K (1980) Cellulose biogenesis: polymerization and crystallization are coupled processes in *Acetobacter xylinum*. Proc Nat Acad Sci USA 77:6678–6682

Colvin JR (1964) The biosynthesis of cellulose. In: Zimmermann MH (ed) The formation of wood in forest trees. Academic Press, New York, pp 189–201

Colvin JR (1972) The structure and biosynthesis of cellulose. Crit Rev Macromol Sci 1:47–81

Colvin JR (1980) The mechanism of formation of cellulose-like microfibrils in a cell-free system from *Acetobacter xylinum*. Planta 149:97–107

Haigler C, Brown RM, Benziman M (1980) Calcofluor white ST alters the *in vitro* assembly of cellulose microfibrils. Science 210:903–906

Herth W (1980) Calcofluor white and congo red inhibit chitin microfibril assembly of *Poterioochromonas*: evidence for a gap between polymerization and microfibril formation. J Cell Biol 87:442–449

Quader H (1981) Interruption of cellulose microfibril crystallization. Naturwissenschaften (in press)

Robinson DG, Quader H (1981) Structure, synthesis and orientation of microfibrils. IX. A freeze-fracture investigation of the *Oocystis* plasma membrane after inhibitor treatments. Eur J Cell Biol (in press)

4 Ultrastructure of the Plant Cell Wall: Biochemical Viewpoint

K. Katō

1 Introduction

The cell walls of higher plants are fundamentally involved in many aspects of plant biology including the morphology, growth, and development of plant cells. Primary cell walls are laid down by undifferentiated cells that are still growing, and it is these primary cell walls that control cell growth. Secondary walls are derived from primary cell walls by cells which have stopped growing and are differentiating. During growth of the cell, polymers of the wall interact and change, and the resulting alteration in the properties of the wall can be correlated with a variation in its function. Cellulose, hemicellulose, pectic polysaccharide, structural protein, and lignin have been identified as the major components of the plant cell wall. A knowledge of their detailed chemical structures and their interconnections should provide a basis for the understanding of both biological function and metabolic interrelationships. Recent work shows that the rather crude description of the primary cell wall as consisting of a network of cellulose microfibrils embedded in an amorphous matrix of pectin and hemicellulose (Northcote 1977, Albersheim 1978, Lamport 1978, Preston 1979, McNeil et al. 1979) is not satisfying.

This chapter will be concerned with recent studies on the chemical structure and interrelationships of the very diverse types of branched and linear polysaccharides that make up plant cell walls.

2 Microfibrillar Component

Crystalline cellulose fibers make up an important part of the framework of the cell walls of all higher plants. The structure of cellulose is characterized by long chains of (1 → 4)-β-linked glucosyl residues. The glucans exist as extended chains with a twofold screw axis, and these are arranged in an ordered manner within the microfibril (Roelofsen 1965, Shafizadeh and McGinnis 1971, Northcote 1972, Preston 1979). The linear glucan molecules of cellulose are bound together by hydrogen bonds. This binding between approximately 40 glucan chains results in 3.5 nm diameter threadlike bodies termed elementary fibrils. These very elongated elementary fibrils are aggregated into 10 to 30 nm diameter ropelike structures, called microfibrils (Frey-Wyssling 1969).

Polymeric material containing sugars other than glucose is always found to be associated with cellulose when it is isolated from the wall. These noncellulosic polysaccharides may be adsorbed onto the surface of the microfibril, or they may be part of individual heteroglucan chains, or they may exist as separate chains intermingled with the outer glucan chains. In any case, because of the crystalline nature of the glucan chains in the microfibrils the noncellulosic polysaccharide chain must be situated at or near the surface, and since they have a structure similar to or even identical with some of the oriented linear polysaccharides of the matrix material, they provide a key material for the entanglement of the microfibrils with the matrix (NORTHCOTE 1972, PRESTON 1979).

The formation of a new cell wall can be investigated as it develops around isolated protoplasts produced from intact cells (NAGATA and TAKEBE 1971, BRIGHT and NORTHCOTE 1974, BURGESS and FLEMING 1974, SHEPARD and TOTTEN 1975, WILLISON and COCKING 1976, FOWKE 1978). During the initial period of cell wall regeneration of soybean protoplasts in liquid medium, pectic polysaccharides are secreted into the media (HANKE and NORTHCOTE 1974), during which time the protoplasts are still sensitive to osmotic shock and have no rigid wall. Glucans are found that can be seen in the electronmicroscope to be microfibrils (FOWKE 1978). The meshwork of microfibrils that develops seems to be able to entrap the continued production of matrix material by the protoplast, and the cells develop a wall that converts the protoplast into a normal cultured cell (HANKE and NORTHCOTE 1974). NORTHCOTE (1977) suggests that the sequence of wall formation is that of synthesis and transport of matrix polysaccharides (pectins and possible hemicelluloses) first followed by cellulose synthesis, laid down as microfibrils at the cell surface and woven into the matrix as in the case of wall formation at telophase.

During the initial stages of cell growth the matrix of the wall is not rigid and the microfibrils may be grouped in bands within matrix. They are oriented at a fairly large angle with respect to the long axis of the cell, and form a loosely interwoven network although some regions exhibit parallel arrangement. The secondary wall is formed by microfibrils that are more closely packed, lie more parallel to one another, and are oriented with a smaller angle to the long axis of the cell (SHAFIZADEH and MCGINNIS 1971).

There are two distinct kinetic stages, corresponding to the formation of primary and secondary walls during the development of cotton fiber. The first stage proceeds very slowly and yields a small amount of "primary" cellulose having a nonuniform degree of polymerization ($\overline{DP}_w$) ranging from 2000 to 6000. The second stage proceeds much faster and provides a large amount of "secondary" cellulose having a $\overline{DP}_w$ of 14,000. During the second stage, the $\overline{DP}_w$ is independent of variations in the kinetics or the rate of synthesis of cellulose (MARX-FIGINI 1966, MARX-FIGINI and SCHULZ 1966).

3 Matrix Noncellulosic Polysaccharides

The matrix noncellulosic polysaccharides have been defined operationally by their presence in fractions obtained by consecutive chemical extraction of isolated cell walls. The two major noncellulosic fractions are the pectic polysaccharides obtained by extracting cell walls with boiling water, EDTA, or dilute acid and the hemicelluloses solubilized by the subsequent extraction of the same wall with alkali. These polymers have been extensively studied, and have been the subject of recent reviews (TIMELL 1964, 1965, WORTH 1967, ASPINALL 1970, 1973, WHISTLER and RICHARDS 1970). Recently, McNEIL et al. (1979) have classified the pectic polysaccharides as those polymers found in covalent association with galacturonosyl-containing polysaccharides. The hemicelluloses are those polysaccharides noncovalently associated with cellulose.

3.1 Hemicelluloses

3.1.1 Xylans

The xylans are laid down throughout the growth of the wall and form the bulk of the hemicellulosic fraction of the angiosperms. The general structure of the xylans in higher plants is that of a main chain of D-xylopyranosyl residues joined by $(1 \rightarrow 4)$-β-links. Attached to the backbone are short side chains. In angiosperms these are 4-*O*-methyl-D-glucopyranosyluronic acid residues which are attached to C-2 of the xylosyl residues of the main chain by α-links, and they are probably distributed along the chain. A xylan, which is a linear unbranched chain of $(1 \rightarrow 4)$-β-linked xylosyl residue, has been isolated from the stalk of tobacco (EDA et al. 1976).

In vivo, about half of the xylosyl residues in the polysaccharides are acetylated. The bulk of the acetylation occurs at C-3 although there are some at C-2, and certain xylosyl residues are acetylated at both C-2 and C-3 (TIMELL 1964). The acetyl groups are numerous enough to prevent alignment of the molecular chains, and molecular aggregation can not take place. The presence of acetyl groups must influence association of these chains with each other and with other polysaccharide complexes within the wall structure (NORTHCOTE 1972, PRESTON 1979).

Side chains of glucuronoarabinoxylans from monocots include L-arabinofuranosyl residues attached to xylosyl residues by $(1 \rightarrow 3)$-α-links in addition to $(1 \rightarrow 2)$-α-linked 4-*O*-methyl-D-glucopyranosyluronic acid residues (ASPINALL 1959, NORTHCOTE 1972, WILKIE 1979).

It has been recently established that a glucuronoarabinoxylan constitutes 5% of the primary walls of suspension-cultured sycamore cells (McNEIL et al. 1979).

An acidic galactoarabinoxylan is also present in the cell walls of Gramineae (WILKIE 1979) such as rye grass (MORRISON 1974), oat plants (REID and WILKIE 1969, BUCHALA et al. 1972), corn leaf and stalk (DUTTON and KABIR 1972) and barley leaf (BUCHALA 1973).

3.1.2 Mixed β-Glucans

Hemicellulosic $(1 \rightarrow 3)$- and $(1 \rightarrow 4)$-β-glucans have been isolated from stem and leaf of oat plants (BUCHALA and WILKIE 1971, FRASER and WILKIE 1971a, b), bamboo (WILKIE and WOO 1976), the stem of barley (MARSHALL 1975), rye endosperm cells (Smith and Stone 1973), wheat endosperm (Mares and Stone 1973a, b), rice endosperm (Shibuya and Misaki 1978), and oat coleoptiles (Wada and Ray 1978, Labavitch and Ray 1978).

The glucan from oat coleoptile has (1→3)- and (1→4)-β-linkages in the ratio of 1:2 (Labavitch and Ray 1978). Apparent molecular weight of the β-glucan from oat coleoptile is 2.0×10^5 (Wada and Ray 1978). The enzymic degradation of this glucan indicates a fairly regular repeating structure composed of (1→3)-β-linked glucosyl alternating with two (1→4)-β-linked glucosyl residues (Labavitch and Ray 1978). There is a decrease in the ratio of (1→3)- to (1→4)-β-D-glucosidic linkages in the β-glucans isolated from stem, leaf, and hull tissues of increasing maturity (Wilkie 1979).

The glucan is a prominent structural wall component of monocots. It is probably the glucan components that have been observed to decrease under auxin treatment, a breakdown that is thought to be involved in cell wall expansion during growth (Loescher and Nevins 1972).

The mixed β-glucan of $\overline{DP}_n$ ca. 80 has been isolated from 2-day-old hypocotyls of *Phaseolus* (Buchala and Franz 1974), but has not been found in older cell walls (Franz 1972). The composition of the cell wall of the cotton fiber changes during development, the most prominent change being a large increase in noncellulosic glucan which occurs just prior to the onset of secondary wall cellulose deposition (Meinert and Delmer 1977). The mixed glucan is also indicated to be present in cultured *Vinca rosea* cell walls as a minor component (Takeuchi and Komamine 1978).

3.1.3 Xyloglucans

Cultured sycamore cell walls contain a xyloglucan polysaccharide that is essentially identical to that isolated from extracellular polysaccharides which are secreted into the culture medium. The xyloglucan binds tightly to purified cellulose by hydrogen bonds (Aspinall et al. 1969, Albersheim 1978).

The glucosyl residues of sycamore xyloglucan are present in the cellulose-like, (1→4)-β-linked glucan backbone, to which xylosyl residues are attached at C-6. The structure of the polymer is based on a repeating heptasaccharide unit which consists of four residues of 4-β-linked glucose and three residues of terminal xylose. A single xylosyl residue is α-linked to C-6 of three glucosyl residues. Some *O*-β-D-galactosyl-(1→2)-α-D-xylosyl side chains are also present. This feature of xyloglucan structure may reduce further lateral associations, giving a monolayer of xyloglucan on the surface of cellulose fibril (Albersheim 1978, McNeil et al. 1979).

The xyloglucan is also isolated from cell walls of etiolated bean (Katō et al. 1977). Endoglucanase treatment of sycamore xyloglucan liberates, in addition to heptasaccharides, a small amount of pentasaccharide which consists of three residues of 4-β-linked glucose and two residues of terminal xylose (Albersheim 1978). The pentasaccharide is obtained as a major structural unit of xyloglucan isolated from mature tobacco leaf midribs, and some of the *O*-α-L-arabinofuranosyl-(1→2)-α-D-xylopyranosyl side chains are present in tobacco arabinoxyloglucan (Eda and Katō 1978a, Mori et al. 1980).

A small amount of xyloglucan is detected in oat coleoptile cell walls (LABAVITCH and RAY 1978) and rice endosperm cell walls (SHIBUYA and MISAKI 1978).

3.1.4 Glucomannans

The glucomannans are common in higher plants and are predominant in hemicellulose of the secondary walls of gymnosperms (TIMELL 1965). Pine glucomannan is essentially a linear polysaccharide composed of both (1 → 4)-β-linked D-mannosyl and D-glucosyl residues. The polysaccharide contains some terminal D-galactosyl residues α-linked to C-6 of the hexosyl residues of the main chan. The mannosyl: glucosyl ratio in the polysaccharide is about 3:1, and the residues are randomly distributed in the chain. About half of the D-mannosyl residues are substituted with *O*-acetyl groups distributed between C-2 and C-3 (LINDBERG et al. 1973).

Glucomannan occurs in very small amounts in the hemicellulose of angiosperms, usually accounting for 3% to 5% of the total cell wall material (TIMELL 1964). The presence of 4-linked mannosyl residues is observed by methylation analyses of the isolated primary cell wall fractions of bean hypocotyl and cultured cells of red kidney bean, soybean and tomato (ALBERSHEIM 1976), and pea epicotyl (GILKES and HALL 1977). This may suggest that these cell walls contain mannan(s) or perhaps a glucomannan.

A small amount of glucomannan is also present in the hemicellulosic fraction of oat coleoptile cell wall (WADA and RAY 1978, LABAVITCH and RAY 1978).

Glucomannan chains have a very similar conformation to that of cellulose. They exist as extended chains with a two-fold screw axis, and are strongly adsorbed at the microfibrillar surface (NORTHCOTE 1972, REES 1972, PRESTON 1979).

3.2 Pectic Polysaccharides

3.2.1 Rhamnogalacturonan

The acidic polysaccharides have a basic structure which consists of a chain of 4-α-linked D-galactopyranosyluronic acid residues in which 2-linked L-rhamnosyl residues are interspersed. A variable number of carboxyl groups are esterified as methyl esters. The rhamnosyl residues are not randomly distributed in the chain, but probably occur as rhamnosyl-(1 → 4)-galactopyranosyluronic acid-(1 → 2)-rhamnosyl units (ASPINALL 1973, MCNEIL et al. 1979). The methylation analyses indicate that approximately 50% of the rhamnosyl residues are branched, having a substituent at C-4 as well as C-2. No aldobiouronic acid with a galactopyranosyluronic acid residue attached to C-4 of a rhamnosyl residue has been isolated. As this is the major, if not the only, branch point of the rhamnogalacturonan chain, this 2,4-linked rhamnosyl residue represents the point of the attachment of at least some of the neutral side chains (MCNEIL et al. 1979).

The rhamnogalacturonan obtained from the walls of suspension-cultured sycamore cells contains, in addition to rhamnosyl and galacturonosyl residues, substantial amounts of arabinosyl and galactosyl residues. The ratio of rhamnosyl to galacturonosyl to arabinosyl to galactosyl residues is 1:1:1.5:1.5. A degree of polymerization is about 10,000 from the results of gel filtration. No information is available on whether the rhamnosidic bonds are in the α or β configuration (McNeil et al. 1979).

Some 4-α-linked homogalacturonans have been isolated from sunflower seeds (Zitko and Bishop 1966) as well as from apple pectin (Barrett and Northcote 1965). A similar galacturonan has been isolated from suspension-cultured sycamore cell walls by treatment with endopolygalacturonase (McNeil et al. 1979).

Apiose-containing rhamnogalacturonan constitutes 3% to 4% of the primary cell walls of suspension-cultured sycamore cells. This galacturonan is a very complex branched polysaccharide yielding, upon hydrolysis, ten different monosaccharides including the rarely observed sugars apiose, 2-*O*-methylxylose, and 2-*O*-methylfucose. In addition, the galacturonan is characterized by the rarely observed glycosyl interconnections of 2-linked galactopyranosyluronic acid, 3,4-linked fucosyl, and 3-linked rhamnosyl residues (Darvill et al. 1978). A component of the cell walls of duckweed has been isolated and identified as an apiogalacturonan, with apiosyl and galacturonosyl residues as the only components (Duff 1965, Beck 1967, Hart and Kindel 1970).

3.2.2 Arabinogalactans

Arabinogalactans are composed of interior chains of $(1\rightarrow3)$-β-linked D-galactopyranosyl residues to which are attached side chains. The side chains are $(1\rightarrow6)$-β-linked D-galacto-oligosaccharides which may carry L-arabinofuranosyl residues linked $(1\rightarrow3)$ or $(1\rightarrow6)$ to the galactosyl residues. Arabinosyl residues are also found attached in a similar manner to galactosyl residues of the main chain. Arabinosyl residues are sometimes present as a disaccharide composed of *O*-β-L-arabinopyranosyl-$(1\rightarrow3)$- and/or $(1\rightarrow5)$-L-arabinofuranosyl residues. Arabinogalactans have been isolated from the tissues and from the extracellular polysaccharides of a variety of plants (Timell 1965, Clarke et al. 1979). However, no arabinogalactan has been isolated from a source known to contain only primary cell walls (McNeil et al. 1979).

3.2.3 Arabinans and Galactans

While pectic arabinogalactan can be separated as a single polymer during electrophoresis of pectic fractions (Barrett and Northcote 1965), the neutral material might contain separate arabinans and galactans. From some plant tissues, degraded material has been used to prepare either an arabinan or a galactan (Hirst and Jones 1974a, b, Eda and Katō 1978b). Works on these degraded materials indicate that L-arabinosyl residues form highly branched polysaccharide by $(1\rightarrow3)$- and $(1\rightarrow5)$-α-linkages, whereas galactosyl residues form a main chain of $(1\rightarrow4)$-β-galactan (Aspinall 1973, Siddiqui and Wood 1974).

Nuclear magnetic resonance analysis of *Rosa glauca* arabinan has provided additional evidence that arabinosyl residues are (1 → 5)-α-linked and in the furanose configuration (JOSELEAU et al. 1977). Arabinans from the bark of *Rosa glauca* have a $\overline{DP}$ of 34 and 100, while an arabinan from willow has a $\overline{DP}$ of 90 (KARÁCSONYI 1975).

A number of complex pectic polysaccharides have been demonstrated to contain arabinosyl residues (NORTHCOTE 1972, ASPINALL 1973, MCNEIL et al. 1979). One investigation, using methylation analysis, has provided evidence that the arabinosyl residues of rapeseed pectic polysaccharides are present as mono- or di-saccharide side chains (ASPINALL and JIANG 1974). On the other hand, glycosyl-linkage analyses of the pectic polysaccharides of primary cell walls suggest the presence of homo-arabinans (GILKES and HALL 1977, ALBERSHEIM 1978).

Galactans have been obtained which contain 6-linked in addition to 4-linked galactosyl residues (TOMAN et al. 1972).

The presence of 4-galactans in the primary cell wall is inferred by the detection of large amounts of 4-linked galactosyl residues upon methylation analysis of total cell walls and of pectic fractions (ALBERSHEIM 1976, GILKES and HALL 1977, RING and SELVENDRAN 1978).

3.3 Glycoprotein of the Walls

The primary cell walls of a wide variety of dicot plants contain a unique hydroxyproline-rich glycoprotein (LAMPORT 1965). Up to 20% of the amino acid residues of this protein are hydroxyproline, and the protein itself accounts for 2% to 10% of plant cell walls (LAMPORT 1965). The amino acid sequence of the wall protein is known only in part. Removal of arabinosyl residues by treatment of tomato cell walls at pH 1 renders the protein tryptic labile. All the tryptic peptides contain at least one unit of the pentapeptide -Ser-Hyp-Hyp-Hyp-Hyp- (LAMPORT 1973). Generally most of the hydroxyproline residues of the protein are glycosylated by a tri- or tetrasaccharide of arabinofuranose. The sequence of tetraarabinoside is Ara*f*(1 → 3)Ara*f*(1 → 2)Ara*f*(1 → 2)Ara*f*-Hyp (LAMPORT 1977, AKIYAMA and KATŌ 1977). Most of the serine residues are galactosylated (LAMPORT et al. 1973).

The covalent attachment of arabinosyl and galactosyl residues to the hydroxyproline-rich proteins is a generally accepted fact. However, the available evidence suggests that the protein is not covalently attached to any of the other cell wall polymers. The evidence does not rule out the possible existence of strong, noncovalent bonding between the protein and the other wall polymers (LAMPORT 1978).

Soluble macromolecules secreted into the medium are thought to be arabinogalactan-glycoproteins (POPE and LAMPORT 1977, HORI and SATO 1977, CLARKE et al. 1979). Some of these glycoproteins show remarkable differences from the wall protein. POPE and LAMPORT (1977) found that 80 mol% of the hydroxyproline arabinosides isolated from the cell wall protein can be accounted for by the tetraarabinoside, while only 4 mol% of the arabinosides obtained from the culture medium glycoprotein can be accounted for by the tetraarabinoside. Perhaps, more importantly, they fail to find any arabinogalactan associated with the cell wall protein, while fully 50% of the glycoprotein of the extracellular macromolecules is accounted for by arabinogalactan.

Two lectin-like protein fractions have been extracted from the cell walls of kidney bean seedlings (KAUSS and GLASER 1974, KAUSS and BOWLES 1976). These lectin-containing fractions bind specifically to galactosyl residues. It has not yet been ascertained whether these lectin-like proteins contain hydroxyproline. The existence of these galactosyl-binding proteins in the wall has led to the suggestion that lectins may be involved in establishing a noncovalent protein-glycan network (KAUSS and BOWLES 1976).

3.4 Lignin

Among noncarbohydrate components of cell walls lignin stands out as the unique aromatic polymer. The products of hydrolysis with dioxan and water, and the catalytic hydrogenolysis of native lignins have given further support to the concept that lignin is a polymeric product arising from an enzyme-initiated dehydrogenative polymerization of cinnamic alcohols (FREUDENBERG and NEISH 1968, SAKAKIBARA 1977). The lignin penetrates the wall from the outside (primary wall) inward at very early stages of secondary thickening (NORTHCOTE 1972). This hydrophobic material replaces the water. As water is displaced, strong hydrogen bonds develop between the polysaccharides both at the microfibrillar matrix interface and between the components of the matrix. In addition, covalent bonds may form between carbohydrates and lignin (HARTLEY 1973, MORRISON 1973, 1974, YAKU et al. 1979). This ensures that the layers and components of the secondary wall will not slip with respect to one another (NORTHCOTE 1972).

4 Polysaccharidic Association Within the Primary Cell Wall

The structures of many of the components of the wall are now documented and the general architecture to which they contribute is known in outline. In order to understand the functional role(s) of the wall, information regarding the relative dispositions of these wall molecules, and the nature and distribution of the covalent and noncovalent bonds among them is needed.

4.1 Dicot Primary Cell Walls

A structural model of the wall of suspension-cultured sycamore cell is presented by ALBERSHEIM (1978) according to the results of methylation analyses of both the isolated cell wall and the defined fragments which are released by the action of specific polysaccharide-degrading enzymes. The preliminary model involves all but one of the wall polysaccharides in a covalently linked network, the exception being xyloglucan hydrogen-bonded to the cellulose microfibrils. The model is further modified by McNeil et al. (1979).

Polymer composition of the walls of suspension-cultured sycamore cells is 34% pectic polysaccharides (consisting of 7% rhamnogalacturonan, 6% homogalacturonan, 9% arabinan, 9% galactan and possible arabinan-galactan, and 3% apiose-containing rhamnogalacturonan), 24% hemicelluloses (consisting of 19% xyloglucan and 5% glucuronoarabinoxylan), 23% cellulose, and 19% hydroxyproline-rich glycoprotein (McNeil et al. 1979).

The neutral pectic polysaccharides, the arabinan and galactan, are covalently attached to the acidic polysaccharide, rhamnogalacturonan. Evidence for the interconnections of these polymers is provided by their permeation- and ion-exchange chromatography and by the β-elimination of the uronosyl residues, which results in a reduction in the molecular size of both the arabinan and galactan (McNeil et al. 1979).

Homogalacturonan has always been assumed to be attached to rhamnogalacturonans. This has been demonstrated by the isolation from sycamore cell walls of oligogalacturonides containing ten or more galacturonosyl residues in which the reducing ends of these oligogalacturonides are covalently attached to single rhamnose residues (McNeil et al. 1979). Barrett and Northcote (1965) have obtained evidence for the interconnection of these polymers in studies of a pectic fraction rich in both neutral sugars and galacturonic acid which migrated as a single acidic component upon electrophoresis. However, a region of neutral sugar-rich pectic polymer can be separated by electrophoresis from a region rich in galacturonic acid after the glycosidic bonds have been cleaved by β-elimination.

Calcium has long been demonstrated to confer rigidity to cell walls. The "egg-box" of Rees and his colleagues (Rees 1972, Grant et al. 1973, Preston 1979) is an attractive model for the manner in which calcium strengthens cell walls. The calcium chelate results not only in increased rigidity of the polyuronans, but also in cross-linking of the chains. The degree of interchain cross-linking due to the presence of calcium ions will be sensitive to the degree of methyl esterification of the galacturonans. Efforts have been made to correlate the degree of calcium cross-linking of galacturonans to the rate of cell wall elongation, but no such correlation has been established (Stoddart et al. 1967).

The possibility of other types of noncovalent interactions between the pectic polysaccharides and other cell wall polymers must be certainly considered (Gould et al. 1965, Dea et al. 1977). The pectic polysaccharides probably interact through noncovalent chemical bonding as well as through covalent bonding. Indeed, noncovalent interactions may provide the most important interconnections between the pectic polysaccharides and the other cell wall polymers (Preston 1979).

The bonding of xyloglucan to cellulose through multiple hydrogen bonds is clearly one of the major interconnections of the cell wall polymers. This bonding may prevent the cellulose fibers from adhering to each other to form enormous aggregates (Albersheim 1978). Xyloglucan chains also offer the possibility of interconnecting the cellulose fibers to other polymers of the primary cell walls. It seems likely that glucuronoarabinoxylans, and perhaps xylogucans, not only bind to cellulose in the cell wall, but also bind themselves. This possibility is increased by the observation that plant arabinoxylans form aggregates in solution (Blake and Richards 1971, McNeil et al. 1979).

Evidence has been provided for a covalent linkage between the xyloglucan chains and the pectic polysaccharides (Albersheim 1978). The attempt to isolate large amounts of solubilized xyloglucan covalently attached to the pectic polysaccharides has been unsuccessful (McNeil et al. 1979), although small amounts of xyloglucan, which has apparent covalent linkage with the pectic polysaccharides, have been isolated (Monro et al. 1976b; Albersheim 1978). The extent of the attachment of xyloglucan to the pectic polysaccharides and the structural importance of this interconnection remain open for further investigation (McNeil et al. 1979).

The position of wall protein is the least characterized link. There is no direct evidence which demonstrates a covalent linkage between the hydroxyproline-rich glycoprotein and polysaccharides of cell walls. Lamport and his colleagues (Mort and Lamport 1977, Lamport 1978) have used anhydrous hydrogen fluoride to deglycosylate glycoproteins. After this treatment of sycamore and tomato cell walls, an insoluble residue which appears microscopically as very thin walls remains, consisting of roughly an equal amount of protein and unidentified material. Thus, carbohydrate removal does not lead to wall protein solubilization. They conclude that the protein is cross-linked through noncovalent interactions to the polysaccharides of the cell wall. Results obtained by alkaline extraction of cell walls (Monro et al. 1976a, Selvendran 1975) also support a lack of covalent linkage between cell wall protein and polysaccharides. It may be that the hydroxyproline-rich glycoprotein or other proteins within the wall, acting as lectins, participate in the cross-linking of the cell wall polymers (Kauss and Bowles 1976).

4.2 Monocot Primary Cell Walls

The hemicellulosic matrix of oat coleoptile cell walls consists of two kinds of polysaccharides, mixed β-glucans and glucuronoarabinoxylans (Wada and Ray 1978, Labavitch and Ray 1978). A minor number of xyloglucan-like polysaccharide chains seem to be present in the oat wall, but the results agree with evidence gathered from other monocot wall substance indicating that xyloglucan is not a prominent component (Albersheim 1976). Polygalacturonan which contains covalently linked rhamnosyl residue is also a minor component of oat cell walls as compared with primary walls of dicots (Wada and Ray 1978).

Evidence obtained by Wada and Ray (1978) shows that the interconnections of the matrix polysaccharides must be labile to relatively gentle extractive treatments such as ammonium oxalate and alkali that do not degrade the major polysaccharides into small fragments, and to which most ordinary glycosidic bonds are stable. This suggests that the different kinds of matrix polysaccharides are not attached one to another as side chains by glycosidic bonds. Alkali-labile bonds such as ester linkage could, however, conceivably cross-link some of the polysaccharides: such linkages could involve either uronic acids or some other organic acid (MARES and STONE 1973b, MARKWALDER and NEUKOM 1976, WHITMORE 1976). The tendency of hemicellulosic components to associate with one another to form gels and water-insoluble complexes suggests that the noncovalent bonding of these matrix polysaccharrides may suffice to create a stable structural material without need of covalent cross-linking (WADA and RAY 1978).

An alkali-soluble hemicellulosic polymer, consisting of xyloglucan, mixed β-glucan, and arabinoxylan moieties, is isolated from rice endosperm cell walls and behaves as a homogeneous polymer in sedimentation analysis and electrophoresis (SHIBUYA and MISAKI 1978). The fact strongly suggests that the three polysaccharide moieties are firmly associated with each other and form a homogeneous architecture of hemicellulose. However, the problems whether these polysaccharide moieties are linked or whether they associate each other by hydrogen bonding or molecular aggregation (BLAKE and RICHARDS 1971, DEA et al. 1973) remain for further investigation (SHIBUYA and MISAKI 1978).

5 Chemical Changes in the Cell Wall During Growth and Differentiation

In many different tissues, the carbohydrate composition of the main components of cell walls changes during development and differentiation of the constituent tissues (NORTHCOTE 1969, 1972, 1974, 1977). During differentiation from a cambial cell to xylem, THORNBER and NORTHCOTE (1961a,b, 1962) have shown that there is a net synthesis of carbohydrate constituents such as glucosyl, xylosyl, and uronosyl residues in several tree species. The increase of uronosyl residue corresponds to that of hemicellulosic polymer consisting of glucuronoxylan. By inducing differentiation in solid callus cultures of bean, JEFFS and NORTHCOTE (1966) have shown that cytological characteristics of differentiation parallel chemical characteristics, and in particular that the ratio of xylose to arabinose in the cell wall increases.

Metabolic relationships of the isolated fractions of actively growing sycamore cells have been demonstrated by pulse labeling with radioactive sugars (STODDART and NORTHCOTE 1967). Glucose can serve as a source of both acidic and neutral precursors, which give rise to a partially methylated galacturonan and to the neutral arabinan-galactan. Arabinose can act only as a neutral precursor and hence arabinan-galactan source. The partially methylated galacturonan can give rise to strongly acidic pectinic acids by receiving neutral blocks of fragments from the arabinan-galactan by transglycosylation. The type of pectin synthesized can also be influenced by application of growth hormone and plasmolysis (RUBERY and NORTHCOTE 1970, BOFFEY and NORTHCOTE 1975). The enzymes which act on synthesis and transglycosylation of the individual polysaccharides of the pectin are important. These enzymic reactions involve control of the texture of the wall and its development (Dalessandro and Northcote 1977a,b,c, Northcote 1977).

It has been shown that pectin in oat coleoptile is synthesized as a methyl galacturonan and that glucose and inositol will serve as uronide precursors (Albersheim 1963, Loewus et al. 1973). The study of pectin in sections of onion root tip has shown the presence of a highly esterified polyuronan in the cell plate and middle lamella, and a certain amount of similar material throughout the primary wall (Albersheim and Killias 1963). Formation of the cell plate at telophase takes place by fusion of vesicles brought up between microtubles to the central region of the mitotic spindle. The vesicles are probably derived from the Golgi bodies. The Golgi body also manufactures and transports material to the cell wall during the secondary thickening of differentiated tissues such as xylem (Wooding and Northcote 1964) and phloem (Northcote and Wooding 1966). Different polysaccharides are used during the formation of the primary and secondary wall and the pectic polysaccharides change during the initial phase of cell growth. Thus, the metabolic function of the Golgi bodies changes continuously during the growth and differentiation of the cell (Northcote 1974, 1977).

Turnover of cell wall polysaccharides of pea stem segment and epicotyl has been followed by gravimetric study and by pulse chase study (Labavitch and Ray 1974a,b, Gilkes and Hall 1977). It is suggested that an important effect of IAA is the promotion of turnover in certain hemicelluloses, notably xyloglucan, moreover, such turnover appears to be inhibited by cytokinin. Substantial loss of galactosyl residues, presumably as galactan, is also noted during growth although the loss is not auxin-dependent. Such losses may present minimal changes in the cell wall which must occur prior to a critical wall-loosening event.

Changes in composition of the cell wall of cotton fiber during development have been studied from the early stages of elongation, 5 DPA (day post anthesis), through the period of secondary wall formation (Meinert and Delmer 1977). The kinetics of the cell wall is relatively constant until about 12 DPA, after this time it markedly increases until secondary wall cellulose deposition is completed. Between 12 and 16 DPA, all components contribute to total wall increase. The deposition of secondary wall cellulose begins at about 16 DPA and continues until about 32 DPA. At the time of onset of secondary wall cellulose deposition, a sharp decline in protein and uronosyl residues occurs. A large increase in noncellulosic 3-linked glucosyl residue occurs just prior to onset of secondary wall cellulose deposition.

On the other hand, it has been shown that a specific decrease in noncellulosic glucosyl residue, which is accompanied by a rough compensating increase in cellulose, occurs in the cell wall of oat coleoptiles elongating in response to auxin in the absence of an external carbon source (Loescher and Nevins 1972) and it has been suggested that changes in noncellulosic wall glucan may be involved in wall loosening in monocots (Nevins 1975).

For the stabilization of the wall from further extension, Lamport (1978) suggests that peroxidases may slow down or limit cell expansion by: (1) oxidizing free auxin (Palmieri et al. 1978); (2) converting ferulic to diferulic acid which can then act as a hemicellulosic crosslink (WHITMORE 1976, MARKWALDER and NEUKOM 1976); (3) generating hydrogen peroxide (via NADH-linked cell wall malate dehydrogenase) for the oxidation of cinnamyl alcohols to their free radical lignin precursors (GROSS et al. 1977); and (4) perhaps using phenolics to crosslink wall protein.

6 Conclusion

The primary cell wall is an amalgam of a dynamic equilibrium of component polysaccharides during growth and differentiation. A network consisting of discrete components seems to be linked together by covalent and noncovalent bonds. It is very important for a more detailed description of the primary cell wall to determine directly not only chemical structure of each of the wall polysaccharides, but also how these polymers are attached to one another or how they are interrelated (MCNEIL et al. 1979).

An important and sometimes overlooked problem is the purification of the cell walls. In order to avoid contamination with intracellular constituents, an improved method for preparing appreciably quantities of cell wall material from potatoes has been developed (RING and SELVENDRAN 1978). The method depends for its success on the selective removal of the contaminants from fresh ball-milled tissue by sequential treatment with 1% aqueous sodium deoxycholate, phenol-acetic acid-water followed by dimethylsulfoxide.

Structural analyses of discrete polysaccharides have been facilitated in reduced sample size by recent advances of analytical methods such as capillary GC-MS and FD-MS spectrometry, ^{1}H- and ^{13}C-NMR, and high pressure liquid chromatography in the field of carbohydrate chemistry. However, the major difficulty in any analysis of cell wall structure is that of finding suitable methods for the isolation of a well-defined interconnecting fragment from the wall. The highly specific hydrolytic enzymes are useful tools for wall disassembly, but they are not so readily available. The increased use of affinity chromatography should help to purify enzymes (MCNEIL et al. 1979).

Alternatively, there are several methods for disassembly of cell wall by direct chemical attack of various linkages of the wall polysaccharides. These include the use of sodium methoxide (WHITMORE 1976, MARKWALDER and NEUKOM 1976, MORRISON 1977), DMSO-$NaBH_4$ cocktail (LAMPORT 1977) and DMSO-paraformaldehyde system (JOHNSON et al. 1976).

The other approaches include studies of biosynthesis and transport of cell wall macromolecules, and regeneration of cell wall from protoplast, and the synchronized differentiation system (KOMAMINE et al. 1978). These approaches will give us valuable information on the size of macromolecular precursors just before their assembly into the wall. It would also tell us which polymers (if any) become covalently linked once they are in the wall or whether there is simply a physical containment of matrix molecules by noncovalent bonds (NORTHCOTE 1977, LAMPORT 1978).

References

Akiyama Y, Katō K (1977) Structure of hydroxyproline-arabinoside from tobacco cells. Agric Biol Chem 41:79–81

Albersheim P (1963) Hormonal control of *myo*-inositol incorporation into pectin. J Biol Chem 238:1608–1610

Albersheim P (1976) The primary cell wall. In: Bonner J, Varner JE (eds) Plant biochemistry. Academic Press, New York, pp 225–274

Albersheim P (1978) Concerning the structure and biosynthesis of the primary cell walls of plant. In: Manners DJ (ed) Biochemistry of carbohydrate II. Int Rev Biochem Vol. 16, Baltimore MD, Univ Park Press, pp 127–150

Albersheim P, Killias U (1963) Histochemical localization at the electron microscope level. Am J Bot 50:732–745

Aspinall GO (1959) Structural chemistry of the hemicelluloses. Adv Carbohydr Chem 14:429–468

Aspinall GO (1970) Pectins, plant gums and other plant polysaccharides. In: Pigman W, Horton D (eds) The carbohydrates. Academic Press, New York, Vol. IIB, pp 515–536

Aspinall GO (1973) Carbohydrate polymers of plant cell walls. In: Loewus F (ed) Biogenesis of plant cell wall polysaccharides. Academic Press, New York, pp 95–115

Aspinall GO, Jiang K-S (1974) Rapeseed hull pectin. Carbohydr Res 38:247–255

Aspinall GO, Molloy JA, Craig JWT (1969) Extracellular polysaccharides from suspension-cultured sycamore cells. Can J Biochem 47:1063–1070

Barrett AJ, Northcote DH (1965) Apple fruit pectic substances. Biochem J 94:617–627

Beck E (1967) Isolierung und Charakterisierung eines Apiogalacturonans aus der Zellwand von *Lemna minor*. Z Pflanzenphysiol 57:444–461

Blake JD, Richards GN (1971) Evidence for molecular aggregation in hemicelluloses. Carbohydr Res 18:11–21

Boffey SA, Northcote DH (1975) Pectin synthesis during the wall regeneration of plasmolysed tobacco leaf cells. Biochem J 150:433–440

Bright SWJ, Northcote DH (1974) Protoplast regeneration from normal and bromodeoxyuridine resistant sycamore cells. J Cell Sci 16:445–463

Buchala AJ (1973) An arabinogalacto(4-O-methylglucurono)xylan from the leaves of *Hordeum vulgare*. Phytochemistry 12:1373–1376

Buchala AJ, Franz G (1974) A hemicellulosic β-glucan from the hypocotyls of *Phaseolus aureus*. Phytochemistry 13:1887–1889

Buchala AJ, Wilkie KCB (1971) The ratio of $\beta(1\to3)$ to $\beta(1\to4)$ glucosidic linkages in non-endospermic hemicellulosic β-glucans from oat plant (*Avena sativa*) tissues at different stages of maturity. Phytochemistry 10:2287–2291

Buchala AJ, Fraser CG, Wilkie KCB (1972) An acidic galactoarabinoxylan from the stem of *Avena sativa*. Phytochemistry 11:2803–2814

Burgess J, Fleming EN (1974) Ultrastructural observations of cell wall regeneration around isolated protoplasts. J Cell Sci 14:439–449

Clarke AE, Anderson RL, Stone BA (1979) Form and function of arabinogalactans and arabinogalactan-proteins. Phytochemistry 18:521–540

Dalessandro G, Northcote DH (1977a) Changes in enzymic activity of nucleoside diphosphate sugar interconversions during differentiation of cambium to xylem in sycamore and poplar. Biochem J 162:267–279

Dalessandro G, Northcote DH (1977b) Possible control site of polysaccharides synthesis during growth and wall expansion of pea seedling (*Pisum sativum* L.). Planta 134:39–44

Dalessandro G, Northcote DH (1977c) Changes in enzymic activities of UDP-D-glucuronate decarboxylase and UDP-D-xylose 4-epimerase during cell division and xylem differentiation in cultured explants of Jerusalem artichoke. Phytochemistry 16:853–859

Darvill AG, McNeil M, Albersheim P (1978) Structure of plant cell walls. VIII. A new pectic polysaccharides. Plant Physiol 62:418–422

Dea ICM, Rees DA, Beveridge RJ, Richards GN (1973) Aggregation with change of conformation in solutions of hemicellulose xylans. Carbohydr Res 29:363–372

Dea ICM, Morris ER, Rees DA, Welsh EJ, Barnes HA, Price J (1977) Associations of like and unlike polysaccharides: Mechanism and specificity in galactomannans, interacting bacterial polysaccharides, and related systems. Carbohydr Res 57:249–272

Duff RB (1965) The occurrence of apiose in *Lemna* (duckweed) and other angiosperms. Biochem J 94:768–772

Dutton CGS, Kabir MS (1972) A comparison of the xylans from corn leaves and stalks. Phytochemistry 11:779–785

Eda S, Ohnishi A, Katō K (1976) Xylan isolated from the stalk of *Nicotiana tabacum*. Agric Biol Chem 40:359–364

Eda S, Katō K (1978a) An arabinoxyloglucan from the midrib of the leaves of *Nicotiana tabacum*. Agric Biol Chem 42:351–357

Eda S, Katō K (1978b) Galactan isolated from the midrib of the leaves of *Nicotiana tabacum*. Agric Biol Chem 42:2253–2257

Fowke LC (1978) Ultrastructure of isolated and cultured protoplasts. In: Thorpe TA (ed) Frontiers of plant tissue culture 1978. Calgary, The Bookstore, Univ Calgary, Canada pp 223–233

Franz G (1972) Polysaccharid-Metabolismus in den Zellwänden wachsender Keimlinge von *Phaseolus aureus*. Planta 102:334–347

Fraser CG, Wilkie KCB (1971a) β-Glucans from oat leaf tissues at different stages of maturity. Phytochemistry 10:1539–1542

Fraser CG, Wilkie KCB (1971b) A hemicellulosic glucan from oat leaf. Phytochemistry 10:199–204

Freudenberg K, Neish AC (1968) Constitution and biosynthesis of lignin. Springer, Berlin, Heidelberg, New York, p 129

Frey-Wyssling A (1969) The ultrastructure and biogenesis of native cellulose. Fortschr Chem Org Naturst 26:1–30

Gilkes NR, Hall MA (1977) The hormonal control of cell wall turnover in *Pisum sativum* L. New Phytol 78:1–15

Gould SEB, Rees DE, Richardson NG, Steele IW (1965) Pectic polysaccharides in the growth of plant cells: Molecular structural factors and their role in the germination of white mustard. Nature 208:876–878

Grant GT, Morris ER, Rees DA, Smith P, Thom D (1973) Biological interactions between polysaccharides and divalent cations: The egg-box model. FEBS Lett 32:195–198

Gross GG, Janse C, Elstner EF (1977) Involvement of malate, monophenols, and the superoxide radical in hydrogen peroxide formation by isolated cell walls from horseradish (*Armoracia lapathifolia* Gilib.). Planta 136:271–276

Hanke DE, Northcote DH (1974) Cell wall formation by soybean callus protoplast. J Cell Sci 14:29–50

Hart DA, Kindel PK (1970) Isolation and partial characterization of apiogalacturonans from the cell wall of *Lemna minor*. Biochem J 116:569–579

Hartley RD (1973) Carbohydrate esters of ferulic acid as components of cell-walls of *Lolium multiflorum*. Phytochemistry 12:661–665

Hirst EL, Jones JKN (1947a) Pectic substances. Part VI. The structure of the araban from *Arachis hypogea*. J Chem Soc 1221–1225

Hirst EL, Jones JKN (1947b) Pectic substances. Part VII. The constitution of the galactan. J Chem Soc 1225–1229

Hori H, Sato S (1977) Extracellular hydroxyproline-rich glycoprotein of suspension-cultured tobacco cells. Phytochemistry 16:1485–1487

Jeffs RA, Northcote DH (1966) Experimental induction of vascular tissue in an undifferentiated plant callus. Biochem J 101:146–152

Johnson DC, Nicholson MD, Haigh FC (1976) Dimethylsulfoxide/paraformaldehyde: A nondegrading solvent for cellulose. Appl Polym Symp 28:931–943

Joseleau JP, Chambat G, Vignon M, Barnoud F (1977) Chemical and ^{13}C-N.M.R. studies on two arabinans from the inner bark of young stems of *Rosa glauca*. Carbohydr Res 58:165–175

Karácsonyi Š, Toman R, Janeček F, Kubačková M (1975) Polysaccharides from the bark of the white willow (*Salix alba* L.).: Structure of arabinan. Carbohydr Res 44:285–290

Katō Y, Asano N, Matsuda K (1977) Isolation of xyloglucans from etiolated *Glycine max* and *Vigna sesquipedalis*. Plant Cell Physiol 18:821–829
Kauss H, Bowles DJ (1976) Some properties of carbohydrate-binding proteins (lectins) solubilized from cell walls of *Phaseolus aureus*. Planta 130:169–174
Kauss M, Glaser C (1974) Carbohydrate-binding proteins from plant cell walls and their possible involvement in extension growth. FEBS Lett 45:304–307
Komamine A, Morigaki T, Fujimura T (1978) Metabolism in synchronous growth and differentiation in plant tissue and cell cultures. In: Thorpe TA (ed) Frontiers of plant tissue culture 1978. Calgary, The Bookstore, Univ Calgary, Canada pp 159–168
Labavitch JM, Ray PM (1974a) Turnover of cell wall polysaccharides in elongating pea stem segments. Plant Physiol 53:669–673
Labavitch JM, Ray PM (1974b) Relationship between promotion of xyloglucan metabolism and induction of elongation by indoleacetic acid. Plant Physiol 54:499–502
Labavitch JM, Ray PM (1978) Structure of hemicellulosic polysaccharides of *Avena sativa* coleoptile cell walls. Phytochemistry 17:933–937
Lamport DTA (1965) The protein component of primary cell walls. Adv Bot Res 2:151–218
Lamport DTA (1973) The glycopeptide linkages of extensin: *O*-D-galactosyl serine and *O*-L-arabinosyl hydroxyproline. In: Loewus F (ed) Biogenesis of plant cell wall polysaccharides. Academic Press, New York, pp 149–164
Lamport DTA (1977) Structure, biogenesis and significance of cell wall glycoproteins. In: Loewus FA, Runeckles VC (eds) The structure, biosynthesis and degradation of wood. Rec Adv Phytochem Vol. 11, Plenum, New York pp 79–115
Lamport DTA (1978) Cell wall carbohydrates in relation to structure and function. In: Thorpe TA (ed) Frontiers of plant tissue culture 1978. The Bookstore, Univ Calgary, Canada, pp 235–244
Lamport DTA, Katona L, Roerig S (1973) Galactosylserine in extensin. Biochem J 133:125–131
Lindberg B, Rosell K-G, Svensson S (1973) Positions of the *O*-acetyl groups in pine glucomannan. Sven Papperstidn 76:383–384
Loescher WH, Nevins DJ (1972) Auxin-induced changes in *Avena* coleoptile cell wall composition. Plant Physiol 50:556–563
Loewus F, Chen MS, Loewus MW (1973) The *myo*-inositol oxidation pathway to cell wall polysaccharides. In: Loewus F (ed) Biogenesis of plant cell wall polysaccharides. Academic Press, New York, pp 1–27
Mares DJ, Stone BA (1973a) Studies on wheat endosperm. I. Chemical composition and ultrastructure of the cell walls. Aust J Biol Sci 26:793–812
Mares DJ, Stone BA (1973b) Studies on wheat endosperm. II. Properties of the wall components and studies on their organization in the wall. Aust J Biol Sci 26:813–830
Markwalder HU, Neukom H (1976) Diferulic acid as possible cross link in hemicelluloses from wheat germ. Phytochemistry 15:836–837
Marshall JJ (1975) Degradation of barley glucan by a purified (1→4)-β-D-glucanase from the snail, *Helix pomatia*. Carbohydr Res 42:203–207
Marx-Figini M (1966) Comparison of the biosynthesis of cellulose *in vitro* and *in vivo* in cotton bolls. Nature 210:754–755
Marx-Figini M, Schulz GV (1966) Zur Biosynthese der Cellulose. Naturwissenschaften 53:466–474
McNeil M, Darvill AG, Albersheim P (1979) The structural polymers of the primary cell walls of dicots. Fortschr Chem Org Naturst 37:191–248
Meinert MC, Delmer DP, (1977) Changes in biochemical composition of the cell wall of the cotton fiber during development. Plant Physiol 59:1088–1097
Monro JA, Bailey RW, Penny D (1976a) Hemicellulose fractions and associated protein of lupin hypocotyl cell walls. Phytochemistry 15:175–181
Monro JA, Penny D, Bailey RW (1976b) The organization and growth of primary cell walls of lupin hypocotyl. Phytochemistry 15:1193–1198
Mori M, Eda S, Katō K (1980) Structural investigation of the arabinoxyloglucan from *Nicotiana tabacum*. Carbohydr Res 84:125–135

Morrison IM (1973) Isolation and analysis of lignin-carbohydrate complexes from *Lolium multiflorum*. Phytochemistry 12:1979–1984
Morrison IM (1974) Lignin-carbohydrate complexes from *Lolium perenne*. Phytochemistry 13:1161–1165
Morrison IM (1977) Extraction of hemicellulose from plant cell-walls with water after preliminary treatment with methanolic methoxide. Carbohydr Res 57:C4–C6
Mort AJ, Lamport DTA (1977) Anhydrous hydrogen fluoride deglycosylates glycoproteins. Anal Biochem 82:289–309
Nagata T, Takebe I (1971) Planting of isolated tobacco mesophyll protoplasts on agar medium. Planta 99:12–20
Nevins DJ (1975) The *in vitro* stimulation of IAA-induced modification of *Avena* cell wall polysaccharides by an exo-glucanase. Plant Cell Physiol 16:495–503
Northcote DH (1969) The synthesis and metabolic control of polysaccharides and lignin during the differentiation of plant cells. Essays Biochem 5:89–137
Northcote DH (1972) Chemistry of the plant cell wall. Annu Rev Plant Physiol 23:113–132
Northcote DH (1974) Site of synthesis of the polysaccharides of the cell wall. In: Pridham JB (ed) Plant carbohydrate chemistry. Academic Press, New York, pp 165–181
Northcote DH (1977) The synthesis and assembly of plant cell walls: Possible control mechanisms. In: Poste G, Nicolson GL (eds) The synthesis, assembly and turnover of cell surface components. North-Holland, Amsterdam Vol. 4, pp 717–739
Northcote DH, Wooding FBP (1966) Development of sieve tubes in *Acer pseudoplatanus*. Proc R Soc Lond Ser B 163:524–537
Palmieri S, Odoardi M, Soressi GP, Salamini F (1978) Indoleacetic acid oxidase activity in two high-peroxidase tomato mutants. Physiol Plant 42:85–90
Pope DG, Lamport DTA (1977) Relationships between hydroxyproline-containing proteins secreted into the cell wall and medium by suspension-cultured *Acer pseudoplatanus* cells. Plant Physiol 59:894–900
Preston RD (1979) Polysaccharide conformation and cell wall function. Annu Rev Plant Physiol 30:55–78
Rees DA (1972) Shapely polysaccharides. Biochem J 126:257–273
Reid JSG, Wilkie KCB (1969) An acidic galactoarabinoxylan and other pure hemicelluloses in oat leaf. Phytochemistry 8:2053–2058
Ring SG, Selvendran RP (1978) Purification and methylation analysis of cell wall material from *Solanum tuberosum*. Phytochemistry 17:745–752
Roelofsen PA (1965) Ultrastructure of the wall in growing cells and its relation to the direction of the growth. Adv Bot Res 2:69–149
Rubery PH, Northcote DH (1970) The effect of auxin (2,4-dichlorophenoxyacetic acid) on the synthesis of cell wall polysaccharides in cultured sycamore cells. Biochim Biophys Acta 222:95–108
Sakakibara A (1977) Degradation products of protolignin and the structure of lignin. In: Loewus F, Runeckles VC (eds) The structure, biosynthesis and degradation of wood. Recent Adv Phytochem Vol. 11, Plenum, New York, pp 117–139
Selvendran RP (1975) Cell wall glycoproteins and polysaccharides of parenchyma of *Phaseolus coccineus*. Phytochemistry 14:2175–2180
Shafizadeh F, McGinnis GD (1971) Morphology and biogenesis of cellulose and plant cell-walls. Adv Carbohydr Chem Biochem 26:297–349
Shepard JF, Totten RE (1975) Isolation and regeneration of tobacco mesophyll cell protoplasts under low osmotic conditions. Plant Physiol 55:689–694
Shibuya N, Misaki A (1978) Structure of hemicellulose isolated from rice endosperm cell wall: Mode of linkages and sequences in xyloglucan, β-glucan and arabinoxylan. Agric Biol Chem 42:2267–2274
Siddiqui IR, Wood P (1974) Structural investigation of oxalate-soluble rapeseed (*Brassica campestris*) polysaccharides. Carbohydr Res 36:35–44
Smith MM, Stone BA (1973) Chemical composition of the cell walls of *Lolium multiflorum* endosperm. Phytochemistry 12:1361–1367
Stoddart RW, Northcote DH (1967) Metabolic relationships of the isolated fractions of the pectic substances of actively growing cells. Biochem J 105:45–59

Stoddart RW, Barrett AJ, Northcote DH (1967) Pectic polysaccharides of growing plant tissue. Biochem J 102:194–204
Takeuchi Y, Komamine A (1978) Changes in the composition of cell wall polysaccharides of suspension-cultured *Vinca rosea* cells during culture. Physiol Plant 42:21–28
Thornber JP, Northcote DH (1961a) Changes in the chemical composition of a cambial cell during its differentiation into xylem and phloem tissues in trees. 1. Main components. Biochem J 81:449–455
Thornber JP, Northcote DH (1961b) Changes in the chemical composition of a cambial cell during its differentiation into xylem and phloem tissues in trees. 2. Carbohydrate constituents of each main component. Biochem J 81:455–464
Thornber JP, Northcote DH (1962) Changes in the chemical composition of a cambial cell during its differentiation into xylem and phloem tissues in trees. 3. Xylan, glucomannan and α-cellulose fractions. Biochem J 82:340–346
Timell TE (1964) Wood hemicelluloses: Part I. Adv Carbohydr Chem 19:247–302
Timell TE (1965) Wood hemiceluloses: Part II, Adv Carbohydr Chem 20:409–483
Toman R, Kárǎcsonyi S, Kubačková M (1972) Polysaccharides from the bark of white willow (*Salix alba* L.): Structure of a galactan. Carbohydr Res 25:371–378
Wada S, Ray PM (1978) Matrix polysaccharides of oat coleoptile cell walls. Phytochemistry 17:923–931
Whistler RL, Richards EL (1970) Hemicelluloses. In: Pigman W, Horton DH (eds) The carbohydrates. Academic Press, New York, Vol. IIA, pp 447–469
Whitmore FW (1976) Binding of ferulic acid to cell walls by peroxidases of *Pinus elliotii.* Phytochemistry 15:375–378
Wilkie KCB (1979) The hemicelluloses of grasses and cereals. Adv Carbohydr Chem Biochem 36:215–264
Wilkie KCB, Woo S-L (1976) Non-cellulosic β-glucans from bamboo, and interpretative problems in the study of cell hemicelluloses. Carbohydr Res 49:399–409
Willison JHM, Cocking EC (1976) Microfibril synthesis at the surface of isolated tobacco mesophyll protoplasts, a freeze-etch study. Protoplasma 88:187–200
Wooding FBP, Northcote DH (1964) The development of the secondary wall of the xylem in *Acer pseudoplatanus*. J Cell Biol 23:327–337
Worth HGJ (1967) The chemistry and biochemistry of pectic substances. Chem Rev 67:465–473
Yaku F, Tsuji S, Koshijima T (1979) Lignin carbohydrate complex. PT. III. Formation of micelles in the aqueous solution of acidic lignin carbohydrate complex. Holzforschung 33:54–59
Zitko V, Bishop CT (1966) Structure of a galacturonan from sunflower pectic acid. Can J Chem 44:1275–1282

5 Biosynthesis and Metabolism of Cellulose and Noncellulosic Cell Wall Glucans

G. FRANZ and U. HEINIGER

1 Introduction: Various Aspects of Cellulose Formation in Vivo and in Vitro

Cellulose is the most plentiful carbohydrate on earth, being widely distributed in the plant kingdom where it serves as skeletal substance in combination with other compounds in the intricate system of the cell wall (see as well Chaps. 1, 2, 3, and 4 this Vol.). During the last decade, several reviews have been published dealing with polysaccharide biosynthesis and cell wall formation (DELMER 1977; ROBINSON 1977; ELBEIN and FORSEE 1973; SHAFIZADEH and McGINNIS 1971; NIKAIDO and HASSID 1971; HASSID 1969). Although there exists considerable information on the chemical structure, physical conformation, and the ultrastructural appearance of cellulose, our understanding of the mechanism of cellulose biosynthesis is still very incomplete.

There are still contradictory opinions on the formation of the glucan macromolecules. The problems of chain association to cellulose microfibrils with a well-defined crystalline structure, and of the deposition of the microfibrillar network to a coherent cell wall, are still poorly understood. Recent publications on cellulose structure (GARDNER and BLACKWELL 1974a, b) present evidence for a parallel arrangement of the glucan chains, so that it is no longer disputed whether two enzyme systems, forming chains by addition of glucose to either the reducing or the nonreducing end of a growing glucan chain, are required. One enzyme system will clearly be sufficient for a parallel arrangement of the glucan chains. Although it is generally accepted that the polysaccharide chains are growing at the nonreducing end, exceptions are known. The glycosylation of proteins in bacterial systems may proceed by addition of sugar units to the reducing end (ROBYT 1979).

The mechanism by which the cellulose chains are initiated and terminated are unknown. MARX-FIGINI and SCHULZ (1966) and MARX-FIGINI (1969) showed that in cotton fibers the cellulose chains are not only differently oriented in the primary and secondary wall, but that they also show different degrees of polymerization. The question is still open whether two different enzyme systems are involved in cellulose biosynthesis of primary and secondary walls and how they are regulated.

Nucleoside diphosphate-sugars (sugar nucleotides) are generally considered to be the active sugar donors for the biosynthesis of oligosaccharides, polysaccharides, glycoproteins, and lipopolysaccharides. As every glucose molecule in the cellulose chain is rotated 180° to its nearest neighbor, some authors suggested an activated cellobiose, the basic repeating unit of the chain, as precursor (DELMER 1977). No evidence has been presented so far.

For several years there have been claims of successful in vitro synthesis of cellulose using different nucleotide-glucoses, but careful product analysis showed that this interpretation was premature since mixed glycans, i.e., glucomannan (ELBEIN and FORSEE 1973), (1→3)-β-glucans, (FEINGOLD et al. 1958) or (1→3), (1→4)-β-glucans (VILLEMEZ et al. 1967) were enzymatically formed.

In the light of data on glycoprotein synthesis (WAECHTER and LENNARZ 1976), a search for lipid-linked intermediates (polyprenols) was undertaken. Even though lipid-linked intermediates were found in glycoprotein-synthesizing plant systems (ELBEIN and FORSEE 1973), such an intermediate has not been clearly identified in cellulose-forming systems. An exception is the alga *Prothotheca zopfii* (HOPP et al. 1978a) in which newly synthesized (1→4)-β-glucan chains are transferred from a lipid-linked intermediate to the protein moiety, suggesting that cellulose in fact may be synthesized as a glycoprotein. Indeed, investigations on cotton fiber cell wall, even after thorough purification, show that some protein is attached to the α-cellulose fraction (NOWAK-OSSORIO et al. 1976, HUWYLER et al. 1978).

The question of possible primers or acceptors for glucan chains, as well as possible activators and other regulating factors of the synthetases, is still being investigated.

Many of the difficulties encountered in the search for "cellulose synthetase" are certainly due to the complexity of the system. Cellulose is apparently not the only glucan found in the cell wall of higher plants. In recent years a series of noncellulosic glucans with varying portions of (1→4)- and (1→3)-β-linkages or with (1→3)-β-linkages alone were demonstrated in cell walls of monocotyledonous (BUCHALA and WILKIE 1971) and dicotyledonous plants (BUCHALA and FRANZ 1974, MEINERT and DELMER 1977, HUWYLER et al. 1978). They seem to be involved in turnover reactions during the process of cell wall differentiation (FRANZ 1972, HUWYLER et al. 1979).

Synthesis of (1→3)-β-glucans in pollen tubes (SOUTHWORTH and DICKINSON 1975), as well as in phloem tissue (BELTRÁN and CARBONELL 1978), uses UDP-glucose as glucose donor. In the phloem tissue this reaction is assumed to lead to the formation of callose. Callose deposition has been known for many years to be an extremely rapid process induced in many plants by wounding (ESCHRICH 1956, 1965) and stress (SMITH and MCCULLY 1977). The wound reaction may well interfere in the assays for cellulose synthesis, competing for the substrate UDP-glucose. Substrate competition seems to be a major problem, as many other enzymic reactions are known to use UDP-glucose as glucosyl donor. Unlike the chitin synthetase of fungal systems (BARTNICKI-GARCIA et al. 1978), the cellulose synthetase does not appear to be zymogenic in nature.

Substrate accessibility also seems to be a problem. EM studies suggest that cellulose synthetase is localized in the plasma membrane, substrate coming from the cytoplasmic side and the product being delivered to the outside of the plasma membrane. (For further information concerning cellular localization of the enzyme system responsible for glucan formation, i.e., cellulose and (1→3)-β-glucan see Chap. 21, this Vol.) Hence, the surface activity detected by RAYMOND et al. (1978) and ANDERSON and RAY (1978), when UDP-glucose was added from the outside, hardly represents the in vivo situation. Similarly, suspension cultures of soybean cells did not incorporate exogenous UDP-glucose, unless they were damaged (BRETT 1978).

Thus the cellulose-generating system is considered to be very fragile and slight disturbance of membrane integrity may cause unexpected side reactions.

2 Possible Substrates for Cellulose Biosynthesis

2.1 Occurrence of NDP-Sugars (Sugar Nucleotides) in Tissues Actively Forming Cell Walls

The discovery of UDP-glucose by LELOIR (1951) opened up a new chapter in carbohydrate biochemistry particularly with regard to complex saccharide synthesis. Since that time sugar nucleotides containing different sugar moieties as well as different bases have been isolated from tissues of all organisms in the plant kingdom, that have been studied in this regard (for further information see FEINGOLD, Chap. 1 Vol. 13A of this series). UDP-glucose was demonstrated to be a direct precursor in vitro for the biosynthesis of trehalose, sucrose, starch, callose, and microbial cellulose (LELOIR 1964). Since that time extensive research has been conducted to identify the naturally occurring sugar nucleotides in tissues which are actively synthesizing cell wall polymers. However, the failure to detect certain sugar nucleotides in any tissue should not be interpreted as signifying that they are of little metabolic importance. The biosynthesis of most sugar nucleotides seems to be under efficient feed-back control.

Glucose-containing sugar nucleotides, which were considered as the natural substrates for cellulose formation, have been reported in a series of publications. Thus UDP-glucose has been detected in growing *Phaseolus aureus* tissue (SOLMS and HASSID 1957), in bamboo tissue (SU 1965), in wheat and oat seedlings (ELNAGHY and NORDIN 1966), in developing cotton fibers (FRANZ 1969), in cambial tissue (CUMMING 1970), in suspension-cultured sycamore cells (BROWN and SHORT 1969), and in a great variety of other plant materials. GDP-glucose on the other hand, which was postulated as a second important substrate for cellulose biosynthesis (BARBER et al. 1964), has not been detected in many cases. Cotton fibers which produce large amounts of cellulose do not contain detectable amounts of GDP-glucose (FRANZ 1969), whereas cambial cells (CUMMING 1970), as well as strawberry leaves, contain small amounts of GDP-glucose inter alia (SELVENDRAN and ISHERWOOD 1967; ISHERWOOD and SELVENDRAN 1970). These rather confusing findings indicate that caution is needed in assessing the physiological function of sugar nucleotides isolated from a certain tissue.

2.2 Sucrose Synthetase and Pyrophosphorylase Activities

The synthesis of nucleotide sugars by pyrophosphorylases and sucrose synthetases are treated in extenso in Chapters 1 and 7 volume 13A, of this series by FEINGOLD and AVIGAD, respectively. A high concentration of a specific sugar nucleotide (i.e., UDP-glucose) does not necessarily mean it is a substrate for specific enzymic reactions. The sugar nucleotides may be immobilized in a pool with low turnover rate or they may serve for specific enzymic reactions due to their enzyme accessibility and the affinity of the enzymes. Thus, the NDP-glucose-synthesizing enzymes, i.e., the corresponding pyrophosphorylases and the sucrose synthetases, may play a key role in the complex process of cell wall glucan synthesis.

NDP-glucose pyrophosphorylases were isolated from several plant tissues. Developing starchy seeds contain specific UDP-glucose and ADP-glucose pyrophosphorylases (SOWOKINOS 1976). VIDRA and LOERCH (1968) also demonstrated GDP-glucose pyrophosphorylase activity in the same plant material. With respect to cell wall synthesis, the activities measured in growing tissue may shed some light on their involvement. AXELOS and PÉAUD-LENOËL (1969) demonstrated UDP-glucose, CDP-glucose, and TDP-glucose pyrophosphorylase activity in pea and bean seedlings, whereas GDP-glucose pyrophosphorylase was only measurable in pea tissue. Both tissues were devoid of ADP-glucose pyrophosphorylase. UDP-glucose pyrophosphorylase is by far the most active pyrophosphorylase, thus explaining the high concentrations of UDP-glucose usually found in plant tissue. The other pyrophosphorylases may not be active at all stages of plant development. Measuring the enzymatic activity along the axis of bean seedlings, UDP-glucose pyrophosphorylase shows highest activity in parts that have already ceased elongation. Other pyrophosphorylases could not be detected (HEINIGER and FRANZ 1980). As the glucan synthetase activity is higher in elongating tissue, HEINIGER and FRANZ (1980) concluded that pyrophosphorylase may be involved in the process of secondary cell wall synthesis, whereas glucan synthetase is linked to the process of cell wall elongation. DELMER and ALBERSHEIM (1970) found that UDP-glucose pyrophosphorylase and sucrose synthetase activities are reduced in photosynthetic tissue. They concluded that UDP-glucose pyrophosphorylase participates in the breakdown of UDP-glucose which was initially produced mainly by sucrose synthetase in nonphotosynthetic tissue.

Sucrose synthetase is unspecific for NDP-glucose (GRIMES et al. 1970). The K_m is similar for all NDP but the v_{max} is considerably higher for UDP (DELMER 1972a) so that again the synthesis of UDP-glucose is favored. This enzyme shows highest activity in the region of elongation (DELMER 1972a), and may thus provide activated sugars which are used for the synthesis of polysaccharides in the elongating cell wall. Sucrose synthetase, measured in the direction of sucrose breakdown, is activated by NADP, IAA, and GA (DELMER 1972b). Hence, factors other than enzyme levels and substrate accessibility could play important regulatory functions.

2.3 Different NDP-Glucoses as Hypothetical Substrates for the in Vitro Biosynthesis of Cellulose

The biosynthesis of cellulose in a cell-free system was first achieved by GLASER (1958), using a particulate enzyme preparation from *Acetobacter xylinum*. Glucose residues were transferred from the substrate UDP-glucose to an alkali-insoluble polysaccharide identified as cellulose. BARBER et al. (1964), repeating the above experiments, and comparing the substrate specificity of different NDP-glucoses, were able to demonstrate that only UDP-glucose, and to a much lesser extent TDP-glucose, could serve as glucosyl donors for cellulose biosynthesis. ADP-glucose and CDP-glucose were completely ineffective.

Using a similar system from higher plants (*Phaseolus aureus*) FEINGOLD et al. (1958) obtained a $(1\rightarrow3)$-β-glucan using UDP-glucose as substrate. In contrast to these findings, cell-free preparations from several higher plants, i.e., pea, corn, squash, and cotton fibers, utilized exclusively GDP-glucose as glucosyl donor for cellulose synthesis (BARBER and HASSID 1965; BARBER et al. 1964; ELBEIN et al. 1964). Addition of GDP-mannose to these in vitro assays greatly stimulated the rate of incorporation of GDP-glucose into the reaction product. This effect was shown by VILLEMEZ (1971) to involve the synthesis of a glucomannan which could only be separated from cellulose with difficulty. BRUMMOND and GIBBONS (1965) obtained different results using UDP-glucose and particulate enzyme preparations from *Lupinus albus*. The alkali-insoluble reaction product possessed exclusively $(1\rightarrow4)$-β-linked glucose residues. These findings were substantiated at least in part by the work of VILLEMEZ et al. (1967) who, using the same substrate and membrane-bound enzyme fractions from *Phaseolus aureus*, obtained a mixed linked glucan containing both $(1\rightarrow3)$- and $(1\rightarrow4)$-β-linkages. FLOWERS et al. (1968), however, disputed these results, demonstrating that only $(1\rightarrow3)$-β-linked glucans were formed by particulate fractions of *Lupinus albus* as well as of *Phaseolus aureus*.

ORDIN and HALL (1967 and 1968), working with particulate preparations from *Avena sativa*, could demonstrate that both GDP-glucose and UDP-glucose produced alkali-insoluble polymers in vitro. Hydrolysis with cellulase of the polysaccharide derived from UDP-glucose gave cellobiose, and to lesser extent a trisaccharide containing both $(1\rightarrow3)$- and $(1\rightarrow4)$-β-glucosidic linkages. Subsequently, several authors using different plant materials as enzyme source for cellulose formation in vitro and trying both hypothetical substrates (GDP-glucose and UDP-glucose) obtained controversial results in examining the chemical nature of the glucan polymers enzymatically synthesized. Table 1 shows some of the results which have been published and aimed at discovering the natural substrate for cellulose synthesis. As can be seen, even after almost 20 years of intensive research, the results remain contradictory and the possibility still exists, that both UDP-glucose and GDP-glucose contribute to the formation of cellulose (HOPP et al. 1978a).

One of the more important findings in recent years was that concerning the influence of substrate concentration upon the type of glycosidic linkage formed in vitro. High levels of UDP-glucose (mM) preferably led to the production of $(1\rightarrow3)$-β-linkages, whereas with low levels (μM) $(1\rightarrow4)$-β-linkages resulted (PEAUD-LENOËL and AXELOS 1970; SMITH and STONE 1973; VAN DER WOUDE et al. 1974). Even the presence or absence of cations seemed to influence the type of linkage formed. Addition of Mg^{2+} to the incubation mixture, containing the substrate UDP-glucose, stimulated the synthesis of $(1\rightarrow4)$-β-glucans and inhibited the synthesis of $(1\rightarrow3)$-β-glucans (TSAI and HASSID 1973; LARSEN and

Table 1. Apparent discrepancies resulting from in vitro assays of cellulose biosynthesis using the two hypothetic glucosyl donors UDP-glucose and GDP-glucose

Author and year of Publication	Substrate used	Source of enzyme (Plant material)	Product of enzymatic reaction (Type of glycosidic linkage) identified as:
GLASER (1958)	UDPG	*Acetobacter xylinum*	Cellulose; (1→4)-β
FEINGOLD et al. (1958)	UDPG	*Phaseolus aureus*	Callose; (1→3)-β
BARBER et al. (1964)	UDPG+TDGP	*Acetobacter xylinum*	Cellulose; (1→4)-β
BARBER et al. (1964)	GDPG	*Phaseolus aureus*	Cellulose; (1→4)-β
BARBER and HASSID (1965)	GDPG	*Gossypium hirsutum*	Cellulose; (1→4)-β
ELBEIN et al. (1964)	GDPG	*Phaseolus aureus*	Cellulose; (1→4)-β
BRUMMOND and GIBBONS (1964, 1965)	UDPG	*Lupinus albus*	Cellulose; (1→4)-β
VILLEMEZ et al. (1967)	UDPG	*Phaseolus aureus*	Alkali-insoluble glucan; 90% (1→4)-β, 10% (1→3)-β
ORDIN and HALL (1968)	UDPG	*Avena sativa*	β-Glucans; one of which is cellulose
SMITH and STANLEY (1969)	UDPG	different higher plants	Glucan; mainly (1→3)-β, some (1→4)-β
FLOWERS et al. (1968)	UDPG	*Phaseolus aureus*	Glucan; (1→3)-β
		Lupinus albus	
RAY et al. (1969)	UDPG/GDPG	*Pisum sativum*	Glucan; (1→4)-β
FRANZ and MEIER (1969b)	UDPG	*Gossypium arboreum*	Glucan; (1→4)-β and (1→3)-β
		Phaseolus aureus	
PÉAUD-LENOËL and AXELOS (1970)	UDPG	*Triticum sativum*	Glucan; (1→4)-β and (1→3)-β
CHAMBERS and ELBEIN (1970)	GDPG	*Phaseolus aureus*	Glucan; (1→4)-β
		Phaseolus aureus	Glucan; (1→3)-β
BATRA and HASSID (1970)	UDPG	*Lupinus albus*	Glucan; (1→3)-β
		Avena sativa	Glucan; (1→3)-β and (1→4)-β
VILLEMEZ and HELLER (1970)	GDPG	*Phaseolus aureus*	Glucomannan, (1→4)-β
STEPANENKO and MAROZOVA (1970)	UDPG	*Gossypium*	Glucan; (1→4)-β
SPENCER et al. (1971)	UDPG	*Pisum sativum*	Glucan; (1→4)-β
TSAI and HASSID (1971)	UDPG	*Avena sativa*	Glucan; (1→4)-β and glucan (1→3)-β
CLARK and VILLEMEZ (1972)	UDPG	*Phaseolus aureus*	Glucan; (1→4)-β, few β-(1→3)-β
SPENCER and MACLACHLAN (1972)	UDPG	*Pisum sativum*	Cellulose; (1→4)-β
ROBINSON and PRESTON (1972)	UDPG/GDPG	*Phaseolus aureus*	Low molecular weight glucans no cellulose
MIYAMOTO and TAMARI (1973)	UDPG	*Phaseolus aureus*	Glucan; (1→3)-β and glucan; (1→3)-β and (1→4)-β
SMITH and STONE (1973)	UDPG	*Lolium multiflorum*	Glucan; (1→4)-β and (1→3)-β

RAY (1973a)	UDPG	*Pisum sativum*	Glucan; (1→4)-β
SHORE and MACLACHLAN (1973)	UDPG/GDPG	*Pisum sativum*	Cellulose
KEMP and LOUGHMAN (1973)	UDPG	*Phaseolus aureus*	Glucan; (1→3)-β inositol-glucoside
ELBEIN and FORSEE (1973)	GDPG	*Phaseolus aureus*	Cellulose
VILLEMEZ (1974)	UDPG	*Phaseolus aureus*	Cellulose; (1→4)-β
LARSEN and BRUMMOND (1974)	UDPG	*Lupinus albus*	Glucan; (1→4)-β, some (1→3)-γβ
VAN DER WOUDE et al. (1974)	UDPG	*Allium cepa*	Glucans; (1→3)-β and (1→4)-β
SOUTHWORTH and DICKINSON (1975)	UDPG	*Lilium longiflorum*	Glucan; (1→3)-β
FRANZ (1976)	UDPG	*Phaseolus aureus*	Glycoprotein; (1→4)-β
BRETT and NORTHCOTE (1975)	UDPG	*Pisum sativum*	Glucan; (1→3)-β and (1→4)-β, glycoprotein
SHORE et al. (1975)	UDPG/GDPG	*Pisum sativum*	Cellulose
TSAI (1975)	UDPG	*Avena sativa*	Glucan; (1→4)-β and glucan (1→3)-β
HELSPER et al. (1977)	UDPG	*Petunia hybrida*	Glucan; (1→4)-β
HEINIGER and DELMER (1977)	UDPG	*Gossypium hirsutum*	Glucan; (1→3)-β
HOPP et al. (1978a)	UDPG/GDPG	*Prototheca zopfii*	Glycoprotein; with (1→4)-β; cellulose; (1→4)-β
BELTRÁN and CARBONELL (1978)	UDPG	*Citrus* phloem	Glucan; (1→3)-β
PHILIPPI (1978)	UDPG	*Polyspondylium pallidum*	Cellulose; (1→4)-β

Table 2. Factors which might influence the type of glycosidic linkage formed in the in vitro assay for cell wall glucan synthesis

1. Plant material (Monocot/Dicot)
2. Stage of development
3. Culture conditions
4. Methods of cell homogenate preparation
5. Fractionation of particulate enzyme
6. Substrate concentration
7. Treatment with detergents
8. Cofactors

Brummond 1974). Table 2 summarizes the different factors which might influence the type of linkage formed in the in vitro assay systems.

Some of the discrepancies mentioned above might result from the inadequate characterization of the reaction product. Careful chemical analysis must be carried out together with enzymatic degradation of the reaction products. Solubility properties, partial acid hydrolysis and product identification, periodate degradation, and methylation analysis should be the essential steps to ascertain the exact structural features of the polysaccharide in question. The DP of the product has also to be analyzed in order to prove true polysaccharide formation and not just a chain-terminating reaction.

2.4 In Vivo Studies on Cellulose Biosynthesis

Early experiments on the biosynthesis of cellulose mainly involved the introduction of glucose, specifically labeled in the different carbon atoms, into biological systems (Edelmann et al. 1954; Neish 1955). The resulting radioactive cellulose was subsequently isolated and distribution of label in the glucose molecules was determined. It was shown that the glucose was in part metabolized, broken down, and resynthesized through the Embden-Meyerhof pathway. In order to find out a more direct precursor of cellulose in the in vivo systems, NDP-sugars were fed to detached cotton fibers and the incorporation into cellulose was studied (Franz and Meier 1969a). In this system UDP-glucose served as a substrate for cellulose formation, whereas glucose was incorporated to a much lesser extent. Similar experiments, also using cotton fibers, compared the capacity for biosynthesis of radioactive, alkali-insoluble products with the hypothetical substrates GDP-glucose and UDP-glucose (Delmer et al. 1974). In these experiments the product formed in vivo from GDP-glucose resembled cellulose. The products formed from UDP-glucose were analyzed as mainly glucolipids, but the yield was very low and could only be detected during primary wall formation.

Delmer et al. (1977) and Heiniger and Delmer (1977), feeding UDP-glucose at different substrate concentrations to the semi-intact fibers of *Gossypium hirsutum*, found exclusively labeled (1→3)-β-glucan after separation from steryl glucosides and sucrose. No incorporation at all was observed into cellulose. These data are in contrast to the findings of Shore et al. (1975) and Raymond et al.

(1978) working with UDP-glucose and thin slices of *Pisum sativum* epicotyl tissue. These authors found a cell surface-located (1→4)-β-glucan synthetase which exclusively used the above substrate for cellulose formation This activity was lost to a great extent by disrupting the tissue. ANDERSON and RAY (1978), working with an identical system and the same glucose donor, could detect no (1→4)-β-linked glucan at all. The radioactive glucan formed in their experiments had (1→3)-β-linkages.

The synthetase system using external UDP-glucose was postulated to be responsible for the formation of wound callose. This finding is substantiated by the fact that intact cultured soybean cells do not incorporate UDP-glucose into β-glucans (BRETT 1978). Only damaged cells produced alkali-insoluble (1→3)-β-glucans.

These contradictory results indicate that the in vivo approach, feeding different labeled NDP-glucoses to the plant material and studying the incorporation into cellulose or other cell wall glucans, cannot decisively show which NDP-glucose is the definite precursor of cellulose.

3 Primer Requirement for Cellulose Biosynthesis

Since the early investigations of polysaccharide biosynthesis, the question of how a polymer chain is initiated has been discussed in many publications. In starch and glycogen synthesis an absolute requirement for a primer or acceptor molecule, shown to be a glycoprotein, has been demonstrated (LAVINTHMAN et al. 1974). In the case of cell wall polysaccharide biosynthesis in the higher plant system and particularly cellulose formation, it is still not clear if and how the glucan chains are initiated and the fibrillar network is formed at the plasma membrane and inserted into the cell wall. From the ultrastructural point of view (for further information see Chaps. 3, 13, and 21, this Vol.) the cellulose microfibrils seem to arise from granular material situated on the outside of the plasma membrane. However, nothing is so far known about the chemical nature of these membrane-bound particles. It was speculated that they contain the multienzyme complex responsible for polymerization reactions (BROWN and MONTEZINOS 1976). When the plasma membrane is removed from the cell wall, as is routinely done in all cell-free assay systems, the enzyme particles are probably disconnected from the growing polysaccharide chains. Regenerating this system in vitro appears to be a very difficult task. A polysaccharide-like acceptor molecule was always assumed to exist in cell-free preparations, since additions of exogenous primer molecules were never necessary for an active in vitro system.

The presence of different carbohydrates associated with particulate enzyme preparations, which inter alia showed (1→4)-β-glucosidic linkages, was demonstrated by BAILEY et al. (1967). On the other hand it could be shown by FRANZ and MEIER (1969b) that endogenous cellodextrins, but not short-chain cello-oligosaccharides, could stimulate the rate of glucose incorporation into alkali-insoluble glucan using the substrate UDP-glucose and mung bean enzyme particles. SPENCER et al. (1971) working with a similar system from peas found that cellobiose activated the reaction. Carboxymethylcellulose and cellodextrins served as competitive acceptor molecules of the unidentified endogenous acceptor molecules.

Cellobiose also stimulated the synthesis of (1→3)-β-glucans from UDP-glucose (FLOWERS et al. 1968). This activation, however, does not seem to be specific, since cellobiose may be replaced by laminaribiose and β-methyl-glucose, although not by α-linked saccharides (DELMER et al. 1977).

Three interesting publications reported the formation and requirement of glucosylinositol as a possible primer for (1→3)-β-glucan synthesis (KEMP and LOUGHMAN 1973, 1974; KEMP et al. 1978) using UDP-glucose as substrate. [The involvement of inositol in the formation of glycosidic linkages has been shown as an essential step for the biosynthesis of oligosaccharides in the raffinose series (LEHLE and TANNER 1973; TANNER and KANDLER 1968), see also KANDLER and HOPF, Chap. 8, Vol. 13A, this series.]

A completely different but very stimulating idea was suggested by MACLACHLAN (1976), based on the findings that only cellulose chains with a high DP were synthesized in vivo. He speculated that a multienzyme cell-surface complex catalyzed coordinated endolysis of existing wall cellulose, producing new chain ends as acceptor sites. Hence, cellulase could have a role in the processes leading to wall-loosening, primer formation, and transglucosylation.

These observations contrast with the results of SATOH et al. (1976a), who isolated a short-chain (1→4)-β-glucan from homogenates of *Phaseolus aureus* seedlings. They believed this might be an intermediate polymer that finally leads to the formation of cellulose in the cell wall. The first observations indicating a membrane-bound component which might function as a primer molecule came from VILLEMEZ (1970). Using detergents, urea, and other protein solvents, he isolated material from *Phaseolus aureus* membranes which appeared to be an oligosaccharide derivative of a low molecular weight membrane protein. FRANZ (1976) working with the same system, was able to catalyze glucosylation of a membrane protein with the substrate UDP-glucose. The presence of (1→4)-β-linkages in the carbohydrate moiety of the in vitro-formed glucoprotein could be established. Further it was shown that the product was at least in part susceptible to β-elimination, demonstrating a linkage to serine or threonine. Turnover experiments indicated that the glycoprotein could act as an acceptor for the initiation and subsequent elongation of glucan chains.

These findings were further substantiated by the fact that purified α-cellulose fractions from cotton fibers always contain tightly bound proteins (NOWAK-OSSORIO et al. 1976), which were regarded as a residue of the synthetic apparatus. PONT LEZICA et al. (1978), working with UDP-glucose and particulate enzyme preparations from *Pisum sativum*, could also demonstrate the formation of a membrane-bound glucoprotein. It is not known, however, whether this glucoprotein can serve as a membrane-bound primer for glucan chain elongation. HOPP et al. (1978a), investigating cellulose biosynthesis in the green algae *Prototheca zopfii*, described the formation of a water-soluble polymer from UDP-glucose which again was analyzed to be a short-chain glucoprotein containing

Fig. 1. Proposed scheme for the reactions leading to the formation of glucoproteins and cellulose. (HOPP et al. 1978a)

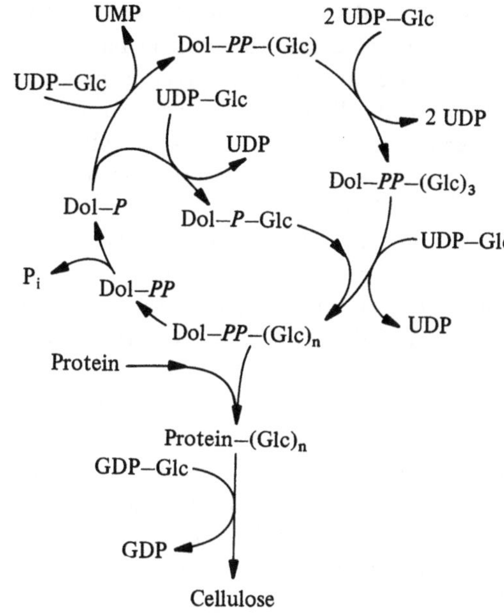

(1→4)-β-linked glucose residues. When GDP-glucose was added to the assay mixture, the soluble glucoprotein became alkali-insoluble and showed properties similar to cellulose. Based on their findings, the authors proposed a reaction sequence, leading to cellulose formation which involves NDP-sugars, UDP-glucose, and GDP-glucose, a hypothetical primer, and further lipid-linked intermediates (Fig. 1). (For further information concerning the biosynthesis of glycoproteins see SELVENDRAN and O'NEILL, Chap. 13 and BOWLES, Chap. 14, Vol. 13A, this series.)

4 Involvement of Lipid-Intermediates in Cellulose Biosynthesis

The role of lipid-linked saccharides in polysaccharide and glycoprotein synthesis is discussed in detail by ELBEIN (Chap. 8, this Vol.). However, some indications of the possible involvement of the so-called lipid-linked intermediates in the field of cellulose biosynthesis will be given, since in this special case of polysaccharide formation, the involvement of such compounds is still a matter of controversy.

There is clear evidence in bacterial, fungal, and mammalian systems that phosphorylated polyprenols function as active carriers of sugars between NDP-sugars and the final cell wall polysaccharides or glycoproteins. A great variety of polyprenols are known to exist in higher plants (WELLBURN and HEMMING 1966) and have been shown to participate in the formation of glycoprotein (ALAM and HEMMING 1971; ERICSON and DELMER 1977, 1978; BEEVERS and MENSE 1977; PONT LEZICA 1979).

KAHN and COLVIN (1961) were the first to postulate a lipid-bound glucose with an intermediate function in bacterial cellulose biosynthesis.

BAILEY et al. (1967), analyzing the composition of particulate enzyme preparations from *Phaseolus aureus* seedlings, found a membrane-bound glucolipid, the exact chemical nature of which was not established. PINSKY and ORDIN (1969) discussed the structural role of lipids in the process of cellulose synthesis in oat seedlings without, however, further characterizing the chemical nature of the lipids involved. Some publications demonstrated the biosynthesis of glucolipids in the in vitro systems; these were shown to be essentially sterol glucosides and were not involved in cellulose biosynthesis (for further information see AXELOS and PÉAUD LENOËL, Chap. 16, Vol. 13A this series). FORSEE and ELBEIN (1972, 1973) demonstrated the synthesis of glucolipids with the biochemical properties of glycosyl phosphoryl-polyprenols by enzyme preparations from cotton fibers. However, the authors were not able to confirm the intermediate role of the lipid derivatives in the process of cellulose formation. Conclusive demonstration of the intermediate function of such a compound is rather difficult, since the known lipid intermediates seem to exhibit a rapid rate of turnover or low steady-state levels. Specific inhibitors are required in order to interrupt the reaction sequence at the glycolipid level.

A further demonstration of the existence and biosynthesis of glucolipids with the properties of a lipid-linked intermediate was presented by BRETT and NORTHCOTE (1975). Their in vitro assay with a particulate pea enzyme preparation and UDP-glucose as substrate-produced glucolipids in which glucose, ranging from one to three glucose residues, was linked through phosphate esters to lipids.

On the other hand, MOROHASHI and BANDURSKI (1976) were not able to detect any labeled glucolipids of the polyprenol-phosphate or polyprenol-pyrophosphate type upon in vivo application of labeled glucose to *Pisum sativum* tissue. HELSPER (1979), working with membrane fractions from pollen tubes of *Petunia hybrida*, found no evidence for the involvement of lipid intermediates. Products synthesized in vitro from UDP-glucose were $(1\rightarrow3)$-β-glucans, sterol glucosides and polyprenol-monophosphate glucose. Kinetic experiments showed a higher initial rate for the formation of $(1\rightarrow3)$-β-glucan than for sterol glucosides or polyprenol monophosphate glucose. Thus, the two glucolipids were not considered to be intermediates for the synthesis of $(1\rightarrow3)$-β-glucans. These findings contrast with the results obtained by PONT LEZICA's group (PONT LEZICA et al. 1975; PONT LEZICA et al. 1976; DALEO and PONT LEZICA 1977; HOPP et al. 1978a; PONT LEZICA et al. 1978; HOPP et al. 1979). Particulate preparations of the chlorophyta *Prototheca zopfii* showed that lipid intermediates are involved in the synthesis of what these authors called a "cellulosic protein". Glucolipids and again oligosaccharide-linked lipids were formed from the substrate UDP-glucose and endogenous lipids. The lipid could be identified as a dolichol-phosphate with a chain length ranging from C_{90} to C_{105}. The oligosaccharides could be transferred from the lipid carrier to a protein acceptor. However, these results were obtained with particulate enzyme preparations from algae and could not be substantiated with higher plant enzymes. A very significant difference seems to be the subcellular localization of the polymerization process.

In the case of higher plants it is generally assumed that the plasma membrane is responsible for the reaction sequence leading to the formation of cellulose (for further information see: Chap. 2, this Vol.). In the case of algae (*Oocystis*), however, there is evidence from the work of ROMANOVICZ and BROWN (1976) that cellulose scales are formed in the dictyosomes. This was substantiated by HOPP et al. (1979) who localized cellulose-synthesizing enzymes in the dictyosomes and ER of chlorophyta.

In summary, one can say that unequivocal evidence for the involvement of lipid-linked intermediates in cellulose synthesis is still lacking.

5 Endogenous and Exogenous Factors Which Might Influence the Biosynthesis of Cellulose

5.1 Hormonal Control of Cellulose Biosynthesis

Since plant hormones have regulatory functions in development, they may also influence the rate and quality of cell wall synthesis. IAA and gibberellin induce elongation growth, a process which is directly linked to cell wall synthesis and turnover.

In *Avena* coleoptiles gibberellic acid treatment results in an increased rate of glucose incorporation into cell wall material (MONTAGUE and IKUMA 1975). The activities of the hemicellulose- and UDP-glucose-β-glucan synthetases are unaffected, whereas the GDP-glucose-β-glucan synthetase is markedly increased (MONTAGUE and IKUMA 1978). Similar effects are attributed to IAA. It promotes turnover reactions of hemicelluloses in pea epicotyls (GILKES and HALL 1977). Although primary wall synthesis is mainly involved in IAA-promoted cell elongation, there is a marked increase in cellulose content and a decrease in noncellulosic glucose polymers in barley coleoptiles (SAKURAI and MASUDA 1978). [^{14}C]-Glucose incorporation is enhanced in oat coleoptiles (HALL and ORDIN 1968) and in pea stems (ABDUL-BAKI and RAY 1971). While decapitated pea stems rapidly lose the particulate UDP-glucose- and GDP-glucose-β-glucan synthetase activities, these activities can be maintained or even restored by IAA application to the cut apex (SPENCER et al. 1972, SHORE and MACLACHLAN 1973). While SPENCER et al. (1972) could inhibit this IAA response by cycloheximide, RAY (1973a) found no cycloheximide inhibition of IAA-induced restoration of UDP-glucose-β-glucan synthetase. He postulated a hormonal induced activation of the enzyme. However, the particulate glucan synthetase from pea and oat could not be activated by IAA in vitro (ORDIN and HALL 1968, RAY 1973b).

In contrast, VAN DER WOUDE et al. (1972), found an increased glucan synthetase activity when 2,4-D was added either to tissue or even to particulate fractions from onion stems. This was discussed as a direct effect of the hormone upon the membrane-bound synthetases. Gibberellin and benzyladenine had no effect on stabilization or restoration of glucan synthetases in pea (SPENCER et al. 1971).

These in part contradictory results do not allow for the moment a conclusive discussion about the different mechanisms of phytohormones involved in the complex of polysaccharide-synthesizing reactions.

5.2 Inhibitors in the Process of Cellulose Formation

Metabolic inhibitors are valuable tools for studying biochemical pathways. Used on the phenomenical and physiological level, they may eventually elucidate biochemical key reactions. Two substances and their derivatives, colchicine and coumarin, have been mainly used in the study of tissue growth, cell expansion, and cell wall formation, processes which may at least in part involve the biosynthesis of cellulose.

Colchicine exerts its effect on microtubules, disturbing mitosis and the deposition of cellulose microfibrils [for the function of microtubules see the review by HEPLER and PALEVITZ (1974) [see also ROBINSON, Chap. 13 this Vol.]. PICKETT-HEAPS (1967) showed that colchicine not only inhibits mitosis, but also leads to irregular wall thickening in xylem cells of wheat seedlings. This was not due to an effect on dictyosomes and their secretory vesicles (MOLLENHAUER and MORRÉ 1976), but rather depends on the disappearance or lack of microtubules (PICKETT-HEAPS 1967; ITOH 1976). Protoplasts of the green alga *Mougeotia* remain spherical when incubated with colchicine. Although microfibrils are formed on the surface of the protoplasts, albeit in a random arrangement, no microtubules are visible (MARCHANT 1979, MARCHANT and HINES 1979). Isopropyl *N*-phenylcarbamate, an inhibitor of microtubule organization (COSS et al. 1975) has the same effects as colchicine (MARCHANT 1979).

Coumarin also has remarkable effects on plant growth. It inhibits root (GOODWIN and TAVES 1950) and seedling growth (HARA et al. 1973), leads to tissue swelling (SVENSSON 1971; HARA et al. 1973), inhibits transverse division of root cells (SVENSSON 1971), and brings about swelling of isolated mesophyll cells (HARADA et al. 1972). In isolated protoplasts both cell wall regeneration and cell division are suppressed by coumarin (BURGESS and LINSTEAD 1977). Coumarin does not seem to act on microtubule arrangement, thus inhibiting the deposition of microfibrils (ITOH 1976; MARCHANT and HINES 1979). However, FASULO et al. (1979) demonstrated a reduced rate of cell division in *Euglena* in the presence of coumarin. SVENSSON (1971) also interpreted his findings as an influence of coumarin on microtubules, thus leading to a random arrangement of microfibrils and eventual inhibition of cell elongation. A reduced rate of cell division in protoplasts in the presence of coumarin (MEYER and HERTH 1978) may not be attributed to a direct coumarin effect on microtubules involved in spindle formation, but rather to a lack of cell wall material, an apparent prerequisite for cell division (MEYER and ABEL 1975).

Coumarin markedly affects ^{14}C-glucose incorporation in sections of bean seedlings. While the incorporation into cellulose is reduced by 40% to 70%, incorporation into other cell wall fractions was barely altered (HARA et al. 1973; HOGETSU et al. 1974a). In the bacterium *Acetobacter xylinum*, cellulose synthesis is also inhibited by coumarin (SATOH et al. 1976b). This might suggest that a general mechanism exists for cellulose biosynthesis.

Dichlorobenzonitrile exhibits similar effects as coumarin, i.e., reduced cell division rates in protoplasts (MEYER and HERTH 1978) and reduced ^{14}C-glucose incorporation into the cellulose fraction by bean seedlings (HOGETSU et al. 1974b).

As coumarin probably acts on the biosynthetic level of cellulose synthesis, one can conclude that the biochemical pathway for cellulose biosynthesis is different at least in some aspects from the pathway of hemicellulose synthesis. Recently it was shown that coumarin inhibits glucose transfer from the poly-prenol-glucoside to the acceptor protein in the course of cellulose biosynthesis in the alga *Prothotheca zopfii* (HOPP et al. 1978b). Although so far no cellulose synthesis via a lipid-linked intermediate has been demonstrated in higher plants, the inhibiting effects of coumarin suggest a similar pathway.

6 Conclusion

The biosynthetic pathways leading to the formation of cellulose macromolecules in the intricate system of the higher plant cell wall still leave major unanswered questions.

The major problem in biochemical studies seems to be the difficult accessibility of the enzyme complex and its fragility. Possibly several enzymic steps have to be coordinated as in the system of glycoprotein synthesis. Therefore the approach via isolated protoplasts which directly exhibit the undisrupted plasma membrane and which are known to produce cellulose microfibrils in in vitro cultures is very promising, although data have been presented that protoplasts do not take up or incorporate UDP-glucose unless they are disrupted. Tracer studies with labeled glucose followed by cell fractionation of the protoplasts and product analysis may finally result in finding a new pathway and precursor.

For biochemical studies new isolation procedures have to be developed in order to preserve the integrity of the membranes and to present the proper side to the artificial substrate. The right milieu for the reactions has also to be created, i.e., concentration of salts, activators, osmoticum etc. A great many prerequisites have to be met to finally solve this intriguing puzzle.

References

Abdul-Baki AA, Ray PM (1971) Regulation by auxin of carbohydrate metabolism involved in cell wall synthesis by pea stem tissue. Plant Physiol 47:537–544

Alam SS, Hemming FW (1971) Betulaprenol phosphate as an acceptor of mannose from GDP-mannose in *Phaseolus aureus* preparations. FEBS Lett 19:60–62

Anderson KL, Ray PM (1978) Labelling of the plasma membrane of pea cells by a surface-localized glucan synthetase. Plant Physiol 61:723–730

Axelos M, Péaud-Lenoël C (1969) Présence et séparation de nucléotide glucose pyrophosphorylases de plantes. Bull Soc Chim Biol 51:261–273

Bailey RW, Haq S, Hassid WZ (1967) Carbohydrate composition of particulate preparations from *Phaseolus aureus* shoots. Phytochemistry 6:293–301
Barber GA, Hassid WZ (1965) Synthesis of cellulose by enzyme preparations from the developing cotton ball. Nature 207:295–296
Barber GA, Elbein AD, Hassid WZ (1964) The synthesis of cellulose by enzyme systems from higher plants. J Biol Chem 239:4056–4061
Bartnicki-Garcia S, Bracker CE, Reys E, Ruiz-Herrera J (1978) Isolation of chitosomes from taxonomically diverse fungi and synthesis of chitin microfibrils *in vitro*. Exp Myocology 2:173–192
Batra KK, Hassid WZ (1970) Determination of linkages of β-D-glucans from *Lupinus albus* and *Avena sativa* by exo-β-(1→3)-D-glucanase. Plant Physiol 45:233–234
Beevers L, Mense RM (1977) Glycoprotein biosynthesis in cotyledons of *Pisum sativum* L. Plant Physiol 60:703–708
Beltrán JP, Carbonell J (1978) (1→3)-β-glucan synthase from *Citrus* phloem. Phytochemistry 17:1531–1532
Brett CT (1978) Synthesis of β-(1→3)-glucan from extracellular uridine diphosphate glucose as a wound response in suspension cultured soybean cells. Plant Physiol 62:377–382
Brett CT, Northcote DH (1975) The formation of oligoglucans linked to lipid during synthesis of β-glucan by characterized membrane fractions isolated from peas. Biochem J 148:107–117
Brown EG, Short KC (1969) The changing nucleotide pattern of sycamore cells during culture in suspension. Phytochemistry 8:1365–1372
Brown RM jr, Montezinos D (1976) Cellulose microfibrils: visualization of biosynthetic and orienting complexes in association with the plasma membrane. Proc Natl Acad Sci USA 73:143–147
Brummond DO, Gibbons AP (1964) The enzymatic synthesis of cellulose by the higher plant *Lupinus albus*. Biochem Biophys Res Commun 17:156–159
Brummond DO, Gibbons AP (1965) Enzymatic cellulose synthesis from UDP-(^{14}C)-glucose by *Lupinus albus*. Biochem Ztg 342:308–318
Buchala AJ, Franz G (1974) A hemicellulosic β-glucan from hypocotyls of *Phaseolus aureus*. Phytochemistry 13:1887–1889
Buchala AJ, Wilkie KCB (1971) The ratio of β-(1→3) to β-(1→4) glucosidic linkages in non-endospermic hemicellulosic β-glucans from oat plant (*Avena sativa*) tissues at different stages of maturity. Phytochemistry 10:2287–2291
Burgess J, Linstead PJ (1977) Coumarin inhibition of microfibril formation at the surface of cultured protoplasts. Planta 133:267–273
Chambers J, Elbein AD (1970) Biosynthesis of glucans in mung bean seedlings. Formation of β-(1→4)-glucans from GDP-glucose and β-(1→3)-glucans from UDP-glucose. Arch Biochem Biophys 138:620–631
Clark AF, Villemez CL (1972) The formation of β-(1→4)-glucan from UDP-α-D-glucose catalyzed by a *Phaseolus aureus* enzyme. Plant Physiol 50:371–374
Coss RA, Bloodgood RA, Brower DL, Pickett-Heaps JD, McIntosh JR (1975) Studies on the mechanism of action of isopropyl N-phenyl carbamate. Exp Cell Res 92:394–398
Cumming DF (1970) Separation and identification of soluble nucleotides in cambial and young xylem tissue of *Larix decidua* Mill. Biochem J 116:189–198
Daleo GR, Pont Lezica R (1977) Synthesis of dolichol phosphate by a cell-free extract from pea. FEBS Lett 74:247–250
Delmer DP (1972a) The purification and properties of sucrose synthetase from etiolated *Phaseolus aureus* seedlings. J Biol Chem 247:3822–3828
Delmer DP (1972b) The regulatory properties of purified *Phaseolus aureus* sucrose synthetase. Plant Physiol 50:469–472
Delmer DP (1977) The biosynthesis of cellulose and other plant cell wall polysaccharides. In: Loewus FA, Runeckles VC (eds) The structure, biosynthesis and degradation of wood. Rec Adv Phytochem Vol 11, Plenum Press, New York, pp 45–77
Delmer DP, Albersheim P (1970) The biosynthesis of sucrose and nucleoside diphosphate glucose in *Phaseolus aureus*. Plant Physiol 45:782–786

Delmer DP, Beasley CA, Ordin L (1974) Utilization of nucleoside diphosphate glucoses in developing cotton fibers. Plant Physiol 53:149–153
Delmer DP, Heininger U, Kulow C (1977) UDP-glucose: glucan synthetase in developing cotton fibers. I. Kinetic and physiological properties. Plant Physiol 59:713–718
Edelman J, Ginsburg V, Hassid WZ (1954) Conversion of monosaccharides to sucrose and cellulose in wheat seedlings. J Biol Chem 213:843–854
Elbein AD, Forsee WT (1973) Studies on the biosynthesis of cellulose. In: Loewus F (ed) Biogenesis of plant cell wall polysaccharides. Academic Press, New York, pp 259–295
Elbein AD, Barber GA, Hassid WZ (1964) The synthesis of cellulose by an enzyme system from a higher plant. J Am Chem Soc 86:309–310
Elnaghy MA, Nordin P (1966) Soluble nucleotides of wheat and oat seedlings. Arch Biochem Biophys 113:72–76
Ericson MC, Delmer DP (1977) Glycoprotein synthesis in plants. I. Role of lipid intermediates. Plant Physiol 59:341–347
Ericson MC, Delmer DP (1978) Glycoprotein synthesis in plants III. Interaction between UDP-N-acetylglucosamine and GDP-mannose as substrates. Plant Physiol 61:819–823
Eschrich W (1956) Kallose (ein kritischer Sammelbericht). Protoplasma 47:487–530
Eschrich W (1965) Physiologie der Siebröhrencallose. Planta 65:280–300
Fasulo MP, Vannini GL, Bruni A, Dal'Olio G (1979) Coumarin as a cytostatic drug for *Euglena gracilis*. A clue to cell cycle study. Z Pflanzenphysiol 93:117–127
Feingold DS, Neufeld EF, Hassid WZ (1958) Synthesis of a β-(1→3)-linked glucan by extracts of *Phaseolus aureus* seedlings. J Biol Chem 233:783–788
Flowers HM, Batra KK, Kemp J, Hassid WZ (1968) Biosynthesis of insoluble glucans from UDP-D-glucose with enzyme preparations from *Phaseolus aureus* and *Lupinus albus*. Plant Physiol 43:1703–1709
Forsee WT, Elbein AD (1972) Biosynthesis of acidic glycolipids in cotton fibers. Possible intermediates in cell wall synthesis. Biochem Biophys Res Commun 49:930–939
Forsee WT, Elbein AD (1973) Biosynthesis of mannosyl- and glucosyl-phosphoryl polyprenols in cotton fibers. J Biol Chem 248:2858–2867
Franz G (1969) Soluble nucleotides in developing cotton hair. Phytochemistry 8:737–741
Franz G (1972) Polysaccharidmetabolismus in den Zellwänden wachsender Keimlinge von *Phaseolus aureus*. Planta 102:334–347
Franz G (1976) The dependence of membrane-bound glucan synthetases on glycoprotein which can act as acceptor molecules. J Polym Sci Part C28:611–621
Franz G, Meier H (1969a) Biosynthesis of cellulose in growing cotton hairs. Phytochemistry 8:597–583
Franz G, Meier H (1969b) Studies of the glucosyl transfer from UDP-glucose to various acceptors using particulate enzyme preparations from different higher plants. XI Int Bot Congr Abstr pp 63–64
Gardner KH, Blackwell J (1974a) The structure of native cellulose. Biopolymers 13:1975–1980
Gardner KH, Blackwell J (1974b) The hydrogen bonding in native cellulose. Biochem Biophys Acta 343:232–242
Gilkes NR, Hall MA (1977) The hormonal control of cell wall turnover in *Pisum sativum*. New Phytologist 78:1–15
Glaser L (1958) The synthesis of cellulose in cell free extracts of *Acetobacter xylinum*. J Biol Chem 232:627–636
Goodwin RH, Taves C (1950) The effect of coumarin derivatives on the growth of *Avena* roots. Am J Bot 37:224–231
Grimes WJ, Jones BL, Albersheim P (1970) Sucrose synthetase from *Phaseolus aureus* seedlings. J Biol Chem 245:188–197
Hall MA, Ordin L (1968) Auxin-induced control of cellulose synthetase activity in *Avena* coleoptile sections. In: Wightman F, Wightman G (eds) Biochem and physiol plant growth substances. Proc 6th Int Conf Plant Subst. Runge Press, Ottawa 1967
Hara M, Umetsu N, Miyamoto C, Tamari K (1973) Inhibition of the biosynthesis of

plant cell wall materials, especially cellulose biosynthesis, by coumarin. Plant Cell Physiol 14:11–28

Harada H, Ohyama K, Chernel J (1972) Effects of coumarin and other factors on the modification of form and growth of isolated mesophyll cells. Z Pflanzenphysiol 66:307–324

Hassid WZ (1969) Biosynthesis of oligosaccharides and polysaccharides in plants. Science 165:137–144

Heiniger U, Delmer DP (1977) UDP-glucose: glucan synthetase in developing cotton fibers II. Structure of the reaction product. Plant Physiol 59:719–723

Heiniger U, Franz G (1980) The role of NDP-glucose pyrophosphorylases in growing mung bean seedlings in relation to cell wall biosynthesis. Plant Sci Lett 17:443–450

Helsper JPF (1979) The possible role of lipid intermediates in the synthesis of β-glucans by a membrane fraction from pollen tubes of *Petunia hybrida*. Planta 144:443–450

Helsper JPF, Veerkamp HH, Sassen MMA (1977) β-glucan synthetase activity in golgi vesicles of *Petunia hybrida*. Planta 133:303–308

Hepler PK, Palevitz BA (1974) Microtubules and microfilaments. Annu Rev Plant Physiol 25:309–362

Hogetsu T, Shibaoka H, Shimokoriyama M (1974a) Involvement of cellulose synthesis in actions of gibberellin and kinetin on cell expansion. Gibberellin-coumarin and kinetin-coumarin interactions on stem elongation. Plant Cell Physiol 15:265–272

Hogetsu T, Shibaoka H, Shimokoriyama M (1974b) Involvement of cellulose synthesis in actions of gibberellin and kinetin on cell expansion. 2,6-Dichlorobenzonitrile as a new cellulose-synthesis inhibitor. Plant Cell Physiol 15:389–393

Hopp EH, Romero PA, Daleo GR, Pont Lezica R (1978a) Synthesis of cellulose precursors. The involvement of lipid-linked sugars. Europ J Biochem 84:561–571

Hopp HE, Romero PA, Pont Lezica R (1978b) On the inhibition of cellulose biosynthesis by coumarin. FEBS Lett 86:259–262

Hopp HE, Romero PA, Daleo GR, Pont Lezica R (1979) Polyprenylphosphate derivatives as intermediates in the biosynthesis of cellulose precursors. Subcellular localisation. In: Appelquist L-A, Liljenberg C (eds) Adv biochem and physiol of plant lipids, Elsevier/North Holland Biomedical Press, Amsterdam, pp 313–318

Huwyler HR, Franz G, Meier H (1978) β-(1→3)-glucans in the cell walls of cotton fibres (*Gossypium arboreum* L). Plant Sci Lett 12:55–62

Huwyler HR, Franz G, Meier H (1979) Changes in the composition of cotton fibre cell walls during development. Planta 146:635–642

Isherwood FA, Selvendran RR (1970) A note of the occurrence of nucleotides in strawberry leaves. Phytochemistry 9:2265–2269

Itoh T (1976) Microfibrillar orientation of radially enlarged cells of coumarin- and colchicine-treated pine seedlings. Plant Cell Physiol 17:385–398

Kahn AW, Colvin IR (1961) Synthesis of bacterial cellulose from labelled precursor. Science 133:2014–2015

Kemp J, Loughman BC (1973) Chain initiation in glucan biosynthesis in mung beans. Biochem Soc Trans 1:446–448

Kemp J, Loughman BC (1974) Cyclitol glucosides and their role in the synthesis of a glucan from uridine diphosphate glucose in *Phaseolus aureus*. Biochem J 142:153–159

Kemp J, Loughman BC, Ephritikhine G (1978) The role of inositol glucosides in the synthesis of β-1,3-glucans. In: Wells WW, Eisenberg F jr (eds) Cyclitoles and Phorphoinositides, Academic Press, New York pp 439–449

Larsen GL, Brummond DO (1974) β-(1→4)-D-glucan synthesis from UDP-(^{14}C)-D-glucose by a solubilized enzyme from *Lupinus albus*. Phytochemistry 13:361–365

Lavinthman N, Tandecarz J, Carceller M, Mendiara S, Cardini CE (1974) Role of Uridine diphosphate glucose in the biosynthesis of starch. Mechanism of formation and enlargement of a glucoproteinic acceptor. Europ J Biochem 50:145–155

Lehle L, Tanner W (1973) The function of myo-inositol in the biosynthesis of raffinose. Purification and characterization of galactinol : sucrose 6-galactosyltransferase from *Vicia faba* seeds. Europ J Biochem 38:103–110

Leloir LF (1951) The enzymatic transformation of uridine diphosphate glucose into a galactose derivate. Arch Biochem Biophys 33:186–190

Leloir LF (1964) Nucleoside diphosphate sugars and saccharide synthesis. Biochem J 91:1–8

Maclachlan GA (1976) A potential role for endo-cellulase in cellulose biosynthesis. J Polym Sci Part C 28:582–596

Marchant HJ (1979) Microtubules, cell wall deposition and the determination of plant cell shape. Nature 278:167–168

Marchant HJ, Hines ER (1979) The role of microtubules and cell-wall deposition in elongation of regenerating protoplasts of *Mougeotia*. Planta 146:41–48

Marx-Figini M (1969) On the biosynthesis of cellulose in higher and lower plants. J Polym Sci C 28:57–67

Marx-Figini M, Schulz GV (1966) Comparison of the biosynthesis of cellulose *in vitro* and *in vivo* in cotton balls. Nature 210:754–755

Meinert MC, Delmer DP (1977) Changes in biochemical compositions of the cell wall of the cotton fiber during development. Plant Physiol 59:1088–1097

Meyer Y, Abel WO (1975) Importance of the wall for cell division and in the activity of the cytoplasma in cultured tobacco protoplasts. Planta 123:33–40

Meyer Y, Herth W (1978) Chemical inhibition of cell wall formation and cytokinesis, but not of nuclear division, in protoplasts of *Nicotiana tabacum* L. cultivated *in vitro*. Planta 142:253–262

Miyamoto C, Tamari K (1973) Synthesis of β-(1→3)-glucan and β-(1→3) and β-(1→4)-glucan from UDP-D-glucose. Agric Biol Chem 37:1253–1260

Mollenhauer HH, Morré DJ (1976) Cytochalasin B, but not colchicine, inhibits migration of secretory vesicles in roots tips of maize. Protoplasma 87:39–48

Montague MJ, Ikuma H (1975) Regulation of cell wall synthesis in *Avena* stem segments by gibberellic acid. Plant Physiol 55:1043–1047

Montague MJ, Ikuma H (1978) Regulation of glucose metabolism and cell wall synthesis in *Avena* stem segments by gibberellic acid. Plant Physiol 62:391–396

Morohashi Y, Bandurski RS (1976) Glucolipids of *Zea mays* and *Pisum sativum*. Plant Physiol 57:846–849

Neish AC (1955) The biosynthesis of cell wall carbohydrates: Formation of cellulose and xylan from labelled monosaccharides in wheat plants. Can J Biochem Physiol 33:658–666

Nikaido H, Hassid WZ (1971) Biosynthesis of saccharides from glycopyranosyl esters of nucleoside phyrophosphates ("sugar nucleotides"). Adv Carbohydr Chem Biochem 26:351–483

Nowak-Ossorio M, Gruber E, Schurz J (1976) Untersuchungen zur Cellulosebildung in Baumwollsamen. Protoplasma 88:255–263

Ordin L, Hall MA (1967) Studies of cellulose synthesis by a cell free oat coleoptile enzyme system: Inactivation by airborne oxidants. Plant Physiol 42:205

Ordin L, Hall MA (1968) Cellulose synthesis in higher plants from UDP-glucose. Plant Physiol 43:473–476

Péaud-Lenoël C, Axelos M (1970) Structural features of the β-glucans enzymatically synthesized from UDP-glucose by wheat seedlings. FEBS Lett 8:224–228

Philippi ML (1978) Enzystierung bei *Polysphondylium* pallidum: Cellulosesynthese, Lokalisation der Cellulosesynthetase und biochemischen Kontrollmechanismen. Diss Univ Zürich, 87 p

Pickett-Heaps JD (1967) The effects of colchicine on the ultrastructure of dividing plant cells, xylem wall differentiation and distribution of cytoplasmic microtubules. Develop Biol 15:206–236

Pinsky A, Ordin L (1969) Role of lipid in the cellulose synthetase enzyme system from oat seedlings. Plant Cell Physiol 10:771–785

Pont Lezica R (1979) Glycosylation of glycoproteins in green plants and algae. Biochem Soc Trans 7:334–337

Pont Lezica R, Brett CT, Martinez PR, Dankert MA (1975) A glucose acceptor in plants with the properties of an α-saturated polyprenyl-monophosphate. Biochem Biophys Res Commun 66:980–987

Pont Lezica R, Romero PA, Dankert MA (1976) Membrane bound UDP-glucose – lipid glucosyltransferases from peas. Plant Physiol 58:675–680
Pont Lezica R, Romero PA, Hopp HE (1978) Glycosylation of membrane bound proteins by lipid linked glucoses. Planta 140:177–183
Ray PM (1973a) Regulation of β-glucan synthetase activity by auxin in pea stem tissue II. Metabolic requirements. Plant Physiol 51:609–614
Ray PM (1973b) Regulation of β-glucan synthetase activity by auxin in pea stem tissue I. Kinetic aspects. Plant Physiol 51:601–608
Ray PM, Shininger MM, Ray MM (1969) Isolation of β-glucan synthetase particles from plant cells and identification with golgi membranes. Proc Natl Acad Sci USA 64:605–612
Raymond Y, Fincher GB, Maclachlan GA (1978) Tissue slice and particulate β-glucan synthetase activities from Pisum epicotyls. Plant Physiol 61:938–942
Robinson DG (1977) Plant cell wall synthesis. Adv Bot Res 5:89–151
Robinson DG, Preston RD (1972) Polysaccharide synthesis in mung mean roots. An X-ray investigation. Biochem Biophys Acta 273:336–345
Robyt JF (1979) Mechanism involved in the biosynthesis of polysaccharides. TIBS Feb 47–49
Romanovicz PK, Brown RM (1976) Biogenesis and structure of golgi derived cellulose scales in *Pleurocrysis*. II Scale composition and supramolecular structure. Appl Polym Symp 28:587–610
Sakurai N, Masuda Y (1978) Auxin-induced extension, cell wall loosening and changes in the wall polysaccharide content of barley coleoptile segments. Plant Cell Physiol 19:1225–1233
Satoh H, Matsuda K, Tamani K (1976a) β-(1→4)-glucan occurring in homogenate of *Phaseolus aureus* seedlings. Possible nascent stage of cellulose biosynthesis in vivo. Plant Cell Physiol 17:1243–1254
Satoh H, Takahara M, Matsuda K (1976b) Inhibition of the synthesis of microbial cellulose by coumarin. Plant Cell Physiol 17:1077–1080
Selvendran RR, Isherwood F (1967) Identification of guanosine diphosphate derivatives of D-xylose, D-mannose, D-glucose and D-galactose in mature strawberry leaves. Biochem J 105:723–728
Shafizadeh F, McGinnis GD (1971) Morphology and biogenesis of cellulose and plant cell walls. Adv Carbohydr Chem Biochem 26:297–349
Shore G, Maclachlan GA (1973) Indoleacetic acid stimulates cellulose deposition and selectively enhances certain β-glucan synthetase activities. Biochem Biophys Acta 329:271–282
Shore G, Raymond Y, Maclachlan GA (1975) The site of cellulose synthesis. Cell surface and intracellular β-(1→4)-glucan (cellulose) synthetase activities in relation to the stage and direction of cell growth. Plant Physiol 56:34–38
Smith MM, McCully ME (1977) Mild temperature "stress" and callose synthesis. Planta 136:65–70
Smith JE, Stanley RG (1969) Stereospecific regulation of plant glucan synthetase *in vitro*. Can J Botany 47:484–496
Smith MM, Stone BA (1973) β-Glucan synthesis by cell-free extracts from *Lolium multiflorum* endosperm. Biochim Biophys Acta 313:72–94
Solms J, Hassid WZ (1957) Isolation of uridine diphosphate N-acetyl-glucosamine and uridine diphosphate glucuronic acid from mung bean seedlings. J Biol Chem 228:357–364
Southworth D, Dickinson DB (1975) β-(1→3)-glucan synthase from *Lilium longiflorum* pollen. Plant Physiol 56:83–87
Sowokinos JR (1976) Pyrophosphorylases in *Solanum tuberosum*. I. Changes in ADP-glucose and UDP-glucose pyrophosphorylase activities associated with starch biogenesis during tuberization, maturation and storage of potatoes. Plant Physiol 57:63–68
Spencer FS, Maclachlan GA (1972) Changes in molecular weight of cellulose in the pea epicotyl during growth. Plant Physiol 49:58–63
Spencer FS, Ziola B, Maclachlan GA (1971) Particulate glucan synthetase activity: Dependence on acceptor, activator and plant growth hormone. Can J Biochem 49:1326–1332
Spencer FS, Shore G, Ziola B, Maclachlan GA (1972) Particulate glucan synthetase activity: generation and inactivation after treatments with indolacetic acid and cycloheximide. Arch Biochem Biophys 152:311–317

Stepanenko BN, Morozova AB (1970) Possibility of cellulose hiosynthesis from UDP-glucose in the cotton plant. Fiziol Rast 17:302–308

Su JS (1965) Carbohydrate metabolism in the shoots of bamboo III. Separation and identification of nucleotides. J Chin Agr Chem Soc:45–54

Svensson S-B (1971) The effect of coumarin on root growth and root histology. Physiol Plant 24:446–470

Tanner W, Kandler O (1968) *Myo*-inositol, a cofactor in the biosynthesis of stachyose. Eur J Biochem 4:233–239

Tsai CM (1975) Isolation of UDP-glucose: β-(1→4)-glucan glucosyltransferase from oat and its properties. J Carbohydr Nucleosid Nucleotid 2:419–431

Tsai CM, Hassid WZ (1971) Solubilisation and separation of UDP-D-glucose: β-1(1→4)-glucan and UDP-D-Glucose: β-(1→3)-glucan glucosyltransferases from coleoptiles of *Avena sativa*. Plant Physiol 47:740–744

Tsai CM, Hassid WZ (1973) Substrate activation of β-(1→3)-glucan synthetase and its effect on the structure of β-glucan obtained from UDP-D-glucose and particulate enzyme of oat coleoptiles. Plant Physiol 51:998–1001

Vidra JD, Loerch JD (1968) A study of pyrophosphorylase activities in maize endosperm. Biochim Biophys Acta 159:551–553

Villemez CL jr (1970) Characterization of intermediates in plant cell wall biosynthesis. Biochem Biophys Res Comm 40:636–641

Villemez CL jr (1971) Rate studies of polysaccharide biosynthesis from GDP-α-D-glucose and GDP-α-D-mannose. Plant Physiol 121:151–157

Villemez CL jr (1974) The relation of plant enzyme-catalyzed β-(1→4)-glucan synthesis to cellulose biosynthesis in vivo. In: Pridham J (ed) Plant carbohydrate biochemistry. Academic Press, New York pp 183–189

Villemez CL jr, Heller JS (1970) Is guanosine diphosphate-D-glucose a precursor of cellulose? Nature 227:80–81

Villemez CL jr, Franz G, Hassid WZ (1967) Biosynthesis of alkali insoluble polysaccharide from UDP-D-glucose with particulate enzyme preparation from *Phaseolus aureus*. Plant Physiol 42:1219–1223

Waechter CJ, Lennarz WJ (1976) The role of polyprenol-linked sugar in glycoprotein synthesis. Annu Rev Biochem 45:95–112

Wellburn AR, Hemming FW (1966) The occurrence and seasonal distribution of higher isoprenoid alcohols in the plant kingdom. Phytochemistry 5:969–975

Van Der Woude WJ, Lembi CA, Morré DJ (1972) Auxin (2,4-D) stimulation (in vivo and in vitro) of polysaccharide synthesis in plasma membrane fragments isolated from onion stems. Biochem Biophys Res Commun 46:245–253

Van Der Woude WJ, Lembi CA, Morré DJ, Kindinger JI, Ordin L (1974) β-glucan synthetase of plasma membrane and golgi apparatus from onion stems. Plant Physiol 54:333–340

6 Metabolism of Noncellulosic Polysaccharides

G.B. FINCHER and B.A. STONE

1 Cell Walls and Cell Wall Metabolism

The cell wall is a polysaccharide-rich, extracellular structure which overlays and encloses the protoplast of higher plant cells. In meristematic cells, the wall is thin and consists of microfibrils embedded in a gel-like matrix. During maturation and specialization, the wall may be thickened and additional wall layers, also composed of microfibrils embedded in a matrix, are often deposited.

In all higher plants the microfibrils are composed of cellulose, an extended, ribbon-like polymer of $(1 \rightarrow 4)$-β-glucopyranosyl residues. These molecules aggregate through extensive inter-molecular hydrogen bonding to form microfibrils. The polysaccharides of the matrix phase, in which the microfibrils are embedded, are heteropolysaccharides. These molecules, like cellulose, generally have extended conformations, but their potential for packing is reduced by kinking of the backbone, as in $(1 \rightarrow 3)$;$(1 \rightarrow 4)$-β-glucans and rhamnogalacturonans, by the presence of side chains, as in heteroxylans, heteroglucans, and heteromannans, or by the nonrepeating arrangement of monomers in the backbone of the molecule, as in glucomannans.

Although these heteropolysaccharides do not pack in a regular way, they may interact through noncovalent forces over limited segments of the molecules to form three-dimensional networks or gel-like matrices. Some local ordering of matrix components in walls is indicated by infra red analysis (MORIKAWA and SENDA 1978, MORIKAWA et al. 1978) and birefringence studies (REIS et al. 1978).

The segments of interaction between polysaccharides are known as "junction zones". Their stability and frequency, which is largely responsible for the strength, resilience, porosity, and plasticity of the matrix (REES and WELSH 1977, DEA 1979), depend on the structure of the interacting polymers. In some cases, their stability approaches that of a crystal lattice. The possibility of covalent association between matrix polysaccharides has been suggested (ALBERSHEIM 1978) but so far linkages of this type have not been specifically identified (MCNEIL et al. 1979).

The hydrated, porous gel matrix of the wall allows apoplastic movement of water, ions, and molecules with diameters up to 5 nm (CARPITA et al. 1979) and also provides some protection of the protoplast from osmotic and mechanical damage. In keeping with the dynamic nature of the wall, the matrix holds a range of hydrolytic and oxidative enzymes involved in post-deposition modification of the wall (Chap. 7, this Vol.). Some of the wall enzymes may also participate in pre-absorptive metabolism of cell nutrients.

Changes in cell walls which accompany growth and differentiation involve metabolism of both matrix and fibrillar phase components. In dividing cells new walls originate at the cell plate by deposition of matrix polysaccharides. The cell plate region can later be recognized as the middle lamella. Further deposition of matrix materials, together with microfibrils, leads to formation of the primary cell wall. This wall may be modified to allow cell expansion during growth and as the surface area increases there is a concomitant deposition of new cell wall material. It is at this stage that cell shape is determined. While the walls of some cells, such as parenchyma and collenchyma, undergo no further observable modification after primary wall deposition has ceased, in other cells, wall composition and organization may be modified to serve a particular function. Thus a secondary wall layer, which is readily distinguished from the primary wall by its morphology and polysaccharide composition, and usually by the presence of the noncarbohydrate polymer lignin, may be deposited. For example, walls of sclerenchyma cells, fibers, and tracheids are thickened and strengthened in this way. During differentiation in other cells, walls may be modified by deposition of waxy cuticles as in the epidermis of leaves, stems and some fruits; by the secretion of mucilaginous slime as in root cap cells; or even by the destruction of the wall fabric as in xylem vessel formation. In senescing tissues such as fruits and leaves, cell walls may be dissolved or softened and in germinating seeds, endosperm cell walls are wholly or partially digested. The wall structure may also be altered in response to environmental changes imposed by water, ionic and temperature stress, to wounding and to invasion by pathogens.

In this chapter the molecular and cellular aspects of noncellulosic, cell wall polysaccharide synthesis and degradation, and their control, are considered.

2 Molecular Aspects of Polysaccharide Synthesis

The biosynthesis of polysaccharides destined for the cell wall may be considered with reference to (1) the establishment of the pool of nucleoside diphosphate sugars (nucleotide sugars), the primary donors for the polymerization process; (2) the assembly of glycosyl residues from nucleotide sugars into polysaccharides, either directly or via intermediary molecules. This process involves chain initiation, copolymerization of different sugars to form heteropolymers, specification of anomeric configuration and linkage position, introduction of branch points and chain termination; (3) the esterification or etherification of glycosyl residues on the polysaccharide; (4) the control of the polymerization process.

2.1 Origins of Monosaccharides and Their Activated Forms

Studies using labeled monosaccharides showed that glucose can be the precursor of all cell wall monosaccharides, although other sugars such as galactose or xylose can serve but are less effective (NEISH 1958).

Each monosaccharide is transferred, directly or indirectly, from a nucleotide sugar to the growing chain. FEINGOLD (Chap. 1, Vol. 13A, this series) has described the routes by which nucleotide sugars are formed from their ultimate sources, glucose, fructose, or sucrose. These sugars are translocated from photosynthetic or storage tissues to the sites of cell wall synthesis and converted there to UDP-Glc, GDP-Glc, etc. (FEINGOLD, Chap. 1, Vol. 13A, this series). Hexose phosphates arising from the catabolism of starch or fructans in the growing tissue may also be a primary monosaccharide source. The metabolic schemes involved in the generation of nucleotide sugars are shown in Fig. 1 of FEINGOLD's Chapter.

2.2 Polysaccharide Assembly

Current methods of analysis allow detailed descriptions of the molecular architecture of noncellulosic cell wall polysaccharides, but the information on how these often complex molecules are assembled is fragmentary. This section describes the present information concerning the various phases of polysaccharide assembly.

2.2.1 Polymerizing Systems

A number of enzyme preparations capable of synthesizing noncellulosic polysaccharides have been examined for their nucleotide sugar specificity and structure of the synthetic product. The data available are collected in Table 1. Except where indicated the sources of the enzymes are mixed-membrane preparations which may bear more than one synthetase, furthermore they may be associated with epimerases and other enzymes able to interconvert nucleotide sugars.

The incorporation of labeled sugars into polymeric material has been interpreted as polysaccharide biosynthesis, but in many cases this conclusion has not been warranted on the basis of the experimental data. Real polysaccharide synthesis involves incorporation of a large number of monomers to form new polymeric chains of high DP (degree of polymerization) rather than the addition of a few residues to pre-existing polymeric acceptors. In most cases this has not been rigorously demonstrated.

2.2.2 Lipid-Saccharide Intermediates

Assembly of saccharide components of glycoproteins involves lipid-glycosyl intermediates of the polyprenol-pyrophosphate or -phosphate type (ERICSON and DELMER 1977, NAGAHASHI and BEEVERS 1978, PONT LEZICA et al. 1978, NORTHCOTE 1979a, ELBEIN 1979, DÜRR et al. 1979). Although there is evidence for the participation of these intermediates in cellulose synthesis (PONT LEZICA et al. 1978, FRANZ and HEINIGER, Chap. 5, this Vol.) their involvement in noncellulosic polysaccharide synthesis has not been conclusively established.

BRETT and NORTHCOTE (1975) found that polyprenol pyrophosphate β-oligoglucosides with $(1\rightarrow3)$- and $(1\rightarrow4)$-linkages in the ratio 3:1 are formed by

Table 1. Product structure and donor specificity of noncellulosic polysaccharide synthetases from higher plant sources

Polymer (Linkage type)	Donor	Source	Sub-cellular fraction	References
Mannan [(1 → 4)-β-mannosyl]	GDP-D-Man	*Orchis morio* tubers		FRANZ (1973)
	GDP-D-Man	*Phaseolus aureus* hypocotyls		HELLER and VILLEMEZ (1972)
	GDP-D-Man	*Acer pseudo-platanus* suspension cultures		SMITH et al. (1976)
Glucomannan [(1 → 4)-β-mannosyl, (1 → 4)-β-glucosyl]	GDP-D-Man GDP-D-Glc	*Pisum sativum* internodes		VILLEMEZ (1974), HINMAN and VILLEMEZ (1975)
Unidentified	GDP-D-Man GDP-D-Glc		Golgi, endo-plasmic reticulum	DÜRR et al. (1979), ELBEIN (1969)
Arabinan	UDP-L-Ara	*Phaseolus aureus* shoots		ODZUK and KAUSS (1972)
Galacturonan [(1 → 4)-α-galacturonyl]	UDP-D-GalU	*Phaseolus aureus* seedlings		VILLEMEZ et al. (1965, 1966)
	TDP-D-GalU	*Phaseolus aureus* seedlings		LIN et al. (1966)
Xylan [(1 → 4)-β-xylosyl]	UDP-D-Xyl	*Phaseolus aureus*		ODZUK and KAUSS (1972)
		Zea mays immature cobs		BAILEY and HASSID (1966)
Galactan [(1 → 4)-β-galactosyl, some (1 → 3)-β-galactosyl]	UDP-D-Gal	*Phaseolus aureus* seedlings		MCNAB et al. (1968), PANAYOTATOS and VILLEMEZ (1973)
Xyloglucan	UDP-D-Glc UDP-D-Xyl	*Phaseolus aureus*		VILLEMEZ and HINMAN (1975)
		Pisum sativum	Golgi-rich	RAY (1975), RAY et al. (1976)
(1 → 3);(1 → 4)-β-glucan [(1 → 3)-β-glucosyl, (1 → 4)-β-glucosyl]	UDP-D-Glc	*Avena sativa* coleoptiles		ORDIN and HALL (1967, 1968)

Table 1 (continued)

Polymer (Linkage type)	Donor	Source	Sub-cellular fraction	References
		Triticum vulgare roots		PÉAUD-LENOËL and AXELOS (1970)
		Lolium multiflorum endosperm suspension cultures		SMITH and STONE (1973a), COOK et al. (1980)
(1→3)-β-glucan [(1→3)-β-glucosyl]	UDP-D-Glc	*Allium cepa* stem	Golgi and plasma membranes	VAN DER WOUDE et al. (1974)
		Lilium longiflorum pollen tubes		SOUTHWORTH and DICKINSON (1975)
		Avena sativa coleoptiles		ORDIN and HALL (1967, 1968), BATRA and HASSID (1970), TSAI and HASSID (1971, 1973)
		Zea mays coleoptiles		HENDRICKS (1978)
		Lolium multiflorum endosperm suspension cultures		SMITH and STONE (1973a), COOK et al. (1980)
		Citrus aurantifolia phloem		BELTRÁN and CARBONELL (1978)
		Glycine max suspension cultures		BRETT (1978)
		Gossypium hirsutum seed hairs		ELBEIN and FORSEE (1973), HEINIGER and DELMER (1977)
		Lupinus albus hypocotyls		FLOWERS et al. (1968), THOMAS and STANLEY (1968), THOMAS et al. (1969), BATRA and HASSID (1970), LARSEN and BRUMMOND (1974)
		Petunia hybrida pollen tubes	Golgi-rich	HELSPER et al. (1977)

Table 1 (continued)

Polymer (Linkage type)	Donor	Source	Sub-cellular fraction	References
		Phaseolus aureus seedlings		FEINGOLD et al. (1958), CHAMBERS and ELBEIN (1970), FLOWERS et al. (1968), MIYAMOTO and TAMARI (1973), CLARK and VILLEMEZ (1972)
		Phaseolus vulgaris		MUSOLAN et al. (1975), MUSOLAN and KINDINGER (1978)
		Vigna sinensis		MUSOLAN et al. (1975)
		Pisum sativum	plasma membrane	ANDERSON and RAY (1978), RAYMOND et al. (1978)
Fucosyl (acid treated root cap mucilage acceptor)	GDP-L-Fucose		smooth endoplasmic reticulum and dictyosomes	JAMES and JONES (1979a, b)
		Zea mays roots	endoplasmic reticulum and dictyosomes	GREEN and NORTHCOTE (1979a, b)

Pisum dictyosome preparations, but their role in biosynthesis of β-glucans was not defined. Specific inhibitors of polyprenol pyrophosphate sugar formation (ELBEIN, Chap. 8, this Vol.) did not prevent $(1 \rightarrow 3)$-β-glucan synthesis by *Phaseolus* preparations (ERICSON et al. 1978) and kinetic and double-labeling experiments with a *Petunia* pollen tube preparation showed that polyprenol monophosphate glucose was not an obligatory intermediate in the formation of $(1 \rightarrow 3)$-β-glucans (HELSPER 1979). Other systems have not been assessed in these ways.

2.2.3 Chain Initiation

There is little information concerning the signal for chain initiation or its control for any noncellulosic wall polysaccharide. WHELAN (1976) has suggested that all nascent polysaccharide chains may be covalently associated with small amounts of protein. In proteoglycan and glycoprotein formation, the glycosyl transferase which initiates chain synthesis recognizes a specific amino acid sequence in the protein (RODÉN and SCHWARTZ 1975, MARSHALL 1979). Oligosaccharide-proteins have been reported to be involved in cellulose synthesis (FRANZ 1976, SATOH et al. 1976, HOPP et al. 1978a, b, PONT LEZICA et al. 1978, FRANZ and HEINIGER, Chap. 5, this Vol.). BRETT and NORTHCOTE (1975) have provided evidence for a protein-linked intermediate in the biosynthesis of *Pisum* cell wall polysaccharides. Glucosyl inositol, which may be protein-bound, has been suggested as an acceptor in (1→3)-β-glucan synthesis (KEMP et al. 1978).

2.2.4 Chain Elongation and Direction of Growth

Elongation of the polysaccharide chain, whether by the addition of monosaccharides or oligosaccharides, could occur at the nonreducing end (tailward growth) or at the reducing end (headward growth) (ROBBINS et al. 1967). Information on this aspect is totally lacking for noncellulosic cell wall polysaccharides. "Tailward growth" operates in amylose biosynthesis, whereas polymerization of certain bacterial cell walls (ROBBINS et al. 1967) and extracellular polysaccharides such as dextran (ROBYT 1979) is by "headward growth".

2.2.5 Sequences of Linkages and Monosaccharides, and Insertion of Side Branches

Noncellulosic cell wall polysaccharides are characteristically heteropolymers with features which enable them to form gel-like assemblies in the matrix (Sect. 1). Thus, in contrast to the strictly repeating sequences in animal connective tissue polysaccharides (RODÉN and SCHWARTZ 1975) and in certain bacterial capsular polysaccharides such as those from *Streptococcus pneumoniae* (LARM and LINDBERG 1976), plant cell wall polysaccharides do not characteristically have repeating units in their backbones or regularly inserted side branches. For example, although sequences of two or three (1→4)-linked residues separated by single (1→3)-linked residues predominate in (1→3);(1→4)-β-glucans (CLARKE and STONE 1962), the sequences are not regularly arranged and account for only 75% to 80% of the molecule. In wheat endosperm heteroxylans the arabinofuranosyl substituents are not evenly distributed; some chain segments are heavily substituted while others are unsubstituted (EWALD and PERLIN 1959, GOLDSCHMID and PERLIN 1963). The partially ordered sequence of linkages or monomers in the main chain or of substituents on the main chain in noncellulosic wall polysaccharides is consonant with the view that their assembly is not a template-directed process, as is the synthesis of proteins and nucleic acids, but is directed by the specificity of the transferases involved. However, the mechanism of assembly of matrix polysaccharides with more than one linkage type or monomer, or with side branches, is not understood. For example, it is not known whether the addition of substituents to the main chains of heteroxylans, heteroglucans or heteromannans is achieved concurrently with main chain synthesis or occurs later. In the fungus *Sclerotium rolfsii* (1→6)-linked glucosyl substituents are added to a (1→3)-β-glucan backbone as synthesis proceeds (BATRA et al. 1969). The synthetases for cell wall polysaccharides are invariably membrane-bound and presumably could exist as a complex which, by concerted action, might control the proportion and sequence of residues, linkage types or branches. Thus for the $(1\rightarrow3);(1\rightarrow4)$-$\beta$-glucan the proportion of $(1\rightarrow3)$- and $(1\rightarrow4)$-linkages could be controlled by two linkage-specific synthetases with different affinities for the terminal saccharide units. In fact, oat coleoptile membrane preparations, which can synthesize $(1\rightarrow3);(1\rightarrow4)$-$\beta$-glucans in addition to $(1\rightarrow3)$- and $(1\rightarrow4)$-β-glucans, can be fractionated to yield separate $(1\rightarrow3)$- and $(1\rightarrow4)$-β-glucan synthetases (TSAI and HASSID 1973) but whether in combination these enzymes synthesize a $(1\rightarrow3);(1\rightarrow4)$-$\beta$-glucan or whether a specific $(1\rightarrow3);(1\rightarrow4)$-$\beta$-glucan synthetase exists is not known. An alternative possibility is that the heteropolymers are assembled from preformed blocks. Thus BRETT and NORTHCOTE (1975) suggested that in the synthesis of mixed-linked polysaccharides, oligosaccharides containing a single linkage type might be first assembled on a lipid or protein intermediate and subsequently joined by a second linkage type to form the completed polysaccharide.

The arabino-$(1\rightarrow3);(1\rightarrow6)$-$\beta$-galactans from *Larix* and *Acacia* spp. appear to be composed of blocks of molecular weight about 6,000 (CHURMS et al. 1977, 1978) which would be consistent with assembly en bloc of preformed saccharides.

2.2.6 Chain Termination

The mechanism of chain termination in polysaccharide biosynthesis is not clear. In fact the matrix polysaccharides of cell walls characteristically exhibit molecular weight polydispersity, which is consistent with enzyme- rather than template-directed polymerization. The glucomannan (VILLEMEZ 1974) and the arabinan (ODZUK and KAUSS 1972) synthesized in vitro by mung bean preparations are also polydisperse with molecular weight averages of 200,000 to 300,000 and 80,000 to 100,000 respectively.

2.3 Biosynthesis of Glycosyl Ethers and Esters

Ester or ether derivatives of monosaccharide residues are frequently found in noncellulosic cell wall polysaccharides (ASPINALL, Chap. 1, this Vol.). It is not known whether these modifications occur before polymerization, concurrently with polymerization, or after the polysaccharide has been assembled. However, VILLEMEZ et al. (1965, 1966) showed that UDP-methyl-galacturonate was not a substrate for polygalacturonate synthetase, indicating that esterification of carboxylic acid groups was achieved after incorporation. *S*-Adenosyl L-methionine is the donor of the methyl groups in a reaction catalyzed by a membrane-bound system (KAUSS and SWANSON 1969). The 4-*O*-methyl groups on glucuronate residues found in other cell wall polysaccharides (Chap. 1, this Vol.) may also be derived from *S*-adenosyl methionine.

The mechanism of acetylation of xylans and pectic polysaccharides has not been documented, although in bacterial systems a membrane-associated enzyme catalyzes the transfer of acyl groups from acyl CoA to lipopolysaccharides (TUNG and BALLOU 1973). Ferulic acid and other phenolic acids occur in ester linkages to cell wall polymers (GEISSMANN and NEUKOM 1973, HARTLEY 1973) but nothing is known of the mechanism for their incorporation. The esterified phenolic acids may function as initiating sites for subsequent polymerization of phenolic-containing macromolecules such as lignin and suberin.

2.4 Control

The polysaccharide composition of the middle lamella, primary wall, and secondary wall layers formed during development show characteristic, qualitative, and quantitative differences (MEIER 1961, THORNBER and NORTHCOTE 1961a, b, 1962, DALESSANDRO and NORTHCOTE 1977a, b, c). The control of polysaccharide synthesis which leads to these differences may be exerted at three points:

1. Through supply of precursors or the intracellular sugar and sugar phosphate pool.
2. Through modulation of the activity of pre-existing enzymes which establish the pool of nucleotide sugars and the enzymes which utilize these as substrates in polymerization reactions.
3. Through control of the amounts of these enzymes and the availability of acceptor molecules.

The stages of monosaccharide interconversion, activation, and polymerization leading to polysaccharide biosynthesis are shown in Fig. 1.

2.4.1 Supply of Monosaccharide Precursors

Little is known about the effect of the supply of glucose, fructose, or sucrose on the amount of wall polysaccharides formed. However, as in starch grain formation, diurnal fluctuations in wall deposition might be expected if photosynthetic sugars were the source of substrate; this has been shown to be the case in cotton hairs, where growth rings are seen in cell walls developing under normal day–night conditions, but are absent in hairs developing in constant light (ANDERSON and KERR 1938), (Fig. 2). Such growth rings are not generally encountered, but have also been observed in stone cells in walnuts (ROTHERT and REINKE, cited in STRASBURGER 1976).

2.4.2 Activation and Interconversion of Monosaccharides

The activation of monosaccharides to the nucleotide form and the reactions involved in the subsequent interconversion of the activated monosaccharides are possible control points in wall biosynthesis (see Fig. 3) since these reactions would control the availability of substrates for the polysaccharide synthetases. The activities of the various enzymes responsible for maintenance of the nucleotide sugar pool varies with the stage of differentiation or growth of the tissue. Thus in the final stages of differentiation of the cambium to xylem in the angiosperms, sycamore (*Acer pseudoplatanus*) and poplar (*Populus robusta*), there

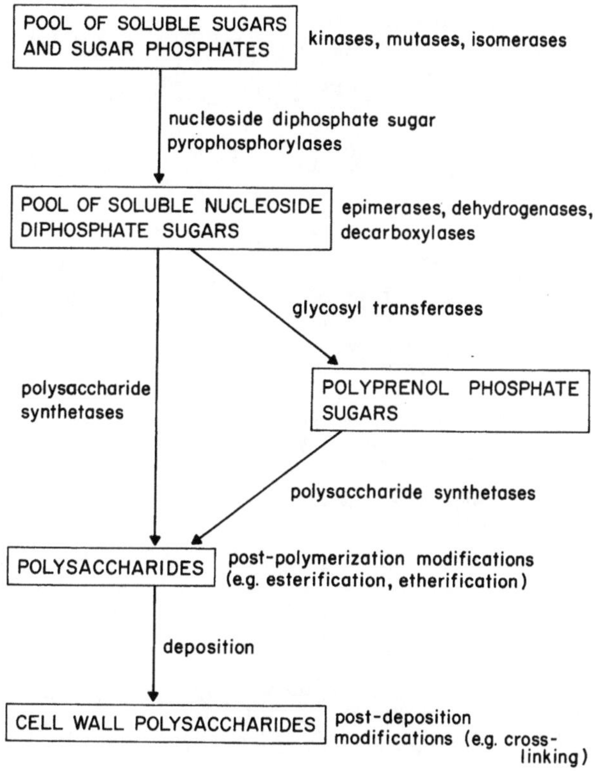

Fig. 1. Stages of synthesis and interconversion of monosaccharides for cell wall polysaccharide formation

is a decrease in the activity of enzymes directly involved in production of soluble precursors of pectin, namely the 4-epimerases for UDP-Gal and UDP-L-Ara, and an increase in the activity of UDP-Glc dehydrogenases and UDP-GlcA decarboxylases which provide precursors for non-cellulosic polysaccharides. These changes correlate well with changes in the chemical composition of the wall during development (DALESSANDRO and NORTHCOTE 1977a). In the gymnosperms, *Pinus silvestris* and *Abies grandis,* the specific activities of these enzymes changed during differentiation of cambium to xylem in a way which was different from the angiosperms and reflected the smaller amounts of xylan and larger amounts of arabinogalactans and galactoglucomannans in the cell walls of gymnosperms (DALESSANDRO and NORTHCOTE 1977b). Similar correlations were observed in cultured explants of Jerusalem artichoke (*Helianthus tuberosus*) during cell division and xylem differentiation (DALESSANDRO and NORTHCOTE 1977d).

In growth and wall expansion in pea seedlings (*Pisum sativum*) an increase occurs in activities of the dehydrogenase and decarboxylase and the epimerases involved in UDP-Glc interconversions. In the conversion of UDP-Glc to UDP-Xyl the activity of the dehydrogenase is much lower than the decarboxylase

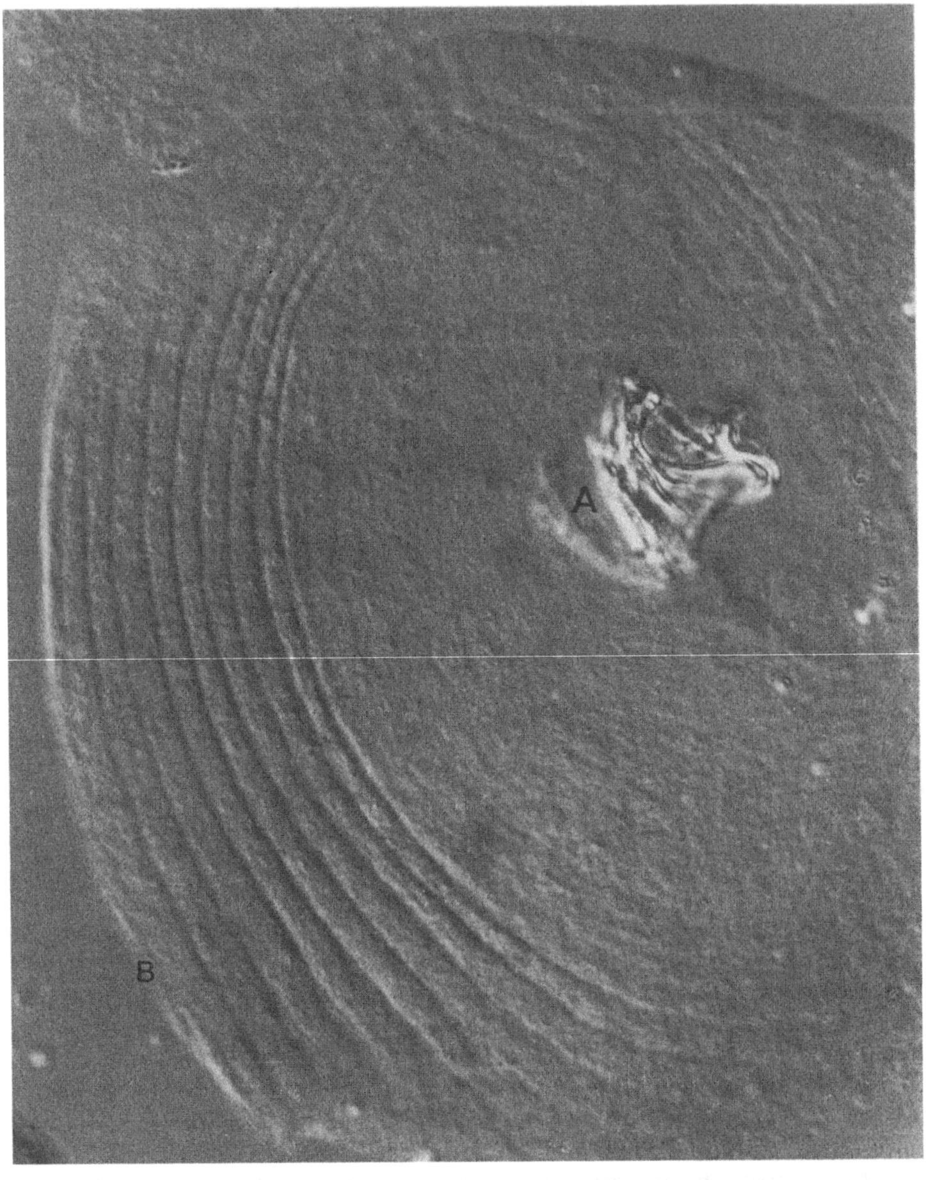

Fig. 2. A transverse section of a cotton fiber from a plant grown at 30° C until anthesis and then subjected to the regime: light: 14 h at 30° C; dark: 10 h at 10° C for 11 days and then grown for an additional 11 days under constant illumination at a temperature of 30° C. After sectioning the fiber was swollen in cupriethylene diamine. Normaski interference optics. ×640. *A* cell lumen; *B* region of primary wall (not visible). (Photograph provided by Professor A.B. Wardrop, Botany Department, La Trobe University, Bundoora, Australia)

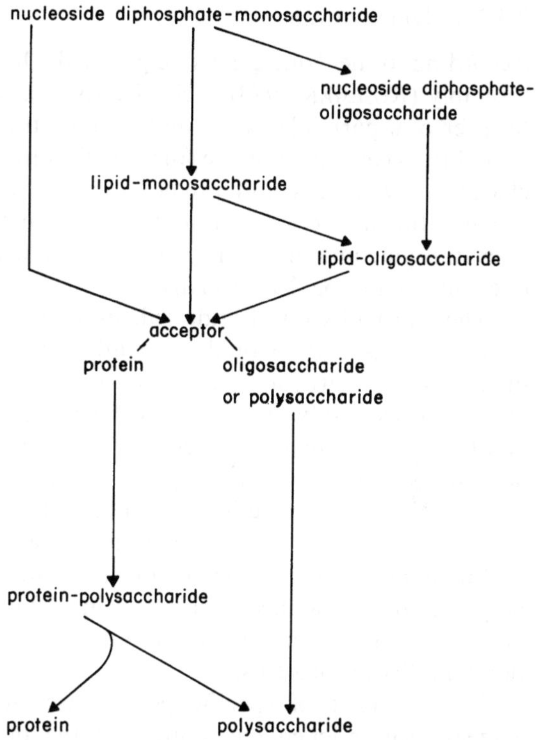

Fig. 3. Possible routes of polymerization from nucleotide sugars

and may represent the rate-controling step in the formation of UDP-Xyl (DALESSANDRO and NORTHCOTE 1977c).

Whether the changes in activity recorded in the experiments of DALESSANDRO and NORTHCOTE (1977a, b, c, d) are due to alterations in the steady state amount of the enzyme or to modulation of the activity of pre-existing enzymes was not investigated. However allosteric modifications of these enzymes are known. UDP-Xyl exerts an inhibitory effect on UDP-Glc dehydrogenase (NEUFELD and HALL 1965, ANKEL et al. 1966, DAVIES and DICKINSON 1972, DALESSANDRO and NORTHCOTE 1977a) and UDP-Glc inhibits UDP-Glc pyrophosphorylases (HOPPER and DICKINSON 1972). Further control possibilities are offered by the formation of nucleotide sugars via the myo-inositol pathway (LOEWUS and DICKINSON, Chap. 6, Vol. 13A, this series).

Although fluctuations in the activities of enzymes involved in the maintenance of the nucleotide sugar pool occur during differentiation and growth, the epimerases, dehydrogenases, and decarboxylases are present even when little or no polysaccharide is being synthesized from some of the nucleotide sugars present in the pool (DALESSANDRO and NORTHCOTE 1977a, b, c). Similarly, the level of UDP-Glc pyrophosphorylase remains high in nonelongating sections of mung bean (*Phaseolus aureus*) hypocotyls at the same time as UDP-Glc glucan synthetase activity is decreasing (HEINIGER and FRANZ 1980). Thus the major control of biosynthesis is probably exerted at the level of the polysaccharide synthetases.

2.4.3 Polymerization

According to the "one gene–one glycosidic linkage" hypothesis for glycoprotein assembly (ROSEMAN 1970) each glycosyl transferase is specific not only for the transfer of a particular sugar but also for the anomeric configuration and position of the glycosidic linkage formed. This hypothesis equally applies to polysaccharide polymerases and predicts for example that there would be separate enzymes for the incorporation of (1→4)-β-mannosyl and (1→4)-β-glucosyl residues into glucomannans or (1→4)-α-galacturonyl and (1→2)-α-rhamnosyl residues into rhamnogalacturonans.

The short oligosaccharide substituents on heteroxylans, heteroglucans etc. (Chap. 1, this Vol.) would be synthesized by successive and specific action of individual synthetases as occurs during glycoprotein biosynthesis (HUGHES 1976). The assembly of a branched β-galactan with (1→6)-inter-residue linkages and (1→3)-interchain linkages may require two polymerases showing different acceptor specificities but using the same donor. It is also possible, however, that a single polymerase is involved but that its specificity is altered by association with a modifier protein, as in the case of lactose synthetase (EBNER 1973). So far none of the enzymes involved in the synthesis of noncellulosic cell wall polysaccharides has been examined in sufficient detail to establish the validity of the "one gene–one glycosidic linkage" hypothesis or the existence of specificity-modifying proteins.

The existence of specific polymerases for polysaccharide assembly offers a means for independent control of the process. This might be achieved by modulating the activity of pre-existing enzymes, but the clear-cut changes in polysaccharide types occurring in the primary to secondary cell wall transition (Sect. 4.5) suggest that the events would probably be controlled by alterations in the steady-state level of specific polymerases, which in turn would be controlled by the rate of enzyme synthesis and degradation, events which presumably are under hormonal influence (NORTHCOTE and WOODING 1966, HADDON and NORTHCOTE 1975, 1976). Another possible control point in polysaccharide biosynthesis is at the level of protein primer availability, which could be regulated at the transcriptional and translational level.

3 Cellular Aspects of Polysaccharide Synthesis and Processing

The sequence of molecular events in polysaccharide synthesis discussed in the previous section is now reviewed in relation to the location of substrates, enzymes, and intermediary molecules in the cytosol and within the endomembrane system (MORRÉ and MOLLENHAUER 1974).

3.1 Location of Enzymes Leading to the Establishment of the Nucleotide Sugar Pool

The enzymes which generate the pools of soluble sugars, sugar phosphates, and nucleotide sugars are generally assumed to be located in the cytosol since they have been detected in buffer-soluble fractions from a wide range of plant tissues (NEUFELD et al. 1957, FAN and FEINGOLD 1969, DALESSANDRO and NORTHCOTE 1977a, b, c, d, CLERMONT and PERCHERON 1979). However, they have also been found in particulate preparations (NEUFELD et al. 1960, FEINGOLD et al. 1960, CARBONELL et al. 1976). For example the UDP-Gal 4-epimerases from wheat (*Triticum sativum*) roots, sycamore (*Acer pseudoplatanus*) cell suspensions (AXELOS and PÉAUD-LENOËL 1978), mung beans (*Phaseolus aureus*) (PANAYOTATOS and VILLEMEZ 1973), and ryegrass (*Lolium multiflorum*) endosperm (MASCARA and FINCHER, unpublished) are associated with membrane fractions. In onion (*Allium cepa*) stem preparations, phosphoglucomutase, and UDP-Glc pyrophosphorylase are associated with smooth membrane, dictyosome, and plasma membrane fractions; the highest specific activity of these enzymes being recorded in the dictyosomes (MORRÉ and MOLLENHAUER 1976). It is not known whether the soluble enzymes are in the cytosol or in the cisternae of the Golgi or endoplasmic reticulum. Immunocytochemistry at the electronmicroscope level (CLARKE and KNOX 1979) could be useful for locating these enzymes.

3.2 Location of Glycosyl Transferases Involving Lipid Intermediates

Glycosyl transferases which catalyze formation of steryl glycosides, polyprenol mono- or diphosphate sugars, and oligosaccharides are associated with membranes (Chap. 8, this Vol.). In soybean (*Glycine max*) the glucosyl transferases which form steryl glucosides are predominantly located in plasma membranes, while the polyprenol monophosphate glucosyl transferases are found both in plasma membrane and Golgi fractions (CHADWICK and NORTHCOTE 1980). In pea (*Pisum sativum*) stems the polyprenol monophosphate glycosyl transferases are mostly associated with the endoplasmic reticulum, whereas the polyprenol diphosphate glycosyl transferases are found in fractions with densities corresponding to either Golgi dictyosomes or plasma membranes (DÜRR et al. 1979).

These observations are consistent with the role of dolichol phosphate sugars as intermediates in glycoprotein synthesis in the endoplasmic reticulum and Golgi but whether they are involved in noncellulosic polysaccharide synthesis in these locations or at plasma membrane sites is unknown at present.

3.3 Location of Polysaccharide Synthetases

Whereas the synthesis of cellulosic microfibrils in higher plants probably occurs exclusively at the plasma membrane (NORTHCOTE and LEWIS 1968, BOWLES and NORTHCOTE 1972, KIERMAYER and DOBBERSTEIN 1973, RAY et al. 1976), noncellulosic wall polysaccharides may be formed throughout the endomembrane system (Sect. 3.4). In a few cases the subcellular locations of synthetases concerned with polysaccharide chain formation or transferases involved in the addition of sugar residues to outer branches of polysaccharides have been examined by fractionating subcellular membrane components on density gradients. The results of these studies are summarized in Table 1.

None of these investigations shows an exclusive association of synthetases or transferases with particular elements of the endomembrane system, although in vivo this may be the case. This may relate to technical difficulties in membrane fractionation after tissue disruption (ROBINSON 1977, QUAIL 1979), since the methods seldom allow complete separation of sub-cellular components. On the other hand a multisite distribution of an enzyme in the endomembrane system may truly reflect the in vivo situation. There is evidence that following synthesis in the rough endoplasmic reticulum, membrane-bound enzymes are moved to their final destination by membrane flow (MORRÉ and MOLLENHAUER 1974) and as a result a particular enzyme may be detected in several endomembrane fractions. Although a polysaccharide synthetase may be found in more than one membrane fraction in vitro, in the absence of corroborative evidence this does not signify that it is active in vivo at that location. Thus, a particular polysaccharide synthetase may become active in vitro after tissue disruption due to changes in accessibility to substrate, acceptor, cofactors, or activators. Conversely, isolation procedures may induce latency in vitro due to vesiculation of membranes (WISHART and FRY 1980) or to masking by other membrane components (VAN DER WOUDE et al. 1971, NORTHCOTE 1979a).

3.4 Subcellular Routes of Polysaccharide Assembly

The biochemical observations discussed in the previous section are now considered in relation to synthesis of noncellulosic polysaccharides in whole tissues. The subcellular sites of cell wall polysaccharide assembly have been the subject of much investigation in the last 15 years and evidence for involvement of endomembrane components has been reviewed on a number of occasions (O'BRIEN 1972, MORRÉ and MOLLENHAUER 1974, 1976, GUNNING and STEER 1975, ROBINSON 1977, NORTHCOTE 1979a, b, MORRÉ et al. 1979).

The main routes or sites of synthesis appear to be (a) via a Golgi dictyosome-secretory vesicle route to the plasma membrane, (b) via an endoplasmic reticulum route, or (c) at the plasma membrane.

The subcellular location of polysaccharide assembly is investigated in intact cell preparations or tissues by light and electronmicroscopy, and by pulse and pulse-chase experiments. Autoradiography, in conjunction with histochemical observations, allows the kinetic and spatial relationships in the events of polysaccharide synthesis within the endomembrane system to be discerned, but does not give precise information about the composition and structure of products formed at individual assembly sites. Such information can, however, be obtained by isolating and identifying specific components of the endomembrane system, followed by analysis of associated polysaccharides and their comparison with those of the cell wall. Although the cytological, autoradiographic, and biochemical approaches all have limitations, taken together they provide significant information on the subcellular events of noncellulosic polysaccharide assembly. The evidence available for the possible routes will now be considered.

3.4.1 Golgi Dictyosome-Golgi Vesicle-Plasma Membrane Route

Cytological investigation of plant cells during their development shows that Golgi activity, as indicated by hypertrophy of cisternae and increased numbers of Golgi-derived vesicles, is correlated with deposition of new cell wall material or mucilage; for example at the developing cell plate (WHALEY and MOLLENHAUER 1963, LEHMANN and SCHULZ 1969, ROBERTS and NORTHCOTE 1970), during primary cell wall synthesis (SIEVERS 1963, PICKETT-HEAPS and NORTHCOTE 1966) during secondary thickening in the xylem (WOODING and NORTHCOTE 1964, ESAU et al. 1966, PICKETT-HEAPS 1967) and during secretion of slimes and mucilages (WOODING and NORTHCOTE 1964, NORTHCOTE and PICKETT-HEAPS 1966, ESAU et al. 1966, PICKETT-HEAPS 1967). These observations and the similar appearance in electronmicrographs of the material of dictyosome cisternae, secretory vesicles, and the cell wall, led to the suggestion that the Golgi and its derived vesicles participate in wall polysaccharide synthesis.

The chemical nature of the material in dictyosome cisternae and vesicles was first investigated using histochemical techniques. Carbohydrate was detected by periodate oxidation in conjunction with silver hexamine (PICKETT-HEAPS 1968) or silver proteinate (THIÉRY 1967, ROUGIER 1971) staining. These methods, however, do not distinguish between carbohydrate in polysaccharides, glycoproteins, or glycolipids. Pectic polysaccharides in dictyosome vesicles have been identified by pectinase treatment (DASHEK and ROSEN 1966) and hydroxylamine-ferric chloride staining (ALBERSHEIM et al. 1960).

A dynamic view of the progress of polysaccharide assembly and transport in the endomembrane system is given by autoradiography of tissue sections following pulse or pulse-chase labeling of cells which are depositing wall polysaccharides or mucilages. These studies suggest that the synthesis of these noncellulosic polysaccharides takes place in the Golgi apparatus. The fully- or partially formed polysaccharides are then carried through the cell in Golgi-derived vesicles which fuse with the plasma membrane, where their contents are deposited in the wall or pass through it to form a surface mucilage. Cells studied by this method include sieve tube cells of wheat (PICKETT-HEAPS 1967), *Cucurbita* meristem cells (COULOMB and COULON 1971), fiber cells of *Marchantia* thalli (FOWKE and PICKETT-HEAPS 1972), developing pollen tubes (MORRÉ and VAN DER WOUDE 1974) and the slime-secreting root cap cells of maize and wheat (MORRÉ and MOLLENHAUER 1976, NORTHCOTE 1979b, Chap. 21, this Vol.). In autoradiographic studies the water-soluble carbohydrate would be lost during the usual fixation procedures, and although it is often assumed that the residual radioactive material is polysaccharide, it might also be glycolipid or glycoprotein.

The monosaccharide composition of the radiochemically labeled material present in Golgi dictyosomes and other endomembrane components has been determined following isolation of the organelle and selective extraction of the putative polysaccharide components. Tissues examined in this way include pea (*Pisum sativum*) seedling shoots (RAY et al. 1976, ROBINSON et al. 1976, ROBINSON 1977, DÜRR et al. 1979) and roots (HARRIS and NORTHCOTE 1971), wheat (*Triticum vulgare*) seedling roots (JILKA et al. 1972) and root tips (MERTZ and NORDIN 1971), maize (*Zea mays*) seedling roots (BOWLES and NORTHCOTE 1972, 1974, 1976) and root tips (PAULL and JONES 1975a, b, 1976) and *Lilium longiflorum* pollen tubes (MORRÉ and VAN DER WOUDE 1974). Direct monosaccharide analyses of contents of isolated Golgi vesicles from *Lilium longiflorum* (VAN DER WOUDE et al. 1971) and *Petunia hybrida* (ENGELS 1974), and of particulate preparations from mung bean (*Phaseolus aureus*) shoots (BAILEY et al. 1967) have also been reported.

In general, the monosaccharide composition of dictyosome material is qualitatively similar to pectic and other noncellulosic polysaccharides of the wall, that is xylose, galactose, arabinose, fucose, glucose and uronic acids are found. The polymeric nature of the radiolabeled carbohydrate has been established by gel filtration chromatography (MERTZ and NORDIN 1971, JILKA et al. 1972,

BOWLES and NORTHCOTE 1976, ROBINSON 1977). BOWLES and NORTHCOTE (1976) also detected low molecular weight sugar phosphates in an aqueous extract of Golgi and rough endoplasmic reticulum, an observation which emphasizes the need to show that membrane-associated carbohydrate in fact represents polysaccharide. In the case of wheat seedlings (JILKA et al. 1972), pea epicotyls (ROBINSON 1977) and maize root cells (BOWLES and NORTHCOTE 1976), the high molecular weight materials were not susceptible to proteolytic digestion.

The secretory vesicles derived from the Golgi dictyosomes may not only be involved in polysaccharide transport but also in modifications in size (BONNETT and NEWCOMB 1966) and in amount of component molecules (PICKETT-HEAPS 1968) during their transit to the plasma membrane. Comparative analyses of pectic polysaccharides in Golgi dictyosomes and secretory vesicles indicate that methyl esterification occurs in these vesicles (DASHEK and ROSEN 1966).

Confirmatory evidence for the role of Golgi dictyosomes and secretory vesicles in deposition of noncellulosic cell wall polysaccharides comes from pulse labeling-inhibitor studies in pea stem segments (ROBINSON and RAY 1977). In the presence of KCN and other respiratory inhibitors, secretion of cell wall polysaccharides through Golgi dictyosomes is slowed and the size of the cisternae increases. When polysaccharide in Golgi dictyosomes was prelabeled with a pulse of ^{14}C-glucose, subsequent polysaccharide synthesis could be inhibited by 2-deoxyglucose and the release of labeled polysaccharide blocked with KCN. On removal of KCN, secretion from dictyosomes occurred and an equivalent amount of radioactivity appeared in the walls (ROBINSON and RAY 1977). In developing *Lilium* pollen tubes, reduction in the number of secretory vesicles observed in the presence of 10 mM $CaCl_2$ was accompanied by reduced incorporation of sugars into matrix polysaccharides (MORRÉ and VAN DER WOUDE 1974).

The observations discussed in this section provide good evidence for the operation of a Golgi dictyosome-vesicle-plasma membrane-cell wall route of polysaccharide assembly and transport.

The time for synthesis and deposition of polysaccharides following the Golgi route is quite short. In *Zea mays* roots (BOWLES and NORTHCOTE 1972, 1974) and pea epicotyls (ROBINSON et al. 1976) the time for passage of polysaccharides through the dictyosomes is about 3 min and in the case of pea epicotyls the entire process of formation and deposition takes only 7 min (ROBINSON 1977). The final event in secretion is fusion of the secretory vesicle with the plasma membrane, which is a Ca^{2+}-dependent process (MORRIS and NORTHCOTE 1977).

Although it is clear that polysaccharide synthesis can occur in the Golgi dictyosomes, it remains uncertain whether synthesis is initiated there or in the endoplasmic reticulum where protein primers, which may be necessary for chain initiation of certain polysaccharides (Sect. 2.2), would be formed.

3.4.2 Endoplasmic Reticulum-Plasma Membrane Route

A number of observations suggest that, in addition to the Golgi route of synthesis and transport of noncellulosic polysaccharides destined for the cell wall, a more direct route involving the endoplasmic reticulum operates. Evidence supporting this alternative route is based almost solely on cytological data. Thus the relative abundance of Golgi and endoplasmic reticulum profiles varies depending on the function and physiological state of the cell. For example the Golgi dictyosomes are plentiful in secretory root cap cells but in central root cap cells where cell wall formation is occurring, the endoplasmic reticulum predominates (CLOWES and JUNIPER 1968). Endoplasmic reticulum as well as Golgi profiles are observed close to the site of cell plate deposition (ROBERTS and NORTHCOTE 1970, JONES and PAYNE 1977) and are often found lying close to and parallel with newly formed walls (HEPLER and NEWCOMB 1964, PICKETT-HEAPS and NORTHCOTE 1966, CLOWES and JUNIPER 1968). A similar association between endoplasmic reticulum and sites of callose deposition has been noted (NORTHCOTE and WOODING 1966, 1968) (Sect. 4.4). In protoplasts which are regenerating cell walls, the endoplasmic reticulum elements are aligned parallel with and close to the plasma membrane (BURGESS and FLEMING 1974, ROBENECK and PEVELING 1977), and in some electronmicrographs, connections between endoplasmic reticulum and plasma membrane are seen (MORRÉ and MOLLENHAUER 1974, ROBENECK and PEVELING 1977). However, there is no evidence in these cases for accumulation of material in the endoplasmic reticulum cisternae. Nonetheless, cytological observations suggest that polysaccharides may sometimes be associated with cisternae. Thus in developing pollen tubes of *Lilium longiflorum* the accumulation of electron translucent material, resembling the material of the pollen tube wall, is associated with complexes of the endoplasmic reticulum, suggesting its possible participation in wall formation (VAN DER WOUDE et al. 1971). In the trichomes of *Pharbitis,* which secrete a protein-carbohydrate mucilage, the endoplasmic reticulum makes frequent direct contact with the plasma membrane and appears to discharge its contents outside the cell (UNZELMAN and HEALY 1974). Numerous dictyosomes are also seen in these secretory cells. In the developing endosperm of *Trigonella foenum-graecum,* a galactomannan forms a thick deposit on the inner surface of the primary cell walls (MEIER and REID 1977). In the cells which are beginning to secrete the galactomannan there are stacks of rough endoplasmic reticulum. The cisternal space, the enchylemma, swells and becomes vacuolated to form a voluminous network. The enchylemma contains material which reacts with Thiéry stain similarly to the galactomannan already in the wall. In these cells, unlike those of *Pharbitis* studied by UNZELMAN and HEALY (1974), dictyosomes were rarely seen.

These few observations provide circumstantial evidence for an endoplasmic reticulum-plasma membrane route for matrix polysaccharide synthesis and transport, but direct evidence is not available. The associations or connections between endoplasmic reticulum and the plasma membrane underlying the walls may also be related to the delivery of polysaccharide synthetases from their sites of synthesis in the rough endoplasmic reticulum to the plasma membrane by membrane flow. Membrane-bound synthetases may also be brought to the plasma membrane by the endoplasmic reticulum-Golgi dictyosome-secretory vesicle route (MORRÉ and MOLLENHAUER 1974).

3.4.3 Assembly at the Plasma Membrane

Although cellulose microfibrils appear to be products of plasma membrane-associated synthetases (Chap. 4, this Vol.), little evidence has been produced for noncellulosic polysaccharide synthesis at this site. ROBINSON et al. (1976)

have argued on the basis of the kinetics of labeling of subcellular organelles in pulse-chase experiments that the synthesis of matrix polysaccharides at the plasma membrane cannot be of major significance in pea cell wall synthesis.

There is, however, good cytological and histochemical evidence that (1→3)-β-glucan (callose) is synthesized at the plasma membrane, for example in sieve plate development (Sect. 4.6) and in wound responses (Sect. 5.3) and in pathogenesis (Sect. 5.4). This is consistent with a plasma membrane location for the (1→3)-β-glucan synthetase of pea epicotyls (RAY et al. 1969, ANDERSON and RAY 1978). Furthermore in growing *Lilium* pollen tubes, Ca^{2+} inhibition of secretory vesicle production only partially inhibits wall polysaccharide formation (MORRÉ and VAN DER WOUDE 1974). Incorporation into the glucose-containing fraction of the wall, which consists of (1→3)- and (1→4)-linked β-glucan (HERTH et al. 1974), is not affected. These results and the apparent decline in importance of the Golgi secretory route of noncellulosic polysaccharide synthesis during secondary wall deposition (Sect. 4.5) suggest that synthetases for noncellulosic polysaccharides may occur at the plasma membrane.

3.4.4 Deposition of Polysaccharides in the Wall

Integration into the wall of polysaccharides formed at the plasma membrane or delivered there in secretory vesicles may occur by a process of self-assembly but, in addition, may involve active metabolism within the wall. A complete understanding of this process requires a thorough knowledge of the noncovalent and covalent interactions, both between matrix and microfibrillar components and between the matrix polysaccharides themselves. The role and origin of the wall protein and other noncarbohydrate components such as phenolic acids and lignin must also be considered.

If covalent associations between matrix polysaccharides occur through glycosidic linkage, and this is by no means certain (McNEIL et al. 1979), then they are presumably generated within the endomembrane system, since glycosidic linkage formation is an energy-dependent process. On the other hand, covalent cross-linking of matrix polysaccharides by ferulic acid dimerization, as suggested by MARKWALDER and NEUKOM (1976), is an oxidative process which would involve wall-bound peroxidases (LAMPORT and CATT, Chap. 7, this Vol.). Similarly, lignification of cell walls, which results in cross-linking of matrix polysaccharides, also occurs by an oxidative process within the wall. The noncovalent association of matrix polysaccharides through junction zone formation may be achieved extracellularly, by self-aggregation or co-gelation (REES and WELSH 1977) of individual molecules or preformed packets or units of noncellulosic polysaccharides (REIS et al. 1978).

Self-assembly could be directed by the presence of seeding or nucleating molecules as occurs in connective tissue deposition (RODÉN and SCHWARTZ 1975) and in the organization of the *Chlamydomonas* cell walls, where nucleating glycoproteins are involved (HILLS 1973, HILLS et al. 1975, CATT et al. 1978). Units of wall might correspond in degree of organization to the mucilage or slime aggregates described by GRANT et al. (1969). A role for carbohydrate-binding proteins and other cell wall proteins as cross-linking (KAUSS and GLASER

1974, KAUSS and BOWLES 1976) or as nucleating molecules is an intriguing possibility (LAMPORT 1980). The observations of PONT LEZICA et al. (1978) suggest that in pea epicotyl preparations a glucose-containing hemagglutinin is an acceptor for growing glucan chains. Proteins of this nature could also direct site-specific wall deposition through their capacity to bind specific carbohydrates.

4 Metabolism During Cell Division, Growth, Differentiation, Senescence and Germination

During growth and morphogenesis of plant cells the cell wall undergoes modifications in both morphology and composition. The changes involved, although occurring in the wall, are under the control of the protoplast. In this section, methodological approaches used to define these modifications at the molecular level are described, the interpretation of results discussed, and the strategies used by the cell to control and coordinate the synthetic and degradative processes outlined. These considerations serve as a background to a description of specific examples of cell wall metabolism.

4.1 Methodological Considerations and Interpretations

Light and electronmicroscopy of cells during differentiation and morphogenesis has provided detailed cytological descriptions of specific alterations in cell walls. In many situations the morphological changes are the only evidence that active metabolism is occurring, but in other cases, this evidence has been supplemented by chemical and biochemical studies.

Direct chemical examination of changes in cell wall polysaccharides is fraught with difficulties. Often cell wall material undergoing metabolism is restricted to limited regions of a tissue or a cell so that isolation of walls or regions of walls in quantities sufficient for analysis presents formidable technical problems. In these circumstances histochemical approaches to wall composition have been used to give qualitative information on the metabolism occurring at localized sites in walls. However, given the uncertain specificity of many histochemical reagents, the results must be interpreted cautiously. Where polysaccharide hydrolases with known purity and specificity are used in conjunction with histochemical techniques the results are more conclusive. Future refinements of sensitive physical methods of analysis such as X-ray (EDAX), infra red, ultra violet, and fluorescence spectroscopy and secondary ion mass spectrometry, and their application to in situ examination of specimens with the dimensions of a cell wall, may make chemical analysis of such material feasible.

Another analytical problem occurs when the cells from a single tissue sample are at various stages of differentiation, or where the tissue is heterogeneous with respect to cell type. A similar analytical problem arises where metabolism in a cell wall is localized at specific regions.

When walls can be isolated in quantities sufficient for analysis by conventional means it is possible to make a complete quantitative inventory of chemical changes occurring in various fractions and to relate these to changes in the physical character of the component polysaccharides, such as degree of polymerization, water-binding capacity, ability to form gels, etc.

Far-reaching physical changes which affect the function of cell walls may be induced by rather subtle alterations in the chemistry of a polysaccharide, for example by alteration in the number and distribution of substituents on main chains, or by changes in the degree of methyl esterification or acetylation.

The important information necessary for understanding wall metabolism is whether a particular polysaccharide component is being synthesized, degraded or both during the differentiation of cell walls.

Metabolic changes in a specific polysaccharide component may follow one of several different patterns:

a) the polysaccharide may be continually synthesized and degraded ("turned over") without change in its absolute amount;
b) the wall polysaccharide is synthesized during part of the developmental period and subsequently: (1) remains metabolically inert, or (2) is wholly or partially depolymerized or debranched, resulting in its loss from the wall or its conversion from a nonextractable to an extractable form or vice versa, or (3) is chemically modified by processes such as chain elongation, acetylation or methyl esterification; by addition of oligosaccharide side chains; by covalent linkage to other wall polysaccharides or proteins; or by cross-linking via noncarbohydrate bridges. Such changes may lead to a redistribution of the polysaccharide between wall fractions.

In assessing which of these metabolic patterns is followed by a particular cell wall component it is necessary to:

a) identify the polysaccharide unequivocally. This depends on a detailed knowledge of its structure and properties. For instance arabinose from an arabinoxylan must be distinguished from arabinose in pectic arabinogalactans or cell wall glycoproteins. Similarly $(1\rightarrow4)$-glucosyl residues in a xyloglucan must be distinguished from those in cellulose, glucomannans or $(1\rightarrow3)$;$(1\rightarrow4)$-β-glucans. In certain cases specific enzymes can be used to identify particular linkage sequences (COOK et al. 1980) and some components of cell walls may be identified through the use of specific carbohydrate-binding proteins (immunoglobulins or lectins). In many situations such detailed information is not available and deductions have to be made from monosaccharide analyses of cell wall fractions, supplemented by quantitative linkage analyses;
b) determine the absolute amount of the specific polysaccharide in the cell wall during the growth period. This information allows a direct conclusion as to whether the amount of a polysaccharide or an individual monosaccharide is increasing or decreasing as wall differentiation proceeds, whereas results reported as relative proportions of polysaccharides, monosaccharides, or linkage types do not;
c) determine whether polysaccharide molecules of a specific type, once deposited in the wall, are removed and subsequently replaced by molecules of the same type. Such information may be obtained from pulse-chase experiments

in which tissues or cells are provided with radioactive precursors and radioactivity measured at intervals in specific wall components. If the total radioactivity in the component remains constant during the chase period, the wall polysaccharide is not subject to degradation. On the other hand, if a decrease in radioactivity of the component is observed, its degradation, or its alteration so that it appears in another wall fraction, is indicated. Such decreases in radioactivity have often been interpreted as "turnover", but in the absence of supplementary data on changes in absolute amounts of a component this conclusion may not be valid, since turnover, sensu stricto, implies continuous synthesis and degradation without any change in the amount of the component per cell. A decrease in specific activity may indicate either that synthesis is continuing without degradation or that both are occurring simultaneously.

4.2 Molecular Strategies in Morphogenesis

There are many instances where changes in cell walls during growth or morphogenesis involve localized alterations, for example in tip growth of cotton hairs or pollen tubes, in sieve plate formation or tracheid differentiation, or in the formation of secondary plasmodesmata. In these situations the question of origin, delivery and localized action of enzymes concerned with biosynthesis or depolymerization must be considered. There appear to be several strategies adopted by the cell to achieve localized or selective wall metabolism:

a) The site of polysaccharide formation or degradation may be determined by the specific positioning of polysaccharide synthetases or hydrolases. Synthetases may be delivered to specific regions of the plasma membrane by other components of the endomembrane system such as the endoplasmic reticulum (Sect. 3.4.2) which has, for example, been observed close to the site of callose deposition during sieve plate formation (Sect. 4.6.3). A similar site-specific change in cell wall composition may be achieved by directed deposition from secretory vesicles of preformed cell wall matrix components, e.g., at pollen tube tips. Endomembrane components may also deliver hydrolytic enzymes involved in cell wall metabolism to specific sites in the cell wall, as may occur during secondary plasmodesmata formation (Sect. 4.6.3). In laticifer wall digestion (Sect. 4.6.1), hydrolases appear to be brought to the site of digestion in membrane-bound vesicles and discharged into the wall, the extent of digestion apparently being controlled by the diffusibility of the enzyme and regulation of its synthesis and secretion.

b) Another strategy which may be adopted is production of synthetases or hydrolases as inactive zymogens and their activation at special sites in the wall. Such a scheme has been proposed to account for local activation of chitin synthetase during yeast septum initiation (Cabib et al. 1974, Chap. 16, this Vol.). Similar molecular mechanisms may be invoked in control of morphogenesis in higher plants, but so far no plant synthetase or hydrolase has been characterized in a zymogen form.

c) Specific regions or layers of cell walls may be rendered resistant to hydrolytic degradation by structural alterations during differentiation, for example by lignification. This permits degradation of unprotected areas or layers by hydrolytic enzymes which may be secreted from the intact protoplast or liberated following its disintegration. Examples of this tactic are the etching of unlignified areas of tracheid cell walls during differentiation (Sect. 4.6.1) or dissolution of outer wall layers of cereal aleurone cells during germination (Sect. 4.8).

d) A different approach to selective wall metabolism is made possible by the production of special cell walls which are chemically distinct from those of adjacent cells. This permits their removal by specific polysaccharide hydrolases secreted by neighboring cells, without damage to the walls of the secreting cells. This strategy is exemplified by the synthesis of a pollen mother cell wall and its subsequent dissolution by specific $(1\rightarrow3)$-β-glucan hydrolases secreted by adjacent tapetal cells (Sect. 4.6.4).

e) At the tissue level, selective wall degradation may occur in discrete regions, such as in abscission zones (Sect. 4.7.1) or in the developing aerenchyma (Sect. 4.6.1). At the appropriate stage of development, cells in these regions secrete or activate polysaccharide hydrolases which partially or completely degrade the cell walls.

4.3 Cell Plate Formation

During cell division the first visible evidence of the new cell wall is the deposition of the cell plate in an equatorial plane between daughter nuclei. The cell plate and its limiting membranes are assembled by fusion of membrane-bound vesicles believed to be derived from Golgi dictyosomes (Sect. 3.4). The cell plate persists during subsequent deposition of primary and secondary cell wall layers and is then recognized as the middle lamella.

There are no definitive chemical analyses of the composition of the cell plate polymers. Histochemical evidence, supported by the observations that pectin hydrolase treatment separates cells in tissue aggregates (COOK and STODDART 1973), suggests that pectic polysaccharides are present (NORTHCOTE 1979b). Arabinogalactans, as evidenced by staining with Yariv artificial antigen, have also been detected in the middle lamella region (CLARKE et al. 1979) and protein was reported by GINZBURG (1958). The cell plate of the dividing meiocyte (WATERKEYN 1961) and of somatic cells (FULCHER et al. 1976, MORRISON and O'BRIEN 1976) is characterized by the bright yellow fluorescence induced by aniline blue. In cell plates and young cell walls the aniline blue fluorescence is evenly distributed, but in older walls only small isolated patches remain (FULCHER et al. 1976, SUTHERLAND and MCCULLY 1976). Whether this is due to removal of the putative $(1\rightarrow3)$-β-glucan or to masking of the fluorescence by newly synthesized wall material is unknown. The intensity of fluorescence in young cell walls of *Zea mays* root hairs was reduced by treatment with the *Bacillus subtilis* $(1\rightarrow3);(1\rightarrow4)$-$\beta$-glucan endohydrolase (EC 3.2.1.73) (SMITH and MCCULLY 1978).

The pectic polysaccharides of the cell plate may be incorporated from the Golgi-derived vesicles (Sect. 3.4), but the observation that these vesicles do not fluoresce with aniline blue (FULCHER et al. 1976) may indicate that $(1 \rightarrow 3)$-β-glucan synthesis occurs at the cell plate membranes. These membranes may originate from the Golgi vesicles or from elements of the endoplasmic reticulum which are found close to the developing cell plate (ROBERTS and NORTHCOTE 1970, JONES and PAYNE 1978).

4.4 Primary Cell Wall Deposition and Expansion

The initial manifestation of primary cell wall formation is the appearance of fibrillar material in the nascent wall. In the transition from cell plate to primary wall deposition, it is likely that changes also occur in the matrix polysaccharides, leading to the formation of the complex association of pectic polysaccharides, heteroglucans, galactans, heteroxylans, $(1 \rightarrow 3);(1 \rightarrow 4)$-$\beta$-glucans and glycoproteins which constitute the predominant noncellulosic matrix polysaccharides of the primary walls (Chap. 1, this Vol.).

During primary wall formation plant cells undergo increases in volume by expansion in a process which is driven by the turgor pressure of the cell. The pattern of expansion of the cell varies; it may occur isodiametrically as in parenchymatous cells; at the tip and side walls as in cotton hairs and in pollen tubes; by widening, as in laticifers; or by elongation as in parenchymatous cells of the apical meristem, such as coleoptiles, epicotyls, and root tips (ROELOFSEN 1965). This phase of cell development determines the final size and shape of the mature cell and it is clear that these various patterns of expansion require differential metabolism in specific regions of the primary cell walls (Sect. 4.2).

The local changes in wall properties which accompany cell expansion may involve loosening of the wall structure by cleavage of intra- or inter-polymer chain linkages and a concomitant synthesis and deposition of new cell wall material (Chap. 11, this Vol.). This mechanism is similar to that proposed by BARTNICKI-GARCIA (1973) for wall expansion in fungi. Notionally it is the alteration of the amount and type of noncellulosic matrix polysaccharides and their relationship with one another and with the cellulosic microfibrils which is most likely to effect loosening of the wall structure. However, our incomplete knowledge of the organization and interactions of the wall polysaccharides makes it difficult to relate physiological observations to cell wall structure.

The expansion process and its control is discussed in detail by CLELAND (Chap. 11, this Vol.) and this section is confined to a description of the metabolic changes in noncellulosic polysaccharides of the primary cell wall matrix during expansion.

Cell wall changes in rapidly elongating epicotyls and coleoptiles, sometimes stimulated by auxins and other phytohormones, have been extensively investigated. In experiments with segments of these tissues, attempts have been made to study wall metabolism related to expansion rather than metabolism concerned with cell wall deposition. This is normally achieved by preincubation of segments in the absence of exogenous carbohydrate sources for a period sufficient to deplete endogenous carbohydrate reserves which might contribute to cell wall biogenesis prior to the addition of auxin or other elongation-stimulating agents.

In pea stem segments elongation results in metabolism of matrix polysaccharides (MACLACHLAN and DUDA 1965, LABAVITCH and RAY 1974a, b, JACOBS and RAY 1975, GILKES and HALL 1977); (1 → 4)-galactans associated with the pectic fraction and arabinose-containing polymers are lost while insoluble xyloglucans are converted to a soluble form.

In *Avena* coleoptile sections pulse-chase experiments show that during expansion there is a transfer of hexoses from water-soluble and dilute acid-soluble fractions into alkali-extractable material (KATZ and ORDIN 1967). Other experiments have suggested that there is a decrease in molecular size and a loss in noncellulosic glucan during expansion (RAY 1962, LOESCHER and NEVINS 1972, SAKURAI et al. 1977, 1979). This glucan is believed to be a (1 → 3);(1 → 4)-β-glucan (NEVINS et al. 1977, LABAVITCH and RAY 1978). In barley coleoptile segments treated with auxin there is also a decrease in noncellulosic glucose content of the cell wall (SAKURAI and MASUDA 1978a, b). Rice coleoptiles undergo similar changes during auxin-induced elongation, but in addition their xylose content increases (ZARRA and MASUDA 1979a, b). In a comprehensive examination of elongating maize coleoptile sections following auxin and pH treatments, DARVILL et al. (1978) showed by methylation analysis of both the wall and released carbohydrates that terminally-linked arabinose, glucuronic acid, 4-*O*-methyl glucuronic acid, galactose, and 3;4-linked xylose, all of which are components of wall heteroxylans, are lost. The changes in noncellulosic glucan reported for maize coleoptiles by LEE et al. (1967) and KIVILAAN et al. (1971) were not observed.

These experiments demonstrate that degradative changes of major noncellulosic cell wall polysaccharides, which in monocotyledons include either (1 → 3); (1 → 4)-β-glucans or heteroxylans and in dicotyledons, heteroglucans or pectic galactans, are important in the expansion phase of primary cell wall development. The exact nature of the alterations involved must await a detailed study of wall chemistry using methods sufficiently sensitive to detect minor, but perhaps significant, changes. Methylation as applied by DARVILL et al. (1978) and its development as described by VALENT et al. (1980) may provide the information needed. LAMPORT (1970) has pointed out the possible involvement of noncarbohydrate molecules, such as phenolic acids or proteins (FORREST and WAINWRIGHT 1977), in associations between wall polymers, but so far no rigorous assessment of their role in expansion has been made.

The metabolism of cell wall polysaccharides of *Vinca rosea* suspension-cultured cells during cell division and expansion has been investigated by pulse-chase techniques (TAKEUCHI et al. 1980). Noncellulosic polysaccharides composed of galacturonic acids, rhamnose, xylose, and glucose were not degraded following their deposition in the wall. However, labeled arabinose and galactose, believed to arise from an arabino-3;6-galactan (TAKEUCHI and KOMAMINE 1978) were lost from the wall preparations during the chase period. This was interpreted as a hydrolytic degradation and release of wall-bound arabinogalactan into the medium (TAKEUCHI et al. 1980). Alternatively the result may simply reflect the secretion of an arabinogalactan through the cell wall, as is believed to occur in *Lolium multiflorum* endosperm (ANDERSON et al. 1977) and *Acer pseudoplatanus* callus (POPE 1977) in tissue culture.

The results of wall analyses and labeling studies discussed in this section imply the involvement of hydrolytic enzymes in the expansion process. This is consistent with the detection of polysaccharide hydrolases in extracts of expanding tissue and with the changes induced in cell walls during autolysis or following the addition of exogenous polysaccharide hydrolases, and provides suggestive but not conclusive evidence to support the "hydrolase hypothesis" (LAMPORT 1970; CLELAND, Chap. 11, this Vol.). In intact tissues, any loosening of walls by hydrolase action will presumably occur concurrently with the synthesis and deposition of new noncellulosic polysaccharides.

In some cells, wall development ceases at the primary wall stage, as in endosperm and in the parenchyma and collenchyma typically found in growing leaves and stems. In other cells, the cessation of primary wall formation gives way to the deposition of secondary wall layers.

4.5 Secondary Cell Wall Deposition and Growth

The change from primary to secondary wall is characterized by alterations in amounts and organization of the microfibrillar phase and is accompanied by changes in composition of noncellulosic polysaccharides. These changes have been followed during the differentiation of angiosperm and gymnosperm cambium cells to xylem and phloem (MEIER 1961, THORNBER and NORTHCOTE 1961 a, b, 1962). From this and other work summarized by TIMELL (1964, 1965), the following general conclusions can be made:

a) pectic polysaccharides are confined to the middle lamella and primary wall;
b) cellulose deposition increases with the formation of secondary cell wall and is accompanied by increases in the content of glucomannan or heteroxylan. Lignification may also occur at this time;
c) heteroxylans formed during secondary thickening have more uronic acid than those of the primary wall. Differences are also seen in the mannose:glucose ratio in the glucomannan fractions.

Developing cotton hairs provide a convenient cell type for investigation of changes in wall composition during the primary to secondary wall transition since they are synchronously growing, single cells whose number per ovule remains constant throughout its development. Using this system HUWYLER et al. (1979) found that the absolute amounts of fucose, galactose, mannose, rhamnose, arabinose, uronic acids, and noncellulosic glucan [$(1 \rightarrow 3)$-β-glucan] all reached a maximum at the end of primary wall formation or beginning of secondary wall and thereafter decreased, implying that degradation of noncellulosic polysaccharides was occurring. In contrast, the amounts of xylose and cellulosic glucan increased until the end of fiber development. In a parallel study, MALTBY et al. (1979) found that both soluble and insoluble forms of $(1 \rightarrow 3)$-β-glucan increased in relative amounts up to the time of onset of secondary wall formation. However pulse-chase experiments provided no evidence for subsequent degradation of the $(1 \rightarrow 3)$-β-glucan.

In *Phaseolus aureus* seedlings relative increases in cellulose and decreases in pectic and other noncellulosic polysaccharides were observed (FRANZ 1972) during the period of growth when secondary cell walls were being deposited. Changes in noncellulosic wall components during maturation, involving secondary wall formation, are also found in tissues of monocotyledons. In oat stems there is a progressive decrease in the ratio of $(1 \rightarrow 3)$- to $(1 \rightarrow 4)$-linked β-glucosyl residues with increasing tissue maturity (BUCHALA and WILKIE 1971). In leaf and stem tissues of barley (BUCHALA and WILKIE 1974) and wheat (BUCHALA and WILKIE 1973) similar changes were observed and in addition, the percentage of xylose increased with maturity, whilst arabinose, glucose, and uronic acids decreased. In wheat stems the xylose:arabinose ratio doubled during maturation.

In the reed, *Arundo donax,* the proportion of an arabinoglucuronoxylan, the major noncellulosic polysaccharide in the cell wall, increases with time of maturation, and although its major structural features were retained, its average DP, uronic acid and acetyl content increased (JOSELEAU and BARNOUD 1976). These authors also noted that the ratio of 4-*O*-methyl glucuronic acid to glucuronic acid decreased with maturation in contrast to the increase found by KAUSS and HASSID (1967) for maize and BUCHALA and WILKIE (1973) for oats.

These studies show that dramatic changes occur in wall composition in the primary to secondary transition. The compositional changes, especially in those properties which govern the strength and permeability of cell walls, relate to development of new wall functions as the cells mature and reflect alterations in the nature, organization, and control of the biosynthetic systems.

The mechanism of alteration in enzymic capacity at the biochemical level has been discussed in Section 3.4 and is probably achieved by removal and replacement of polysaccharide synthetases. Cytologically, secondary wall formation may be characterized by fewer Golgi dictyosomes and vesicles, and by an increase in the occurrence of endoplasmic reticulum profiles in the regions of wall synthesis. Thus in cotton seed hairs WESTAFER and BROWN (1976) suggest that endoplasmic reticulum plays a major role in the synthesis of secondary wall polysaccharides, and although they report swelling of the endoplasmic reticulum cisternae and the presence of microvesicles believed to be of endoplasmic reticulum origin, RYSER (1979) could detect no vesicles containing periodate-oxidizable carbohydrate during secondary wall formation. Again, the main function of the endoplasmic reticulum might be to deliver to the plasma membrane newly formed polysaccharide synthetases required for secondary wall formation. Synthetases for cellulose are believed to occur as transmembrane assemblies at the plasma membrane (PRESTON 1979, MUELLER and BROWN 1980, GIDDINGS et al. 1980) and the tenacious association of xylans and glucomannans with cellulosic microfibrils suggests that synthetases for noncellulosic polysaccharides may also be part of a complex at the plasma membrane. However, apart from $(1\rightarrow3)$-β-glucan synthetase, no evidence for other noncellulosic polysaccharide synthetases at the plasma membrane is available (Sect. 3.4).

In the thick-walled sclerenchyma fibre cells from the flowering stem of *Lolium temulentum* dictyosomes are found not only during elongation but also in the post-elongation phase when fibre walls are thickening (JUNIPER et al. 1981).

4.6 Cell Wall Metabolism During Differentiation in Specific Tissues

4.6.1 Tracheids, Laticifers, Aerenchyma and Lateral Root Initiation

Tracheids. The formation of xylem tracheids or tracheary elements has been followed cytologically (BUVAT 1964a, b, c, WARDROP 1964, ESAU et al. 1966, O'BRIEN and THIMANN 1967, O'BRIEN 1970, 1974). Tracheid differentiation is characterized initially by deposition of patterned, lignified secondary wall thickenings and by localized increases in the thickness of primary end walls, followed by lysis of the protoplast (WODZICKI and BROWN 1973) and loss of noncellulosic wall polysaccharides in all unlignified areas.

In the end walls, single or multiple pores are developed by complete removal of unlignified regions and these pores interconnect adjacent cells of mature water-conducting tracheary elements (Fig. 4a, b). Microfibril-rich remnants of the unlignified lateral walls persist as "pit membranes" (SASSEN 1965, LIESE 1965, O'BRIEN 1970, BUTTERFIELD and MEYLAN 1972, MEYLAN and BUTTERFIELD 1972, CZANINSKI 1972, SCHEIRER 1973, O'BRIEN 1974). At the tracheid-parenchyma interface the dissolution proceeds only as far as the middle lamella region (O'BRIEN and THIMANN 1967, O'BRIEN 1970).

The hydrolases responsible for the dissolution of the noncellulosic polysaccharides of the primary wall possibly originate from cytoplasmic vesicles or the vacuole itself, which appears to vesiculate during the final stages of differentiation (BUVAT 1964a, b, c, SASSEN 1965). There have been no systematic studies of the polysaccharide hydrolases in developing tracheids, although the viscosity of carboxymethyl-cellulose is lowered by xylem extracts (SASSEN 1965, SHELDRAKE 1970). In *Acer pseudoplatanus* xylem extracts, the cellulase is predominantly membrane-bound (SHELDRAKE 1970) and this enzyme may play a role in wall dissolution.

In quiescent root meristems of *Allium cepa*, localized cell wall dissolution occurs in cortical and provascular elements, and is accompanied by elaboration of massive amounts of rough endoplasmic reticulum. Vesicles formed from the endoplasmic reticulum can be seen in close proximity to the cell walls and appear to participate in their dissolution (BAL and PAYNE 1972).

Laticifers. Laticifers are latex-containing cells or series of cells associated with the phloem of some tissues. Compound laticifers are derived from a group of cells which become united by dissolution of intervening walls (ESAU 1977). This process has been followed cytologically in *Achras sapota*. The breakdown process starts in the center of the transverse cell wall, which in young laticifers is thin, and gradually extends centrifugally to the lateral walls. In contrast to tracheid wall dissolution these events occur in living cells. Electronmicroscopy reveals the close association of the plasma membrane at the sites of wall resorption and the occurrence of periplasmic bodies, especially at the rim of the dissolving wall (SASSEN 1965). Extracts of latex from *Hevea* spp. and other plants with articulated laticifers hydrolyze carboxymethyl-cellulose (SHELDRAKE and MOIR 1970, SHELDRAKE 1970).

Aerenchyma. These highly porous, lacunate tissues are associated with gas transport in parts of plants having no direct access to the atmosphere. They are found in fruits, seeds, stems, leaf mesophyll and notably in the roots of aquatic plants. They develop in response to oxygen deficiency in roots of plants which have become waterlogged (CLOWES and JUNIPER 1968) and can be experimentally induced by ethylene treatment (KAWASE 1974).

Their development involves selective cell dissolution (lysogeny) or cell separation (schizogeny) or may involve both processes. Using *Helianthus* stems, KAWASE (1979) found an increase in carboxymethyl-cellulose hydrolyzing capacity in extracts of stems which had been water-logged or treated with ethylene.

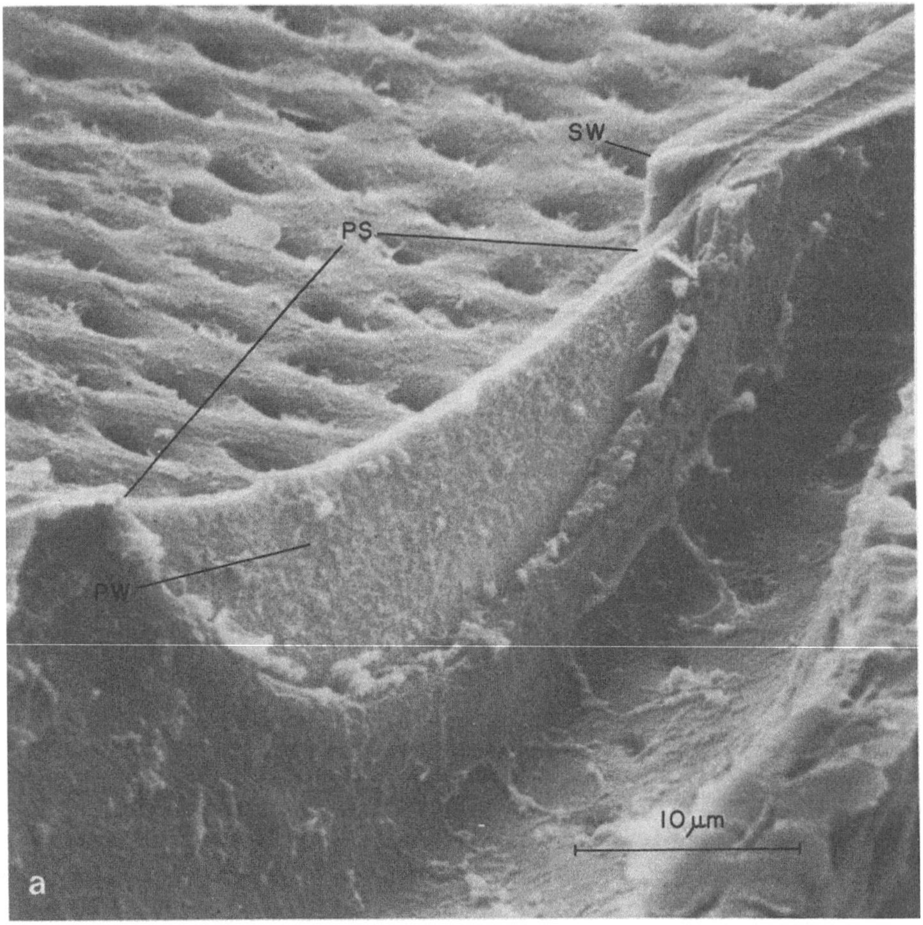

Fig. 4a, b. Scanning electron micrographs of half a perforation plate in a xylem vessel from wood of *Knightia excelsa*. **a** prior to removal of primary wall (*PW*) at the pore site (*PS*). *SW* secondary wall at rim of perforation plate. **b** after removal of primary wall from the perforation plate. Note that primary wall (*PW*) remains between secondary walls (*SW*) at the rim of the perforation plate. (Courtesy of Dr. B.A. Meylan, from Meylan and Butterfield, 1972)

Lateral Root Initiation. Changes in root pericycle cells in which lateral root meristems are initiated involve thinning of the originally thick secondary walls. This is accompanied by delignification, the appearance (or unmasking) of aniline blue fluorescent material and the insertion of plasmodesmata (BELL and MCCULLY 1970, SUTHERLAND and MCCULLY 1976), indicating that the process may involve biosynthetic and degradative changes in the walls.

4.6.2 Tyloses

Parenchyma cells adjacent to tracheids sometimes form balloon-like outgrowths, or tyloses, which extend through pits into the lumen of the abutting tracheid.

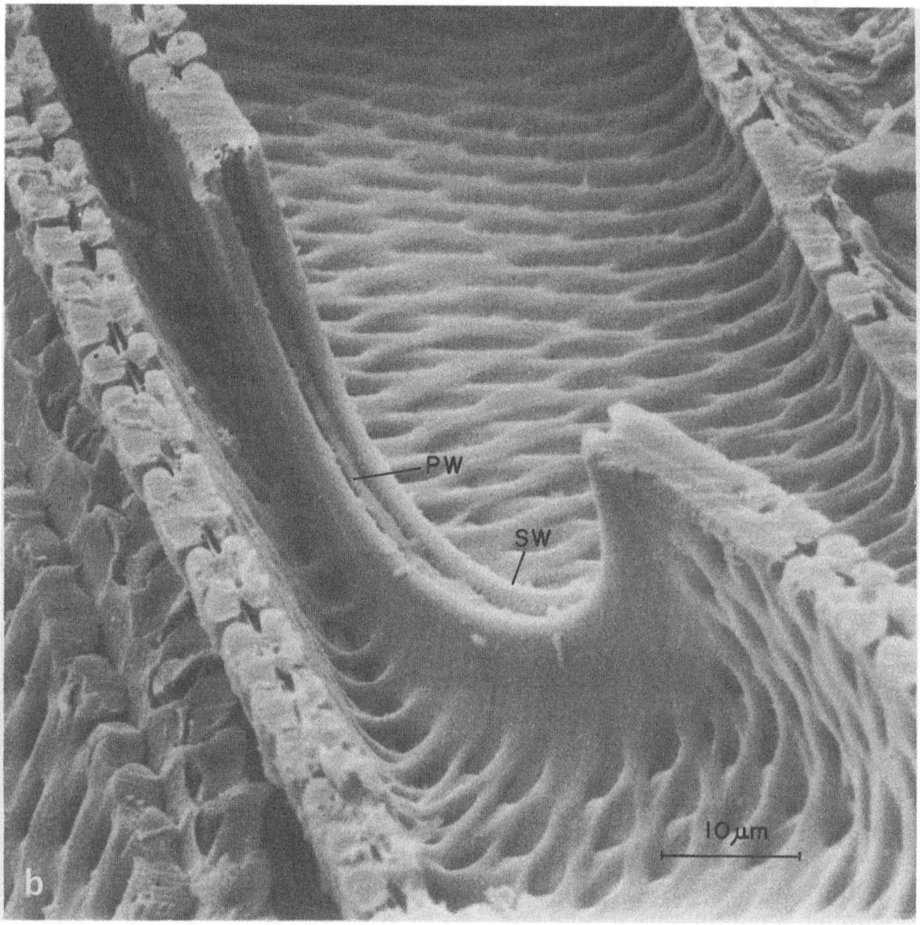

Fig. 4b

Surfaces of the tyloses are covered by a "protective" cell wall layer which also extends over the entire parenchyma wall surface adjacent to the tracheid (O'BRIEN 1974, ESAU 1977).

The protective layer stains with periodic acid-silver proteinate (CZANINSKI 1973) and has been shown by histochemical methods to be rich in acidic polysaccharides (O'BRIEN 1970). The deposition of this unusual wall layer has not been investigated.

4.6.3 Formation of Plasmodesmata and Differentiation of Sieve Plates

Plasmodesmata Formation. The protoplasts of living plant cells are connected through the cell wall by plasmodesmata, which are tubular channels lined with plasma membrane and which often contain endoplasmic reticulum elements. The primary plasmodesmata are seen to develop in the cell plate at the time of wall formation but secondary plasmodesmatal connections may be established between mature cells, for example between sieve tube elements and companion cells, during the initiation of new lateral roots (SUTHERLAND and MCCULLY 1976) and even between cells of different genetic origin as in the junction zone between graft hybrids and between parasitic plants and their hosts (ROBARDS 1975, JONES 1976). The origin and nature of the hydrolases involved in wall dissolution or thinning during the formation of these symplastic connections is not understood, nor is the control of the insertion of plasmodesmata at specific sites. The close proximity of endoplasmic reticulum to sites of plasmodesmata formation (NORTHCOTE and WOODING 1966, CLOWES and JUNIPER 1968) suggest their involvement with the localized dissolution process, which may be mediated by membrane-associated hydrolases.

Sieve Plate Differentiation. In developing phloem, perforated, sieve-like plates are found either in lateral walls or in end walls of adjoining sieve elements. The pores in the sieve plate are established following a complex differentiation program which has been established by cytological examination. The pores are initiated on each side of the developing sieve plate, at the openings of the plasmodesmatal connections. Early in the development of the sieve plate, elements of the endoplasmic reticulum become closely applied to the plasma membrane at the pore site region and remain there at all stages of development. An electron translucent substance, believed to be callose, is deposited outside the plasma membrane at the pore site region. The callose continues to be deposited and replaces previously deposited wall material to form cone-shaped pads which penetrate deeper into the wall as development proceeds.

Callose deposition continues until the opposing callose masses fuse and form cylinders surrounding the plasmodesmata. At this stage, the plasmodesmata begin to enlarge by dissolution of the surrounding callose until a large intercellular pore is formed. As in the original plasmodesmatal connection,the plasma membrane is continuous through the pore. Residual callose is nearly always observed as a thin-walled cylinder around the pore.

Pore formation is thus a complex series of morphogenetic events involving hydrolytic and synthetic operations in a strictly controlled sequence. Moreover, normal wall thickening is occurring coincidentally in sites adjacent to those of pore differentiation. Again the origin of the enzymes for callose formation and depolymerization and for the dissolution of the original wall fabric is not clear, nor is the mechanism by which these enzymes are directed to their specific sites of action. The endoplasmic reticulum appears to be involved (NORTHCOTE and WOODING 1966, 1968, HEPLER and NEWCOMB 1967, ROBERTS and NORTHCOTE 1970, NORTHCOTE 1974a, EVERT 1977) as judged by its proximity to the developing pores. However, ESCHRICH et al. (1972) and ESCHRICH (1976) have presented evidence suggesting that in *Yucca flaccida* and *Cucurbita maxima,* the filamentous P-protein which occurs in and around sieve pores in mature sieve tubes, is able to synthesize callose in the presence of UDP-Glc. The P-protein itself is apparently synthesised in companion cells.

In a few plants, such as *Vitis* and *Rosa,* where the sieve element functions for two seasons, a pad of "dormancy" callose is deposited on each face of the sieve plate at the end of the first season and is removed at the beginning of the next growing season. CLARKE and STONE (1962) could detect no increases in (1 → 3)-β-glucan hydrolase in *Vitis* phloem extracts at the break of dormancy but local increases at the sieve plate during callose dissolution would have gone undetected in the background of activity from other sources of the enzyme in phloem tissue. At the end of the operational life of a phloem element a pad of definitive callose is deposited on each face of the sieve plate (ESAU 1965).

Although the origin of the synthetases for dormancy or definitive callose formation is uncertain, in the mature sieve element the plasma membrane remains intact and elements of the endoplasmic reticulum, which are also present, could participate in the process (EVERT 1977, but cf. NORTHCOTE 1974b).

4.6.4 Gametogenic Tissues

During pollen formation the sporogenous cells, located in the anther, are surrounded by a multilayered tissue. The innermost layer, the tapetum, is involved with the nurture of the developing pollen grain. The morphogenetic events of pollen formation are complex and have been followed cytologically (WATERKEYN 1961, 1964). Prior to meiosis the primary wall of the sporogenous cell (meiocyte), which is rich in aniline blue-positive material (callose), is deposited as a thin layer inside the existing primary wall, apparently by membrane-bound cytoplasmic inclusions whose contents have the same appearance in the electronmicroscope as the callose wall (ESCHRICH 1962). Following meiotic division in angiosperms the tetrad of microspores may also be separated by special callose-rich walls which form in the cell plate region. These walls are at first penetrated by wide plasmodesmata but later become continuous around the meiocytes and are of varying thickness and form depending on the species. Both the pollen mother cell wall and the special microspore callosic walls have a transitory appearance and during pollen development are dissolved to release the developing pollen grain (Fig. 5).

The cells of the parental tapetum which are in close contact with the pollen mother cell wall are the probable source of hydrolases for callose wall dissolution (ESCHRICH 1966, ECHLIN and GODWIN 1968, MEPHAM and LANE 1969). Dictyosome vesicles become particularly abundant in the tapetal cells at this time (ECHLIN and GODWIN 1968). Examination of *Lilium longiflorum* anthers (STIEGLITZ and STERN 1973, STIEGLITZ 1974, 1977), which have a relatively high degree of meiotic synchrony, has allowed a definitive account of the role of (1 → 3)-β-glucan hydrolases in microspore release. There is a peak of hydrolase activity at the time of the rapid degradation of the callose wall. (1 → 3)-β-Glucan endo- and exo-hydrolases were found in the tapetal extract and were separated and characterized. Only the endo-hydrolase was effective in removing the callose walls and it was found in the immediate surroundings of the meiotic cells, whereas the exo-hydrolase, at the time of breakdown of the callose wall, was found only in the outer somatic layers of the anther. It was speculated that the exo-hydrolases were involved in the intra-cellular depolymerization of

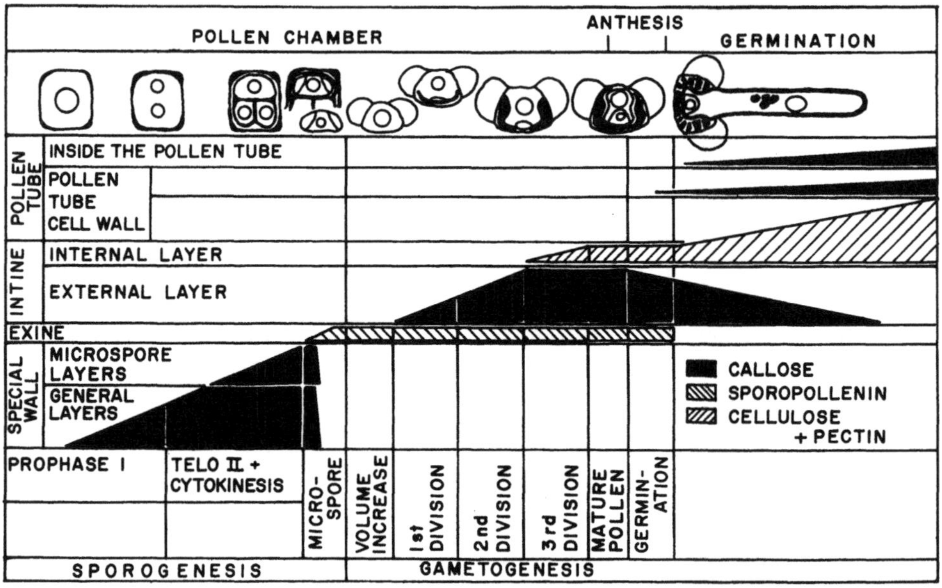

Fig. 5. Stages of appearance and subsequent fate of various wall deposits during microsporogenesis and gametogenesis in *Pinus*. (After WATERKEYN 1964)

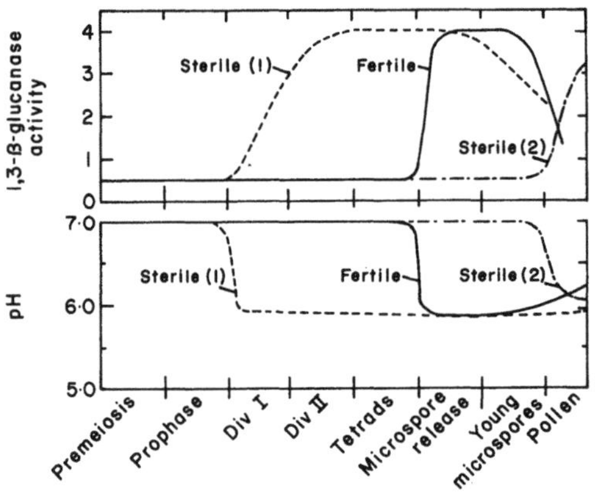

Fig. 6. Development of (1→3)-β-glucan hydrolase activity and changes in pH during microsporogenesis in anthers of a fertile and two cytoplasmic male sterile *Petunia* lines. (Redrawn from IZHAR and FRANKEL 1971)

(1→3)-β-oligoglucosides released from the callose by the endohydrolase. These oligomers had been observed by ESCHRICH (1961) in anthers of *Cucurbita ficifolia*.

The crucial role of the (1→3)-β-glucan endo-hydrolase in microspore cell wall dissolution is shown by studies on cytoplasmic male sterile and male fertile lines of *Petunia* (FRANKEL et al. 1969). In fertile lines, (Fig. 6) the pH in the anthers drops from about 6.9 to 6.0 just prior to the rapid increase in endohydrolase activity which precedes microspore release. In some sterile lines, (sterile 1, Fig. 6) the pH drops and the (1→3)-β-glucan hydrolase develops precociously, just after prophase. In these lines, callose is absent from the microspore mother cell wall and the mother cells have a distorted appearance. In another sterile line (sterile 2, Fig. 6) the pH remained high, no increase in hydrolase was detected and the callose walls remained intact at the tetrad stage. Callose wall dissolution and the associated drop in pH and increase in hydrolase activity occurred much later (IZHAR and FRANKEL 1971).

Precocious removal of special walls has been observed in cytoplasmic male sterile lines in *Sorghum bicolor* (OVERMAN and WARMKE 1972) while in a male sterile line of *Capsicum annuum* the callosic meiocyte walls are not dissolved (HORNER and ROGERS 1974).

In megasporogenesis in angiosperms and gymnosperms, callose is found in the cell wall around the megasporocyte at meiotic prophase and later in the transverse walls of the dyads and tetrads. As in microsporogenesis the callose has only a transitory existence and eventually disappears from the tetrad when the active megaspore differentiates into the embryo sac (see KAPIL and TIWARI 1978 for a review).

4.6.5 Mucilage (Slime)- and Gum-secreting Cells

Mucilage-secreting cells are found in a variety of tissues, for example in ducts of stems, petioles and leaf mid-veins, in the epidermis of seed coats, as trichomes on leaf surfaces and in root caps (FAHN 1979; Chap. 21, this Vol.). Mucilages are polysaccharides or polysaccharide-proteins of a complex type and have been characterized in mustard seeds (GRANT et al. 1969) and *Zea mays* root cap (WRIGHT and NORTHCOTE 1974, 1976). The assembly of the mucilage occurs in the Golgi apparatus, although it may be initiated in the endoplasmic reticulum (GREEN and NORTHCOTE 1978; WILLISON, Chap. 21, this Vol.).

Root cap mucilage and other mucilages are secreted into the periplasmic space following fusion of Golgi-derived vesicles with the plasma membrane (FAHN 1979). Enzymes are present in root-tip homogenates which are capable of detaching polysaccharide from protein (GREEN and NORTHCOTE 1978), but it is not known whether slime polysaccharides are released into the periplasmic space still attached to protein, or whether the polysaccharides are released in the Golgi.

The mucilage appears on the outer surface of the root cap cell wall in the form of a droplet (MORRÉ et al. 1967). The transfer of these relatively large slime polysaccharide complexes through the cell wall is not well understood. CARPITA et al. (1979) have shown that the cell walls of *Raphanus sativa* root hairs have an average pore size of 3.5 to 3.8 nm. They calculate that such a pore size would exclude dextrans of molecular weights in excess of 6,500 and conclude that the cell wall must act as a significant permeability barrier for large secreted polysaccharides such as root-cap slimes. PAULL and JONES (1976) have microscopical evidence that the slime accumulates initially in the periplasmic space, but as the root cap cells approach the surface of the root tip the slime accumulation is no longer observed, suggesting that the wall no longer serves as a barrier to secretion. This may be due to partial dissolution of the cell wall matrix. Hydrolases for noncellulosic polysaccharides have been observed in the apical regions of the root and $(1 \rightarrow 3)$-β-glucan hydrolases were particularly active in the terminal root tip cells (PAULL and JONES 1976).

Gum ducts are found in many plants and apparently arise following mechanical, physiological, or pathological injury. Few critical cytological studies have been performed on the cellular origin of polysaccharide gums, however in *Commiphora mukul,* the gum (guggul) is secreted by thin-walled cells lining the ducts (SETIA et al. 1977).

4.7 Senescence

This section deals with changes in cell walls which occur during the senescence of tissues, for which abscission and fruit senescence provide two well-documented examples. In addition, cell wall changes have been reported in senescing flowers and leaves.

4.7.1 Abscission

The shedding of leaves, flowers, petals, fruits and stems from plants occurs by a process known as abscission. This process is controlled by auxin, abscisic acid, and ethylene, and may be influenced by gibberellin and cytokinin (ADDICOTT and WIATR 1977).

Cytological studies show that changes leading to abscission occur in a localized region known as the separation layer or abscission zone. In leaf abscission, dissolution of the middle lamella has been observed in the cells of the abscission zone and this breakdown may be sufficient for separation. In some cases, however, disruption of both middle lamella and primary cell wall occurs, but cellulosic microfibrils remain (FACEY 1950, RASMUSSEN 1965, WEBSTER 1973, SEXTON 1976, ADDICOTT and WIATR 1977).

During abscission there is a loss of wall material staining with ruthenium red or hydroxylamine-ferric chloride (FACEY 1950, BORNMAN 1967, WEBSTER and LEOPOLD 1972, WEBSTER 1973) suggesting that pectic polysaccharides are removed. This is supported by analyses of cell walls in the abscission zone, which show an increase in cold- and hot water-soluble polysaccharide and a decrease in acid-extractable pectic polysaccharides (FACEY 1950, MORRÉ 1968). Although these observations suggest the involvement of pectin hydrolases and possibly pectin esterases, no clear correlations between the activity of these enzymes and the progress of abscission have been demonstrated (HÄNISH TEN CATE et al. 1975, BERGER and REID 1979). Furthermore, no correlation was found between strength of abscission tissue and the activities of polygalacturonase, pectin transeliminase, "arabinase", or "hemicellulase" (ABELES and LEATHER 1971).

Another hydrolase which may participate in abscission, and which has been detected in abscission zones by viscometric assays using carboxymethyl-cellulose as substrate, is cellulase (HORTON and OSBORNE 1967, ABELES 1969, SHELDRAKE 1970, RASSMUSSEN 1973, LEW and LEWIS 1974, REID et al. 1974, HUBERMAN et al. 1975, HÄNISH TEN CATE et al. 1975, HUBERMAN and GOREN 1979). In *Phaseolus vulgaris,* the cellulase appears in three locations; intracellularly as a buffer-soluble form (pI 4.5) which is present at all times, extracellularly as a wall-bound form (pI 9.5) which can be extracted with 1 M NaCl and which increases during abscission (LINKINS et al. 1973), and as a plasma membrane-bound form. The salt-extractable cellulase (pI 9.5) is formed de novo during abscission, whereas the soluble cellulase (pI 4.5) is not (LEWIS and VARNER 1970).

The assumption that cellulase is directly involved in wall separation has been questioned since increases in cellulase activity do not always coincide with abscission (HÄNISH TEN CATE et al. 1975, HUBERMAN and GOREN 1979). However, using an anti-serum prepared to the pI 9.5 cellulase from *Phaseolus vulgaris* it has been shown immunocytologically that this enzyme is located in the 2 to 3 rows of cortical cells in the separation layers and in the vascular bundle of the petiole. Significantly, injection of anti-cellulase serum into the fracture plane successfully prevented loss of break strength of abscission zones (SEXTON et al. 1980). These results point to the direct involvement of the pI 9.5 cellulase in abscission. However, until more precise information on changes in wall composition during abscission and on the substrate specificity of the cellulase is available, it will not be clear whether fibrillar cellulose or a noncellulosic polysaccharide such as xyloglucan is the substrate for these hydrolases.

4.7.2 Fruit Senescence

When fruit have matured a series of transformations occur which collectively constitute senescence. The pattern of fruit senescence follows one of two courses. In climacteric fruits (apples, pears, peaches, plums, avocados) there is a sudden upsurge in respiration when development is complete and before senescence begins. In nonclimacteric fruits, such as strawberries, pineapples, oranges, lemons and grapes, development and senescence proceed successively without a sudden increase in respiration. The process is under hormonal control and in the initial stages, at least, involves active metabolism.

The changes encountered include depolymerization of storage materials, changes in pigments, flavor- and aroma-producing compounds, as well as alterations to cell walls, which contribute to fruit softening and ultimately lead to its disintegration. In this section cell wall changes which have been established cytologically are compared and related to alterations in wall chemistry. The correlations between cell wall changes and polysaccharide hydrolases found in senescing fruits are also summarized.

Most studies of fruit senescence have concentrated on changes leading to the stage when the fruit is acceptable for consumption, that is, the fruit is ripe. This point is judged by texture, flavor, color etc. and varies for each fruit. The texture-determining factors relate chiefly to the thickness and physical state of the cell wall, but cell size and turgor also contribute.

Cytological examination of walls of fruit parenchyma and collenchyma during senescence reveals that the middle lamella becomes hydrated, swells and is then dissolved. There are also indications of disaggregation of the primary wall, which later may lose its structural integrity and lead to cellular collapse. Fruits which have been shown to follow this pattern during senescence include pears (STERLING 1954, BEN ARIE et al. 1979), dates (COGGINS et al. 1967), strawberries (KNEE et al. 1977), peaches (REEVE 1959) and avocados (SCOTT et al. 1963, PESIS et al. 1978).

Changes in the composition of the pectic complex have been recorded as major and early events in wall senescence. Thus the degree of esterification of peach pectins decreases during senescence (POSTLMAYR et al. 1956) and losses of galactose, arabinose, and galacturonic acid from pectic fractions have been observed in apples (NELMES and PRESTON 1968, KNEE 1973, 1975, SEIPP 1978), pears (JERMYN and ISHERWOOD 1956, YAMAKI et al. 1979), peaches (POSTLMAYR et al. 1956, REEVE 1959, STERLING and KALB 1959, PRESSEY et al. 1971, WHAN, unpublished results), and tomatoes (WALLNER and BLOOM 1977, GROSS and WALLNER 1979). These changes are accompanied by increases in water-extractable polyuronide in tomatoes (GROSS and WALLNER 1979), apples (KNEE 1973, BARTLEY 1974, 1976), pears (YAMAKI et al. 1979) and peaches (POSTLMAYR et al. 1956). In apples, alkali-soluble glucans are also lost (KNEE 1973).

The cytological and chemical evidence for changes in cell wall components, especially the pectic polysaccharides, has led to extensive examination of the occurrence and properties of polysaccharide and glycoside hydrolases and their relationship to wall alteration during senescence. Polygalacturonases have been characterized from many senescent fruits (FOGARTY and WARD 1974, PILNIK and ROMBOUTS 1979) and two basic types distinguished; polygalacturonate endo-hydrolases (EC 3.2.1.15) and exo-hydrolases (EC 3.2.1.67). Both are found in pears (PRESSEY and AVANTS 1976), tomatoes (PRESSEY and AVANTS 1973a), and peaches (PRESSEY and AVANTS 1973b) but only the exo-hydrolase is present in apples (PILNIK and VORAGEN 1970, BARTLEY 1978).

There are good correlations between rises in polygalacturonase activity and the softening of peaches (PRESSEY and AVANTS 1973b), pears (BARTLEY 1974, YAMAKI and MATSUDA 1977, YAMAKI and KAKIUCHI 1979), avocados (AWAD and YOUNG 1979), tomatoes (BESFORD and HOBSON 1972, WALLNER and BLOOM 1977, GROSS and WALLNER 1979) and dates (HASEGAWA et al. 1969). The implication that cleavage of $(1 \rightarrow 4)$-α-galacturonosyl linkages in pectic polysaccharides has a critical role in wall modification is supported by the observation that the mutant tomatoes *rin* and *nor*, which fail to soften, do not show the normal increase in polygalacturonase activity (BUESCHER and TIGCHELAAR 1975, POOVAIAH and NUKAYA 1979). Immunochemical studies also show that in normal tomatoes there is a massive increase in polygalacturonase during senescence but that in the mutants this protein is absent or present in greatly reduced amounts. The increase in polygalacturonase is thus likely to be due to de novo synthesis rather than to relief of inhibition of a preformed enzyme (BRADY et al. 1980).

Pectin esterases (EC 3.1.1.11) have been found in mature and senescent tomatoes (HOBSON 1963, BUESCHER and TIGCHELAAR 1975), bananas (YOUNG 1965), cherries (AL-DELAIMY et al. 1966, DAVIGNON 1961), lemons (ROUSE and KNORR 1969), oranges (VERSTEEG et al. 1978), and Japanese pears (YAMAKI and MATSUDA 1977), but their activities do not appear to increase during senescence in the same way as polygalacturonate hydrolases. In fact, in avocados (AWAD and YOUNG 1979), pears (NAGEL and PATTERSON 1967), tomatoes (TAHIR and ELAHI 1978), and capsicum (THIRUPATHAIAH and SUBRAMANIAN 1977) the pectin esterase activity decreases during senescence. The pectin esterase from fruits is partly wall-bound and occurs in multiple forms in tomatoes (PRESSEY and AVANTS 1972) and bananas (HULTIN et al. 1966, BRADY 1976). Changes in pectic polysaccharides involving de-esterification, cleavage of galacturonyl linkages, release of galactose- and arabinose-containing polymers, and soluble polygalacturonates occur during senescence. It has been inferred from these observations and from the specificity of the polygalacturonases that preliminary de-esterification of polygalacturonate esters and removal of arabinose- and galactose-containing side branches is necessary for subsequent depolymerization by the exo- and endo-polygalacturonase. However, these interpretations must remain tentative pending further information on the specificities of the enzymes and more detailed structural analyses of changes in pectic polysaccharides occurring in the different species.

Cellulases are also found in senescing fruit such as tomatoes (LEWIS et al. 1974, BUESCHER and TIGCHELAAR 1975, POOVAIAH and NUKAYA 1979) in both normal and *rin* types (HOBSON 1968, PHARR and DICKINSON 1973, SOBOTKA and STELZIG 1974), pears (YAMAKI and MATSUDA 1977, YAMAKI and KAKIUCHI 1979), freestone peaches (HINTON and PRESSEY 1974, LEWIS et al. 1974), dates (HASEGAWA et al. 1969), pineapples (SUZUKI et al. 1971), and capsicum (chilli) (THIRUPATHAIAH and SUBRAMANIAN 1977). In ripe avocados (LEWIS et al. 1974, AWAD and YOUNG 1979) the cellulase activity reaches prodigious levels.

The increase in cellulase activity observed in senescing fruits is in accord with the disaggregation of the primary wall which follows dissolution of the middle lamella. Whether the cellulases are capable of digesting microfibrillar cellulose is known only for tomato, from which a multi-enzyme complex which degrades microcrystalline cellulose has been described (SOBOTKA and STELZIG 1974). The loss in cellulose from walls of Japanese pears in the later stages of senescence (over-ripening) (YAMAKI et al. 1979) might be attributed to the action of such an enzyme complex.

4.7.3 Flower and Leaf Senescence

Among the catabolic changes occurring in senescent flowers and leaves is the mobilization of wall polysaccharides. Thus during senescence of the ephemeral flowers of *Ipomea tricolor,* 43% of noncellulosic and 40% of cellulosic polysaccharides of the corolla are degraded. This resorption of wall polysaccharides is accompanied by an increase in the activity of a number of glycoside and glycan hydrolases (WIEMKEN-GEHRIG et al. 1974). Similarly in senescing tobacco leaf discs, there is extensive degradation of noncellulosic wall polysaccharides (MATILE 1974). The enzymes responsible were not examined but in senescing *Nicotiana glutinosa* leaves there is a 14-fold increase in $(1 \rightarrow 3)$-β-glucan hydrolase activity (MOORE and STONE 1972).

4.8 Germination

During seed germination reserve proteins, lipids and polysaccharides are mobilized to support the growth of the developing seedling. These reserves are found in two major storage tissues: in the cotyledons of the embryo and in the extra-embryonic endosperm. In grass and cereal seeds (monocotyledons), the endosperm is the major storage tissue, whereas in some dicotyledonous seeds (e.g., in peas) the endosperm has a transient existence during maturation and the cotyledons are the major storage organ in the mature seed. In other dicotyledons, for example in fenugreek, lucerne, locust, guar, carrots, soybean, and lettuce, the endosperm is retained at maturity and both endosperm and cotyledons act as storage tissue although the quantitative contribution from each is variable. Other seed parts may also have storage functions; thus in *Yucca* the nucellus persists as a storage tissue (perisperm), and in gymnosperms the storage tissue is the gametophyte.

In the seeds of grasses the endosperm is composed of relatively thin-walled cells tightly packed with starch granules and protein. The endosperm is surrounded by the aleurone layer which consists of cells with thick, bi-layered walls and characteristic cytoplasmic inclusions. Endosperm walls in wheat (MARES and STONE 1973a), barley (FINCHER 1975) and *Lolium* (MARES and STONE 1973b) have a low content (<5%) of microfibrillar cellulose. Heteroxylans and (1→3);(1→4)-β-glucans, together with some glucomannan, are the predominant wall components in wheat (MARES and STONE 1973a, BACIC and STONE 1980), barley (FINCHER 1975) and rye grass *Lolium multiflorum* (SMITH and STONE 1973b, MARES and STONE 1973b, ANDERSON and STONE 1978). The water-residue of rice endosperm cell walls has been reported to have an unusually high cellulose content (36.3%) and, in addition to the heteroxylan and (1→3); (1→4)-β-glucan, pectic polysaccharides, xyloglucan, and a hydroxyproline-rich glycoprotein were also present (SHIBUYA and IWASAKI 1978, SHIBUYA and MISAKI 1978).

The starchy endosperm and aleurone cells are ontogenetically related, the aleurone cells being differentiated from the outer layer of the developing starchy endosperm. In wheat and barley the overall composition of the aleurone cell walls is very similar to that of the starchy endosperm (BACIC and STONE 1981a, b), but it is not known how the cell wall components are distributed between the thin inner and the thick outer layers of the aleurone walls. This bilayered structure is retained following water and 8 M urea extraction, but dilute alkali entirely disaggregates the wall.

An early event in the germination of grass and cereal seeds is the dissolution of endosperm cell walls (BROWN and MORRIS 1890, DICKSON and SHANDS 1941, POMERANZ 1972, FINCHER and STONE 1974, PALMER and BATHGATE 1976). Hydrolases capable of depolymerizing the wall polysaccharides have been extensively investigated (PREECE 1957, ANDERSON et al. 1976). They include (1→3)-β-glucan hydrolase (EC 3.2.1.39), (1→4)-β-glucan hydrolase (EC 3.2.1.4), (1→3); (1→4)-β-glucan hydrolase (EC 3.2.1.73), β-xylan hydrolase (EC 3.2.1.32), arabinofuranosidase (EC 3.2.1.55), and β-glucosidase (EC 3.2.1.21). The enzymes arise from the cells of the aleurone layer but there is evidence that some may also originate from the scutellum (BALLANCE et al. 1976, GIBBONS 1980) and the endosperm itself (BALLANCE et al. 1976). Gibberellic acid arising from the embryo is responsible for the initiation of de novo synthesis or release of some of these enzymes from aleurone cells (HALMER and BEWLEY, Chap. 21, Vol. 13A, this series).

Degradation of the endosperm walls and cell contents begins in the endosperm underlying the aleurone near the scutellar epithelium-aleurone interface and proceeds centripetally across the grain and towards its distal end (GIBBONS 1980, OKAMOTO et al. 1980). In some grasses, such as *Bromus*, endosperm cell walls are relatively thick, so that their digestion products could be quantitatively important sources of carbohydrate for the embryo. MORRALL and BRIGGS (1978) have prepared a balance sheet for the utilization of aleurone and starchy endosperm polysaccharides in germinating barley and found that the nonstarchy polysaccharides of the endosperm and aleurone cell walls were able to provide a significant 18.5% of the carbohydrate degraded and used for seedling development.

Cytological studies show that during germination the secretion of hydrolytic enzymes by the aleurone cells is accompanied by a progressive digestion of the outer aleurone wall layer while the inner layer remains intact (TAIZ and JONES 1970, ASHFORD and JACOBSEN 1974). In this way the aleurone cell is maintained as an active metabolic unit throughout the period of endosperm dissolution. The chemical or physical basis for the selective resistance of the inner aleurone wall is not understood. In this regard, a role for ferulic acid, which is covalently linked to both aleurone and endosperm polysaccharides, has been suggested (FULCHER et al. 1972). Its content in aleurone walls is 10 to 20 times higher on a weight basis than in endosperm walls (BACIC 1979). However, such a role for ferulic acid appears unlikely since the degree of substitution of wall polysaccharides with ferulic acid residues would be too low to prevent enzymic hydrolysis (BACIC 1979).

Other quantitatively important constituents of the aleurone wall which could be involved in the differential resistance are proteins and the acetyl substituents to polysaccharides. The latter comprise 4.5% in wheat and 2.1% in barley aleurone cell walls and could be significant in preventing hydrolysis, as has been demonstrated by the depressed rumen digestion of acetylated forage polysaccharides (BACON et al. 1975).

In leguminous seeds such as fenugreek, guar, locust, and honey locust, where cotyledons and endosperm co-exist at maturity, the carbohydrate reserve in the endosperm is not starch but galactomannan (MEIER and REID 1977). In fenugreek this polysaccharide is found as an inner, thick deposit on the endosperm wall (MEIER and REID 1977) and, in addition to its role as a carbohydrate reserve, may function through its water-binding capacity to protect the germinating seed from desiccation (REID and BEWLEY 1979). During germination the galactomannan is removed by hydrolases secreted from the aleurone (REID 1971, REID and MEIER 1972, 1973, MCLENDON et al. 1976, MEIER and REID 1977, REID et al. 1977), while the thin outer layer of the endosperm cell wall persists (REID and MEIER 1972). The enzymes involved in galactomannan depolymerization include (1→4)-β-mannan hydrolase (EC 3.2.1.78), β-mannosidase (EC 3.2.1.25), and α-galactosidase (EC 3.2.1.22) (REID and MEIER 1973, MCCLEARY and MATHESON 1974, 1975, SEILER 1977), and in contrast to cereals their synthesis and secretion by the aleurone cells appears to occur without embryo participation (REID and MEIER 1972). As in germinating cereals, the aleurone cell walls themselves are partially but not completely dissolved (REID and MEIER 1972; see also Chap. 11, Vol. 13A, this series).

In lettuce (*Lactuca sativa*) seeds the endosperm completely envelops the cotyledons, is two cells thick, and is characterized by thick cell walls which extend numerous peg-like projections into the cytoplasm (JONES 1974). The polysaccharides of the cell wall have a high mannose content (58%) and galactose (9.5%), glucose (10%), arabinose (9.5%), and uronic acids (10%) are the other main monosaccharide components (HALMER et al. 1976). In germinated seeds the endosperm cell walls, except those adjacent to the integument, are extensively degraded, as shown by loss of periodate-Schiff staining. Digestion proceeds from the plasma membrane outward, suggesting that the enzymes arise from the cells themselves (JONES 1974). Among the enzymes involved is a (1→4)-β-mannan hydrolase (EC 3.2.1.78) which increases about 100-fold within 15 h of the start of germination (HALMER et al. 1978). The increased enzyme activity is inhibited by cycloheximide but stimulated by gibberellin (HALMER et al. 1976).

The endosperm of celery (*Apium graveolens*) forms the bulk of the seed and completely encloses the embryo. The endosperm cells are large, thick-walled, and angular, containing many aleurone grains, an abundance of lipids, but no starch. The cell walls stain blue with toluidine blue, are periodate-Schiff-positive but do not give a positive hydroxylamine-ferric chloride reaction for pectic polysaccharides. During germination the outer portion of the endosperm walls is digested, leaving a thin layer of cell wall surrounding the protoplast, which remains viable throughout. The persistent layer is periodate-Schiff-positive and stains purple with toluidine blue. This selective wall dissolution transforms the endosperm into a mass of separated cells. The enzymes concerned with wall digestion appear to arise from the endosperm and can be induced by gibberellic acid in the absence of the embryo (JACOBSEN et al. 1976, JACOBSEN and PRESSMAN 1979).

In *Yucca* seeds the nucellus persists as the storage tissue (perisperm) surrounding the embryo. Cytological and histochemical studies have shown that noncellulosic components of the thick perisperm cell walls are degraded during germination to leave a cellulosic framework (HORNER and ARNOTT 1966).

In date (*Phoenix dactylifera*) seeds the mannan of the endosperm cell wall (MEIER 1958) is degraded by enzymes in a "dissolution zone" surrounding the haustorium (KEUSCH 1968), which is the source of the hydrolases. In this region two sub-zones can be distinguished; a primary zone consisting of several cell layers where the secondary wall becomes swollen but does not lose its form, and a much smaller secondary zone where the mannan is degraded to mannose residues. The cellulose of the primary wall of secondary zone cells resists decomposition and remains as thread-like elements.

In contrast to endospermic seeds, cell walls of cotyledonary seeds are not extensively degraded during germination, although some preliminary evidence suggests that cellulose, pectic polysaccharides, but not other noncellulosic polysaccharides undergo limited metabolism during the mobilization of protein and lipid reserves in the cotyledons of white mustard seeds (GOULD and REES 1965).

5 Metabolism in Response to Environmental Changes and Pathogenesis

In addition to the changing patterns of polysaccharide structure and composition which occur during normal growth and development of cells, qualitative and quantitative changes in noncellulosic components of cell walls have been observed in the face of changing environmental conditions and in the presence of pathogens. The chemistry and biochemistry of these alterations and their control have seldom been rigorously examined but the available information will be discussed in the following sections.

5.1 Tropisms

5.1.1 Geotropism and Phototropism

Plant shoots and roots bend in response to light or gravity due to differential growth on one side of the tissue. This response may be regarded as a site-specific elongation of cells in the extending zone of the tissue and presumably occurs by a mechanism comparable to normal primary wall extension (Sect. 4.4). No comparative studies on wall composition of cells in the various zones in which differential growth occurs have been reported.

5.1.2 Reaction Wood

Nonvertical branches and trunks of leaning trees are characterized by the presence of reaction wood, which is formed as a result of attempts by the plant to counteract the force inducing the nonvertical position. In gymnosperms, reaction wood is located on the underside of the limb or trunk and is termed compression wood, whereas in arborescent angiosperms reaction wood is located on the upper side and is termed tension wood. Reaction woods are morphologically distinct from normal wood (WARDROP 1964, SCURFIELD 1973) and their cell wall composition is also different. In both compression and tension wood the relative amount of $(1\rightarrow4)$-β-galactan is increased compared with normal wood (HÄGGLUND 1951, BOUVENG and MEIER 1959, MEIER 1962). These β-galactans appear to be structurally related to the pectic galactans rather than to the soluble (1→3);(1→6)-β-galactans found in the lumen of gymnosperm tracheids (CLARKE et al. 1979).

Gymnosperm compression wood also contains a (1→3)-β-glucan, which has associated uronic acid residues. This polysaccharide has been extracted from reaction woods of *Pinus resinosa,* (red pine, HOFFMANN and TIMELL 1972a), *Pinus sylvestris* (Scot's pine, FU et al. 1972) and *Larix laricina* (tamarack, HOFFMANN and TIMELL 1970, 1972b) but also appears in small amounts in the normal wood of red pine and other conifers (HOFFMANN and TIMELL 1972c). This (1→3)-β-glucan may arise from the helical bands of aniline-blue fluorescent material found in the S_2 layer of gymnosperm compression wood tracheids (BRODZKI 1972, WLOCH 1975) where it may form as a wound response induced by the schizogenous development of the helical cavities.

Collectively these observations indicate that alterations in the relative activities of polysaccharide synthetases occur as part of the metabolic changes associated with reaction wood formation.

5.1.3 Hypogravity

In marigold (*Tagetes patula*) plants grown under conditions of hypogravity, increases in the proportion of alkali-soluble, noncellulosic polysaccharides (SIEGEL et al. 1977) and salt-extractable proteins (OPUTA and MAZELIS 1977) have been reported. Sunflower seedlings subjected to hypogravity showed an increased extensibility which was correlated with an increase in auxin-stimulated cellulase activity (BARA and GORDON 1972). The functional significance of these changes remains unexplained.

5.2 Environmental Stress

Several types of stress may be imposed on a plant by changes in the environment. These include alterations in water availability, changes in ambient temperature, increases in salinity, and the presence of toxic molecules or metal ions.

Cell wall metabolism is very sensitive to lowered water availability and is affected by as little as a 5 bar reduction in tissue water potential (HSAIO 1973). In *Avena* coleoptiles (ORDIN 1960, CLELAND 1967), and in sunflower and almond leaves (PLAUT and ORDIN 1961, 1964) subjected to low water potential, cell elongation or expansion is completely inhibited, but cell wall synthesis, as measured by ^{14}C-glucose incorporation, continues at a depressed rate. The decrease in wall synthesis could be due to inactivation of polysaccharide synthetases or to physical perturbation of the plasma membrane-cell wall interface.

Continued exposure to a low water potential results in "softening" of the coleoptile cell wall (CLELAND 1967), an observation which implies that the hydrolase action necessary for wall loosening continues in the cell wall after synthesis or deposition of wall polysaccharides is inhibited, and that the normal balance of synthetase and hydrolase activity in elongating cells is tipped in favor of wall degradation.

Environmental stress due to chilling, to high temperatures, or to raised salt concentrations, may also result in physical alteration of the plasma membrane-cell wall interface due to changes in membrane structure (LYONS 1973) or to lowered tissue water potential. These changes might be expected to affect cell wall metabolism.

In chilling-sensitive cucumbers, bananas and sweet potatoes held at low temperatures there is an increase in the proportion of low methoxy pectic polysaccharides, but in chilling-resistant peas and turnips, the amount of low methoxy pectic polysaccharides decreases at low temperatures (FUKUSHIMA and YAMAZAKI 1978). The increase in low methoxy pectic polysaccharide in chilling-sensitive plants is consistent with the observation that levels of pectin esterases increase in peaches subjected to low temperatures (BEN-ARIE and LAVEE 1971).

Another response to physiological stresses which is commonly encountered is the formation of callose deposits on the plasma membrane (Table 2). The mechanism of the induction of this response is discussed in Section 5.3 and appears to be due to alteration in the environment of plasma membrane-bound $(1\rightarrow3)$-β-glucan synthetases. Callose deposition might therefore be useful as an indicator of changes at the plasma membrane-wall interface.

5.3 Wounding Responses

Wounding of tissues produces changes in cell walls which can be followed cytologically. In mechanically damaged tissues, cells respond by depositing a suberin layer on their inner wall surface. In some tissues, wounding induces cell division and a suberin layer is found between primary and secondary wall layers in newly formed cells (BARCKHAUSEN 1978).

A well-documented response to chemical and physical trauma is the appearance of callose, which is identified on the basis of its aniline blue fluorescence and is found at the plasma membrane-wall interface, in pit fields and on sieve plates. Because of technical difficulties, none of these deposits have been isolated and analyzed, but those formed in the stigma papillae as a response to self-incompatible mating (Chap. 23, this Vol.), which can be considered as a type of wound response, have been shown by chemical and enzymic techniques to contain both $(1\rightarrow3)$- and $(1\rightarrow4)$-β-glucosidic linkages, some of which occur in the same polymer (VITHANAGE et al. 1980).

Callose deposition can occur within minutes of injury (ESCHRICH 1956, 1975, CRAFTS and CURRIER 1963) but whether this results from the high capacity of latent, plasma membrane-bound $(1\rightarrow3)$-β-glucan synthetases or from deposition of preformed polysaccharides is not known. A latent capacity to synthesize callose at the sieve plate is shown by the rapid deposition of radioactive callose when *Acer pseudoplatanus* stems are sectioned in the presence of ^{3}H-glucose (NORTHCOTE and WOODING 1966, 1968). Further support for the idea that the $(1\rightarrow3)$-β-glucan synthetase is activated is given by the observation that callose deposited as a response to heat shock in *Phaseolus vulgaris* and *Vigna sinensis* is paralleled by an increase in the capacity of cell-free extracts to synthesize $(1\rightarrow3)$-β-glucosidic linkages in polysaccharides (MUSOLAN et al. 1975).

The synthesis of (1 → 3)-β-glucans from exogenously supplied UDP-^{14}C-Glc has been followed in experimentally damaged tissues. In tissue-cultured soybean cells damaged by stirring (BRETT 1978) and in pea epicotyls damaged by slicing (RAYMOND et al. 1978, ANDERSON and RAY 1978) there is rapid synthesis of (1 → 3)-β-glucan. The UDP-Glc is utilized directly without prior metabolism. These observations may be interpreted in terms of an activation of the (1 → 3)-β-glucan synthetase on wounding, or may simply reflect increased accessibility of the synthetase to its substrate after membranes have been exposed. It is known that (1 → 3)-β-glucan synthesis is enhanced by the high concentration of UDP-Glc (>0.5 mM) used in these experiments. This raises the possibility that release of substrate from a sub-cellular compartment, which would normally isolate it from the enzyme, could account for the rapid deposition of callose on wounding. The triggering of callose deposition on wounding in vivo may involve an hormonal signal. Thus in decapitated *Salix viminalis* shoots callose is deposited on sieve plates (THOMAS and HALL 1975) and ethylene, which is known to be liberated on wounding (LIEBERMAN 1979), increases callose deposition. Indole acetic acid and gibberellin, but not abscissic acid or kinetin, inhibit its formation (THOMAS and HALL 1975).

Provided that wounding does not result in irreversible damage, as in osmotic (ESCHRICH 1957) or ultrasonic (CURRIER and WEBSTER 1964) treatments, the wound callose is slowly resorbed after removal of the stimulus. This implies local operation of a (1 → 3)-β-glucan hydrolase known to be widely distributed in higher plants (CLARKE and STONE 1962).

5.4 Pathogenesis

Infecting micro-organisms initiate a number of resistance mechanisms in plants. These appear to depend on the plant's ability to recognize the pathogen as "nonself", a characteristic also expressed in plant mating systems and somatic cell interactions (CLARKE and KNOX 1979). Among the responses to infection are production of phytoalexins, hydrolytic and other enzymes, and chemical modifications to cell walls, especially those in the region of contact with the infecting agent (WOODWARD et al. 1980). Changes in cell walls include thickening (HIRUKI and TU 1972), increases in phenolic acids (SWAIN 1977, FRIEND 1979, HARTLEY et al. 1978), polyphenolics (MAYER and HAREL 1979), lignin (KIMMINS and BROWN 1973), and hydroxyproline-rich glycoproteins (ESQUERRÉ-TUGAYÉ et al. 1979). Furthermore, in tissues infected by fungi, nematodes, and viruses, local deposits variously known as papillae, callosities, callus, or lignitubers are seen on cell walls (AIST 1976, GARCIA-ARENAL and SAGASTA 1977, SPEAKMAN and LEWIS 1978). The deposits appear to be similar to those produced by physical or chemical wounding in that they react with the aniline blue fluorochrome (Table 2). Some also give positive tests for lignin (AIST 1976). The availability of techniques for examination of papillae in situ by acoustic microscopy (ISRAEL et al. 1980) and for their isolation (KUNOH et al. 1979) will assist in a critical assessment of their development and composition.

Table 2. Observed deposition of aniline blue staining of fluorescing deposits (callose) in plant tissues following physiological stress, wounding, or infection

Physiological stress	
Plasmolysis	CURRIER and STRUGGER (1956), CURRIER (1957), ESCHRICH (1957), PRAT and ROLAND (1971)
Heating	WEBSTER and CURRIER (1965), SHIH and CURRIER (1969), DEKAZOS (1972), MCNAIRN (1972), SMITH and MCCULLY (1977)
Low temperature stress	MAJUMDER and LEOPOLD (1967), SMITH and MCCULLY (1977), LANG et al. (1979)
Wounding	
Chemical	CURRIER (1957), SCHUSTER (1960), ULLRICH (1963), MCNAIRN and CURRIER (1965), ESCHRICH et al. (1965), HUGHES and GUNNING (1980), PETERSON and RAUSER (1979)
Mechanical	ITÔ (1949), CURRIER and STRUGGER (1956), CURRIER (1957), ENGLEMAN (1965), DEKAZOS and WORLEY (1967), DEKAZOS (1972) THOMAS and HALL (1975)
Ultrasound	CURRIER and WEBSTER (1964)
Infection	
Fungal infection	MANGIN (1899), KUSANO (1911), ITÔ (1949), AIST (1976), GARCIA-ARENAL and SAGASTA (1977), see AIST (1976) for further references
Virus infection	MOERICKE (1955), SPRAU (1955), SCHUSTER and BYHAN (1958), SCHUSTER (1960), DE BOKX (1967), ESAU and CRONSHAW (1967), WU et al. (1969), WU and DIMITMAN (1970), HIRUKI and TU (1972), WU (1973), ALLISON and SHALLA (1974), SHIMOMURA and DIJKSTRA (1975), SCHUSTER and FLEMING (1976), STOBBS and MANOCHA (1977), PENNAZIO et al. (1978)
Mycoplasma-like bodies	HIRUKI and SHUKLA (1973), COUSIN 1979

Fig. 7. Callose deposition in roots of *Zea mays* infected with *Phytophthora cinnamomi*. Section prepared 4 h after initial contact of zoospores with the root, stained with aniline blue, and examined by fluorescence microscopy. The germ tubes have grown through the thick slime layer, and penetrated the epidermis. Fluorescent callose deposits are seen in epidermal cells. (Scale 1 cm = 26 µm). (Photograph provided by Dr. A.E. Clarke, Botany School, Melbourne University, Australia)

Another response to infection is the deposition of "gums" or "gels" in xylem vessels. The gels are of plant origin and possibly arise by swelling of preformed wall structures, but their composition is unknown (VANDERMOLEN et al. 1977).

In most cases the chemical changes in host cell walls are limited to cells close to the point of infection. The altered cell wall may in some cases encapsulate the micro-organism in the intercellular spaces (MENDGEN 1978). In other situations, as in local necrotic lesions found in some viral, fungal, or bacterial infections, the altered walls may function to arrest the spread of the pathogen. In these lesions the changes appear to be general throughout the wall, however localized deposits (papillae) (Fig. 7) may also restrict the spread of the pathogen (SPEAKMAN and LEWIS 1978) or may serve to plug or seal the wound, restricting interchange of ions, metabolites, and toxins between injured and healthy host cells or between the pathogen and host cells.

The chemical changes in host walls following infection all involve active metabolism and may be induced or elicited by specific molecules originating from the pathogen, as in the phytoalexin response (ALBERSHEIM and VALENT 1978), or by hormones such as ethylene produced on wounding, or by a combination of these effects (WOODWARD et al. 1980).

Acknowledgments. We are indebted to Dr. Adrienne E. Clarke for helpful suggestions concerning the manuscript, to Professor A.B. Wardrop and Dr. Clarke for providing Figs. 2 and 7, respectively, to Ms. Jo Cook for the compilation of references, and to Mrs. Pat Janicke for typing the manuscript.

References

Abeles FB (1969) Abscission: role of cellulase. Plant Physiol 44:447–452
Abeles FB, Leather GR (1971) Abscission: Control of cellulase secretion by ethylene. Planta 97:87–91
Addicott FT, Wiatr SM (1977) Hormonal controls of abscission: Biochemical and ultrastructural aspects. In: Pilet PE (ed) Plant growth regulation. Springer, Berlin, Heidelberg, New York, pp 249–258
Aist JR (1976) Papillae and related wound plugs of plant cells. Annu Rev Phytopathol 14:145–163
Albersheim P (1978) Concerning the structure and biosynthesis of the primary cell walls of plants. In: Manners DJ (ed) Biochemistry of carbohydrates II. Vol. 16 Park Press, Baltimore, pp 127–150
Albersheim P, Valent BS (1978) Host-pathogen interactions in plants. Plants, when exposed to oligosaccharides of fungal origin, defend themselves by accumulating antibiotics. J Cell Biol 78:627–643
Albersheim P, Mühlethaler K, Frey-Wyssling A (1960) Stained pectin as seen in the electron microscope. J Cell Biol 8:501–506
Al-Delaimy KA, Borgstrom G, Bedford CL (1966) Pectic substances and pectic enzymes of fresh and processed Montmorency cherries. Q Bull Michigan State Univ Agric Exp St 49:164–171
Allison AV, Shalla TA (1974) The ultrastructure of local lesions induced by potato virus X: a sequence of cytological events in the course of infection. Phytopathology 64:784–793
Anderson DB, Kerr T (1938) Growth and structure of cotton fiber. Ind Eng Chem 30:48–54
Anderson RL, Ray PM (1978) Labeling of the plasma membrane of pea cells by a surface-localized glucan synthetase. Plant Physiol 61:723–730
Anderson RL, Stone BA (1978) Studies on *Lolium multiflorum* endosperm in tissue culture. III. Structural studies on the cell walls. Aust J Biol Sci 31:573–586
Anderson RL, Anderson MA, Stone BA (1976) Some structural and enzymological aspects of non-starchy polysaccharides in cereal grains. Proc 14th Convention Inst Brewing (Aust N Z Sect), Melbourne, 153–159
Anderson RL, Clarke AE, Jermyn MA, Knox RB, Stone BA (1977) A carbohydrate-binding arabinogalactan-protein from liquid suspension cultures of endosperm from *Lolium multiflorum*. Aust J Plant Physiol 4:143–158
Ankel H, Ankel E, Feingold DS (1966) Biosynthesis of uridine diphosphate D-xylose. III. Uridine diphosphate D-glucose dehydrogenase of *Cryptococcus laurentii*. Biochemistry 5:1864–1869
Ashford AE, Jacobsen JV (1974) Cytochemical localization of phosphatase in barley aleurone cells: The pathway of gibberellic acid-induced enzyme release. Planta 120:81–105
Awad M, Young RE (1979) Postharvest variation in cellulase, polygalacturonase, and pectinmethylesterase in avocado (*Persea americana* Mill, cv. Fuerte) fruits in relation to respiration and ethylene production. Plant Physiol 64:306–308
Axelos M, Péaud-Lenoël C (1978) Glycosyl transfers from UDP-sugars to lipids of plant membranes: identification and specificity of the transferases. Biochimie 60:35–44
Bacic A (1979) Biochemical and ultrastructural studies on endosperm cell walls. Ph. D. Thesis, Biochem Dept, La Trobe Univ
Bacic A, Stone BA (1980) A (1→3) and (1→4)-linked β-D-glucan in the endosperm cell-walls of wheat. Carbohydr Res 82:372–377
Bacic A, Stone BA (1981a) Isolation and ultrastructure of aleurone cell walls from wheat and barley. Aust J Plant Physiol, in press
Bacic A, Stone BA (1981b) Chemistry and organization of aleurone cell wall components from wheat and barley. Aust J Plant Physiol, in press
Bacon JSD, Gordon AH, Morris EJ, Farmer VC (1975) Acetyl groups in cell-wall preparations from higher plants. Biochem J 149:485–487
Bailey RW, Hassid WZ (1966) Xylan synthesis from uridine-diphosphate-D-xylose by particulate preparations from immature corncobs. Proc Natl Acad Sci USA 56:1586–1593

Bailey RW, Haq S, Hassid WZ (1967) Carbohydrate composition of particulate preparations from mung bean (*Phaseolus aureus*) shoots. Phytochemistry 6:293–301
Bal AK, Payne JF (1972) Endoplasmic reticulum activity and cell wall breakdown in quiescent root meristems of *Allium cepa* L. Z Pflanzenphysiol 66:265–272
Ballance GM, Meredith WOS, Laberge DE (1976) Distribution and development of endo-β-glucanase activities in barley tissues during germination. Can J Plant Sci 56:459–466
Bara M, Gordon SA (1972) The effect of gravity compensation on growth and cell wall-loosening enzymes in *Helianthus annuus* hypocotyls. Physiol Plant 27:277–280
Barckhausen R (1978) Ultrastructural changes in wounded plant storage tissue cells. In: Kahl G (ed) Biochemistry of wounded plant tissues. Walter de Gruyter, Berlin, New York, pp 1–42
Bartley IM (1974) β-galactosidase activity in ripening apples. Phytochemistry 13:2107–2111
Bartley IM (1976) Changes in the glucans of ripening apples. Phytochemistry 15:625–626
Bartley IM (1978) Exo-polygalacturonase of apple. Phytochemistry 17:213–216
Bartnicki-Garcia S (1973) Fundamental aspects of hyphal morphogenesis. Symp Soc Gen Microbiol 23:245–267
Batra KK, Hassid WZ (1970) Determination of linkages of β-D-glucans from *Lupinus albus* and *Avena sativa* by exo-β-(1→3)-D-glucanase. Plant Physiol 45:233–234
Batra KK, Nordin JH, Kirkwood S (1969) Biosynthesis of the β-D-glucan of *Sclerotium rolfsii* Sacc. Direction of chain propogation and the insertion of the branch residues. Carbohydr Res 9:221–229
Bell JK, McCully ME (1970) A histological study of lateral root initiation and development in *Zea mays*. Protoplasma 70:179–205
Beltrán JP, Carbonell J (1978) 1,3-β-Glucan synthase from *Citrus* phloem. Phytochemistry 17:1531–1532
Ben-Arie R, Lavee S (1971) Pectic changes occurring in Elberta peaches suffering from woolly breakdown. Phytochemistry 10:531–538
Ben-Arie R, Kislev N, Frenkel C (1979) Ultrastructural changes in the cell walls of ripening apple and pear fruit. Plant Physiol 64:197–202
Berger RK, Reid PD (1979) Role of polygalacturonase in bean leaf abscission. Plant Physiol 63:1133–1137
Besford RT, Hobson GE (1972) Pectic enzymes associated with the softening of tomato fruit. Phytochemistry 11:2201–2205
Bonnett HT Jr, Newcomb EH (1966) Coated vesicles and other cytoplasmic components of growing root hairs of radish. Protoplasma 62:59–75
Bornman CH (1967) Some ultrastructural aspects of abscission in *Coleus* and *Gossypium*. S Afr J Sci 63:325–331
Bouveng HO, Meier H (1959) Studies on a galactan from Norwegian spruce compression wood (*Picea abies* Karst.). Acta Chem Scand 13:1884–1889
Bowles DJ, Northcote DH (1972) The sites of synthesis and transport of extracellular polysaccharides in the root tissues of maize. Biochem J 130:1133–1145
Bowles DJ, Northcote DH (1974) The amounts and rates of export of polysaccharides found within the membrane system of maize root cells. Biochem J 142:139–144
Bowles DJ, Northcote DH (1976) The size and distribution of polysaccharides during their synthesis within the membrane system of maize root cells. Planta 128:101–106
Brady CJ (1976) The pectinesterase of the pulp of the banana fruit. Aust J Plant Physiol 3:163–172
Brady CJ, Ali ZM, MacAlpine GA, McGlasson WB (1980) Polygalacturonase and pectinesterase in tomato fruits. Abst Aust Soc Plant Physiol, 21st Gen Meet, p 29
Brett CT (1978) Synthesis of β-(1→3)-glucan from extracellular uridine diphosphate glucose as a wound response in suspension-cultured soybean cells. Plant Physiol 62:377–382
Brett CT, Northcote DH (1975) The formation of oligoglucans linked to lipid during synthesis of β-glucan by characterized membrane fractions isolated from peas. Biochem J 148:107–117
Brodzki P (1972) Callose in compression wood tracheids. Acta Soc Bot Pol 41:321–327
Brown HT, Morris GH (1890) XXX – Researches on the germination of some of the Gramineae. Part 1. J Chem Soc 458–528
Buchala AJ, Wilkie KCB (1971) The ratio of β(1→3) to β(1→4) glucosidic linkages in

non-endospermic hemicellulosic β-glucans from oat plant (*Avena sativa*) tissues at different stages of maturity. Phytochemistry 10:2287–2291
Buchala AJ, Wilkie KCB (1973) Total hemicelluloses from wheat at different stages of growth. Phytochemistry 12:499–505
Buchala AJ, Wilkie KCB (1974) Total hemicelluloses from *Hordeum vulgare* plants at different stages of maturity. Phytochemistry 13:1347–1351
Buescher RW, Tigchelaar EC (1975) Pectinesterase, polygalacturonase, Cx-cellulase activities and softening of the *rin* tomato mutant. Hort Sci 10:624–625
Burgess J, Fleming EN (1974) Ultrastructural observations of cell wall regeneration around isolated tobacco protoplasts. J Cell Sci 14:439–449
Butterfield BG, Meylan BA (1972) Scalariform performation plate development in *Laurelia novae-zelandiae* A. Cunn: a scanning electron microscope study. Aust J Bot 20:253–259
Buvat R (1964a) Infrastructures protoplasmiques des vaisseaux du métaxylème de *Cucurbita pepo* au cours de leur différenciation. Compt Rend 258:5243–5246
Buvat R (1964b) Comportement des membranes plasmiques lors de la différenciation des parois latérales des vaisseaux (métaxylème de *Cucurbita pepo*). Compt Rend 258:5511–5514
Buvat R (1964c) Observations infrastructurales sur les parois transversales des éléments de vaisseaux (métaxylème de *Cucurbita pepo*) avant leur perforation. Compt Rend 258:6210–6212
Cabib E, Ulane R, Bowers B (1974) A molecular model for morphogenesis: the primary septum of yeast. In: Horecher BL, Stadtman ER (eds) Current topics in cellular regulation, vol 8. Academic Press, New York, pp 1–32
Carbonell J, Beltrán JP, Conejero V (1976) Activity, extraction and stability of enzymes involved in polysaccharide biosynthesis in *Citrus*. Phytochemistry 15:1873–1876
Carpita N, Sabularse D, Montezinos D, Delmer DP (1979) Determination of the pore size of cell walls of living plant cells. Science 205:1144–1147
Catt JW, Hills GJ, Roberts K (1978) Cell wall glycoproteins from *Chlamydomonas reinhardii*, and their self-assembly. Planta 138:91–98
Chadwick CM, Northcote DH (1980) Glucosylation of phosphorylpolyisoprenol and sterol at the plasma membrane of soya-bean (*Glycine max*) protoplasts. Biochem J 186:411–421
Chambers J, Elbein AD (1970) Biosynthesis of glucans in mung bean seedlings. Formation of β-(1→4)-glucans from GDP-glucose and β-(1→3)-glucans from UDP-glucose. Arch Biochem Biophys 138:620–631
Churms SC, Merrifield EH, Stephen AM (1977) Structural features of the gum exudates from some *Acacia* species of the series *Phyllodineae* and *Botryocephalae*. Carbohydr Res 55:3–10
Churms SC, Merrifield EH, Stephen AM (1978) Regularity within the molecular structure of arabinogalactan from Western larch (*Larix occidentalis*). Carbohydr Res 64:C1–C2
Clark AF, Villemez CL (1972) The formation of β-1→4-glucan from UDP-α-D-glucose catalyzed by a *Phaseolus aureus* enzyme. Plant Physiol 50:371–374
Clarke AE, Knox RB (1979) Plants and immunity. Developmental and comparative immunology 3:571–589
Clarke AE, Stone BA (1962) β-1,3-Glucan hydrolases from the grape vine (*Vitis vinifera*) and other plants. Phytochemistry 1:175–188
Clarke AE, Anderson RL, Stone BA (1979) Form and function of arabinogalactans and arabinogalactan-proteins. Phytochemistry 18:521–540
Cleland R (1967) A dual role of turgor pressure in auxin-induced cell elongation in *Avena* coleoptiles. Planta 77:182–191
Clermont S, Percheron F (1979) Uridine diphosphogalactose-4 épimérase de fenugrec: essais de purification et quelques propriétés. Phytochemistry 18:1963–1965
Clowes FAL, Juniper BE (eds) (1968) Plant cells. Blackwell Scientific Publications, Oxford-Edinburgh
Coggins CW Jr, Lewis LN, Knapp JCF (1967) Progress report: chemical and histological studies of tough and tender Deglet Noor dates. Date Growers' Inst Annu Rept 44:15
Cook GMW, Stoddart RW (1973) Surface carbohydrates of the eukaryotic cell, p 179. Academic Press, London New York
Cook JA, Fincher GB, Keller F, Stone BA (1980) The use of specific β-glucan hydrolases

in the characterization of β-glucan synthetase products. In: Marshall JJ (ed) Mechanisms of saccharide polymerization and depolymerization. Academic Press, New York, pp 301–315

Coulomb P, Coulon J (1971) Fonctions de l'appareil de Golgi dans les méristèmes radiculaires de la courge (*Cucurbita pepo* L. Cucurbitacée). J Microsc Paris 10:203–214

Cousin MT (1979) Appearance of a neosynthesized callosic substance inside the intine of pollen grains of *Vinca rosea* (Apocynaceae) infected with Stolbur disease (mycoplasma disease). Compt Rend 289:1339–1342

Crafts AS, Currier HB (1963) On sieve tube function. Protoplasma 57:188–202

Currier HB (1957) Callose substance in plant cells. Amer J Bot 44:478–488

Currier HB, Strugger S (1956) Aniline blue and fluorescence microscopy of callose in bulb scales of *Allium cepa* L. Protoplasma 45:552–559

Currier HB, Webster DH (1964) Callose formation and subsequent disappearance: studies in ultrasound stimulation. Plant Physiol 39:843–847

Czaninski Y (1972) Observations ultrastructurales sur l'hydrolyse des parois primaires des vaisseaux chez le *Robinia pseudo-acacia* L. et l'*Acer pseudoplatanus* L. Compt Rend 275:361–363

Czaninski Y (1973) Observations on a new wall layer in vessel associated cells of *Robinia* and *Acer*. Protoplasma 77:221–229

Dalessandro G, Northcote DH (1977a) Changes in enzymic activities of nucleoside diphosphate sugar interconversions during differentiation of cambium to xylem in sycamore and poplar. Biochem J 162:267–279

Dalessandro G, Northcote DH (1977b) Changes in enzymic activities of nucleoside diphosphate sugar interconversions during differentiation of cambium to xylem in pine and fir. Biochem J 162:281–288

Dalessandro G, Northcote DH (1977c) Possible control sites of polysaccharide synthesis during cell growth and wall expansion of pea seedlings (*Pisum sativum* L.) Planta 134:39–44

Dalessandro G, Northcote DH (1977d) Changes in enzymic activities of UDP-D-glucuronate decarboxylase and UDP-D-xylose 4-epimerase during cell division and xylem differentiation in cultured explants of Jerusalem artichoke. Phytochemistry 16:853–859

Darvill AG, Smith CJ, Hall MA (1978) Cell wall structure and elongation growth in *Zea mays* coleoptile tissue. New Phytol 80:503–516

Dashek WV, Rosen WG (1966) Electron microscopical localization of chemical components in the growth zone of lily pollen tubes. Protoplasma 61:192–204

Davies MD, Dickinson DB (1972) Properties of uridine diphosphoglucose dehydrogenase from pollen of *Lilium longiflorum*. Arch Biochem Biophys 152:53–61

Davignon ML (1961) Contribution a l'étude de l'évolution chimique des substances pectiques au cours de la croissance, de la maturation et de la sénescence des fruits. CR Acad Agri Fr 47:62–66

Dea ICM (1979) Interactions of ordered polysaccharide structures – synergism and freeze-thaw phenomena. In: Blanshard JMV, Mitchell JR (eds) Polysaccharides in food. Butterworths, London, Boston, pp 229–247

De Bokx JA (1967) The callose test for the detection of leafroll virus in potato tubers. Europ Potato J 10:221–234

Dekazos ED (1972) Callose formation by bruising and heating of tomatoes and its presence in processed products. J Food Sci 37:562–567

Dekazos ED, Worley JF (1967) Induction of callose formation by bruising and aging of red tart cherries. J Food Sci 32:287–289

Dickson JG, Shands HL (1941) Cellular modification of the barley kernel during malting. Amer Soc Brew Chem Proc 4th Ann Gen Meet, 1–10

Dürr M, Bailey DS, Maclachlan G (1979) Subcellular distribution of membrane-bound glycosyltransferases from pea stems. Eur J Biochem 97:445–453

Ebner KE (1973) Lactose synthetase. In: Boyer PD (ed) The enzymes, Vol IX, 3rd edn. Academic Press, New York, pp 363–377

Echlin P, Godwin H (1968) The ultrastructure and ontogeny of pollen in *Helleborus foetidus* L. I. The development of the tapetum and Ubisch bodies. J Cell Sci 3:161–174

Elbein AD (1969) Biosynthesis of a cell wall glucomannan in mung bean seedlings. J Biol Chem 244:1608–1616
Elbein AD (1979) The role of lipid-linked saccharides in the biosynthesis of complex carbohydrates. Annu Rev Plant Physiol 30:239–272
Elbein AD, Forsee WT (1973) Studies on the biosynthesis of cellulose. In: Loewus F (ed) Biogenesis of plant cell wall polysaccharides. Academic Press, New York, pp 259–295
Engels FM (1974) Function of golgi vesicles in relation to cell wall synthesis in germinating *Petunia* pollen. II. Chemical composition of Golgi vesicles and pollen tube wall. Acta Bot Neerl 23:81–89
Engleman EM (1965) Sieve element of *Impatiens sultanii*. 1. Wound reaction. Ann Bot 29:83–101
Ericson MC, Delmer DP (1977) Glycoprotein synthesis in plants. 1. Role of lipid intermediates. Plant Physiol 59:341–347
Ericson MC, Gafford J, Elbein AD (1978) Bacitracin inhibits the synthesis of lipid-linked saccharides and glycoproteins in plants. Plant Physiol 62:373–376
Esau K (1965) Anatomy and cytology of *Vitis* phloem. Hilgardia 37:17–72
Esau K (1977) Anatomy of Seed Plants. 2nd edn. Wiley and Sons, New York
Esau K, Cheadle VI, Gill RH (1966) Cytology of differentiating tracheary elements. I. Organelles and membrane systems. Am J Bot 53:756–764
Esau K, Cronshaw J (1967) Relation of tobacco mosaic virus to the host cells. J Cell Biol 33:665–678
Eschrich W (1956) Callose. Protoplasma 47:487–530
Eschrich W (1957) Kallosebildung in plasmolysierten *Allium-cepa*-epidermen Planta 48:578–586
Eschrich W (1961) Untersuchungen über den Ab- und Aufbau der Callose. III. Mitteilung über Callose. Z Bot 49:153–218
Eschrich W (1962) Elektronenmikroskopische Untersuchungen an Pollen-Mutterzellen-Callose. IV. Mitteilung über Callose. Protoplasma 55:419–422
Eschrich W (1966) Der Calloseabbau bei den Pollentetraden von *Cucurbita ficifolia*. Z Pflanzenphysiol 54:463–471
Eschrich W (1975) Sealing systems in phloem. In: Zimmermann MH, Milburn JA (eds) Transport in plants I. Phloem transport, Encyclopedia of Plant Physiology, Vol I, Springer, Berlin, Heidelberg, New York, pp 39–56
Eschrich W (1976) Phloemprotein bei *Cucurbita maxima*. Ber Deut Bot Ges 89:515–523
Eschrich W, Currier HB, Yamaguchi S, McNairn RB (1965) The influence of increased callose formation upon the transport of assimilates in sieve tubes. Planta 65:49–64
Eschrich W, Hüttermann A, Heyser W, Tammes PML, Van Die J (1972) Evidence for the synthesis of callose in sieve-tube exudate of *Yucca flaccida*. Z Pflanzenphysiol 67:468–470
Esquerré-Tugayé MT, Lafitte C, Mazau D, Toppan A, Touzé A (1979) Cell surfaces in plant-microorganism interactions. II. Evidence for the accumulation of hydroxyproline-rich glycoproteins in the cell wall of diseased plants as a defense mechanism. Plant Physiol 64:320–326
Evert RF (1977) Phloem structure and histochemistry. Annu Rev Plant Physiol 28:199–222
Ewald CM, Perlin AS (1959) The arrangement of branching in an arabino-xylan from wheat flour. Can J Chem 37:1254–1259
Facey V (1950) Abscission of leaves in *Fraxinus americana* L. New Phytol 49:103–116
Fahn A (1979) Secretory tissues in plants. Academic Press, New York
Fan D-F, Feingold DS (1969) Nucleoside diphosphate-sugar 4-epimerases. I. Uridine diphosphate glucose 4-epimerase of wheat germ. Plant Physiol 44:599–604
Feingold DS, Neufeld EF, Hassid WZ (1958) Synthesis of a β-1,3-linked glucan by extracts of *Phaseolus aureus* seedlings. J Biol Chem 233:783–788
Feingold DS, Neufeld EF, Hassid WZ (1960) The 4-epimerization and decarboxylation of uridine diphosphate D-glucuronic acid by extracts from *Phaseolus aureus* seedlings. J Biol Chem 235:910–913

Fincher GB (1975) Morphology and chemical composition of barley endosperm cell walls. J Inst Brew 81:116–122
Fincher GB, Stone BA (1974) Some chemical and morphological changes induced by gibberellic acid in embryo-free wheat grain. Aust J Plant Physiol 1:297–311
Flowers HM, Batra KK, Kemp J, Hassid WZ (1968) Biosynthesis of insoluble glucans from uridine-diphosphate-D-glucose with enzyme preparations from *Phaseolus aureus* and *Lupinus albus*. Plant Physiol 43:1703–1709
Fogarty WM, Ward OP (1974) Pectinases and pectic polysaccharides. Prog Ind Microbiol 13:59–119
Forrest IS, Wainwright T (1977) The mode of binding of β-glucans and pentosans in barley endosperm cell walls. J Inst Brew 83:279–286
Fowke HC, Pickett-Heaps JD (1972) A cytochemical and autoradiographic investigation of cell wall deposition in fiber cells of *Marchantia berteroana*. Protoplasma 74:19–32
Frankel R, Izhar S, Nitsan J (1969) Timing of callase activity and cytoplasmic male sterility in *Petunia*. Biochem Genet 3:451–455
Franz G (1972) Polysaccharide metabolism in the cell walls of growing *Phaseolus aureus* seedlings. Planta 102:334–347
Franz G (1973) Biosynthesis of salep mannan. Phytochemistry 12:2369–2373
Franz G (1976) The dependence of membrane-bound glucan synthetases on glycoprotein which can act as acceptor molecules. App Polym Symp 28:611–621
Friend J (1979) Phenolic substances and plant disease. In: Swain T, Harborne JB, Van Sumere CF (eds) Biochemistry of plant phenolics. Recent advances in phytochemistry, Vol 12, Plenum Press, New York pp. 557–588
Fu Y-L, Gutmann PJ, Timell TE (1972) Polysaccharides in the secondary phloem of Scots pine (*Pinus sylvestris* L.). I. Isolation and characterization of callose. Cellulose Chem Technol 6:507–512
Fukushima T, Yamazaki M (1978) Chilling injury in cucumbers. V. Polysaccharide changes in cell walls. Sci Hortic 8:219–227
Fulcher RG, O'Brien TP, Lee JW (1972) Studies on the aleurone layer. I. Conventional and fluorescence microscopy of the cell wall with emphasis on phenol-carbohydrate complexes in wheat. Aust J Biol Sci 25:23–34
Fulcher RG, McCully ME, Setterfield G, Sutherland J (1976) β-1,3-Glucans may be associated with cell plate formation during cytokinesis. Can J Bot 54:539–542
Garcia-Arenal F, Sagasta EM (1977) Callose deposition and phytoalexin accumulation in *Botrytis cinerea*-infected bean (*Phaseolus vulgaris*) Plant Sci Lett 10:305–312
Gibbons GC (1980) On the sequential determination of α-amylase transport and cell wall breakdown in germinating seeds of *Hordeum vulgare*. Carlsberg Res Commun 45:177–184
Geissmann T, Neukom H (1973) A note on ferulic acid as a constituent of the water-insoluble pentosans of wheat flour. Cereal Chem 50:414–416
Giddings TH Jr, Brower DL, Staehelin LA (1980) Visualization of particle complexes in the plasma membrane of *Micrasterias denticulata* associated with the formation of cellulose fibrils in primary and secondary cell walls. J Cell Biol 84:327–339
Gilkes NR, Hall MA (1977) The hormonal control of cell wall turnover in *Pisum sativum* L. New Phytol 78:1–15
Ginzburg BZ (1958) Evidence for a protein component in the middle lamella of plant tissue: a possible site for indolylacetic acid action. Nature 181:398–400
Goldschmid HR, Perlin AS (1963) Interbranch sequences in the wheat arabinoxylan. Selective enzymolysis studies. Can J Chem 41:2272–2277
Gould SEB, Rees DA (1965) Polysaccharides and germination: some chemical changes that occur during the germination of white mustard. J Sci Food Agric 16:702–709
Grant GT, McNab C, Rees DA, Skerrett RJ (1969) Seed mucilages as examples of polysaccharide denaturation. Chem Commun 805–806
Green JR, Northcote DH (1978) The structure and function of glycoproteins synthesized during slime-polysaccharide production by membranes of the root-cap cells of maize (*Zea mays*). Biochem J 170:599–608
Green JR, Northcote DH (1979a) Polyprenol phosphate sugars synthesized during slime-

polysaccharide production by membranes of the root-cap cells of maize (*Zea mays*). Biochem J 178:661–671
Green JR, Northcote DH (1979b) Location of fucosyl transferases in the membrane system of maize root cells. J Cell Sci 40:235–244
Gross KC, Wallner SJ (1979) Degradation of cell wall polysaccharides during tomato fruit ripening. Plant Physiol 63:117–120
Gunning BES, Steer MW (1975) Ultrastructure and the biology of plant cells. Edward Arnold, London
Haddon LE, Northcote DH (1975) Quantitative measurement of the course of bean callus differentiation. J Cell Sci 17:11–26
Haddon LE, Northcote DH (1976) The influence of gibberellic acid and abscisic acid on cell and tissue differentiation of bean callus. J Cell Sci 20:47–56
Hägglund E (1951) Chemistry of wood. Academic Press, New York
Halmer P, Bewley JD, Thorpe TA (1976) An enzyme to degrade lettuce endosperm cell walls. Appearance of a mannanase following phytochrome- and gibberellin-induced germination. Planta 130:189–196
Halmer P, Bewley JD, Thorpe TA (1978) Degradation of the endosperm cell walls of *Lactuca sativa* L. cv Grand Rapids. Timing of mobilization of soluble sugars, lipid and phytate. Planta 139:1–8
Hänish Ten Cate ChH, Van Netten J, Dortland JF, Bruinsma J (1975) Cell wall solubilization in pedicel abscission of *Begonia* flower buds. Physiol Plant 33:276–279
Harris PJ, Northcote DH (1971) Polysaccharide formation in plant Golgi bodies. Biochim Biophys Acta 237:56–64
Hartley RD (1973) Carbohydrate esters of ferulic acid as components of cell walls of *Lolium multiflorum*. Phytochemistry 12:661–665
Hartley RD, Harris PJ, Russell GE (1978) Degradability and phenolic components of cell walls of wheat in relation to susceptibility to *Puccinia striiformis*. Ann Appl Biol 88:153–158
Hasegawa S, Maier VP, Kaszychi HP, Crawford JK (1969) Polygalacturonase content of dates and its relation to maturity and softness. J Food Sci 34:527–529
Heiniger U, Delmer DP (1977) UDP-glucose: glucan synthetase in developing cotton fibers. II. Structure of the reaction product. Plant Physiol 59:719–723
Heiniger U, Franz G (1980) The role of NDP-glucose pyrophosphorylases in growing mung bean seedlings in relation to cell wall biosynthesis. Plant Sci Lett 17:443–450
Heller JS, Villemez CL (1972) Solubilization of a mannose-polymerizing enzyme from *Phaseolus aureus*. Biochem J 128:243–252
Helsper JPFG (1979) The possible role of lipid intermediates in the synthesis of β-glucans by a membrane fraction from pollen tubes of *Petunia hybrida*. Planta 144:443–450
Helsper JPFG, Veerkamp JH, Sassen MMA (1977) β-Glucan synthetase activity in Golgi vesicles of *Petunia hybrida*. Planta 133:303–308
Hendriks T (1978) The distribution of glucan synthetase in maize coleoptiles: a comparison with K-ATPase. Plant Sci Lett 11:261–274
Hepler PK, Newcomb EH (1964) Microtubules and fibrils in the cytoplasm of *Coleus* cells undergoing secondary wall deposition. J Cell Biol 20:529–533
Hepler PK, Newcomb EH (1967) Fine structure of cell plate formation in the apical meristem of *Phaseolus* roots. J Ultrastruct Res 19:498–513
Herth W, Franke WW, Bittiger H, Kuppel A, Keilich G (1974) Alkali-resistant fibrils of β-1,3- and β-1,4-glucans: structural polysaccharides in the pollen tube wall of *Lilium longiflorum*. Cytobiologie 9:344–367
Hills GJ (1973) Cell wall assembly *in vitro* from *Chlamydomonas reinhardii*. Planta 115:17–23
Hills GJ, Phillips JM, Gay MR, Roberts K (1975) Self-assembly of a plant cell wall *in vitro*. J Mol Biol 96:431–441
Hinman MB, Villemez CL (1975) Glucomannan biosynthesis catalyzed by *Pisum sativum* enzymes. Plant Physiol 56:608–612
Hinton DM, Pressey R (1974) Cellulase activity in peaches during ripening. J Food Sci 39:783–785

Hiruki C, Shukla P (1973) Mycoplasma-like bodies associated with witches'-broom of bleeding heart. Phytopathology 63:88–92
Hiruki C, Tu JC (1972) Light and electron microscopy of potato virus M lesions and marginal tissue in red kidney bean. Phytopathology 62:77–85
Hobson GE (1963) Pectinesterase in normal and abnormal tomato fruit. Biochem J 86:358–365
Hobson GE (1968) Cellulase activity during the maturation and ripening of tomato fruit. J Food Sci 33:588–592
Hoffmann GC, Timell TE (1970) Isolation of a β-1,3-glucan (laricinan) from compression wood of *Larix laricina*. Wood Sci Technol 4:159–162
Hoffmann GC, Timell TE (1972a) Polysaccharides in ray cells of compression wood of red pine (*Pinus resinosa*). Tappi 55:871–873
Hoffmann GC, Timell TE (1972b) Polysaccharides in compression wood of tamarack (*Larix laricina*). 1. Isolation and characterization of laricinan, an acidic glucan. Svensk Papperstidn 75:135–142
Hoffmann GC, Timell TE (1972c) Polysaccharides in ray cells of normal wood of red pine (*Pinus resinosa*). Tappi 55:733–736
Hopp HE, Romero PA, Daleo GR, Pont Lezica R (1978a) Synthesis of cellulose precursors. The involvement of lipid-linked sugars. Europ J Biochem 84:561–571
Hopp HE, Romero PA, Pont Lezica R (1978b) On the inhibition of cellulose biosynthesis by coumarin. FEBS Lett 86:259–262
Hopper JE, Dickinson DB (1972) Partial purification and sugar nucleotide inhibition of UDP-glucose pyrophosphorylase from *Lilium longiflorum* pollen. Arch Biochem Biophys 148:523–535
Horner HT Jr, Arnott HJ (1966) A histochemical and ultrastructural study of pre- and post-germinated *Yucca* seeds. Bot Gaz 127:48–64
Horner HT, Rogers MA (1974) A comparative light and electron microscopic study of microsporogenesis in male-fertile and cytoplasmic male-sterile pepper (*Capsicum annuum*). Can J Bot 52:435–441
Horton RF, Osborne DJ (1967) Senescence, abscission and cellulase activity in *Phaseolus vulgaris*. Nature 214:1086–1088
Hsaio TC (1973) Plant responses to water stress. Annu Rev Plant Physiol 24:519–570
Huberman M, Goren R (1979) Exo- and endo-cellular cellulase and polygalacturonase in abscission zones of developing orange fruits. Physiol Plant 45:189–196
Huberman M, Goren R, Birk Y (1975) The effects of pH, ionic strength and ethylene on the extraction of cellulase from abscission zones of citrus leaf explants. Plant Physiol 55:941–945
Hughes JE, Gunning BES (1980) Glutaraldehyde-induced deposition of callose. Can J Bot 58:250–258
Hughes RC (1976) Membrane glycoproteins. A review of structure and function. Butterworths, London
Hultin HO, Sun B, Bulger J (1966) Pectin methylesterases of the banana. Purification and properties. J Food Sci 31:320–327
Huwyler HR, Franz G, Meier H (1979) Changes in the composition of cotton fibre cell walls during development. Planta 146:635–642
Israel HW, Wilson RG, Aist JR, Kunoh H (1980) Cell wall appositions and plant disease resistance: acoustic microscopy of papillae that block fungal ingress. Proc Natl Acad Sci USA 77:2046–2049
Itô K (1949) Studies on "Murasaki-monpa" disease caused by *Helicobasidium mompa* Tanaka. Bull Governm For Exp Sta (Tokyo) 43:1–26
Izhar S, Frankel R (1971) Mechanism of male sterility in *Petunia*: the relationship between pH, callase activity in the anthers, and the breakdown of the microsporogenesis. Theor Appl Genet 41:104–108
Jacobs M, Ray PM (1975) Promotion of xyloglucan metabolism by acid pH. Plant Physiol 56:373–376
Jacobsen JV, Pressman E (1979) A structural study of germination in celery (*Apium graveolens* L.) seed with emphasis on endosperm breakdown. Planta 144:241–248

Jacobsen JV, Pressman E, Pyliotis NA (1976) Gibberellin-induced separation of cells in isolated endosperm of celery seed. Planta 129:113–122
James DW Jr, Jones RL (1979a) Characterization of GDP-fucose. Polysaccharide fucosyl transferase in corn roots (*Zea mays* L.). Plant Physiol 64:909–913
James DW Jr, Jones RL (1979b) Intracellular localization of GDP-fucose. Polysaccharide fucosyl transferase in corn roots (*Zea mays* L.). Plant Physiol 64:914–918
Jermyn MA, Isherwood FA (1956) Changes in the cell wall of the pear during ripening. Biochem J 64:123–132
Jilka R, Brown O, Nordin P (1972) Uptake of [U-^{14}C]glucose into subcellular particles of wheat seedling roots. Arch Biochem Biophys 152:702–711
Jones MGK (1976) The origin and development of plasmodesmata. In: Gunning BES, Robards AW (eds) Intercellular communication in plants: studies on plasmodesmata. Springer, Berlin, Heidelberg, New York, pp 81–105
Jones MGK, Payne HL (1978) Cytokinesis in *Impatiens balsamina* and the effect of caffeine. Cytobios 20:79–91
Jones RL (1974) The structure of the lettuce endosperm. Planta 121:133–146
Joseleau J-P, Barnoud F (1976) Cell wall carbohydrates and structural studies of xylans in relation to growth in the reed *Arundo donax*. J Appl Polym Sci Appl Polym Symp 28:983–992
Juniper BE, Lawton JR, Harris PJ (1981) Cellular organelles and cell wall formation in fibres from the flowering stem of *Lolium temulentum* L. New Phytol (in press)
Kapil RN, Tiwari SC (1978) Embryological investigations and fluorescence microscopy – an assessment of integration. Int Rev Cytol 53:291–331
Katz M, Ordin L (1967) Metabolic turnover in cell wall constituents of *Avena sativa* L. coleoptile sections. Biochim Biophys Acta 141:118–125
Kauss H, Bowles DJ (1976) Some properties of carbohydrate-binding proteins (lectins) solubilized from cell walls of *Phaseolus aureus*. Planta 130:169–174
Kauss H, Glaser C (1974) Carbohydrate-binding proteins from plant cell walls and their possible involvement in extension growth. FEBS Lett 45:304–307
Kauss H, Hassid WZ (1967) Biosynthesis of the 4-*O*-methyl-D-glucuronic acid unit of hemicellulose B by transmethylation from *S*-adenosyl-L-methionine. J Biol Chem 242:1680–1684
Kauss H, Swanson AL (1969) Cooperation of enzymes responsible for polymerisation and methylation in pectin biosynthesis. Z Naturforsch 24B:28–33
Kawase M (1974) Role of ethylene in induction of flooding damage in sunflower. Physiol Plant 31:29–38
Kawase M (1979) Role of cellulase in aerenchyma development in sunflower. Am J Bot 66:183–190
Kemp J, Loughman BC, Ephritikhine G (1978) The role of inositol glucosides in the synthesis of β-1,3-glucans. In: Wells WW, Eisenberg F (eds) Cyclitols and phosphoinositides. Academic Press, New York, pp 439–449
Keusch L (1968) Mobilization of reserve mannan in germinating date seeds. Planta 78:321–350
Kiermayer O, Dobberstein B (1973) Dictyosome-derived membrane complexes as templates for the extraplasmatic synthesis and orientation of microfibrils. Protoplasma 77:437–451
Kivilaan A, Bandurski RS, Schulze A (1971) A partial characterization of an autolytically solubilized cell wall glucan. Plant Physiol 48:389–393
Kimmins WC, Brown RG (1973) Hypersensitive resistance. The role of cell wall glycoproteins in virus localization. Can J Bot 51:1923–1929
Knee M (1973) Polysaccharide changes in cell walls of ripening apples. Phytochemistry 12:1543–1549
Knee M (1975) Changes in structural polysaccharides of apples ripening during storage. Colloques internationaux. C N R S 238:341–345
Knee M, Sargent JA, Osborne DJ (1977) Cell wall metabolism in developing strawberry fruits. J Exp Bot 28:377–396
Kunoh H, Aist JR, Israel HW (1979) Microsurgical isolation of intact plant cell wall appositions for microanalyses. Can J Bot 57:1349–1353

Kusano S (1911) *Gastrodia elata* and its symbiotic association with *Armillaria mellea.* J Coll Agric Tokyo Imp Univ 4:1–65

Labavitch JM, Ray PM (1974a) Turnover of cell wall polysaccharides in elongating pea stem segments. Plant Physiol 53:669–673

Labavitch JM, Ray PM (1974b) Relationship between promotion of xyloglucan metabolism and induction of elongation by indoleacetic acid. Plant Physiol 54:499–502

Labavitch JM, Ray PM (1978) Structure of hemicellulosic polysaccharides of *Avena sativa* coleoptile cell walls. Phytochemistry 17:933–937

Lamport DTA (1970) Cell wall metabolism. Annu Rev Plant Physiol 21:235–270

Lamport DTA (1980) Structure and function of plant glycoproteins. In: Stumpf PK, Conn EE (eds) The biochemistry of plants. Vol 3. Preiss J (ed) Carbohydrates: Structure and function. Academic Press, New York, pp 501–541

Lang A, Canney M, Whitty R (1979) Chilling effects upon sieve tube ultrastructure. Aust Soc Plant Physiol, 20th Gen Meet, p 44

Larm O, Lindberg B (1976) The pneumococcal polysaccharides: a re-examination. Advan Carbohydr Chem Biochem 33:295–322

Larsen GL, Brummond DO (1974) β-(1→4)-D-Glucan synthesis from UDP-[^{14}C]-D-glucose by a solubilized enzyme from *Lupinus albus*. Phytochemistry 13:361–365

Lee S, Kivilaan A, Bandurski RS (1967) *In vitro* autolysis of plant cell walls. Plant Physiol 42:968–972

Lehmann H, Schulz D (1969) Elektronenmikroskopische Untersuchungen von Differenzierungsvorgängen bei Moosen. II. Die Zellplatten- und Zellwandbildung. Planta 85:313–325

Lew FT, Lewis LN (1974) Purification and properties of cellulase from *Phaseolus vulgaris.* Phytochemistry 13:1359–1366

Lewis LN, Varner JE (1970) Synthesis of cellulase during abscission of *Phaseolus vulgaris* leaf explants. Plant Physiol 46:194–199

Lewis LN, Linkins AE, O'Sullivan S, Reid PD (1974) Two forms of cellulase in bean plants. 8th Int Conf Plant Growth Substan, Hirokawa, Tokyo pp 708–718

Lieberman M (1979) Biosynthesis and action of ethylene. Annu Rev Plant Physiol 30:533–591

Liese W (1965) The fine structure of bordered pits in softwoods. In: Côté WA Jr (ed) Cellular ultrastructure of woody plants. Syracuse Univ Press pp 271–290

Lin T-Y, Elbein AD, Su J-C (1966) Substrate specificity in pectin synthesis. Biochem Biophys Res Commun 22:650–657

Linkins AE, Lewis LN, Palmer RL (1973) Hormonally induced changes in the stem and petiole anatomy and cellulase enzyme patterns in *Phaseolus vulgaris* L. Plant Physiol 52:554–560

Loescher W, Nevins DJ (1972) Auxin-induced changes in *Avena* coleoptile cell wall composition. Plant Physiol 50:556–563

Lyons JM (1973) Chilling injury in plants. Annu Rev Plant Physiol 24:445–466

Maclachlan GA, Duda CT (1965) Changes in concentration of polymeric components in excised pea-epicotyl tissue during growth. Biochim Biophys Acta 97:288–299

Majumder SK, Leopold AC (1967) Callose formation in response to low temperature. Plant Cell Physiol 8:775–778

Maltby D, Carpita NC, Montezinos D, Kulow C, Delmer DP (1979) β-1,3-Glucan in developing cotton fibers: structure, localization, and relationship of synthesis to that of secondary wall cellulose. Plant Physiol 63:1158–1164

Mangin ML (1899) Sur le Piétin ou maladie du pied du blé. Bull Soc Mycol France 15:210–239

Mares DJ, Stone BA (1973a) Studies on wheat endosperm. I. Chemical composition and ultrastructure of the cell walls. Aust J Biol Sci 26:793–812

Mares DJ, Stone BA (1973b) Studies on *Lolium multiflorum* endosperm in tissue culture. II. Fine structure of cells and cell walls and the development of cell walls. Aust J Biol Sci 26:135–150

Markwalder HU, Neukom H (1976) Diferulic acid as a possible crosslink in hemicelluloses from wheat germ. Phytochemistry 15:836–837

Marshall RD (1979) Some observations on why many proteins are glycosylated. Biochem Soc Trans 7:800–805
Matile P (1974) Cell wall degradation in senescing tobacco leaf discs. Experientia 30:98–99
Mayer AM, Harel E (1979) Polyphenol oxidases in plants. Phytochemistry 18:192–215
McCleary BV, Matheson NK (1974) α-D-Galactosidase activity and galactomannan and galactosylsucrose oligosaccharide depletion in germinating legume seeds. Phytochemistry 13:1747–1757
McCleary BV, Matheson NK (1975) Galactomannan structure and β-mannanase and β-mannosidase activity in germinating legume seeds. Phytochemistry 14:1187–1194
McClendon JH, Nolan WG, Wenzler HF (1976) The role of the endosperm in the germination of legumes: galactomannan, nitrogen, and phosphorus changes in the germination of guar (*Cyamopsis tetragonoloba*; Leguminosae) Am J Bot 63:790–797
McNab JM, Villemez CL, Albersheim P (1968) Biosynthesis of galactan by a particulate enzyme preparation from *Phaseolus aureus* seedlings. Biochem J 106:355–360
McNairn RB (1972) Phloem translocation and heat-induced callose formation in field-grown *Gossypium hirsutum* L. Plant Physiol 50:366–370
McNairn RB, Currier HB (1965) The influence of boron on callose formation in primary leaves of *Phaseolus vulgaris* L. Phyton 22:153–158
McNeil M, Darvill AG, Albersheim P (1979) The structural polymers of the primary walls of dicots. Prog Chem Org Natl Prod 37:191–249
Meier H (1958) On the structure of cell walls and cell wall mannans from ivory nuts and from dates. Biochim Biophys Acta 28:229–240
Meier H (1961) The distribution of polysaccharides in wood fibers. J Polym Sci 51:11–18
Meier H (1962) Studies on a galactan from tension wood of beech (*Fagus silvatica* L.). Acta Chem Scand 16:2275–2283
Meier H, Reid JSG (1977) Morphological aspects of the galactomannan formation in the endosperm of *Trigonella foenum-graecum* L. (Leguminosae). Planta 133:243–248
Mendgen K (1978) Attachment of bean rust cell wall material to host and non-host plant tissue. Arch Microbiol 119:113–117
Mepham RH, Lane GR (1969) Formation and development of the tapetal periplasmodium in *Tradescantia bracteata*. Protoplasma 68:175–192
Mertz J, Nordin P (1971) Uptake of labeled glucose by root tips from etiolated wheat seedlings. Phytochemistry 10:1223–1227
Meylan BA, Butterfield BG (1972) Perforation plate development in *Knightia excelsa* R. Br: a scanning electron microscope study. Aust J Bot 20:79–86
Miyamoto C, Tamari K (1973) Synthesis of β-1,3-glucan and β-1,3 and β-1,4-mixed glucan from UDP-α-D-glucose. Agric Biol Chem 37:1253–1260
Moericke V (1955) Über den Nachweis der Blattrollkrankheit in Kartoffelknollen durch den Resorzintest. Phytopathol Z 24:462–464
Moore AE, Stone BA (1972) Effect of senescence and hormone treatment on the activity of a β-1,3-glucan hydrolase in *Nicotiana glutinosa* leaves. Planta 104:93–109
Morikawa H, Senda M (1978) Infrared analysis of oat coleoptile cell walls and oriented structure of matrix polysaccharides in the walls. Plant Cell Physiol 19:327–336
Morikawa H, Hayashi R, Senda M (1978) Infrared analysis of pea stem cell walls and oriented structure of matrix polysaccharides in them. Plant Cell Physiol 19:1151–1159
Morrall P, Briggs DE (1978) Changes in cell wall polysaccharides of germinating barley grains. Phytochemistry 17:1495–1502
Morré DJ (1968) Cell wall dissolution and enzyme secretion during leaf abscission. Plant Physiol 43:1545–1559
Morré DJ, Mollenhauer HH (1974) The endomembrane concept: a functional integration of endoplasmic reticulum and Golgi apparatus. In: Robards AW (ed) Dynamic aspects of plant ultrastructure. McGraw Hill, New York, pp 84–137
Morré DJ, Mollenhauer HH (1976) Interactions among cytoplasm, endomembranes, and the cell surface. In: Stocking CR, Heber U (eds) Transport in plants III, Encyclopedia of plant physiology, New Series, Volume 3, Springer, Berlin, Heidelberg, New York, pp 288–344
Morré DJ, Van Der Woude WJ (1974) Origin and growth of cell surface components.

In: Hay ED, King TJ, Papaconstantinou J (eds) Marcromolecules regulating growth and development. 30th Symp Soc Dev Biol Academic Press, New York, pp 81–111
Morré DJ, Jones DD, Mollenhauer HH (1967) Golgi apparatus mediated polysaccharide secretion by outer root cap cells of *Zea mays*. I. Kinetics and secretory pathways. Planta 74:286–301
Morré DJ, Kartenbeck J, Franke WW (1979) Membrane flow and interconversions among endomembranes. Biochim Biophys Acta 559:71–152
Morris MR, Northcote DH (1977) Influence of cations at the plasma membrane in controlling polysaccharide secretion from sycamore suspension cells. Biochem J 166:603–618
Morrison IN, O'Brien TP (1976) Cytokinesis in the developing wheat grain; division with and without a phragmoplast. Planta 130:57–67
Mueller SC, Brown RM Jr (1980) Evidence for an intramembrane component associated with a cellulose microfibril-synthesizing complex in higher plants. J Cell Biol 84:315–326
Musolan C, Kindinger J (1978) The synthesis of β-1,3- and β-1,4-glucosyl glucosidic linkages at different concentrations of UDP-α-D-glucose by *Phaseolus vulgaris* leaves. Rev Roum Biochim 15:53–61
Musolan C, Ordin L, Kindinger JE (1975) Effects of heat shock on growth and on lipid and β-glucan synthetases in leaves of *Phaseolus vulgaris* and *Vigna sinensis*. Plant Physiol 55:328–332
Nagahashi J, Beevers L (1978) Subcellular localization of glycosyl transferases involved in glycoprotein biosynthesis in the cotyledons of *Pisum sativum* L. Plant Physiol 61:451–459
Nagel CW, Patterson ME (1967) Pectic enzymes and development of the pear (*Pyrus communis*). J Food Sci 32:294–297
Neish AC (1958) The biosynthesis of cell wall carbohydrates. IV. Further studies on cellulose and xylan in wheat. Can J Biochem Physiol 36:187–193
Nelmes BJ, Preston RD (1968) Wall development in apple fruits: a study of the life history of a parenchyma cell. J Exp Bot 19:496–518
Neufeld EF, Hall CW (1965) Inhibition of UDP-D-glucose dehydrogenase by UDP-D-xylose: a possible regulatory mechanism. Biochem Biophys Res Commun 19:456–461
Neufeld EF, Feingold DS, Hassied WZ (1960) Phosphorylation of D-galactose and L-arabinose by extracts from *Phaseolus aureus* seedlings. J Biol Chem 235:906–909
Neufeld EF, Ginsburg V, Putman EW, Fanshier D, Hassid WZ (1957) Formation and interconversion of sugar nucleotides by plant extracts. Arch Biochem Biophys 69:602–616
Nevins DJ, Huber DJ, Yamamoto R, Loescher WH (1977) β-D-Glucan of *Avena* coleoptile cell walls. Plant Physiol 60:617–621
Northcote DH (1974a) Sites of synthesis of the polysaccharides of the cell wall. In: Bridham JB (ed) Plant carbohydrate biochemistry, Academic Press, New York, pp 165–181
Northcote DH (1974b) Complex envelope system. Membrane systems of plant cells. Phil Trans R Soc Lond Ser B 268:119–128
Northcote DH (1979a) The involvement of the Golgi apparatus in the biosynthesis and secretion of glycoproteins and polysaccharides. Biomembranes 10:51–76
Northcote DH (1979b) Polysaccharides of the plant cell during its growth. In: Blanshard JMV, Mitchell JR (eds) Polysaccharides in food. Butterworths, London, pp 3–13
Northcote DH, Lewis DR, (1968) Freeze-etched surfaces of membranes and organelles in the cells of pea root tips. J Cell Sci 3:199–206
Northcote DH, Pickett-Heaps JD (1966) A function of the Golgi apparatus in polysaccharide synthesis and transport in the root-cap cells of wheat. Biochem J 98:159–167
Northcote DH, Wooding FBP (1966) Development of sieve tubes in *Acer pseudoplatanus*. Proc R Soc London Ser B 163:524–535
Northcote DH, Wooding FBP (1968) The structure and function of phloem tissue. Sci Prog 56:35–58
O'Brien TP (1970) Further observations on hydrolysis of the cell wall in the xylem. Protoplasma 69:1–14
O'Brien TP (1972) The cytology of cell-wall formation in some eukaryotic cells. Bot Rev 38:87–118

O'Brien TP (1974) Primary vascular tissues. In: Robards AW (ed) Dynamic aspects of plant ultrastructure. McGraw-Hill, New York, pp 414–440
O'Brien TP, Thimann KV (1967) Observations on the fine structure of the oat coleoptile. III. Correlated light and electron microscopy of the vascular tissues. Protoplasma 63:443–478
Odzuck W, Kauss H (1972) Biosynthesis of pure araban and xylan. Phytochemistry 11:2489–2494
Okamoto K, Kitano H, Akazawa T (1980) Biosynthesis and excretion of hydrolases in germinating cereal seeds. Plant Cell Physiol 21:201–204
Oputa CO, Mazelis M (1977) Simulated hypogravity and proline incorporation into salt-extractable macromolecules from cell walls. Phytochemistry 16:673–675
Ordin L (1960) Effect of water stress on cell wall metabolism of *Avena* coleoptile tissue. Plant Physiol 35:443–450
Ordin L, Hall MA (1967) Studies on cellulose synthesis by a cell-free oat coleoptile enzyme system: inactivation by airborne oxidants. Plant Physiol 42:205–212
Ordin L, Hall MA (1968) Cellulose synthesis in higher plants from UDP-glucose. Plant Physiol 43:473–476
Overman MA, Warmke HE (1972) Cytoplasmic male sterility in sorghum. II. Tapetal behavior in fertile and sterile anthers. J Heredity 63:227–234
Palmer GH, Bathgate GN (1976) Malting and Brewing. In: Pomeranz Y (ed) Advances in Cereal Science and Technology Chap 5. Amer Assoc Cereal Chem, St Paul, Minnesota
Panayotatos N, Villemez CL (1973) The formation of a β-(1→4)-D-galactan chain catalysed by a *Phaseolus aureus* enzyme. Biochem J 133:263–271
Paull RE, Jones RL (1975a) Studies on the secretion of maize root cap slime. II. Localization of slime production. Plant Physiol 56:307–312
Paull RE, Jones RL (1975b) Studies on the secretion of maize root-cap slime. III. Histochemical and autoradiographic localization of incorporated fucose. Planta 127:97–110
Paull RE, Jones RL (1976) Studies on the secretion of maize root cap slime. V. The cell wall as a barrier to secretion. Z Pflanzenphysiol 79:154–164
Péaud-Lenoël C, Axelos M (1970) Structural features of the β-glucans enzymatically synthesized from uridine diphosphate glucose by wheat seedlings. FEBS Lett 8:224–228
Pennazio S, D'Agostino G, Appiano A, Redolfi P (1978) Ultrastructure and histochemistry of the resistant tissue surrounding lesions of tomato bushy stunt virus in *Gomphrena globosa* leaves. Physiol Plant Pathol 13:165–171
Pesis E, Fuchs Y, Zauberman G (1978) Cellulase activity and fruit softening in avocado. Plant Physiol 61:416–419
Peterson CA, Rauser WE (1979) Callose deposition and photoassimilate export in *Phaseolus vulgaris* exposed to excess cobalt, nickel and zinc. Plant Physiol 63:1170–1174
Pharr DM, Dickinson DB (1973) Partial characterization of Cx cellulase and cellobiase from ripening tomato fruits. Plant Physiol 51:577–583
Pickett-Heaps JD (1967) The use of radioautography for investigating wall secretion in plant cells. Protoplasma 64:49–66
Pickett-Heaps JD (1968) Further ultrastructural observations on polysaccharide localization in plant cells. J Cell Sci 3:55–64
Pickett-Heaps JD, Northcote DH (1966) Relationship of cellular organelles to the formation and development of the plant cell wall. J Exp Bot 17:20–26
Pilnik W, Rombouts FM (1979) Pectic enzymes. In: Blanshard JMV, Mitchell JR (eds) Polysaccharides in food. Butterworths, London, pp 109–126
Pilnik W, Voragen AGJ (1970) Pectic substances and other uronides. In: Hulme AC (ed) The biochemistry of fruits and their products. Academic Press, London, pp 53–87
Plaut Z, Ordin L (1961) Effect of soil moisture content on the cell wall metabolism of sunflower and almond leaves. Physiol Plant 14:646–658
Plaut Z, Ordin L (1964) The effect of moisture tension and nitrogen supply on cell wall metabolism of sunflower leaves. Physiol Plant 17:279–286
Pomeranz Y (1972) Scanning electron microscopy of the endosperm of malted barley. Cereal Chem 49:5–19

Pont Lezica R, Romero PA, Hopp HE (1978) Glucosylation of membrane-bound proteins by lipid-linked glucose. Planta 140:177–183

Poovaiah BW, Nukaya A (1979) Polygalacturonase and cellulase enzymes in the normal Rutgers and mutant *rin* tomato fruits and their relationship to the respiratory climacteric. Plant Physiol 64:534–537

Pope DG (1977) Relationships between hydroxyproline-containing proteins secreted into the cell wall and medium by suspension-cultured *Acer pseudoplatanus* cells. Plant Physiol 59:894–900

Postlmayr HL, Luh BS, Leonard SJ (1956) Characterization of pectin changes in freestone and clingstone peaches during ripening and processing. Food Technol 10:618–625

Prat R, Roland J-C (1971) Étude ultrastructurale des premiers stades de néoformation d'une enveloppe par les protoplastes végétaux séparés mécaniquement de leur paroi. Compt Rend 273:165–168

Preece IA (1957) Malting relationships of barley polysaccharides – an account of original work on the enzymolysis of β-glucosan and araboxylan. Wallerstein Lab Commun 20:147–160

Pressey R, Avants JK (1972) Multiple forms of pectinesterase in tomatoes. Phytochemistry 11:3139–3142

Pressey R, Avants JK (1973a) Two forms of polygalacturonase in tomatoes. Biochim Biophys Acta 309:363–369

Pressey R, Avants JK (1973b) Separation and characterization of endopolygalacturonase and exopolygalacturonase from peaches. Plant Physiol 52:252–256

Pressey R, Avants JK (1976) Pear polygalacturonases. Phytochemistry 15:1349–1351

Pressey R, Hinton DM, Avants JK (1971) Development of polygalacturonase activity and solubilization of pectin in peaches during ripening. J Food Sci 36:1070–1073

Preston RD (1979) Polysaccharide conformation and cell wall function. Annu Rev Plant Physiol 30:55–78

Quail PH (1979) Plant cell fractionation. Annu Rev Plant Physiol 30:425–484

Rasmussen HP (1965) Chemical and physiological changes associated with abscission layer formation in the bean (*Phaseolus vulgaris* L. cv. Contender). Ph D thesis, Michigan State Univ

Rasmussen GK (1973) Changes in cellulase and pectinase activities in fruit tissues and separation zones of citrus treated with cycloheximide. Plant Physiol 51:626–628

Ray PM (1962) Cell wall synthesis and cell elongation in oat coleoptile tissue. Am J Bot 49:928–939

Ray PM (1975) Golgi membranes form xyloglucan from UDPG and UDP-xylose. Plant Physiol 56 Suppl:16

Ray PM, Shininger TL, Ray MM (1969) Isolation of β-glucan synthetase particles from plant cells and identification with Golgi membranes. Proc Natl Acad Sci USA 64:605–612

Ray PM, Eisinger WR, Robinson DG (1976) Organelles involved in cell wall polysaccharide formation and transport in pea cells. Ber Dtsch Bot Ges 89:121–146

Raymond Y, Fincher GB, Maclachlan GA (1978) Tissue slice and particulate β-glucan synthetase activities from *Pisum* epicotyls. Plant Physiol 61:938–942

Rees DA, Welsh EJ (1977) Secondary and tertiary structure of polysaccharides in solutions and gels. Angew Chem 16:214–224

Reeve RM (1959) Histological and histochemical changes in developing and ripening peaches. II. The cell walls and pectins. Am J Bot 46:241–248

Reid JSG (1971) Reserve carbohydrate metabolism in germinating seeds of *Trigonella foenum-graecum* L. (Leguminosae). Planta 100:131–142

Reid JSG, Bewley JD (1979) A dual role for the endosperm and its galactomannan reserves in the germinative physiology of fenugreek (*Trigonella foenum-graecum* L.), an endospermic leguminous seed. Planta 147:145–150

Reid JSG, Meier H (1972) The function of the aleurone layer during galactomannan mobilisation in germinating seeds of fenugreek (*Trigonella foenum-graecum* L.), crimson clover (*Trifolium incarnatum* L.) and lucerne (*Medicago sativa* L.): a correlative biochemical and ultrastructural study. Planta 106:44–60

Reid JSG, Meier H (1973) Enzymic activities and galactomannan mobilisation in germinating seeds of fenugreek (*Trigonella foenum-graecum* L. Leguminosae). Planta 112:301–308

Reid JSG, Davies C, Meier H (1977) Endo-β-mannanase, the legunimous aleurone layer and the storage galactomannan in germinating seeds of *Trigonella foenum-graecum* L. Planta 133:219–222

Reid PD, Strong HG, Lew F, Lewis LN (1974) Cellulase and abscission in the red kidney bean (*Phaseolus vulgaris*). Plant Physiol 53:732–737

Reis D, Vian B, Roland JC (1978) *In vitro* and *in vivo* polysaccharides assembly. Ultrastructural and cytochemical study of growing plant cell wall components. In: Sturgess JM (ed) Ninth Int Cong Electron Microsc Vol II, Microsc Soc Canada, pp 434–435

Robards AW (1975) Plasmodesmata. Annu Rev Plant Physiol 26:13–29

Robenek H, Peveling E (1977) Ultrastructure of the cell wall regeneration of isolated protoplasts of *Skimmia japonica* thunb. Planta 136:135–145

Roberts K, Northcote DH (1970) The structure of sycamore callus cells during division in a partially synchronised suspension culture. J Cell Sci 6:299–321

Robbins PW, Bray D, Dankert M, Wright A (1967) Direction of chain growth in polysaccharide synthesis. Science 158:1536–1542

Robinson DG (1977) Plant cell wall synthesis. Adv Bot Res 5:89–151

Robinson DG, Ray PM (1977) The reversible cyanide inhibition of Golgi secretion in pea cells. Cytobiologie 15:65–77

Robinson DG, Eisinger WR, Ray PM (1976) Dynamics of the Golgi system in wall matrix polysaccharide synthesis and secretion by pea cells. Ber Dtsch Bot Ges 89:147–161

Robyt JF (1979) Mechanisms involved in the biosynthesis of polysaccharides. Trends Biochem Sci 2:47–49

Rodén L, Schwartz NB (1975) Biosynthesis of connective tissue proteoglycans. In: Whelan WJ (ed) Biochemistry of carbohydrates. MTP Int Rev Sci Biochem Ser One, Vol 5, Butterworths, Univ Park Press, pp 95–152

Roelofsen PA (1965) Ultrastructure of the wall in growing cells and its relation to the direction of the growth. Adv Bot Res 2:69–149

Roseman S (1970) The synthesis of complex carbohydrates by multiglycosyltransferase systems and their potential function in intercellular adhesion. Chem Phys Lipids 5:270–297

Rougier M (1971) Étude cytochimique de la secrétion des polysaccharides végétaux à l-aide d'un matériel de choix: les cellules de la coiffe de *Zea mays*. J Microscopie 10:67–82

Rouse AH, Knorr LC (1969) Seasonal changes in pectin esterase activity, pectin and citric acid of Florida lemons. Food Technol 23:829–831

Ryser U (1979) Cotton fibre differentiation: occurrence and distribution of coated and smooth vesicles during primary and secondary wall formation. Protoplasma 98:223–239

Sakurai N, Masuda Y (1978a) Auxin-induced changes in barley coleoptile cell wall composition. Plant Cell Physiol 19:1217–1223

Sakurai N, Masuda Y (1978b) Auxin-induced extension, cell wall loosening and changes in the wall polysaccharide content of barley coleoptile segments. Plant Cell Physiol 19:1225–1233

Sakurai N, Nevins DJ, Masuda Y (1977) Auxin- and hydrogen ion-induced cell wall loosening and cell extension in *Avena* coleoptile segments. Plant Cell Physiol 18:371–380

Sakurai N, Nishitani K, Masuda Y (1979) Auxin-induced changes in the molecular weight of hemicellulosic polysaccharides of the *Avena* coleoptile cell wall. Plant Cell Physiol 20:1349–1357

Sassen MMA (1965) Breakdown of the plant cell wall during the cell-fusion process. Acta Bot Neerl 14:165–196

Satoh S, Matsuda K, Tamari K (1976) β-1,4-Glucan occurring in homogenate of *Phaseolus aureus* seedlings. Possible nascent stage of cellulose biosynthesis in vivo. Plant Cell Physiol 17:1243–1254

Scheirer DC (1973) Hydrolysed walls in the water-conducting cells of *Dendroligotrichum* (Bryophyta): histochemistry and ultrastructure. Planta 115:37–46

Schuster G (1960) Auslösung von Kallosebildung in den Siebröhren der Kartoffelknolle durch Applikation von Gerbsäure und Berberinsulfat. Naturwissenschaften 47:427

Schuster G, Byhan O (1958) Zur Präzisierung des Kallosetestes für den Nachweis der Blattrollkrankheit bei Kartoffeln. Z Landwirtsch Vers Untersuchungswes 4:37–49

Schuster G, Flemming M (1976) Studies on the formation of diffusion barriers in hypersensitive hosts of tobacco mosaic virus and the role of necrotization in the formation of diffusion barriers as well as in the localization of virus infections. Phytopathol Z 87:345–352

Scott FM, Bystrom BG, Bowler E (1963) *Persea americana*, mesocarp cell structure, light and electron microscope study. Bot Gaz 124:423–428

Scurfield G (1973) Reaction wood: its structure and function. Science 179:647–655

Seiler A (1977) Galactomannan breakdown in germinating carob seeds (*Ceratonia siliqua* L.). Planta 134:209–221

Seipp D (1978) Pectin decomposition in early and late ripening apple cultivars. II. Early ripening cultivars. Gartenbauwissenschaft 43:248–253

Setia RC, Parthasarathy MV, Shah JJ (1977) Development, histochemistry and ultrastructure of gum-resin ducts in *Commiphora mukul* Engl. Ann Bot 41:999–1004

Sexton R (1976) Some ultrastructural observations on the nature of foliar abscission in *Impatiens sultani*. Planta 128:49–58

Sexton R, Durbin ML, Lewis LN, Thomson WW (1980) Use of cellulase antibodies to study leaf abscission. Nature 283:873–874

Sheldrake AR (1970) Cellulase and cell differentiation in *Acer pseudoplatanus*. Planta 95:167–178

Sheldrake AR, Moir GFJ (1970) A cellulase in *Hevea* latex. Physiol Plant 23:267–277

Shibuya N, Iswasaki T (1978) Polysaccharides and glycoproteins in the rice endosperm cell wall. Agric Biol Chem 42:2259–2266

Shibuya N, Misaki A (1978) Structure of hemicellulose isolated from rice endosperm cell wall: mode of linkages and sequences in xyloglucan, β-glucan and arabinoxylan. Agric Biol Chem 42:2267–2274

Shih CY, Currier HB (1969) Fine structure of phloem cells in relation to translocation in the cotton seedling. Am J Bot 56:464–472

Shimomura T, Dijkstra J (1975) The occurrence of callose during the process of local lesion formation. Neth J Plant Pathol 81:107–121

Siegel S, Speitel T, Shiraki D, Fukumoto J (1977) Effects of experimental hypogravity on peroxidase and cell wall constituents in the dwarf marigold. Life Sci Space Res 16:105–109

Sievers A (1963) Beteiligung des Golgi-Apparates bei der Bildung der Zellwand von Wurzelhaaren. Protoplasma 56:188–192

Smith MM, McCully ME (1977) Mild temperature "stress" and callose synthesis. Planta 136:65–70

Smith MM, McCully ME (1978) A critical evaluation of the specificity of aniline blue induced fluorescence. Protoplasma 95:229–254

Smith MM, Stone BA (1973a) β-Glucan synthesis by cell-free extracts from *Lolium multiflorum* endosperm. Biochem Biophys Acta 313:72–94

Smith MM, Stone BA (1973b) Chemical composition of the cell walls of *Lolium multiflorum* endosperm. Phytochemistry 12:1361–1367

Smith MM, Axelos M, Péaud-Lenoël C (1976) Biosynthesis of mannan and mannolipids from GDP-Man by membrane fractions of sycamore cell cultures. Biochimie 58:1195–1211

Sobotka FE, Stelzig DA (1974) An apparent cellulase complex in tomato (*Lycopersicon esculentum* L.) fruit. Plant Physiol 53:759–763

Southworth D, Dickinson DB (1975) β-1,3-Glucan synthase from *Lilium longiflorum* pollen. Plant Physiol 56:83–87

Speakman JB, Lewis BG (1978) Limitation of *Gaeumannomyces graminis* by wheat root responses to *Philophora radicicola*. New Phytol 80:373–380

Sprau F (1955) Pathologische Gewebeveränderungen durch das Blattrollvirus bei der Kartoffel und ihr färbetechnischer Nachweis. Ber Dtsch Bot Ges 68:239–246

Sterling C (1954) Sclereid development and the texture of bartlett pears. J Food Res 19:433–443
Sterling C, Kalb AJ (1959) Pectic changes in peach during ripening. Bot Gaz 121:111–113
Stieglitz HB (1974) Somatic regulation of microspore release: an analysis of meiocyte wall breakdown. Ph D Thesis, Univ of California, San Diego
Stieglitz H (1977) Role of β-1,3-glucanase in postmeiotic microspore release. Develop Biol 57:87–97
Stieglitz H, Stern H (1973) Regulation of β-1,3-glucanase activity in developing anthers of *Lilium*. Dev Biol 34:169–173
Stobbs LW, Manocha MS, Dias HF (1977) Histological changes associated with virus localization in TMV-infected Pinto bean leaves. Physiol Plant Pathol 11:87–94
Strasburger's Textbook of Botany (1976) Longman, London, New York
Sutherland J, McCully ME (1976) A note on the structural changes in the walls of pericycle cells initiating lateral root meristems in *Zea mays*. Can J Bot 54:2083–2087
Suzuki H, Imai S, Nisizawa K, Murachi T (1971) Some properties of cellulases from pineapple stem. Bot Mag 84:389–397
Swain T (1977) Secondary compounds as protective agents. Annu Rev Plant Physiol 28:479–501
Tahir MA, Elahi M (1978) Effect of growth on pectinesterase, vitamin C and protein contents of tomato. Pakistan Sci Res 30:8–10
Taiz L, Jones RL (1970) Gibberellic acid, β-1,3-glucanase and the cell walls of barley aleurone layers. Planta 92:73–84
Takeuchi Y, Komamine A (1978) Changes in the composition of cell wall polysaccharides of suspension-cultured *Vinca rosea* cells during culture. Physiol Plant 42:21–28
Takeuchi Y, Komamine A, Saito T, Watanabe K, Morikawa N (1980) Turnover of cell wall polysaccharides of a *Vinca rosea* suspension culture. Radio gas chromatographical analyses. Physiol Plant 48:536–541
Thiéry J-P (1967) Mise en évidence des polysaccharides sur coupes fines en microscopie électronique. J Microsc Paris 6:987–1018
Thirupathaiah V, Subramanian D (1977) Physiological changes in chilli (*Capsicum annuum* L.) fruits during maturation and ripening. Indian J Exp Biol 15:683–685
Thomas B, Hall MA (1975) The effect of growth regulators on wound-stimulated callose formation in *Salix viminalis* L. Plant Sci Lett 4:9–15
Thomas D, Des S, Stanley RG (1968) Glycerol activation of glucan synthesis in cell-free preparations from *Lupinus albus*. Biochem Biophys Res Commun 30:292–296
Thomas D, Des S, Smith JE, Stanley RG (1969) Stereospecific regulation of plant glucan synthetase in vitro. Can J Bot 47:489–496
Thornber JP, Northcote DH (1961a) Changes in the chemical composition of a cambial cell during its differentiation into xylem and phloem tissues in trees. 2. Carbohydrate constituents of each main component. Biochem J 81:455–464
Thornber JP, Northcote DH (1961b) Changes in the chemical composition of a cambial cell during its differentiation into xylem and phloem tissues in trees. I. Main components. Biochem J 81:449–455
Thornber JP, Northcote DH (1962) Changes in the chemical composition of a cambial cell during its differentiation into xylem and phloem tissues in trees. 3. Xylan, glucomannan and α-cellulose fractions. Biochem J 82:340–346
Timell TE (1964) Wood hemicelluloses: part I. Adv Carbohydr Chem 19:247–302
Timell TE (1965) Wood hemicelluloses: part II. Adv Carbohydr Chem 20:409–483
Tsai CM, Hassid WZ (1971) Solubilization and separation of uridine diphospho-D-glucose: β-(1→4) glucan and uridine diphospho-D-glucose: β-(1→3) glucan glucosyltransferases from coleoptiles of *Avena sativa*. Plant Physiol 47:740–744
Tsai CM, Hassid WZ (1973) Substrate activation of β-(1→3) glucan synthetase and its effect on the structure of β-glucan obtained from UDP-D-glucose and particulate enzyme of oat coleoptiles. Plant Physiol 51:998–1001
Tung K-K, Ballou CE (1973) Biosynthesis of a mycobacterial lipopolysaccharide. Properties of the polysaccharide:acyl coenzyme A acyltransferase reaction. J Biol Chem 248:7126–7133

Ullrich W (1963) Über die Bildung von Kallose bei einer Hemmung des Transportes in den Siebröhren durch Cyanid. Planta 59:387–390
Unzelman JM, Healey PL (1974) Development, structure and occurrence of secretory trichomes of *Pharbitis*. Protoplasma 80:285–303
Valent BS, Darvill AG, McNeil M, Robertsen BK, Albersheim P (1980) A general and sensitive chemical method for sequencing the glycosyl residues of complex carbohydrates. Carbohydr Res 79:165–192
VanderMolen GE, Beckman CH, Rodehorst E (1977) Vascular gelation: a general response phenomenon following infection. Physiol Plant Pathol 11:95–100
Van der Woude WJ, Morré DJ, Bracker CE (1971) Isolation and characterization of secretory vesicles in germinated pollen of *Lilium longiflorum*. J Cell Sci 8:331–351
Van der Woude WJ, Lembi CA, Morré DJ, Kindinger JI, Ordin L (1974) β-Glucan synthetases of plasma membrane and Golgi apparatus from onion stem. Plant Physiol 54:333–340
Versteeg C, Rombouts FM, Pilnik W (1978) Purification and some characteristics of two pectinesterase isoenzymes from orange. Lebensm Wiss Technol 11:267–274
Villemez CL (1974) Molecular sieve chromatography of biosynthetic cell [^{14}C] glucomannan chains. Arch Biochem Biophys 165:407–412
Villemez CL, Hinman M (1975) UDP-glucose stimulated formation of xylose-containing polysaccharides. Plant Physiology 56:Suppl 15
Villemez CL, Lin T-Y, Hassid WZ (1965) Biosynthesis of the polygalacturonic acid chain of pectin by a particulate enzyme preparation from *Phaseolus aureus* seedlings. Proc Natl Acad Aci USA 54:1626–1632
Villemez CL, Swanson AL, Hassid WZ (1966) Properties of a polygalacturonic acid-synthesizing enzyme system from *Phaseolus aureus* seedlings. Arch Biochem Biophys 116:446–452
Vithanage HIMV, Gleeson PA, Clarke AE (1980) The nature of callose produced during self-pollination in *Secale cereale*. Planta 148:498–509
Wallner SJ, Bloom HL (1977) Characteristics of tomato cell wall degradation *in vitro*. Implications for the study of fruit-softening enzymes. Plant Physiol 60:207–210
Wardrop AB (1964) The structure and formation of the cell wall in xylem. In: Zimmermann MH (ed) The formation of wood in forest trees. Academic Press, New York, pp 87–134
Waterkeyn L (1961) Étude des dépôts de callose au niveau des parois sporocytaires au moyen de la microscopie de fluorescence. Compt Rend 252:4025–4027
Waterkeyn L (1964) Callose microsporocytaire et callose pollinique. In: Linskens HF (ed) Pollen physiology and fertilization. North Holland, Amsterdam, pp 52–58
Webster BD (1973) Anatomical and histochemical changes in leaf abscission. In: Kozlowski TT (ed) Physiological Ecology. Academic Press, New York, pp 45–83
Webster BD, Leopold AC (1972) Stem abscission in *Phaseolus vulgaris* explants. Bot Gaz 133:292–298
Webster DH, Currier HB (1965) Callose: lateral movement of assimilates from phloem. Science 150:1610–1611
Westafer JM, Brown RM Jr (1976) Electron microscopy of the cotton fibre: new observations on cell wall formation. Cytobios 15:111–138
Whaley WG, Mollenhauer HH (1963) The Golgi apparatus and cell plate formation – a postulate. J Cell Biol 17:216–225
Whelan WJ (1976) On the origin of primer for glycogen synthesis. Trends Biochem Sci 1:13–15
Wiemken-Gehrig V, Wiemken A, Matile P (1974) Cell wall breakdown in wilting flowers of *Ipomoea tricolor* Cav. Planta 115:297–307
Wishart GJ, Fry DJ (1980) Evidence from rat liver nuclear preparations that latency of microsomal UDP-glucuronosyltransferase is associated with vesiculation. Biochem J 186:687–691
Wloch W (1975) Longitudinal shrinkage of compression wood in dependence on water content and cell wall structure. Acta Soc Bot Polon 44:217–229
Wodzicki TJ, Brown CL (1973) Organization and breakdown of the protoplast during maturation of pine tracheids. Am J Bot 60:631–640

Wooding FBP, Northcote DH (1964) The development of the secondary wall of the xylem in *Acer pseudoplatanus*. J Cell Biol 23:327–337

Woodward JR, Keane PJ, Stone BA (1980) β-Glucans and β-glucan hydrolases in plant pathogenesis with special reference to wilt-inducing toxins from *Phytophthora* species. In: Sandford PA, Matsuda K (eds) Fungal Polysaccharides. ACS Symp Ser No. 126. Amer Chem Soc Washington DC, pp 113–141

Wright K, Northcote DH (1974) The relationship of root-cap slimes to pectins. Biochem J 139:525–534

Wright K, Northcote DH (1976) Identification of β-1→4-glucan chains as part of a fraction of slime synthesized within the dictyosomes of maize root caps. Protoplasma 88:225–239

Wu JH (1973) Wound-healing as a factor in limiting the size of lesions in *Nicotiana glutinosa* leaves infected by the very mild strain of tobacco mosaic virus (TMV-VM). Virology 51:474–484

Wu JH, Dimitman JE (1970) Leaf structure and callose formation as determinants of TMV movement in bean leaves as revealed by UV irradiation studies. Virology 40:820–827

Wu JH, Blakely LM, Dimitman JE (1969) Inactivation of a host resistance mechanism as an explanation for heat activation of TMV-infected bean leaves. Virology 37:658–666

Yamaki S, Kakiuchi N (1979) Changes in hemicellulose-degrading enzymes during development and ripening of Japanese pear fruit. Plant Cell Physiol 20:301–309

Yamaki S, Matsuda K (1977) Changes in the activities of some cell wall-degrading enzymes during development and ripening of Japanese pear fruit (*Pyrus serotina* Rehder *var.* culta Rehder). Plant Cell Physiol 18:81–93

Yamaki S, Machida Y, Kakiuchi N (1979) Changes in cell wall polysaccharides and monosaccharides during development and ripening of Japanese pear fruit. Plant Cell Physiol 20:311–321

Young RE (1965) Extraction of enzymes from tannin-bearing tissue. Arch Biochem Biophys 111:174–180

Zarra I, Masuda Y (1979a) Growth and cell wall changes in rice coleoptiles growing under different conditions. I. Changes in turgor pressure and cell wall polysaccharides during intact growth. Plant Cell Physiol 20:1117–1124

Zarra I, Masuda Y (1979b) Growth and cell wall changes in rice coleoptiles growing under different conditions. II. Auxin-induced growth in coleoptile segments. Plant Cell Physiol 20:1125–1133

7 Glycoproteins and Enzymes of the Cell Wall

D.T.A. LAMPORT and J.W. CATT

1 Introduction

Morphogenetic machines perform according to preprogrammed design by assembling a vastly intricate pattern woven out of cell secretions. Thus eukaryotic cells, both plant and animal, embed themselves in a surprisingly similar extracellular matrix containing hydroxyproline-rich glycoproteins. Mention any other amino acid and the comparison would be trivial; but hydroxyproline occurs in few proteins, has no codon, and must be synthesized by post-translational modification of prolyl residues via prolyl hydroxylase, a remarkable enzyme catalyzing the direct fixation of molecular oxygen and a concomitant stoichiometric decarboxylation of α-ketoglutaric acid with ascorbic acid as a cofactor. Furthermore, hydroxyproline-rich glycoproteins have been with us virtually since the origin of the *Chlamydomonas* type where they comprise the major, if not the only component of a cellulose-free cell wall. Thus from an evolutionary point of view we can regard cell surface glycoproteins as exceedingly primitive features, probably first elaborated by wall-less *Archaebacteria* as cell surface protection (LAMPORT 1980). The role of plant cell wall proteins is less clear (LAMPORT 1965, 1970, 1977). This chapter will summarize recent results and major conclusions.

The major questions concern structure and function. There is slow but steady progress but no unambiguous resolution of the precise role played by hydroxyproline-rich proteins whose diversity gradually becomes apparent. Currently, three major types or classes of extracellular hydroxyproline-containing glycoproteins are recognized: (1) covalently bound wall protein "extensin", (2) freely soluble arabinogalactan proteins (AGP's or "β-lectins"), and (3) classical lectins, such as potato lectin described by ALLEN and NEUBERGER (1973) and dealt with in Chapter 13, Volume 13 A of this series by SELVENDRAN and O'NEILL. As a working hypothesis we can summarize the respective roles of these glycoproteins as structural, cell recognition, and disease resistance, but these may be rather convenient points of departure than correct points of view. A major question which cannot be answered but which is well worth keeping in mind throughout this chapter concerns the significance of the hydroxyproline-rich domains (which contain several contiguous hydroxyproline residues) present in each of the glycoproteins. Does the hydroxyproline-rich region of each protein correspond to a common structural feature, and if so does this feature presuppose some common functional element? And if so what is it?

Finally one should keep in mind that there is still no adequate model of the primary cell wall. The earlier Albersheim model (KEEGSTRA et al. 1973) has undergone extensive revision; replacing earlier details with broad generalities: "Our present picture of the cell wall remains one in which the cellulose fibers are interconnected by the pectic polysaccharides ... The extent of covalent attachment between the hemicelluloses and the pectic polysaccharides is unknown." (McNEIL et al. 1979).

2 Structural or Matrix Proteins

2.1 Higher and Lower Plants

The protein content of secondary cell walls is negligible; our discussion then is of the primary cell wall, that first formed thin (approx. 1000 nm) extensible sheath around the turgid protoplast. This extensible sheath is sufficiently similar in structure, composition, and function throughout higher and lower plants for us to regard it as homologous throughout the plant kingdom, and hence describe it using only the one term: primary cell wall. Adopting the five-kingdom classification (MARGULIS 1974), in which algae comprise part of the protists, makes sense if for no other reason than that in algae the presence of a true primary cell wall (i.e., similar to that of plants) is at best debatable, though, as we shall see, there are fascinating similarities. First evidence for a distinct wall protein characterized by its rich hydroxyproline content came in 1960 (DOUGAL and SHIMBAYASHI 1960, LAMPORT and NORTHCOTE 1960a). There was some earlier evidence based on chromatographic demonstration of the common protein amino acids in acid hydrolysates of isolated cotton primary cell walls. (TRIP et al. 1951). However, other evidence continues to accumulate up to the most recent, which involves the synthesis of putative cell wall protein by its messenger RNA in a cell-free translation system (SMITH 1981a). STEWARD et al. (1967, 1974); ASHFORD and NEUBERGER (1980) provide useful criticism and alternative viewpoints.

2.1.1 Glycopeptide Linkages

The insolubility of cell wall protein is a major obstacle to its study. However, enzymic degradation of primary cell walls isolated from tomato cell suspension cultures gave a mixture of glycopeptides rich in hydroxyproline, serine, arabinose, and lesser amounts of galactose (LAMPORT 1969). Alkaline hydrolysis of glycopeptides and cell walls released a series of low molecular weight compounds chemically characterized as hydroxyproline-*O*-arabinosides (LAMPORT 1967, 1969), showing that an arabinose residue was glycosidically linked to the trans C4 position of hydroxyproline.

The arabinofuranosyl hydroxyproline linkage is very labile to acid. At pH 1 complete cleavage occurs within 1 h at 100° C. This is a simple but effective method for the deglycosylation (or at least de-arabinosylation) of wall protein. Deglycosylation greatly enhances subsequent tryptic degradation (LAMPORT

1974, CHO and CHRISPEELS 1976). The isolated tryptic peptides were microheterogeneous due to the presence of galactosylated serine (LAMPORT et al. 1973).

While the glycosidic linkages of hydroxyproline arabinosides (involving the hydroxyproline γ carbon) are fairly stable to alkali, glycosidic linkages involving the β carbon of the peptide-linked hydroxyamino acids serine and threonine are unstable in dilute alkali, undergoing β-elimination to yield the free sugar and the corresponding dehydroamino acid.

The galactosylserine linkages in tryptic peptides of wall protein are somewhat more resistant to alkali-induced β-elimination than comparable linkages such as xylosylserine in the chondroitin sulfate of animal tissues. Even more resistant to β-elimination are the galactosylserine linkages of undegraded wall protein actually in the wall (LAMPORT 1977). This is most likely due to the ionized hydroxyl groups of arabinosyl hydroxyproline residues at high pH acting as an electrostatic screen against hydroxyl ion attack (cf. ALLEN et al. 1978). Based on enzymic evidence the anomeric configuration is α-D-galactosylserine (O'NEILL and SELVENDRAN 1980).

2.1.2 Hydroxyproline Glycosides

Alkaline hydrolysis of isolated cell walls releases a series of hydroxyproline arabinosides containing from one to four arabinose residues separable by ion exchange chromatography (LAMPORT and MILLER 1971) or high resolution gel filtration on Biogel P2 (KLIS and EELTINK 1979). The hydroxyproline arabinoside profile is a species-specific characteristic (LAMPORT and MILLER 1971), though this profile may change slightly as a function of growth state (KLIS and EELTINK 1979) or in response to infection (ESQUERRÉ-TUGAYÉ and LAMPORT 1979).

The anomeric configuration of hydroxyproline arabinosides, earlier regarded as probably all β (LAMPORT 1974, 1977, AKIYAMA and KATO 1977, ALLEN et al. 1978), is correct with the exception of the hydroxyproline tetra-arabinoside where the nonreducing terminal arabinose is α-linked (AKIYAMA et al. 1980). Thus the tetra-arabinoside structure is: α-L-Ara*f*(1 → 3)-β-L-Ara*f*(1 → 2)-β-L-Ara*f*(1 → 2)-β-L-Ara*f*(1 → 4)Hyp. Other hydroxyproline glycosides such as hydroxyproline-*O*-galactose also occur in algal cell walls (LAMPORT and MILLER 1971) and in the arabinogalactan proteins (FINCHER et al. 1974, POPE 1977). There is also some evidence for the existence of hydroxyproline-*O*-glucose (MANI et al. 1979a).

2.1.3 Peptide Sequence and Conformation

The entire amino acid sequence of bound wall protein is not yet known, even the amount sequenced (about a third) is an approximation, but there are encouraging signs that sequencing of precursor or perhaps loosely bound wall protein (STUART et al. 1980, STUART and VARNER 1980) will remedy that deficiency. In the meantime we only know that all tryptic peptides so far sequenced from cultured tomato or sycamore (LAMPORT 1973) and from melon hypocotyls (ESQUERRÉ-TUGAYÉ and LAMPORT 1979) contain at least one or more repeats of the pentapeptide Ser-Hyp-Hyp-Hyp-Hyp (i.e., Ser-Hyp_4). Although at first glance it seems unlikely that other proteins would share this pentapeptide, work with potato lectin (ALLEN et al. 1978) and hydroxyproline-rich arabinogalactan proteins suggests that they too may turn out to contain either Ser-Hyp_4 regions, or perhaps X-Hyp_4 regions, where X is another amino acid such as alanine or aspartic acid. What peculiar properties does this pentapeptide confer on the protein? Earlier suggestions (HEATH and NORTHCOTE 1971, LAMPORT 1974) that the structure might have a conformation very similar to that of polyhydroxyproline are supported by circular dichroism spectra of peptides from tomato cells walls (LAMPORT 1977) and *Chlamydomonas* walls (HOMER and ROBERTS 1979). These data imply that the hydroxyproline-rich regions exist in a "polyproline II" type conformation, that is, a three residue per turn left-handed helix with a pitch of approximately 0.94 nm. This is of course much steeper than an alpha helix (pitch 0.54 nm, 3.7 residues per turn) and also much more rigid because the pyrollidine ring prevents rotation around the φ link (α carbon to N). This gives a rod-like molecule, which, combined with the striking amino acid periodicity typifies a structural rather than enzymatic macromolecule.

2.1.4 Role of Glycosylation

In most plants, nearly all of the wall hydroxyproline residues are glycosylated, with the interesting exception of the Gramineae (LAMPORT and MILLER 1971), where the majority of hydroxyproline residues are not glycosylated. The size of the oligosaccharide attached to hydroxyproline varies from one to four residues, and we have the impression that the size of the oligoarabinoside increases as we ascend the plant kingdom. Thus dicotyledons usually have three or four arabinofuranosyl residues.

What is the significance of these oligoarabinosides? The linkage analysis first determined by Dr. Arthur Karr in our laboratory fits the "crumpled sheet" description of REES (1977). Molecular models (LAMPORT 1980) show that the β-linked oligosaccharide can hydrogen bond with and "nest" along the peptide helix created by four contiguous hydroxyproline residues, (LAMPORT 1980) as exclusively α-linked arabinofuranosides cannot do so. We can rationalize this possible interaction in at least two ways: (1) close association of oligosaccharide and peptide may provide mutual defense as each tends to shield the other's critical linkages from enzymatic attack, (note also the β-linkages; virtually all arabinosidases isolated are specific for the α-linked sugar). Moreover, enzymes cleaving the hydroxyprolyl-hydroxyproline linkage are few; (2) there is also the possibility that a hydrogen-bonded oligoarabinoside may do for extensin what a triple-stranded polypeptide helix does for collagen, namely provide a fibrous molecule with tensile strength. It also seems significant that glycosylation is highest in herbaceous dicotyledonous plants that rely heavily on cell turgor for support, while hydroxyproline glycosylation is lowest in the Gramineae which have alternative means of support such as silica deposits.

The role of serine galactosylation is discussed in Sections 2.1.5 and 2.1.6.2.

2.1.5 Assembly

There is surprisingly little biochemical work dealing with assembly of the primary cell wall macromolecules because it is so difficult to characterize microamounts of complex precursors chemically. What is the size of these macromolecular precursors to the wall? Where and how are they synthesized, packaged, exported, and incorporated into the total wall structure? Answers to these questions are generally lacking or at best rudimentary. The extensin precursor problem is no exception, even though work showing the presence of precursor in smooth vesicles has appeared (DASHEK 1970). Dashek's data indicate that extensin precursor is not readily solubilized from the vesicles and this, together with the high rate of turnover and hence small amount of material present, plus the difficulty of distinguishing precursor extensin from arabinogalactan or other hydroxyproline protein precursor, makes for a most formidable problem. However, the problem is not intractable, and there are encouraging signs of future substantial progress in this area: microanalytical methods have greatly improved over the past 10 years, an immunochemical method of detecting wall protein now exists (SMITH 1981b). Methods for discriminating between extensin and AGP precursors are close at hand, and there is clear evidence that, at least under some conditions, a lesser or greater proportion of extensin is only loosely bound to the wall and can be extracted with aqueous salt solutions (BRYSK and CHRISPEELS 1972, SMITH 1981a, STUART and VARNER 1980) or chaotropic agents such as guanidine thiocyanate (MONRO et al. 1972, 1974). In lupin hypocotyls (BAILEY and KAUSS 1974) there even seems to be a positive correlation between growth rate and easily extractable wall protein. Nevertheless the actual route of extensin precursor from cytoplasm to its covalent site in the wall remains somewhat debatable. While the initial site of synthesis is almost certainly rough endoplasmic reticulum, there is no definitive proof, nor is there overwhelming evidence, that further packaging processing and transport involves the classical Golgi apparatus, although the evidence of GARDINER and CHRISPEELS (1975) and KAWASAKI and SATO (1979) is attractive. Unfortunately the crucial marker enzyme, inosine diphosphatase, is not necessarily a unique marker for Golgi (QUAIL 1979), while the autoradiographic work of ROBERTS and NORTHCOTE (1972) does not support classical Golgi involvement in extensin transport and secretion.

We have known for quite some time (LAMPORT 1965) that trace amounts of newly synthesized wall protein are only loosely bound to the wall and can be washed off with salt solutions. At present it is not clear whether this pool of loosely bound wall protein represents a true precursor to covalently bound wall proteins or whether perhaps *loosely* bound wall protein serves some other function. BRYSK and CHRISPEELS (1972) claim that loosely bound cell surface wall protein exhibits pulse-chase kinetics, being progressively incorporated into bound wall protein, but the actual data (cf. Fig. 3 of their paper) are not compelling while other workers (POPE 1977) concluded that ionically bound wall protein is *not* a precursor to extensin per se. The glycosylation state of the serine residues may provide a clue. According to BRYSK and CHRISPEELS (1972), serine residues of ionically bound wall protein are, unlike covalently bound wall protein, not galactosylated. This suggests that galactosylation is a necessary part of covalently assembled cell wall protein. When and where does this galactosylation occur? By analogy with normal eukaryotic glycoprotein biosynthesis, galactosylation would occur as the translated product enters the lumen of the endoplasmic reticulum, implying that galactose-free extensin at the cell surface has undergone "processing" (degalactosylation) during transport, presumably to serve some other than a purely structural function.

2.1.6 Possible Roles for the Matrix Protein Extensin

2.1.6.1 The Extensin Hypothesis

The extensin hypothesis was formulated some years ago as a working hypothesis for approaching the problem of cell extension (LAMPORT 1963, 1965). There is now general agreement that the control of cell extension involves the control of cell wall plasticity presumably by regulating the cross-linking of a cellulose network, as originally suggested by HEYN (1940). The molecular details remain speculative. The extensin hypothesis fills this void conceptually, as extensin has the merit of being a substance rather than a state of mind. Extensin is not an "effect"! Thus the hypothesis predicts intermolecular linkages or "cross-links" between extensin and other wall components, and that cell wall extensibility depends on the amount of extensin (see Sect. 2.1.6.4) and the state of its cross-links. The next section examines the possible chemical identity of such cross-links.

2.1.6.2 Possible Intermolecular Linkages – "Cross-Links"

Conceivably extensin could bond to itself, to cell wall polysaccharides, or to cell wall phenolic components. The linkages involved could be static or dynamic labile cross-links under metabolic control. We shall consider first evidence for the role of specific linkages, followed by more generalized evidence for interaction between extensin and other cell wall components.

a) Hydroxyproline Arabinosides. These are natural contenders for a cross-link role, but direct methylation analysis of cell walls (TALMADGE et al. 1973) indicates that all the hydroxyproline arabinosides exist as short arabino-oligosaccharide side chains with the nonreducing end unsubstituted and therefore *not* involved in further attachment to polysaccharide. However, methylation analysis does not rule out the possibility of an alkaline labile, e.g., ester linkage.

b) Galactosyl Serine and Cell Wall Galactosidases. The occurrence of galactosylserine in extensin is well established (LAMPORT et al. 1973, CHO and CHRISPEELS 1976) and is therefore another possible cross-link site under metabolic control. We shall therefore briefly discuss evidence for: (1) the lability of the galactosylserine linkage in vivo, (2) attachment of oligo- or polysaccharide to extensin via galactosylserine and (3) specific enzymatic cleavage in the oligosaccharide serine region.

1. If the galactosyl serine linkage turns over during growth, this should be detectable by measuring the steady-state level of galactosylserine. Unfortunately conventional alkaline-induced β-elimination of serine saccharide residues is not stoichiometric, presumably because of the electrostatic screening mentioned earlier (Sect. 2.1.1). An alternative assay of galactosylserine linkages based on hydrazinolysis depends on the empirical observation that serine residues involved in *O*-glycosidic linkages do not survive hydrazinolysis and post-hydrolysis to regenerate parent amino acids from the hydrazides (LAMPORT et al. 1973), presumably because β-elimination occurs yielding the acid-labile dehydroamino acid. Hydrazine-labile serine is therefore a likely indication of glycosyl serine residues. There are three reports that hydrazine-labile serine, as a percentage of total wall serine, increases as the rate of cell extension decreases or as the tissue ages (LAMPORT 1973, KLIS 1976, and WINTER et al. 1978).

2. Turnover of the galactosylserine linkage implies the existence of further polysaccharide attachment at that site. The best evidence obtained so far is the actual isolation of a glycopeptide *fraction* (by alkaline degradation of sycamore maple cell walls) which contains a galactose oligosaccharide of about ten residues, attached to serine (LAMPORT 1977), but further characterization is necessary.

3. If galactan attachment to serine is involved in the growth control mechanism then we should expect both direct and indirect evidence of appropriate enzymic activities in the wall, and evidence for their control. Thus auxin-activated cell wall β-galactosidase occurs in oat coleoptiles (JOHNSON et al. 1974) and pea stems (TANIMOTO and IGARI 1976), but D-galactono-1,4-lactone a specific inhibitor of β-galactosidase did not bring about the inhibition of cell extension predicted by the hypothesis above (EVANS 1974, PERLEY and PENNY 1974). However, this simply indicates the noninvolvement of β-galactosyl linkages while, as mentioned in Section 2.1.1.1, galactosylserine of extensin is probably α-linked. α-Galactosidases are very widespread in plants (e.g., DEY and PRIDHAM 1968, PETEK et al. 1969, THOMAS and WEBB 1977, SHARMA and SHARMA 1977, WALLNER and WALKER 1975) and they are *not* inhibited by D-galactono-1,4-lactone (PERLEY and PENNY 1974), but as yet there is no report of a direct correlation between α-galactosidase activity and growth (NEVINS 1970). However, the broader specificity of α-D-galactosidases is a possible significant sidelight as some can also cleave β-L-arabinosides (DEY 1973, MALHOTRA and SINGH 1976), which *might* include the β-linked hydroxyproline arabinofuranosides, although information concerning the conformational specificity (pyranoside vs. furanoside) of these enzymes is lacking. Highly specific β-arabinosidases appear to be extremely rare in nature; there is only one report (DEY 1973). α-L-arabinofuranosidases by contrast are extremely common, especially in pathogenic fungi (BATEMAN 1976). Activity is even reported of α-L-arabinofuranosidases in the culture medium of rapidly growing *Scopolia japonica* cultures (TANAKA and UCHIDA 1978).

c) Macroscopic Evidence from Cellulase and HF Extraction. The striking amino acid periodicity of extensin attests to its structural role (Sect. 2.1.3). Like many other structural proteins the bulk of extensin is not soluble in normal protein solvents. Even after severe enzymic attack by hemicellulase/cellulase/protease mixtures much of the extensin remained as an insoluble wall-shaped residue (LAMPORT 1965), implying that extensin, like other structural proteins, might form a coherent network. Extensin remains insoluble even after treatment with anhydrous hydrogen fluoride, an extremely powerful protein solvent which also cleaves glycosidic linkages (MORT and LAMPORT 1977). Evidently something more than glycosidic linkages holds extensin together as a coherent network!

d) Interaction with Phenolics Including Lignin. Some of the extensin network cross-links are evidently stable in hydrogen fluoride. Thus the involvement of phenolic carbon-carbon cross-links, which are quite stable to HF, is an attractive possibility and consistent with the following observations:

1. some extensin glycopeptides contain an unidentified tyrosine derivative (LAMPORT 1974) which may perhaps have arisen through peroxidatic coupling (see Sect. 3.2.1.1), and in the wall may even involve further linkage to lignin (WHITMORE 1978a, b, VANCE et al. 1980) or other polyphenolic substances;
2. insoluble cell wall residues remaining after HF treatment contain extensin together with an approximately equal amount of another substance which remains chemically undefined (MORT and LAMPORT 1977);
3. solubilization of hydroxyproline-rich glycoprotein fractions by delignification procedures involving oxidation with sodium chlorite (MORT 1978, O'NEILL and SELVENDRAN 1980).

It follows that there may be two interacting but semi-independent wall networks (LAMPORT 1978), glycoprotein-phenolic and polysaccharide. This interesting possibility should be compared with some algal cell walls which are composed entirely of hydroxyproline-rich glycoproteins (see Sect. 2.2.2).

e) Correlation of Growth with Extensin Extractability. Usually *nondegradative* reagents, such as high ionic strength buffers, extract only a relatively small but variable proportion of hydroxyproline-rich protein from the wall (LAMPORT 1965, BRYSK and CHRISPEELS 1972, MONRO et al. 1972, 1974, BAILEY and KAUSS 1974, MONRO et al. 1976, STUART and VARNER 1980), and "the proportion of material bound tightly to the wall was definitely greater in the non-growing parts" (BAILEY and KAUSS 1974). This indication of greater cross-linking in less rapidly growing regions is also consistent with the much higher alkali-lability (fragmentation into dialyzable fragments) of extensin from elongating rather than nonelongating regions of lupin hypocotyls (MONRO et al. 1974).

There is also the interesting observation that within 24 h of excision, carrot root discs secrete hydroxyproline-rich material (BRYSK and CHRISPEELS 1972) which is mostly ionically bound and has a composition typical of extensin (STUART and VARNER 1980) yet does *not* rapidly become covalently attached to other wall-bound extensin, i.e., ionically bound extensin is not a precursor to covalently bound extensin (POPE 1977). Possibly this is because ionically bound extensin contains little or no galactose (BRYSK and CHRISPEELS 1972). Thus serine galactosylation could be an essential requirement for covalent attachment (SMITH 1981a. cf. DARVILL et al. 1978) consistent with WINTERBURN and PHELPS' (1972) dictum that "... glycosylation codes for topographical location."

Codes for noncovalent attachment are also possible. For example KAUSS and GLASER (1974) used 0.5 M potassium phosphate (pH 7.1) to elute carbohydrate-binding proteins from mung bean cell walls, (i.e., these extracts agglutinate trypsinized rabbit erythrocytes) and suggested that "Their function in the wall could take the form of a non-covalent 'glueing' substance" noting that "extracts from nongrowing walls exhibit higher binding capacity." Whether or not these "carbohydrate binding proteins" contain hydroxyproline remains a moot point, but there is the speculation that "... extensins may be lectins whose role in cellular metabolism remains to be discovered" (LANG 1976).

2.1.6.3 Disease Resistance

Hydroxyproline-rich proteins often increase in direct response to pathogens (ESQUERRÉ-TUGAYÉ and MAZAU 1974) or at the site of strictly mechanical wounding such as the cut ends of epicotyls (STUART and VARNER 1980).

In the classic example of melon pathogenesis by *Colletotrichum*, wall-bound hydroxyproline increased dramatically after infection (ESQUERRÉ-TUGAYÉ and MAZAU 1974) and the extensin tryptic peptide Ser-Hyp-Hyp-Hyp-Hyp-Lys was isolated in good yield from the infected hypocotyls (ESQUERRÉ-TUGAYÉ and LAMPORT 1979). As short partial sequences of this size can be used to identify proteins fairly unambiguously (DAYHOFF and ORCUTT 1979), it is reasonable to conclude that the increased wall-bound hydroxyproline of infected melon hypocotyls *is* extensin.

Ethylene treatment enhances the accumulation of wall-bound hydroxyproline in pea stems (RIDGE and OSBORNE 1971) and in hypocotyls infected with *Colletotrichium* (ESQUERRÉ-TUGAYÉ et al. 1979). In view of this and also because "... inhibiting the synthesis of this glycoprotein in diseased plants is strictly correlated with an accelerated and more intense colonization of the host by the pathogen", ESQUERRÉ-TUGAYÉ et al. (1979) postulated that *wall protein may act as a defense mechanism against microbial attack*. This hypothesis is consistent with the following observations:

1. increased glycosylation of the hydroxyproline residues of melon hypocotyls after infection by *Colletotrichum* (ESQUERRÉ-TUGAYÉ and LAMPORT 1979). Glycosylation tends to enhance the resistance of glycoproteins to proteolysis (LAMPORT 1980);
2. increased amounts of wall-bound hydroxyproline corresponding with enhanced resistance of potato and tomato to nematode attack (GIEBEL and STOBIECKA 1974, ZACHEO et al. 1977);
3. cytoplasmic "hydroxyproline-poor glycoproteins" involved in the *suppression* of systemic virus infections (KIMMINS and BROWN 1975);
4. sequestration of an invading bacterial pathogen such as *Pseudomonas solanacearum* by potato lectin (SEQUEIRA 1978), another hydroxyproline-rich glycoprotein containing hydroxyproline arabinosides (MURAY and NORTHCOTE 1978);
5. malformin, a plant teratogen produced by *Aspergillus niger* enhances by about fourfold wall-bound hydroxyproline in bean internodes. At least one of the wall receptors for malformin is "... probably a glycoprotein which

contains hydroxyproline and sulfhydryl groups" (CIARLANTE and CURTIS 1977);

6. a rapid (beginning within 24 h) increase in the wall-bound hydroxyproline of primary wheat leaves on infection with *Erysiphe graminis* (CLARKE et al. 1981).

Extensin may perhaps also act as a defense mechanism by providing a template for initial lignin deposition as suggested by WHITMORE (1978a). This interaction with lignin might involve covalent interaction (cf. Sect. 2.1.6.2. d and 3.2.3) or hydrogen bonding between phenolic hydroxyl and protein carbonyl groups. These hydrogen bonds are, according to LOOMIS (1974) the strongest ones known. The involvement of lignin itself in disease resistance mechanisms now seems well established (VANCE et al. 1980).

2.1.6.4 Physiological Control of Extensin Levels

Discussion in the previous section shows that the level of wall-bound hydroxyproline is not a chance event but well regulated and co-ordinated. We shall now briefly survey physiological aspects of this regulation by considering such physiological parameters as oxygen tension, ethylene levels, light quality, phytochrome involvement, and hydroxylation inhibitors.

a) Oxygen Effects. Plant and animal prolyl hydroxylases are remarkably similar enzymes (cf. SADAVA and CHRISPEELS 1971), hence the K_m of O_2 for plant prolyl hydroxylase is probably similar to that of the animal enzyme namely, about 5% O_2 tension. Thus FUJII and SHIMOKORIYAMA (1976) confirmed the predictable effect of low oxygen tensions (8%) in decreasing the wall hydroxyproline content. Their additional data also support the hypothesis (CLELAND and KARLSNES 1967) of *an inverse relationship between the level of wall-bound hydroxyproline and the potential for rapid cell extension,* as low oxygen tensions greatly *accelerated* growth of *Avena* coleoptile sections despite a *decrease* in overall ATP production (an effect they term "oxygen-sensitive" growth). The cell wall pH *increased* during "oxygen-sensitive" growth, and this conflicts with Hager's "acid-induced growth" hypothesis (HAGER et al. 1971).

If oxygen becomes limiting at such relatively high oxygen tensions (8%) how can wall protein play a significant role in the regulation of cell expansion at the very low oxygen tensions encountered by some plants, for example under conditions of flooding? VARNER (private communication) suggests ingeniously that the cyanide-insensitive respiratory pathway may function by decreasing the competition for oxygen between prolyl hydroxylase and the cytochrome system. He suggests that low oxygen tensions somewhat paradoxically switch off the "high oxygen affinity" cytochrome system, replacing it with the "low oxygen affinity" cyanide-insensitive respiratory pathway which then competes on a much more equal basis with prolyl hydroxylase for the limited oxygen available. There is indeed a good correlation between the ability to synthesize hydroxyproline and the activity of the cyanide-insensitive respiratory pathway (ARRIGONI et al. 1977b).

b) Ethylene Effects. While low oxygen tensions may show fairly small effects on levels of wall-bound hydroxyproline, ethylene induces a striking increase (approx. 200%) in the pea epicotyl, combined with enhanced lateral expansion and an inhibition of cell elongation (RIDGE and OSBORNE 1970). Ethylene may turn out to be the common factor involved in several different "effects" leading to enhanced levels of wall hydroxyproline. For example there are the pathological increases of hydroxyproline described in Section 2.1.6.3, while PEGG's (1976) work showing that ethylene-treated plants became more resistant to *Verticillium* is also relevant here. In addition JOTTERAND-DOLIVO and PILET (1976) report that roots have less hydroxyproline in the upper (extending) half of the georeactive region. This can perhaps be ascribed to auxin-induced ethylene production (YANG et al. 1980) on the lower side leading to an increased level of hydroxyproline, which would be consistent with the postulated inverse relationship between hydroxyproline and cell extension just mentioned. Geotropic responses, especially of roots, are of course highly complex (cf. WILKINS 1977).

c) Phytochrome Involvement. In some instances ethylene production may involve the action of phytochrome (GOESCHL et al. 1967), but the relationship is not clear-cut, as we see in the following scene:

In the pea epicotyls, "Red light is one of the factors which suppresses ethylene production ..." (LIEBERMAN 1979) and as ethylene increases wall-bound hydroxyproline (RIDGE and OSBORNE 1970) we should expect red light to decrease wall-bound hydroxyproline, but on the contrary red light increases it (PIKE et al. 1979), a paradox awaiting resolution. Nevertheless when red light inhibits the growth of intact pea internodes there is a corresponding increase in deposition of wall-bound hydroxyproline, and "... since other growth inhibitors also increased hydroyproline deposition, there is evidence for a common mechanism of action", i.e., increased hydroxyproline deposition inhibits growth (PIKE et al. 1979).

d) Inhibition of Proline Hydroxylation. All these reports of an inverse relationship between growth rate and wall-bound hydroxyproline levels lead to the attractive but debatable proposition that growth rates decline as a direct result of increased hydroxyproline deposition. However, while some (KLIS 1976) report that cessation of cell elongation coincides with attainment of the maximum level of wall-bound hydroxyproline, others (CLELAND and KARLSNES 1967, SADAVA et al. 1973, LANG 1976) report that wall-bound hydroxyproline increases for some time even after cell elongation ceases. What happens, however, if we make a defective wall protein, for example by inhibiting proline hydroxylation? The ferrous iron chelator α,α'-dipyridyl inhibits proline hydroxylation (HOLLEMAN 1967) and also enhances elongation (HEATH and CLARK 1956, BARNETT 1970). However, this does not establish a causal relationship (LANG 1976), and it might be worthwhile to examine the effects of other inhibitors of proline hydroxylation such as: palladium ions (RAPAKA et al. 1976); lycorine (ARRIGONI et al. 1977a); nitroblue tetrazolium and copper chelates (MYLLYLA et al. 1979); polyproline (MYLLYLA et al. 1977); other peptides such as (Gly-Pro-Ala), (RAO and ADAMS 1978), and dehydroascorbate (MYLLYLA et al. 1978).

e) Morphoregulation. The preceding sections have summarized much suggestive evidence for extensin as a structural component involved in growth regulation. BASILE (1973) raises the intriguing possibility that hydroxyproline-rich wall proteins play a direct role in the morphogenesis of leafy liverworts, where inhibition of proline hydroxylation or the use of a proline analog produces phenovariants possessing a primitive leaf arrangement. Hence BASILE (1979) speculates that "hydroxyproline proteins regulate several aspects of morphogenesis in leafy liverworts by suppressing growth and development at critical times in highly localized areas." There are strong echoes of this at the higher plant level with the emphasis placed on the inverse relationship between growth rate and wall protein content. Hence we suggest that wall protein falls nicely into the vacant morphogenetic niche described by ROLAND and VIAN (1979): "turgor pressure is a nondirectional force, while the response is directional. One of the problems is to find the sites and the mechanism of this polarized response."

2.2 Algae

2.2.1 Occurrence of Matrix Proteins

Difficulties have arisen over the definition of the cell wall in higher plants (see for example LAMPORT 1970, p. 235). In the algae the semantic problem is even more difficult because of the extreme diversity in cell wall structure. (see DODGE 1973 for a general discussion of "cell coverings"). Many algae continuously slough material off the surface of the cell perhaps as protection against mechanical damage and desiccation. For example, the "naked" cells of *Rhodella maculata* are embedded in a mucilaginous sheath (EVANS 1970). Others have scales (as in *Pleurochrysis*). Some, such as *Chlamydomonas fimbriata,* embed themselves in a matrix, which can in turn be surrounded by a sheath as in *Eudorina* and *Volvox,* while some algae such as *Derbesia-Halicystis* have different cell walls at different stages of their life cycle (WUTZ and ZETSCHE 1976). For purposes of this discussion we regard the cell wall in a broad sense as any material which may even be transiently associated with the cell surface.

Proteins are invariably associated with the complex polysaccharides of algal cell walls (see Chap. 12, this Vol.) (NORTHCOTE et al. 1960, JONES 1962, THOMPSON and PRESTON 1967, COOMBS and VOLCANI 1968, NEVO and SHARON 1969, BROWN et al. 1970, HERTH et al. 1972). Critical analysis of this protein in the mucilaginous sheath of *Porphyridium cruentum* reveals the glycopeptide linkages, xylosyl serine and xylosyl threonine (HEANEY-KIERAS et al. 1977). In this instance the protein component of the purified material is so small (3% to 10%) that it is probably more correct to regard this cell surface material as a proteoglycan rather than a glycoprotein in the strict sense (cf. LAMPORT 1980). However, the most common cell surface glycoproteins contain hydroxyproline and they occur throughout the algae particularly in the Volvocales, (PUNNETT and DERRENBACKER 1966, THOMPSON and PRESTON 1967, GOTELLI and CLELAND 1968, ARONSON et al. 1969, TAYLOR 1969, AARONSON 1970) and even in diatoms which also contain dihydroxyproline (SADAVA and VOLCANI 1977). Again, we know few glycopeptide linkage types, namely arabinosyl hydroxyproline, and galactosyl hydroxyproline, and so far only in the Chlorophyceae: *Chlorococcum oleofaciens* (MILLER 1978), *Chlamydomonas gymnogama* (MILLER et al. 1974) *Eudorina* (TAUTVYDAS 1978) and *Volvox carteri* (LAMPORT 1974, KOCHERT 1975, MITCHELL 1980). There is also histochemical evidence of these glycoproteins by light and electronmicroscopy in *Pandorina* (FULTON 1978), *Eudorina* (TAUTVYDAS 1978) and by co-fractionation of protein and carbohydrate, in Chlamydomonas (SCHLÖSSER 1966). Coomassie blue and PAS staining after polyacrylamide gel electrophoresis of various Chlamydomonads (ROBERTS 1974) shows that these glycoproteins comprise the bulk components of Chlorophycean, especially Volvocalian cell walls. Indeed the highly ordered glycoprotein component of Volvocalian cell walls is crystalline (ROBERTS 1974) and the cell walls of this group of algae fall into a number of classes which may be of taxonomic importance. ETTL (1971) has also suggested from a survey of 73 *Chlamydomonas* species that all contained periodic structures of unknown composition in the cell wall.

2.2.2 The Chlamydomonas Type Wall

Chlamydomonas reinhardtii is the most thoroughly investigated of the algae with glycoprotein cell walls and an examination of its structure, composition, and morphogenesis can be used as a basic type for the genera and perhaps even for the entire Volvocalian family. The cell wall consists of a number of discrete, hydroxyproline-rich glycoproteins (ROBERTS et al. 1972, ROBERTS 1974, HILLS et al. 1975, CATT et al. 1976) although some investigators have been unable to find evidence of these (GUNNISON and ALEXANDER 1975, JIANG and BARBER 1975, RAY et al. 1978). Fixation of cells with glutaraldehyde supplemented with osmic acid or low molecular weight tannic acid followed by sectioning shows a wall consisting of an inner fibrillar layer (30–200 nm) covered by a regular tripartite layer (28 nm). Negative staining (in ammonium molybdate or methylamine tungstate) of cell wall preparations shows that the regular layer is in fact, crystalline. The repeating unit in this crystalline layer is a parallelogram 28.5 nm by 23.6 nm, at an angle of 80°. Fixation artifacts can be ruled out because glancing sections through the cell wall reveal the same crystalline structure. The amorphous inner wall layer (5%–10% by weight of the cell wall) is insoluble in chaotropic agents, detergents, and denaturants, but has a gross chemical composition very similar to the outer crystalline layer. Its apparent amorphous state and stability is probably due to more subtle interactions (covalent cross-linking?) than compositional data indicate. The crystalline layer, which is soluble in chaotropes (such as molar perchlorate isocyanate or 8 M lithium chloride) has been extensively investigated by electronmicroscopy and image analysis (ROBERTS et al. 1972, HILLS et al. 1975, CATT et al. 1976, 1978). Chaotrope-solubilized material reassembles into crystalline structures upon removal of the salt across a surface (such as dialysis or by placing drops of digest on agar or floating millipore filters). When the inner wall layer is also present, the cell walls formed are identical to native cell walls. Absence of the inner wall layer during salt removal gives rise to the formation of crystalline sheets rather than discrete cell walls. (HILLS 1973, HILLS et al. 1975, CATT et al. 1978).

This implies that the crystalline layer is inherently capable of forming itself and that directed assembly by the inner wall layer does not occur.

The chaotrope-soluble material separates into two components on exclusion chromatography via Sepharose 2B (2BI and 2BII). Component 2BII has a molecular weight, from ultracentrifugation, of about 278,000, and is inherently capable of forming crystalline sheets, but the presence of the relatively minor component 2BI, (about 20% of the soluble material) may promote the reassembly process (CATT et al. 1978). Chemical analyses of 2BI and 2BII show similar but not identical compositions. Both are hydroxyproline-rich glycoproteins, containing arabinose, galactose, and mannose (with a trace of glucose). There are small quantitative differences in the sugar compositions (CATT et al. 1976, ROBERTS 1979) and some are sulfated (ROBERTS et al. 1980). Component 2BII can be further split into four subunits by the action of sodium dodecyl sulfate and mercaptoethanol, followed by polyacrylamide gel electrophoresis. Electronmicroscopy of the components has been difficult because of the high salt concentrations needed for dissociation; if the salt is absent the components are aggregated. Recent work by HILLS and ROBERTS (personal communication) indicates that both 2BI and 2BII are very elongated molecules, as predicted from ultracentrifugation. Further investigation of the 2BII fraction (ROBERTS 1979, HOMER and ROBERTS 1979) shows hydroxyproline-rich highly glycosylated, and hydroxyproline-poor low glycosylation domains. The hydroxyproline-rich domain (isolated by thermolytic digestion of 2BII) gives a circular dichroism spectrum typical of a polyproline type II helix, discussed further below. The precise chemical identity and role of the salt-soluble components and their relation to the inner wall layer in the cell wall await further investigation.

A genetic approach to the morphogenesis of the cell wall is also invaluable (reviewed by DAVIES and ROBERTS 1976). Nearly 80 cell wall-less mutants were isolated (20 of the most stable are now in the Cambridge Culture Collection for Algae and Protozoa). Extensive mating analysis revealed a complex situation. Wild-type phenotype can be restored by passage through the zygospore stage (the zygospore cell wall is unaffected in these mutants). Diploid, vegetative strains often have wild-type phenotypes and the mutants often do not show Mendelian segregation. To account for this DAVIES and LYALL (1973) proposed that extranuclear control of cell wall synthesis is involved. Further evidence for this is based on a relationship between the cell wall-less mutant loci and the "yellow" locus (y1) which is not found on any of the 16 chromosomes. Although much knowledge has accumulated about the cell wall of *C. reinhardtii*, some crucial questions remain; for example the relationship of the components to each other and how they are assembled in vivo (does a self-assembly process occur?), as well as the biological role of the cell wall itself. The presence of contractile vacuoles and viable wall-less mutants argues against a purely osmotic role.

2.2.3 Phylogenetic Considerations

The major structural extracellular matrix glycoprotein of eukaryotic cells is hydroxyproline-rich. In the evolutionary line classed as Animalia this glycoprotein is collagen: a triple-stranded glycine-rich polypeptide based on a tripeptide periodicity of high fidelity (Gly-X-Y). In the plant evolutionary line sequence data are so far only available from the higher plants where a pentapeptide periodicity (of relatively low fidelity being interspersed with other short sequences) occurs: Ser-Hyp-Hyp-Hyp-Hyp. Apart from the presence of hydroxyproline there appears to be no reason to regard these proteins as homologous, especially considering that glycosylated hydroxyproline occurs in plants but not in animals. However, both proteins have a similar tertiary rod-like structure consisting of a polyproline II type helix (a left-handed three residue per turn helix of pitch approx. 0.94 nm). This is clear both from molecular models and circular dichroism spectra obtained from *Chlamydomonas* (HOMER and ROBERTS 1979) and higher plants (LAMPORT 1977). However, one cannot argue for homology based on shape alone – there is often more than one solution to a single problem and a saltatory origin of periodic proteins (YCAS 1976) could account for the separate origins of both hydroxyproline-rich tripeptide and pentapeptide periodicities. This would side-step the difficult problem of accounting for the origin of glycine-rich tripeptide periodic structure from a glycine-poor structure of a different (although probably pentapeptide) periodicity. But it seems very strange that, despite its ubiquity in the metazoa, collagen has yet to be detected among the protists, while the "plant-type" glycosylated hydroxyproline-rich glycoproteins occur in nearly all the photosynthetic protists. Indeed the correlation between photosynthesis and presence of hydroxyproline-rich glycoprotein is so strong that we (DELMER and LAMPORT 1976, LAMPORT 1980) suggested earlier that prolyl hydroxylase probably evolved in the first oxygen producers (Cyanobacteria), some of which do in fact contain hydroxyproline (PUNNETT and DERRENBACKER 1966, GOTELLI and CLELAND 1968) so that essentially the same enzyme (SADAVA and CHRISPEELS 1971) probably hydroxylates proline residues of all hydroxyproline-rich proteins, no doubt including the arabinogalactan proteins (β-lectins) and potato-type lectins. Although at the present time we cannot distinguish between a common or separate origin for collagen and extensin-like proteins, there are three possibilities worthy of consideration in future work along these evolutionary lines. First there are the arabinogalactan proteins (AGP's) which exhibit a wide range of hydroxyproline levels while being generally rich in alanine. The ubiquity of the AGP's does not yet extend to the algae, but that may simply be a problem of looking for them there. If AGP's are present in the algae then there is the possibility that AGP's represent the archetypal hydroxyproline-rich protein. Second there is the possibility that the early evolution of hydroxyproline-rich glycoproteins created a protein with two distinct domains ultimately destined to become extensin and collagen respectively. We base this speculation on the fact that autolysis of *Chlamydomonas* cell walls releases autolytic peptides rich in glycine (CATT and LAMPORT 1980, unpublished data). Third, there is the very distinct possibility suggested earlier (DELMER and LAMPORT 1976) that the evolutionary precursor to hydroxyproline-rich glycoproteins such as extensin and collagen was a prokaryotic cell surface glycoprotein. Glycoproteins are generally rare in prokaryotes except for the Archaebacteria which usually have a cell surface coat or "cell wall" of glycoprotein rather than peptidoglycan, tempting the speculation that the Archaebacteria contained the original "pro-eukaryote" (LAMPORT 1980). The recent discovery that the cell wall protein of the Archaebacterium *Thermomicrobium roseum* contains 34 mol % glycine and 14 mol % proline (MERKEL et al. 1980) will therefore be of great significance if future analysis indicates a structure similar to procollagen!

While the extensin-collagen evolutionary connection is highly speculative, we feel on safer ground in considering that the hydroxyproline-rich glycoproteins of algae are homologous with those of higher plants (despite the frustrating absence of explicit amino acid sequence data from the algae) because one can infer that algae and higher plant extensin share similar sequences: they have remarkably similar gross amino acid compositions, and they probably both share the same unique sequon (amino acid sequence) coding for hydroxyproline arabinosylation.

If we regard evolution as a process of message refinement (REICHERT et al. 1976) how refined has cell wall protein become over the aeons? There is a clear trend in two directions: namely reduction and cross-linking. Thus we regard the primitive condition extant in *Chlamydomonas* which has a cellulose-free cell wall built virtually exclusively from a glycoprotein arranged as a crystalline lattice. (From this view point the glycoprotein is a more primitive feature than cellulose!). Other cell wall components soon appear along the Chlorophycean evolutionary pathway, notably the $(1 \rightarrow 4)$-β-polymers, chitin and cellulose, combined with other polysaccharides which are no doubt the forerunners of hemicellulose, followed later by the true pectic substances. Judged from difficulty of extraction, wall protein changed fairly early from a crystalline lattice to a network structure by the acquisition of covalent cross-links (i.e., in the Chlorophyceae e.g., *Chlorella vulgaris*) although the beginning of this network may perhaps be represented by the small amount of chaotrope-insoluble cell wall residue of *Chlamydomonas reinhardtii*. From then on major evolutionary progress involved a rapid reduction of the protein component from a major to a relatively minor component. For example the wall hydroxyproline levels of some algae and bryophytes are remarkably low and this continues, with interesting exceptions, such as herbaceous dicots where high levels of wall protein may enhance wall tensile strength (LAMPORT 1965), right up to the Gramineae. We consider that this trend represents a change in function – from a bulk structural component to that of a controlling element in cell extension, while its role in cell surface defense may have been retained throughout evolution.

But there is also a minor evolutionary line where the trend was toward an increase in the amount of cell wall protein by enormous hypertrophy creating an extended extracellular matrix. This hydroxyproline-rich matrix forms the morphogenetic substrate for the entire order Volvocales (MITCHELL 1980), culminating in the creation of a true multicellular blastula-like organism. *Volvox* is not a colony; the vegetative cells are terminally differentiated! We conclude that "... the evolution of the eukaryotic cell and the evolution of glycoproteins go hand in hand" (DELMER and LAMPORT 1976).

3 Cell Wall Enzymes

3.1 Introduction

Complex and sophisticated eukaryotic genomes ultimately translate into form. This form is extracellular matrix, a supramolecular structure created both by self-assembly and directed assembly. Evidently reactions which occur *outside* cells are just as important as reactions which occur inside cells! The catalysts must also be extracellular. Besides assembly other extracellular recognition, signalling, and metabolic processes occur. The primary cell wall is definitely not a passive inert dead structure (LAMPORT 1965, 1970)! This brief section provides definitions and criteria for cell wall enzymes, their detection and assay, their release and purification, and most importantly their possible biological role. Any enzyme associated with the cell wall can be regarded as a cell wall enzyme. However our operational definition refers to enzymes "associated" with the mechanically isolated cell wall, and this definition tends to exclude enzymes of the plasma membrane, periplasmic space at the wall-plasma membrane junction, and freely soluble enzymes which may just happen to leak through the plasma membrane.

Detection of cell wall enzymes via histochemical techniques gives semiquantitative results at best. It also selects for the easiest methods: such as phosphatases and peroxidase. By contrast, isolated primary cell wall preparations allow quantitative assays and determination of kinetic constants (LAMPORT and NORTHCOTE 1960b, LAMPORT 1965). Model systems indicate that the matrix-bound enzyme behaves kinetically very much like its free counterpart. Theoretically this can be confirmed by releasing enzymes "associated" with the cell wall either by elution with strong salt solutions (ionic binding) or by enzymic degradation of the wall (covalent binding or perhaps just a physical entrapment or entanglement within the wall). Cytoplasmic contamination is not usually a serious problem provided the cell walls and enzymic activity in question meet criteria offered earlier (LAMPORT 1965).

3.2 Biological Role

3.2.1 Involvement in Cell Wall Assembly and Cell Extension

3.2.1.1 Peroxidases

The peroxidase system tends to increase the number of cell wall cross-links while glycosidases tend to decrease cross-linking. Therefore the former slows while the latter enhance cell extension. Peroxidases have various specificities and may therefore operate in several different ways (LAMPORT 1978, FRY 1979) as follows:

a) Some may be auxin oxidases (MORITA et al. 1967, PALMIERI et al. 1978, KOKKINAKIS and BROOKS 1979).

b) Others may function as lignification catalysts involving peroxidatic coupling of cinnamyl alcohols (GROSS 1977, GRISEBACH 1977). Remarkably, NADH-driven reversal of the general peroxidase reaction generates the required hydrogen peroxide (ELSTNER and HEUPEL 1976) while a wall-bound malic dehydrogenase (STEPHENS and WOOD 1974) generates the NADH (GROSS and JANSE 1977, GROSS et al. 1977, GROSS 1977).

c) Ferulic acid ester-linked to hemicelluloses is of widespread occurrence (HARRIS and HARTLEY 1976), and thus may also undergo peroxidatic coupling to form diferuloyl bridges between polysaccharide chains (MARKWALDER and NEUKOM 1976, WHITMORE 1976).

d) Exclusion of water at specific wall sites by adsorption of phenolic oxidation products may lower the effective concentration of cell wall water, strengthen hydrogen bonding between adjacent polymers, and/or restrict glycosidase access, thereby leading to an overall stiffening of the wall (FRY 1979).

e) Peroxidatic coupling of extensin subunits (LAMPORT 1978, 1980) may occur, perhaps involving the tyrosine derivative mentioned earlier or a peroxidatic deamination of lysyl residues to lysyl aldehyde which may then react with, and become cross-linked by, quinones also generated by peroxidase (STAHMANN and SPENCER 1977). Although peroxidase will catalyze (in very low yield) hydroxylation of proline in model systems (YIPP 1964) there is, despite earlier suggestions (RIDGE and OSBORNE 1970, 1971), no evidence at all of this role for peroxidase in plants (cf. Sect. 1).

Histochemical evidence in support of a role for peroxidase as an agent that increases cross-linking includes the observation that peroxidase tends *not* to occur in regions of highest growth rate such as the tips of pollen tubes (DASHEK et al. 1979) and protoplasts during initial resynthesis of primary cell walls (MADER et al. 1976). Apparently this observation does not apply in root hair tips which, unlike pollen, have hydroxyproline-rich cell walls (ZAAR 1979).

3.2.1.2 Role of Glycosidases and Other Enzymes in Cell Wall Autolysis

We have already discussed in Section 2.1.6.2 b the possible role of galactosidases in cell extension. Other wall-bound carbohydrases include cellobiase (CHANG and BANDURSKI 1964), cellulase (FAN and MACLACHLAN 1967), polygalacturonase strongly bound to fruit cell walls (WALLNER and WALKER 1975) but much less firmly associated with *Avena* cell walls (PRESSEY and AVANTS 1977), pectinesterase (PRESSEY and AVANTS 1977), β-glucosidases (NEVINS 1970, TANIMOTO and PILET 1978, JOHNSON et al. 1974), (1→3)-β-glucanase (WALLNER and WALKER 1975, HEYN 1969) and xylosidases (NEVINS 1970).

Some or all of these enzymes are involved in cell wall autolysis (i.e., autolysins; see Chap. 14, this Vol. also LAMPORT 1970) which even occurs in vitro (LEE et al. 1967) and in corn coleoptiles involves at very least degradation of a (1→3) (1→4)-β-linked glucan (HUBER and NEVINS 1979). Auxin enhances this autolysis in vivo, while a β-glucosidase inhibitor (5-amino 5-deoxyglucose) blocks both auxin-induced growth and autolysis (NEVINS and LOESCHER 1974), implying that in oat coleoptile (hardly a typical plant tissue; it does not show auxin-induced lateral growth response, BURG and BURG 1966), extension growth probably involves degradation of the mixed linkage glucan. In dicotyledons (and probably nongraminaceous monocots; MANKARIOS et al. 1980) the situation seems to be more complex; the primary cell wall has a battery of glycosidases (KEEGSTRA and ALBERSHEIM 1970). Glucose, galactose (substantial), arabinose, xylose, and mannose residues of pea epicotyl cells walls turn over (LABAVITCH and RAY 1974, GILKES and HALL 1977), presumably a reflection of complex cell extension chemistry and consistent with the view that cell wall glycosidases are indeed genuine autolysins. Curiously there is as yet no report of cell wall proteases, but in pepper fruits the appearance of free hydroxyproline corresponds to the period of most rapid cell expansion (FUKUDA and YAMAGUCHI 1979) implying the existence of proteases which can degrade wall proteins (cf. CRASNIER et al. 1980; Sect. 3.2.3).

3.2.2 Involvement in Transport

Cell wall enzymes including those in the periplasmic space contribute to transport by hydrolyzing organic molecules in the extracellular fluid. Thus invertase hydrolyzes sucrose to fructose and glucose which can then, like sucrose itself, enter the cell by active transport (COOMBE 1976, BAKER 1978). Similarly, the periplasmic β-galactosidase of immature *Cucurbita pepo* leaves may hydrolyze imported galactosylsucrose oligosaccharides to facilitate transport of a free sugar (THOMAS and WEBB 1979).

Cell wall phosphatases are virtually ubiquitous and were among the first wall enzymes to be rigorously demonstrated and studied kinetically (LAMPORT and NORTHCOTE 1960b, LAMPORT 1965) and purified recently by CRASNIER et al. (1980). Despite the enormous literature dealing with these enzymes (largely histochemical), their role is not yet clearly defined. However, CHANG and BANDURSKI (1964) pointed out that these enzymes make it "... possible for corn roots to enzymically hydrolyze and solubilize organic soil macromolecules independently of soil microbial activity," and most significantly BIELESKI (1964) demonstrated a derepressible cell surface alkaline phosphatase in the water plant *Spirodela oligorrhiza.* This plant responds to phosphate deficiency by producing very high activities (400-fold increase) of the cell surface enzyme. How can we account for cell wall hydrolases in nonroot tissues? The substrate source is the plant itself arising from cell autolysis or secretion, and there is evidence for the occurrence of phosphorylated compounds such as UDP glucose (BRETT 1978) and NADPH (GROSS and JANSE 1977) in the extracellular space.

3.2.3 Involvement in Recognition Phenomena and Disease Resistance

While cell wall enzymes are probably not informational macromolecules or specific effectors per se, they may well play a support role in recognition phenomena at the metabolic level. Intuitively one expects them to play a part in cell–cell interactions such as pollination including fertilization, epiphytism, symbiosis, pathogenesis, and induced disease resistance, yet the undoubted enzymic component is largely unknown. Even a recent excellent review of cell recognition in flowering plants (CLARKE and KNOX 1978) makes only a passing reference to enzymatic (esterase) activity in the sticky stigma receptive layer, although there is appreciable evidence (reviewed by MASCARENHAS 1975) for many pollen wall enzymes, mainly hydrolases, including a protease.

Symbionts such as *Rhizobium* undergo an initial lectin-mediated attachment apparently through a cross-reactive antigen which is present on the *Rhizobium* capsule (DAZZO 1980). Subsequent entry and infection, almost certainly involves enzymatic cell wall degradation, the source, microbial one presumes (DAZZO 1980), but a triggered autolysis of the host cell wall is an interesting possibility.

Pathogens elicit many different responses from the host (MUSSELL and STRAND 1977) but the role of enhanced wall peroxidase levels (RAA et al. 1977) in disease resistance hinges on at least two peroxidatic reactions: (1) formation of bactericidal quinones from phenolics, and (2) lignification (reviewed by VANCE et al. 1980). Infection of primary *Cucumis sativa* leaves with *Colletotrichum lagenarium,* led to enhanced peroxidase levels in subsequent leaves, and this correlated well with the reduction in number and size of necrotic lesions produced when those leaves were challenged with the fungus (HAMMERSCHMIDT and KUC 1980).

4 Arabinogalactan Proteins (AGP's) and β-Lectins

4.1 General Properties

All "β-lectins" are AGP's, but not all AGP's are "β-lectins" (cf. JERMYN 1974). Arabinogalactan proteins reviewed comprehensively by CLARKE et al. (1979a) are both intracellular (LAMPORT 1970) and extracellular (JERMYN and YEOW 1975), throughout the plant kingdom, and invariably contain lesser or greater amounts of hydroxyproline *O*-glycosidically linked to an arabinogalactan (FINCHER et al. 1974, POPE 1977, POPE and LAMPORT 1974, cf. YAMAGISHI et al. 1976). *Most* AGP's behave as β-lectins with a broad specificity directed towards β-D-glycopyranosyl linkages, effectively shown by precipitation with the highly colored artificial antigens of YARIV et al. (1962), such as Tris-(4β-D-glucopyranosyloxyphenylazo)-phloroglucinol which is also useful as a method of isolation. However, β-lectins differ markedly from classical lectins in their composition, specificity, and physical properties. The carbohydrate content of β-lectins often exceeds 90%, which may explain their thermal stability and general resistance to denaturation (JERMYN and YEOW 1975). At low concentrations they do not agglutinate erythrocytes, presumably because each molecule has a single site which reacts with only a single molecule of Yariv antigen (JERMYN and YEOW 1975). Precipitation occurs because the Yariv antigens themselves are very highly aggregated (WOODS et al. 1978), but we know very little about the location or chemistry of the site which actually binds β-glycosyl residues. Curiously, conventional glycoside haptens do not compete for the binding site, although some flavonol glycosides present in the original tissue extract may prevent initial binding between AGP and Yariv antigen (JERMYN 1978). There is a strong hint that the binding site is in the hydroxyproline-rich domain: crude enzymic degradation (helicase and subtilisin) of several different AGP's removed about 50% of the carbohydrate and left a hydroxyproline-rich core which still retained binding activity (JERMYN and YEOW 1975, GLEESON and JERMYN 1979). Presumably the binding site is protein, but there is a strong possibility of carbohydrate involvement (M.A. Jermyn, private communication).

4.2 Chemistry, Physical Properties, and Biosynthesis

AGP heterogeneity is almost legendary: "Choice of methods for following the progress of isolation and assessment of homogeneity of the final preparation of arabinogalactan or arabinogalactan proteins is difficult because often the preparation apparently contains a continuous spectrum of closely related molecular species ..." (CLARKE et al. 1979a). What is the source of this heterogeneity, carbohydrate, protein, or both? The major carbohydrate component is typically arabino-3,6-galactan, of D.P. approximately 50 residues attached *O*-glycosidically to hydroxyproline residues of the protein; thus the major AGP glycopeptide linkage is *O*-galactosyl hydroxyproline. The galactan backbone consists of $(1 \rightarrow 3)\beta$-linked D-galactose residues with galactosyl (oligosaccharide) side-chains at C6. Either $(1 \rightarrow 6)$- or $(1 \rightarrow 3)$-β-linked L-arabinofuranosyl residues are substituents on most if not all the terminal galactose residues, as well as many of the side chain galacto-oligosaccharide residues. AGP heterogeneity (polydispersity) is almost certainly due to variability in the number of arabinogalactan polysaccharide units attached to the protein core, but in addition we should be aware of the possibility that arabinogalactan chains themselves are heterogeneous with respect to size, branching, degree of arabinosylation, and even perhaps the presence of other sugars or uronic acids in small amounts. AGP heterogeneity may also arise through variation in the undoubted "minor" AGP carbohydrate substituents, such as oligoarabinosides attached to hydroxyproline, mannan oligosaccharides attached via Asn-GlcNAc glycopeptide linkages (inferred from the presence of both GlcNAc and mannose in most AGP's), and the less well documented *O*-glycosidic substituents of serine or threonine, involving *N*-acetyl galactosamine (HORI 1978) or galactose. With the possibility of seven different glycopeptide linkages, Hyp-Gal, Hyp-Ara, Asn-GlcNAc, Ser/Thr-GalNAc, and Ser/Thr-Gal, there is evident scope for variation!

Analysis of the protein component of AGP's rich in Hyp, Ala, Ser, and Asp+Glu presents a challenging problem. Structure is the key to function, but this key is so well buried in a saccharide matrix, that until recently there seemed little hope of obtaining primary amino acid sequence data. Anhydrous hydrogen fluoride deglycosylates glycoproteins; *O*-glycosidic linkages of neutral sugars cleave within 1 h at 0° while those of *N*-acetylated amino sugars cleave within 3 h at room temperature. Under these conditions *there is no evidence of significant peptide bond cleavage* (MORT and LAMPORT 1977). However HF-deglycosylation of sycamore-maple AGP's (MANI et al. 1979b) gave material which showed apparent heterogeneity on SDS-polyacrylamide disc gel electrophoresis, even after initial gel filtration and cation exchange chromatography of the deglycosylated material (MANI et al. in preparation). At present the source of this heterogeneity remains obscure, but the high level of acidic amino acids

(Asp+Glu approximates 20 mol %) raises the interesting possibility of highly variable amidation, and therefore gross charge heterogeneity. As a very rough estimate the molecular weight of the Hyp-rich AGP protein component approximates only 15,000 or about 150 residues, based on 5% protein in a hypothetical AGP of 300,000 (cf. JERMYN and YEOW 1975). Ultracentrifugation data suggest that the molecular weight may be somewhat lower; thus the protein may turn out to be less than 100 residues in length containing 20 to 30 Hyp residues and 20 to 30 Asp+Glu residues. The possibility of some sequence homology with extensin (pentapeptide periodicity?) is intriguing. These proteins certainly share much else in common including probable biosynthetic route, involving prolyl hydroxylase. However the site and mechanism of glycosylation together with the mode of secretion (maximal in log phase cells, PAULL and JONES 1978) remain unknown. Does each arabinogalactan chain "grow" while attached to the protein or is there a block transfer of entire polysaccharide chains? Presumably the entire synthesis is cytoplasmic, but how then do AGP's cross cell walls whose average pore size is only 4 nm (CARPITA et al. 1979)? Possibly via slow permeation through a few relatively large pores as histochemical staining with Yariv antigens suggests that AGP's are not actually in walls, (ANDERSON et al. 1977), just very closely associated. Plasma membrane, periplasm, intercellular spaces and "ducts" seem to be the major sites of extracellular AGP's (ANDERSON et al. 1977, LARKIN 1978). Possibly some AGP release occurs in suspension cultures as daughter cells expand and rupture the old wall (cf. LAMPORT 1964).

4.3 Biological Role

A search for the AGP role is akin to the quest for the proverbial riddle covered with enigma and wrapped in mystery. AGP's are too widespread and conservative in composition for a trivial role, yet so fundamental that we probably overlook it! We perceive three possible major physiological involvements: water relations, mechanical cell–cell interaction, and involvement as informational macromolecules in cell recognition (e.g., recognition of self vs. non-self).

4.3.1 Water Relations

If xylem sap AGP is significant, where is it going? It may end up in young expanding leaves or at exudation sites (guttation), where the AGP's could plausibly function as antitranspirants or even contribute to frost hardiness as an anti-freeze.

4.3.2 Mechanical Cell–Cell Interactions

Suspension cultures secrete appreciable amounts of AGP into the growth medium (POPE 1977) which becomes in a sense an extended middle lamella. Hence AGP, which constitutes 12% of cambial wall tissue (SIMSON and TIMELL 1978) may be a prominent component of the middle lamella, but is it lubricant or glue? Intrusive growth typical of cambia requires a middle lamellar lubricant.

On the other hand AGP's account for nearly half of the *Gladiolus* style and stigma mucilage (GLEESON and CLARKE 1979) implying an important role in pollination and fertilization. CLARKE et al. (1979b) argue that stigma AGP's, having the molecular properties of a good adhesive, can trap pollen (anemophilous species should have very AGP-rich stigmas). How can AGP's possess the seemingly opposed properties of *both* glue and lubricant? The answer may lie in the degree of AGP arabinosylation as the factor determining sol-gel transformation. De-arabinosylation drastically reduces AGP solubility (FINCHER et al. 1974, GLEESON and CLARKE 1979). AGP's can then bind with each other via unblocked terminal galactose residues. There is evidence for de-arabinosylation in vivo (ROSENFIELD and LOEWUS 1975) while further AGP degradation may also permit stylar AGP's to function as a carbon source for pollen tube growth (LOEWUS and LABARCA 1973).

4.3.3 Cell Recognition

As of yet there is no report of AGP's showing biological activity, but their apparent mobility, their plasma membrane location, their chemical properties and their ability to be bound by other lectins, make them attractive contenders for a role in cell recognition, perhaps even as receptor sites for molecules such as β-glucan elicitors! (CLARKE et al. 1979a).

Apparent AGP heterogeneity (iso-β-lectins?), even after deglycosylation, may indicate a family of closely related macromolecules.

5 Concluding Remarks

There is great scope for further work, ranging from simple chemical identification to the complex three dimensional interaction of macromolecules during cell wall assembly. Nothing in science ever seems to be finally settled. For example there is probably only one major covalently bound hydroxyproline-rich glycoprotein in the wall, i.e. extensin. Almost certainly less easily recognized (structural?) proteins rich in glycine exist (especially in the Gramineae) but remain to be characterized.

The role of extensin is bound up with its structure. We need to know much more of its cross-link chemistry, including its possible attachment to other phenolics including lignin. We cannot now claim a purely structural role for extensin; its involvement in disease resistance is also likely. Then there are the problems of extensin biosynthesis and assembly into the cell wall, including the task of describing the control mechanisms. In comparison the evolutionary origin of extensin might almost seem to be a sidelight were it not for the possibility that such knowledge might be relevant to other biological problems such as the origin of collagen and the metazoa.

The undoubtedly complex chemistry occurring in a dynamic extracellular matrix helps to rationalize the presence of cell wall enzymes, but again the role of some is only just beginning to be understood. Others, such as ascorbic acid oxidase, we have ignored because their role is a complete mystery; but we did not ignore the arabinogalactan proteins because of their intriguing similarities to extensin and the interest shown them by many workers: surely the best sign of vitality for any field of endeavor.

Acknowledgments. We wish to thank Joann Lamport for her advice and help with all stages of the manuscript and to Dr. Michael Jermyn for his help and valuable comments. This work is supported by DOE contract DE-ACO2-76ERO1338.

Note Added in Proof. An important distinction between prolyl hydroxylase from plants and animals can now be made on the basis of synthetic peptide substrates. Thus poly-L-proline is a good substrate for plant prolyl hydroxylase, but inhibits the enzyme from animals (cf. Sect. 2.1.6.4d; TANAKA et al. 1980).

References

Aaronson S (1970) Molecular evidence for evolution in algae: a possible affinity between plant cell walls and animal skeletons. Ann NY Acad Sci 175:531–540

Akiyama Y, Katō K (1977) Structure of hydroxyproline-arabinoside from tobacco cells. Agric Biol Chem 41:79–81

Akiyama Y, Mori M, Katō K (1980) ^{13}C-NMR Analysis of hydroxyproline arabinosides from *Nicotiana tabacum*. Agric Biol Chem 44:2487–2489

Allen AK, Neuberger A (1973) The purification and properties of the lectin from potato tubers, a hydroxyproline-containing glycoprotein. Biochem J 135:307–314

Allen AK, Desai NN, Neuberger A, Creeth JM (1978) Properties of potato lectin and the nature of its glycoprotein linkages. Biochem J 171:665–674

Anderson RL, Clarke AE, Jermyn MA, Knox RB, Stone BA (1977) A carbohydrate-binding arabinogalactan-protein from liquid suspension cultures of endosperm from *Lolium multiflorum*. Aust J Plant Physiol 4:143–158

Aronson JM, Klapprott JA, Lin CC (1969) Chemical composition of the cell walls of some green algae. XIth Int Bot Congr Abs p 5 Seattle

Arrigoni O, Arrigoni-Liso R, Calabrese G (1977a) Ascorbic acid requirement for biosynthesis of hydroxyproline-containing proteins in plants. FEBS Lett 82:135–138

Arrigoni O, DeSantis A, Arrigoni-Liso R, Calabrese G (1977b) The increase of hydroxyproline-containing proteins in Jerusalem artichoke mitochondria during the development of cyanide-insensitive respiration. Biochem Biophys Res Commun 74:1637–1641

Ashford D, Neuberger A (1980) 4-Hydroxy-L-proline in plant glycoproteins. Where does it come from and what is it doing there? Trends Biochem Sci 5:245–248

Bailey RW, Kauss H (1974) Extraction of hydroxyproline-containing proteins and pectic substances from cell walls of growing and non-growing mung bean hypocotyl segments. Planta 119:233–245

Baker DA (1978) Proton co-transport of organic solutes by plant cells. New Phytol 81:485–497

Barnett NM (1970) Dipyridyl-induced cell elongation and inhibition of cell wall hydroxyproline biosynthesis. Plant Physiol 45:188–191

Basile DV (1973) Hydroxy-L-proline and 2,2′-dipyridyl-induced phenovariation in the liverwort *Jungermannia lanceolata*. Bull Torrey Bot Club 100:350–352

Basile DV (1979) Hydroxyproline-induced changes in form, apical development, and cell wall protein in the liverwort *Plagiochilla arctica*. Am J Bot 66:776–783

Bateman DF (1976) Plant cell wall hydrolysis by pathogens. In: Friend J, Threlfall DR

(eds) Biochemical aspects of plant-parasitic relationships Ann Proc Phytochem Soc 13:70–103
Bieleski RL (1974) Development of an externally-located alkaline phosphatase as a response to phosphorus deficiency. In: Ferguson AR, Cresswell MM (eds) Mechanism of regulation of plant growth. R Soc NZ Bull pp 165–170
Brett CT (1978) Synthesis of beta-(1→3)-glucan from extracellular uridine diphosphate glucose as a wound response in suspension-cultured soybean cells. Plant Physiol 62:377–382
Brown RM Jr, Franke WW, Kleinig M, Sitte P (1970) Scale formation in Chrysophycean algae I. Cellulosic and non-cellulosic wall components made by the Golgi apparatus. J Cell Biol 45:246–271
Brysk MM, Chrispeels MJ (1972) Isolation and partial characterization of a hydroxyproline-rich cell wall glycoprotein and its cytoplasmic precursor. Biochim Biophys Acta 257:421–432
Burg SP, Burg EA (1966) The interaction between auxin and ethylene and its role in plant growth. Proc Natl Acad Sci USA 55:262–269
Carpita N, Sabularse D, Montezinos D, Delmer DP (1979) Determination of the pore size of cell walls of living plant cells. Science 205:1144–1147
Catt JW, Hills GJ, Roberts K (1976) A structural glycoprotein containing hydroxyproline, isolated from the cell walls of *Chlamydomonas reinhardtii*. Planta 131:165–171
Catt JW, Hills GJ, Roberts K (1978) Cell wall glycoproteins from *Chlamydomonas reinhardtii*, and their self assembly. Planta 138:91–98
Chang CW, Bandurski RS (1964) Exocellular enzymes of corn roots. Plant Physiol 39:60–64
Cho YP, Chrispeels MJ (1976) Serine-*O*-galactosyl linkages in glycopeptides from carrot cell walls. Phytochemistry 15:165–169
Ciarlante D, Curtis RW (1977) Isolation and characterization of the cell wall receptor of ^{14}C-malformin in *Phaseolus vulgaris* L. Plant Cell Physiol 18:225–234
Clarke AE, Knox RB (1978) Cell recognition in flowering plants. Quart Rev Biol 53:3–28
Clarke AE, Anderson RL, Stone BA (1979a) Form and function of arabinogalactans and arabinogalactan-proteins. Phytochemistry 18:521–540
Clarke A, Gleeson P, Harrison S, Knox RB (1979b) Pollen-stigma interactions: identification and characterization of surface components with recognition potential. Proc Natl Acad Sci USA 76:3358–3362
Clarke JA, Lisker N, Ellingboe AH, Lamport DTA (1981) Hydroxyproline enhancement as a primary event in the successful development of Erysiphe in wheat. Plant Physiol 67:188–189
Cleland R, Karlsnes AM (1967) A possible role of hydroxyproline-containing proteins in the cessation of cell elongation. Plant Physiol 42:669–671
Coombe BG (1976) The development of fleshy fruits. Annu Rev Plant Physiol 27:507–28
Coombs J, Volcani BE (1968) Studies on the biochemistry and fine structure of silica shell formation in diatoms. Chemical changes in the wall of *Navicula pelliculosa* during its formation. Planta 82:280–292
Crasnier M, Noat G, Ricard J (1980) Purification and molecular properties of acid phosphatase from sycamore cell walls. Plant Cell Environ 3:217–224
Darvill AG, Smith CJ, Hall MA (1978) Cell wall structure and elongation growth in *Zea mays* coleoptile tissue. New Phytol 80:503–516
Dashek WV (1970) Synthesis and transport of hydroxyproline-rich components in suspension cultures of sycamore-maple cells. Plant Physiol 46:831–838
Dashek WV, Erickson SS, Hayward DM, Lindbeck G, Mills RR (1979) Peroxidase in cytoplasm and cell wall of germinating lily pollen. Bot Gaz 140:261–265
Davies DR, Lyall V (1973) The role of heritable factors and of physical structure. Molec Gen Genet 124:21–34
Davies DR, Roberts K (1976) Genetics of cell wall synthesis in *Chlamydomonas reinhardtii*. In: Levin RA (ed) Genetics of the algae botanical monographs, Vol 12, Univ California Press, pp 63–68
Dayhoff MO, Orcutt BC (1979) Methods for identifying proteins by using partial sequences. Proc Natl Acad Sci USA 76:2170–2174

Dazzo FB (1980) Adsorption of microorganisms to roots and other plants surfaces. In: Britton G, Marshall KC (eds) Adsorption of microorganisms to surfaces. Willey and Sons, New York, pp 1–51

Delmer DP, Lamport DTA (1976) The origin and significance of plant glycoproteins. In: Solheim B, Raa J (eds) Cell wall biochemistry related to specificity in host-plant pathogen interactions. Columbia Univ Press, New York, pp 85–104

Dey PM (1973) β-L-arabinosidase from Cajanus indicus: a new enzyme. Biochim Biophys Acta 302:393–398

Dey PM, Pridham JB (1968) Multiple forms of α-galactosidase in *Vicia faba* seeds. Phytochemistry 7:1737–1739

Dodge JD (1973) The cell covering. In: The fine structure of algal cells, 1–251, Academic Press, London New York, pp 21–25

Dougal DK, Shimbayashi K (1960) Factors affecting growth of tobacco callus tissue and its incorporation of tyrosine. Plant Physiol 35:396–404

Elstner EF, Heupel A (1976) Formation of hydrogen peroxide by isolated cell walls from horseradish (*Armoracia lapathifolia* Gilib). Planta 130:175–180

Esquerré-Tugayé MT, Lamport DTA (1979) Cell surfaces in plant microorganism interactions. I A structural investigation of cell wall hydroxyproline-rich glycoproteins which accumulate in fungus-infected plants. Plant Physiol 64:314–319

Esquerré-Tugayé MT, Mazau D (1974) Effect of a fungal disease on extensin, the plant cell wall glycoprotein. J Exp Bot 25:509–513

Esquerré-Tugayé MT, Lafitte C, Mazau D, Toppan A, Touzé A (1979) Cell surfaces in plant-microorganism interactions. II. Evidence for the accumulation of hydroxyproline-rich glycoproteins in the cell wall of diseased plants as a defense mechanism. Plant Physiol 64:320–326

Ettl v H (1971) *Chlamydomonas,* geeigneter Modellorganismen für vergleichende cytomorphologische Untersuchungen. Arch Hydrobiol Suppl 39, Algal Stud 5:259–300

Evans LV (1970) Electron microscopical observations on a new red algal unicell *Rhodella maculata* gen nov sp no. Br Phycol J 5:1–13

Evans ML (1974) Rapid responses to plant hormones. Annu Rev Plant Physiol 25:195–223

Fan DF, Maclachlan GA (1967) Studies on the regulation of cellulase activity and growth in excised pea epicotyl sections. Can J Bot 45:1837–1844

Fincher GB, Sawyer WH, Stone BA (1974) Chemical and physical properties of an arabinoglactan-peptide from wheat endosperm. Biochem J 139:535–545

Fry SC (1979) Phenolic components of the primary cell wall and their possible role in the hormonal regulation of growth. Planta 146:343–351

Fujii T, Shimokoriyama M (1976) A possible role of hydroxyproline-protein in oxygen-sensitive growth. Plant Cell Physiol 17:483–492

Fukuda M, Yamaguchi T (1979) Quantitative changes and distribution of free hydroxyproline in sweet pepper fruits during maturation. Agric Biol Chem 43:1145–1146

Fulton AB (1978) Colonial development in *Pandorina morum* I. Structure and composition of the extracellular matrix. Dev Biol 64:224–235

Gardiner M, Chrispeels MJ (1975) Involvement of the Golgi apparatus in the synthesis and secretion of hydroxyproline-rich cell wall glycoproteins. Plant Physiol 55:536–541

Giebel J, Stobiecka M (1974) Role of amino acids in plant tissue response to *Heterodera rostochiensis*. I. Protein-proline and hydroxyproline content in roots of susceptible and resistant solanaceous plants. Nematologica 20:407–414

Gilkes NR, Hall MA (1977) The hormonal control of cell wall turnover in *Pisum sativum* L. New Phytol 78:1–15

Gleeson PA, Clarke AE (1979) Structural studies on the major component of *Gladiolus* style mucilage, an arabinogalactan protein. Biochem J 181:607–621

Gleeson PA, Jermyn MA (1979) Alteration in the composition of β-lectins caused by chemical and enzymic attack. Aust J Plant Physiol 6:25–38

Goeschl JD, Pratt HK, Bonner BA (1967) An effect of light on the production of ethylene and the growth of the plumular portion of etiolated pea seedlings. Plant Physiol 42:1077–1088

Gotelli IB, Cleland R (1968) Differences in the occurrence and distribution of hydroxyproline-proteins among the algae. Am J Bot 55:907–914

Griesbach H (1977) Biochemistry of lignification. Naturwissenschaften 64:619–625

Gross GG (1977) Cell wall-bound malate dehydrogenase from horseradish. Phytochemistry 16:319–321

Gross GG, Janse C (1977) Formation of NADH and hydrogen peroxide by cell wall-associated enzymes from *Forsythia* xylem. Z Pflanzenphysiol 84:447–452

Gross GG, Janse C, Elstner EF (1977) Involvement of malate, monophenols, and the superoxide radical in hydrogen peroxide formation by isolated cell walls from horseradish (*Armoracia lapathifolia* Gilib.). Planta 136:271–276

Gunnison D, Alexander M (1975) Basis for the susceptibility of several algae to microbial decomposition. Can J Microbiol 21:619–628

Hager A, Menzel H, Krauss A (1971) Versuche und Hypothese zur Primärwirkung des Auxins beim Streckungswachstum. Planta 100:47–75

Hammerschmidt R, Kuc J (1980) Enhanced peroxidase activity and lignification in the induced systemic protection of cucumber. Phytopathol 70:689

Harris PJ, Hartley RD (1976) Detection of bound ferulic acid in cell walls of the Gramineae by ultraviolet fluorescence microscopy. Nature 259:508–510

Heaney-Kieras J, Roden L, Chapman DJ (1977) The covalent linkage of protein to carbohydrate in the extracellular protein-polysaccharide from the red alga *Porphyridium cruentum*. Biochem J 165:1–9

Heath MF, Northcote DH (1971) Glycoprotein of the wall of sycamore tissue-cultured cells. Biochem J 125:953–961

Heath OVS, Clark JE (1956) Chelating agents as plant growth substances. A possible clue to the mode of action of auxin. Nature 177:118–1121

Herth W, Franke WW, Stadler J, Bittiger H, Keilick G, Brown RM Jr (1972) Further characterization of the alkali stable material from the scales of *Pleurochrysis scherffelii:* a cellulosic glycoprotein. Planta 105:79–92

Heyn ANJ (1940) The physiology of cell elongation. Bot Rev 6:515–574

Heyn ANJ (1969) Glucanase activity in coleoptiles of *Avena*. Arch Biochem Biophys 132:442–449

Hills GJ (1973) Cell walll assembly in vitro from *Chlamydomonas reinhardtii*. Planta 115:17–23

Hills GJ, Phillips JM, Gay MR, Roberts K (1975) Self assembly of a plant cell wall *in vitro*. J Mol Biol 96:431–441

Holleman J (1967) Direct incorporation of hydroxyproline into protein of sycamore cells incubated at growth-inhibitory levels of hydroxyproline. Proc Natl Acad Sci USA 57:50–54

Homer RB, Roberts K (1979) Glycoprotein conformation in plant cell walls. Circular dichroism reveals a polyproline II structure. Planta 146:217–222

Hori H (1978) The demonstration of galactosamine in extracellular hydroxyproline-rich macromolecule purified from the culture media of tobacco cells. Plant Cell Physiol 19:501–505

Huber DJ, Nevins DJ (1979) Autolysis of the cell wall β-D-glucan in corn coleoptiles. Plant Cell Physiol 20:201–212

Jermyn MA (1974) A class of lectins widespread in the flowering plants. Proc Aust Biochem Soc 7:32

Jermyn MA (1978) Isolation from the flowers of *Dryandra praemorsa* of a flavonol glycoside that reacts with β-lectins. Aust J Plant Physiol 5:697–705

Jermyn MA, Yeow YM (1975) A class of lectins present in the tissues of seed plants. Aust J Plant Physiol 2:501–531

Jiang K, Barber GA (1975) Polysaccharide from cell walls of *Chlamydomonas reinhardtii*. Phytochemistry 14:2459–2461

Johnson KD, Daniels D, Dowler MJ, Rayle DL (1974) Activation of *Avena* coleoptile cell wall glycosidases by hydrogen ions and auxin. Plant Physiol 53:224–228

Jones RF (1962) Extracellular mucilage of the red algae *Porphyridium cruentum*. J Cell Comp Physiol 60:61–64

Jotterand-Dolivo MC, Pilet PE (1976) Wall hydroxyproline and growth of georeactive roots (*Zea mays* L.) Experientia 32:874–875

Kauss H, Glaser C (1974) Carbohydrate-binding proteins from plant cell walls and their possible involvement in extension growth. FEBS Letts 45:304–307

Kawasaki S, Sato S (1979) Isolation of the Golgi apparatus from suspension cultured tobacco cells and preliminary observation on the intracellular transport of extensin-precursor. Bot Mag 92:305–314

Keegstra K, Albersheim P (1970) The involvement of glycosidases in the cell wall metabolism of suspension cultured *Acer pseudoplatanus* cells. Plant Physiol 45:675–678

Keegstra K, Talmadge KW, Bauer WD, Albersheim P (1973) The structure of plant cell walls. III. A model of the walls of suspension-cultured sycamore cells based on the interconnections of the macromolecular components. Plant Physiol 51:188–196

Kimmins WC, Brown RG (1975) Effect of a non-localized infection by southern bean mosaic virus on a cell wall glycoprotein from bean leaves. Phytopathology 65:1350–1351

Klis FM (1976) Glycosylated seryl residues in wall protein of elongating pea stems. Plant Physiol 57:224–226

Klis FM, Eeltink H (1979) Changing arabinosylation patterns of wall-bound hydroxyproline in bean cell cultures. Planta 144:479–484

Kochert G (1975) Developmental mechanisms of *Volvox* reproduction. In: Markert CL, Papaconstantinou J (eds) The developmental biology of reproduction (33rd Symp Soc Dev Biol) Academic Press, New York, pp 55–90

Kokkinakis DM, Brooks JL (1979) Hydrogen peroxide-mediated oxidation of indole-3-acetic acid by tomato peroxidase and molecular oxygen. Plant Physiol 64:220–223

Labavitch JM, Ray PM (1974) Turnover of cell wall polysaccharides in elongating pea stem segments. Plant Physiol 53:669–673

Lamport DTA (1963) Oxygen fixation into hydroxyproline of plant cell wall protein. J Biol Chem 238:1438–1440

Lamport DTA (1964) Cell suspension cultures of higher plants: isolation and growth energetics. Exp Cell Res 33:195–206

Lamport DTA (1965) The protein component of primary cell walls. In: Preston RD (ed) Advances in Botanical Research, Vol 2 Academic Press, New York, pp 151–218

Lamport DTA (1967) Hydroxyproline-*O*-glycosidic linkage of the plant cell wall glycoprotein extensin. Nature 216:1322–1324

Lamport DTA (1969) The isolation and partial characterization of hydroxyproline-rich glycopeptides obtained by enzymic degradation of primary cell walls. Biochemistry 8:1155–1163

Lamport DTA (1970) Cell wall metabolism. Annu Rev Plant Physiol 21:235–270

Lamport DTA (1973) Is the primary cell wall a protein-glycan network? Colloq Int CNRS 212:26–31

Lamport DTA (1974) The role of hydroxyproline-rich proteins in the extracellular matrix of plants. In: Hay ED, King TJ, Papaconstantinou J (eds) Macromolecules regulating growth and development. 30th Symp Soc Dev Biol, pp 113–130

Lamport DTA (1977) Structure, biosynthesis and significance of cell wall glycoproteins. In: Loewus FA, Runeckles VC (eds) The structure, biosynthesis and degradation of wood. (Recent Adv Phytochem, Vol 11) Plenum Press, New York, pp 79–115

Lamport DTA (1978) Cell wall carbohydrates in relation to structure and function. In: Thorpe TA (ed) Frontiers of plant tissue culture 1978, Proc 4th Int Cong Plant Tissue Cell Culture, Calgary, Canada, pp 235–244

Lamport DTA (1980) Structure and function of plant glycoproteins. In: Stumpf PK, Conn EE (eds) The biochemistry of plants. Vol 3 Preiss J (ed) Carbohydrates: structure and function. Academic Press, New York, pp 501–541

Lamport DTA, Miller DH (1971) Hydroxyproline arabinosides in the plant kingdom. Plant Physiol 48:454–456

Lamport DTA, Northcote DH (1960a) Hydroxyproline in primary cell walls of higher plants. Nature 188:665–666

Lamport DTA, Northcote DH (1960b) The use of tissue cultures for the study of plant cell walls. Biochem J 76:52P

Lamport DTA, Katona L, Roerig S (1973) Galactosylserine in extensin. Biochem J 133:125–131

Larkin PJ (1978) Plant protoplast agglutination by artificial carbohydrate antigens. J Cell Sci 30:283–292

Lang W (1976) Biosynthesis of extensin during normal and light-inhibited elongation of radish hypocotyls. Z Pflanzenphysiol 78:228–235

Lee SH, Kivilaan A, Bandurski RS (1967) In vitro autolysis of plant cell walls. Plant Physiol 42:968–972

Lieberman M (1979) Biosynthesis and action of ethylene. Annu Rev Plant Physiol 30:533 91

Loewus F, Labarca C (1973) Pistil secretion product and pollen tube wall formation. In: Loewus F (ed) Biogenesis of plant cell wall polysaccharides. Academic Press, New York, pp 175–193

Loomis WD (1974) Overcoming problems of phenolics and quinones in the isolation of plant enzymes and organelles. Methods Enz 31:528–544

Mader M, Meyer Y, Bopp M (1976) Zellwandregeneration und Peroxidase-Isoenzym-Synthese isolierter Protoplasten von *Nicotiana tabacum* L. Planta 129:33–38

Malhotra OP, Singh H (1976) Enzymic hydrolysis of α-galactosides. Ind J Biochem Biophys 13:208–212

Mani UV, Akiyama Y, Mohrlok S, Lamport DTA (1979a) Evidence for a new glycopeptide linkage, glucosyl hydroxyproline, in arabinogalactan proteins. Plant Physiol 63 Suppl:31, Abstr 167

Mani UV, Akiyama Y, Mohrlok S, Lamport DTA (1979b) Deglycosylation via HF solvolysis of arabinogalactan proteins (AGPs) from sycamore cultures. Fed Am Soc Exp Biol 38:Part 1, 418 Abstr 995

Mankarios AT, Hall MA, Jarvis MC, Threlfall DR, Friend J (1980) Cell wall polysaccharides from onions. Phytochemistry 19:1731–1733

Margulis L (1974) The classification and evolution of prokaryotes and eukaryotes. In: King RC (ed) Handbook of Genetics. Plenum Press, New York, Vol 1, pp 1–41

Markwalder HU, Neukom H (1976) Diferulic acid as a possible crosslink in hemicelluloses from wheat germ. Phytochemistry 15:836–837

Mascarenhas JP (1975) The biochemistry of angiosperm pollen development. Bot Rev 41:259–314

McNeil M, Darvill AG, Albersheim P (1979) The structural polymers of the primary cell walls of dicots. Prog Chem Org Nat Prod 37:191–249

Merkel GJ, Durham DR, Perry JJ (1980) The atypical cell wall composition of *Thermomicrobium roseum*. Can J Microbiol 26:556–559

Miller DH (1978) Cell wall chemistry and ultrastructure of *Chlorococcum oleofaciens* (Chlorophyceae). J Phycol 14:189–194

Miller DH, Mellman IS, Lamport DTA, Miller M (1974) The chemical composition of the cell wall of *Chlamydomonas gymnogama* and the concept of cell wall protein. J Cell Biol 63:420–429

Mitchell B (1980) Polyhydroxyproline in the extracellular matrix of Volvox. Masters Thesis: Michigan State Univ, East Lansing, pp 1–73

Monro JA, Bailey RW, Penny D (1972) Polysaccharide composition in relation to extensibility and possible peptide linked arabino-galactan of lupin hypocotyl cell walls. Phytochemistry 11:1597–1602

Monro JA, Bailey RW, Penny D (1974) Cell wall hydroxyproline-polysaccharide associations in *Lupinus* hypocotyls. Phytochemistry 13:375–382

Monro JA, Bailey RW, Penny D (1976) Hemicellulose fractions and associated protein of *Lupinus* hypocotyl cell walls. Phytochemistry 15:175–181

Morita Y, Kominato Y, Shimizu K (1967) Studies on phyto-peroxidase Part XIX. Some further aspects of oxidation of indole-3-acetic acid by peroxidase. Res Inst Food Sci Kyoto Univ 28:1–17

Mort AJ (1978) Partial characterization of extensin by selective degradation of cell walls. Ph D dissertation, Michigan State Univ, East Lansing, pp 1–169
Mort AJ, Lamport DTA (1977) Anhydrous hydrogen fluoride degycosylates glycoproteins. Anal Biochem 82:289–309
Muray RHA, Northcote DH (1978) Oligoarabinosides of hydroxyproline isolated from potato lectin. Phytochemistry 17:623–629
Mussell H, Strand LL (1977) Pectic enzymes: involvement in pathogenesis and possible relevance to tolerance and specificity. In: Solheim B, Raa J (eds) Cell wall biochemistry related to specificity in host-plant pathogen interactions. Columbia Univ Press, New York, pp 31–70
Myllyla R, Tuderman L, Kivirikko KI (1977) Mechanism of the prolyl hydroxylase reaction. 2 Kinetic analysis of the reaction sequence. Eur J Biochem 80:349–357
Myllyla R, Kuutti-Savolainen ER, Kivirikko KI (1978) The role of ascorbate in the prolyl hydroxylase reaction. Biochem Biophys Res Commun 83:441–448
Myllyla R, Schubotz LM, Weser U, Kivirikko KI (1979) Involvement of superoxide in the prolyl and lysyl hydroxylase reactions. Biochem Biophys Res Commun 89:98–102
Nevins DJ (1970) Relation of glycosidases to bean hypocotyl growth. Plant Physiol 46:458–462
Nevins D, Loescher W (1974) Nature of the glucan released from Avena walls during auxin-induced growth. In: Plant Growth Subst, Proc Int Conf 8th, Hirokawa Publ Co, Tokyo pp 828–837
Nevo Z, Sharon N (1969) The cell wall of *Peridinium westii*, a non cellulosic glucan. Biochim Biophys Acta 173:161–175
Northcote DH, Goulding KJ, Horne RW (1960) The chemical composition and structure of the cell wall of *Hydrodictyon africanum* (Yaman). Biochem J 77:503–508
O'Neill MA, Selvendran RR (1980) Glycoproteins from the cell wall of *Phaseolus coccineus*. Biochem J 187:53–63
Palmieri S, Odoardi M, Soressi GP, Salamini F (1978) Indoleacetic acid oxidase activity in two high-peroxidase tomato mutants. Physiol Plant 42:85–90
Paull RE, Jones RL (1978) Regulation of synthesis and secretion of fucose-containing polysaccharides in cultured sycamore cells. Aust J Plant Physiol 5:457–67
Pegg GF (1976) The response of ethylene-treated tomato plants to infection by *Verticillium albo-atrum*. Physiol Plant Pathol 9:215–226
Perley JE, Penny D (1974) The effect of some aldonolactones on the IAA-induced growth of lupin hypocotyl segments. J Bot 12:503–12
Petek F, Villarroya E, Courtois JE (1969) Purification proprietes de l'alpha-galactosidase des graines germees de *Vicia sativa*. Eur J Biochem 8:395–402
Pike CS, Un H, Lystash JC, Showalter AM (1979) Phytochrome control of cell wall-bound hydroxyproline content in etiolated pea epicotyl. Plant Physiol 63:444–449
Pope DG (1977) Relationships between hydroxyproline-containing proteins secreted into the cell wall and medium by suspension-cultured *Acer pseudoplatanus* cells. Plant Physiol 59:894–900
Pope DG, Lamport DTA (1974) Hydroxyproline-rich material secreted by cultured *Acer pseudoplatanus* cells: Evidence for polysaccharide attached directly to hydroxyproline. Am Soc Plant Physiol 54:Suppl p 15
Pressey R, Avants JK (1977) Occurrence and properties of polygalacturonase in *Avena* and other plants. Plant Physiol 60:548–553
Punnett T, Derrenbacker EC (1966) The amino acid composition of algal cell walls. J Gen Microbiol 44:105–114
Quail PH (1979) Plant cell fractionation. Annu Rev Plant Physiol 30:425–84
Raa J, Robertsen B, Solheim B, Tronsmo A (1977) Cell surface biochemistry related to specificity of pathogenesis and virulence of microorganisms. In: Solheim B, Raa J (eds) Cell wall biochemistry related to specificity in host-plant pathogen interactions. Columbia Univ Press, New York, pp 11–30
Rao NV, Adams E (1978) Partial reaction of prolyl hydroxylase. $(Gly\text{-}Pro\text{-}Ala)_n$ stimulates

alpha-ketoglutarate decarboxylation without prolyl hydroxylation. J Biol Chem 253:6327–6330
Rapaka RS, Sorensen KR, Lee SD, Bhatnagar RS (1976) Inhibition of hydroxyproline synthesis by palladium ions. Biochim Biophys Acta 429:63–71
Ray DA, Solter KM, Gibor (1978) Flagellar surface differentiation: evidence for multiple sites involved in mating of *Chamydomonas reinhardtii.* Exptl Cell Res 114:185–189
Rees DA (1977) Polysaccharides shapes. In: Ashworth JM (ed) Polysaccharide shapes. Chapman and Hall, Wiley and Sons, New York, pp 6–80
Reichert TA, Yu JMC, Christensen RA (1976) Molecular evolution as a process of message refinement. J Mol Evol 8:41–54
Ridge I, Osborne DJ (1970) Hydroxyproline and peroxidases in cell walls of *Pisum sativum*: regulation by ethylene. J Exp Bot 21:843–856
Ridge I, Osborne DJ (1971) Role of peroxidase when hydroxyproline-rich protein in plant cell walls is increased by ethylene. Nature London New Biol 229:205–208
Roberts K (1974) Crystalline glycoprotein cell walls of algae: their structure, composition and assembly. Philos Trans R Soc London Ser B 268:129–146
Roberts K (1979) Hydroxyproline its asymmetric distribution in a cell wall glycoprotein. Planta 146:275–280
Roberts K, Northcote DH (1972) Hydroxyproline: Observations on its chemical and autoradiographical localization in plant cell wall protein. Planta 107:43–51
Roberts K, Gurney-Smith M, Hills GJ (1972) Structure, composition and morphogenesis of the cell wall of *Chlamydomonas reinhardtii* I. Ultrastructure and preliminary chemical analysis. J Ultrastruct Res 40:599–613
Roberts K, Gay MR, Hills GJ (1980) Cell wall glycoproteins from *Chlamydomonas reinhardtii* are sulfated. Physiol Plant 49:421–424
Roland JC, Vian B (1979) The wall of the growing plant cell: Its three-dimensional organization. Int Rev Cytobiol 61:129–166
Rosenfield CL, Loewus FA (1975) Carbohydrate interconversions in pollen-pistil interactions of the lily. In: Mulcahy DL (ed) Gamete competition in plants and animals. North-Holland, Amsterdam, pp 151–160
Sadava D, Chrispeels MJ (1971) Hydroxyproline biosyntheis in plant cells peptidyl proline hydroxylase from carrot disks. Biochim Biophys Acta 227:278–287
Sadava D, Volcani BE (1977) Studies on the biochemistry and fine structure of silica shell formation in diatoms: formation of hydroxyproline and dihydroxyproline in *Nitzchia angularis.* Planta 135:7–11
Sadava D, Walker F, Chrispeels MJ (1973) Hydroxyproline-rich cell wall protein (extensin): biosynthesis and accumulation in growing pea epicotyls. Dev Biol 30:42–48
Schlösser UG (1966) Enzymatisch gesteuerte Freisetzung von Zoosporen bei *Chlamydomonas reinhardtii* (Dangeard) in Synchronkultur. Arch Mikrobiol 54:129–159
Sequeira L (1978) Lectins and their role in host-pathogen specificity. Annu Rev Phytopathol 16:453–481
Sharma CB, Sharma TN (1977) Multiple forms of α-galactosidase in chick pea seedlings. Phytochemistry 16:1053–1054
Simson BW, Timell TE (1978) Polysaccharides in cambial tissues of *Populus tremuloides* and *Tilia americana.* III. Isolation and constitution of an arabinogalactan. Cellulose Chem Technol 12:63–77
Smith M (1981 a) Characterization of carrot cell wall protein. I. The effect of α-α'-dipyridyl on cell wall protein synthesis and secretion in incubated carrot disks. Plant Physiol (in press)
Smith M (1981 b) Characterization of carrot cell wall protein. II. Immunological study of cell wall protein. Plant Physiol (in press)
Stahmann MA, Spencer AK (1977) Deamination of protein lysyl epsilon amino groups by peroxidase *in vitro*. Biopolymers 16:1299–1306
Stephens GJ, Woods RKS (1974) Release of enzymes from cell walls by an endopectate-trans-eliminase. Nature 251:358

Steward FC, Israel HW, Salpeter MM (1967) The labeling of carrot cells with ^{3}H-Proline: Is there a cell-wall protein? Proc Natl Acad Sci USA 58:541–544
Steward FC, Israel HW, Salpeter MM (1974) The labeling in cultured cells of Acer with [^{14}C] proline and its significance. J Cell Biol 60:695–701
Stuart DA, Varner JE (1980) Purification and characterization of a salt-extractable hydroxyproline-rich glycoprotein from aerated carrot discs. Plant Physiol 66:787–792
Stuart DA, Mozer TV, Varner JE (1980) Induction of a hydroxyproline-rich glycoprotein and isolation of poly C-rich, poly A RNA from aerated carrot root discs. Plant Physiol 65 Suppl:106, Abst 583
Talmadge KW, Keegstra K, Bauer WD, Albersheim P (1973) The structure of plant cell walls. I. The macromolecular components of the wall of suspension-cultured sycamore cells with a detailed analysis of the pectic polysaccharides. Plant Physiol 51:158–173
Tanaka M, Uchida T (1978) Purification and properties of α-L-arabinofuranosidase from plant *Scopolia japonica* calluses. Biochim Biophys Acta 522:531–540
Tanaka M, Shibata H, Uchida T (1980) A new prolyl hydroxylase acting on poly-L-proline, from suspension cultured cells of *Vinca rosea.* Biochim Biophys Acta 616:188–198
Tanimoto E, Igari M (1976) Correlation between β-galactosidase and auxin-induced elongation growth in etiolated pea stems. Plant Cell Physiol 17:673–682
Tanimoto E, Pilet PE (1978) α- and β-glycosidases in maize roots. Planta 138:119–122
Tautvydas KJ (1978) Isolation and characterization of an extracellular hydroxyproline-rich glycoprotein and a mannose-rich polysaccharide from *Eudorina californica* (Shaw). Planta 140:213–220
Taylor IEP (1969) Hydroxyproline containing protein in wall preparations of *Chlamydomonas moewusii.* XIth Int Bot Congr Abs p 125
Thomas B, Webb JA (1977) Multiple forms of α-galactosidase in mature leaves of *Cucurbita pepo.* Phytochemistry 16:203–206
Thomas B, Webb JA (1979) Intracellular distribution of α-galactosidase in leaves of *Cucurbita pepo.* Can J Bot 57:1904–1911
Thompson RD, Preston EW (1967) Proteins in cell walls of some green algae. Nature 212:684–685
Tripp VW, Moore AT, Rollins ML (1951) Some observations on the primary wall of the cotton fiber. Textile Res 21:886–894
Vance CP, Kirk TK, Sherwood RT (1980) Lignification as a mechanism of disease resistance. Annu Rev Phytopathol 18:259–288
Wallner SJ, Walker JE (1975) Glycosidases in cell wall-degrading extracts of ripening tomato fruits. Plant Physiol 55:94–98
Whitmore FW (1976) Binding of ferulic acid to cell walls by peroxidases of *Pinus elliottii.* Phytochemistry 15:375–378
Whitmore FW (1978a) Lignin-protein complex catalyzed by peroxidase. Plant Sci Letts 13:241–245
Whitmore FW (1978b) Lignin-carbohydrate complex formed in isolated cell walls of callus. Phytochemistry 17:421–425
Wilkins MB (1977) Gravity and light-sensing guidance systems in primary roots and shoots. In: Jennings DH (ed) Integration of activity in the higher plant. Soc Exp Biol Symp 31, Cambridge Univ Press, pp 310–321
Winter H, Wiersema ICM, Walbrecht DT, Buffinga H (1978) The role of glycosylated amino acids of wall-bound protein in cell extension of stems of *Pisum sativum* L. plants. Acta Bot Neerl 27:405–415
Winterburn PJ, Phelps CF (1972) The significance of glycosylated proteins. Nature 236:147–151
Woods EF, Lilley GG, Jermyn MA (1978) The self association of glycosyl phenylazo dyes (Yariv antigens). Aust J Chem 31:2225–2238
Wutz M, Zetsche K (1976) Zur Biochemie und Regulation des heteromorphen Generationswechsels der Grünalge *Derbesia-Halicystis.* Planta 129:211–216
Yamagishi T, Matsuda K, Watanabe T (1976) Characterization of the fragments obtained by enzymic and alkaline degradation of rice-bran proteoglycans. Carbohydr Res 50:63–74

Yang SF, Adams DO, Lizada C, Yu Y, Bradford KJ, Cameron AC, Hoffman NE (1980) Mechanism and regulation of ethylene biosynthesis. Skoog F (ed) Plant growth substances 1979. Springer, Berlin Heidelberg New York, pp 219–229

Yariv J, Rapport MM, Graf L (1962) The interaction of glycosides and saccharides with antibody to the corresponding phenylazo glycosides. Biochem J 85:383–388

Ycas M (1976) Origin of periodic proteins. Fed Proc Fed Am Soc Exp Biol 35:2139–2140

Yipp C (1964) The hydroxylation of proline by horseradish peroxidase Biochim Biophys Acta 92:395–396

Zaar K (1979) Peroxidase activity in root hairs of Cress (*Lepidium sativum* L.) Cytochemical localization and radioactive labelling of wall bound peroxidase. Protoplasma 99:263–274

Zacheo G, Lamberti F, Arrigoni-Liso R, Arrigoni O (1977) Mitochondrial protein-hydroxyproline content of susceptible and resistant tomatoes infected by *Meloidogyne incognita*. Nematologica 23:471–476

8 The Role of Lipid-Linked Saccharides in the Biosynthesis of Complex Carbohydrates

A.D. ELBEIN

1 Introduction

Lipid intermediates, or lipid-linked saccharides, were first discovered in bacteria where they were shown to be involved in the biosynthesis of cell wall polymers such as peptidoglycan, lipopolysaccharide, teichoic acid and capsular polysaccharides (OSBORN 1969, HEMMING 1974). Those studies set the stage for later experiments in eukaryotic systems which also showed the synthesis and utilization of similar types of lipid intermediates in the biosynthesis of complex carbohydrates (WAECHTER and LENNARZ 1976). In this chapter, the lipid intermediates that have been identified in higher plants will be discussed and their role in biosynthetic reactions will be considered. The plant systems will be compared to those studies that have been done in animal and yeast systems. Figure 1 outlines a series of reactions that have been proposed to account for the formation of the lipid-linked oligosaccharide that serves as the final donor in the glycosylation of protein. These reactions are considered in detail in the ensuing pages.

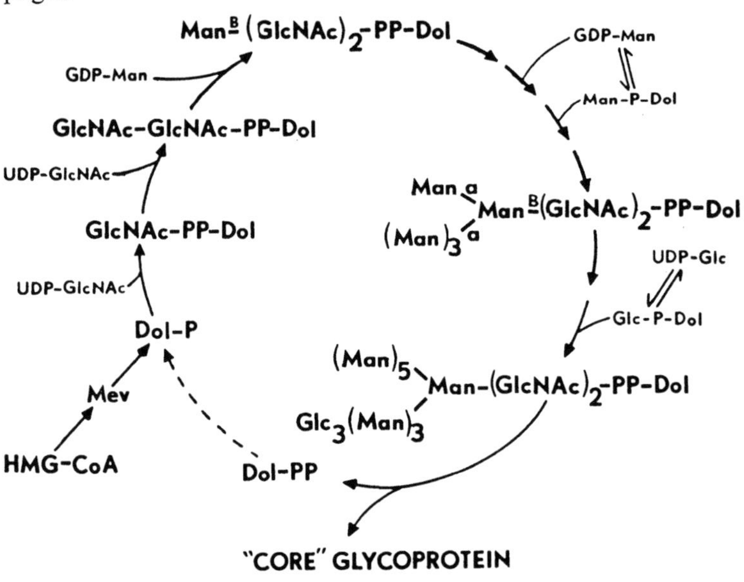

Fig. 1. Series of reactions in the lipid-linked saccharide pathway. In this pathway *N*-acetylglucosamine (GlcNAc)-1-P is first added to dolichyl-P, followed by the addition of a second GlcNAc. A number of mannose residues are then added to the lipid intermediate and in some systems glucose may also be transferred to the oligosaccharide. Finally, the oligosaccharide is transferred en bloc to the polypeptide chain

2 Nature of the Lipid Carrier

The lipid portion of the lipid-linked saccharides that serve as intermediates in glycosylation reactions are all of the polyisoprenyl type with the general structure shown in Fig. 2A. However, the lipids that act as carriers in the eukaryotic systems differ from those found in bacterial cells in at least two ways: (1) eukaryotic polyprenols are generally composed of 14 to 24 isoprene units whereas bacterial prenols are smaller and contain 10 to 12 isoprene units, and (2) eukaryotic polyprenols have a saturated α-isoprene unit while bacterial prenols have all unsaturated isoprene units (HEMMING 1974). Apparently these differences in structure confer unique properties on these lipid carriers. The name dolichol has been used to designate those lipids that have an α-saturated isoprene unit. Figure 2B presents the structure of dolichyl-phosphoryl-mannose (also referred to as mannosyl-phosphoryl-dolichol), the lipid intermediate which functions as a mannosyl donor in various reactions. This lipid-linked saccharide and others are discussed in more detail in other sections of this chapter.

Dolichols are found in most tissues as components of the membrane. Most of the dolichol in animal tissues is esterified to fatty acids, but some is also found as the free dolichol, and a small amount is present as dolichyl-phosphate and as lipid-linked saccharides. In yeast, the phosphorylated dolichols represented about 10 to 20% of the total dolichols of the cell (JUNG and TANNER 1973). Dolichols and other polyprenols are synthesized from acetyl CoA as shown in Fig. 3 (MARTIN and THORNE 1974). The synthesis proceeds along the pathway of cholesterol biosynthesis through hydroxymethylglutaryl CoA (HMG-CoA) and mevalonic acid to farnesyl-pyrophosphate (FPP). At this point there is a bifurcation in which FPP can be utilized either by squalene synthetase to form squalene and then cholesterol, or it can be used by a dolichyl-phosphate synthetase that condenses isopentyl-pyrophosphate and farnesyl-pyrophosphate to form 2,3-dehydrodolichyl-pyrophosphate. This compound then undergoes reduction and dephosphorylation to yield dolichyl-phosphate (GRANGE and

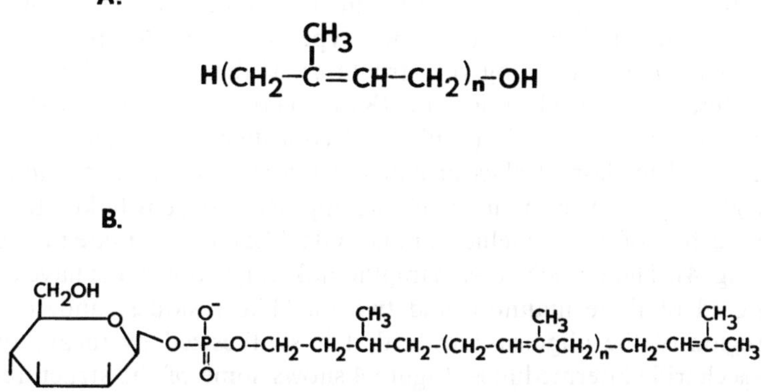

Fig. 2. A Structure of the polyisoprenyl-P (dolichyl-P) which serves as a carrier of sugars for the reactions in Fig. 1. **B** Structure of mannosyl-phosphoryl-dolichol (dolichyl-phosphoryl-mannose)

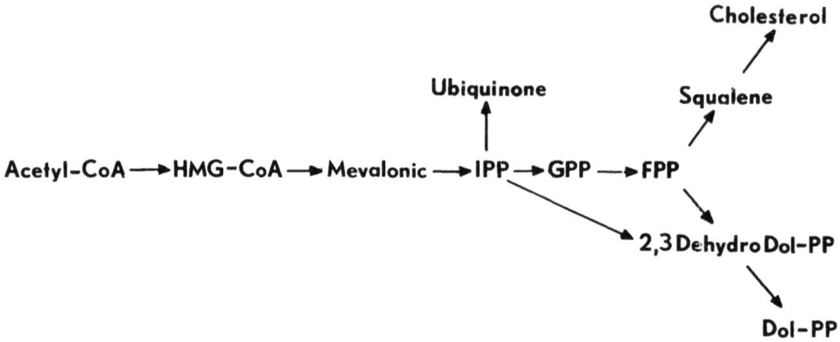

Fig. 3. Pathway for the synthesis of dolichyl-P from acetyl CoA

ADAIR 1977). The synthesis of cholesterol is prevented by feedback inhibition at the level of HMG-CoA reductase and also at the level of the squalene synthetase (GOLDSTEIN and BROWN 1977). There is some evidence to indicate that the synthesis of dolichyl-phosphate may be regulated at the level of HMG-CoA reductase or dolichyl-phosphate synthetase, or both (MILLS and ADAMANY 1978, KELLER et al. 1979). The reactions for the synthesis of dolichyl-phosphate and its regulation have not been as well documented in plant systems, but some studies indicate that they are probably similar (DALEO et al. 1977).

In addition to the de novo pathway, dolichyl-phosphate may also be derived from the phosphorylation of free dolichol. Thus, a kinase which utilizes CTP for the phosphorylation of free dolichol has been reported (ALLEN et al. 1978, WEDGEWOOD and STROMINGER 1980). Whether this reaction represents a salvage pathway, or what its exact function is, remains to be determined.

3 Glycoproteins or Other Complex Carbohydrates

Glycoproteins are widely distributed in nature and play important structural and functional roles in various types of cells. A great deal of information is available concerning the structures of the oligosaccharide portion of many different animal glycoproteins (KORNFELD and KORNFELD 1976), but considerably less is known about plant glycoproteins (SHARON and LIS 1978, ELBEIN 1979). The class of glycoproteins of interest in this discussion are those having a glycosylamine bond in which the oligosaccharide is linked through the anomeric carbon of *N*-acetylglucosamine (GlcNAc) to the amide nitrogen of asparagine (Fig. 4). These GlcNAc-asparagine-linked glycoproteins have a core region composed of three mannose and two GlcNAc residues, and it is apparently this region of the oligosaccharide that is synthesized by means of the lipid-linked saccharide intermediates. Figure 4 shows some of the structures of the oligosaccharides that are found in these glycoproteins. These structures are derived from animal cell glycoproteins, but they are probably similar in plant cells. Both the "high-mannose" and the "complex" type have the common core

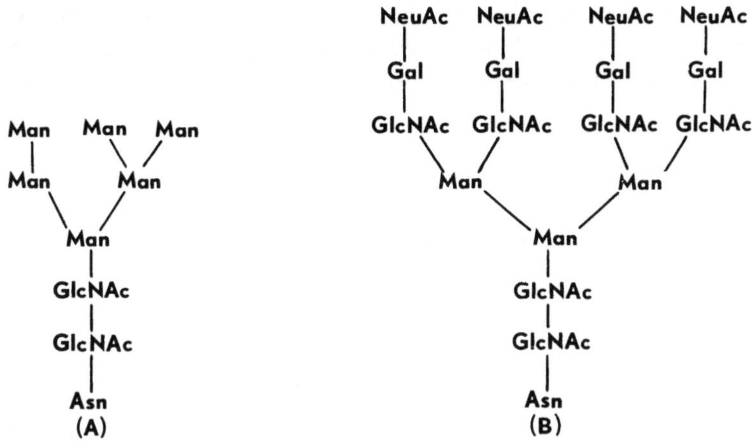

Fig. 4A, B. General structures of the oligosaccharide chains of GlcNAc→asparagine-linked glycoproteins. Both the "high mannose" and "complex" types of structures may be present in these proteins

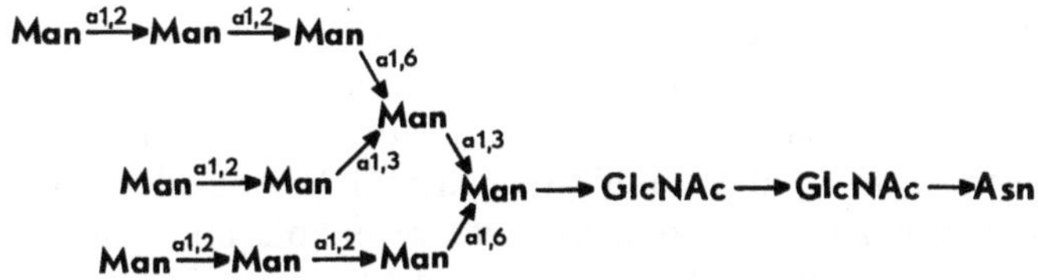

Fig. 5. Structure of the oligosaccharide portion of soybean lectin. (LIS and SHARON 1978)

of two GlcNAc residues in an *N,N'*-diacetylchititobiose unit to which is attached the branched mannose-trisaccharide. The first mannose is attached to the GlcNAc in a (1→4)-β-linkage while the next two mannoses are attached in (1→3)-α and (1→6)-α bonds. Beyond this point the oligosaccharide may vary greatly in structure, depending on the specific glycoprotein. The "high mannose" type may have varying amounts of mannose, whereas the complex type has the trisaccharide sequence; sialic acid-galactose-GlcNAc. Fucose or additional GlcNAc residues may also be attached to these complex oligosaccharides.

Although the complex type of oligosaccharide has not been reported in plant tissues, plants have been shown to have the "high mannose" type of glycoproteins. For example, soybean lectin has recently been characterized as a "high mannose" glycoprotein (LIS and SHARON 1978). The structure of the oligosaccharide portion of this glycoprotein was elucidated by the elegant studies of LIS and SHARON, and is shown in Fig. 5. The similarity between the first five sugars in this oligosaccharide (i.e., Manα1-3(Manα1-6)Manβ1-4GlcNAcβ1-4GlcNAc) and those presented in Fig. 4 is evident. Vicillin (ERICSON and CHRISPEELS 1973) and legumin (BASHA and BEEVERS 1976) may be other examples of this type of structure. In addition to glycoproteins, plant cells contain many other types of complex carbohydrates including cellulose, glucans, mannans, glucomannans, galactomannans, arabinogalactans, and so on. It is not known at this time whether the synthesis of these polymers also involves the participation of lipid-linked saccharide intermediates. However, these kinds of intermediates have been implicated in the synthesis of cellulose (HOPP et al. 1978a, b).

4 Lipid-Linked Monosaccharides

4.1 Mannose

The first indication that there might be transient types of sugar-containing lipids in eukaryotic cells came almost simultaneously in yeast (TANNER 1969), animals (CACCAM et al. 1969, ZATZ and BARONDES 1969), and plants (VILLAMEZ and CLARK 1969, KAUSS 1969). Basically these observations showed that upon incubation of GDP-^{14}C-mannose with membrane preparations of the various tissues, ^{14}C-mannose was transferred to compounds that were soluble in organic solvents such as chloroform:methanol (2:1 or 1:1). In those studies where the organic solvent-soluble material was examined in some detail, the mannolipid was found to be labile to mild acid hydrolysis but is was resistant to alkaline conditions that are generally used to hydrolyze acyl glycerides. The formation of this glycolipid was inhibited when GDP was added to the incubation mixtures, suggesting that the reaction was reversible and that the mannose was present in the mannolipid in an "activated" linkage. Furthermore, the kinetics of mannose incorporation showed that the transfer of radioactivity to lipid reached a maximum very rapidly and then declined, indicating that the mannolipid was being "turned over". These properties would be in keeping with a role as an intermediate in glycosylation reactions.

More detailed studies in animal (RICHARDS and HEMMING 1972) and plant (FORSEE and ELBEIN 1973) systems presented considerable evidence that these lipids were chemically related to the polyisoprenol types of intermediates that participate in the biosynthesis of peptidoglycan and lipopolysaccharide in bacteria (OSBORN 1969). The isolation and purification of sufficient amounts of these mannolipids from animals (BAYNES et al. 1973), yeast (JUNG and TANNER 1973), and plants (DELMER et al. 1978) allowed their definitive characterization by mass spectrometry as well as by infra red and nuclear magnetic resonance spectroscopy (EVANS and HEMMING 1973). The lipid was shown to be a dolichol and the mannolipid was characterized as dolichyl-phosphoryl-mannose (BAYNES et al. 1973, JUNG and TANNER 1973, DELMER et al. 1978, EVANS and HEMMING 1973). The structure of this compound is shown in Fig. 2B. Chemical characterization as well as chemical synthesis of this mannolipid showed that the mannose was attached in a β-linkage (WARREN and JEANLOZ 1973).

The mannosyl transferase that catalyzes the reaction:

$$\text{GDP-mannose} + \text{dolichyl-phosphate} \underset{}{\overset{Mn^{2+}}{\rightleftharpoons}} \text{dolichyl-P-mannose} + \text{GDP}$$

has been solubilized from microsomal preparations of aorta (HEIFETZ and ELBEIN 1977a) and from Acanthamoeba (CARLO and VILLAMEZ 1979). The aorta-solubilized enzyme required polyprenyl-phosphate as a mannose acceptor as well as divalent cation (Mn^{2+} or Mg^{2+}). Ficaprenyl-phosphate, dolichyl-phosphate or other polyprenyl-phosphates could act as mannose acceptors, but as indicated above, the natural mannose carrier is apparently dolichyl-phosphate. This is interesting in view of the fact that plants contain large amounts of the unsaturated polyprenol, ficaprenol. Perhaps this polyprenol functions as a carrier of other sugars or in other as yet undetermined reactions.

4.2 Glucose

Shortly after the demonstration of mannose incorporation into lipid, the incorporation of glucose from UDP-glucose into a glucolipid was reported using a particulate enzyme preparation from rat liver (BEHRENS and LELOIR 1970). The formation of this glucolipid was inhibited when either EDTA or UDP were added to the incubation mixtures. When [β-^{32}P]UDP-glucose was used as the substrate, no ^{32}P was found in the glucolipid, indicating that glucose and not glucose-1-P was being transferred. The lipid portion of the glucolipid was identified as a polyisoprenol-P of the dolichol type and the linkage of sugar to phospholipid was shown to be a β-linkage. Thus, this enzyme catalyzes the following reaction:

$$\text{UDP-glucose} + \text{dolichyl-P} \overset{Mn^{2+}}{\rightleftharpoons} \text{glucosyl-P-dolichol} + \text{UDP}$$

The structure of this glycolipid is identical to that shown in Fig. 2B except that the sugar moiety is glucose rather than mannose. The enzyme catalyzing this reaction was most prevalent in smooth I microsomal fractions and then in order of decreasing activity: outer mitochondrial membranes, smooth II microsomal fractions, rough microsomal fractions, nuclear membranes, Golgi apparatus, inner mitochondrial membranes, and finally plasma membranes (DALLNER et al. 1972).

Glucose is also transferred to lipid in plants (FORSEE and ELBEIN 1973). Particulate preparations from peas have been reported to catalyze the transfer of glucose-1-P from UDP-glucose to endogenous acceptors to form glucosyl-pyrophosphoryl-polyprenol (HOPP et al. 1978b). When the UDP-^{14}C-glucose was labeled with ^{32}P in the β phosphate, the glycolipid became doubly labeled with ^{32}P and ^{14}C. These same enzyme preparations also transferred glucose to form dolichyl-phosphoryl-glucose (PONT-LEZICA et al. 1976). Endogenous lipids were isolated from peas by solvent extraction and these lipids were purified on DEAE cellulose columns. The purified lipids acted as glucose acceptors with enzyme preparations from either animal or plant tissues. For characterization, the glucolipid was chromatographed on thin layer plates against standards such as glucosyl-phosphoryl-ficaprenol or glucosyl-phosphoryl-dolichol, and the enzymatically synthesized glucolipid was also subjected to catalytic hydrogenation and phenol treatment. The data indicated that the acceptor lipid from peas was an α-saturated polyprenol of about the same molecular size as dolichol (80–100 carbons) and different from ficaprenol (PONT-LEZICA et al. 1976).

A word of caution should be indicated at this point. Many plant tissues incorporate glucose into lipid, but much if not all of the glucose incorporated is into steryl glucosides (LAINE and ELBEIN 1971). These steryl glucosides are fairly common in plants, and most plant tissues will catalyze their cell-free synthesis. Therefore it is necessary to demonstrate that the biosynthetic product is charged (i.e., binds to DEAE cellulose because of the presence of phosphate), and has the properties of a polyisoprenol derivative.

4.3 N-Acetylglucosamine

During the time when these mannolipids and glucolipids were being discovered, another study showed that liver microsomes catalyzed the incorporation of *N*-acetylglucosamine (GlcNAc) from UDP-^{14}C-GlcNAc into lipid. However, in this case the evidence suggested that GlcNAc-1-P was being transferred to the lipid rather than GlcNAc alone. Thus, when [β-^{32}P]UDP-^{14}C-GlcNAc was used as the substrate, the GlcNAc-lipid became labeled with both ^{14}C and ^{32}P (MOLNAR et al. 1971). Also the fact that the GlcNAc-lipid bound more firmly to DEAE cellulose than the glucolipid or the mannolipid, as well as the fact that its synthesis was inhibited by UMP rather than UDP, suggested that it contained a pyrophosphoryl linkage instead of a phosphoryl linkage. Thus, the synthesis of the GlcNAc-lipid proceeds by the reaction shown in Fig. 6. This lipid has been characterized as GlcNAc-pyrophosphoryl-dolichol. The GlcNAc-1-P transferase was solubilized from aorta membrane fractions by treatment with the nonionic detergent Nonidet P-40, and the enzyme was

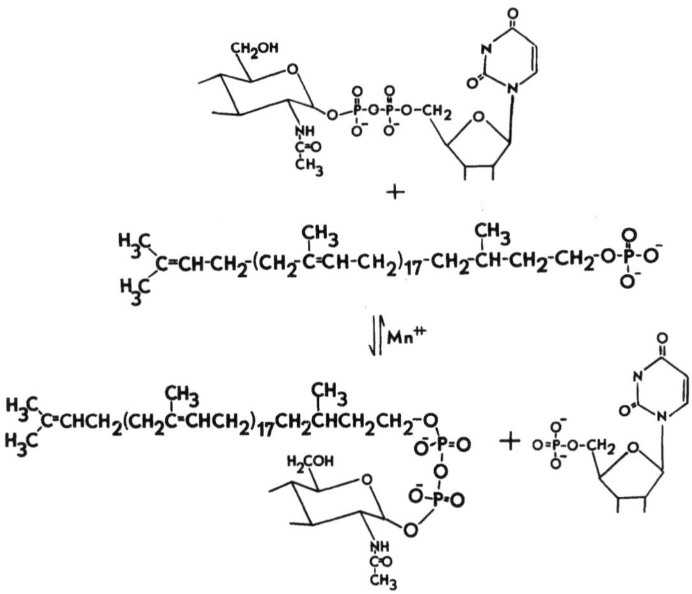

Fig. 6. Biosynthetic reaction for the synthesis of the first lipid intermediate, GlcNAc-pyrophosphoryl-dolichol

partially purified (Heifetz and Elbein 1977a). The enzyme catalyzed the reversible reaction shown in Fig. 6, and required polyprenyl-phosphate for activity. This enzyme is the site of action of the tunicamycin antibiotics (see Sect. 8).

Hypocotyls from *Phaseolus aureus* were also shown to incorporate glucosamine into lipid, but in those initial experiments the lipids were not characterized (Roberts 1975). Cell-free studies with membrane preparations from cotton fibers, mung bean hypocotyls or other plant tissues showed that radioactivity from UDP-^{3}H-GlcNAc was transferred to a lipid which had the following properties: (a) it bound more firmly to DEAE cellulose than did the mannolipid indicating the presence of a pyrophosphoryl-bond, (b) the sugar was found to be attached to the lipid in a linkage that was susceptible to mild acid hydrolysis, and (c) the mobility of the GlcNAc-lipid on thin layer chromatography in a neutral, acidic, and basic solvent was indicative of a polyisoprenol type of lipid (Forsee et al. 1976). However, in the cotton system, the GlcNAc was found to be present mostly in a *N,N'*-diacetylchitobiose-lipid and only small amounts of the mono-GlcNAc-lipid were detected. However, in other plant systems, the major GlcNAc-lipid may be the mono-GlcNAc or it may be a mixture of mono- and di-GlcNAc-lipids (Brett and Leloir 1977, Forsee and Elbein 1975, Lehle et al. 1976). It seems likely that the specific conditions of the incubation mixture, and/or the state of the tissue with regard to the type of endogenous lipid acceptor and the activities of the various glycosyl transferases determine whether one or two GlcNAc residues are transferred in these cell-free systems.

In the case of GlcNAc transfer in plants, it is still not clear whether the lipid carrier is a dolichol. Some studies indicate, however, that the lipid carrier for GlcNAc is different from that for mannose. For example, GDP-mannose and UDP-GlcNAc do not compete with each other for lipid acceptor when the particulate enzyme preparation from mung beans is used. These results would suggest either that there are different pools of polyprenol acceptors for these two sugars, or that the polyprenols are different. In addition, the polyprenyl-phosphates obtained by mild acid hydrolysis of the GlcNAc-lipid stimulated GlcNAc incorporation into lipid but had no effect on mannose incorporation (Ericson et al. 1978a). Likewise, dolichyl-phosphate stimulated mannose incorporation but did not stimulate the incorporation of GlcNAc into lipid. Thus the GlcNAc-lipid may be a dolichol, but if so, it must have different properties than the mannolipid.

4.4 Other Monosaccharide Lipids

Several other monosaccharides have been found to be linked to polyprenols. In oviduct, a xylose-containing lipid is formed from UDP-xylose and endogenous lipid. The xylosyl-lipid is apparently xylosyl-phosphoryl-polyprenol (Waechter et al. 1974). Early reports in plants suggested that monosaccharides from UDP-xylose, UDP-arabinose, UDP-galactose, UDP-glucuronic acid, and GDP-glucose were incorporated into lipid (Villamez and Clark 1969). However, these studies have not been confirmed, nor have the products been identified. Glucuronic acid from UDP-glucuronic acid was found to be incorporated into lipid by microsomal preparations of lung, but this lipid proved to be a disaccharide with the structure, glucuronosyl-GlcNAc-pyrophosphoryl-dolichol (Turco and Heath 1977).

5 Oligosaccharide Derivatives

The role of the lipid-linked saccharides became much clearer with the discovery of lipid-linked oligosaccharides that contained mannose and GlcNAc. As shown in Fig. 1, the lipid-linked monosaccharides are precursors for the formation of the lipid-linked oligosaccharides. The lipid-linked oligosaccharides were first discovered by virtue of the fact that their solubility in organic solvents is different from that of the lipid-linked monosaccharides (Behrens et al. 1971). Thus, most incubation mixtures containing particulate enzyme, sugar nucleotide (i.e., either GDP-^{14}C-mannose, UDP-^{3}H-GlcNAc or UDP-^{3}H-glucose), and divalent cation Mn^{2+} or Mg^{2+}) are treated in the following way after incubation: The reaction is terminated by the addition of chloroform: methanol (1:1) in sufficient amounts to give a final mixture of chloroform: methanol:water (1:1:1). Following a thorough mixing, the phases are separated by centrifugation and the lower, chloroform phase that contains the lipid-linked monosaccharides is removed and saved. The lipid-linked oligosaccharides are apparently too polar to be extracted by this solvent and therefore they remain associated with the particulate enzyme at the interface. Thus, after removal of the first chloroform extract, the particulate material is isolated by centrifugation, and after several washings with 50% methanol to remove any remaining sugar nucleotide, the lipid-linked oligosaccharides are extracted by suspending the particles in chloroform:methanol:water (10:10:3). This procedure is generally useful for separating and isolating the different classes of lipids, but the separation is not absolute. Thus under some conditions, disaccharide-, trisaccharide-, and even tetrasaccharide-lipid may be partially extracted into the first chloroform extraction.

As shown in Fig. 1, the lipid-linked saccharide pathway is initiated by the formation of GlcNAc-pyrophosphoryl-dolichol. The synthesis of this lipid was discussed in the previous section. The particulate enzyme from most tissues also catalyzes the addition of a second GlcNAc from UDP-GlcNAc to form *N,N'*-diacetylchitobiosyl-pyrophosphoryl-dolichol. In some tissues, the disaccharide-lipid is the major product, whereas in others the mono-GlcNAc-lipid is the major product. In these latter cases, a second incubation with an excess of UDP-GlcNAc is necessary to form the di-GlcNAc-lipid.

The third step in the pathway is the addition of mannose to the *N,N'*-diacetylchitobiosyl-pyrophosphoryl-dolichol to form the trisaccharide-lipid, Manβ-GlcNAc-GlcNAc-pyrophosphoryl-dolichol. The mannosyl donor for this first mannose, which is β-linked, was shown to be GDP-mannose rather then dolichyl-phosphoryl-mannose (Levy et al. 1974, Chen and Lennarz 1976). The enzyme that transfers mannose to the di-GlcNAc-lipid was solubilized from

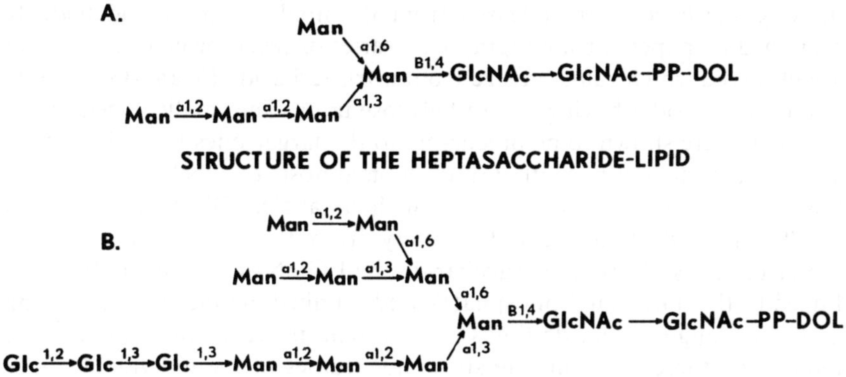

Fig. 7. Structure for several of the oligosaccharides isolated from lipid-linked oligosaccharide intermediates. **A** Structure of the heptasaccharide, $Man_5GlcNAc_2$. **B** Structure of the large oligosaccharide, $Glc_3Man_5GlcNAc_2$. (Li and Kornfeld 1979, Spiro et al. 1976, Robbins et al. 1977)

aorta membranes with the detergent Nonidet P-40. This solubilized enzyme utilized GDP-mannose but not dolichyl-phosphoryl-mannose as the mannosyl donor. This mannosyl transferase also showed an absolute reqzirement for the *N,N'*-diacetylchitobiose-lipid as the mannose acceptor (Heifetz and Elbein, 1977b). In cotton fibers, the trisaccharide-lipid was synthesized from GDP-^{14}C-mannose as part of a mixture of oligosaccharide-lipids and this oligosaccharide was partially characterized as Manβ-GlcNAc-GlcNAc (Forsee and Elbein, 1975).

Historically, the next lipid-linked oligosaccharide that was isolated and characterized was a heptasaccharide-lipid that contained five mannose and two GlcNAc residues. Although the individual steps in the synthesis of this oligosaccharide-lipid were not demonstrated, the ^{14}C-mannose-labeled-heptasaccharide-lipid accumulated when particulate enzymes from oviduct (Lucas et al. 1975), mouse myeloma (Hsu et al. 1974), and liver (Behrens et al. 1973: Oliver and Hemming, 1975) were incubated with GDP-^{14}C-mannose. The last four mannose residues in this heptasaccharide are all α-linked and are released by treatment of the heptasaccharide with α-mannosidase. Apparently, addition of mannose residues to the trisaccharide-lipid proceeds so rapidly that little or no tetrasaccharide-, pentasaccharide- or hexasaccharide-lipids accumulate. However, small amounts of these other lipid-linked oligosaccharides have been detected in large-scale incubations and some of these intermediates have been characterized (Chapman et al. 1979). The heptasaccharide-lipid that accumulates in Chinese hamster ovary cells has been characterized using a variety of chemical techniques and shown to have the structure presented in Fig. 7A. This oligosaccharide is branched at the two α-linked mannoses, one of them being in a 1-6 and the other a 1-3 bond. The next two α-linked mannoses are attached in 1-2 linkages (Li and Kornfeld 1979).

In plants, a series of lipid-linked oligosaccharides are formed when GDP-^{14}C-mannose is incubated with the particulate enzyme from cotton fibers or mung bean seedlings (Forsee and Elbein 1975, Ericson and Delmer 1977). When the oligosaccharides are released from the lipids by mild acid hydrolysis and examined by paper chromatography, a series of seven or eight radioactive peaks are observed which range in size from a trisaccharide (Manβ-GlcNAc-GlcNAc) to oligosaccharides having 10 to 12 glycose units. The smaller-sized oligosaccharide-lipids were shown to be precursors to the larger ones by a chase experiment. Thus, after incubating with GDP-^{14}C-mannose to label the oligosaccharide-lipids, the incubations were chased with unlabeled GDP-mannose. This chase resulted in a shift in the radioactivity from the smaller oligosaccharides to the larger ones. These oligosaccharides had GlcNAc at their reducing end (i.e., linked to the polyprenyl-phosphate), and α-linked mannose at their nonreducing end. The oligosaccharide-lipids are present in very small amounts in these tissues and therefore complete structural studies of the individual oligosaccharides have not yet been accomplished.

Much more detailed studies have been done on the oligosaccharides (isolated from the lipid-linked oligosaccharides) from animal cells. The major advantage to using these cells is that many animal cells, such as Chinese hamster ovary cells, can be grown in tissue culture so that radioisotopes such as ^{3}H-mannose or ^{3}H-leucine can be included in the growth medium. This enables one to label all of the sugar residues in the lipid-linked oligosaccharides, and thereby allows the structural characterization of oligosaccharides with small amounts of material. In addition, many animal cells in culture can be infected with viruses that have glycoprotein coats, so that one can look specifically at the synthesis of the viral glycoprotein. From the ^{3}H-oligosaccharide-lipids obtained from Chinese hamster ovary cells, it was possible to isolate small amounts of tetrasaccharide-, pentasaccharide- and hexasaccharide, as well as the major oligosaccharide, the heptasaccharide. Since only one isomer of each of these oligosaccharides was found, it was reasonable to assume that there was a definite sequence of addition of sugars from trisaccharide to heptasaccharide as is shown in Fig. 8. The structures of these compounds and the sequence of formation was deduced by methylation of the ^{3}H-oligosaccharides, and identification of the ^{3}H-methyl-mannoses by thin layer chromatography. The sequence shown in Fig. 8 indicates that the first mannose that is added to the Manβ-GlcNAc-GlcNAc is an (1→3)-α-linked mannose followed by an (1→6)-α-linked mannose and then two (1→2)-α-linked mannoses (CHAPMAN et al. 1979). Although these reactions are still somewhat hypothetical, the available evidence does support this series of reactions. It will, however, probably require the stepwise synthesis with soluble enzymes to verify this pathway.

The four α-linked mannose residues in the heptasaccharide-lipid have been postulated to come from dolichyl-phosphoryl-mannose, This postulation is based on the assumption that the transfer of sugars from sugar donor to acceptor involves an inversion of the configuration of the anomeric carbon atom, but several lines of evidence indicate that at least some of these α-linked mannose units are transferred directly from GDP-mannose. First of all, using the particulate enzyme preparation from aorta, EDTA inhibits the transfer of mannose from GDP-mannose to dolichyl-phosphate preventing the formation of dolichyl-phosphoryl-mannose. Even in the presence of EDTA, radioactive mannose is still transferred from GDP-mannose into the lipid-linked oligosaccharides. The

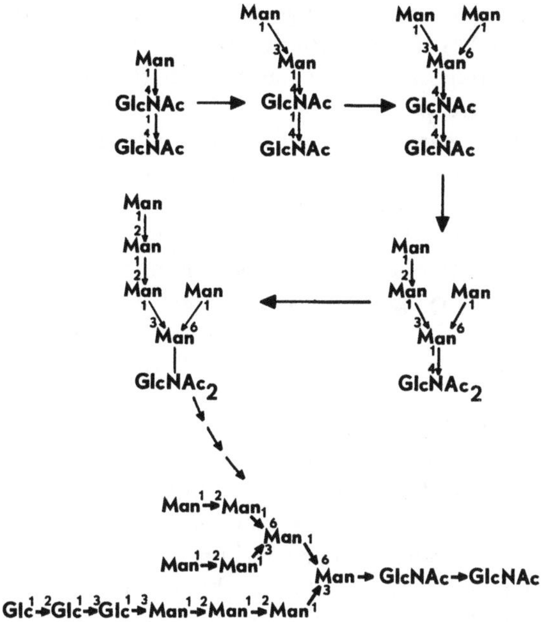

Fig. 8. Postulated reactions for the addition of mannose residues to the trisaccharide (Man-GlcNAc-GlcNAc)-lipid to form the heptasaccharide (Man_5GlcNA_2)-lipid (LI and KORNFELD 1979)

major oligosaccharide present in these lipid-linked oligosaccharides is a heptasaccharide (CHAMBERS et al. 1977). The synthesis of dolichyl-phosphoryl-mannose is also inhibited by the antibiotic, amphomycin. In the presence of high enough concentrations of amphomycin to inhibit the formation of dolichyl-phosphoryl-mannose more than 90%, ^{14}C-mannose is still transferred from GDP-^{14}C-mannose to lipid-linked oligosaccharides, and again the major oligosaccharide is a heptasaccharide containing five mannose and two GlcNAc residues (KANG et al. 1978). Finally some of the mannosyl transferases that catalyze the addition of α-linked mannoses to the acceptor lipid to form the heptasaccharide-lipid have been solubilized from the particulate enzyme preparation. These solubilized enzymes utilize GDP-mannose as the mannosyl donor rather than dolichyl-phosphoryl-mannose (SPENCER and ELBEIN 1980), but it is not known which of the α-linked mannoses come directly from GDP-mannose and it will require more extensive studies with purified enzymes to characterize each of these reactions. However, recent studies with a partially purified mannosyl transferase which adds (1→2)-α-mannosyl linkages suggest that the two (1→2)-α-linked mannoses come directly from GDP-mannose (SCHUTZBACH et al. 1980). In addition, a mutant lymphoma cell line which cannot synthesize dolichyl-phosphoryl-mannose was recently reported. This mutant accumulated the heptasaccharide-lipid (i.e., $Man_5GlcNAc_2$), but apparently could not elongate it further (CHAPMAN et al. 1980). This latter study suggests that the first five mannoses come from a substrate other than dolichyl-phosphoryl-mannose. These kinds of studies has not yet been done in plants, but it will be important in these systems to determine the mannosyl donors of the various mannose residues.

In several different animal systems, it has been found that the heptasaccharide-lipid can be lengthened by the addition of four mannose and three glucose units to give an oligosaccharide-lipid having the composition, Glc_3-Man_9-$GlcNAc_2$ (LI and KORNFELD 1978, 1979, SPIRO et al. 1976, ROBBINS et al. 1977). This oligosaccharide has been partially characterized and shown to have the structure presented in Fig. 7B. The first seven sugars of this oligosaccharide are the same as those in the heptasaccharide shown in Fig. 7A, and indeed the evidence indicates that the heptasaccharide-lipid is the precursor for the large oligosaccharide-lipid, but exactly how the mannoses and glucoses are added and in what sequence is not known. It seems likely that the mannose donor is dolichyl-phosphoryl-mannose and that the glucosyl donor is dolichyl-phosphoryl-glucose. As indicated in the next section, after transfer of this large oligosaccharide to the protein, all of the glucoses and some of the mannoses are removed from the oligosaccharide before the protein reaches its final destination.

An oligosaccharide-lipid such as that shown in Fig. 7B has not been demonstrated in plants. That is, particulate enzymes from plants do incorporate glucose from UDP-glucose into lipid-linked oligosaccharides, but it is not clear whether these oligosaccharides also contain mannose and GlcNAc (HOPP et al. 1978b). For example, in the bacterium *Acetobacter xylinum*, which produces an extracellular cellulose, lipid-linked glucose, lipid-linked cellobiose, and lipid-linked cellotriose have been implicated in cellulose synthesis (KJOBAKKEN and COLVIN 1973). Particulate enzymes from the algae *Prototheca zopfii* also catalyzed the incorporation of glucose from UDP-glucose into lipid-linked oligosaccharides (PONT-LEZICA et al. 1978). However, these oligosaccharides were apparently quite different from the oligosaccharides of animal cells since no GlcNAc was found in them, and the sugar at the reducing end was apparently glucose.

6 Protein Glycosylation

The ultimate goal of the lipid-linked oligosaccharides that have been described in preceding sections is to donate the oligosaccharide portion of these molecules to a polypeptide chain to form the glycoprotein. Early studies in the field of glycoprotein biosynthesis suggested that radioactive glucosamine was incorporated into protein while the polypeptide chain was still attached to polysomes (i.e., on nascent chains) (ROBINSON 1969, MELCHERS 1973). More recently, oviduct slices that were synthesizing the glycoprotein ovalbumin were shown to incorporate both radioactive mannose and radioactive glucosamine into protein. These sugars were found to be linked to protein while it was still bound to polyribosomes attached to the endoplasmic reticulum. The protein was identified as ovalbumin by immunoprecipitation (KIELY et al. 1976).

In some of the initial studies on the formation and utilization of the lipid-linked oligosaccharides, GDP-^{14}C-mannose or UDP-^{3}H-glucose were incubated ed with membrane fractions from various tissues and the incorporation of radioactivity into the various dolichol intermediates was followed. In most of these experiments, radioactivity was also incorporated into a residue that was insoluble in organic solvents, in trichloroacetic acid and in water. However, this water-insoluble material could be rendered water-soluble upon digestion with a proteolytic enzyme such as Pronase. Pronase-digested material behaved like a mixture of glycopeptides when subjected to high voltage paper electrophoresis or to gel filtration. In addition, SDS gel electrophoresis (PAGE) of the intact insoluble material revealed the presence of several radioactive protein bands. Similar kinds of radioactive products have also been identified in several plant systems. In these initial studies, the kinetics of transfer of radioactive mannose from GDP-^{14}C-mannose was compatible with the following series of reactions:

$$\text{GDP-mannose} \rightleftharpoons \text{Dolichyl-P-mannose} \rightarrow$$
$$\rightarrow \text{Lipid-linked oligosaccharides} \rightarrow \text{Protein}$$

In order to prove conclusively that these lipid-linked oligosaccharides were involved in the transfer of sugars to protein, it was necessary to isolate sufficient quantities of the radioactive lipid-linked saccharides so that these compounds could be used as substrates in cell-free enzyme experiments. Several early studies gave promising results. For example, the oligosaccharide moiety of the lipid-linked oligosaccharide could be incorporated into an exogenous, carbohydrate-free form of a κ-type immunoglobulin light chain, produced in the presence of 2-deoxyglucose (EAGON and HEATH 1977). The enzymatic transfer of the oligosaccharide portion of the lipid-linked oligosaccharides to denatured forms of three secretory proteins, ovalbumin, α-lactalbumin and ribonuclease A, was demonstrated using a membrane fraction from hen oviduct (PLESS and LENNARZ 1977). Based on a survey of ten proteins denatured by sulfitolysis, the presence of the tripeptide sequence, asn-X-ser (thr) appeared to be necessary but not sufficient for the protein to serve as an acceptor of carbohydrate in vitro. These studies also indicated that unfolding of the peptide chain was required in order to expose sites for carbohydrate attachment.

Additional studies along these lines have led to some interesting and rather unexpected results. In animal systems, the lipid-linked oligosaccharide that contains glucose is a better substrate for the transfer of oligosaccharide to protein than is the lipid-linked oligosaccharide that only contains mannose and GlcNAc. Thus in one set of experiments, the ^{14}C-mannose-labeled lipid-linked oligosaccharide was synthesized from GDP-^{14}C-mannose in cell-free extracts and then unlabeled UDP-glucose was added to one half of the incubation to further glucosylate these oligosaccharides (TURCO et al. 1977). The glucose-free and glucose-containing lipid-linked oligosaccharides were then compared for their ability to glycosylate proteins in a cell-free enzyme preparation. As much as 40% of the glucose-containing oligosaccharide was transferred to protein, as compared to only 5% of the glucose-free oligosaccharide. In another type of experiment, the lipid-linked oligosaccharide that contained 2 GlcNAc, 11 mannose and 1 or 2 glucose residues was synthesized in thyroid slices in the presence of radioactive sugars (SPIRO et al. 1978, 1979). Approximately 50% transfer of this oligosaccharide from its lipid carrier to protein was achieved with the thyroid particulate enzyme. When 40% of the mannose residues were removed from the oligosaccharide-lipid with jack bean α-mannosidase there was no change in the rate of transfer. However, removal of 80% of the glucose residues with a thyroid glucosidase reduced the donor activity of this oligosaccharide-lipid by 80%. Thus, these experiments indicate that the glucose in these oligosaccharide-lipids plays some critical role in directing or facilitating the transfer of this oligosaccharide to protein. Further reactions of the glucose-containing glycoproteins are discussed in the next section.

Particulate enzyme preparations from *Pisum sativum* were reported to incorporate glucose from UDP-^{14}C-glucose into an oligosaccharide-lipid which contained 7 to 8 glucose units linked to dolichol (PONT-LEZICA et al. 1978). Thus this lipid was quite different in composition than that described above in the animal systems. The glycoprotein formed in this plant system was a membrane-bound glucoprotein that was suggested to be a pea lectin. An oligosaccharide-lipid was shown to act as donor of oligosaccharide to a sulfitolyzed form of ribonuclease A using a membrane fraction prepared from the endosperm tissue of 3-day-old castor beans. This membrane fraction was composed of endoplasmic reticulum membranes (MELLOR et al. 1978). However, in these studies it was not clear what the length of the oligosaccharide was (i.e., how many sugar residues it contained), nor whether it contained glucose.

A number of studies have been done to examine the temporal and topological relationship between protein biosynthesis and core glucosylation in cell-free systems. In one of these studies, the mRNA for the mouse MOPC-46B κ chain was prepared and this mRNA was translated in an Ehrlich ascites cell-free system. The polypeptide was identified as the κ chain by its antigenic properties and by the analysis of the products obtained by trypsin digestion. This polypeptide could be glycosylated in this cell-free system after translation (TUCKER and PESKA 1977).

Probably the best system for examining protein glycolysation both in vivo and in vitro has been that utilizing mammalian cells that have been infected with a glycoprotein-containing virus such as vesicular stomatitis virus (VSV). Detailed studies on the VSV glycoproteins are covered by D. Bowles in chapter 14, Volume 13A, this Series, and will be only briefly summarized here. VSV is a budding virus which has an envelope glycoprotein called the G protein. G has a molecular weight of 69,000 and contains an asparagine-linked oligosaccharide that accounts for 10% of the mass. A number of studies have shown that G is synthesized on membrane-bound polysomes as a polypeptide containing a leader sequence (i.e., a signal peptide). This signal peptide is involved in interaction with the membrane and the insertion of G protein into the membrane. The signal peptide becomes inserted into the membrane as the polypeptide is being synthesized, and membrane insertion and glycosylation are cotranslational events (TONEGUZZO and GHOSH 1977, ROTHMAN and LODISH 1977, KATZ et al. 1977, IRVING et al. 1979).

Similar studies have been done on the synthesis of the secretory protein, α-lactalbumin. In this case, mRNA from rat mammary glands was translated in a cell-free wheat germ system in the presence or absence of membranes prepared from rough microsomes of dog pancreas. In the absence of membranes, the synthesized protein was apparently unglycosylated since it did not bind to concanavalin A-sepharose or to ricin-agarose, but it did have a signal peptide of 19 amino acids. However, in the presence of membranes, a processed form of α-lactalbumin was synthesized that lacked the signal peptide, was segregated within the microsomal vesicles, and was glycosylated. Thus, in these studies also, membranes had to be present during the translation for processing, segregation, and core glycosylation to occur (LINGAPPA et al. 1978).

Although similar detailed studies have not been done in plants, several preliminary reports suggest that plant systems will probably be similar. For example, in one study, oocytes from *Xenopus laevis* were injected with mRNA from either normal or opaque-2 maize. The zein proteins synthesized in these oocytes were about 2000 smaller in molecular weight than those synthesized in wheat germ cell-free systems (HURKAMN et al. 1979). These results suggest that the zein proteins are synthesized with a signal peptide that is removed in the oocyte system. In another study, the synthesis of the major seed globulin of *Phaseolus vulgaris* was translated in a cell-free system using a wheat germ extract supplemented with bean polysomes. The major seed protein, G_1, is a glycoprotein and each of the G_1 subunits are glycosylated. However, when total poly-(A) containing RNA from developing cotyledons was used, only two major polypeptides were formed and these appeared to be unglycosylated forms of the two subunits. These same forms of the proteins were synthesized in cotyledon slices incubated with tunicamycin (SUN et al. 1979). Perhaps these proteins were not glycosylated because the cell-free system did not contain membranes.

7 Further Reactions – Processing of Proteins

As indicated in the preceeding section, in animal cells the $Glc_3Man_9GlcNAc_2$ oligosaccharide is assembled on the lipid carrier and this oligosaccharide is then transferred en bloc to the polypeptide chain. Once this transfer has occurred, the oligosaccharide undergoes a number of processing or "trimming" reactions whereby all of the glucose residues and four to six or more of the mannose residues are removed from the protein. The experiments that have been designed to show these trimming reactions have used pulse-chase techniques to determine the fate of the oligosaccharide after transfer to protein in vivo. Thus various mammalian cells were pulse-labeled with a radioactive sugar and the radioactive lipid-linked oligosaccharides and glycoproteins were isolated at various times following the pulse label, and were characterized. After transfer, the outermost glucose residue (i.e., at the nonreducing end of the oligosaccharide) is rapidly removed with a half-time of about 2 min. The second glucose is removed more slowly with a half-time of about 5 min and the third glucose is removed most slowly (TURCO and ROBBINS 1979, HUBBARD and ROBBINS 1979).

A glucosidase was identified in thyroid microsomes that selectively releases glucose from the dolichol-linked oligosaccharides or from the protein-linked oligosaccharide (SPIRO et al. 1979). This enzyme was less active on free oligosaccharide or on the oligosaccharide from which some of the mannoses had been removed by α-mannosidase digestion. The glucosidase was solubilized by treatment of the microsomes with Triton X-100 and this enzyme had a pH optimum between 6 and 7 and was unaffected by EDTA. The pH optima indicate that this is not a lysosomal enzyme. Since the glucose residues in the oligosaccharide could not be released by other glucosidases of known anomeric specificities, the configuration of the glucose in the oligosaccharide was not clear. However, chromic acid oxidation of the oligosaccharide indicated that the glucose residues were in the α-configuration.

Following the removal of the glucose residues, some of the mannose residues are also excised. It is tempting to speculate that for some glycoproteins trimming stops after the removal of glucose to give a high mannose type of glycoprotein. For example, the $Man_9GlcNAc_2$ oligosaccharide that would remain after removal of glucose is similar to one of the components found in a glycopeptide from calf thyroglobulin, and the branching patterns of these oligosaccharides is similar. However, in other cases, processing may continue with the removal of 1 or more mannose residue. This could give rise to a family of high mannose oligosaccharides, or the oligosaccharide could be trimmed to the $Man_3GlcNAc_2$ core structure by the removal of 6 mannose residues. This core structure (attached to protein) is believed to be the precursor for the formation of the complex types of oligosaccharides shown in Fig. 4. Thus after removal of the mannose, the core glycoprotein serves as the acceptor for the addition of the common trisaccharide structure: *N*-acetylneuraminic acid (sialic acid)-galactose-GlcNAc. Each of these three sugars is added to the protein by means of specific glycosyltransferases, each utilizing the appropriate sugar nucleotide donor.

In terms of the removal of mannose residues from the glycoprotein, a specific α-mannosidase has recently been isolated from the Golgi apparatus of rat liver (TABAS and KORNFELD, 1978). This enzyme was purified about 3000 to 6000-fold by subcellular fractionation, Triton X-100 solubilization, and ion-exchange and hydroxylapatite chromatography. The enzyme had a pH optimum of 6 to 6.5, and showed specificity for (1→2)-α-linked mannose units.

Based on the above studies, the formation of the complex type of glycoprotein can probably be summarized as follows: The $Glc_3Man_9GlcNAc_2$ oligosaccharide is added to the polypeptide chain in the rough endoplasmic reticulum while the peptide is being synthesized on membrane-bound polysomes, and insertion of the peptide into membrane vesicles is occurring. After transfer of the oligosaccharide and perhaps after completion of protein synthesis, the glucoses are removed by membrane-bound glucosidases in the endoplasmic reticulum. The $Man_9GlcNAc_2$-protein is then transported to the Golgi apparatus while enclosed in membrane vesicles (ROTHMAN and FENE 1980) and there, a number of mannose residues are removed by the Golgi α-mannosidase. Finally, the sugars, GlcNAc, galactose, and sialic acid are added to this protein in a stepwise fashion from their respective sugar nucleotide derivatives utilizing Golgi glycosyltransferases to yield the complex type of glycoprotein. Whether any types of processing or trimming reactions of the oligosaccharide portions of plant glycoproteins occur is still unknown, but studies in this direction should be forthcoming.

8 Effect of Antibiotics and Other Inhibitors

During the past 5 or 10 years, a number of inhibitors of protein glycosylation have been described, and these compounds have proven to be valuable tools for examining the role of glycosylation in various cell functions. Probably the most useful of these compounds is the antibiotic tunicamycin which is produced by *Streptomyces lysosuperificus* (TAKATSUKI et al. 1971). Tunicamycin is a uracil nucleoside antibiotic containing glucosamine, an 11-carbon sugar called tunicamine, and fatty acids, and its structure is shown in Fig. 9 (TAKATSUKI et al. 1977, ITO et al. 1980). The antibiotic is produced as a family of closely related compounds which differ in the length and structure of the fatty acid component. In cell-free extracts of several different tissues from animals, tunicamycin was shown to inhibit the first enzyme in the lipid-linked saccharide pathway, the GlcNAc-1-P transferase, thereby inhibiting the formation of dolichyl-pyrophosphoryl-GlcNAc (TKACZ and LAMPEN 1975, TAKATSUKI et al. 1975, STRUCK and LENNARZ 1977). This antibiotic also inhibits the synthesis of GlcNAc-pyrophosphoryl-polyprenol in cell-free extracts of several plant tissues (ERICSON et al. 1977). In yeast membrane fractions, tunicamycin was shown to inhibit the transfer of the first GlcNAc to dolichyl-P but to have no effect on the transfer of the second GlcNAc (LEHLE and TANNER 1976).

In intact cells, tunicamycin inhibits the incorporation of ^{3}H-mannose into those proteins whose oligosaccharide is formed via the lipid-linked saccharide pathway. Since the antibiotic prevents the formation of the first lipid intermediate, it inhibits the synthesis of lipid-linked oligosaccharides. However, under these conditions the proteins are still synthesized but cannot become glycosylated. Most tunicamycin preparations do have some inhibitory activity toward protein synthesis but this activity is usually much lower than that toward protein glycosylation. For example, under conditions where mannose incorporation into protein is inhibited 90% or more (usually 1 to 5 μg/ml of antibiotic), ^{3}H-leucine incorporation into protein might be inhibited about 10 to 20%. Much of the inhibitory activity toward protein synthesis can be removed when tunicamycin is purified by high performance liquid chromatography (MAHONEY and DUSKIN 1979).

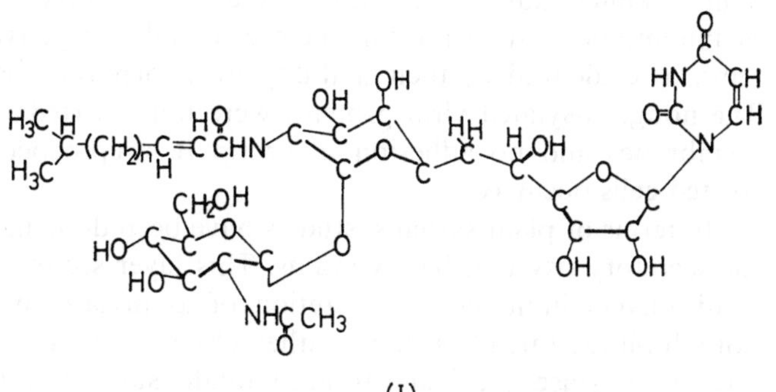

(I) n = 8,9,10,11

Fig. 9. Structure of tunicamycin (TAKATSUKI et al. 1977)

When tunicamycin is used with whole cells, various effects have been observed as a result of the inhibition of glycosylation, but at this point it is difficult to make generalizations regarding the role of the carbohydrate in glycoprotein function. For example, this antibiotic blocked the conversion of procollagen to collagen in fibroblasts, probably by preventing the glycosylation of a procollagen peptidase (DUSKIN and BORNSTEIN 1977). Tunicamycin also inhibited the incorporation of glucosamine into immunoglobulins and the secretion of these proteins in a MOPC 315 cell line (HICKMANN et al. 1977). However, in hepatocytes where glycosylation was blocked by this antibiotic, both the unglycosylated transferrin and unglycosylated very low density lipoprotein (VLDL) were still secreted (STRUCK et al. 1979). A number of other studies in a variety of animal systems have also given conflicting results in terms of the function of the carbohydrate in secretion or membrane insertion, making it difficult to pinpoint a role for the oligosaccharide moiety. It may be that the carbohydrate confers certain conformational properties on the protein and that it is protein conformation that is critical in secretion. Likewise, as shown by some of the studies discussed below, there is also some controversy as to whether the unglycosylated proteins undergo more rapid degradation than do the normal proteins. Again these differences may be due to alterations in protein conformation.

Tunicamycin prevents the glycosylation of viral coat proteins in influenza virus-infected mammalian cell lines, and also blocks the insertion of viral glycoproteins into the membrane. In this case, the unglycosylated proteins remained in the cytoplasm of the infected cells. With some strains of influenza virus, the unglycosylated proteins underwent much more rapid degradation in the cytoplasm than did the normally glycosylated proteins, or other proteins (SCHWARZ et al. 1976, 1978). However, in another virus system, somewhat different results were observed. In the case of cells infected with Semliki Forest virus, tunicamycin also prevented the glycosylation of proteins, but in this system glycosylation was not a prerequisite for correct insertion and cleavage of such membrane ectoproteins (GAROFF and SCHWARZ 1978). The influenza viral glycoproteins that are present in the mammalian cell membrane have been shown to be receptor sites for group B streptococci (PAN et al. 1979). When canine kidney cells are infected with influenza virus in the presence of the antibiotic tunicamycin, these cells fail to glycosylate the viral glycoproteins, as evidenced by their inability to incorporate ^{3}H-mannose into protein. The nonglycosylated viral proteins were not inserted into the mammalian cell membranes and no adhesion of group B streptococci to these tunicamycin-treated cells occurred.

In terms of plant systems, studies have been done in isolated aleurone layers, the secretory tissue of barley grains. This layer secretes the enzyme, α-amylase. Tunicamycin inhibited the secretion of α-amylase by 60 to 80%, but it did not inhibit the rate of protein synthesis by more than 10%. However, intracellularly no amylase was found to accumulate, suggesting the possibility that when glycosylation is prevented, the synthesis of the protein portion of the glycoprotein is specifically inhibited (SCHWAIGER and TANNER 1979).

Bakers' yeast produce carboxypeptidase Y, a glycoprotein enzyme of molecular weight 61,000 which has been proposed to have four asparagine-linked oligosaccharides with the general structure ($Man_{13}GlcNAc_2$). The enzyme produced in the presence of tunicamycin has a molecular weight of 51,000 as expected for the protein devoid of carbohydrate. Tunicamycin also specifically reduced the amount of carbohydrate-free enzyme present in the cells. Since the product was found to be metabolically stable, the authors suggest that the reduced amount of carboxypeptidase Y must be due to a reduced rate of synthesis and propose that there is a regulatory link between the glycosylation of the protein moiety and its biosynthesis (HASILIK and TANNER 1978).

In sea urchin embryos, tunicamycin also blocks the formation of the lipid intermediate, dolichyl-pyrophosphoryl-GlcNAc. When this antibiotic is added to developing embryos as early as 5 h after fertilization, no morphological effects on development are observed until the early gastrula stage where development is arrested (SCHNEIDER et al. 1978). However, treatment of the embryos with tunicamycin after gastrulation results in the arrest of spicule formation and arm growth. These results suggest a role for glycoproteins in embryonic development.

Several other antibiotics have been described that are related to tunicamycin. These antibiotics are streptovirudin (ECKHARDT et al. 1975), mycospocidin (NAKAMURA et al. 1957), and antibiotic 24010 (MIZUNO et al. 1971). Both streptovirudin and antibiotic 24010 were found to have the same mechanism of action as tunicamycin, i.e., they both inhibited the formation of dolichyl-pyrophosphoryl-GlcNAc in cell-free particulate fractions prepared from aorta tissue (ELBEIN et al. 1979). Both of these antibiotics were also effective inhibitors of this reaction in a cell-free particulate extract prepared from mung bean seedlings (JAMES and ELBEIN 1980). Since the activities of streptovirudin and antibiotic 24010 were destroyed by treatment with periodic acid under conditions which also destroy the activity of tunicamycin, they probably all contain a similar periodate-sensitive structure. Streptovirudin and antibiotic 24010 are also similar to tunicamycin in inhibiting the incorporation of mannose into glycoprotein in suspension cultures of soybean cells (JAMES and ELBEIN 1980).

In addition to the tunicamycin group of antibiotics, several other inhibitors of protein glycosylation or of the lipid-linked saccharide pathway have been reported. For example, the addition of deoxyglucose, glucosamine, or one of the fluorosugars to the culture medium of mammalian cells decreases the incorporation of radioactive sugars into the lipid-linked oligosaccharides and into glycoproteins. Based on a kinetic analysis, it appeared that the lipid-linked oligosaccharides were the first to be inhibited and then, after a lag, protein glycosylation was inhibited (SCHWARZ et al. 1978). Deoxyglucose not only affected the total amount of the lipid-linked oligosaccharides formed, but it also affected the size of the oligosaccharide moiety. Thus, in the presence of the inhibitor, the lipid-linked oligosaccharides were heterogeneous and of decreased molecular weight. In addition, these smaller-sized lipid-linked oligosaccharides were not transferred to protein. The deoxyglucose is incorporated into both GDP-deoxyglucose and into dolichyl-phosphoryl-deoxyglucose and even into some small-sized lipid-linked saccharides such as dolichyl-pyrophosphoryl-$(GlcNAc)_2$-deoxyglucose and dolichyl-pyrophosphoryl-$(GlcNAc)_2$-Man-deoxyglucose. Apparently these compounds cannot be further elongated. However, it is still not clear exactly how these inhibitors block protein glycosylation, but in a number of cell culture system, 2-deoxyglucose and glucosamine inhibit in vivo protein glycosylation (SCHWARZ and DATEMA 1980). Thus, the synthesis of the glycoprotein enzymes, invertase and acid phosphatase, by protoplasts of a *Saccharomyces* mutant is inhibited by 2-deoxyglucose following a 20- to 30-min lag period (KUO and LAMPEN 1972). Glycosylation of the κ-46 chain of the immunoglobulins by mouse myeloma cells was also inhibited by 2-deoxyglucose, but in these cells the nonglycosylated protein was still secreted. However, intracellular transport and secretion of these proteins was retarded (EAGON and HEATH 1977).

Several peptide antibiotics, bacitracin, amphomycin, and tsushimycin, also inhibit the synthesis of lipid-linked saccharides in both plant and animal systems. In bacteria, bacitracin was initially shown to inhibit the dephosphorylation of the C_{55} polyisoprenyl-pyrophosphate, which is a necessary step in the reutilization of polyprenyl-phosphate for cell wall polysaccharide synthesis (STONE and STROMINGER 1971). In cell-free extracts of mung bean seedlings, bacitracin inhibited the formation of both dolichyl-phosphoryl-mannose and dolichyl-pyrophosphoryl-GlcNAc with 50% inhibition requiring about 5 mM antibiotic (ERICSON et al. 1978b). Bacitracin also blocked the incorporation of mannose into lipid and into glycoprotein in carrot slices. Although bacitracin also inhibited the individual reactions of the lipid-linked saccharide pathway in extracts from animal tissues, this antibiotic was not effective in whole cells probably because it is not able to enter the cells (SPENCER et al. 1978). Amphomycin and tsushimycin are two other peptide antibiotics that inhibit the formation of dolichyl-phosphoryl-mannose and dolichyl-pyrophosphoryl-GlcNAc in cell-free extracts of animal tissue (KANG et al. 1978), and plant tissue (ERICSON et al. 1978c). While both of these antibiotics completely inhibited the transfer of mannose from GDP-^{14}C-mannose to dolichyl-phosphoryl-mannose, they only partially inhibited the incorporation of mannose into the lipid-linked oligosaccharides. These studies indicated that some of the mannose residues in the lipid-linked oligosaccharides come directly from GDP-mannose without the participation of dolichyl-phosphoryl-mannose (see also Sect. 5). The major oligosaccharide which accumulated in these experiments with amphomycin and tsushimycin was a heptasaccharide, probably containing five mannose and two GlcNAc residues (KANG et al. 1978). This oligosaccharide has the same general structure as the oligosaccharide which accumulates in a mutant cell line that is unable to synthesize dolichyl-phosphoryl-mannose (CHAPMAN et al. 1980).

One rather interesting group of compounds, which may have some important effects on the lipid-linked saccharide pathways, are inhibitors of dolichol synthesis. As indicated earlier in this chapter, dolichol is synthesized by way of the cholesterol pathway and involves the regulatory enzyme, hydroxymethylglutaryl-CoA. This enzyme is a major regulatory enzyme in the biosynthesis of cholesterol and is inhibited by high concentrations of cholesterol. Some studies have shown that 25-hydroxycholesterol, which suppresses the HMG-CoA reductase in cultured cells, inhibited the incorporation of acetate into dolichyl-pyrophosphoryl-oligosaccharides and into cholesterol to the same degree (MILLS and ADAMANY 1978). These results indicated that this enzyme is also a rate-limiting enzyme for dolichol synthesis. However, another study (JAMES and KANDUTSCH 1979) suggested that HMG-CoA reductase affects the rate of dolichol synthesis by influencing the pool size of one or more of the intermediates in the sterol pathway that are also intermediates of dolichol. Their results also showed that large fluctuations in the activity of the HMG-CoA reductase and in the rate of cholesterol biosynthesis can occur simultaneously with relatively minor changes in the rate of dolichol synthesis. Compactin is another inhibitor of cholesterol biosynthesis which should prove to be useful for studies on dolichol synthesis and its relation to protein glycosylation. Compactin is produced by several strains of *Penicillium* and its structure includes a lactonized ring that apparently resembles the lactone form of mevalonic acid (ENDO et al. 1976). Thus, compactin is a potent and reversible, competitive inhibitor of the HMG-CoA reductase (GOLDSTEIN et al. 1979). In sea urchins, compactin induced abnormal gastrulation at concentrations of antibiotic that had no effect on earlier embryonic development. The results suggested that the lesion resulted from the inhibition of dolichol biosynthesis with a corresponding inhibition in the synthesis of the oligosaccharide portion of the glycoproteins (CARSON and LENNARZ 1979).

9 Subcellular Location of the Enzymes of the Dolichol Pathway

As indicated in other sections of this chapter, the enzymes involved in the formation of the lipid-linked saccharides are located in the membranous fraction of the cell. However, it is not absolutely clear what membrane component is responsible for these various transfer reactions, nor is it clear whether the end products of these reactions are membrane components or secretory substances, or both. For example, in rat liver the highest specific activity for glycosyl transfer to dolichyl-P was found in the outer mitochondrial membrane and in the smooth microsomal fraction (DALLNER et al. 1972). However, other studies have suggested that the highest activity was located in the rough endoplasmic reticulum (MARTIN and THORNE 1974), but activity for these reactions has also been found in the plasma membrane (STRUCK and LENNARZ 1976).

In plant systems several studies have been done to pinpoint the location of these various activities. For example, membranes isolated from hypocotyls of *Phaseolus aureus* were separated by continuous sucrose gradient centrifugation. The highest specific activity of the enzyme which synthesizes dolichyl-phosphoryl-mannose was found in the fraction enriched in membranes of the endoplasmic reticulum (LEHLE et al. 1978). However, the highest specific activity for the enzymes that incorporate mannose into ethanol-insoluble material (or perhaps the highest concentration of endogenous acceptors) was found in the fraction that was enriched in Golgi membranes. In castor bean endosperm, the highest specific activity, as well as the highest total activity of the enzymes involved in the formation of the polyprenyl-linked sugars, was found in the endoplasmic reticulum fraction (MARRIOTT and TANNER 1979). On the other hand, the activities were very low in glyoxysomes and in mitochondria. This is interesting in view of the fact that some glyoxysomal enzymes have been found to be glycoproteins. The activities of the various reactions involved in lipid-linked saccharide synthesis increased several-fold during the first few days of germination which is the time during which rapid formation of glyoxysomes occurs (MARRIOTT and TANNER 1979).

10 Conclusions

It seems clear from the evidence accumulated thus far that lipid-linked saccharides play an important role in the glycosylation of those proteins that have an oligosaccharide chain attached to protein in a GlcNAc-asparagine linkage. Although much of the initial work on the lipid-linked saccharide pathway has come from studies in animal systems, many similar types of reactions have been demonstrated in plant systems. Therefore it appears likely that the pathway of assembly of the oligosaccharide moiety in plants is generally similar to that in animal systems. It is, however, not known whether the plant lipid-linked oligosaccharides also contain glucose, in addition to mannose and GlcNAc. It is also not known whether the oligosaccharide portions of plant glycoproteins undergo any type of processing reactions such as those described in animal tissues. Hopefully, as the cell-free protein synthesizing systems are coupled with membrane glycosylating enzymes, the nature of the initial "glycoprotein", as well as any later forms will become more apparent. Perhaps also the use of glycosylation inhibitors or other compounds that inhibit glycoprotein assembly at various steps will help in the understanding of the mechanism and function of glycosylation.

References

Allen CM, Kalin JR, Sack J, Veruzzo D (1978) CTP-dependent dolichol phosphorylation by mammalian cell homogenates. Biochemistry 17:5020–5026

Basha SM, Beevers L (1976) Glycoprotein metabolism in the cotyledons of *Pisum sativum* during development and germination. Plant Physiol 57:93–97

Baynes JW, Hsu AF, Heath EC (1973) The role of mannosyl-phosphoryl-dihydropolyisoprenol in the synthesis of mammalian glycoproteins. J Biol Chem 248:5693–5704

Behrens NH, Leloir LF (1970) Dolichol monophosphate glucose: an intermediate in glucose transfer in liver. Proc Natl Acad Sci USA 66:153–159

Behrens NF, Parodi AI, Leloir LF (1971) Glucose transfer from dolichol monophosphate glucose. Proc Natl Acad Sci USA 68:2857–2860

Behrens NF, Carminatti H, Staneloni RJ, Leloir LF, Cantarella AI (1973) Formation of lipid-bond oligosaccharides containing mannose. Their role in glycoprotein synthesis. Proc Natl Acad Sci USA 70:3390–3394

Brett CT, Leloir LF (1977) Dolichol monophosphate and its sugar derivatives in plants. Biochem J 161:93–101

Caccam JF, Jackson JJ, Eyler EH (1969) The biosynthesis of mannose containing glycoproteins. Biochem Biophys Res Commun 35:505–511

Carlo PL, Villamez CL (1979) Solubilization and properties of polyprenyl phosphate: GDP-mannose mannosyl transferase. Arch Biochem Biophys 198:117–123

Carson DD, Lennarz WJ (1979) Inhibition of polyisoprenoid and glycoprotein biosynthesis causes abnormal embryonic development. Proc Natl Acad Sci USA 76:5709–5713

Chambers J, Forsee WT, Elbein AD (1977) Transfer of mannose from mannosyl-phosphoryl-polyisoprenol to lipid-linked oligosaccharides by extracts of pig aorta. J Biol Chem 252:2498–2506

Chapman A, Li E, Kornfeld S (1979) The biosynthesis of the major lipid-linked oligosaccharide of chinese hamster ovary cells occurs by the ordered addition of mannose residues. J Biol Chem 254:10243–10249

Chapman A, Fujimoto K, Kornfeld S (1980) The primary glycosylation defect in class E thy-l-negative mutant mouse lymphoma cells is an inability to synthesize dolichol-P-mannose. J Biol Chem 255:4441–4446

Chen WW, Lennarz WJ (1976) Participation of a trisaccharide-lipid in glycosylation of oviduct membrane glycoproteins. J Biol Chem 251:7802–7809

Daleo CR, Pont-Lezica R (1977) Synthesis of dolichol phosphate by a cell free extract from peas. FEBS Lett 74:247–250

Daleo GR, Hopp HE, Romero PA, Pont-Lezica R (1977) Biosynthesis of dolichol phosphate by subcellular fractions from liver. FEBS Lett 81:411–414

Dallner G, Behrens NH, Parodi A, Leloir LF (1972) Subcellular distribution of dolichol phosphate. FEBS Lett 24:315–317

Delmer DP, Kulow C, Ericson MC (1978) Glycoprotein synthesis in plants. II. Structure of mannolipid intermediates. Plant Physiol 61:25–29

Duskin D, Bornstein P (1977) The role of glycosylation in the enzymatic conversion of procollagen to collagen. Studies using Tunicamycin and conconavalin A. Arch Biochem Biophys 185:326–332

Eagon PC, Heath EC (1977) Glycoprotein biosynthesis in myeloma cells. Characterization of the nonglycosylated immunoglobulin light chain secreted in the presence of 2-deoxy-glucose. J Biol Chem 252:2372–2383

Eckhardt K, Thrum H, Bradler G, Tonew E, Tonew M (1975) Streptovirudin: new antibiotics with antibacterial and antiviral activity. J Antibiot Ser A 28:274–279

Elbein AD (1979) The role of lipid-linked saccharides in the biosynthesis of complex carbohydrates. Annu Rev Plant Physiol 30:239–272

Elbein AD, Gafford J, Kang MS (1979) Inhibition of lipid-linked saccharides. Comparison of tunicamycin, streptovirudin, and antibiotic 24010. Arch Biochem Biophys 196:311–318

Endo A, Kuroda M, Tanzawa K (1976) Competitive inhibition of 3-hydroxymethylglutaryl CoA reductase by ML-236B and ML-236A. Fungal metabolites having hypercholesterolenic activity. FEBS Lett 72:323–326

Ericson MC, Chrispeels MJ (1973) Isolation and characterization of glucosamine-containing glycoproteins from cotyledons of *Phaseolus aureus*. Plant Physiol 52:98–104

Ericson MC, Delmer DP (1977) Glycoprotein synthesis in plants. I. Role of lipid intermediates. Plant Physiol 59:341–347

Ericson MC, Gafford J, Elbein AD (1977) Tunicamycin inhibits GlcNAc-lipid formation in plants. J Biol Chem 252:7431–7433

Ericson MC, Gafford J, Elbein AD (1978a) Evidence that the lipid carrier for GlcNAc is different from that for mannose in mung beans and cotton fibers. Plant Physiol 61:274–277

Ericson MC, Gafford J, Elbein AD (1978b) Bacitracin inhibits the synthesis of lipid-linked saccharides and glycoproteins in plants. Plant Physiol 62:373–376

Ericson MC, Gafford J, Elbein AD (1978c) *In vivo* and *in vitro* inhibition of lipid-linked saccharides and glycoprotein synthesis in plants by amphomycin. Arch Biochem Biophys 191:698–704

Evans PJ, Hemming FW (1973) The unambiguous characterization of dolichol phosphate mannose as a product of mannosyl transferase in pig liver endoplasmic reticulum. FEBS Lett 31:335–338
Forsee WT, Elbein AD (1973) Biosynthesis of glucosyl and mannosyl-phosphoryl-polyprenols. J Biol Chem 248:2858–2867
Forsee WT, Elbein AD (1975) Glycoprotein biosynthesis in plants. Demonstration of lipid-linked oligosaccharides. J Biol Chem 250:9283–9293
Forsee WT, Valkovich G, Elbein AD (1976) Glycoprotein biosynthesis in plants. Formation of lipid-linked oligosaccharides of mannose and GlcNAc by mung bean seedlings. Arch Biochem Biophys 174:469–479
Garoff H, Schwarz RT (1978) Glycosylation is not necessary for membrane insertion and cleavage of semliki forest virus membrane proteins. Nature 274:487–489
Goldstein JL, Brown MS (1977) The low density lipoprotein pathway and its relation to atherosclerosis. Annu Rev Biochem 46:897–930
Goldstein JL, Hegelson JAS, Brown MS (1979) Inhibition of cholesterol synthesis on the low density lipoprotein receptor. J Biol Chem 254:2403–2409
Grange DK, Adair WL Jr (1977) Studies on the biosynthesis of dolichyl-phosphate. Evidence for *in vitro* formation of 2,3-dehydrodolichyl-phosphate. Biochem Biophys Res Commun 79:734–740
Hasilik A, Tanner W (1978) Carbohydrate moiety of carboxypeptidase Y and its perturbation of its biosynthesis. Eur J Biochem 91:567–575
Heifetz A, Elbein AD (1977a) Solubilization and properties of mannose and GlcNAc transferases involved in the formation of polyprenyl-sugar intermediates. J Biol Chem 252:3057–3063
Heifetz A, Elbein AD (1977b) Biosynthesis of a man-beta-GlcNAc-GlcNAc-pyrophosphoryl-polyprenol by a solubilized enzyme from aorta. Biochem Biophys Res Commun 75:20–28
Hemming FW (1974) Lipids in glycan biosynthesis. In: Goodwin TW (ed) Biochemistry of lipids. 4, Butterworth, London, pp 39–58
Hickman S, Kulczycki A Jr, Lynch RG, Kornfeld S (1977) Studies on the mechanism of tunicamycin inhibition of IgA and IgE secretion by plasma cells. J Biol Chem 252:4402–4408
Hopp HE, Romero PA, Pont-Lezica R (1978a) On the inhibition of cellulose by coumarin. FEBS Lett 86:259–262
Hopp HE, Romero PA, Pont-Lezica R (1978b) Synthesis of cellulose precursors. Eur J Biochem 84:259–262
Hsu A-F, Baynes JW, Heath EC (1974) The role of a dolichol oligosaccharide as an intermediate in glycoprotein biosynthesis. Proc Natl Acad Sci USA 71:2391–2395
Hubbard SC, Robbins PW (1979) Synthesis and processing of protein-linked oligosaccharides *in vivo*. J Biol Chem 254:4568–4576
Hurkman WJ, Pedersen K, Smith LD, Larkins BA (1979) Synthesis and processing of maize storage proteins in *Xenopus* oocytes infected with maize mRNA. Plant Physiol 63:524A
Irving RA, Toneguzzo F, Rhee SH, Hofmann T, Ghosh AP (1979) Biosynthesis and assembly of membrane glycoproteins. Presence of leader peptide in nonglycosylated precursor of membrane glycoproteins of vesicular stomatitis virus. Proc Natl Acad Sci USA 76:570–574
Ito T, Takatsuki A, Kawamura K, Sato K, Tamura G (1980) Isolation and structures of components of tunicamycin. Agric Biol Chem 44(3):695–698
James DW Jr, Elbein AD (1980) Effects of several tunicamycin like antibiotics on glycoprotein biosynthesis in mung beans and suspension-cultured soybean cells. Plant Physiol 65:460–464
James MJ, Kandutsch AA (1979) Interrelationships between dolichol and sterol synthesis in mammalian cell cultures. J Biol Chem 254:8442–8446
Jung P, Tanner W (1973) Identification of the lipid intermediate in yeast mannan biosynthesis. Eur J Biochem 37:1–6
Kang MS, Spencer JP, Elbein AD (1978) Amphomycin inhibition of mannose and GlcNAc incorporation into lipid-linked saccharides. J Biol Chem 254:8860–8866

Katz FN, Rothman JE, Lingappa VR, Blobel G, Lodish HF (1977) Membrane assembly, *in vitro* synthesis, glycosylation, and assymetric insertion of a transmembrane protein. Proc Natl Acad Sci USA 74:3278–3282

Kauss H (1969) A plant mannosyl-lipid acting in reversible transfer of mannose. FEBS Lett 5:81–84

Keller RL, Adair WL Jr, Ness G (1979) Studies on the regulation of glycoprotein synthesis. J Biol Chem 254:9966–9969

Kiely ML, McKnight GS, Schimke RT (1976) Studies on the attachment of carbohydrate to ovalbumin nascent chains in hen oviduct. J Biol Chem 251:5490–5495

Kjobakken J, Colvin JR (1973) Biosynthesis of cellulose by a particulate enzyme system from *Acetobacter xylinum*. In: Loewus F (ed) Biogenesis of plant cell wall polysaccharides, Academic Press, New York, pp 361–371

Kornfeld R, Kornfeld S (1976) Comparative aspects of glycoprotein structure. Annu Rev Biochem 45:217–237

Kuo S-C, Lampen JO (1972) Inhibition by 2-deoxyglucose of synthesis of a glycoprotein enzyme by protoplasts of saccharomyces. J Bacteriol 111:419–429

Laine R, Elbein AD (1971) Steryl glucosides in *Phaseolus aureus*. Biochemistry 10:2547–2553

Lehle L, Tanner W (1976) The specific site of tunicamycin inhibition in the formation of dolichol bound *N*-acetylglucosamine derivatives. FEBS Lett 71:167–170

Lehle L, Fartaczek F, Tanner W, Kauss H (1976) Formation of polyprenyl-linked mono- and oligosaccharides in *Phaseolus aureus*. Arch Biochem Biophys 175:419–426

Lehle L, Bowles DJ, Tanner W (1978) Subcellular site of mannosyl transfer to dolichyl phosphate in *Phaseolus aureus*. Plant Sci Lett 11:27–34

Levy JA, Carminatti H, Cantarella AI, Behrens NH, Leloir LF, Tabora E (1974) Mannose transfer to lipid-linked Di-*N*-Acetylchitobiose. Biochem Biophys Res Commun 60:118–125

Li E, Kornfeld S (1978) Structure of the altered oligosaccharide present in glycoprotein from a clone of Chinese hamster ovary cells deficient in *N*-acetylglucosaminyltransferase activity. J Biol Chem 253:6426–6431

Li E, Kornfeld S (1979) Structural studies on the major high mannose oligosaccharide units from Chinese hamster ovary cell glycoproteins. J Biol Chem 254:1600–1606

Lingappa VR, Lingappa JR, Prasad R, Ebner KE, Blobel G (1978) Coupled cell free synthesis, segregation and core glycosylation of a secretory protein. Proc Natl Acad Sci USA 75:2338–2342

Lis H, Sharon N (1978) Soybean agglutinin, a plant glycoprotein. J Biol Chem 253:3468–3476

Lucas JJ, Waechter CW, Lennarz WJ (1975) The participation of lipid-linked oligosaccharides in the synthesis of membrane glycoproteins. J Biol Chem 250:1992–2002

Mahoney WC, Duskin D (1979) Biological activities of the two major components of tunicamycin. J Biol Chem 254:6572–6576

Marriott KM, Tanner W (1979) Dolichyl-phosphate dependent glycosyl transfer reactions in the endoplasmic reticulum of castor bean endosperm. Plant Physiol 64:445–449

Martin HG, Thorne KJ (1974) The involvement of endogenous dolichol in the formation of lipid-linked precursors of glycoproteins in rat liver. Biochem J 138:281–289

Melchers F (1973) Biosynthesis, intracellular transport and secretion of immunoglobulins. Biochemistry 12:1471–1475

Mellor RB, Roberts LM, Lord JM (1978) Glycosylation of endogenous proteins by endoplasmic reticulum membranes from castor beans. Biochem J 182:629–631

Mills JJ, Adamany AM (1978) Impairment of dolichyl saccharide synthesis and dolichol mediated glycoprotein assembly in aorta smooth muscle cells in culture by inhibitors of cholesterol biosynthesis. J Biol Chem 253:5270–5273

Mizuno M, Shimojima Y, Sugawara T, Takeda J (1971) On antibiotic 24010. J Antibiotics 24:896–899

Molner J, Chao H, Ikehara Y (1971) Phosphoryl-*N*-acetylglucosamine transfer to a lipid acceptor of liver microsomal fractions. Biochim Biophys Acta 239:401–411

Nakamura S, Arai M, Karasawa K, Yonehara H (1957) On the antibiotic, mycospocidin. J Antibiotics Ser A 10:248–253

Oliver GJA, Hemming FW (1975) The transfer of mannose to dolichol diphosphate oligosaccharides in pig liver endoplasmic reticulum. Biochem J 152:191–199
Osborn MJ (1969) Structure and biosynthesis of the bacterial cell wall. Annu Rev Biochem 38:501–538
Pan YT, Schmitt JW, Sanford BA, Elbein AD (1979) Adherence of bacteria to mammalian cells: inhibition by tunicamycin and streptovirudin. J Bacteriol 139:507–514
Pless DD, Lennarz WJ (1977) Enzymatic conversion of proteins to glycoproteins. Proc Natl Acad Sci USA 74:134–138
Pont-Lezica R, Romero PA, Dankert MA (1976) Membrane-bound UDP-glucose: lipid glucosyl transferases from peas. Plant Physiol 58:675–680
Pont-Lezica R, Romero PA, Hopp HE (1978) Glucosylation of membrane-bound proteins by lipid-linked glucose. Planta 140:177–183
Richards JB, Hemming FW (1972) The transfer of mannose from GDP-mannose to dolichyl-phosphate and protein by pig liver endoplasmic reticulum. Biochem J 130:77–93
Robbins PW, Krag SS, Liu T (1977) Effect of UDP-glucose addition on the synthesis of mannosyl-lipid-linked oligosaccharides by cell-free fibroblast preparations. J Biol Chem 252:1780–1785
Roberts RM (1975) The incorporation of glucosamine into glycolipids and glycoproteins of membrane preparations from *Phaseolus aureus*. Plant Physiol 55:431–436
Robinson GB (1969) The role of polyribosomes in the biosynthesis of glycoproteins. Biochem J 115:1077–1078
Rothman JE, Fene RE (1980) Coated vesicles transport newly synthesized membrane glycoproteins from endoplasmic reticulum to plasma membranes in two successive stages. Proc Natl Acad Sci USA 77:780–784
Rothman JE, Lodish HF (1977) Synchronized transmembrane insertion and glycosylation of a nascent membrane protein. Nature 269:775–780
Schneider EG, Nguyen HT, Lennarz WJ (1978) The effect of tunicamycin, an inhibitor of protein glycosylation on embryonic development in the sea urchin. J Biol Chem 253:2348–2355
Schwaiger H, Tanner W (1979) Effects of gibberellic acid and of tunicamycin on glycosyl transferase activities and on amylase secretion in barley. Eur J Biochem 102:375–381
Schutzbach JS, Springfield JD, Jensen JW (1980) The biosynthesis of oligosaccharide-lipids. Formation of an α-1,2-mannosyl-mannose linkage. J Biol Chem 255:4170–4175
Schwarz RJ, Datema R (1980) Inhibitors of protein glycosylation. Trends Biochem Sci 5:65–68
Schwarz RT, Rohrschneider JM, Schmidt MFG (1976) Suppression of glycoprotein formation of semiliki forest, influenza and avian sarcoma virus by tunicamycin. J Virol 19:782–791
Schwarz RT, Schmidt MFG, Lehle L (1978) Glycosylation *in vitro* of Semliki forest virus and influenza virus glycoproteins and its suppression by nucleotide-2-deoxyhexose. Eur J Biochem 85:163–172
Sharon N, Lis H (1978) Comparative biochemistry of plant glycoproteins. Biochem Soc Trans 7:783–805
Spencer JP, Elbein AD (1980) Transfer of mannose from GDP-mannose to lipid-linked oligosaccharides by soluble mannosyl transferase. Proc Natl Acad Sci USA 77:2524–2527
Spencer JP, Kang MS, Elbein AD (1978) Inhibition of lipid-linked saccharide synthesis by bacitracin. Arch Biochem Biophys 190:829–837
Spiro MJ, Spiro RJ, Bhoyroo VD (1976) Lipid-saccharide intermediates in glycoprotein biosynthesis. J Biol Chem 251:6420–6425
Spiro MJ, Spiro RJ, Bhoyroo VD (1978) Utilization of oligosaccharide-lipids in glycoprotein biosynthesis by thyroid enzyme. Fed Proc Fed Am Soc Exp Biol 37:2285
Spiro MJ, Spiro RJ, Bhoyroo VD (1979) Glycosylation of proteins by oligosaccharide-lipids. J Biol Chem 254:7668–7674
Stone KJ, Strominger JL (1971) Mechanism of action of bacitracin complexation with metal ions and C_{55}-isoprenyl pyrophosphate. Proc Natl Acad Sci USA 68:3223–3227

Struck DK, Lennarz WJ (1976) Utilization of exogenous GDP-mannose for the synthesis of mannose-containing lipids and glycoproteins by oviduct cells. J Biol Chem 251:2511–2519
Struck DK, Lennarz WJ (1977) Evidence for the participation of saccharide-lipids in the synthesis of the oligosaccharide chain of ovalbumin. J Biol Chem 252:1007–1013
Struck DK, Suita PB, Lane MD, Lennarz WJ (1979) Effect of tunicamycin on the secretion of serum proteins by primary cultures of rat and chick hepatocytes. J Biol Chem 254:5334–5337
Sun SM, Ma Y, Buchbinder BO, Hall TC (1979) Comparison of Gl polypeptide synthesized *in vitro* and *in vivo* in the presence and absence of glycosylation inhibitors. Plant Physiol 63:529A
Tabas I, Kornfeld S (1978) The synthesis of complex type oligosaccharides. J Biol Chem 253:7779–7786
Takatsuki A, Arima K, Tamura G (1971) Tunicamycin, a new antibiotic. J Antiobiotics 24:215–223
Takatsuki A, Kohno K, Tamura G (1975) Inhibition of biosynthesis of polyisoprenol sugars in chick embryo microsomes by tunicamycin. Agric Biol Chem 39:2089–2091
Takatsuki A, Kawamura K, Okina M, Kodama Y, Ito T, Tamura G (1977) The structure of tunicamycin. Agric Biol Chem 41(11):2307–2309
Tanner W (1969) A lipid intermediate in mannan biosynthesis in yeast. Biochem Biophys Res Commun 35:144–150
Tkacz JS, Lampen JO (1975) Tunicamycin inhibition of polyisoprenyl-*N*-acetylglucosamine-pyrophosphate formation in calf liver microsomes. Biochem Biophys Res Commun 65:248–257
Toneguzzo F, Ghosh HP (1977) Synthesis and glycosylation *in vitro* of glycoprotein of vesicular stomatitis virus. Proc Natl Acad Sci USA 75:1516–1520
Tucker P, Peska S (1977) *De novo* synthesis and glycosylation of MOPC-46B mouse immunoglobulin light chains in cell free extracts. J Biol Chem 252:4474–4486
Turco SJ, Heath EC (1977) Glucuronosyl-*N*-acetylglucosaminyl-pyrophosphoryl-dolichol. J Biol Chem 252:2918–2928
Turco SJ, Robbins PW (1979) The initial stages of processing of protein bound oligosaccharide *in vitro* J. Biol Chem 254:4560–4567
Turco SJ, Stetson B, Robbins PW (1977) Comparative rates of transfer of lipid-linked oligosaccharides to endogenous glycoprotein acceptors *in vitro*. Proc Natl Acad Sci USA 74:4411–4414
Villamez CL, Clark CF (1969) A particle bound intermediate in the biosynthesis of plant cell wall polysaccharides. Biochem Biophys Res Commun 36:57–63
Waechter CJ, Lennarz WJ (1976) The role of polyprenol-linked sugars in glycoprotein biosynthesis. Annu Rev Biochem 45:95–112
Waechter CJ, Lucas JJ, Lennarz WJ (1974) Evidence for xylosyl-lipids as intermediates in xylosyl transfer in hen oviduct membranes. Biochem Biophys Res Commun 56:343–350
Warren CD, Jeanloz RW (1973) Chemical synthesis of P^1-dolichyl-P^2-α-mannopyranosyl-pyrophosphate. Biochemistry 14:412–419
Wedgewood JF, Strominger JL (1980) Enzymatic activities in cultured human lymphocytes that dephosphorylate dolichyl-pyrophosphate and dolichyl-phosphate. J Biol Chem 255:1120–1123
Zatz M, Barondes SH (1969) Incorporation of mannose into mouse brain lipid. Biochem Biophys Res Commun 36:511–517

9 Biosynthesis of Lignin

T. Higuchi

1 Occurrence of Lignin in Plants

The origin of lignins in plant tissues is closely related to plant evolution. Land plants, which are believed to have been derived from the Psilophytales during the Silurian to Devonian eras, are always subjected to the mechanical stresses of gravity, winds, and rains, and they have acquired strong supporting organs, such as vascular bundles composed of phloem and xylem. Xylem is comprised of wood fibers, tracheids, and vessels reinforced with lignins. The roles of lignins, which are aromatic polymers, in xylem tissues are: (1) to impart rigidity to the cell walls and enable terrestrial plants to develop large upright forms resistant to various stresses, (2) to assist in the smooth transportation of solutes by decreasing the permeability of cell walls in the conductive xylem tissues, and (3) to resist attack by microorganisms.

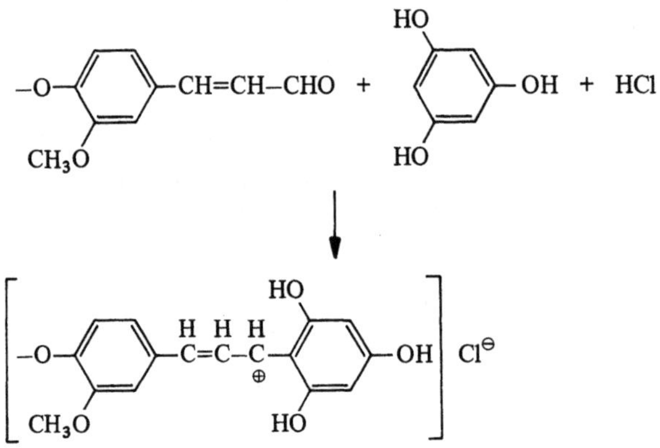

Fig. 1. Mechanism of formation of the chromogen responsible for the red-purple color of the phloroglucinol reaction of lignins

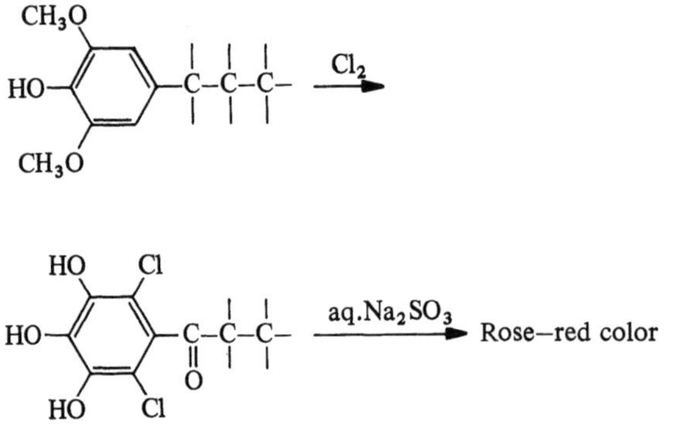

Fig. 2. Suggested mechanism of the Cross and Bevan color reaction of syringyl lignins

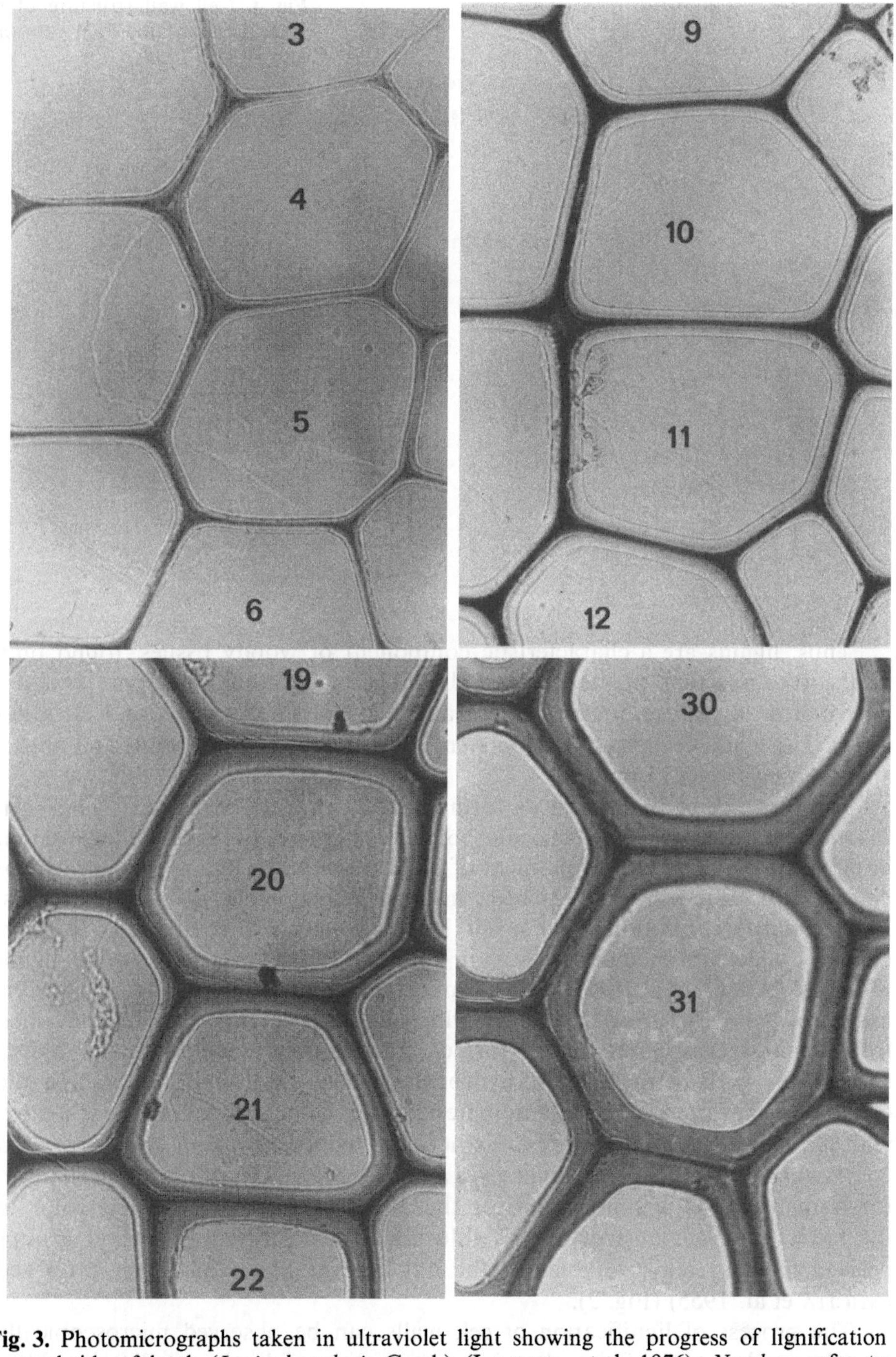

Fig. 3. Photomicrographs taken in ultraviolet light showing the progress of lignification in tracheids of larch (*Larix leptolepis* Gord.) (IMAGAWA et al. 1976). *Numbers* refer to the cell number counted from the cambium

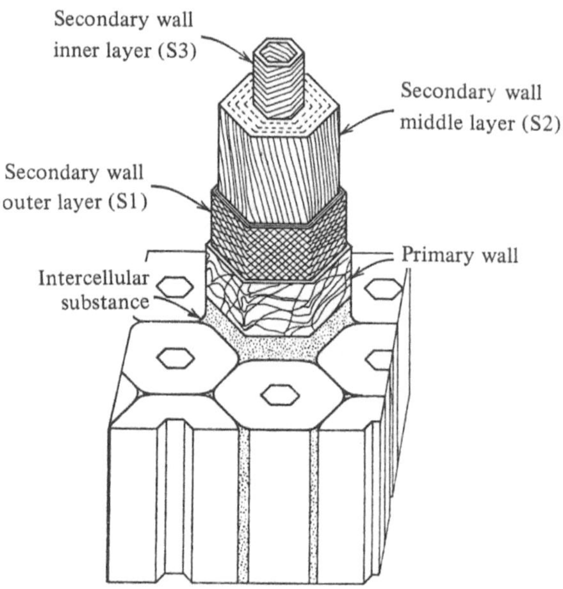

Fig. 4. Cell wall structure of a typical wood fiber. (WARDROP and BLAND 1959)

Thus, lignins are a characteristic component of woody tissues, found only in the true vascular plants, ferns and plants higher than the fern; and are absent from the algae, mosses, fungi, and bacteria (KAWAMURA and HIGUCHI 1964). The lignin content of woody stems of arboreus gymnosperms and angiosperms range from 15 to 36%.

The presence of lignin and its formation in plant cell walls have long been studied by botanists using specific color reactions. For example, formation of a red-purple color upon treatment of plant tissues with phloroglucinol-hydrochloric acid (Wiesner test) has been used to detect lignin in plant materials. The red-purple coloration (mechanism, Fig. 1) has been attributed to p-hydroxycinnamyl aldehyde moieties such as in coniferyl aldehyde groups in the lignin molecule (ADLER et al. 1948). Other important color reactions are the Mäule, and Cross and Bevan reactions, by which angiosperm lignin is distinguished from gymnosperm lignin. In the former reaction, plant materials are treated successively with permanganate, hydrochloric acid, and ammonia, and in the Cross and Bevan reaction plant materials are chlorinated and then treated with sodium sulfite solution. In both reactions angiosperm lignin gives a rose-red color, whereas gymnosperm lignin gives a brown color. The principal mechanism for both reactions is similar and for the direct precursor of the red color in the latter reaction, a structure of chlorinated 3,4,5-trihydroxyphenyl groups derived from syringyl residues in the angiosperm lignin has been proposed (MIGITA et al. 1955) (Fig. 2).

The process of lignification of cell walls can be observed microscopically by taking advantage of the ultraviolet absorption of lignins at 280 nm. Thus ultraviolet microscopy (Fig. 3) has been used to show the progress of lignification in cell walls of larch tracheids (*Larix leptolepis* Gord.) (IMAGAWA et al. 1976).

Lignification was shown to be initiated in the primary walls adjacent to the corners of the cells undergoing lignin deposition in the S1 layer near the cambium (cell no. 3–6), then to proceed in the intercellular layers, primary walls (cell no. 9–12) and secondary walls (cell no. 19–31). In the secondary walls, three layers having different orientation of cellulose microfibrils are referred to as the outer (S1), middle (S2), and inner (S3) layers, and are usually distinguishable by electronmicroscopy (WARDROP and BLAND 1959) (Fig. 4).

A radioautographic study of cottonwood (SALEH et al. 1967) has also shown three different stages of lignification in the development of cells. Administered tritiated ferulic acid was incorporated first into the lignin of cell corners in xylem tissues nearest the cambium. Lignification extended along the middle lamella, first proceeding along the radial walls. Intercellular regions of the cell corners completed lignification last. A recent study (FUJITA and HARADA 1979) of the lignification of compression wood of *Cryptomeria japonica* D. Don has also shown that tritiated phenylalanine and ferulic acid are first incorporated into the intercellular and cell corner regions of the cells which are depositing S1, and the cells in transitional development from S1 to S2, and then into the outer region of the S2 layer of the cells during a rapid S2 thickening and lignification of secondary walls.

2 Morphological Distribution of Lignins in Plant Cell Walls

Recent ultraviolet microscopy studies (SCOTT et al. 1969) on the distribution of lignin in completely lignified cell walls of tracheids of black spruce (*Picea mariana* Mill.) have shown that lignin concentrations are 22, 50, and 85% for the secondary wall, compound middle lamella, and cell corners in earlywood, and 22, 60, and 100% for the corresponding layers in latewood, respectively. The fraction of total cell lignin present in secondary walls, calculated from the lignin concentration and volume fraction, is 72% in earlywood and 81% in latewood. The fraction of lignin in middle lamellae, whose lignin concentration is 85 to 100%, on the other hand, is only 28 and 19% of the lignin in earlywood and latewood, reflecting the considerably smaller volume of the middle lamella.

Distribution of lignin in cell walls of birch wood (*Betula papyrifera* Marsh.) has also been investigated by ultraviolet microscopy (MUSHA and GORING 1975). The cell walls and middle lamellae of vessels gave a characteristic ultraviolet spectrum with a maximum at 276 nm, and were suggested to contain guaiacyl lignin. The middle lamellae of fibers gave a spectrum which has a maximum at a lower wavelength (275 to 276 nm), suggesting that the lignin is composed of guaiacyl-syringyl units. The secondary walls of fibers and ray cells, on the other hand, exhibited a flat maximum at about 270 nm characteristic of syringyl lignin. These results suggest that in hardwoods the localization of the enzyme systems in the biosynthesis of lignin precursors is different in the respective cells and their cell wall layers, and that the lignin in the secondary walls of fibers and ray cells is composed mainly of syringyl units, that the cell corner

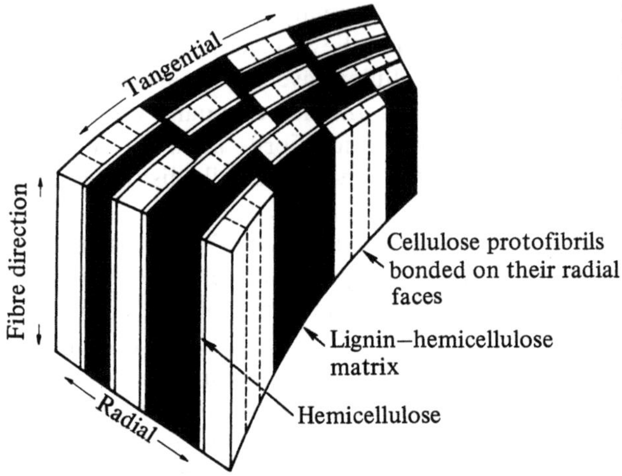

Fig. 5. Interrupted lamella model for the ultrastructural arrangement of lignin, cellulose and hemicellulose in the wood cell wall. (RUEL et al. 1978)

and middle lamella lignins of these cells are of the guaiacyl-syringyl type, and that vessel lignin is mainly of a guaiacyl type.

The pattern of the fine distribution of lignin in cell walls of black spruce and silver fir (*Abies alba* Mill.) has been observed recently by scanning transmission electronmicroscopy (RUEL et al. 1978); ultrathin cross-sections of tracheids treated with permanganate showed that the S2 layer is lamellated at the level of the individual microfibrils, and that the lignin in each lamella is arranged as dark spots or stripes aligned in approximately parallel rows and that the direction of alignment is parallel to the middle lamella (Fig. 5).

Lignin always occurs in intimate association with the cell wall polysaccharides, and even lignins extracted by mild conditions, as is the case with dioxane lignin, suffer to some extent from condensations, oxidations, additions, and substitutions.

However, when wood powder is milled in toluene by a vibrational ball mill, cellulose microfibrils in the cell walls are degraded extensively, and thus subsequent extraction of the milled wood with aqueous dioxane yields about 30% of the original lignin. The lignin is called milled wood lignin (MWL) and was first prepared by BJÖRKMAN (1956). Analytical and degradation studies of MWL have shown that the lignin contains free phenolic hydroxyl groups, methoxyl groups, p-hydroxybenzyl alcohol moieties, carbonyl groups conjugated or nonconjugated with aromatic rings, and ethylenic double bonds, and that the original structure of lignin is preserved without significant secondary reactions. Since BJÖRKMAN's investigation MWL has been used as a standard lignin.

Chemical studies of spruce MWL (ADLER 1977) have indicated that the lignin is an aromatic polymer in which the monomeric guaiacyl propane units are joined by both ether- and carbon-to-carbon linkages; the ether linkages include guaiacylglycerol-β-aryl ether (**1**) which is the commonest inter-phenylpropane unit linkage in lignin, and guaiacylglycerol-α-aryl ether (**2**). Guaiacyl propane units involving both ether and carbon-to-carbon bonds, such as phenylcoumaran (**3**) and the pinoresinol-type structure (**4**) have also been found in the lignin. With regard to carbon-to-carbon bonds, the biphenyl-type linkage (**5**), which is predominant in lignin, β-5′ (**6**), and α-6′ (**7**) combinations have been found. Also, the 1,2-diguaiaylpropane-1,3-diol substructure (**8**), and diphenyl ether substructure (**9**) have recently been established to be present as main substructures in lignin.

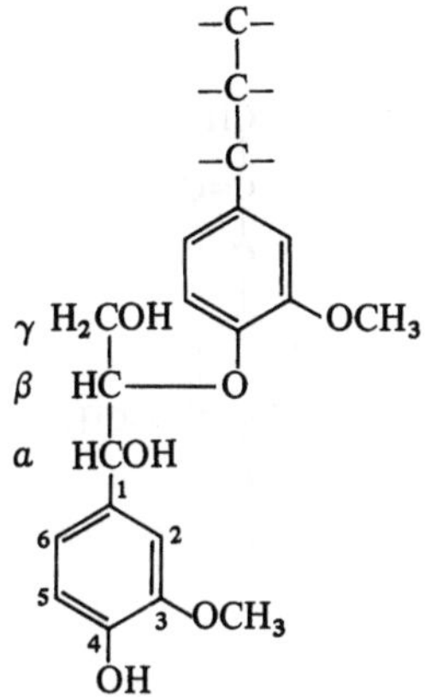

(1)

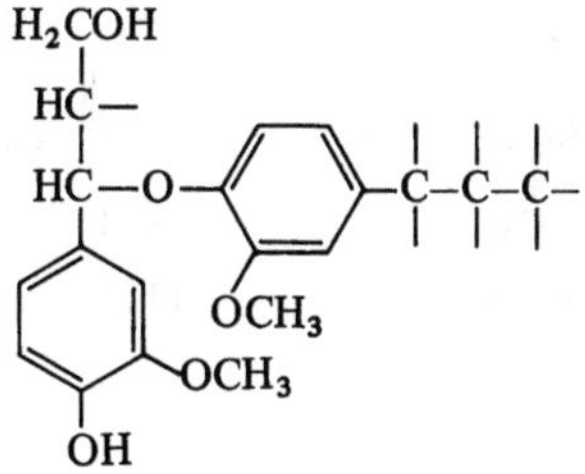

(2)

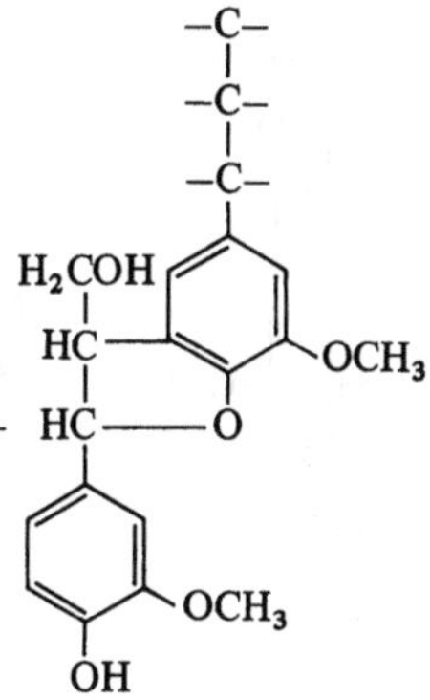

(3)

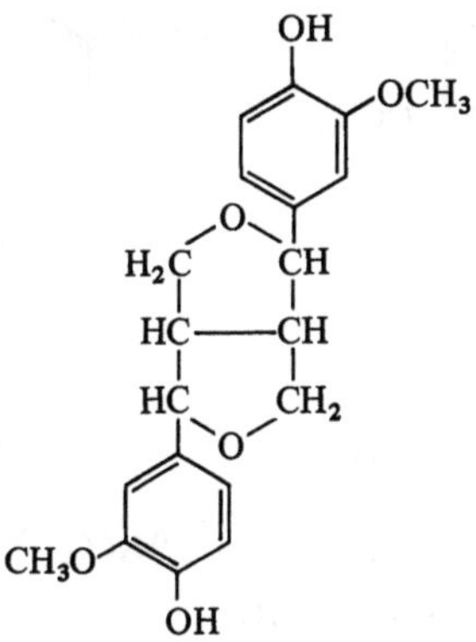

(4)

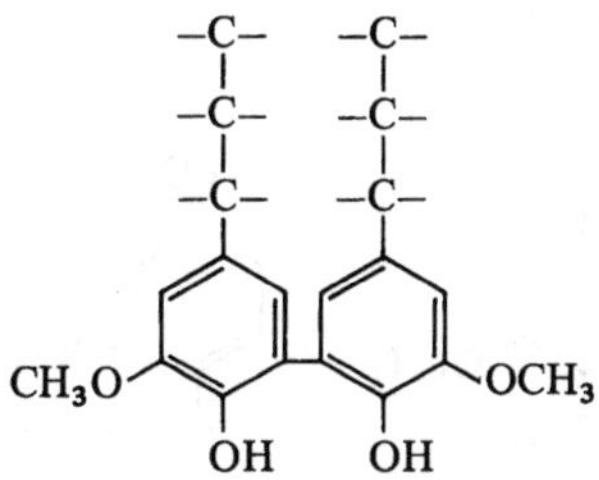

(5)

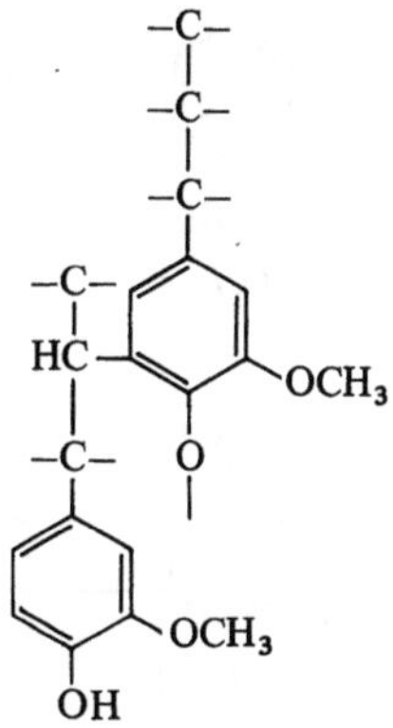

(6)

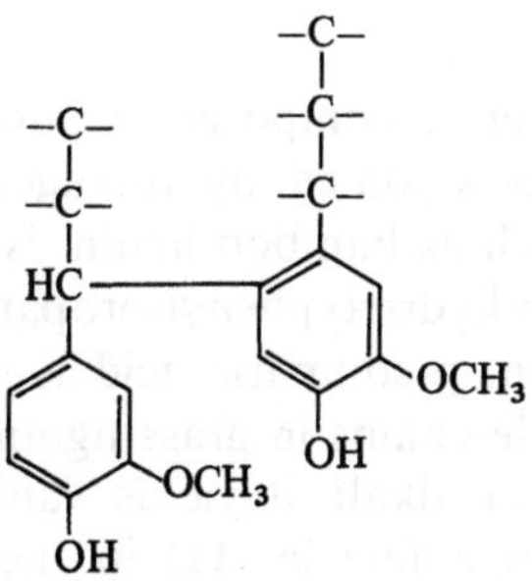

(7)

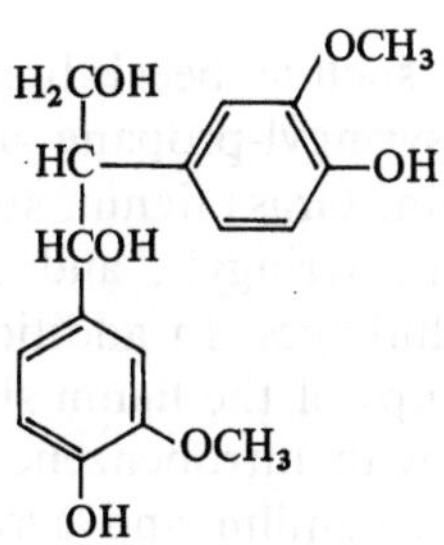

(8)

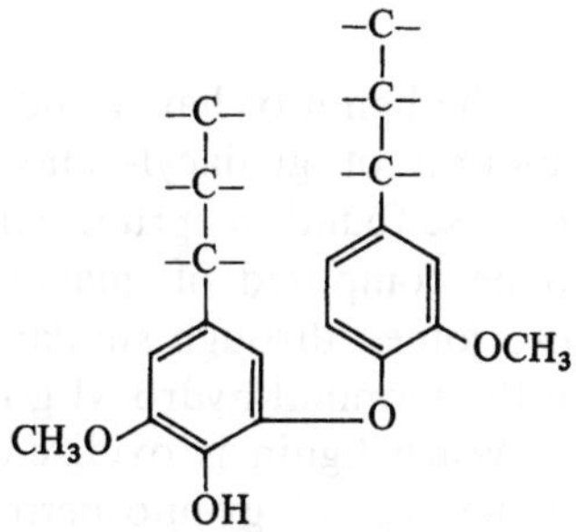

(9)

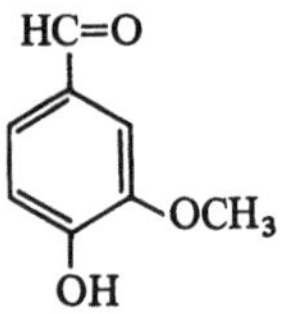

(10)

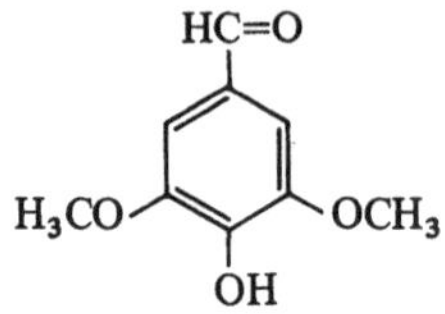

(11)

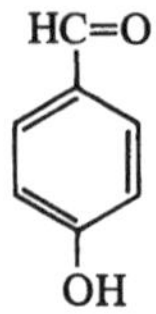

(12)

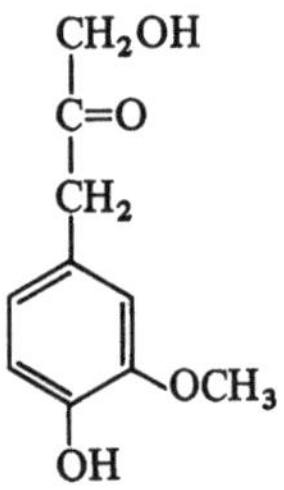

(13)

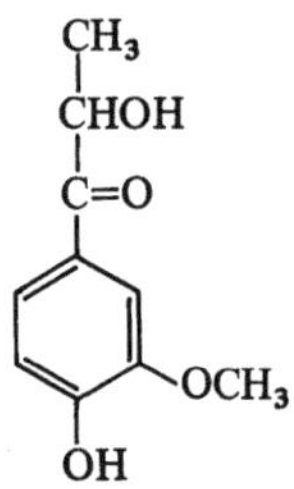

(14)

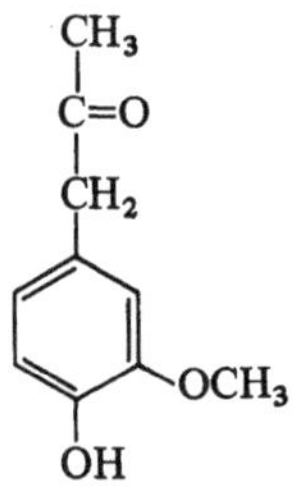

(15)

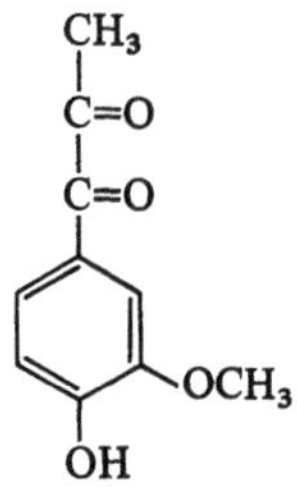

(16)

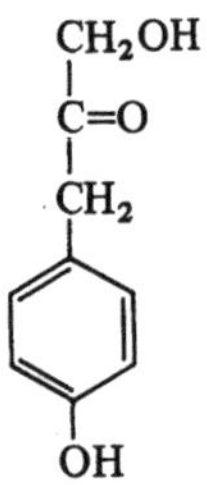

(17)

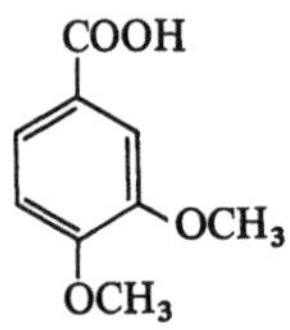

(18)

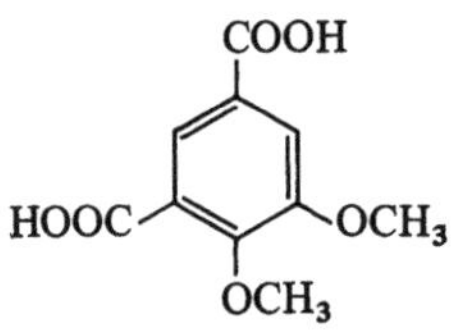

(19)

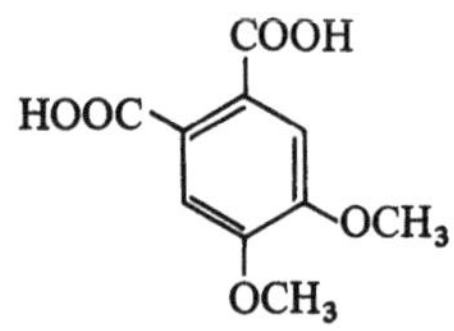

(20)

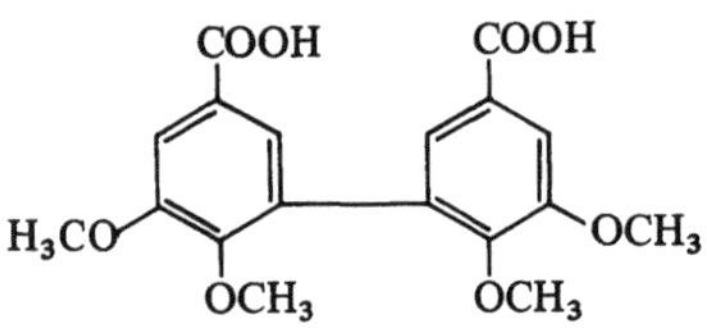

(21)

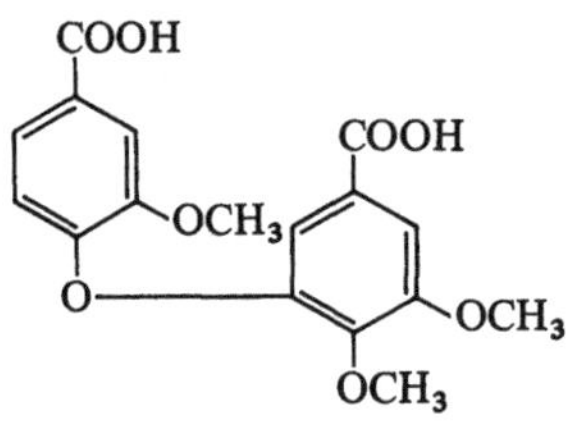

(22)

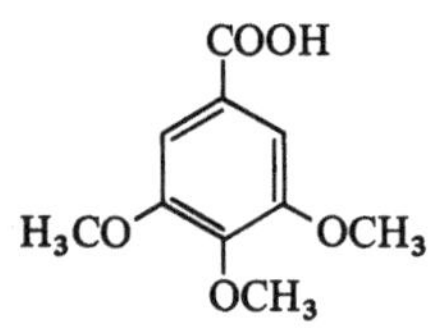

(23)

The lignin of hardwoods, such as beech lignin, is composed of about equal amounts of guaiacyl- and syringyl-propane units joined by linkages similar to those found in spruce lignin. Grass lignin, such as bamboo lignin, is thought to be composed of guaiacyl-, syringyl-, and p-hydroxyphenyl-propane units, also joined through similar linkages. In addition, p-coumaric acid is esterified to the terminal hydroxyl groups of the lignin side chains in grass lignins.

When lignin is oxidized with nitrobenzene in alkali, it yields vanillin (**10**) in the case of gymnosperms; vanillin and syringaldehyde (**11**) in the case of angiosperms; and p-hydroxybenzaldehyde (**12**), vanillin, and syringaldehyde in the case of grasses. When lignin is refluxed with dioxane-water (9:1, v/v) containing 0.2 N hydrochloric acid, it gives a mixture of "acidolysis monomers," including β-hydroxyconiferyl alcohol, (**13**) α-hydroxypropioguaiacone (**14**), guaiacyl acetone (**15**), vanilloyl methyl ketone (**16**) and β-hydroxy-*p*-coumaryl alcohol (**17**) in the case of gymnosperms, and, in addition, the corresponding syringyl derivatives in the case of angiosperms and grasses. High pressure catalytic hydrogenolysis of gymnosperm lignins gives guaiacylpropanes, cyclohexylpropanes, or both, and that of angiosperm lignins gives the corresponding syringylpropanes, as well as guaiacyl derivatives.

Upon permanganate oxidation, methylated gymnosperm lignin gives methoxybenzoic acids, including veratric (**18**), isohemipinic (**19**), metahemipinic (**20**), and dehydrodiveratric acids (**21**), and 2,5′,6′-trimethoxy-4,3′-dicarboxy-diphenyl ether (**22**). Angiosperm lignins give 3,4,5-trimethoxybenzoic acid (**23**) and other products in addition to 18–22 (ADLER 1977).

3 Biogenesis of Lignin Precursors

It was established (ACERBO et al. 1960, KRATZL and FAIGLE 1958) that the distribution of incorporated radioactivity in the aromatic ring of the vanillin obtained by nitrobenzene oxidation of D-1-^{14}C-glucose-fed plants was similar to that observed in the microbial biosynthesis of aromatic amino acids via shikimic acid from glucose.

These results and an efficient incorporation of $^{14}CO_2$ into lignins indicate that lignins are derived from sugars produced by photosynthesis, and that biosynthesis of phenylpropanoids may follow similar pathways in both plants and microorganisms.

3.1 Shikimic Acid – Phenylalanine Pathway

The pathway for synthesis of the aromatic amino acids, L-phenylalanine, L-tyrosine, and L-tryptophan in microorganisms is now well established as a result of the work of DAVIS et al. (DAVIS 1955, SPRINSON 1960), who worked with auxotrophic mutants of *Escherichia coli*. The auxotrophic mutants could convert shikimic acid but not carbohydrates to the aromatic amino acids, thus establishing the acid as an obligate intermediate (see HARBORNE, Chap. 6, Vol. 8, this Series).

Efficient incorporation of shikimic acid into lignins was well demonstrated by tracer experiments (BROWN and NEISH 1955, HASEGAWA et al. 1960, GAMBORG 1967), and it was found by EBERHARDT and SCHUBERT (1956) that [2,6-^{14}C]-shikimic acid is incorporated into the corresponding carbons of the aromatic ring of the vanillin derived from sugarcane lignin.

Shikimic acid, without any rearrangement of the carbon atoms of its cyclohexene ring, was converted to the aromatic rings of the lignin. The conversion

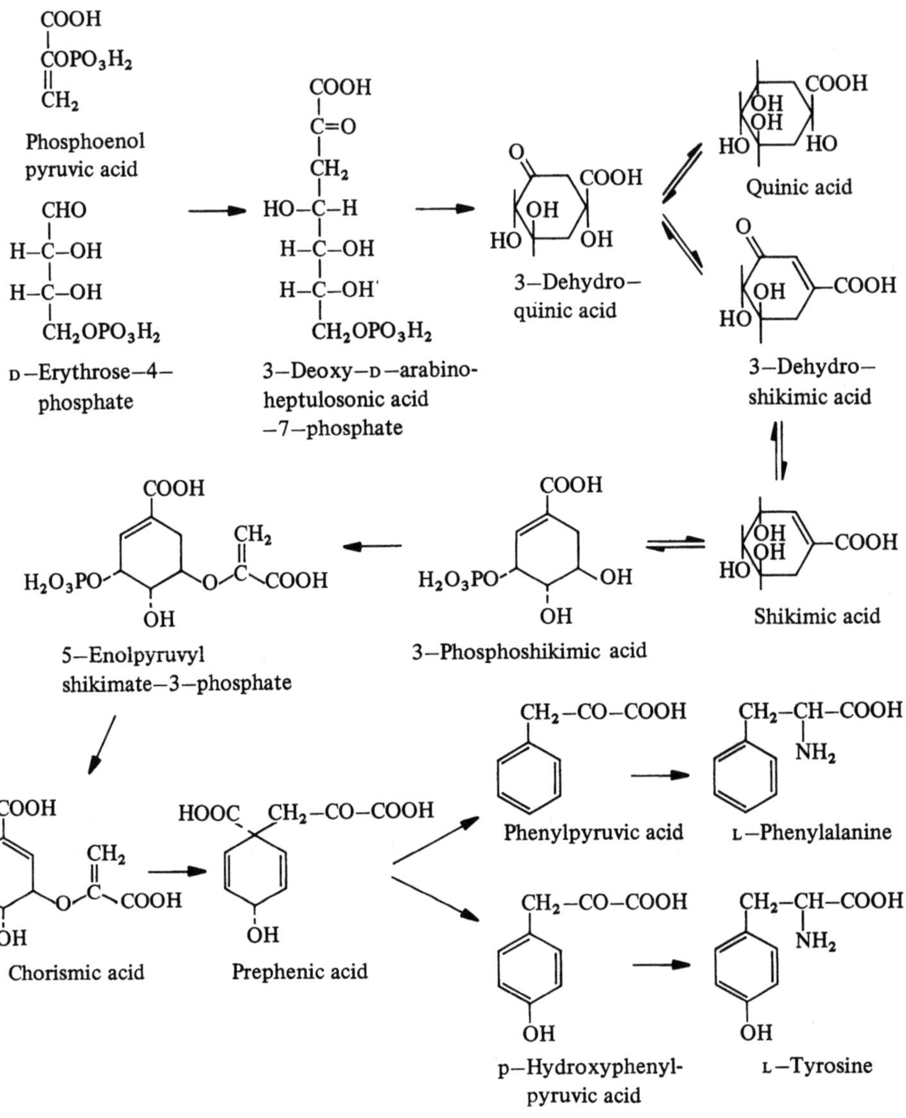

Fig. 6. Biosynthesis of the phenylpropanoid amino acids, L-phenylalanine, and L-tyrosine from carbohydrates

was in agreement with the mode of formation of the aromatic acids in microorganisms. Incorporation of shikimic acid into both L-phenylalanine and L-tyrosine has also been established in several plants (MCCALLA and NEISH 1959, GAMBORG and NEISH 1959). In view of the established shikimic acid pathway in microorganisms, and the wide distribution of shikimic acid in plants, aromatic amino acids are considered to be intermediates between shikimic acid and lignins (Fig. 6).

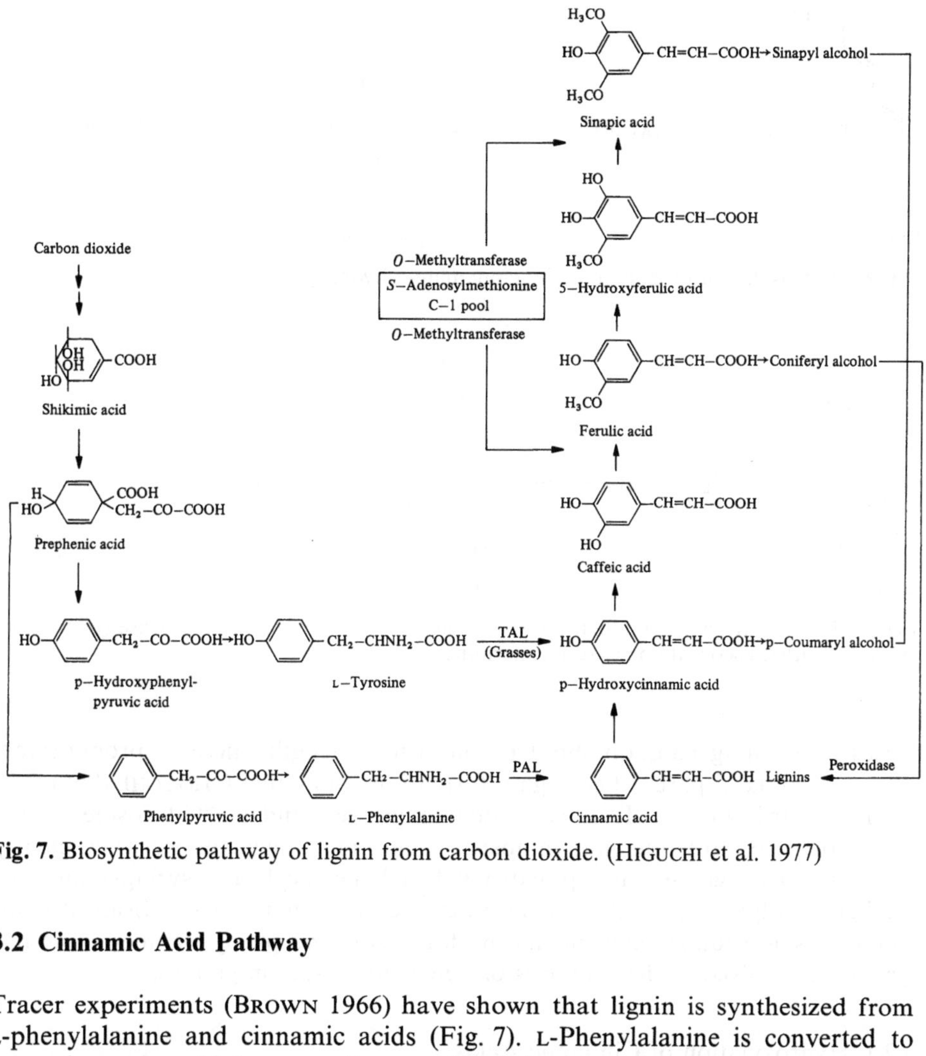

Fig. 7. Biosynthetic pathway of lignin from carbon dioxide. (HIGUCHI et al. 1977)

3.2 Cinnamic Acid Pathway

Tracer experiments (BROWN 1966) have shown that lignin is synthesized from L-phenylalanine and cinnamic acids (Fig. 7). L-Phenylalanine is converted to trans-cinnamic acid in a reaction catalyzed by phenylalanine ammonia-lyase (PAL) (EC 4.3.1.5) (KOUKOL and CONN 1961), which is a widely distributed and key enzyme in the synthesis of various phenolic compounds including lignin in higher plants. Taxonomic surveys (HIGUCHI and BARNOUD 1964, YOUNG et al. 1966) of the distribution of the ammonia-lyase showed that the enzyme is present only in organisms which can synthesize lignin or other cinnamic acid-derived products.

The enzyme activity is relatively high in lignifying tissues of various plant stems (HIGUCHI 1966, RUBERY and NORTHCOTE 1968), but considerably lower in young non-differentiated, or in completely lignified, tissues.

Tyrosine ammonia-lyase (TAL) (NEISH 1961), which catalyzes the formation of p-coumaric acid from L-tyrosine, is characteristically found in grasses, the lignin of which contains p-coumaryl alcohol as an additional lignin monomer, as well as esterified p-coumaric acid residues. The activities of both ammonia-

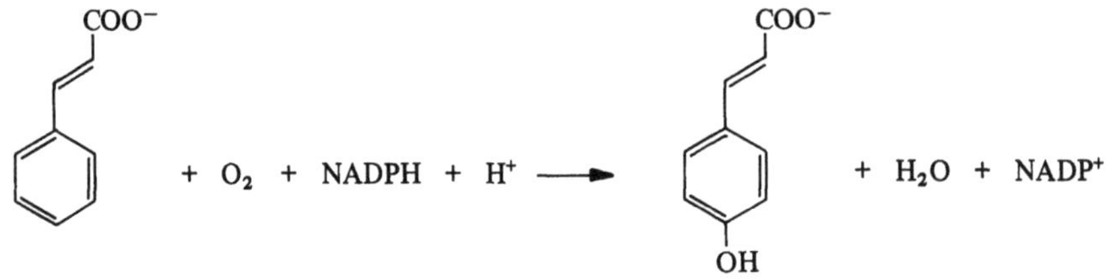

Cinnamate p–Coumarate

Fig. 8. Hydroxylation of cinnamate by cinnamate-4-hydroxylase

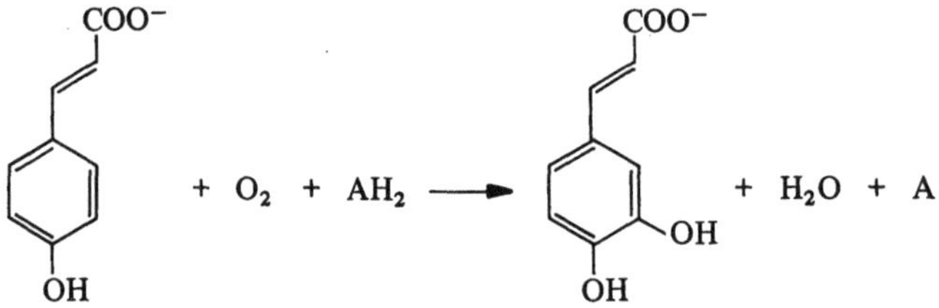

p–Coumarate Caffeate

Fig. 9. Hydroxylation of p-coumarate to caffeate by p-coumarate 3-hydroxylase (phenolase) AH_2: NADPH, ascorbate or reduced pteridines

lyases in a young bamboo shoot (about 1 m in length) increase progressively from top to basal parts where lignin formation is initiated (HIGUCHI 1966).

The distribution of TAL is limited to grasses, and L-^{14}C-tyrosine, which is not converted to lignins in gymnosperms and most angiosperms, is efficiently transformed by grasses into p-hydroxyphenyl, guaiacyl, and syringyl units of the lignin polymer, and also to the esterified p-coumaric acid (HIGUCHI et al. 1967). It is accordingly evident that both tyrosine and phenylalanine ammonia-lyases are involved in the synthesis of lignin precursors in grasses.

3.2.1 Hydroxylation of Cinnamic Acids

Hydroxylation of cinnamic acid to p-coumaric acid (RUSSEL 1971, HILL and RHODES 1975) in the cinnamic acid pathway is meditated by cinnamate-4-hydroxylase, (EC 1.14.13.11). The hydroxylase has been studied extensively, and is known to be located in the microsomal fraction of plant tissues. The enzyme is a monooxygenase of the p-450 type and requires cinnamate, NADPH and oxygen (Fig. 8).

An enzyme which mediates the hydroxylation of p-coumaric acid to caffeic acid has been isolated from spinach beet (VAUGHAN and BUTT 1970). It is a phenolase and requires NADPH, ascorbate, or reduced pteridines as electron donors (Fig. 9). The microsomal fraction from the root of *Quercus* seedlings was found to mediate the formation of caffeic acid from L-phenylalanine via p-coumaric acid (ALIBERT et al. 1972), which suggests the occurrence of p-coumarate-3-hydroxylase as a membrane-associated activity.

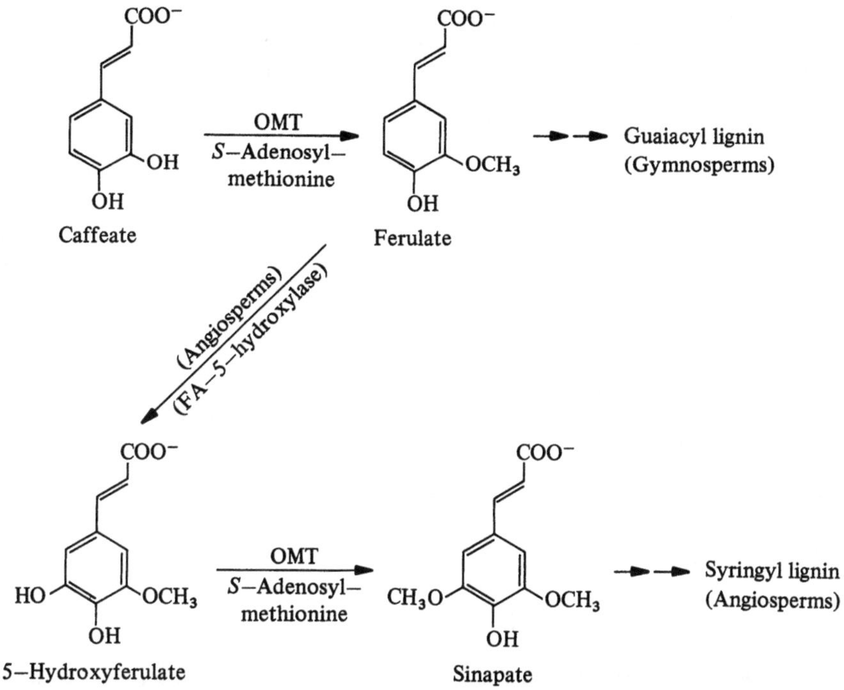

Fig. 10. Formation of sinapate from ferulate in angiosperms

3.2.2 Methylation of Hydroxycinnamic Acids

The conversion of caffeic acid to ferulic acid which is catalyzed by *S*-adenosyl L-methionine: catechol *O*-methyltransferase (OMT) (EC 2.1.1.6) was first shown to occur in plants by FINKLE and NELSON (1963), and FINKLE and MASRI (1964), and has since been found to be distributed widely in higher plants. Ferulic acid thus formed presumably is hydroxylated to 5-hydroxyferulic acid, and 5-hydroxyferulic acid is methylated to yield sinapic acid in angiosperms (Fig. 10). Tracer experiments (HIGUCHI and BROWN 1963) support this, but 5-hydroxyferulic acid has not been found in nature.

In investigations on the substrate specificities of OMT's from gymnosperms and angiosperms, it was found that both caffeic and 5-hydroxyferulic acids are very good substrates and that 3,4,5-trihydroxycinnamic acid, 5-hydroxyvanillin, protocatechuic aldehyde and chlorogenic acid are also fairly good substrates for angiosperm OMT's (SHIMADA et al. 1970). The substrate specificity of the gymnosperm OMT's (SHIMADA et al. 1973), on the other hand, is completely different, and caffeic acid is the most favorable substrate, followed by protocatechuic aldehyde and 3,4-dihydroxyphenyl acetate, whereas 5-hydroxyferulic acid is a poor substrate. Catechylglycerol-β-guaiacyl ether, which was used as a dimeric lignin model compound, is not methylated by Japanese black pine (*Pinus thunbergii* Parl.) OMT, suggesting that the methylation of lignin takes place at the monomeric stage prior to polymerization of coniferyl and sinapyl alcohols (Table 1).

Table 1. Substrate specificity of various plant OMT's[a]

Substrate	Relative methylation		
	Pine %	Poplar %	Bamboo %
Caffeic acid	100	100	100
5-Hydroxyferulic acid	10	320	124
3,4,5-Trihydroxycinnamic acid	25	60	50
Chlorogenic acid	10	46	3
iso-Ferulic acid	0	0	5
m-Coumaric acid	0	0	0
p-Coumaric acid	0	0	0
3,4-Dihydroxyphenyl acetic acid	54	0	0
3,4-Dihydroxymandelic acid	0	0	0
Protocatechuic aldehyde	68	46	45
5-Hydroxyvanillin	0	190	59
Protocatechuic acid	20	0	28
Gallic acid	0	0	0
Pyrocatechol	3	30	0
Pinosylvin	10	–	–
D-Catechin	0	0	0
Catechylglycerol-β-guaiacyl ether	0	–	–

[a] The reaction mixture contained the following components: 0.1 ml each of 0.01 M substrate, 0.04 M $MgCl_2$, 0.04 M sodium ascorbate, 0.005 M *S*-adenosylmethionine, 0.2 ml of 1 M Tris buffer (pH 8.0) and 1 ml of the enzyme solution. The reaction mixture was incubated at 30° C for 30 min. After the enzyme reaction was stopped by addition of HCl, the mixture was extracted with ether. The ether extract was chromatographed on paper in toluene-AcOH-H_2O (4:1:5, upper phase) and $CHCl_3$-AcOH-H_2O (2:1:1, lower phase) and identified by comparison of the fluorescent band with that of authentic compound. The band was then extracted with ethanol, and the product was determined spectrophotometrically at the wave length of its maximal absorption

The relationships between the ratio of sinapic acid to ferulic acid (SA/FA) formed by OMT's, the ratio of syringaldehyde to vanillin (S/V) on nitrobenzene oxidation of the lignins and the Mäule reaction are given in Table 2. Plants possessing higher SA/FA ratio, such as angiosperms, give the greater S/V ratio and a positive Mäule reaction, whereas plants with lower SA/FA ratio, such as conifers and ferns, give a lower S/V ratio and negative Mäule reaction. This result points to an intimate correlation between the evolution of lignin and the two types of OMT's. Some gymnosperms, such as Ephedra, Gnetum and Podocarpus, which have been found to give a positive Mäule reaction due to syringyl groups, might have OMT's with higher SA/FA ratios similar to angiosperm enzymes (SHIMADA et al. 1973, HIGUCHI et al. 1977).

The SA/FA ratio of bamboo OMT was found to remain constant during purification by chromatography on DEAE-cellulose, Sephadex gel filtration and polyacrylamide gel electrophoresis, and the two methylating activities for FA and SA formation were established to belong to a single enzyme protein. The enzymatic methylation of caffeic acid to ferulic acid was competitively

Table 2. Relationship between the SA/FA ratio, the S/V ratio and Mäule reaction of lignins[a]

Plant species	SA/FA (OMT)	S/V (lignin)	Mäule reaction
Pteridophyta			
Psilotum nudum	0.2	0	–
Angiopteris lygodifolia	0.3	–	–
Gymnospermae			
Ginkgo biloba	0.1	0	–
Pinus thunbergii	0.1	0	–
Podocarpus macrophylla	0.0	0	–
Taxus cuspidata	0.3	0	–
Angiospermae (Dicotyledoneae)			
Magnolia grandiflora	3.0	2.2	+
Cercidophyllum japonicum	3.2	2.9	+
Populus nigra	3.0	2.0	+
Pueraria thunbergiana	2.5	–	+
Angiospermae (Monocotyledoneae)			
Oryza sativa	0.9	1.0	+
Triticum aestivum	1.0	1.0	+
Phyllostachys pubescens	1.3	1.1	+

[a] SA, FA = activity against sinapic and ferulic acids, respectively. S = syringaldehyde, V = vanillin

inhibited by 5-hydroxyferulic acid, suggesting a preferential formation of syringyl lignin in angiosperms.

The purified OMT from Japanese black pine seedlings (KURODA et al. 1975), on the other hand, mainly catalyzed the formation of FA. K_m values for FA and SA were $5.1 \cdot 10^{-5}$ M and $2.8 \cdot 10^{-4}$ M, respectively, and SA formation was competitively inhibited by caffeic acid, indicating a preferential formation of guaiacyl lignin in conifers.

It is remarkable in relation to plant phylogeny that the angiosperm OMT is a difunctional enzyme which catalyzes the formation of both SA and FA from caffeic and 5-hydroxyferulic acids and is favorable to the formation of syringyl lignin, whereas gymnosperm and fern OMTs are essentially monofunctional and catalyze the formation of FA leading to give guaiacyl lignin.

3.2.3 Reduction of Hydroxycinnamic Acids

Ferulic and sinapic acids are reduced to the corresponding cinnamyl alcohols by successive mediation of three enzymes, hydroxycinnamate: CoA ligase, hydroxycinnamoyl-CoA reductase and hydroxycinnamyl alcohol oxidoreductase, which were isolated from *Salix* and *Forsythia* by MANSELL et al. (1972), and from cell suspension culture of *Glycine max* by EBEL and GRISEBACH (1973), respectively. The same enzyme system has recently been found in *Brassica* (RHODES and WOOLTORTON 1974) and many other plants (GROSS 1977).

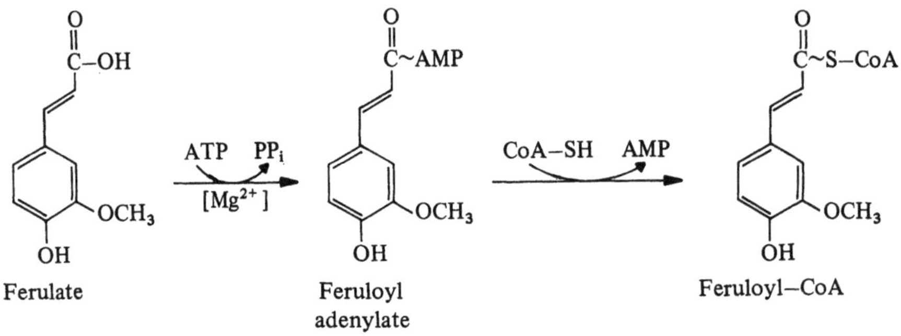

Fig. 11. Activation of ferulate by hydroxycinnamate:CoA ligase

The first step in the reduction of ferulic acid is activation of the carboxyl group via formation of a CoA ester. Ferulic acid is converted to feruloyl adenylate in the presence of ATP, and feruloyl adenylate is subsequently converted to feryloyl-CoA and AMP by CoA (Fig. 11). Cinnamate:CoA ligase (EC 6.2.1.X), which catalyzes this reaction, is distributed in various higher plants, especially in young lignifying stems. The enzymes isolated from *Forsythia* and *Brassica* have similar substrate specificities, and hydroxycinnamic acids, such as p-coumaric and ferulic acids, are effective substrates. Methylated or glycosylated ferulate is not effective , and sinapic acid is also not effective (GROSS 1977).

HAHLBROCK and GRISEBACH (1979) have further found that the ligase activity of soybean cell cultures could be separated into two isoenzymes, each with characteristic properties. The isoenzyme 1 has relatively low K_m and high V/K_m values only for the three typical lignin precursors, p-coumarate, ferulate, and sinapate. The isoenzyme 2, on the other hand, has relatively high affinities for p-coumarate and caffeate and does not activate sinapate. They suggested that the isoenzyme 1 is involved in lignin biosynthesis, whereas the isoenzyme 2 in flavonoid biosynthesis.

Thus, the isoenzyme 1 of soybean culture is the only enzyme shown to date to be effective with sinapic acid. Our recent study (KUTSUKI H, SHIMADA M and HIGUCHI T, unpublished data) has indicated, however, that sinapic acid, as well as ferulic and p-coumaric acids, are effectively converted to their CoA esters by bamboo shoot enzymes, and that the activity for sinapic acid is labile and is lost during storage of the enzyme preparation in a refrigerator.

Hydroxycinnamoyl-CoA is reduced to the corresponding aldehyde by cinnamoyl-CoA reductase (EC 1.2.1.44) (EBEL and GRISEBACH 1973, STÖCKIGT et al. 1973, RHODES and WOOLTORTON 1975) (Fig. 12). The enzyme requires NADPH as hydrogen donor, and feruloyl-CoA is found to be the best substrate followed by p-coumaroyl, sinapoyl, and 5-hydroxyferuloyl-CoA's. The enzyme is specific to cinnamoyl-CoA, other aromatic or aliphatic CoA esters are not effective, and therefore cinnamoyl-CoA:NADP oxydoreductase is proposed as a systematic name.

The last step in the formation of cinnamyl alcohol is the conversion of cinnamyl aldehyde to the corresponding alcohol mediated by an aromatic alcohol dehydrogenase (Fig. 13). Hydroxycinnamyl alcohol oxidoreductases from *For-*

Fig. 12. Reduction of feruloyl-CoA by cinnamoyl-CoA reductase

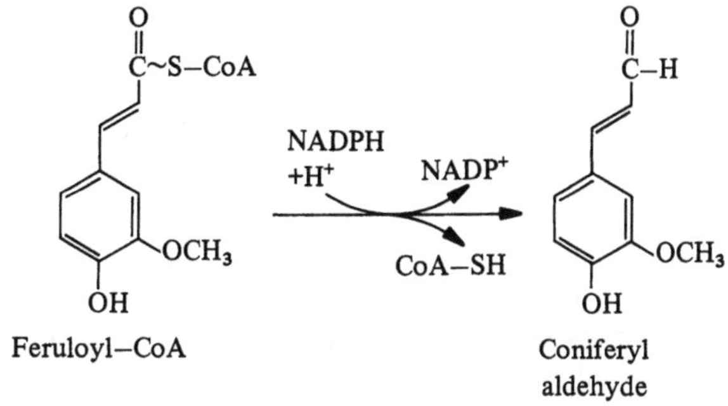

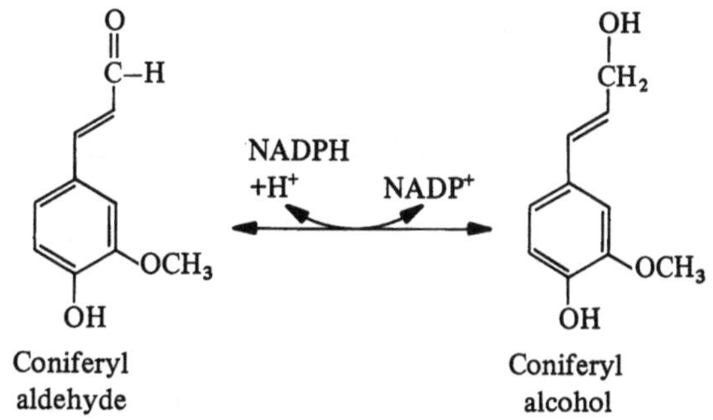

Fig. 13. Reduction of coniferyl aldehyde by cinnamyl alcohol dehydrogenase

sythia and *Glycine* have been extensively investigated (MANSELL et al. 1974). The enzymes have rather broad substrate specificities for p-coumaraldehyde, coniferyl aldehyde, sinapyl aldehyde, and their methyl derivatives, and require NADPH as hydrogen donor.

The enzyme is distributed in several organs of vascular plants, and is especially active in the cambial zone, which suggests that it plays a role in lignification (GROSS 1977). The enzyme is specific for cinnamyl aldehydes; aliphatic aldehydes are ineffective.

Thus, enzymes involved in the formation of the precursors of guaiacyl, syringyl, and p-hydroxyphenyl lignins have been characterized during the past few years; only the putative ferulate-5-hydroxylase remains uncharacterized.

4 Dehydrogenative Polymerization of Hydroxycinnamyl Alcohols to Lignins

FREUDENBERG (1965) found that a lignin-like dehydrogenation polymer was produced in vitro by treating coniferyl alcohol under aerobic conditions with a crude mushroom phenol oxidase, which was later characterized by HIGUCHI (1958) as a laccase. This polymer is closely related to spruce milled wood lignin (MWL) in many aspects, such as in functional groups, ultraviolet, infrared,

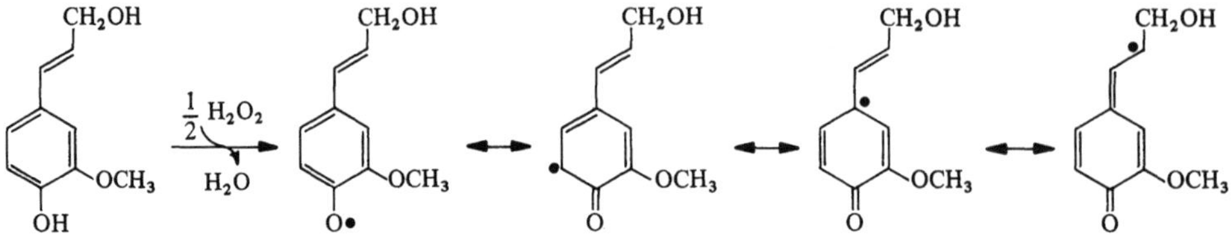

Fig. 14. Dehydrogenation of coniferyl alcohol by peroxidase showing the formation of phenoxy radicals

proton- and ^{13}C-nuclear magnetic resonance spectra, and in degradation products formed on nitrobenzene oxidation, on permanganate oxidation after methylation, and on acidolysis.

Further study has shown that coniferyl alcohol is dehydrogenated by either laccase/O_2 or peroxidase/H_2O_2; peroxidase was later shown to be the actual enzyme involved in lignification (HIGUCHI 1959, HARKIN and OBST 1973). Oxidation of coniferyl alcohol by these enzymes results in the mesomeric free radicals shown in Fig. 14. The radicals of coniferyl alcohol formed couples nonenzymatically in a random fashion to give dimers, trimers, and higher oligomers as racemic mixtures. Figure 15 shows examples of the coupling of the radicals to give quinone methides, which result in guaiacylglycerol-β-coniferyl ether (**24**), dehydro-diconiferyl alcohol (**25**) and *dl*-pinoresinol (**26**) by the addition of water or intramolecular nucleophilic attack on the benzyl carbons by hydroxyl groups. Dimers thus formed are further dehydrogenated to their radicals which give polymers via reactions such as the following (FREUDENBERG 1965) (Fig. 16).

1. Addition of water to the quinone methides formed by the coupling of monomer- and dimer radicals. (Example, guaiacylglycerol-β-pinoresinol ether (**27**)).
2. Addition of coniferyl alcohol to the dimeric quinone methides. (Example, guaiacylglycerol-α,β-bisconiferyl ether (**28**)).
3. Formation of benzyl aryl ether linkages by polymerization of quinone methides
4. Rearrangement of benzyl aryl ether linkages to α-5′ (**29**) and α-6′ (**30**) linkages.
5. Addition of sugars to quinone methides to result in lignin-carbohydrate complexes (**31**).

The lignin formed by these processes would be optically inactive, as are natural lignins.

Figure 17 shows a schematic constitution of spruce lignin based on dehydrogenation experiments of coniferyl alcohol and on analytical and degradative investigations of spruce lignin (ADLER et al. 1969).

FREUDENBERG (FREUDENBERG and NEISH 1968) found that a solution comprised of a mixture of coniferyl and sinapyl alcohols in approximately equal amounts gave a mixed dehydrogenation polymer by laccase/O_2 treatment, and that the polymer was closely related to beech MWL in both analytical and degradative features.

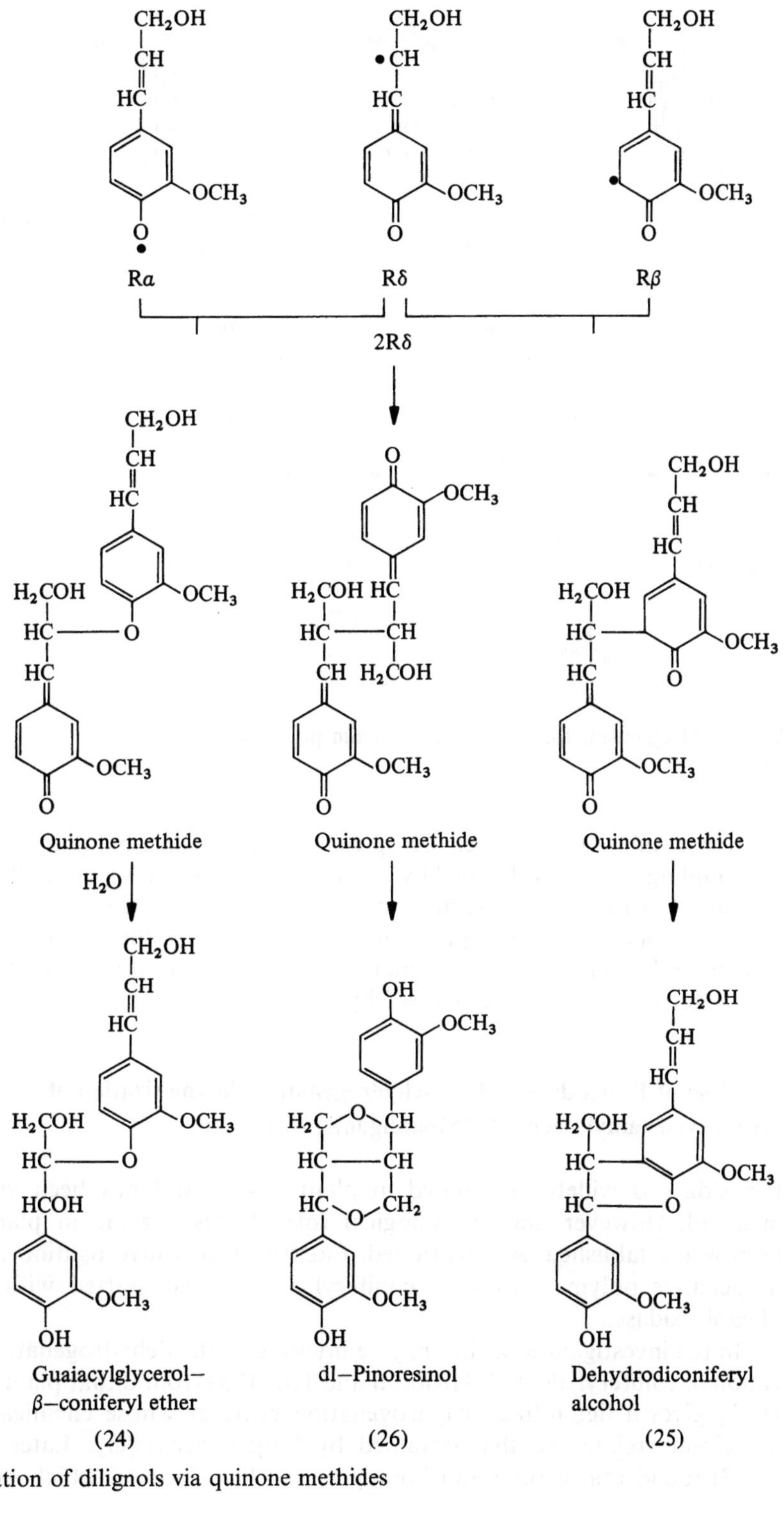

Fig. 15. Formation of dilignols via quinone methides

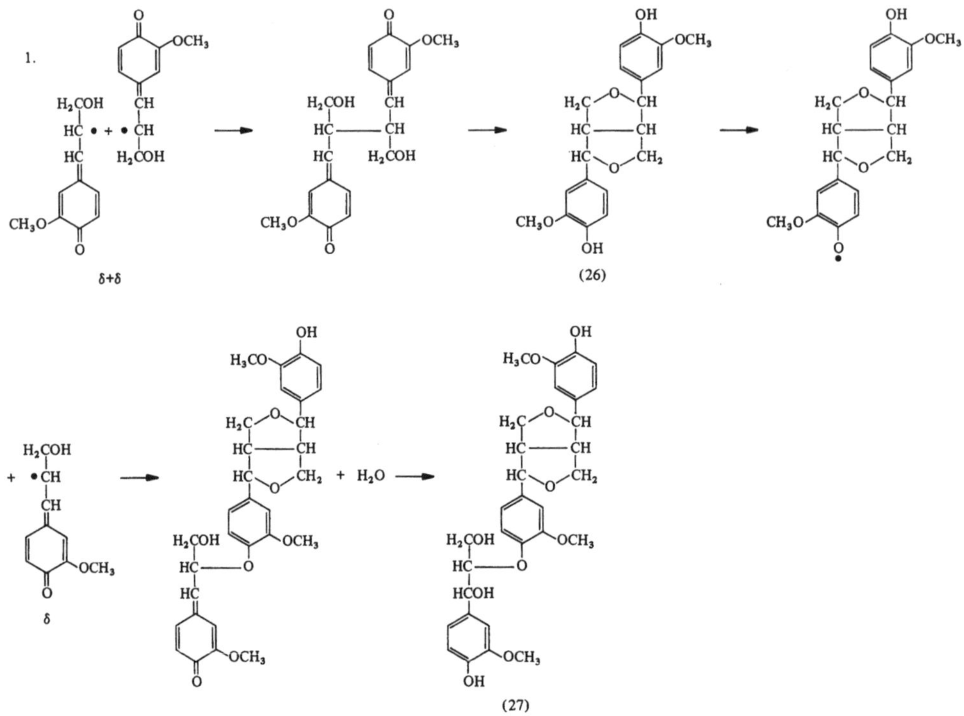

Fig. 16. Mechanisms for growth of the lignin polymer

It has subsequently been accepted that gymnosperm lignin is formed by the coupling of radicals formed by enzymic dehydrogenation of coniferyl alcohol, that angiosperm lignin is formed by the coupling of radicals of coniferyl and sinapyl alcohols, and that grass lignin is formed by the coupling of radicals of coniferyl, sinapyl, and p-coumaryl alcohols. These three cinnamyl alcohols were referred to as "monolignols" by FREUDENBERG.

4.1 Role of Peroxidase in the Dehydrogenative Polymerization of Hydroxycinnamyl Alcohols (Monolignols) to Lignins

Peroxidase is widely distributed in plant tissues and has been characterized in detail. However, the physiological role of this enzyme in plants has not been well established. As mentioned, FREUDENBERG's investigation of the dehydrogenative polymerization of coniferyl alcohol was started with mushroom phenol oxidase.

In the investigation of the enzyme involved in the dehydrogenative polymerization of coniferyl alcohol, HIGUCHI and ITO (1958) found that plant peroxidase/H_2O_2 gives a lignin-like dehydrogenation polymer whose chemical properties are closely related to that obtained by fungal laccase/O_2. Later NAKAMURA (1967) found that a pure and homogeneous *Rhus* laccase, which was obtained

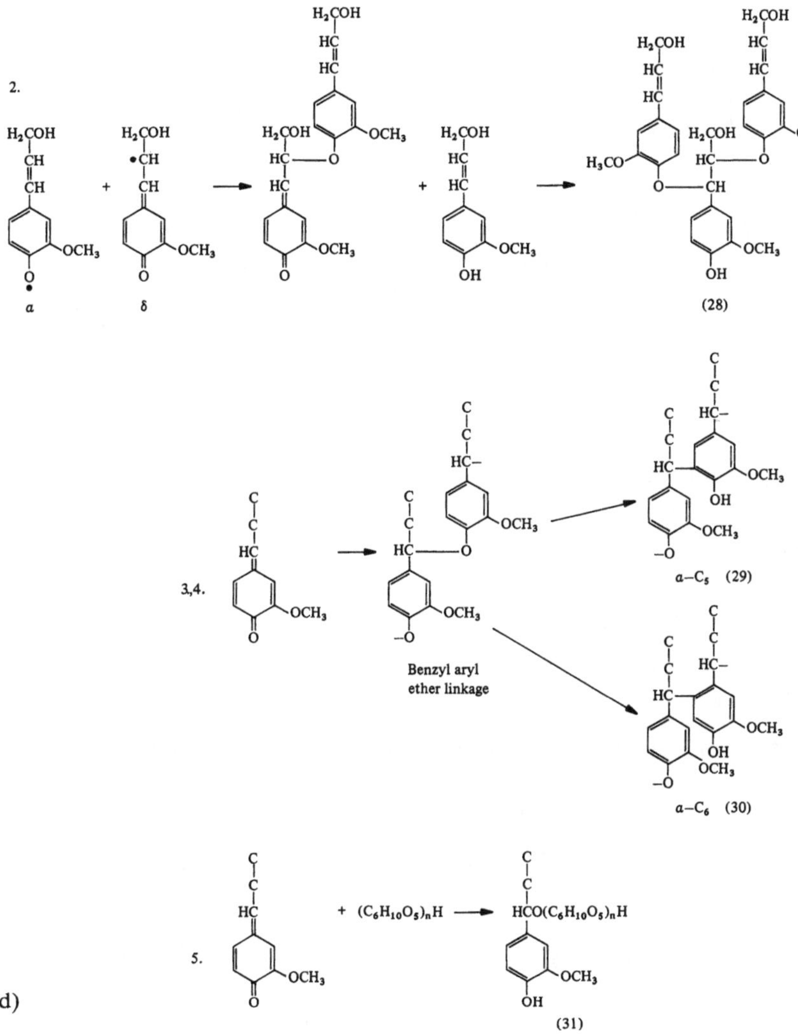

Fig. 16
(continued)

from fractional precipitations, ion-exchange column chromatography, etc., is incapable of catalyzing the oxidation of coniferyl alcohol, but that pure bamboo shoot peroxidases oxidize coniferyl alcohol efficiently (NAKAMURA and NOZU 1967). It is now accepted that peroxidase is actually involved in dehydrogenative polymerization of monolignols to lignins in higher plants (HARKIN and OBST 1973).

As to the origin of hydrogen peroxide required as substrate in dehydrogenative polymerization of monolignols, ELSTNER and HEUPEL (1976) recently demonstrated the formation of H_2O_2 by isolated cell walls from horseradish at the expense of NAD(P)H. This complex reaction involves the superoxide radical ($O_2^{-\cdot}$), which is formed by the reduction of O_2 by NAD$^\cdot$, as intermediate. The formation of NAD$^\cdot$ is thought to be initiated through oxidation of NADH

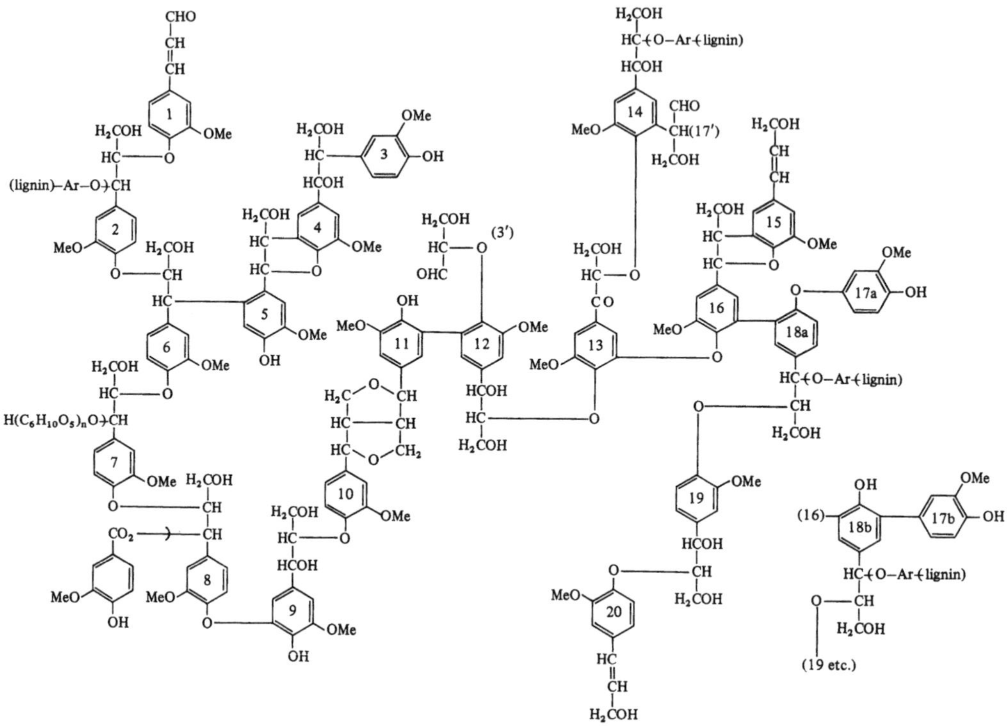

Fig. 17. A schematic constitution of spruce lignin. (ADLER et al. 1969)

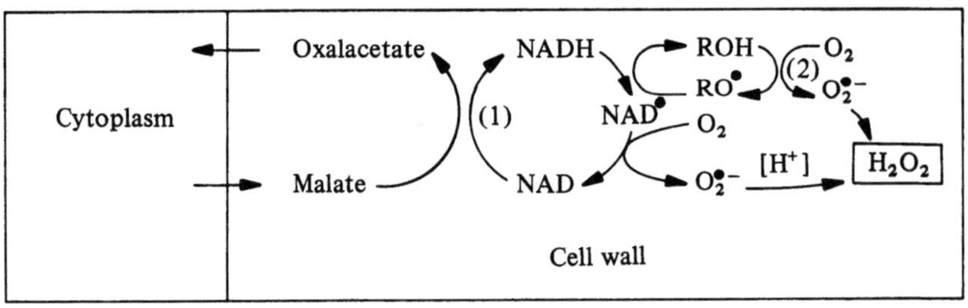

$$2\,\text{Malate} + 2\,O_2 \longrightarrow 2\,\text{Oxalacetate} + 2\,H_2O_2$$

Fig. 18. Hypothetical scheme for the formation of hydrogen peroxide in the plant cell wall. Enzymes: *1* Malate dehydrogenase; *2* Peroxidase+Mn^{2+}. (Redrawn from GROSS et al. 1977)

by phenoxy radicals formed in a peroxidase reaction in the presence of Mn^{2+}. GROSS et al. (1977) further showed that horseradish cell walls contain a bound malate dehydrogenase which forms sufficient NADH to allow the subsequent synthesis of H_2O_2 by the peroxidase (Fig. 18). This scheme can explain how H_2O_2 can be synthesized directly at the proper site without a requirement for transportation of toxic H_2O_2 from the cytoplasm to the cell wall. GROSS et al. (1977) have actually shown that a lignin-like dehydrogenation polymer of coniferyl alcohol is formed in the presence of added malate by isolated lignifying cell walls of *Forsythia,* which contain bound malate dehydrogenase and peroxidase, without addition of H_2O_2.

4.2 Structural Variation in Dehydrogenation Polymers

FREUDENBERG (FREUDENBERG and NEISH 1968) found that the amounts of dilignols, such as dehydrodiconiferyl alcohol and *dl*-pinoresinol, are considerably higher than the amount of guaiacylglycerol-β-coniferyl ether when coniferyl alcohol solution is added at once to the peroxidase solution ("Zulaufverfahren"), whereas the amount of guaiacylglycerol-β-coniferyl ether increases when substrate is added dropwise over long periods of time to the enzyme solution ("Zutropfverfahren").

SARKANEN and LUDWIG (1971) discussed in detail the polymerization mechanism of coniferyl alcohol in both methods, and concluded that in the former method large amounts of dilignols would be formed because of the higher

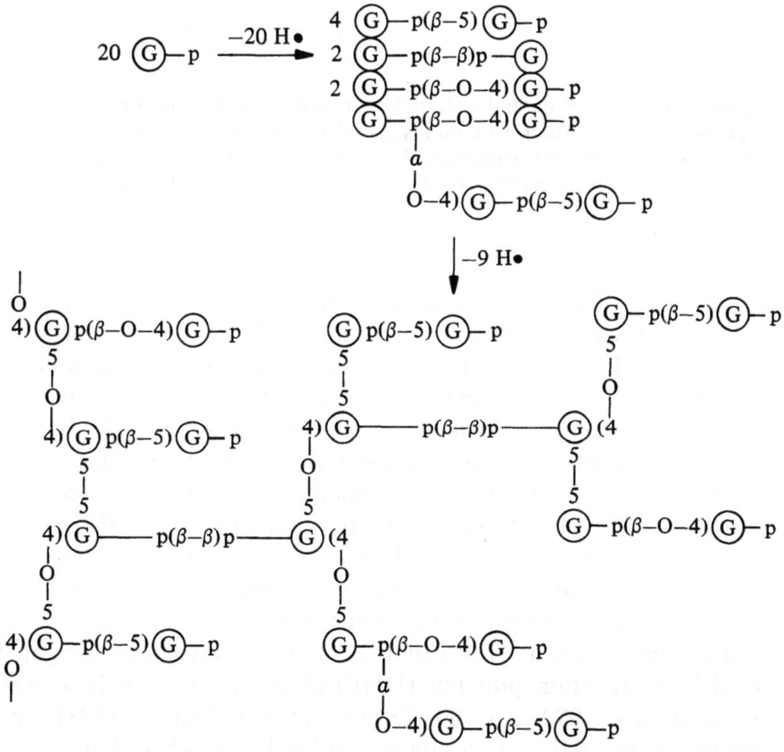

Fig. 19. Formation of a bulk polymer of coniferyl alcohol by the "Zulauf" method, from SARKANEN 1971. The radicals of coniferyl alcohol (20 molecules) couple to form dimers, which are dehydrogenatively polymerized to tetramers, and finally to a large molecule. *G* is the symbol of guaiacyl group and *p* represents a generalized propyl side chain structure, including both the allyl alcohol side chains in precursor molecules as well as structures resulting from free radical coupling and quinone methides

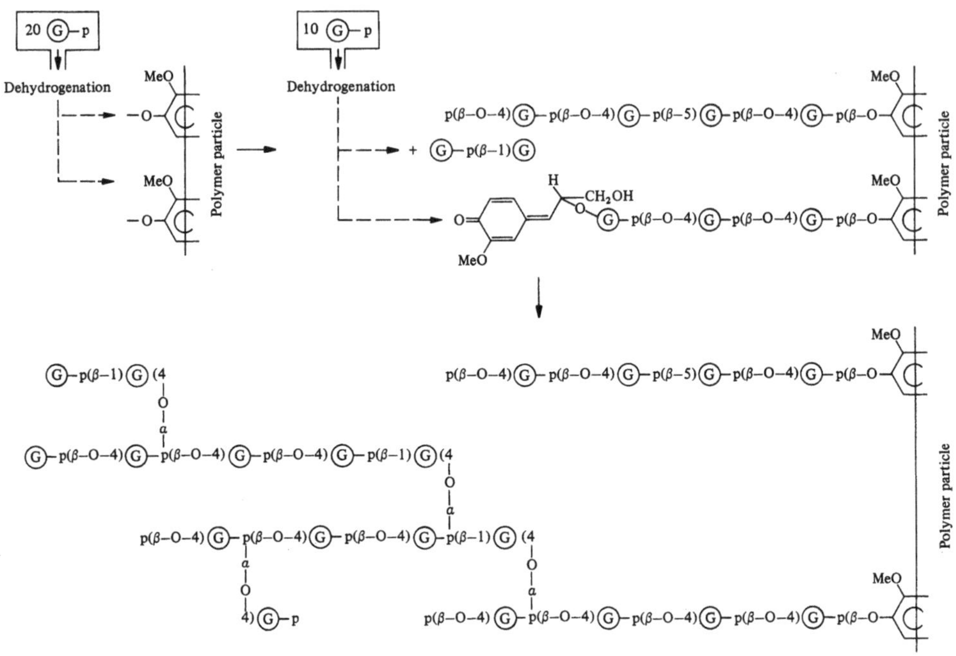

Fig. 20. Formation of an end-wise polymer of coniferyl alcohol by the "Zutropf" method. (SARKANEN 1971). Coniferyl alcohol is added dropwise for long periods to a system consisting of polymer particles and peroxidase/H_2O_2. Coniferyl alcohol radicals couple with phenoxy radicals present on the surface of the polymer particles. Abbreviations as in Fig. 19

concentration of monolignol radicals, and the dilignols thus formed are dehydrogenatively polymerized to tetramers and finally to a large molecule via C4-*O*-C5 and C5-C5 linkages. The "bulk polymer" thus formed would contain considerable amounts of double bonds in the side chains (Fig. 19). In the slower method, however, the concentration of monomer radicals would be lower and the monomer radicals would be preferentially coupled to oligolignol radicals previously formed, via mainly *β*-*O*-4 linkages to produce an "end-wise polymer" containing relatively few side chain double bonds. (Fig. 20)

It is likely that in the actual lignification of plant cell walls lignin synthesis depends on the amount of monolignols supplied to cell walls and probably proceeds via a "Zutropfverfahren-type mode".

It has been shown (HIGUCHI 1971) that the molecular weight of the lignin-like dehydrogenation polymer (DHP) of coniferyl alcohol formed by peroxidase is usually 1,000 to 1,100. TANAHASHI and HIGUCHI (1981) recently found, however, that DHP's with higher molecular weights than usual DHP's are formed when a cellulosic dialysis tube which contains horseradish peroxidase is added into a beaker containing H_2O_2 and coniferyl alcohol solutions. The results suggest that the concentration of DHP radicals in the tube is high; smaller molecules of oligolignols diffuse through the cellulose membrane. Coupling between DHP radicals in the tube and monolignol radicals would proceed in preference to monolignol-monolignol coupling, resulting in higher polymers. In the lignification of plant cell walls, polymerization of lignins may proceed in a similar way, involving diffusion across cell membranes and cell walls.

5 Formation and Distribution of Syringyl Lignin in Angiosperm Woods

In the investigation of the dehydrogenative polymerization of monolignols, FREUDENBERG (FREUDENBERG and NEISH 1968) reported that sinapyl alcohol alone does not give a lignin-like polymer, but yields mainly syringaresinol and dimethoxybenzoquinone, and he suggested that there is no syringyl lignin in nature. However, YAMASAKI et al. (1976) found that considerable amounts of DHP are formed from sinapyl alcohol alone with peroxidase and H_2O_2 by use of the "Zutropf" method. Ultraviolet, infrared, proton- and ^{13}C-nuclear magnetic resonance spectra and functional group analysis of the polymer showed characteristic features of syringyl lignin. Acidolysis of the polymer gave syringa-

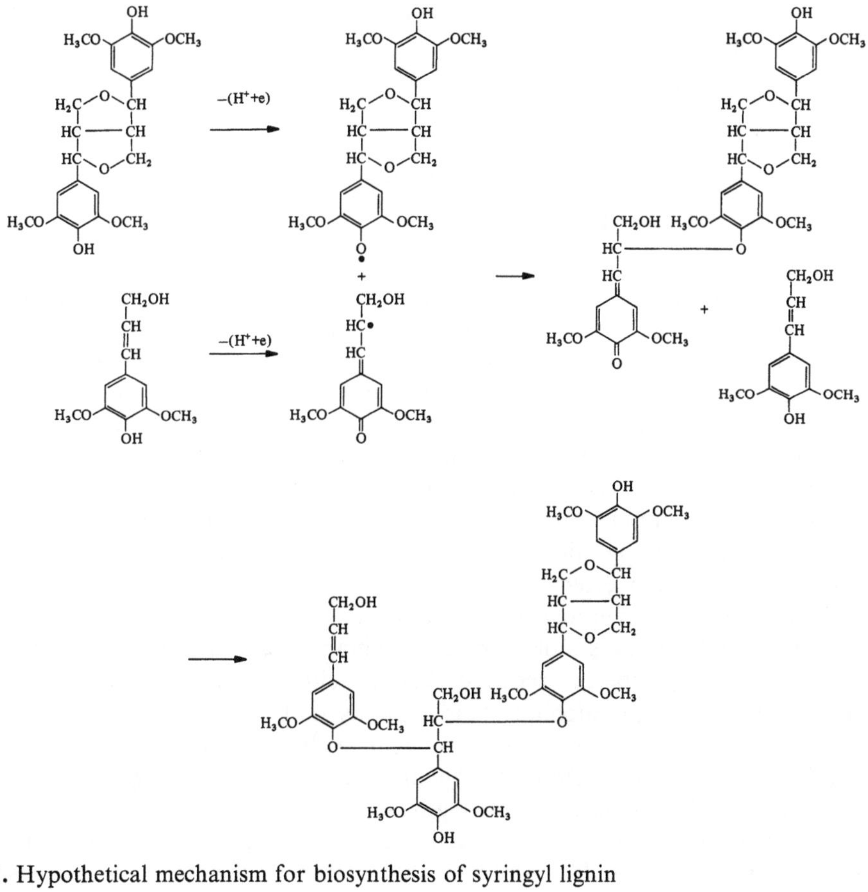

Fig. 21. Hypothetical mechanism for biosynthesis of syringyl lignin

resinol, with considerable amounts of typical acidolysis ketols, indicating the occurrence of the *β-O*-4 linkage, which is the most important structural unit of lignin polymers. The result indicates that the phenoxy radicals of sinapyl alcohol, formed enzymically, are coupled not only by the *β-β* mode to form syringaresinol, but also by the *β-O*-4 mode to produce a growing, polymeric syringyl lignin as shown in Fig. 21.

Chemical evidence for the occurrence of syringyl lignin in hardwoods has also been obtained. Advantage was taken of the solubility difference between mercurated derivatives of guaiacyl and syringyl lignins in acetic acid to isolate a syringyl lignin. The solubility of acetoxymercurated DHP of sinapyl alcohol in acetic acid is considerably higher than that of coniferyl alcohol (YAMASAKI et al. 1978a). This solubility difference was thus applied to the isolation of syringyl lignin from hardwood lignins (YAMASAKI et al. 1978b); MWL of *Myrica rubra* Sieb et Zucc. could be separated into syringyl-rich lignin containing 25.2% OCH_3, which corresponds to a syringyl unit content of about 85% in the lignin. Beech dioxane lignin also gave a syringyl-rich fraction whose methoxyl content, 27.8%, corresponded to a content of 95% syringyl units. These results indicate that hardwood lignins are not a uniformly copolymerized guaiacyl-syringyl lignin, but are heterogeneously comprised in part by a syringyl unit-rich fraction, and in part by a guaiacyl unit-rich fraction, in addition to syringyl-guaiacyl copolymer.

6 Differences Between Gymnosperms and Angiosperms in Lignin Biosynthesis

When ^{14}C-ferulic acid is administered to gymnosperms it is converted to only guaiacyl lignin but when it is administered to angiosperms, to both guaiacyl and syringyl lignins. Reasons for the preferential formation of guaiacyl lignin in gymnosperms include the absence of ferulate-5-hydroxylase and the substrate specificities of *O*-methyltransferases (OMT) as discussed in Sections 3.2.1 and 3.2.2 above. Another reason is the lack of activation and/or reduction of sinapic acid in gymnosperms (NAKAMURA et al. 1974, HIGUCHI et al. 1977).

Sinapic acid administered to seedlings of Japanese black pine was not converted to either sinapyl aldehyde or sinapyl alcohol; sinapyl aldehyde, on the other hand, was converted to syringyl lignin via sinapyl alcohol which is not ordinarily found in gymnosperms (NAKAMURA et al. 1974).

Poplar and cherry shoots converted sinapic acid to sinapyl alcohol, but the callus of angiosperms reduced ferulic acid to coniferyl alcohol but not sinapic acid to sinapyl alcohol. This suggests that the occurrence of activating and/or reducing enzymes for sinapic acid such as sinapate: CoA ligase are closely related to the differentiation and lignification of angiosperm tissues. Angiosperm callus makes primarily guaiacyl lignin.

GROSS (1977) found that the activities of the enzymes for the reduction of hydroxycinnamic acids are remarkably higher in the lignifying tissues than

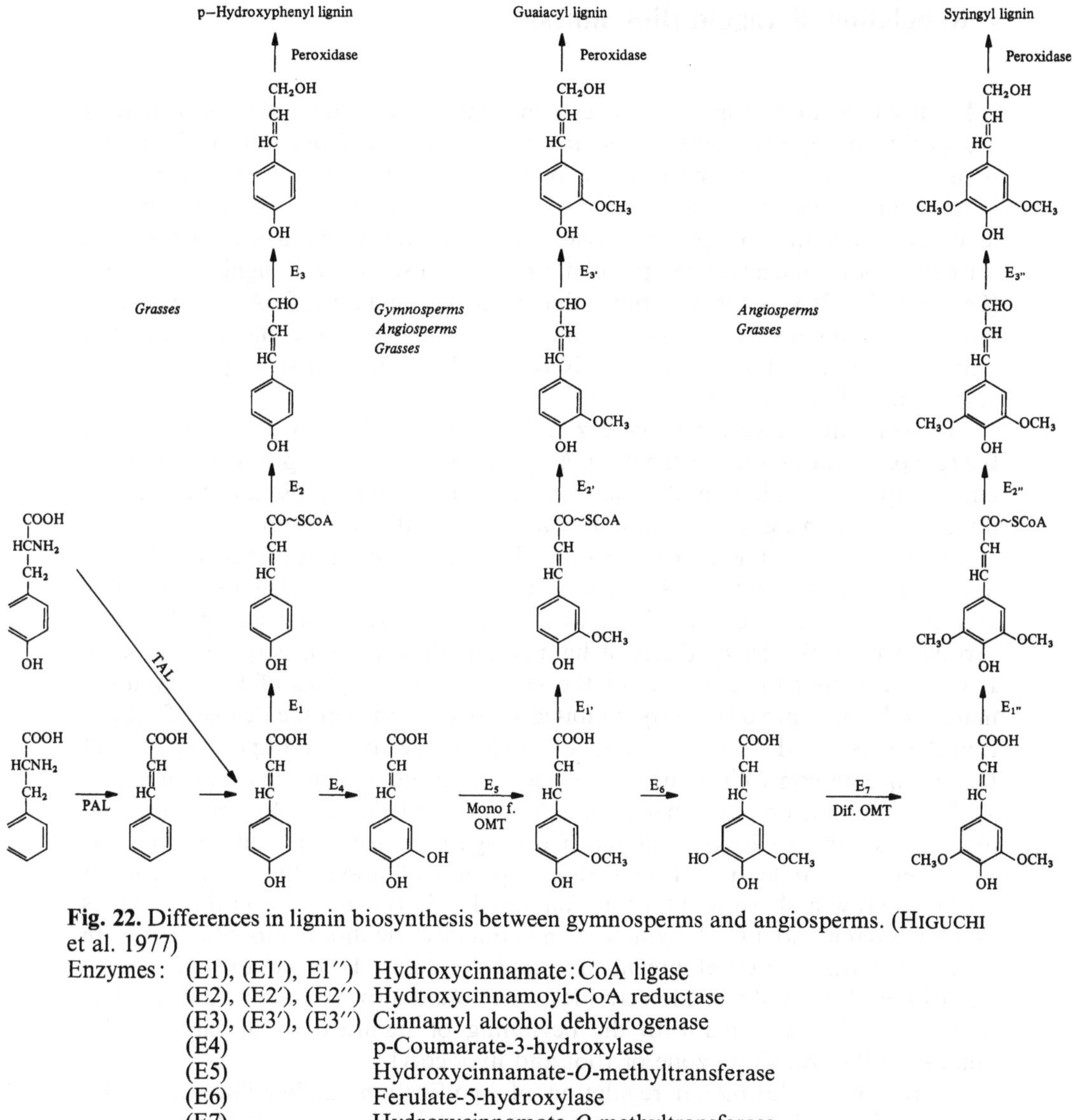

Fig. 22. Differences in lignin biosynthesis between gymnosperms and angiosperms. (HIGUCHI et al. 1977)

Enzymes:	(E1), (E1′), E1″)	Hydroxycinnamate:CoA ligase
	(E2), (E2′), (E2″)	Hydroxycinnamoyl-CoA reductase
	(E3), (E3′), (E3″)	Cinnamyl alcohol dehydrogenase
	(E4)	p-Coumarate-3-hydroxylase
	(E5)	Hydroxycinnamate-*O*-methyltransferase
	(E6)	Ferulate-5-hydroxylase
	(E7)	Hydroxycinnamate-*O*-methyltransferase

in callus tissues. Thus it is concluded that lignin synthesis by gymnosperms and angiosperms differs at the reactions catalyzed by ferulate-5-hydroxylase, OMT, cinnamate:CoA ligase, and cinnamoyl-CoA reductase, but is comparatively similar in the reduction of cinnamyl aldehydes to the corresponding alcohols, and in their dehydrogenation to lignins. Whereas in grasses the pool of p-coumaric acid, formed by TAL as well as by PAL, is considerably larger than the pools of ferulic and sinapic acids, the former acid is converted both to p-coumaroyl-CoA and incorporated into lignin via p-coumaryl alcohol, and to p-coumaric acid esters in the lignin side chains (Fig. 22).

7 Regulation of Lignin Biosynthesis

It has been found that most of the enzymes involved in the synthesis of lignins are specific for lignin precursors and are extracted mainly from actively lignifying tissues. PAL has been found to occur mainly in xylem tissues of various plants; it was found that the activity of PAL of an aspen tree is 30 times higher in outer xylem than in phloem tissue, and that the enzyme involved in the formation of ferulic acid from p-coumaric acid is also active in lignifying tissues (GROSS 1977). It was further found that hydroxycinnamate:CoA ligase, cinnamoyl-CoA reductase and cinnamyl alcohol dehydrogenase could be extracted only from vascular bundles, especially from xylem, and not from parenchymatous tissues of celery (GROSS et al. 1975).

These results suggest that the enzymatic controls, the substrate specificities, the rate of synthesis and degradation of appropriate enzymes, product inhibition, and compartmentation of the enzymes and substrates in tissues, all play an important role in lignin biosynthesis during xylem differentiation.

With respect to regulation via the substrate specificity of enzymes, HAHLBROCK and GRISEBACH (1979) showed the occurrence of isoenzymes of OMT and hydroxycinnamate:CoA ligase with properties suggesting their preferential involvement in the biosynthesis of lignins and flavonoids in soybean cultures. It was found that the activities of the respective isoenzymes of hydroxycinnamate:CoA ligase are differently inhibited by AMP and that the degree of inhibition depends on the concentration of ATP, suggesting an important role of the enzymes in controlling the rate of the formation of lignin and flavonoids.

It was found, on the other hand, that the activities of PAL, cinnamate-4-hydroxylase and hydroxycinnamate:CoA ligase in soybean cultures increase simultaneously at least tenfold within a period of about 20 h at the end of the linear growth phase, and then decline rapidly. Large increases in the activities of hydroxycinnamoyl-CoA reductase and cinnamyl alcohol oxidoreductase were observed during the development of spruce seedlings (HAHLBROCK and GRISEBACH 1979). Maximal enzyme activities were found approximately 12 days after sowing, which correlated with the beginning of lignification in the vascular bundles at the transition zone of root and hypocotyl.

In relation to hormonal regulation of lignification, gibberellin was found to increase both lignin biosynthesis and PAL activity (CHENG and MARSH 1968). Kinetin stimulated lignin formation and the activities of PAL (RUBERY and FOSKET 1969) and OMT (YAMADA and KUBOI 1976). Ethylene was shown to stimulate lignin biosynthesis in swede roots (RHODES and WOOLTORTON 1973a, b), with a dramatic increase of the activities of PAL, cinnamate hydroxylase, and cinnamate:CoA ligase. Shikimate dehydrogenase, OMT, and feruloyl-CoA reductase were moderately stimulated, but the activities of cinnamyl alcohol dehydrogenase and peroxidase were not stimulated in the swede roots (RHODES et al. 1976). SIEGEL (1954) showed that polymerization of lignin precursors by *Elodea densa* is inhibited by IAA, which competitively inhibits oxidations catalyzed by peroxidase. He suggested that high levels of IAA in meristematic tissues would suppress peroxidase activity, and hence lignin deposition, whereas with the decline of IAA concentration that accompanies maturity, lignification would increase. HIGUCHI (1957) suggested that a reducing system such as glutathione-ascorbic acid is involved in controlling the dehydrogenative polymerization of coniferyl alcohol, and with decline of the concentration of the reducing compounds lignification increases.

Recent investigations (KOMAMINE 1978) of xylogenesis in tissue cultures of carrot roots have indicated that xylem formation is induced by the cooperation of auxin and zeatin, the formation of which is mediated by the levels of cylic AMP, which in turn is affected by illumination. GRISEBACH and HAHLBROCK (1974) also found that the activities of the enzymes in both lignin and flavonoid biosynthesis are induced by illumination of cultured cells of parsley and soybean.

Many factors are involved in the initiation of lignification. Further systematic investigations are needed on the regulating mechanism for induction of enzyme synthesis, on compartmentation, and on genetic controls of phenylpropanoid metabolism during differentiation of plant tissues.

In relation to the organelles involved in lignification, PICKETT-HEAPS (1968) found that tritiated cinnamic acid, as well as tyrosine, phenylalanine, and methionine, are incorporated well into xylem cell walls of a wheat plant. Cinnamic acid was markedly concentrated in the xylem thickenings. Electronmicroscopic observation showed that in developing xylem cells, cinnamic acid was first incorporated into both endoplasmic reticulum and Golgi bodies, and that these generally aggregated in the cytoplasm near the bands of wall microtubules. Aromatic aldehydes, such as p-hydroxybenzaldehyde, obtained by alkaline nitrobenzene oxidation of cinnamic acid-fed plants, were heavily radioactive. PICKETT-HEAPS (1968) suggested that the role of the endoplasmic reticulum is not very clear, but that it probably acts as a transport system in the cell, as well as being one of the pools of metabolites within the cell. He suggested further that Golgi bodies are concerned with the actual synthesis of the wall materials, whose properties change as secondary thickening commences, to accommodate synthesis of different types of polysaccharides and lignin.

STAFFORD (1974) indicated that the microsomal fraction of green shoots of *Sorghum vulgare* contains an enzyme system converting aromatic amino acids to cinnamic acids; this would include enzymes such as PAL, TAL, cinnamate-4-hydroxylase, p-coumarate-3-hydroxylase, and OMT. Her results seem consistent with PICKETT-HEAPS' experiment and appear to support the idea that lignin precursors are also synthesized in Golgi bodies or in endoplasmic reticulum, transported to the cell wall via vesicles, and dehydrogenatively polymerized to lignin by peroxidase bound onto the cell walls.

References

Acerbo SN, Schubert WJ, Nord FF (1960) Investigation on lignins and lignification. XXII. The conversion of D-glucose into lignin in Norway spruce. J Am Chem Soc 82:735–739
Adler E (1977) Lignin chemistry-past, present and future. Wood Sci Technol 11:169–218
Adler E, Björkvist KJ, Haggroth S (1948) Über die Ursache der Farbreaktion des Holzes. Acta Chem Scand 2:93–94

Adler E, Larsson S, Lundquist K, Miksche GE (1969) Acidolytic, alkaline, and oxidative degradation of lignin. Abst Int Wood Chem Symp 1

Alibert G, Ranjeva R, Boudet A (1972) Recherches sur les enzymes catalysant la biosynthèse des acides phenolique chez *Quercus pedunculata* III. Formation sequentielle, à partir de la phenylalanine, des acid cinnamique, *p*-coumarique et caffeique, par des organites cellulaires isolés. Physiol Plant 27:240–243

Björkman A (1956) Studies on finely divided wood I. Extraction of lignin with neutral solvents. Svensk Papperstidn 59:477–485

Brown SA (1966) Lignins. Annu Rev Plant Physiol 17:223–244

Brown SA, Neish AC (1955) Shikimic acid as a precursor in lignin biosynthesis. Nature 175:688–690

Cheng CKC, Marsh HV (1968) Gibberellic acid-promoted lignification and phenylalanine ammonia-lyase activity in a dwarf pea (*Pisum sativum*). Plant Physiol 43:1755–1759

Davis BD (1955) Intermediates in amino acid biosynthesis. Advan Enzymol 16:247–312

Ebel J, Grisebach H (1973) Reduction of cinnamic acids to cinnamyl alcohols with an enzyme preparation from cell suspension cultures of soybean (*Glycine max*). FEBS Lett 30:141–143

Eberhardt G, Schubert WJ (1956) Investigations on lignin and lignification. XVII. Evidence for the mediation of shikimic acid in the biogenesis of lignin building stones. J Am Chem Soc 78:2835–2837

Elstner EF, Heupel A (1976) Formation of hydrogen peroxide by isolated cell walls from horseradish (*Armoracia lapathifolia* Gilib.) Planta 130:175–180

Finkle BJ, Masri MS (1964) Methylation of polyhydroxyaromatic compounds by pampas grass *O*-methyltransferase. Biochim Biophys Acta 85:167–169

Finkle BJ, Nelson RF (1963) Enzyme reactions with phenolic compounds a meta-*O*-methyltransferase in plants. Biochim Biophys Acta 78:747–749

Freudenberg K (1965) Lignin: Its constitution and formation from *p*-hydroxycinnamyl alcohols. Science 148:595–600

Freudenberg K, Neish AC (1968) Constitution and biosynthesis of lignin. Springer-Verlag, Berlin, Göttingen, Heidelberg 78–122

Fujita M, Harada H (1979) Autoradiographic investigations of cell wall development. II. Tritiated phenylalanine and ferulic acid assimilation in relation to lignification. Mokuzai Gakkaishi 25:89–94

Gamborg OL (1967) Aromatic metabolism in plants. V. The biosynthesis of chlorogenic acid and lignin in potato cell cultures. Can J Biochem 45:1451–1457

Gamborg OL, Neish AC (1959) Biosynthesis of phenylalanine and tyrosine in young wheat and buckwheat plants. Can J Biochem Physiol 37:1277–1285

Grisebach H, Hahlbrock K (1974) Enzymology and regulation of flavonoid and lignin biosynthesis in plants and plant cell suspension cultures. Recent Adv Phytochem 8:21–52

Gross GG (1977) Biosynthesis of lignin and related monomers. Recent Adv in Phytochem 11:141–184

Gross GG, Janse C, Elstner EF (1977) Involvement of malate, monophenols and superoxide radical in hydrogen peroxide formation by isolated cell walls from horseradish (*Aromoracia lapathifolia* Gilib.). Planta 136:271–276

Gross GG, Mansell RL, Zenk MH (1975) Hydroxycinnamate: CoA ligase from lignifying tissue of higher plant. Some properties and taxonomic distribution. Biochem Physiol Pflanz 168:41–51

Hahlbrock K, Grisebach H (1979) Enzymic controls in the biosynthesis of lignin and flavonoids. Annu Rev Plant Physiol 30:105–130

Harkin JM, Obst TR (1973) Lignification in trees: indication of exclusive peroxidase participation. Science 180:296–298

Hasegawa M, Higuchi T, Ishikawa H (1960) Formation of lignin in tissue culture of *Pinus strobus*. Plant Cell Physiol 1:173–182

Higuchi T (1957) Biochemical studies of lignin formation II. Physiol Plant 10:621–632

Higuchi T (1958) Further studies on phenol oxidase related to the lignin biosynthesis. J Biochem (Tokyo) 45:515–528

Higuchi T (1959) Studies on the biosynthesis of lignin. In: Kratzl K, Billek G (eds) Biochemistry of Wood. Pergamon Press, New York, pp 161–188

Higuchi T (1966) Role of phenylalanine deaminase and tyrase in the lignification of bamboo. Agr Biol Chem 30:667–673

Higuchi T (1971) Formation and biological degradation of lignins. Advan Enzymol 34:207–283

Higuchi T, Barnoud F (1964) Les lignins dans les tissus végétaux cultivés in vitro 1. Chimie Biochimie de la Lignine, Cellulose, Hemicelluloses. Les Imprimeries Réunies de Chambery, 255–274

Higuchi T, Brown SA (1963) Studies of lignin biosynthesis using isotopic carbon. XII. The biosynthesis and metabolism of sinapic acid. Can J Biochem Physiol 41:613–620

Higuchi T, Ito Y (1958) Dehydrogenation products of coniferyl alcohol formed by the action of mushroom phenol oxidase, Rhus-laccase and radish peroxidase. J Biochem (Tokyo) 45:575–579

Higuchi T, Ito Y, Kawamura I (1967) *p*-Hydroxyphenylpropane component of grass lignin and role of tyrosine ammonia-lyase in its formation. Phytochemistry 6:875–881

Higuchi T, Shimada M, Nakatsubo F, Tanahashi M (1977) Differences in biosyntheses of guaiacyl and syringyl lignins in woods. Wood Sci Technol 11:153–167

Hill AC, Rhodes MJC (1975) The properties of cinnamic-4-hydroxylase of aged swede root disks. Phytochemistry 14:2387–2391

Imagawa H, Fukazawa K, Ishida S (1976) Study on the lignification in tracheids of Japanese larch, *Larix leptolepis* Gord. Res Bull Col Exp Forests, Hokkaido Univ 33:No 1, 127–138

Kawamura I, Higuchi T (1964) Comparative studies of milled wood lignins from different taxonomical origins by IR spectrometry. Chimie Biochimie de la Lignine, Cellulose, Hemicelluloses. Les Imprimeries Reunies de Chambery, 439–456

Komamine A (1978) Biochemical aspects on the growth and differentiation of cultured cells. Abst. 14th Symp Plant Chem (Japan) 23

Koukol J, Conn EE (1961) Purification and properties of the phenylalanine deaminase of *Hordeum vulgare*. J Biol Chem 236:2692–2698

Kratzl K, Faigle H (1958) Der Einbau von D-Glucose-[1-^{14}C] in das Phenylpropanegerüst des Fichtenlignins. Monatsh Chem 89:708–715

Kuroda H, Shimada M, Higuchi T (1975) Purification and properties of *O*-methyltransferase involved in the biosynthesis of gymnosperm lignin. Phytochemistry 14:1759–1763

Mansell RL, Gross GG, Stöckigt J, Franke H, Zenk MH (1974) Purification and properties of cinnamyl alcohol dehydrogenase from higher plants involved in lignin biosynthesis. Phytochemistry 13:2427–2435

Mansell RL, Stöckigt T, Zenk MH (1972) Reduction of ferulic acid to coniferyl alcohol in a cell free system from a higher plant. Z Pflanzenphysiol 68:286–288

McCalla DR, Neish AC (1959) Metabolism of phenylpropanoid compounds in *Salvia* I. Biosynthesis of phenylalanine and tyrosine. Can J Biochem Physiol 37:531–536

Migita N, Nakano J, Okada T, Ohashi G, Noguchi M (1955) On the mechanism of color reaction of lignin (IV). Cross-Bevan color reaction of hardwood lignin. J Jpn Forest Soc 37:26–34

Musha Y, Goring DAI (1975) Distribution of syringyl and guaiacyl moieties in hardwoods as indicated by ultraviolet microscopy. Wood Sci Technol 9:45–58

Nakumura W (1967) Studies on the biosynthesis of lignin I. Disproof against the catalytic activity of laccase in the oxidation of coniferyl alcohol. J Biochem (Tokyo) 62: 54–60

Nakamura W, Nozu Y (1967) Studies on the biosynthesis of lignin II. Purification and properties of peroxidase from bamboo shoot. J Biochem (Tokyo) 62:308–314

Nakamura Y, Fushiki H, Higuchi T (1974) Metabolic differences between gymnosperms and angiosperms in the formation of syringyl lignin. Phytochemistry 13:1777–1784

Neish AC (1961) Formation of *m*- and *p*-coumaric acids by enzymatic deamination of the corresponding isomers of tyrosine. Phytochemistry 1:1–24

Pickett-Heaps JD (1968) Xylem wall deposition autoradiographic investigations using lignin precursors. Protoplasma 65:181–205

Rhodes MJC, Wooltorton LSC (1973a) Stimulation of phenolic acid and lignin biosynthesis in swede root tissue by ethylene. Phytochemistry 12:107–118

Rhodes MJC, Wooltorton LSC (1973b) Changes in phenolic acid and lignin biosynthesis

in response to treatment of root tissue of the swedish turnip (*Brassica napo-brassica*) with ethylene. Qualitas Plant 23:145–155
Rhodes MJC, Wooltorton LSC (1974) Reduction of the CoA thioesters of *p*-coumaric and ferulic acids by extract of aged Brassica *napo-brassica* root tissue. Phytochemistry 13:107–110
Rhodes MJC, Wooltorton LSC (1975) Enzymes involved in the reduction of ferulic acid to coniferyl alcohol during the aging of disks of swede root tissues. Phytochemistry 14:1235–1240
Rhodes MJC, Hill RC, Wooltorton LSC (1976) Activity of enzymes involved in lignin biosynthesis in swede root disks. Phytochemistry 15:707–710
Rubery PH, Fosket DE (1969) Changes in phenylalanine ammonia-lyase activity during xylem differentiation in *Coleus* and soybean. Planta 87:54–62
Rubery PH, Northcote DH (1968) Site of phenylalanine ammonia-lyase activity and synthesis of lignin during xylem differentiation. Nature 219:1230–1234
Ruel K, Barnoud F, Goring DAI (1978) Lamellation in the S2 layer of softwood tracheids as demonstrated by scanning transmission electron microscopy. Wood Sci Technol 12:287–291
Russel DW (1971) The metabolism of aromatic compounds in higher plant. J Biol Chem 246:3870–3878
Saleh TM, Leney L, Sarkanen KV (1967) Radioautographic studies of cotton wood, Douglas fir and wheat plants. Holzforschung 21:116–120
Sarkanen KV, Ludwig CH (1971) Lignins: Occurrence, Formation, Structure and Reactions. Wiley-Interscience, New York, pp 150–155
Scott JAN, Procter AR, Fergus BJ, Goring DAI (1969) The application of ultraviolet microscopy to the distribution of lignins in wood. Description and validity of the technique. Wood Sci Technol 3:73–92
Shimada M, Ohashi H, Higuchi T (1970) *O*-Methyltransferase involved in the biosynthesis of lignins. Phytochemistry 9:2463–2470
Shimada M, Kuroda H, Higuchi T (1973) Evidence for the formation of methoxyl groups of ferulic and sinapic acids in Bambusa by the same *O*-methyltransferase. Phytochemistry 12:2873–2875
Siegel SM (1954) Studies on the biosynthesis of lignins. Physiol Plant 7:41–49
Stafford HA (1974) Possible multienzyme complexes regulating the formation of C6-C3 phenolic compounds and lignins in higher plants. Recent Adv Phytochem 8:53–79
Sprinson DB (1960) The biosynthesis of aromatic compounds from D-glucose. Adv Carbohydr Chem 15:235–270
Stöckigt J, Mansell RL, Gross GG, Zenk MH (1973) Enzymic reduction of *p*-coumaric acid via *p*-coumaroyl-CoA to *p*-coumaryl alcohol by a cell free system from *Forsythia* sp Z Pflanzenphysiol 70:305–307
Tanahashi M, Higuchi T (1981) Dehydrogenative polymerization of monolignols by peroxidase and H_2O_2 in a dialysis tube I preparation of highly polymerized DHPs. Wood Research No. 67:29–42
Vaughan PFT, Butt VS (1970) The action of *o*-dihydric phenols in the hydroxylation of *p*-coumaric acid by a phenolase from leaves of spinach beet (*Beta vulgaris* L.). Biochem J 119:89–94
Wardrop AB, Bland DE (1959) The process of lignification in woody plants. In: Kratzl K, Billek G (eds) Biochemistry of woods. Pergamon Press, New York, pp 92–116
Yamada Y, Kuboi T (1976) Significance of caffeic acid *O*-methyltransferase in lignification of cultured tobacco cells. Phytochemistry 15:395–396
Yamasaki T, Hata K, Higuchi T (1976) Dehydrogenation polymer of sinapyl alcohol by peroxidase and hydrogen peroxide. Mokuzai Gakkaishi 22:582–588
Yamasaki T, Hata K, Higuchi T (1978a) Separation of s-DHP from a mixture of c- and s-DHP with special reference to the isolation of syringyl lignin. Holzforschung 32:20–23
Yamasaki T, Hata K, Higuchi T (1978a) Isolation and characterization of syringyl component rich lignin. Holzforschung 32:44–47
Young MR, Towers GHN, Neish AC (1966) Taxonomic distribution of ammonia-lyase for L-phenylalanine and L-tyrosine in relation to lignification. Can J Bot 44:341–349

10 Hydrophobic Layers Attached to Cell Walls. Cutin, Suberin and Associated Waxes

P.E. KOLATTUKUDY, K.E. ESPELIE, and C.L. SOLIDAY

1 Introduction

The polysaccharidic cell walls do not constitute a significant diffusion barrier, and, therefore, plants attach a hydrophobic layer to the wall to erect such a barrier. Such diffusion barriers consist of an insoluble polymeric structural component and a complex mixture of extractable lipids, collectively called wax. We shall briefly summarize the general characteristics, structure, composition, metabolism, and function of the waxes and the polymeric materials. Other more detailed reviews on the subject are available (MARTIN and JUNIPER 1970, KOLATTUKUDY 1980b) and the function of the components is discussed by J. SCHÖNHERR, Chapter 1.6, in Volume 12B of this series.

2 Location and Ultrastructure of Cutin, Suberin, and Waxes

The aerial organs are covered by a barrier called the cuticle which consists of the polymer, cutin, embedded in waxes and attached to epidermal cell walls via pectinaceous materials. On the other hand, underground organs and periderms in all organs are protected from diffusion by the polymer, suberin, and associated waxes. In this case, the barrier layer is in the plasma membrane side of the cell wall and in some cases suberization extends throughout the wall. A suberin-type barrier appears to be formed in cases where a diffusion barrier is to be erected at some time during a normal developmental process or as a result of injury caused by environmental factors. Sites of suberization during the developmental process include the endodermis (Casparian bands) (ROBARDS et al. 1973), epidermis, and hypodermis of roots (FERGUSON and CLARKSON 1976, OLESEN 1978, PETERSON et al. 1978), the bundle sheaths of grasses (O'BRIEN and CARR 1970, O'BRIEN and KUO 1975), the sheaths around idioblasts (WATTENDORF 1969, 1976), the boundary between secretory organs (such as glands and trichomes) and the rest of the plant (THOMSON et al. 1976, OLESEN 1979), the pigment strands of grains (ZEE and O'BRIEN 1970), and the connection between seed coats and vascular tissue (ESPELIE et al. 1980b). Suberization occurs as a result of injury caused by fungi, insects, or mechanical means, and by disorders such as hollow heart in potato tubers (DEAN and KOLATTUKUDY 1976, MULLICK 1977, DEAN et al. 1977, KRÄHMER 1980).

Cuticular surfaces show characteristic wax crystals in most cases (Fig. 1 A and B). Upon removal of the wax from isolated cuticles, the outer surface

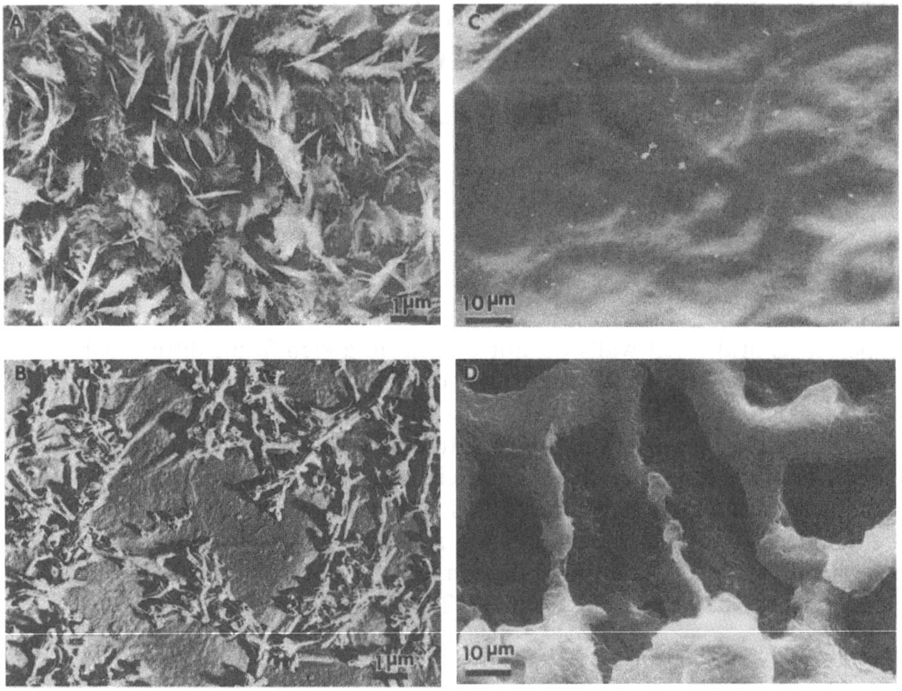

Fig. 1. Electronmicrographs of the leaf surface wax of **A** *Eucalyptus papuana*. (Courtesy of Dr. N.D. Hallam) and **B** *Picea abies*. (Courtesy of Dr. B.E. Juniper). Scanning electronmicrographs of the isolated cutin polymer (WALTON and KOLATTUKUDY, 1972a) from the fruit of *Lycopersicon esculentum:* **C** the external surface of the cutin membrane and **D** the internal surface of the polymer snowing intercellular grooves. (Courtesy of R.W. Davis)

of the polymer is rather smooth (Fig. 1C), whereas the internal surface attached to the epidermis shows ridges which fit into the intracellular grooves of the epidermal layer of cells (Fig. 1D). Cross-sections of cuticles usually appear amorphous with some reticulate intrusions of unknown composition (Fig. 2A). In some cases lamellar appearance has been observed (Fig. 2B); in one such case, *Atriplex,* the chemical composition of the polymer appeared to be quite typical of cutin from plants with amorphorus appearance. All suberized layers thus far examined showed lamellar structure composed of light bands of waxes and dark bands which presumably consist of phenolic materials (Fig. 2C).

Angiosperms and gymnosperms have cutin and this polymer is also present in primitive plants such as liverworts (HOLLOWAY et al. 1972a, HUNNEMAN and EGLINTON 1972, CALDICOTT and EGLINTON 1976) and a moss (P.E. KOLATTUKUDY, unpublished results). All aerial organs including stems, petioles, leaves (including substomatal cavities), flowers, fruits, and seed coats are covered by cutin (MARTIN and JUNIPER 1970, KOLATTUKUDY 1980b, ESPELIE et al. 1980b). The amount of cutin (<0.02 to $1.5\ \mathrm{mg\ cm^{-2}}$) and wax ($<0.01$ to $>1.0\ \mathrm{mg\ cm^{-2}}$) per surface area varies a great deal among different species and among different organs within the same species (MARTIN and JUNIPER 1970, HOLLOWAY and BAKER 1970).

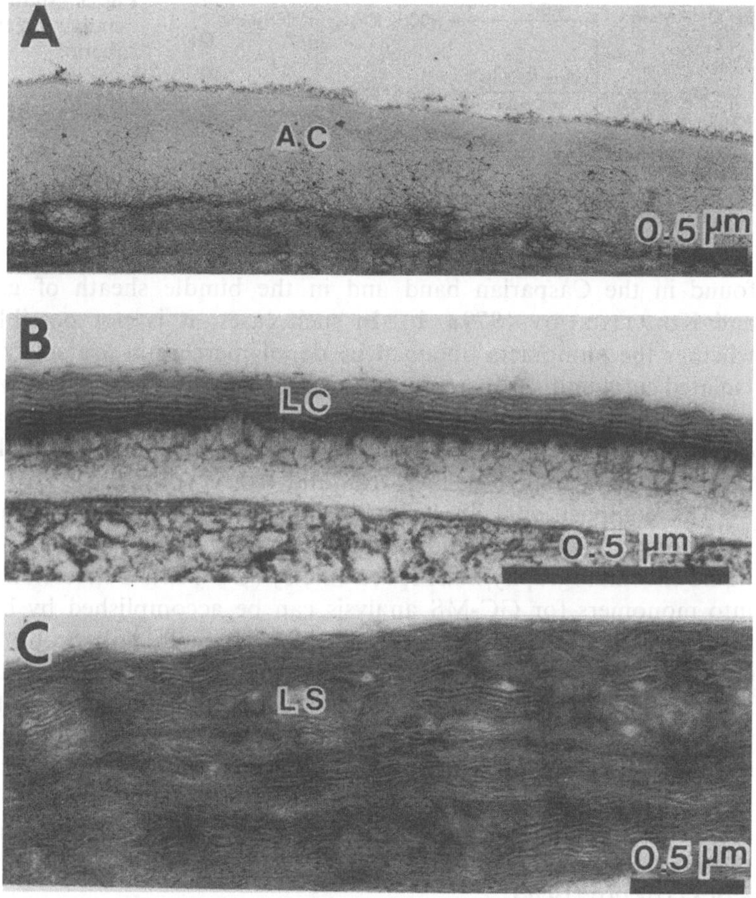

Fig. 2. Electronmicrographs of **A** amorphous cuticle (*AC*) from the leaf of *Frankenia grandifolia*. (Courtesy of Dr. W.W. Thomson); **B** lamellar cuticle (*LC*) from a developing leaf of *Phormium tenax*. (Courtesy of Dr. A.B. Wardrop); **C** lamellar suberin (*LS*) in the wall of a cell in the chalazal region of the inner seed coat of *Citrus paradisi*. (Courtesy of R.W. Davis)

3 Composition and Structure of Cutin and Suberin

3.1 Isolation and Depolymerization of Cutin and Suberin

Many plant cuticles can be isolated as transparent membranes by treatment of the plant tissue with pectin-disrupting reagents such as ammonium oxalate/oxalic acid, $ZnCl_2/HCl$, cuprammonium ion, or by enzymatic digestion (ORGELL 1955, MATIC 1956, EGLINTON and HUNNEMANN 1968, WALTON and KOLATTUKUDY 1972a). Suberin, on the other hand, can be isolated only as enriched preparations because of its location and tight attachment to the cell wall (SITTE 1962, 1975, HOLLOWAY 1972a, b, KOLATTUKUDY et al. 1975). This problem is especially evident when dealing with internal suberin polymers such as that

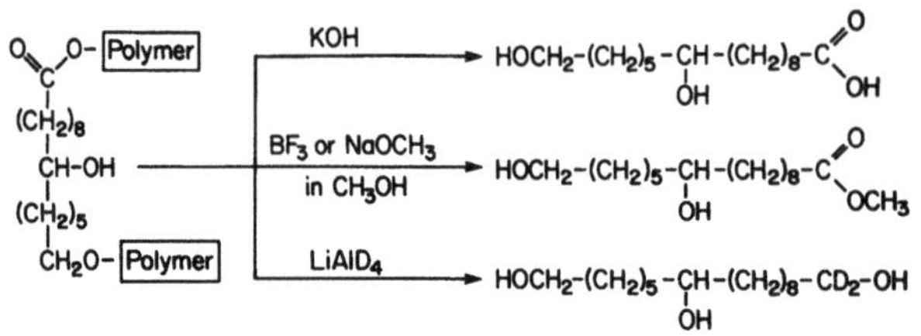

Fig. 3. Methods for chemical depolymerization of cutin and suberin

found in the Casparian band and in the bundle sheath of grasses (ESPELIE and KOLATTUKUDY 1979a, b). In such cases, it is not possible to ascertain whether the monomers obtained by depolymerization are solely from suberin. Isolated cutin and suberin preparations are usually subjected to extensive extractions with organic solvent to remove nonpolymeric materials and enzymatic digestion with cellulase and pectinase to remove any residual carbohydrate prior to chemical analysis (WALTON and KOLATTUKUDY 1972a).

The compositional analysis of cutin and suberin has been greatly aided by chromatographic techniques and especially by the advent of combined gas-liquid chromatography/mass spectrometry (GC-MS). Conversion of the polymer into monomers for GC-MS analysis can be accomplished by base hydrolysis, transesterification with BF_3 or $NaOCH_3$ in methanol, or by hydrogenolysis ($LiAlH_4$) generating free acids, methyl esters, or alkanols, respectively (Fig. 3). Each technique has its own merits and the most complete compositional information about a sample is usually obtained by a combination of depolymerization techniques. Reduction with $LiAlD_4$ is particularly advantageous because labile functional groups present in the polymer (e.g., aldehydes and epoxides) are converted to stable alcohols with the incorporation of deuterium to indicate the original structure of each monomer (WALTON and KOLATTUKUDY 1972a, KOLATTUKUDY 1974).

3.2 Composition of the Monomers of Cutin and Suberin

Most cutin samples which have been chemically examined are composed primarily of a C_{16} family and a C_{18} family of acids (Fig. 4). The most common C_{16} monomers are palmitic acid, 16-hydroxypalmitic acid, 9,16-dihydroxy- and 10,16-dihydroxypalmitic acid and among them the dihydroxy acids are usually the most dominant ones. The common members of the C_{18} family of acids are stearic, oleic, linoleic, 18-hydroxyoleic, 18-hydroxy-9,10-epoxystearic, 9,10,18-trihydroxystearic and the corresponding Δ^{12} unsaturated analogs. Monomers other than these have been reported as rare or minor cutin components (HUNNEMAN and EGLINTON 1972, DEAS et al. 1974, KOLATTUKUDY 1974, CALDICOTT et al. 1975). These less common components have either a greater or lesser degree of substitution and oxidation and/or shorter or longer chain lengths. The plants from which most (>90%) of the aliphatic components of cutin have been identified are listed in Table 1. The most characteristic cutin component, dihydroxypalmitate, is present in almost all cases and frequently it is a major, if not the most dominant component. This monomer has a mid-chain hydroxyl moiety of L configuration (ESPELIE and KOLATTUKUDY 1978) and occurs as a mixture of positional isomers in which the mid-chain hydroxyl group is at the 7-, 8-, 9- or 10-position. The composition varies with species, anatomical location, age of the tissue and possibly environmental factors such as light and temperature (HOLLOWAY and DEAS 1971, WALTON and KOLATTUKUDY 1972a, ESPELIE et al. 1979). In rapidly growing plant organs cutin is composed of chiefly the C_{16} family of monomers whereas cutin from slowly growing organs often has a high proportion of C_{18} acids.

Table 1. Plants from which the aliphatic components of cutin or suberin polymer(s) have been analyzed in detail

Source		Reference	Major fatty acid percentage in parenthesis
Cutin			
Agave americana	Leaf	1	9,10,18-trihydroxy C_{18} (50)
Amygdalus persica	Fruit	2	10,16-dihydroxy C_{16} (33)
Anemia phyllitidis	Leaf	3	ω-hydroxy C_{16} (47)
Citrus deliciosa	Fruit	4	ω-hydroxy-10-oxo C_{16} (34)
	Leaf	4	10,16-dihydroxy C_{16} (47)
Citrus limon	Fruit	4	ω-hydroxy-10-oxo C_{16} (34)
	Leaf	4	10,16-dihydroxy C_{16} (49)
Citrus paradisi	Fruit	5	ω-hydroxy-10-oxo C_{16} (43)
	Juice sac	5	10,16-dihydroxy C_{16} (19)
	Leaf	5	10,16-dihydroxy C_{16} (52)
	Seed coat	5	ω-hydroxy-9,10-epoxy C_{18} (37)
Citrus reticulata	Fruit	4	10,16-dihydroxy C_{16} (37)
	Leaf	4	10,16-dihydroxy C_{16} (47)
Citrus sinensis	Fruit	4	10,16-dihydroxy C_{16} (31)
	Leaf	4	10,16-dihydroxy C_{16} (49)
Coffea arabica	Leaf	6	9,16-dihydroxy C_{16} (32)
Ginkgo biloba	Leaf	3	10,16-dihydroxy C_{16} (45)
Gnetum gnemon	Leaf	3	10,16-dihydroxy C_{16} (73)
Hordeum vulgare	Leaf	7	ω-hydroxy-9,10-epoxy C_{18} (34)
	Seed coat	7	ω-hydroxy-9,10-epoxy C_{18} (30)
Lactuca sativa	Leaf	7	9,16-dihydroxy C_{16} (18)
	Seed coat	7	ω-hydroxy-9,10-epoxy C_{18} (19)
Malabar papaiarnasium	Fruit	2	9,16-dihydroxy C_{16} (73)
Malus pumila	Fruit	8	ω-hydroxy-9,10-epoxy $C_{18:1}$ (22)
	Leaf	7	10,16-dihydroxy C_{16} (41)
	Petal	7	10,16-dihydroxy C_{16} (54)
	Stigma	7	10,16-dihydroxy C_{16} (57)
Pinus sylvestris	Leaf	3	9,16-dihydroxy C_{16} (49)
Pisum sativum	Leaf	7	9,16-dihydroxy C_{16} (26)
	Seed coat	7	9,16-dihydroxy C_{16} (24)
Psilotum nudum	Stem	9	ω-hydroxy C_{16} (53)
Pyrus communis	Fruit	2	10,16-dihydroxy C_{16} (40)
Ribes grossularia	Fruit	3	10,16-dihydroxy C_{16} (83)
Rosmarinus officinalis	Leaf	10	10,16-dihydroxy C_{16} (47)
Sapindus saponaria	Leaf	3	10,16-dihydroxy C_{16} (64)
Solanum tuberosum	Leaf	11	dihydroxy C_{16} (51)

Table 1 (continued)

Source		Reference	Major fatty acid percentage in parenthesis
Sorghum bicolor	Stem	12	9,16-dihydroxy C_{16} (11)
Spinacia oleracea	Leaf	13	ω-hydroxy-9,10-epoxy C_{18} (63)
Tmesipteris viellardii	Leaf	10	ω-hydroxy C_{14} (29)
Tropaeolum majus	Stigma	14	10,16-dihydroxy C_{16} (45)
Vaccinium macrocarpon	Fruit	15	9,10,18-trihydroxy C_{18} (44)
Vicia faba	Flower	16	10,16-dihydroxy C_{16} (80)
	Leaf	17	10,16-dihydroxy C_{16} (78)
	Stem	7	10,16-dihydroxy C_{16} (63)
Vitis vinifera	Fruit	2	ω-hydroxy-9,10-epoxy C_{18} (32)
Zea mays	Leaf	7	ω-hydroxy $C_{18:1}$ (32)
	Seed coat	7	ω-hydroxy $C_{18:1}$ (51)
Suberin			
Beta vulgaris	Storage organ	18	ω-hydroxy $C_{18:1}$ (21)
Betula pendula	Bark	19	9,10,18-trihydroxy C_{18} (43)
Brassica napobrassica	Storage organ	18	ω-hydroxy $C_{18:1}$ (15)
Brassica rapa	Storage organ	18	ω-hydroxy $C_{18:1}$ (26)
Citrus paradisi	Inner seed coat	5	ω-hydroxy $C_{18:1}$ (20)
Daucus carota	Storage organ	18	ω-hydroxy $C_{18:1}$ (17)
Ipomoea batatas	Storage organ	18	ω-hydroxy $C_{18:1}$ (33)
Pastinaca sativa	Storage organ	18	ω-hydroxy $C_{18:1}$ (16)
Quercus suber	Bark	19	ω-hydroxy C_{22} (25)
Ribes grossularia	Bark	20	ω-hydroxy $C_{18:1}$ (26)
Ribes nigrum	Bark	20	ω-hydroxy C_{20} (15)
Solanum tuberosum	Storage organ	11	ω-hydroxy $C_{18:1}$ (33)
Sorghum bicolor	Stele	12	ω-hydroxy C_{16} (17)
Zea mays	Bundle sheath	21	9,16-dihydroxy C_{16} (15)

References (1) CRISP 1965, (2) WALTON and KOLATTUKUDY 1972a, (3) HUNNEMAN and EGLINTON 1972, (4) BAKER and PROCOPIOU 1975, (5) ESPELIE et al. 1980b; (5) HOLLOWAY et al. 1972b, (7) ESPELIE et al. 1979, (8) HOLLOWAY 1973, (9) CALDICOTT et al. 1975, (10) BRIESKORN and KABELITZ 1971, (11) BRIESKORN and BINNEMAN 1975, (12) ESPELIE and KOLATTUKUDY 1979a, (13) HOLLOWAY 1974, (14) SHAYKH et al. 1977b, (15) CROTEAU and FAGERSON 1972, (16) KOLATTUKUDY et al. 1974, (17) KOLATTUKUDY and WALTON 1972, (18) KOLATTUKUDY et al. 1975, (19) HOLLOWAY 1972a, (20) HOLLOWAY 1972b, (21) ESPELIE and KOLATTUKUDY 1979b

Chemical analysis of suberin has been more limited than that of cutin in: (a) the number of species examined, (b) the variety of anatomical locations of the polymer, and (c) the proportion of monomers identified. The composition of only the aliphatic components has been studied in detail (Fig. 4), and the plants for which such information is available are listed in Table 1. The aliphatic monomers, which comprise only 10 to 40% of the suberin polymers (KOLATTUKUDY et al. 1975, KOLATTUKUDY 1978) consist of ω-hydroxy even-chain acids (C_{16}-C_{24}) and α,ω-dicarboxylic acids (mainly C_{16} and C_{18}), and very long-chain (C_{20}-C_{30}) alcohols and acids. Although most suberin polymers have only a small proportion of the mid-chain substituted acids which are prevalent in cutin, a fairly high proportion of such polyhydroxylated acids has been found in some suberin polymers (HOLLOWAY 1972a, HOLLOWAY et al. 1972a, ESPELIE and KOLATTUKUDY 1979b).

Phenolic materials apparently comprise a large portion (20–60%) of the suberin polymer (HERGENT 1958, SWAN 1968, KOLATTUKUDY et al. 1975). The suberin-enriched polymers from several storage organs contain small amounts of esterified ferulic acid (RILEY and KOLATTUKUDY 1975), and nitrobenzene oxidation of suberin-enriched preparations gives rise to vanillin and *p*-hydroxybenzaldehyde. Although these results, together with the staining properties of suberin layers (SCOTT and PETERSON 1979), confirm the presence of phenolics in suberin, the nature and composition of the major portion of the phenolic matrix is largely unknown.

3.3 Intermolecular Linkages in Cutin and Suberin

X-ray diffraction studies indicated that cutin is an amorphorus polymer (V.P. AGRAWAL and P.E. KOLATTUKUDY, unpublished results, 1974, WILSON and STERLING 1976). Only a limited amount of information is available concerning the intermolecular linkages in cutin. Since the polymer is held together largely by ester linkages, selective partial degradation by chemical techniques is extremely difficult. Chemical modification of free hydroxyl groups in the polymer followed by analyses of the depolymerization products showed that most of the primary hydroxyl groups are esterified in the polymer. Although the majority of the free hydroxyl groups are secondary, a significant portion of the mid-chain hydroxyl groups are involved in ester linkages (DEAS and HOLLOWAY 1977, KOLATTUKUDY 1977). In general agreement with these conclusions pancreatic lipase, which is known to be specific for primary alcohol esters, caused extensive depolymerization of cutin (BROWN and KOLATTUKUDY 1978a) and released soluble secondary alcohol esters (I. MAITI and P.E. KOLATTUKUDY, unpublished results). Cutins which contain predominantly ω-hydroxy-9,10-epoxy C_{18} acids (e.g., *Vitis vinifera* fruit) (WALTON and KOLATTUKUDY, 1972a) or ω-hydroxy-8,9 or 10-oxo C_{16} acids (e.g., *Citrus* fruit) (DEAS et al. 1974, BAKER and PROCOPIOU 1975, ESPELIE et al. 1980b) must be predominantly a linear polymer because of the absence of functional groups which can participate in cross-linking. In tomato fruit the degree of cross-linking was reported to be independent of the stage of growth (DEAS and HOLLOWAY 1977), suggesting that the secondary structure may not change with the physiological state of the fruit. However, the functional significance of the degree of cross-linking remains unknown. On the basis of the available limited structural information the tentative model shown in Fig. 4 was proposed for cutin containing only the C_{16} family of monomers.

The secondary structure of suberin remains largely unknown. On the basis of the very limited amount of available information concerning the suberization process and the composition of suberin, the tentative model shown in Fig. 4 was proposed as a working hypothesis (KOLATTUKUDY 1977). According to this model, a phenolic matrix is attached to the cell wall possibly in a manner similar to that found in lignification. This matrix is probably somewhat similar

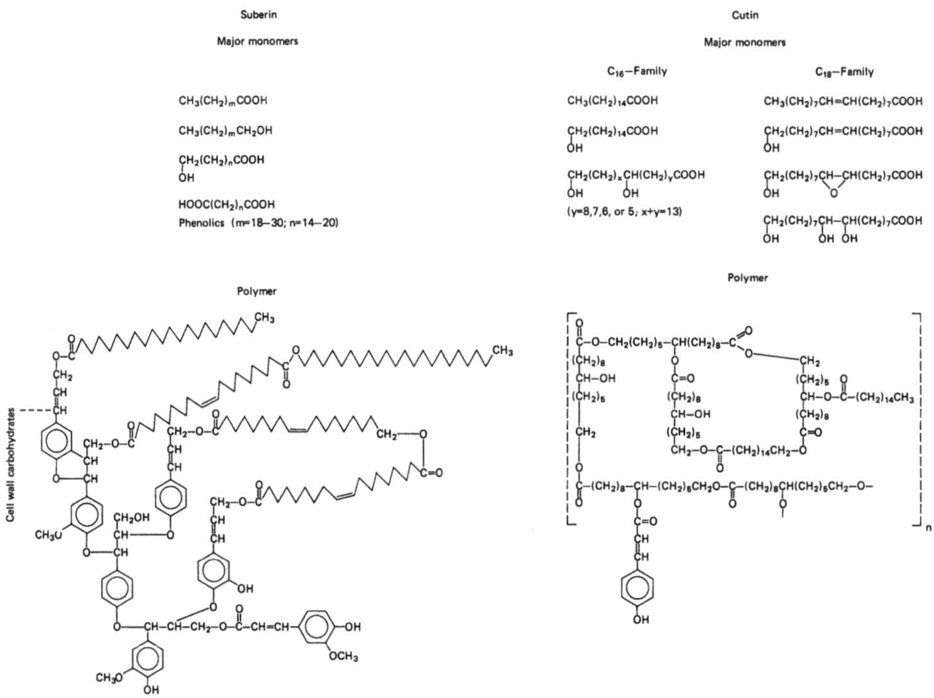

Fig. 4. The major monomers of cutin and suberin and proposed models of the two polymers

to lignin but contains less methoxy groups and is probably different in other ways, as suggested by recent specific staining tests (BRISSON et al. 1976). The aliphatic components, which by themselves obviously cannot form an extensive polymer, probably serve to cross-link the phenolic domains and provide a hydrophobic environment which accommodates the soluble waxes giving rise to the dark and light lamellar appearance in electronmicrographs. Evidence supporting such a model has been recently summarized (KOLATTUKUDY 1980a).

4 Biosynthesis of Cutin and Suberin

4.1 Biosynthesis of the C_{16} Family of Monomers

The results from the radiotracer studies with [1-^{14}C]palmitate and 16-hydroxy [1-^{14}C]palmitate, suggested a biosynthetic pathway for the C_{16} family of cutin acids (Fig. 5). The enzymatic activities required to catalyze the postulated reactions have been demonstrated in cell-free preparations. ω-Hydroxylation, the first step involved in the biosynthesis of most of the aliphatic components of cutin, is catalyzed by enzymes located in the endoplasmic reticulum (SOLIDAY and KOLATTUKUDY 1977). This ω-hydroxylation requires NADPH and O_2 as cofactors; however, involvement of a thioester of the fatty acid in this reaction

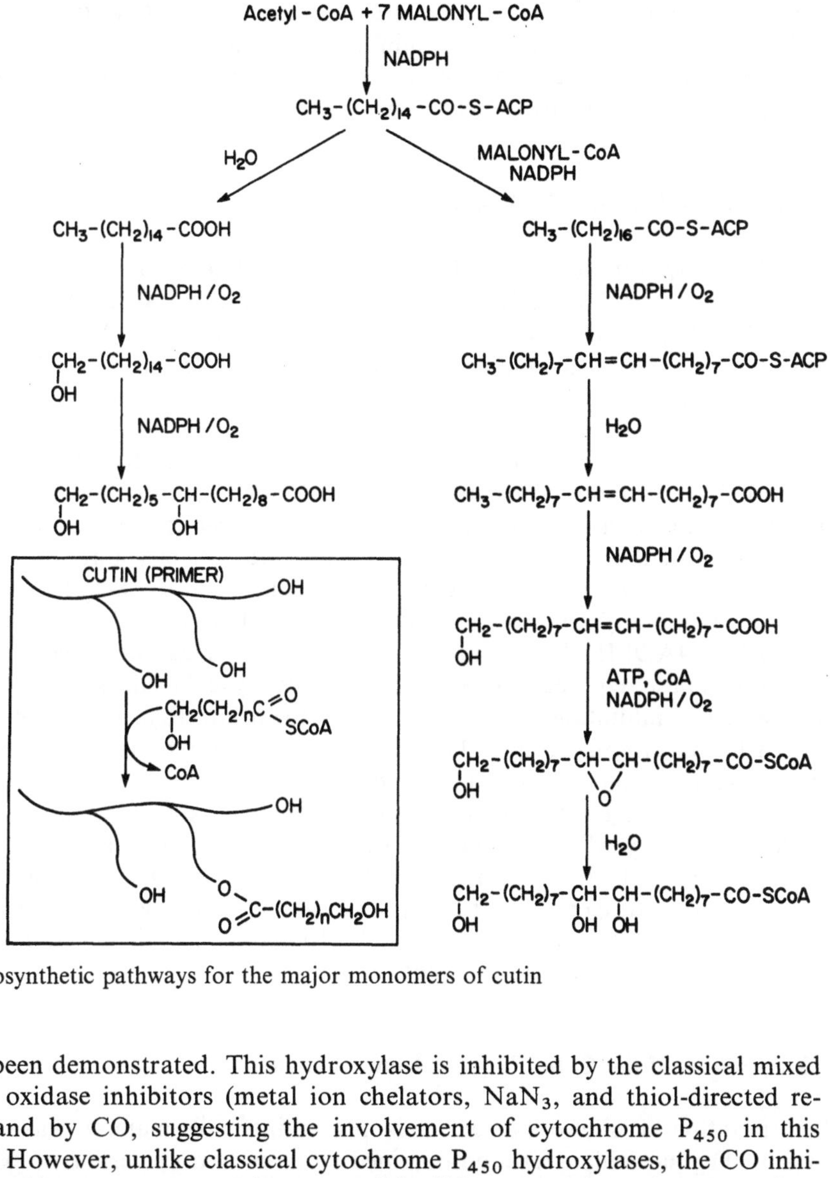

Fig. 5. Biosynthetic pathways for the major monomers of cutin

has not been demonstrated. This hydroxylase is inhibited by the classical mixed function oxidase inhibitors (metal ion chelators, NaN_3, and thiol-directed reagents) and by CO, suggesting the involvement of cytochrome P_{450} in this reaction. However, unlike classical cytochrome P_{450} hydroxylases, the CO inhibition in this case was not photoreversible (450 nm) perhaps because of an unusually high affinity of the ω-hydroxylase for CO (30% CO resulted in complete inhibition). The mid-chain hydroxylation of ω-hydroxypalmitate to the dihydroxypalmitate of cutin involves a direct hydroxylation rather than hydration of a double bond (KOLATTUKUDY and WALTON 1972). A crude cell-free preparation from the excised epidermis of *V. faba* catalyzed C-10 hydroxylation of ω-hydroxypalmitate (WALTON and KOLATTUKUDY 1972b). Further characterization of this hydroxylation was accomplished with an endoplasmic reticulum fraction from embryonic shoots of *V. faba*. Mid-chain hydroxylation by this preparation required O_2 and NADPH and was inhibited by the typical mixed function oxidase inhibitors as well as CO (SOLIDAY and KOLATTUKUDY 1978). Unlike the ω-hydroxylation, CO inhibition of mid-chain hydroxylation was photoreversible as expected of a typical cytochrome P_{450} hydroxylase. Whether the CoA ester of the hydroxy acid is involved in this reaction remains doubtful.

4.2 Biosynthesis of the C_{18} Family of Monomers

Specific incorporation of [1-^{14}C]oleic acid, [1-^{14}C]linoleic acid and [1-^{14}C]linolenic acid into the C_{18} family of cutin monomers in tissue slices (KOLATTUKUDY et al. 1973a) led to the postulation of the pathway shown in Fig. 5. The direct conversion of exogenous synthetic 18-hydroxy[18-^{3}H]oleic acid into 18-hydroxy-9,10-epoxystearic and 9,10,18-trihydroxystearic acids and conversion of exogenous 18-hydroxy-9,10-epoxy [18-^{3}H]stearic acid into 9,10,18-trihydroxystearic acid of cutin in plant tissues (CROTEAU and KOLATTUKUDY 1974a) strongly supported the proposed biosynthetic pathway. The reactions postulated to be involved in this pathway have been demonstrated in cell-free preparations. Thus, a 3,000 g particulate fraction from cell-free preparations of young spinach leaves catalyzed the conversion of 18-hydroxy[18-^{3}H]oleic acid to 18-hydroxy-cis-9,10-epoxystearic acid (CROTEAU and KOLATTUKUDY 1975a). This epoxidation required O_2, NADPH, ATP, and CoA as cofactors, suggesting that this mixed function oxidase required a thioester of the hydroxyacid as the substrate. Photoreversible CO inhibition suggested that a cytochrome P_{450} type protein is involved in this enzymatic epoxidation. This epoxidase required a free hydroxyl at the ω-carbon and a cis-double bond in the substrate. A 3,000 g particulate fraction, containing cuticular membranes, from the skin of rapidly expanding apple fruit catalyzed hydration of 18-hydroxy-9,10-epoxystearic acid without requiring any cofactors (CROTEAU and KOLATTUKUDY 1975b). This enzyme showed fairly stringent substrate specificity and generated *threo*-9,10,18-trihydroxystearic acid from 18-hydroxy-cis-9,10-epoxystearate reflecting the composition of the apple cutin monomers.

4.3 Synthesis of the Cutin Polymer from Monomers

Incorporation of labeled hydroxyacids into an insoluble material was catalyzed by a 3,000 g particulate fraction obtained from excised epidermis of expanding *V. faba* leaves (CROTEAU and KOLATTUKUDY 1973). This incorporation represented cutin biosynthesis because the label incorporated into the insoluble material could be released only by cutinase and not by other hydrolases. Incorporation of hydroxyacids into cutin required ATP and CoA, strongly suggesting that the CoA esters of the hydroxy-acids are the true substrates. The enzyme, upon ultrasonic dissociation from the endogenous polymer contained in such preparations, revealed a requirement for a primer (CROTEAU and KOLATTUKUDY 1974b). While cutin from different plants could serve as primer, other possible acyl acceptors such as cellulose, glycerol, hexadecanol, cholesterol, and polyethylene glycol were ineffective acceptors for the hydroxyacyl groups. The most efficient primer for the *V. faba* enzyme was cutin isolated from very young leaves, while cutin from older leaves was significantly less effective. Studies with derivatives of the hydroxyacid substrates and chemically modified primers showed that the enzyme catalyzes transfer of an hydroxyacyl moiety from the CoA ester preferentially to the free primary hydroxyl groups of the primer (Fig. 5).

4.4 Biosynthesis of the Aliphatic Components of Suberin

The finding that the composition of the aliphatic components of suberin of the periderm formed on wound-healing potato tuber discs is identical to that of the intact potato tuber (Kolattukudy and Dean 1974) permitted the use of wound-healing tissue slices for biosynthetic studies. On the basis of incorporation of labeled substrates into suberin components in such tissues, a pathway for the biosynthesis of the aliphatic components of suberin (Fig. 6) was proposed (Kolattukudy 1978). The unique reaction involved in this pathway is the conversion of ω-hydroxy acids to dicarboxylic acids, which are characteristically major aliphatic components of suberin. This conversion is a two-step process catalyzed by two separable enzymes, an ω-hydroxyacid dehydrogenase and an ω-oxoacid dehydrogenase (Agrawal and Kolattukudy 1978a). Since only the former, which catalyzes the first step in this process, is induced in the wound-healing tissue, this enzyme was purified to homogeniety. This NADP-specific enzyme, which is a dimer of 30,000 molecular weight protomers, catalyzes the reversible oxidation with an equilibrium constant of 1.4×10^{-9} M at pH 9.5 and 30° C. An unusual feature of this enzyme is that hydride from the A-side of the nicotinamide is transferred to the ω-oxoacid at nearly the same rate as that from the B-side. Mechanistic studies indicated that an arginine

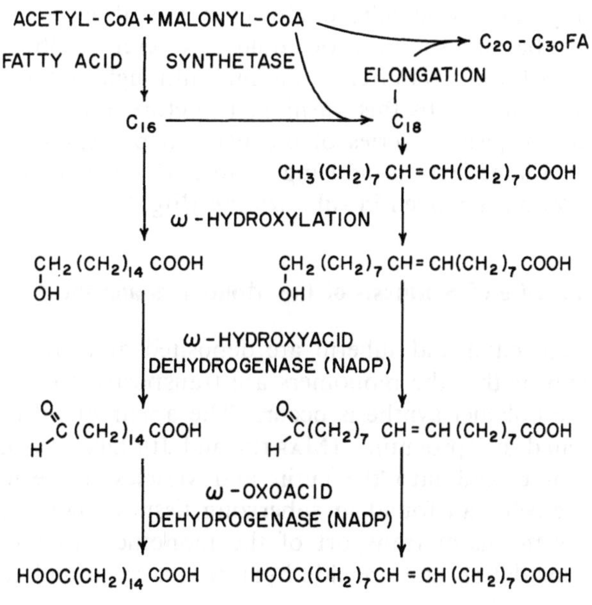

Fig. 6. Proposed biosynthetic pathways for the major aliphatic components of suberin

residue participates in the binding of the nucleotide while the ε-amino group of a lysine interacts with the ω-carboxyl group of the substrate and a histidine residue functions as a proton/acceptor (AGRAWAL and KOLATTUKUDY 1978b). These results and substrate specificity studies strongly suggest that this enzyme induced for suberization, is similar to other dehydrogenases except that it has evolved a specific substrate binding site for the ω-hydroxy acid. This site is such that it cannot readily accommodate very long ($>C_{20}$) chains and this specificity explains the observation that dicarboxylic acids of suberin contain much less very long-chain components than are found in the ω-hydroxy acid fraction of suberin (HOLLOWAY 1972a, b, KOLATTUKUDY et al. 1975, ESPELIE and KOLATTUKUDY 1979a, b).

4.5 Regulation of Suberization

Wound healing involves suberization of two layers of cells beneath the wound surface in both the aerial and underground organs of plants (DEAN and KOLATTUKUDY 1976). In potato tuber, suberization began after a lag period of about 3 days at 23° C and the process was completed within 7 days (KOLATTUKUDY and DEAN 1974). Experiments with inhibitors of protein and nucleic acid synthesis showed that the transcriptional and translational processes responsible for the production of the aliphatic components of suberin occurred between 72 and 96 h after wounding, resulting in the appearance of maximal levels of ω-hydroxy fatty acid dehydrogenase activity and subsequent deposition of the aliphatic components of suberin (AGRAWAL and KOLATTUKUDY 1977). It would appear that some substance generated by the wound triggers this chain of events. In fact, thorough washing of the wound prior to this induction period (72–96 h) inhibited suberization (SOLIDAY et al. 1978). Exogenous abscisic acid partially reversed this inhibition and abscisic acid induced suberization in potato tissue culture. However, this abscisic acid effect might be an indirect one because suberization could be severely inhibited by washing the tissue slices even three days after wounding, although neither free nor bound abscisic acid was removed by this washing. Therefore, it is tentatively concluded that abscisic acid triggers a series of biochemical processes, which result in the formation of a suberization-inducing factor, which in turn induces the synthesis of the enzymes involved in suberization (Fig. 6).

4.6 Site of Synthesis of the Monomers and the Polymers

Both cutin and suberin are deposited at extracellular locations and it would appear that the monomers are transported to this extracellular location where the polymer synthesis occurs. The apparent oil droplets and/or vesicles designated as "procutin" (MARTIN and JUNIPER 1970), and the reticular structures that extend into the cutin and vesicles between the plasma membrane and the cell wall found in suberizing tissues (BARCKHAUSEN 1978) could represent the means of transport of the monomers to the extracellular location. Since epoxidation and epoxide hydration were found to be catalyzed by particulate fractions containing cuticular membranes, it is possible that these two reactions occur at an extracellular location. However, such a possibility would have to assume that the required cofactors are available at the extracellular location. Ectodesmata and/or the reticulate structures often found in the interior regions of cutin might represent a means to get the required enzymes and cofactors to the extracellular location. In any case, ω-hydroxylation and mid-chain hydroxylation of the ω-hydroxyacid have been shown to be catalyzed by endoplasmic reticulum and these hydroxyacids are presumably transported via vesicles which fuse with plasma membrane and thus empty their contents at an extracellular location.

5 Biodegradation of Cutin and Suberin

Vast quantities of cutin and suberin generated by plants must be degraded by microbes present in the soil such as those which have been shown to grow on cutin as the sole source of carbon (HEINEN and DE VRIES 1966, HANKIN and KOLATTUKUDY 1971, PURDY and KOLATTUKUDY 1973). The fate of the large amounts of cutin and suberin ingested by insects and animals was unknown until recently when it was shown that pancreatic lipase hydrolyzed biosynthetically labeled cutin ingested by animals, and metabolized the resulting hydroxyacids (BROWN and KOLATTUKUDY 1978b). Plants themselves probably do not catalyze metabolic turnover of these extracellular barriers. However, in specialized situations some degradation of cutin probably does occur. For example, localized cleavages might occur during expansion of the organ so that a "break and extend" type of mechanism can accomodate the polymer synthesis needed to cover the increased surface area. Cutinase was recently suggested to be responsible for creating the opening in the cuticle above the newly differentiated sunken stomata (CARR and CARR 1978). Another example of cutin degradation by plants is during penetration of the pollen tube into the stigma with an intact cuticle (CHRIST 1959, KNOX et al. 1976). Cutinases have been purified from phytopathogenic fungi and from pollen and these two classes of enzymes are discussed below.

5.1 Fungal Cutinase

Cutinase was first purified from the extracellular fluid of *Fusarium solani f. pisi* grown on apple cutin, by the use of gel filtration and ion exchange column chromatography (PURDY and KOLATTUKUDY 1975a, b). Using the same procedures cutinases have been isolated from other pathogenic fungi such as *Fusarium roseum culmorum* (SOLIDAY and KOLATTUKUDY 1976), *F. roseum sambucinum, Heminthosporum sativum, Ulocladium consortiale,* and *Streptomyces scabies* (LIN and KOLATTUKUDY 1980b). If the fungal culture was initially grown on glucose and the glucose supply allowed to become exhausted, the production of cutinase could be induced by the addition of small quantities of cutin hydrolysate (LIN and KOLATTUKUDY 1978). The minimum structural features needed to induce cutinase production under these conditions were an aliphatic chain of about 16 carbons and a primary hydroxyl group at one end of the aliphatic chain.

All fungal cutinases thus far isolated are very similar in molecular weight (25,000), amino acid composition, and catalytic properties such as pH optimum (pH 10), substrate specificity, and the nature of the active site (Table 2). In addition to cutin, they hydrolyze p-nitrophenyl esters of short-chain acids (e.g., butyrate) but not of long-chain acids (e.g., palmitate). The fungal cutinases show specificity for the hydrolysis of primary alcohol esters and they also catalyze the hydrolysis of some triglycerides (T.-S. LIN and P.E. KOLATTUKUDY, unpublished). Ouchterlony double diffusion analyses showed that the enzymes from the different fungal sources are immunologically quite different but cutinase from Fusarium species showed some similarities (SOLIDAY and KOLATTUKUDY 1976, LIN and KOLATTUKUDY 1980b).

The fungal cutinases thus far isolated from six phytopathogens are glycoproteins containing 3.5 to 6% carbohydrates (LIN and KOLATTUKUDY 1976, 1977). Of the seven enzymes examined, five contain alkali-labile *O*-glycosidic linkages, whereas the enzymes from *H. sativum* and *S. scabies* contain alkali-resistant carbohydrate linkages (Table 2) presumably via asparaginyl residues (LIN and KOLATTUKUDY 1980b). Cutinase I from *F. solani pisi* has mannose, arabinose, glucosamine (*N*-acetyl), and glucuronic acid attached *O*-glycosidically to four different amino acids, two of which, serine and threonine, are commonly found to be involved in this type of linkage in other glycoproteins (LIN and KOLATTUKUDY 1980a). However, the other two residues, identified as β-hydroxyphenylalanine and β-hydroxytyrosine, have not been found heretofore in any other protein (LIN and KOLATTUKUDY 1976, 1979). Of the other cutinases containing *O*-glycosidic linkages, all of them contained *O*-glycosidically linked serine, while *O*-glycosidically linked β-hydroxyphenylalanine was present in four, threonine in three, and β-hydroxytyrosine in two. All five fungal cutinases which contained *O*-glycosidically linked carbohydrates also contained an N-terminal glycine attached to glucuronic acid by an amide linkage, another novel structural feature not previously found in any other protein (LIN and KOLATTUKUDY 1977).

Growth of the fungi on cutin also resulted in the production of small quantities of a nonspecific esterase which catalyzed the hydrolysis of a variety of small esters including *p*-nitrophenyl esters of the longer fatty acids (e.g., palmitate), but not the hydrolysis of cutin. Several areas of similarity of this nonspecific esterase with cutinase, such as amino acid composition, immunological cross-reactivity, involvement of active serine in catalysis, and common *O*-glycosidic linkages, have suggested that this glycoprotein may be a "pro-cutinase", representing incomplete processing (KOLATTUKUDY 1977).

5.2 Pollen Cutinase

Pollen cutinase has been isolated only from nasturtium (*Tropaeolum majus*), although the presence of a cutinase activity has been suggested from experimental results obtained with pollen from other plants (HEINEN and LINSKENS 1961, LINSKENS and HEINEN 1962). Soaking of the nasturtium pollen in a buffer results

Table 2. Comparison of the properties of cutinases from fungi and pollen

Properties	Fungus[a,b]	Pollen[c]
Molecular weight	25,000	40,000
Subunit composition	Single peptide	Single peptide
Carbohydrate content	4%–6%	7%
Acidic amino acid content	17.5%	28.5%
Free SH groups	Absent	Present
Inhibition by SH directed reagents	None	Severe
Inhibition by active serine-directed reagents (DFP)	Severe	None
pH optimum	10	6.8
Effect of chelators	None	None
Hydrolysis of p-nitrophenyl esters	$<C_{10}$	C_2-C_{18}

[a] PURDY and KOLATTUKUDY (1975a, b), [b] LIN and KOLATTUKUDY (1980a, b), [c] MAITI et al. (1979)

in release of the bulk of the cutinase into the medium because mature pollen already contains maximal levels of cutinase and synthesis of new enzyme does not occur during pollen germination (SHAYKH et al. 1977b). From this medium cutinase was isolated in homogeneous form using Sephadex G-100 gel filtration, QAE-Sephadex chromatography and isoelectric focusing (MAITI et al. 1979). This enzyme is a single polypeptide of about 40,000 molecular weight and contains 7% carbohydrate. It catalyzes the hydrolysis of *p*-nitrophenyl esters of C_2 to C_{18} fatty acids with comparable K_m and V_{max}. This enzyme, unlike that from the fungi, has a low pH optimum and uses a thiol moiety rather than an "active" serine in catalysis. In general, molecular and catalytic properties of the fungal and pollen cutinases are quite different (Table 2).

6 Isolation and Analysis of Waxes

Cuticular waxes can be readily extracted with little contamination from the internal lipids by dipping the plant organ in organic solvents such as chloroform for short periods of time (10 s to 1 min). However, waxes deeply embedded in thick cuticles and suberin-associated waxes are more difficult to extract (TULLOCH 1976, ESPELIE et al. 1980a). Thin-layer chromatography on silica gel separates waxes into classes of lipids which can then be subjected to combined gas-liquid chromatography and mass spectrometry (KOLATTUKUDY and WALTON 1973, TULLOCH 1975, 1976).

7 Composition of Waxes

The waxes associated with cutin and suberin are usually a complex mixture of several classes of aliphatic compounds each of which may contain many homologous series of compounds (MARTIN and JUNIPER 1970, KOLATTUKUDY

Table 3. Cuticular and suberin-associated wax components[a]

Compound type	General structure	Range
Hydrocarbons		
n-Alkanes	$CH_3(CH_2)_nCH_3$	C_{21} to C_{35}
iso-Alkanes	$CH_3CH(CH_3)R$	C_{25} to C_{35}
anteiso-Alkanes	$CH_3CH_2CH(CH_3)R$	C_{24} to C_{36}
Alkenes	$R_1CH{=}CHR_2$	C_{17} to C_{33}
Terpenoid hydrocarbons	Farnasene, Pristane, and cyclic hydrocarbons, phyllocladene, isophyllocladene (+)kaurene, and isokaurene etc.	
Aromatic hydrocarbons	Anthracene and phenanthrene type with an attached alkane chain	
Ketones	R_1COR_2	C_{25} to C_{33}
α-Ketols	$R_1CH(OH)COR_2$	
Secondary alcohols	$R_1CH(OH)R_2$	C_9 to C_{33}
β-Diketones	$R_1COCH_2COR_2$	
Hydroxy- or oxo-β-diketones	$R_1CH(OH)(CH_2)_nCOCH_2COR_2$	
Monoesters	R_1COOR_2	C_{30} to C_{60}
Phenolic esters	1,16-Dioxo, 1-hydroxy-16-oxo, and 1,16-dihydroxy-hexadecan-7-yl-p-coumarate. Ferulic acid esters of C_{18} to C_{28} primary alcohols.	
Diesters	$R_1-\overset{O}{\overset{\parallel}{C}}-OCH_2(CH_2)_nCH_2-O\overset{O}{\overset{\parallel}{C}}-R_2$	Diol C_9 to C_{12}, Acids C_{22} and C_{24} trans Δ^2
Polyesters	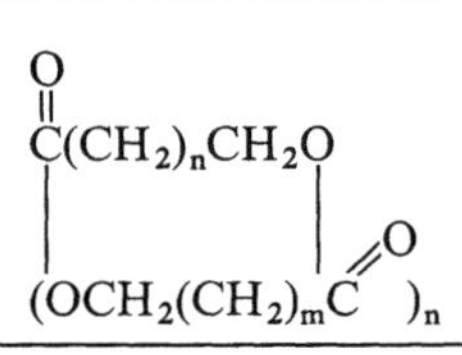	

Most common	Source (example)	Comments
C_{29}, C_{31}	Most plants	Most common component. High proportion of even-chain length and shorter chain length in suberin-associated waxes
C_{27}, C_{29}, C_{31}, C_{33}	Lavender blossom, poplar	Not as widespread as n-alkanes
C_{28}, C_{30}, C_{32}	leaf, tobacco	
C_{26}, C_{29}, C_{31}, C_{33}	Sugar cane, rose petal, algae, *Citrus*	In algae dienes and trienes also found. Branched alkenes might also be minor components
	Apple, Podocarpaceae	
	Banana leaf	
C_{29}, C_{31}	*Brassica*, roses, *Leptochloa digitata*, etc.	Not as common as alkanes
C_{29}	*Brassica*	Minor and rare component
C_{29}, C_{31}	Pea leaves, Rosaceae, *Brassica*, apple	Hydroxyl usually near the middle but towards the end also found; 2-ols found in suberin-associated waxes and in cuticular waxes which have β-diketones
C_{29}, C_{31}, C_{33}	*Eucalyptus*, *Poa colenasia*, grasses	Carbonyls at 12,14-, 14,16-, 16,18-positions; in some cases major wax components
(C_{31}) Hentriacontan-9-ol-14,16-dione	grasses	Usually minor components
C_{44}, C_{46}, C_{48}, C_{50}	Most plants	Composed of even chain saturated acids and alcohols. Esters of aromatic acids found in carnauba
	Pinus densiflora pollen	Commonly found in suberin-associated wax
	S. tuberosum, *Pseudosuga menziesii* cork	
	Wheat, rye	Rare and minor components
C_{12} to C_{16} ω-hydroxy acids 5 to 7/mole	Gymnosperms	Cyclic and noncyclic. Major components only in gymnosperms

Table 3 (continued)

Compound type	General structure	Range
Primary alcohols	RCH_2OH	C_{12} to C_{36}
Aldehydes	RCHO	C_{14} to C_{34}
Acids		
Alkanoic acids	RCOOH	C_{12} to C_{36}
Dicarboxylic acids	$HOOC(CH_2)_nCOOH$	C_{16} to C_{30}
γ-Lactones	$RCHCH_2CH_2CO$	C_{22} to C_{28}
ω-Hydroxy acids	$CH_2(OH)(CH_2)_nCOOH$	C_{10} to C_{34}
Diols		
α,ω-Diols	$CH_2(OH)(CH_2)_nCH_2OH$	C_{20} to C_{32}
α,β-Diols	$CH_2(OH)CH(OH)(CH_2)_nCH_3$	C_{14} to C_{32}
Terpenes, and Steroids	Ursolic acid, oleanolic acid, betulin, β-sitosterol	

[a] Adapted from Kolattukudy (1980b)

and Walton 1973, Tulloch 1976, Kolattukudy 1980b). There are wide variations in the amount and type of components present in different cuticular waxes; the most common wax components are listed in Table 3. There is a high degree of species specificity in the composition of cutin-associated waxes although anatomical location, age, and environmental conditions can all affect the composition. Hydrocarbons are the most ubiquitous wax components, although their percentage in wax varies widely. Cuticular hydrocarbons are predominantly odd-chain length alkanes with C_{29} and C_{31} frequently being the dominant components; olefins are rare and minor components. Esters and fatty acids are common but usually minor components, whereas fatty alcohols more often comprise a major portion of the wax.

In contrast to cuticular waxes, suberin-associated waxes have not been examined thoroughly. A survey of waxes from the periderm of underground storage organs indicates that suberin-associated waxes contain the same classes of components as cuticular waxes, although there may be some characteristic differences (Espelie et al. 1980a). The hydrocarbons, for example, seem to have a shorter average chain length, a broader distribution of homologs, and a larger proportion of even-chain length homologs than cuticular alkanes. The recent report that suberin-associated waxes serve as the major barrier to diffusion (Soliday et al. 1979), makes the examination of a wider variety of suberin-

Most common	Source (example)	Comments
C_{26}, C_{28}	Most plants	Even chains predominate
C_{26}, C_{28}, C_{30}	Grapes, apple, *B. oleracea*	Not as common as alcohols, polymeric forms also found. Even chain and saturated
C_{24}, C_{26}, C_{28}	Most plants	Very common. Even chain saturated chains predominate.
	Barks, spores of *Equisetum*.	Uncommon. Found in suberin-associated wax.
C_{26}	Roses	Rare and minor
C_{24}, C_{26}, C_{28}	Carnauba, apple	Uncommon. Found in suberin-associated wax
C_{22}, C_{24}, C_{28}, C_{30}	Carnauba, apple	Probably occur as esters (see diester above), not very common in waxes
C_{18}, C_{20}, C_{22}, C_{24}	Apple	Rare and minor
–	Bark and roots	Probably common in suberin-associated wax

associated waxes more relevant. There has been as yet, no correlation made between the chemical composition and the function of either cutin-associated or suberin-associated waxes.

8 Biosynthesis of Waxes

8.1 Biosynthesis of Very Long Fatty Acids

The very long aliphatic chains (C_{20}–C_{30}) characteristic of plant waxes are formed by chain elongation of fatty acids generated by fatty acid synthetase. Chain elongation was demonstrated by incorporation of C_2 to C_{24} acids into long acids in several plant tissues (KOLATTUKUDY and WALTON 1973) and by synthesis of such acids in cell-free preparations from different plants (MACEY and STUMPF 1968, KOLATTUKUDY and BUCKNER 1972, LESSIRE 1973, CASSAGNE and LESSIRE 1979). These studies showed that chain elongation required malonyl-CoA, NADPH, and an endogenous acyl moiety which appears to be CoA ester rather than an acyl carrier protein derivative (CASSAGNE and LESSIRE 1978). Differential effects of inhibitors and cofactors on chain elongation (LIU 1972, KOLATTUKUDY and BROWN 1975, KOLATTUKUDY 1975, KOLATTUKUDY et al. 1976, CASSAGNE and LESSIRE 1978, VON WETTSTEIN-KNOWLES 1979), and certain genetic evidence (VON WETTSTEIN-KNOWLES 1974) suggest that several chain-elongating enzyme systems with different specificities are involved in the generation of the wide variety of chain length classes found in plant waxes.

8.2 Biosynthesis of Fatty Alcohols

Labeled C_2 to C_{20} fatty acids were incorporated into very long fatty alcohols in plant tissues, suggesting that the exogenous fatty acids were elongated and subsequently reduced (MAZLIAK 1963, KOLATTUKUDY 1966, 1980b, CASSAGNE 1970). A cell-free preparation which catalyzed fatty acid reduction to alcohol was obtained from *Euglena gracilis* (KOLATTUKUDY 1970). In this organism acyl-CoA reductase is located mainly in the microsomal membranes (KHAN and KOLATTUKUDY 1973). Since in the membranes acyl-CoA reduction is coupled to aldehyde reduction, free aldehydes are not found. On the other hand, acyl-CoA reductase activity contained in an acetone powder extract of young *Brassica oleracea* leaves was fractionated into two components; one catalyzed NADPH-dependent reduction of the acyl-CoA to an aldehyde and the other protein fraction catalyzed an NADPH-dependent reduction of the aldehyde to an alcohol (KOLATTUKUDY 1971). Such enzyme systems are probably present in plants which have free aldehydes in their wax.

8.3 Biosynthesis of Wax Esters

Biosynthesis of wax ester would be expected to involve an acyl transfer mechanism. A partially purified protein fraction from the acetone powder extract of *B. oleracea* leaves catalyzed esterification of fatty alcohols with acyl-CoA as the acyl donor (KOLATTUKUDY et al. 1967). The acetone powder also contained enzymes which catalyzed esterification of fatty alcohols with free fatty acids and phospholipids as acyl donors. However, acyl-CoA fatty alcohol transacylase probably is the physiologically most significant route for wax ester synthesis. On the basis of the available experimental evidence the biosynthesis of wax esters can be summarized as shown in Fig. 7.

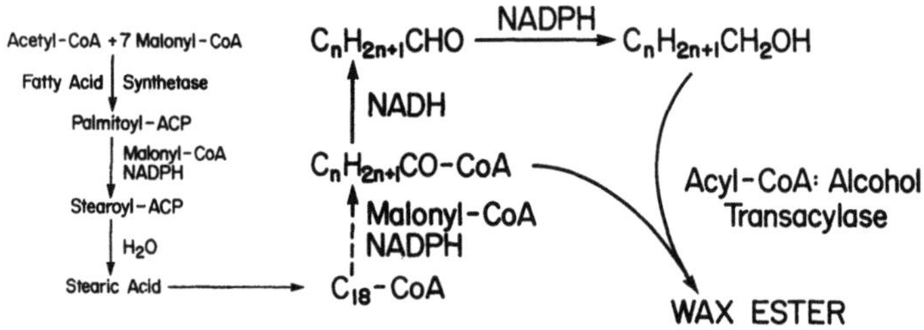

Fig. 7. Biosynthetic pathways for the production of very long chain fatty acids, aldehydes, alcohols and wax esters

8.4 Biosynthesis of Hydrocarbons and Derivatives

Incorporation of specifically labeled precursors into alkanes by leaf tissue slices ruled out a head-to-head condensation between two fatty acids as a possible mechanism for the biosynthesis of hydrocarbons as discussed elsewhere (KOLATTUKUDY and WALTON 1973, KOLATTUKUDY 1975). On the basis of such experimental results an elongation-decarboxylation mechanism was proposed for the biosynthesis of hydrocarbons (KOLATTUKUDY 1966, 1967). According to this hypothesis, a C_{16} or C_{18} fatty acid is elongated to appropriate chain length followed by decarboxylation (Fig. 8). Subsequently, a large amount of experimental evidence favoring this hypothesis has been obtained with higher plant tissues as summarized in other reviews (KOLATTUKUDY et al. 1976, VON WETTSTEIN-KNOWLES 1979, KOLATTUKUDY 1980b). This experimental evidence includes direct conversions of exogenous very long acids into alkanes with one less carbon than the precursor acid (KOLATTUKUDY et al. 1972, CASSAGNE and LESSIRE 1974, KOLATTUKUDY et al. 1974). Perhaps the most convincing evidence for the elongation-decarboxylation mechanism was the demonstration that cell-free preparations from *Pisum sativum* catalyzed the conversion of [9,10,11-^{3}H]C_{32} acid to C_{31} and C_{30} alkanes (KHAN and KOLATTUKUDY 1974). This conversion required O_2 and ascorbate and was inhibited by thiol compounds such as dithioerythritol. More recent experimental results strongly suggest that the elongating enzyme system is localized in the cytoplasmic side of the endoplasmic reticulum, whereas the decarboxylating enzymes are in the luminal side of the membrane (A. BOGNAR and P.E. KOLATTUKUDY, unpublished observations). The hydrocarbons thus generated in the lumen might be excreted via vesicles which fuse with the plasma membrane.

The α-oxidation activity contained in the microsomes generated odd-chain fatty acids from the even-chain fatty acid substrates and the decarboxylating enzyme preferentially used these odd-chain acids. Detergent-solubilized decarboxylase could not be resolved from α-oxidation activity. The decarboxylase from *P. sativum* decarboxylated C_{18}, C_{22}, C_{24} and C_{32} fatty acids, strongly

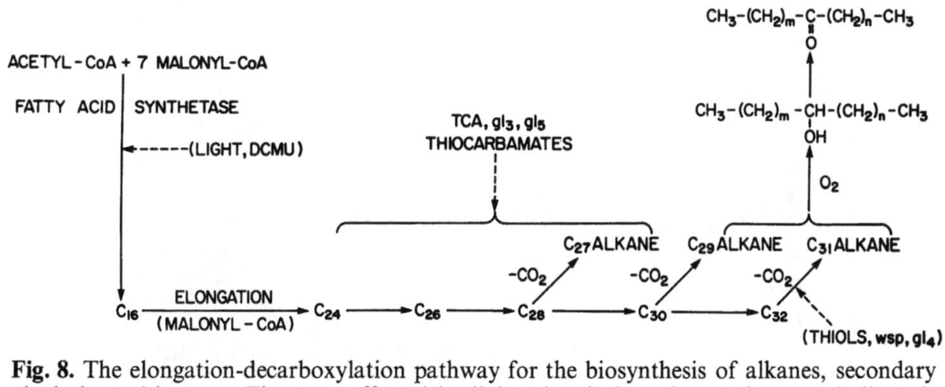

Fig. 8. The elongation-decarboxylation pathway for the biosynthesis of alkanes, secondary alcohols, and ketones. The steps affected by light, chemicals and mutations are indicated: gl_3, gl_4, and gl_5 are mutants of *B. oleracea* and *wsp* is a mutant of *P. sativum*

suggesting that the specificity of the chain elongation process determines the chain length of the alkane generated by an organism. The mechanism of decarboxylation and the nature of the intermediate involved remain obscure.

The secondary alcohols and ketones found in plant waxes are generated by hydroxylation of alkanes followed by oxidation of the secondary alcohol (Fig. 8). This conclusion was supported by several lines of evidence summarized elsewhere (KOLATTUKUDY 1975). A series of experiments demonstrated that exogenous alkane was converted to the secondary alcohol and that exogenous secondary alcohol was oxidized to the ketone by *B. oleracea* leaves providing direct evidence for the above hypothesis (KOLATTUKUDY and LIU 1970, KOLATTUKUDY et al. 1973b).

8.5 Biosynthesis of β-Diketones

β-Diketones are probably generated by a modified elongation-decarboxylation mechanism (Fig. 9). The distribution of label in hentriacontan-14,16-dione generated from specifically ^{14}C-labeled acetate and the observation that exogenous labeled C_{12}, C_{14}, and C_{16} acids but not C_{18} were incorporated into the aliphatic chains of the β-diketone on both sides of the diketone function support the proposed mechanism (VON WETTSTEIN-KNOWLES 1979, MIKKELSON 1979). Genetic evidence and differential inhibition of incorporation of labeled precursors into various classes of wax components are consistent with this proposed pathway. Hydroxy-β-diketones are probably derived from β-diketones and alkan-2-ols, found in plants which contain β-diketones, are probably derived from the same β-keto intermediates as the β-diketones.

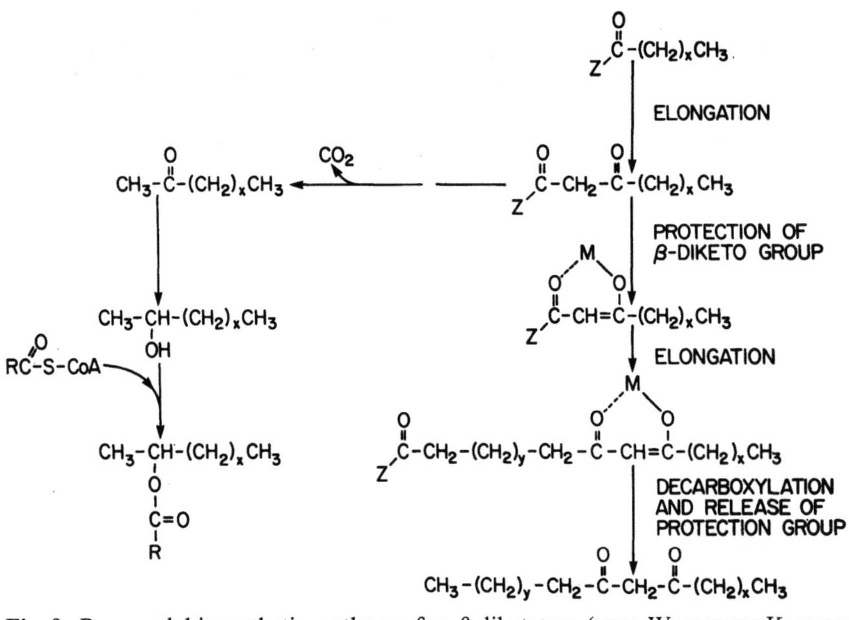

Fig. 9. Proposed biosynthetic pathway for β-diketones (VON WETTSTEIN-KNOWLES 1979) and a possible mode of synthesis of esters of alkan-2-ols

9 Function of Cutin, Suberin and Associated Waxes

The major function of the cuticle and of suberized walls is to serve as a physical barrier to prevent diffusion of moisture and/or other molecules. It is possible to regulate transpiration with the dynamic stomata because the cuticle is nearly waterproof due to the presence of wax. Experiments with isolated cuticle showed that waxes associated with cutin provide the major diffusion barrier (see SCHÖNHERR, Chapter 1.6, Volume 12B, this series). Similar experiments with suberized periderm indicated that waxes might not play as dominant a role as in the cuticle in preventing water loss (SCHÖNHERR and ZIEGLER 1980, ESPELIE et al. 1980a). However, specific inhibition of wax synthesis, without significantly affecting the polymer synthesis, prevented development of diffusion resistance of periderm in wound-healing potato tuber slices, demonstrating the central role played by waxes in waterproofing the wound periderm (SOLIDAY et al. 1979). Cutin and suberin provide the structural matrix which becomes waterproof upon deposition of the waxes.

Control of solute movement is another function usually associated with the cuticle and suberized walls. Suberized walls perform important functions in the root (TORREY and CLARKSON 1975). For example, suberization of the hypodermis probably reduces the radial movement of ions into the stele. The suberized Casparin band of endodermis helps to minimize apoplastic transport of water and solutes. Similarly, barrier layers in secretory glands prevent the influx of secreted chemicals as well as protecting the plant against moisture loss. In grass leaves the vascular bundles are encased in bundle sheaths with suberized walls which reduce apoplastic exchange of water and solutes between the mesophyll and vascular tissue. The presence of weak ion exchange sites in cuticles and suberized walls is thought to play a role in ion uptake as well as in the leaching of ions from plants.

Cuticles and suberized walls also perform an important function as physical barriers against entry of pathogens and possibly incompatible pollen tubes into the stigma. Pathogenic fungi, which break the cutin barrier, do so either mechanically or enzymatically (VAN DEN ENDE and LINSKENS 1974). The isolation of cutinase from *F. solani pisi* enabled the use of ferritin-labeled antibodies to show that during infection of its host (*Pisum sativum*) this fungus excretes cutinase (SHAYHK et al. 1977a). Recent demonstration (MAITI and KOLATTUKUDY 1979) that inhibition of the enzyme prevents infection shows that enzymatic penetration is involved in this case. Apparent suberization triggered by fungal infection probably helps to prevent the spread of invading pathogens in the host (ROYLE 1975, AIST 1976) and suberization caused by mechanical wounding helps to guard against entry of pathogens into the plants. Enzymatic degradation of suberin might be important in pathogenesis involving roots and barks and some recent ultrastructural studies are beginning to support such a notion (PARAMESWARAN and WILHELM 1979, GRÜNWALD and SEEMÜLLER 1979).

Chemical communication between the plant and other organisms in the environment is a probable but poorly understood role of the extracellular barrier layers. Cuticular components might play a role in host selection of insects by eliciting or retarding biting response (LIN et al. 1971, LEVY et al. 1974). The similarity of some cuticular components to compounds involved in chemical communication in insects (JACKSON and BLOMQUIST 1976), suggests that such plant products might elicit other responses of biological significance in insects. Plant cuticular components probably influence the growth and differentiation of fungi and thus play an important role in the interaction among plants and fungi. Numerous scattered reports suggest that some cuticular components are both inhibitors and promoters of fungal spore germination, growth, and differentiation of infective structures (MARTIN 1964, LAMPARD and CARTER 1973, PARBERY and BLAKEMAN 1978). There is suggestive evidence that some components, such as the phenolics covalently attached to either cutin or suberin polymers, might be released during infection and as a result play a defensive role or in some cases such compounds might promote differentiation of the infection structures of the fungi (GRAMBOW and GRAMBOW 1978). However, these potentially important roles of the extracellular barriers remain largely unexplored.

Acknowledgments. The work from the authors' laboratory discussed in this chapter was supported by grants from the U.S. Public Health Service, GM 18278, The National Science Foundation, PCM 77-00927, and from the Washington State Tree Fruit Research Commission.

References

Agrawal VP, Kolattukudy PE (1977) Biochemistry of suberization; ω-hydroxyacid oxidation in enzyme preparations from suberizing potato tuber disks. Plant Physiol 59:667–672

Agrawal VP, Kolattukudy PE (1978a) Purification and characterization of a wound-induced ω-hydroxyfatty acid: NADP oxidoreductase from potato tuber disks. Arch Biochem Biophys 191:452–465

Agrawal VP, Kolattukudy PE (1978b) Mechanism of action of a wound-induced ω-hydroxyfatty acid: NADP oxidoreductase isolated from potato tubers (*Solanum tuberosum* L). Arch Biochem Biophys 191:466–478

Aist JR (1976) Cytology of penetration and infection-fungi. In: Heitefuss R, Williams PH (eds) Encyclopedia plant physiology, new series Vol 4, Physiological plant pathology. Springer, Berlin Heidelberg New York, pp 197–221

Baker EA, Procopiou J (1975) The cuticles of *Citrus* species. Composition of the intracuticular lipids of leaves and fruits. J Sci Food Agric 26:1347–1352

Barckhausen R (1978) Ultrastructural changes in wounded plant storage tissue cells. In: Kahl G (ed) Biochemistry of wounded plant tissues. de Gruyter, Berlin New York, pp 1–42

Brieskorn CH, Binneman PH (1975) Carbonsäuren und Alkanole des Cutins und Suberins von *Solanum tuberosum*. Phytochemistry 14:1363–1367

Brieskorn CH, Kabelitz L (1971) Hydroxyfettsäuren aus dem Cutin des Blattes von *Rosmarinus officinalis*. Phytochemistry 10:3195–3204

Brisson JD, Robb J, Peterson RL (1976) Phenolic localization by ferric chloride and other iron compounds. Microsc Soc Can 3:174–175

Brown AJ, Kolattukudy PE (1978a) Evidence that pancreatic lipase is responsible for the hydrolysis of cutin. Arch Biochem Biophys 190:17–26

Brown AJ, Kolattukudy PE (1978b) Mammalian utilization of cutin, the cuticular polyester of plants. J Agric Food Chem 26:1263–1266

Caldicott AB, Eglinton G (1976) Cutin acids from Bryophytes: an ω-1 hydroxy alkanoic acid in two liverwort species. Phytochemistry 15:1139–1143

Caldicott AB, Simoneit BRT, Eglinton G (1975) Alkane triols in Psilotophyte cutins. Phytochemistry 14:2223–2228

Carr DJ, Carr SGM (1978) Origin and development of stomatal microanatomy in two species of *Eucalyptus*. Protoplasma 96:127–148
Cassagne C (1970) Les hydrocarbures végétaux: biosynthesis et localisation cellulaire. Ph D Thesis, Univ Bordeaux, France
Cassagne C, Lessire R (1974) Studies on alkane biosynthesis in epidermis of *Allium porrum* L. leaves. Arch Biochem Biophys 165:274–280
Cassagne C, Lessire R (1978) Biosynthesis of saturated very long chain fatty acids by purified membrane fractions from leek epidermal cells. Arch Biochem Biophys 191:146–152
Cassagne C, Lessire R (1979) Biosynthesis of the very long chain fatty acids in higher plants from exogenous substrates. In: Appelqvist L-Å, Liljenberg C (eds) Advances in the biochemistry and physiology of plant lipids. Elsevier, New York, pp 393–398
Christ B (1959) Entwicklungsgeschichtliche und physiologische Untersuchungen über die Selbststerilität von *Cardamine pratensis* L. Z Bot 47:88–112
Crisp CE (1965) The biopolymer cutin. Ph D Thesis, Univ California, Davis
Croteau R, Fagerson IS (1972) The constituent cutin acids of cranberry cuticle. Phytochemistry 11:353–363
Croteau R, Kolattukudy PE (1973) Enzymatic synthesis of a hydroxy fatty acid polymer, cutin, by a particulate preparation from *Vicia faba* epidermis. Biochem Biophys Res Commun 52:863–869
Croteau R, Kolattukudy PE (1974a) Direct evidence for the involvement of epoxide intermediates in the biosynthesis of the C_{18} family of cutin acids. Arch Biochem Biophys 162:471–480
Croteau R, Kolattukudy PE (1974b) Biosynthesis of hydroxy fatty acid polymers. Enzymatic synthesis of cutin from monomer acids by cell-free preparations from the epidermis of *Vicia faba* leaves. Biochemistry 13:3193–3202
Croteau R, Kolattukudy PE (1975a) Biosynthesis of hydroxy fatty acid polymers. Enzymatic epoxidation of 18-hydroxy oleic acid to 18-hydroxy-*cis*-9,10-epoxystearic acid by a particulate preparation from spinach (*Spinacia oleracea*). Arch Biochem Biophys 170:61–72
Croteau R, Kolattukudy PE (1975b) Biosynthesis of hydroxy fatty acid polymers. Enzymatic hydration of 18-hydroxy-*cis*-9,10-epoxystearic acid to *threo*-9,10,18-trihydroxystearic acid by a particulate preparation from apple (*Malus pumila*). Arch Biochem Biophys 170:73–81
Dean BB, Kolattukudy PE (1976) Synthesis of suberin during wound-healing in jade leaves, tomato fruit, and bean pods. Plant Physiol 58:411–416
Dean BB, Kolattukudy PE, Davis RW (1977) Chemical composition and ultrastructure of suberin from hollow heart tissue of potato tubers (*Solanum tuberosum*). Plant Physiol 59:1008–1010
Deas AHB, Holloway PJ (1977) The intermolecular structure of some plant cutins. In: Tevini M, Lichtenthaler HK (eds) Lipids and lipid polymers in higher plants. Springer, Berlin Heidelberg New York, pp 293–299
Deas AHB, Baker EA, Holloway PJ (1974) Identification of 16-hydroxyoxohexadecanoic acid monomers in plant cutins. Phytochemistry 13:1901–1905
Eglinton G, Hunneman DH (1968) Gas chromatographic-mass spectrometric studies of long chain hydroxy acids. I. The constituent cutin acids of apple cuticle. Phytochemistry 7:313–322
Espelie KE, Kolattukudy PE (1978) The optical rotation of a major component of plant cutin. Lipids 13:832–833
Espelie KE, Kolattukudy PE (1979a) Composition of the aliphatic components of suberin of the endodermal fraction from the first internode of etiolated Sorghum seedlings. Plant Physiol 63:433–435
Espelie KE, Kolattukudy PE (1979b) Composition of the aliphatic components of 'suberin' from the bundle sheaths of *Zea mays* leaves. Plant Sci Lett 15:225–230
Espelie KE, Dean BB, Kolattukudy PE (1979) Composition of lipid-derived polymers from different anatomical regions of several plant species. Plant Physiol 64:1089–1093
Espelie KE, Sadek NZ, Kolattukudy PE (1980a) Composition of suberin-associated waxes

from the subterranean storage organs of seven plants [parsnip, carrot, rutabaga, turnip, red beet, sweet potato and potato]. Planta, 148:468–476
Espelie KE, Davis RW, Kolattukudy PE (1980b) Composition, ultrastructure and function of the cutin and suberin-containing layers in the leaf, fruit peel, juice-sac and inner seed coat of grapefruit (*Citrus paradisi* Macfed). Planta, 149:498–511
Ferguson IB, Clarkson DT (1976) Ion uptake in relation to the development of a root hypodermis. New Phytol 77:11–14
Grambow HJ, Grambow GE (1978) The involvement of epicuticular and cell wall phenols of the host plant in the *in vitro* development of *Puccinia graminis* F. sp. *tritici*. Z Pflanzenphysiol 90:1–9
Grünwald J, Seemüller E (1979) Zerstörung der Resistenzeigenschaften des Himbeerrutenperiderms als Folge des Abbaus von Suberin und Zellwandpolysacchariden durch die Himbeerrutengallmücke *Thomasiniana theobaldi* Barnes (Dipt., Cecidomyiidae). Z Pflanzenkr Pflanzenschutz 86:305–314
Hankin L, Kolattukudy PE (1971) Utilization of cutin by a Pseudomonad isolated from soil. Plant Soil 34:525–529
Heinen W, Linskens HF (1961) Enzymic breakdown of stigmatic cuticula of flowers. Nature 191:1416
Heinen W, de Vries H (1966) Stages during the breakdown of plant cutin by soil microorganisms. Arch Mikrobiol 54:331–338
Hergent HL (1958) Chemical composition of cork from white fir bark. Forest Prod J 8:335–339
Holloway PJ (1972a) The composition of suberin from the corks of *Quercus suber* L and *Betula pendula* Roth. Chem Phys Lipids 9:158–170
Holloway PJ (1972b) The suberin composition of the cork layers from some *Ribes* species. Chem Phys Lipids 9:171–179
Holloway PJ (1973) Cutins of *Malus pumila* fruits and leaves. Phytochemistry 12:2913–2920
Holloway PJ (1974) Intracuticular lipids of spinach leaves. Phytochemistry 13:2201–2207
Holloway PJ, Baker EA (1970) The cuticles of some angiosperm leaves and fruits. Ann Appl Biol 66:145–154
Holloway PJ, Deas AHB (1971) The occurrence of positional isomers of dihydroxyhexadecanoic acid in plant cutins and suberins. Phytochemistry 10:2781–2785
Holloway PJ, Baker EA, Martin JT (1972a) The chemistry of plant cutins and suberins. An Quim 68:905–916
Holloway PJ, Deas AHB, Kabarra AM (1972b) Composition of cutin from coffee leaves. Phytochemistry 11:1443–1447
Hunneman DH, Eglinton G (1972) The constituent acids of gymnosperm cutins. Phytochemistry 11:1989–2001
Jackson LL, Blomquist GJ (1976) Insect waxes. In: Kolattukudy PE (ed) Chemistry and biochemistry of natural waxes. Elsevier, New York, pp 201–235
Khan AA, Kolattukudy PE (1973) Control of synthesis and distribution of acyl moieties in etiolated *Euglena gracilis*. Biochemistry 12:1939–1948
Khan AA, Kolattukudy PE (1974) Decarboxylation of long chain fatty acids to alkanes by cell-free preparations of pea leaves (*Pisum sativum*). Biochem Biophys Res Commun 61:1379–1386
Knox RB, Clarke A, Harrison S, Smith P, Marchalonis JJ (1976) Cell recognition in plants: determination of the stigma surface and their pollen interactions. Proc Natl Acad Sci USA 73:2788–2792
Kolattukudy PE (1966) Biosynthesis of wax in *Brassica oleracea*. Relation of fatty acids to wax. Biochemistry 5:2265–2275
Kolattukudy PE (1967) Biosynthesis of paraffins in *Brassica oleracea*: Fatty acid elongation-decarboxylation as a plausible pathway. Phytochemistry 6:963–975
Kolattukudy PE (1970) Reduction of fatty acids to alcohols by cell-free preparations of *Euglena gracilis*. Biochemistry 9:1095–1102
Kolattukudy PE (1971) Enzymatic synthesis of fatty alcohols in *Brassica oleracea*. Arch Biochem Biophys 142:701–709
Kolattukudy PE (1974) Biosynthesis of a hydroxy fatty acid polymer, cutin. Identification

and biosynthesis of 16-oxo-9- or 10-hydroxy palmitic acid, a novel compound in *Vicia faba*. Biochemistry 13:1354–1363

Kolattukudy PE (1975) Biochemistry of cutin, suberin and waxes, the lipid barriers on plants. In: Galliard T, Mercer EI (eds) Recent advances in the chemistry and biochemistry of plant lipids. Academic Press, New York, pp 203–246

Kolattukudy PE (1977) Lipid polymers, and associated phenols, their chemistry, biosynthesis, and role in pathogenesis. In: Loewus FA, Runeckles VC (eds) The structure, biosynthesis, and degradation of wood. Plenum Press, New York, pp 185–246

Kolattukudy PE (1978) Chemistry and biochemistry of the aliphatic components of suberin. In: Kahl G (ed) Biochemistry of wounded plant tissues. de Gruyter, Berlin New York, pp 43–84

Kolattukudy PE (1980a) Biopolyester membranes of plants: cutin and suberin. Science 208:990–1000

Kolattukudy PE (1980b) Cutin, suberin and waxes. In: Stumpf PK, Conn EE (eds) The biochemistry of plants Vol 4. Stumpf PK (ed) Lipids: structure and function. Academic Press, New York, pp 571–645

Kolattukudy PE, Brown L (1975) Fate of naturally occurring epoxy acids: a soluble epoxide hydrase, which catalyzes *cis* hydration from *Fusarium solani pisi*. Arch Biochem Biophys 166:599–607

Kolattukudy PE, Buckner JS (1972) Chain elongation of fatty acids by cell-free extracts of epidermis from pea leaves (*Pisum sativum*). Biochem Biophys Res Commun 46:801–807

Kolattukudy PE, Dean BB (1974) Structure, gas chromatographic measurement, and function of suberin synthesized by potato tuber tissue slices. Plant Physiol 54:116–121

Kolattukudy PE, Liu TJ (1970) Direct evidence for biosynthetic relationships among hydrocarbons, secondary alcohols and ketones in *Brassica oleracea*. Biochem Biophys Res Commun 41:1369–1374

Kolattukudy PE, Walton TJ (1972) Structure and biosynthesis of the hydroxy fatty acids of cutin in *Vicia faba* leaves. Biochemistry 11:1897–1907

Kolattukudy PE, Walton TJ (1973) The biochemistry of plant cuticular lipids. Progr Chem Fats Other Lipids 13:119–175

Kolattukudy PE, Buckner JS, Brown L (1972) Direct evidence for a decarboxylation mechanism in the biosynthesis of alkanes in *B. oleracea*. Biochem Biophys Res Commun 47:1306–1313

Kolattukudy PE, Walton TJ, Kushwaha RPS (1973a) Biosynthesis of the C_{18} family of cutin acids: ω-hydroxyoleic acid, ω-hydroxy-9,10-epoxystearic acid, 9,10,18-trihydroxystearic acid, and their Δ^{12}-unsaturated analogs. Biochemistry 12:4488–4498

Kolattukudy PE, Buckner JS, Liu TYJ (1973b) Biosynthesis of secondary alcohols and ketones from alkanes. Arch Biochem Biophys 156:613–620

Kolattukudy PE, Croteau R, Brown L (1974) Hydroxylation of palmitic acid and decarboxylation of C_{28}, C_{30} and C_{32} acids in *Vicia faba* flowers. Plant Physiol 54:670–677

Kolattukudy PE, Kronman K, Poulose AJ (1975) Determination of structure and composition of suberin from the roots of carrot, parsnip, rutabaga, turnip, red beet and sweet potato by combined gas-liquid chromatography and mass spectrometry. Plant Physiol 55:567–573

Kolattukudy PE, Croteau R, Buckner JS (1976) The biochemistry of plant waxes. In: Kolattukudy PE (ed) The chemistry and biochemistry of natural waxes. Elsevier, New York, pp 289–347

Krähmer H (1980) Wundreaktionen von Apfelbäumen und ihr Einfluß auf Infektionen mit *Nectria galligena*. Z Pflanzenkr Pflanzenschutz 87:97–112

Lampard JF, Carter GA (1973) Chemical investigations on resistance to coffee berry disease in *Coffea arabica*. An antifungal compound in coffee cuticular wax. Ann Appl Biol 73:31–37

Lessire R (1973) Etude de la biosynthèse des acides gras a tres longue chaine et leurs relations avec les alcanes dans la cellule d'epiderme d'*Allium porrum*. PhD Thesis, Univ Bordeaux, France

Levy EC, Ishaaya I, Gurevite E, Cooper R, Lavie D (1974) Isolation and identification

of host compounds eliciting attraction and bite stimuli in the fruit tree bark beetle, *Scolytus mediterraneus*. J Agr Food Chem 22:376–379
Lin K, Yamada H, Kato M (1971) Free fatty acids promote feeding behavior of silkworm, *Bombyx mori* L. Mem Fac Sci Kyoto Univ Ser Biol 4:108–115
Lin T-S, Kolattukudy PE (1976) Evidence for novel linkages in a glycoprotein involving β-hydroxyphenylalanine and β-hydroxytyrosine. Biochem Biophys Res Commun 72:243–250
Lin T-S, Kolattukudy PE (1977) Glucuronyl glycine, a novel N-terminus in a glycoprotein. Biochem Biophys Res Commun 75:87–93
Lin T-S, Kolattukudy PE (1978) Induction of a biopolyester hydrolase (cutinase) by low levels of cutin monomers in *Fusarium solani f. sp. pisi*. J Bacteriol 133:942–951
Lin T-S, Kolattukudy PE (1979) Direct evidence for the presence of β-hydroxyphenylalanine and β-hydroxytyrosine in cutinase from *Fusarium solani pisi*. Arch Biochem Biophys 196:225–264
Lin T-S, Kolattukudy PE (1980a) Structural studies on cutinase, a glycoprotein containing novel amino acids and glucuronic acid amide at the N-terminus. Europ J Biochem 106:341–351
Lin T-S, Kolattukudy PE (1980b) Isolation and characterization of a cuticular polyester (cutin) hydrolyzing enzyme from phytopathogenic fungi. Physiol Plant Pathol 17: 1–15
Linskens HF, Heinen W (1962) Cutinase-Nachweis in Pollen. Z Bot 50:338–347
Liu TYJ (1972) M.S. Thesis, Washington State University, Pullman, Washington, USA
Macey MJK, Stumpf PK (1968) Fat metabolism in higher plants XXXVI: long chain fatty acid synthesis in germinating peas. Plant Physiol 43:1637–1647
Maiti IB, Kolattukudy PE (1979) Prevention of fungal infection of plants by specific inhibition of cutinase. Science 205:507–508
Maiti IB, Kolattukudy PE, Shaykh M (1979) Purification and characterization of a novel cutinase from nasturtium (*Tropaeolum majus*) pollen. Arch Biochem Biophys 196:412–423
Martin JT (1964) Role of cuticle in the defense against plant disease. Annu Rev Phytopathol 2:81–100
Martin JT, Juniper BE (1970) The cuticles of plants. St Martins Press, New York
Matic M (1956) The chemistry of plant cuticle. A study of cutin from *Agave americana* L. Biochem J 63:168–178
Mazliak P (1963) La cire cuticulaire des pommes (*Malus pumila* L.) Ph D Thesis, Univ Paris, France
Mikkelsen JD (1979) Structure and biosynthesis of β-diketones in barley spike epicuticular wax. Carlsberg Res Commun 44:133–147
Mullick DB (1977) The non-specific nature of defense in bark and wood during wounding, insect and pathogen attack. In: Loewus FA, Runeckles VC (eds) The structure, biosynthesis and degradation of wood. Plenum Press, New York, pp 395–442
O'Brien TP, Carr DJ (1970) A suberized layer in the cell walls of the bundle sheath of grasses. Aust J Biol Sci 23:275–287
O'Brien TP, Kuo J (1975) Development of the suberized lamella in the mestome sheath of wheat leaves. Aust J Bot 23:783–794
Olesen P (1978) Studies on the physiological sheaths in roots. I. Ultrastructure of the exodermis in *Hoya carnosa* L. Protoplasma 94:325–340
Olesen P (1979) Ultrastructural observations on the cuticular envelope in salt glands of *Frankenia pauciflora*. Protoplasma 99:1–9
Orgell WH (1955) The isolation of plant cuticles with pectic enzymes. Plant Physiol 30:78–80
Parameswaren N, Wilhelm GE (1979) Micromorphology of naturally degraded beech and spruce barks. Eur J For Pathol 9:103–112
Parbery DG, Blakeman JP (1978) Effect of substances associated with leaf surfaces on appressorium formation by *Colletotrichum acutatum*. Trans Br Mycol Soc 70:7–19
Peterson CA, Peterson RL, Robards AW (1978) A correlated histochemical and ultrastructural study of the epidermis and hypodermis of onion roots. Protoplasma 96:1–21
Purdy RE, Kolattukudy PE (1973) Depolymerization of a hydroxy fatty acid biopolymer,

cutin, by an extracellular enzyme from *Fusarium solani f. pisi:* isolation and some properties of the enzyme. Arch Biochem Biophys 159:61–69

Purdy RE, Kolattukudy PE (1975a) Hydrolysis of plant cuticle by plant pathogens. Purification, amino acid composition, and molecular weight of two isozymes of cutinase and a nonspecific esterase from *Fusarium solani f. pisi*. Biochemistry 14:2824–2831

Purdy RE, Kolattukudy PE (1975b) Hydrolysis of plant cuticle by plant pathogens. Properties of cutinase I, cutinase II, and a nonspecific esterase isolated from *Fusarium solani pisi*. Biochemistry 14:2832–2840

Riley RG, Kolattukudy PE (1975) Evidence for covalently attached *p*-coumaric and ferulic acid in cutins and suberins. Plant Physiol 56:650–654

Robards AW, Jackson SM, Clarkson DT, Sanderson J (1973) The structure of barley roots in relation to the transport of ions into the stele. Protoplasma 77:291–311

Royle DJ (1975) Structural features of resistance to plant diseases. In: Friend J, Threlfall DR (eds) Biochemical aspects of plant parasite relationships. Academic Press, New York, pp 161–194

Schönherr J, Ziegler H (1980) Water permeability of *Betula* periderm. Planta 147:345–354

Scott MG, Peterson RL (1979) The root endodermis in *Ranunculus acris*. II. Histochemistry of the endodermis and the synthesis of phenolic compounds in roots. Can J Bot 57:1063–1077

Shaykh M, Soliday C, Kolattukudy PE (1977a) Proof for the production of cutinase by *Fusarium solani f. pisi* during penetration into its host, *Pisum sativum*. Plant Physiol 60:170–172

Shaykh M, Kolattukudy PE, Davis R (1977b) Production of a novel extracellular cutinase by the pollen and the chemical composition and ultrastructure of the stigma cuticle of nasturtium (*Tropaeolum majus*). Plant Physiol 60:907–915

Sitte P (1962) Zum Feinbau der Suberinschichten in Flaschenkork. Protoplasma 54:555–559

Sitte P (1975) Die Bedeutung der molekularen Lamellenbauweise von Korkzellwänden. Biochem Physiol Pflanz 168:287–297

Soliday CL, Kolattukudy PE (1976) Isolation and characterization of a cutinase from *Fusarium roseum culmorum* and its immunological comparison with cutinases from *F. solani pisi*. Arch Biochem Biophys 176:334–343

Soliday CL, Kolattukudy PE (1977) Biosynthesis of cutin. ω-Hydroxylation of fatty acids by a microsomal preparation from germinating *Vicia faba*. Plant Physiol 59:1116–1121

Soliday CL, Kolattukudy PE (1978) Midchain hydroxylation of 16-hydroxypalmitic acid by the endoplasmic reticulum fraction from germinating *Vicia faba*. Arch Biochem Biophys 188:338–347

Soliday CL, Dean BB, Kolattukudy PE (1978) Suberization: inhibition by washing and stimulation by abscisic acid in potato disks and tissue culture. Plant Physiol 61:170–174

Soliday CL, Kolattukudy PE, Davis RW (1979) Chemical and ultrastructural evidence that waxes associated with the suberin polymer constitute the major diffusion barrier to water vapor in potato tuber (*Solanum tuberosum* L) Planta 146:607–614

Swan EP (1968) Alkaline ethanolysis of extractive-free western red cedar bark. Tappi 51:301–304

Thomson WW, Platt-Aloia KA, Endress AG (1976) Ultrastructure of oil gland development in the leaf of *Citrus sinensis* L. Bot Gaz 137:330–340

Torrey JG, Clarkson DT (eds) (1975) The development and function of roots. Academic Press, New York

Tulloch AP (1975) Chromatographic analysis of natural waxes. J Chromatograph Sci 13:403–407

Tulloch AP (1976) Chemistry of waxes of higher plants. In: Kolattukudy PE (ed) Chemistry and biochemistry of natural waxes. Elsevier, New York, pp 235–287

van den Ende G, Linskens HF (1974) Cutinolytic enzymes in relation to pathogenesis. Annu Rev Phytopathol 12:247–258

von Wettstein-Knowles P (1974) Gene mutation in barley inhibiting the production and use of C_{26} chains in epicuticular wax formation. FEBS Lett 42:187–191

von Wettstein-Knowles P (1979) Genetics and biosynthesis of plant epicuticular waxes.

In: Appleqvist L-Å, Liljenberg C (eds) Advances in the biochemistry and physiology of plant lipids. Elsevier, New York, pp 1–26
Walton TJ, Kolattukudy PE (1972a) Determination of the structures of cutin monomers by a novel depolymerization procedure and combined gas chromatography and mass spectrometry. Biochemistry 11:1885–1897
Walton TJ, Kolattukudy PE (1972b) Enzymatic conversion of 16-hydroxypalmitic acid into 10,16-dihydroxypalmitic acid in *Vicia faba* epidermal extracts. Biochem Biophys Res Commun 46:16–21
Wattendorf J (1969) Feinbau und Entwicklung der verkorkten Calciumoxalat-Kristallzellen in der Rinde von *Larix decidua* Mill. Z Pflanzenphysiol 60:307–347
Wattendorf J (1976) Ultrastructure of the suberized styloid cells in *Agave* leaves. Planta 128:163–165
Wilson LA, Sterling C (1976) Studies on the cuticle of tomato fruit I. Fine structure of the cuticle. Z Pflanzenphysiol 77:359–371
Zee S-Y, O'Brien TP (1970) Studies on the ontogeny of the pigment strand in the caryopsis of wheat. Aust J Biol Sci 23:1153–1171

11 Wall Extensibility: Hormones and Wall Extension

R.E. CLELAND

1 Introduction

Plant cells can undergo striking amounts of cell elongation. For example, a cell initially 20 to 30 μ in length can end up over 2000 times as long (BANNON 1964). During elongation there must be a proportional increase in wall area. This increase occurs in one of three patterns. In algal rhizoids, fungal hyphae, root hairs, and pollen tubes (GREEN 1969) the wall increases in area only at the tip (tip growth). In bacteria (FIEDLER and GLAZER 1973), yeast (GOODAY and TRINCI 1980) and the red alga *Griffithsia pacifica* (WAALAND et al. 1972) growth is restricted to only a part of the lateral wall (band growth); but in the majority of cells growth occurs throughout the whole lateral surface (ROELOFSEN 1965, ROLAND and VIAN 1979). In this case (surface growth), growth involves an extension of wall already present as well as synthesis of new wall. As it is primarily surface growth which is controlled by plant hormones, further discussion of cell elongation will be restricted to cells which undergo this type of extension.

2 Cellular Parameters Which Control Cell Elongation

The rate at which a cell enlarges (dV/dt) is a function of four parameters; the hydraulic conductivity (Lp), the wall extensibility (WEx), the wall yield stress (Y) and the osmotic potential gradient between the cell and the outside solution ($\Delta\pi$), as described in Eq. (1) (LOCKHART 1965, RAY et al. 1972). This equation is derived by combining a theoretically derived equation relating the

$$dV/dt = Lp \cdot WEx \cdot (\Delta\pi - Y)/(Lp + WEx) \tag{1}$$

extension rate to Lp and to the water potential gradient ($\Delta\Psi$) [Eq. (2)] with an empirically derived equation [Eq. (3)] which relates the extension rate to WEx, Y, and to the turgor pressure (P).

$$dV/dt = Lp \cdot \Delta\Psi, \tag{2}$$

$$dV/dt = WEx \cdot (P - Y). \tag{3}$$

The importance of Eq. (1) is that it indicates that when a hormone increases the rate of cell elongation, it can do so only by changing one or more of these four parameters. For example, there must be an increase in Lp, WEx, $\Delta\pi$, or a decrease in Y in order for the extension rate to increase. The problem is to determine which of these parameters is altered by a particular hormone so as to lead to an increase in the extension rate (RAY 1974, CLELAND 1977, PENNY and PENNY 1978). The ability of hormones to control WEx is the aspect of this problem which is considered here.

3 Do Hormones Control Wall Extension via Changes in Wall Extensibility?

HEYN (1931) first presented evidence that changes in WEx can control the rate of cell elongation. While subsequent investigators have accepted this conclusion, they have not agreed as to what WEx is or how it should be measured (BURSTRÖM et al. 1967, CLELAND 1971a, PILET 1971, PRESTON 1974, MASUDA 1978, PENNY and PENNY 1978). Therefore a discussion of WEx and how it is measured must precede any consideration of whether hormones alter WEx.

3.1 WEx and the Mechanical Properties of Cell Walls

3.1.1 What is WEx?

WEx, as used in Eq. (1), is the ability of cell walls to undergo irreversible extension when under constant stress. At a particular moment walls will have mechanical properties which permit a certain, limited amount of extension whenever the wall is under stress (CLELAND 1971a). If continued extension is to occur, there must be, in addition, biochemical modifications of the walls so as to restore the physical capacity of extension. This is shown by the rapid inhibition of growth by metabolic inhibitors (e.g., RAY and RUESINK 1962, HAGER et al. 1971) and by the inability of deproteinized walls to undergo continued extension when under constant applied stress (CLELAND and RAYLE 1978). These modifications, or *wall loosening,* may consist of cleavage of already existing load-bearing bonds in the wall and/or the synthesis of new cell wall. The rate of wall extension (and thus the amount of WEx) depends on the rate of wall loosening events and the amount of physical extension which occurs as a result of each wall loosening event.

The rate of wall loosening, in turn, must depend upon at least two factors. Since wall loosening is an extracellular event, while the site of hormone action appears to be intracellular, there must be some factor(s) (wall loosening factor, or WLF) which is released from the cytoplasm which initiates wall loosening. Regardless of the nature of the WLF, the rate of wall loosening will have to be a function of the rate of release of the WLF. In addition, the walls must possess the capacity to respond to the WLF by undergoing the actual wall loosening. This wall loosening capacity is probably dependent on the composition and structure of the wall, and on the presence of enzymes needed for the wall loosening.

3.1.2 Mechanical Properties of Cell Walls

The mechanical properties of walls can be assessed by determining the response of isolated walls to applied stress. When a wall is subjected to a constant applied stress it extends in the time-dependent manner shown in Fig. 1a. A plot of extension vs. log time (Fig. 1b) gives a nearly straight line which can be extrapolated back to zero extension by extending the line into the short time periods. This type of extension indicates that the wall is a viscoelastic material (FERRY 1970). The viscoelastic extension has several properties worth noting for comparison with the wall extension which occurs during growth (CLELAND 1971b). First, it can give only a limited amount of extension within a reasonable time period. Secondly, it requires neither metabolism nor even proteins in the wall. Finally, the Q_{10} of this extension is about 1.05.

If after a time the stress is removed, the walls contract in a time-dependent viscoelastic manner (Fig. 1a). The amount of contraction depends on the past extension history of the walls; in particular, on the direction of the force vectors (i.e., the directions in which the stress acts) and on the amount of prior extension. If a particular force vector is applied for the first time, or if the applied stress is greater than that applied previously with the same force vectors, extension will be only partly reversible (elastic) and irreversible (plastic) extension will have occurred. However, if the tissue has already been subjected previously to the same or greater stress with the same vectors, the extension will be almost entirely elastic. It should be noted that a wall can undergo repeated elastic viscoelastic extension but can undergo plastic extension only once unless further biochemical modification occurs.

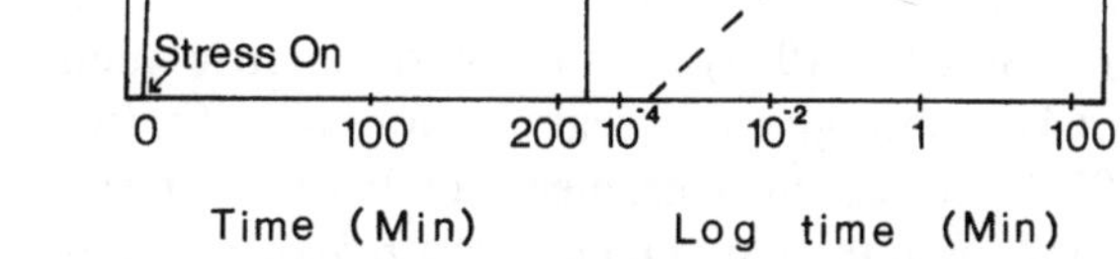

Fig. 1a, b. Extension of isolated cell walls in response to applied stress. **a** Extension and contraction of a wall as a function of time. Wall first subjected to applied stress for 100 min, then stress removed. Extension and contraction are effectively completed within 100 min. **b** Extension from **a** between 1 and 100 min as a function of log time (*solid line*). The curve can be extrapolated back to zero extension by using the short-time portion of the time scale (*dashed line*)

3.1.3 Changes in WEx in Vivo

At any particular moment, while the walls of a turgid, intact cell will be elastically extended by the turgor, they will have almost no potential for irreversible extension (and thus WEx) because any irreversible extension which could occur will already have taken place. In order for the walls to now develop WEx, one of four things must happen. First, an increase in turgor (increase in stress with the same force vectors) will result in a burst of viscoelastic extension. Secondly, an applied stress, because it will have different force vectors than turgor, will also result in some irreversible viscoelastic extension. Neither of these methods of generating WEx will lead to continued wall extension since only a single burst of irreversible extension will occur in each case.

If load-bearing bonds in the wall are cleaved, WEx will be generated and the wall can undergo additional irreversible extention, the amount of which depends upon the stress and the number of bonds broken. Continued reoccurrence of the wall loosening events will result in continued extension of the wall. Finally, the elastic extension of the wall may be converted into irreversible extension by apposition of new cell wall material; this must be followed by regeneration of the elasticity by cleavage of bonds in the wall. This type of extension, in contrast to the previous one, can only occur in the presence of new wall synthesis.

Normal cell extension must involve one or perhaps both of these two bond-cleaving mechanisms. The requirement for respiratory metabolism (BONNER 1933, RAY and RUESINK 1962) and protein synthesis (NOODÉN and THIMANN 1963), the necessity for undenaturated wall proteins (HAGER et al. 1971, RAYLE and CLELAND 1972) and the high Q_{10} (RAYLE and CLELAND 1972), all indicate that bond cleavage is involved in wall extension. We may conclude, therefore, that WEx involves two components; the rate of biochemical wall loosening events, and the amount of irreversible viscoelastic extension which can result from each wall loosening event.

3.2 Measurement of WEx

3.2.1 Measurement of the Mechanical Properties of Isolated Walls

In order to determine whether a hormone causes a change in WEx, one must measure WEx before and after hormone treatment. How should WEx be measured? Indirect methods such as the bending technique (HEYN 1931, TAGAWA and BONNER 1957) or the plasmometric method (CLELAND 1959, BURSTRÖM 1964) cannot give any real measure of WEx (for further discussion, see CLELAND 1971a); direct measurements of WEx are needed, instead. The three most commonly used tests are *creep, stress relaxation,* and the *Instron technique*. Characteristics of these tests are listed in Table 1.

The creep test is a direct measure of the viscoelastic properties of walls. After removal of enzymatic activity by boiling, walls are subjected to a constant applied stress and the extension is measured with time. The data are usually plotted as extension vs. log time, and give a line which is often nearly straight

Table 1. Characteristics of tests for measuring WEx

Test	Material used	Parameter held constant	Parameters measured	Type of WEx measured
Creep	Isolated walls	Stress	Increased strain vs. time	Past WEx
Stress relaxation	Isolated walls	Strain	Decreased stress vs. time	Past WEx(?)
Instron	Isolated walls	Strain rate	Stress vs. strain	Past WEx
Constant-stress	Live tissue	Stress	Increased strain vs. time	Present WEx

(PROBINE and PRESTON 1962, CLELAND 1971b). Extension per decade of log time (e.g., from 3–30 min) is usually used as a measure of extensibility.

In stress relaxation the walls are extended and held at a constant strain while stress along the walls is measured as a function of time. Although the wall cannot actually extend during the relaxation period, wall polymers can rearrange and slip past each other with the result that stress decays with time. A plot of stress vs. log time gives a nearly straight line (HAUGHTON et al. 1968, CLELAND and HAUGHTON 1971, YAMAMOTO et al. 1970).

Stress relaxation curves (and creep curves) can be subjected to conventional rheological analysis (e.g., FERRY 1970), but this has been performed only on algal cell walls (HAUGHTON et al. 1968, HAUGHTON and SELLEN 1969). Masuda and coworkers (reviewed in MASUDA 1978) have analyzed their relaxation curves by fitting them to the equation:

$$S = b \cdot \log \frac{(t+Tm)}{(t+To)} + C \tag{4}$$

where S is the stress, C is a constant, b gives the relative relaxation rate, and To and Tm are, respectively, the times when relaxation starts and stops. To and Tm have been related to the amount or molecular weights (viscosity) of the hemicellulosic and cellulosic parts of the wall, respectively. It should be noted, however, that the values of To and Tm are very sensitive to the rate at which the walls are extended at the start of stress relaxation (FUJIHARA et al. 1978a) and differences in To which occur at one strain rate may disappear at a different extension rate.

The most widely used technique for measuring the mechanical properties of cell walls is the Instron technique, pioneered by OLSON et al. (1965). Deproteinized walls are subjected to two successive extensions at a constant rate of strain, and the stress along the walls is recorded as a function of extension. The first extension involves both a reversible and an irreversible component, while the second extension is entirely reversible. The slope of each curve (% extension/unit stress) is measured at a particular stress; the slope of the second curve gives the elastic extensibility (DE) while the difference in slopes between the two extensions is the plastic extensibility (DP) (for methods of determining DE and DP, see CLELAND 1967a).

The Instron technique has two advantages over creep and stress relaxation. First, the rapidity with which a measurement can be made means that a much larger number of samples can be analyzed conveniently. Secondly, in the Instron technique one obtains a separate measure of the plastic and elastic extensibility, while in creep and stress relaxation only the total, combined extensibility is measured.

Theoretical analysis (FUJIHARA et al. 1978b) and experimental results (CLELAND 1971a, 1971b, CLELAND and HAUGHTON 1971) show that the Instron technique, creep tests, and stress relaxation give the same qualitative information about the mechanical properties of the walls; i.e., a change detected by one test will be mirrored by similar changes in the other two tests. Thus any of these three tests can be employed.

How do the results of these tests relate to WEx? They cannot measure WEx directly, since the force vectors used in these tests are different from those imposed by turgor pressure, and since they give no indication of the rate at which wall loosening events will occur in the future. CLELAND (1967a, 1971a, unpublished data) believes that the DP values from an Instron test are proportional to the average WEx which existed in the hour or so preceeding the time the tissue was harvested. As bonds are cleaved during wall loosening, those with the correct orientation will be converted into extension, but the others which are broken but not converted into extension give the extensibility measured with the Instron. MASUDA (1978) argues that the To parameter in stress relaxation is inversely proportional to the capacity of the wall for *future* extension, although the data are mostly compatible with the idea that To is related to past WEx. In either case, although these mechanical tests do not measure WEx directly, they can be used to determine whether hormone-induced changes in the growth rate involve changes in WEx. For example, a tissue is given a treatment which results is an increased growth rate. Samples are collected just before the treatment and 1 h after the new growth rate is attained, and either DP is determined with the Instron or To is measured by stress relaxation. If the increased growth rate was due to an increase in WEx, DP or To will show a change, but if the change in growth rate was due simple to a change in Lp, $\Delta\pi$ or Y, DP and To will be unchanged.

3.2.2 Direct Measurement of WEx

WEx can be measured directly if live tissues are subjected to a constant applied force, without disrupting normal cellular metabolism (constant-stress method). Because the applied stress replaces the driving force of osmotic water uptake, Lp, $\Delta\pi$, and Y no longer influence the extension rate. Therefore if addition of a hormone causes a change in the extension rate, it can only be due to a change in WEx (Table 1). Oddly, this technique has rarely been used.

3.2.3 Other Mechanical Testing Procedures

The resonance frequency technique has been used as a method for measuring elasticity (BURSTRÖM et al. 1967, MORRÉ and EISENGER 1968). Here an elongate piece of tissue is fixed at one end and a piece of iron is attached to the free end. That end is made to resonate by application of a magnetic field of variable frequency. The frequency which causes resonance of maximum amplitude is a function of tissue elasticity and cell turgor (FALK et al. 1958). The problem with this technique is that when a change in frequency is found, there is no independent way to determine whether it reflects a change in elasticity or only a change in turgor.

3.3 Cases Where Hormones Affect WEx

3.3.1 Auxin

In every case so far examined where auxin increases the growth rate, it also causes an increase in WEx, whether measured by the Instron technique, by stress relaxation or by the creep test. Table 2 lists the plant material which has been examined so far. In each case the effect of auxin is to increase DP in the Instron test, with only a lesser and more variable increase in DE, to cause a decrease in To in stress relaxation with a variable effect on b and Tm, or to increase the creep rate in the creep test.

The striking correlation between changes in the growth rate, induced by metabolic inhibitors or by variation in the auxin concentration, and changes in WEx (CLELAND 1967a, COURTNEY and MORRÉ 1980a, MASUDA et al. 1974) suggests that changes in WEx are the primary determinant of changes in the growth rate of auxin-sensitive tissues. However, changes in WEx do not always lead to parallel changes in the growth rate. For example, when the turgor is lowered below Y by application of an external osmoticum, even though auxin has no effect on the growth rate it can still cause an increase in WEx (CLELAND 1967b, MASUDA et al. 1974).

Direct measurement of WEx by means of the constant applied stress technique has only been attempted with pea epicotyl section (YODA and ASHIDA 1960), pea epicotyl epidermal strips (YAMAGATA and MASUDA 1976) and *Avena* coleoptiles (R.E. CLELAND, unpublished data). In each case auxin caused a large increase in WEx.

If hydrogen ions are the wall loosening factor in auxin-induced growth, as has been proposed (reviewed in CLELAND and RAYLE 1978), they should also induce an increase in WEx. Acid-induced increases in WEx have been detected in *Avena* coleoptiles by the Instron technique (RAYLE and CLELAND 1970), in cucumber hypocotyls by both stress relaxation and the Instron technique (IWAMI and MASUDA 1973) and in pea epicotyl epidermal strips by stress relaxation (YAMAMOTO et al. 1974) and by the constant-stress method (YAMAGATA et al. 1974). It is interesting to note that acid *during* stress relaxation caused just the opposite effect on To (an increase) and Tm (a decrease), as did a pretreatment of the tissues with acid prior to the isolation of the walls (YAMAMOTO et al. 1974). Hydrogen ions also mimic auxin in decreasing the tissue elasticity of pea epicotyl sections as measured by the resonance frequency method (UHRSTRÖM 1974).

Table 2. Plant systems where auxin causes an increase in WEx

Plant system	Technique	Reference
Avena coleoptile	Instron	HEYN 1933 OLSON et al. 1965 CLELAND 1967a, 1967b MASUDA 1969 COURTNEY and MORRÉ 1980a
	Stress relaxation	YAMAMOTO et al. 1970 CLELAND and HAUGHTON 1971
	Creep	CLELAND 1971b
Barley coleoptile	Stress relaxation	SAKURAI and MASUDA 1978
Rice coleoptile	Stress relaxation	ZARRA and MASUDA 1979
Zea mays mesocotyl	Instron	COURTNEY et al. 1967
Pea epicotyl	Instron	OSBORNE 1977 COURTNEY and MORRÉ 1980a
	Stress relaxation	TANIMOTO and MASUDA 1971 MASUDA et al. 1974
	Constant stress	YODA and ASHIDA 1960 YAMAGATA and MASUDA 1975
Vigna angularis epicotyl	Stress relaxation	NISHITANI et al. 1979
Lupin hypocotyl	Instron	PENNY et al. 1972
	Creep	PENNY et al. 1974
Soybean hypocotyl	Instron	COURTNEY and MORRÉ 1980a
Cucumber hypocotyl	Instron	CLELAND et al. 1968 COURTNEY and MORRÉ 1980a
	Stress relaxation	KATSUMI and KAZAMA 1978
Mung bean hypocotyl	Instron	LOCKHART 1967
Japanese black pine hypocotyl	Stress relaxation	SUMIYA and YAMADA 1974
Celery petiole	Instron	PILET and ROLAND 1974

3.3.2 Gibberellin

The effect of gibberellin on WEx depends upon the tissue (Table 3). An increase in WEx after gibberellin treatment has been recorded for lettuce hypocotyls using stress relaxation (KAMISAKA et al. 1972, KAWAMURA et al. 1976), for *Avena* stem internodes as measured by the Instron technique (ADAMS et al. 1975), for pea epicotyl tissues by stress relaxation (NAKAMURA et al. 1975), and by a bending technique (LOCKHART 1960). On the other hand, no effect of gibberellin on WEx could be detected in gibberellic acid-treated cucumber hypocotyl tissues by either the Instron technique (CLELAND et al. 1968) or by stress relaxation

Table 3. Effect of gibberellin on WEx

Plant system	Increase in WEx?	Technique	Reference
Lettuce hypocotyl	Yes	Stress relaxation	KAMISAKA et al. 1972
	Yes	Stress relaxation	KAWAMURA et al. 1976
Avena stem internode	Yes	Instron	ADAMS et al. 1975
Pea epicotyl	Yes	Stress relaxation	NAKAMURA et al. 1975
	Yes	Bending	LOCKHART 1960
	No	Constant-stress	YODA and ASHIDA 1960
Cucumber hypocotyl	No	Instron	CLELAND et al. 1968
	No	Stress relaxation	KATSUMI and KAZAMA 1978

(KATSUMI and KAZAMA 1978); in this tissue gibberellin appears to induce growth by increasing $\Delta\pi$.

3.3.3 Other Hormones

There have been only a few attempts to determine the effects of other plant hormones on WEx. Cytokinins increase the WEx of radish cotyledons as measured with the Instron, in parallel with the induction of cell expansion (ROSS CW, unpublished data). The phytotoxin fusicoccin, which mimics auxin in causing tissues to excrete protons and to grow (MARRÉ 1979), increases WEx of celery petioles as measured by creep (JACCARD and PILET 1979) and of pea epicotyls as measured by stress relaxation (YAMAGATA and MASUDA 1975). Ethylene did not affect the Instron extensibility of auxin-treated pea epicotyl sections (MORRÉ and EISINGER 1968) but did decrease the extensibility of intact pea epicotyls (OSBORNE 1977). Both growth and Instron extensibility of *Regnillidium* petioles were increased by an ethylene treatment (OSBORNE 1977).

3.3.4 Conclusions

In all tissues in which auxin induces rapid cell enlargement, it appears to do so by increasing the WEx parameter. In some cases gibberellin promotes growth in a similar manner, but in other cases its effect must be on another of the four cellular growth parameters. There is not enough information available about other hormones to make conclusions at the present concerning their effects on WEx.

4 The Mechanism of Auxin-Induced Wall Loosening

4.1 Wall Structure, Wall Synthesis and Wall Loosening

The mechanism of hormone-induced cell wall loosening has been examined primarily in connection with auxin-induced growth. Progress in this area has been hampered by lack of exact knowledge on the composition and structure of the primary cell wall; without such knowledge, it is difficult to determine which component of the wall might be altered so as to lead to wall loosening.

KEEGSTRA et al. (1973) proposed that dicotyledonous walls consist of a single cross-linked lattice containing rhamnogalacturonan covalently linked to an arabinogalactan which, in turn, is covalently bonded to a xyloglucan. The xyloglucan is then hydrogen-bonded to the cellulose to complete the lattice. Evidence for the wall structure has been obtained from a variety of dicotyledonous walls (ALBERSHEIM 1976). However, MONRO et al. (1976) have concluded from differential extraction procedures that lupin walls contain at least two networks, and that the uronic acids are not cross-linked to either the xyloglucans or to the cellulose. The walls of monocotyledonous grasses contain an arabinoxylan instead of the xyloglucan (BURKE et al. 1974, MCNEIL et al. 1975, WADA and RAY 1978). The controversy concerns the presence of mixed-link $(1 \rightarrow 3),(1 \rightarrow 4)$-$\beta$-glucans. BURKE et al. (1974) and MCNEIL et al. (1975) indicated that they did not exist as part of the cell wall, but WADA and RAY (1978), DARVILL et al. (1978), and NEVINS et al. (1977), have all found the glucan as part of the walls of *Avena* and *Zea mays* coleoptiles. Further information about the structure and composition of cell walls can be found in Chapters 1–10 of this Volume.

There has been a difference of opinion as to whether cell wall loosening involves wall synthesis (e.g., intussusception of wall polysaccharides) or only wall degradation (e.g., cleavage of load-bearing wall bonds). The wall synthesis mechanism, which is similar to that of bacteria (FIEDLER and GLAZER 1973) and fungi (GOODAY and TRINCI 1980), has been championed by BURSTRÖM (e.g., 1975) in particular. Auxin does stimulate the synthesis of new cell walls (e.g., BAKER and RAY 1965) but the amount of wall synthesis is not necessarily correlated with the amount of wall loosening (RAY 1962, LOESCHER and NEVINS 1972, COURTNEY and MORRÉ 1980b), and stimulation of wall synthesis is not apparent until 1 h after addition of auxin (BAKER and RAY 1975) while growth is induced after only a 10-min lag. In addition, wall loosening of isolated, frozen-thawed *Avena* coleoptile walls can be induced by protons (RAYLE et al. 1970) with characteristics similar to those of auxin-induced growth of live sections (RAYLE and CLELAND 1972) such as the requirement for tension for wall loosening, and a similar Q_{10}. This suggests that wall loosening can be initiated without wall synthesis, although wall synthesis is certainly required in vivo over longer time periods, if only to maintain the structure of the wall. It is now generally accepted that the primary step in wall loosening must involve a modification of the wall already present.

4.2 Possible Mechanisms for Wall Loosening

4.2.1 Breakage of Hydrogen Bonds

KEEGSTRA et al. (1973) suggested that wall loosening might consist of a rupturing of the hydrogen bonds between the xyloglucans and cellulose whenever the wall solution becomes acidic. This attractive possibility seems unlikely, since acid was unable to break the hydrogen-bond links between cellulose and xyloglucan in a model system (VALENT and ALBERSHEIM 1974) and agents capable of breaking hydrogen bonds such as 8 mol urea failed to cause any significant wall loosening in isolated *Avena* coleoptile cell walls (TEPFER and CELAND 1979).

4.2.2 Calcium Cross-Links

The ability of calcium ions to inhibit auxin-induced growth (COOIL and BONNER 1957) and decrease the extensibility of live *Avena* coleoptile tissues (TAGAWA and BONNER 1957) led to the suggestion that the strength of walls is imparted by calcium cross-links between pectic chains, and that auxin loosens walls by causing the release of calcium from the walls. However, CLELAND (1960) was unable to detect any release of calcium from *Avena* coleoptile walls during auxin-induced growth, and more recent measurements of the effect of calcium ions on the mechanical properties of isolated cell walls (CLELAND and RAYLE 1977, TEPFER and CLELAND 1979, COURTNEY and MORRÉ 1980a) indicate that calcium bridges are not the load-bearing bonds in the wall, and thus their removal cannot cause wall loosening. In the alga *Valonia ventricosa*, on the other hand, calcium bridges may play a role in wall strength since removal of calcium increases wall extensibility (TEPFER and CLELAND 1979).

4.2.3 Nonenzymatic Wall Loosening

RAYLE and CLELAND (1970) suggested that the acid-induced wall loosening, and by analogy the auxin-induced wall loosening, occurred by a nonenzymatic cleavage of wall polysaccharides, on the basis of the fact that wall loosening only appeared to occur when the wall pH was 4 or below. RAYLE (1973) subsequently showed that the requirement for such a low pH was an artifact of the impermeability of the cuticle to protons, and that maximum wall loosening of *Avena* coleoptiles occurred at 4.7, a pH more compatible with an enzymatic mechanism. Furthermore, removal of proteins from *Avena* coleoptile walls eliminated the ability of these walls to undergo acid-induced wall loosening (CLELAND and RAYLE 1978, TEPFER and CLELAND 1979). It is now generally accepted that auxin-induced wall loosening is an enzymatically mediated process. Nonenzymatic wall loosening can occur in algae, however, as boiling of *Valonia* walls (TEPFER and CLELAND 1979) and *Nitella* walls (MÉTRAUX and TAIZ, 1977) failed to eliminate H^+-induced wall loosening.

4.2.4 Enzymatic Wall Loosening

4.2.4.1 Possible Mechanisms; Transglycosylation Versus Hydrolysis

Two general mechanisms of enzymatic cell wall loosening have been considered (Fig. 2). First, wall loosening may be mediated by a transglycosylase which breaks polysaccharide cross-links and then reforms the ends with new partners (CLELAND and RAYLE 1972, ALBERSHEIM 1976); this mechanism has been called "chemical creep". Such a mechanism would explain why tension is required

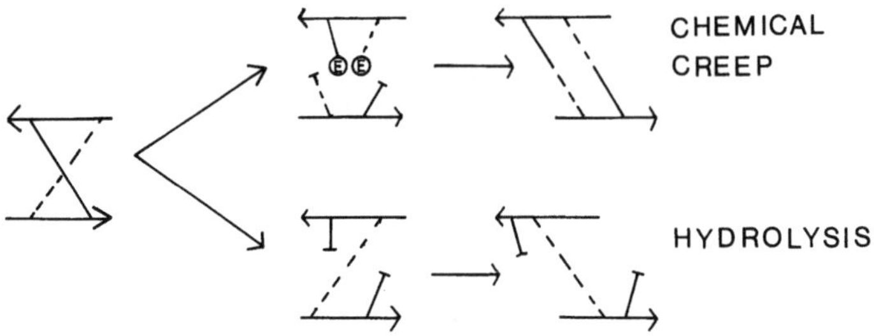

Fig. 2. Two mechanisms for cell wall loosening. Wall polymers are restrained from moving by cross-links (*left*). When links are broken by a transglycosylase (*upper*), extension occurs until restrained by new cross-links formed by the transglycosylase (chemical creep). When links are broken by a hydrolase (*lower*) extension occurs until limited by already existing cross-links (hydrolysis)

for wall loosening (CLELAND and RAYLE 1972, RAYLE and CLELAND 1972), since in the absence of tension the bonds would simply reform in their original position resulting in an unchanged wall, while if the wall is under tension some extension would occur before the bonds reform with new partners. The problem is that such a transglycosylation mechanism will be difficult to detect after extension because the wall ends up being unchanged chemically. The solution is to find a transglycosylase which will cause wall loosening when added to isolated walls under tension. YAMAMOTO and NEVINS (1979) report that the $(1 \rightarrow 3)$-β-glucanase from the fungus *Sclerotinia libertiana,* which can cause limited extension of *Avena* coleoptile (MASUDA and YAMAMOTO 1970), has some transglycosylase activity when tested against model substrates.

The second mechanism involves irreversible breakage of load-bearing bonds in the wall by polysaccharide hydrolases. This mechanism has been examined in two ways; by looking for wall hydrolases which either induce wall loosening when added back to walls, or whose activity is stimulated by auxin, and by looking for either solubilization of cell wall components or changes in the molecular weight of wall polymers during auxin-induced growth.

4.2.4.2 Changes in Wall Hydrolases in Relation to Auxin-Induced Wall Loosening

Several of enzymes capable of degrading cell wall polysaccharides have been found to occur in cell walls (see Chap. 7, this Vol.). In certain tissues, auxin activates already existing hydrolases or initiates the synthesis of additional ones. For example, JOHNSON et al. (1974) found that *Avena* coleoptile walls contain a variety of exo-glycosidases including β-glucosidase, β-galactosidase, and α-mannosidase, and that β-galactosidase is activated during auxin-induced growth, apparently by the lower wall pH which occurs under those conditions. It seems doubtful, however, that either β-glucosidase or β-galactosidase are responsible for wall loosening, since gluconolactone and galactonolactone, potent inhibitors of these two enzymes, failed to block auxin-induced growth (EVANS 1974, GOLDBERG 1977, see discussion in RAYLE and CLELAND 1977). Auxin also causes an increase in $(1 \rightarrow 6)$-α-glucosidase of *Avena* coleoptiles (HEYN 1970), and the β-galactosidase of pea epicotyls (TANIMOTO and IGARI 1976).

Auxin induces a massive increase in cellulase activity in pea stems (FAN and MACLACHLAN 1967) which is due to increased synthesis of the mRNA for cellulase (VERMA et al. 1975). The increase in cellulase cannot be the cause of the growth, however, since it occurs after the start of auxin-induced growth (DATKO and MACLACHLAN 1968). MACLACHLAN (1977) has concluded that the cellulase promotes the synthesis of cellulose but does not participate in wall loosening. The latter conclusion is reinforced by RUESINK (1969), who found that direct addition of cellulase to *Avena* coleoptiles did not promote growth, and by TANIMOTO and MASUDA (1968), who found that auxin causes no increase in cellulase in the barley coleoptile.

Auxin stimulates the activity of (1 → 3)-β-glucanase in both barley coleoptiles (TANIMOTO and MASUDA 1968) and *Avena* coleoptiles (MASUDA and YAMAMOTO 1970), where the increase could be detected 10 min after addition of auxin. The possibility that this enzyme might participate in wall loosening is suggested by the fact that nojirimycin, an inhibitor of exo-glucanase activity, effectively blocks auxin-induced growth (NEVINS 1975a, SAKURAI et al. 1977). Addition of an exo-(1 → 3)-β-glucanase from *Sclerotinia libertiana* to *Avena* coleoptiles resulted in a limited amount of cell elongation in the absence of auxin (YAMAMOTO and MASUDA 1971). The importance of these results was uncertain, however, because CLELAND (1971a) and RUESINK (1969) were unable to duplicate the growth promotion when they used endo- and exo-(1 → 3)-β-glucanases from other sources. YAMAMOTO and NEVINS (1979) may have resolved these differences by finding that the *Sclerotinia* (1 → 3)-β-glucanase also possesses transglucosylase activity as well, while other exo-(1 → 3)-β-glucanases lack this activity. Correlation between the (1 → 3)-β-glucanase activity and growth is striking enough to indicate that this enzyme deserves further attention.

4.2.4.3 Effects on Cell Wall Polymers

Auxin-induced growth is accompanied by a massive release of glucose from the mixed-linked glucans of the *Avena* coleoptile (KATZ and ORDIN 1967, LOESCHER and NEVINS 1972, WADA et al. 1968), corn coleoptiles (DARVILL et al. 1978, HUBER and NEVINS 1979) and barley coleoptiles (SAKURAI and MASUDA 1978). This glucan release is inhibited when growth is inhibited by agents such as cycloheximide (SAKURAI and MASUDA 1979). Since exogenous exo-(1 → 3)-β-glucanase will cause a similar glucan release (NEVINS 1975b), and since auxin increases the activity of this enzyme (see Sect. 4.2.4.2), it is not surprising that glucan degradation has been linked to wall loosening (MASUDA 1978). In addition, there is a close correlation between glucan release and the change in the mechanical properties of walls as measured by stress relaxation (MASUDA 1978, SAKURAI and MASUDA 1978).

The role of glucan degradation is further suggested by the ability of nojirimycin to inhibit glucan release in parallel with its inhibition of growth (NEVINS 1975a, SAKURAI et al. 1977). Although nojirimycin is an effective inhibitor of exo-β-glucanases, caution must be used in interpreting its effects. Nojirimycin is not a specific inhibitor (DIGBY and FIRN 1977) and it may be acting also to inhibit the release of wall loosening factors from the cell (e.g., it blocks auxin-induced H^+-excretion in *Avena* coleoptiles, R.E. CLELAND, unpublished data).

These are three reasons to question whether glucan release is directly connected with wall loosening. First, glucan release can be detected only 1–3 h after addition of auxin (LOESCHER and NEVINS 1972, SAKURAI et al. 1977, SAKURAI and MASUDA 1978), while growth begins after only a 10-min lag. SAKURAI et al. (1979), who extracted the glucans and then measured their molecular weights by gel permeation and viscosity, were able to detect a small decrease in molecular weight within the first half hour after addition of auxin. COURTNEY and MORRÉ (1980b) could find no evidence for polysaccharide chain cleavage in auxin-treated *Avena* or *Zea mays* coleoptile walls, but changes in the β-glucans may have been masked by the other wall polysaccharides. Thus wall loosening might simply require cleavage of bonds within the glucans which would start at the time growth starts, while glucan release may be detectable only after sufficient bond cleavage has occurred to solubilize the glucans.

Secondly, glucan release does not occur under one set of conditions known to cause wall loosening and growth (CLELAND and RAYLE 1978); namely, the treatment of *Avena* coleoptile sections with acidic solutions (SAKURAI et al. 1977). Finally, glucan release occurs under conditions which apparently do not permit wall loosening. When growth of *Avena* coleoptiles is inhibited by osmotic reduction of turgor (e.g., 0.2 M mannitol), little wall loosening occurs, as indicated by the small potential for future expansion which builds up during such a period (CLELAND and RAYLE 1972, RAYLE and CLELAND 1972, GÖRING et al. 1978). If glucan release was the cause of wall loosening, it should not occur in the presence of 0.2 M mannitol, yet the glucan release under these conditions is as great as it is under full turgor (LOESCHER and NEVINS 1973, SAKURAI and MASUDA 1978). Thus at the present time it seems premature to conclude that glucan release is the mechanism of wall loosening in coleoptile walls.

In *Zea mays* coleoptiles, in addition to the glucose release, there is an auxin-induced release of arabinose and xylose which is better correlated with wall loosening than is the glucose release (DARVILL et al. 1978). No such comparable release or arabinose and xylose occurs in *Avena* coleoptiles, however (LOESCHER and NEVINS 1972, MASUDA 1978).

Dicotyledonous walls do not possess the β-glucans [although GOLDBERG (1977) has reported that auxin induces glucan release from *Phaseolus aureus* walls]. In the pea epicotyl, the component of the wall which is solubilized during auxin-induced growth is the xyloglucan (LABOVITCH and RAY 1974a). This xyloglucan release is well correlated with growth (LABOVITCH and RAY 1974b) and occurs in response to acid as well (JACOBS and RAY 1975). It can be detected by homogenization of the tissue (LABOVITCH and RAY 1974a) or by centrifuging the wall solution from the section (TERRY et al. 1980). Xyloglucan solubilization can be detected during acid treatment of isolated pea walls (BATES and RAY 1979), although in this case other components of the walls also appear to be solubilized. In *Vigna angularis* walls, on the other hand, there is no apparent loss of xyloglucans during auxin-induced growth, but there is a marked change in the turnover of wall galactose (NISHITANI and MASUDA

1980). Thus xyloglucan release does not seem to be a common feature of dicot walls which are undergoing wall loosening.

4.2.5 Conclusions

Although glucan release and the presence of $(1 \rightarrow 3)$-β-glucanases in coleoptiles, and xyloglucan release in peas are both correlated with auxin-induced growth, it is not certain that these effects are directly connected with wall loosening. The possibility that wall loosening occurs by transglycosylation rather than by a hydrolytic mechanism deserves more attention.

References

Adams PA, Montague MJ, Tepfer M, Rayle DL, Ikuma H, Kaufman PB (1975) Effect of gibberellic acid on the plasticity and elasticity of *Avena* stem segments. Plant Physiol 56:757–760

Albersheim P (1976) The primary cell wall. In: Bonner J, Varner JE (eds) Plant Biochemistry, 3rd edn. Academic Press, New York, pp 225–274

Baker DB, Ray PM (1965) Direct and indirect effects of auxin on cell wall synthesis in oat coleoptile tissue. Plant Physiol 40:345–352

Bannon MW (1964) Tracheid size and anticlinal division in the cambium of *Pseudotsuga*. Can J Bot 42:603–631

Bates GW, Ray PM (1979) pH dependent release of polymers from isolated cell walls. Plant Physiol 63:S20

Bonner J (1933) The action of plant growth hormones. J Gen Physiol 17:63–76

Burke D, Kaufman P, McNeil M, Albersheim P (1974) The structure of plant cell walls. VI. A survey of the walls of suspension-cultured monocots. Plant Physiol 54:109–115

Burström H (1964) Calcium, water conditions and growth of pea seedling stems. Physiol Plant 17:207–219

Burström H (1975) Growth, solute and water fluxes in the etiolated *Pisum* stem. Z Pflanzenphysiol 76:339–352

Burström H, Uhrström I, Wurscher R (1967) Growth, turgor, water potential and Young's modulus in pea internodes. Physiol Plant 20:213–231

Cleland RE (1959) Effect of osmotic concentration on auxin-action and on irreversible and reversible expansion of the *Avena* coleoptile. Physiol Plant 12:809–825

Cleland RE (1960) Effect of auxin on loss of calcium from cell walls. Nature 185:44

Cleland RE (1967a) Extensibility of isolated cell walls: measurement and changes during cell elongation. Planta 74:197–209

Cleland RE (1967b) A dual role of turgor pressure in auxin-induced cell elongation in *Avena* coleoptiles. Planta 77:182–191

Cleland RE (1971a) Cell wall extension. Annu Rev Plant Physiol 22:197–222

Cleland RE (1971b) Mechanical behaviour of isolated *Avena* coleoptile walls subjected to constant stress. Plant Physiol 47:805–811

Cleland RE (1977) The control of cell enlargement. In: Jennings DH (ed) Integration of activity in the higher plant. Soc Exp Biol Symp 31. Cambridge Press, Cambridge pp 101–115

Cleland RE, Haughton PM (1971) The effect of auxin on stress relaxation of isolated *Avena* coleoptiles. Plant Physiol 47:812–815

Cleland RE, Rayle DL (1972) Absence of auxin-induced stored growth in *Avena* coleoptiles and its implications concerning the mechanism of wall extension. Planta 106:61–71

Cleland RE, Rayle DL (1977) Reevaluation of the effect of calcium ions on auxin-induced elongation. Plant Physiol 60:709–712

Cleland RE, Rayle DL (1978) Auxin, H^+-excretion and cell elongation. Bot Mag Tokyo Spec Issue 1:125–139
Cleland RE, Thompson M, Rayle DL, Purves WK (1968) Differences in the effects of auxins and gibberellins on wall extensibility of cucumber hypocotyls. Nature 219:510–511
Cooil B, Bonner J (1957) Effects of calcium and potassium ions on the auxin-induced growth of *Avena* coleoptile section. Planta 48:696–723
Courtney JS, Morré DJ (1980a) Studies on the role of wall extensibility in the control of cell expansion. Bot Gaz 141:56–62
Courtney JS, Morré DJ (1980b) Studies on the chemical basis of auxin-induced cell wall loosening. Bot Gaz 141:63–68
Courtney JS, Morré DJ, Key JL (1967) Inhibition of RNA synthesis and auxin-induced cell wall extensibility and growth by actinomycin D. Plant Physiol 42:434–439
Darvill AG, Smith CJ, Hall MA (1978) Cell wall structure and elongation growth in *Zea mays* coleoptile tissue. New Phytol 80:503–516
Datko AH, Maclachlan GA (1968) IAA and synthesis of glucanases and pectic enzymes. Plant Physiol 43:735–742
Digby J, Firn RD (1977) Some criticisms of the use of nojirimycin as a specific inhibitor of auxin-induced growth. Z Pflanzenphysiol 82:355–362
Evans ML (1974) Evidence against the involvement of galactosidase and glucosidase in auxin- or acid-stimulated growth. Plant Physiol 54:213–215
Falk SO, Hertz CH, Virgin HI (1958) On the relation between turgor pressure and tissue rigidity. I. Experiments on resonance frequency and tissue rigidity. Physiol Plant 11:802–817
Fan DF, Maclachlan GA (1967) Massive synthesis of RNA and cellulase in the pea epicotyl in response to IAA, with or without concurrent cell division. Plant Physiol 42:1114–1122
Ferry JD (1970) Viscoelastic properties of polymers, 2nd ed. Wiley, New York, p 671
Fiedler F, Glazer L (1973) Assembly of bacterial cell walls. Biochem Biophys Acta 300:467–485
Fujihara S, Yamamoto R, Masuda Y (1978a) Viscoelastic properties of plant cell walls. II. Effect of pre-extension rate on stress relaxation. Biorheology 15:77–85
Fujihara S, Yamamoto R, Masuda Y (1978b) Viscoelastic properties of plant cell walls. III. Hysteresis loop in the stress-strain curve at constant strain rate. Biorheology 15:87–97
Goldberg R (1977) On possible connections between auxin induced growth and cell wall glucanase activities. Plant Sci Lett 8:233–242
Gooday GW, Trinci APJ (1980) Wall structure and biosynthesis in fungi. Symp Soc Gen Microbiol 30:207–252
Göring H, Bleiss W, Schenk D, Kretschmer H (1978) Dependence of the detectability of stored growth on the elongation rates in IAA- and acid-induced elongation of wheat coleoptile section. Plant Cell Physiol 19:833–838
Green PB (1969) Cell morphogenesis. Annu Rev Plant Physiol 20:365–394
Hager A, Menzel H, Krauss A (1971) Versuche und Hypothese zur Primärwirkung des Auxins beim Streckungswachstum. Planta 100:47–75
Haughton PM, Sellen DB (1969) Dynamic mechanical properties of the cell walls of some green algae. J Exp Bot 20:516–535
Haughton PM, Sellen DB, Preston RD (1968) Dynamic mechanical properties of the cell walls of *Nitella opaca*. J Exp Bot 19:1–12
Heyn ANJ (1931) Der Mechanismus der Zellstreckung. Rec Trav Bot Néerl 28:113–244
Heyn ANJ (1933) Further investigations on the mechanism of cell elongation and the properties of the cell wall in connection with elongation. Protoplasma 19:78–96
Heyn ANJ (1970) Dextranase activity in coleoptiles of *Avena*. Science 167:874–875
Huber DJ, Nevins DJ (1979) Autolysis of cell wall β-D-glucan in corn coleoptile. Plant Cell Physiol 20:201–212
Iwami S, Masuda Y (1973) Hydrogen-ion induced curvature in cucumber hypocotyls. Plant Cell Physiol 14:757–762
Jaccard M, Pilet PE (1979) Growth and rheological changes of collenchyma cells: the fusicoccin effect. Plant Cell Physiol 20:1–7

Jacobs M, Ray PM (1975) Promotion of xyloglucan metabolism by acid pH. Plant Physiol 56:373–376
Johnson KD, Daniels D, Dowler MJ, Rayle DL (1974) Activiation of *Avena* coleoptile cell wall glycosidases by hydrogen ions and auxin. Plant Physiol 53:224–228
Kamisaka S, Sano H, Katsumi M, Masuda Y (1972) Effects of cyclic-AMP and gibberellic acid on lettuce hypocotyl elongation and mechanical properties of its cell wall. Plant Cell Physiol 12:167–174
Katsumi M, Kazama H (1978) Gibberellin control of cell elongation in cucumber hypocotyl sections. Bot Mag Tokyo Spec Issue 1:141–158
Katz M, Ordin L (1967) A cell wall polysaccharide-hydrolyzing enzyme system in *Avena sativa* L coleoptiles. Biochem Biophys Acta 141:126–134
Kawamura H, Kamisaka S, Masuda Y (1976) Regulation of lettuce hypocotyl elongation by gibberellic acid. Correlation between cell elongation, stress-relaxation properties of the cell walls and wall polysaccharide content. Plant Cell Physiol 17:23–34
Keegstra K, Talmadge KW, Bauer WD, Albersheim P (1973) The structure of plant cell walls. III. A model of the walls of suspension-cultured sycamore cells based on the interconnections of the macromolecular components. Plant Physiol 51:188–197
Labovitch JM Ray PM (1974a) Turnover of cell wall polysaccharide in elongating pea stem sections. Plant Physiol 53:669–673
Labovitch JM, Ray PM (1974b) Relationship between promotion of xyloglucan metabolism and induction of elongation by IAA. Plant Physiol 54:499–502
Lockhart JA (1960) Intracellular mechanisms of growth inhibition by radiant energy. Plant Physiol 35:129–135
Lockhart JA (1965) An analysis of irreversible plant cell elongation. J Theor Biol 8:264–275
Lockhart JA (1967) Physical nature of irreversible deformation of plant cells. Plant Physiol 42:1545–1552
Loescher W, Nevins DJ (1972) Auxin-induced changes in *Avena* coleoptile cell wall composition. Plant Physiol 50:556–563
Loescher WH, Nevins DJ (1973) Turgor-dependent changes in *Avena* coleoptile cell wall composition. Plant Physiol 52:248–251
Maclachlan GA (1977) Cellulose metabolism and cell growth. In: Pilet PE (ed) Plant growth regulation. Springer, Berlin Heidelberg New York, pp 13–20
McNeil M, Albersheim P, Taiz L, Jones RL (1975) The structure of plant cell walls. VII. Barley aleurone cells. Plant Physiol 55:64–68
Marré E (1979) Fusicoccin: a tool in plant physiology. Annu Rev Plant Physiol 30:273–288
Masuda Y (1969) Auxin-induced cell expansion in relation to cell wall extensibility. Plant Cell Physiol 10:1–9
Masuda Y (1968) Auxin-induced cell wall loosening. Bot Mag Tokyo Spec Issue 1:103–123
Masuda Y, Yamamoto R (1970) Effect of auxin on β-1,3-glucanase activity in *Avena* coleoptile. Dev Growth Differ 11:287–296
Masuda Y, Yamamoto R, Kawamura H, Yamagata Y (1974) Stress relaxation properties of the cell wall of tissue segments under different growth conditions. Plant Cell Physiol 15:1083–1092
Métraux JP, Taiz L (1977) Cell wall extension in *Nitella* as influenced by acid and ions. Proc Natl Acad Sci USA 74:1565–1569
Monro JA, Penny D, Bailey RW (1976) The organization and growth of primary cell walls of lupin hypocotyls. Phytochemistry 15:1193–1198
Morré DJ, Eisinger WR (1968) Cell wall extensibility; its control by auxin and relationship to cell elongation. In: Wightman F, Setterfield G (eds) Biochemistry and physiology of plant growth substances. Runge Press, Ottawa, pp 625–645
Nakamura T, Sekine S, Arai K, Takahashi N (1975) Effects of gibberellic acid and IAA on stress-relaxation properties of pea hook cell wall. Plant Cell Physiol 16:127–138
Nevins DJ (1975a) The effect of nojirimycin on plant growth and its implications concerning a role of exo-β-glucanases in auxin-induced cell expansion. Plant Cell Physiol 16:347–356
Nevins DJ (1975b) The in vitro simulation of IAA-induced modification of *Avena* cell wall polysaccharides by an exo-glucanase. Plant Cell Physiol 16:495–503
Nevins DJ, Huber DJ, Yamamoto R, Loescher W (1977) β-D-glucan of *Avena* coleoptile cell wall. Plant Physiol 60:617–620

Nishitani K, Masuda Y (1980) Modifications of cell wall polysaccharides during auxin-induced growth in azuki bean epicotyl segments. Plant Cell Physiol 21:169–181
Nishitani K, Shibaoka H, Masuda Y (1979) Growth and cell changes in azuki bean epicotyls. II. Changes in wall polysaccharides during auxin-induced growth of excised segments. Plant Cell Physiol 20:463–472
Noodén LD, Thimann KV (1963) Evidence for a requirement for protein synthesis for auxin-induced cell enlargement. Proc Natl Acad Sci USA 50:194–200
Olson AC, Bonner J, Morré DJ (1965) Force extension analysis of *Avena* coleoptile cell walls. Planta 66:127–133
Osborne DJ (1977) Auxin and ethylene and the control of cell growth. Identification of three classes of target cells. In: Pilet PE (ed) Plant growth regulation. Springer, Berlin Heidelberg New York, pp 161–171
Penny P, Penny D (1978) Rapid responses to phytohormones. In: Letham DS, Goodwin PB, Higgins TJV (eds) Phytohormones and related compounds. Vol II. North Holland, Amsterdam, pp 537–597
Penny P, Penny D, Marshall D, Heyes JK (1972) Early responses of excised stem segments to auxins. J Exp Bot 23:23–36
Penny D, Penny P, Marshall DC (1974) High resolution measurement of plant growth. Can J Bot 52:959–969
Pilet PE (1971) Les Parois Cellulaires. Doin, Paris, p 172
Pilet PE, Roland J-C (1974) Growth and extensibility of collenchyma cells. Plant Sci Lett 2:203–207
Preston RD (1974) The physical biology of plant cell walls. Chapman and Hall, London, p 491
Probine MC, Preston RD (1962) Cell growth and the structure and mechanical properties of the wall in internodal cells of *Nitella opaca*. II. Mechanical properties of the walls. J Exp Bot 13:111–127
Ray PM (1962) Cell wall synthesis and cell elongation in oat coleoptile tissues. Am J Bot 49:928–939
Ray PM (1974) The biochemistry of the action of IAA on plant growth. In: Runeckles VC, Sondheimer E (eds) The chemistry and biochemistry of plant hormones. Academic Press, New York, pp 93–122
Ray PM, Ruesink AW (1962) Kinetic experiments on the nature of the growth mechanism in oat coleoptile cells. Dev Biol 4:377–397
Ray PM, Green PB, Cleland RE (1972) Role of turgor in plant cell growth. Nature 239:163–164
Rayle DL (1973) Auxin-induced hydrogen-ion excretion in *Avena* coleoptiles and its implications. Planta 114:63–73
Rayle DL, Cleland RE (1970) Enhancement of wall loosening and elongation by acid solutions. Plant Physiol 46:250–253
Rayle DL, Cleland RE (1972) The in-vitro acid-growth response: relation to in-vivo growth responses and auxin action. Planta 104:282–296
Rayle DL, Cleland RE (1977) Control of plant cell enlargement by hydrogen ions. Curr Top Dev Biol 11:187–214
Rayle DL, Haughton PM, Cleland RE (1970) An in vitro system that simulates plant cell extension growth. Proc Natl Acad Sci USA 67:1814–1817
Roelofsen PA (1965) Ultrastructure of the wall in growing cells and its relation to the direction of growth. Adv Bot Res 2:69–149
Roland J-C, Vian B (1979) The wall of the growing cell: its three dimensional organization. Int Rev Cytol 61:129–166
Ruesink AW (1969) Polysaccharidases and the control of cell wall elongation. Planta 89:95–107
Sakurai N, Masuda Y (1978) Auxin-induced extensibility, cell wall loosening and changes in the wall polysaccharide content of barley coleoptile segments. Plant Cell Physiol 19:1225–1233
Sakurai N, Masuda Y (1979) Effect of cycloheximide and cordycepin on auxin-induced elongation and β-glucan degredation of non-cellulosic polysaccharides of *Avena* coleoptile cell walls. Plant Cell Physiol 20:593–603

Sakurai N, Nevins DJ, Masuda Y (1977) Auxin and hydrogen ion-induced cell wall loosening and cell extension in *Avena* coleoptile segments. Plant Cell Physiol 18:371–380
Sakurai N, Nishitani K, Masuda Y (1979) Auxin-induced changes in the molecular weight of hemicellulosic polysaccharides of the *Avena* coleoptile cell wall. Plant Cell Physiol 20:1349–1357
Sumiya K, Yamada T (1974) Effect of IAA on stress relaxation of Japanese black pine seedling. Wood Res 56:13–20
Tagawa T, Bonner J (1957) Mechanical properties of the *Avena* coleoptile as related to auxin and to ionic interactions. Plant Physiol 32:207–212
Tanimoto E, Igari M (1976) Correlation between β-galactosidase and auxin-induced elongation growth in etiolated pea stems. Plant Cell Physiol 17:673–682
Tanimoto E, Masuda Y (1968) Effect of auxin on cell wall degrading enzymes. Physiol Plant 21:820–826
Tanimoto E, Masuda Y (1971) Role of the epidermis in auxin-induced elongation of light-grown pea stem segments. Plant Cell Physiol 12:663–673
Tepfer M, Cleland RE (1979) A comparison of acid-induced cell wall loosening in *Valonia ventricosa* and in oat coleoptiles. Plant Physiol 63:898–902
Terry M, Rubinstein B, Jones RL (1980) Changes in soluble cell wall polysaccharides and growth. Plant Physiol 65:S23
Uhrström I (1974) The effect of auxin and low pH on Young's modulus in *Pisum* stems and on water permeability in potato parenchyma. Physiol Plant 30:97–102
Valent BS, Albersheim P (1974) The structure of plant cell walls. V. On the binding of xyloglucan to cellulose fibers. Plant Physiol 54:105–108
Verma DPS, Maclachlan GA, Byrne H, Ewings D (1975) Regulation and in vitro translation of messenger RNA for cellulose from auxin-treated pea epicotyls. J Biol Chem 250:1019–1026
Waaland SD, Waaland JR, Cleland RE (1972) A new pattern of plant cell elongation: bipolar band growth. J Cell Biol 54:184–190
Wada S, Ray PM (1978) Matric polysaccharides of oat coleoptile cell walls. Phytochemistry 17:923–931
Wada S, Tanimoto E, Masuda Y (1968) Cell elongation and metabolic turnover of the cell wall as affected by auxin and cell wall degrading enzymes. Plant Cell Physiol 9:269–276
Yamagata Y, Masuda Y (1975) Comparative studies on auxin and fusicoccin actions on plant growth. Plant Cell Physiol 16:41–52
Yamagata Y, Masuda Y (1976) Auxin-induced extension of the isolated epidermis of light-grown pea epicotyls. Plant Cell Physiol 17:1235–1242
Yamagata Y, Yamamoto R, Masuda Y (1974) Auxin and hydrogen ion actions on light-grown pea epicotyl sections. II. Effect of hydrogen ions on extension of isolated epidermis. Plant Cell Physiol 15:833–841
Yamamoto R, Masuda Y (1971) Stress-relaxation properties of the *Avena* coleoptile cell wall. Physiol Plant 25:330–335
Yamamoto R, Nevins DJ (1979) A transglucosylase from *Sclerotinia libertiana*. Plant Physiol 64: 193–196
Yamamoto R, Shinozyki K, Masuda Y (1970) Stress-relaxation properties of plant cell walls with special reference to auxin action. Plant Cell Physiol 11:947–956
Yamamoto R, Makai K, Masuda Y (1974) Auxin and hydrogen ion actions on light-grown pea epicotyl segments. III. Effect of auxin and hydrogen ions on stress-relaxation properties. Plant Cell Physiol 15:1027–1038
Yoda S, Ashida J (1960) Effect of gibberellin and auxin on the extensibility of the pea stem. Plant Cell Physiol 1:99–105
Zarra I, Masuda Y (1979) Growth and cell wall changes in rice coleoptiles growing under different conditions. II. Auxin-induced growth in coleoptile segments. Plant Cell Physiol 20:1125–1133

II. Cell Walls of Algae and Fungi

12 Algal Walls – Composition and Biosynthesis

E. PERCIVAL and R.H. MCDOWELL

1 Introduction

Although the cell walls of algae have structural features in common with those of land plants they differ from them in many respects, as would be expected in plants which live in water instead of in air. Cell walls are the main components of plant tissues which are responsible for their mechanical properties and must be suitable for withstanding the forces to which they are subjected. Macroscopic algae do not require rigid structures which can withstand gravitational forces but must be strong enough to resist damage by swiftly moving and turbulent water. The constituents of the cell walls and intercellular regions are therefore those which give tensile strength and flexibility to the plant. Typically the proportion of fibrillar material is less than and that of the matrix and intercellular material greater than in land plants, and in addition many species exude polysaccharide material onto their outer surfaces.

As might be expected considering the wide range of types included in the algae, many different polysaccharides are found in their cell walls, a number of them being found only in one class of algae, while others are very widespread. Some, for example cellulose, are similar to those found in land plants, while polysaccharides substituted with sulfate ester groups, universal in the algae, are not found in land plants, although they are similar to sulfate esters present in animal tissue polysaccharides.

Detailed chemical studies on the polysaccharides of algae have been made on comparatively few species. Alginates from Phaeophyceae and galactans from Rhodophyceae have received the most attention in view of their economic value. A general survey of algal carbohydrates has been given by PERCIVAL and MCDOWELL (1967). Algal cell wall polysaccharides have been reviewed by MACKIE and PRESTON (1974) and HAUG (1974), and the biochemistry of algal polysaccharides by TURVEY (1978). In this chapter therefore particular attention has been paid to the more recent work, and more reference to earlier studies can be found in the publications mentioned above.

Following the determination of the constituent monosaccharide residues and their sequence in some of the polysaccharides, it has been possible to study the shapes of the macromolecules (REES 1972, REES and WELSH 1977), and how these control their physical properties. While there is little doubt that fibrillar polysaccharides are located in the cell wall proper, it is not always certain which of several polysaccharides in an alga constitutes the crystalline material observed by optical or X-ray methods (FREI and PRESTON 1961a) although cellulose when present is always considered to form at least part of the fibrillar structure. To determine whether nonorientated materials are present as matrix substances in the cell wall or are located in the intercellular region presents even more difficulty.

Stains for light microscopy which have been used most successfully in recent years are toluidine blue, which gives a pink color with carboxyl or sulfate groups, and periodic acid-Schiff (PAS), which is specific for polysaccharides with adjacent free hydroxyl groups in the sugar residues (MCCULLY 1966, EVANS and HOLLIGAN 1972). Alcian blue and Alcian yellow dyes are also useful for detecting acid polysaccharides in tissues (PARKER and DIBOLL 1966). Unfortunately, these methods are not sufficiently specific to distinguish between the different polysaccharides in the algae.

A more specific method of distinguishing between the different polysaccharides has been developed by VREELAND (1970, 1972). She prepared antibodies from the serum of rabbits sensitized by injection of alginate, coupled this with a fluorescent dye and applied it to algal sections for microscopic examination. GORDON and MCCANDLESS (1973) applied this method to the detection of κ-carrageenan and GORDON-MILLS and MCCANDLESS (1975) to λ-carrageenan.

A polysaccharide containing an element not present in other substances in the plant can be located by X-ray microanalysis. CALLOW and EVANS (1976) used the sulfur in fucans to locate it in tissues of *Laminaria.* The various methods used agree that the mucilaginous constituents of algae are to be found in both the cell wall and intercellular regions.

In nearly all the chemical studies on the polysaccharides of algae, extracts have been prepared from whole plants, although in many cases a sequential method of extraction has been used with a series of reagents or physical conditions bringing out the more easily extracted materials in the early stages (PERCIVAL and MCDOWELL 1967). Thus the whole plant, including cell contents, cell wall, and intercellular regions, is obtained in a series of solutions and an insoluble residue from which the carbohydrates can be separated. It has sometimes been assumed that cell contents, intercellular material and cell wall constituents would be extracted in that order, but this is certainly not always true.

In this chapter all the polysaccharides obtained from the algae are classed as cell wall materials except those which are generally considered to be food reserve materials, e.g., starch or laminaran. HANIC and CRAIGIE (1969) isolated a chemically resistant cuticle from a number of algae. As in those cases where a chemical analysis was made it was found to be almost entirely protein, it will not be considered further in this chapter. This very thin cuticle should not be confused with the outer layer of mannan also referred to as cuticle isolated from *Porphyra umbilicalis* by FREI and PRESTON (1964b).

In multicellular plants it is very difficult to separate cell walls from intercellular regions, and unicellular algae are frequently enclosed in a layer of extracellular mucilage. Isolated cell walls of *Fucus* sp. free from such contaminants have been studied from their earliest stages of development (QUATRANO and STEVENS 1976) and been found to consist of cellulose, alginate, and fucans, substances which are the structural polysaccharides of mature whole plants.

Each stage of sequential extraction of an alga contains a mixture of polysaccharides: they can be separated to some extent by various fractionation techniques but after even a series of such separations it cannot be expected that the product will consist of molecules of identical size and composition, and analysis will give results which are averages with little information on the degree of polydispersivity. With several species variations have been found in the fine structure of polysaccharides from different stages of their life cycle.

In plant cell walls the various constituents form a complex structure in which different polysaccharides may be held together by covalent links to protein (THOMPSON and PRESTON 1967). Extraction methods used to obtain "pure" compounds must therefore break down much of their structure, but preparations of, for example, sulfated polysaccharides of the Chlorophyceae and fucans from the brown seaweeds commonly contain some protein. The nature of the carbohydrate protein linkages awaits further study.

Although much of the work on the primary stages of carbon fixation in photosynthesis was carried out on *Chlorella,* very little is known about the steps involved in the biosynthesis of algal polysaccharides (cf. TURVEY 1978). A few nucleotides have been extracted from algae and, as with higher plants, it is probable that they are the precursors of polysaccharides, but extraction of the enzymes which bring about the synthesis has proved extremely difficult. The work that has been done is considered in the sections on individual polysaccharides.

Little in the way of chemical investigation, apart from a determination of the constituents (cf. PERCIVAL and TURVEY 1974), has been carried out on the cell wall polysaccharides of members of the Pyrrophyta, Euglenophyta, Chrysophyta, Xanthophyta, and Charophyta.

Where possible the classification of PARKE and DIXON (1976) or otherwise that of SILVA (1962) will be used in this chapter.

2 Cellulose and Other Glucans

Cellulose is very widely distributed in the different classes of algae, and when present forms all or part of the microfibrils which make up the skeleton of the cell wall. The amount present, however, is frequently small compared with other polysaccharides and in only a few cases have microfibrils consisting of $(1\rightarrow4)$-β-glucan been isolated free from other sugars.

The methods used for freeing cellulose from other constituents, and the nomenclature for the products obtained were worked out by the industries using the cellulose from higher plants and can in some cases be misleading when applied to algal cellulose. The residue remaining after extraction with hot dilute acid and alkali was referred to as cellulose in a survey of its quantity in red seaweeds (ROSS 1953), while a chlorine treatment was included by BLACK (1950) in his figures for the Phaeophyceae.

In preparing the material which they termed α-cellulose CRONSHAW et al. (1958) included treatment with cold 4N-alkali followed by chlorite. With few exceptions α-cellulose, which retained the microfibrillar structure, contained minor amounts of other sugars as well as glucose. Treatment with 2.5N-sulfuric acid at 100 °C for 24 h (DENNIS and PRESTON 1961), which removed the other sugars, destroyed the microfibrillar structure but left cellullose as rodlets which gave the cellulose I X-ray diagram.

In the Phaeophyceae cellulose appears to be a universal wall constituent, although in some cases the amount is very small, *Pelvetia canaliculata* having only 0.6 to 1.5% (BLACK 1950). On the other hand, in both the Chlorophyceae and the Rhodophyceae cellulose is absent in some species, the fibrillar components of the cell walls being made up of highly crystalline xylans or mannans.

2.1 Chlorophyceae

The most highly organized algal cellulose is found in *Valonia* and species of *Cladophora* and *Chaetomorpha*. The organization of the microfibrils in the cell wall has been extensively studied and is discussed in Chapters 3 and 13, this Volume. The cell walls of these algae contain a high proportion of highly crystalline cellulose (FREI and PRESTON 1961a) and it has been used for detailed X-ray crystallographic studies. There has been some difficulty in reconciling the results with the Meyer-Misch unit cell. It was suggested by HONJO and WATANABE (1958), using *Valonia* sp., that the unit cell should be made up of eight chains and this was supported by observations on *Chaetomorpha melagonium* (FREI and PRESTON 1961b). A re-examination of the cell walls of this alga (NIEDUSZYNSKI and ATKINS 1970) revealed some uncertainties which could be resolved only with richer X-ray diagrams. On the other hand, SARKO and MUGGLI (1974) concluded that for *Valonia* cellulose a two-chain unit cell would account for all the reflections that had been observed. On the basis of the energy of packing of the chains they consider that these are parallel, although an antiparallel arrangement is not definitely ruled out, as energy differences are small.

The unicellular alga *Glaucocystis* (SCHNEPF 1965, ROBINSON and PRESTON 1971) has similarly well organized cellulose microfibrils.

In *Ulva* and *Enteromorpha* the residues after sequential extraction give, on hydrolysis, glucose with smaller quantities of xylose (CRONSHAW et al. 1958, MCKINNELL and PERCIVAL 1962b). That from *Urospora penicilliformis* (BOURNE et al. 1974) was shown by methylation analysis and periodate oxidation to contain (1→4)-linked glucose. The residues from *Urospora wormskioldii* and its *Codiolum pusillum* phase both contained major amounts of glucose, but the latter also had an appreciable amount of mannose. In the alga *Derbesia marina-Halicystis ovalis* the cell walls of the *Derbesia* phase give mainly mannose on hydrolysis, while those of *Halicystis* contain glucose and xylose (HUIZING and RIETEMA 1975, WURTZ and ZETSCHE 1976). The latter is probably a mixture of cellulose with a glucoxylan (FREI and PRESTON 1961a).

NORTHCOTE et al. (1958) separated from the cell walls of *Chlorella pyrenoidosa* a component made up mainly of glucose and galactose. CONTE and PORE (1973) concluded that these sugars were present as heteroglycans in the cell walls of that and other species of *Chlorella* and *Prototheca*. Glucosamine was detected in their material by NORTHCOTE et al. (1958) and also as a major constituent by TAKEDA and HIROKAWA (1978) in *C. ellipsoidea*. The latter authors were unable to find any cellulose.

2.2 Rhodophyceae

The red algae, apart from the Bangiales, contain cellulose in their cell walls (FREI and PRESTON 1961 a), but the microfibrils are built up of other sugars as well as glucose.

WHYTE and ENGLAR (1971) purified the cellulose from the residue after sequential extraction of *Rhodymenia pertusa* by dissolution in cuprammonium and reprecipitation. This was methylated and shown by gas-liquid chromatography, mass spectrometry and proton magnetic resonance to consist of (1 → 4)-β-D-glucopyranose residues.

2.3 Phaeophyceae

PERCIVAL and ROSS (1948, 1949) purified the materials remaining after sequential extraction of *Laminaria digitata, L. hyperborea* and *Fucus vesiculosus* by dissolving in cuprammonium solution and reprecipitating. They then found that they gave only glucose on hydrolysis and were able to prepare cellobiose octaacetate from each of them. Cellulose was shown to be present in *F. vesiculosus* zygotes within 30 min of fertilization (STEVENS and QUATRANO 1978), but the isolated walls, after removal of other polysaccharides, retained birefringence only from cells more than 4 h after fertilization.

2.4 Biosynthesis

The formation of cellulose from simple sugars has not been studied in detail for an alga, but there is no evidence to suggest that the intermediates involved are any different from those in higher plants (NIKAIDO and HASSID 1971, DELMER 1977, FRANZ and HEINIGER, Chap. 5, this Vol.).

In the biosynthesis of cellulose in the cell walls of *Fucus* species QUATRANO and STEVENS (1976) suggest that the glucose is derived from laminaran present in the unfertilized egg and demonstrated the release of glucose from laminaran by a particulate enzyme preparation from *F. distichus*.

2.5 (1 → 3)-β-glucan

Although the (1 → 3)-β-glucan of the Phaeophyceae is the storage product laminaran, it appears that a similar polysaccharide may at times play a structural role in some algae. When plants of *Caulerpa simpliciuscula* are wounded, a plug rapidly forms at the damaged surface. It has been shown (HOWARD et al. 1976, DREHER et al. 1978, HAWTHORNE et al. 1979) that this is a gelled form of a low molecular weight (1 → 3)-β-glucan normally held in solution in the vacuole which occupies the greater part of the volume of this alga.

3 Xylans

Although xylose has been reported as a constituent of fucose-containing polysaccharides of the Phaeophyceae, no one has isolated a pure xylan from the brown seaweeds.

3.1 Rhodophyceae

A major polysaccharide of *Rhodymenia palmata* (dulse) is a water-soluble essentially linear xylan made up of (1→3)- and (1→4)-units in purely random sequence (BJORNDAHL et al. 1965) with an average one branch point in each molecule linked at C-3 and C-4. More recent work, by extraction of the residual weed of *Rhodymenia* with dilute alkali, gave a (1→4)-xylan (TURVEY and WILLIAMS 1970). Extracts from *Rhodochorton floridulum* (TURVEY and WILLIAMS 1970) contain a xylan similar in many respects to the water-soluble extract of *Rhodymenia* but with a higher proportion of branch points. Since these extracts also contain glucose units, it is tentatively suggested by the authors that the polysaccharide from this weed may be a heteropolymer in which the xylose occurs as short branched chains on a glucan backbone. Other (1→3)- and (1→4)-xylans have been separated from *Porphyra umbilicalis, Laurencia pinnatifida* (TURVEY and WILLIAMS 1970) and from *Chaetangium fastigiatum* CEREZO et al. 1971, CEREZO 1972). In contrast to *Rhodymenia* further extraction with alkali of *Porphyra umbilicalis* gave a pure (1→3)-xylan (TURVEY and WILLIAMS 1970). This confirms earlier work of FREI and PRESTON (1964b) who mechanically separated the cell walls of *P. umbilicalis* and *Bangia fuscopurpurea* and showed from X-ray diffraction and electronmicroscope studies that they consisted of chains of (1→3)-xylose in a random network of microfibrils.

3.2 Chlorophyceae

Many of the Chlorophyceae appear to be devoid of cellulose and synthesize instead a xylan and/or a mannan as the cell wall polysaccharide. The xylans of *Caulerpa filiformis* (MACKIE and PERCIVAL 1959), *C. brachypus, C. racemosa, C. anceps, Halimeda cuneata, Udotea orientalis, Chlorodesmis formosana* and *Pseudodichotomosiphon constricta* (MIWA et al. 1961, IRIKI et al. 1960) consist of chains of 50 (1→3)-β-xylose units. Light-scattering osmometry, gel permeation chromatography and viscosity techniques on xylan from the cell wall of *Penicillus dumetosus* isolated under very mild conditions indicated degrees of polymerization above 10,000 (MACKIE and SELLEN 1971). From X-ray studies it is deduced that the (1→3)-β-xylans consist of hexagonally packed double-stranded helices present as microfibrils (FREI and PRESTON 1964a, ATKINS et al. 1969). Reconsideration (ATKINS and PARKER 1969) of these results supplemented by polarized infra red observations have shown that these polysaccharides exist in right-handed helices with three xylan chains froming a triple helix, thus giving a strong fibrous structure.

Xylose and glucose are the dominant monosaccharides in the cell walls of the gametophytic macrothallus of *Bryopsis maxima, B. plumosa, Bryopsidella neglecta* and *Derbesia tenuissima* (HUIZING et al. 1979).

4 Mannans

Mannans have never been detected in the Phaeophyceae, although mannose has been found as a constituent of some "fucans". Mannans have only been reported as a cuticle in the Rhodophycean algae, *Porphyra umbilicalis* and *Bangia fuscopurpurea* (FREI and PRESTON 1964a).

4.1 Chlorophyceae

Codium fragile (LOVE and PERCIVAL 1964b) and *Acetabularia crenulata* (BOURNE et al. 1972) synthesize a (1→4)-β-mannan. Electronmicrographs of the walls of these two algae (MACKIE and PRESTON 1968) show fibrils with a heavily encrusted appearance. The mannan from *C. fragile* has degrees of polymerization which range from below 20 to 10,000 (MACKIE and SELLEN 1969). Similar partially crystalline mannans constitute the cell walls of *Codium latum, C. intricata, C. adherens, Derbesia lamourouxii, Acetabularia calyculus, Halycoryne wrightii* (MIWA et al. 1961, IRIKI and MIWA 1960), *Dasycladus clavaeformis, D. vermicularis, Batophora oerstedi, Neomeris annulata, Cymopolia barbata, Acetabularia mediterranea, A. crenulata* (FREI and PRESTON 1968). *Codiolum pusillum,* the erect stage in the life history of *Urospora wormskioldii* synthesizes a (1→4)-β-mannan branched at C-6 and carrying half ester sulfate groups at C-2 on some of the units (CARLBERG and PERCIVAL 1977). Mannose and glucose are the dominant monosaccharides in the sporophytic microthallus of *Bryopsis plumosa* and of several *Derbesia* species (HUIZING et al. 1979).

Xylomannan constitutes the major part of the cell wall of *Prasiola japonica* Yatabe (TAKEDA et al. 1968). It consists on average of about 1 xylose and 13 mannose residues, most of which appear to be connected through (1→4)-β-linkages.

A water-soluble (1→3)-α-mannan is synthesized by *Urospora penicilliformis* (BOURNE et al. 1974).

4.2 Bacillariophyceae

The cell walls of the diatom, *Phaeodactylum tricornutum* consist of chains of (1→3)-β-mannose units carrying side chains of *O*-D-glucopyranosyluronic acid-(1→3)-*O*-D-mannopyranosyl-(1→2)-*O*-D-mannopyranose (FORD and PERCIVAL 1965).

5 Alginic Acid

Alginic acid has been found in all species of the Phaeophyceae that have been examined, but is not present in any other plant tissues. However, the extracellular mucilage formed by certain bacteria (LINKER and JONES 1964, GORIN and SPENCER 1966) differs from algal alginic acid only in being acetylated.

The proportion of alginate shows wide variations from one species of alga to another, and in some species changes with the seasons and location of growth (PERCIVAL and MCDOWELL 1967). An indication of the variation is the nearly 50% of dry weight found in some samples of *Durvillea antarctica* (SOUTH 1979) and 15% in *Padina pavonia* (JABBAR MIAN and PERCIVAL 1973a).

The alginate can be extracted from the algae by ion exchange to convert it into the soluble sodium alginate and dissolution in water: the product thus obtained always contains small amounts of sulfated polysaccharide and protein, but can be purified by precipitation as the calcium salt, from which the acid is recovered by leaching out the calcium. The fact that the alginate can be extracted by such relatively mild methods indicates that it is not covalently linked to insoluble cell wall constituents. It must be expected, however, that some degradation of the polymer takes place during extraction, and as preparations with degrees of polymerization (DP) in the range 1,000 to 10,000 are commonly obtained the alginate must exist as a very large molecule in the plant.

5.1 Constitution and Structure

Alginic acid is a linear polymer of (1 → 4)-β-D-mannuronic acid and α-L-guluronic acid the proportions of the two units varying widely in the alginates in different algal species (PERCIVAL and MCDOWELL 1967). It has been shown that only these units linked in this way are present (REES and SAMUEL 1967).

HAUG et al. (1966, 1967) used partial hydrolysis with dilute acid to determine the sequence of the units. They found that part of the alginate was solubilized and that the insoluble residue could be fractionated into a mannuronic-rich fraction (M-blocks) soluble at pH 2.85 and a guluronic-rich fraction (G-blocks) insoluble at that pH. The fraction solubilized by partial hydrolysis contained roughly equal amounts of the two uronides (MG-blocks) and was at first thought to be built up of alternating mannuronic and guluronic units. SIMONESCU et al. (1976), from a kinetic study of acid hydrolysis, concluded that residues in the MG-blocks are arranged in a more random fashion and further work using other methods has confirmed this.

MIN et al. (1977) using poly-G degrading lyases derived from *Pseudomonas* sp. and BOYD and TURVEY (1978) using a similar lyase from *Klebsiella aerogenes* found by analysis of the oligosaccharides obtained by digestion of MG-block preparations with the lyases that this material has in it both mannuronic followed by mannuronic residues (MM sequences) and guluronic followed by guluronic (GG sequences) as well as linkages between the two different units. All the

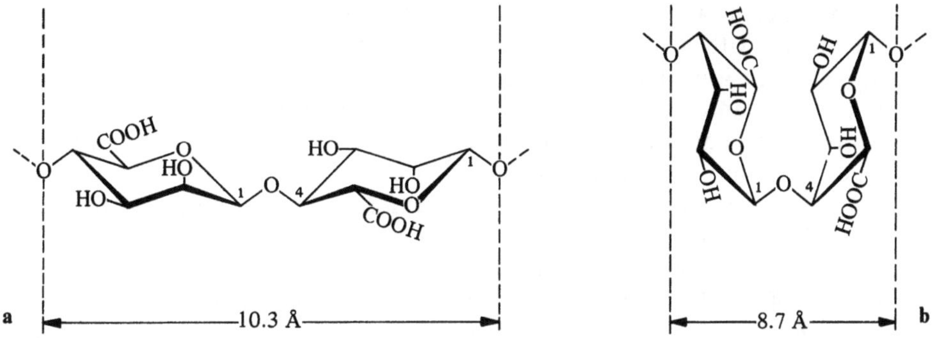

Fig. 1a, b. Parts of alginic acid molecule. **a** Mannuronic acid residues; **b** Guluronic acid residues

enzymes known to degrade alginates are eliminases so that the nonreducing end of a fragment liberated by the enzymes is a 4,5-unsaturated uronic acid, the structure being the same whether derived from mannuronic or guluronic residue, and complete details of the order of residues in the polysaccharide remain uncertain.

In a study of methods of determining the proportions of M and G residues in an alginate PENMAN and SANDERSON (1972) found that using proton magnetic resonance they could distinguish the signals from H-1 and H-5 in the guluronic residues and from H-1 in mannuronic acid when they examined the homopolymeric blocks obtained by the partial hydrolysis of alginates. By the use of high resolution proton magnetic resonance GRASDALEN et al. (1979b) were able to use slightly degraded whole alginate, as well as fractions from partial hydrolysis, to distinguish signals of H-5 of guluronic residues with a mannuronic neighbor (GM sequence) from those with a guluronic neighbor (GG sequence). In conjunction with the assignments for H-1 they were able to obtain the proportion of each of the four sequences: MM, GG, MG, and GM. They (GRASDALEN et al. 1979a) also used ^{13}C nuclear magnetic resonance to obtain these sequences and the triplet sequences with M units in the middle.

Their results agree with those from the enzymic work in showing that there are some MM and GG sequences in the alginate fraction solubilized in acid hydrolysis and also that there are some MM, MG, and GM sequences in G-blocks and GG, MG, and GM sequences in the M-blocks. Although these methods yield new information about alginate structure, it is not yet possible to determine the lengths of uninterrupted sequences of one monomer. However, BOYD and TURVEY (1978) found that M blocks remaining after the attack by their poly-G lyase on whole alginate and M-rich preparations all had a chain length of about 24 residues.

5.2 Ion Exchange

The relative affinities of a number of bases for alginates from different sources have been investigated and expressed as selectivity coefficients (HAUG 1964).

Those for calcium and strontium against potassium, obtained by equilibrating an alginate gel with a solution containing various concentrations of potassium and the divalent ions, have a high correlation with the guluronic acid content of the alginate, and it was shown later that the high affinity for calcium and strontium was a characteristic of the G-blocks in the alginate (HAUG and SMIDSROD 1967, SMIDSROD and HAUG 1968). Measurements on solutions of calcium polymannuronate and polyguluronate of different chain lengths (KOHN and LARSEN 1972) gave considerably lower calcium ion activities in calcium polyguluronate samples with more than 20 units, than of those of lower molecular weight and of all the calcium polymannuronates. This, together with further ion exchange experiments (SMIDSROD and HAUG 1972b), suggested an autocooperative binding in which aggregation of the polyguluronide chains binds more firmly the calcium ions which initiate the aggregation. Circular dichroism measurements in the wavelength range 190 to 250 nm confirmed the idea that calcium ions introduced into an alginate solution combined preferentially with the G-blocks (MORRIS et al. 1973). Further studies using the same techniques (MORRIS et al. 1978) suggest that the primary process in cooperative binding is a dimerization of G-blocks.

5.3 Conformation

The difference in selectivity for calcium of the M- and G-blocks is a consequence of their stereochemistry. X-ray studies (ATKINS et al. 1970, 1971) of fibers of polymannuronic acid derived from *Fucus* receptacles and others from alginic acid with a high guluronic content have shown spacings along the fiber axis of 10.35 Å for mannuronic and 8.72 Å for the guluronic polymers, both having a twofold screw axis. Fibers could not be prepared from G-blocks, but comparison of powder X-ray diffraction patterns showed that in an alginic acid with about 60% G-blocks the polyguluronic pattern was obtained with the mannuronic reflections completely absent (ATKINS et al. 1970, 1971). Conformations of the uronic acid units in agreement with these spacings are C1 for the (1 → 4)-β-D-mannuronic units giving an extended diequatorial structure similar to that of cellulose, and 1C for the (1 → 4)-α-L-guluronic units in which all the glycosidic links are diaxial (Fig. 1).

MACKIE (1971) found that in the salts of polymannuronic acid the repeating unit was made up of three residues still in the extended conformation to give a spacing of 15 Å, while the polyguluronic salts retained the twofold repeating sequence with the 8.7 Å spacing. Using a combination of viscosity and light scattering measurements with conformational energy calculations, SMIDSROD et al. (1973) concluded that the same conformation was retained in solution. In their assignment of proton magnetic resonance peaks PENMAN and SANDERSON (1972) found that these were satisfactorily explained by the M and G residues having these conformations in solution.

5.4 Gel Formation

A consequence of this conformation is that there are cavities between residues in the G-blocks in which calcium ions can coordinate with carboxyl groups and ring oxygen atoms in each of two parallel blocks. Thus structures ranging from dimers of polyguluronate chains to extended sheets of such chains are stabilized with calcium ions and have been described in terms of an "egg box model" (GRANT et al. 1973). Strontium ions have an even better fit in the cavities and will be retained in preference if there is competition between calcium and strontium.

An alginate gel is therefore envisaged as a three-dimensional network of long chain molecules held together by junction zones which are formed from G-block sections of the molecules and calcium ions. In the gels prepared by diffusion of calcium chloride into sodium alginate solutions (SMIDSROD et al. 1972, SMIDSROD and HAUG 1972a) the counterions would all be calcium. In the plant, however, the alginate is in equilibrium with seawater and is combined principally with calcium, magnesium, and sodium (HAUG 1964, HAUG and SMIDSROD 1967), and it is logical to suppose that it is in the form of a gel with the calcium ions being concentrated in G-blocks, largely in junction zones, the other ions being associated with other parts of the alginate molecules.

It is clear that alginate gels can contribute a wide range of mechanical properties to algae depending on the proportion and arrangements of their constituent polymers. As at least part of the alginate is orientated (THIELE and ANDERSEN 1955), the degree and direction of anisotropy could also be significant.

5.5 Variations in Structure

The earlier work on variation between species and in different tissues of the same plant (FISCHER and DÖRFEL 1955, HAUG 1964) was confined to M/G ratios, but in view of the importance of block structure the more recent publications which take this into account are of greater significance. HAUG et al. (1969, 1974) give figures based on partial hydrolysis and GRASDALEN et al. (1979a) on ^{13}C-nuclear magnetic resonance. Except in those cases where a high proportion of the alginate was solubilized by partial hydrolysis (e.g., old tissues from *Ascophyllum*), there is good agreement between the two methods.

A few of the results are given in Table 1. Outstanding are the high proportion of G-blocks in *Laminaria hyperborea* stipes and of mannuronic acid in the intercellular mucilage which can be squeezed out of the receptacles of *Ascophyllum*. These alginates with markedly different properties represent the extremes of whole alginates which have been used in the work on ion selectivity and gel formation.

In the living plant alginates have been shown to be present both in the cell wall and in the intercellular mucilage (VREELAND 1970, 1972, EVANS and HOLLIGAN 1972). The very considerable variation in the alginate composition from different parts of algae could be partly a result of the tissues extracted having different proportions of intercellular material. The separation of alginate from the apparently homogeneous tissues of new fronds of *L. digitata* into two parts with M/G ratios of 1.65 and 2.9 by fractionation with manganese sulfate and potassium chloride (HAUG 1964) could perhaps be a consequence of the difference between cell wall and intercellular alginates.

Table 1. Block distribution and doublet frequency of alginate samples

Source of sample		MG+GM[a]	MM	GG	M/G
Laminaria digitata					
new fronds	b	0.26	0.54	0.20	2.16
	c	0.37	0.50	0.14	2.10
L. hyperborea,					
new fronds	d	0.22	0.53	0.25	1.8
old fronds	d	0.33	0.34	0.33	1.0
L. hyperborea,					
old stipe	b	0.22	0.15	0.62	0.36
	c	0.25	0.0	0.75	0.38
Macrocystis pyrifera,	b	0.40	0.40	0.20	1.50
stipe	c	0.44	0.43	0.13	1.20
Ascophyllum nodosum	b	0.29	0.53	0.18	2.05
tip (new growth)	c	0.34	0.54	0.12	2.38
A. nodosum,					
base (old tissue)	b	0.44	0.32	0.24	1.32
	c	0.72	0.20	0.08	1.23
A. nodosum, receptacles intercellular mucilage	d	0.09	0.84	0.06	7.7

[a] In partial hydrolysis the proportion solubilized
[b] Doublet frequency by ^{13}C-nuclear magnetic resonance (GRASDALEN et al. 1979a)
[c] Block distribution by partial hydrolysis (GRASDALEN et al. 1979b)
[d] Block distribution by partial hydrolysis (HAUG et al. 1974)

5.6 Biosynthesis

Although knowledge of the biosynthesis of alginate is very far from complete, a number of possible steps have been elucidated. LIN and HASSID (1966a) isolated the nucleotides GDP-mannuronic acid and GDP-guluronic acid from *Fucus gardneri* (=*distichus*) and also showed the presence of the enzymes (LIN and HASSID 1966b) catalyzing the sequence of reactions: D-mannose → D-mannose-6-P → D-mannose-l-P → GDP-D-mannose → GDP-D-mannuronate, and also the incorporation of ^{14}C-labeled GDP-D-mannose into sodium alginate. They also showed that slices of *F. gardneri* were able to synthesize alginic acid from mannose at a greater rate than from glucose.

At the present time no pathway has been proved for the formation of mannose from the primary products of photosynthesis. Mannitol is the main product obtained after several hours of photosynthesis (BIDWELL et al. 1972) by *Fucus vesiculosus*, but there has been no demonstation of conversion of this sugar alcohol into mannose or other sugars by algal enzymes. On the other hand, HELLEBUST and HAUG (1972) found that only 16% of newly synthesized carbon was present as mannitol in *Laminaria digitata* blades, whereas about 50% was present in the form of amino acids. As moreover amino acids disappeared much more rapidly in the dark than mannitol, during which time alginate was being synthesized, they suggested amino acids as possible precursors of alginate. On the basis of chase experiments using $^{14}CO_2$, BIDWELL et al. (1972) suggested that a compound which can be extracted from tissues of *Fucus vesiculosus* by acid might be a stage in alginic acid synthesis.

In a study of the development of zygotes starting with fertilization of the eggs of several species of *Fucus*, QUATRANO and STEVENS (1976) suggested that laminaran present in the unfertilized eggs was the source of carbon for the cell wall polysaccharides, as well as providing energy.

In view of the presence of GDP-guluronic acid in *F. gardneri* tissues LIN and HASSID (1966b) suggested that there should be a reaction

GDP-L-guluronic acid + alginate acceptor $\xrightarrow{\text{transferase}}$ alginic acid with additional guluronic residues,

but they were unable to detect it.

On the other hand it has been shown (MADGWICK et al. 1973) that *Pelvetia canaliculata* contains an enzyme which can convert D-mannuronic residues into L-guluronic residues in the polymer. Previous studies (HAUG and LARSEN 1971, LARSEN and HAUG 1971) had shown that an enzyme isolated from the alginate synthesizing organism *Azotobacter vinelandii* was capable of converting polymannuronic acid into an alginate with both heteropolymeric and guluronic blocks the extent of conversion depending on the concentration of calcium in the medium. Using tritiated water it was found with the enzyme preparation from both *Pelvetia* and *Azotobacter* that tritium was incorporated into the guluronic units formed by enzyme action, suggesting that a hydrogen atom adjacent to the carboxyl group is removed and replaced with inversion. Working with samples of degraded polymannuronic acid LARSEN and HAUG (1973) found that a product with a minimum DP of between 10 and 15 was necessary for this enzymic epimerization to take place.

It is possible therefore that in the plant polymannuronic acid is first synthesized and that some of the mannuronic residues are then isomerized to guluronic, the number and position in the polymer being controlled in a manner which gives alginic acid with the properties required in the particular tissue. This must take place at an early stage in the development of an alga, as QUATRANO and STEVENS (1976) found that alginic acid, having an excess of guluronic over mannuronic units, was the major polysaccharide in the cell walls of 16- to 24-h *Fucus* zygotes.

6 Galactans

The galactans are the major polysaccharides of the majority of the Rhodophyceae (cf. PERCIVAL 1972, 1978b, 1978c). They all appear to consist of essentially linear chains of alternating $(1 \rightarrow 3)$-β-galactose (A) and $(1 \rightarrow 4)$-α-galactose (B) (REES 1969). Many of these residues are masked by substitution with half ester sulfate with methoxyl groups and with pyruvic acid groups (as the 4,6-*O*-1-

carboxyl-ethylidene derivative), and by modification to the 3,6-anhydro sugar (Fig. 2). Extracts from particular genera differ in the proportion of D- and L-galactose and in the degree of substitution and modification of the residues, and it is these differences which determine the overall shape or conformation of the macromolecules and hence the physical properties of the extracts. With the large number of possible variations it is clear that wide differences in the properties of extracts from different seaweeds are possible. There are extracts which give stiff gels in dilute solution and others that have no gelling properties at all. The presence of the 3,6-anhydro derivative induces gelling. Where this is replaced by galactose 6-sulfate the gelling power is considerably less and gelation appears to be completely prevented by the presence of D-galactose 2,6-disulfate in place of the 3,6-anhydro sugar, or by the presence of 2-sulfate on the (1 → 3)-linked galactose units. It is in these variations that differences in the galactans from different species have appeared.

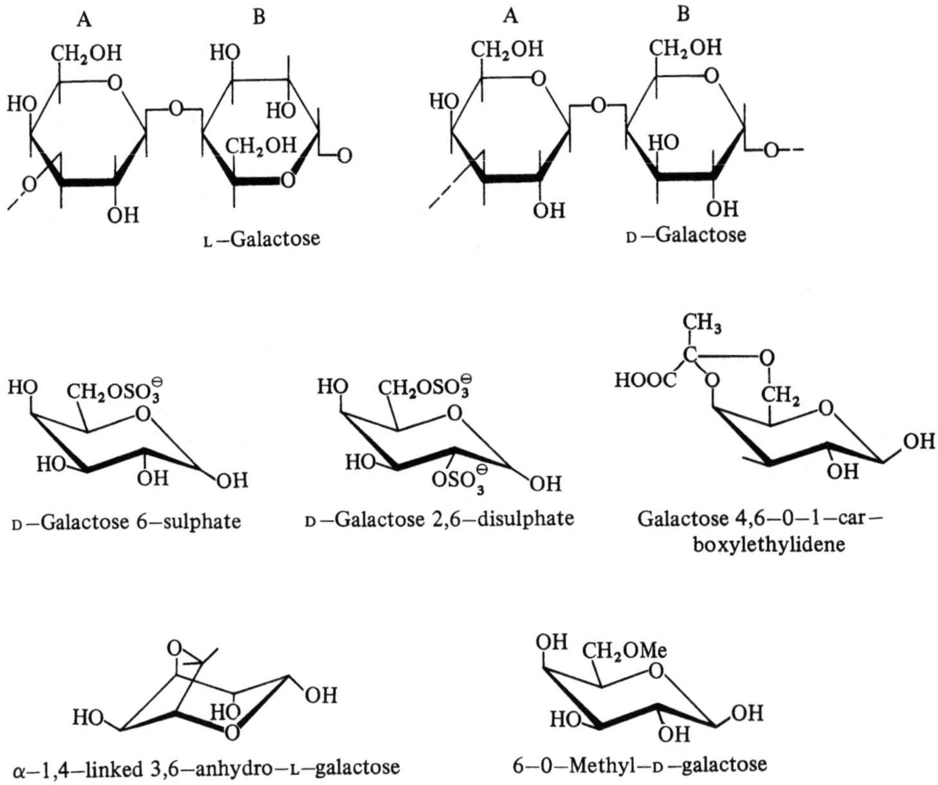

Fig. 2. Galactose residues found in the Rhodophyceae

The galactans can be divided into those from genera which synthesize agar-type molecules – alternating (1 → 3)-D-galactose units and (1 → 4)-L-galactose units – and others that metabolize carrageenan-type molecules in which all the units are D-galactose and galactans which show affinities both to agar and to carrageenan (PERCIVAL 1972). In every case the extracts consist of a family of molecules which differ in the fine details of structure and it is the average of these molecules which the chemist determines.

6.1 Agar and Related Molecules

It has been possible in some cases to separate the extremes of the structure. The separation of agarose (Fig. 3) (DUCKWORTH and YAPHE 1971) a nonsulfated fraction from agar, leaving behind agaropectin, a mixture of variously sulfated molecules (IZUMI 1971 a, b, 1972), is an example of this fractionation. 6-*O*-Methyl-D-galactose may also be present in agarose to the extent of about 1% to 20%, depending upon the algal species (ARAKI 1966) and 4-*O*-methyl-L-galactose has been found in agar from *Gelidium amansii* (ARAKI et al. 1967). Agaropectin contains 3% to 10% sulfate, glucuronic acid and in some species a small proportion of pyruvic acid (HIRASE 1957; YOUNG et al. 1971) linked in acetal linkage (Fig. 2).

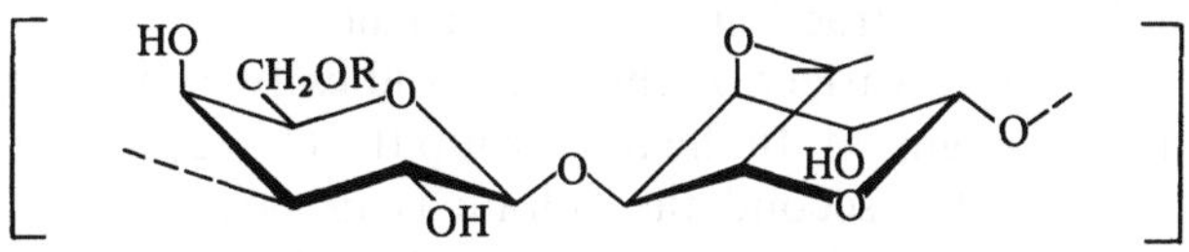

Fig. 3. Agarose

Agarose R=H or OMe

Agar is made commercially from a number of species of which the most important are from the genera *Gelidium* and *Gracilaria*. Some seaweeds which are not at present used commercially yield extracts which are similar to agar in containing the L-isomer of 3,6-anhydrogalactose or galactose 6-sulfate. Examples of these are galactans from the *Porphyra* species, but here the repeating pattern is masked not only by 6-*O*-methyl on the D-galactose units but as much as 50% of the anhydro-L-units are replaced by L-galactose 6-sulfate (cf. PERCIVAL and McDOWELL 1967). *Laurencia* (BOWKER and TURVEY 1968) *Ceramium rubrum* (TURVEY and WILLIAMS 1976) and a number of genera belonging to the Grateloupiaceae (ALLSOBROOK et al. 1969, NUNN and PAROLIS 1969) have C-2 of the D-galactose units substituted by methoxyl. In the extract from *Aeodes orbitosa* (NUNN and PAROLIS 1968) this is the major methyl sugar comprising one in every six of the galactose residues. 4- and 6-*O*-Methyl residues are also present, the former on the L-galactose units. The last-named extract also contains glycerol. In *Anatheca dentata* (NUNN et al. 1971), a member of the Solieriaceae, 3-*O*-methylgalactose has been found as a constituent of the galactan. The only effect recorded of methoxyl groups on the physical properties of the agaroids appears to be an increase of gelling temperature with increase of methoxyl content, as much as 12 °C for an 8% increase of methoxyl (GUISELEY 1970).

Agarase, an enzyme separated from the bacterium *Pseudomonas atlantica*, has been used to characterize agar in marine algae (YAPHE 1957, YOUNG et al. 1972, GROLEAU and YAPHE 1977). ^{13}C-nuclear magnetic resonance analysis (HAMER et al. 1977) established the structure of the enzymic hydrolysis products of agarose, and characteristic ^{13}C-nuclear magnetic resonance spectra have been obtained for agars from different genera and for a variety of carrageenans (BHATTACHARJEE et al. 1979).

6.2 Carrageenan and Related Polysaccharides

The precipitation by *K* ions of κ-carrageenan (SMITH and COOK 1953, STANCIOFF 1965) (Fig. 4) a sulfated gelling fraction from a whole extract of *Chondrus crispus* or *Gigartina* species (CEREZO 1973) are examples of the fractionation of carrageenan. The material left in solution, the nongelling fraction known as λ-carrageenan, is a mixture of variously substituted molecules, which can be further fractionated (Table 2) and which may also contain acetal-linked pyruvic acid (HIRASE and WATANABE 1972). REES (1969) suggests that the term λ-carrageenan should be restricted to one fraction (Table 2) (Fig. 5) of this material. Unfortunately many of the reports in the literature refer to the whole KCl-soluble fraction as λ-carrageenan.

From a variety of chemical studies two major groups of carrageenans have been recognized. In the first group the (1 → 3)-units are sulfated in the 4-position, while in the second the sulfate is in position 2. Both groups are subdivided according to the nature of the (1 → 4)-residues (Table 2). In nature the 3,6-anhydrides of κ- and ι-carrageenan are formed by enzymic elimination of the 6-sulfate from the μ- and ν-forms, but the conversion is not always complete. For some seaweeds these carrageenan types can be isolated in almost pure forms while in others they exist as copolymers.

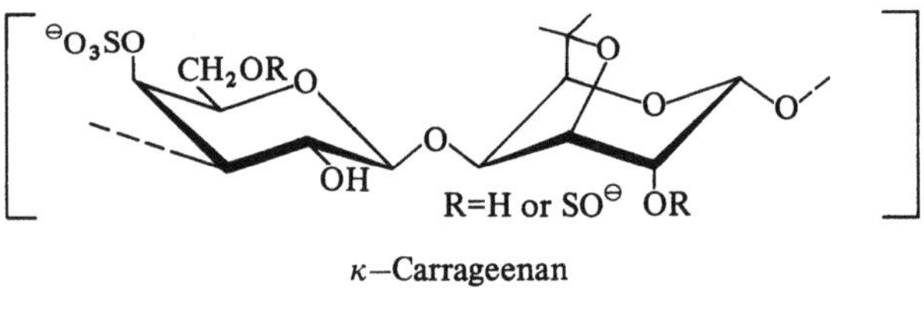

Fig. 4. κ-Carrageenan

Table 2. Repeating units of carrageenans

$—^{3}A^{1\beta}—^{4}B^{1\alpha}—^{3}A^{1\beta}—^{4}B^{1\alpha}—^{3}A^{1\beta}—$

A units	B units	Name
D-Galactose 4-sulfate	D-Galactose 6-sulfate	μ
	D-Galactose 2,6-disulfate	ν
	3,6-Anhydro-D-galactose	κ
	3,6-Anhydro-D-galactose 2-sulfate	ι
D-Galactose 2-sulfate	D-Galactose 2-sulfate	ξ
	D-Galactose 2,6-disulfate	λ
	3,6-Anhydro-D-galactose 2-sulfate	θ

Fig. 5. λ-Carrageenan

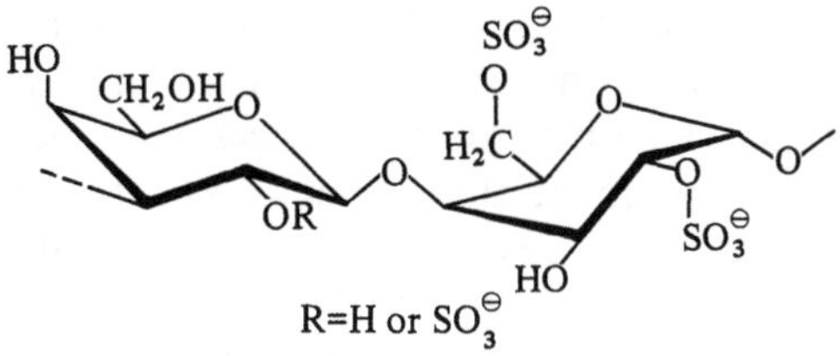

Furcellaran or Danish agar, extracted from *Furcellaria fastigiata*, has many similarities with the κ-fraction of carrageenan (YAPHE 1959, PAINTER 1966). *Eucheuma spinosum* and *Agardheilla tenera* extracts consist of ι-carrageenan (Fig. 6). DAWES (1979) reports that at least two types of ι-carrageenan can be found in different species of *Eucheuma*. *Gigartina atropurpurea* synthesizes ξ-carrageenan (LAWSON et al. 1973, PENMAN and REES 1973). SANTOS and DOTY (1979) have investigated the carrageenan content of four species of red algae from Hawaii, and SHIMAHARA and SUGIYAMA (1974) that of *Hypnea charoides* from the coasts of Japan and Korea. For structural analysis with ^{13}C-nuclear magnetic resonance spectroscopy of carrageenan see BHATTACHARJEE et al. (1979) and YAROTSKII et al. (1978). Both the carrageenans and the furcellarans appear to be devoid of methoxyl residues.

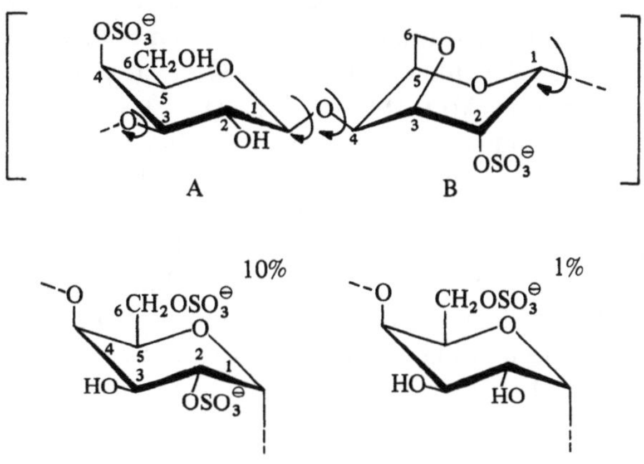

Fig. 6. ι-Carrageenan

Enzymes specific for the κ- and λ-fractions of carrageenan and a glycosulfatase has been obtained by WEIGL and YAPHE (1966) from *Pseudomonas carrageenovora*.

6.3 Conformation (see REES and WELSH 1977)

It was found from the X-ray diffraction pattern (REES 1969, ARNOTT et al. 1974a) that a single right-handed helix could be recognized for chains of κ- and ι-carrageenan having a pitch of 26 Å. The presence of specific hydrogen bonding between the helices has been confirmed by infra red and deuteration-dichroism studies (ANDERSON et al. 1968), making the conformation very stable. In natural ι-carrageenan the regular alternation is replaced by a λ-like segment (10% of the 3,6-anhydro units are replaced by 2,6-disulfate units) (Fig. 6), which results in a change in direction in the chain. Thus a single molecule may have a number of "kinks" in it and each bit of the helix between the kinks can form double helices with helices from different molecules. In this way a complete three-dimensional network can be built up. The solution that exists before gelation is a typical solution of polymer molecules in random coil formation, and gel formation involves the association of chain segments resulting in a three-dimensional framework with junction zones between the molecules and with water held in the interstices of the framework.

X-ray diffraction studies by ARNOTT et al. (1974b) on fibers of agarose showed the presence of threefold helices, but the presence of the anhydro sugar in the L-configuration reverses the screw sense from that of carrageenans, giving left-handed helices. The pitch of the helix is shorter than that for ι- and κ-carrageenan but this is explained by the authors as following the trend of changing sulfate content.

Changes in molecular conformation and interaction of agarose during the sol-gel transition have been studied (LIANG et al. 1979) by vacuum-ultraviolet circular dichroism (VUCD).

6.4 Gametophyte and Sporophyte Carrageenans

The majority of the chemical studies of the galactans have been made on extracts from a mixture of whole plants at different stages of growth. In 1973 extracts of plants of *Chondrus crispus* at particular stages of the life cycle were examined by MCCANDLESS et al. (1973), who were able to show that sporophytic plants contained λ-carrageenan and the gametophytic plants mainly κ-carrageenan with as much as 25% of KCl soluble material. PICKMERE et al. (1973) and PARSONS et al. (1977) studying a number of species of *Gigartina* also found that the sporophyte contained only λ-carrageenan and that the male and female gametophytic plants had ratios of κ:λ ranging from 1.0 to 4.0 and *G. atropurpurea* 0.8. Similar results were obtained from three species of *Iridaea* and two more species of *Gigartina* (WAALAND 1975) and from *Iridaea cordata* (MCCANDLESS et al. 1975).

6.5 Immunochemistry

JOHNSTON and MCCANDLESS (1968) injected λ- and κ-carrageenan from *Chondrus crispus* into rabbits and were able to isolate specific precipitating antibodies which could be used to distinguish different types of carrageenans from a variety of red algae at different stages of their life cycle (HORSFORD and MCCANDLESS 1975, DININNO and MCCANDLESS 1978a, b 1979). Antibodies produced by injection of λ-carrageenan into goats were used (MCCANDLESS et al. 1979) to distinguish two classes of λ-carrageenans.

6.6 Cell Wall Localization

κ- and λ-carrageenans occurring in the cell walls of the appropriate stages of the life cycle were identified by fluorescent antibody technique (GORDON-

MILLS and MCCANDLESS 1975). Mild physical separation of the cell walls of *Porphyra umbilicalis* (GUNAWARDENA and WILLIAMSON 1975) located the agar-like polysaccharide in the intercellular matrix.

Energy-dispersive X-ray analysis (EDX) of carposporic and tetrasporic plants of *C. crispus* revealed the presence of sulfur primarily in the intercellular matrix and cell walls (MCCANDLESS et al. 1977). The microfibrils visible with transmission electron microscopy in cell walls and in the intercellular matrix of both generations of *C. crispus* appeared to run parallel to the cell surface, i.e., circumferentially. After extraction of the sulfated polysaccharide the microfibrils, in a disorganized state, still persisted (MCCANDLESS et al. 1977).

6.7 Biosynthesis

Pieces of freshly harvested *C. crispus* were incubated in the presence of ^{35}S Na_2SO_4 (LOEWUS et al. 1971). Both κ- and λ-carrageenan subsequently isolated contained radioactive half-ester sulfate groups. From autoradiographic and histochemical studies on *Eucheuma nudum* (CLAIRE and DAWES 1976) it was found that ι-carrageenan was localized in the middle lamella of epidermal, cortical, and medullary cells. $^{35}SO_4$ studies indicated that the label was first incorporated in the inner wall and ultimately deposited in the middle lamella of all cells, and in an outer wall layer of the epidermal cells. It is considered that a nonsulfated polygalactan and sulfatransferase enzyme(s) are secreted by the Golgi apparatus and endoplasmic reticulum respectively. Free sulfate in the wall would be bound to the precursor polysaccharide with much of the resulting carrageenan migrating to the middle lamella.

Enzyme preparations isolated from *Porphyra umbilicalis* (REES 1961) and from *Gigartina stellata* (LAWSON and REES 1970) converted L-galactose 6-sulfate and D-galactose 6-sulfate into the corresponding 3,6-anhydrogalactose in the respective polysaccharides. WONG and CRAIGIE (1978) separated similar enzymes, sulfohydrolases, which catalyzed this reaction in both haploid and diploid plants of *C. crispus* and *G. stellata* at the polymer level. CRAIGIE and WONG (1979) have assembled biological and chemical information which enabled them to formulate likely pathways and identify potential control points in the biosynthesis of the different carrageenans. They consider that these must arise by specific sulfotransferase action on the precursor, with transformation by sulfohydrolase to give the less heavily sulfated carrageenans containing anhydrogalactose.

7 Fucans

The sulfated polysaccharides of brown algae have been described under such names as fucoidin, fucoidan, ascophyllan, sargassan, pelvetian, glucuronoxylofucans, and fucans. They all comprise a family of polydisperse, heteromolecules based on fucose, xylose, glucuronic acid, and half ester sulfate. In some species of seaweeds, galactose and/or mannose may also be major constituents (PERCIVAL

1979). A small amount of protein is found with all fucans, provided they have not been subjected to extensive purification, indicating that they may be proteoglycans. Tentative evidence of linkage through serine and threonine (MEDCALF and LARSEN 1977) has been advanced for the fucan from *Ascophyllum nodosum*. LARSEN (1978) suggests that the name fucoidan be restricted to the polymeric material extractable with acid, but this is misleading since it depends on acid strength and temperature (BLACK et al. 1952), and QUATRANO et al. (1979) confuse the names fucoidan and fucan still further. According to polysaccharide nomenclature rules the correct name would be L-fuco-D-galacto-D-glucurono-D-manno-D-xyloglycan, but for simplicity in this article such polymers will be called fucans on the understanding that they contain other sugars in addition to fucose.

In the early studies extensive purification of extracts was carried out in an attempt to isolate a fucan containing only fucose residues. This was never achieved completely and even in the so-called pure sample from *Himanthalia lorea* small proportions of galactose, xylose, and uronic acid persisted (PERCIVAL and ROSS 1950).

The proportion of fucan in a brown seaweed varies, not only seasonally but also with the particular species. The fucan from *Pelvetia canaliculata* comprises from 18 to 24% of the dry weight of the weed, whereas in *Durvillea potatorum* the proportion of fucan is very small (0 to 2%) (MADGWICK and RALPH 1969). The isolation of fucans is not restricted to water or acid extracts of the algae. Subsequent extractions with chelating agents, like EDTA, alkali, or alkali at elevated temperatures will remove additional amounts of fucans. Sequential extraction in aqueous solution of three genera of weeds acid, alkali, ammonium oxalate and chlorite gave in every case fucans (JABBAR MIAN and PERCIVAL, 1973a) which contained the same sugars with perhaps a small increase in the proportion of uronic acid in the alkaline and chlorite extracts.

Sulfated polysaccharides have been isolated from the following seaweeds: *Ascophyllum nodosum* (LARSEN et al. 1966, 1970, MEDCALF and LARSEN 1977), *Fucus vesiculosus* (BERNARDI and SPRINGER 1962, MEDCALF and LARSEN 1977), *Fucus spiralis* and *Pelvetia canaliculata* (QUILLET 1959), *Himanthalia lorea, Bifurcaria bifurcata, Padina pavonia* (JABBAR MIAN and PERCIVAL 1973a): *Sargassum pallidum* (OVODOV et al. 1973), *Sargassum linifolium* (ABDEL-FATTAH et al. 1973), *Pelvetia wrightii* (ANNO et al. 1966, PAVLENKO and OVODOV 1974), *Lessonia flavicans* (VILLARROEL and ZANLUNGO 1975), *Dictyota dichotoma* (ABDEL-FATTAH et al. 1978, HUSSEIN et al. 1979). The yield, and fucose, uronic acid, and sulfate content of hot water extracts of 21 different species (including 3 species of *Laminaria*) of brown seaweeds is detailed by FUJIKAWA and NAKASHIMA (1975) and a more extensive study of *Nemacystus decipiens* has been carried out by FUJIKAWA and WADA (1975). The proportions of the sugars present in the hydrolysates of fucans from 11 different species of brown seaweeds is given by QUILLET (1961). The methods of isolation, purification and fractionation vary considerably and in some instances only a small fraction of the native glycan has been examined (MEDCALF et al. 1978).

Free-boundary electrophoresis experiments (O'NEILL 1954, LARSEN and HAUG 1963) gave evidence of the heterogeneity of fucans. Fractionation on a column of DE-52-cellulose of extracts from *Himanthalia lorea, Bifurcaria bifurcata* and *Padina pavonia* (JABBAR MIAN and PERCIVAL 1973a), *Desmarestia aculeata* (PERCIVAL and YOUNG 1974), *Desmarestia ligulata* and *D. firma* (CARLBERG et al. 1978) by elution with increasing concentration of KCl resulted in every case in the separation with dilute eluant of a fraction with a high proportion of glucuronic acid and a relatively low fucose and sulfate content. As the concentration of the eluting solution rose so the proportion of uronic acid decreased and the fucose and sulfate increased. Similar heterogeneity of acid-extracted fucans from *Ascophyllum nodosum* and *Fucus vesiculosus* was demonstrated by fractional precipitation with ethanol containing $MgCl_2$ or $CaCl_2$ (MEDCALF and LARSEN 1977, MEDCALF et al. 1978).

7.1 Structure

In spite of all these variations, where structural studies have been carried out, the linkages between the individual sugars appear to be the same. Methylation, periodate oxidation, and partial hydrolysis studies have revealed the essential structural similarity of the fucans separated and fractionated from *Himanthalia lorea* and *Bifurcaria bifurcata* (JABBAR MIAN and PERCIVAL 1973b). The high fucose-, high sulfate-containing material more closely resembles the structure of the fucan from *Fucus vesiculosus* reported by CONCHIE and PERCIVAL (1950), O'NEILL (1954), COTE (1959), and by PERCIVAL (1971) for the fucan from *Ascophyllum nodosum;* namely (1→2)- and (1→3)-fucose residues with sulfate at C-4. The glucuronic acid and xylose residues are not sulfated and appear to be on the periphery of highly branched molecules. Partial hydrolysis revealed D-glucuronosyl-(1→3)-L-fucose as a structural feature of the macromolecule (PERCIVAL 1968, JABBAR MIAN and PERCIVAL 1973b). The galactose present in the fucan from *Desmarestia aculeata* appears to be present as end group and (1→3)-units (PERCIVAL and YOUNG 1974). The fucans are clearly a family of complex highly branched polysaccharides and various macromolecular structures have been suggested LARSEN et al. (1966), MEDCALF and LARSEN (1977), MEDCALF et al. (1978), ABDEL-FATTAH et al. (1974), but other interpretations of their results are possible and further study is necessary before these structures can be accepted.

7.2 Site of Sulfate

Confirmation that C-4 did indeed carry half ester sulfate was confirmed by the isolation of fucose 4-sulfate from a partial acid hydrolyzate of the fucan from *Pelvetia wrightii* by ANNO et al. (1969). Partial hydrolysis of the acid-extracted fucan from *Sargassum linifolium* by ABDEL-FATTAH et al. (1974) gave a number of sulfated fucose and galactose oligosaccharides and residual material consisting of glucuronic acid and mannose. PAVLENKO et al. (1976) report the presence of sulfated mannose, galactose, and xylose in the fucans from *Sargassum pallidum* and *Pelvetia wrightii.* However, no other workers have reported any sulfated residues other than fucose and galactose and attempts to apply the Russian methods to other fucans (E. PERCIVAL and M.A. RAHMAN, unpublished work) revealed only sulfated fucose; the interpretation of their results by PAVLENKO et al. must therefore be viewed with caution.

7.3 Biosynthesis

Little is known about the synthesis of fucans. BIDWELL and GHOSH (1963) postulate that sulfate is rapidly exchanged without complete breakdown or resynthesis of the fucan. This is supported by LESTANG and QUILLET (1973, 1974), who found that living fronds of *Pelvetia canaliculata* immersed in seawater from which sulfate ions had been removed secreted sulfate into the water and that fronds which had been partially depleted of sulfate absorbed sulfate from the seawater.

QUATRANO and CRAYTON (1973) and QUATRANO and STEVENS (1976) have studied the zygotes and young embryos of *Fucus distichus* (=*gardneri*), *F. vesiculosus* and *F. inflatus* (for a full discussion see QUATRANO et al. 1979). In the early stages after fertilization a fucan low in fucose and sulfate (F_1) is present at the same time as the acquisition of structural integrity and birefringence of the cell wall. Four hours after fertilization the cell wall consists of 60% alginate 20% cellulose and 20% fucan which proportions remain unchanged during the next 20 h although more polysaccharides are synthesized. At 16 to 24 h a further more highly sulfated component (F_2) can be separated from (F_1) by electrophoresis of the cell wall extract and the composition of the whole fucan is then fucose:xylose:glucose:mannose:galactose=8:4:2:2:1 together with some glucuronic acid. In normal conditions when (F_2) becomes sulfated during development it concentrates in the region of the rhizoid, but when grown in a sulfate-free medium the fucan is uniformly distributed over the zygote cell wall. In the latter conditions the rhizoids develop but fail to adhere to the substratum (CRAYTON et al. 1974, QUATRANO et al. 1979a).

EVANS et al. (1979), working with zygotes of *Fucus serratus*, found from experiments with ^{35}S-sulfate and autoradiography that sulfation occurs in the Golgi bodies after 3 min and can be detected in the cell wall fraction after 10 min. QUATRANO et al. (1979b) agree that the Golgi bodies are the site of sulfation.

7.4 Location and Function (see PERCIVAL 1979)

Fucans are present in the intercellular tissues and in the mucilage that exudes from the surface of fronds of *Laminaria digitata, Ascophyllum nodosum* and *Macrocystis pyrifera* (DONER and WHISTLER 1973). This mucilage contains in addition small amounts of algin and proteins. Methods necessary for the removal of these impurities or for the extraction of fucans from the seaweeds destroy the colloidal properties.

Light microscope histochemistry (EVANS and CALLOW 1974) and electronmicroscope X-ray (CALLOW and EVANS 1976) show that fucan is located in secretory canals, the middle lamella of cell walls, and on the thallus surface. Sulfation occurs in the Golgi bodies of specialized secretory cells.

The studies of McCULLY (1966) and VREELAND (1971) on Fucus spp. indicated that in general the immediate cell walls are composed of alginate with some fucan in inner layers while the matrix regions are composed mainly of fucans.

It has long been thought that by their hygroscopic mucilaginous nature fucans aid the plant to resist desiccation when exposed at low tide. The French workers LESTANG and QUILLET (1973) have shown that there is a much higher proportion of magnesium than sodium associated with the fucan of *Pelvetia canaliculata* although the seawater in contact with it has a great excess of sodium over magnesium. They point out that the magnesium ion is highly hydrated and suggest that this affinity for magnesium protects the plant against desiccation and high salt concentrations. This is supported by the very low fucan content of weeds which are permanently submerged.

8 Sulfated Polysaccharides of the Chlorophyceae

The major polysaccharides of many of the green seaweeds are complex water-soluble heteropolysaccharides containing a wide variety of sugars and in some genera glucuronic acid (PERCIVAL and McDOWELL 1967, PERCIVAL 1978a). In all of them half ester sulfate groups are linked to hydroxyl groups of one or more of the sugars. Chemical investigation has been on extracts of whole weeds at different stages of growth and the results consequently are the average of all the molecules present in the plant. It has proved possible to separate food reserve materials, starch-type polysaccharides from the extracts of the majority of genera and inulin-type polysaccharides from extracts of the Dasycladales. Fractionation techniques applied to the sulfated polysaccharides have resulted in separation of similar type molecules indicating the presence of a family of polydisperse molecules. Complete removal of protein has proved impossible (FISHER and PERCIVAL 1957), indicating the presence of a proteoglycan. Those from *Cladophora, Chaetomorpha* and *Rhizoclonium* species (Cladophorales) have a positive specific rotation and contain varying proportions of arabinose, galactose, and xylose (PERCIVAL and YOUNG 1971), and in some species, rhamnose and glucose are also constituents.

Studies on the polysaccharide from *Cladophora rupestris* (FISHER and PERCIVAL 1957, JOHNSON and PERCIVAL 1969) established the presence of arabinose, galactose, xylose, rhamnose, and glucose in the approximate molar proportions of 3.7:2.8:1.0:0.4:0.2 (+8% protein) and 19.6% half ester sulfate. Application of degradative studies provided evidence of a highly branched structure, with xylose and galactose units at the ends of the branches and galactose, arabinose, and rhamnose residues occurring in the inner part of the molecules. Evidence of (1→6)- and/or 6-sulfated galactofuranose units was obtained. Partial hydrolysis experiments (HIRST et al. 1965, BOURNE et al. 1970) led to the separation and characterization of the following fragments: L-arabinose 3-sulfate, D-galactose 6-sulfate, (1→3)-linked and (1→6)-D-galactobioses, (1→4)- or (1→5)-Larabinobiose 3-sulfate, (1 → 4)-D-xylobiose, a mixture of trisaccharides containing sulfated galactose and arabinose, and a mixture of pentasaccharides in which the molar ratio of arabinose to galactose was 4:1.

Parallel studies on *Chaetomorpha linum* and *C. capillaris* water-soluble polysaccharides (HIRST et al. 1965) showed their essential similarity to the corresponding polysaccharide of *Cladophora rupestris*.

Samples of *Codium fragile* (Codiales) harvested from South Africa and from near Biarritz (France) at different seasons gave aqueous extracts which resembled those from the Cladophorales in their main monosaccharide content, except that mannose was also a constituent. However by fractionation on a cellulose column LOVE and PERCIVAL (1964a) separated from this extract 27% of material containing only galactose, arabinose, and 29% half ester sulfate, and 7% of material devoid of these two sugars and containing glucose, mannose, xylose, and rhamnose. After partial hydrolysis in addition to the monosaccharides, 3-*O*-β-L-arabinopyranosyl-L-arabinose, 3-*O*-β-D-galactopyranosyl-D-galactose, together with galactose 4- and 6-sulfates, were separated and characterized.

A somewhat similar type of polysaccharide, also with a positive specific rotation, is synthesized by *Caulerpa filiformis*, *C. racemosa* and *C. sertularoides*. Mannose is again a constituent, but the proportion of arabinose varies with the season of harvesting and was absent from some samples (MACKIE and PERCIVAL 1961). Fractional precipitation of the free acid polysaccharide with saturated barium hydroxide gave material with increasing proportion of galactose and arabinose and decreasing xylose and mannose. From the fact that the major structural polysaccharide synthesized by this seaweed is a (1 → 3)-β-xylan (Sect. 3) it is possible that the xylose-containing material in the aqueous extracts is part of the xylan subsequently extracted from the residual weed with alkali.

In contrast to these water-soluble polysaccharides, those from *Ulva lactuca*, *Enteromorpha compressa* (Ulvales), *Acrosiphonia centralis*, *Urospora penicilliformis*, *Urospora wormskioldii*, and *Codiolum pusillum* (Acrosiphoniales) all have negative specific rotations, and contain rhamnose, xylose, glucuronic acid, and half ester sulfate. That from *A. centralis* (*Spongomorpha arcta*) (12% of the dry weight of the weed) comprised rhamnose and xylose in the molar proportions of 1.4:1.6 together with traces of galactose and mannose, 19% glucuronic acid and 7.8% of half ester sulfate (O'DONNELL and PERCIVAL 1959). The aldobiouronic acid 4-*O*-β-D-glucopyruronosyl-L-rhamnose was shown to be a major structural feature of this polysaccharide. End-group xylose, (1 → 4)-linked xylose and rhamnose, and a relatively large proportion of triply-linked rhamnose are also features of the polymer.

The acid polysaccharide from *Ulva lactuca* contained a similar polysaccharide (BRADING et al. 1954, MCKINNELL and PERCIVAL 1962a). After removal of a starch-type polysaccharide from the aqueous extract, the residual polysaccharide contained L-rhamnose, D-xylose, and D-glucose in the molar proportion of about 4.2:1.3:1.0 with a trace of mannose and 24% of D-glucuronic acid and 19% of ester sulfate and about 5% of protein. ABDEL-FATTAH and EDREES (1972) report the presence of arabinose as well as the other constituents in the water-soluble polysaccharide from *U. lactuca* collected at Alexandria, Egypt.

Methylation, desulfation, partial hydrolysis, and degradative studies established the presence of the following structural units in this polysaccharide (HAQ and PERCIVAL 1966a, b):

$$\text{GA-(1}\rightarrow\text{4)-}\underset{2S}{\text{R}}\text{, (major);} \quad \text{GA-(1}\rightarrow\text{3)-Xy;} \quad \text{GA-(1}\rightarrow\text{4)-Xy;}$$

$$\text{GA-(1}\rightarrow\text{4)-R-(1}\rightarrow\text{3)-Xy;} \quad \text{G-(1}\rightarrow\text{3)-Xy;} \quad \text{R-(1}\rightarrow\text{4)-Xy-(1}\rightarrow\text{3)-G/GA.}$$

It was tentatively suggested that the following heptasaccharide constituted a repeating unit in the macromolecule:

$$\text{GA-(1}\rightarrow\text{4)-}\underset{2S}{\text{R}}\text{-(1}\rightarrow\text{4)-GA-(1}\rightarrow\text{3)-Xy-(1}\rightarrow\text{4)-}\underset{2S}{\text{R}}\text{-(1}\rightarrow\text{3)-}\underset{6S}{\text{G}}\text{-(1}\rightarrow\text{4)-Xy}$$

Key: GA = D-glucopyranuronic acid; R = L-rhamnopyranose; Xy = D-xylopyranose; G = D-glucopyranose: S = 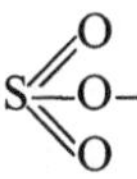

MCKINNELL and PRCIVAL (1962b) found that *Enteromorpha compressa* synthesized a similar water-soluble polysaccharide in which 4-*O*-D-glucopyranuronosyl-L-rhamnopyranose was again a major structural feature.

From *Urospora penicilliformis* aqueous extraction gave a mixture of a mannan (Sect. 4) and a glucuronosylxylorhamnan $[\alpha]_D$-80° (BOURNE et al. 1974). Subsequent extraction of the weed residues with acid (pH 2, 70 °C) and with sodium chlorite gave additional quantities of these two polysaccharides. Final extraction with 13% sodium hydroxide gave material which precipitated at pH5 and consisted mainly of glucose with traces of rhamnose and xylose, and the final residue contained only glucose (Sect. 2). Methylation of the rhamnan established the linkages between the units and from autohydrolyzates of the free acid polysaccharide, α-L-rhamnose(1 → 4)-xylose and β-D-glucopyranuronosyl(1 → 3)-α-L-rhamnopyranosyl(1 → 4)-D-xylopyranose (with rhamnose sulfated at C-2 and C-4 in both) were separated and characterized.

Glucuronoxylorhamnans are also synthesized by *Urospora wormskioldii* and *Codium pusillum* (different stages in the life history of a single alga) (CARLBERG and PERCIVAL 1977). These polysaccharides have many features in common with those already described. They consist of a family of polydisperse heteropolysaccharides built up on the same general plan but with varying proportions of the individual sugars, extent of branching and sulfation. From the various results a possible formula and scheme for the glucuronoxylorhamnan was advanced.

The water-soluble polysaccharide from *Acetabularia crenulata* ($[\alpha]_D$)-7 to -22° contains variable proportions of galactose, xylose, rhamnose, glucuronic acid, and 4-*O*-methylgalactose with half ester sulfate groups, depending upon whether the extraction was with cold or hot water of the seaweed stalks or caps (PERCIVAL and SMESTAD 1972). Structural studies established (1 → 3)-linked D-galactose 4-sulfate (major) and 6-sulfate, and (1 → 2)-linked L-rhamnose as the main structural features of the macromolecules. Glucuronic acid, galactose, and rhamnose are all present as end groups, indicating highly branched molecules. Glucuronic acid is linked to both rhamnose and galactose, and galactose residues are mutually linked in the polysaccharide.

Although these polysaccharides are water-soluble, in many algae it is impossible to remove them completely from the plant even by hot water and acid extraction. Only after chlorite treatment and alkaline extraction is the residual material free from these polysaccharides. They appear to comprise the matrix material of the cell wall and are present in the plant as a gel.

HAUG (1976) found that the soft gel of 1% of the glucuronoxylorhamnan from *U. lactuca* was insoluble in seawater. He showed also that only at the pH 8 of seawater containing the amount of borate and calcium normally found therein was the polysaccharide retained in the live fronds when immersed. He postulated that the cis hydroxyl groups on C-2 and C-3 of the rhamnose units complexed with the borate ion and that the Ca ions stabilized the complex. However, a number of the rhamnose hydroxyl groups on C-2 are replaced by sulfate groups (PERCIVAL and WOLD 1963) and would be unable to complex with borate. It seems to the author that by the extent of this sulfation the alga regulates the amount of complexing with borate and hence the strength of its gel. *Cladophora rupestris* polysaccharide only needs the presence of Ca ions for the water-soluble polysaccharide to form a stiff gel in a 1% solution.

Biosynthesis

BIDWELL (1958) and CRAIGIE et al. (1966) growing *Ulva lactuca, Cladophora* sp., *Chaetamorpha melagonium, Monostroma fuscum,* and *Enteromorpha intestinalis* and members of the Volvocales with $^{14}CO_2$ as the carbon source for 2 h found that sucrose was the most highly labeled low molecular weight carbohydrate and that a considerable amount of radioactivity was also incorporated into the constituents insoluble in 80% ethanol. Further studies (PERCIVAL and SMESTAD 1972) on *U. lactuca* cultured in $^{14}CO_2$ showed that sucrose was again the most highly labeled sugar, although glucose, fructose, xylose and *myo*-inositol also incorporated radioactivity. The radioactivity of the sulfated polysaccharide increased 24 times in the 3 h culture compared with a 10 min culture. The activity in the individual sugars in this glucuronoxylorhamnan was also measured and from the results it was tentatively concluded that a rhamnan is built up first and that the glucuronic acid and xylose units are added later. The residual algal material after aqueous extraction contained 37% of the total radioactivity after 10 min and 49% after 3 h. Glucose was the major constituent of this residue and had the highest radioactivity both after 10 min and 3 h.

Oligosaccharides containing L-rhamnose separated from the ethanolic extracts of *Codiolum pusillum,* and neutral polysaccharides separated from the aqueous extracts of *Urospora wormskioldii* and *C. pusillum* are thought (CARLBERG and PERCIVAL 1977) to be the precursors of the glucuronoxylorhamnans synthesized by these algae.

9 Polysaccharides of the Bacillariophyceae

For *Phaeodactylum tricornutum* see Section 4. Analyses by PARSONS et al. (1961) on a series of residues of diatom species and on *Skeletonema costatum* by HANDA (1969) indicate that mannose is the dominant monomer of this type of polymer.

ALLAN et al. (1972) isolated polysaccharides from diatom residues of *Nitzschia frustulum, N. angularis, Asterionella socialis* and *Cylindrotheca fusiformis* after extraction with hot water and found that they comprised different proportions of rhamnose, fucose, galactose, mannose and xylose. Glucuronic acid was reported to be a constituent. HAUG and MYKLESTAD (1976) examined the alkali-soluble polysaccharides of the cell wall from seven different marine diatoms, mainly Chaetoceras sp., and showed that they contained different proportions of the same neutral sugars.

10 Extracellular Polysaccharides

10.1 From Phaeophyceae

Many brown seaweeds exude mucilaginous material thought to be a fucan (Sect. 7) (PERCIVAL and McDOWELL 1967); that from *Macrocystis pyrifera* was examined in detail (SCHWEIGER 1962).

10.2 From Rhodophyceae

The microcscopic unicellular red alga *Rhodella maculata* produces a soluble extracellular mucilage which freeze-dries to a white fibrous solid, soluble in water only in dilute solution: 0.3% gives a viscous opaque solution (SHEIK-FAREED 1975). Chemical studies (EVANS et al. 1974) showed that the mucilage comprises D-xylose, D-glucuronic acid (both major), 3-*O*-methylxylose (SHEIK-FAREED 1975), rhamnose, galactose, glucose, and about 16% half ester sulfate. Protein to the extent of 16 to 20% is also present (15 amino acids were identified). Gel electrophoresis experiments indicated that this is not linked to the bulk of the polysaccharide. Methylation, periodate oxidation and partial hydrolysis (EVANS et al. 1974) established the linkages present and the characterization of several oligouronic acids confirmed the heterogeneity of the polysaccharide and the mutual linkage of glucuronic acid, xylose and galactose.

Electron microscope autoradiography using ^{35}S-sulfate (EVANS et al. 1974) showed that the mucilage is packaged in the Golgi bodies, passing to the plasmalemma in large vesicles. Sulfation of the mucilage occurs in the Golgi cisternae.

Porphyridium cruentum, another unicellular red alga, produces an extracellular mucilage which contains D- and L-galactose, D-glucose, xylose, glucuronic acid, and a wide variety of amino acids (JONES 1962, MEDCALF et al. 1975). The mucilage contains 6% to 7% protein and 10% of ester bound sulfate. Viscous solutions of the polymer are obtained at concentrations of 1% or less, which show good pH and temperature stability. Further work (HEANEY-KIERAS and CHAPMAN 1976, HEANEY-KIERAS et al. 1977, KIERAS et al. 1976) confirmed some of the earlier results and showed that protein is linked to carbohydrate via serine-xylose and threonine-xylose linkages. The polysaccharide was reported to comprise D- and L-galactose, D-glucose, xylose, D-glucuronic acid and its 2-*O*-methyl derivative, and sulfate in the molar proportions of 2.12:1.0:2.42:1.22:2.61 respectively, and the aldobiouronic acids 3-*O*-(α-D-glucopyranosyluronic acid)-L-galactose, 3-*O*-(2-*O*-methyl-α-D-glucopyranosyluronic acid)-D-galactose and 3-*O*-(2-*O*-methyl-α-D-glucopyranosyluronic acid)-D-glucose were isolated from partial hydrolysates.

PERCIVAL and FOYLE (1979) confirmed much of the above work on this mucilage but also found 3-*O*-methylxylose, 3- and 4-*O*-methylgalactoses to be constituents. In their mucilage the 2-*O*-methylglucuronic acid is linked to C-4 of L-galactose. Molecular weight determinations on Sepharose 4B indicate 4×10^6 for the mucilage.

The extracellular mucilage from the unicellular freshwater red alga *Porphyridium aerugineum* has many similarities with that of *P. cruentum*. RAMUS (1972) showed that it contained xylose, galactose, glucose, and 7.6% half ester sulfate, together with several minor unidentified constituents. PERCIVAL and FOYLE (1979) continued the investigations on this mucilage and found both D- and L-galactose, D-glucose, D-glucuronic acid, 2,4-di-*O*-methyl-, 3- and 4-*O*-methylgalactoses and 3-*O*-methylxylose together with approximately 10% half ester sulfate. The aldobiouronic acid 3-*O*-(β-D-glucopyranosyluronic acid)-D-galactose and a number of tentatively identified neutral oligosaccharides were also separated from hydrolysates. Methylation and periodate oxidation studies established that the linkages between the sugars are similar to those found for *P. cruentum*. Elution from a column of Sepharose 4B gave a single peak corresponding to molecular weight of 5×10^6 in agreement with the earlier results of RAMUS (1972).

Electronmicroscopy of *Porphyridium aerugineum* (RAMUS 1972) revealed that Golgi vesicles transport the mucilage to and through the cell membrane. ^{14}C-bicarbonate experiments resulted in all the constituents in the mucilage being labeled. Rates of excretion of the mucilage were found to follow a cyclic pattern correlated with the division cycle of the cell. Evidence of rapid incorporation of sulfate in the solubilized capsular polysaccharide was demonstrated with ^{35}S sulfate by RAMUS and GROVES (1972).

10.3 From the Xanthophyceae

From the filamentous yellow-green alga *Tribonema aquale* (Tribonemataceae) an extracellular mucilage can be isolated containing 79% carbohydrate comprising glucose plus galactose, rhamnose and xylose in the molar proportions of 1.0:0.5:0.7. Uronic acid and a trace of fucose were also detected (CLEARE and PERCIVAL 1972).

10.4 From Bacillariophyceae

The production of soluble extracellular polysaccharides from several diatom species has been reported by ALLAN et al. (1972), although only *Nitzschia frustulum* produced any quantity. It comprised rhamnose (24%), mannose (34%), galactose (8%) and two unidentified components (14% and 20%) together with an unspecified amount of glucuronic acid.

Chaetoceros affinis, C. curvisetus, and *C. desipiens* produce extracellular polysaccharides comprising rhamnose, fucose, galactose, and half ester sulfate (SMESTAD et al. 1975). Structural studies revealed similarities and differences between the three species.

Investigation of the extracellular mucilage of the diatom, *Coscinodiscus nobilis* by PERCIVAL et al. (1980) established some of the structural features of a complex heteropolysaccharide comprising fucose, rhamnose, mannose, xylose, galactose (trace), glucuronic acid, and half ester sulfate.

11 Conclusions

11.1 Functions

As the polysaccharides of comparatively few species of algae have been examined in detail there is not yet sufficient information to reach sound conclusions which are generally applicable. It seems likely, however, that structural variations in the different groups of substances discussed in this chapter enable them to serve a number of functions in the living plant.

As polysaccharides constitute a high proportion of the cell wall constituents they clearly have a structural function, whether as fibrillar or matrix components. It appears that some algae have an enzymatic regulatory system which will modify polysaccharides to give them the physical properties most suitable for particular conditions. Examples are the conversion of mannuronic to guluronic acid units in alginic acid (MADGWICK et al. 1973), and the conversion of galactose 6-sulfate to 3,6-anhydrogalactose in agar and carrageen (REES 1961, LAWSON and REES 1970, WONG and CRAIGIE 1978) in each giving the possibility of stiffer gels. The gel formation by borate complexing of rhamnose in the glucuronoxylorhamnan from *Ulva lactuca* may be controlled by the degree of sulfation (Sect. 8).

These instances come from widely different genera and no doubt a search would reveal many more examples. It is known that the alginates from different parts of some algae show variation in composition (Table 1). As most investigations have been carried out on extracts from whole plants, any such differences in polysaccharide composition would generally have been undetected.

A somewhat similar modification of cell wall properties is found in some species of red and green algae, where different polysaccharides are found in different phases of the plants' life cycle (Sects. 4 and 6).

The major polysaccharides are acidic due to the presence of uronic acid units or half ester sulfate groups and are therefore associated with the cations present in seawater. They show ion selectivity, extensively studied in the case of alginates (Sect. 5). This complements the enzymic control of the physical properties, but probably also plays some part in ion transport in the plant. Ion selectivity is also a factor in the protection of intertidal brown algae against desiccation, the highly hydrated magnesium ions being concentrated by fucans (LESTANG and QUILLET 1973).

The role of fucans in the development of *Fucus* zygotes has been studied by QUATRANO et al. (1979) and no doubt acidic polysaccharides play some part in the reproductive cycle of other algae.

A polysaccharide has also been shown to be involved in wound healing in at least one algal species (DREHER et al. 1978).

11.2 Taxonomy

Here the lack of wide-ranging studies is an even greater drawback, but some limited conclusions can be drawn. There is a clear distinction between the polysaccharides of the three important classes of algae, Chlorophyceae, Rhodophyceae, and Phaeophyceae, on which most of the studies have been made, but only a few correlations can be found between polysaccharides and genera and species.

The taxonomic implications of the polysaccharides of red and brown algae have recently been discussed (PERCIVAL 1978b), and there appear to be no clear-cut distinctions which are associated with their presence in different genera. Although all the green seaweeds contain sulfated heteropolysaccharides, a distinction can be made between those from the Cladophorales and those from the Ulvales (Sect. 8). The close relationship of the Acrosiphoniales to the Ulvales is confirmed by the similarity of their polysaccharides.

It is possible that detailed studies on the polysaccharides of more species will help to resolve some of the uncertainties in the classification of the algae. However, a deeper understanding of the mechanism of the variation of polysaccharides between different generations in some algae would be necessary.

References

Abdel-Fattah AF, Edrees M (1972) A study of the polysaccharide content of *Ulva lactuca* L. Qual Plant Mater Veg 22:15–22

Abdel-Fattah AF, Hussein MM-D, Salem HM (1973) Sargassan, a sulfated heteropolysaccharide from *Sargassum linifolium*. Phytochemistry 12:1995–1998

Abdel-Fattah AF, Hussein MM-D, Salem HM (1974) Some structural features of sargassan, a sulfated heteropolysaccharide from *Sargassum linifolium*. Carbohydr Res 33:19–24

Abdel-Fattah AF, Hussein MM-D, Fouad ST (1978) Carbohydrates of the brown seaweed *Dictyota dichotoma*. Phytochemistry 17:741–743

Allan GG, Lewin J, Johnson PG (1972) Marine polymers IV. Diatom polysaccharides. Bot Mar 15:102–108

Allsobrook AJR, Nunn JR, Parolis H (1969) Some Grateloupiaceae polysaccharides. In:

Margalef R (ed) Proc 6th Int Seaweed Symp. Direccion General de Pesca Maritima, Madrid, pp 417–420

Anderson NS, Dolan TCS, Lawson CJ, Penman A, Rees DA (1968) The masked repeating structures of lambda- and mu-carrageenans. Carbohydr Res 7:468–473

Anno K, Terahata H, Hayashi Y, Seno N (1966) Isolation and purification of fucoidin from brown seaweed *Pelvetia wrightii*. Agric Biol Chem 30:495–499

Anno K, Seno N, Ota M (1969) Structural studies on fucoidan from *Pelvetia wrightii*. In: Margalef R (ed) Proc Int 6th Seaweed Symp. Direccion General de Pesca Maritima, Madrid, pp 421–426

Araki C (1966) Some recent studies on the polysaccharides of agarophytes. In: Young EG, McLachlan JL (eds) Proc 5th Int Seaweed Symp. Pergamon Press, Oxford New York, pp 3–17

Araki C, Arai K, Hirase S (1967) Studies on the chemical constitution of agar-agar XXIII. Isolation of D-xylose, 6-*O*-methyl-D-galactose, 4-*O*-methyl-L-galactose and O-methylpentose. Bull Chem Soc Jpn 40:959–962

Arnott S, Scott WE, Rees DA, McNab CGA (1974a) iota-Carrageenan: Molecular structure and packing of polysaccharide double helices in oriented fibres of divalent cation salts. J Mol Biol 90:253–267

Arnott S, Fulmer A, Scott WE, Moorhouse R, Rees DA (1974b) The agarose double helix and its function in agarose gel structure. J Mol Biol 90:269–284

Atkins EDT, Parker KD (1969) The helical structure of a β-D-1,3-xylan. J Polym Sci Part C 28:69–81

Atkins EDT, Parker KD, Preston RD (1969) The helical structure of the β-1,3-linked xylan in some siphoneous green algae. Proc R Soc London Ser B 173:209–221

Atkins EDT, Mackie W, Smolko EE (1970) Crystalline structures of alginic acid. Nature (London) 225:626–628

Atkins EDT, Mackie W, Parker KD, Smolko EE (1971) Crystalline structures of poly-D-mannuronic and poly-L-guluronic acids. Polym Lett 9:311–316

Bernardi G, Springer GF (1962) Properties of highly purified fucan. J Biol Chem 237:75–80

Bhattacharjee SS, Yaphe W, Hamer GK (1979) Study of agar and carrageenan by ^{13}C-nuclear magnetic resonance spectroscopy. In: Jensen A, Stein JR (eds) Proc 9th Int Seaweed Symp. Princeton Science Press, Princeton, pp 379–385

Bidwell RGS (1958) Photosynthesis and metabolism of marine algae. II. A survey of rates and products of photosynthesis in $C^{14}O_2$. Can J Bot 36:337–349

Bidwell RGS, Ghosh NR (1963) Photosynthesis and metabolism in marine algae. VI. Uptake and incorporation of ^{35}S in *Fucus vesiculosus*. Can J Bot 41:209–220

Bidwell RGS, Percival Elizabeth, Smestad B (1972) Photosynthesis and metabolism of marine algae. VIII. Incorporation of ^{14}C into the polysaccharides metabolised by *Fucus vesiculosus* during pulse labeling experiments. Can J Bot 50:191–197

Bjorndahl H, Eriksson K-E, Garegg PJ, Lindberg B, Swan B (1965) Studies on the xylan from the red seaweed *Rhodymenia palmata*. Acta Chem Scand 19:2309–2315

Black WAP (1950) The seasonal variation in the cellulose content of the common Scottish Laminariaceae and Fucaceae. J Mar Biol Assoc UK 29:379–387

Black WAP, Dewar ET, Woodward FN (1952) Manufacture of algal chemicals. IV. Laboratory scale isolation of fucoidin from brown marine algae. J Sci Food Agric 3:122–129

Bourne EJ, Johnson PG, Percival Elizabeth (1970) The water-soluble polysaccharides of *Cladophora rupestris*. Part IV. Autohydrolysis, methylation of the partly desulphated material and correlation with the results of Smith degradation. J Chem Soc C:1561–1569

Bourne EJ, Percival Elizabeth, Smestad B (1972) Carbohydrates of *Acetabularia* species, Part 1. *A. crenulata*. Carbohydr Res 22:75–82

Bourne EJ, Megarry ML, Percival Elizabeth (1974) Carbohydrates metabolised by the green seaweed *Urospora penicilliformis*. J Carbohydr Nucleosid Nucleotid 1:235–264

Bowker DM, Turvey JR (1968) Water-soluble polysaccharides of the red alga *Laurencia pinnatifida*. Part I. Constituent units. Part. II Methylation analysis of the galactan sulphate. J Chem Soc C 983–988, 989–992

Boyd J, Turvey JR (1978) Structural studies of alginic acid using a bacterial poly-α-L-guluronate lyase. Carbohydr Res 66:187–194

Brading JWE, Georg-Plant MMT, Hardy DM (1954) The polysaccharide from the alga *Ulva lactuca*, purification, hydrolysis, and methylation. J Chem Soc 319–324
Callow ME, Evans LV (1976) Localization of sulfated polysaccharides by X-ray microanalysis in *Laminaria saccharina*. Planta 131:155–157
Carlberg GE, Percival Elizabeth, (1977) The carbohydrates of the green seaweeds *Urospora wormskioldii* and *Codiolum pusillum*. Carbohydr Res 57:223–234
Carlberg GE, Percival Elizabeth, Rahman MA (1978) Carbohydrates of the seaweeds *Desmarestia ligulata* and *D firma*. Phytochemistry 17:1289–1292
Cerezo AS (1972) The fine structure of *Chaetangium fastigiatum* xylan: Studies of the sequence and configuration of the 1,3-linkages. Carbohydr Res 22:209–211
Cerezo AS (1973) The carrageenans of *Gigartina skottsbergii* S et G. Pt III. Methylation analysis of the fraction precipitated with 0.3-0.4-M-KCl. Carbohydr Res 26:335–340
Cerezo AS, Lezerovich A, Labriola R, Rees DA (1971) A xylan from the red seaweed *Chaetangium fastigiatum*. Carbohydr Res 19:289–296
Claire JWL, Dawes CJ (1976) An auto-radiographic and histochemical localization of sulfated polysaccharides of *Eucheuma nudum*. J Phycol 12:368–375
Cleare M, Percival Elizabeth (1972) Carbohydrates of the fresh-water alga *Tribonema aequale* I. Low molecular weight and polysaccharides. Br Phycol J 7:185–193
Conchie J, Percival EGV (1950) Fucoidin Pt. II. The hydrolysis of a methylated fucoidin from *Fucus vesiculosus*. J Chem Soc C 827–832
Conte MV, Pore RS (1973) Taxonomic implications of *Prototheca* and *Chlorella* cell wall polysaccharide characterisation. Arch Mikrobiol 92:227–233
Cote RH (1959) Disaccharides from fucoidin. J Chem Soc C:2248–2254
Craigie JS, Wong KF (1979) Carrageenan biosynthesis. In: Jensen A, Stein JR (eds) Proc 9th Int Seaweed Symp. Princeton Science Press, Princeton, pp 369–377
Craigie JS, McLachlan J, Majak W, Ackman RG, Tocher CS (1966) Photosynthesis in Algae. II. Green algae with special reference to *Dunaliella* spp and *Tetraselmis* spp. Can J Bot 44:1247–1254
Crayton MA, Wilson E, Quatrano RS (1974) Sulfation of fucoidan in *Fucus* embryos II. Separation from initiation of polar growth. Dev Biol 39:164–167
Cronshaw J, Myers A, Preston RD (1958) A chemical and physical investigation of the cell walls of some marine algae. Biochim Biophys Acta 27:89–103
Dawes CJ (1979) Physiological and biochemical comparisons of *Eucheuma* spp. (Florideophyceae) yielding iota-carrageenan. In: Jensen A, Stein JR (eds) Proc 9th Int Seaweed Symp. Princeton Science Press, Princeton, pp 199–207
Delmer DP (1977) Biosynthesis of cellulose and other plant cell wall polysaccharides. In: Loewus FA, Runeckles VC (eds) Structure, biosynthesis and degradation of wood. Recent Adv Phytochem, Vol XI Plenum Press, New York London, pp 45–77
Dennis DT, Preston RD (1961) Constitution of cellulose microfibrils. Nature (London) 191:667–668
DiNinno V, McCandless EL (1978a) The chemistry and immunochemistry of carrageenans from *Eucheuma* and related algal species. Carbohydr Res 66:85–93
DiNinno V, McCandless EL (1978b) The immunochemistry of λ-type carrageenans from certain red algae. Carbohydr Res 67:235–241
DiNinno V, McCandless EL (1979) Immunochemistry of kappa-type carrageenans from certain red algae. Carbohydr Res 72:157–163
Doner LW, Whistler RL (1973) Fucoidan In: Whistler RL, Bemiller J (eds) Industrial gums 2nd edn. Academic Press, London New York, pp 115–121
Dreher TW, Grant BR, Wetherbee R (1978) The wound response in the siphonous alga *Caulerpa simpliciuscula* C. Ag Fine structure and cytology. Protoplasma 96:189–203
Duckworth M, Yaphe W (1971) The structure of agar Part 1. Fractionation of a complex mixture of polysaccharides. Carbohydr Res 16:189–197 and refs cited therein
Evans LV, Callow ME (1974) Polysaccharide sulfation of *Laminaria*. Planta 117:93–95
Evans LV, Holligan MS (1972) Correlated light and electron microscope studies on brown algae. I. Localization of alginic acid and sulfated polysaccharide in *Dictyota*. New Phytol 71:1161–1172
Evans LV, Callow ME, Percival Elizabeth, Sheik-Fareed V (1974) Studies on the synthesis

and composition of extracellular mucilage in the unicellular red alga *Rhodella*. J Cell Sci 16:1–21

Evans LV, Callow ME, Coughlan SJ (1979) Sulfated polysaccharide synthesis in *Fucus* (Phaeophyceae) zygotes. In: Jensen A, Stein JR (eds) Proc 9th Int Seaweed Symp, Princeton Science Press, Princeton, pp 329–335

Fischer FG, Dörfel H (1955) Die Polyuronsäuren der Braunalgen. Hoppe-Seyler's Z Physiol Chem 302:186–203

Fisher IS, Percival Elizabeth (1957) The water-soluble polysaccharides of *Cladophora rupestris*. J Chem Soc:2666–2675

Ford CW, Percival Elizabeth (1965) Carbohydrates of *Phaeodactylum tricornutum*. Pt II. A sulphated glucuronomannan. J Chem Soc:7042–7046

Frei E, Preston RD (1961a) Variants in the structural polysaccharides of algal cell walls. Nature (London) 192:939–943

Frei E, Preston RD (1961b) Cell wall organization and wall growth in the filamentous green algae *Cladophora* and *Chaetomorpha* I. The basic structure and its formation. Proc R Soc London Ser B 154:70–94

Frei E, Preston RD (1964a) Non-cellulosic structural polysaccharides in algal walls. I. Xylan in siphoneous green algae. Proc R Soc London Ser B 160:293–313

Frei E, Preston RD (1964b) Non-cellulosic structural polysaccharides in algal cell walls. II. Association of xylan and mannan in *Porphyra umbilicalis*. Proc R Soc London Ser B 160:314–327

Frei E, Preston RD (1968) Non-cellulosic structural polysaccharides in algal cell walls. III. Mannan in siphoneous green algae. Proc R Soc London Ser B 169:127–145

Fujikawa T, Nakashima K (1975) The occurrence of fucoidan and fucoidan analogues in brown seaweed. J Agric Chem Soc Jpn 49:455–461

Fujikawa T, Wada M (1975) Mucilage from the brown seaweed. *Nemacystus decipiens*. I. Constituents of the mucilage and their properties. Agric Biol Chem 39:1109–1114

Gordon EM, McCandless EL (1973) Ultrastructure and histochemistry of *Chondrus crispus* Stackhouse. Proc N S Inst Sci 27; Suppl 111–133

Gordon-Mills EM, McCandless EL (1975) Carrageenans in the cell walls of *Chondrus crispus* Stack (Rhodophyceae Gigartinales) I. Localisation with fluorescent antibody. Phycologia 14:275–281

Gorin PAJ, Spencer JFT (1966) Exocellular alginic acid from *Azotobacter vinelandii*. Can J Chem 44:993–998

Grant T, Morris ER, Rees DA, Smith JC, Thom D (1973) Biological interactions between polysaccharides and divalent cations: The egg box model. FEBS Lett 32:195–198

Grasdalen H, Larsen B, Smidsrod O (1979a) Uronic acid sequence in alginates by ^{13}C-NMR. In: Jensen A, Stein JR (eds) Proc 9th Int Seaweed Symp. Princeton Science Press, Princeton, pp 309–317

Grasdalen H, Larsen B, Smidsrod O (1979b) A PMR study of the composition and sequence of uronate residues in alginates. Carbohydr Res 68:23–31

Groleau D, Yaphe W (1977) Enzymic hydrolysis of agar. Purification and characterization of *β-neo*agarotetraose hydrolase from *Pseudomonas atlantica*. Can J Microbiol 23:672–679

Guiseley KB (1970) The relationship between methoxyl content and gelling temperature of agarose. Carbohydr Res 13:247–256

Gunawardena P, Williamson FB (1975) Structure and composition of the cell wall, intercellular matrix and outer cuticle of *Porphyra umbilicalis* (L.). In: Proc 8th Int Seaweed Symp. Bangor, Wales, in press

Hamer GK, Bhattacharjee SS, Yaphe W (1977) Analysis of the enzymic hydrolysis products of agarose by ^{13}carbon-NMR. Carbohydr Res 54:C7–C10

Handa N (1969) Carbohydrate metabolism in the marine diatom *Skeletonema costatum*. Mar Biol 4:208–214

Hanic LA, Craigie JS (1969) Studies on the algal cuticle. J Phycol 5:89–102

Haq QN, Percival Elizabeth (1966a) Structural studies on the water-soluble polysaccharide from the green seaweed *Ulva lactuca*. In: Barnes H (ed) Some contemporary studies in marine science. Georg Allen and Unwin Ltd, London, pp 355–368

Haq QN, Percival Elizabeth (1966b) Structural studies on the water-soluble polysaccharide of the green seaweed *Ulva lactuca*. Part IV Smith degradation. In: Young EG, McLachlan JL (eds)Proc 5th Int Seaweed Symp. Pergamon Press, Oxford New York, pp 261–270
Haug A (1964) Composition and properties of alginates. Rep No 30. Norw Inst Seaweed Res, Trondheim
Haug A (1974) Chemistry and biochemistry of algal cell wall polysaccharides. In: Northcote DH (ed) Int Rev Sci Biochem Ser I, vol 11 Plant Biochemistry, Butterworths, London and Baltimore, Univ Park Press, pp 51–88
Haug A (1976) The influence of borate and calcium on the gel formation of a sulfated polysaccharide from *Ulva lactuca*. Acta Chem Scand 30B:562–566
Haug A, Larsen B (1971) Biosynthesis of alginate, Pt. II. Polymannuronic acid C-5-epimerase from *Azotobacter vinelandii* (Lipman). Carbohydr Res 17:297–308
Haug A, Myklestad S (1976) Polysaccharides of marine diatoms with special reference to *Chaetoceros* species. Mar Biol 34:217–222
Haug A, Smidsrod O (1967) Strontium, calcium and magnesium in brown algae. Nature (London) 215:1167–1168
Haug A, Larsen B, Smidsrod O (1966) A study of the constitution of alginic acid by partial acid hydrolysis. Acta Chem Scand 20:183–190
Haug A, Larsen B, Smidsrod O (1967) Studies on the sequence of uronic acid residues in alginic acid. Acta Chem Scand 21:691–704
Haug A, Larsen B, Baardseth E (1969) Comparison of the constitution of alginates from different sources. In: Margalef R (ed) Proc 6th Int Seaweed Symp. Direccion General de Pesca Maritima, Madrid, pp 443–451
Haug A, Larsen B, Smidsrod O (1974) Uronic acid sequence in alginate from different sources. Carbohydr Res 32:217–225
Hawthorne DB, Sawyer WH, Grant BR (1979) The structure of the low molecular weight glucans isolated from the siphonous green alga *Caulerpa simpliciuscula*. Carbohydr Res 77:157–167
Heaney-Kieras J, Chapman DJ (1976) Structural studies on the extracellular polysaccharide of the red alga *Porphyridium cruentum*. Carbohydr Res 52:169–177
Heaney-Kieras J, Roden L, Chapman DJ (1977) The covalent linkage of protein to carbohydrate in the extracellular protein-polysaccharide from the red alga *Porphyridium cruentum*. Biochem J 165:1–9
Hellebust JA, Haug A (1972) In situ studies on alginic acid synthesis and other aspects of the metabolism of *Laminaria digitata*. Can J Bot 50:177–184
Hirase S (1957) Chemical constitution of agaragar. XIX Bull Chem Soc Jpn 30:68–79
Hirase S, Watanabe K (1972) The presence of pyruvate residues in lambda-carrageenan and a similar polysaccharide. Bull Inst Chem Res Kyoto Univ 50:332–336
Hirst Sir E, Mackie W, Percival Elizabeth (1965) The water-soluble polysaccharides of *Cladophora rupestris* and of *Chaetomorpha* spp. Part II. The site of ester sulphate groups and the linkage between the galactose residues. J Chem Soc:2958–2967
Honjo G, Watanabe M (1958) Examination of cellulose fibre by the low temperature specimen method of electron diffraction and electron microscopy. Nature (London) 181:326–328
Horsford SPC, McCandless EL (1975) Immunochemistry of carrageenans from gametophytes and sporophytes of certain red algae. Can J Bot 53:2835–2841
Howard RJ, Wright SW, Grant BR (1976) Structure and some properties of soluble 1,3-β-glucan isolated from the green alga *Caulerpa simpliciuscula*. Plant Physiol 58:459–463
Huizing HJ, Rietema H (1975) Xylan and mannan as cell wall constituents of different stages in the life histories of some siphoneous green algae. Br Phycol J 10:13–16
Huizing HJ, Rietema H, Sietsma JH (1979) Cell wall constituents of several siphoneous green algae in relation to morphology and taxonomy. Br Phycol J 14:25–32
Hussein MM-D, Fouad ST, Abdel-Fattah AF (1979) Structural features of a sulfated fucose-containing polysaccharide from the brown seaweed *Dictyota dichotoma*. Carbohydr Res 72:177–181
Iriki Y, Miwa T (1960) Chemical nature of the cell wall of the green algae *Codium*, *Acetabularia* and *Halicoryne*. Nature (London) 185:178–179

Iriki Y, Suzuki T, Nisizawa K, Miwa T (1960) Xylan of siphonaceous green algae *Bryopsis maxima, Caulerpa anceps, Halimeda cuneata, Chlorodesmis formosana*. Nature (London) 187:82–83

Izumi K (1971a) Chemical heterogeneity of the agar from *Gelidium amansii*. Carbohydr Res 17:227–230

Izumi K (1971b) Chemical heterogeneity in *Gloiopeltis furcata*. Agric Biol Chem 35:651–655

Izumi K (1972) Chemical heterogeneity of the agar from *Gracilaria verrucosa*. J Biochem (Tokyo) 72:135–140

Jabbar Mian A, Percival Elizabeth (1973a) Carbohydrates of the brown seaweeds, *Himanthalia lorea, Bifurcaria bifurcata* and *Padina pavonia*. Pt 1. Extraction and fractionation. Carbohydr Res 26:133–146

Jabbar Mian A, Percival Elizabeth (1973b) Carbohydrates of the brown seaweeds, *Himanthalia lorea, Birfurcaria bifurcata*. Part 2. Structural studies of the "fucans". Carbohydr Res 26:147–161

Johnson PG, Percival Elizabeth (1969) Water-soluble polysaccharides of *Cladophora rupestris*. Part III. Smith degradation. J Chem Soc C:906–909

Johnston KH, McCandless EL (1968) The immunologic response of rabbits to carrageenans, sulphated galactans extracted from marine algae. J Immunol 101:556–562

Jones RF (1962) Extracellular mucilage of the red alga *Porphyridium cruentum*. J Cell Comp Physiol 60:61–64

Kieras JH, Kieras FJ, Bowen DV (1976) 2-*O*-Methyl-D-glucuronic acid, a new hexuronic acid of biological origin. Biochem J 155:181–185

Kohn R, Larsen B (1972) Preparation of water-soluble polyuronic acids and their calcium salts and the determination of calcium ion activity in relation to the degree of polymerization. Acta Chem Scand 26:2455–2468

Larsen B (1978) Fucoidan. In: Hellebust JA, Craigie JS (eds) Physiological and biochemical methods. Cambridge Univ Press, Cambridge, pp 152–156

Larsen B, Haug A (1963) Free-boundary electrophoresis of acidic polysaccharides from the marine alga *Ascophyllum nodosum* (L) Le Jol. Acta Chem Scand 17:1646–1652

Larsen B, Haug A (1971) Biosynthesis of alginate. Pt III. Tritium incorporated with polymannuronic acid 5-epimerase from *Azotobacter vinelandii*. Carbohydr Res 20:225–232

Larsen B, Haug A (1973) Biosynthesis of alginate. In: Nisizawa K (ed) Proc 7th Int Seaweed Symp. Univ Tokyo Press, Tokyo, pp 491–495

Larsen B, Haug A, Painter TL (1966) Sulfated polysaccharides in brown algae. I. Isolation and preliminary characterization of three sulfated polysaccharides from *Ascophyllum nodosum*. Acta Chem Scand 20:219–230

Larsen B, Haug A, Painter T (1970) Sulfated polysaccharides in brown algae III. The native state of fucoidan in *Ascophyllum nodosum* and *Fucus vesiculosus*. Acta Chem Scand 24:3339–3352

Lawson CJ, Rees DA (1970) An enzyme for the metabolic control of polysaccharide conformation and function. Nature (London) 227:193

Lawson CJ, Rees DA, Stancioff DJ, Stanley NF (1973) Carrageenans, Pt VIII. Repeating structures of galactan sulfates from *Furcellaria fastigiata, Gigartina canaliculata, Gigartina chamissoi, Gigartina atropurpurea, Ahnfeltia durvillaei, Gymnogongrus furcellatus, Eucheuma cottonii, Eucheuma spinosum, Eucheuma isiforme, Eucheuma uncinatum, Aghardhiella tenera, Pachymenia hymantophora* and *Gloiopeltis cervicornis*. J Chem Soc C: 2177–2182

Lestang G De, Quillet M (1973) Renouvellement rapide des esters sulfuriques de la fucoidine des frondes intactes de *Pelvetia canaliculata* (Dcne et Thur) pendent leur immersion en mer. C R Acad Sci Paris Ser D:277:2005–2008

Lestang G De, Quillet M (1974) Comportement du fucoidane sulfuryré de *Pelvetia canaliculata* (Dcne and Thur) vis à vis des cations de la mer: propriétés d'échange renouvellement des radicaux sulfuriques, coenzyme d'activation des sulfates. Intérêt fonctionnel. Physiol Vég 12:199–227

Liang JN, Stevens ES, Morris ER, Rees DA (1979) Spectroscopic origin of conformation-sensitive contributions to polysaccharide optical activity: vacuum-ultraviolet circular dichroism of agarose. Biopolymers 18:327–333

Lin TY, Hassid WZ (1966a) Isolation of guanosine diphosphate uronic acids from a marine brown alga. *Fucus gardneri* Silva. J Biol Chem 241:3283–3293
Lin TY, Hassid WZ (1966b) Pathway of alginic acid synthesis in the marine brown alga *Fucus gardneri* Silva. J Biol Chem 241:5284–5297
Linker A, Jones RS (1964) A polysaccharide resembling alginic acid from a *Pseudomonas* microorganism. Nature (London) 204:187–188
Loewus F, Wagner G, Schiff JA, Weistrop J (1971) The incorporation of ^{35}S-labelled sulfate into carrageenan in *Chondrus crispus*. Plant Physiol 48:373–375
Love J, Percival Elizabeth (1964a) The polysaccharides of the green seaweed *Codium fragile*. Part II. The water-soluble sulphated polysaccharides. J Chem Soc:3338–3345
Love J, Percival Elizabeth (1964b) The polysaccharides of the green seaweed *Codium fragile* Pt. III. A β-1,4-linked mannan. J Chem Soc:3345–3350
Mackie IM, Percival Elizabeth (1959) The constitution of xylan from the green seaweed *Caulerpa filiformis*. J Chem Soc:1151–1156
Mackie IM, Percival Elizabeth (1961) Polysaccharides from the green seaweeds of *Caulerpa* spp. Part III. Detailed study of the water-soluble polysaccharide of *C. filiformis*. Comparison with the polysaccharides synthesised by *C. racemosa* and *C. sertularioides*. J Chem Soc:3010–3015
Mackie W (1971) Conformation of crystalline alginic acids and their salts. Biochem J 125:89P
Mackie W, Preston RD (1968) The occurrence of mannan microfibrils in the green algae *Codium fragile* and *Acetabularia crenulata*. Planta 79:249–253
Mackie W, Preston RD (1974) Cell wall and intercellular region polysaccharides. In: Stewart WD (ed) Algal physiology and biochemistry. Blackwell, Oxford and Univ Calif Press, Berkeley, pp 40–85
Mackie W, Sellen DB (1969) The degree of polymerization and polydispersity of mannan from the cell wall of the green seaweed *Codium fragile*. Polymer 10:621–632
Mackie W, Sellen DB (1971) Degreee of polymerization and polydispersity of xylan from the cell wall of the green seaweed *Penicillus dumetosus*. Biopolymers 10:1–9
Madgwick JC, Ralph BJ (1969) Chemical constituents of Australian bull kelp, *Durvillea potatorum*. In: Margalef R (ed) Proc 6th Int Seaweed Symp. Direccion General de Pesca Maritima, Madrid, pp 539–544
Madgwick J, Haug A, Larsen B (1973) Polymannuronic acid 5-epimerase from the marine alga *Pelvetia canaliculata* (L) Dcne et Thur. Acta Chem Scand 27:3592–3594
McCandless EL, Craigie JS, Walter JA (1973) Carrageenans in the gametophytic and sporophytic stages of *Chondrus crispus*. Planta 112:201–212
McCandless EL, Craigie JS, Hansen JE (1975) Carrageenans of gametangial and tetrasporangial stages of *Iridaea cordata* (Gigartinaceae). Can J Bot 53:2315–2318
McCandless EL, Okada WT, Lott JNA, Volmer CM, Gordon-Mills EM (1977) Structural studies of *Chondrus crispus*: the effect of extraction of carrageenan. Can J Bot 55:2053–2064
McCandless EL, Evelegh MJ, Vollmer CM, DiNinno VL (1979) Immunological characterisation of lambda-carrageenans. In: Jensen A, Stein JR (eds) Proc 9th Int Seaweed Symp. Princeton Science Press, Princeton, pp 347–352
McCully ME (1966) Histological studies on the genus *Fucus*. 1. Light microscopy of the mature vegetative plant. Phytoplasma 62:287–305
McKinnell JP, Percival Elizabeth (1962a) The acid polysaccharide from the green seaweed *Ulva lactuca*. J Chem Soc C:2082–2083
McKinnell JP, Percival Elizabeth (1962b) Structural investigations on the water-soluble polysaccharide of the green seaweed *Enteromorpha compressa*. J Chem Soc C:3141–3148
Medcalf DG, Larsen B (1977) Fucose containing polysaccharides in the brown algae *Ascophyllum nodosum* and *Fucus vesiculosus*. Carbohydr Res 59:531–537, 539–546
Medcalf DG, Scott JR, Brannon JH, Hemerick GA, Cunningham RL, Chessen JH, Shah J (1975) Structural features and visosometric properties of the extracellular polysaccharide from *Porphyridium cruentum*. Carbohydr Res 44:87–96
Medcalf DG, Schneider TL, Barnett RW (1978) Structural features of a novel glucuronogalactofucan from *Ascophyllum nodosum*. Carbohydr Res 66:167–171

Min KH, Sasaki SF, Kashiwabara Y, Umekawa M, Nisizawa K (1977) Fine structure of SMG alginate fragment in the light of its degradation by alginate lyases of *Pseudomonas* sp. J Biochem (Tokyo) 81:555–562
Miwa T, Iriki Y, Suzuki T (1961) Mannan and xylan as essential cell wall constituents of some siphonous green algae. Colloq Int CNRS 103:135–143
Morris ER, Rees DA, Thom D (1973) Characterisation of polysaccharide structure and interactions by circular dichroism: order-disorder transition in the calcium alginate system. Chem Commun 245–246
Morris ER, Rees DA, Thom D, Boyd J (1978) Chiroptical and stoichiometric evidence of a specific primary dimerization process in alginate gelation. Carbohydr Res 66:145–154
Nieduszynski IA, Atkins EDT (1970) Preliminary investigation of algal cellulose I. X-ray intensity data. Biochem Biophys Acta 222:109–118
Nikaido H, Hassid WZ (1971) Biosynthesis of saccharides from glycopyranosyl esters of nucleoside pyrophosphates (sugar nucleotides) 3. Cellulose. In: Tipson RS, Horton D (eds) Adv Carbohydr Chem Biochem, vol 26, Academic Press, London New York, pp 386–391
Northcote DH, Goulding KJ, Horne RW (1958) The chemical composition and structure of the cell wall of *Chlorella pyrenoidosa*. Biochem J 70:391–397
Nunn JR, Parolis H (1968) A polysaccharide from *Aeodes orbitosa*. Carbohydr Res 6:1–11, 8:361–362
Nunn JR, Parolis H (1969) A polysaccharide from *Phyllomenia cornea*. Carbohydr Res 9:265–276
Nunn JR, Parolis H, Russell I (1971) Sulfated polysaccharides of the Solieriaceae family. Part 1. A polysaccharide from *Anatheca dentata*. Carbohydr Res 20:205–215
O'Donnell JJ, Percival Elizabeth (1959) Structural investigations on the water-soluble polysaccharides from the green seaweed *Acrosiphonia centralis (Spongomorpha arcta)*. J Chem Soc C:2168–2178
O'Neill AN (1954) Degradative studies on fucoidin. J Am Chem Soc 76:5074–5076
Ovodov YS, Khomenko VA, Guseva TF (1973) Polysaccharides of brown algae, VIII. Sargassan from *Sargassum pallidum* Khim Prir Soedin 9:107–108
Painter TJ (1966) The location of the sulfate half-ester groups in furcellaran and kappa-carrageenan. In: Young EG, McLachlan JL (eds) Proc 5th Int Seaweed Symp. Pergamon Press, London Oxford New York, pp 305–313
Parke M, Dixon PS (1976) Check-list of British marine algae – third revision. J Mar Biol Assoc UK 56:527–594
Parker BC, Diboll AG (1966) Alcian stains for histochemical localisation of acid and sulphated polysaccharides in algae. Phycologia 6:37–46
Parsons RT, Stephens K, Strickland JDH (1961) Chemical composition of eleven species of marine phytoplankters. J Fish Res Board Can 18:1001–1013
Parsons MJ, Pickmere SE, Bailey RW (1977) Carrageenan composition in New Zealand species of *Gigartina* (Rhodophyta). Geographic variation and interspecific differences. N Z J Bot 15:589–595
Pavlenko AF, O'Vodov YS (1974) Polysaccharides of brown algae. X. Structure of side chains of pelvetian molecule. Khim Prir Soedin 697–699
Pavlenko AF, Belogortseve NI, Kalinovskiai, Ovodov YS (1976) Determination of the positions of the sulfate groups in sulfated polysaccharides, Khim Prir Soedin 573–577
Penman A, Rees DA (1973) Carrageenans Part. IX. Methylation analysis of galactan sulfates from *Furcellaria fastigiata, Gigartina canaliculata, Gigartina chamissoi, Gigartina atropurpurea, Ahnfeltia durvillaei, Gymnogongrus furcellatus, Eucheuma isiforme, Eucheuma uncinatum, Aghardhiella tenera, Pachymenia hymantophora* and *Gloiopeltis cervicornis*. Structure of xi-carrageenan. J Chem Soc C:2182–2187
Penman A, Sanderson GR (1972) A method for the determination of uronic acid sequence in alginates. Carbohydr Res 25:273–282
Percival EGV, Ross AG (1948) The cellulose of marine algae. Nature (London) 162:895–897
Percival EGV, Ross AG (1949) Marine algal cellulose. J Chem Soc:3041–3043

Percival EGV, Ross AG (1950) Fucoidin I. The isolation and purification of fucoidin from brown seaweeds. J Chem Soc: 717–720
Percival Elizabeth (1968) Glucuronoxylofucan, a cell-wall component of *Ascophyllum nodosum*. Part I. Carbohydr Res 7: 272–283
Percival Elizabeth (1971) Glucuronoxylofucan. A cell wall component of *Ascophyllum nodosum*, Part II. Methylation. Carbohydr Res 17: 121–126
Percival Elizabeth (1972) Chemistry of agaroids, carrageenans and furcellarans. J Sci Food Agric 23: 933–940
Percival Elizabeth (1978a) Sulphated polysaccharides metabolised by the marine Chlorophyceae – A review, In: Schweiger RG (ed) ACS Symp Ser 77. Am Chem Soc, Washington DC, pp 203–212
Percival Elizabeth (1978b) Sulphated polysaccharides of the Rhodophyceae – A review. In: Schweiger RG (ed) ACS Symp Ser 77. Am Chem Soc, Washington DC, pp 213–224
Percival Elizabeth (1978c) Do the polysaccharides of brown and red seaweeds ignore taxonomy? In: Irvine DEG, Price JH (eds) Systematics association special, vol X. Modern approaches to the taxonomy of red and brown algae. Academic Press, London New York, pp 47–62
Percival Elizabeth (1979) The polysaccharides of green, red and brown seaweeds. Their basic structure, biosynthesis and function. Br Phycol J 14: 103–117
Percival Elizabeth, Foyle RAJ (1979) The extracellular polysaccharides of *Porphyridium cruentum* and *Porphyridium aerugineum*. Carbohyd Res 72: 165–176
Percival Elizabeth, McDowell RH (1967) Chemistry and enzymology of marine algal polysaccharides. Academic Press, London New York
Percival Elizabeth, Smestad B (1972) Photosynthetic studies on *Ulva lactuca*. Phytochemistry 11: 1967–1972
Percival Elizabeth, Turvey JR (1974) Polysaccharides of algae. In: Laskin AI, Lechevalier HA (eds) CRC Handbook of microbiology. CRC Press Inc, Cleveland Ohio, pp 532–550
Percival Elizabeth, Wold JK (1963) The acid polysaccharide from the green seaweed *Ulva lactuca*. Part II. The site of ester sulphate. J Chem Soc C: 5459–5468
Percival Elizabeth, Young M (1971) Low molecular weight carbohydrates and water-soluble polysaccharide metabolized by the Cladophorales. Phytochemistry 10: 807–812
Percival Elizabeth, Young M (1974) Carbohydrates of the brown seaweeds. Pt III. *Desmarestia aculeata*. Carbohydr Res 32: 195–201
Percival Elizabeth, Rahman MA, Weigel H (1980) Aspects of the chemistry of the polysaccharides of the diatom *Coscinodiscus nobilis* Phytochemistry 9: 809–811
Pickmere SE, Parsons MJ, Bailey RW (1973) Composition of *Gigartina* carrageenan in relation to sporophyte and gametophyte stages of the life cycle. Phytochemistry 12: 2441–2444
Quatrano RS, Crayton MA (1973) Sulfation of fucoidan in *Fucus* embryos. I. Possible role in localisation. Dev Biol 30: 29–41
Quatrano RS, Stevens P (1976) Cell wall assembly in *Fucus* zygotes. I. Characterization of the polysaccharide components. Plant Physiol 58: 224–231
Quatrano RS, Hogsett WS, Roberts M (1979a) Localization of a sulfated polysaccharide in the rhizoid wall of *Fucus distichus* (Phaeophyceae) zygotes. In: Jensen A, Stein JR (eds) Proc 9th Int Seaweed Symp. Princeton Science Press, Princeton, pp 113–123
Quatrano RS, Brawley SH, Hogsett WE (1979b) The control of the polar deposition of a sulfated polysaccharide in *Fucus* zygotes. In: Subtelny, Konisberg (eds) Determinants of spatial organization. Academic Press London New York, pp 77–95
Quillet M (1959) Sur la composition chimique des mucilage de *Fucus spiralis* et *Pelvetia canaliculata*. Bull Lab Marit Dinard 45: 63–68
Quillet M (1961) Sur la composition chimique de la fucoidine des algues brunes I. Les oses non esterifies. Colloq Int CNR 103: 145–158
Ramus J (1972) Production of extracellular polysaccharide by the unicellular red alga *Porphyridium aerugineum*. J Phycol 8: 97–111
Ramus J, Groves ST (1972) Incorporation of sulfate into the capsular polysaccharide of the red alga *Porphyridium*. J Cell Biol 54: 399–407

Rees DA (1961) Enzymic synthesis of 3,6-anhydrogalactose within Porphyran from L-galactose 6-sulfate units. Biochem J 81:347–352
Rees DA (1969) Structure, conformation and mechanism in the formation of polysaccharide gels and networks. In: Wolfrom ML, Tipson RS (eds) Adv Carbohydr Chem Biochem, vol 24, Academic Press, London New York, pp 267–332
Rees DA (1972) Shapely polysaccharides. Biochem J 126:257–273
Rees DA, Samuel JWB (1967) The structure of alginic acid. Part VI. Minor features and structural variations. J Chem Soc C:2295–2298
Rees DA, Welsh EJ (1977) Secondary and tertiary structure of polysaccharides in solutions and gels. Angew Chem Int Ed Engl 16:214–224
Robinson DG, Preston RD (1971) Studies on the fine structure of *Glaucocystis nostochinearum* Itzigs I. Wall structure. J Exp Bot 22:635–643
Ross AG (1953) Some typical analyses of red seaweeds. J Sci Food Agric 4:333–335
Santos GA, Doty MS (1979) Carrageenans from some Hawaiian red algae. In: Jensen A, Stein JR (eds) Proc 9th Int Seaweed Symp. Princeton Science Press, Princeton, pp 361–367
Sarko A, Muggli R (1974) Packing analysis of carbohydrates and polysaccharides III. *Valonia* cellulose and cellulose II. Macromolecules 7:486–494
Schnepf E (1965) Struktur der Zellwände und cellulöse Fibrillen bei *Glaucocystis*. Planta 67:213–224
Schweiger RG (1962) Methanolysis of fucoidan I, II. J Org Chem 27:4267–4269, 4270–4272
Sheik-Fareed V (1975) Structural investigation of the extracellular polysaccharides metabolized by S19 *Xanthomonas* type bacterium and by the red alga *Rhodella maculata*. Ph D Thesis, London
Shimahara H, Sugiyama N (1974) A sulfated galactan from the red seaweed *Hypnea charoides*. Agric Biol Chem 38:2569–2570
Silva PC (1962) Appendix A. Classification. In: Lewin RA (ed) Physiol Biochem Algae. Academic Press, London New York, pp 828–837
Simonescu CI, Popa VI, Rusan V, Liga A (1976) The influence of structural units distribution on macromolecules conformation of alginic acid. Cellul Chem Technol 10:587–594
Smestad B, Haug A, Myklestad M (1975) Structural studies of the extracellular polysaccharide produced by the diatom *Chaetoceros curvisetas* Cleve. Acta Chem Scand B 29:337–340
Smidrod O, Haug A (1968) Dependence upon uronic acid composition of some ion-exchange properties of alginates. Acta Chem Scand 22:1989–1997
Smidsrod O, Haug A (1972a) Properties of poly (1,4-hexuronates) in the gel state II. Comparison of gels of different chemical composition. Acta Chem Scand 26:79–88
Smidsrod O, Haug A (1972b) Dependence upon the gel-solstate of the ion-exchange properties of alginates. Acta Chem Scand 26:2063–2074
Smidsrod O, Haug A, Lian B (1972) Properties of poly (1,4-polyhexuronates) in the gel state I. Evaluation of a method for the determination of stiffness. Acta Chem Scand 26:71–78
Smidsrod O, Glover R, Whittington SG (1973) The relative extension of alginates having different chemical compositions. Carbohydr Res 27:107–118
Smith DB, Cook WH (1953) Fractionation of carrageenin. Arch Biochem Biophys 45:232–233
South GR (1979) Alginate levels in New Zealand *Durvillaea* (Phaeophyceae) with particular reference to age variations in *D. antarctica*. In: Jensen A, Stein JR (eds) Proc 9th Int Seaweed Symp. Princeton Science Press, Princeton, pp 133–142
Stancioff DJ (1965) Extraction of hydrocolloid fractions from sea plants. (to Marine Colloids Inc) US Pat 3:176,003
Stevens PT, Quatrano RS (1978) Cell wall assembly in *Fucus* zygotes II. Cellulose synthesis and deposition is controlled at the post transitional level. Dev Biol 62:518–525
Takeda H, Hirokawa T (1978) Studies on the cell wall of *Chlorella*. I Quantitative changes in cell wall polysaccharides during the cell cycle of *Chlorella ellipsoidea*. Plant Cell Physiol 19:591–598

Takeda H, Nisizawa K, Miwa T (1968) A xylomannan from the cell wall of *Prasiola japonica* Yatabe. Sci Rep Tokyo Kyoiku Daigaku Sect B No 198. 13:183–198
Thiele H, Andersen G (1955) Ionotrope Gels von Polyuronsauren. III Alginat und Pektin in nativen organisierten Gelen. Kolloid Z 143:21–31
Thompson EW, Preston RD (1967) Proteins in the cell walls of some green algae. Nature (London) 213:684–685
Turvey JR (1978) Biochemistry of algal polysaccharides. In: Manners DJ (ed) Int Rev Biochem Biochem Carbohydr II, vol 16. Univ Park Press, Baltimore, pp 151–177
Turvey JR, Williams EL (1970) The structure of some xylans from red algae. Phytochemistry 9:2383–2388
Turvey JR, Williams EL (1976) The agar-type polysaccharide from the red alga *Ceramium rubrum*. Carbohydr Res 49:419–425
Villarroel LH, Zanlungo AB (1975) Polisacaridos solubles de algas chilenas. I. el fucoidano de la *Lessonia flavicans*. Rev Latinoam Quim 6:127–130
Vreeland V (1970) Localization of a cell wall polysaccharide in a brown alga with labeled antibody. J Histochem Cytochem 18:371–373
Vreeland V (1971) An immunological study of extracellular polysaccharides in the brown alga *Fucus distichus*. Ph D Thesis, Standford Univ California
Vreeland V (1972) Immunocytochemical localization of the extracellular polysaccharide alginic acid in the brown seaweed *Fucus distichus*. J Histochem Cytochem 20:358–367
Waaland JR (1975) Differences in carrageenan in gametophytes and tetrasporophytes of red algae. Phytochemistry 14:1359–1362
Weigl J, Yaphe W (1966) The enzymic hydrolysis of carrageenan by *Pseudomonas carrageenovora*. Can J Microbiol 12:874–876, 939–947
Whyte JNC, Englar JR (1971) Polysaccharides of the red alga *Rhodymenia pertusa*. Part II. Cell wall glucan. Proton magnetic resonance studies on permethylated polysaccharides. Can J Chem 49:1302–1305
Wong F, Craigie JS (1978) Sulfohydrolase activity and carrageenan biosynthesis. Plant Physiol 61:663–666
Wurtz M, Zetsche K (1976) Zur Biochemie und Regulation des heteromorphen Generationswechsels der Grünalge *Derbesia-Halicystis*. Planta 129:211–216
Yaphe W (1957) The use of agarase from *Pseudomonas atlantica* in the identification of agar in marine algae (Rhodophyceae). Can J Microbiol 3:987–993
Yaphe W (1959) The determination of kappa-carrageenan as a factor in the classification of the Rhodophyceae. Can J Bot 37:751–757
Yarotskii SV, Shashkov AS, Usov AS (1978) Polysaccharides of algae. XXV. Use of carbon-13NMR spectrosopy for the structural analysis of kappa-carrageenan group polysaccharides. Biorg Khun 41:745–751
Young K, Duckworth M, Yaphe W (1971) The structure of agar. Part III. Pyruvic acid a common feature of agars from different agarophytes. Carbohydr Res 16:446–448
Young K, Hong KC, Duckworth M, Yaphe W (1972) Enzymic hydrolyses of agar and properties of bacterial agarases. In: Nisizawa K (ed) Proc 7th Int Seaweed Symp. Univ Tokyo Press, Tokyo, pp 469–472

13 Algal Walls – Cytology of Formation

D.G. Robinson

1 Matrix Polysaccharide and Slime Production

Although the production of nonfibrillar extracellular polysaccharides is a feature of nearly all algal cells and indeed attains industrial importance in some instances (kelps), comparatively few cytological studies have been carried out with respect to their sites of synthesis and modes of liberation. Nevertheless it is clear that elements of the endomembrane system (endoplasmic reticulum, ER; Golgi apparatus, GA) are principally involved. Autoradiography (Evans et al. 1974) and comparative micromorphometry of secretory and nonsecretory stages (Ramus 1972, Ramus and Robins 1975) have implicated the GA in red algal mucilage production. The participation of ER and GA in the production of the glycoprotein adhesive secreted by the ship-fouling green alga *Enteromorpha* has been demonstrated by cytochemical (for the carbohydrate moiety) and autoradiographic (for the protein moiety) means (Callow and Evans 1977). Ultrastructural, autoradiographical, and microanalytical (for sulfur detection in sulfated polysaccharides) evidence pertaining to GA participation in slime-polysaccharide secretion has also been provided for various representatives of the brown algae (Evans and Holligan 1972, Evans et al. 1973, Evans and Callow 1974, Callow and Evans 1973, 1976). Analysis of isolated dictyosomal fractions from *Fucus* zygotes (Callow et al. 1978, Coughlan and Evans 1978) has furthermore shown this fraction to be the first labeled when ^{35}S-sulfate is applied exogenously. Moreover this fraction contains an enzyme responsible for the transfer of galactose from UDP-galactose to fucose and fucoidan.

2 Microfibril Synthesis and Orientation

2.1 The Formation and Secretion of Scales

2.1.1 Scale Structure

Although recorded in a few other protists (Darley et al. 1969, Furtado and Olive 1971), the existence of a cell covering composed of discrete subunits or scales is a feature usually confined to marine members of three algal families: Chrysophyceae, Prasinophyceae, and Haptophyceae (or Prymnesiophyceae sensu Hibberd 1976). All these scales have an organic base upon which either silica (Chrysophyceae) or calcite (Haptophyceae, the scales being then referred to as coccoliths) may be deposited. With the exception of the demonstration that the scale complexes (thecae) of *Platymonas tetrahele* (Prasinophyceae) are of a pectin-like nature (Gooday 1971) a detailed chemical analysis of scales has only been carried out on two members of the Haptophyceae. The nonmineralized scales of both *Pleurochrysis scherffelii* and *Chrysochromulina chiton* have been demonstrated to contain cellulose (BROWN et al. 1970, ALLEN and NORTHCOTE 1975). Interesting, however, is the protein content of these scale preparations: on a dry weight basis this amounts to 9% for *Pleurochrysis* (HERTH et al. 1972) but is as high as 65% in the case of *Chrysochromulina*. Even after alkali treatment (10% NaOH for 24 h at room temperature) a large proportion of the protein remains bound to the polysaccharide portion of the scale. The amino acid composition of this alkali-resistant glycoprotein is characterized by approximately 10% (mol %) each of alanine, leucine, glutamine and aspargine for *Chrysochromulina* and by serine (32%) and glycine (21%) for *Pleurochrysis* (HERTH et al. 1972). A further feature of the chemistry of these scales is the high proportion of ribose residues in hydrolysates of the alkali-soluble fractions.

The cellulose component of these scales is present as two sets of thin microfibrils (ca. 2×4 nm in cross-section) which intersect more or less at right angles (see Fig. 1). In one set, usually that facing the medium, the microfibrils are concentrically or spirally arranged, and in the other set, facing the cell surface, radially arranged. In some scales the spiral set may be reduced to one or two microfibrils at the rim of the scale. ROMANOVICZ and BROWN (1976) have shown for *Pleurochrysis* scales that is possible to dissolve the radial microfibrils while leaving the spiral microfibrils intact, and have presented evidence indicating that the radial microfibrils are probably noncellulosic in nature. Scales may be circular, 0.5 to 3 µm in diameter, or elliptical, in which case the radial microfibrils are arranged in four similar quarters.

A difficulty with these flagellates is that not only several different scale types may be produced at the same time (even mineralized and nonmineralized scales may be produced together, e.g. *Hymenomonas carterae* (FRANKE and BROWN 1971), but that an alteration of generations may take place with correspondingly different scale types. Thus the nonmotile vegetative cells of the well-known, noncoccolith forming *Pleurochrysis scherffelii*, which forms pseudofilamentous colonies on agar plates, are believed to be the haploid generation while the diploid generation exists as the motile, coccolith-forming *Hymenomonas (Cricosphaera) carterae* (see LEADBEATER 1971, for appropriate references). To complicate matters even further, each generation may reproduce itself through motile unscaled zoospores. Perhaps corresponding to the well-known production of gametes by cultivation in a nitrogen-free medium (see Chap. 14, this Vol.) is the observation of BROWN and ROMANOVICZ (1976) that the *Hymenomonas* stage of *Pleurochrysis* may be induced by deletion of two of the three amino acids from the culture medium.

2.1.2 The Golgi Apparatus and Scale Production

With the exception of the silica scales of the Chrysophyceae, whose synthesis occurs in special flattened vesicles lying directly adpressed on periplastidal ER (SCHNEPF and DEICHGRÄBER 1969), both the microfibrillar scales of the Haptophyceae and the nonmicrofibrillar scales of the Prasinophyceae are produced in the cisternae of the Golgi apparatus. In the Haptophyceae there is usually only one dictyosome present whose cisternae lie at an angle to both the nuclear envelope and the plasmalemma; in the Prasinophyceae there may be several dictyosomes present. A forming face, opposing an arm of the ER, is clearly seen, and, particularly in the Haptophyceae, the cisternae at the maturing face of the dictyosome are very dilated and contain the completed scales (Fig. 2). A range of possibilities for cisternal scale synthesis exists, e.g., one or several scales per cisterna may be produced: where only one is produced all cisternae may be the same (e.g., *Pleurochrysis*) or different cisternae may synthesize different scales (e.g., *Chrysochromulina chiton*, MANTON 1967); where several are made the scales may be identical (e.g., *Pyramimonas amylifera*, MANTON 1966)

1
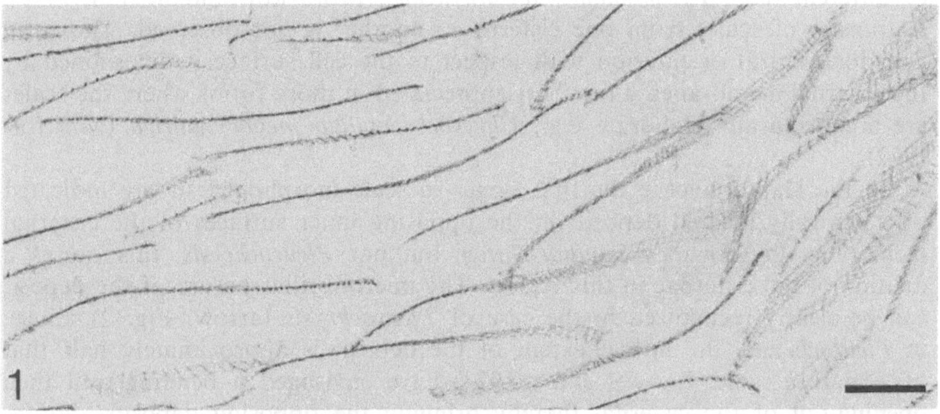

2
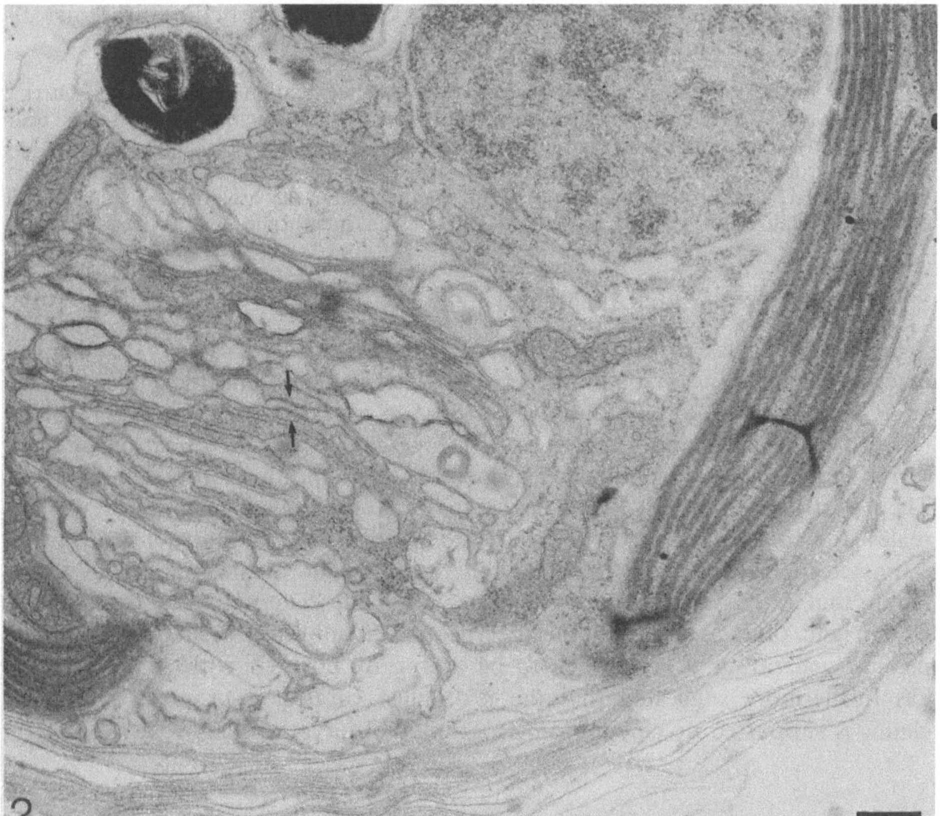

Fig. 1. Sectioned cellulosic scales of *Pleurochrysis scherffelii*. ×55,000. *Bar* 0.2 µm. (Micrograph W. Herth)

Fig. 2. The Golgi apparatus and scale secretion in *Pleurochrysis*. Fibrillar elements are first observed as opposing scale-halves (*arrows*). ×40,000. *Bar* 0.2 µm. (Micrograph W. Herth) or different (e.g., *Pyramimonas tetrarhynchus,* Moestrup and Walne 1979). A transfer of scales from one cisterna to another is not observed. The form and dorsiventral orientation with respect to the cell surface is determined by the cisternal membranes, a fact best appreciated in those forms where the scales are architecturally elaborate e.g., *Chrysochromulina megacylindrica* (Manton 1972).

In the Haptophyceae the first stages in scale morphogenesis are indicated by a centrally located deposit on the opposing inner surfaces of the cisternal membrane. In *Chrysochromulina chiton,* but not *Pleurochrysis,* this causes a dilation of the cisternae in this region. The microfibrillar nature of the deposit can be clearly recognized in the case of *Pleurochrysis* (arrow, Fig. 2). Since, in *Pleurochrysis,* the lateral extent of the deposit is approximately half that of a mature scale, Brown et al. (1973) have envisaged a bending and then opening-out of the cisterna, thereby bringing the opposing deposits side by side and creating the radial microfibril network. Subsequently the spiral microfibrils are added to this framework.

An interesting complication in these systems is the calcification of such scales. Outka and Williams (1971) have demonstrated for *Hymenomonas carterae* whose scales bear calcified rims, that osmiophilic granules ("coccolithosomes", 5–8 nm diameter) collect together in groups in cisternal pockets at the scale periphery. These granules may then be seen clustered round the developing calcite plaques. Thus the organization and control of scale mineralization is also a function of the cisternal membrane.

2.1.3 Scale Transport and Liberation

Whereas the scales of Haptophyceaen algae are released directly to the cell exterior through fusion of mature Golgi cisternae with the plasmalemma, in some Prasinophyceae the scales are retained in a vesicle reservoir before being released (Moestrup and Thomsen 1974). In *Pleurochrysis* the direction of scale secretion appears to be controlled by two sets of flanking microtubule bundles. Their destruction by colchicine leads to an autophagic consumption of the scales due to their retarded release (Brown et al. 1973). Brown (1969) has estimated for *Pleurochrysis* that scale liberation occurs at the rate of one every 2 min, that scale synthesis and transport within the cisternal stack takes about 60 min and that, as a result, the dictyosomal membranes turn over 80 times per cell generation. For the cell to be uniformly covered with scales Brown has furthermore suggested that the protoplast rotates, a proposal which he claims is supported by cinephotomicrographic observations.

2.2 Microfibril Deposition in Cellulosic Algae

2.2.1 Cladophorales, Siphonocladales

Several species of these two groups of coenocytic, mainly marine, algae have become classic objects for studies on cell wall structure, particularly through the work of Preston and colleagues (see Preston 1974 for references). The walls are rich in cellulose (in *Valonia* spp. ca. 70%), which is arranged in layers of parallel microfibrils lying at right angles to one another and possessing either slow or fast helical distributions with respect to the main cell axis. Occasionally a third microfibril orientation is to be seen, intersecting the other two at 45°. As a result of the observation that on the innermost, i.e., most recently synthesized, layer of such microfibrils (prepared through stripping or plasmolysis) groups of granular bodies are to be seen in which microfibrils appeared to begin/end, PRESTON put forward (1964) the "ordered granule hypothesis". According to this hypothesis, the synthesis of regularly alternating layers of parallel oriented microfibrils is achieved through a regular arrangement, in three dimensions, of enzyme complexes approx. 50 nm diameter which lie at the protoplast surface. Since each microfibril is thought to grow through the individual complexes by addition of glucose/cellobiose units the PRESTON model presupposes that extensive areas of the protoplast are covered with such granules. (See Chaps. 2, 3, and 5, this Vol. for further discussion.)

Using zoospores of *Cladophora rupestris* and *Chaetomorpha melagonium*, attempts have been made to examine this model using freeze-etching.

After release from the parent plant the zoospores remain motile for several hours before settling and producing a cell wall. The first stage in wall development is the production of the so-called "fibrous layer", a layer consisting of a mosaic of parallel arranged filaments approximately 23 nm diameter (NICOLAI and PRESTON 1959). A mass of granular material is then deposited between this layer and the plasmalemma. Microfibrils arise in this granular region. BARNETT and PRESTON (1970) have published pictures of this stage in which the granular material does indeed appear cubically close-packed. However, the unusual nature of these pictures, which were not repeatable in later investigations (ROBINSON and PRESTON 1971, ROBINSON et al. 1972) cannot be regarded as convincing support for PRESTON's model, particularly since the first layers of microfibrils are randomly oriented and therefore probably do not need an organized template of this nature. Nevertheless views of the plasmalemma have been obtained from which a role for granular or particulate elements in microfibril synthesis and orientation may be inferred.

Strings, up to 4 μm in length, of 20 nm particles in the EF plasmalemma fracture plane (nomenclature of BRANTON et al. 1975) and corresponding grooves in the PF plane are to be seen during the first stages of microfibril production around *Cladophora* zoospores, (ROBINSON et al. 1972). Similar strings of particles, though shorter (ca. 0.2 μm) and smaller (11 nm diameter) have also been recorded at the beginning of wall development round *Pelvetia* embryos (PENG and JAFFE 1976). These authors have shown that, as the cells grow, turgor pressure is developed which results in the impression of the microfibrils on the plasmalemma. Imprints of the microfibrils are thus registered as ridges or grooves in the EF or PF fracture planes respectively. As the wall develops the particle strings line up along these grooves. Particle-associated microfibril imprints in the plasmalemma may also be recognized in *Cladophora* during rhizoid production, when the microfibrils are synthesized transversely with respect to the cell axis (reinterpretation of Figs. 15 and 16 of ROBINSON et al. 1972).

Independently of the type of plasmalemma particle arrangement associated with microfibril synthesis, be they strings of particles or a more complex arrangement as envisaged by PRESTON, the question remains as to what controls the orientation of the synthetic complexes. HEATH (1974) has suggested that cortical microtubules are responsible for this event, however in germinating zoospores of *Cladophora* microtubules are absent from the cell cortex during the periods of random and transverse microfibril production. They do, however, reappear longitudinally oriented prior to the synthesis of the first longitudinally oriented microfibrils (ROBINSON et al. 1972). The relevance of this observation, however, must be evaluated in terms of the fact that in the cylindrical cells of the growing filament, where the microfibrils are helically arranged, cortical microtubules are also found longitudinally oriented (MACDONALD and PICKETT-HEAPS 1976). It would be interesting to see these observations extended to *Chaetomorpha* and *Valonia*.

2.2.2 Chlorococcales

Representative of this group may be uninucleate or coenocytic, unicellular or colonial. Because many of them reproduce through the production of autospores, which are initially naked, many important observations on cell wall formation have been made with members of this algal order, particularly since they are usually freshwater organisms and easier to cultivate in the laboratory than the marine Cladophorales. The first step in wall deposition is the production of a layer which presumably serves to protect the autospore during enzymatic dissolution of the mother cell wall. This layer is usually less electron-dense than the cellulosic wall which subsequently develops between it and the plasmalemma. In *Oocystis solitaria* it has a thickness of 50 to 70 nm and, when tangentially sectioned, is seen to contain interlocking fibrillar material (SACHS et al. 1976). In some species of *Chlorella* and *Scenedesmus* the layer is thinner (approx. 20 nm) but is more electron-dense at its inner and outer surfaces, thus giving rise to a trilaminar appearance. ATKINSON et al. (1972) have shown that this trilaminar layer contains sporopollenin and develops first as small plaques which then grow together (see Fig. 3a and b).

The majority of those species investigated develop a cellulosic wall in which the microfibrils are randomly oriented. During this period cortical microtubules are notably absent (ATKINSON et al. 1971). This is particularly apparent in *Hydrodictyon:* net production is accomplished by the adhesion of neighboring zoospores at sites where many cortical microtubules are located; as soon as wall production is begun at these contact regions the microtubules disappear (MARCHANT and PICKETT-HEAPS 1972).

Fig. 3a, b. First stages in wall development around autospores of *Scenedesmus obliquus*. A trilaminar layer (*TLL*) is deposited, first as plaques, outside the plasmalemma. ×90,000. *Bar* 0.2 μm

Fig. 4. Sandwich wall plus cortical microtubules (*arrows*) in *Oocystis solitaria* produced through addition and then removal of microtubule depolymerizing agents. ×55,000. *Bar* 0.2 μm. F. QUADER et al. (1978)

Fig. 5. *Oocystis solitaria:* PF-fracture face. Granule bands (*GB*) running from left to right are recognizable. ×60,000. *Bar* 0.2 μm

Fig. 6. *Oocystis solitaria:* EF-fracture face. Terminal complexes (*TC*) with associated microfibril imprints (MI) are seen. ×70,000. *Bar* 0.2 μm

Fig. 7. Microfibrils and associated particles associated with the trilaminar layer (*TLL*) in *Scenedesmus*. (*PL* Plasmalemma EF-face). ×70,000. *Bar* 0.2 μm

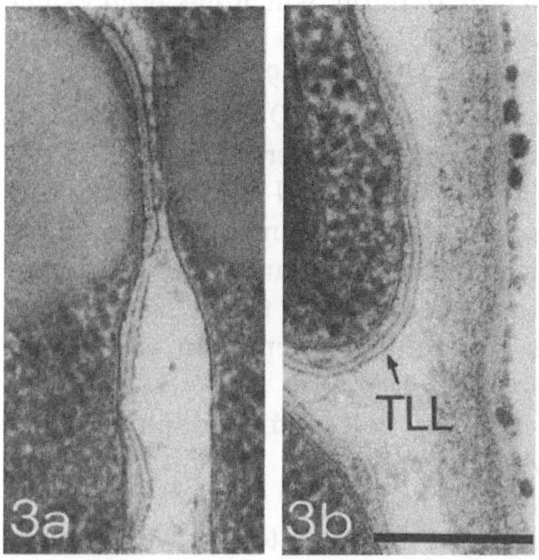
3a
TLL
3b

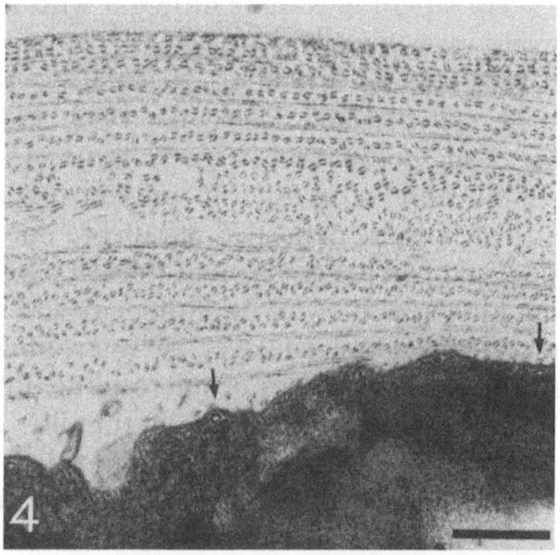
4

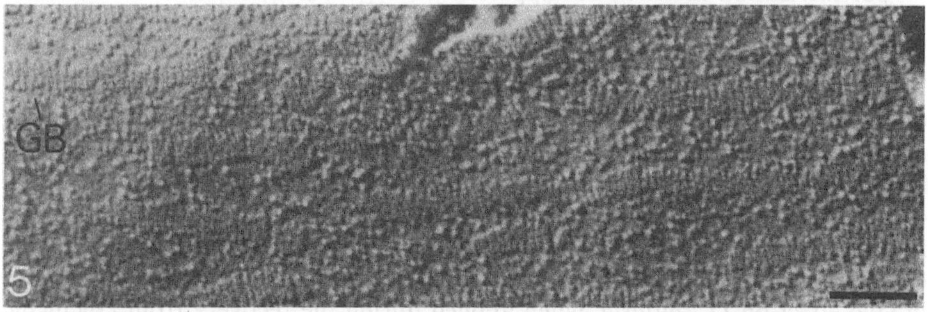
GB
5

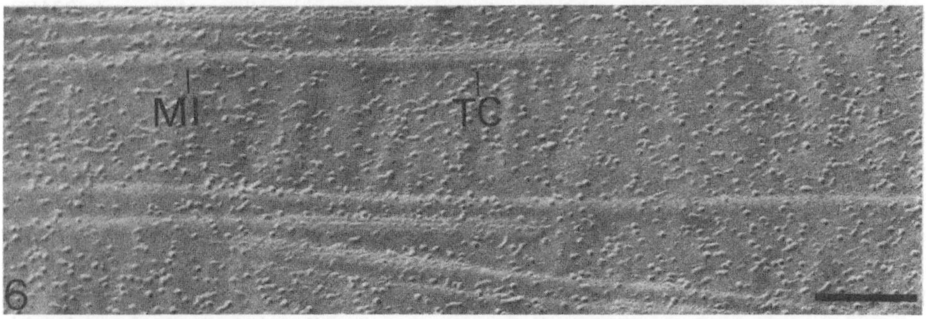
MI
TC
6

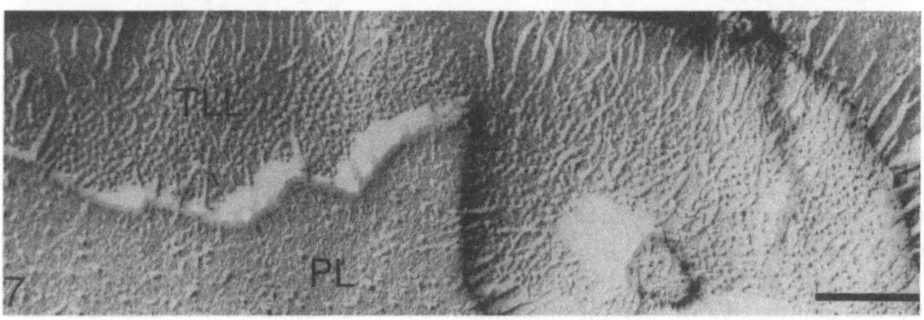
TLL
PL
7

The situation is quite different with *Oocystis solitaria,* which has been extensively investigated by Robinson and coworkers (Robinson et al. 1976, Sachs et al. 1976, Grimm et al. 1976, Robinson and Herzog 1977, Quader et al. 1978, Quader and Robinson 1979, 1980, Robinson and Quader 1980). This organism possesses extremely large (16 nm diameter) microfibrils which are readily observable in thin sections. The cells are elliptical in shape and the microfibrils are deposited parallel to one another in alternating layers describing slow and fast helices around the cell. Cortical microtubules are present throughout all stages of wall development and are aligned parallel to the last or most recently synthesized layer of microfibrils. Microfibril orientation can be influenced by two types of microtubule inhibitors:

1. Substances which bind to tubulin: colchicine, colcemid, podophyllotoxin, griseofulvin, methylbenzimidazole-2-yl-carbamate (see Dustin 1978 for necessary background on biochemistry of microtubules).
2. Substances which do not bind to tubulin and do not interfere with tubulin polymerization in vitro; but apparently deregulate the levels of intracellular calcium: the herbicides amiprophosmethyl (APM), trifluralin and oryzalin (Hertel et al. 1980).

When, through using these agents, the microtubules are depolymerized in vivo, the microfibrils are still synthesized parallel to one another, but without the regular change of orientation from layer to layer. These effects are fully reversible, resulting in the production of "sandwich walls" (Fig. 4), and the reassembly of cortical microtubules. Although the reappearance of the microtubules is not blocked by cycloheximide application, continual protein synthesis is apparently necessary for microfibril production. Similar to cycloheximide in their in vivo effect on *O. solitaria,* when applied during the recovery from colchicine treatment are D_2O and several ion deregulatory agents (ionophores A 23187, X-537 A, dianemycin; chelators EDTA, CTC; cryptates 211, 221). Among other possible effects, all of these agents considerably reduce protein synthesis. Although it is extremely unlikely that cortical microtubules are not involved in the orientation of cellulose synthesis in *O. solitaria,* observations on the similar walled algae *Glaucocystis nostochinearum, Eremosphaera viridis* and *O. lacustris* (Robinson 1977 and unpublished observations) do not allow a generalization of this type for all plant cells.

Freeze-etch results have been particularly illuminating with respect to microfibril synthesis. In *Chlorella* and *Scenedesmus,* where microfibrils are randomly oriented, particles 8 to 11 nm diameter, occasionally close-packed, and with attached microfibrils, are seen associated with the trilaminar outer layer (Fig. 7) rather than the plasmalemma (Sassen et al. 1970, Pickett-Heaps and Staehelin 1975). In an earlier paper on *Chlorella* Staehelin (1966) proposed that these particles originate at the plasmalemma and migrate to the inner surface of the outer layer, but in a later paper on *Scenedesmus* (Staehelin and Pickett-Heaps 1975) it is claimed that the outer layer behaves as a lipid bilayer and that the microfibril-associated particles are exposed on the inner face of the outer leaflet thereof.

The highly ordered synthesis of microfibrils in *Oocystis solitaria* is mirrored by the complexity of the plasmalemma. First investigated by Robinson and

PRESTON (1972), who demonstrated "granule bands" and mistakenly allocated them to the surface of the plasmalemma, this organism has received much publicity due to additional information provided by BROWN and MONTEZINOS (1976) and MONTEZINOS and BROWN (1976). In addition to confirming the existence of the granule bands which consist of stacked rows of eight paired 8.5 nm particles (Fig. 5), these authors have described a further structure: so-called "terminal complexes" which consist of three rows of 30 to 40 7-nm particles (Fig. 6). Although originally claiming therefore to have visualized the site(s) of cellulose synthesis, these authors concede later that "there is presently no *direct* evidence for a biosynthetic role for terminal complexes in *Oocystis*" (MONTEZINOS and BROWN 1978). Granule bands are located on the PF fracture plane of the plasmalemma and organized helically (like the microfibrils) with respect to the cell axis, terminal complexes are arranged antiparallel and are found on the EF plane. Destruction of cortical microtubules through colchicine causes a disturbance in arrangement of both granule and terminal complexes; inhibition of cellulose synthesis through cycloheximide results in a disappearance of the terminal complexes (ROBINSON, unpublished observation). Although biochemical investigations have not yet been carried out, there can be no doubt that *Oocystis* is one of the most important study objects for cellulose synthesis.

2.2.3 Conjugales (Placoderm Desmids)

In this algal group the cell wall is made up of three parts: an outer slime layer, a thin "primary" wall which may or may not be discarded during growth of the young semi-cell and a more permanent "secondary" wall. The primary wall contains randomly oriented microfibrils and the secondary wall layers of microfibrils which are oriented into 0.15 to 0.3 μm bands of up to 15 microfibrils each (MIX 1972). In *Micrasterias* similar-sized bands of 5 to 6 interconnected microtubules are also visible in the cortical cytoplasm during secondary wall formation (KIERMAYER 1968). Is this similarity in size a coincidence or do the microtubules control the orientation of the microfibril bands? Unfortunately inhibitor experiments of the type carried out with *Oocystis* have not yet been undertaken. Similar-sized matrices have recently been detected as an integral part of the plasmalemma through freeze-fracturing (KIERMAYER and SLEYTR 1979, GIDDINGS et al. 1980). These matrices take the form of diamond-shaped collections of approximately 175 "rosettes". Each rosette consists of five to six 7-nm particles in the PF-plane. Complementary to the hollow center of the rosettes are hexagonally arranged single particles in the EF plane which terminate in bands of microfibril impressions. An analogy to *Oocystis* is clear: granule bands ≡ rosettes; terminal complexes ≡ single particles. Proof of the participation of cortical microtubules in microfibril orientation in *Micrasterias* would make this analogy with *Oocystis* almost perfect.

The dictyosomal origin of the plasmalemma matrices in *Micrasterias* has been suggested from thin sections (DOBBERSTEIN and KIERMAYER 1972, KIERMAYER and DOBBERSTEIN 1973, KIERMAYER 1977). Special Golgi vesicles, so-called "F-vesicles" the lumen side of the membrane of which bears rows of particles

20 nm in diameter (Fig. 8) appear during secondary wall formation. These vesicles have now been demonstrated by Giddings et al. (1980) to bear rosette particle complexes and pictures are presented showing their fusion with the plasmalemma.

Another most unusual kind of vesicle is that seen in species of *Penium* (Mix and Manshard 1977). The vesicles are large (2.7 × 0.5 µm) and are situated in the cell cortex, particularly in the polar regions. One portion of the vesicle matrix is electron-dense and the other part contains a bundle of up to 12 microfibrils (Fig. 9a and b). Detailed information on the origin and fate of these vesicles would be most interesting.

2.3 The Production of Chitin Microfibrils

2.3.1 Poterioochromonas

Work by Schnepf and Herth (Schnepf et al. 1975, Herth et al. 1977, Schnepf et al. 1977) has established beyond question the existence of chitinous microfibrils around this chrysophycean flagellate. The fibrils constitute a lorica which initially forms at the posterior end after attachment to a substrate. Cortical microtubules, which arise from the outermost of a series of concentrically arranged flagellar root fibers ("apical platform"), progress helically into the tail and are mirrored, 1:1, by the orientation of the fasciated primary fibrils (Fig. 10). Depolymerization of the microtubules through colchicine application leads to disoriented primary fibrils, indicating the microtubular control over organization of synthesis. Although freeze-etching data is not yet available, microfibril synthesis is clearly a property of the plasmalemma and is very fast (1–2 glycosidic bonds · μm^{-2} membrane · min^{-1}).

2.3.2 Centric Diatoms

Although the chitinous nature of the fibrillar appendages of certain pelagic diatoms has been known for some 15 years (Maclachlan et al. 1965) the

→

Fig. 8. "F-Vesicles" in *Micrasterias denticulata* during secondary wall formation. × 63,000. (Kiermayer 1977)

Fig. 9a, b. Microfibril-containing vesicles **a** longitudinally × 22,000, **b** transversely × 37,000 sectioned, in *Penium spirostriolatum*. *Bar* 0.2 µm. (Mix and Manshard 1977)

Fig. 10. Negatively stained whole mount of *Poterioochromonas stipitata*. Chitin microfibrils run parallel to helically arranged microtubules. × 42,000. *Bar* 0.2 µm. (Schnepf et al. 1975)

Fig. 11. Marginal pore origin of chitin microfibrils in the diatom *Cyclotella meneghiniana*, *CF* Chitin fibril; *SP* strutted process. × 40,000. *Bar* 0.2 µm (Herth 1979)

Fig. 12. Chitin microfibril synthesis complex in *Thalassiosira fluviatilis*. *pm* plasma membrane; *cc* cytoplasmic coat; *v* vesicles which fuse with the plasmalemma. × 107,000. *Bar* 0.2 µm (Herth 1979)

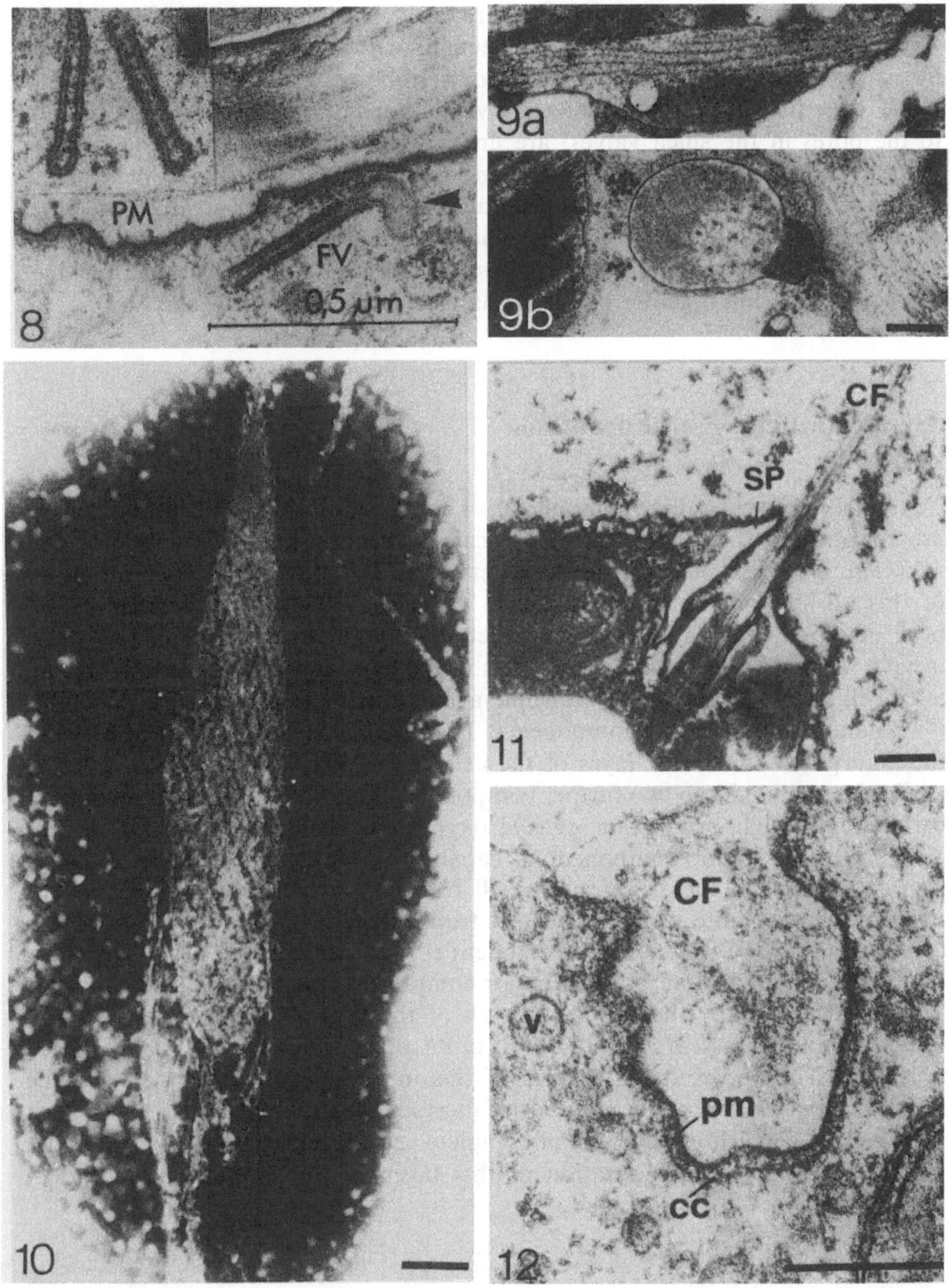
8
PM
FV
0,5 µm
9a
9b
10
11
SP
CF
12
CF
v
pm
cc

cytology of their formation has only recently been investigated (HERTH and BARTHLOTT 1979, HERTH 1979). In keeping with the unusual nature of the fibrils (extremely crystalline 50 nm broad ribbons; HERTH and ZUGENMAIER 1977), their sites of synthesis (conical invaginations of the plasmalemma interior to the pores in the silica valves) are unique structures (Fig. 11). The peg-like tip of the fibril has amorphous appendages which appear attached to the plasmalemma whose cytoplasmic side is decorated with hexagonally arranged material of unknown composition (Fig. 12). These observations have been interpreted as favoring a two-step microfibril synthesis, i.e., individually synthesized polymer chains crystallizing together by lateral association (see Chap. 3, this Vol).

3 Glycoprotein Wall Formation

Cell walls whose crystalline component is neither cellulosic nor fibrillar in nature are confined to the algal group Volvocales. The best-known example is that of *Chlamydomonas reinhardii* (ROBERTS et al. 1972), though four other variations of this basic type are also recognized (ROBERTS 1974). Wall formation has been investigated both in vitro (HILLS et al. 1975, CATT et al. 1976, 1978) and in vivo around regenerating protoplasts (ROBINSON and SCHLÖSSER 1978). Isolated cell walls treated with chaotropic agents can be fractionated into soluble (94%) and insoluble (6%) components. The latter is structurally identical with the amorphous innermost layer of the cell wall ("W1"). The soluble fraction can be fractionated into two further components both of which are glycoproteins containing a high proportion of hydroxyproline residues (see Chap. 7, this Vol). Only the major component is capable of self-assembly when dialyzed against water, and then only into small fragments which are identical to the stainable parts of the so-called central triplet (layers "W2 and W6") of the cell wall (Fig. 13b). The minor component is itself not capable of self-assembly but increases the yield and size of major component fragments; this component is believed to occupy the central portion of the central triplet (layer "W4"). For a complete reassembly of the wall in vitro the inner layer is also believed necessary. The apparent importance of this inner layer as a template for the assembly process in vivo is reduced by the fact that mutants exist which lack this layer, but which are capable of complete central triplet production (DAVIES 1972); moreover the visualization of this layer in thin section is dependent

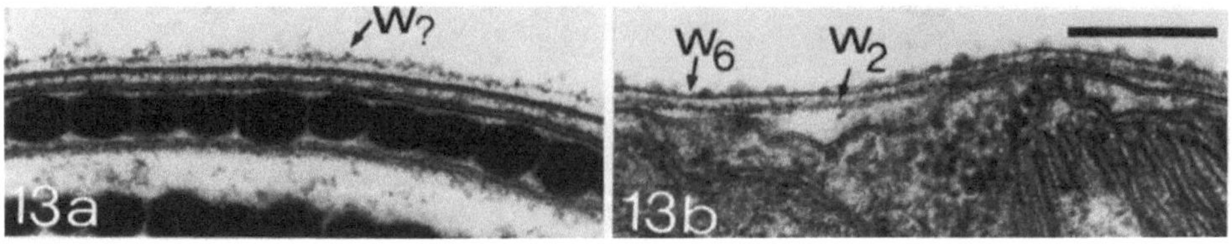

Fig. 13a, b. Wall regeneration around *Chlamydomonas smithii* protoplasts. **a** after 1 h; **b** after 2 h. ×75,000. *Bar* 0.2 μm

upon species and culture conditions. In regenerating protoplasts an amorphous layer external to the plasmalemma is observed after 1 h of incubation, but it is not clear whether this represents the inner layer or part of the central triplet or both (Fig. 13a). The full cell wall is completed after 2 h. A role for elements of the endomembrane system (ER, GA) in this regeneration is not apparent, in contrast with the results of MINAMI and GOODENOUGH (1978) on zygote wall formation, where fibrous elements, similar to those in the wall, are observable in secretory vesicles budding off from the GA.

Acknowledgments. The author would like to thank Drs. Mix and Herth for generously providing micrographs. Work carried out in my laboratory and reported in this article was supported by the Deutsche Forschungsgemeinschaft.

References

Allen DM, Northcote DH (1975) The scales of *Chrysochromulina chiton*. Protoplasma 83:389–412

Atkinson AW, Gunning BES, John PCL, McCullough W (1971) Centrioles and microtubules in *Chlorella*. Nature (London) New Biol 234:24–25

Atkinson AW, Gunning BES, John PCL (1972) Sporopollenin in the cell wall of *Chlorella* and other algae: Ultrastructure, chemistry and incorporation of ^{14}C-acetate, studied in synchronous cultures. Planta 107:1–32

Barnett JR, Preston RD (1970) Arrays of granules associated with the plasmalemma in swarmers of *Cladophora*. Ann Bot 34:1011–1017

Branton D, Bullivant S, Gilula NB, Karnovsky MJ, Moor H, Mühlethaler K, Northcote DH, Packer L, Satir B, Satir P, Speth V, Staehelin LA, Weinstein RS (1975) Freeze-etching nomenclature. Science 190:54–56

Brown RM (1969) Observations on the relationship of the Golgi apparatus to wall formation in the marine chrysophycean alga *Pleurochrysis scherffelii* Pringsheim. J Cell Biol 41:109–123

Brown RM, Montezinos D (1976) Cellulose microfibrils: visualization of biosynthetic and orienting complexes in association with the plasma membrane. Proc Natl Acad Sci USA 73:143–147

Brown RM, Romanovicz DK (1976) Biogenesis and structure of Golgi-derived cellulosic scales in *Pleurochrysis*. I. Role of the endomembrane system in scale assembly and exocytosis. Appl Polym Symp 28:537–585

Brown RM, Franke WW, Kleinig H, Falk H, Sitte P (1970) Scale formation in chrysophycean algae. I. Cellulosic and non-cellulosic wall components made by the Golgi apparatus. J Cell Biol 45:246–271

Brown RM, Herth W, Franke WW, Romanovicz D (1973) The role of the Golgi apparatus in the biosynthesis and secretion of a cellulosic glycoprotein in *Pleurochrysis:* A model system for the synthesis of structural polysaccharides. In: Loewus F (ed) Biogenesis of Plant Cell Wall Polysaccharides. Academic Press, New York, pp 207–257

Callow ME, Evans LV (1973) Studies on the ship-fouling alga *Enteromorpha*. III. Cytochemistry and autoradiography of adhesive production. Protoplasma 80:15–27

Callow ME, Evans LV (1976) Localization of sulfated polysaccharides by X-Ray microanalysis in *Laminaria saccharina*. Planta 131:155–157

Callow ME, Evans LV (1977) Studies on the ship-fouling alga *Enteromorpha* (Chlorophyceae, Ulvales). IV. Polysaccharide and nucleoside diphosphatase localization. Protoplasma 16:313–320

Callow ME, Coughlan SJ, Evans LV (1978) The role of Golgi bodies in polysaccharide sulfation in *Fucus* zygotes. J Cell Sci 32:337–356

Catt JW, Hills GJ, Roberts K (1976) A structural glycoprotein, containing hydroxyproline, isolated from the cell wall of *Chlamydomonas reinhardii*. Planta 131:165–171

Catt JW, Hills GJ, Roberts K (1978) Cell wall glycoproteins from *Chlamydomonas reinhardii*, and their self assembly. Planta 138:91–98

Coughlan S, Evans LV (1978) Isolation and characterization of Golgi bodies from vegetative tissue of the brown alga *Fucus serratus*. J Exp Bot 29:55–68

Darley WM, Porter D, Fuller MS (1969) Cell wall composition and synthesis via Golgi-directed scale formation in the marine eucaryote, *Schizochytrium aggregatum*, with a note on *Thraustochytrium* sp. Arch Mikrobiol 90:89–106

Darley WM, Porter D, Fuller MS (1969) Cell wall composition and synthesis via Golgi-directed scale formation in the marine eucaryote, *Schizochytrium aggregatum*, with a note on *Thraustochytrium* sp. Arch Mikrobiol 90:89–106

Davies DR (1972) Electrophoretic analyses of wall glycoproteins in normal and mutant cells. Exp Cell Res 73:512–516

Dobberstein B, Kiermayer O (1972) Das Auftreten eines besonderen Typs von Golgivesikeln während der Sekundärwandbildung vom *Micrasterias denticulata* Bréb. Protoplasma 75:185–194

Dustin P (1978) Microtubules. Springer, Berlin Heidelberg New York

Evans LV, Callow ME (1974) Polysaccharide sulphation in *Laminaria*. Planta 117:93–95

Evans LV, Holligan MS (1972) Correlated light and electron microscope studies on brown algae. I. Localization of alginic acid and sulphated polysaccharides in *Dictyota*. New Phytol 71:1161–1172

Evans LV, Simpson M, Callow ME (1973) Sulphated polysaccharide synthesis in brown algae. Planta 110:237–252

Evans LV, Callow ME, Percival E, Fareed V (1974) Studies on the synthesis and composition of extracellular mucilage in the unicellular red alga *Rhodella*. J Cell Sci 16:1–21

Franke WW, Brown RM (1971) Scale formation in chrysophycean algae. III. Negatively stained scales of the coccolithophorid *Hymenomonas*. Arch Mikrobiol 77:12–19

Furtado JS, Olive LS (1971) Ultrastructure of the protostelid *Ceratiomyxella tahitiensis* including scale formation. Nova Hedwigia 21:537–576

Giddings TH, Brower DL, Staehelin LA (1980) Visualization of particle complexes in the plasma membrane of *Micrasterias denticulata* associated with the formation of cellulose fibrils in primary and secondary walls. J Cell Biol 84:327–339

Gooday GW (1971) A biochemical and autoradiographic study of the role of the Golgi bodies in thecal formation in *Platymonas tetrahele*. J Exp Bot 22:959–971

Grimm I, Sachs H, Robinson DG (1976) Structure, synthesis and orientation of microfibrils. II. The effect of colchicine on the wall of *Oocystis solitaria*. Cytobiologie 14:61–74

Heath IB (1974) A unified hypothesis for the role of membrane bound enzyme complexes and microtubules in plant cell wall synthesis. J Theoret Biol 48:445–449

Hertel C, Quader H, Robinson DG, Marmé D (1980) Antimicrotubule herbicides and fungicides affect Ca^{2+} – transport in plant mitochondria. Planta 149:336–340

Herth W (1979) The site of β-chitin fibril formation in centric diatoms. II. The chitin-forming cytoplasmic structures. J Ultr Res 68:16–27

Herth W, Barthlott W (1979) The site of β-chitin fibril formation in centric diatoms. I. Pores and fibril formation. J Ultr Res 68:6–15

Herth W, Zugenmaier P (1977) Ultrastructure of the chitin microfibrils of the centric diatom *Cyclotella cryptica*. J Ultr Res 61:230–239

Herth W, Franke WW, Stadler J, Bittiger H, Keilich G, Brown RM (1972) Further characterization of the alkali-stable material from the scales of *Pleurochrysis scherfellii*: a cellulosic glycoprotein. Planta 105:72–92

Herth W, Kuppel A, Schnepf E (1977) Chitinous fibrils in the lorica of the flagellate chrysophyte *Poterioochromonas stipitata* (syn. *Ochromonas malhamensis*). J Cell Biol 73:311–321

Hibberd DJ (1976) The ultrastructure and taxonomy of the Chrysophyceae and Prymnesiophyceae (Haptophyceae): a survey with some new observations on the ultrastructure of the Chrysophyceae. Bot J Linn Soc 72:55–80

Hills GJ, Philips JM, Gay MR, Roberts K (1975) Self-assembly of a plant cell wall *in vitro*. J Mol Biol 96:431–441

Kiermayer O (1968) The distribution of microtubules in differentiating cells of *Micrasterias denticulata* Bréb. Planta 83:223–236

Kiermayer O (1977) Biomembran als Träger morphogenetischer Information. Untersuchungen bei der Grünalge *Microasterias*. Naturwiss Rundsch 30:161–165
Kiermayer O, Dobberstein B (1973) Membrankomplex dictyosomaler Herkunft als „Matritzen" für die extraplasmatische Synthese und Orientierung von Mikrofibrillen. Protoplasma 77:437–451
Kiermayer O, Sleytr UB (1979) Hexagonally ordered "rosettes" of particles in the plasma membrane of *Micrasterias denticulata* Bréb. and their significance for microfibril formation and orientation. Protoplasma 101:133–138
Leadbeater BSC (1971) Observations on the life history of the haptophycean alga *Pleurochrysis scherffelii* with special references to the microanatomy of the different types of motile cells. Ann Bot 35:429–439
MacDonald K, Pickett-Heaps JD (1976) Ultrastructure and differentiation in *Cladophora glomerata*. I. Cell division. Am J Bot 63:592–601
MacLachlan J, McInnes AG, Falk M (1965) Studies on the chitin (chitin-poly-N-acetylglucosamine) fibers of the diatom *Thalasiosira fluviatilis* Hustedt. I. Production and isolation of chitin fibers. Can J Bot 43:707–713
Manton I (1966) Observations on scale production in *Pyramimonas amylifera* Conrad. J Cell Sci 1:429–438
Manton I (1967) Further observations on the fine structure of *Chrysochromulina chiton* with special reference to the haptonema, "peculiar" Golgi structure and scale production. J Cell Sci 2:265–272
Manton I (1972) Observations on the biology and micro-anatomy of *Chrysochromulina megacylindrica*. Br Phycol J 7:235–248
Marchant HJ, Pickett-Heaps JD (1972) Ultrastructure and differentiation of *Hydrodictyon reticulatum*. III. Formation of the vegetative daughter net. Aust J Biol Sci 25:265–278
Minami SA, Goodenough UW (1978) Novel glycopolypeptide synthesis induced by gametic cell fusion in *Chlamydomonas reinhardii*. J Cell Biol 77:165–181
Mix M (1972) Die Feinstruktur der Zellwände bei Mesotaeniceae und Gonatozygaceae mit einer vergleichenden Betrachtung der verschiedenen Wandtypen der Conjugatophyceae und deren systematischen Wert. Arch Mikrobiol 81:197–220
Mix M, Manshard E (1977) Über Mikrofibrillen-Aggregate in langgestreckten Vesikeln und ihre Bedeutung für die Zellwandbildung bei einem Stamm von *Penium* (Desmidiales). Ber Dtsch Bot Ges 90:517–526
Moestrup O, Thomsen HA (1974) An ultrastructural study of the flagellate *Pyramimonas orientalis* with particular emphasis on Golgi apparatus activity and the flagellar apparatus. Protoplasma 81:247–269
Moestrup Ø, Walne PL (1979) Studies on scale morphogenesis in the Golgi apparatus of *Pyraminonas tetrahynchus* (Prasinophyceae). J Cell Sci 36:437–459
Montezinos D, Brown RM (1976) Surface architecture of the plant cell: biogenesis of the cell wall, with special emphasis on the role of the plasma membrane in cellulose biosynthesis. J Supramol Struct 5:277–290
Montezinos D, Brown RM (1978) Cell wall biogenesis in *Oocystis*: experimental alteration of microfibril assembly and orientation. Cytobios 23:119–139
Nicolai E, Preston RD (1959) Cell wall studies in the Chlorophyceae. III. Differences in structure and development in the Cladophoraceae. Proc Roy Soc London Ser B 151:244–255
Outka DE, Williams DC (1971) Sequential coccolith morphogenesis in *Hymenomonas carterae*. J Protozool 18:285–297
Peng HB, Jaffe LF (1976) Cell wall formation in *Pelvetia* embryos. A freeze-fracture study. Planta 133:57–71
Pickett-Heaps JD, Staehelin LA (1975) The ultrastructure of *Scenedesmus* (Chlorophyceae). II. Cell division and colony formation J Phycol 11:186–202
Preston RD (1964) Structural and mechanical aspects of plant cell walls with particular reference to synthesis and growth. In: Zimmermann MH (ed) The formation of wood in forest trees. Academic Press, New York, pp 169–188
Preston RD (1974) The Physical Biology of Plant Cell Walls. Chapman and Hall, London
Quader H (1981) Effects of D_2O on microtubules and microfibrils in *Oocystis*. In: Robinson DG, Quader H (eds) Cell walls 81. Wiss Verlagsges, Stuttgart, pp. 198–205

Quader H, Robinson DG (1979) Structure, synthesis and orientation of microfibrils. VI. The role of ions in microfibril deposition in *Oocystis solitaria*. Eur J Cell Biol 20:51–56

Quader H, Wagenbreth I, Robinson DG (1978) Structure, synthesis and orientation of microfibrils. V. On the recovery of *Oocystis solitaria* from microtubule inhibitor treatment. Cytobiologie 18:39–51

Ramus J (1972) The production of extracellular polysaccharide by the unicellular red alga *Porphyridium aerugineum*. J Phycol 8:97–111

Ramus J, Robins DM (1975) The correlation of Golgi activity and polysaccharide secretion in *Porphyridium*. J Phycol 11:70–74

Roberts K (1974) Crystalline glycoprotein cell walls of algae: their structure, composition and assembly. Philos Trans R Soc London Sev B 268:129–146

Roberts K, Gurney-Smith M, Hills GS (1972) Structure, composition and morphogenesis of the cell wall of *Chlamydomonas reinhardii*. I. Ultrastructure and preliminary chemical analysis. J Ultr Res 40:599–613

Robinson DG (1977) Structure, synthesis and orientation of microfibrils. IV. Microtubules and microfibrils in *Glaucocystis*. Cytobiologie 15:475–484

Robinson DG, Herzog W (1977) Structure, synthesis and orientation of microfibrils. III. A survey of the action of microtubule inhibitors on microtubules and microfibril orientation in *Oocystis solitaria*. Cytobiologie 15:463–474

Robinson DG, Preston RD (1971) Fine structure of swarmers of *Cladophora* and *Chaetomorpha*. I. The plasmalemma and Golgi apparatus in naked swarmers. J Cell Sci 9:581–601

Robinson DG, Preston RD (1972) Plasmalemma structure in relation to microfibril biosynthesis in *Oocystis*. Planta 104:234–246

Robinson DG, Quader H (1980) Structure, synthesis and orientation of microfibrils. VII. Microtubule reassembly *in vivo* after cold treatment in *Oocystis* and its relevance to microfibril orientation. Eur J Cell Biol 21:229–230

Robinson DG, Schlösser UG (1978) Cell wall regeneration by protoplasts of *Chlamydomonas*. Planta 141:83–92

Robinson DG, White RK, Preston RD (1972) Fine structure of swarmers of *Cladophora* and *Chaetomorpha*. III. Wall synthesis and development. Planta 107:131–144

Robinson DG, Grimm I, Sachs H (1976) Colchicine and microfibril orientation. Protoplasma 89:375–380

Romanovicz DK, Brown RM (1976) Biogenesis and structure of Golgi-derived cellulosic scales in *Pleurochrysis*. II. Scale composition and supramolecular structure. Appl Polym Symp 28:587–610

Sachs H, Grimm I, Robinson DG (1976) Structure synthesis and orientation of microfibrils. I. Architecture and development of the wall of *Oocystis solitaria*. Cytobiologie 14:49–60

Sassen A, Eyden-Emons van A, Lamers A, Wanka F (1970) Cell wall formation in *Chlorella pyrenoidosa:* a freeze etching study. Cytobiologie 1:373–382

Schnepf E, Deichgräber G (1969) Über die Feinstrukturen von *Synura petersenii* unter besonderer Berücksichtigung der Morphogenese ihrer Kieselschuppen. Protoplasma 68:85–106

Schnepf E, Röderer G, Herth W (1975) The formation of the fibrils in the lorica of *Poterioochromonas stipitata*: tip growth, kinetics, site, orientation. Planta 125:45–62

Schnepf E, Deichgräber G, Röderer G, Herth W (1977) The flagellar root apparatus, the microtubular system and associated organelles in the chrysophycean flagellate *Poterioochromonas malhamensis* Peterfi (syn *Poterioochromonas stipitata* Scherffel and *Ochromonas malhamensis* Pringsheim). Protoplasma 92:87–107

Staehelin LA (1966) Die Ultrastruktur der Zellwand und des Chloroplasten von Chlorella. Z Zellforsch 74:325–350

Staehelin LA, Pickett-Heaps JD (1975) The ultrastructure of *Scenedesmus* (Chlorophyceae). I. Species with the "Reticulate" or "warty" type of ornamental layer. J Phycol 11:163–185

14 Algal Wall-Degrading Enzymes – Autolysines

U.G. Schlösser

1 Introduction

The structure of the walls of cells which grow, differentiate, and fuse with other cells must be changeable during ontogenesis. An obvious expression of structural change is lysis of the cell wall. Normal (nonpathological) cell wall autolysis seems to appear during different events in the life cycle of plants:

- *active growth and vegetative differentiation*
  (e.g., during extending growth; tip growth of fungal hyphae; differentiation of xylem vessels, of pores in sieve vessels, of milk vessels, and of lysigenic excretion bodies)
- *formation and liberation of reproduction units*
  (e.g., from sporangia and gametangia; lysis of tapetum cells into periplasmodium with Pteridophytes and Spermatophytes; transitory formation of pits between pollen mother cells; autolysis of fungal fruit bodies)
- *cell fusions*
  (e.g., of sexual reproduction cells; dissolution of pistil tissue during lysigenic pollen tube growth; formation of anastosomes, clamp connections, and dissolution of septa with hyphae of fungi; fusion of carposporogenic threads with auxiliar cells, and formation of secondary pits in red algae)
- *transition from resting stage to growth*
  (e.g., germination of cysts, zygotes, spores, pollen, and seed)
- *mobilization of wall material during senescence*
  (e.g., in whithering flowers and leaves; lysis of pectic substances in separating zones in stalks of leaves and fruit, and in ripened fruit tissue).

Natural cell wall autolysis is always specific to a stage of development and, in multicellular plants, locally confined to special cells or tissues.

Such autolytic phenomena have been observed microscopically and very often with nearly all divisions of algae. Usually, the operating wall-degrading enzymes are not known. Enzymic factors of unknown or known catalytic action dissolving cell walls of the secreting species totally or in part are called "autolysines" in the following text.

The release of reproduction units from sporangia and gametangia is connected very often with particularly striking and short-term changes of the walls, e.g., gelatinization, partial or total lysis, differentiation of pores, channels, or lids (Fig. 1). This chapter refers to a simple method for the demonstration and isolation of autolysines by use of synchronized release of reproduction units. It requires the following steps:

- selection of suitable cell systems with wall autolysis;

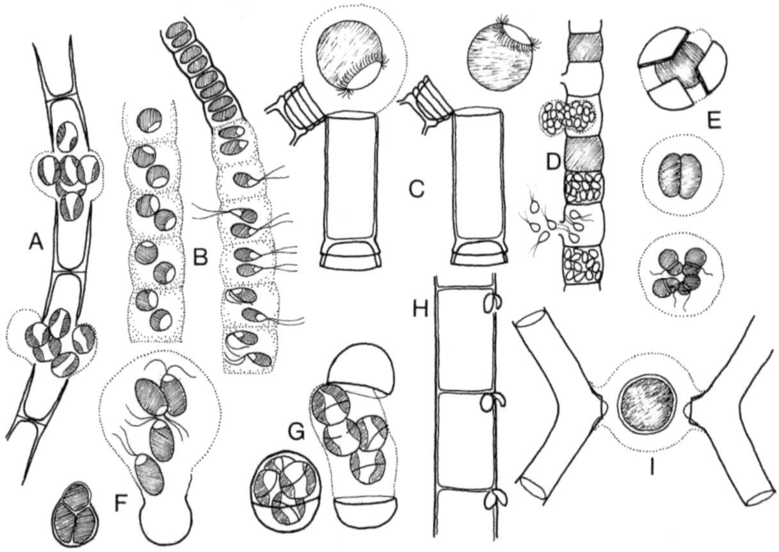

Fig. 1 A–I. Release of reproduction units from sporangia and gametangia connected with changes of the walls. **A** Aplanospores in *Tribonema* sp., swelling jelly. **B** Aplano- and zoospores in *Microspora quadrata,* unlimited swelling of the sporangium walls. **C, D, E, F** Release in a transitory vesicle of the zoospore in *Oedogonium concatenatum,* gametes in *Ulothrix zonata,* germination of zygotes in *Glenodinium lubiniensiforme* and *Hydrodictyon reticulatum.* **G** Aplanospores in *Chlorothecium inaequale,* swelling jelly which lifts an operculum. **H** Gametangia of *Acrosiphonia* sp., formation of operculum. **I** Zygote of *Mougeotia oedogonioides,* separated from the empty gamete walls by swelling jelly. (Schematically redrawn after A PASCHER 1939, B SKUJA 1956, C HIRN 1900, D DODEL 1876, E DIWALD 1938, F SCHLÖSSER, original, G PASCHER 1932, H KORNMANN 1965, I CZURDA 1931)

- synchronization of both events, formation and release, in high cell density for the purpose of concentration of the autolysines and their special substrates in the walls;
- bioassays in order to prove and control the enzymic activities.

Synchronized cultures of unicellular or multicellular microalgae, where all cells are forced to pass an autolytic reproduction stage at the same time have proved themselves to be a suitable material. Until now the method has been shown successful for the green algae *Chlamydomonas, Volvox, Chlorella,* and *Geminella.*

2 Autolysines Found in Algae

2.1 Chlamydomonas (Chlorophyceae, Volvocales)

2.1.1 Evidence

In this genus cell wall autolysis may be observed microscopically in at least three stages of the life cycle, e.g., in *C. reinhardii* (Fig. 2), two of which are now known to be mediated by the action of a special autolysine.

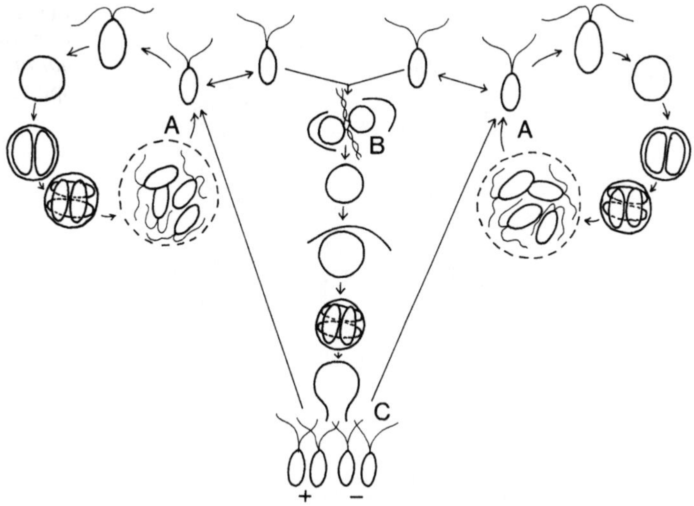

Fig. 2A–C. Scheme of the life cycle of *Chlamydomonas reinhardii*. **A** total dissolution of the sporangium wall during zoospore release. **B** discharge of gametes connected with partial dissolution of their walls. **C** partial dissolution of the zygote wall during release of meiospores. (SCHLÖSSER 1976)

A *"sporangium autolysine"* can be demonstrated in *C. reinhardii* according to the following method (SCHLÖSSER 1966, 1976, confirmed by MIHARA and HASE 1975): Cells were synchronized in aerated liquid mineral medium by light–dark changes every day and dilution to a constant cell number at the end of each dark period (light–dark = 14:10, 15,000 lux, compressed air + 2% CO_2, dilution to 1.5×10^6 cells ml^{-1}). Medium and method of cell synchronization in a culture chamber are the same as developed for *Chlorella* (KUHL and LORENZEN 1964). When cell divisions are completed cells accumulate quantitatively in the sporangium stage (Fig. 3) during the dark phase for several hours. This is because release of zoospores is slightly inhibited in the aerated culture conditions. Concentrated by centrifugation to one-tenth of its volume, a sporangium suspension can be induced to release zoospores quantitatively within a short period by incubation in illuminated dishes without aeration. A total dissolution of the sporangium walls and secretion of autolysine into the medium thus ensues. Autolysine action is demonstrable in this "sporulation medium" after separation from the zoospores by centrifugation in a *bioassay:* Heat-fixed sporangia (5 min 55 °C) from 2 ml of a standard synchronized culture were incubated at 30 °C with 2 ml of the sporulation medium. Activity of the autolysine was measured microscopically by the time required for quantitative release of the killed zoospores ("sporangiolysis"). Normally this requires 5 to 20 min.

Wall dissolving ability of the sporangium autolysine was proved with an isolated and purified fraction of the sporangium wall (Fig. 4).

A "gamete autolysine" was demonstrated by CLAES (1971) in *C. reinhardii* which becomes active and secreted into the medium after flagellar contact of the + and − gametes, and induces shedding of the gamete walls. With this autolysine the gametes temporarily become protoplasts. The conditions for the production and proof of the gamete autolysine were improved by application

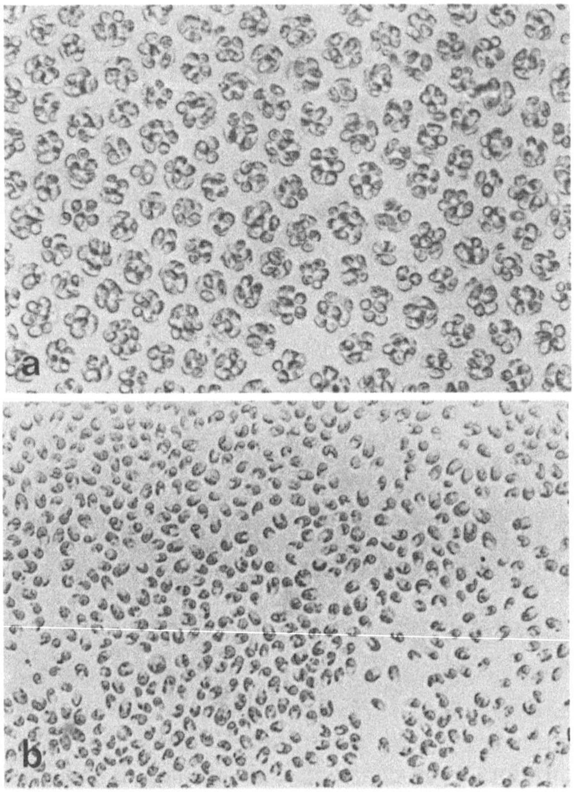

Fig. 3a, b. Synchronized culture of *Chlamydomonas reinhardii*. **a** Sporangia. **b** Zoospores

of suitable strains in the following way (SCHLÖSSER 1976): Synchronized cultures of *C. reinhardii* 11–32b and 11–32c (strains of the Sammlung von Algenkulturen, Universität Göttingen) were transferred into a "gametogenesis medium" (10^{-3} M $MgSO_4$, 10^{-4} M $CaCl_2$, 5×10^{-3} M Na-phosphate buffer pH 6.0 in distilled water) after centrifugation 9 h after the start of the light period, that is just before cell divisions begin, and aerated again in the culture chamber. Nine hours later the cells, which are in the sporangium or released daughter cell stage, were tranferred after centrifugation into new gametogenesis medium concentrated to 1.5×10^7 cells ml^{-1}. Sixty ml of these dense cell suspensions were exposed to light (4–10 klux) in dishes of 20 cm diameter for 3 h at room temperature. Then the gamete suspensions were mixed 1:1. At once their flagella agglutinate, and within 30 to 60 min all cells shed their walls which are totally dissolved shortly after. The autolysine accumulates in this "zygote medium" and can be separated from the cells by centrifugation. The *bioassay* is based upon the liberation of protoplasts from living vegetative cells under the influence of the gamete autolysine. Cells during the 11th h after start of the light period of 4 ml of a synchronized culture of *C. smithii* 54.72, a species closely related to *C. reinhardii*, were incubated in 2 ml of the zygote medium at 30 °C. The activity was controlled microscopically with India ink preparations and measured as the time required for quantitative protoplast formation. Normally all cells

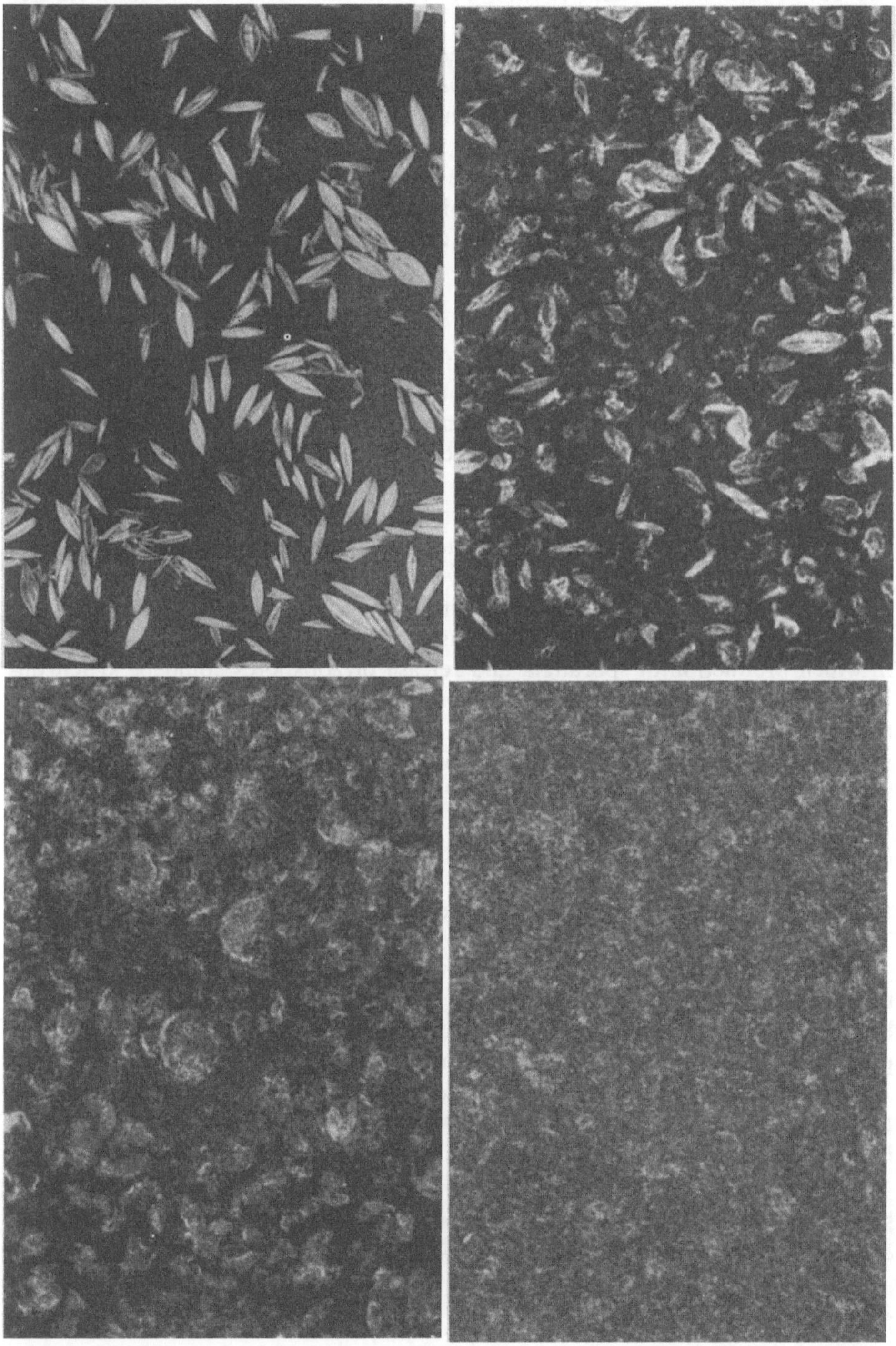

Fig. 4. Lysis of an isolated and purified fraction of the sporangium wall of *Chlamydomonas reinhardii* by the action of the sporangium autolysine. (SCHLÖSSER 1976)

slip out of their walls within 10 min, and round themselves off within a further 15 min.

HEIMKE and STARR (1979) described the action of a "sexual hormone" in the oogamous *Chlamydomonas zimbabwiensis* which might be regarded as a gamete autolysine. This is secreted into the medium during release of the male gametes by partial lysis of their mother cell wall. It causes vegetative cells to cast off their cell walls by partial lysis and function as female gametes. Its activity is proved by a quantitative bioassay based upon liberation of protoplasts = female gametes from vegetative cells.

2.1.2 Specificity of Action

The autolysines of *C. reinhardii* differ from each other in their *developmental stage specificity* of action which could be demonstrated in the bioassays (CLAES 1971, SCHLÖSSER 1976). The sporangium autolysine dissolves only the sporangium wall, i.e., the mother cell wall which has lifted from the protoplast during the first division step. With sporangium autolysines from species whose walls contain a broad peripheral jelly layer, this jelly in vegetative cells can be dissolved but not the proper cell wall. In contrast to this limited action of the sporangium-autolysine the gamete autolysine dissolves the wall of every cell stage (gamete, zoospore, vegetative cell, and sporangium) except that of the zygote.

A sporangium autolysine was demonstrable in all species of *Chlamydomonas* in which production of sporangia and release of zoospores could be synchronized. A comparative investigation where the sporulation media of all these active strains were cross-tested against their sporangia resulted in a *species specificity,* i.e., the sporangium autolysines act only between morphologically similar species and in some cases only upon the sporangium of the producer strain (SCHLÖSSER 1976). From the 56 active *Chlamydomonas* strains 15 autolysine groups could be distinguished.

A lytic action between these groups has been proven in several cases. The autolysines from four groups each acted on sporangia of one other group. This action was nonreciprocal (e.g., Table 1).

Both the sporangium and the gamete autolysine of *C. reinhardii* agree in their species specificity: production of protoplasts as well as sporangiolysis by the gamete autolysine occur only with those strains of *Chlamydomonas* which possess, together with *C. reinhardii,* a mutual sporangium autolysine.

The action of the gamete autolysine of *C. zimbabwiensis* is also specific to developmental stage and species (HEIMKE and STARR 1979). Only living flagellated vegetative cells were induced to cast off their walls, but not nonflagellated dividing cells or sporangia. Its action is confined to the similar oogamous species *C. capensis* and *C. pseudogigantea*. *C. eugametos, C. gigantea, C. gymnogama,* and *C. reinhardii* were not susceptible.

2.1.3 Isolation and Properties

Sporangium autolysines from eight species of different sporangium autolysine groups and the gamete autolysine of *C. reinhardii* have some characteristics

Table 1. Nonreciprocal action of sporangium autolysines from different *Chlamydomonas* strains on their heat-fixed sporangia in the bioassay: autolysines from the strains 4.72, 26.72, and 14.72 dissolve the sporangium walls of the strains 81.72, 7.73, 11–32a, 11–32b, 11–32c, 73.72, but not vice versa. Explanation of symbols: quantitative release of spores within +++ 5–20 min ++ 60 min + 24 h – no release

Autolysines from:	Sporangia of									
	81.72 *C. glob.*	7.73 *C. inc.*	11–32a *C. reinh.*	11–32b *C. reinh.*	11–32c *C. reinh.*	73.72 *C. reinh.*	54.72 *S. smith.*	4.72 *C. ang.*	26.72 *C. kom.*	14.72 *C. deb.*
81.72 *C. globosa*	+++	+++	+++	+++	+++	+++	+++	–	–	–
7.73 *C. incerta*	+++	++	+++	+++	+++	+++	+++	–	–	–
11–32a *C. reinhardii*	+++	++	+++	+++	+++	+++	+++	–	–	–
11–32b *C. reinhardii*	++	++	+++	+++	+++	+++	+++	–	–	–
11–32c *C. reinhardii*	++	++	+++	+++	+++	+++	+++	–	–	–
73.72 *C. reinhardii*	++	++	++	++	+++	+++	+++	–	–	–
54.72 *C. smithii*	++	++	+++	+++	+++	+++	+++	–	–	–
4.72 *C. angulosa*	+	+	+	+	+	+	+	++	++	+
26.72 *C. komma*	+	+	+	+	+	+	+	++	++	+
14.72 *C. debaryana*	+	+	+	+	+	+	+	+	+	+

in common (SCHLÖSSER 1976): they can be salted out with $(NH_4)_2SO_4$, were nondialyzable, temperature-labile (irreversible inactivated after 10 min at 70 °C), and inactivated with 5×10^{-4} M $HgCl_2$, with the proteinase papain, or with 10^{-2} M EDTA. Inactivation with EDTA of the gamete autolysine can be overcome in the presence of divalent cations (Ca^{2+}, Mg^{2+}, Mn^{2+}, Ba^{2+}, Sr^{2+}, Co^{2+}, Zn^{2+}).

The gamete autolysine of *C. reinhardii* was characterized as a neutral tryptic serine protease by JAENICKE and WAFFENSCHMIDT (1981). After lyophilization and precipitation with 40% to 70% $(NH_4)_2SO_4$ it was purified and enriched about 90-fold by ion exchange on CM-cellulose and passage through Sephadex G-75 to an electrophoretically uniform fraction. This purified enzyme has a molecular weight between 40,000 and 37,000 and is a single peptide chain. It hydrolyzes proteins like casein, hemoglobin, or insulin at basic residues. It is a weak basic protein with a pH optimum about 7.5. Different serine specific and trypsin inhibitors are effective, indicating serine in its active center and tryptic properties. The heat inactivation kinetics is first-order, pointing to only one active species of the lysis enzyme.

The properties of the gamete autolysine of *C. zimbabwiensis* (HEIMKE and STARR 1979) are similar to those of the *C. reinhardii* autolysine: nondialysable, concentrated by adsorption to CM-cellulose or lyophilization, molecular weight about 35,500, high tolerance to pH (active at pH 4.0–10.0), temperature-labile (inactive after 15 min 55 °C), inactivated by pronase.

A protease nature of the autolysines makes sense because in the genus *Chlamydomonas* and generally in the Volvocales there is a special type of cell envelope: it is a glycoprotein without cellulose (see Chap. 7, this Vol.). Probably the Volvocales are generally characterized by proteases acting as specific cell envelope autolysines. The high specificity of action of the *Chlamydomonas* autolysines reminds one of comparable circumstances in the cell walls of bacteria (STROMINGER and GHUYSEN 1967). Here, a number of specific proteases are known which hydrolyze the peptide bridges connecting poly-disaccharide chains of murein sacculus each at a specific site.

2.1.4 Regulation

The activation and secretion of all autolysines are strictly developmental stage-specific. Only flagellated cell stages activate autolysines. The sporangium autolysines are secreted by zoospores and young vegetative cells (SCHLÖSSER 1976).

The gamete autolysine of the oogamous *C. zimbabwiensis* (HEIMKE and STARR 1979) is shed by the male gametes in absence of the females. The gamete autolysine of the isogamous *C. reinhardii* is secreted only after preceding flagellar contact of + and − gametes. Gametes of both mating types produce the autolysine, which was demonstrated after agglutination of living gametes of one mating type with glutaraldehyde-fixed gametes, or with isolated flagella, or with antibodies against isolated flagella of vegetative cells of the other mating type and vice versa (CLAES 1971, 1977, GOODENOUGH and WEISS 1975, WEISS et al. 1977). In contrast to this MATSUDA et al. (1978) stated with other strains of *C. reinhardii* that only gametes of the + mating type secrete the autolysine during agglutination, but not that of the − mating type. They agglutinated glutaraldehyde-fixed gametes or isolated flagella with intact gametes. This contradiction may be due to different methods of the fixation of gametes or amputation of the flagella, or depend on strain-specific differences within this species.

Obviously the flagella, besides their function in motility, are recipient and transfer organelles in the regulation of gamete autolysine activation. Gamete walls are not shed and no autolysine is secreted if the flagellar agglutination is inhibited, e.g., with trypsin (CLAES 1971). Instead of flagellar agglutination the secretion of the autolysine can be stimulated unspecifically in the − mating type but not in the + mating type of *C. reinhardii* by immobilized concanavalin A, or by adhesion of the flagella to polystyrene (plastic petri dishes), or by iso-agglutination of the flagella with antibodies against isolated flagella of vegetative cells (CLAES 1977). The ionophore A 23187 induced in the presence of Ca^{2+} shedding of the walls and secretion of autolysine in gametes of both mating types and also in vegetative cells (CLAES 1980). The ionophore induces autolysine secretion only in cells with intact flagella, deflagellated cells do not respond. A common activation mechanism of the autolysine is presumed therefore for both gamete mating types and is triggered by the increase of free Ca^{2+} concentration in the cells.

A hypothesis of RAY and GIBOR (1977) assumes different flagellar membrane sites for agglutination and autolysine activation, because deflagellated gametes after regeneration of 10% of the original flagellar length are able to agglutinate, while soluble cell wall carbohydrates accumulate in the medium only when more than 50% of the flagellar length is regenerated. The autolysine release site seems to be more sensitive to trypsin than the agglutination site (SOLTER and GIBOR 1977).

Secretion and action in the bioassay of the gamete autolysine are totally inhibited in the presence of 50 $\mu g\ ml^{-1}$ concanavalin A (CLAES 1975), which does not disturb the gamete flagellar agglutination. This lectin inhibition can be reversed by the addition of D-glucose or methly-α-D-mannopyranoside. This may be interpreted as an action on glucose or mannose residues in the active site of the autolysine which is supposed to be a glycoprotein itself.

Specificity to developmental stage of both secretion and action of the autolysines suggests that in natural cell wall lysis two requirements have to be met: the supply of the lytic enzyme and the achievement of a dissolvable stage of the cell wall, which changes qualitatively during the cell cycle.

Time of synthesis and site of deposition of the autolysines in the cell are not known. The active sporangium autolysine seems to be formed in the young zoospores during ripening of the sporangia. MIHARA and HASE (1975) with their synchronization conditions obtain the release of zoospores from the sporangia in the 23rd h of the 24-h developmental cycle. Zoospores may be liberated by sonification before this, and then they secrete autolysine into the medium if liberated in the 17th h, but not when liberated in the 14th h. Gametes from *C. reinhardii* broken in a French press yielded 8,000 *g* supernatants which contained the autolysine in an inactive and sedimentable state (CLAES 1977). The preparations could be activated by sonification. These preliminary observations indicate that the gamete autolysine might be localized in vesicles and that its

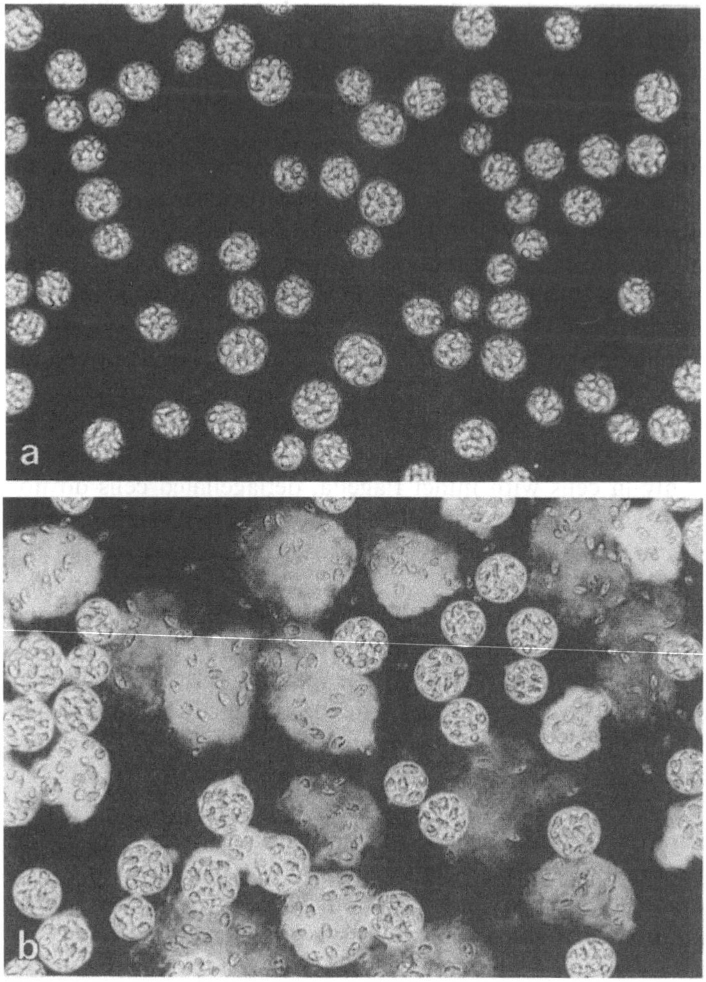

Fig. 5 a, b. Release of spores in *Chlamydomonas reinhardii* from heat-fixed sporangia, contrasted with India ink. **a** Sporangia. **b** Swelling of sporangium jelly during autolysine-induced dissolution of the sporangium walls. (SCHLÖSSER 1976)

release might be connected with membrane fusions as in other secretory systems.

As an accumulation of autolysine in the medium could disturb the development of later cell stages, regulation of autolysine activity outside the cells is required. All autolysines are thermo-labile, and thus inactivation may be due to a quick thermal denaturation, at least under culture conditions: Sporangium autolysine activity was demonstrable in the medium of synchronized cultures after release of zoospores at 34 °C only during 4 h, at 25 °C during 12 h, and at 0 °C during more than 23 h (SCHLÖSSER 1966). Possibly special inhibitors which are secreted simultaneously with the autolysine or afterward contribute to this inactivation. According to JAENICKE and WAFFENSCHMIDT (1981) the

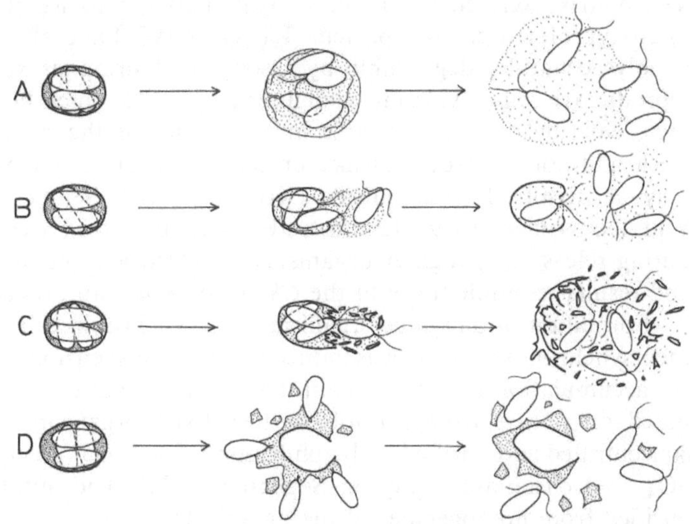

Fig. 6A–D. Different modes of autolysine-mediated zoospore release in the genus *Chlamydomonas*. **A** Total dissolution of the sporangium wall, unlimited swelling of the sporangium jelly. **B, C** Partial dissolution of the sporangium wall resulting in a hole or in fragmentation in pieces respectively, unlimited swelling of the sporangium jelly. **D** Partial dissolution of the sporangium wall connected with explosive turning inside out, no swelling of the sporangium jelly. (SCHLÖSSER 1976)

zygote medium of *C. reinhardii* contains, together with the autolysine, an inhibitor of protein nature which could be separated by salting out with 70% to 100% $(NH_4)_2SO_4$ and after addition to the autolysine fraction was able to reduce its activity.

2.1.5 Reproduction Cell Release Mechanism

The autolysine seems to be only one component of the reproduction cell release mechanism. In the sporangia the zoospores are embedded in a jelly which swells and dissolves at once in living as well as heat-killed sporangia when the limiting sporangium wall is removed mechanically or by autolysine action. Thus cooperation of this "swelling body" with the autolysine results in spore liberation (Fig. 5). Different modes of this event have been observed (Fig. 6).

Both the gamete autolysines of *C. reinhardii* and *C. zimbabwiensis* induce shedding of the wall only in living cells, but not in heat- or formol-fixed ones (except sporangia of *C. reinhardii,* where the mother cell wall is lifted from the protoplast when divisions start). Perhaps turgor pressure is an essential part of the liberation process of cells from their own walls.

2.2 Volvox (Chlorophyceae, Volvocales)

This genus comprises simple organisms which differentiate mostly somatic flagellated cells and few reproductive cells, all situated at the periphery of a globular jelly matrix excreted by the cells. This matrix consists of an outer sheath of hydroxyproline-rich glycoprotein (LAMPORT 1974) and an inner gel, presumably a polysaccharide, degradable by lysozyme. During asexual reproduction the reproductive cells divide into daughter organisms which become flagellated and then leave by dissolving a hole in the sheath of the dying mother organism. With addition of proteases like pronase, trypsin, or chymotrypsin the sheath may be dissolved from outside (KOCHERT 1968, STARR 1969). JAENICKE and WAFFENSCHMIDT (1979) demonstrated that in *Volvox carteri* the natural lysis during release of daughter organisms is indeed a proteolytic process. Methods and results resemble those of the *Chlamydomonas* autolysine.

Formation of daughter organisms was synchronized in aerated cultures and their synchronous release was induced in dense suspensions by a stop in aeration. An accumulation of autolysine in the medium resulted. A quantitative bioassay based on release from formol- or heat-fixed organisms. The autolysine was concentrated and purified by lyophilization, fractional precipitation with $(NH_4)_2$ SO_4, and chromatography on Sephadex 200 S. The autolysine could also be purified from homogenates of the organisms.

The purified autolysine proved to be a protease characterized by the following properties. It is a neutral (glyco)protein with tryptic properties and serine and histidine residues in the active site, with pH-optimum at 7.4 to 7.6. It hydrolyzes casein, hemoglobin, or insulin at basic residues, shows heat-lability with a half-time of 30 min at 50 °C, has a molecular weight between 30,000 and 26,000, and is stabilized by Ca^{2+}, and inhibited by PO_4^{3-} or EDTA. The latter can be overcome by excess of Ca^{2+} or Mg^{2+}. The heat inactivation kinetics are first order, pointing to only one active species of the lysis enzyme. The highly purified protease is able to disintegrate only the matrix of killed but not of living *Volvox* organisms. Lysis of the matrix and daughter organism release is blocked in vivo by: arginine, lysine, benzamidine, and by lectins, particularly concanavalin A. On the other hand, myristyl lecithin causes premature liberation of the daughter organisms.

The lysis enzyme has limited species specificity. Sensitive are, e.g., *V. pocockiae, V. aureus, V. spermatioides,* and *V. africanus,* but not *V. dissipatrix* and *V. gigas*. There is no mutual cross-reacting with the *Chlamydomonas reinhardii* gamete autolysine. The spermatozoids of *Volvox* have to dissolve the matrix at least twice, i.e., during release from the male and penetration of the female organism. Preliminary investigations in *V. carteri* revealed proteolytic activity in vitro (sonification extracts from sperms) and in vivo (daughter organism release from fixed organisms by living sperm). This activity was Ca^{2+}-dependent and inhibited by EDTA or reagents blocking serine-hydroxyl-groups (JAENICKE and WAFFENSCHMIDT 1981). The autolysines of the asexual and sexual reproduction phase here are possibly identical.

2.3 Chlorella (Chlorophyceae, Chlorococcales)

In synchronized cultures of *Chlorella fusca* the sporangium walls seem to be dissolved during spore release. It is, however, not possible to demonstrate autoly-

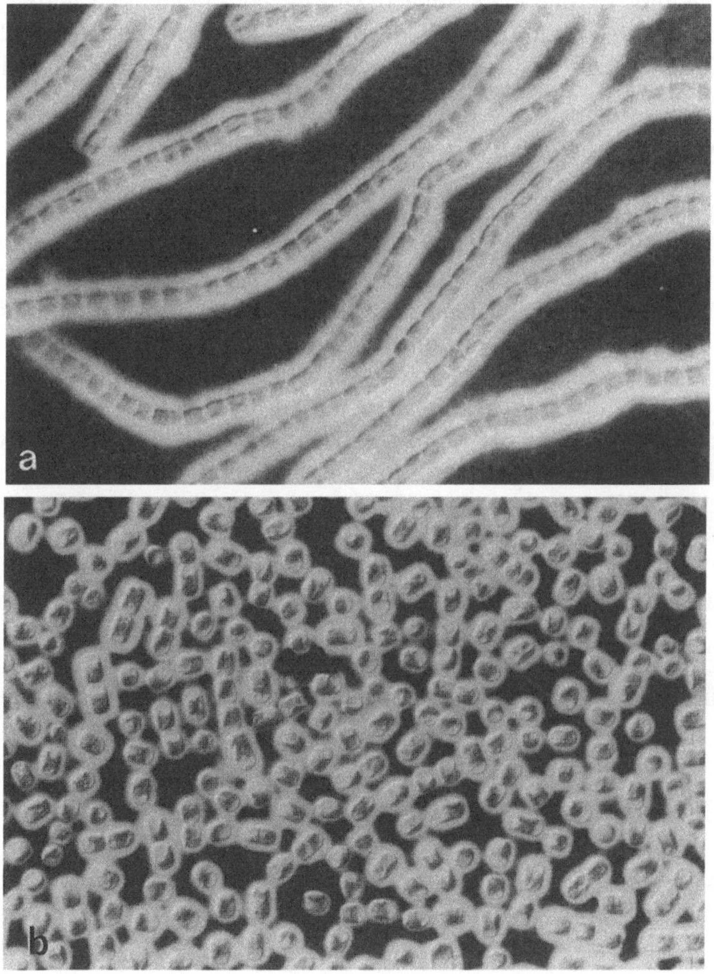

Fig. 7a, b. Asexual reproduction in *Geminella* sp. (Ulotrichales) by fragmentation. **a** Filaments. **b** Induced fragmentation into single cells. Jelly contrasted with India ink

sine activity by incubation of fixed sporangia in the sporulation medium as in *Chlamydomonas* (U.G. SCHLÖSSER, unpublished). This might be due to a protecting sporopollenin layer in the outer part of the wall (ATKINSON et al. 1972) which is present in many species of the Chlorococcales. An inner layer of the sporangium wall containing microfibrils is dissolved during spore release, as demonstrated in electronmicrographs, and this layer can also be removed from isolated sporangium walls by snail digestive juice (a mixture of carbohydrases) (ATKINSON et al. 1972).

Subsequently TOUET and AACH (1979) demonstrated the activity of the natural autolysine in *C. fusca* by incubating isolated dyed sporangium walls, from which dyed particles dissolved into the medium under the influence of concentrated sporulation medium, which was then measured photometrically. The medium of the synchronized culture contained lytic activity only during the spore release phase. The autolysine is thermo-labile and stabile at pH 5 to 7. It hydrolyzes (1→4)-β-xylan, indicating (1→4)-β-xylanase activity.

2.4 Geminella (Chlorophyceae, Ulotrichales)

This filamentous genus reproduces asexually by fragmentation of the longer filaments into short filaments or single cells. A *Geminella* sp. 49.80 was cultivated in an aerated culture chamber like *Chlamydomonas* (25 °C, light–dark = 14:10, 5 klux, nutrient solution of KUHL and LORENZEN 1964) and induced to fragment into single cells quantitatively (Fig. 7) by impoverishment of the medium and cessation of aeration (SCHLÖSSER 1981). Dense-growing cultures were transferred into an induction medium (10^{-3} M $MgSO_4$, 10^{-4} M $CaCl_2$, 5×10^{-3} M Na-phosphate buffer pH 6.0 in distilled water) for 24 h and afterwards into distilled water in large dishes for 12 to 24 h, where fragmentation occurs. The fragmentation medium contains autolytic activity as demonstrated in a bioassay of heat-fixed filaments (5 min, 55 °C) against the fragmentation medium. In this assay the killed filaments were fragmented totally into single cells within 12 h. The autolysine is heat-labile (inactivated at 5 min, 100 °C) and does not fragment filaments of other genera like *Klebsormidium* or *Ulothrix*. It seems to be synthetized during the induction phase in distilled water because presence of cycloheximide (1 mg/ml) during this phase totally prevents fragmentation.

3 Indication of Autolysine Action in Other Algae

Many observations, mostly accidental, indicate autolysine action during reproduction in other algae. The following selection draws attention to three different aspects.

3.1 Cell Wall Changes in the Reproduction Phase

Zoospore or gamete release seems rarely to occur by means of total dissolution of the mother cell wall, e.g., in *Spondylomorum* (SCHILLER 1927), *Chlorogonium* and *Hyalogonium* (PRINGSHEIM 1969), *Radiosphaera* (OCAMPO-PAUS and FRIEDMAN 1966), and *Pleurochrysis* (BROWN et al. 1969, 1970), but occurs instead more frequently by partial dissolution. The latter very often consists of a gelatinization of inner mother cell wall layers, e.g., the cellulose-containing layers in the Chlorococcales such as *Chlorella* (SOEDER 1965, OSCHMAN 1967, WANKA 1968, GRIFFITHS and GRIFFITHS 1969, ATKINSON et al. 1972), *Chlorococcum* (HEUSSLER 1972), *Coelastrum* (REYMOND 1975), *Hydrodictyon* (BIERBACH 1972), *Oocystis* (SCHWERTNER et al. 1972), *Pediastrum* (GAWLIK and MILLINGTON 1969), *Scenedesmus* (BISALPUTRA and WEIER 1963, BURCZYK et al. 1971, HEGEWALD and SCHNEPF 1974, PICKETT-HEAPS and STAEHELIN 1975), *Scotiella* (HEUSSLER 1972), and *Spongiochloris* (MCLEAN 1969) or Dinophyceae such as *Gymnodinium* and *Woloszynskia* (v. STOSCH 1973). The gelatinized inner mother cell wall layer probably causes as a "swelling body" which bursts the outer cell wall layers, and in several genera transports the offspring outside in a transitory vesicle, e.g., in *Ulothrix* (DODEL 1876), *Pediastrum* (GAWLIK and MILLINGTON 1969), *Sorastrum* (GEITLER 1924), *Pteromonas* (ETTL 1964), *Halosphaera* (PARKE and DEN HARTOG-ADAMS 1965), and zygotes of *Hydrodictyon* (PIRSON 1972b).

The details of cell wall changes during the release process may vary with the species and even differ within the life cycle of one species, as revealed by a film on *Hydrodictyon* (PIRSON 1972a, b): here the mother cell wall dissolves slowly during daughter colony release except for a very thin outer layer which bursts; gamete release is connected with the dissolution of inner layers of the mother cell wall which as a swelling body drive the gamete masses outside through a preformed hole; meiospores were released from germinating zygotes by dissolution of inner wall layers which extend outside as a transitory vesicle (Fig. 1).

The jelly of the colonies of *Pandorina morum* dissolves during release of gametes when colonies of compatible clones contact with their flagella. Addition of cell-free nutrient solution of a compatible mating type induces iso-agglutination, swelling of the jelly, and release of gametes (COLEMAN 1959).

Mature female gametophytes of six laminarian brown algae (*Laminaria digitata, L. hyperborea, L. saccharina, Alaria esculenta, Pterygophora californica, Macrocystis pyrifera*) produce one or more substances which cause explosive opening of the male gametangia and release of spermatozoids as well as their attraction in all 36 combinations of the male and female gametophytes (LÜNING and MÜLLER 1978). It was possible to attribute spermatozoid-releasing and -attracting activity to one specific compound in *Laminaria digitata,* a highly volatile olefinic hydrocarbon (MÜLLER et al. 1979). This hydrocarbon possibly triggers activation of an autolysine.

3.2 Lysis by Exogenous Enzymes

Cell wall lysis under the influence of enzyme preparations from other organisms may be a hint to the decisive cell wall binding sites as well as to the possible nature of the natural autolysines. Proteolytic enzymes dissolve the matrix jelly of *Volvox* (KOCHERT 1968, STARR 1969), and the cell envelope of *Euglena* (KLEBS 1883, KIRK 1964, PRICE and BOURKE 1966). Lysozyme may be used for the preparation of spheroplasts in different unicellular and filamentous species of Cyanophyceae (review of ADAMICH and HEMMINGSEN 1980) by the method of BIGGINS (1967, 1971). Cellulase preparations dissolve the inner sporangium wall layer of *Chlorella* (BRAUN and AACH 1975, FEYEN 1977, BERLINER 1977), and *Scenedesmus* (BURCZYK et al. 1971), and have been used for protoplast isolation in *Micrasterias* (BERLINER and WENC 1976), *Mougeotia* (MARCHANT and FOWKE 1977), *Spirogyra, Zygnema* (OHIWA 1977), *Klebsormidium, Stigeoclonium,* and *Ulothrix* (MARCHANT and FOWKE 1977, GABRIEL 1970). Cellulase was used to free septal plugs in the red alga *Griffithsia pacifica* from a clean cell wall fraction (RAMUS 1971).

3.3 Dependency on Divalent Cations

Stabilization with Ca^{2+} or Mg^{2+} of the autolysines from *Chlamydomonas* and *Volvox* and their inhibition with EDTA points to an important role of divalent cations in the autolytic process and its regulation. Dependency of the zoospore release on Ca^{2+}, which in these cases could not be replaced by other divalent cations, was also shown in *Protosiphon botryoides* (O'KELLEY and HERNDON 1959, 1961, O'KELLEY and DEASON 1962) and in *Chlorococcum echinozygotum* (GILBERT and O'KELLEY 1964). Spore release in *Chlorella fusca* requires Ca^{2+} or Sr^{2+}, and perhaps Mg^{2+} as in Mg^{2+}-deficient cultures sporangium walls accumulate in the medium (GALLING 1963). Participation of Ca^{2+} in the sexual flagellar agglutination of isogametes in *Chlamydomonas* (LEWIN 1954, JONES and WIESE 1962, WIESE and JONES 1963), in *Volvulina steinii* (CAREFOOT 1966), and *Pandorina unicocca* (RAYBURN 1974) suggest its function in the regulation of this preceding step to autolysine activation.

4 Application of Autolysines

Beside their role in release mechanisms, the autolysines offer different possibilities of other applications. They may be used as special tools for the analysis of cell wall structure at the molecular level by hydrolysis into fragments of natural composition. The gamete autolysine of *Chlamydomonas reinhardii* was employed in the preparation of protoplasts (SCHLÖSSER et al. 1976); in contrast to other methods of protoplast isolation this "natural" method is without any damage to the cells. The formation of a new cell wall may be investigated in homogenous suspensions of autolysine-isolated protoplasts (ROBINSON and SCHLÖSSER 1978). Species-specific sporangium autolysines can be used as chemotaxonomic markers characterizing groups of related species; this might permit a more natural classification in the large genus *Chlamydomonas*, of which more than 500 species are described. Strains of colony-forming green algae like *Coelastrum, Pediastrum,* and *Scenedesmus* under cultural conditions often fragment into single cells. This might be due to the action of autolysines. Perhaps the regulation of autolysine activity decides also in nature whether a species is unicellular or colony-forming.

Acknowledgments. The support of the Deutsche Forschungsgemeinschaft is gratefully acknowledged. I thank Prof. Dr. D.G. Robinson for correcting the English.

References

Adamich M, Hemmingsen BB (1980) Protoplast and spheroplast production. In: Gantt E (ed) Handbook of phycological methods. Developmental and cytological methods. Cambridge Univ Press, Cambridge, pp 153–169

Atkinson AW, Gunning BE, John PCL (1972) Sporopollenin in the cell wall of *Chlorella* and other algae: ultrastructure, chemistry and incorporation of ^{14}C-acetate, studied in synchronous cultures. Planta 107:1–32
Berliner MD (1977) Protoplast induction in *Chlorella vulgaris*. Plant Sci Lett 9:201–204
Berliner MD, Wenc KA (1976) Protoplast induction in *Micrasterias* and *Cosmarium*. Protoplasma 89:389–393
Bierbach M (1972) Lebenszyklus und Kultur von *Hydrodictyon reticulatum*. Staatsexamensarb, Pflanzenphysiol Inst, Univ Göttingen
Biggins J (1967) Preparation of metabolically active protoplasts from the blue-green alga, *Phormidium luridum*. Plant Physiol 42:1442–1446
Biggins J (1971) Protoplasts of algal cells. In: San Pietro A (ed) Methods in enzymology, vol 23A. Academic Press, London New York, pp 209–211
Bisalputra T, Weier TE (1963) The cell wall of *Scenedesmus quadricauda*. Am J Bot 50:1011–1019
Braun E, Aach HG (1975) Enzymic degradation of the cell wall of *Chlorella*. Planta 126:181–185
Brown RM, Franke WW, Kleinig H, Falk H, Sitte P (1969) Cellulosic wall component produced by the golgi apparatus of *Pleurochrysis scherffelii*. Science 166:894–896
Brown RM, Franke WW, Kleinig H, Falk H, Sitte P (1970) Scale formation in chrysophycean algae. J Cell Biol 45:246–271
Burczyk J, Grzybek H, Banás J, Banás E (1971) Presence of cellulase in the alga *Scenedesmus*. Exp Cell Res 63:451–453
Carefoot JR (1966) Sexual reproduction and intercrossing in *Volvulina steinii*. J Phycol 2:150–156
Claes H (1971) Autolyse der Zellwand bei den Gameten von *Chlamydomonas reinhardii*. Arch Mikrobiol 78:180–188
Claes H (1975) Influence of concanavalin A on autolysis of gametes from *Chlamydomonas reinhardii*. Arch Mikrobiol 103:225–230
Claes H (1977) Non-specific stimulation of the autolytic system in gametes from *Chlamydomonas reinhardii*. Exp Cell Res 108:221–229
Claes H (1980) Calcium ionophore-induced stimulation of secretory activity in *Chlamydomonas reinhardii*. Arch Microbiol 124:81–86
Coleman A (1959) Sexual isolation in *Pandorina morum*. J Protozool 6:249–264
Czurda V (1931) Ein neuer, eigenartiger Kapulationsablauf bei einer Mougeotia (*M. oedogonioides* Czurda) Beih Bot Centralbl 48:286–290
Diwald K (1938) Die ungeschlechtliche und geschlechtliche Fortpflanzung von *Glenodinium lubiniensiforme* sp. nov. Flora 132:174–192
Dodel A (1876) Die Kraushaar-Alge *Ulothrix zonata*. Engelmann, Leipzig
Ettl H (1964) Die Morphologie einiger *Pteromonas*-Arten. Nova Hedwigia Z Kryptogamenkd 8:323–331
Feyen V (1977) Versuche zur somatischen Hybridisation von *Chlorella saccharophila*. Ph D Thesis, Tech Hochschule Aachen, Germany
Gabriel M (1970) Formation, growth, and regeneration of protoplasts of the green alga, *Uronema gigas*. Protoplasma 70:135–138
Galling G (1963) Analyse des Magnesium-Mangels bei synchronisierten Chlorellen. Arch Mikrobiol 46:150–184
Gawlik SR, Millington WF (1969) Pattern formation and the fine structure of the developing cell wall in colonies of *Pediastrum boryanum*. Am J Bot 56:1084–1093
Geitler L (1924) Die Entwicklungsgeschichte von *Sorastrum spinulosum* und die Phylogenie der Protococcales. Arch Protistenkd 47:440–447
Gilbert WA, O'Kelley JC (1964) The effects of replacement of calcium by strontium on the reproduction of *Chlorococcum echinozygotum*. Am J Bot 51:866–869
Goodenough UW, Weiss RL (1975) Gametic differentiation in *Chlamydomonas reinhardtii*. III. Cell wall lysis and microfilament-associated mating structure activation in wild-type and mutant strains. J Cell Biol 67:623–637
Griffiths DA, Griffiths DJ (1969) The fine structure of autotrophic and heterotrophic cells of *Chlorella vulgaris* (Emerson strain). Plant Cell Physiol 10:11–19

Hegewald E, Schnepf E (1974) Beitrag zur Kenntnis der Grünalgenart *Scenedesmus verrucosus* Roll. Arch Hydrobiol Suppl 46:151–162
Heimke JW, Starr RC (1979) The sexual process in several heterogamous *Chlamydomonas* strains in the subgenus *Pleiochloris*. Arch Protistenkd 122:20–42
Heussler P (1972) Licht- und elektronenmikroskopische Untersuchungen zur Entwicklungsmorphologie zoosporenbildender und coccaler Grünalgen. Ph D Thesis, Univ München, Germany
Hirn KE (1900) Monographie und Iconographie der Oedogoniaceen. Acta Soc Sci Fennicae 27:1–394
Jaenicke L, Waffenschmidt S (1979) Matrix-lysis and release of daughter spheroids in *Volvox carteri* – a proteolytic process. FEBS Lett 107:250–253
Jaenicke L, Waffenschmidt S (1981) Liberation of reproduction units in *Volvox* and *Chlamydomonas*: Proteolytic processes. Ber Dtsch Bot Ges, in press
Jones RF, Wiese L (1962) Studies on the mating reaction in *Chlamydomonas*. J Gen Physiol 46:362A
Kirk JTO (1964) The effect of trypsin on the pellicle of *Euglena gracilis*. J Protozool 11:435–437
Klebs G (1883) Über die Organisation einiger Flagellatengruppen und ihre Beziehungen zu Algen und Infusorien. Unters Bot Inst Tübingen 1:233–262
Kochert G (1968) Differentiation of reproductive cells in *Volvox carteri*. J Protozool 15:438–452
Kornmann P (1965) Was ist *Acrosiphomia arcta*? Helgol Wiss Meeresunters, 12:40–51
Kuhl A, Lorenzen H (1964) Handling and cultering of *Chlorella*. In: Prescott DM (ed) Methods in cell physiology, vol I, Academic Press, London New York, pp 159–187
Lamport DTA (1974) The role of hydroxyproline rich proteins in the extracellular matrix of plants. In: Hay ED, King TJ, Papaconstantinou J (eds) Macromolecules regulating growth and development. 30th Symp Soc Dev Biol. Academic Press. New York, pp 113–129
Lewin RA (1954) Sex in unicellular algae. In: Wenrich DH (ed) Sex in microorganisms. Am Assoc Adv Sci, Washington DC, pp 100–133
Lüning K, Müller DG (1978) Chemical interaction in sexual reproduction of several Laminariales (Phaeophyceae): Release and attraction of spermatozoids. Z Pflanzenphysiol 89:333–341
Marchant HJ, Fowke LC (1977) Preparation, culture, and regeneration of protoplasts from filamentous green algae. Can J Bot 55:3080–3086
Matsuda Y, Tamaki S, Tsubo Y (1978) Mating type specific induction of cell wall lytic factor by agglutination of gametes in *Chlamydomonas reinhardtii*. Plant Cell Physiol 19:1253–1261
McLean RJ (1969) Rejuvenation of senescent cells of *Spongiochloris typica*. J Phycol 5:32–37
Mihara S, Hase E (1975) Studies on the vegetative life cycle of *Chlamydomonas reinhardi* Dangeard in synchronous culture. III. Some notes on the process of zoospore liberation. Plant Cell Physiol 16:371–375
Müller DG, Gassmann G, Lüning K (1979) Isolation of a spermatozoid-releasing and -attracting substance from female gametophytes of *Laminaria digitata*. Nature 279:430–431
Ocampo-Paus R, Friedmann I (1966) *Radiosphaera negevensis* sp. n., a new chlorococcalean desert alga. Am J Bot 53:663–671
Ohiwa T (1977) Preparation and culture of *Spirogyra* and *Zygnema* protoplasts. Cell Struct Funct 2:249–255
O'Kelley JC, Deason TR (1962) Effect of nitrogen, sulfur and other factors on zoospore production by *Protosiphon botryoides*. Am J Bot 49:771–777
O'Kelley JC, Herndon WR (1959) Effect of strontium replacement for calcium on production of motile cells in *Protosiphon*. Science 130:718
O'Kelley JC, Herndon WR (1961) Alkaline earth elements and zoospore release and development in *Protosiphon botryoides*. Am J Bot 48:796–802
Oschman JL (1967) Structure and reproduction of the algal symbionts of *Hydra viridis*. J Phycol 3:221–228

Parke M, den Hartog-Adams I (1965) Three species of *Halosphaera*. J Mar Biol Assoc UK 45:537–557

Pascher A (1939) Heterokonten. In: Kolkwitz R (Herausg.) Rabenhorsts Kryptogamenflora Bd 11 Akadem Verlagsges Leipzig

Pickett-Heaps JD, Staehelin LA (1975) The structure of *Scenedesmus* (Chlorophyceae). II. Cell division and colony formation. J Phycol 11:186–202

Pirson A (1972a) Ungeschlechtliche Fortpflanzung der Grünalge *Hydrodictyon reticulatum*. Wiss Film C 1042. Inst Wiss Film, Göttingen

Pirson A (1972b) Geschlechtliche Fortpflanzung der Grünalge *Hydrodictyon reticulatum*. Wiss Film C 1043. Inst Wiss Film, Götttingen

Price CA, Bourke ME (1966) "Spheroplasts" prepared from *Euglena gracilis* by proteolysis. J Protozool 13:474–477

Pringsheim EG (1969) Die Gattungen *Chlorogonium* und *Hyalogonium*. Nova Hedwigia Z Kryptogamenkd 18:831–867

Ramus J (1971) Properties of septal plugs from the red alga *Griffithsia pacifica*. Phycologia 10:99–103

Ray DA, Gibor A (1977) The role of flagella in mating of *Chlamydomonas reinhardi* – evidence for two binding sites by use of deflagellation. Plant Physiol Suppl 59:111

Rayburn WR (1974) Sexual reproduction in *Pandorina unicocca*. J Phycol 10:258–265

Reymond O (1975) La paroi cellulaire de *Coelastrum* (Chlorophyceés). Arch Microbiol 102:95–101

Robinson DG, Schlösser UG (1978) Cell wall regeneration by protoplasts of *Chlamydomonas*. Planta 141:83–92

Schiller J (1927) Über *Spondylomorum caudatum* sp. n., seine Fortpflanzung und Lebensweise. Jahrb Wiss Bot 66:274–284

Schlösser UG (1966) Enzymatisch gesteuerte Freisetzung von Zoosporen bei *Chlamydomonas reinhardii* Dangeard in Synchronkultur. Arch Mikrobiol 54:129–159

Schlösser UG (1976) Entwicklungsstadien- und sippenspezifische Zellwand-Autolysine bei der Freisetzung von Fortpflanzungszellen in der Gattung *Chlamydomonas*. Ber Dtsch Bot Ges 89:1–56

Schlösser UG (1981) Release of reproduction cells by action of cell wall autolytic factors in *Chlamydomonas* and *Geminella*. Ber Dtsch Bot Ges, in press

Schlösser UG, Sachs H, Robinson DG (1976) Isolation of protoplasts by means of a "species-specific" autolysine in *Chlamydomonas*. Protoplasma 88:51–64

Schwertner HA, Butcher WI, Richardson B (1972) Structure and composition of *Oocystis polymorpha* cell walls from the culture medium. J Phycol 8:144–146

Skuja H (1956) Taxonomische und biologische Studien über das Phytoplankton schwedischer Binnengewässer. Nova Acta Regiae Soc Sci Ups IV 16 (3):1–404

Soeder CJ (1965) Elektronenmikroskopische Untersuchung der Protoplastenteilung bei *Chlorella fusca* Shihira et Krauss. Arch Mikrobiol 50:368–377

Solter KM, Gibor A (1977) The role of flagella in the mating of *Chlamydomonas reinhardi* – evidence for two binding sites by use of trypsin. Plant Physiol 59 Suppl:111

Starr RC (1969) Structure, reproduction and differentiation in *Volvox carteri f. nagariensis* Iyengar, strains HK 9 & 10. Arch Protistenkd 111:204–222

Strominger JL, Ghuysen JM (1967) Mechanisms of enzymatic bacteriolysis. Science 156:213–221

Stosch HA v (1973) Observations on vegetative reproduction and sexual life cycles of two freshwater Dinoflagellates, *Gymnodinium pseudopalustre* Schiller and *Woloszynskia apiculata* sp. nov. Br Phycol J 8:105–134

Touet G, Aach HG (1979) Isolation and first characterization of a cell-wall-degrading agent of *Chlorella*. Naturwissenschaften 66:525–526

Wanka F (1968) Ultrastructural changes during normal and colchicine-inhibited cell division of *Chlorella*. Protoplasma 66:105–130

Weiss RL, Goodenough DA, Goodenough UW (1977) Membrane differentiations at sites specialized for cell fusion. J Cell Biol 72:144–160

Wiese L, Jones RF (1963) Studies on gamete copulation in heterothallic Chlamydomonads. J Cell Comp Physiol 61:265–274

15 Fungal Cell Walls: A Survey

J.G.H. Wessels and J.H. Sietsma

1 Introduction

The cell wall is commonly regarded as an assemblage of polymers, mainly polysaccharides, that occurs outside the plasma membrane of cells of plants, fungi, and bacteria. Because of its rigidity it maintains the shape of the cell and offers resistance to unlimited influx of water into the cell. In addition it offers protection and, being the outermost cover of cells, often contains molecules involved in interaction between cells.

Cell walls of fungi have received much attention (1) because of the general accepted view that different groups of fungi differ widely in the nature of the polymers that make up their cell walls and (2) because of the realization that morphogenesis of fungal cells is intricately associated with cell wall metabolism.

Because this volume contains detailed accounts of certain classes of fungal wall polymers (Chaps. 16, 19), this review will concentrate on general principles to try to integrate what is known about the chemical architecture of fungal walls and how the dynamics of these walls could be involved in growth and morphogenesis.

1.1 Methodological Difficulties

Often a distinction is made between the cell wall proper and capsular or extracellular polymers (mucilage, slime), the latter being soluble in water. This appears justified if rigidity is considered an essential feature of cell walls and it offers the methodological advantage that the cell wall fraction can be obtained by mechanical breaking of the cells, low speed centrifugation, and extensive washings to remove cytoplasmic contamination. This procedure not only removes more or less water-soluble polymers, but also involves the risk of slouching off functional polymers loosely adhering to the wall. The water-soluble polymers often bear a close resemblance to insoluble wall polymers (Gorin and Spencer 1968, Phaff 1971, Sietsma et al. 1977, Sietsma and Wessels 1979). Therefore it is possible that the procedure of obtaining cell walls often results in breaking up of a functional entity. On the other hand it is not always certain that the extensive washing procedure does remove all cytoplasmic contamination. For instance, the high lipid content of wall fractions obtained from members of the Mucorales (Chenouda 1972) suggests contamination with plasma membranes. Also, the wall fraction may become contaminated during breaking of the cells. In *Neurospora crassa,* Harold (1962) has shown that intracellular polyphosphates can bind to ionic groups in the wall during fragmentation of the mycelium. Although it is now clear that proteins (glycoproteins) can be an intrinsic constituent of fungal walls (Ballou 1974, Gander 1974) it often remains difficult to decide whether proteins found in the wall fraction result from cytoplasmic contamination or represent enzymes that normally occur in the periplasmic space rather than proteins which play a structural role in the wall (and may have enzymic activity at the same time).

To circumvent these problems, some workers have used detergents, salt solutions, or urea to break noncovalent bonds and to remove presumptive contaminating materials (MAHADEVAN and TATUM 1965, MITCHELL and TAYLOR 1969). However, such treatments may also remove genuine wall polymers held insoluble by noncovalent bonds to other wall components (DATEMA et al. 1977a, b).

Because of the insoluble nature of at least a large part of the wall it is extremely difficult to obtain a comprehensive picture of its chemical and morphological organization. Solubilization of polymers from the insoluble complex mostly involves the use of alkali or acid, treatments likely to break covalent bonds and to modify polymers to a state different from the native condition. This not only concerns the chemical structure of the polymers in which labile linkages may be cleaved and monomers may become modified; it also refers to possible changes in the aggregation of polymers or segments thereof resulting in crystallites and fibers not originally present. The present review will emphasize the importance of these considerations in understanding the native structure of the wall. Other aspects of methodology related to wall analysis have been critically evaluated by TAYLOR and CAMERON (1973).

Although much information about the composition and ultrastructure of fungal walls has been obtained (for reviews see ARONSON 1965, BARTNICKI-GARCIA 1968, PHAFF 1971, LINDBERG and SVENSON 1973, ROSENBERGER 1976, FARKAS 1979), the relationship between structure and function of the various identified components is not yet clear. An understanding of function, e.g., in relation to mechanical strength, permeability, growth and morphogenesis, would require knowledge of the exact positions, physical states, and interactions between various components identified after isolation from the wall. It is now generally held that the hyphal wall consists of coaxial layers of which at least the inner layer contains a fibrillar component, chitin, and sometimes cellulose. The fibrils mesh together to form a net; the spaces in the net being filled by a matrix of other polymers such as $(1\rightarrow3)$-β-D-glucans (HAWKER 1965, MAHADEVAN and TATUM 1967, HUNSLEY and BURNETT 1970, VAN DER VALK et al. 1977). The fibrils would resist stretching, the matrix would resist compression of the wall. However, as will be emphasized in this review, doubt exists whether such an architecture, as revealed by subsequent extractions of the wall, always reflects the real structural organization as it exists in the native wall.

The basic importance of knowledge of the structural organization of the native wall is paramount in studies concerned with lysis and biosynthesis of the wall in relation to growth and morphogenesis. The study of degradative processes meets with difficulties because of our ignorance of the relative importance of the various chemical bonds in determining the physical properties of the wall. Studies on biosynthesis now proceed at the level of individual polymers such as chitin, β-glucan and mannan, but our understanding of the integration of all biosynthetic activities in producing a complex wall is still very incomplete.

2 Survey of Wall Polymers

2.1 Distribution Among Fungi

Cell wall analyses of different taxonomic groups of fungi have revealed a remarkable heterogeneity with respect to polymers or combinations of polymers present, as first pointed out clearly by BARTNICKI-GARCIA (1968). On the other hand, ultrastructural studies (BRACKER 1967, TROY and KOFFLER 1969, HUNSLEY and BURNETT 1970, BURNETT 1976) have shown a general similarity in construction of the wall, i.e., an inner layer containing chitin or cellulose embedded in

Table 1. Various polymers occurring in the cell wall of fungi

Taxonomic groups	Cell wall polymers	
	Alkali-soluble	Alkali-insoluble
Basidiomycetes	Xylo-manno-protein (1 → 3)-α-D-glucan [a]	(1 → 3)-β/(1 → 6)-β-D-Glucan chitin
Ascomycetes	(Galacto)-manno-protein (1 → 3)-α-D-glucan [a]	(1 → 3)-β/(1 → 6)-β-D-Glucan chitin
Zygomycetes	Glucuronomanno-protein polyphosphate	Polyglucuronic acid chitosan chitin
Chytridiomycetes	Glucan [b]	Glucan [b] chitin
Hyphochytridiomycetes	not determined	Chitin cellulose
Oomycetes	(1 → 3)-β/(1 → 6)-β-D-Glucan	(1 → 3)-β/(1 → 6)-β-D-Glucan cellulose

[a] In a number of cases the alkali-soluble fraction also contains part of the (1 → 3)-β/(1 → 6)-β-D-glucan

[b] Incompletely characterized; probably (1 → 3)-β and (1 → 6)-β-linked

other polymers and an outer layer. As a rule the outer layer is soluble in dilute alkali leaving the inner layer as an insoluble residue. Table 1 lists the various types of alkali-soluble and alkali-insoluble polymers as they occur among the fungi.

It appears that within the fungi convergent evolution has provided for the occurrence of a variety of different polymers fulfilling similar functional requirements of the walls. Among these are rigidity to perform a skeletal function, flexibility to absorb physical stress, regulated patterns of synthesis and degradation to accommodate growth and morphogenesis, and in a number of cases, the possibility of wall components to be used as a reserve substance when a shortage of external nutrients occurs. On the molecular level the convergent evolution indicates that the molecular conformations of the different polymers must exhibit similarities which enable them to form similar secondary or tertiary structures resulting in, for example, microfibrillar crystallites or gel-like assemblies (KIRKWOOD 1974, REES 1977).

In previous publications much emphasis has been put on the taxonomic value of cell wall composition in fungi (BARTNICKI-GARCIA 1968). At present, more data are known on the wall composition of fungal species and when compared, one is struck by the similarities rather than by the differences (see Table 1). Chitin is almost ubiquitously present in all fungi except in some Oomycete species, where it is replaced by cellulose (BARTNICKI-GARCIA 1968, SIETSMA et al. 1969). In addition some yeasts, e.g., *Schizosaccharomyces* species (KREGER 1954), appear to lack chitin. Its content varies from 1%–2% in *Saccharomyces* species (PHAFF 1971) to about 60% in *Allomyces macrogynus* (ARONSON and MACHLIS 1959). Chitin is usually closely associated with (1→3)-β-D/(1→6)-β-D-glucan, first described in yeast as yeast glucan; again in the Oomycetes this glucan is associated with cellulose (ZEVENHUIZEN and BARTNICKI-GARCIA 1969, SIETSMA et al. 1975a). Hyphae of Zygomycetes are exceptional in not containing any glucan but polysaccharides rich in D-glucuronic acid (BARTNICKI-GARCIA and REYES 1968a, b, KREGER 1970, DATEMA et al. 1977a, b), in addition the deacetylation ratio of chitin is very high, which means that chitosan can be isolated from these species (KREGER 1954). The only wall entities which could be species-specific are the glycoproteins, usually containing mannose and/or galactose in the Ascomycete species and xylose and mannose in the Basidiomycetes (see Sect. 2.2.6).

The simultaneous occurrence of chitin and cellulose in fungal walls has been a matter of dispute since the early studies of wall composition (ARONSON 1965, LIN and ARONSON 1970). Convincing evidence for the presence of both substances has been given for the wall of *Rhizidiomyces* sp. (FULLER and BARSHAD 1960, FULLER 1960) so that only the order of the Hyphochytridiomycetes is presently considered as including organisms having both cellulose and chitin in their walls. Recently LIN et al. (1976) have recorded X-ray diffraction patterns of chitin and cellulose I in a preparation obtained after drastic chemical treatment of the wall of *Apodachlya* sp. (Oomycetes), indicating that the wall of this organism contains long stretches of poly-(1→4)-β-D-linked *N*-acetylglucosamine and poly-(1→4)-β-D-linked glucose. On the other hand SIETSMA et al. (1969) reported that *Apodachlya brachynema* walls were not able to induce chitinase in a *Streptomyces* species capable of producing this enzyme in the presence of only 0.01% of chitin. Also no chitin X-ray diffraction pattern was observed after boiling these walls in 2% hydrochloric acid for 1 h (J.H. SIETSMA unpublished).

2.2 Individual Polymers

2.2.1 (1→4)-β-D-Glycosaminoglycans (Chitin, Chitosan)

Chitin is usually considered as a homopolymer of *N*-acetylglucosamine [(1→4)-2-acetamido-2-deoxy-β-D-glucan] but even in crystalline chitin a number of non-acetylated glucosamine residues may occur (HACKMAN and GOLDBERG 1965, RUDALL and KENCHINTON 1973). Because chitin will be comprehensively treated in the next chapter (Chap. 16, this Vol.) a few remarks pertaining to its possible association with other polymers will suffice here.

Evidence for the occurrence of chitin in walls is mostly obtained by chemical analysis, infra red spectroscopy and X-ray diffraction analysis of wall residues obtained after consecutive extractions with alkali and acid. After such treatments electronmicroscopical observations mostly reveal long interweaving microfibrils of varying width in random orientation (TROY and KOFFLER 1969, MANOCHA and COLVIN 1967, WANG and BARTNICKI-GARCIA 1970, HUNSLEY and BURNETT 1970, BURNETT 1976, VAN DER VALK et al. 1977), sometimes short rodlets (KITAZIMA et al. 1972) or granules (HOUWINK and KREGER 1953). X-ray diffraction of such preparations shows a microcrystalline configuration known as α-chitin, a crystal form which also occurs when chains of chitin crystallize from solution, in contrast to the polymorphs β and γ which also occur in nature (cf. MUZZARELLI 1977).

Usually nontreated walls do not show clear X-ray diffraction lines of chitin (HOUWINK and KREGER 1953, KREGER 1954, REID and BARTNICKI-GARCIA 1976, SIETSMA and WESSELS 1977). For the walls of *Schizophyllum commune* it was concluded that in native and in alkali-extracted walls most of the chitin exists in a noncrystalline condition because the percentage of chitin was high enough (18% in alkali-extracted walls) to produce good X-ray reflections if it were microcrystalline. Only after treatments of the alkali-insoluble residue with hot mineral acid, periodic acid (Smith degradation) or (1→3)-β-glucanase, each of which removed most of the β-glucan present, did sharp X-ray reflections of chitin (SIETSMA and WESSELS 1977) and abundant microfibrils (VAN DER VALK et al. 1977) occur. It would thus appear that only a small fraction of the chitin in the native wall, if any, exists in microcrystalline microfibrillar condition. During extraction of the glucan dispersed chitin chains may be brought into contact and crystallize spontaneously to form fibrillar structures. A similar situation seems to hold for chitin in the yeast *Saccharomyces cerevisiae*. In this organism the chitin content is low and chitin seems to be confined to the bud scars (BACON et al. 1966, BERAN et al. 1972). Again, alkali-extracted wall preparations do not show chitin reflections by X-ray diffraction analysis but such reflections appear after treatment with hot HCl together with chitin granules in the bud scars (HOUWINK and KREGER 1953).

In contrast, when protoplasts of *S. commune* regenerate a wall, chitin synthesis initially proceeds without β-glucan deposition (DE VRIES and WESSELS 1975) and this chitin is microcrystalline and microfibrillar in the native wall (VAN DER VALK and WESSELS 1976). Apparently, the absence of β-glucan at this stage allows for the alignment and crystallization of the chitin chains. Also in regenerating yeast protoplasts the chitin is synthesized in a crystalline microfibrillar form along with an unbranched crystalline (1→3)-β-D-glucan (KREGER and KOPECKÁ 1976).

Apart from dispersed chitin chains, the regular occurrence of embedded microfibrils on the inner face of shadowed wall preparations (MARET 1972, REID and BARTNICKI-GARCIA 1976, VAN DER VALK and WESSELS 1977) suggests that at least some of the chitin exists in the native wall in microcrystalline condition. However, some caution is necessary because these walls may have become modified by drying for electronmicroscopy. On the other hand it is difficult to envisage that the nonrandom orientation of chitin microfibrils seen in rare cases such as in sporangiophore walls of *Phycomyces* (ROELOFSEN 1951), rhizoid walls of *Allomyces* (ARONSON and PRESTON 1960), walls from the stipe of fruiting bodies of *Coprinus* (GOODAY 1975), the wall at the germ tube apex of *Phytophthora palmivora* (BARTNICKI-GARCIA 1973), and in walls of *Polyporus mylittae* (SCURFIELD 1967) and *Choanephora cucurbitarum* (LETOURNEAU et al. 1976), arose as aggregation artifacts during preparation of the walls. Thus, although both crystalline and disperse chitin may occur in fungal walls, at the moment it is difficult to assess their quantitative contribution to the structure of the wall.

With respect to the association of chitin and glucan in one alkali-insoluble complex the suggestion has often been made that covalent linkages may be involved. For instance, after enzymic or chemical removal of the β-glucan, part of the *N*-acetylglucosamine dissolves indicating that *N*-acetylglucosamine is not only present as chitin (TROY and KOFFLER 1969, SIETSMA and WESSELS 1977). Also amino acids have been found which remain associated with the residue otherwise containing only (*N*-acetyl)glucosamine (WANG and BARTNICKI-GARCIA 1970, MARET 1972, SIETSMA and WESSELS 1977). In *Schizophyllum commune* covalent interaction between chitin and β-D-glucan via amino acids was strongly indicated by a marked change in solubility characteristics of the β-glucan after selective depolymerization of the chitin and by the isolation of a fragment containing (*N*-acetyl)glucosamine, amino acids and glucose (SIETSMA and WESSELS 1979).

Also in fungi not containing β-glucan, *N*-acetylglucosamine appears to exist in more than one polymeric form (DATEMA et al. 1977b). Using nitrous acid that specifically attacks nonacetylated glucosamine residues in the polymers, it was found that three fractions could be distinguished in the wall of *Mucor mucedo:* One fraction which was solubilized with nitrous acid contained *N*-acetylglucosamine interspersed with glucosamine. Another fraction became nitrous acid-soluble after treatment of the walls with pronase or alkali, indicating a polymer containing *N*-acetylglucosamine and glucosamine to which peptides were linked. The remaining fraction appeared to consist of a homopolymer of *N*-acetylglucosamine with an X-ray diffraction pattern of α-chitin.

In contrast to *Mucor mucedo,* other Zygomycetes, e.g., *Phycomyces blakesleeanus* (KREGER 1954) and *Mucor rouxii* (BARTNICKI-GARCIA and NICKERSON 1962, BARTNICKI-GARCIA and REYES 1968a) contain a homopolymer of glucosamine [(1 → 4)-2-amino-2-deoxy-β-D-glucan)] called chitosan. Unlike the acetylated glycan (chitin) it can be extracted from the wall with dilute acid. The cationic property of polymers containing glucosamine requires the presence of a counterion, a role probably fulfilled by inorganic polyphosphate and glycuronans (BARTNICKI-GARCIA and REYES 1968a, b, DATEMA et al. 1977a, b). How chitosan exists in the native wall is unknown. A significant observation is that the crystalline conformation of chitosan is only detected after extraction of the substance from the wall with dilute acid, a treatment likely to break covalent bonds.

In conclusion, it is questionable whether the simple model of chitin microfibrils physically embedded in a matrix of amorphous material is always tenable. A large fraction of the poly-(*N*-acetyl)glucosamine segments seems to be part of heteropolymers containing other sugars and amino acids. If microfibrils of microcrystalline chitin do exist in native walls, substituted chains of poly-(*N*-acetyl)glucosamine might be hydrogen-bonded to the outer surface of such microfibrils, linking individual microfibrils. The molecular architecture would then be not unlike the cellulose/hemicellulose/pectin complex envisioned in primary walls of higher plant cells (KEEGSTRA et al. 1973).

2.2.2 (1 → 4)-β-D-Glucan (Cellulose)

In contrast to the ubiquitous occurrence of cellulose in plant cells, (1 → 4)-β-D-glucan is found in the walls of only a limited number of fungi belonging to the Oomycetes (BARTNICKI-GARCIA 1968, GORIN and SPENCER 1968). In these walls it occurs together with a large amount of branched (1 → 3)-β-D/(1 → 6)-β-D-glucan which shares an extreme insolubility with the cellulose (ZEVENHUIZEN and BARTNICKI-GARCIA 1969, SIETSMA et al. 1969). However, a large part of this branched glucan can be solubilized in alkali after treatment with dilute acid. Also most of the proteinous material present in these walls appears to be acid-soluble (BARTNICKI-GARCIA 1966, NOVEAES-LEDIEU and JIMENEZ-MARTINEZ 1968). After treatment of a wall preparation with Schweitzer's reagent to remove cellulose, the residual glucan still contains (1 → 4)-linkages, and also after chemical or enzymatical removal of the (1 → 3)-β-D/(1 → 6)-β-D-glucan, a fibrillar preparation is obtained still containing (1 → 3) and (1 → 6) linkages (NOVAES-LEDIEU and JIMENEZ-MARTINEZ 1968, ZEVENHUIZEN and BARTNICKI-GARCIA 1969, SIETSMA et al. 1975a, b). Native walls do not show an X-ray diffraction pattern, only after very drastic chemical treatments could a poorly crystalline cellulose I pattern be obtained (PARKER et al. 1963, ARONSON and FULLER 1969, COOPER and ARONSON 1967, SIETSMA et al. 1975a). One is thus tempted to interpret these facts by assuming a chemical association between both types of glucans, similar to the association between chitin and glucan as inferred in Section 2.2.1.

A puzzling fact remains that after treating the walls with peroxide and acid, which undoubtedly cleaves covalent linkages, an X-ray diffraction pattern is obtained of cellulose I which differs from that of regenerated cellulose obtained when cellulose crystallizes spontaneously (cf. STÖCKMANN 1972). Also microfibrils have been observed at the inside of the native wall (TOKUNAGA and BARTNICKI-GARCIA 1971, SIETSMA et al. 1975b) and after enzymic treatment to remove the noncellulosic glucan (HUNSLEY and BURNETT 1970).

2.2.3 (1 → 3)-β-D/(1 → 6)-β-D-Glucan

With the exception of the hyphal walls of Zygomycetes, glucans with (1 → 3)-β and (1 → 6)-β linkages seem to occur in the walls of all fungi (BARTNICKI-GARCIA 1968, GORIN and SPENCER 1968, ROSENBERGER 1976). They may occur as more or less water-soluble glucans often forming a gel-like layer around the hyphae (GORIN and SPENCER 1968, BUCK et al. 1968, WESSELS et al. 1972) or as a genuine wall component often existing in an alkali-insoluble condition. The extreme insolubility of these glucans makes it difficult to establish whether or not only one glucan species is present. Attempts to isolate homogenous glucans nearly always involve the use of alkali or acid with the risk of breaking covalent bonds.

The difficulties in analysis of these glucans are well illustrated by the fact that exposure of alkali-insoluble (1 → 3)-β-D/(1 → 6)-β-D-glucan to hot dilute HCl leads to the formation of a highly crystalline (1 → 3)-β-D-glucan (hydroglucan) which is soluble in alkali (HOUWINK and KREGER 1953, KREGER 1967). Even conditions as mild as treatment with sodium acetate buffer at pH 5.0 for 3 h at 75° C have been reported to make most of the alkali-insoluble glucan from *Saccharomyces cerevisiae* soluble in dilute alkali (BACON et al. 1969).

Most efforts have gone into determining the structure of yeast (*S. cerevisiae*) glucan with conflicting results (cf. PHAFF 1963, 1971). The most recent results seem to indicate that the wall of yeast contains three different glucan fractions: (1) an alkali-soluble glucan mainly (1→3)-β-linked with a small number of (1→6)-β-linked branches (FLEET and MANNERS 1976), (2) an alkali-insoluble but acetic acid-soluble branched (1→6)-β-linked glucan with a high number of glucose residues triply linked at C-1, C-3 and C-6 (BACON et al. 1969, MANNERS et al. 1973a), and (3) a major component (85%) consisting of a high molecular weight (1→3)-β-linked glucan with a few (1→6)-β-linked branches which may connect the (1→3)-β-linked chains (MISAKI et al. 1968, MANNERS et al. 1973b).

The reason why most of the β-glucan in the native wall of *S. cerevisiae* is alkali-insoluble is still not fully understood. The possibility remains that the various chemical or enzymic treatments used to extract distinct glucans break covalent bonds, destroying a continuous network of β-glucan around the cell. To explain the insolubility of the glucan the existence of such a continuous network has been inferred (NORTHCOTE and HORNE 1952, LAMPEN 1968, MACWILLIAM 1970).

The most insoluble part of the β-glucan in the wall of *S. cerevisiae* appears to be located in and around the bud scars, which also contain most of the chitin of the wall (HOUWINK and KREGER 1953, BACON et al. 1966, BERAN et al. 1972). Upon boiling with dilute HCl granules of crystalline chitin and fibrils of crystalline (1→3)-β-D-glucan (hydroglucan) appear in these bud scar regions (HOUWINK and KREGER 1953). Therefore the possibility should be considered that in these regions there originally existed a highly insoluble glycosaminoglycan containing homopolymeric stretches of both (1→3)-β-linked glucose residues and (1→4)-β-linked *N*-acetylglucosamine residues. Upon limited hydrolysis these stretches would be freed and give rise to the crystalline hydroglucan and crystalline chitin, the former being soluble in alkali.

More direct evidence for covalent linkage between (1→3)-β/(1→6)-β-linked glucan chains and chitin chains has been obtained for the highly insoluble R-glucan/chitin complex from the wall of the basidiomycete *Schizophyllum commune* (SIETSMA and WESSELS 1977, 1979). The glucan could not be dissolved in alkali; it even resisted treatment with 40% KOH at 100° C (a treatment that removes most of the β-glucan in yeast). However, after specific hydrolysis of the chitin chains with purified chitinase or with nitrous acid after deacetylation of *N*-acetylglucosamine with alkali, part of the glucan became water-soluble, the rest alkali-soluble. The alkali-soluble glucan very much resembled a glucan also found in the culture medium, a (1→3)-β-D-glucan with single glucose branches attached by (1→6)-β-linkages. The water-soluble glucan was (1→3)-β-linked but also contained longer (1→6)-β-linked branches. Again, boiling of the R-glucan/chitin complex with dilute mineral acid generated sharp X-ray reflections of hydroglucan and chitin (WESSELS 1965), indicating breaking of linkages in the R-glucan/chitin complex followed by alignment and crystallization of homopolymeric segments of (1→3)-β-linked glucose residues and of homopolymeric segments of (1→4)-β-linked *N*-acetylglucosamine residues.

The finding that acid-treated pieces of rhizomorph of *Armillaria mellea* showed oriented X-ray reflections of hydroglucan enabled JELSMA and KREGER (1975) to show that hydroglucan has a conformation of three intertwining helices each containing six glucose residues per turn of the helix, in accordance with a suggestion of REES (1973). In native walls or alkali-extracted residues, which only give very diffuse hydroglucan reflections, such ordered structures are probably frequently disrupted by side branches, giving the glucan gel-like properties.

Unlike normal cells, regenerating protoplasts of *Saccharomyces cerevisiae* form fibrils of well-crystallized (1 → 3)-β-D-glucan (KREGER and KOPECKÁ 1976). The x-ray reflections were identical to those of paramylon A, the granular (1 → 3)-β-D-glucan of the Euglenophyta, which differ from those of hydroglucan in a small shift in two spacings. These results were interpreted to mean that on these regenerating protoplasts the (1 → 3)-β-D-glucan synthesizing enzymes were active without the chain-interrupting activity of enzymes that normally are involved in branching.

2.2.4 (1 → 3)-α-D-Glucan with Variable Amounts of (1 → 4)-α-Linkages

The occurrence in fungal walls of an alkali-soluble glucan containing a preponderance of (1 → 3)-α-linkages was firmly established only recently. The glucan was first tentatively identified in *Polyporus* (*Piptoporus*) *betulinus* by DUFF (1952) and rigorously proved in *Aspergillus niger* by JOHNSTON (1965) using chemical methods. BACON et al. (1968) then showed that the X-ray reflections of this glucan were identical to those previously reported for a glucan in the walls of a variety of fungi (KREGER 1954, WESSELS 1965). Ever since the number of fungi with (1 → 3)-α-D-glucan in their walls has grown steadily (cf. ROSENBERGER 1976, PHAFF 1977). Even in *Neurospora crassa* the additional presence of (1 → 3)-α-D-glucan was overlooked by earlier investigators (DE VRIES 1974).

The different (1 → 3)-α-D-glucans studied contain various amounts of (1 → 4)-α-linkages; they vary from a (1 → 3)-α-D-glucan (S-glucan, pseudo-nigeran) apparently free of (1 → 4)-α-linkages in *Schizophyllum commune* (SIETSMA and WESSELS 1977) to a glucan with alternating (1 → 3)-α- and (1 → 4)-α-linkages (nigeran) isolated as a hot water-soluble glucan from a few species of *Penicillum* and *Aspergillus* (BOBBITT et al. 1977). Most (1 → 3)-α-D-glucans studied are not entirely free of (1 → 4)-α-linkages (SHIDA et al. 1978), although one should be aware of contamination with glycogen.

(1 → 3)-α-D-glucan clearly occurs in the wall in a microcrystalline condition, e.g., in *Schizophyllum commune* (WESSELS et al. 1972, VAN DER VALK and WESSELS 1976) and in *Piptoporus betulinus* and *Laetiporus sulphureus* (JELSMA and KREGER 1978). The X-ray reflections recorded with the native wall were similar to those of the isolated glucan precipitated from alkali but with the latter two species a small shift was recorded in the major reflections of the precipitated glucan and a third polymorph of the glucan was found after rigorous drying of the sample (JELSMA and KREGER 1979). Deviating X-ray diffraction patterns of (1 → 3)-α-D-glucan recorded for *Fusicoccum amygdali* (OBAIDAH and BUCK 1971) and *Tremella mesenterica* (REID and BARTNICKI-GARCIA 1976) can now be interpreted as representing the dehydrated polymorph of the glucan (JELSMA and KREGER 1978). On the basis of X-ray fiber patterns obtained with trama tissue of *P. betulinus*, JELSMA (1979) also derived the unit cells of the three polymorphs and a chain conformation representing a stretched, ribbon-like structure, the same in all polymorphs, and confirming earlier theoretical models (SUNDARALINGAM 1968, REES and SCOTT 1971, SATHANARAYANA and RAO 1972).

A detailed study on the crystal structure of isolated nigeran has also been carried out (PÉREZ et al. 1979), revealing a slight difference in X-ray diffraction pattern between freshly precipitated and dehydrated material.

As to the electronmicroscopical appearance, (1 → 3)-α-D-glucan, both in native walls and in precipitated samples, consists of thick microfibrils (20 to 30 nm) with an irregular outline (CARBONELL et al. 1970, SAN BLASS and CARBONELL 1974) which show cross-striations after negative staining (VAN DER VALK and WESSELS 1976).

2.2.5 Homo- and Hetero-Glucuronans

Acidic polysaccharides are major wall constituents in the hyphal walls of Zygomycetes in which polymers containing the cationic monomer glucosamine are also abundantly present. In the alkali-soluble fractions of the walls of these fungi heteropolysaccharides are found containing fucose, glucuronic acid, galatose, and/or mannose (BARTNICKI-GARCIA and REYES 1968a, b, MIYAZAKI and IRINO 1970, 1971, BALLESTA and ALEXANDER 1971, DATEMA et al. 1977a).

The heteroglucuronan isolated with alkali from walls of *Mucor rouxii* (mucoran) was water-soluble and contained fucose, mannose, and glucuronic acid in a 2:3:5 ratio. It had a molecular weight of about 100,000 and 60 to 70% consisted of (1 → 4)-linked aldobiuronic acids of the type GlcA-(1 → 4)-α-D-Man (BARTNICKI-GARCIA and LINDBERG 1972). Another acidic polysaccharide resisted both alkali and acid extraction but could be solubilized by alkali after acid treatment. The water-insoluble polysaccharide extracted in this way (mucoric acid) was microcrystalline and appeared to contain mainly glucuronic acid. The X-ray diffraction pattern showed that this substance is identical to a substance previously shown to be present in acid-treated walls of *Phycomyces blakesleeanus* (KREGER 1954, 1970). Although isolated as two distinct polysaccharides, the possibility was considered that mucoran and mucoric acid might have been derived from a single polysaccharide (BARTNICKI-GARCIA and REYES 1968a).

By depolymerizing the glucosamine-containing polymers by nitrous acid, DATEMA et al. (1977a) isolated from the wall of *M. mucedo* a single water-soluble heteroglucuronan containing all the neutral and acidic sugars of the wall (fucose, mannose, galactose, and glucuronic acid in a molar ratio of 5:1:1:6). This heteroglucuronan could also be extracted quantitatively with salt solutions of high ionic strength and partially with alkali. This was interpreted as evidence that the acidic polysaccharide was kept in the wall in an insoluble form by ionic binding to the insoluble polymers containing glucosamine. By treatment of the isolated water-soluble heteroglucuronan with 1M HCl (100 °C) it was partly converted into a water-insoluble crystalline glucuronan, containing only glucuronic acid, with the properties of mucoric acid. This indicates that in *M. mucedo* mucoric acid, which can be extracted from the wall by alkali after acid treatment, is not a genuine wall component, but arises by partial acid hydrolysis of a single heteroglucuronan and subsequent crystallization of homopolymeric segments containing glucuronic acid.

Uronic acids have also been detected in small amounts in the walls of fungi not belonging to the Zygomycetes (GANCEDO et al. 1966). How these uronic acids are present is unknown. In addition, heteroglycuronans have been isolated by water extraction from fruit-bodies of basidiomycetes (UKAI et al. 1974, YOSHIOKA et al. 1975), but their relationship to the wall is not clear.

2.2.6 Glycoproteins

The occurrence of proteins in the cell wall of fungi has been a matter of controversy. It is often difficult to establish whether protein material is a genuine wall component or a cytoplasmic contaminant. Also, secreted enzymes may stay for some time in the periplasmatic space or even in the wall itself; usually they consist of glycoproteins in which the polysaccharide closely resembles genuine wall components (LAMPEN 1968, VAN RIJN et al. 1972).

It has now been firmly established that glycoproteins form a structural part of the fungal wall. Most of these glycoproteins can be extracted with dilute alkali, a small part of the polymeric amino acids, however, resists this treatment. The latter material is usually very closely associated with the glucan/chitin complex. The amount of peptide material in this fraction is low and the largest part of it is soluble in dilute acid, the rest becomes soluble in alkali after this acid treatment (MARET 1972, REID and BARTNICKI-GARCIA 1976, SIETSMA and WESSELS 1977). This indicates special properties for this kind of material and probably a special role in the fungal wall. SIETSMA and WESSELS (1977) have shown in *Schizophyllum commune* that these peptides have a function in the linkage between the glucan and chitin.

The glycoprotein material soluble in dilute alkali often contains mannose and therefore can be separated from other alkali-soluble material, e.g., the $(1\rightarrow3)$-α-D-glucan, by formation of an insoluble copper-complex (GORIN and SPENCER 1970). The best-characterized are the alkali-soluble glycoproteins (the mannans and galactomannans) of yeast and yeast-like organisms (BALLOU 1976). In *Saccharomyces cerevisiae* the polysaccharide moities consist exclusively of mannose linked by $(1\rightarrow3)$, $(1\rightarrow6)$ and $(1\rightarrow2)$-α-linkages. In other yeast species galactose molecules are also present and the linkage type may differ, even β-linkages may occur (GORIN and SPENCER 1968). In *Neurospora, Aspergillus,* and *Penicillium* mannose and galactose occur in the glycoprotein fraction (CARDEMIL and PINCHEIRA 1979, ZONNEVELD 1971, GANDER 1974, TROY and KOFFLER 1969). Analysis of walls of several other fungi show traces of sugars other than glucose, probably also occurring in glycoproteins located at the outer periphery of the wall. In Basidiomycetes xylose and mannose are often present in trace amounts, suggesting that the polysaccharide moities of the glycoproteins in these organism are xylomannans (ANGYAL et al. 1974, REID and BARTNICKI-GARCIA 1976, WESSELS et al. 1972).

Since in cell wall analysis most wall preparations have been treated with alkali which involves the risk of separating polysaccharides and peptides by a β-elimination reaction, more careful studies of cell walls could reveal that materials now identified as pure polysaccharides are linked to peptide material in the native wall.

3 Ultrastructural Localization of Wall Polymers

3.1 Methodological Difficulties

Since fungal hyphae generally grow by extension of a small apical region, most of the wall material isolated from growing cultures originates from the larger part of the hyphae no longer capable of growth (mature walls). Such wall preparations, and sometimes intact hyphae, have been examined with the electronmicroscope before and after various chemical or enzymic extractions in order to probe the location of different polymers in the wall. However, many observations thus obtained do not allow straightforward conclusions and have contributed much confusion. Among the methodological difficulties are the following: (1) An apparent layering is often observed in sections. However, it is possible that the stains used for electronmicroscopy are taken up by minor components of the wall, e.g., proteins, and that the removal of these components bears no relationship to the removal of major components as intended by the extractions. (2) Mostly the shadowing technique is used to view the surfaces of the wall specimens before and after extractions. During the drying process the wall fabric is flattened and it becomes difficult to judge whether the appearance of structures, e.g., microfibrils, was due to removal of an overlaying layer or to the removal of (matrix) material filling the spaces between the structures uncovered. The only possibility to avoid this problem is to use the shadowing technique on sectioned material (SCHNEIDER and WARDROP 1979). (3) Fixatives such as glutaraldehyde/OsO_4 fix wall material poorly so that during dehydration a phase separation may be observed that was not originally present (VAN DER VALK et al. 1977). (4) The most specific agents to remove particular wall components are thought to be enzymes. However, commercial enzyme preparations often contain a variety of contaminating enzymes, and a check for purity is seldom made (DAVIS and DOMER 1977). (5) Because of cross-linking between wall polymers degradation of one polymer may lead to solubilization of another polymer. For instance, complete depolymerization of chitin may render previously alkali-insoluble β-glucan soluble in water or alkali (SIETSMA and WESSELS 1979) and depolymerization of glucosamine-containing polymers can lead to release of a hetero-glycuronan (DATEMA et al. 1977a). (6) The removal of a wall component from the wall may lead to rearrangement of remaining polymer chains and thus produce an artifactual image in the electronmicroscope (see Sect. 2.2.1 and 2.2.3). (7) Wall components in the intact wall may be inaccessible to enzymes (see Sect. 5.1) and failure of an enzyme to produce a visible change does not necessarily mean that the substrate for the enzyme is missing. A polymer may even occupy several positions in the wall with a differential susceptibility to enzymes (BOBBITT et al. 1977).

These considerations make it extremely difficult to evaluate the existing literature and therefore only a generalized view will be presented here based on much interpretation by the present reviewers.

3.2 Mature Walls

3.2.1 Filamentous Fungi

Electronmicroscopic observations combined with the use of more or less specific enzymic or chemical extractions have been made on the walls of a variety of filamentous fungi (ARONSON 1965, MAHADEVAN and TATUM 1967, MANOCHA and COLVIN 1967, 1968, TROY and KOFFLER 1969, HUNSLEY and BURNETT 1970, MARET 1972, WESSELS et al. 1972, BARTNICKI-GARCIA 1973, HUNSLEY and GOODAY 1974, SIETSMA et al. 1975a, LETOURNEAU et al. 1976, MICHALENKO et al. 1976, BOBBITT et al. 1977, POLACHEK and ROSENBERGER 1977, VAN DER VALK et al. 1977, SCHNEIDER and WARDROP 1979, WESSELS and SIETSMA 1979). Al-

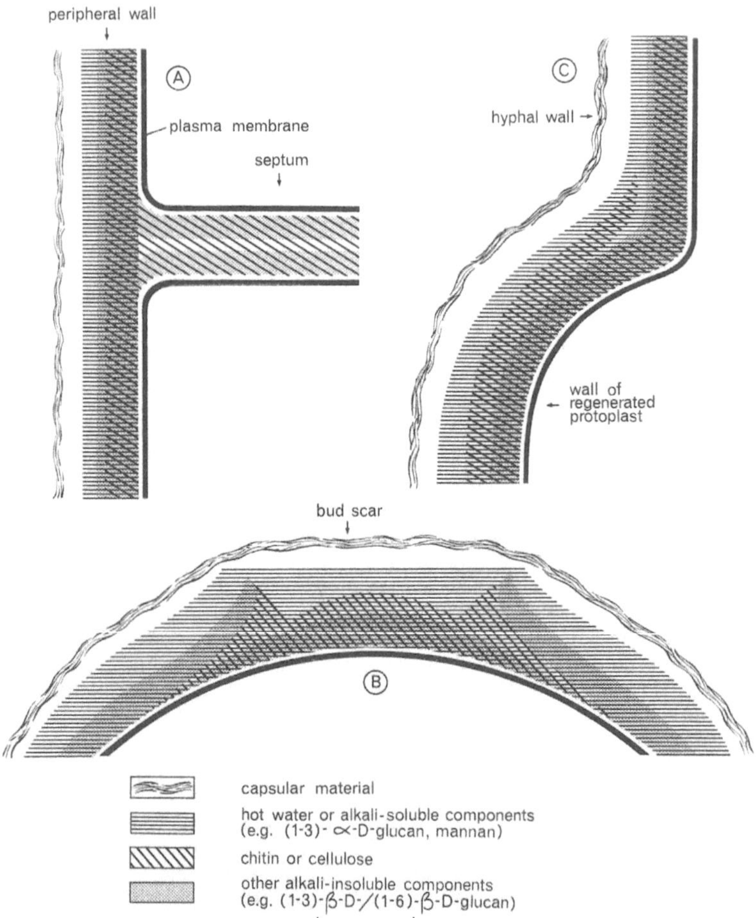

Fig. 1A–C. Generalized models of fungal walls and septa. **A** Peripheral wall and septum of filamentous fungi. **B** Wall and bud scar (septum) of yeast. **C** Transition zone between wall of regenerated protoplast (or encysted zoospore) and emerging hyphal tube

though in some of these studies coaxial layering of the wall is explicitly inferred (e.g., HUNSLEY and BURNETT 1970) most studies appear compatible with a model as depicted in Fig. 1A. In this model the wall consists of essentially one layer, some components of which extend through the inner wall portion to the surface where they accumulate and can be found in a rather homogeneous state. This does not exclude the occurrence of distinct outer wall layers in certain specialized cells such as aerial hyphae, sporangiophores, and spores, but most of the studies cited above were aimed at elucidating the structure of the wall of undifferentiated hyphae growing in submerged condition.

The capsular material accumulating at the outside of the hyphae and sometimes found in abundance in the culture medium is very variable in composition. Often it consists of simple glucans (Sect. 2.2.3) which may be consumed by the fungus during carbon starvation (WESSELS et al. 1972). In other cases they are highly specific glycopeptides or glycoproteins (GANDER 1974) which may be determinants of specificity in host–pathogen interactions (ALBERSHEIM and ANDERSON-PROUTY 1975, CALLOW 1977, ALVIANO et al. 1979) or mating interactions (CRANDALL 1976).

The water-insoluble polymers that accumulate at the outer surface of the wall can sometimes be removed by hot water (nigeran) but mostly with dilute alkali. Such substances include (1→3)-α-D-glucan (S-glucan) in *Schizophyllum commune* (WESSELS et al. 1972, VAN DER VALK et al. 1977). *Agaricus bisporus* (MICHALENKO et al. 1976) and *Aspergillus nidulans* (POLACHEK and ROSENBERGER 1977) and (1→3)-α/(1→4)-α-D-glucan (nigeran) in *Aspergillus niger* and *A. awamori* (BOBBITT et al. 1977). The conclusion that the glucans are present not only at the outside but also in the inner portion of the walls is based on the susceptibility of other wall components in wall preparations before and after extraction of the α-D-glucans (WESSELS and DE VRIES 1973, BOBBITT et al. 1977) or in wall preparations from normal and mutant strains lacking (1→3)-α-D-glucan (POLACHEK and ROSENBERGER 1977).

Although the α-D-glucans accumulated at the surface of the walls are mostly called "amorphous" in electronmicroscopic studies, it is important to note that in fact these glucans exist in the wall (at least partly) in a microcrystalline condition (WESSELS et al. 1972, BOBBITT et al. 1977). When the packing is not too dense, microfibrils of (1→3)-α-D-glucan can also be observed (CARBONELL et al. 1970, VAN DER VALK and WESSELS 1976), especially after negative staining.

In freeze-etched surfaces of hyphae of *Schizophyllum commune* small parallel arrays of rodlets with a periodicity of about 10 nm have been observed (WESSELS et al. 1972) which were also attributed to crystalline (1→3)-α-D-glucan. Similar structures have been observed at the wall surface of many fungi, particularly on spores (see BURNETT 1976). However, quite similar structures in the wall of microconidia of *Trichophyton mentagrophytes* have been related to the presence of a distinct glycoprotein layer very resistant to proteolytic digestion (HASHIMOTO et al. 1976). This has been confirmed in a recent study (COLE et al. 1979) for rodlet layers on the surface of conidia of a variety of fungi. It is possible that this layer is typical for aerial structures because it is often found associated with lipids and thus could confer hydrophobic properties to the surface of the wall.

Glycoproteins forming reticulate fibers are present at the outside of walls of *Neurospora crassa* (MAHADEVAN and TATUM 1967, HUNSLEY and BURNETT 1970) and *Chaetomium globosum* (MARET 1972). Whether the glycoproteins are also present in the inner portion of the wall is not known.

Apart from loosely adhering capsular material, the wall of Oomycetes is rather insoluble in alkali, as judged from studies on *Pythium debaryanum* (COOPER and ARONSON 1967, MANOCHA and COLVIN 1968, YAMADA and MIYAZAKI 1976), *P. acanthicum* (SIETSMA et al. 1975a) and *Phytophthora palmivora* (*parasitica*) (HUNSLEY and BURNETT 1970, TOKUNAGA and BARTNICKI-GARCIA 1971, HEGNAUER and HOHL 1973). Here the alkali-insoluble (1→3)-β/(1→6)-β-D-glucan, present in the inner portion of the wall together with (1→4)-β-D-glucan (cellulose), seems to extend beyond the fibrillar phase obscuring microfibrils when the walls are viewed from the outside.

In Chytridiomycetes the construction of the wall is similar to that of Ascomycetes and Basidiomycetes; chemical extraction of the wall reveals microfibrils composed of crystalline chitin (ARONSON and PRESTON 1960).

In the Zygomycetes, which do not contain glucan in their hyphal walls, little is known about the location of polymers in the wall. One would expect, however, that the cationic polymers containing glucosamine and the anionic glycuronans, which together with chitin make up the most of the wall (BARTNICKI-GARCIA and REYES 1968a, b, DATEMA et al. 1977a, b), occur in close association.

Apart from the Zygomycetes, most authors mentioned above consider the alkali-insoluble portion of the wall as consisting of microfibrils of chitin (Basidiomycetes, Ascomycetes, and Chytridiomycetes) or cellulose (Oomycetes) embedded in a matrix of (1 → 3)-β/(1 → 6)-β-D-glucan. However, the concept of microfibrils physically embedded in a matrix of β-glucan may not always be entirely correct. As has been pointed out in Sections 2.2.1 and 2.2.3, apart from chitin and (1 → 3)-β/(1 → 6)-β-D-glucan occurring as individual polymers, covalent binding between these polymers is likely to occur and at least some of the microfibrils observed may be artifacts of preparation of the wall residues. How these findings will eventually influence our views of the microfibril-matrix concept is too early to say, but at least they suggest that the alkali-insoluble portion of the wall is a much more intricate covalently linked molecular network than originally thought.

Some confusion exists as to the occurrence of more than one wall layer in fungi. As indicated above, we consider most fungal walls as consisting of essentially one layer, some components of which accumulate at the outer surface. Even very thick walls such as occur in hyphae in fruit-bodies of *Schizophyllum commune* apparently have the same architecture as the walls of thin-walled hyphae (VAN DER VALK and MARCHANT 1978). SCHNEIDER and WARDROP (1979) have proposed that a clear case of multiple layers, comparable to primary and secondary wall layers in higher plants, would require the demonstration of different orientations of microfibrils in successive layers. An example in case would be the wall of the sporangiophore of *Phycomyces* (ROELOFSEN 1951, MIDDLEBROOK and PRESTON 1952). Other cases of different orientations of microfibrils in successive layers have been reported (ARONSON and PRESTON 1960, MARCHANT 1966, SCURFIELD 1967, MANOCHA and COLVIN 1968, BARTNICKI-GARCIA 1973, GOODAY 1975, LETOURNEAU et al. 1976). However, it is possible that some of these cases arose due to reorientation of microfibrils by shearing forces during preparation of the walls (SCHNEIDER and WARDROP 1979).

The ultrastructure of septa in filamentous fungi was recently reviewed by GULL (1978). In Ascomycetes and Basidiomycetes the septum is typically trilamellate: a central layer that remains translucent after various staining procedures bounded at both sides by a layer that is continuous with the inner portion of the peripheral wall. The location of polymers as depicted in Fig. 1A is mainly deduced from a study with *Schizophyllum commune* (WESSELS and MARCHANT 1974, VAN DER VALK et al. 1977). In this organism it was concluded that the central plate consists of a rather homogeneous assembly of chitin microfibrils bordered on both sides by a layer consisting of a chitin/β-glucan complex as present in the peripheral wall. However, while in addition to the β-glucan the peripheral wall also contained $(1 \rightarrow 3)$-α-D-glucan, the latter substance was apparently missing from the septa. Similarly, HUNSLEY and GOODAY (1974) have concluded that glycoproteins present in the peripheral wall are absent in the septa in *Neurospora crassa*. Not shown in Fig. 1A is the septal swelling around the central pore in the septum of Basidiomycetes. This swelling appears to consist mainly of $(1 \rightarrow 3)$-β/$(1 \rightarrow 6)$-β-D-glucan (VAN DER VALK et al. 1977).

The apparent homogeneity of chitin and the concentric orientation of chitin microfibrils often observed around the central pore (SCURFIELD 1967, MARET 1972, VAN DER VALK et al. 1977) indicate that the microfibrils in the central plate are not an artifact. It rather suggests that the first materials synthesized on the infolding plasma membrane during ontogeny of the septum are only chains of $(1 \rightarrow 4)$-β-D-aminoglycan which crystallize to give chitin microfibrils. This would then be followed by coordinate synthesis of chitin and β-glucan as occurring in the peripheral wall. Such a chain of events would not be unlike that observed on the plasma membrane of naked protoplasts which regenerate a new wall (Sect. 3.2.3).

3.2.2 Yeasts

Most wall components found in filamentous fungi have also been detected in yeasts but many yeast walls are particularly rich in manno-proteins (PHAFF 1971, BARTNICKI-GARCIA and MCMURROUGH 1971). Because yeasts do not constitute a natural taxonomic unit, the variations in wall composition encountered are partly explicable on the basis of taxonomic diversity (MANNERS et al. 1974). For instance, the ascomycetous yeasts, such as *Saccharomyces cerevisiae* and *Candida albicans*, are rich in $(1 \rightarrow 3)$-β/$(1 \rightarrow 6)$-β-D-glucan and mannan with only a small amount of chitin. In contrast, basidiomycetous yeasts (e.g., *Sporobolomyces*) contain mainly chitin and galacto-mannan. Although a high mannan content is typical for many yeasts, it appears to be very low in some species. For instance, little galacto-mannan and no chitin was found in the fission yeast *Schizosaccharomyces* which does contain a large amount of $(1 \rightarrow 3)$-α-D-glucan and $(1 \rightarrow 3)$-β-D-glucan in its cell wall (KREGER 1954, BACON et al. 1968, GORIN and SPENCER 1970).

Only the walls of *S. cerevisiae* and *C. albicans* have been studied in enough detail to construct models of the molecular architecture of their walls. Such models depict the $(1 \rightarrow 3)$-β/$(1 \rightarrow 6)$-β-D-glucan (yeast glucan) as an inner layer and the manno-proteins as forming a continuous layer at the outside (LAMPEN 1968, KIDBY and DAVIES 1970). The model of LAMPEN emphasizes the cross-linking of manno-proteins by phospho-diester bonds between the mannan moieties; in the model of KIDBY and DAVIS disulfide linkages between the protein moieties are also considered of major importance.

As in filamentous fungi, the concept of more than one layer in the wall may not be entirely correct. Cytochemical staining procedures selective for manno-proteins (LINNEMANS et al. 1977, HORISBERGER and VONLANTHEN, 1977, POULAIN et al. 1978) have shown that manno-proteins are present throughout the wall and the apparent layering of the wall may thus merely reflect an accumulation of manno-proteins at the outside of the wall.

In *S. cerevisiae,* chitin seems to be confined to the bud scars (BACON et al. 1966, CABIB and BOWERS 1971, BERAN et al. 1972) which also contain β-glucan (HOUWINK and KREGER 1953) and mannan (BAUER et al. 1972). Figure 1B shows a model of the wall and bud scar based on this information and on published electronmicrographs of sections by KREGER-VAN RIJ and VEENHUIS (1969) and CABIB and BOWERS (1971). Of particular interest is the structure of the bud scar which marks the area on the mother cell where the daughter cell separated along the original septum. As in the filamentous fungi, the original septum separating the two cells was trilamellate with the central plate (i.e., the first wall material formed between the invaginating plasma membrane) consisting of rather homogeneous chitin. Not only in ascomycetous yeasts but also in basidiomycetous yeasts a nonstainable central plate in the septum has been observed (KREGER-VAN RIJ and VEENHUIS 1971), probably consisting of rather pure chitin.

3.3 Newly Formed Walls

3.3.1 Reverted Protoplasts and Germinated Zoospores

Protoplasts of fungi are readily formed in osmotically stabilized media by using wall-lytic enzymes from various sources (VILLANUEVA and GARCIA ACHA 1971). When placed in an osmotically stabilized growth medium lacking the wall-lytic enzymes, the protoplasts quickly manufacture a new wall around them (regeneration) which may or may not contain all the components found in the normal wall. The walled protoplasts (primary cells), obtained from filamentous fungi, may then first go through a stage of abnormal budding or normal hyphae emerge (reversion) directly from the primary cells (for reviews see: NECAS 1971, WESSELS et al. 1976, PEBERDY 1978). At this point it is relevant to discuss the architecture of the wall of the primary cell and the transition in architecture when this wall merges into that of the forming hyphal tube. Figure 1C interprets the changes as seen in a number of cases.

Both in *Pythium acanthicum* (SIETSMA et al. 1975b) and *Schizophyllum commune* (DE VRIES and WESSELS 1975, VAN DER VALK and WESSELS 1976) protoplast regeneration starts with the synthesis of microfibrillar components and the subsequent architectural change that occurs during completion of regeneration and reversion appears much the same, although the chemical nature of some of the wall components is very different. In *S. commune,* the first components formed are a network of highly crystalline chitin microfibrils on the plasma membrane covered by a mass of crystalline $(1\rightarrow3)$-α-D-glucan. The synthesis of the alkali-insoluble $(1\rightarrow3)$-β/$(1\rightarrow6)$-β-D-glucan starts later and from that point on all newly synthesised chitin microfibrils appear embedded in this glucan. Consequently, the uncoupling of chitin synthesis from β-glucan synthesis in the beginning results in the appearance of free chitin microfibrils at the outside of the regenerated protoplast, seen after removal of the $(1\rightarrow3)$-α-D-glucan. If the regenerated protoplast grows by budding, which probably involves the regeneration of a new protoplast emerging through a weak spot in the wall of the primary cell (RAMOS and GARCIA-ACHA 1975), the bud attains the same wall architecture as the primary cell.

If, however, a hyphal tube is initiated, chitin and β-glucan synthesis occur coordinately underneath the already existent microfibrillar chitin layer; at a certain point the layer containing the chitin/glucan complex breaks through the microfibrillar layer and a hyphal tube emerges (VAN DER VALK and WESSELS 1976) in which chitin and β-glucan synthesis occur coordinately. This results in a sharp demarcation of wall architecture at the site where the hyphal tube emerges (Fig. 1C).

A similar change in wall architecture is found at the site of germination of encysted zoospores of *Phytophthora palmivora* (TOKUNAGA and BARTNICKI-GARCIA 1971, HEGNAUER and HOHL 1973) and *Allomyces macrogynus* (FULTZ and WOOLF 1972). As in reverting protoplasts, the change in wall architecture may be explained by assuming an unbalanced synthesis on the originally naked cell, followed by a coordinate synthesis of the fibrillar and matrix components at the hyphal apex.

In regenerating protoplasts of *Saccharomyces cerevisiae* walls are produced not only with an aberrant architecture, but also with microfibrils of crystalline (1→3)-β-D-glucan, not found in the normal wall (NECAS 1971, KREGER and KOPECKÁ 1976). This has been explained by assuming that the biosynthetic machinery for wall synthesis is somehow deranged due to difficulties in retaining diffusible polymers and enzymes by the regenerating protoplasts.

3.3.2 Growing Areas of the Wall

Intercalary elongation of the wall is a rare event in fungi. It occurs in growing sporangiophores of *Phycomyces* and the study of this system has actually led to the formulation of the multi-net growth theory (ROELOFSEN 1959) also used to explain the change from a predominantly transverse orientation of cellulose microfibrils at the inside to a predominantly longitudinal orientation at the outside in cell walls of elongated cells of higher plants. Intercalary growth also seems to occur in elongating stipes of Agarics and GOODAY (1975) has shown a predominant transverse orientation of chitin microfibrils at the inside of the walls of the elongating stipe cells of *Coprinus cinereus* and demonstrated the intercalary deposition of wall material during elongation. Such striking examples of intercalary growth should not divert attention from the fact that in general all fungal hyphae grow by extension of a small area at their apices. Because of its small size and minute contribution to the material in wall preparations, it is not surprising that little is known about the molecular architecture of the wall in this most important part of the hypha.

After chemical treatments to visualize microfibrils, it is generally observed that the hyphal tip is completely covered with a random meshwork of microfibrils similar to and continuous with that of the subapical region. Earlier claims as to discontinuities at the very apex (STRUNK 1963, MARCHANT 1966) have been contradicted (cf. HUNSLEY and BURNETT 1970, BARTNICKI-GARCIA 1973, SCHNEIDER and WARDROP 1979). In all the cases examined the microfibrils appear to be embedded in a matrix as they are in subapical regions and only become visible in abundance after removal of this matrix material.

After chemical treatments to visualize microfibrils, these are seen to increase in thickness toward the subapical region (HUNSLEY and BURNETT 1968) and in *Neurospora crassa* the chitin microfibrils have also been seen to increase in density of packing toward the base of the tip (BURNETT 1976). In view of the uncertainties about the generation of microfibrils of chitin by the chemical treatments used to visualize the microfibrils, it is difficult to interpret these results. Also, because the length of the extension zone of the hyphae examined is not known, it is difficult to know whether the change in microfibril diameter occurred within or outside the extension zone. However, it appears clear that these interesting changes cannot be related to an increase in wall thickness. TRINCI and COLLINGE (1975) have established that in growing hyphae of *N. crassa* wall thickness remains uniform in the extension zone and even in the zone of rigidification where the hyphal diameter attains its maximum value. Increase in wall thickness did not start within 10 to 50 μm from the apex (the larger distance was found in leading hyphae), whereas the measurements of BURNETT (1976) fell within the first 15 μm from the apex. Eventually the wall thickness increased up to five times that in the apical part, probably due to accumulation of glycoprotein at the outside of the wall.

The accumulation of certain wall polymers at the outer surface some distance from the tip may also account for differences found between apical and subapical parts of the walls in staining by fluorescent antibodies (FULTZ and SUSSMAN 1966, MARCHANT and SMITH 1968) or fluorescent brightners such as calcofluor (GULL and TRINCI 1974). With the fluorescent brightener not only the apices but also the septa, known to lack some of the surface polymer (HUNSLEY and GOODAY 1974, VAN DER VALK et al. 1977), were stained. The absence of abundant quantities of such polymers at the apex and in septa could expose stainable polymers ubiquitously present in the wall. In any case, no certainty exists that the boundary between stainable and nonstainable wall marks the transition between extensible and nonextensible wall. Therefore any chemical or architectural changes occurring in this area remain obscure at the moment.

4 Wall Composition and Cellular Morphology

4.1 Does Wall Composition Determine Cellular Morphology?

Wall preparations from fungal cells retain the shape of the original cells. On account of this, the statement that the wall determines the shape of the fungal cell is often made in the literature. This may be true in a general way but the simplicity of this statement has led some students of morphogenesis to propose that the wall in itself contains the determinants of cellular shape and therefore has resulted in many studies intended to show a causal relationship between wall composition and cellular form (e.g., yeast-like form vs. hyphal form) or mycelial morphology (e.g., branching pattern of hyphae).

As outlined before, in the fungal kingdom both the hyphal and the yeast-like form can be constructed on the basis of quite different wall components. In addition, large variations in wall composition may occur during growth of a single species, often in relation to changing environmental conditions. For instance, GOLD et al. (1973) have shown that in *Aspergillus aculeatus* the wall component nigeran increases dramatically during nitrogen starvation and SIETSMA et al. (1977) have shown that growth of *Schizophyllum commune* under elevated levels of carbon dioxide leads to the production of a double amount of water-soluble (1 → 3)-β-D-glucan and less than half the amount of alkali-insoluble wall glucan containing the same type of linkages. Yet no changes in cellular shape occurred. Also it has never been demonstrated that any one of the individual wall components is responsible for maintaining the integrity of the wall. Chemical or enzymic removal of individual wall components may lead to thinning of the wall, but the original shape is conserved until the last component is removed (cf. HUNSLEY and BURNETT 1970, WESSELS and MARCHANT 1974).

The best evidence to show that the presence of an individual wall component does not determine shape comes from experiments in which mutations or environmental conditions prevent synthesis of such a wall component without interfering with cellular morphology. KATZ and ROSENBERGER (1970a, 1971a) isolated a temperature-sensitive mutant of *Aspergillus nidulans* that had a greatly reduced level of chitin when grown at the nonpermissive temperature (41° C). This phenotype was due to a single recessive mutation and could be reversed by glucosamine or *N*-acetylglucosamine, suggesting the loss of an enzyme involved in amino-sugar synthesis. Yet, when grown at the nonpermissive temperature in the presence of an osmotic stabilizer, the hyphae had a normal morphology; when transferred to a dilute buffer these hyphae did not burst. These findings suggest that in this case the synthesis of chitin is not necessary to achieve hyphal morphology nor to produce a rigid wall. A puzzling phenomenon remains that the hyphae did burst when they were transferred to a growth medium lacking osmotic stabilizer. ZONNEVELD (1974) and POLACHECK and ROSENBERGER (1977) studied mutants of *A. nidulans* containing little or no (1→3)-α-D-glucan in their walls. Although no detailed information was given on hyphal morphology, the mutants grew on normal media and apparently had rigid walls. J.H. SIETSMA (unpublished) has shown that regeneration and reversion of protoplasts from *Schizophyllum commune* proceeds normally in the presence of 2-deoxyglucose, although this compound completely inhibited the synthesis of (1→3)-α-D-glucan. Again it appears that this major wall polysaccharide is of no consequence to hyphal morphology. On the other hand, it was found in this system that inhibition of chitin synthesis with polyoxin D, which also inhibited synthesis of the insoluble (1→3)-β/(1→6)-β-D-glucan in the wall, prevented the generation of hyphal tubes (DE VRIES and WESSELS 1975). Under these conditions, however, protoplasts regenerated a wall essentially consisting of only (1→3)-α-D-glucan which conferred osmotic stability to these protoplasts.

Although it is unlikely that a direct causal relationship exists between wall composition and cellular morphology, it is clear that during synthesis of the wall the shape of the cell comes into being and is subsequently maintained by the wall due to its rigidity. The issue of spatially regulated synthesis of wall components giving rise to cellular form will be considered in Section 6.3. Here a few cases will be mentioned in which the establishment of a relationship between wall composition and cellular shape has been attempted, because they may contain clues pertaining to the processes that generate cellular form.

4.2 Changes in Wall Composition Accompanying Changes in Cell Morphology

Among the earliest recognized differences in wall composition related to cellular form is the presence of mannan in many yeasts compared to the scarcity of this polymer in hyphal fungi (GARZULY-JANKE 1940). Particular interest has been devoted to cases where environmental manipulation induces the transition from the hyphal to the yeast-like form and vice versa (dimorphism). In some cases the appearance of the yeast-like form was indeed accompanied by the appearance of mannan in the wall, e.g., in *Mucor rouxii* (BARTNICKI-GARCIA 1963), but in other cases an inverse relationship has been observed (BARTNICKI-GARCIA 1968). In *Paracoccidioides brasiliensis* it has been found that the yeast-like form contains (1 → 3)-α-D-glucan as the only glucan in the wall but the hyphal wall contains (1 → 3)-β/(1 → 6)-β-D-glucan in addition, stressing the importance of the β-glucan for hyphal morphology (KANETSUNA et al. 1972). In *Histoplasma capsulatum*, however, both forms contained the β-glucan but the (1 → 3)-α-D-glucan was missing from the hyphal walls (KANETSUNA et al. 1974). On the other hand, PREVIATO et al. (1979), working with *Sporothrix schenckii*, found approximately the same ratios of alkali-soluble to alkali-insoluble glucans in the yeast-like and hyphal forms, but they emphasized the importance of quantitative and qualitative differences in wall proteins.

Perhaps the most promising results along these lines have come from a long-term study of the effects of mutations on hyphal morphology in *Neurospora crassa* (cf. BRODY 1973, MISHRA 1977), especially since they seem to point at mutational changes in membranes which probably are intimately involved in synthesis of the wall. Approximately 120 of the known 500 loci of *Neurospora* cause morphological aberrations, the best-studied of which are those that give colonial growth of some kind. These are characterized by compact colonies in which the hyphae are highly branched. Most of these mutants (*colonial, balloon, frost, ragged*) have no nutritional deficiencies but some (*inos, chol*) only produce the aberrant phenotype when grown on suboptimal levels of inositol or choline, respectively.

The walls of the colonial mutants mostly show an increase in chitin, the alkali-insoluble β-glucan can be increased or decreased, and although they contain more of an alkali-soluble fraction (glycoproteins) less peptide material is found in this fraction (MAHADEVAN and TATUM 1965, WRATHALL and TATUM 1974). However, in the inositol requiring mutant suboptimal levels of inositol, giving rise to the abnormal morphology, lower the glucosamine content (chitin) but increase the amount of galactosamine (derived from the alkali-soluble fraction) in the wall (HANSON and BRODY 1979). Again, at the level of wall composition, there appears little consistency relating the mutational lesion to morphology.

Where the primary lesions in the prototrophic colonial mutants have been pinpointed, they all appear to occur in enzymes of carbohydrate metabolism viz phosphoglucomutase (PGM), glucose-6-phosphate dehydrogenase (G6PH), and 6-phosphogluconate dehydrogenase (6PGD). In the PGM mutants the specific activity of PGM is greatly reduced. The effect on morphology is thought to occur via a reduction in glucose-1-phosphate resulting in a reduction in UDPG. A shortage of UDPG would lower the synthesis of (1 → 3)-β-D-glucan which indeed is found in lower quantities in the walls of these mutants. The possibility that abnormal levels of phosphorylated sugars (such as glucose-6-phosphate) could modify the biosynthesis of glucan was also considered. Inhibition of β-glucan synthesis has also been inferred to explain the effect of sorbose which, when added to the medium of growing wild-type strains, produced phenocopies of the colonial mutants (MISHRA and TATUM 1972).

The effects of defective G6PD and 6PGD have been related to morphology in a less direct way. These mutations do not influence the specific activities, but rather cause a decrease in the substrate affinities of these enzymes. This would lead to the decreased NADPH levels observed in these mutants which in turn would result in a decrease in the amount of unsaturated fatty acids, such as linoleic acid, in the cellular membranes. Considering that wall-synthesizing enzymes are membrane-bound, this could lead to changes in the activities of these enzymes. In this way wall synthesis and morphology could be modified. Such effects on membrane composition have also been proposed in choline and inositol auxotrophs that exhibit the aberrant morphology. In the case of an inositol mutant growing on suboptimal levels of inositol a dramatic decrease in inositol-containing lipids was observed (HANSON and BRODY 1979).

5 Wall-Degrading Enzymes

5.1 Wall Components as Substrates for Degrading Enzymes

Fungi produce a variety of extracellular enzymes involved in degradation of biopolymers in the external milieu, the breakdown products of which can then be taken up by the fungus for nutrition. Such enzymes include chitinase (STIRLING et al. 1979), $(1\rightarrow3)$-β-glucanases and $(1\rightarrow6)$-β-glucanases (VILLANUEVA et al. 1976), and cellulases (ERICKSON and HAMP 1978). Among these polysaccharidases, $(1\rightarrow3)$-β-glucanases and $(1\rightarrow6)$-β-glucanases are constitutively produced, but chitinase and cellulase require inducers while the formation of all these enzymes is reduced in the presence of a readily usable carbon source such as glucose (REESE 1972, 1977).

Such enzymes which appear to function primarily in the degradation of extracellular biopolymers can be potentially active against the wall polymers of the fungus that produces these enzymes. For instance, concentrated extracellular hydrolases produced by *Trichoderma harzianum* (*viride*) can be used to liberate protoplasts from the same fungus (DE VRIES and WESSELS 1973a).

At least two explanations can be given why potentially active wall-degrading enzymes do not generally disrupt the wall structure: (1) The excreted enzymes are rapidly diluted in the environment and never reach concentrations high enough to attack the wall components of the living fungus. (2) Although individual wall components after extraction from the wall may be susceptible, their physical or chemical status in the intact wall may prevent action of the hydrolytic enzymes. Special enzymes may be necessary to first disrupt the integrity of the wall in order to allow general hydrolases to act on individual wall components.

That native walls are quite resistant to general hydrolases has often been noted, but our ignorance of the mechanisms of resistance reflects our incomplete understanding of the molecular interactions between polymers in the wall. One factor which has been involved is the presence of melanin in the walls of differentiated hyphae (BLOOMBIELD and ALEXANDER 1967, BULL 1970a, DURRELL 1964, HOROWITZ et al. 1961, LEONARD 1971, ROWLEY and PIRT 1972, GARCIA MENDOZA et al. 1979). Melanin has been reported to inhibit wall-lytic enzymes (BULL 1970b, KUO and ALEXANDER 1967). It is mostly produced in older cultures when the absence of catabolic repression promotes the production of potentially wall-lytic enzymes and thus may be a very effective agent. However, resistance to degradation is also noted in unmelanized walls of young hyphae. In such cases several possibilities can be considered.

One possibility to be considered is that the presence of certain wall components may hinder access of lytic enzymes to susceptible wall components. In *Schizophyllum commune* the removal from the wall of (1 → 3)-α-D-glucan by alkali (VAN DER VALK et al. 1977) or by (1 → 3)-α-glucanase in living hyphae (DE VRIES and WESSELS 1972, 1973b) greatly enhance the degradation of (1 → 3)-β/(1 → 6)-β-D-glucan and chitin by added enzymes. By employing a mutant of *Aspergillus nidulans* unable to synthesize (1 → 3)-α-D-glucan, POLACHECK and ROSENBERGER (1977) showed that in unmelanized walls this polymer seems to protect against degradation of the other wall components by wall-bounded autolysins. The synergistic effects of mixtures of lytic enzymes leads to the same conclusion. For instance, it has been found that preliminary treatment with β-glucanases of the β-glucan/chitin complex of the wall renders the chitin more susceptible to chitinase, but the reverse is not true (SKUJINS et al. 1965, WESSELS and MARCHANT 1974). This has been interpreted as indicating that the chitin is buried in β-glucan, making the chitin inaccesible to chitinase. Similarly, the effect of sulfhydryl reagents in making yeast walls more susceptible to hydrolases has been attributed to a loosening of glycoproteins, permitting the hydrolases to reach their substrates (NICKERSON 1963, DOOIJEWAARD-KLOOSTERZIEL et al. 1973, BASTIDE et al. 1975).

Perhaps an even more important factor conferring resistance of the wall against general hydrolases is the intricate linkage between polymer segments by covalent and noncovalent bonds. The resulting molecular structures may prevent penetration of lytic enzymes and/or sterically hinder their activity. For instance, the (1 → 3)-β/(1 → 6)-β-D-glucan (R-glucan) of *Schizophyllum commune,* containing 35% (1 → 3)-β-linkages, is rather resistant to exo- and endo-(1 → 3)-β-glucanase. However, the structure is easily attacked by an endo (1 → 6)-β-glucanase (R-glucanase) which releases large segments of this glucan which are now easily hydrolyzed by the (1 → 3)-β-glucanases (WESSELS 1969b). In crystalline polymers, such as cellulose or chitin, special enzymes may be needed to disrupt hydrogen bonds before hydrolytic enzymes can gain access to the polysaccharide chains. The importance of such primary "wall-loosening" enzymes in determining whether or not a wall polymer will be degraded by general hydrolytic enzymes has been emphasized by REESE (1972, 1977).

Apart from control at the level of wall-degrading enzymes, the in vivo regulation of wall degradation may also involve a control at the level of substrate susceptibility. In other words, during synthesis of a polymer in the wall its susceptibility to degradation may be determined. In the case of the R-glucan/R-glucanase system of *Schizophyllum commune* it has been shown that the R-glucan in the septa is more susceptible to degradation than R-glucan in peripheral walls (WESSELS and MARCHANT 1974) and that mutations (WESSELS and KOLTIN 1972) and environmental conditions prevailing during wall synthesis (SIETSMA et al. 1977) can change the susceptibility of the glucan to degradation.

5.2 Net Degradation of Wall Components in Relation to Development

An important role of wall-degrading enzymes becomes manifest during prolonged growth of fungi in submerged culture. The type of autolysis that ensues is probably initiated by a lack of nutrients and not only concerns the wall but also other cell constituents (cf. FENCL 1978). An important view relating to filamentous fungi is that autolysis may proceed in subapical parts and that breakdown products are re-used for synthesis in apical parts. Autolysis can thus be considered as a mechanism to survive under starvation conditions.

A special case of controlled autolysis is the rapid decay of fruiting bodies of Agarics after spore release. In *Coprinus lagopus* ITEN and MATILE (1970) found a sharp rise in chitinase activity just before autolysis and they argued that chitinase is one of a series of hydrolases contained in vacuoles which would function as lysosomes (MATILE 1971).

It has long been known that carbon starvation enhances the formation of fruiting bodies and turnover of cellular material is therefore necessary to sustain development of these fruiting bodies. Since fungi invest a large amount of polymers in the construction of their cell walls, it is not surprising that certain wall polymers can serve as a reserve when the external supply of carbohydrates falls short. For instance, the outgrowth of fruiting bodies (the formation of pilei) in *Schizophyllum commune* can occur in the absence of a carbon and a nitrogen supply in the medium and is accompanied by a net breakdown of $(1\rightarrow 3)$-β/$(1\rightarrow 6)$-β-D-glucans (WESSELS 1965, NIEDERPRUEM and WESSELS 1969, WESSELS and SIETSMA 1979). In the culture medium a water-soluble glucan with these linkage types is broken down by an exo-$(1\rightarrow 3)$-β-glucanase. In the walls, however, glucans with these linkages occur in a highly insoluble complex with chitin, the so-called R-glucan/chitin complex (SIETSMA and WESSELS 1977, 1979) and here breakdown appears to be initiated by a special endo-$(1\rightarrow 6)$-β-glucanase (R-glucanase) before $(1\rightarrow 3)$-β-glucanases can proceed to degrade the resulting large soluble products (WESSELS 1969b). R-glucanase appears in the dikaryon, and not in the monokaryon, as a response to glucose depletion in the medium (WESSELS 1966, WESSELS and NIEDERPRUEM 1967) and R-glucan degradation may then follow in vegetative mycelium and stunted fruiting bodies, the breakdown products being partly used for the construction of growing pilei.

Degradation of R-glucan, however, is not only dependent on the presence of R-glucanase. A mutation (WESSELS 1965, 1966) or environmental conditions prevailing during synthesis of the walls of the dikaryon, such as high temperature (WESSELS 1965) or high carbon dioxide concentration (SIETSMA et al. 1977) may prevent R-glucan degradation from occurring in vivo notwithstanding the presence of R-glucanase. In these cases a decrease in susceptibility of R-glucan in the wall to R-glucanase was observed and the failure to degrade R-glucan was correlated with the absence of pilei.

This raises the question of a possible causal relationship between R-glucan degradation and the development of pilei. It was found that in the absence of an exogenous nitrogen supply the development of pilei is suppressed by exogenous glucose. On the other hand, a mutant unable to degrade R-glucan and to produce pilei could be induced to form pilei by maintaining a steady low concentration of glucose in the medium. Therefore the theory was advanced that the slow degradation of a wall polymer, such as R-glucan, would provide for a steady flow of carbon compounds at low concentration toward the developing pilei. This would ensure sufficient building blocks for the developing pilei, yet suppression of pileus development by too-high concentrations of carbon compounds, as generated by exogenous glucose, would not occur (WESSELS 1965).

Re-utilization of wall components has also been implicated in the process of fruit-body formation in *Flammulina velutipes* (KITAMOTO and GRUEN 1976) and *Coprinus congregatus* (ROBERT 1977a, b). Of particular interest is the formation of cleistothecia in *Aspergillus nidulans*. Here, not only the β-glucan but especially the (1 → 3)-α-D-glucan is subject to degradation by an exo-splitting (1 → 3)-α-glucanase during carbon starvation (ZONNEVELD 1972a, b). The significance of this process for cleistothecium development is indicated by the fact that mutations or environmental manipulations that prevent the accumulation of (1 → 3)-α-D-glucan also prevent the development of cleistothecia (ZONNEVELD 1974, POLACHECK and ROSENBERGER 1977).

Wall material in different parts of the cell may be differentially susceptible to wall-lytic enzymes. This can be illustrated by considering septal dissolution accompanying nuclear migration in basidiomycetes. While nuclear migration is a transient process during matings involving monokaryons with different *A*- and *B* factors, resulting in a dikaryon, it occurs continuously in heterokaryons containing nuclei differing only in *B* factor alleles. Continuous nuclear migration also occurs in homokaryons with a particular mutation in the *B* factor (RAPER and RAPER 1973). In such cases septa are normally synthesized but degradation of these septa may start within the hour (NIEDERPRUEM 1971). During septal dissolution an abundance of vesicles is observed in the cytoplasma, at least some of which carry acid phosphatase (RAUDASKOSKI 1976). Such vesicles apparently undergo fusion with the plasma membrane covering the cross walls, suggesting extrusion of lytic enzymes (MARCHANT and WESSELS 1973).

Of the lytic enzymes probably involved, the activity of R-glucanase is sharply increased when septal dissolution occurs (WESSELS and NIEDERPRUEM 1967, WESSELS 1969a, WESSELS and KOLTIN 1972). Little, if any, increase is found in the activities of (1 → 3)-β-glucanases and chitinase. By incubating hyphal wall preparations, still containing the cross walls, with lytic enzymes, it could be shown that a combination of R-glucanase and chitinase completely dissolves the cross walls while leaving the peripheral walls structurally intact (WESSELS and MARCHANT 1974). Chitinase alone was quite ineffective and R-glucanase alone affected dissolution of the R-glucan component but left the central chitinous plate of the cross wall (cf. Fig. 1 A) intact. Simultaneous determination of the degree of cross wall dissolution and of the breakdown of wall polymers in the whole wall preparation revealed that the R-glucan/chitin complex in the cross walls is much more susceptible to the R-glucanase-chitinase mixture than the complex in the peripheral walls. This is possibly due to the absence of (1 → 3)-α-D-glucan in the cross walls which seems to protect the R-glucan/chitin complex in the peripheral walls against degradation. Yet some degradation of R-glucan in the peripheral walls occurs and this corresponds to the situation in vivo where hyphae in which continuous septal dissolution occurs are characterized by irregularly shaped outlines and a lowered amount of R-glucan in the wall (WESSELS 1969a). In contrast to the septa of monokaryons the septa of dikaryons remain entire when high activities of R-glucanase develop. It was found that in this case the susceptibility of the R-glucan/chitin complex of the peripheral walls is similar to that in the monokaryon, but in the cross walls this complex is somehow protected against enzymic attack (WESSELS and MARCHANT 1974).

Apparently during synthesis the cross walls are prepared to yield or to resist to lytic enzymes. This is quite reminiscent of the processes involved in vessel differentiation in higher plants (cf. BRYANT 1976). Here unthickened end walls between vessel elements are dissolved and pores in the end walls of sieve cells are formed by (1 → 3)-β-glucanase in those areas where previously callose was deposited.

5.3 "Wall-Loosening" Enzymes

"Wall-loosening" enzymes are thought to be involved in plasticizing the wall so that it yields to the turgor, e.g., in lateral branch formation, bud emergence in yeast, and hyphal tip extension.

The involvement of lytic enzymes in branch initiation appears obvious because a previously rigid area of the wall has to become plastic. Support for this concept has been drawn from a positive correlation between the activity of lytic enzymes and the degree of branching (THOMAS and MULLINS 1967, MAHADEVAN and MAHADKAR 1970, FÈVRE 1972, MULLINS and ELLIS 1974). An increase was not only found in cytoplasmic and secreted enzymes but also wall preparations from more highly branched mycelia showed a higher autolytic activity. A difficulty with the observed correlation is that it is not immediately apparent that localized softening of the wall would be manifested in a general increase in lytic activity. In addition, the increased lytic activities may be a result rather than a cause of increased branching because hyphal apices may be major sources of extracellular enzyme production (CHANG and TREVITHICK 1974).

Lytic enzymes have also been related to budding, conjugation, and sporulation in yeasts. BROCK (1965) did not find changes in β-glucanases specifically related to conjugation in *Hansenula wingii* but peaks of wall-autolytic activities were found in *Schizosaccharomyces pombe* during conjugation and sporulation (KRÖNIG and EGEL 1974). Budding in *Saccharomyces cerevisiae* has been related to an increase in total exo-(1 → 3)-β-glucanase activity (CORTAT et al. 1972). However, SANTOS et al. (1979) isolated a mutant of *S. cerevisiae* deficient in the production of exo-(1 → 3)-β-glucanase and this mutant shows normal budding and mating (cited in DEL REY et al. 1979). In addition, DEL REY et al. (1979) showed that in normal strains of *S. cerevisiae* the increase of exo- and endo-(1 → 3)-β-glucanase is restricted in the cell cycle to the transition from the S to the G_2 phase and therefore is not related in time to bud emergence. In addition they could not detect an increase in enzyme during mating when the cells are arrested in G_1; only during sporulation a significant increase in (1 → 3)-β-glucanase occurs.

Some authors have emphasized the role of soluble lytic enzymes packaged in vesicles and released at the site of wall loosening (e.g., CORTAT et al. 1972, FARKAS et al. 1973). Others have drawn attention to the fact that lytic enzymes sometimes appear tightly bound to isolated walls which display autolytic activity (MAHADEVAN and MAHADKAR 1970, BARRAS 1972, POLACHECK and ROSENBERGER 1975, 1978). In *Aspergillus nidulans* POLACHECK and ROSENBERGER (1975) found that isolated walls rapidly release material during the first 3 to 5 h of incubation; thereafter the release is much slower, a total of approximately 3% of the dry weight being lost after 48 h. They showed that the slowing down of autolysis is not due to inactivation of enzymes but to depletion of an easily degradable part of the wall. By using walls from a mycelium briefly exposed to ^{14}C-glucose (to label apices) they could show that this easily degradable part can be equated with apical wall material. Similarly, FÈVRE (1977) observed that hyphal tips of *Saprolegnia monoica* are most susceptible to degradation. In *A. nidulans* POLACHECK and ROSENBERGER (1977) found that although mutants lacking (1 → 3)-α-D-glucan and melanin show a more rapid release of wall materials, the greater susceptibility of the apical wall is maintained. They also found (POLACHECK and ROSENBERGER 1978) that autolysins can be detached from the wall by the cationic detergent cetyltrimethylammoniumbromide, but not by other detergents and LiCl, and once released do not readsorb to the wall. Therefore they inferred that binding of the autolysins is mediated by lipids, probably involving lipid vesicles entrapped in the wall during growth. Since turnover of wall components in vivo is hardly detectable (POLACHECK and ROSENBERGER 1977) shearing forces during wall isolation probably release some of the autolysins resulting in wall autolysis in vitro. The conclusion that autolysins are distributed along the hyphal wall and are not localized in specific parts, e.g., at the sites of wall synthesis as in the case of autolysins of streptococci (SHOCKMAN et al. 1967), is strengthened by the observations of KRITZMAN et al. (1978). These authors were able to detect an apparent localization of (1 → 3)-β-glucanase at hyphal apices, branching points, clamp connections and new septa in *Sclerotium rolfsii* using immunoflurescence. With diethylpyrocarbonate, a powerful irreversible inactivator of (1 → 3)-β-glucanase, the immunofluorescence disappears. However, after treatment with the inactivator about 70% of the active enzyme could still be extracted from the walls by the detergent Triton X-100 and thus was present in a "masked" form. With respect to the occurrence of "free" enzyme in growing wall areas it is possible that this enzyme is actively involved in growth processes, but such growth areas are probably also channels through which lytic enzymes are extruded into the extracellular milieu (CHANG and TREVITHICK 1974). Immunofluorescence may therefore detect the presence of lytic enzymes in these areas.

The idea that autolytic enzymes are actually involved in the process of wall growth was first advanced by JOHNSON (1968), who observed that inhibition of wall synthesis in *Schizosaccharomyces* by 2-deoxy-glucose leads to lysis of the cells in the growing areas. Unfortunately, a search for an enzyme that hydrolyzes the major $(1 \rightarrow 3)$-α-glucan component in this fission yeast has been unsuccessful (FLEET and PHAFF 1975). In hyphal fungi, too, it has been observed that interference with wall synthesis, e.g., by inhibiting chitin synthesis by polyoxin D, leads to bursting of hyphal tips (BARTNICKI-GARCIA and LIPPMAN 1972a, b). On the basis of such observations BARTNICKI-GARCIA (1973) has proposed a unitary model of cell wall growth in which growth at the apex is only possible by continuous hydrolytic cleavage of existing wall polymers creating new starting points for synthesizing enzymes which by insertion of new polymer can enlarge the surface area. In the author's words this would require a delicate balance between wall synthesis and wall lysis. This points to the major problem attached to this model, namely how the cell can control both the activities of synthesizing and lytic enzymes in such a way that synthesis can proceed without the danger of losing the integrity of the wall fabric at the apex. Also the continuity of microfibrils at the apex is difficult to reconcile with this model.

6 Synthesis of the Wall

6.1 Introduction

Knowledge on the process of wall formation in fungi has recently been advanced by studying regenerating protoplasts (WESSELS et al. 1976, PEBERDY 1978) and cell-free systems (GOODAY 1977, FARKAS 1979). Aspects of wall formation will also receive detailed attention in Chapters 16, 17, 18 and 19 of this Volume dealing with individual wall components. The present part of this survey will therefore be confined to a brief outline of the subject in order to indicate some of the problems on wall growth and morphogenesis in fungi.

6.2 Biosynthesis of Individual Wall Components

6.2.1 Chitin

The enzyme chitin synthase (uridine-5′-diphosphate-2-acetamido-2-deoxy-D-glucose: chitin-4-β-acetamidodeoxyglucosyltransferase; EC 2.4.1.16) that synthesizes chitin from uridine-5′-diphosphate-*N*-acetyl-D-Glucosamine (UDP-GlcNAc) has been isolated from many chitin-containing fungi (BURNETT 1976). It is always present in particulate fractions of cell homogenates from which it can be solubilized with agents such as butanol (GLASER and BROWN 1957) or digitonin (DE ROUSSET-HALL and GOODAY 1975).

Results obtained both with particulate and solubilized enzyme preparations indicate that chitin synthase displays cooperative kinetics, UDP-GlcNAc being both a substrate and activator while *N*-acetyl-glucosamine (GlcNAc) and diacetylchitobiose are allosteric activators at low substrate concentrations (DE ROUSSET-HALL and GOODAY 1975, MCMURROUGH and BARTNICKI-GARCIA 1973). The product uridine-5′-diphosphate inhibits the enzyme. Although this provides a number of possible controls of enzyme activity, it is unknown how these are exercised in vivo.

Cellular localization of chitin synthase has been a matter of some controversy. Upon differential centrifugation of a cell homogenate all fractions contain at least some activity. In the Mucorales most of the activity is associated with the low-speed fraction containing cell walls, e.g., in *Mucor rouxii* (MCMURROUGH and BARTNICKI-GARCIA 1971), *Phycomyces blakesleeanus* (JAN 1974, VAN LAERE and CARLIER 1978), and *Cunninghamella elegans* (MOORE and PEBERDY 1975). This has led to the hypothesis that the enzyme is actually secreted into the wall to perform synthesis of chitin (cf. BARTNICKI-GARCIA 1973). However, JAN (1974) proposed that in *Phycomyces* the plasma membrane is the principal location of chitin synthase, the high activity in the wall would then be due to contamination with plasma membranes. In Ascomycetes, such as *Aspergillus nidulans* (RYDER and PEBERDY 1977) and *A. flavus* (LOPEZ-ROMERO and RUIZ-HERRERA 1976), and Basidiomycetes, such as *Coprinus cinereus* (GOODAY 1977) and *Schizophyllum commune* (this laboratory, unpublished), little activity is associated with the wall and most activity is recovered from higher speed fractions.

Substantial evidence for the attachment of chitin synthase to the plasma membrane has been reported by DURÁN et al. (1975) for *Saccharomyces cerevisiae*, BRAUN and CALDERONE (1978) for *Candida albicans*, and VERMEULEN et al. (1979) for *Schizophyllum commune*. In all cases plasma membranes were isolated from concanavalin-A coated protoplasts. DURÁN et al. (1975) found that most of the chitin synthase in the plasma membrane of *Saccharomyces cerevisiae* is present in an inactive state, possibly as a zymogen, which could be activated by limited proteolytic digestion. This was related to the localized synthesis of chitin in this yeast at the site of the septum (CABIB and FARKAS 1971). In contrast, the enzyme in the plasma membrane of *Schizophyllum commune* protoplasts, which are very active in synthesizing chitin all over the plasma membrane (VAN DER VALK and WESSELS 1977), is mainly present in an active state (VERMEULEN et al. 1979). In fact, about 90% of the active enzyme in these protoplasts was found associated with the plasma membrane.

Activation of chitin synthase by limited proteolytic digestion has been found in enzyme preparations from many fungi (RUIZ-HERRERA and BARTNICKI-GARCIA 1976, LOPEZ-ROMERO and RUIZ-HERRERA 1976, ARCHER 1977, RYDER and PEBERDY 1977, VAN LAERE and CARLIER 1978, ARROYO-BEGOVICH and RUIZ-HERRERA 1979, VERMEULEN et al. 1979). The recognition that the enzyme can occur in an inactive state also led to the isolation of small discrete particles containing chitin synthase from the 54,000 *g* supernantant of a cell-free extract from the yeast form of *Mucor rouxii* (RUIZ-HERRERA et al. 1975, BRACKER et al. 1976). Such particles, called chitosomes, were also isolated from a number of other taxonomically diverse fungi (BARTNICKI-GARCIA et al. 1978). Similar particles were also found when chitin synthase was detached from crude membrane preparations by incubating the membranes with UDP-GlcNAc and GlcNAc at 0° C (RUIZ-HERRERA and BARTNICKI-GARCIA 1974, RUIZ-HERRERA et al. 1975).

Whereas membrane preparations contain mostly active chitin synthase (except in yeast), post-membrane high-speed fractions may contain the enzyme in an inactive form (VERMEULEN et al. 1979, ARROYA-BEGOVICH and RUIZ-HERRERA 1979). Failure to detect such a difference may be accounted for by the presence in extracts of proteolytic activities (RUIZ-HERRERA and BARTNICKI-GARCIA 1976). With respect to the site of chitin synthesis in vivo, electronmicroscopic autoradiography of chitin synthesis on protoplasts of *Schizophyllum commune* strongly suggests that chitin synthesis only occurs in very close association with the plasma membrane (VAN DER VALK and WESSELS 1977). VERMEULEN et al. (1979) have shown that the chitin synthesized in vitro on the plasma membranes of these protoplasts also remains attached to the membranes. These results are best explained by assuming that chitin synthase is formed in the cell in an inactive state, precluding the synthesis of chitin within the cell, and is transported as such to the plasma membrane. During or after incorporation into the membrane the enzyme becomes activated, possibly by limited proteolytic digestion although other mechanisms of activation cannot be excluded at the moment.

Both activated chitosomal preparations (RUIZ-HERRERA and BARTNICKI-GARCIA 1974) and isolated plasma membranes (VERMEULEN et al. 1979) synthesize chitin in a highly crystalline microfibrillar form. Similarly, young regenerating protoplasts not yet fully equipped to produce all wall components make microcrystalline microfibrillar chitin on their plasma membrane surfaces. The formation of such a highly crystalline product in these cases may reflect the absence of other enzyme activities which could cross-link the chitin chains to other wall polymers, thus increasing the chance of perfect crystallization (KREGER and KOPECKÁ 1976, SIETSMA and WESSELS 1979).

6.2.2 Glucan

In spite of many attempts, the demonstration of synthesis of α- and β-glucans in vitro from nucleoside-diphosphate-glucose has met with more difficulties than the synthesis of chitin. FARKAS (1979) lists several studies on β-glucan synthesis but mostly the products were either poorly characterized or the products were clearly not identical to glucans present in the wall. There seem to be no reports on the synthesis of α-glucan in a cell-free system from fungi.

Of special interest is a report of WANG and BARTNICKI-GARCIA (1976), who showed that a mixed membrane fraction from *Phytophthora cinnamomi* synthesized massive amounts of an alkali-soluble (1→3)-β-D-glucan from uridine-5′-diphosphate-D-glucose. The product had a typical microfibrillar appearance and an X-ray diffraction diagram similar to that of hydroglucan. Although such a glucan is not present in the wall of this fungus, a similar glucan is synthesized on regenerating protoplasts of *Saccharomyces cerevisiae* (KREGER and KOPECKÁ 1976). A reasonable conclusion is that the (1→3)-β-D-glucan synthase is functioning in isolation in both systems and is not accompanied by other transferases normally involved in branching of the β-D-glucan. This would result in the production of a linear glucan that can crystallize, a process normally prevented by the attachment of side chains. Another possibility that should be considered is that special enzymes may be needed to link the glucans to other wall polymers, e.g., chitin (SIETSMA and WESSELS 1979). The absence of such enzymes in enzyme preparations could result in the formation of products with solubility characteristics vastly different from those in the cell wall. The construction of a functional wall polymer would thus require the coordinate activity of many enzymes. Unbalanced activity of such enzymes would lead to over-production of some components of the polymer complex resulting in secretion of these components. This may be the case in *Schizophyllum commune* where strains secrete variable amounts of a soluble (1→3)-β-(1→6)-β-D-glucan which very much resembles one of the chain types in the insoluble R-glucan/chitin complex (SIETSMA and WESSELS 1979). Elevated carbon dioxide concentrations seem to enhance secretion of these chains and to lower their incorporation into the R-glucan/chitin complex (SIETSMA et al. 1977).

6.2.3 Glycoprotein

Polysaccharide–protein complexes probably occur in all fungal walls and are especially abundant in yeast walls. Structurally the fungal glycoproteins show many similarities with polysaccharide–protein complexes of higher organisms and also the biosynthetic pathways are nearly the same (GANDER 1974, KORNFELD and KORNFELD 1976). The *N*-glycosidic or *O*-glycosidic bonds which link the carbohydrate moiety to the peptide chain are the main structural features of glycoproteins. In the synthesis of these bonds a terpene-like intermediate, polyprenylphosphate, plays an important role. This has been very well described for the biosynthesis of manno-protein in *Saccharomyces cerevisiae* (BALLOU 1976, CABIB 1975, WAECHTER and LENNARZ 1976, FARKAS 1979) and will also be discussed in Chap. 19 of this Vol.).

6.3 Wall Synthesis and Morphogenesis

A well-studied case of spatial and temporal regulation of synthesis of a wall component concerns the deposition of chitin at the site of septum formation during a specific period in the cell cycle of *Saccharomyces cerevisiae*. This system will be described in detail in Chapter 16, this Volume. Therefore it suffices to remark here that the emerged model involves the presence of inactive chitin synthase uniformly distributed in the plasma membrane and the activation of this supposed zymogen by proteolysis at the time and site of septum formation.

Mutations affecting only the spatial component have been described (SLOAT and PRINGLE 1978) and, in the absence of a division septum, α-factor-treated cells accumulate chitin in the area of pheromone-stimulated growth, i.e., in the walls of the conjugation tubes (SCHEKMAN and BRAWLEY 1979).

As indicated in Sections 3.2.1 and 3.2.2, there seems to be a similarity between the septa of yeasts and those of filamentous Ascomycetes and Basidiomycetes in that the first material deposited between the invaginating plasma membranes (PATTON and MARCHAT 1978) appears to be rather homogeneous chitin. Also in regenerating protoplasts one of the first components to appear on the naked plasma membrane is often pure chitin (Sect. 3.2.3). It is thus remarkable that in all these cases the formation of a wall on a naked plasma membrane appears to start with the synthesis of chitin, later followed by the simultaneous deposition of other wall components.

In contrast to septum formation, hyphal wall extension is a continuous process and only a spatial element has to be considered. Autoradiography shows that wall synthesis is at a maximum at the very apex of the hypha and declines toward the area where the hyphal tube assumes its maximum diameter (BARTNICKI-GARCIA and LIPPMAN 1969, KATZ and ROSENBERGER 1970b, 1971b, GOODAY 1971, VAN DER VALK and WESSELS 1977). Such a gradient of wall deposition is in general agreement with models developed by Green (cf. GREEN 1974) in which a hemispherical growth zone displays nondirectional expansion with a gradient in growth rate of any point on the hemisphere declining as the cosine of the angle between the longitudinal axis and the considered point on the hemispherical tip. Since the shapes of tips of rapidly growing hyphae approximates more closely to half ellipsoids of revolutions than to hemispheres, TRINCI and SAUNDERS (1977) have derived that better quantitative fits to observed data are given by cotangent rather than by cosine curves. The driving force for expansion is the turgor pressure. Therefore, whatever the exact mathematical description in a particular case may be, both the plasticity of the wall and the wall synthetic activity must decline from the very tip toward the area where the cylindrical shape is attained. Since the thickness of the wall in the growing tip area does not seem to change (TRINCI and COLLINGE 1975) the rate of expansion must equal the rate of addition of wall material at any point of the growing tip.

With regard to the mechanisms involved in tip growth, great importance can be attributed to the concept of a continuous flow of vesicles from subapical parts to the apex of the hypha. At the apex these vesicles are thought to fuse with the plasma membrane, extending the surface area of the plasma membrane and extruding their contents (possibly wall precursors and enzymes) into the periplasmic space (MCCLURE et al. 1968, GIRBARDT 1969, GROVE and BRACKER 1970, HEATH et al. 1971). In accordance with the view of a flow at a constant rate of vesicles from subapical parts towards the apex, COLLINGE and TRINCI (1974) showed that in *Neurospora crassa* the total number of vesicles in transverse sections slightly increased toward the tip, resulting in a steep increase in the number of vesicles per unit protoplasm in the tapered apex. They also established that the steep increase in vesicles parallelled the steep increase in rate of wall synthesis as determined by autoradiography in these hyphae (GOODAY 1971), strongly suggesting a relationship between the two phenomena. Recently these phenomena have been incorporated into a mathematical model which predicts changes in hyphal lengths and numbers and position of branches and septa on the basis of changes in vesicle and nuclear concentrations (PROSSER and TRINCI 1979). With respect to the motive forces that drive vesicles to the tip it has been suggested that an electrical potential difference or an osmotic flow, caused by a decline of potassium pumps (ATPase) toward the tip plays a role (cf. JENNINGS 1979).

The molecular mechanisms that operate during hyphal wall extension remain largely unknown although several hypotheses have been advanced. GREEN (1974) has proposed that the units of wall material deposited near the tip can be expanded by stretching, due to the turgor pressure in the hypha, but that the resistance to stretch is a function of stretch. At the base of the tip where the hypha becomes cylindrical, the added wall material would be maximally stretched and resist the hydrostatic pressure in the hypha. The whole process would thus be self-regulatory. On the other hand, mainly on the basis of experiments in which bursting of the tips is observed after inhibition of wall synthesis, the concept has arisen that hydrolytic enzymes continuously loosen the wall at the tip and permit the insertion of new wall material (BARTNICKI-GARCIA 1973). Such hydrolytic enzymes could be carried to the tip, together with wall-synthesizing enzymes, in the vesicles that fuse with the plasma membrane. Given the known stability of such hydrolytic enzymes one would then expect a gradient in the resistance of the wall material in the tip increasing toward the base of the tip. This generally agrees with the finding that the newly deposited wall is more susceptible to hydrolytic enzymes than the older wall material (see Sect. 5.3). One could also assume that the newly deposited wall material has plastic properties of its own, as suggested by GREEN (1974), but that progressive cross-linking, such as between chitin and β-glucan (SIETSMA and WESSELS 1979) results in a stiff nondeformable wall structure at the base of the tip. It has been suggested for plant and animal cells that the degree of interlinking of the matrix material with the fibrous phase constitutes the mechanical properties of the wall (KIRKWOOD 1974). This could also be true for fungal walls where the degree of cross-linking between the β-glucan and chitin determines the flexibility or rigidity of the wall. Wall-loosening enzymes would then only be required to induce plasticity again, e.g., during the formation of branches.

Also the gradient in wall synthetic activity in hyphal tips is not yet satisfactorily explained. For instance, if chitin synthase is present in the plasma membrane, possibly inserted by fusion of the plasma membrane with vesicles or particles containing inactive chitin synthase (VERMEULEN et al. 1979), what determines the steep gradient in synthetic activity? If the activity depends on a continuous flow of inactive chitin synthase to the growing tip this would require a continuous process of sequential activation and inactivation of the enzyme. Since incubation of cell-free preparations containing inactive chitin synthase with proteolytic enzymes first leads to activation and then, on continued incubation, to inactivation of the chitin synthase, such a mechanism has also been proposed to explain the steep gradient of synthetic activity at the hyphal apex (RUIZ-HERRERA and BARTNICKI-GARCIA 1976). If inactivation is irreversible, this would mean that the active enzyme must have a very short half-life. Experiments with cycloheximide using regenerating protoplasts (de Vries and Wessels 1975) or growing hyphae (Katz and Rosenberger 1971 b, Sternlicht et al. 1973) do not support the notion of such a short half-life. In hyphae the main effect of cycloheximide or osmotic shock is a displacement of wall synthetic activities from the apex to a more subapical region of the hypha (cf. Park and Robinson 1966). A short half-life of chitin synthase would not be required in an hypothesis involving reversible inactivation of the enzyme by binding with an inactivating protein (Lopez-Romero et al. 1978). Another possibility, not considered so far, is that active chitin synthase is a rather stable membrane protein that can move laterally within the membrane together with the advancing tip, e.g., under influence of an electrical field. The abundance of hypotheses that can be advanced serves to illustrate our ignorance of even a "simple" morphogenetic process as the construction of a hyphal tube.

References

Albersheim P, Anderson-Prouty AJ (1975) Carbohydrate, proteins, cell surface, and the biochemistry of pathogenesis. Annu Rev Plant Physiol 26:31–52

Alviano CS, Gorin PAJ, Travassos LR (1979) Surface polysaccharides of phytopathogenic strains of *Ceratocystis paradoxa* and *Ceratocystis fimbriata* isolated from different hosts. Exp Mycol 3:174–187

Angyal SJ, Bender VJ, Ralph BJ (1974) Structure of polysaccharides from the *Polyporus tumulosus* cell wall. Biochim Biophys Acta 362:175–187

Archer DB (1977) Chitin biosynthesis in protoplasts and subcellular fractions of *Aspergillus fumigatus*. Biochem J 164:653–658

Aronson JM (1965) The cell wall. In: Ainsworth GC, Sussman AS (eds) The fungi, vol I. Academic Press, London New York, pp 49–76

Aronson JM, Fuller MS (1969) Cell wall structure of the marine fungus *Atkinsiella dubia*. Arch Mikrobiol 68:295–305

Aronson JM, Machlis L (1959) The chemical composition of the hyphal walls of the fungus *Allomyces*. Am J Bot 46:292–300

Aronson JM, Preston RD (1960) An electron microscopic and X-ray analysis of the cell wall of the filamentous fungus *Allomyces*. Proc R Soc London Ser B 152:346–352

Arroyo-Begovich A, Ruiz-Herrera J (1979) Proteolytic activation and inactivation of chitin synthese from *Neurospora crassa*. J Gen Microbiol 113:339–345

Bacon JSD, Davidson ED, Jones D, Taylor IF (1966) The location of chitin in the yeast cell wall. Biochem J. 101:36C–38C

Bacon JSD, Jones D, Farmer VC, Webley DM (1968) The occurrence of α-(1–3)-glucan in *Cryptococcus, Schizosaccharomyces* and *Polyporus* species and its hydrolysis by a streptomycete culture filtrate lysing cell walls of *Cryptococcus*. Biochim Biophys Acta 158:313–315

Bacon JSD, Farmer VC, Jones D, Taylor IF (1969) The glucan components of the cell wall of baker's yeast (*Saccharomyces cerevisiae*) considered in relation to its ultrastructure. Biochem J 114:557–567

Ballesta J-PG, Alexander M (1971) Resistance of *Zychorynchus* species to lysis. J. Bacteriol 106:938–945

Ballou C (1974) Some aspects of the structure, immuno-chemistry and genetic control of yeast mannans. Adv Enzymol 40:239–270

Ballou C (1976) Structure and biosynthesis of the mannan component of the yeast cell envelope. Adv Microbiol Physiol 14:93–158

Barras DR (1972) A β-glucan endohydrolase from *Schizosaccharomyces pombe* and its role in cell wall growth. Antonie van Leeuwenhoek J Microbiol Serol 38:65–80
Bartnicki-Garcia S (1963) Symposium on the biochemical bases of morphogenesis in fungi III. Mold-yeast dimorphism of *Mucor*. Bacteriol Rev 27:293–304
Bartnicki-Garcia S (1966) Chemistry of hyphal walls of *Phytophthora*. J Gen Microbiol 42:57–69
Bartnicki-Garcia S (1968) Cell wall chemistry morphogenesis, and taxonomy of fungi. Annu Rev Microbiol 22:87–108
Bartnicki-Garcia S (1973) Fundamental aspects of hyphal morphogenesis. In: Microbial differentiation. 23rd Symp Soc Gen Microbiol. Univ Press, Cambridge, pp 245–267
Bartnicki-Garcia S, Lindberg B (1972) Partial characterization of mucoran: the glucoromannan component. Carbohydr Res 23:75–85
Bartnicki-Garcia S, Lippman E (1969) Fungal morphogenesis: Cell wall construction in *Mucor rouxii*. Science 165:302–304
Bartnicki-Garcia S, Lippman E (1972a) Inhibition of *Mucor rouxii* by polyoxin D: effects on chitin synthetase and morphological development. J Gen Microbiol 71:301–309
Bartnicki-Garcia S, Lippman E (1972b) The bursting tendency of hyphal tips of fungi: Presumptive evidence for a delicate balance between wall synthesis and wall lysis in apical growth. J Gen Microbiol 73:487–500
Bartnicki-Garcia S, McMurrough I (1971) Biochemistry of morphogenesis in yeasts. In: Rose AH, Harrison JH (eds) The yeasts, vol II. Academic Press, New York, pp 441–491
Bartnicki-Garcia S, Nickerson WJ (1962) Isolation, composition, and structure of cell walls of filamentous and yeast-like forms of *Mucor rouxii*. Biochim Biophys Acta 58:102–119
Bartnicki-Garcia S, Reyes E (1968a) Chemical composition of sporoangiophore walls of *Mucor rouxii*. Biochim Biophys Acta 165:32–42
Bartnicki-Garcia S, Reyes E (1968b) Polyuronides in the cell walls of *Mucor rouxii*. Biochim Biophys Acta 170:54–62
Bartnicki-Garcia S, Bracker CE, Reyes E, Ruiz Herrera J (1978) Isolation of chitosomes from taxonomically diverse fungi and synthesis of chitin microfibrils in vitro. Exp Mycol 2:173–192
Bastide M, Trave P, Bastide JM (1975) L'hydrolyse enzymatique de la paroi appliqué a la classification des levures. Ann Microbiol 126:275–294
Bauer H, Horisberger M, Bush DA, Sigarlakie E (1972) Mannan as a major component of the bud scars of *Saccharomyces cerevisiae*. Arch Mikrobiol 85:202–208
Beran K, Holan Z, Baldrian J (1972) The chitin-glucan complex in *Saccharomyces cerevisiae*. I. IR and X-ray observations. Folia Microbiol (Prague) 17:322–330
Bloomfield BJ, Alexander M (1967) Melanins and resistance of fungi to lysis. J Bacteriol 93:1276–1280
Bobitt TF, Nordin JH, Roux M, Revol JF, Marchessault RH (1977) Distribution and conformation of crystalline nigeran in hyphal walls of *Aspergillus niger* and *Aspergillus awamori*. J Bacteriol 132:691–703
Bracker CE (1967) Ultrastructure of fungi. Annu Rev Phytopathol 5:343–374
Bracker CE, Ruiz-Herrera J, Bartnicki-Garcia S (1976) Structure and transformation of chitin synthase particles (chitosomes) during microfibril synthesis in vitro. Proc Natl Acad Sci USA 73:4570–4574
Braun PC, Calderone RA (1978) Chitin synthesis in *Candida albicans*: comparison of yeast and hyphal forms. J Bacteriol 135:1472–1477
Brock TD (1965) Biochemical and cellular changes during conjugation in *Hansenula wingei*. J Bacteriol 90:1019–1025
Brody S (1973) Metabolism, cell walls, and morphogenesis. In: Coward SX, (ed) Developmental regulation, aspects of cell differentiation. Academic Press, London New York, pp 107–154
Bryant JA (1976) Molecular aspects of differentiation. In: Bryant JA (ed) Molecular aspects of gene expression in plants. Academic Press, London New York, pp 217–248
Buck KW, Chen AW, Dickerson AG, Chain EB (1968) Formation and structure of extracellular glucans produced by *Claviceps* species. J Gen Microbiol 51:337–352

Bull AT (1970a) Chemical composition of wild-type and mutant *Aspergillus nidulans* cell walls. The nature of polysaccharide and melanin constituents. J Gen Microbiol 63:75–94
Bull AT (1970b) Inhibition of polysaccharases by melanin: enzyme inhibition in relation to mycolysis. Arch Biochem Biophys 137:345–356
Burnett JH (1976) Fundamentals of mycology, 2nd edn. Edward Arnold, London, pp 1–664
Cabib E (1975) Molecular aspects of yeast morphogenesis. Annu Rev Microbiol 29:191–214
Cabib E, Bowers B (1971) Chitin and yeast budding. Localization of chitin in yeast bud scars. J Biol Chem 246:152–159
Cabib E, Farkas V (1971) The control of morphogenesis: an enzymatic mechanism for the initiation of septum formation in yeast. Proc Natl Acad Sci USA 68:2052–2056
Callow JA (1977) Recognition, resistance and the role of plant lectins in host-parasite interactions. In: Preston RD, Woolhouse HW (eds) Advances in botanical research vol IV. Academic Press, London New York, pp 1–49
Carbonell LM, Kanetsuna F, Gil F (1970) Chemical morphology of glucan and chitin in the cell wall of the yeast phase of *Paracoccidioides brasiliensis*. J Bacteriol 101:636–642
Cardemil L, Pincheira G (1979) Characterization of the carbohydrate component of fraction I in the *Neurospora crassa* cell wall. J Bacteriol 137:1067–1072
Chang PLY, Trevithick JR (1974) How inportant is secretion of exoenzymes through apical cell walls of fungi? Arch Microbiol 101:281–293
Chenouda MS (1972) Lipid and phospholipid composition of non-pigmented and pigmented hyphal cell walls of (–) strain of *Blakeslea trispora*. J Gen Appl Microbiol 18:155–163
Cole GT, Sekiya T, Kasai R, Yokoyama T, Nozawa Y (1979) Surface ultrastructure and chemical composition of cell walls of conidial fungi. Exp Mycol 3:132–156
Collinge AJ, Trinci APJ (1974) Hyphal tips of wild-type and spreading colonial mutants of *Neurospora*. Arch Microbiol 99:353–368
Cooper BA, Aronson JM (1967) Cell wall structure of *Pythium debaryanum*. Mycologia 59:658–670
Cortat M, Matile P, Wiemken A (1972) Isolation of glucanase-containing vesicles from budding yeast. Arch Mikrobiol 82:189–205
Crandall M (1976) Mechanisms of fusion in yeast cells. In: Peberdy JF, Rose AH, Rogers HJ, Cocking EC (eds) Microbial and plant protoplasts. Academic Press, London New York, pp 161–175
Datema R, Ende H van den, Wessels JGH (1977a) The hyphal wall of *Mucor mucedo* 1. Polyanionic polymers. Eur J Biochem 80:611–619
Datema R, Wessels JGH, Ende H van den (1977b) The hyphal wall of *Mucor mucedo* 2. Hexosamine-containing polymers. Eur J Biochem 80:621–626
Davis TE, Domer JE (1977) Glycohydrolase contamination of commercial enzymes frequently used in the preparation of fungal cell walls. Anal Biochem 80:593–600
Del Rey F, Santos T, Garcia-Acha I, Nombela C (1979) Synthesis of 1,3-β-glucanases in *Saccharomyces cerevisiae* during the mitotic cycle, mating, and sporulation. J Bacteriol 139:924–931
Dooijewaard-Kloosterziel AMP, Sietsma JH, Wouters JTM (1973) Formation and regeneration of *Geotrichum candidum* protoplasts. J Gen Microbiol 74:205–209
Duff RB (1952) The constitution of a glucosan from the fungus *Polyporus betulinus*. J Chem Soc III:2592–2594
Durân A, Bowers B, Cabib E (1975) Chitin synthetase zymogen is attached to the yeast plasma membrane. Proc Natl Acad Sci USA 72:3952–3955
Durell LW (1964) The composition and structure of walls of dark fungus spores. Mycopathol Mycol Appl 23:339–345
Erikson KE, Hamp SG (1978) Regulation of endo-1,4-β-glucanase production in *Sporotrichum pulverulentum*. Eur J Biochem 90:183–190
Farkas V (1979) Biosynthesis of cell walls of fungi. Microbiol Rev 43:117–144
Farkas V, Biely P, Bauer S (1973) Extracellular β-glucanases of the yeast *Saccharomyces cerevisiae*. Biochim Biophys Acta 321:246–255
Fencl Z (1978) Cell ageing and autolysis. In: Smith JE, Berry DR (eds) The filamentous fungi, vol III. Developmental mycology. Edward Arnold, London, pp 389–405

Fèvre M (1972) Contribution to the study of the determination of mycelium branching of *Saprolegina monoica* Pringsheim. Z Pflanzenphysiol 68:1–10

Fèvre M (1977) Subcellular localization of glucanase and cellulase in *Saprolegnia monoica*. J Gen Microbiol 103:287–295

Fleet GH, Phaff HJ (1975) Glucanases in *Schizosaccharomyces*. Isolation and properties of an exo-β-glucanase from the cell extracts and culture fluid of *Schizosaccharomyces japonicus* var. *versatilis*. Biochim Biophys Acta 401:318–332

Fleet GH, Manners DJ (1976) Isolation and composition of an alkali-soluble glucan from the cell walls of *Saccharomyces cerevisiae*. J Gen Microbiol 941:180–192

Fuller MS (1960) Biochemical and microchemical study of the cell walls of *Rhizidiomyces sp*. Am J Bot 47:838–842

Fuller MS, Barshad I (1960) Chitin and cellulose in the cell walls of *Rhizidiomyces* sp. Am J Bot 47:105–109

Fultz SA, Sussman AS (1966) Antigenic differences in the surface of hyphae and rhizoids in *Allomyces*. Science 152:785–787

Fultz SA, Woolf RA (1972) Surface structure in *Allomyces* during germination and growth. Mycologia 64:212–218

Gancedo JM, Gancedo C, Asensio C (1966) Uronic acids in fungal cell walls. Biochem Z 346:328–332

Gander JE (1974) Fungal cell wall glycoproteins and peptido-polysaccharides. Annu Rev Microbiol 28:103–119

Garcia Mendoza C, Leal JA, Novaes-Ledieu M (1979) Studies of the spore walls of *Agaricus bisporus* and *Agaricus campestris*. Can J Microbiol 25:32–39

Garzuly-Janke R (1940) Über das Vorkommen von Mannan bei Hyphen- und Sprosspilzen. Zentralbl Bakteriol Parasitenkd Infektionskr 102:361–365

Girbardt M (1969) Die Ultrastruktur der Apicalregion von Pilzhyphen. Protoplasma 67:413–441

Glaser L, Brown DH (1957) The synthesis of chitin in cell-free extracts of *Neurospora crassa*. J Biol Chem 228:729–742

Gold MH, Mitzel D, Segel I (1973) Regulation of nigeran accumulation by *Aspergillus aculeatus*. J Bacteriol 113:856–862

Gooday GW (1971) An autoradiographic study of hyphal growth of some fungi. J Gen Microbiol 67:125–133

Gooday GW (1975) The control of differentiation in fruit bodies of *Coprinus cinereus*. Rep Tottori Mycol Inst 12:151–160

Gooday GW (1977) Biosynthesis of the fungal wall – Mechanism and implications. J Gen Microbiol 99:1–11

Gorin PAJ, Spencer JFT (1968) Structural chemistry of fungal polysaccharides. Adv Carbohydr Chem 23:367–417

Gorin PAJ, Spencer JFT (1970) Proton magnetic resonance spectroscopy. An aid in identification and chemotaxonomy of yeasts. Adv Appl Microbiol 13:25–89

Green PB (1974) Morphogenesis of the cell and organ axis – Biophysical models. Brookhaven Symp Biol 25:166–190

Grove SN, Bracker CE (1970) Protoplasmic organization of hyphal tips among fungi: Vesicles and Spitzenkörper. J Bacteriol 104:989–1009

Gull K (1978) Form and function of septa in filamentous fungi. In: Smith JE, Berry DR (eds) The filamentous fungi, vol III. Developmental mycology. Edward Arnold, London, pp 78–93

Gull K, Trinci PJ (1974) Detection of areas of wall differentiation in fungi using fluorescent staining. Arch Microbiol 96:53–57

Hackman RH, Goldberg M (1965) Studies on chitin. VI. The nature of α- and β-chitins. Aust J Biol Sci 18:935–942

Hanson B, Brody S (1979) Lipid and cell wall changes in an inositol-requiring mutant of *Neurospora crassa*. J Bacteriol 138:461–466

Harold FM (1962) Binding of inorganic polyphosphate to the cell wall of *Neurospora crassa*. Biochim Biophys Acta 57:59–65

Hashimoto T, Wu-Yuan CF, Blumenthal HJ (1976) Isolation and characterization of the rodlet layer of *Trichophyton mentagrophytes* microcondidial wall. J Bacteriol 127:1543–154

Hawker LW (1965) Fine structure of fungi as revealed by electron microscopy. Biol Rev 40:52–92

Heath IB, Gay JL, Greenwood AD (1971) Cell wall formation in the Saprolegniales: cytoplasmic vesicles underlaying developing walls. J Gen Microbiol 65:225–232

Hegnauer H, Hohl HR (1973) A structural comparison of cyst and germ tube wall in *Phytophthora palmivora*. Protoplasma 77:151–163

Houwink AL, Kreger DR (1953) Observations on the cell wall of yeasts. An electron microscope and X-ray diffraction study. Antonie van Leeuwenhoek J Microbiol Serol 19:1–24

Horisberger M, Vonlanthen M (1977) Location of mannan and chitin on thin sections of budding yeast with gold markers. Arch Microbiol 115:1–7

Horowitz NH, Fling M, Maclead H, Watanabe Y (1961) Structural and regulatory genes controlling tyrosinase synthesis in *Neurospora*. Cold Spring Harbor Symp Quant Biol 26:233–238

Hunsley D, Burnett JH (1968) Dimensions of microfibrillar elements in fungal walls. Nature (London) 218:462–463

Hunsley D, Burnett JH (1970) The ultrastructural architecture of the walls of some fungi. J Gen Microbiol 62:203–218

Hunsley D, Gooday GW (1974) The structure and development of septa in *Neurospora crassa*. Protoplasma 82:125–146

Iten W, Matile P (1970) Role of chitinase and other lysosomal enzymes of *Coprinus lagopus* in the autolysis of fruiting bodies. J Gen Microbiol 61:301–303

Jan YN (1974) Properties and cellular localization of chitin synthase in *Phycomyces blakesleeanus*. J Biol Chem 249:1973–1979

Jennings DH (1979) Membrane transport and hyphal growth. In: Burnett JH, Trinci APJ (eds) Fungal walls and hyphal growth. Cambridge Univ Press, London New York Melbourne, pp 279–294

Jelsma J (1979) Ultrastructure of glucans in fungal cell walls. Thesis, Univ Groningen

Jelsma J, Kreger DR (1975) Ultrastructural observation on (1–3)-β-D-glucan from fungal cell walls. Carbohydr Res 43:200–203

Jelsma J, Kreger DR (1978) Observations on the cell wall composition of the bracket fungi *Laetiporus sulphureus* and *Piptoporus betulinus*. Arch Microbiol 119:249–253

Jelsma J, Kreger DR (1979) Polymorphism in crystalline (1–3)-β-D-glucan from fungal cell-walls. Carbohydr Res 71:51–64

Johnson B (1968) Lysis of yeast cell walls induced by 2-deoxyglucose at their sites of glucan synthesis. J Bacteriol 95:1169–1172

Johnston IR (1965) The partial acid hydrolysis of a highly dextrorotatory fragment of the cell wall of *Aspergillus niger*. Isolation of the α-(1–3)-linked dextrin series. Biochem J 96:659–664

Kanetsuna F, Carbonell LM, Azuma I, Yamamura Y (1972) Biochemical studies on the thermal dimorphism of *Paracoccidioides brasiliensis*. J Bacteriol 110:208–218

Kanetsuna F, Carbonell LM, Gil F, Azuma I (1974) Chemical and ultrastructural studies on the cell walls of the yeastlike and mycelial forms of *Histoplasma capsulatum*. Mycopathol Mycol Appl 54:1–13

Katz D, Rosenberger RF (1970a) A mutation in *Aspergillus nidulans* producing hyphal walls which lack chitin. Biochim Biophys Acta 208:452–460

Katz D, Rosenberger RF (1970b) The utilization of galactose by an *Aspergillus nidulans* mutant lacking galactose phosphate-UDP glucose transferase and its relation to cell wall synthesis. Arch Mikrobiol 74:41–51

Katz D, Rosenberger RF (1971a) Lysis of an *Aspergillus nidulans* mutant blocked in chitin synthesis and its relation to wall assembly and wall metabolism. Arch Mikrobiol 80:284–292

Katz D, Rosenberger RF (1971b) Hyphal wall synthesis in *Aspergillus nidulans*: Effect of protein synthesis inhibition and osmotic shock on chitin insertion and morphogenesis. J Bacteriol 108:184–190

Katz D, Goldstein D, Rosenberger RF (1972) Model for branch initiation in *Aspergillus nidulans* based on measurements of growth parameters J Bacteriol 109:1097–1100

Keegstra K, Talmadge KW, Bauer WD, Albersheim P (1973) The structure of plant cell walls. Plant Physiol 51:188–196

Kidby DK, Davies R (1970) Invertase and disulphide bridges in the yeast wall. J Gen Microbiol 61:327–333

Kirkwood S (1974) Unusual polysaccharides. Annu Rev Biochem 43:401–417

Kitamoto Y, Gruen HE (1976) Distribution of cellular carbohydrates during development of the mycelium and fruit bodies of *Flammulina velutipes*. Plant Physiol 58:485–491

Kitazima Y, Banno Y, Noguchi T, Nozawa Y, Ito Y (1972) Effects of chemical modification of structural polymer upon cell wall integrity of *Trichophyton*. Arch Biochem Biophys 152:811–820

Kornfeld RD, Kornfeld S (1976) Comparative aspects of glycoprotein structure. Annu Rev Biochem 45:217–237

Kreger DR (1954) Observations on cell walls of yeasts and some other fungi by X-ray diffraction and solubility tests. Biochim Biophys Acta 13:1–9

Kreger DR (1967) Röntgenographical aspects of yeast cell walls with particular reference to the glucan components (Baker's yeast, *Saccharomycopsis guttulata, Schizosaccharomyces octosporus* and *Candida albicans*). Abh Dtsch Akad Wiss Berlin Kl Med 1966:81–89, 349

Kreger DR (1970) Polyuronides as structural components of cell walls of fungi and green algae. Nature (London) 227:81–82

Kreger DR, Kopecká M (1976) On the nature and formation of the fibrillar nets produced by protoplasts of *Saccharomyces cerevisiae* in liquid media: an electronmicroscopic, X-ray diffraction and chemical study. J Gen Microbiol 92:207–220

Kreger-Van Rij NJW, Veenhuis M (1969) A study of vegetative reproduction in *Endomycopsis platypodis* by electronmicroscopy. J Gen Microbiol 58:341–346

Kreger-Van Rij NJW, Veenhuis M (1971) A comparative study of the cell wall structure of basidiomycetous and related yeasts. J Gen Microbiol 68:87–95

Kritzman G, Chet I, Henis Y (1978) Localization of β-(1,3)-glucanase in the mycelium of *Sclerotium rolfsii*. J Bacteriol 134:470–475

Krönig A, Egel R (1974) Autolytic activities associated with conjugation and sporulation in fission yeast. Arch Microbiol 99:241–249

Kuo MJ, Alexander M (1967) Inhibition of the lysis of fungi by melanins. J Bacteriol 94:624–629

Laere A van, Carlier AR (1978) Synthesis and proteolytic activation of chitin synthase in *Phycomyces blakesleeanus* Burgeff. Arch Microbiol 116:181–184

Lampen JO (1968) External enzymes of yeast: their nature and formation. Antonie van Leeuwenhoek J Microbiol Serol 34:1–18

Leonard TJ (1971) Phenoloxydase and fruiting body formation in *Schizophyllum commune*. J Bacteriol 106:162–167

Letourneau DR, Deven JM, Manocha MS (1976) Structure and composition of the cell wall of *Choanephora cucurbitanum*. Can J Microbiol 22:486–494

Lin CC, Aronson JM (1970) Chitin and cellulose in the cell walls of the Oomycete *Apodachlya* sp. Arch Mikrobiol 72:111–114

Lin CC, Sicher Jr RC, Aronson JM (1976) Hyphal wall chemistry in *Apodachlya* Arch Microbiol 108:85–91

Lindberg B, Svenson S (1973) Microbial polysaccharides. In: Aspinal GO (ed) Carbohydrates. MTP Int Rev Sci Org Chem Ser 1, vol VII. Butterworth London, pp 319–331

Linnemans WAM, Boer P, Elbers PF (1977) Localization of acid phosphatase in *Saccharomyces cerevisiae*: a clue to cell wall formation. J Bacteriol 131:638–644

Lopez-Romero E, Ruiz-Herrera J (1976) Synthesis of chitin by particulate preparations from *Aspergillus flavus*. Antonie van Leeuwenhoek J Microbiol Serol 42:261–276

Lopez-Romero E, Ruiz-Herrera J, Bartnicki-Garcia S (1978) Purification and properties of an inhibitory protein of chitin synthesis from *Mucor rouxii*. Biochim Biophys Acta 525:338–345

MacWilliam IC (1970) The structure, synthesis and functions of the yeast cell wall. J Inst Brew 76:524–536
Mahadevan PR, Mahadkar VR (1970) Role of enzymes in growth and morphology of *Neurospora crassa*: Cell-wall-bound enzymes and their possible role in branching. J Bacteriol 101:941–947
Mahadevan PR, Tatum EL (1965) Relationship of the major constituents of the *Neurospora crassa* cell wall to wild-type and colonial morphology. J Bacteriol 90:1073–1081
Mahadevan PR, Tatum EL (1967) Localization of structural polymers in the cell wall of *Neurospora crassa*. J Cell Biol 35:295–302
Manners DJ, Masson AJ, Patterson JC (1973a) The structure of a β-(1-6)-D-glucan from yeast cell walls. Biochem J 135:31–36
Manners DJ, Masson AJ, Patterson JC (1973b) The structure of a β-(1–3)-D-glucan from yeast cell walls. Biochem J 135:19–30
Manners DJ, Masson AJ, Patterson JC (1974) The heterogeneity of glucan preparations from the walls of various yeasts. J Gen Microbiol 80:411–417
Manocha MS, Colvin JR (1967) Structure and composition of the cell wall of *Neurospora crassa*. J Bacteriol 94:202–212
Manocha MS, Colvin JS (1968) Structure of the cell wall of *Pythium debaryanum*. J Bacteriol 95:1140–1152
Marchant R (1966) Wall structure and spore germination in *Fusarium culmorum*. Ann Bot 30:821–830
Marchant R, Smith DG (1968) A serological investigation of hyphal growth in *Fusarium culmorum*. Arch Mikrobiol 63:85–94
Marchant R, Wessels JGH (1973) Septal structure in normal and modified strains of *Schizophyllum commune* carrying mutations affecting septal dissolution. Arch Mikrobiol 90:35–45
Marchant R, Wessels JGH (1974) An ultrastructural study of spetal dissolution in *Schizophyllum commune*. Arch Mikrobiol 96:175–182
Maret R (1972) Chemie et morphologie submicroscopique des parois cellulaires de l'Ascomycète *Chaetomium globosum*. Arch Mikrobiol 81:68–90
Matile P (1971) Vacuoles, lysosomes of *Neurospora*. Cytobiology 3:324–330
McClure K, Park D, Robinson PM (1968) Apical organization in the somatic hyphae of fungi. J Gen Microbiol 50:177–182
McMurrough I, Bartnicki-Garcia S (1971) Properties of a particulate chitin synthetase from *Mucor rouxii*. J Biol Chem 246:4008–4016
McMurrough I, Bartnicki-Garcia S (1973) Inhibition and activation of chitin synthesis by *Mucor rouxii* cell extracts. Arch Biochem Biophys 158:812–816
Michalenko GO, Hohl HR, Rast D (1976) Chemistry and architecture of the mycelial wall of *Agaricus bisporus*. J Gen Microbiol 92:251–262
Middlebrook MJ, Preston RD (1952) Spiral growth and spiral structure. III. Wall structure in the growth zone of *Phycomyces*. Biochim Biophys Acta 9:32–48
Misaki A, Johnson Jr J, Kirkwood S, Scaletti JV, Smith F (1968) Structure of the cell wall glucan of yeast. Carbohydr Res 6:150–164
Mishra NC (1977) Genetics and biochemistry of morphogenesis in *Neurospora*. Adv Genet 19:341–405
Mishra NC, Tatum EL (1972) Effects of L-sorbose in polysaccharide synthetase of *Neurospora*. Proc Natl Acad Sci USA 69:313–317
Mitchell A, Taylor IF (1969) Cell wall protein in *Aspergillus niger* and *Chaetomium globosum*. J Gen Microbiol 59:103–108
Miyazaki T, Irino T (1970) Acidic polysaccharides from the cell wall of *Absidia cylindrospora, Mucor mucedo* and *Rhizopus nigricans*. Chem Pharm Bull 18:1930–1934
Miyazaki T, Irino T (1971) Fungal polysaccharides. IX. Acidic polysaccharides from the cell wall of *Rhizopus nigricans*. Chem Pharm Bull 19:2545–2549
Moore PM, Peberdy JF (1975) Biosynthesis of chitin by particulate fractions from *Cunninghamella elegans*. Microbios 12:29–39
Mullins JT, Ellias EA (1974) Sexual morphogenesis in *Achlya*: Ultrastructural basis for the hormonal induction of antheridial hyphae. Proc Natl Acad Sci USA 71:1347–1350
Muzzarelli RAA (1977) Chitin. Pergamin Press, Oxford, pp 1–254

Necas O (1971) Cell wall synthesis in yeast protoplasts. Bacteriol Rev 35:149–170
Nickerson JW (1963) Symposium on the biochemical bases of morphogenesis in fungi. IV. Molecular bases of form in yeasts. Bacteriol Rev 27:305–324
Niederpruem DJ (1971) Kinetic studies of septum synthesis, erosion and nuclear migration in a growing B-mutant of *Schizophyllum commune*. Arch Mikrobiol 75:189–196
Niederpruem DJ, Wessels JGH (1969) Cytodifferentiation and morphogenesis in *Schizophyllum commune*. Bacteriol Rev 33: 505–535
Northcote DH, Horne RW (1952) The chemical composition and structure of the yeast cell wall. Biochem J 51:232–236
Novaes-Ledieu M, Jimenez-Martinez A (1968) The structure of cell walls of Phycomycetes. J Gen Microbiol 54:407–415
Obaidah MA, Buck KW (1971) Characterization of two cell-wall polysaccharides from *Fusicoccum amygdali*. Biochem J 125:473–480
Park D, Robinson PM (1966) Internal pressure of hyphal tips of fungi and its significance in morphogenesis. Ann Bot 30:425–439
Parker BC, Preston RD, Fogg GE (1963) Studies of the structure and chemical composition of the cell walls of *Vaucheriaceae* and *Saprolegniaceae*. Proc Soc London Ser, B 158:435–445
Patton AM, Marchant R (1978) An ultrastructural study of septal development in hyphae of *Polyporus biennis*. Arch Microbiol 118:271–277
Peberdy JF (1978) Protoplasts and their development. In: Smith JE, Berry DR (eds) The filamentous fungi, vol III. Developmental mycology. Edward Arnold, London, pp 119–131
Pérez S, Roux M, Revol JF, Marchessault RH (1979) Dehydration of nigeran crystals: crystal structure and morphological aspects. J Mol Biol 129:113–133
Phaff HJ (1963) Cell wall of yeasts. Annu Rev Microbiol 17:15–30
Phaff HJ (1971) Structure and biosynthesis of the yeast cell envelope. In: Rose AH, Harrison JH (eds) The yeasts, vol II. Academic Press, London New York, pp 135–210
Phaff HJ (1977) Enzymatic yeast cell wall degradation: In: Feeney RE, Whitaker JR (eds) Advances in chemistry. Ser no 160. Am Chem Soc, pp 244–282
Polacheck Y, Rosenberger RF (1975) Autolytic enzymes in hyphae of *Aspergillus nidulans*: their action on old and newly formed walls. J Bacteriol 121:332–337
Polacheck I, Rosenberger RF (1977) *Aspergillus nidulans* mutant lacking (1–3)-α-D-glucan, melanin, and cleistothecia. J Bacteriol 132:650–656
Polacheck I, Rosenberger RF (1978) Distribution of autolysis in hyphae of *Aspergillus nidulans*: Evidence for a lipid-mediated attachment to hyphal walls. J Bacteriol 135:741–747
Poulain DG, Tronchin G, Dubremetz JF, Biquet J (1978) Ultrastructure of the cell wall of *Candida albicans* blastospores: study of its constitutive layer by the use of a cytochemical technique revealing polysaccharides. Ann Microbiol 129:141–153
Previato JO, Gorin PAJ, Travassos LR (1979) Cell wall composition in different cell types of the dimorphic species *Sporotrix schenckii*. Exp Mycol 3:83–91
Prosser JI, Trinci AJP (1979) A model for hyphal growth and branching. J Gen Microbiol 111:153–164
Ramos S, Garciá-Acha I (1975) Cell wall enzymatic lysis of the yeast form of *Pullularia pullulans* and wall regeneration by protoplasts. Arch Microbiol 104:271–277
Raper JR, Raper CA (1973) Incompatibility factors: regulatory genes for sexual morphogenesis in higher fungi. Brookhaven Symp Biol 25:19–38
Raudaskoski M (1976) Acid phosphatase activity in the wild-type and B-mutant hyphae of *Schizophyllum commune*. J Gen Microbiol 94:373–379
Rees DA (1973) Polysaccharide conformation. In: Aspinal GO (ed) MTP Int Rev Sci Org Chem, ser 1, vol VII. Carbohydrates. Butterworth, London, pp 270–293
Rees DA (1977) Polysaccharide shapes. Chapman and Hall, London, pp 1–152
Rees DA, Scott WE (1971) Polysaccharide conformation Part VI. J Chem Soc B:469–479
Reese ET (1972) Enzyme production from insoluble substrates. Biotechnol Bioeng Symp 3:43–62
Reese ET (1977) Degradation of polymeric carbohydrates by microbial enzymes. In: Loewus F, Runeckles VC (eds) The structure, biosynthesis and degradation of wood. Recent Adv Phytochem, vol XI. Plenum Press, New York London, pp 311–367

Reid ID, Bartnicki-Garcia S (1976) Cell-wall composition and structure of yeast cells and conjugation tubes of *Tremella mesenterica*. J Gen Microbiol 96:35–50
Rijn HJ van, Boer P, Steyn-Parvé EP (1972) Biosynthesis of acid phosphatase of baker's yeast. Factors influencing its production by protoplasts and characterization of the secreted enzyme. Biochim Biophys Acta 268:431–441
Robert JC (1977a) Fruiting of *Coprinus congregatus:* biochemical changes in fruit-bodies during morphogenesis. Trans Br Mycol Soc 68:379–387
Robert JC (1977b) Fruiting of *Coprinus congregatus:* relationships to biochemical changes in the whole culture. Trans Br Mycol Soc 68:389–395
Roelofson PA (1951) Cell wall structure in the growth zone of *Phycomyces* sporangiophores. II. Double refraction and electron microscopy. Biochim Biophys Acta 6:357–373
Roelofsen PA (1959) The plant cell wall. Encyclopedia of plant anatomy, vol. III, part 4, Bornträger, Berlin-Nikolassee
Rosenberger RF (1976) The cell wall. In: Smith JE, Berry DR (eds) The filamentous fungi, vol II. Biosynthesis and metabolism, Edward Arnold, London, pp 328–344
Rousset-Hall A de, Gooday GW (1975) A kinetic study of sobulized chitin synthetase preparation from *Coprinus cinereus*. J Gen Microbiol 89:146–154
Rowley BI, Pirt SJ (1972) Melanin production by *Aspergillus nidulans* in batch and chemostat cultures. J Gen Microbiol 72:553–563
Rudall KM, Kenchinton W (1973) The chitin system. Biol Rev 48:597–636
Ruiz-Herrera J, Bartnicki-Garcia S (1974) Synthesis of cell wall microfibrils in vitro by a "soluble" chitin synthetase from *Mucor rouxii*. Science 186:357–359
Ruiz-Herrera J, Bartnicki-Garcia S (1976) Proteolytic activation and inactivation of chitin synthase from *Mucor rouxii*. J Gen Microbiol 97:241–249
Ruiz-Herrera J, Sing VO, Woude van der WJ, Bartnicki-Garcia S (1975) Microfibril assembly by granules of chitin synthetase. Proc Natl Acad Sci USA 72:2706–2710
Ryder NS, Peberdy JF (1977) Chitin synthetase in *Aspergillus nidulans:* properties and proetolytic activation. J Gen Microbiol 99:69–76
San Blass G, Carbonel LM (1974) Chemical and ultrastructural studies on the cell walls of the yeast like and mycelial forms of *Histoplasma farcinosum*. J Bacteriol 119:602–611
Santos T, Del Rey F, Conde J, Villanueva JR, Nombela C (1979) *Saccharomyces cerevisiae* mutant defective in exo-1,3-β-glucanase production. J Bacteriol 139:333–338
Sathyanarayana BK, Rao VSR (1972) Conformational studies of α-glucans. Biopolymers 11:1379–1394
Schekman R, Brawley V (1979) Localized deposition of chitin on the yeast cell surface in response to mating pheromone. Proc Natl Acad Sci USA 76:645–649
Schneider EF, Wardrop AB (1979) Ultrastructural studies on the cell walls in *Fusarium sulphureum*. Can J Microbiol 25:75–85
Scurfield G (1967) Fine structure of the cell walls of *Polyporus myllitae*. J Linn Soc London Bot 60:159–166
Shida M, Uchida T, Matsuda K (1978) A (1-3)α-D-glucan isolated from the fruit bodies of *Lentinus edodes*. Carbohydr Res 60:117–127
Shockman GD, Pooley HM, Thompson JS (1967) Autolytic enzyme system of *Streptococcus faecalis*. III Localization of autolysin at the side of wall synthesis. J Bacteriol 94:1525–1530
Sietsma JH, Wessels JGH (1977) Chemical analysis of the hyphal wall of *Schizophyllum commune*. Biochim Biophys Acta 496:225–239
Sietsma JH, Wessels JGH (1979) Evidence for covalent linkages between chitin and β-glucan in a fungal wall. J Gen Microbiol 114:99–108
Sietsma JH, Eveleigh DE, Haskins RH (1969) Cell wall composition and protoplast formation of some Oomycete species. Biochim Biophys Acta 184:306–317
Sietsma JH, Child JJ, Nesbitt LR, Haskins RH (1975a) Chemistry and ultrastructure of the hyphal walls of *Pythium acanthicum*. J Gen Microbiol 86:29–38
Sietsma JH, Child JJ, Nesbitt LR, Haskins RH (1975b) Ultrastructural aspects of wall regeneration by *Pythium* protoplasts. Antonie van Leeuwenhoek J Microbiol Serol 41:17–23
Sietsma JH, Rast D, Wessels JGH (1977) The effect of carbon dioxide on fruiting and

on the degradation of a cell-wall glucan in *Schizophyllum commune*. J Gen Microbiol 102:385–389
Skujins JJ, Potgieter HJ, Alexander M (1965) Dissolution of fungal cell walls by a streptomycete chitinase and β-(1–3)glucanase. Arch Biochem Biophys 111:358–364
Sloat BF, Pringle JR (1978) A mutant of yeast defective in cellular morphogenesis. Science 200:1171–1173
Sternlicht E, Katz D, Rosenberger RF (1973) Subapical wall synthesis and wall thickening induced by cycloheximide in hyphae of *Aspergillus nidulans*. J Bacteriol 114:819–823
Stirling JL, Cook GA, Pope AMS (1979) Chitin and its degradation. In: Burnett JH, Trinci APJ (eds) Fungal walls and hyphal growth. Cambridge Univ Press, London, pp 169–188
Stöckmann VE (1972) Developing a hypothesis: Native cellulose elementary fibrils are formed with metastable structure. Biopolymers 11:251–256
Strunk C (1963) Über die Substruktur der Hyphenspitzen von *Polystictus versicolor*. Allg Mikrobiol 3:265–274
Sundaralingam M (1968) Some aspects of stereochemistry and hydrogen bonding of carbohydrates related to polysaccharide conformations. Biopolymers 6:189–213
Taylor IEP, Cameron DS (1973) Preparation and quantitative analysis of fungal cell walls: strategy and tactics. Annu Rev Microbiol 27:243–260
Thomas D des S, Mullins JT (1967) Role of enzymatic wall-softening in plant morphogenesis: Hormonal induction in *Achlya*. Science 156:84–85
Tokunaga J, Bartnicki-Garcia S (1971) Structure and differentiation of the cell wall of *Phytophthora palmivora*: cysts, hyphae and sporangia. Arch Mikrobiol 79:293–310
Trinci APJ, Collinge A (1975) Hyphal wall growth in *Neurospora crassa* and *Geotrichum candidum*. J Gen Microbiol 91:355–361
Trinci APJ, Saunders PT (1977) Tip growth of fungal hyphae. J Gen Microbiol 103:243–248
Troy FA, Koffler H (1969) The chemistry and molecular architecture of the cell walls of *Penicillium chrysogenum*. J Biol Chem 244:5563–5576
Ukai S, Hirose K, Kiho T, Hara C (1974) Polysaccharides in fungi I Purification and characterization of acidic heteroglycans from aqueous extracts of *Tremella fuciformis* Berk. Chem Pharm Bull 22:1102–1107
Valk P van der, Marchant R (1978) Hyphal ultrastructure in fruit-body primordia of the basidiomycetes *Schizophyllum commune* and *Coprinus cinereus*. Protoplasma 95:57–72
Valk P van der, Wessels JGH (1976) Ultrastructure and localization of wall polymers during regeneration and reversion of protoplasts of *Schizophyllum commune*. Protoplasma 90:65–87
Valk P van der, Wessels JGH (1977) Light and electron microscopic autoradiography of cell wall regeneration by *Schizophyllum commune* protoplasts. Acta Bot Neerl 26:43–52
Valk P van der, Marchant R, Wessels JGH (1977) Ultrastructural localization of polysaccharides in the wall and septum of the basidiomycete *Schizophyllum commune*. Exp Mycol 1:69–82
Vermeulen CA, Raeven MBJM, Wessels JGH (1979) Localization of chitin synthase activity in subcellular fractions of *Schizophyllum commune* protoplasts. J Gen Microbiol 114:87–97
Villanueva JR, Garcia Acha I (1971) Production and use of fungal protoplasts. In: Booth (ed) Methods in microbiology, vol IV. Academic Press, London New York, pp 665–718
Villanueva JR, Notario V, Santos T, Villa TG (1976) β-glucanases in nature. In: Peberdy JF, Rose AH, Rogers HJ, Cocking EC (eds) Microbial and plant protoplasts. Academic Press, London New York, pp 323–358
Vries OMH de (1974) Formation and cell wall regeneration of protoplasts from *Schizophyllum commune*. Thesis, Univ Groningen
Vries OMH de, Wessels JGH (1972) Release of protoplasts from *Schizophyllum commune* by a lytic enzyme preparation from *Trichoderma viride*. J Gen Microbiol 73:13–22
Vries OMH de, Wessels JGH (1973a) Effectiveness of a lytic enzyme preparation from *Trichoderma viride* in releasing spheroplasts from fungi, particularly basidiomycetes. Antonie van Leeuwenhoek J Microbiol Serol 39:397–400
Vries OMH de, Wessels JGH (1973b) Release of protoplasts from *Schizophyllum commune* by combined action of purified α-1,3-glucanase and chitinase derived from *Trichoderma viride*. J Gen Microbiol 76:319–330

Vries OMH de, Wessels JGH (1975) Chemical analysis of cell wall regeneration and reversion of protoplasts from *Schizophyllum commune*. Arch Microbiol 102:209–218

Waechter CJ, Lennarz WJ (1976) The role of polyprenol-linked sugars in glycoprotein synthesis. Annu Rev Biochem 45:95–112

Wang MC, Bartnicki-Garcia S (1970) Structure and composition of walls of the yeast form of *Verticillium alboatrum*. J Gen Microbiol 64:41–54

Wang MC, Bartnicki-Garcia S (1976) Synthesis of β-1,3-glucan microfibrils by a cell-free extract from *Phytophthora cinnamomi*. Arch Biochem Biophys 175:351–354

Wessels JGH (1965) Morphogenesis and biochemistry of *Schizophyllum commune*. Wentia 13:1–113

Wessels JGH (1966) Control of cell-wall glucan degradation in *Schizophyllum commune*. Antonie van Leeuwenhoek J Microbiol Serol 32:341–355

Wessels JGH (1969a) Biochemistry of sexual morphogenesis in *Schizophyllum*: Effect of mutations affecting the incompatibility system on cell wall metabolism. J Bacteriol 98:697–704

Wessels JGH (1969b) A (1–6)-β-D-glucan glucanohydrolase involved in hydrolysis of cell-wall glucan in *Schizophyllum commune*. Biochim Biophys Acta 178:191–193

Wessels JGH, Koltin Y (1972) R-glucanase activity and susceptibility of hyphal walls to degradation in mutants of *Schizophyllum* with disrupted nuclear migration. J Gen Microbiol 71:471–475

Wessels JGH, Marchant R (1974) Enzymic degradation of septa in hyphal wall preparations from a monokaryon and a dikaryon of *Schizophyllum commune*. J Gen Microbiol 83:359–368

Wessels JGH, Niederpruem DJ (1967) Role of a cell-wall glucan-degrading enzyme in mating of *Schizophyllum commune*. J Bacteriol 94:1594–1602

Wessels JGH, Sietsma JH (1979) Wall structure and growth in *Schizophyllum commune*. In: Burnett JH, Trinci APJ (eds) Fungal walls and hyphal growth. Cambridge Univ Press, London, pp 27–48

Wessels JGH, Vries OMH de (1973) Wall structure, wall degradation protoplast liberation and wall regeneration in *Schizophyllum commune*. In: Villanueva JR, Garciá-Acha, Gascón S, Uruburu F (eds) Yeast, mould and plant protoplasts. Academic Press, London New York, pp 295–306

Wessels JGH, Kreger DR, Marchant R, Regensburg BA, Vries OMH de (1972) Chemical and morphological characterization of the hyphal surface of the basidiomycete *Schizophyllum commune*. Biochim Biophys Acta 273:346–358

Wessels JGH, Valk P van der, Vries OMH de (1976) Wall synthesis by fungal protoplasts. In: Peberdy JF, Rose AH, Rogers HJ, Cocking EC (eds) Microbial and plant protoplasts. Academic Press, London New York, pp 267–281

Wrathall CR, Tatum EL (1974) Hyphal wall peptides and colonial morphology in *Neurospora crassa*. Biochem Genet 12:59–68

Yamada M, Miyazaki (1976) Ultrastructure and chemical analysis of the cell wall of *Pythium debaryanum*. Jpn J Microbiol 20:83–91

Yoshioka Y, Emori M, Ikekawa T, Fukuoka F (1975) Isolation, purification, and structure of components from acidic polysaccharides of *Pleurotus ostreatus* (Fr.) Quél Carbohydr Res 43:305–320

Zevenhuizen LPTM, Bartnicki-Garcia S (1969) Chemical structure of the insoluble hyphal wall glucan of *Phytophthora cinnamomi*. Biochemistry 8:1496–1502

Zonneveld BJM (1971) Biochemical analysis of the cell wall of *Aspergillus nidulans*. Biochim Biophys Acta 249:506–514

Zonneveld BJM (1972a) A new type of enzyme, an exo-splitting α-1,3-glucanase from non-induced cultures of *Aspergillus nidulans*. Biochim Biophys Acta 258:541–547

Zonneveld BJM (1972b) Morphogenesis in *Aspergillus nidulans*. The significance of α-1,3-glucan of the cell wall and α-1,3-glucanase for cleistothecium development. Biochim Biophys Acta 273:174–187

Zonneveld BJM (1974) α-1,3-glucan synthesis correlated with α-1,3-glucanase synthesis, conidiation and fructification in morphogenetic mutants of *Aspergillus nidulans*. J Gen Microbiol 81:445–451

16 Chitin: Structure, Metabolism, and Regulation of Biosynthesis

E. CABIB

1 Introduction

Chitin is an ubiquitous component of fungal cell walls. Only a few classes of fungi, such as the Oomycetes and Trichomycetes (BARTNICKI-GARCIA 1968) appear to lack chitin. Even in species of the Saccharomycetaceae, which were classified as devoid of chitin by BARTNICKI-GARCIA, small amounts of the polysaccharide are found in the cell wall, with a preferential localization in the septum (CABIB et al. 1974). Because of its function in the generation and maintenance of cell shape and in the formation of septa, chitin is of great importance in fungal growth and physiology. For the same reasons, it constitutes a very useful model for the study of morphogenetic processes. Hence the interest of many investigators in this polysaccharide, which may be considered the equivalent, for fungi, of cellulose in the plant kingdom.

2 Chemical and Physical Structure

Chitin is a linear polysaccharide composed of $(1 \rightarrow 4)$-β-linked *N*-acetylglucosamine units (Fig. 1). The chain length for chitin from the crab *Scylla serrata* varies, with different methods of measurement, between 5,000 and 8,000 (HACKMAN and GOLDBERG 1974). There is no information available to this reviewer on the molecular weight of fungal chitin. The chitin chains associate by forming very strong hydrogen bonds between the $>N-H$ groups of one chain and the $>C=O$ groups of the adjacent chain (CARLSTRÖM 1957, MINKE and BLACKWELL 1978). This accounts for the extreme insolubility of chitin in most solvents, with the exception of hexafluoroisopropanol, hexafluoroacetone sesquihydrate, some chloroalcohols, and concentrated acids (MUZZARELLI 1977). With the last type of solvent some degradation is known to occur.

X-ray diffraction studies of chitin samples from various sources document the occurrence in nature of different conformations of the polysaccharide. In the structure prevalent among fungi, termed α-chitin, the sugar chains appear to run antiparallel to each other, i.e., with the reducing groups of adjacent chains on opposite ends (CARLSTRÖM 1957). In β-chitin the chains are parallel (all reducing groups on the same side) and in γ-chitin it appears that of every three chains two are parallel and one antiparallel (MUZZARELLI 1977). It should be kept in mind that a configuration equivalent to the antiparallel one can also be achieved by a chain folding back on itself.

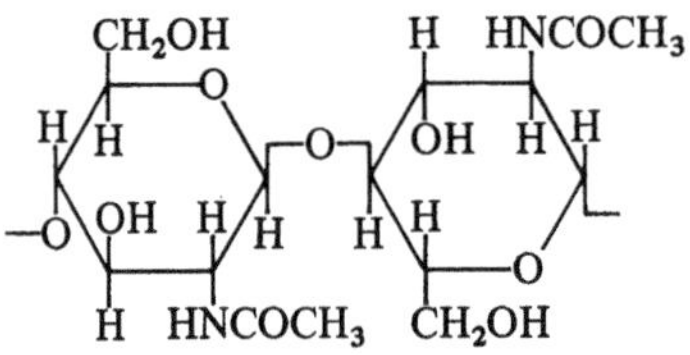

Fig. 1. Repeating unit of chitin

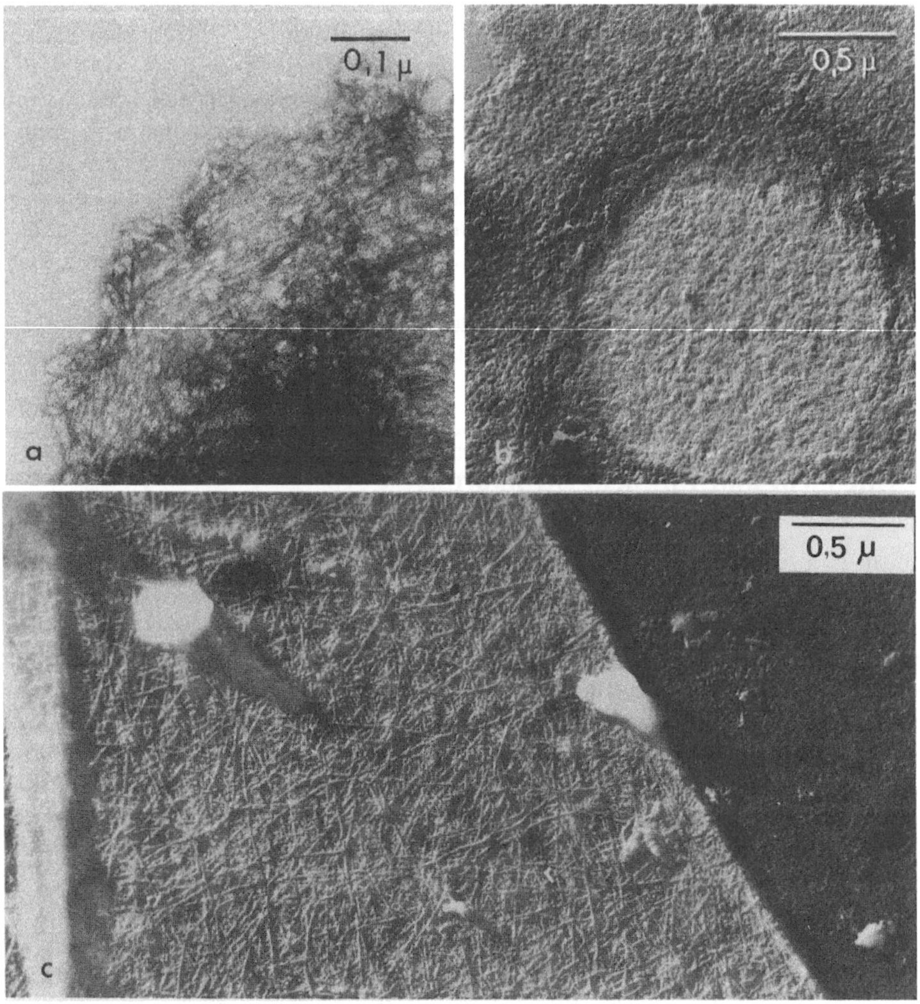

Fig. 2. **a** Chitin of *Saccharomyces cerevisiae* primary septum, as observed by negative staining with 0.5% uranyl acetate. (Reprinted from CABIB et al. 1979, with permission of the publisher.) **b** Chitin of *Saccharomyces cerevisiae* primary septum, in a metal-shadowed preparation. **c** Chitin fibrils of *Neurospora* cell wall, revealed by enzymatic treatment, as described by HUNSLEY and BURNETT (1970). (Electronmicrograph kindly contributed by Prof. J.H. BURNETT)

In the cell walls of fungi, chitin chains associate to form fibrils which can be several μm long (Fig. 2c). The fibrils are probably not artifacts of the preparation because they can be exposed after digesting the external wall layers under very mild conditions, i.e., by enzymatic digestion (HUNSLEY and BURNETT 1970). In contrast, the chitin of yeast primary septa seems to be organized in short rod-like units, about 60×5 nm in dimensions (Fig. 2a and b and CABIB et al. 1979).

3 Distribution and Localization

As mentioned in the Introduction, chitin is a major component of the cell wall of most fungi, except for Oomycetes and Trichomycetes (BARTNICKI-GARCIA 1968, ROSENBERGER 1976). Where its location has been studied in detail, chitin has been found in the inner layers of the cell wall, close to the plasmalemma (HUNSLEY and BURNETT 1970, VAN DER VALK et al. 1977). In the apical region of a hypha, where elongation occurs, chitin appears to be the principal component (HUNSLEY and BURNETT 1970), therefore it seems logical that a study of its formation should be essential to an understanding of the process of hyphal growth.

In budding yeasts chitin is found almost exclusively in the primary septa (CABIB and BOWERS 1971, CABIB et al. 1974) although a small amount (less than 10%) is also uniformly dispersed over the whole cell surface (HORISBERGER and VONLANTHEN 1977, MOLANO et al. 1980).

Because, after cell division, the primary septum remains embedded in the bud scar on the mother cell, the amount of chitin present in any given cell is proportional to the number of bud scars, i.e., to the number of times the cell has budded (BERAN et al. 1972).

Of special interest is the cell wall formed in *Blastocladiella emersonii* during the conversion of zoospores into "round cells" (see Sect. 6.4). This cell wall is said to consist of chitin (SELITRENNIKOFF et al. 1976). The best evidence for this conclusion is apparently the finding that the insoluble hexosamine-containing material in round cells is recovered in the low-speed pellet of a cell sonicate, which presumably includes the cell walls (SELITRENNIKOFF et al. 1976). The hexosamine is released by acid hydrolysis but not by alkali treatment. This reviewer, however, has been unable to find direct data on wall composition of the round cell in the literature. Better information on this point is highly desirable, especially because the formation of cell wall in the transition from zoospore to round cell has been used as a model for morphogenesis (see Sect. 6.4).

4 The Enzymatic Synthesis of Chitin and its Regulation

4.1 Biosynthesis of Precursors

The first specific step in the pathway from glucose to chitin is the formation of glucosamine-6-phosphate from fructose-6-phosphate and glutamine (Fig. 3).

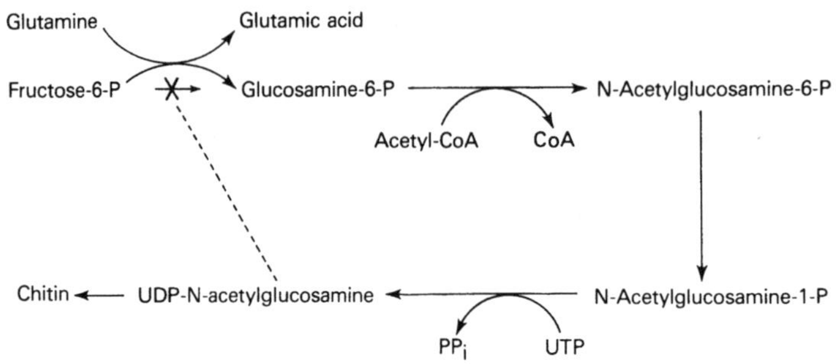

Fig. 3. Metabolic pathway leading from fructose-6-P to chitin. The feedback inhibition of hexosamine formation by UDP-*N*-acetylglucosamine is symbolized by the *broken line* and the *crossed arrow*

This reaction was first reported in *Neurospora* (LELOIR and CARDINI 1953), and later in bacteria and in mammalian tissues (GHOSH et al. 1960). As is the case for the mammalian enzyme (KORNFELD et al. 1964, KORNFELD 1967), those from both yeast (MORIGUCHI et al. 1976) and *Blastocladiella* (NORMAN et al. 1975, SELITRENNIKOFF et al. 1976, SELITRENNIKOFF and SONNEBORN 1976b) were found to be inhibited by UDP-*N*-acetylglucosamine (UDP-GlcNAc), the immediate precursor of chitin. This feedback inhibition would prevent the accumulation of UDP-GlcNAc in the cell, without interfering with the last step of chitin synthesis.

In the presence of acetyl-CoA, glucosamine-6-phosphate is transformed into the *N*-acetyl derivative (Fig. 3) by an acetylase (LELOIR and CARDINI 1953, DAVIDSON et al. 1957). The next step, conversion of *N*-acetyl-glucosamine-6-phosphate into the corresponding 1-phosphate is catalyzed by a mutase, which was first detected in *Neurospora* (REISSIG 1956) and is also present in *Blastocladiella* (SELITRENNIKOFF and SONNEBORN 1976b). Finally, *N*-acetylglucosamine-1-phosphate and UTP react to yield UDP-*N*-acetylglucosamine (Fig. 3). The pyrophosphorylase that catalyzes this reaction has been studied in yeast (YAMAMOTO et al. 1976). The reader interested in a more comprehensive account of early work on all these enzymatic reactions is directed to DAVIDSON (1966).

4.2 Chitin Synthetase and its Regulation

4.2.1 General Properties of Chitin Synthetase

Since the initial report of GLASER and BROWN (1957), who detected chitin synthetase in an extract from *Neurospora,* the presence of this enzyme has been reported in many fungi (JAWORSKI et al. 1965, PORTER and JAWORSKI 1966, PLESSMANN CAMARGO et al. 1967, OHTA et al. 1970, KELLER and CABIB 1971, MCMURROUGH and BARTNICKI-GARCIA 1971, MCMURROUGH et al. 1971, HORI et al. 1974a, b, c, JAN 1974, GOODAY and ROUSSET-HALL 1975, MOORE and PEBERDY 1975, PEBERDY and MOORE 1975, ROUSSET-HALL and GOODAY 1975, MOORE and PEBERDY 1976, LOPEZ-ROMERO and RUIZ-HERRERA 1976,

RYDER and PEBERDY 1977a, RYDER and PEBERDY 1977b, BARTNICKI-GARCIA et al. 1978, BRAUN and CALDERONE 1978). The substrate is UDP-*N*-acetylglucosamine, and the reaction may be described by the equation:

$$2n \text{ UDP-GlcNAc} \rightarrow 2n \text{ UDP} + [\text{GlcNAc } \beta(1 \rightarrow 4) \text{ GlcNAc}]_n \quad (1)$$

The stoichiometry has been measured only with the enzyme from *Saccharomyces carlsbergensis* (KELLER and CABIB 1971). In Eq. (1) a primer is not included because need for a primer has not yet been demonstrated (see below).

Most preparations of the enzyme show requirement for a divalent cation, usually Mg^{2+} (GLASER and BROWN 1957, PLESSMANN CAMARGO et al. 1967, KELLER and CABIB 1971, JAN 1974, GOODAY and ROUSSET-HALL 1975, PEBERDY and MOORE 1975, LOPEZ-ROMERO and RUIZ-HERRERA 1976, RYDER and PEBERDY 1977b) and for free *N*-acetylglucosamine or chitodextrins (GLASER and BROWN 1957, PORTER and JAWORSKI 1966, PLESSMANN CAMARGO et al. 1967, KELLER and CABIB 1971, MCMURROUGH and BARTNICKI-GARCIA 1971, JAN 1974, GOODAY and ROUSSET-HALL 1975, ROUSSET-HALL and GOODAY 1975, LOPEZ-ROMERO and RUIZ-HERRERA 1976, MOORE and PEBERDY 1976, RYDER and PEBERDY 1977b). Only with the enzyme from *Blastocladiella* is *N*-acetylglucosamine incorporated into the final product (PLESSMANN CAMARGO et al. 1967), a result that suggests the possible function of the amino sugar as a primer in the reaction. In some cases the curve of activity as a function of UDP-GlcNAc concentration is sigmoid in shape when measured in the absence of GlcNAc (MCMURROUGH and BARTNICKI-GARCIA 1971, ROUSSET-HALL and GOODAY 1975, RUIZ-HERRERA et al. 1977, RYDER and PEBERDY 1977b) but is partially or totally converted to a hyperbola by addition of the acetylamino sugar (ROUSSET-HALL and GOODAY 1975, RUIZ-HERRERA et al. 1977, RYDER and PEBERDY 1977b).

The physiological significance of the stimulation by *N*-acetylglucosamine or oligosaccharides is not well understood. One would not expect to find significant amounts of these as free sugars in the cell, but a high local concentration might be built up at the site of chitin formation, especially if some chitinase works in concert with the synthetase (see Sect. 5).

Since the synthetase is particulate (Sect. 4.2.3) the first step toward purification must be solubilization. Some enzyme was apparently brought into solution by GLASER and BROWN (1957) by the use of butanol, but digitonin proved to be much more effective with the synthetases from *Coprinus* (GOODAY and ROUSSET-HALL 1975) and from *Saccharomyces cerevisiae* (DURAN and CABIB 1978). After further purification by gel filtration, the synthetase from *S. cerevisiae* manifested a requirement for an acidic phospholipid, whereas the stimulation by free *N*-acetylglucosamine was greatly decreased with respect to the particulate preparation. Both solubilized enzymes, from *Coprinus* and *Saccharomyces,* showed no requirement for an added primer. Similar results were recently reported with the synthetase from *Candida albicans* (BRAUN and CALDERONE 1979).

In chitin synthesis, there is no clear indication of a lipid intermediate of the type involved in the formation of bacterial cell wall polysaccharides (WRIGHT and KANEGASAKI 1971) or of glycoproteins (PARODI and LELOIR 1979). Some stimulation of chitin synthetase of *Mucor* by a crude lipid extract from the

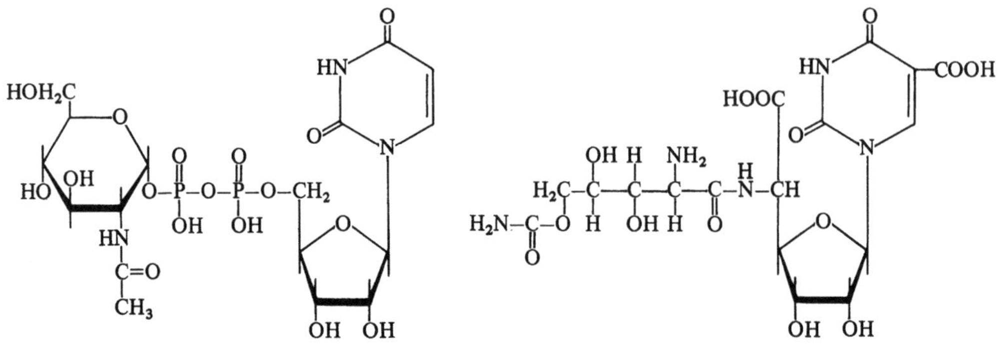

Fig. 4. Structures of polyoxin D and UDP-*N*-acetylglucosamine

same organism has been reported (MCMURROUGH and BARTNICKI-GARCIA 1971). Preliminary evidence on the involvement of a glycolipid with the properties of a glycosyldiglyceride in the formation of chitin in *Blastocladiella* has also been presented (MILLS and CANTINO 1978). Nevertheless, the crucial experiment, i.e., transfer of hexosamine from enzymatically formed putative intermediate to chitin, was not performed. On the other hand, addition of dolichyl phosphate to the solubilized yeast enzyme failed to enhance the activity and tunicamycin, a potent inhibitor of dolichyl pyrophosphoryl *N*-acetylglucosamine formation (LEHLE and TANNER 1976), was almost without effect (DURAN and CABIB 1978).

The polyoxins, a family of antibiotics from *Streptomyces cacoi* (ISONO et al. 1969) are very effective competitive inhibitors ($K_i \sim 10^{-6}$ M) of chitin synthetase (ENDO et al. 1970, KELLER and CABIB 1971, HORI et al. 1974a, b, c, JAN 1974, LOPEZ-ROMERO and RUIZ-HERRERA 1976). Their action appears to be due to their structural resemblance to the substrate UDP-GlcNAc (Fig. 4 and HORI et al. 1971, 1974a, b). Thus far, polyoxins appear to be quite specific for chitin synthetase, therefore they have been used to stop chitin synthesis either in vitro or in vivo and to study the resulting effects (ENDO et al. 1970, OHTA et al. 1970, BARTNICKI-GARCIA and LIPPMAN 1972, GOODAY 1972, BOWERS et al. 1974, HORI et al. 1974a, b, c, CABIB and BOWERS 1975, BENITEZ et al. 1976, GOODAY et al. 1976, WOOD and HAMMOND 1977). Polyoxins are also used in the field as fungicides (SUHADOLNIK 1970). One drawback to the use of these antibiotics in vivo is the great resistance manifested by some species, such as for instance *S. cerevisiae* (BOWERS et al. 1974), which has forced the use of very high concentrations of the inhibitor in the medium. A curious observation for which no explanation has been given is that peptides, when present in the medium, prevent the effect of the antibiotic on intact cells (MITANI and INOUE 1968, BOWERS et al. 1974).

The product of the synthetase reaction has been identified as chitin because of its content of *N*-acetylglucosamine, its insolubility and sensitivity to chitinase (KELLER and CABIB 1971). The chain length of the polysaccharide has not, however, been determined. The chitin synthesized by the solubilized synthetase from yeast has been observed in the electronmicroscope with negative staining

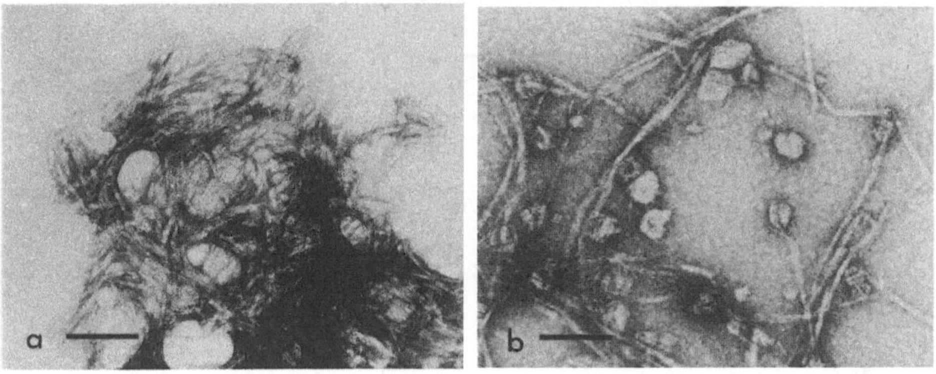

Fig. 5. a Chitin formed in vitro by solubilized chitin synthetase from *S. cerevisiae*. **b** Chitin fibrils formed in vitro by a chitosome preparation from *Mucor rouxii*. (Photograph kindly contributed by Dr. S. BARTNICKI-GARCIA, and reprinted with permission of the publisher.) *Bars*=0.1 μm

(DURAN and CABIB 1978) and found to consist of rod-like particles about 60 nm long and 9 nm wide (Fig. 5a). Intact membranes give rise to similar material (CABIB et al. 1979). On the other hand, a preparation of "chitosomes" (see Sect. 4.2.3) forms much longer fibrils (Fig. 5b). Thus, although an ordered arrangement of the chains can apparently arise even from a nonparticulate enzyme, the production of long fibrils as usually found in cell walls may require some organization of the synthetase in an organelle or membrane.

4.2.2 Regulation of Chitin Synthetase Activity

Modulation of chitin synthetase activity by divalent cations and *N*-acetylglucosamine or chito-oligosaccharides may be of some importance in vivo. Nevertheless, more far-reaching and permanent changes can be brought about by irreversible modification of the enzyme. In 1971, CABIB and FARKAS reported that the synthetase from *S. cerevisiae* or *S. carlsbergensis* could be prepared in an inactive form, which gave rise to active enzyme upon incubation either with an appropriate soluble fraction from the same yeast or with trypsin. It was postulated that the enzyme could exist in a zymogen or pro-enzyme form which could be converted into active enzyme, presumably by proteolytic action. These initial observations were later extended. The activating factor from yeast was identified as a proteinase (CABIB and ULANE 1973) which was subsequently purified to homogeneity (ULANE and CABIB 1976) and found to have identical properties to the so-called proteinase B which had been earlier detected in yeast (LENNEY and DALBEC 1967, HATA et al. 1967). A heat-stable proteinaceous inhibitor of proteinase B (LENNEY and DALBEC 1969) was also purified to homogeneity from *Saccharomyces* (CABIB and KELLER 1971, BETZ et al. 1974, ULANE and CABIB 1974) and recently sequenced (MAIER et al. 1979). The proteinase was found to be sequestered in the yeast cell vacuoles (CABIB et al. 1973, HASILIK et al. 1974, LENNEY et al. 1974), whereas the inhibitor appeared to be in soluble

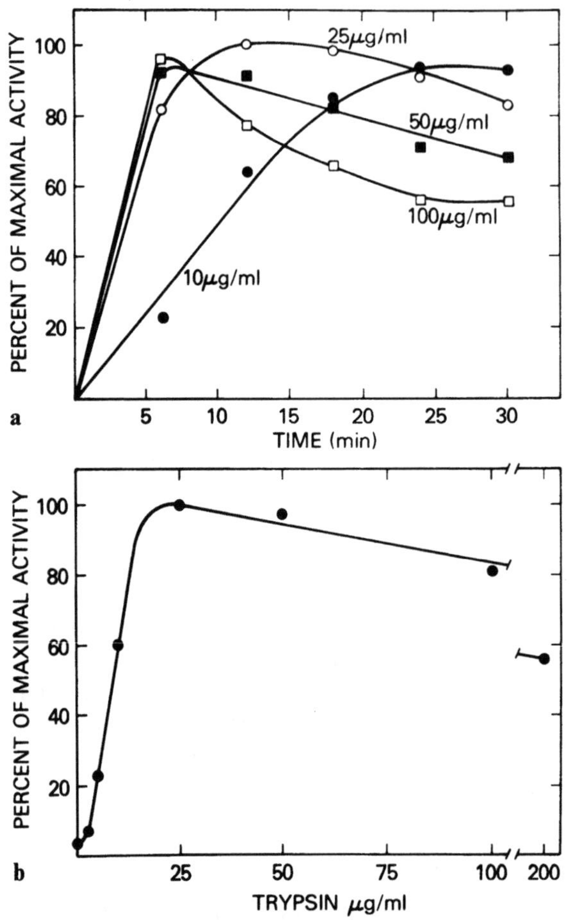

Fig. 6. Effect of trypsin concentration (**a**) and incubation time (**b**) on activation of chitin synthetase zymogen from *S. cerevisiae*. In **b**, the *numbers* on each line represent the trypsin concentration. (Reprinted from DURAN and CABIB 1978, with permission of the publisher)

form in the cytoplasm (CABIB and FARKAS 1971, HASILIK et al. 1974, LENNEY et al. 1974).

Proteolytic activation is not an exclusive property of the yeast enzyme. On the contrary, most of the synthetases from other fungi exhibit to a greater or lesser degree the same behavior (MCMURROUGH and BARTNICKI-GARCIA 1973, LOPEZ-ROMERO and RUIZ-HERRERA 1976, RUIZ-HERRERA and BARTNICKI-GARCIA 1976, ARCHER 1977, RUIZ-HERRERA et al. 1977, RYDER and PEBERDY 1977a, 1977b, BARTNICKI-GARCIA et al. 1978, BRAUN and CALDERONE 1978, ISAAC et al. 1978, VAN LAERE and CARLIER 1978, ARROYO-BEGOVICH and HERRERA 1979), with the possible exception of *Coprinus* (GOODAY 1979, see Sect. 6.3). Although commercial proteases have been used most often for convenience, in some cases proteases from the same organism have been found to function as activators of the homologous chitin synthetase (CABIB and FARKAS 1971, MCMURROUGH and BARTNICKI-GARCIA 1973, ULANE and CABIB 1976, CAMPBELL and PEBERDY 1979).

Chitin synthetase continues to manifest its zymogen characteristics, i.e., requires a proteinase for activation, after being solubilized (Fig. 6 and DURAN and CABIB 1978), an indication that the latent state of the enzyme is not caused by physical masking in a membrane. Therefore, direct activation of a zymogen, as observed in classical cases, still appears to be the simplest explanation for the action of proteases on chitin synthetase. However, other mechanisms are equally possible. If the inactive form consisted of an enzyme-inhibitor complex, the protease might act either by digesting the (proteinaceous) inhibitor or by splitting off that part of the enzyme that contains the binding site for the inhibitor, in this case not necessarily a protein. The latter type of mechanism seems to be involved in the activation of yeast glycogen synthetase, D form, by proteases (HUANG and CABIB 1974). It is unlikely that this problem can be solved until chitin synthetase is purified to homogeneity in reasonable amounts. In the meantime it is interesting to mention that a proteinaceous noncompetitive inhibitor of the synthetase was partially purified from *Mucor* (LOPEZ-ROMERO et al. 1978).

With the preparations that show proteolytic activation the basal activity of enzyme varies widely, from almost zero in yeast (DURAN and CABIB 1978) to about 60% in plasma membranes of *Schizophyllum* (VERMEULEN et al. 1979). According to HASILIK (1974), in *Saccharomyces* the enzyme from log phase cells exhibited about 5 to 10% of the maximal activity, whereas that from stationary phase was practically inactive. Great caution should be exercised in interpreting these results, because it is difficult to evaluate how much activation of the enzyme, by proteolysis or otherwise, might have occurred in vivo or during extraction. Furthermore, treatment with protease in large amounts and for a long time usually results in partial or total inactivation of the synthetase (HASILIK 1974, RUIZ-HERRERA and BARTNICKI-GARCIA 1976, DURAN and CABIB 1978). Clearly, post-translational modifications can greatly change the enzymatic activities. This fact complicates the evaluation and comparison of the different preparations described in the previous section (Sect. 4.2.1), since it is not known how much their properties might have been artifactually changed from those prevalent in vivo.

Despite the widespread occurrence of proteolytic activation among chitin synthetases, there are serious doubts about the physiological significance of this phenomenon. Recently, two different laboratories reported the isolation of mutants of *S. cerevisiae* that are defective in proteinase B (WOLF and EHMANN 1978, 1979, ZUBENKO et al. 1979). These mutants were not affected in their growth or in their ability to synthesize chitin in vivo. It is possible that in vivo activation of chitin synthetase results from the action of another, hitherto undetected, protease, or from an altogether different type of process.

4.2.3 Subcellular Distribution of Chitin Synthetase

In all cell-free extracts that catalyze the formation of chitin, the synthetase has been found to be in the particulate form. It is of obvious importance to determine to what organelle or membrane the enzyme is bound in order to understand how it functions in vivo. By coating yeast protoplasts with concanavalin A, a method successfully applied by SCARBOROUGH (1975) to *Neurospora*, DURAN et al. (1975) were able to reinforce the plasma membranes and subse-

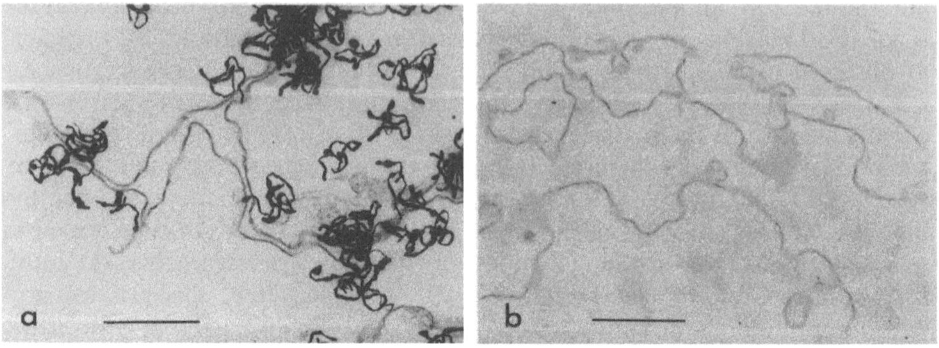

Fig. 7a, b. Autoradiograph of chitin synthesized by yeast plasma membrane from UDP-^{3}H-*N*-acetylglucosamine. **a** normal incubation mixture; **b** incubation in the presence of polyoxin D (120 μg/ml), to inhibit chitin synthesis. (DURAN et al. 1979). *Bars*=0.5 μm

quently isolate them in intact form by density-gradient centrifugation. They found almost all of the chitin synthetase of the lysate in the plasma membrane fraction, although a small amount sometimes appeared in a lighter fraction. BRAUN and CALDERONE (1978) reported a similar distribution for the synthetase from the dimorphic fungus *Candida albicans,* both in the yeast and in the mycelial phase. In contrast, SCHEKMAN and BRAWLEY (1979), who used a similar method but without pretreatment of the protoplasts, found only 60% of the synthetase activity in the plasma membrane fraction (as identified by ^{125}I-labeling), and the remainder in a lighter fraction. Somewhat similar results were reported by VERMEULEN et al. (1979) for the basidiomycete *Schizophyllum.* In this case, the plasma membrane fraction, which was obtained from concanavalin A-coated protoplasts, contained almost fully active enzyme, whereas in the lighter fraction the synthetase was recovered in the zymogen form.

The distribution of chitin synthetase on yeast plasma membranes previously treated with protease, was studied by an indirect method by DURAN et al. (1979). These authors determined the localization of nascent chains of the polysaccharide manufactured by the intact membrane preparation, either by fluorescence microscopy in the presence of Calcofluor White M2R or by electronmicroscope autoradiography (Fig. 7). In both cases the chitin was seen to emerge at many different locations and it was concluded that the synthetase was also widely disseminated on the membrane surface. With regard to the orientation of the enzyme in the plane of the membrane, it was shown that the synthetase was not affected by applying glutaraldehyde to the external face, i.e., to intact protoplasts, but was irreversibly inactivated when the treatment with glutaraldehyde was carried out after exposure of the inner face of the plasma membrane (DURAN et al. 1975). It was concluded that the synthetase faces the cytoplasm, a not unexpected finding, if one considers that the substrate UDP-GlcNAc is synthesized inside the cell.

A completely different approach to the determination of synthetase localization was taken by BARTNICKI-GARCIA, BRACKER and their co-workers (RUIZ-HERRERA et al. 1975, BRACKER et al. 1976, RUIZ-HERRERA et al. 1977, BARTNICKI-

GARCIA et al. 1978). Rather than resorting to gentle lysis of protoplasts in order to conserve the integrity of subcellular structures, these investigators used mechanical breakage of intact cells of several fungi with glass beads. By a series of subsequent steps, they isolated a fraction enriched for particles with a high content of chitin synthetase, in the zymogen state, which they name "chitosomes". When chitosomes, which are of roundish shape and 40 to 70 nm in diameter, are incubated in the presence of UDP-GlcNAc and protease, they give rise to chitin fibrils. Initially, the fibrils appear to form inside the particle; then the membrane surrounding the particle seems to open, and the fibrils grow and elongate into long strands. These findings were initially reported for *Mucor,* but later extended to many other organisms, including *S. cerevisiae* (BARTNICKI-GARCIA et al. 1978). Because of the drastic procedures employed in their isolation, it is difficult to decide whether chitosomes are autonomous organelles or small vesiculated fragments of plasma membrane. They might well be, however, a transport form of chitin synthetase from the site of enzyme formation to its final destination, presumably the plasma membrane.

5 Chitin Degradation

Although data on the possible turnover of chitin in vivo are lacking, there are reasons to believe that some degradation of the polysaccharide may be a normal feature of cell growth. One obvious case is branching in filamentous fungi, where some lysis of the preexisting polysaccharide may be required to allow branch emergence. Indeed, POLACHECK and ROSENBERGER (1978) have found chitinase in the cell wall of *Aspergillus nidulans*. Hydrolysis of chitin may also accompany its synthesis during normal wall growth. For instance, the polysaccharide chains formed by the synthetase may need to be cut after reaching a certain length.

In this respect, it is interesting to notice that chitinases appear to be much more active on nascent than on preformed chains of their substrate (MOLANO et al. 1979). Their activity should therefore be reassessed under conditions closer to those existing in vivo. The possible physiological function of these enzymes merits further scrutiny.

6 Regulation and Localization of Chitin Synthesis in Vivo

6.1 Introduction

As noted in Section 1, chitin is a major component of cell walls and septa. Its contribution to these structural cell components is interesting not only for the physiology of the individual organisms but also, more generally, as a model for morphogenetic processes. Here we will consider in some detail those systems involving chitin formation that have attracted a considerable number of studies in recent times, i.e., the formation of the primary septum in budding yeasts, the apical growth of hyphae and the formation of cell wall in *Blastocladiella* during the transition from the zoospore to the "round cell" stage.

6.2 Primary Septum Formation in Budding Yeasts

As mentioned in Section 3, the primary septum of *Saccharomyces* consists of chitin. The first appearance of septal material, as a ring at the base of an emerging bud, has been detected either by electronmicroscopy (MARCHANT and SMITH 1968) or by fluorescence microscopy (HAYASHIBE and KATOHDA 1973, SEICHERTOVA et al. 1973, CABIB and BOWERS 1975) in the presence of either primulin or Calcofluor White MR2 (Fig. 8a, b). In later stages, the septum grows concentrically and becomes a disk that seals the connection between mother and daughter cell (Fig. 8c, d). After the plasma membranes have been pinched off, new wall material is excreted from both sides, to form secondary septa (Fig. 8e), which contain both glucan (CABIB and BOWERS 1971) and mannan (BAUER et al. 1972). After cell division is completed, the chitinous primary septum remains in the bud scar, on the surface of the mother cell (Fig. 8f).

The validity of this scheme has been confirmed by the use of polyoxin D as an inhibitor of chitin synthesis in vivo (BOWERS et al. 1974, CABIB and BOWERS 1975). The antibiotic prevents formation of the initial chitin ring without apparently interfering with bud emergence and growth. The primary septum is not synthesized and, as a consequence, secondary septal material when formed at all has an aberrant aspect and is oriented parallel to the channel between mother and daughter cell rather than transversally to it. Finally, most cells burst at the junction, perhaps by the action of "runaway" glucanases, which under normal conditions presumably facilitate cell separation. Growth of a glucosamine auxotroph of *S. cerevisiae* in the absence of amino sugar leads to a somewhat similar morphology (BALLOU et al. 1977).

The formation of a chitin ring prior to septum completion was confirmed recently by VRSANSKA et al. (1979), who isolated virgin cells and allowed them to bud synchronously. Cell walls were isolated and digested with a chitinase-free lytic enzyme from *Arthrobacter*. The resistant material left after enzyme action on walls from early budding cells had the shape of a ring; in later stages the ring was filled in by the chitinous material.

From the scheme presented in Fig. 8 it is clear that the cell must be endowed with a mechanism that triggers chitin synthesis at a precise time in the cell cycle (bud emergence) and at a well-defined location in the cell (presumably the connection between mother cell and emerging bud). How is this process started? Because, in yeast, chitin synthetase appears to be distributed uniformly over the surface of the plasma membrane (Sect. 4.2.3) and is found mainly in the latent or zymogen state, it seems reasonable to assume that chitin synthesis is initiated by localized activation of the zymogen. It has been proposed that the activator may be proteinase B (CABIB and FARKAS 1971, CABIB et al. 1974, ULANE and CABIB 1976), which would be carried inside the vacuoles in which

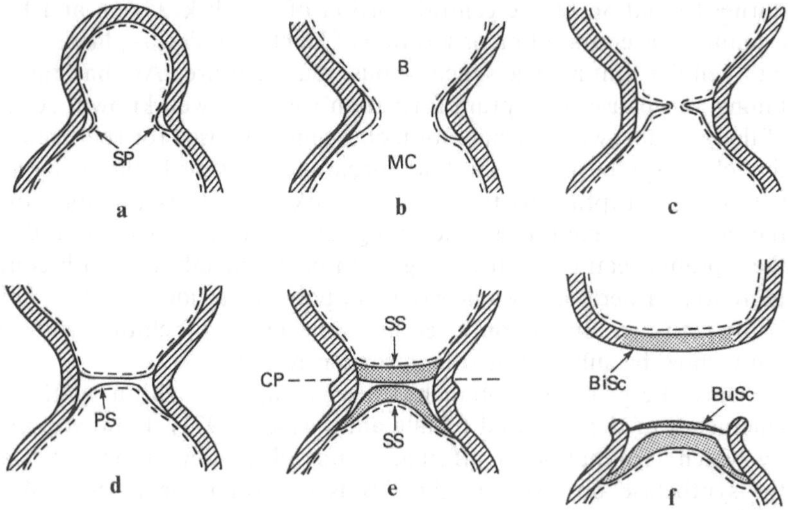

Fig. 8a–f. Scheme of septum formation in *S. cerevisiae*. *SP* septal primordia; *B* bud; *MC* mother cell; *PS* primary septum; *SS* secondary septa; *CP* cleavage plane; *BiSc* birth scar; *BuSc* bud scar

it has been found (see Sect. 4.2.2) to the sites where chitin synthesis must begin. However, the recent finding of mutants lacking proteinase B but showing normal growth and cell division (see Sect. 4.2.2) casts serious doubts on the function of this proteinase as the physiological activator. Nevertheless, the basic hypothesis on the mechanism of chitin synthesis initiation has not been invalidated. It seems very probable that the activator, whether it is another protease or has an altogether different nature, is carried in a particulate form to insure precise localization of the activation. In this respect, it is worthwhile noting that small vesicles have been observed in buds during the early stages of their formation (SENTANDREU and NORTHCOTE 1969).

In certain cases, a delocalization of chitin formation has been observed. Thus, in *S. cerevisiae* mutant *cdc* 24, which is temperature-sensitive for bud formation, incubation at the nonpermissive temperature results in deposition of chitin all over the wall of the much enlarged cells (SLOAT and PRINGLE 1978). Similarly, chitin synthesis continues when budding is inhibited in cells exposed to α mating factor. In these cells, the polysaccharide is found in a wide area in the region of the cell where "shmoo" formation takes place (SCHEKMAN and BRAWLEY 1979), and an increase in the active form of chitin synthetase is also detected. The mechanism underlying this behavior is not clear. It is possible, however, that when budding is artificially prevented, as in the two cases above, the system that normally directs the activator to the site of bud emergence becomes partially or totally disorganized and activation occurs in a wider area.

From the scheme presented in Fig. 8 it seems possible that chitin synthesis might occur in two stages or bursts, the first at bud initiation and the second during formation of the central portion of the disk. CABIB and FARKAS (1971) examined the course of incorporation of label, supplied as glucose in the medium, into cellular chitin in a synchronous yeast culture. At that time, the precise timing and phases of septum formation were not well known. A reexamination of these results, with consideration of the method used for cell counting, indicates that chitin synthesis starts at bud appearance or shortly thereafter and continues without interruption to the time of cytokinesis. It is possible, therefore, that the chitin ring continues thickening during cell growth and that closing of the septum is caused by inward growth of the membrane, with continued chitin excretion, rather than by a new burst of polysaccharide synthesis. This hypothesis is attractive in that it only requires one event of chitin synthesis initiation, but it must be substantiated by stronger evidence.

Once the primary septum has been completed, chitin synthesis stops until emergence of the next bud (CABIB and FARKAS 1971). However, the mechanism by which termination is affected is not clear. Based on the inactivation of the synthetase observed at high levels of proteinase B (Sect. 4.2.2), HASILIK (1974) proposed that the proteinase might be the terminating agent. Because the involvement of proteinase B in chitin synthesis now seems unlikely, this possibility appears remote.

6.3 Hyphal Growth

At the growing tip of a hypha, in at least some fungi, chitin appears to be the main component of the cell wall, whereas other polysaccharides are added later in subapical areas (HUNSLEY and BURNETT 1970, TRINCI 1978). Therefore, chitin probably plays a fundamental role in the ontogeny of the mycelial cell wall and it would be of great importance to determine how its synthesis is regulated. Unfortunately, little is known on this topic.

The active synthesis and accessibility of chitin at hyphal tips has been documented by autoradiography after administration of radioactive hexosamine (BARTNICKI-GARCIA and LIPPMAN 1969, 1977, GOODAY 1971, GALUN 1972, GALUN et al. 1976) and by observation of fluorescence in the presence of either brighteners (DARKEN 1961) or a fluorescein conjugate of wheat germ agglutinin (GALUN et al. 1976). The importance of chitin in cell wall maintenance was explored, as in yeast, by inhibiting chitin synthesis either with polyoxin D or by the use of a hexosamine auxotroph. Polyoxin D led in several cases either to weakening of the cell wall or to lysis (ENDO et al. 1970, BARTNICKI-GARCIA and LIPPMAN 1972, BENITEZ et al. 1976, GOODAY et al. 1976). Growth of a temperature-sensitive glucosamine-requiring *Aspergillus nidulans* mutant at the nonpermissive temperature in the absence of amino sugar results in lysis of the hyphae (KATZ and ROSENBERGER 1971a). Lysis can be prevented by inclusion of an osmotic stabilizer in the medium; chitin-free walls formed under such conditions withstand subsequent submersion into water. If, however, the mycelia are allowed to grow again in low osmolality medium, lysis takes place. The lysis occurs not only at growing tips, but also along the lateral portion of hyphae. This suggests an activation of wall lytic enzymes under the conditions of the experiment, but the interpretation is not clear.

With respect to the distribution of chitin synthetase, results in the literature are somewhat contradictory. As mentioned in Section 4.2.3, in *Candida albicans* the synthetase has been found in the plasma membrane fraction (BRAUN and CALDERONE 1978), whereas in *Schizophyllum commune* about half of it appears to be in some other particles (VERMEULEN et al. 1979). The percentage of basal activity (activity measured before proteolytic activation) also varies; in *Mucor rouxii* it tends to be higher in the mycelial than in the yeast form (RUIZ-HERRERA and BARTNICKI-GARCIA 1976). In *Schizophyllum* the plasma membrane-bound enzyme is almost fully active, whereas that found in lighter particles has a total requirement for proteolytic activation (VERMEULEN et al. 1979). Some authors (ARCHER 1977, ISAAC et al. 1978), using *Aspergillus* report that the membranes liberated during early stages of protoplast formation, i.e., originating in hyphal tips, have a higher basal activity than those arising from lateral walls, a result that would be in keeping with the synthesis of chitin at the hyphal apex. Nevertheless, VERMEULEN et al. (1979) were unable to confirm this finding.

An unusual situation is found in the enzyme from *Coprinus* stipes, which is completely active as obtained and is not affected by proteolysis (GOODAY and ROUSSET-HALL 1975, GOODAY 1979). It is interesting to note that in cells from stipes of another Basidiomycete, *Agaricus bisporus,* chitin synthesis takes place simultaneously all over the wall, as observed by autoradiography (CRAIG et al. 1979). In stipe cells, therefore, there may be no need for a zymogenic form of the synthetase.

The presence of a latent form of the enzyme under in vivo conditions may be inferred by the results of KATZ and ROSENBERGER (1970, 1971b). By the use of autoradiography, these authors found in *Aspergillus nidulans* that chitin deposition normally occurs only at hyphal tips. Treatment with cycloheximide or osmotic shock, however, results in labeling of chitin all along the hypha and in an increase in branching and septation. It seems probable that these effects were produced by activation of preexisting zymogen, especially because they were observed during inhibition of protein synthesis.

Despite all these observations, the regulation of chitin synthesis in growing hyphae is still not clear. It is possible that the enzyme (in chitosomes?) is delivered to the plasma membrane at the hyphal apex in latent form, then activated, and finally again deactivated as growth leaves it behind in subapical positions. There is, however, no clear evidence on this point. FARKAS (1979) takes the opposite position, i.e., that the enzyme reaches the growing tips in active form and is later inactivated by putative inhibitors that are somehow directed to subapical positions. Although this is an interesting hypothesis, there is no experimental basis for it. One note of caution should be introduced here. If the physiological regulation of chitin synthetase is not by a proteolytic process, the so-called "basal" and "total" activities of the enzyme (as measured before and after protease treatment, respectively) may not reflect precisely the situation in vivo. Therefore, interpretations based on this type of result may be quite erroneous.

6.4 Chitin Synthesis During Round Cell Formation in *Blastocladiella emersonii*

One of the stages in the development of the water mold, *Blastocladiella emersonii*, is the flagellate zoospore (LOVETT 1975). Zoospores may be kept indefinitely in a medium containing Ca^{2+} but, if transferred to a K^+ solution, rapidly undergo a series of irreversible changes, which result in the formation of germlings (LOVETT 1975). Some of these changes, i.e., retraction of the flagellum, rounding up of the cell and formation of a chitinous (see Sect. 3) cell wall can take place in the apparent absence of protein synthesis (LOVETT 1975). It seems, therefore, that activation of preexisting enzymes is involved in these events, including the formation of chitin. Indirect evidence of an activation of the hexosamine biosynthetic pathway during "round cell" formation was presented by SELITRENNIKOFF et al. (1976). These authors showed that the hexosamine present in the zoospore is only about 10% of that required for cell wall formation. They also found that the first enzyme of the hexosamine pathway (glutamine-fructose-6-phosphate amidotransferase, see Sect. 4.1) is, as in other organisms, subject to feedback inhibition by UDP-GlcNAc. UTP and UDP partially relieve the inhibition and were considered as possible effectors in the activation of the pathway. Other enzymes involved in the formation of UDP-GlcNAc, such as *N*-acetylglucosamine-phosphate mutase and UDP-GlcNAc pyrophosphorylase, also present in zoospores, are not inhibited by UDP-GlcNAc, a finding that supports the amidotransferase as the critical control point (SELITRENNIKOFF et al. 1976, SELITRENNIKOFF and SONNEBORN 1976a, b).

With respect to chitin synthetase itself, the enzyme appears to be present both in the zoospore and in the encysted cells. No obvious difference in its properties that depended on the state of development was observed (PLESSMANN CAMARGO et al. 1967) with the exception of the effect of Ca^{2+}, which stimulates the enzyme from cells but not that from spores. According to CANTINO and MYERS (1972, 1974) chitin synthetase is sequestered in the so-called γ-particles present in the zoospore stage and thereby isolated from its substrate. In the early phases of encystment γ-particles undergo great morphological changes and finally disappear. According to CANTINO's interpretation, the opening of the particles would bring about contact of the synthetase with UDP-GlcNAc and subsequent formation of chitin. The polysaccharide would be incorporated into the cell wall by fusion of the modified γ-particles with the plasma membrane and exocytosis. It should be kept in mind, however, that the exclusive localization of chitin synthetase in γ-particles has not been clearly established and that most of this scheme is in the speculative state. Clearly, more work is needed to understand the regulation of wall formation in *Blastocladiella*.

Acknowledgments. I am indebted to Drs. B. BOWERS, J. CORREA, P. MCPHIE, V. NOTARIO and R. ROBERTS for useful criticism. I also thank Drs. S. BARTNICKI-GARCIA and J.H. BURNETT for supplying electronmicrographs and Dr. B. BOWERS for help with the illustrations.

References

Archer DB (1977) Chitin biosynthesis in protoplasts and subcellular fractions of *Aspergillus fumigatus*. Biochem J 164:653–658

Arroyo-Begovich A, Herrera JR (1979) Proteolytic activation and inactivation of chitin synthase from *Neurospora crassa*. J Gen Microbiol 113:339–345

Ballou CE, Maitra SK, Walker JW, Whelan WL (1977) Developmental defects associated with glucosamine auxotrophy in *Saccharomyces cerevisiae*. Proc Natl Acad Sci USA 74:4351–4355

Bartnicki-García S (1968) Cell wall chemistry, morphogenesis and taxonomy of fungi. Annu Rev Microbiol 22:87–108

Bartnicki-García S, Lippman E (1969) Cell wall construction in *Mucor rouxii*. Science 165:302–304

Bartnicki-García S, Lippman E (1972) Inhibition of *Mucor rouxii* by polyoxin D: effects on chitin synthetase and morphological development. J Gen Microbiol 71:301–309

Bartnicki-García S, Lippman E (1977) Polarization of cell wall synthesis during spore-germination of *Mucor rouxii*. Exp Mycol 1:230–240

Bartnicki-García S, Bracker CE, Reyes E, Ruiz-Herrera J (1978) Isolation of chitosomes from taxonomically diverse fungi and synthesis of chitin microfibrils in vitro. Exp Mycol 2:173–192

Bauer H, Horisberger M, Bush DA, Sigarlakie E (1972) Mannan as a major component of the bud scars of *Saccharomyces cerevisiae*. Arch Mikrobiol 85:202–208

Benitez T, Villa TG, García Acha I (1976) Effect of polyoxin D on germination, morphological development and biosynthesis of the cell wall of *Trichoderma viride*. Arch Microbiol 108:183–188

Beran K, Holan Z, Baldrian J (1972) The chitin-glucan complex in *Saccharomyces cerevisiae* I. IR and X-ray observations. Folia Microbiol 17:322–330

Betz H, Hinze H, Holzer H (1974) Isolation and properties of two inhibitors of proteinase B from yeast. J Biol Chem 249:4515–4521

Bowers B, Levin G, Cabib E (1974) Effect of polyoxin D on chitin synthesis and septum formation in *Saccharomyces cerevisiae*. J Bacteriol 119:564–575

Bracker CE, Ruiz-Herrera J, Bartnicki-García S (1976) Structure and transformation of chitin synthetase particles (chitosomes) during microfibril synthesis *in vitro*. Proc Natl Acad Sci USA 73:4570–4574

Braun PC, Calderone RA (1978) Chitin synthesis in *Candida albicans*: comparison of yeast and hyphal forms. J Bacteriol 133:1472–1477

Braun PC, Calderone RA (1979) Regulation and solubilization of *Candida albicans* chitin synthetase. J Bacteriol 140:666–670

Cabib E, Bowers B (1971) Chitin and yeast budding. Localization of chitin in yeast bud scars. J Biol Chem 246:152–159

Cabib E, Bowers B (1975) Timing and function of chitin synthesis in yeast. J Bacteriol 124:1586–1593

Cabib E, Farkas V (1971) The control of morphogenesis: an enzymatic mechanism for the initiation of septum formation in yeast. Proc Natl Acad Sci USA 68:2052–2056

Cabib E, Keller FA (1971) Chitin and yeast budding. Allosteric inhibition of chitin synthetase by a heat-stable protein from yeast. J Biol Chem 246:167–173

Cabib E, Ulane R (1973) Chitin synthetase activating factor from yeast, a protease. Biochem Biophys Res Commun 50:186–191

Cabib E, Ulane R, Bowers B (1973) Yeast chitin synthetase. Separation of the zymogen from its activating factor and recovery of the latter in the vacuole fraction. J Biol Chem 248:1451–1458

Cabib E, Ulane R, Bowers B (1974) A molecular model for morphogenesis: the primary septum of yeast. In: Horecker BL, Stadtman ER (eds) Current topics in cellular regulation, vol VIII. Academic Press, London New York, pp 1–32

Cabib E, Durán A, Bowers B (1979) Localized activation of chitin synthetase in the initiation of yeast septum formation. In: Burnett JH, Trinci APJ (eds) Fungal walls and hyphal growth. Cambridge Univ Press, Cambridge, pp 189–201

Campbell JMcA, Peberdy JF (1979) Proteases of *Aspergillus nidulans* and the possible role of a neutral component in the activation of chitin synthase zymogen. FEMS Microbiol Lett 6:65–69
Cantino EC, Myers RB (1972) Concurrent effect of visible light on γ-particles, chitin synthetase, and encystment capacity in zoospores in *Blastocladiella emersonii*. Arch Mikrobiol 83:203–215
Cantino EC, Myers RB (1974) Basic mechanisms in plant morphogenesis. Brookhaven Symp Biol 25:51–74
Carlström D (1957) The crystal structure of α-chitin. J Biophys Biochem Cytol 3:669–683
Craig GD, Newsam RJ, Gull K, Wood DA (1979) An ultrastructural and autoradiographic study of stipe elongation in *Agaricus bisporus*. Protoplasma 98:15–29
Darken MA (1961) Application of fluorescent brighteners in biological techniques. Science 133:1704–1705
Davidson EA (1966) Metabolism of amino sugars. In: Balazs EA, Jeanloz RW (eds) The amino sugars, vol IIB. Academic Press, London New York, pp 1–44
Davidson EA, Blumenthal HJ, Roseman S (1957) Glucosamine metabolism II Studies on glucosamine 6-phosphate N-acetylase. J Biol Chem 226:125–133
Durán A, Cabib E (1978) Solubilization and partial purification of yeast chitin synthetase. Confirmation of the zymogenic nature of the enzyme. J Biol Chem 253:4419–4425
Durán A, Bowers B, Cabib E (1975) Chitin synthetase zymogen is attached to the yeast plasma membrane. Proc Natl Acad Sci USA 72:3952–3955
Durán A, Cabib E, Bowers B (1979) Chitin synthetase distribution on the yeast plasma membrane. Science 203:363–365
Endo A, Kakiki K, Misato T (1970) Mechanism of action of the antifungal agent polyoxin D. J Bacteriol 104:189–196
Farkas V (1979) Biosynthesis of cell walls of fungi. Microbiol Rev 43:117–144
Galun E (1972) Morphogenesis of *Trichoderma*: autoradiography of intact colonies labeled by [^{3}H]N-acetylglucosamine as a marker of new cell wall biosynthesis. Arch Mikrobiol 86:305–314
Galun M, Braun A, Frensdorff A, Galun E (1976) Hyphal walls of isolated lichen fungi. Autoradiographic localization of precursor incorporation and binding of fluorescein-conjugated lectins. Arch Microbiol 108:9–16
Ghosh S, Blumenthal HJ, Davidson E, Roseman S (1960) Glucosamine metabolism V. Enzymatic synthesis of glucosamine 6-phosphate. J Biol Chem 235:1265–1273
Glaser L, Brown DH (1957) The synthesis of chitin in cell-free extracts of *Neurospora crassa*. J Biol Chem 228:729–742
Gooday GW (1971) An autoradiographic study of hyphal growth of some fungi. J Gen Microbiol 67:125–133
Gooday GW (1972) The effect of polyoxin D on morphogenesis in *Coprinus cinereus*. Biochem J 129:17P
Gooday GW (1979) Chitin synthesis and differentiation in *Coprinus cinereus*. In: Burnett JH, Trinci APJ (eds) Fungal walls and hyphal growth. Cambridge Univ Press, Cambridge, pp 203–223
Gooday GW, Rousset-Hall A (1975) Properties of chitin synthetase from *Coprinus cinereus*. J Gen Microbiol 89:137–145
Gooday GW, Rousset-Hall A, Hunsley D (1976) Effect of polyoxin D on chitin synthesis in *Coprinus cinereus*. Trans Br Mycol Soc 67:193–200
Hackman RH, Goldberg M (1974) Light-scattering and infrared-spectrophotometric studies of chitin and chitin derivatives. Carbohydr Res 38:35–45
Hasilik A (1974) Inactivation of chitin synthase in *Saccharomyces cerevisiae*. Arch Microbiol 101:295–301
Hasilik A, Muller H, Holzer H (1974) Compartmentation of the tryptophan-synthase-proteolyzing system in *Saccharomyces cerevisiae*. Eur J Biochem 48:111–117
Hata T, Hayashi R, Doi E (1967) Purification of yeast proteinases. I. Fractionation and some properties of the proteinases. Agric Biol Chem 31:150–159
Hayashibe M, Katohda S (1973) Initiation of budding and chitin ring. J Gen Appl Microbiol 19:23–39

Hori M, Kakiki K, Suzuki S, Misato T (1971) Studies on the mode of action of polyoxins. Part III. Relation of polyoxin structure to chitin synthetase inhibition. Agric Biol Chem 35:1280–1291

Hori M, Eguchi J, Kakiki K, Misato T (1974a) Studies on the mode of action of polyoxins. VI. Effect of polyoxin B on chitin synthesis in polyoxin-sensitive and resistant strains of *Alternaria kikuchiana*. J Antibiot 27:260–266

Hori M, Kakiki K, Misato T (1974b) Further study on the relation of polyoxin structure to chitin synthetase inhibition. Agric Biol Chem 38:691–698

Hori M, Kakiki K, Misato T (1974c) Interaction between polyoxin and active center of chitin synthetase. Agric Biol Chem 38:699–705

Horisberger M, Vonlanthen M (1977) Location of mannan and chitin on thin sections of budding yeasts with gold markers. Arch Microbiol 115:1–7

Huang KP, Cabib E (1974) Yeast glycogen synthetase in the glucose 6-phosphate-dependent form. II. The effect of proteolysis. J Biol Chem 249:3858–3861

Hunsley D, Burnett JH (1970) The ultrastructural architecture of the walls of some hyphal fungi. J Gen Microbiol 62:203–218

Isaac S, Ryder NS, Peberdy JF (1978) Distribution and activation of chitin synthase in protoplast fractions released during the lytic digestion of *Aspergillus nidulans* hyphae. J Gen Microbiol 105:45–50

Isono K, Asahi K, Suzuki S (1969) Studies on polyoxins, antifungal antibiotics. XIII The structure of polyoxins. J Am Chem Soc 91:7490–7505

Jan YN (1974) Properties and cellular localization of chitin synthetase in *Phycomyces blakesleeanus* J Biol Chem 249:1973–1979

Jaworski EG, Wang LC, Carpenter WD (1965) Biosynthesis of chitin in cell-free extracts of *Venturia inaequalis*. Phytopathology 55:1309–1312

Katz D, Rosenberger RF (1970) A mutation in *Aspergillus nidulans* producing hyphal walls which lack chitin. Biochim Biophys Acta 208:452–460

Katz D, Rosenberger RF (1971a) Lysis of an *Aspergillus nidulans* mutant blocked in chitin synthesis and its relation to wall assembly and wall metabolism. Arch Mikrobiol 80:284–292

Katz D, Rosenberger RF (1971b) Hyphal wall synthesis in *Aspergillus nidulans*: effect of protein synthesis inhibition and osmotic shock on chitin insertion and morphogenesis. J Bacteriol 108:184–190

Keller FA, Cabib E (1971) Chitin and yeast budding. Properties of chitin synthetase from *Saccharomyces cerevisiae*. J Biol Chem 246:160–166

Kornfeld R (1967) Studies on L-glutamine D-fructose 6-phosphate amidotransferase. J Biol Chem 242:3135–3141

Kornfeld S, Kornfeld R, Neufeld EF, O'Brien PJ (1964) The feedback control of sugar nucleotide biosynthesis in liver. Proc Natl Acad Sci USA 52:371–379

Lehle L, Tanner W (1976) The specific site of tunicamycin inhibition in the formation of dolichol-bound N-acetylglucosamine derivatives. FEBS Lett 71:167–170

Leloir LF, Cardini CE (1953) The biosynthesis of glucosamine. Biochim Biophys Acta 12:15–22

Lenney JF, Dalbec JM (1967) Purification and properties of two proteinases from *Saccharomyces cerevisiae*. Arch Biochem Biophys 120:42–48

Lenney JF, Dalbec JM (1969) Yeast proteinase B: identification of the inactive form as an enzyme-inhibitor complex. Arch Biochem Biophys 129:407–409

Lenney JF, Matile P, Wiemken A, Schellenberg M, Meyer J (1974) Activities and cellular localization of yeast proteases and their inhibitors. Biochem Biophys Res Commun 60:1378–1383

López-Romero E, Ruiz-Herrera J (1976) Synthesis of chitin by particulate preparations from *Aspergillus flavus*. Antonie van Leeuwenhoek J Microbiol Serol 42:261–276

López-Romero E, Ruiz-Herrera J, Bartnicki-García S (1978) Purification and properties of an inhibitor protein of chitin synthetase from *Mucor rouxii*. Biochim Biophys Acta 525:338–345

Lovett JS (1975) Growth and differentiation of the water mold *Blastocladiella emersonii*: cytodifferentiation and the role of ribonucleic acid and protein synthesis. Bacteriol Rev 39:345–404

Maier K, Müller H, Tesch R, Trolp R, Witt I, Holzer H (1979) Primary structure of yeast proteinase B inhibitor 2. J Biol Chem 254:12555–12561
Marchant R, Smith DG (1968) Bud formation in *Saccharomyces cerevisiae* and a comparison with the mechanism of cell division in other yeasts. J Gen Microbiol 53:163–169
Mc Murrough I, Bartnicki-Garcia S (1971) Properties of a particulate chitin synthetase from *Mucor rouxii*. J Biol Chem 246:4008–4016
Mc Murrough I, Bartnicki-Garcia S (1973) Inhibition and activation of chitin synthesis by *Mucor rouxii* cell extracts. Arch Biochem Biophys 158:812–816
Mc Murrough I, Flores-Carreon A, Bartnicki-Garcia S (1971) Pathway of chitin synthesis and cellular localization of chitin synthetase in *Mucor rouxii*. J Biol Chem 246:3999–4007
Mills GL, Cantino EC (1978) The lipid composition of the *Blastocladiella emersonii* γ-particle and the function of γ-particle lipid in chitin formation. Exp Mycol 2:99–109
Minke R, Blackwell J (1978) The structure of α-chitin. J Mol Biol 120:167–181
Mitani M, Inoue Y (1968) Antagonists of antifungal substance polyoxin. J Antibiot 21:492–496
Molano J, Polachek I, Durán A, Cabib E (1979) An endochitinase from wheat germ. Activity on nascent and preformed chitin. J Biol Chem 254:4901–4907
Molano J, Bowers B, Cabib E (1980) Distribution of chitin in the yeast cell wall. An ultrastructural and chemical study. J Cell Biol 85:199–212
Moore PM, Peberdy JF (1975) Biosynthesis of chitin by particulate fractions from *Cunninghamella elegans*. Microbios 12:29–39
Moore PM, Peberdy JF (1976) A particulate chitin synthase from *Aspergillus flavus* Link: the properties, location and levels of activity in mycelium and regenerating protoplast preparation. Can J Microbiol 22:915–921
Moriguchi M, Yamamoto K, Kawai H, Tochikura T (1976) The partial purification of L-glutamine D-fructose 6-phosphate amidotransferase from baker's yeast. Agric Biol Chem 40:1655–1656
Muzzarelli RA (1977) Chitin. Pergamon Press, Oxford
Norman J, Giddings TH, Cantino EC (1975) Partial purification and properties of L-glutamine: D-fructose 6-phosphate amidotransferase from zoospores of *Blastocladiella emersonii*. Phytochemistry 14:1271–1274
Ohta N, Kakiki K, Misato T (1970) Studies on the mode of action of polyoxin D. II Effect of polyoxin D on the synthesis of fungal cell wall chitin. Agric Biol Chem 34:1224–1234
Parodi AJ, Leloir LF (1979) The role of lipid intermediates in the glycosylation of proteins in the eucaryotic cell. Biochim Biophys Acta 559:1–37
Peberdy JF, Moore PM (1975) Chitin synthase in *Mortierella vinacea*: properties, cellular location and synthesis in growing cultures. J Gen Microbiol 90:228–236
Plessmann Camargo E, Dietrich CP, Sonneborn D, Strominger JL (1967) Chitin synthesis in spores and vegetative cells of *Blastocladiella emersonii*. J Biol Chem 242:3121–3128
Polacheck I, Rosenberger RF (1978) Distribution of autolysins in hyphae of *Aspergillus nidulans*: evidence for a lipid mediated attachment to hyphal walls. J Bacteriol 135:741–747
Porter CA, Jaworski EG (1966) The synthesis of chitin by particulate preparations of *Allomyces macrogynus*. Biochemistry 5:1149–1154
Reissig JL (1956) Phosphoacetylglucosamine mutase of Neurospora. J Biol Chem 219:753–767
Rosenberger RF (1976) The cell wall. In: Smith JE, Berry DR (eds) The filamentous fungi, vol II. Edward Arnold, London, pp 328–344
Rousset-Hall A, Gooday GW (1975) A kinetic study of a solubilized chitin synthetase preparation from *Coprinus cinereus*. J Gen Microbiol 89:146–154
Ruiz-Herrera J, Bartnicki-García S (1976) Proteolytic activation and inactivation of chitin synthetase from *Mucor rouxii*. J Gen Microbiol 97:241–249
Ruiz-Herrera J, Sing VO, Van der Woude WJ, Bartnicki-García S (1975) Microfibril assembly by granules of chitin synthetase. Proc Natl Acad Sci USA 72:2706–2710
Ruiz-Herrera J, Lopez-Romero E, Bartnicki-García S (1977) Properties of chitin synthetase in isolated chitosomes from yeast cells of *Mucor rouxii*. J Biol Chem 252:3338–3343

Ryder NS, Peberdy JF (1977a) Chitin synthase activity in dormant conidia of *Aspergillus nidulans*. FEMS Microbiol Lett 2:199–201

Ryder NS, Peberdy JF (1977b) Chitin synthase in *Aspergillus nidulans*: properties and proteolytic activation. J Gen Microbiol 99:69–76

Scarborough GA (1975) Isolation and characterization of *Neurospora crassa* plasma membranes. J Biol Chem 250:1106–1111

Schekman R, Brawley V (1979) Localized deposition of chitin on the yeast cell surface in response to mating pheromone. Proc Natl Acad Sci USA 76:645–649

Seichertová O, Beran K, Holan Z, Pokorny V (1973) The chitin-glucan complex of *Saccharomyces cerevisiae*. II Location of the complex in the encircling region of the bud scar. Folia Microbiol 18:207–211

Selitrennikoff CP, Sonneborn DR (1976a) Posttranslational control of *de novo* cell wall formation during *Blastocladiella emersonii* zoospore germination. Dev Biol 54:37–51

Selitrennikoff CP, Sonneborn DR (1976b) The last two pathway-specific enzyme activities of hexosamine biosynthesis are present in *Blastocladiella emersonii* zoospores prior to germination. Biochim Biophys Acta 451:408–416

Selitrennikoff CP, Allin D, Sonneborn DR (1976) Chitin biosynthesis during *Blastocladiella* zoospore germination: evidence that the hexosamine biosynthetic pathway is post translationally activated during cell differentiation. Proc Natl Acad Sci USA 73:534–538

Sentandreu R, Northcote DH (1969) The formation of buds in yeast. J Gen Microbiol 55:393–398

Sloat BF, Pringle JT (1978) A mutant of yeast defective in cellular morphogenesis. Science 200:1171–1173

Suhadolnik RJ (1970) Nucleoside antibiotics. J Wiley, New York

Trinci APJ (1978) Wall and hyphal growth. Sci Prog (New Haven) 65:75–99

Ulane RE, Cabib E (1974) The activating system of chitin synthetase from *Saccharomyces cerevisiae*. Purification and properties of an inhibitor of the activating factor. J Biol Chem 249:3418–3422

Ulane RE, Cabib E (1976) The activating system of chitin synthetase from *Saccharomyces cerevisiae*. Purification and properties of the activating factor. J Biol Chem 251:3367–3374

Van der Valk P, Marchant R, Wessels JGH (1977) Ultrastructural localization of polysaccharides in the wall and septum of the basidiomycete *Schizophyllum commune*. Exp Mycol 1:69–82

Van Laere AJ, Carlier AR (1978) Synthesis and proteolytic activation of chitin synthetase in *Phycomyces blakesleeanus* Burgeff. Arch Microbiol 116:181–184

Vermeulen CA, Raeven MBJM, Wessels JGH (1979) Localization of chitin synthase activity in subcellular fractions of *Schizophyllum commune* protoplasts J Gen Microbiol 114:87–97

Vrsanska M, Kratky Z, Biely P, Machala S (1979) Chitin structures of the cell walls of synchronously grown virgin cells of *Saccharomyces cerevisiae*. Z Allg Mikrobiol 19:357–362

Wolf DH, Ehmann C (1978) Isolation of yeast mutants lacking proteinase B activity. FEBS Lett 92:121–124

Wolf DH, Ehmann C (1979) Studies on a proteinase B mutant of yeast. Eur J Biochem 98:375–384

Wood DA, Hammond JBW (1977) Inhibition of growth and development of *Agaricus bisporus* by polyoxin D. J Gen Microbiol 98:625–628

Wright A, Kanegasaki S (1971) Molecular aspects of lipopolysaccharides. Physiol Rev 51:748–784

Yamamoto K, Kawai H, Moriguchi M, Tochikura T (1976) Purification and characterization of yeast UDP-N-acetylglucosamine pyrophosphorylase. Agric Biol Chem 40:2275–2281

Zubenko GS, Mitchell AP, Jones EW (1979) Septum formation, cell division and sporulation in mutants of yeast deficient in proteinase B. Proc Natl Acad Sci USA 76:2395–2399

17 Fungal Glucans – Structure and Metabolism

G.H. FLEET and H.J. PHAFF

1 Introduction

Glucans constitute important structural or skeletal components of the cell envelope of yeasts and filamentous fungi. Isolation of individual glucan components and elucidation of their chemical structures have presented enormous difficulties. Early investigators generally subjected whole cells of baker's or brewer's yeast to more or less drastic treatments with alkali and acid to obtain cell wall glucan residues. It is now recognized that some glucans are extracted by these treatments and their contribution to the cell wall structure was overlooked. In addition, chemical degradation of some polysaccharides occurred as a result of heating with alkali and acids. Thirdly, as will be discussed later on, it was not recognized until the last decade that the alkali-insoluble glucan of baker's yeast actually consists of two different polysaccharides that are difficult to separate. As a consequence, the value of the early structural investigations on yeast "glucan" was greatly diminished.

With the advent of procedures for cell rupture and the preparation of purified cell walls, much progress has been made in the analysis of such walls. In the preparation of pure cell walls attention must be paid to the control of endogenous, wall-associated β-glucanases, which could cause autolysis and modify the chemical configuration significantly. Keeping the cell walls in Tris buffer at pH 8.5 during the preparative steps controls the action of endogenous glucanases (FLEET and PHAFF 1973). Another factor that influences cell wall composition significantly is the conditions under which the cells are grown, as well as culture age (TAYLOR and CAMERON 1973). This is especially striking in several dimorphic fungi that change from a mycelial to a yeast phase.

From purified cell walls a number of different glucans have been isolated and studied in considerable detail. Of the numerous species of yeast and filamentous fungi, relatively few have been studied in depth. Enzymatic evidence indicates both qualitative and quantitative differences in cell wall glucans among yeast species (TANAKA et al. 1966) and thus much work remains to be done in this area.

The present chapter deals with the various glucans that occur in the cell wall of yeasts and filamentous fungi, their chemical structure, enzymatic degradation, and biosynthesis. Previous reviews on these subjects are cited in the text that follows.

2 Glucans from Yeasts

2.1 Cell Envelope Glucans

The historical development of our knowledge of yeast cell walls has been given by PHAFF (1963, 1971). Most information has come from studies with *Saccharomyces cerevisiae,* although, more recently, other species have also been studied. The principal components of the cell walls of *Sacch. cerevisiae* and most other ascomycetous yeasts are the polysaccharides glucan and mannan which, together, account for approximately 80 to 90% of the dry weight of cell walls. Protein, chitin, phosphorus, and lipids are also part of the wall complex. The glucan component is responsible for the tensile strength, rigidity, and shape of the cell.

The reported glucan content of cell walls of *Sacch. cerevisiae* has ranged from 30 to 60% depending on the study (NORTHCOTE and HORNE 1952, ROELOFSEN 1953, MILL 1966, BOWDEN and HODGSON 1970, FLEET and MANNERS 1976). This large variation seems related to yeast cultural conditions (MCMURROUGH and ROSE 1967, RAMSAY and DOUGLAS 1979), methods of cell wall preparation, and glucan extraction and fractionation (BACON et al. 1969, FLEET and MANNERS 1976). It is now evident that the in situ glucan of *Sacch. cerevisiae* walls consists of at least three molecular types, namely an alkali-insoluble, acetic acid-insoluble $(1 \rightarrow 3)$-β-D-glucan, an alkali-soluble $(1 \rightarrow 3)$-β-D-glucan and an alkali-insoluble, acetic acid-soluble $(1 \rightarrow 6)$-β-D-glucan.

2.1.1 Structure

Early studies (reviewed by PHAFF 1963) focused on the alkali-insoluble, acetic acid-insoluble glucan. This glucan was prepared by hot alkali extraction of whole yeast cells, sometimes followed by hot acid extraction of the residue. The alkali treatment served to extract the mannan and the acid treatment to remove storage glycogen, leaving an insoluble residue of glucan which still retained the cell shape. Methylation and partial acid hydrolysis of these preparations established the predominance of $(1 \rightarrow 3)$-β-D-linked glucose residues (BELL and NORTHCOTE 1950, PEAT et al. 1958a, b). Later studies by MISAKI et al. (1968) showed this glucan to be branched and to contain a small proportion of unsubstituted $(1 \rightarrow 6)$-β-D-linked residues. MANNERS et al. (1973a) questioned the purity of these earlier glucan preparations and demonstrated that, as generally prepared, the alkali-insoluble residue was contaminated with a $(1 \rightarrow 6)$-β-linked glucan which could only be removed by exhaustive acetic acid extraction and specific enzymatic digestion steps. Their highly purified, alkali-insoluble, acetic acid-insoluble glucan had a degree of polymerization (DP) of ca. 1,500 and accounted for 85% of the original alkali-insoluble material. The molecule consisted exclusively of $(1 \rightarrow 3)$-β-D-linked glucose residues, 3% of which were branched through $(1 \rightarrow 6)$-β-interchain linkages.

The association of a $(1 \rightarrow 6)$-β-linked glucan with the cell wall of *Sacch. cerevisiae* was first reported by BACON et al. (1969). This glucan was obtained by acid extraction of isolated cell wall preparations but was not structurally studied. MANNERS et al. (1973b) isolated a $(1 \rightarrow 6)$-β-glucan from *Sacch. cerevisiae* by acetic acid extraction of the alkali-insoluble glucan. The $(1 \rightarrow 6)$-β-glucan (ca. 15% of the original alkali-insoluble material), which was water-soluble, was freed of glycogen contamination by iodine precipitation followed by amylase digestion. The purified glucan had a DP of 130–140 and was highly branched. About one half of the residues were linked through C-1 and C-6, about one quarter were triply linked through C-1, C-3 and C-6, and about 11% were linked through C-1 and C-3. MANNERS et al. (1974) have also noted alkali-insoluble, acid-soluble $(1 \rightarrow 6)$-β-D-glucans in *Kloeckera apiculata, Kluyveromyces fragilis,* and *Sacch. fermentati.* TANAKA et al. (1966) obtained evidence by enzymatic analysis that *Hansenula anomala* cell walls contain a high percentage of $(1 \rightarrow 6)$-β-linked glucans.

Over the years there have been several reports on the occurrence of an alkali-soluble glucan in *Sacch. cerevisiae* cell walls (ROELOFSEN 1953, EDDY and WOODHEAD 1968) and in some studies this glucan appeared to be complexed with the wall mannan component (KESSLER and NICKERSON 1959, KORN and NORTHCOTE 1960). FLEET and MANNERS (1976) obtained this glucan by extraction of isolated cell walls of *Sacch. cerevisiae* with 3% NaOH at 4° C and neutralization of the extract with acetic acid. On neutralization the glucan precipitated as a gel and could be separated from the soluble mannan by centrifugation. The alkali-soluble glucan, which represented about 20% of the cell wall dry weight, had a DP of about 1,500 and contained 80 to 85% $(1 \rightarrow 3)$-β-D-linked residues, 8 to 12% unsubstituted $(1 \rightarrow 6)$-β-D-linked residues and 3 to 4% branched residues linked through C-1, C-3, and C-6. This glucan molecule, after purification, contained about 15 to 25 mannose residues, that, on the basis of specific enzymatic degradation studies, appeared to be attached to the $(1 \rightarrow 6)$-β-D-linked portion of the glucan (FLEET and MANNERS 1977). After examining the hydrolytic action of specific $(1 \rightarrow 3)$-β-D- and $(1 \rightarrow 6)$-β-D-glucanases on isolated *Sacch. cerevisiae* cell walls, the latter authors suggested that the cell wall mannan is probably attached through the alkali-soluble glucan component, although, as yet, there has been no direct chemical evidence of mannose-glucose oligosaccharides in cell wall hydrolysates.

Combined enzymatic and electronmicroscopy studies suggest that the cell walls of *Sacch. cerevisiae* consist of an outer layer of mannan embedded in an amorphous matrix of alkali-soluble glucan under which lies the more rigid, fibrillar, alkali-insoluble $(1 \rightarrow 3)$-β-glucan (BACON et al. 1969, KOPECKA et al. 1974, FLEET and MANNERS 1976, 1977). On the other hand, the existence of a fibrillar network of β-glucan could not be detected by BOWDEN and HODGSON (1970) in either native cell walls or in cell walls after treatment with various agents.

The cell wall structure of *Candida albicans* has received some recent attention (CASSONE et al. 1979) because of the pathogenic nature of this yeast and its occasional increase in resistance to polyene antibiotics. Glucans account for about 75% of its cell carbohydrate. Three glucan types have been described (1) an alkali-soluble glucan, (2) an alkali-insoluble, acid-insoluble glucan, and (3) an alkali-insoluble, acid-soluble glucan. The relative proportions of these glucans in exponentially growing *C. albicans* cells were 5%, 20%, and 52% respectively, based on total cell carbohydrate. In starved cells the proportion of alkali-insoluble, acid-insoluble glucan is increased at the expense of the other two glucans. The alkali-soluble glucan contained 63 to 73% unsubstituted (1 → 6)-β-linkages. (BISHOP et al. 1960, YU et al. 1967). This molecule is branched through about 6% residues triply linked by C-1, C-3, and C-6 and there are about 16% of unsubstituted (1 → 3)-β-D-linkages. The other two glucans from *Candida albicans* have not been studied chemically, but enzymatic and cytochemical observations suggest that the alkali-insoluble, acid-insoluble glucan contains a predominance of (1 → 3)-β-linked residues (CHATTAWAY et al. 1976, POULAIN et al. 1978).

X-ray diffraction and infra red spectroscopy studies have revealed another type of alkali-soluble glucan in the cell walls of some species of *Cryptococcus, Endomyces,* and *Schizosaccharomyces* (KREGER 1954, BACON et al. 1968). Unlike the alkali-soluble glucans of *Sacch. cerevisiae* or *C. albicans* this glucan contains (1 → 3)-α-D-linked residues (pseudonigeran). In the case of *Schizosaccharomyces pombe* the alkali-soluble α-D-glucan comprises between 18 and 28% of the cell wall material, has a DP of about 200, is unbranched, and contains a small percentage of (1 → 4)-α-D-linked residues (BUSH et al. 1974, MANNERS and MEYER 1977). In addition to pseudonigeran, alkali-soluble β-D-glucans, an alkali-insoluble, acid-soluble glucan, and an alkali-insoluble, acid-insoluble glucan have been isolated from *S. pombe* cell walls; these show much the same basic structures as those from *Sacch. cerevisiae* (MANNERS and MEYER 1977).

Recently, a glucomannan complex was described from the cell walls of the pigmented yeast *Rhodotorula glutinis* (ARAI et al. 1978, ARAI and MURAO 1978). This polysaccharide appears to be responsible for the tensile strength and rigidity of the cell wall but further structural studies are required.

Although glycogen is usually considered primarily an intracellular storage polysaccharide (NORTHCOTE 1953, MUNDKUR 1960), the exclusive intracellular location has recently been questioned. Glycogen is present in yeast as an alkali-soluble and an acid-soluble fraction. EVANS and MANNERS (1971) recognized that acid-soluble glycogen preparations from baker's yeast may be contaminated with some of the cell wall (1 → 6)-β-glucan, and such contamination should be taken into consideration when interpreting studies on glycogen yields and chemical properties. In a comparison of glycogen isolated from intact cells and protoplasts of *Sacch. cerevisiae,* GUNJA-SMITH and SMITH (1974) demonstrated that the acid-soluble glycogen was located external to the cytoplasmic membrane in the periplasmic space. It was later suggested that this glycogen was associated with cell wall polysaccharides (GUNJA-SMITH et al. 1977). These workers have proposed the existence of two different pools of storage glycogen each having specific roles in cell development and differentiation.

Flocculation of *Sacch. carlsbergensis* and *Sacch. cerevisiae* cells has been correlated with the build-up of acid-soluble glycogen reserves. Glycogen-less mutants fail to flocculate and factors which induce glycogen degradation are accompanied by a loss in flocculence (PATEL and INGELDEW 1975a, b).

Some species of yeasts synthesize exocellular polysaccharides which give rise to capsules of varying thickness.

Extracellular amylose-like glucans are produced by *Cryptococcus laurentii* and most other *Cryptococcus* species when they are cultured in media where the pH drops to less than 3.0 (FODA and PHAFF 1970, 1978). The starches from these yeasts have been isolated and are linear amylose polymers of (1 → 4)-α-linked D-glucose residues with molecular weights of ca 10^6 (GORIN et al. 1966). The strain designated *Rhodotorula peneaus* that was used by these authors has been reclassified as *Cryptococcus laurentii* var. *flavescens.* A nonstarch, α-glucan has been isolated from culture fluids of *Cr. laurentii* grown at pH values above 5.0 (ABERCROMBIE et al. 1960) but this may be cell wall originated (1 → 3)-α-D-glucan (PHAFF 1971). Other yeast species which form starch-like compounds in acidic media include *Bullera alba, Candida curvata. C. humicola, Trichosporon cutaneum,* and *Lipomyces* spp. (LODDER 1970). Some species, such as *Cryptococcus macerans, Cr. informo-miniatus* (LODDER 1970) and *Phaffia rhodozyma* (MILLER et al. 1976), produce extracellular starches under conditions that are not dependent on a low pH of the medium. The chemical nature of these starches has not been examined.

It should be pointed out, however, that the capsular polysaccharides of yeasts are generally not glucans, but phosphorylated polysaccharides or heteropolysaccharides.

2.1.2 Degradation

Because of the role of glucans in maintaining structural strength and rigidity considerable attention has been given to enzymes which degrade these molecules. Such degradation may be achieved by (1) endogenous yeast glucanases, and (2) exogenous glucanases from microbial and other sources. The enzymatic degradation of yeast cell walls has been reviewed recently (PHAFF 1977, 1979).

2.1.2.1 Endogenous Yeast Glucanases

Types of Enzymes. The earliest evidence for the existence of glucanases in yeasts was the ability of some yeasts to grow on laminarin, a (1 → 3)-β-linked glucan, as carbon source (MORRIS 1955, CHESTERS and BULL 1963a, b).

An exo-(1 → 3)-β-D-glucanase was first isolated from intracellular extracts of *Sacch. cerevisiae* by BROCK (1965). This enzyme hydrolyzed laminarin; the (1 → 6)-β-D-glucan, pustulan; and p-nitrophenyl-β-D-glucoside (pNPG). A similar enzyme occurs in *Hansenula wingei* (BROCK 1964). Subsequently, exo-(1 → 3)-β-glucanases of high purity were isolated from extracts and culture fluids of *Sacch. cerevisiae, Kluyveromyces fragilis,* and *Hansenula anomala* (ABD-EL-AL and PHAFF 1968). These enzymes too were able to hydrolyze both (1 → 3)- and (1 → 6)-β-D-glucosidic linkages as well as p-nitrophenyl-β-D-glucoside. Such nonspecific exoglucanases have now been described in *Kluyveromyces lactis* (TINGLE and HALVORSON 1971), *Pichia polymorpha* (VILLA et al. 1975), in cell extracts of *Candida utilis* (NOTARIO et al. 1976), *Cryptococcus albidus* (NOTARIO et al. 1975), *Kluyveromyces aestuarii* (LACHANCE et al. 1977), *K. phaseolosporus* (VILLA et al. 1978), and in the fission yeast *Schizosaccharomyces versatilis* (FLEET and PHAFF 1975).

In addition to these nonspecific exo-β-glucanases, some yeasts, such as *Pichia polymorpha* and *C. utilis,* produce an exo-enzyme that is specific for (1 → 3)-β-D-linkages and that does not hydrolyze pustulan (VILLA et al. 1975, 1976). A third type of exo-β-glucanase, with hydrolytic activity on periodate-oxidized laminarin and pustulan but without activity on p-nitrophenyl-β-D-glucoside, was reported in *Kluyveromyces phaseolosporus* (VILLA et al. 1978) and in the extracellular broth of *Sacch. cerevisiae* (VILLA and PHAFF 1980). Finally, FARKAŠ et al. (1973) isolated still another exo-β-glucanase from *Sacch. cerevisiae* with hydrolytic activity on laminarin, periodate-oxidized laminarin, pustulan, and pNPG.

Endo-(1 → 3)-β-glucanases have been isolated from the culture supernatants, cell extracts and cell walls of numerous yeast species. Endo-(1 → 3)-β-D-glucanases in yeasts were first described by ABD-EL-AL and PHAFF (1969) in the bipolarly budding species *Hanseniaspora valbyensis* and *H. uvarum*. Later, FLEET and PHAFF (1974a) characterized an endo-(1 → 3)-β-glucanase from the cell walls of the fission yeast, *Schizosaccharomyces japonicus* var. *versatilis;* they gave qualitative evidence for the occurrence of similar enzymes in the walls of *Sacch. rosei, Kluyveromyces fragilis, Schiz. pombe, Schiz. octosporus, Schiz. malidevorans, Hansenula anomala,* and *Pichia pastoris.* Endo-(1 → 3)-β-glucanases have been demonstrated in *Sacch. cerevisiae* (MADDOX and HOUGH 1971, ARNOLD 1972, FARKAŠ et al. 1973), but enzymes from this yeast have not been isolated or characterized. More recently, endo-(1 → 3)-β-glucanases have been isolated from cell extracts and culture supernatants of *Pichia polymorpha* (VILLA et al. 1975), culture supernatants of *Candida utilis* (VILLA et al. 1979) and cell extracts of *Kluyveromyces phaseolosporus* (VILLA et al. 1978). A comparison of the properties of the endo-(1 → 3)-β-glucanases so far described shows two basic groups: those which produce principally oligosaccharides during laminarin hydrolysis and those which rapidly produce glucose and laminaribiose as end products of laminarin hydrolysis. The functional significance of the different glucanase types is not yet understood.

(1 → 6)-β-D-glucanase (ARNOLD 1972) and (1 → 3)-α-D-glucanase activities (MEYER and PHAFF 1977) have been found in some yeasts, but very little is known about these enzymes. Interestingly, (1 → 3)-α-glucanase activity has been found not only in some yeasts where (1 → 3)-α-glucan is a component of the cell wall (e.g., *Cryptococcus albidus*) but also in yeasts where this glucan is apparently absent from the wall structure (e.g., *Rhodotorula minuta*) (MEYER and PHAFF 1977).

Function and Regulation of Activity. Because of the potential of endogenous yeast glucanases to hydrolyze cell wall rigid components, much speculation has been given as to the function and control of these enzymes. Possible functions include cell wall expansion during budding, cell separation (bud abscission), the formation of conjugation tubes and cellular fusion during cell mating (BARTNICKI-GARCIA and MCMURROUGH 1971, CRANDALL et al. 1977), release of ascopores from asci, and cell autolysis (PHAFF 1977).

BROCK (1964) noted that the level of (1 → 3)-β-glucanase in cell extracts of conjugating cells of *H. wingei* was much higher than in extracts of vegetatively growing cells, and proposed that such an enzyme might be responsible for cell wall softening and degradation during cell conjugation and fusion. Similarly, sharp increases in the (1 → 3)-β-glucanase activity of *Schiz. versatilis* during the early stages of cell agglutination and conjugation were noted by FLEET and PHAFF (1975), supporting this postulated function.

Yeasts, such as *Kluyveromyces fragilis, Hansenula anomala, Hanseniaspora valbyensis,* and *Schiz. versatilis,* where the spores are rapidly liberated (dehisced) from the ascus on spore maturation, exhibit much higher levels of endogenous $(1 \rightarrow 3)$-β-glucanases than those species where the spores are liberated more slowly by a swelling process. This has led to the conclusion that endogenous $(1 \rightarrow 3)$-β-glucanases may be involved in spore release from the ascus (ABD-EL-AL and PHAFF 1968, 1969, FLEET and PHAFF 1975).

In budding yeasts, such as *Sacch. cerevisiae,* cell division involves modification of cell wall properties at the location of the new bud. The sudden extrusion of buds seems to depend on the local softening of the existing cell wall (BERAN 1968). Cytological studies with *Sacch. cerevisiae* have shown that the budding process is initiated by a localized vesiculation of the endoplasmic reticulum (MOOR 1967, MARCHANT and SMITH 1968, SENTANDREU and NORTHCOTE 1969). The small vesicles fuse with the cell wall at the site of the prospective bud. $(1 \rightarrow 3)$-β-Glucanase activity has been found associated with these vesicles (MATILE et al. 1971, CORTAT et al. 1972). Endogenous glucanase activity in actively budding cells is significantly higher than in nonbudding cells, supporting the concept that glucanases are involved in cell wall plasticizing and degradation during the budding process (MADDOX and HOUGH 1971, CORTAT et al. 1972).

The control and regulation of endogenous yeast glucanase activity is little understood. The plant hormone, auxin, causes cellular elongation in certain strains of *Sacch. ellipsoideus.* Since auxin does not affect sphaeroplasts of this yeast, the primary effect of this compound appears to be in the cell wall (YANAGISHIMA 1963, SHIMODA et al. 1967, SHIMODA and YANAGISHIMA 1971). Addition of a fungal exo-$(1 \rightarrow 3)$-β-D-glucanase to cultures of auxin-sensitive strains also caused cell elongation (SHIMODA and YANAGISHIMA 1968). Endogenous levels of $(1 \rightarrow 3)$-β-D-glucanases also were found to be higher in auxin-treated yeasts and inhibition of this glucanase activity with glucono-δ-lactone prevented cell elongation (SHIMODA and YANAGISHIMA 1971). It was concluded that auxin-induced cell wall expansion is due to the increased activity of wall-degrading enzymes such as $(1 \rightarrow 3)$-β-D-glucanases.

Uncontrolled glucanase activity and cell wall degradation is probably one cause of yeast autolysis. On isolation and subsequent incubation in buffer, yeast cell walls undergo self-degradation or autolysis. This is characterized by a breakdown of cell wall glucans and a solubilization of wall-associated $(1 \rightarrow 3)$-β-glucanases (BARRAS 1972, FLEET and PHAFF 1974a, KRÖNING and EGEL 1974). In *Schiz. versatilis* and *Schiz. pombe,* the glucanases responsible for this degradation have been isolated and extensively studied (FLEET and PHAFF 1974a, REICHELT 1978). In *Kluyveromyces phaseolosporus,* which contains two endo-$(1 \rightarrow 3)$-β-glucanases and two exo-β-glucanases (VILLA et al. 1978), both endoglucanases were lytic on yeast cell walls. In addition, an atypical exo-β-glucanase with activity on periodate-oxidized laminarin also produced lysis and the three enzymes together acted synergistically.

Vesiculation as described by CORTAT et al. (1972) is one means of compartmentalizing and localizing endogenous glucanase activity. However, numerous studies indicate that the bulk of yeast glucanase activity is external to the cytoplasmic membrane and is located in the periplasmic space and, further, may be tightly associated with its cell wall substrate (BARRAS 1972, CORTAT et al. 1972, FARKAŠ et al. 1973, FLEET and PHAFF 1974a, VILLA et al. 1975). It has been suggested that isosteric effectors (such as laminaribiose) may control this activity (FLEET and PHAFF 1974a). These last authors also suggested that endogenous glucanases may exist in an inactive or zymogen form and may require controlled activation by proteolysis as occurs in yeast chitin synthetase (CABIB et al. 1974). REICHELT (1978) has obtained evidence showing a proteolytic activation of wall-associated endo-(1→3)-β-D-glucanases of *Schiz. pombe*.

A number of recent studies have shown that certain yeasts (and perhaps many others) may contain a complex of three or four exo- and endo-(1→3)-β-glucanases, with the different glucanases possibly serving different hydrolytic functions in the cell cycle (see the preceding section for the nature of these enzymes). Although total glucanase activity is known to vary during the cell cycle, the contributions of the different glucanase types to this activity may also change, so that one form of control would operate at the transcriptional or biosynthetic level (NOTARIO et al. 1975, DEL REY et al. 1979).

In connection with specific cellular functions it is noteworthy that the purified endo-(1→3)-β-glucanase from *Sch. versatilis* is very lytic on isolated cell walls of yeasts, thereby suggesting an in vivo wall-degradative function (FLEET and PHAFF 1974a). In contrast, the nonspecific exo-(1→3)-β-glucanase from this yeast shows negligible hydrolytic activity on the β-glucans of the cell wall of this yeast (FLEET and PHAFF 1975) and this enzyme may only have an auxiliary function by acting on products of the endoglucanase action. The discovery by SANTOS et al. (1979b) of a mutant strain of *Sacch. cerevisiae* that was defective in nonspecific exo-(1→3)-β-glucanase production, but grew normally, suggests that this type of enzyme is not essential for cell wall modification during the events of morphogenesis and growth. The fact that no exo-β-glucanase activity could be detected in *Schiz. pombe* lends support for this nonessential role (REICHELT 1978). This species, however, produces two endo-(1→3)-β-glucanases; but only one of these enzymes was found to be lytic on isolated cell walls. The lytic glucanase was not apparent in cells dividing vegetatively, but evidently was produced only during ascosporulation where ascus walls undergo lysis (REICHELT 1978). In *Kluyveromyces phaseolosporus* two lytic endo-(1→3)-β-glucanases were found together with a nonlytic, nonspecific exo-β-glucanase and a lytic exo-β-glucanase. The last enzyme differs from the nonlytic exoglucanase by its ability to hydrolyze periodate-oxidized laminarin and inability to hydrolyze pNPG (VILLA et al. 1978). The functional significance of the lytic exoglucanase in the cell cycle events is not clear. Further mutational studies similar to those reported by SANTOS et al. (1979b) may help to resolve this question.

2.1.2.2 Exogenous Glucanases

Research on the enzymatic digestion of yeast cell wall glucans by exogenous glucanases has been stimulated for three main reasons; (a) the preparation of protoplasts or spheroplasts for physiological studies; (b) studies on the biosynthesis and regeneration of cell walls from protoplasts; and (c) analysis of native and isolated cell walls and cell wall glucans. Wall and glucan analysis with selected and specific enzymes avoids the often used harsh, analytical treatments with strong acids and alkalis, but great care must be taken of using only highly purified enzyme preparations to avoid false conclusions (MARSHALL 1974).

Early observations showed that yeast cell walls were degraded by the complex of enzymes found in the digestive juices of the snail, *Helix pomatia*, and that this complex of enzymes could be used to prepare yeast protoplasts (BROWN 1971; (see also reviews by PHAFF 1963, 1971, 1977). Snail digestive fluid contains some 30 enzymes including β-glucanases. Chromatographic fractionation and testing of the fractions clearly showed that the yeast cell wall digestive properties resided in those fractions containing endo-(1→3)-β-glucanase and endo-(1→6)-β-glucanase activities (ANDERSON and MILLBANK 1966).

Yeast cell wall hydrolytic enzymes are also found in the extracellular culture fluids of numerous bacterial and fungal species (PHAFF 1971, 1977). Such enzymes are produced by these organisms when grown in a medium containing autoclaved yeasts or isolated yeast cell walls as sole carbon source. The complex of enzymes found in these culture fluids is capable of producing protoplasts when incubated with the respective whole yeast cells and of complete lysis of isolated yeast cell walls. Biochemical studies have identified various β-glucanases as the main hydrolytic enzymes within these culture fluids. PHAFF (1971, 1977) has reviewed the many microbial species that produce wall hydrolytic enzymes. Some of the more comprehensively studied organisms include *Bacillus circulans* (TANAKA and PHAFF 1965, TANAKA et al. 1966, HORIKOSHI 1973, FLEET and PHAFF 1974b, ROMBOUTS and PHAFF 1976a, b, TANAKA et al. 1978). *Cytophaga johnsonii* (BACON et al. 1970, BACON 1973), *Arthrobacter* spp. (KITAMURA et al. 1972, KITAMURA and YAMAMOTO 1972, DOI et al. 1976), *Streptomyces* (ELORZA et al. 1966), *Rhizopus chinensis* (YAMAMOTO and NAGASAKI 1975), and *Oerskovia* spp. (MANN et al. 1972, OBATA et al. 1977).

At least five endo-β-glucanases were found in the culture fluids of *Bacillus circulans* grown on baker's yeast cell walls, four of which have been extensively characterized. The predominant enzymes, an endo-(1→3)-β-glucanase and an endo-(1→6)-β-glucanase, produce typical oligosaccharide products upon hydrolysis of their respective substrates, laminarin and pustulan, but show negligible activity on isolated yeast cell walls and their glucans (FLEET and PHAFF 1974b, ROMBOUTS et al. 1978). The remaining two endo-(1→3)-β-glucanases and an endo-(1→6)-β-glucanase also show typical endo-cleavage of their substrates but they are very lytic on isolated yeast cell walls (ROMBOUTS and PHAFF 1976a, b). The lytic endo-(1→6)-β-glucanase appears capable of also hydrolyzing the (1→3)-β bonds surrounding a (1→6)-β-linked branch point. This debranching ability may explain its unusual lytic activity (ROMBOUTS et al. 1978, VILLA et al. 1977).

Upon chromatographic fractionation, the lytic culture fluids of *Cytophaga johnsonii* show the presence of two types of endo-(1→3)-β-glucanases and several (1→6)-β-glucanases (BACON et al. 1970). The predominant enzyme, an endo-

(1 → 3)-β-glucanase, showed typical endohydrolysis of laminarin but had insignificant activity on cell walls of *Sacch. cerevisiae*. However, the other endo-(1 → 3)-β-glucanase, which was present in the culture fluid in only small amounts, caused extensive solubilization of yeast cell walls and glucans, producing as end products mainly oligosaccharides with five or more residues. Laminarin hydrolysis by this enzyme was also characterized by the production of pentasaccharide units (Bacon et al. 1970). Lytic enzymes producing mainly pentasaccharide units as hydrolysis products of yeast cell walls and wall glucans have also been described in *Arthrobacter* species (Kitamura and Yamamoto 1972, Doi et al. 1973) and in *Rhizopus* (Yamamoto et al. 1972). These organisms simultaneously produce high amounts of other (1 → 3)-β-glucanases (which lack extensive hydrolytic acitivity on yeast walls) and, therefore, pentasaccharides are not found in the culture fluid of such organisms grown on cell walls.

An exo-(1 → 3)-β-glucanase excreted by the basidiomycetous fungus designated as QM 806 has the ability to produce protoplasts from several yeasts (Bauer et al. 1972). The purification (Huotari et al. 1968) action pattern and specificity (Nelson et al. 1969) and hydrolytic mechanism (Nelson 1970) of this enzyme have been thoroughly studied. Cell walls of *Sacch. cerevisiae* and *Wickerhamia fluorescence* are solubilized by the action of this enzyme (Bauer et al. 1972). The ability of this enzyme to by-pass (1 → 6)-β-linked side chains or branch points in (1 → 3)-β-linked glucans may be related to its lytic properties. We wish to point out, however, that in our opinion the enzyme preparation used by Bauer et al. (1972) for the preparation of protoplasts was not tested sufficiently for homogeneity, and thus the possible presence of traces of a lytic (1 → 3)-β-glucanase contaminant working synergistically with the exo-β-glucanase cannot be definitely excluded (Fleet and Phaff 1975).

It is evident that the predominant enzymes required for yeast cell wall solubilization are (1 → 3)-β-glucanases – an observation that reinforces the role of the (1 → 3)-β-linkage in maintaining cell rigidity and integrity. However, enzymes which show very high (1 → 3)-β-glucanase activity as usually measured with laminarin as substrate are not necessarily endowed with lytic activity on yeast cell walls or on wall (1 → 3)-β-glucans. The reasons why some endo-(1 → 3)-β-glucanases and endo-(1 → 6)-β-glucanases are lytic while others are nonlytic are not fully understood. This difference may relate to enzyme affinity for substrates of various size, degree of branching of substrates, and the ability to cleave residues surrounding branch points, and to the tertiary structure or conformation of the polysaccharide molecule (Fleet and Manners 1977, Phaff 1977, Rombouts et al. 1978).

As indicated by the enzymic complexity of microbial culture fluids, complete hydrolysis of yeast cell walls and wall glucans by exogenous enzymes appears greatly facilitated by the synergistic action of a number of different glucanases (Rombouts and Phaff 1976b, Tanaka et al. 1978, Obata et al. 1977), thereby underscoring the overall complexity of yeast cell wall structure. Selective and limited hydrolysis of isolated cell walls and wall glucan components with specific, purified, exogenous enzymes can contribute significantly to an understanding of this structure (Kopecká et al. 1974, Fleet and Manners 1977).

3 Glucans from Filamentous Fungi

Fungi may produce glucans as part of their cell walls, as extracellular products that are often loosely associated with the cell wall, and as cytoplasmic reserve materials. Glucans have been found in many fungal species and it is now recognized that the presence and structure of these glucans may vary with the environmental conditions of growth, the stage of the fungal life cycle, and the species (BARTNICKI-GARCIA 1973). Many of the fungal glucans described have not received the detailed chemical study as has been done with *Sacch. cerevisiae* glucans and, furthermore, structures have often been reported for preparations whose homogeneity has not been adequately defined. Early observations on the occurrence and structure of fungal glucans have been reviewed by BARTNICKI-GARCIA (1968) and GORIN and SPENCER (1968).

3.1 Structure

3.1.1 β-Linked Glucans

(1→4)-β-Glucan. Cellulose is a linear polysaccharide composed of (1→4)-β-linked glucose residues which may achieve chain lengths of up to 400 residues (ARONSON 1965). The techniques of X-ray diffraction, solubility in aqueous cupra-ammonium hydroxide (Schweizer reagent), and susceptibility to the action of cellulase have been used to identify cellulose as a component in the cell walls of numerous fungal species. Fungi containing cellulose in their walls have been cited by ARONSON (1965), BARTNICKI-GARCIA (1968) and GORIN and SPENCER (1968), and include species of *Phytophthora, Pythium, Saprolegnia, Sapromyces,* and *Dictyostelium.* More recently, methylation studies have confirmed the presence of (1→4)-β-linked glucose residues in the cell walls of *Phytophthora cinnamomi* (ZEVENHUIZEN and BARTNICKI-GARCIA 1969), *Phytophthora palmivora* (TOKUNAGA and BARTNICKI-GARCIA 1971), Basidiomycete QM 806 (BUSH and HORISBERGER 1972) and *Pythium acanthicum* (SIETSMA et al. 1975).

In the case of the slime mold *Dictyostelium discoideum,* cellulose is formed only during the fruiting stage and becomes part of the sheath and spore walls. Both alkali-soluble and alkali-insoluble forms of cellulose have been described in this species (HEMMES et al. 1972, ROSNESS and WRIGHT 1974) but, as with other fungal species, a structural analysis of these celluloses has not been presented.

(1→3)-β-Linked Glucans. The chemistry of glucans containing (1→3)-β-linkages has been reviewed by CLARKE and STONE (1963), BULL and CHESTERS (1966), and BARRAS et al. (1969). Although the (1→3)-β-glucosidic linkage occurs widely in fungi, it is seldom present as linear, homopolymer type glucans, and the molecules are generally branched through (1→6)-β-glucosidic bonds.

Poria cocos, a tree rot fungus, has sclerotia composed of pachyman, an alkali-soluble, water-insoluble (1→3)-β-glucan of approximately 250 to 700 glucose residues and three to six branch points per molecule (SAITO et al. 1968, HOFFMANN et al. 1971). Cytoplasmic reserves of a water-soluble, (1 → 3)-β-linked glucan, termed mycolaminaran, are found in *Phytophthora cinnamomi* and *P. palmivora*. These glucans contain only 30 to 40 glucose residues with one or two branch points per molecule (ZEVENHUIZEN and BARTNICKI-GARCIA 1970, WANG and BARTNICKI-GARCIA 1972, 1974).

A novel type of (1 → 3)-β-glucan is found in the extracellular culture fluids of *Claviceps purpurea* (PERLIN and TABER 1963), *Sclerotium rolfsii* (JOHNSON et al. 1963), *Pullularia pullulans* (BOUVENG et al. 1963), *Claviceps fusiformis* (BUCK et al. 1968), *Schizophyllum commune* (WESSELS et al. 1972, SIETSMA and WESSELS 1977) and *Monilinia fructigena* (SANTAMARIA et al. 1978). In each case the glucan consists of a main chain of (1 → 3)-β-linked residues, to which single glucose residues are attached by (1 → 6)-β-bonds. Depending upon the fungal species and stage of the life cycle, the side groups are regularly attached to either every second, every third, or every fourth glucose residue and the DP of these glucans may range from 10 to 400. In the species of *Claviceps, Schizophyllum,* and *Monilinia,* the extent of glucan production decreases with culture age and it has been suggested that this glucan serves as an extracellular reserve, being remetabolized upon depletion of nutrient levels in the growth medium.

(1 → 6)-β-Linked Glucans. Predominantly (1 → 6)-β-linked glucans, such as those found in the cell walls of *Sacch. cerevisiae* (MANNERS et al. 1973b), have not been described in fungi. However, systematic studies for the presence of such glucans in fungi have not been conducted.

Luteic acid and islandic acid are water-soluble, extracellular, acidic polysaccharides produced by *Penicillium luteum* and *P. islandicum,* respectively. These linear molecules consist of 80 to 90 glucose residues connected by (1 → 6)-β-bonds. Malonic acid is attached to the polysaccharide as a hemiester at a frequency of about one malonic acid residue to every one or two glucose residues (NAKAMURA and TANABE 1963, EBERT and ZENK 1967). A linear, water-soluble (1 → 6)-β-D-glucan (DP 120), termed pustulan, can be extracted from the lichen *Umbilicaria pustulata* and other *Umbilicaria* species (HELLERQVIST et al. 1968, NISHIKAWA et al. 1970).

Mixed (1 → 3)-β- and (1 → 6)-β-Linked Glucans. Glucans containing blocks of unsubstituted (1 → 3)-β- and (1 → 6)-β-linked residues, as well as glucose residues branched through carbon atoms 1, 3, and 6, are found in the cell walls of numerous fungal species (BARTNICKI-GARCIA 1968, GORIN and SPENCER 1968). Both alkali-soluble and alkali-insoluble (1 → 3)-(1 → 6)-β-glucans have been isolated, and in a few cases some structural features have been determined.

The alkali-insoluble glucan of *Phytophthora cinnamomi* is predominantly (1 → 3)-β-linked but is highly branched through residues triply bonded at carbons 1, 3, and 6 (ZEVENHUIZEN and BARTNICKI-GARCIA 1969). The alkali-insoluble glucan (termed R-glucan) from *Schizophyllum commune* represents about 55 to 60% of the cell wall and is closely associated with the wall chitin. This glucan is also highly branched and contains approximately equal proportions of (1 → 3)-β- and (1 → 6)-β-linked residues (WESSELS et al. 1972, SIETSMA and WES-

SELS 1977). In *Pythium acanthicum,* the alkali-insoluble glucan comprises about 70% of the cell wall. Combined chemical and enzymatic studies have revealed a predominantly (1→3)-β-linked main chain occasionally substituted in the C-6 position with very short side chains. About 17% of the glucose residues are triply linked through carbons 1, 3, and 6 and there are about 4% of linear (1→6)-β-linked residues (SIETSMA et al. 1975). Alkali-insoluble glucans with similar structures have been examined from the cell walls of *Piricularia oryzae* and *Tremella fuciformis.* The polysaccharide from *P. oryzae* has a molecular weight of 38,500 and comprises 60% of the cell wall (NAKAJIMA et al. 1970, 1972). For *T. fuciformis,* the glucan may be covalently linked to a glucurono-xylo-mannan (SONE and MISAKI 1977). The alkali-insoluble glucan extracted from the cell walls of chlamydospores from *Aureobasidium pullulans* differs from the above glucans in that there is a higher proportion of (1→6)-β-linkages compared to (1→3)-β-linkages (BROWN et al. 1973, DOMINGUEZ et al. 1978).

Although not well characterized, the (1→3)-β-glucans reported in *Aspergillus nidulans* (BULL 1970, ZONNEVELD 1971), *Neurospora crassa* (MAHADEVAN and TATUM 1965), *Penicillium chrysogenum* (TROY and KOFFLER 1969), *Paracoccidioides brasiliensis* (KANETSUNA and CARBONELL 1970), *Blastomyces dermatitides* (KANETSUNA and CARBONELL 1971) and *Histoplasma* species (SAN-BLAS and CARBONELL 1974) probably contain mixed (1→3)-β- and (1→6)-β-linkages. Yeast-mycelial dimorphism occurs in the last three genera and higher β-glucan levels are usually associated with the cell walls of the mycelial form (KANETSUNA et al. 1972).

Alkali-soluble glucans containing both (1→3)-β- and (1→6)-β-linked residues have been isolated from the fruiting bodies of *Polyporus fomentarius* and *P. igniarius* (BJORNDAL and LINDBERG 1970) and the mycelium of Basidiomycete QM 806 (BUSH and HORISBERGER 1972). The glucans of *Polyporus* species contain a greater preponderance of (1→6)-β-linkages; and chains of four to five (1→4)-β-linked glucuronic acid residues are β-linked to the 3-position of some glucose residues in the backbone. This basidiomycete glucan is branched and contains (1→4)-β-linked residues in addition to the (1→3)- and (1→6)-β-linkages.

3.1.2 α-Linked Glucans

Starch and Glycogen. As determined by iodine staining, glycogen is found in all and starch in some fungal species; these polysaccharides are thought to act as energy reserve materials. GORIN and SPENCER (1968) have summarized these early observations on starches in *Aspergillus* and *Penicillium* species and ZALOKAR (1959) of glycogens in *Neurospora crassa* and BHAVANANDAN et al. (1964) in *Polyporus giganteus.* The glycogen of *P. giganteus* has a unit chain length of eight to nine residues. Glycogen constitutes about 35% of the total carbohydrate in aggregated cells of *Dictyostelium discoideum* but is rapidly degraded during subsequent cellular differentiation (ROSNESS and WRIGHT 1974).

Pullulan. Aureobasidium pullulans (formerly *Pullularia pullulans*) produces an abundance of a viscous, water-soluble, extracellular polysaccharide when grown in the presence of a suitable carbohydrate source (YUEN 1974, JEANES 1977).

Because of potential industrial applications, the chemistry and biochemistry of this polysaccharide, commonly referred to as pullulan, has recently been reviewed in detail (CATLEY 1979). Pullulan is a linear molecule containing mainly maltotriose residues connected by (1 → 6)-α-linkages. Its molecular weight varies between 10^5 and 10^6 depending upon the strain of the fungus and stage of its life cycle. *A. pullulans* shows yeast-mycelial dimorphism with polysaccharide production being restricted to the yeast phase. Nitrogen limitation in the growth medium triggers pullulan production (CATLEY and KELLY 1975).

Mycodextran. Mycodextran, or nigeran, is a linear molecule with alternating (1 → 3)-α- and (1 → 4)-α-linked D-glucopyranosyl residues. Its production has been associated with a number of *Aspergillus* and *Penicillium* species (GORIN and SPENCER 1968). It forms about 5% of the cell wall of *A. niger* from which it may be recovered by hot water extraction. However, growth conditions and culture age may affect this percentage in the cell wall (JOHNSTON 1965a).

Pseudonigeran. Pseudonigeran is a water-insoluble, alkali-soluble glucan consisting of a linear arrangement of (1 → 3)-α-linked glucose residues. It has been obtained by alkali extraction of *Polyporus tumulosus* (RALPH and BENDER 1965), *Aspergillus niger* (JOHNSTON 1965b, HORISBERGER et al. 1972), *A. nidulans* (BULL 1970, ZONNEVELD 1971), *Paracoccidioides brasiliensis* (KANETSUNA and CARBONELL 1970), *Histoplasma capsulatum* (KANETSUNA et al. 1974), *H. farciminosum* (SAN-BLAS and CARBONELL 1974), *Blastomyces dermatitides* (KANETSUNA and CARBONELL 1971), and *Schizophyllum commune* (WESSELS et al. 1972, SIETSMA and WESSELS 1977). Pseudonigeran is a wall-associated polysaccharide and in the case of *Aspergillus* spp. it may represent up to 30% of the dry weight of the cell wall. With *A. nidulans* the content of wall-associated pseudonigeran decreases as fructification or cleistothecium formation commences, and it was postulated that this glucan might act as a carbon and energy reserve for this purpose (ZONNEVELD 1972a). The species of *Histoplasma, Paracoccidioides,* and *Blastomyces* exhibit yeast-mycelial dimorphism and by far the highest levels of this glucan are associated with the cell walls of the yeast form. Chemical analyses of the glucan obtained from *A. niger* have revealed a linear structure and the presence of a small percentage (<10%) of (1 → 4)-α-linked residues. The molecule exhibited a DP of ca. 330 (JOHNSTON 1965b, HORISBERGER et al. 1972). The "S" or (1 → 3)-α-glucan of *Schizophyllum commune* constitutes about 28% of the cell wall; it is nonbranched and consists of about 90 glucose residues (SIETSMA and WESSELS 1977).

Elsinan. Recently, a novel type of glucan was isolated from the culture fluids of *Elsinoe leucospila,* a pathogenic fungus of tea plants. The glucan, which is water-soluble, is essentially linear and consists of blocks of (1 → 4)-α-linked maltotriose units connected by (1 → 3)-α-bonds (MISAKI et al. 1978).

3.2 Fungal Glucanases and Glucan Degradation

Because of their potential industrial and scientific uses, considerable attention has been given to the many extracellular glucanases that are produced by fungi when grown on polysaccharides. Generally, these enzymes are not directly involved in fungal glucan metabolism and will not be discussed further. Reviews of their production and properties may be found in BARRAS et al. (1969), MANDELS (1975) and REESE (1977). Various glucanases have been discovered in cell extracts and cell walls of some fungal species and, more recently, attention has been directed toward the possible roles of these enzymes in fungal morphogenesis (BARTNICKI-GARCIA 1973). However, unlike the endogenous yeast glucanases, the fungal glucanases have not received extensive biochemical characterization.

ROSNESS (1968) reported the appearance of cellulase activity in extracts of *Dictyostelium discoideum* at the onset of fruiting and spore formation. The evidence suggested the presence of at least one endocellulase which was converted into an active form when the degradation of cell wall cellulose was required. JONES et al. (1979) confirmed the presence of cellulase in spores of *D. discoideum* and showed that this activity was released into the medium on spore germination. On chromatographic fractionation of this activity two cellulases with molecular weights of 136,000 and 69,000 were separated. Their modes of action on cellulose were not reported. Cellulase activity has also been reported in extracts of *Saprolegnia monoica,* where it is involved in cell wall autolysis (FÈVRE 1977).

An enzyme has been found in extracts of *Schizophyllum commune* that degrades the β-(1 $\rightarrow$ 3)(1 $\rightarrow$ 6)-linked R-glucan of the cell wall. Although the mechanism and pattern of hydrolysis of the wall glucan are not known, the enzyme has a specificity for hydrolyzing (1 $\rightarrow$ 6)-β-glucosidic linkages (WESSELS and NIEDERPRUEM 1967, WESSELS 1969, WESSELS and KOLTIN 1972). (1 $\rightarrow$ 6)-β-Glucanase activity also occurs in extracts of *Neurospora crassa* and *Trichoderma viride* (DEL REY et al. 1979), but it is possible that this may represent one activity of a nonspecific, exo-glucanase type enzyme as is found in yeasts (PHAFF 1977).

In view of the widespread existence of the (1 $\rightarrow$ 3)-β-glucosidic linkage in fungal cell walls, it is not surprising to find the frequent occurrence of (1 $\rightarrow$ 3)-β-glucanases in fungal species. Isolated cell walls of *N. crassa* undergo autolysis, and this is associated with the presence of wall-bound (1 $\rightarrow$ 3)-β-glucanase activity. The level of wall-glucanase varies with culture age and is postulated to function in hyphal branching (MAHADEVAN and MAHADKAR 1970). The autolysis complex associated with hyphal walls of *Aspergillus nidulans* contains (1 $\rightarrow$ 3)-β-glucanase activity. This activity may be solubilized through wall auto-digestion or by extraction with a cationic detergent (POLACHECK and ROSENBERGER 1978). Triton X-100 was used to solubilize the (1 $\rightarrow$ 3)-β-glucanase from the cell walls of *Sclerotium rolfsii* (KRITZMAN et al. 1978). Although this enzyme was subsequently purified through DEAE-cellulose and Sephadex G-100 columns, its properties were not reported. Fluorescent-antibody studies showed the enzyme to be localized in hyphal tips, clamp connections, and newly formed septa. Three (1 $\rightarrow$ 3)-β-glucanases, I, II, and III, have been found in extracts of *Penicillium italicum* (SANTOS et al. 1978a, b). These enzymes have been tentatively characterized as an endo-glucanase, an exo-glucanase, and a mixed exo- and endo-glucanase, respectively. Only small amounts of the enzymes II and III are produced when the fungus is grown in the presence of excess glucose. On transfer to a medium low in glucose, this glucose repression is overcome, and increases in the production of enzymes II and III occur. In addition, glucanase I is now produced (SANTOS et al. 1978a). It appears that the glucose repressive effect on (1→3)-β-glucanase production is exerted at a pre-translational level (SANTOS et al. 1978b). Although glucanases I, II, and III may be recovered from cytoplasmic extracts and extra-cellular culture fluids of *P. italicum*, glucanases II and III are also found attached to isolated cell walls and cause wall autolysis. It is postulated that these two glucanases function in the mobilization of cell wall glucans in vivo (SANTOS et al. 1979a). The (1→3)-β-glucanase found in extracts of *N. crassa* is also subject to glucose repression, but this was not the case for the enzyme found in *Trichoderma viride* (DEL REY et al. 1979). (1→3)-β-Glucanases also occur in the dimorphic fungi *Histoplasma capsulatum*, *Blastomyces dermatitidis* (DAVIS et al. 1977) and *Paracoccidioides brasiliensis* (FLORES-CARREON et al. 1979). In the latter species higher glucanase levels in extracts and cell walls were associated with the transition to the mycelial phase.

Extracts of *Aspergillus nidulans* exhibit (1→3)-α-glucanase activity. High levels of this enzyme are found only after glucose in the growth medium has been depleted and after (1→3)-α-glucan has become a major wall component. During fructification, which occurs after glucose depletion, the (1→3)-α-glucan content of the cell wall decreases and this is attributable to the development of (1→3)-α-glucanase activity (ZONNEVELD 1972a). The enzyme has been isolated and characterized as an exo-(1→3)-α-glucanase (ZONNEVELD 1972b). Low levels of (1→3)-α-glucanase have also been detected in extracts of *Paracoccidioides brasiliensis* (FLORES-CARREON et al. 1979).

4 Biosynthesis

Little information on the biosynthesis of yeast cell wall glucans is available from the literature as compared to that on the biosynthesis of mannans and chitin. One reason for this limited information may be a greater sensitivity of the glucan-synthesizing enzymes to isolation procedures than those responsible for mannan and chitin synthesis (FARKAŠ 1979). Factors responsible for negative results could include disturbance of a delicate balance among biosynthetic enzymes during the isolation steps and hydrolysis of biosynthetic products by insufficiently separated endogenous enzymes (cf. Sect. 2.1.2.1).

In only a limited number of attempts was in vitro biosynthesis of glucan demonstrated. From a homogenized cell mass of *Phytophthora cinnamomi* a particulate fraction was isolated that catalyzed the incorporation of radioactive glucose from UDPG into a β-glucan containing both (1→3)-β- and (1→6)-β-bonds (WANG and BARTNICKI-GARCIA 1966). Most of the activity appeared to be associated with the cell walls. However, in later experiments (WANG and BARTNICKI-GARCIA 1976) a high glucan synthetase activity was demonstrated in a cell wall-free membrane fraction of this fungus that formed β-glucan microfibrils from UDPG. Activation of the enzyme system by trypsin suggested that at least some of the enzyme components were present in a zymogen form.

This reaction product appeared to consist entirely of (1 → 3)-β-linked glucose residues.

NAMBA and KURODA (1974a, b) used a cellular homogenate of *Cochliobolus miyabeanus* and also demonstrated incorporation of UDPG [^{14}C] into a polysaccharide with both (1 → 3)-β- and (1 → 6)-β-glucosidic bonds. Biosynthesis of a β-glucan from UDPG by a cell wall fraction of *Neurospora* was demonstrated by MISHRA and TATUM (1972), but the product was not characterized in detail. A cellulose-type of polysaccharide was synthesized by cell-free extracts of *Saprolegnia monoica* (FÈVRE and DUMAS 1977) from UDPG. Mg^{2+} and cellobiose stimulated the biosynthesis of β-glucan.

Glucan biosynthesis in yeast has been studied only in *Sacch. cerevisiae.* SENTANDREU et al. (1975) treated whole cells of this species with toluene plus ethanol and such cells were able to incorporate labeled glucose from UDPG into (1 → 3)-β-glucan. The synthesized glucan was recovered from the cell homogenate in a membrane fraction, apparently prior to its transport into the cell wall.

BÁLINT et al. (1976) isolated a particulate membrane fraction from *Sacch. cerevisiae* that catalyzed the incorporation of radioactive glucose from both UDPG and GDPG into β-glucans. This suggested the presence of two separate glucosyltransferases in the enzyme preparation. Digestion of the products with partially purified β-glucanases indicated that the glucan synthesized from UDPG contained a high proportion of (1 → 3)-β-bonds, whereas the product synthesized from GDPG contained more (1 → 6)-β-bonds. Both products were alkali-soluble.

LÓPEZ-ROMERO and RUIZ-HERRERA (1977), on the other hand, in a similar system, observed only negligible incorporation of labeled glucose from GDPG, whereas from UDPG a glucan was synthesized with both (1 → 3)-β-and (1 → 6)-β-linkages. These authors found the highest activity in the cell wall fraction of *Sacch. cerevisiae.*

References

Abd-El-Al ATH, Phaff HJ (1968) Exo-β-glucanases in yeast. Biochem J 109:347–360

Abd-El-Al ATH, Phaff HJ (1969) Purification and properties of endo-β-glucanases in the yeast *Hanseniaspora valbyensis.* Can J Microbiol 15:697–701

Abercrombie M, Jones J, Lock M, Perry M, Stoodley R (1960) The polysaccharides of *Cryptococcus laurentii* (NRRL Y-1401). Can J Chem 38:1617–1624

Anderson FB, Millbank JW (1966) Protoplast formation and yeast cell wall structure. The action of enzymes of the snail *Helix pomatia.* Biochem J 99:682–687

Arai M, Murao S (1978) Characterisation of oligosaccharides from an enzymatic hydrolysate of red yeast cell walls by lytic enzyme. Agric Biol Chem 42:1651–1659

Arai M, Lee TH, Murao S (1978) Substrate specificity of the *Penicillium lilacinum* enzyme lytic to the cell wall of *Rhodotorula glutinis* and the structure of the *Rhodotorula* cell wall glucomannan. Curr Microbiol 1:185–188

Arnold WN (1972) The structure of the yeast cell wall. Solubilization of a marker enzyme, β-fructofuranosidase, by the autolytic system. J Biol Chem 247:1161–1167

Aronson JM (1965) The cell wall. In: Ainsworth GC, Sussman AS (eds) The fungi, vol I. Academic Press, London New York, pp 49–76

Bacon JSD (1973) In: Villanueva JR, Garcia-Acha I, Gascon S, Uruburu F (eds) Yeast mould and plant protoplasts. Academic Press, London New York, pp 61–103

Bacon JSD, Jones D, Farmer VC, Webley DM (1968) The occurrence of α-(1→3)-glucan in *Cryptococcus, Schizosaccharomyces* and *Polyporus* species, and its hydrolysis by a streptomycete culture filtrate lysing cell walls of *Cryptococcus*. Biochim Biophys Acta 158:313–315

Bacon JSD, Farmer VC, Jones D, Taylor IF (1969) The glucan components of the cell wall of baker's yeast (*Saccharomyces cerevisiae*) considered in relation to its ultrastructure. Biochem J 114:557–567

Bacon JSD, Gordon AH, Jones D, Taylor IF, Webley DM (1970) The separation of β-glucanases produced by *Cytophaga johnsonii* and their role in the lysis of yeast cell walls. Biochem J 120:67–78

Bálint Š, Farkaš V, Bauer Š (1976) Biosynthesis of β-glucans catalysed by a particulate enzyme preparation from yeast. FEBS Lett 64:44–47

Barras DR (1972) A β-glucan endo-hydrolase from *Schizosaccharomyces pombe* and its role in cell wall growth. Antonie van Leeuwenhoek J Microbiol Serol 38:65–80

Barras DR, Moore AE, Stone BA (1969) Enzyme–substrate relationships among β-glucan hydrolyases. Adv Chem Ser 95:105–138

Bartnicki-Garcia S (1968) Cell wall chemistry, morphogenesis and taxonomy of fungi. Annu Rev Microbiol 22:87–107

Bartnicki-Garcia S (1973) Fundamental aspects of hyphal morphogenesis. In: Ashworth JM, Smith JE (eds) *Microbial differentiation*. Univ Press, Cambridge, pp 245–267

Bartnicki-Garcia S, McMurrough I (1971) Biochemistry of morphogenesis in yeasts. In: Rose AH, Harrison JS (eds) The yeasts – physiology and biochemistry of yeasts, vol II. Academic Press, London New York, pp 441–491

Bauer H, Bush DA, Horisberger M (1972) Use of the exo-β-1,3-glucanase from Basidiomycete QM 806 in studies on yeast. Experientia 28:11–13

Bell DJ, Northcote DH (1950) The structure of a cell wall polysaccharide of baker's yeast. J Chem Soc 1944–1947

Beran F (1968) Budding of yeast cells. Their scars and aging. Adv Microb Physiol 2:143–171

Bhavanandan VP, Bouveng HO, Lindberg B (1964) Polysaccharides from *Polyporus giganteus*. Acta Chem Scand 18:504–512

Bishop CT, Blank F, Gardner PE (1960) The cell wall polysaccharides of *Candida albicans:* glucan, mannan and chitin. Can J Chem 38:869–881

Bjorndal H, Lindberg B (1970) Polysaccharides elaborated by *Polyporus fomentarius* and *Polyporus igniarius*. Carbohydr Res 12:29–35

Bouveng HO, Kiessling H, Lindberg B, McKay J (1963) Polysaccharides elaborated by *Pullularia pullulans*. III Polysaccharides synthesised from xylose solutions. Acta Chem Scand 17:1351–1356

Bowden JK, Hodgson B (1970) Evidence against the presence of fibers or chemically distinct layers in the cell wall of *Saccharomyces*. Antonie von Leeuwenhoek J Microbiol Serol 36:81–108

Brock TD (1964) Enzyme synthesis during conjugation in the yeast *Hansenula wingei*. J Cell Biol 23:15A

Brock TD (1965) β-Glucanase of yeast. Biochem Biophys Res Commun 19:623–629

Brown JP (1971) Susceptibility of the cell walls of some yeasts to lysis by the enzymes of *Helix pomatia*. Can J Microbiol 17:205–208

Brown RG, Hanic LA, Hsiao M (1973) Structure and chemical composition of yeast chlamydospores of *Aureobasidium pullulans*. Can J Microbiol 19:163–168

Buck KW, Chen AW, Dickerson AG, Chain EB (1968) Formation and structure of extracellular glucans produced by *Claviceps* species J Gen Microbiol 51:337–352

Bull AT (1970) Chemical composition of wild-type and mutant *Aspergillus nidulans* cell walls. The nature of polysaccharide and melanin constituents. J Gen Microbiol 63:75–94

Bull AT, Chesters CGC (1966) The biochemistry of laminarin and the nature of laminarinase. Adv Enzymol 28:325–364

Bush DA, Horisberger M (1972) Structure of a β-D-glucan from the mycelial wall of *Basidiomycete* QM *806*. Carbohydr Res 22:361–367

Bush DA, Horisberger M, Horman I, Wursch P (1974) The wall structure of *Schizosaccharomyces pombe*. J Gen Microbiol 81:199–206

Cabib E, Ulane R, Bowers B (1974) A molecular model for morphogenesis: the primary septum of yeast. Cun Top Cell Regul 6:1–32
Cassone A, Kerridge D, Gale EF (1979) Ultrastructural changes in the cell wall of *Candida albicans* following cessation of growth and their possible relationship to the development of polyene resistance. J Gen Microbiol 110:339–349
Catley BJ (1979) Pullulan synthesis by *Aureobasidium pullulans*. In: Berkeley RCW, Gooday GW, Elwood DC (eds) Microbial polysaccharides and polysaccharidases. Soc Gen Microbiol. Academic Press, London New York, pp 69–84
Catley BJ, Kelly DJ (1975) Metabolism of trehalose and pullulan during the growth cycle of *Aureobasidium pullulans*. Biochem Soc Trans 3:1079–1081
Chattaway FW, Chenolikar S, O'Reilly J, Barlow AJE (1976) Changes in the cell surface of the dimorphic forms of *Candida albicans* by treatment with hydrolytic enzymes. J Gen Microbiol 95:335–347
Chesters CGC, Bull AT (1963a) The enzymic degradation of laminarin. I. The distribution of laminarinase among microorganisms. Biochem J 86:28–31
Chesters CGC, Bull AT (1963b) The enzymic degradation of laminarin. 2. The multicomponent nature of fungal laminarinases. Biochem J 86:31–38
Clarke AE, Stone BA (1963) Chemistry and biochemistry of β-(1→3)-glucans. Rev Pure Appl Chem 13:134–156
Cortat M, Matile P, Wiemken A (1972) The isolation of glucanase-containing vesicles from budding yeast. Arch Microbiol 82:189–205
Crandall M, Egel R, Mackay VL (1977) Physiology of mating in three yeasts. Adv Microb Physiol 15:307–398
Davis TE, Domer JE, Li YT (1977) Cell wall studies of *Histoplasma capsulatum* and *Blastomyces dermatitidis* using autologous and heterologous enzymes. Infect Immun 15:978–987
Del Rey FD, Garcia-Acha I, Nombela C (1979) The regulation of β-glucanase synthesis in fungi and yeast. J Gen Microbiol 110:83–89
de Vries OMH, Wessels JGH (1972) Release of protoplasts from *Schizophyllum commune* by a lytic enzyme from *Trichoderma viride*. J Gen Microbiol 73:13–22
Doi K, Doi A, Fukui T (1973) Purification and properties of a lytic β-glucanase from an Arthrobacter bacterium. Agric Biol Chem 37:1619–1627
Doi K, Doi A, Ozaki T, Fukui T (1976) Further studies on the heterogeneity of the lytic activity for isolated yeast cell wall of the components of an *Arthrobacter* glucanase system: Properties of the two components of a β-1,3-glucanase. Agric Biol Chem 40:1355–1362
Dominguez JB, Goni FM, Uruburu F (1978) The transition from yeast-like to chlamydospore cell in *Pullularia pullulans*. J Gen Microbiol 108:111–117
Ebert E, Zenk MH (1967) Luteic acid and islandic acid, composition and structure. Phytochemistry 6:309–312
Eddy AA, Woodhead JS (1968) An alkali-soluble glucan fraction from the cell walls of the yeast *Saccharomyces carlsbergensis*. Fed Eur Biochem Soc Let 1:67–68
Elorza MV, Ruiz EM, Villanueva JR (1966) Production of yeast cell wall lytic enzymes on a semi-defined medium by a *Streptomyces*. Nature (London) 210:442–443
Evans RB, Manners DJ (1971) Observations on the purity of some glycogen preparations. Biochem J 125:31
Farkaš V (1979) Biosynthesis of cell walls of fungi. Microbiol Rev 43:117–144
Farkaš V, Biely P, Bauer Š (1973) Extracellular β-glucanases of the yeast *Saccharomyces cerevisiae*. Biochim Biophys Acta 321:246–255
Fèvre M (1977) Subcellular localisation of glucanase and cellulase in *Saprolegnia monoica* Pringsheim. J Gen Microbiol 103:287–295
Fèvre M, Dumas C (1977) β-Glucan synthetases from *Saprolegnia monoica*. J Gen Microbiol 103:297–306
Fleet GH, Manners DJ (1976) Isolation and composition of an alkali-soluble glucan from the cell walls of *Saccharomyces cerevisiae*. J Gen Microbiol 94:180–192
Fleet GH, Manners DJ (1977) The enzymic degradation of an alkali-soluble glucan from the cell walls of *Saccharomyces cerevisiae*. J Gen Microbiol 98:315–327

Fleet GH, Phaff HJ (1973) Effect of glucanases of yeast and bacterial origin on cell walls of *Schizosaccharomyces* species. In: Villanueva JR, Garcia-Acha I, Gascon S, Uruburu F (eds) Yeast, mould and plant protoplasts. Academic Press, London New York, pp 33–59

Fleet GH, Phaff HJ (1974a) Glucanases in *Schizosaccharomyces.* Isolation and properties of the cell wall-associated β-(1→3)-glucanases. J Biol Chem 249:1717–1728

Fleet GH, Phaff HJ (1974b) Lysis of yeast cell walls: Glucanases from *Bacillus circulans* WL-12. J Bacteriol 119:207–219

Fleet GH, Phaff HJ (1975) Glucanases in *Schizosaccharomyces.* Isolation and properties of an exo-β-glucanase from the cell extracts and culture fluid of *Schizosaccharomyces japonicus* var. *versatilis.* Biochim Biophys Acta 401:318–332

Flores-Carreon A, Gomez-Villaneuva A, San-Blas G (1979) β-(1→3)-Glucanase and dimorphism in *Paracoccidioides brasiliensis.* Antonie van Leeuwenhoek J Microbiol Serol 45:265–274

Foda MS, Phaff HJ (1970) The synthesis of starch-like compounds by *Cryptococcus laurentii.* In: Ahearn DH (ed) Recent trends in yeast research-Spectrum, vol I. Georgia State Univ, Atlanta Georgia pp 181–198

Foda MS, Phaff HJ (1978) Properties and kinetics of glucan phosphorylase of the amylase-forming yeast *Cryptococcus laurentii.* Z Allg Mikrobiol 18:95–106

Gorin PAJ, Spencer JFT (1968) Structural chemistry of fungal polysaccharides. Adv Carbohydr Chem 23:367–414

Gorin PAJ, Spencer JFT, MacKenzie S (1966) Estimated molecular weight of amylose from the yeast *Rhodotorula peneaus.* Can J Chem 44:2087–2090

Gunja-Smith Z, Smith EE (1974) Evidence for the periplasmic location of glycogen in *Saccharomyces.* Biochem Biophys Res Commun 56:588–592

Gunja-Smith Z, Patil NB, Smith EE (1977) Two pools of glycogen in *Saccharomyces.* J Bacteriol 130:818–825

Hellerqvist CG, Lindberg B, Samuelsson K (1968) Methylation analysis of pustulan. Acta Chem Scand 22:2376–2377

Hemmes DE, Kojima-Buddenhagen ES, Hohl HH (1972) Structure and enzymatic analysis of the spore wall layers in *Dictyostelium discoideum.* J Ultrastruct Res 41:406–417

Hoffmann GC, Simson BW, Timell TE (1971) Structure and molecular size of pachyman. Carbohydr Res 20:185–188

Horikoshi K (1973) In: Villanueva JR, Garcia-Acha I, Gascon S, Uruburu F (eds) Yeast, mould and plant protoplasts. Academic Press, London New York, pp 25–32

Horisberger M, Lewis BA, Smith F (1972) Structure of a (1→3)-α-D-glucan (pseudonigeran) of *Aspergillus niger* NRRL 326 cell wall. Carbohydr Res 23:183–188

Huotari F, Nelson T, Smith F, Kirkwood S (1968) Purification of an exo-β-D-1,3-glucanase from Basidiomycete species QM 806. J Biol Chem 243:952–956

Jeanes A (1977) Dextrans and pullulans: Industrially significant α-D-glucans. In: Sandford PA, Laskin A (eds) Am Chem Soc Symp 45:284–297

Johnson J, Kirkwood S, Misaki A, Nelson TE, Scaletti JV, Smith F (1963) Structure of a new glucan. Chem Ind 20:820–822

Johnston IR (1965a) The composition of the celll wall of *Aspergillus niger.* Biochem J 96:651–658

Johnston IR (1965b) The partial acid hydrolysis of a highly dextrorotary fragment of the cell wall of *Aspergillus niger.* Isolation of the α-(1→3)-linked dextrin series. Biochem J 96:659–664

Jones THD, de Renobales M, Pon N (1979) Cellulases released during the germination of *Dictyostelium discoideum* spores. J Bacteriol 137:752–757

Kanetsuna F, Carbonell LM (1970) Cell wall glucans of the yeast and mycelial forms of *Paracoccidioides brasiliensis.* J Bacteriol 101:675–680

Kanetsuna F, Carbonell LM (1971) Cell wall composition of the yeast-like and mycelial forms of *Blastomyces dermatitidis.* J Bacteriol 106:946–948

Kanetsuna FC, Carbonell M, Azuma I, Yamamura Y (1972) Biochemical studies on the thermal dimorphism of *Paracoccidioides brasiliensis.* J Bacteriol 110:208–218

Kanetsuna F, Carbonell CM, Gill F, Azuma I (1974) Chemical and ultrastructural studies

on the cell walls of the yeast-like and mycelial forms of *Histoplasma capsulatum*. Mycopathol Mycol Appl 54:1–13
Kessler G, Nickerson WJ (1959) Glucomannan-protein complexes from cell walls of yeasts. J Biol Chem 234:2281–2283
Kitamura K, Yamamoto Y (1972) Purification and properties of an enzyme, zymolyase, which lyses viable yeast cells. Arch Biochem Biophys 153:403–406
Kitamura K, Kaneko T, Yamamoto Y (1972) Lysis of viable yeast cells by enzymes of *Arthrobacter luteus*. I. Isolation of lytic strain and studies of its lytic activity. J Gen Appl Microbiol 18:57–71
Kopecká M, Phaff HJ, Fleet GH (1974) Demonstration of a fibrillar component in the cell wall of the yeast *Saccharomyces cerevisiae* and its chemical nature. J Cell Biol 62:66–76
Korn ED, Northcote DH (1960) Physical and chemical properties of polysaccharides and glucoproteins of the yeast cell wall. Biochem J 75:12–17
Kreger DR (1954) Observation on cell walls of yeast and some other fungi by X-ray diffraction and solubility tests. Biochim Biophys Acta 13:1–9
Kritzman G, Chet I, Henis Y (1978) Localisation of β-(1→3)-glucanase in the mycelium of *Sclerotium rolfsii*. J Bacteriol 134:470–475
Kröning A, Egel R (1974) Autolytic activities associated with conjugation and sporulation in fission yeast. Arch Microbiol 99:241–249
Lachance MA, Villa TG, Phaff HJ (1977) Purification and partial characterisation of an exo-β-glucanase from the yeast *Kluyveromyces aestuarii*. Can J Biochem 55:1001–1006
Lodder J (ed) (1970) The yeasts – a taxonomic study. North Holland Publ Co, Amsterdam, pp 1–1385
López-Romero E, Ruiz-Herrera J (1977) Biosynthesis of β-glucans by cell free extracts from *Saccharomyces cerevisiae*. Biochim Biophys Acta 500:372–384
Maddox IS, Hough JS (1971) Yeast glucanase and mannanase. J Inst Brew London 77:44–47
Mahadevan PR, Mahadkar UR (1970) Role of enzymes in growth and morphology of *Neurospora crassa:* cell-wall-bound enzymes and their possible role in branching. J Bacteriol 101:941–947
Mahadevan PR, Tatum EL (1965) Relationship of the major constituents of the *Neurospora crassa* cell wall to wild-type and colonial morphology. J Bacteriol 90:1073–1081
Mandels M (1975) Microbial sources of cellulases. In: Wilke CR (ed) Cellulose as a chemical and energy source. Wiley and Sons, New York, pp 81–105
Mann JW, Heinz CE, Macmillan JD (1972) Yeast spheroplasts formed by cell wall degrading enzymes from *Oerskovia* sp. J Bacteriol 111:821–824
Manners DJ, Meyer MT (1977) The molecular structure of some glucans from the cell walls of *Schizosaccharomyces pombe*. Carbohydr Res 57:189–203
Manners DJ, Masson AJ, Patterson JC (1973a) The structure of a β-(1→3)-D-glucan from yeast cell walls. Biochem J 135:19–30
Manners DJ, Masson AJ, Patterson JC, Bjorndal H, Lindberg B (1973b) The structure of a β-(1→6)-D-glucan from yeast cell walls. Biochem J 135:31–36
Manners DJ, Masson AJ, Patterson JC (1974) The heterogeneity of glucan preparations from the walls of various yeasts. J Gen Microbiol 80:411–417
Marchant R, Smith DG (1968) Bud formation in *S. cerevisiae* and a comparison of the mechanism of cell division in other yeasts. J Gen Microbiol 53:163–169
Marshall JJ (1974) Application of enzymic methods to the structural analysis of polysaccharides. Adv Carbohydr Chem Biochem 30:257–353
Matile P, Cortat M, Wiemken A, Frey-Wyssling A (1971) Isolation of glucanase-containing particles from budding *Saccharomyces cerevisiae*. Proc Natl Acad Sci USA 68:636–640
McMurrough I, Rose AH (1967) Effect of growth rate and substrate limitation on the composition and structure of the cell wall of *Saccharomyces cerevisiae*. Biochem J 105:189–203
Meyer MT, Phaff HJ (1977) Survey for α-(1→3)-glucanase activity among yeasts. J Bacteriol 131:702–706
Mill PJ (1966) Phosphomannans and other components of flocculent and non-flocculent walls of *Saccharomyces cerevisiae*. J Gen Microbiol 44:329–341

Miller MW, Yoneyama M, Soneda M (1976) *Phaffia*, a new yeast genus in the *Deuteromycotina (Blastomycetes)*. Int J Syst Bacteriol 26:286–291

Misaki A, Johnson J, Kirkwood S, Scaletti JV, Smith F (1968) Structure of the cell wall glucan of yeast (*Saccharomyces cerevisiae*). Carbohydr Res 6:150–164

Misaki A, Tsumuraya Y, Takaya S (1978) A new fungal α-D-glucan, elsinan, elaborated by *Elsinoe leucospila*. Agric Biol Chem 42:491–493

Mishra NC, Tatum EL (1972) Effect of L-sorbose on polysaccharide synthetases of *Neurospora crassa*. Proc Natl Acad Sci USA 69:313–317

Moor H (1967) Endoplasmic reticulum as the initiation of bud formation in yeast (*S. cerevisiae*). Arch Mikrobiol 57:135–146

Morris EO (1955) Seaweed as a source of yeast food. J Sci Food Agric 6:611–621

Mundkur B (1960) Electron microscopical studies of frozen-dried yeast (1) Localisation of polysaccharides. Exp Cell Res 20:28–42

Nakajima T, Tamari K, Matsuda K, Tanaka H, Ogasawara N (1970) Studies on the cell wall of *Piricularia oryzae*. Part II. The chemical constituents of the cell wall. Agric Biol Chem 34:553–560

Nakajima T, Tamari K, Matsuda K, Tanaka H, Ogasawara N (1972) Studies on the cell wall of *Piricularia oryzae*. Part III. The chemical structure of the β-D-glucan. Agric Biol Chem 36:11–17

Nakamura N, Tanabe O (1963) Studies on an enzyme capable of splitting β-D-(1→6)-glucosidic linkage. Isolation of a lutease-producing microorganism and some properties of lutease. Agric Biol Chem 27:80–87

Namba H, Kuroda H (1974a) Studies on fungicides. X. Biosyntheses of β-glucan and chitin-like substance of cell wall from *Cochliobolus miyabeamus*. Chem Pharm Bull 22:610–616

Namba H, Kuroda H (1974b) Studies on fungicides. XII. Biosyntheses of β-glucan and chitin-like substance of cell wall from *Cochliobolus miyabeamus*. Chem Pharm Bull 22:1895–1901

Nelson TE (1970) The hydrolytic mechanism of an exo-β-1,3-glucanase. J Biol Chem 245:869–872

Nelson TE, Johnson J, Jantzen E, Kirkwood S (1969) Action pattern and specificity of an exo-β-(1→3)-D-glucanase from Basidiomycete species QM 806. J Biol Chem 244:5972–5980

Nishikawa Y, Tanaka M, Shibata S, Fukuoka F (1970). Polysaccharides of lichens and fungi. IV. Antitumour active o-acetylated pustulan type glucans from the lichens of *Umbilicaria* species. Chem Pharm Bull 18:1431–1434

Northcote DH (1953) The molecular structure and shapes of yeast glycogen. Biochem J 53:348–352

Northcote DH, Horne RW (1952) The chemical composition and structure of the yeast cell wall. Biochem J 51:232–236

Notario V, Villa TG, Benitez T, Villanueva JR (1975) β-Glucanases in the yeast *Cryptococcus albidus* var *aerius*. Production and separation of β-glucanases in a synchronous culture. Can J Microbiol 22:261–268

Notario V, Villa TG, Villanueva JR (1976) Purification of an exo-β-D-glucanase from cell-free extracts of *Candida utilis*. Biochem J 159:555–562

Obata T, Fujioka K, Hara S, Namba Y (1977) The synergistic effects among β-1,3-glucanases from *Oerskovia* sp. CK on lysis of viable yeast cells. Agric Biol Chem 41:671–677

Patel GB, Ingledew WM (1975a) The relationship of acid-soluble glycogen to yeast flocculation. Can J Microbiol 21:1608–1613

Patel GB, Ingledew WM (1975b) Glycogen – a physiological determinant of yeast flocculation. Can J Microbiol 21:1614–1621

Peat S, Whelan WJ, Edwards TE (1958a) Polysaccharides of baker's yeast. Part II. Yeast glucan. J Chem Soc 3862–3868

Peat S, Turvey JR, Evans JM (1958b) Polysaccharides of baker's yeast. Part III. The presence of 1:6-linkages in yeast glucan. J Chem Soc 3868–3870

Perlin AS, Taber WA (1963) A glucan produced by *Claviceps purpurea*. Can J Chem 41:2278–2282

Phaff HJ (1963) Cell wall of yeasts. Annu Rev Microbiol 17:15–30
Phaff HJ (1971) Structure and biosynthesis of the yeast cell envelope. In: Rose AH, Harrison JS (eds) The yeasts – physiology and biochemistry of yeast, vol II. Academic Press, London New York, pp 135–210
Phaff HJ (1977) Enzymatic yeast cell wall degradation. Adv Chem Ser No 160:244–282
Phaff HJ (1979) A retrospective and current view on endogenous β-glucanases in yeast. Advances in protoplast research. Proc 5th Int Protoplast Symp, Szeged, Hungary, July 9–14, 1979. Publ House Hung Acad Sci, pp 171–182
Polacheck I, Rosenberger RF (1978) Distribution of autolysins in hyphae of *Aspergillus nidulans*: evidence for a lipid-mediated attachment to hyphal walls. J Bacteriol 135:741–747
Poulain D, Tronchin G, Dubremetz JF, Biguet J (1978) Ultrastructure of the cell wall of *Candida albicans* blastospores: study of its constitutive layers by the use of a cytochemical technique revealing polysaccharides. Ann Microbiol (Paris) 129:141–153
Ralph BJ, Bender J (1965) Isolation of two new polysaccharides from the cell wall of *Polyporus tumulosus*. Chem Ind 1181
Ramsay AM, Douglas LJ (1979) Effects of phosphate limitation of growth on the cell wall and lipid composition of *Saccharomyces cerevisiae*. J Gen Microbiol 110:185–191
Reese ET (1977) Degradation of polymeric carbohydrates by microbial enzymes. Recent Adv Phytochem 11:311–367
Reichelt B (1978) β-Glucanases in *Schizosaccharomyces pombe*. MSci Thesis, Univ New South Wales, Australia
Roelofsen PA (1953) Yeast mannan, a cell wall constituent of baker's yeast. Biochim Biophys Acta 10:477–478
Rombouts FM, Phaff HJ (1976a) Lysis of yeast cell walls. Lytic β-(1→6)-glucanase from *Bacillus circulans* WL-12. Eur J Biochem 63:109–120
Rombouts FM, Phaff HJ (1976b) Lysis of yeast cell walls. Lytic β-(1→3)-glucanases from *Bacillus circulans* WL-12. Eur J Biochem 63:121–130
Rombouts FM, Fleet GH, Manners DJ, Phaff HJ (1978) Lysis of yeast cell walls: non-lytic and lytic (1→6)-β-D-glucanases from *Bacillus circulans* WL-12. Carbohydr Res 64:237–249
Rosness PA (1968) Cellulolytic enzymes during morphogenesis in *Dictyostelium discoideum*. J Bacteriol 96:639–645
Rosness PA, Wright BE (1974) In vivo changes of cellulose, trehalose and glycogen during differentiation of *Dictyostelium discoideum*. Arch Biochem Biophys 164:60–72
Saito H, Misaki A, Harada T (1968) A comparison of the structure of curdlan and pachyman. Agric Biol Chem 32:1261–1269
San-Blas G, Carbonell LM (1974) Chemical and ultrastructural studies on the cell walls of the yeast-like and mycelial forms of *Histoplasma farciminosum*. J Bacteriol 119:602–611
Santamaria F, Fuensanta R, Lahoz R (1978) Extracellular glucan containing (1→3)-β- and (1→6)-β-linkages isolated from *Monilinia fructigena*. J Gen Microbiol 109:287–293
Santos T, Sanchez M, Villanueva JR, Nombela C (1978a) Regulation of the β-(1→3)-glucanase system in *Penicillium italicum*: glucose repression of the various enzymes. J Bacteriol 133:465–471
Santos T, Villanueva JR, Nombela C (1978b) Regulation of β-(1→3)-glucanase synthesis in *Penicillium italicum*. J Bacteriol 133:542–548
Santos T, Sanchez M, Villanueva JR, Nombela C (1979a) Derepression of β-(1→3)-glucanases in *Penicillium italicum*: localisation of the various enzymes and correlation with cell wall glucan mobilisation and autolysis. J Bacteriol 137:6–12
Santos T, del Rey F, Code J, Villanueva JR, Nombela C (1979b) *Saccharomyces cerevisiae* mutant defective in exo-1,3-β-glucanase production. J Bacteriol 139:333–338
Sentandreu R, Northcote DH (1969) The formation of buds in yeast. J Gen Microbiol 55:383–398
Sentandreu R, Elorza MV, Villaneuva JR (1975) Synthesis of yeast wall glucan. J Gen Microbiol 90:13–20

Shimoda C, Yanagishima N (1968) Strain dependence of the cell-expanding effect of β-1,3-glucanase in yeast. Physiol Plant 21:1163–1169
Shimoda C, Nasada Y, Yanagishima N (1967) Nucleic acid metabolism involved in auxin-induced elongation of yeast cells. Physiol Plant 20:299–305
Shimoda C, Yanagishima N (1971) Role of wall degrading enzymes in auxin-induced cell wall expansion in yeast. Physiol Plant 24:46–50
Sietsma JH, Wessels JGH (1977) Chemical of the hyphal wall of *Schizophyllum commune*. Biochim Biophys Acta 496:225–239
Sietsma JH, Child JJ, Nesbitt LR, Haskins RH (1975) Chemistry and ultrastructure of the hyphal walls of *Pythium acanthicum*. J Gen Microbiol 86:29–38
Sone Y, Misaki A (1977) Structures of the cell wall polysaccharides of *Tremella fuciformis*. Agric Biol Chem 42:763–777
Tanaka H, Phaff HJ (1965) Enzymic hydrolysis of yeast cell wall. I. Isolation of wall decomposing organisms and separation and purification of lytic enzymes. J Bacteriol 89:1570–1580
Tanaka H, Phaff HJ, Higgins LW (1966) Enzymatic hydrolysis of yeast cell walls. II. Susceptibilities of isolated cell walls and ascus walls of various yeasts to the actions of bacteriol endo-β-glucanases. Symp Yeast Protoplasts. Abh Dtsch Akad Wiss Berlin Kl Med 6:113–129, 353–357
Tanaka H, Itakura K, Toda K (1978) Concerted-induction of β-glucanases of *Bacillus circulans* WL-12 in response to various yeast glucans. Agric Biol Chem 42:1631–1636
Taylor IEP, Cameron DS (1973) Preparation and quantitative analysis of fungal cell wall: strategy and tactics. Annu Rev Microbiol 27:243–259
Tingle MA, Halvorson HO (1971) A comparison of β-glucanase and β-glucosidase in *Saccharomyces lactis*. Biochim Biophys Acta 250:165–171
Tokunaga J, Bartnicki-Garcia S (1971) Cyst wall formation and endogenous carbohydrate ultilisation during synchronous encystment of *Phytophthora palmivora* zoospores. Arch Mikrobiol 70:283–292
Troy FA, Koffler H (1969) The chemistry and molecular architecture of the cell walls of *Penicillium chrysogenum*. J Biol Chem 244:5563–5576
Villa TG, Phaff HJ (1980) Recovery of invertase and laminarinases from industrial waste broths of baker's yeast. Eur J Appl Microbiol Biotechnol 9:9–14
Villa TG, Notario V, Villanueva JR (1975) β-Glucanases of the yeast *Pichia polymorpha*. Arch Microbiol 104:201–206
Villa TG, Notario V, Benitez T, Villanueva JR (1976) Purification of an exo-(1→3)-β-glucanase from *Candida utilis*. Can J Biochem 54:927–934
Villa TG, Lachance MA, Phaff HJ (1977) On the structure of the β-(1→3)-glucan component of the cell wall of baker's yeast. FEMS Microbiol Lett 1:317–319
Villa TG, Lachance MA, Phaff HJ (1978) β-Glucanases of the yeast *Kluyveromyces phaseolosporus*: partial purification and characterisation. Exp Mycol 2:12–25
Villa TG, Notario V, Villanueva JR (1979) Occurrence of an endo-(1→3)-β-glucanase in culture fluids of the yeast *Candida utilis*. Biochem J 177:107–114
Wang MC, Bartnicki-Garcia S (1966) Biosynthesis of β-1,3 and β-1,6-linked glucan by *Phytophthora cinnamomi* hyphal walls. Biochem Biophys Res Commun 24:832–837
Wang MC, Bartnicki-Garcia S (1973) Novel phosphoglucans from the cytoplasm of *Phytophthora palmivora* and their selective occurrence in certain life cycle stages. J Biol Chem 248:4112–4118
Wang MC, Bartnicki-Garcia S (1974) Mycolaminarans: storage (1→3)-β-D-glucans from the cytoplasm of the fungus *Phytophthora palmivora*. Carbohydr Res 37:331–338
Wang MC, Bartnicki-Garcia S (1976) Synthesis of β-1,3-glucan microfibrils by a cell-free extract from *Phytophthora cinnamomi*. Arch Biochem Biophys 175:351–354
Wessels JGH (1969) A β-(1→6)-glucan glucanohydrolase involved in hydrolysis of cell-wall glucan in *Schizophyllum commune*. Biochim Biophys Acta 178:191–193
Wessels JGH, Koltin Y (1972) R-glucanase activity and susceptibility of hyphal walls to degradation in mutants of *Schizophyllum* with disrupted nuclear migration. J Gen Microbiol 71:471–475

Wessels JGH, Niederpruem DJ (1967) Role of a cell-wall glucan-degrading enzyme in mating of *Schizophyllum commune*. J Bacteriol 94:1594–1602
Wessels JGH, Kreger DR, Marchant R, Regensburg BA, de Vries OMH (1972) Chemical and morphological characterisation of the hyphal wall surface of the Basidiomycete *Schizophyllum commune*. Biochim Biophys Acta 273:346–358
Yamamoto S, Nagasaki S (1975) Purification, crystallisation and properties of an endo-β-(1→3)-glucanase from *Rhizopus chinensis* R-69 Agric Biol Chem 39:2163–2169
Yamamoto S, Shiraishi T, Nagasaki S (1972) A crystalline enzyme which degrades the cell wall of living yeast. Biochem Biophys Res Commun 46:1802–1809
Yanagishima N (1963) Effect of auxin and anti-auxin on cell elongation in yeast. Plant Cell Physiol 4:257–264
Yu RJ, Bishop CT, Cooper FP, Blank F, Hasenclever HF (1967) Glucans from *Candida albicans* (serotype B) and from *Candida parapsilosis*. Can J Chem 45:2264–2267
Yuen S (1974) Pullullan and its application. Process Biochem 22:7–9
Zalokar M (1959) Growth and differentiation of *Neurospora hyphae*. Am J Bot 46:602–609
Zevenhuizen L, Bartnicki-Garcia S (1969) Chemical structure of the insoluble hyphal wall glucan of *Phytophthora cinnamomi*. Biochemistry 8:1496–1501
Zevenhuizen L, Bartnicki-Garcia S (1970) Structure and role of a soluble cytoplasmic glucan from *Phytophthora cinnamomi*. J Gen Microbiol 61:183–188
Zonneveld BJM (1971) Biochemical analysis of the cell wall of *Aspergillus nidulans*. Biochim Biophys Acta 249:506–514
Zonneveld BJM (1972a) Morphogenesis in *Aspergillus nidulans*. The significance of α-(1→3)-glucan of the cell wall and α-(1→3)-glucanase for cleistothecium development. Biochim Biophys Acta 273:174–187
Zonneveld BJM (1972b) A new type of enzyme, an exo-splitting α-(1→3)-glucanase from non-induced cultures of *Aspergillus nidulans*. Biochim Biophys Acta 258:541–547

18 Mannoproteins: Structure

R.E. COHEN and C.E. BALLOU

1 Definition and Occurrence

The bulk of this chapter will be concerned with mannoproteins of the yeast *Saccharomyces cerevisiae*. For this review, mannoproteins are defined as glycoproteins that contain 15 to 90% mannose by weight. Although this volume emphasizes extracellular carbohydrates, the discussion will include the cell wall and periplasmic mannoproteins and, for comparison, intracellular glycoproteins found in the yeast vacuole. Other reviews of mannoprotein structure have appeared (BALLOU 1974, 1976, FARKAŠ 1979, PHAFF 1971, SPENCER and GORIN 1973), and this article will aim at summarizing the central facts and describing some newer developments.

We now recognize four kinds of mannoproteins, delineated by their structure, location, and function. In the cell wall, intermeshed with the glucan network, is a structural mannoprotein containing up to 90% mannose (PEAT et al. 1961). In the periplasm of the cell are the mannoprotein enzymes, such as invertase and acid phosphatase, that contain about 50% mannose (LAMPEN 1968). Inside the cell, localized predominantly in the vacuole, are hydrolytic enzymes, such as carboxypeptidase, with about 15% mannose (HAYASHI 1976). In each of these three types, most of the carbohydrate is attached to the protein by way of di-*N*-acetylchitobiose units linked to asparagine (NAKAJIMA and BALLOU 1974b). In contrast, the fourth type of mannoprotein, represented by the sexual agglutinin carried on the surface of *Hansenula wingei* (CRANDALL and BROCK 1968) and other strongly agglutinative yeasts (BURKE et al. 1980), has about 85% mannose, most or all of which is linked to serine and threonine residues in the protein (YEN and BALLOU 1974b). About 10% of the mannose in the periplasmic and cell wall mannoproteins mentioned above is linked to protein in a similar fashion (NAKAJIMA and BALLOU 1974a).

2 Isolation and Criteria of Homogeneity

The classical isolation of cell wall and periplasmic mannoproteins (PEAT et al. 1961) makes use of their solubility in high concentrations of ammonium sulfate and their ease of precipitation by complexation with alkaline cupric ion, both a consequence of the high mannose content. Mannoproteins with less than 40% mannose may be precipitated by 75% ammonium sulfate (LEHLE et al.

1979), and the slightly alkaline cetyltrimethylammonium borate complexation (LLOYD 1970) is a milder procedure than Fehling's solution for mannoprotein precipitation. As investigations of mannoprotein structure become more sophisticated, there is need to develop methods for preparing homogeneous materials without denaturating the protein component. Standard methods of protein purification involving ion exchange chromatography and gel filtration can lead to fairly homogeneous invertase preparations (NEUMANN and LAMPEN 1967), and specialized affinity adsorbents may be found, such as that for isolation of carboxypeptidase Y (JOHANSEN et al. 1976). Lectins, particularly concanavalin A, can also be useful as affinity adsorbents.

Homogeneity of mannoproteins is not easily established or, perhaps, even easily defined. An invertase preparation might be considered homogeneous if it were purified to a maximum and constant catalytic activity. To establish that the protein component is indeed homogeneous requires that the carbohydrate part be removed and the protein be analyzed by ion exchange chromatography or gel electrophoresis. Even with the specific endo-*N*-acetyl-β-D-glucosaminidases now available, however, this is not easily accomplished (TARENTINO et al. 1974). Should the protein be demonstrated to be homogeneous, there is still the question of the nature of the carbohydrate component. As illustrated in this review and elsewhere (TRIMBLE and MALEY 1977a), purified yeast invertase possesses a variety of carbohydrate chains, and even the more easily purified carboxypeptidase Y yields at least four different oligosaccharide chains when digested with an endo-*N*-acetyl-β-D-glucosaminidase (HASHIMOTO et al. 1981).

3 Methods for Structural Analysis

3.1 Selective Chemical Degradations

The chemical dissection of mannoproteins has played an important role in their characterization. Because they are released by β-elimination during mildly alkaline treatment, the oligosaccharides attached to hydroxyamino acids can be separated from the chains linked to asparagine (NAKAJIMA and BALLOU 1974a). The latter attachment can be split under mild conditions by an endo-*N*-acetyl-β-D-glucosaminidase (NAKAJIMA and BALLOU 1974b), which leaves a single *N*-acetylglucosamine still linked to the protein; or the polysaccharide chain can be obtained intact by hydrazinolysis (HASILIK and TANNER 1978b), which cleaves the amide bonds of the protein and of the mannan.

Some mannoproteins contain sugar units attached to the polysaccharide side-chains in phosphodiester linkage (THIEME and BALLOU 1971), and such sugar units are released by hydrolysis in 0.1 M HCl at 100° C for 15 min, a treatment that does not significantly affect other glycosidic bonds in the molecule. These other bonds differ from each other in their susceptibility to acetolysis, a procedure that allows selective cleavage of (1→6)-linkages (ROSENFELD and BALLOU 1974a). Because *S. cerevisiae* mannan has an (1→6)-α-linked backbone, partial acetolysis yields a mixture of oligosaccharides that represent the side-chain units of the polymer (GORIN and PERLIN 1956).

The various chemical degradation procedures differ in their severity, so the order of a sequential fragmentation should be chosen that will yield the desired information. Thus, the alkaline conditions for β-elimination may lead to some hydrolysis of phosphodiester bonds, whereas the acid hydrolysis of the latter can be carried out without degradation of the *O*-glycosidic bonds. Similarly, β-elimination should precede acetolysis if the aim is to distinguish *O*-linked from *N*-linked side-chains.

3.2 Selective Enzymic Degradations

Two D-mannanases of bacterial origin have been isolated, one an exo-α-mannanase that will remove the terminal mannose units (JONES and BALLOU 1969), and the other an endo-(1→6)-α-mannanase that will degrade an unsubstituted (1→6)-α-mannan chain (NAKAJIMA et al. 1976). The exomannase acts preferentially to remove the (1→2)-α- and (1→3)-α-linked side-chains, which allows isolation of an almost intact backbone, although the enzyme does slowly attack even the backbone. The exomannanase is inhibited by the presence of phosphate in the mannan, so the removal of the side-chains is most successful if the mannan is from a mutant that is unable to phosphorylate the glycoprotein. For some reason, the exomannanase is unable to remove the last mannose unit attached to serine and threonine in *H. wingei* 5-agglutinin (SING et al. 1976). Whereas the endomannanase degrades an unsubstituted (1→6)-α-mannan to a mixture of mannose and mannobiose, it will also attack an (1→6)-α-mannan that has occasional side-chains to yield larger oligosaccharides with (1→6)-α-linkages and an (1→2)-α-linked side-chain mannose (COHEN et al. 1980). Jack bean α-mannosidase is not very active on yeast mannans (LI 1967), although it will degrade the isolated oligosaccharides.

The endo-*N*-acetyl-β-D-glucosaminidase H has been used with variable success to remove the asparagine-linked polymannose chains of yeast mannoproteins (TARENTINO et al. 1974, TRIMBLE and MALEY 1977b). It appears to remove the larger polymer chains from invertase preferentially, perhaps because they are most exposed (LEHLE et al. 1979), and complete deglycosylation occurs only after carboxymethylation of the protein. Yeast mannoproteins do not contain the "complex" type of carbohydrate chains found in mammalian glycoproteins, so there is no interference from *N*-acetylneuraminic acid or fucose, but other structural variations might interfere with the action of this enzyme on mannoproteins.

Yeast mannoproteins are susceptible to proteases (THIEME and BALLOU 1972), but the 5-agglutinin of *H. wingei* is uncommonly resistant (YEN and BALLOU 1974b), undoubtedly due to the extensive glycosylation of the molecule, two-thirds of the amino acids being substituted in this manner.

3.3 Mutant Analysis

Simplification of the carbohydrate component of *S. cerevisiae* mannoproteins through "mutational surgery" has played an important role in structure determination (RASCHKE et al. 1973). Because the mannoproteins are exposed on the cell surface and are immunogenic, it has been possible to obtain mannan-specific antisera, and these in turn can be used to select for cells with altered surface determinants from a mutagenized culture, thus enriching for mutants. A bank of antisera against all available mannan mutants can be used to identify or score the different mutations, and such sera are useful in selecting for still "deeper" mutations. Recently, the technique has been improved by using fluorescent antibodies or lectins that will stick to the cell wall and distinguish wild-type from mutant cells; then, by processing the culture in a fluorescence-activated cell sorter, fractions with up to 5% mutant cells can be obtained (DOUGLAS 1979, DOUGLAS and BALLOU 1979). An important advantage of this technique is that it permits the selection of slow-growing mutants, whereas enrichment for mutants by antibody agglutination (RASCHKE et al. 1973) requires several generations of cell growth, with the result that the percentage of slow-growing cells in the culture is reduced. Thus, the fluorescence-activated cell sorter has been used to select several mutants of *S. cerevisiae* that have shortened mannoprotein outer chains (BALLOU et al. 1980). All of these mutants grow slowly, and previous attempts to isolate them by the antibody agglutination method were unsuccessful. These mutants are discussed more fully in Section 4.2.

In addition to their value for structure determination, mannan mutants have been useful for defining the biosynthetic pathway, the genetic control of biosynthesis and the function of mannoproteins (BALLOU 1974). Some mutants point to a distinction between the biosynthesis of the core and outer chain, whereas others show that these two parts utilize some enzymes in common. Mutations that appear to have a regulatory function can be obtained (KARSON and BALLOU 1978), which suggests that synthesis of mannoproteins, like that of other macromolecules, is controlled during cell proliferation. Finally, the ability of cells to function normally with extensive alterations in the mannan sets some limit on the physiological role served by the carbohydrate component of wild-type mannoproteins.

3.4 Immunochemical Methods

Yeast mannoproteins are the immunodominant structures of the cell surface (BALLOU and RASCHKE 1974). Terminal $(1\rightarrow3)$-α-linked mannose units (SUZUKI et al. 1968) and mannosylphosphate units (RASCHKE and BALLOU 1971) appear more immunogenic than terminal $(1\rightarrow2)$-α-linked mannose, but the terminal $(1\rightarrow2)$-α-linked *N*-acetylglucosamine of *Kluyveromyces lactis* is a good immunogen (RASCHKE and BALLOU 1972). Why these distinctions should exist is not clear, but they serve as a useful warning that all parts of the mannoprotein structure cannot be studied with equal ease by this technique. Moreover, the loss of an immunochemical determinant may not always reflect the loss of the structure, but instead can be due to modification of the antigen as exampled by the change from a mannosylphosphate unit to a mannobiosylphosphate unit in a diploid obtained by crossing *S. cerevisiae* A364A and X2180 (ROSENFELD and BALLOU 1974b).

An antiserum raised in rabbits against whole heat-killed yeast cells appears to precipitate all mannoproteins of the particular cell type, a reflection of the common carbohydrate structures (SMITH and BALLOU 1974). If prepared against an isolated mannoprotein such as invertase, however, the resulting antiserum can be adsorbed with cell wall mannan and made specific for the protein component of the invertase. Such a serum can be useful for studies on the enzymic synthesis and processing of a mannoprotein molecule (HASILIK and TANNER 1978a, b).

3.5 Physical Methods

Physical studies of mannoproteins are often complicated because of the high carbohydrate content. Analysis of bulk mannan by gel filtration and by sedimentation velocity usually reveals extensive size heterogeneity, and fractionation by ion exchange chromatography shows that charge heterogeneity is also present (THIEME and BALLOU 1972). Purified mannoproteins, such as invertase, give broad diffuse bands on gel electrophoresis in sodium dodecyl sulfate that are somewhat sharper for the enzyme from mutants with a simplified carbohydrate structure (LEHLE et al. 1979). High performance liquid chromatography is an excellent procedure for fractionating the acetolysis fragments (SEYMOUR et al. 1976) and the core fragments (LEHLE et al. 1979) from mannoproteins. Optical rotation and circular dichroism measurements have been useful in solving some specialized structural problems (RASCHKE and BALLOU 1972).

^{1}H (LEE and BALLOU 1965, GORIN et al. 1968, 1969) and ^{13}C (GORIN 1973) nuclear magnetic resonance have been used with considerable success in studies of mannoproteins and their degradation products. The mannose anomeric proton chemical shifts of intact mannans allow direct structural comparisons of taxonomic value, whereas the characterization of acetolysis fragments is greatly facilitated by [^{1}H]NMR. A recent detailed analysis of mannooligosaccharide reference compounds has led to a catalog of ^{1}H chemical shift data (COHEN and BALLOU 1980) that can be used to assign structures to more complex core fragments obtained from mannoproteins by endo-*N*-acetyl-*β*-D-glucosaminidase H digestion.

4 *Saccharomyces cerevisiae* Mannoprotein

4.1 Bulk Cell Wall Mannoprotein

The present, most refined structure for the carbohydrate component of *S. cerevisiae* cell wall mannoprotein is shown in Fig. 1, and the following experimental facts document this structure. (1) The composition reveals a mannose: phosphate ratio of about 75, that varies somewhat between preparations, and mild acid hydrolysis releases mainly (1→3)-α-mannobiose in an amount equimolar to the phosphate content. During this hydrolysis, the previously diesterified phos-

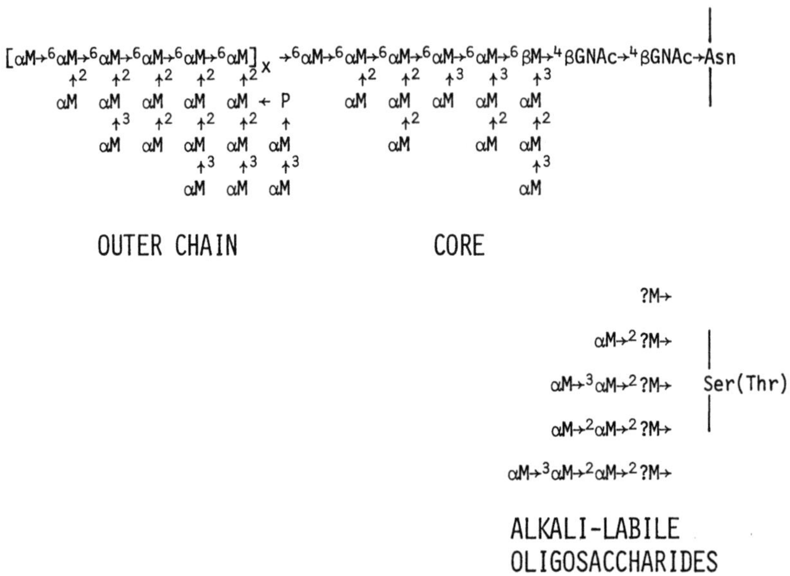

Fig. 1. Generalized structure of the carbohydrate components of *Saccharomyces cerevisiae* X2180 wild-type cell wall mannoprotein. No sequence of side-chains is implied, although it has been established that 3,6-di-*O*-substituted mannose is present only in the core. It is not known whether the alkali-labile oligosaccharides and asparagine-linked polysaccharide are attached to the same protein, and the anomeric configuration of the mannose-hydroxy-amino acid linkage has not been determined (see however, Chap. 19, this Vol.). Phosphate is attached to position 6 of the side-chain mannose. The *bracketed fragment* is not a repeating unit, but it is meant to indicate the assortment of side-chains found in the outer chain. The value of the subscript x is 10 to 15. *M* D-mannose; *P* phosphate; *GNAc* *N*-acetyl-D-glucosamine; *Asn, Ser,* and *Thr* amino acids in the polypeptide chain(s)

phate becomes monoesterified and remains with the large molecular weight mannoprotein residue. (2) The β-elimination products contain mannose, mannobiose, mannotriose, and mannotetraose of the indicated structures, which usually represent about 10% of the total mannose of the mannoprotein. (3) The acetolysis fragments of the β-eliminated mannan consist of mannose, mannobiose, mannotriose, mannotetraose, and mannotetraose phosphate. (4) Exhaustive digestion of the mannan by the bacterial exo-α-mannanase yields predominantly an unbranched (1→6)-α-polymer of mannose. (5) Digestion of the *mnn2* mutant mannoprotein of *S. cerevisiae* with endo-(1→6)-α-mannanase and endo-*N*-acetyl-β-D-glucosaminidase yields mannooligosaccharides terminated at the reducing ends by *N*-acetylglucosamine. The compositions of these oligosaccharides range from Man_{12}-GlcNAc to Man_{17}-GlcNAc. (6) Acetolysis of these oligosaccharides yields the pentasaccharide αMan→^{3}αMan→^{2}αMan→^{3}βMan→^{4}GlcNAc and other fragments that are mainly terminated at the reducing end by (1→3)-α-linked mannose.

Demonstration of the core structure of a new mannoprotein generally depends on isolation of the mannan from a strain carrying the *mnn2* mutation (NAKAJIMA and BALLOU 1974b, RASCHKE et al. 1973). This is easily done if the new yeast is interfertile with *S. cerevisiae* X2180, because such a strain can be constructed by genetic techniques (FINK 1970). Alternatively, the analogous mutant must be isolated in the new strain, which can require considerable effort. Moreover, it is not certain that all mannoproteins will prove to have a structure similar to *S. cerevisiae*. In fact, the (1→6)-linked backbone appears to be altered in other yeasts by interspersion of (1→2) and (1→3) linkages, and such a polymer would be more difficult to characterize than that found in *S. cerevisiae* (YEN and BALLOU 1974a).

4.2 Characteristics of Mannan Mutants

The mannan mutants of *S. cerevisiae* selected on the basis of altered surface antigens show the following changes. The *mnn1* defect is the most wide-ranging (BALLOU 1974) in that the mannan lacks terminal (1→3)-linked mannose in the core and *O*-linked oligosaccharides as well on the outer chain, and the interpretation of the associated changes is that a single (1→3)-α-mannosyltransferase of broad specificity is involved. The *mnn2* strain lacks the side-chains of the outer chain, a result explainable by a defect in an (1→2)-α-mannosyltransferase that adds the first mannose to the backbone. The *mnn5* locus appears to involve the activity of a second (1→2)-α-mannosyltransferase because this mutant carries single (1→2)-α-mannosyl units attached to the (1→6)-α-backbone. Both the *mnn2* and *mnn5* mutants lack phosphate in the mannan because the outer chain does not possess the acceptors for addition of the mannosylphosphate units. The *mnn3* locus was originally thought (RASCHKE et al. 1973) to have the *mnn5* phenotype, but it is now recognized to affect (1→2)-α-and (1→3)-α-linked mannose units on both the *O*-linked oligosaccharides and the outer chain. These multiple structural changes cannot be rationalized by the loss of discrete mannosyltransferase activities. Instead, the *mnn3* phenotype must be a pleiotropic affect, possibly resulting from limited availability of substrates necessary for mannoprotein biosynthesis (COHEN et al. 1980). The *mnn4* and *mnn6* loci (KARSON and BALLOU 1978) regulate the addition of mannosylphosphate to the mannan, but they differ in that the former is dominant and the latter is recessive. Possibly one is the structural gene for the transferase and the other is a regulatory gene.

In an attempt to isolate mutants that lack the outer chain, the *mnn2* strain has been mutagenized and cells unable to react with antiserum directed against the unbranched (1→6)-α-polymannose chain were selected (BALLOU et al. 1980). Genetic analysis proved that the mutants were not revertants to the wild type, and clones representing at least four complementation groups (designated *mnn7, mnn8, mnn9,* and *mnn10*) were isolated. Such mutants acquire the ability to react with antiserum directed against terminal (1→3)-α-mannose units, as though the core were more exposed, and the ratio of total mannose in the mannoprotein to that released by β-elimination is much reduced compared to the parent strain, as expected if the outer chain had been reduced in size. Structural studies indicate that the mutant mannoproteins retain a normal core but have only about 45 mannoses in the outer chain, compared with 260 in the wild type. This number is reduced to 15 or less for the double mutants containing the

Fig. 2. Structures of the asparagine-linked polysaccharides of the *Saccharomyces cerevisiae* mutant mannoproteins. The effects of the mutations upon the alkali-labile oligosaccharides are discussed in the text (Sect. 4.2). In comparison with the wild-type structure (Fig. 1) where x=10 to 15, the principal effect of the *mnn7, mnn8, mnn9,* or *mnn10* mutations is a dramatic shortening of the outer chain such that y=0 to 2. Other subtle differences may exist between the four complementation groups. Symbols are explained in the legend to Fig. 1

mnn2 lesion, which is to be compared with 125 for the *mnn2* outer chain. A surprising observation is that the mannoprotein from such mutants contains phosphate, an anomalous result because previous studies suggest that the core is not phosphorylated in the wild-type yeast. Finally, the mutants are of particular interest because all exhibit an altered cellular physiology. The cells grow slowly and tend to lyse, fail to divide cleanly, and have distorted cell walls. How these properties are associated with the structural alterations in the mannoprotein remains to be determined.

Figure 2 shows generalized structures for the asparagine-linked carbohydrate chains of the *S. cerevisiae* mannoprotein mutants. Not shown are the serine- and threonine-linked oligosaccharides (see Fig. 1), which are affected only by the *mnn1* and *mnn3* lesions. In the *mnn1* mutant, the tetrasaccharides are absent due to the lack of the terminal (1 → 3)-α-mannosyltransferase, whereas in the *mnn3* mutant the amounts of both tri- and tetrasaccharide *O*-linked oligosaccharides are diminished.

The mannoproteins of several interfertile *Saccharomyces* strains have been analyzed by acetolysis, and it is notable that the wild-type variants differ from each other by the presence or absence of one or both of two (1 → 3)-α-linked mannose units (BALLOU et al. 1974). Presumably, the genes controlling two (1 → 3)-transferases and the corresponding null loci are segregating in these yeast strains, thus leading to the observed polymorphism. What the natural selection pressure is that retains this polymorphism is unknown, but the presence of variable surface determinants is reminiscent of the pattern observed in many other systems, such as the blood group substances (WATKINS 1972).

4.3 Invertase, an Example of a "Homogeneous" Extracellular Glycoprotein

Yeasts contain a variety of extracellular enzymes, mostly hydrolases that are secreted in response to the growth conditions (LAMPEN 1968). In *S. cerevisiae*, invertase (β-fructofuranosidase), acid phosphatase, and asparaginase II have been shown to be mannoproteins (NEUMANN and LAMPEN 1967, BOER and STEYN-PARVÉ 1966, DUNLOP et al. 1978), whereas α-galactosidase (BUCKHOLZ and ADAMS 1978) and (1 → 3)-β-glucanase (BIELY et al. 1976) are also glycosylated and may prove to be mannoproteins.

Of the several genetic loci that encode invertase in yeast, only one of these, *SUC2*, is active in *S. cerevisiae* X2180 (WINGE and ROBERTS 1957, OTTOLENGHI 1971). External invertase can be produced on a large scale either by adjusting the growth medium to optimize secretion of the enzyme (LEHLE et al. 1979) or by using strains that are genetically derepressed for external invertase production (NEUMANN and LAMPEN 1969).

The carbohydrate components of external invertases from wild-type and *mnn* mutants of *S. cerevisiae* are immunochemically indistinguishable from the bulk cell wall mannoproteins of the respective strains (SMITH and BALLOU 1974). Thus, polysaccharide chains of these molecules must have similar structures and are probably assembled by the same enzymes. The carbohydrate associated with *S. cerevisiae* invertase is linked to asparagine (NEUMANN and LAMPEN 1969, TARENTINO et al. 1974, TRIMBLE and MALEY 1977a, LEHLE et al. 1979), although hydroxyamino acid-linked mannooligosaccharides have been reported as the major carbohydrate for an invertase from an unidentified yeast strain (GREILING et al. 1969).

S. cerevisiae external invertase contains 50% mannose and has a molecular weight of 270,000 (Neumann and Lampen 1967). When examined by polyacrylamide gel electrophoresis in sodium dodecyl sulfate, the enzyme gives a single, diffuse band between 90,000 and 160,000 molecular weight (Trimble and Maley 1977a). Nearly all of the carbohydrate can be removed from the native enzyme by digestion with endo-β-*N*-acetylglucosaminidase H (Tarentino et al. 1974), and the products appear as three or four discrete glycoprotein bands between 63,000 and 69,000 molecular weight on gel electrophoresis. Complete deglycosylation, which requires denaturation of the invertase by reduction and carboxymethylation, yields a single protein of 60,000 molecular weight. The studies suggest that the native external invertase is a dimer of identical polypeptides, each of which has an average of nine asparagine-linked carbohydrate chains (Trimble and Maley 1977a).

Recently, Lehle et al. (1979) have extended the findings of Maley and his coworkers, making use of invertase from the *S. cerevisiae* X2180 *mnn2* mutant to facilitate the distinction of the core from the outer chain carbohydrate. Only the *mnn2* outer chain is degraded by endo-(1→6)-α-mannanase or recognized by anti-(1→6)-α-mannosyl sera, whereas anti-(1→3)-α-mannosyl sera will recognize only the core. Two forms of *mnn2* invertase were obtained from homogenized cells by fractionation with 75% saturated ammonium sulfate. The soluble fraction (S75 invertase) contained 53% carbohydrate, and the precipitated fraction (P75 invertase) had 36% carbohydrate. Anti-(1→6)-α-mannosyl sera reacted only with the S75 invertase, whereas the anti-(1→3)-α-mannosyl sera reacted with both. Digestion of the S75 invertase with endo-(1→6)-α-mannanase removed the outer chain and converted the enzyme to a form similar to the P75 invertase, which itself was resistant to the endomannanase.

Polyacrylamide gel electrophoresis of endo-β-*N*-acetylglucosaminidase digests of either the S75 or P75 invertase gave a set of 63,000 to 69,000 molecular weight proteins similar to that obtained from the wild-type invertase. The bands differed by a molecular weight of about 3,000, which is approximately the size of a core oligosaccharide. Whereas the untreated S75 invertase had a mobility on gel electrophoresis similar to the wild-type protein, the P75 fraction appeared smaller (75,000 to 100,000 molecular weight) and was similar to S75 invertase after its digestion with endo-(1→6)-α-mannanase. These results are summarized in Fig. 3.

The carbohydrate released by endo-β-*N*-acetylglucosaminidase digestion of both S75 and wild-type invertases contained products of two size classes. The smaller to these had an average composition of Man_{14}-GlcNAc, and it was similar in structure to the core oligosaccharides obtained from the *mnn2* bulk cell wall mannoprotein. The larger polysaccharide corresponded to the mannan outer chain and, for the S75-derived material, was converted to core oligosaccharides by endo-(1→6)-α-mannanase digestion. Only two or three of the nine carbohydrate chains per invertase monomer possessed an outer chain structure. In the S75 fraction, these had an average length of 115 sugar units, whereas the core fragments averaged 15 sugars. In contrast, the P75 form of the enzyme contained core oligosaccharides with little or no outer chain.

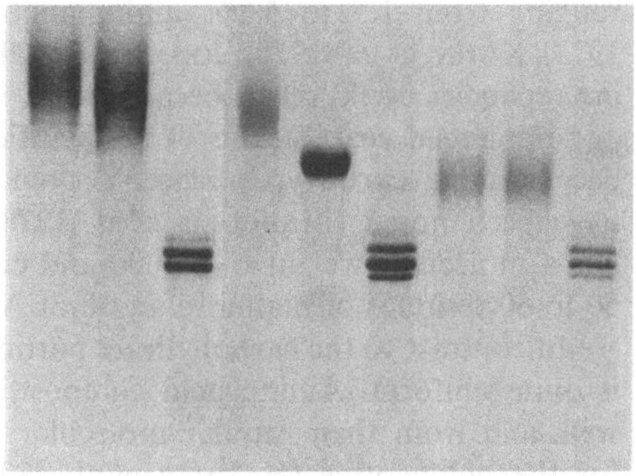

Fig. 3. Polyacrylamide gel electrophoresis in sodium dodecyl sulfate of *S. cerevisiae* invertase preparations. Lanes *1* to *3* are wild-type invertases, lanes *4* to *6* are *mnn2* S75 invertases, and lanes *7* to *9* are *mnn2* P75 invertases. The first sample in each set is the untreated enzyme, the second is after endo-(1 → 6)-α-mannanase digestion, and the third is after endo-β-*N*-acetylglucosaminidase digestion. (LEHLE et al. 1979)

That two classes of carbohydrate chains are released from wild-type and *mnn2* S75 invertase by endo-β-*N*-acetylglucosaminidase was first reported by TARENTINO et al. (1974). They observed, however, that the oligosaccharides had intermediate sizes of 26 and 54 mannose units, respectively. This discrepancy may reflect incomplete separation of the two populations, with the consequence that cross-contamination skewed their analysis toward the median of the chain lengths.

The discovery of two forms of glycosylated *mnn2* invertase, one with only core oligosaccharides and the other with additional polymannose chains, suggests a precursor-product relationship. Preliminary experiments indicate that the P75 form is membrane-associated, whereas the S75 invertase is probably from the periplasm. The idea that invertase biosynthesis occurs in two stages by transfer of a core oligosaccharide to the polypeptide and subsequent elaboration of the outer chain is compatible with current models of mannoprotein biosynthesis (reviewed in Chap. 19, this Vol.).

4.4 Carboxypeptidase Y, an Example of a "Homogeneous" Intracellular Mannoprotein

Glycosylated intracellular proteins are known to occur in the yeast vacuole. Two proteases, carboxypeptidase Y (formerly termed proteinase C) and proteinase A, were shown to contain carbohydrate by HATA et al. (1967), and subsequent work established that carboxypeptidase Y is a mannoprotein with four carbohydrate chains. More recently, *S. cerevisiae* alkaline phosphatase (ONISHI et al. 1979) and α-mannosidase (OPHEIM 1978) have been identified as glycoproteins, and both are found associated with the vacuolar membrane (BAUER and SIGARLAKIE 1975, OPHEIM 1978).

The carbohydrate component of carboxypeptidase Y consists of mannose and *N*-acetylglucosamine (HASILIK and TANNER 1976, MARGOLIS et al. 1978). Reports of the compositions vary considerably, with the hexosamine content ranging from 1.5 to 5.3% and mannose from 7.4 to 22% (HAYASHI et al. 1973, KUHN et al. 1974, JOHANSEN et al. 1976, MARGOLIS et al. 1978). These discrepancies could result because most investigations have employed a variety of commercial yeast strains of unidentified backgrounds, and systematic differences among carboxypeptidase Y preparations from different baker's yeasts have been noted (MARGOLIS et al. 1978). An "average" carboxypeptidase Y has a molecular weight of 61,000 and contains about 15% carbohydrate, with 50 to 60 residues of mannose and 8 of *N*-acetylglucosamine.

In contrast to the carbohydrate portion, the carboxypeptidase Y polypeptide is quite uniform. Amino acid compositions from different preparations agree well, and from them similar molecular weights (ca. 50,000) can be calculated for the polypeptide (HAYASHI et al. 1973, KUHN et al. 1974, JOHANSEN et al. 1976, MARGOLIS et al. 1978). A value of 51,000 molecular weight was observed for the deglycosylated enzyme (TRIMBLE and MALEY 1977b) and for an unglycosylated carboxypeptidase Y synthesized in tunicamycin-treated cells (HASILIK and TANNER 1978b). *S. cerevisiae* mutants have been selected that lack carboxypeptidase Y activity (WOLF and FINK 1975, JONES 1977). Because these mutations are recessive and affect single alleles, it is unlikely that the parent yeasts produce isozymes of carboxypeptidase Y. Thus, the differences observed between preparations are probably a consequence of differences in carbohydrate structure.

As with invertase (TRIMBLE and MALEY 1977a), endo-β-*N*-acetylglucosaminidase H treatment of native carboxypeptidase Y releases only part of the carbohydrate, and digestion of the denatured enzyme is required to release the remainder (TRIMBLE and MALEY 1977b). The residual protein contains no mannose and half of the original 8 *N*-acetylglucosamine residues, indicating a stoichiometry of four carbohydrate chains per carboxypeptidase Y molecule. As expected, the oligosaccharides contain *N*-acetylglucosamine at the reducing end, and the average composition is Man_{14}-GlcNAc. Similar results were obtained by HASILIK and TANNER (1978b), who analyzed the compositions of carbohydrate-containing fragments produced by protease digestion and hydrazinolysis of the enzyme. These studies indicate that the oligosaccharides are linked to the protein exclusively through di-*N*-acetylchitobiose units to asparagines; no carbohydrate susceptible to β-elimination has been found in carboxypeptidase Y.

The size, composition, and linkage of the carboxypeptidase Y carbohydrate chains suggest a structural resemblance to the core fragments of the cell wall mannoprotein (NAKAJIMA and BALLOU 1975) and external invertase (LEHLE et al. 1979). This was noted by HASILIK and TANNER (1976), who demonstrated that these three mannoproteins bear similar immunochemical determinants. Identification by [^{1}H]NMR of βMan(1 $\rightarrow$ 4)GlcNAc and 3,6-di-*O*-substituted α-mannose in the oligosaccharides released from carboxypeptidase Y by endo-β-*N*-acetylglucosaminidase H confirm their core-like structure (HASHIMOTO et al. 1981). The same workers have shown that some of the oligosaccharides are phosphorylated.

That carboxypeptidase Y possesses carbohydrate structures similar to those found in the core of extracellular mannoproteins implies that at least part of the machinery responsible for glycosylation is shared by these different proteins. By examining the effects of various *mnn* mutations, it may be possible to determine where the processing of intracellular (vacuolar) and extracellular (secreted) mannoproteins diverge and, simultaneously, provide some insight regarding the localization of various mannosyltransferases within the cell.

5 *Hansenula wingei* Sexual Agglutination Factors

5.1 Description of the Mating Reaction in Yeasts

The initiation of sexual mating in yeasts involves two distinct steps, a specific agglutination between haploid cells of opposite mating type that increases the efficiency of conjugation and an exchange of diffusible pheromones that induces the overall developmental process leading to zygote formation (FOWELL 1969, MANNEY and MEADE 1977). Weakly agglutinative strains, such as *S. cerevisiae* X2180, show an enhanced agglutinability as a result of pheromone action (FEHRENBACHER et al. 1978), and this change is associated with a visible reorganization of the cell surface (LIPKE et al. 1976, TKACZ and MACKAY 1979). Some yeasts show a strong constitutive agglutination, a property that is highly specific and is absent in the heterozygous diploid. Studies on the yeast *H. wingei* have shown that the agglutination is due to mannoproteins located on the cell surface (CRANDALL and BROCK 1968), and similar factors appear to be present on other yeasts such as *Pichia amethionina* and *Saccharomyces kluyveri* (BURKE et al. 1980). The following discussion deals with the isolation and properties of some of these agglutination factors.

5.2 Isolation and Structure of the 5-Cell Agglutinin

Subtilisin digestion of *H. wingei* 5-cells releases a soluble macromolecule with ability to agglutinate *H. wingei* 21-cells (CRANDALL and BROCK 1968). The material can be purified by affinity adsorption to 21-cells at pH 4 and elution at pH 1.8, followed by gel filtration on Bio-Gel A5m (YEN and BALLOU 1974b). The pure active 5-agglutinin thus obtained has a molecular weight of 960,000, and is composed of 85% mannose, 10% protein, and 5% phosphate. The phosphate is not required for biological activity because 5-agglutinin, isolated from a mutant that is unable to phosphorylate its mannoproteins, is fully active (SING et al. 1976). The protein contains 55% serine and 9% threonine, and β-elimination of the 5-agglutinin destroys 85% of these two amino acids and releases all of the carbohydrate as oligosaccharides with 1 to 16 mannose units. Thus, the molecule is composed of an assortment of mannooligosaccharides in *O*-glycosidic linkage to about half of the amino acids of the protein chain.

The activity of 5-agglutinin is stable to heat, but it is lost on treatment with reducing agents, which releases small glycopeptide fragments that retain the ability to bind to 21-cells and must represent the recognition sites of the agglutinin (TAYLOR and ORTON 1971). Agglutinative activity can be partly reconstituted by reoxidation of a mixture of core and binding fragments. The reduced mixture can be separated by gel filtration into a large (molecular weight 900,000) inactive core and the small (molecular weight 12,000) binding fragments. The latter is a glycopeptide with a lower serine and higher threonine content than the core glycoprotein, although it still contains about 75% mannose. The activity of the binding fragment is destroyed by exomannanase or protease digestion.

These studies suggest that the 5-agglutinin is a long, brush-like glycoprotein, with several small glycopeptide units attached at various points along the backbone through disulfide linkages involving cysteine residues (TAYLOR and ORTON 1971, YEN and BALLOU 1974b). The molecule solubilized by subtilisin presumably represents a degraded form, and an apparently intact 5-agglutinin has been prepared by β-glucanase digestion of the yeast cell wall. Although it is clearly larger, structural analysis of this material has not been pursued, and the chemical nature of the additional component that may serve to anchor the molecule in the cell wall is unknown.

5.3 Isolation and Structure of 21-Factor

Trypsin digestion of 21-cells destroys their agglutinability by 5-agglutinin, but it releases in soluble form a small, heat-labile protein that will inhibit the agglutination reaction although it does not agglutinate 5-cells (CRANDALL and BROCK 1968). Thus, this 21-factor appears to be monovalent. It has been purified by ion exchange chromatography, by affinity adsorption to 5-cells and by gel filtration (BURKE et al. 1980). The active material has a molecular weight of 27,000 and a pI of 3.5, which is correlated with a high content of glutamic and aspartic acids. About 5% by weight of mannose appears to be attached to the protein through di-*N*-acetylchitobiose linked to asparagine.

The calculated equivalent combining weights of 5-agglutinin and 21-factor are about 5:1, and in the agglutination inhibition assay the reaction is reversed at a ratio of about 3:1. Since this depends on the reversal of a multivalent interaction between 5-agglutinin and 21-cells by the monovalent 21-factor, it suggests that the 21-factor preparation has retained essentially full biological activity.

As with the 5-agglutinin preparation, a larger form of 21-factor is obtained if the cell wall is solubilized by β-glucanase digestion, and it is converted to a material with the properties of trypsin-released 21-factor if subjected to controlled proteolysis. Thus, the active component of 21-factor is also apparently anchored in the cell wall by some additional structure.

5.4 Comparison with Agglutination Factors in Other Yeasts

Pichia amethionina and *Saccharomyces kluyveri* show a constitutive agglutination that is similar to that of *H. wingei,* and the properties of the different factors suggest that they are similar to each other (BURKE et al. 1980, TAYLOR 1964). The agglutinative ability of one haploid cell type from each pair (*H. wingei* 5-cell, *P. amethionina* α-cell, and *S. kluyveri* 16-cell) is stable to heat but is destroyed by treatment with mercaptoethanol, whereas the other haploid cell type from each pair (*H. wingei* 21-cell, *P. amethionina* a-cell, and *S. kluyveri* 17-cell) is stable to mercaptoethanol but is inactivated by heat. This similarity is enhanced by the observation that solubilized *P. amethionina* α-cell factor is a large glycoprotein and acts like an agglutinin, whereas the factor from the a-cell type is a small, monovalent protein.

If the generalization proves to hold, it is possible to relate the mating types of widely divergent yeasts by a comparison of their agglutination factors. This is so because *S. kluyveri* 16-cells respond to α-factor isolated from *S. cerevisiae* (McCullough and Herskowitz 1979). Thus, we conclude that the a mating type (as represented in the nomenclature of *S. cerevisiae*) controls formation of the multivalent, heat-stable agglutinin.

References

Ballou CE (1974) Some aspects of the structure, biosynthesis and genetic control of yeast mannans. In: Meister A (ed) Advances in enzymology, vol 40. John Wiley & Sons, New York, pp 239–270

Ballou CE (1976) Structure and biosynthesis of the mannan component of the yeast cell envelope. In: Rose AH, Tempest DW (eds) Advances in microbial physiology, vol XIV. Academic Press, London New York, pp 93–158

Ballou CE, Raschke WC (1974) Polymorphism of the somatic antigen of yeast. Science 184:127–134

Ballou CE, Lipke PN, Raschke WC (1974) Structure and immunochemistry of the cell wall mannans from *Saccharomyces chevalieri, S. italicus, S. diastaticus*, and *S. carlsbergensis*. J Bacteriol 117:461–467

Ballou DL, Cohen RE, Ballou CE (1980) *Saccharomyces cerevisiae* mutants that make mannoproteins with a truncated carbohydrate outer chain. J Biol Chem 255:5986–5981

Bauer H, Sigarlakie E (1975) Localization of alkaline phosphatase in *Saccharomyces cerevisiae* by means of ultrathin frozen sections. J Ultrastruct Res 50:208–215

Biely P, Krátky Z, Bauer Š (1976) Interaction of concanavalin A with external mannan proteins of *Saccharomyces cerevisiae: glycoprotein nature of β-glucanases.* Eur J Biochem 70:75–81

Boer P, Steyn-Parvé EP (1966) Isolation and purification of an acid phosphatase from baker's yeast (*Saccharomyces cerevisiae*). Biochim Biophys Acta 128:400–402

Buckholz RG, Adams BG (1978) Synthesis and secretion of yeast α-galactosidase: possible processing and precursor relationships. Abstr 9th Int Conf Yeast Genet Mol Biol, p 100

Burke D, Previato LM, Ballou CE (1980) Cell-cell recognition in yeast. Isolation of the 21-cell sexual agglutination factor from *Hansenula wingei* and comparison of the factors from three genera. Proc Natl Acad Sci USA 77:318–322

Cohen RE (1980) Structure of glycoprotein core oligosaccharides. Doc Diss, Univ California, Berkeley CA

Cohen RE, Ballou CE (1979) Characterization of glycoprotein mannooligosaccharides by PMR. Abstr ACS/CJS Chem Congr, Honolulu Hawaii, April 1–6

Cohen RE, Ballou CE (1980) Linkage and sequence analysis of mannose-rich glycoprotein core oligosaccharides by proton nuclear magnetic resonance spectroscopy. Biochemistry 19:4345–4358

Cohen RE, Ballou DL, Ballou CE (1980) *Saccharomyces cerevisiae* mannoprotein mutants: isolation of the *mnn5* mutant and comparison with the *mnn3* strain. J Biol Chem 225:7700–7707

Crandall MA, Brock TD (1968) Molecular basis of mating in the yeast *Hansenula wingei.* Bacteriol Rev 32:139–163

Douglas RH (1979) Mannan biosynthesis in *Kluyveromyces lactis.* Doct Diss, Univ California, Berkeley CA

Douglas RH, Ballou CE (1979) Mannan biosynthesis in *Kluyveromyces lactis*. Abstr ACS/CJS Chem Congr, Honolulu Hawaii, April 1–6
Dunlop PC, Meyer GM, Ban D, Roon RJ (1978) Characterization of two forms of asparaginase in *Saccharomyces cerevisiae*. J Biol Chem 253:1297–1304
Farkaš V (1979) Biosynthesis of cell walls of fungi. Microbiol Rev 43:117–144
Fehrenbacher G, Perry K, Thorner J (1978) Cell-cell recognition in *Saccharomyces cerevisiae*: Regulation of mating-specific adhesion. J Bacteriol 134:893–901
Fink GR (1970) The biochemical genetics of yeast. In: Tabor H, Tabor CW (eds) Methods in enzymology, vol XVIIA. Academic Press, London New York, pp 59–78
Fowell R (1969) Life cycles in yeasts. In: Rose AH, Harrison JS (eds) The yeasts, vol I. Biology of yeasts. Academic Press, London New York, pp 461–471
Gorin PAJ (1973) The position of phosphate groups in phosphomannan of *Hansenula capsulata*, as determined by carbon-13 magnetic resonance spectroscopy. Can J Chem 51:2105–2109
Gorin PAJ, Perlin AS (1956) A mannan produced by *Saccharomyces rouxii*. Can J Chem 34:1796–1803
Gorin PAJ, Mazurek M, Spencer JFT (1968) Proton magnetic resonance spectra of *Trichosporon aculeatum* mannan and its borate complex and their relationship to chemical structure. Can J Chem 46:2305–2310
Gorin PAJ, Spencer JFT, Bhattacharjee SS (1969) Structures of yeast mannans containing both α- and β-linked D-mannypyranose units. Can J Chem 47:1499–1505
Greiling H, Vogele P, Kisters R, Ohlenbusch HD (1969) A carbohydrate-protein bond in β-fructofuranosidase. Hoppe-Seyler's Z Physiol Chem 350:517–518
Hashimoto C, Cohen RE, Zhang W-J, Ballou CE (1981) The carbohydrate chains on yeast carboxypeptidase Y are phosphorylated. Proc Natl Acad Sci USA 78:2244–2248
Hasilik A, Tanner W (1976) Inhibition of the apparent rate of synthesis of the vacuolar glycoprotein carboxypeptidase Y and its protein antigen by tunicamycin in *Saccharomyces cerevisiae*. Antimicrob Agents Chemother 10:402–410
Hasilik A, Tanner W (1978a) Biosynthesis of the vacuolar yeast glycoprotein carboxypeptidase Y. Conversion of precursor into the enzyme. Eur J Biochem 85:599–608
Hasilik A, Tanner W (1978b) Carbohydrate moiety of carboxypeptidase Y and perturbation of its biosynthesis. Eur J Biochem 91:567–575
Hata T, Hayashi R, Doi E (1967) Purification of yeast proteinases. Part III. Isolation and physicochemical properties of yeast proteinases A and C. Agric Biol Chem 31:357–367
Hayashi R (1976) Carboxypeptidase Y. In: Lorand L (ed) Methods in enzymology, vol XLVB. Academic Press, London New York, pp 568–587
Hayashi R, Moore S, Stein WH (1973) Carboxypeptidase from yeast. Large scale preparation and the application of COOH-terminal analysis of peptides and proteins. J Biol Chem 248:2296–2302
Johansen JT, Breddam K, Ottesen M (1976) Isolation of carboxypeptidase Y by affinity chromatography. Carlsberg Res Commun 41:1–14
Jones EW (1977) Proteinase mutants of *Saccharomyces cerevisiae*. Genetics 85:23–33
Jones GH, Ballou CE (1969) Studies on the structure of yeast mannan. I. Purification and some properties of an α-mannosidase from an Arthrobacter species. J Biol Chem 244:1043–1051
Karson EM, Ballou CE (1978) Biosynthesis of yeast mannan. Properties of a mannosylphosphate transferase in *Saccharomyces cerevisiae*. J Biol Chem 253:6484–6492
Kuhn RW, Walsh KA, Neurath H (1974) Isolation and partial characterization of an acid carboxypeptidase from yeast. Biochemistry 13:3871–3877
Lampen JO (1968) External enzymes of yeast: their nature and formation. Antonie von Leeuwenhoek J Microbiol Serol 34:1–18
Lee YC, Ballou CE (1965) Preparation of mannobiose, mannotriose, and a new mannotetraose from *Saccharomyces cerevisiae* mannan. Biochemistry 4:257–264
Lehle L, Cohen RE, Ballou CE (1979) Carbohydrate structure of yeast invertase. Demonstration of a form with only core oligosaccharides and a form with completed polysaccharide chains. J Biol Chem 254:12209–12218

Li Y-T (1967) Studies on the glycosidases in jack bean meal. I. Isolation and properties of α-mannosidase. J Biol Chem 242:5474–5480

Lipke PN, Taylor A, Ballou CE (1976) Morphogenic effects of α-factor on *Saccharomyces cerevisiae* a-cells. J Bacteriol 127:610–618

Lloyd KO (1970) Isolation, characterization, and partial structure of peptido galactomannans from the yeast form of *Cladosporium werneckii*. Biochemistry 9:3446–3453

Manney TR, Meade JH (1977) Cell-cell interactions during mating in *Saccharomyces cerevisiae*. In: Reissig JL (ed) Receptors and recognition, series B, vol III. Microbial interactions. Chapman and Hall, London, pp 283–321

Margolis HC, Nakagawa Y, Douglas KT, Kaiser ET (1978) Multiple forms of carboxypeptidase Y from *Saccharomyces cerevisiae*. Kinetic demonstration of effects of carbohydrate residues on the catalytic mechanism of a glycoenzyme. J Biol Chem 253:7891–7897

McCullough J, Herskowitz I (1979) Mating pheromones of *Saccharomyces kluyveri*: pheromone interactions betweeen *Saccharomyces kluyveri* and *Saccharomyces cerevisiae*. J Bacteriol 138:146–154

Nakajima T, Ballou CE (1974a) Characterization of the carbohydrate fragments obtained from *Saccharomyces cerevisiae* mannan by alkaline degradation. J Biol Chem 249:7679–7684

Nakajima T, Ballou CE (1974b) Structure of the linkage region between the polysaccharide and protein parts of *Saccharomyces cerevisiae* mannan. J Biol Chem 249:7685–7694

Nakajima T, Ballou CE (1975) Microheterogeneity of the inner core region of yeast mannoprotein. Biochem Biophys Res Commun 66:870–879

Nakajima T, Maitra SK, Ballou CE (1976) An endo-α-1→6-D-mannanase from a soil bacterium. Purification, properties and mode of action. J Biol Chem 251:174–181

Neumann NP, Lampen JO (1967) Purification and properties of yeast invertase. Biochemistry 6:468–475

Neumann NP, Lampen JO (1969) The glycoprotein structure of yeast invertase. Biochemistry 8:3552–3556

Onishi HR, Tkacz JS, Lampen JO (1979) Glycoprotein nature of yeast alkaline phosphatase. Formation of active enzyme in the presence of tunicamycin. J Biol Chem 254:11943–11952

Opheim DJ (1978) The regulation, purification, and characterization of the α-mannosidase of *Saccharomyces cerevisiae*. Abstr 9th Int Conf Yeast Genet Mol Biol, p 97

Ottolenghi P (1971) Properties of five non-allelic β-D-fructofuranosidases (invertases) of *Saccharomyces*. C R Trav Lab Carlsberg 38:213–221

Peat S, Whelan WJ, Edwards TE (1961) Polysaccharides of baker's yeast. Part IV. Mannan. Biochem J 29–34

Phaff HJ (1971) Structure and biosynthesis of the yeast cell envelope. In: Rose AH, Harrison JS (eds) The yeasts, vol II. Physiology and biochemistry of yeasts. Academic Press, London New York, pp 135–210

Raschke WC, Ballou CE (1971) Immunochemistry of the phosphomannan of the yeast *Kloeckera brevis*. Biochemistry 10:4130–4135

Raschke WC, Ballou CE (1972) Characterization of a yeast mannan containing N-acetyl-D-glucosamine as an immunochemical determinant. Biochemistry 11:3807–3816

Raschke WC, Kern KA, Antalis C, Ballou CE (1973) Genetic control of yeast mannan structure. Isolation and characterization of mannan mutants. J Biol Chem 248:4660–4666

Rosenfeld L, Ballou CE (1974a) Acetolysis of disaccharides. Comparative kinetics and mechanism. Carbohydr Res 32:287–298

Rosenfeld L, Ballou CE (1974b) Genetic control of yeast mannan structure. Biochemical basis for the transformation of *Saccharomyces cerevisiae* somatic antigen. J Biol Chem 249:2319–2321

Seymour FR, Slodki ME, Plattner RD, Stodola RM (1976) Methylation and acetolysis of extracellular D- mannans from yeast. Carbohydr Res 48:225–237

Sing V, Yeh YF, Ballou CE (1976) Isolation of a *Hansenula wingei* mutant with an altered sexual agglutinin. In: Bradshaw RA, Frazier WA, Merrell RC, Gottlieb DI, Hogue-Angeletti RA (eds) Surface membrane receptors. Plenum Publ Co, New York London, pp 87–97

Smith WL, Ballou CE (1974) Immunochemical characterization of the mannan component of the external invertase of *Saccharomyces cerevisiae*. Biochemistry 13:355–361

Spencer JFT, Gorin PAJ (1973) Mannose-containing polysaccharides of yeasts. Biotechnol Bioeng 15:1–12

Suzuki S, Sunayama H, Saito T (1968) Studies on the antigenic activity of yeasts. I. Analysis of the determinant groups of the mannan of *Saccharomyces cerevisiae*. Jpn J Microbiol 12:19–24

Tarentino AL, Plummer TH, Maley F (1974) The release of intact oligosaccharides from specific glycoproteins by endo-β-*N*-acetylglucosaminidase H. J Biol Chem 249:818–824

Taylor NW (1964) Inactivation of sexual agglutination in *Hansenula wingei* and *Saccharomyces kluyveri* by disulfide-cleaving agents. J Bacteriol 88:929–936

Taylor NW, Orton WL (1971) Cooperation among the active binding sites in the sex-specific agglutinin from the yeast *Hansenula wingei*. Biochemistry 10:2043–2049

Thieme TR, Ballou CE (1971) Nature of phosphodiester linkage of the phosphomannan from the yeast *Kloeckera brevis*. Biochemistry 10:4121–4129

Thieme TR, Ballou CE (1972) Subunit structure of the phosphomannan from *Kloeckera brevis* yeast cell wall. Biochemistry 11:1115–1119

Tkacz JS, MacKay VL (1979) Sexual conjugation in yeast. Cell surface changes in response to the action of mating hormones. J Cell Biol 80:326–333

Trimble RB, Maley F (1977a) Subunit structure of external invertase from *Saccharomyces cerevisiae*. J Biol Chem 252:4409–4412

Trimble RB, Maley F (1977b) The use of endo-β-*N*-acetylglucosaminidase H in characterizing the structure and function of glycoproteins. Biochem Biophys Res Commun 78:935–944

Watkins WM (1972) Blood-group specific substances. In: Gottschalk A (ed) Glycoproteins, part B. Elsevier Publ Co, Amsterdam New York, pp 830–891

Winge O, Roberts C (1957) A genetic analysis of melibiose and raffinose fermentation. C R Trav Lab Carlsberg 25:419–459

Wolf DH, Fink GR (1975) Proteinase C (carboxypeptidase Y) mutant of yeast. J Bacteriol 123:1150–1156

Yen PH, Ballou CE (1974a) Structure and immunochemistry of *Hansenula wingei* Y-2340 mannan. Biochemistry 13:2420–2427

Yen PH, Ballou CE (1974b) Partial characterization of the sexual agglutination factor from *Hansenula wingei* Y-2340 type 5 cells. Biochemistry 13:2428–2437

19 Biosynthesis of Mannoproteins in Fungi

L. LEHLE

1 Introduction

Among the various fungal glycoproteins, yeast mannans have been studied most systematically, not only from a structural point of view, but also in terms of their biosynthesis. Although considerable progress has been made since earlier reviews by BALLOU (1976), CABIB (1975) and FARKAŠ (1979), we are still in a state of having more open questions than clear answers. Several aspects of mannoprotein synthesis have been investigated in the past. For example, one has been directed to the overall process of mannoprotein synthesis by labeling intact cells and analyzing them by autoradiography. Another has been concerned with the formation and secretion of mannoproteins in protoplasts. Finally, intensive efforts have been made to dissect the numerous enzymatic steps involved in the assembly of the carbohydrate portion, as well as to study the cellular localization and regulation of the individual reactions. These aspects will be the subject of discussion of this article. Since yeast mannoproteins are glycoproteins and in many respects similar to the mannose-rich glycoproteins of animal and plant origin, it is not surprising that many parallels can be seen. Therefore, in order to be familiar with our present understanding on protein glycosylation in eukaryotic cells in general, the reader may also consult reviews related to this topic (HEMMING 1974, WAECHTER and LENNARZ 1976, PARODI and LELOIR 1979, see also Chap. 8, this Vol.).

2 Biosynthesis of Mannoproteins in *Saccharomyces* Species

2.1 Early in Vitro Studies with Guanosine Diphosphate Mannose

Most of our knowledge on mannan biosynthesis comes from studies using baker's yeast and dates back to the discovery of the nucleoside diphosphate sugars by LELOIR more than 20 years ago. When guanosine diphosphate mannose (GDP) was first isolated from yeast, it was proposed that it might be a precursor of mannan (CABIB and LELOIR 1954), but the direct evidence for this hypothesis was presented by ALGRANATI et al. in 1963 and later by BEHRENS and CABIB in 1968. They showed mannose transfer from the sugar nucleotide into a product with characteristics of mannan. The particulate system isolated from *S. carlsbergensis* was highly specific for GDP-Man, had an almost absolute requirement for Mn^{2+}, and an optimum pH between 5.5 and 7.2. Detailed degradation studies of the product further revealed that mannose was incorporated into all positions of the fragments obtained by selective acetolysis. This means that the preparation had synthesized a branched polymer that was similar to that formed in vivo. For some reason no tetrasaccharide was present in the hydrolysate from the mannan obtained in vitro. It is not yet clear whether the (1→3)-α-transferase is missing in the system used, or whether this is due to an overdegradation during acetolysis. The enzymic activity in the membrane system accounts for only about 1% of the rate in growing cells.

Another important contribution leading to the elucidation of the steps involved in the glycosylation of yeast mannoproteins and glycoproteins in general was made by TANNER in 1969. He showed that a novel mannosylphosphoryl-lipid was formed, when GDP-Man was incubated with a membrane fraction from yeast, and from the time dependence of the reaction a typical precursor-product relationship for mannan synthesis was evident. Formation of the glycolipid needed Mg^{2+} or Mn^{2+}, whereas the further transfer into the polymer fraction was dependent on Mn^{2+}. The reaction was easily reversible by GDP, indicating that the transfer potential of the sugar nucleotide is retained in the mannosyl-lipid. The radioactivity could be chased into a polymer fraction by addition of unlabeled GDP-Man. At the same time the existence of sugar-lipid derivatives with the possible function as intermediates in the synthesis of glycoproteins or polysaccharides was also reported in animal and plant systems (CACCAM et al. 1969, ZATZ and BARONDES 1969, KAUSS 1969, BEHRENS and LELOIR 1970), thus paving the way for a new development in this field.

2.2 Participation of Lipid-Linked Intermediates

2.2.1 Identification of the Glycosylated Lipid Intermediate

The mannolipid formed was different from known lipids in being degraded by mild acid (80% decomposition in 15 min with 0.1 N HCl at 20° C) and resistant to alkaline saponification. A hint on the polyprenol nature of the lipid came from experiments, in which it was shown that exogenous dolichyl monophosphate from yeast or liver enhanced the formation of the mannolipid (TANNER et al. 1971). The participation of a lipid intermediate in mannan biosynthesis was confirmed by SENTANDREU and LAMPEN (1971). JUNG and TANNER (1973) isolated the endogenous lipid acceptor and showed it to be dolichyl monophosphate. After dephosphorylation the mass spectrum of the compound was identical to that of authentic yeast dolichol. The lipid fraction was a mixture of dolichols with 14 to 18 isoprene units, somewhat shorter than the dolichols in animal systems, but longer than the bacterial polyprenols involved in the synthesis of various polysaccharides. Only about 10 to 20% of the free dolichol found in yeast cells is phosphorylated. Indirect evidence for the polyprenol nature of the lipid was also obtained by SENTANDREU and LAMPEN (1972).

It was speculated whether the different chain lengths of dolichol serve different functions, in a way that dolichol of a certain size carries *N*-acetylglucosamine, while others serve as acceptor for mannose or glucose (see below). However, REUVERS et al. (1978), who examined the lipid moiety of Dol-PP-GlcNAc from yeast by high pressure liquid chromatography, concluded that no particular homolog was preferred in the enzymic transfer.

2.2.2 Role of Dolichyl Monophosphate in the Formation of Mannooligosaccharide Chains Linked Through the Hydroxyl Group of Serine and Threonine

Further studies on the function of Dol-P-Man have revealed that it is involved both in the formation of the *O*-glycosidic and *N*-glycosidic linkage. The first

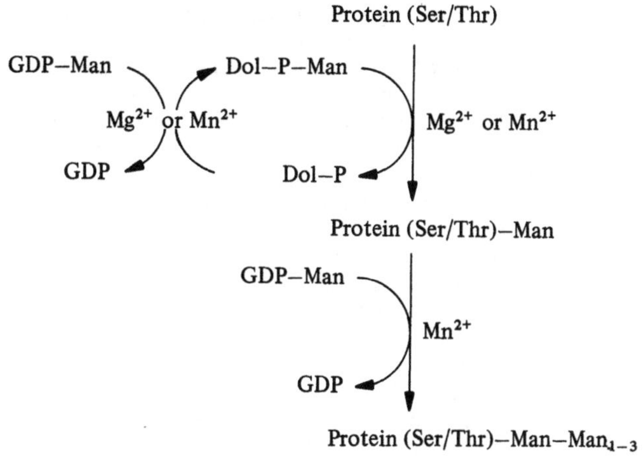

Fig. 1. Formation of fungal mannooligosaccharide chains linked through the hydroxyl group of serine or threonine

case has been clarified in studies by BABCZINKSI and TANNER (1973) and SHARMA et al. (1974). It was found that after incubation of *S. cerevisiae* membrane fraction with GDP-^{14}C-Man about 80% of the total radioactivity, incorporated into the glycoprotein fraction, could be released under mild alkaline conditions, a standard procedure to break glycosidic bonds to serine and threonine (β-elimination). The products obtained were mannose, mannobiose, and mannotriose. Mannosyl transfer was also achieved using Dol-P-^{14}C-Man as glycosyl donor. The oligosaccharide pattern after β-elimination, however, was different in this case: only free mannose was detectable. However, if Dol-P-^{14}C-Man and in addition unlabeled GDP-Man were incubated, again the formation of oligosacccharides was observed with restriction of the radioactivity to the reducing end. To rationalize the results the scheme in Fig. 1 has been proposed.

In support of this reaction sequence is the observation that aging of the membrane fraction leads to a loss of mannosyltransfer activity to Dol-P and consequently to a decrease in the transfer of the first mannose to serine or threonine. The amount of β-eliminable mannobiose and mannotriose is much less affected. Moreover, a different ion requirement of the various enzymes involved could be demonstrated. Formation of Dol-P-Man and also subsequent transfer of its mannose to protein proceeds both in the presence of Mg^{2+} and Mn^{2+}, whereas the elongation reactions occur only in the presence of Mn^{2+}. The configuration of mannose in Dol-P-Man was determined to be β (LEHLE and TANNER 1978a) and inversed during transfer to form an α-mannosyl-ser/thr-linkage (BAUSE and LEHLE 1979). Exogenously added yeast cell wall mannoprotein serves as acceptor for mannosyl groups, both with GDP-Man or Dol-P-Man as mannosyl donor (LEHLE and TANNER 1978a). The very specific role of Dol-P for the *O*-glycosylation in *S. cerevisiae* seems to be restricted to fungi so far (see below). No evidence is available as yet to indicate that this mechanism applies also to animal systems for the transfer, e.g., of *N*-acetylgalactosamine, or galactose to serine or threonine residues (BABCZINKSI 1980, HANOVER et al. 1980).

2.2.3 Role of Dolichyl Diphosphate-Linked Oligosaccharides in the Formation of the Core Structure Containing the *N*-Acetylglucosaminyl-Asparagine Linkage

Studies by LELOIR and coworkers in the early 1970's (BEHRENS et al. 1971, BEHRENS et al. 1973) using liver microsomes have demonstrated the formation of lipid-linked oligosaccharides containing *N*-acetylglucosamine, mannose, and also glucose, and it was implicated that they are potential intermediates in the synthesis of asparagine-linked carbohydrate chains. With yeast it has been shown (LEHLE and TANNER 1975, 1976) that a lipid extract obtained upon incubation of membranes with UDP-GlcNAc separated into four compounds on silica gel thin layer plates. Two of them have been identified as dolichyl pyrophosphate derivatives, Dol-PP-*N*-acetylglucosamine and Dol-PP-chitobiose. The results further indicated that Dol-PP-GlcNAc is the precursor for the chitobiosyl-lipid and that UDP-GlcNAc is the immediate donor for the latter compound. The same particulate preparation transferred mannose from GDP-Man to Dol-PP-GlcNAc$_2$, giving rise to Dol-PP-GlcNAc$_2$-Man. The oligosaccharide residue was first thought to be of the size of a tetrasaccharide due to its elution behavior on gel filtration. However, later work showed that it is a trisaccharide having an unexpected high chromatographic mobility caused by the two acetyl groups (LEHLE et al. 1976). The synthesis of Dol-PP-GlcNAc, Dol-PP-GlcNAc$_2$ and Dol-PP-GlcNAc$_2$-Man in yeast was subsequently also described by others (REUVERS et al. 1977, NAKAYAMA et al. 1976, PALAMARCZYK and JANCZURA 1977).

Further elongation of exogenous Dol-PP-GlcNAc$_2$ with GDP-Man to a larger oligosaccharide has been difficult to demonstrate in vitro (LEHLE and TANNER 1978b). In addition, Dol-P-Man seems to be the immediate donor for some of the mannose units of the oligosaccharide (see below), whereas at least the first mannose attached to the nonreducing end of chitobiose is transferred directly from GDP-Man. With somewhat more success formation of a mannose-containing large oligosaccharide-lipid in yeast was achieved using endogenous, membrane-bound acceptor (LEHLE and TANNER 1978b, PARODI 1978). The oligosaccharide residue formed with GDP-Man could be released from the lipid by mild acid and was shown to have *N*-acetylglucosamine at its reducing end. It is linked via a pyrophosphate bridge to the lipid moiety, as indicated by the chromatographic behavior of the glycolipid on DEAE-cellulose. The dolichol nature was verified indirectly. It is based on the observation that the lipid moiety from nonradioactive lipid-oligosaccharide obtained after mild acid treatment stimulated incorporation of mannose into Dol-P-Man (LEHLE and TANNER 1978b).

In recent studies (LEHLE 1980) with a membrane preparation from yeast protoplasts lipid-oligosaccharide was synthesized in high yield, allowing a precise determination of its size. It was found that the major compound formed with GDP-Man has the composition Man$_9$GlcNAc$_2$ and minor amounts of the radioactivity were incorporated into the species Man$_8$GlcNAc$_2$, Man$_7$GlcNAc$_2$, and Man$_6$GlcNAc$_2$, obviously representing precursor forms for the Man$_9$GlcNAc$_2$ compound. A similar oligosaccharide pattern was obtained after incubation with Dol-P-Man as donor, indicating that beside its function in *O*-glycosylation described above, this compound also plays an intermediatory role in the assembly, probably in the "late" steps of oligosaccharide formation. The size of the lipid-linked oligosaccharide consisting of 2 *N*-acetylglucosamine and 9 mannose has to be seen in relation to the fact that the protein-bound core region of native mannoproteins is larger, having the composition Man$_{12-17}$GlcNAc$_2$ (NAKAJIMA and BALLOU 1975, LEHLE et al. 1979). This indicates that apparently not the whole core structure is synthesized via the lipid intermediate. In support of this conclusion is the observation that the mutant strain mnn1, which lacks terminal (1→3)-α-linkages, both in outer chain and core region (NAKAJIMA and BALLOU 1975), has the same sized lipid-oligosaccharide fraction as wild-type yeast. Moreover, it could be demonstrated that the Man$_9$GlcNAc$_2$-oligosaccharide fraction can be elongated with GDP-Man to the size of the protein-bound core. Dol-P-Man does not function as glycosyl donor (LEHLE 1980).

Studies dealing with glycoprotein synthesis in animal cells have provided evidence that asparagine-linked oligosaccharides are derived from a common lipid-bound precursor, containing also glucose beside *N*-acetylglucosamine and mannose (ROBBINS et al. 1977, HUNT et al. 1978, TABAS et al. 1978, SPIRO et al. 1976). First evidence for a glucose-containing oligosaccharide in yeast was obtained by PARODI (1976 and 1977) and was later confirmed and investigated further by LEHLE (1980). Three species of the composition $Glc_{1-3}Man_9GlcNAc_2$ were formed upon incubation of yeast membranes with UDP-Glc. The same oligosaccharides were found to occur also in vivo after labeling yeast cells with glucose (LEHLE et al. 1980). The largest oligosaccharide (Glc_3Man_9-$GlcNAc_2$) generated in yeast is identical in size with the largest corresponding oligosaccharide found in animal cells (LI et al. 1978). In vitro, all the glucose residues were incorporated via Dol-P-Glc (LEHLE 1980). Moreover, the glucose-oligosaccharide could be transferred to endogenous protein acceptor followed by enzymic excision of the glucose residues by a glucosidase activity associated with the membranes. In these studies also transfer of the glucose-free oligosaccharide-lipid was detected, however, with a lower rate of incorporation.

In earlier studies NAKAYAMA et al. (1976) and REUVERS et al. (1977) demonstrated that Dol-PP-$GlcNAc_2$ led to glycosylation of endogenous acceptor. LEHLE and TANNER (1978a) compared the rate of transfer of *N*-acetylglucosamine, chitobiose, and mannosyl-chitobiose from the dolichyl diphosphate-bound form to endogenous glycoprotein. It was found that Dol-PP-GlcNAc was a poor substrate, and transfer becomes more efficient with increasing chain length. Attempts to build up the core region on the trisaccharide residue after its transfer to protein have failed; no elongation was achieved neither using GDP-Man (LEHLE and TANNER 1978a) nor Dol-P-Man (L. LEHLE unpublished results) as donor. One can assume, therefore, that the reactions above simulate in vitro glycosyl transfer from lipid-linked oligosaccharide rather than being of physiological relevance for mannoprotein synthesis. However, other possibilities, such as an involvement of the small glycolipids in glycosylation of different yet unknown acceptors, e.g., with only short carbohydrate chains, cannot be entirely ruled out.

2.3 Mannosyl Transfer Reactions Involved in the Assembly of the Outer Chain Structure

Biochemical and genetic evidence shows that synthesis of the core and outer chain is catalyzed, at least in part, by different mannosyl transferases. Once the core is established on the polypeptide chain the elaboration of the outer chain occurs apparently by a direct transfer of mannosyl units from GDP-Man (LEHLE and TANNER 1978a, PARODI 1979). The complexity of the carbohydrate components suggests that several different glycosyl transferases are involved, perhaps in some kind of a multienzyme complex.

A logical step in order to define the process has been to utilize endogenous acceptors and to solubilize and separate individual enzymes in combination with appropriate mutants having an altered mannan structure. Membrane-bound enzyme activities catalyzing Mn^{2+}-dependent mannosyl transfer from GDP-Man to free mannose, mannobiose, and mannotriose to yield the next highest homolog have been demonstrated (LEHLE and TANNER 1974). The mannobiose formed was almost exclusively $(1\rightarrow 2)$-α-linked. From the results no definite conclusion could be drawn, whether the activities were due to enzymes involved in *O*-glycosylation or in the formation of the outer chain. The latter was first shown by NAKAJIMA and BALLOU (1975). Selective assays of four mannosyltransferases have been devised. (1 → 6)-α-Mannoseoligosaccharides, resembling that of the polymannose backbone, can serve as acceptor for both the (1 → 6)-α-mannosyltransferase and the (1 → 2)-α-mannosyltransferase, while the mannotetraose αMan1 → 3Man1 → 2Man1 → 2Man is a specific acceptor for the (1 → 6)transferase. Assay of the (1 → 3)-α-transferase uses reduced (1 → 2)-α-mannotriose. Cell-free biosynthesis of "mannan" was achieved with a Triton-solubilized and 100-fold purified enzyme fraction. (1 → 6)-α-Mannopentaitol was used as acceptor and elongated to give a product reaching at least 55 mannoses. Analysis of acetolysis fragments indicated the action of (1 → 6)-α- and (1 → 2)-α-transferases. Diversity of mannosyltransferases using exogenous acceptors was also studied by FARKAŠ et al. (1976). It was shown that the enzymes catalyzing the formation of different types of glycosidic bonds differed beside in their acceptor specificity also in pH activity curves and rates of thermal inactivation.

High molecular weight polysaccharide as an acceptor to study outer chain formation has been used by LEHLE and TANNER (1978a) and in a more detailed investigation by PARODI (1979). Addition of wild-type mannan to a solubilized enzyme extract resulted in a stimulation of mannosyltransfer from GDP-Man to methanol insoluble material. Acetolysis of the product revealed Man_1 to Man_4-saccharides using wild-type extract; no tetrasaccharide is formed with the extract from the *mnn1* mutant. This is in accordance with the established mannan structure of this strain, indicating a defect in the terminal (1 → 3)-α-transferase. In the study above indirect evidence also suggested that no lipid intermediates are involved in the synthesis of the outer chain and that the latter and Dol-PP-derivatives are synthesized by different mannosyltransferases (PARODI 1979).

Depending on the strain, *S. cerevisiae* cell wall mannoproteins contain more or less mannosylphosphate and mannobiosylphosphate groups attached by a phosphate diester linkage to oligosaccharide side chains in the outer chain. KARSON and BALLOU (1978) have solubilized and purified a membrane-bound mannosylphosphate transferase. It catalyzes the transfer of mannosylphosphate units from GDP-Man to reduced (1 → 2)-α-mannotetraose to yield mannosylphosphoryl-mannotetraose. The enzyme has a preference for oligosaccharides with an (1 → 2)-α-linkage at the nonreducing end of the chain, and a much lower activity with acceptors modified by the (1 → 3)-α-mannosyl transferase. The product is analogous in structure to the phosphorylated mannan side chains. Exogenous mannan from mutants lacking phosphate, which would be expected to have acceptor sites, did not compete significantly with the reduced mannotetraose. This indicates that the enzyme cannot act on large mannan molecules and could mean that the phosphate branches are introduced during the stepwise elaboration of the outer chain and not as a subsequent modification.

3 Biosynthesis of Mannoproteins in Other Species

3.1 *Hansenula* Species

Different types of mannan are made in these strains, thus making the system more complex. There is a phosphomannan with a teichoic acid-like structure (SLODKI 1963). KOZAK and BRETTHAUER (1970) have described a neutral mannan similar to that of *S. cerevisiae*. A species of mannoprotein containing only *O*-glycosidically linked carbohydrate residues has been described in *Hansenula wingei* (YEN and BALLOU 1973).

The enzymatic synthesis of phosphomannan using particulate preparations from *Hansenula holstii* and *Hansenula capsulata* has been studied. BRETTHAUER et al. (1969) demonstrated Mn^{2+}-dependent transfer of mannosyl-1-phosphate from GDP-Man to form mannosyl-6-phosphoryl mannan and GMP as products. This was concluded from results using double labeled GDP-Man. ^{14}C-Mannose was released by mild acid hydrolysis and mannose-6-phosphate under stronger acid conditions. Thus, the phosphoryl group in *H. holstii* is incorporated in the same way as in *S. cerevisiae* and this reaction may represent a general way by which phosphoryl groups are incorporated into phosphomannans and peptidophosphomannans. Further studies by KOZAK and BRETTHAUER (1970) showed that the product consists of a phosphomannan as well as of a neutral mannan fraction giving rise to mannose, mannobiose, mannotriose, and mannotetraose on acetolysis. Enzymatic transfer of mannose to serine and threonine was also demonstrated. Glycopeptides were isolated from an in vitro-labeled endogenous acceptor and shown to release mannose and oligosaccharides after β-elimination.

The participation of Dol-P-Man as an intermediate in the formation of the *O*-glycosidic carbohydrate part was demonstrated. The results confirm the pathway that has been worked out by Tanner and coworkers for *S. cerevisiae* (see above): Dol-P-Man is the immediate donor for mannose residues directly linked to the hydroxy amino acids and additional mannoses are introduced from the sugar nucleotide (BRETTHAUER et al. 1973, BRETTHAUER and WU 1975). MAYER (1971) reported the enzymatic synthesis of the phosphomannan in *H. capsulata*. A particulate enzyme has been shown to catalyze mannosylation of endogeneous acceptor with GDP-Man as substrate. Mannose, mannosyl-6-phosphate, and mannobiosylphosphate were obtained by mild acid hydrolysis as the expected products of the proposed acid phosphomannan structure.

3.2 *Kluyveromyces lactis*

The structure of *K. lactis* mannan is very similar to that of *S. cerevisiae*, but it has one additional kind of side chain in which *N*-acetylglucosamine is attached in $(1 \rightarrow 2)$-α-linkage to the penultimate sugar of the mannotetraose side branch unit. SMITH et al. (1975) have studied its incorporation what might be considered a terminal step in mannan biosynthesis. Particles catalyzed a Mn^{2+}-dependent transfer of *N*-acetylglucosamine from UDP-GlcNAc to the mannotetraose side chains of endogenous acceptor yielding $Man_4GlcNAc$ on acetolysis. Preincubation with GDP-Man did not increase acceptor activity for UDP-GlcNAc, suggesting that new mannan synthesis is not required for GlcNAc incorporation. Moreover, structural analysis of the mannan of a mutant designated as *mnn1* has shown that the polymer lacks both the $Man_4GlcNAc$ and Man_4 side chains. This may indicate that in vivo Man_4 must be formed to serve as acceptor before the (1 → 2)-α-N-acetylglucosaminyltransferase can act. Exogenous mannotetraose and related oligosaccharides also served as an acceptor. The most important requirement is a terminal (1 → 3)-α-mannobiosyl unit, but a pentasaccharide with two terminal (1 → 3)-α-linkages was inactive. Man_4 yielded $Man_4GlcNAc$ and the product from the reaction with αMan1 → 3Man as acceptor had *N*-acetylglucosamine attached to the mannose unit at the reducing end. This supports the conclusion that the cell-free glycosyl transferase activity is identical with that involved in mannan synthesis.

3.3 *Cryptococcus laurentii*

This organism is classified among the fungi imperfecti. Its cell envelope contains a β-linked glucan, an acidic capsular polysaccharide, composed of mannose, xylose, and glucuronic acid, and a neutral glycoprotein having three different types of *O*-glycosidically linked oligosaccharide side chains probably on the same protein (RAIZADA et al. 1975). One is a pentasaccharide with the structure αMan1 → 2αMan1 → 6αMan1 → 3(βXyl1 → 2)Man and the others are a dodecasaccharide with the composition $(\alpha Gal1 \rightarrow 6)_{10}\beta Gal \rightarrow Man$ and a trisaccharide of the structure αMan1 → 2αMan1 → 2Man.

Starting with mannose, a particulate enzyme fraction was shown (ANKEL et al. 1970, SCHUTZBACH and ANKEL 1971, SCHUTZBACH et al. 1974) to catalyze a stepwise de novo synthesis of the branched pentasaccharide in the presence of GDP-Man, UDP-Xyl, and the divalent cations Mn^{2+} and Mg^{2+}. Three distinct mannosyltransferases have been detected, each specific for the formation of one type of bond (SCHUTZBACH and ANKEL 1972). Biosynthesis of the oligogalactosyl side chain was demonstrated in the same particulate fraction (RAIZADA et al. 1974). The data suggest addition of multiple galactosyl residues involving two different galactosyl transferases to an endogenous glycoprotein acceptor. Formation of a mannosyl-lipid in the presence of GDP-Man was observed (ANKEL et al. 1970). The preliminary characterization of this compound, however, does not support a possible role as an intermediate in the glycosylation of the protein-bound mannose residue.

3.4 *Aspergillus* Species

The hyphal cell wall of *Aspergillus* species contains galactomannan and peptidogalactomannans (AZUMA et al. 1971, SAKAGUCHI et al. 1968, BARDALAYE and NORDIN 1977). The chemical structure has been determined from that of *A. niger* by BARDALAYE and NORDIN (1977); it was found to consist of approximately equal numbers of galactose and mannose units, but no biosynthetic studies were described in their paper. However, Letoublon and coworkers have reported on mannosyl transfer from GDP-Man to endogenous glycoprotein acceptor in microsomal preparations from *A. niger* (LETOUBLON et al. 1971) and *A. oryzae* (RICHARD et al. 1971). Subsequent studies in the *A. niger* system revealed that polyprenyl phosphate mannose is involved in the enzymic transfer to β-eliminable positions of the peptide acceptor. No further characterization of the product was made (LETOUBLON et al. 1973, LETOUBLON and GOT 1974, 1977). BARR and HEMMING (1972a, b) isolated a polyprenyl phosphate (exo methylene hexahydroprenyl phosphate) from *A. niger* and showed its mannosylation and possible function in glycosylation of endogenous polymer. Besides structural components, several glycosidases secreted by *Aspergillus* species have been shown to be of glycoprotein nature containing oligosaccharide chains of mannose and glucosamine (ADYA and ELBEIN 1977, ELBEIN et al. 1977, RUDICK and ELBEIN 1973, 1974, 1975). A smooth membrane fraction of *A. niger* catalyzed transfer of mannose from GDP-Man to endogenous lipid and protein acceptor. Structural analysis of the oligosaccharide chain of the in vitro polymer is very similar to the α-glucosidase of this organism and which in turn is almost identical to the carbohydrate chain from ovalbumin (RUDICK 1979). The mannolipid synthesized appears to be a polyprenyl derivative.

3.5 *Penicillium* Species

GANDER et al. (1974) have found that *Penicillium charlesii* secretes a peptidophosphogalactomannan, which contains a phosphogalactomannan region and 10 to 12 mannose-containing low molecular weight saccharides each *O*-glycosidically linked to a polypeptide consisting of about 30 amino acid residues. The glycopeptide presumably is not derived from the cell wall but from a membrane-bound lipopeptidophosphogalactomannan (GANDER and FANG 1976). A membrane fraction (GANDER et al. 1977) requiring Mn^{2+} was shown to catalyze incorporation of mannose from GDP-Man independently into both the phosphogalactomannan (to about 10%) and mannose-oligosaccharide region of the glycopeptide. Analysis of the product demonstrated that the system mainly transfers only one mannose residues to endogenous acceptor. The main product of the oligosaccharide region following treatment with mild alkaline was mannobiose with the radioactivity at the nonreducing end indicating mannosylation of serine/threonine-mannose residues. The factors limiting incorporation into the galactomannan part or further elongation of mannobiose are not known. No evidence was obtained for the formation and participation of a lipid-mannose intermediate. Since no free hydroxy-amino acid residues appear to be mannosylated, this enzyme activity could be missing in the preparation. The data also suggest that separate transferases are responsible for mannosyl transfer to oligosaccharide and polysaccharide region. The latter could be solubilized, purified, and shown to incorporate mannose into exogenous peptidophosphogalactomannan. SAMUEL and NORDIN (1971) studied the biosynthesis of mycodextranase, an inducible, secreted glycoprotein from *P. melinii*. A microsomal mannosyltransferase activity utilizing an endogenous acceptor was found to be directly correlated with the induction of mycodextranase. The glycosylated product has chemical and enzymatic properties of mycodextranase. The incorporated mannose was linked *O*-glycosidically.

3.6 Other Species

Formation of glycoproteins in *Neurospora crassa* was studied by GOLD and HAHN (1976, 1979) and in *Fusarium solani* f. *pisi* by SOLIDAY and KOLATTUKUDY (1979). In both cases introduction of *O*-glycosidically linked mannose into endogenous protein acceptor via Dol-P-Man was demonstrated. In the case of *Fusarium* the endogenous lipid acceptor was characterized and shown to contain dolichols varying in chain length from C_{90} to C_{110}. Since *A. niger* was reported to contain C_{90} to C_{120} dolichols (BARR and HEMMING 1972a), whereas the acceptor lipid from *S. cerevisiae* contains shorter species, C_{70} to C_{90}, filamentous fungi possibly utilize longer dolichols than those used by yeast. In addition attempts have been made to characterize the end product. SDS-electrophoresis indicates that one of the in vitro products might be cutinase, a glycoprotein with *O*-glycosidically linked carbohydrates. However, exogenous cutinase was not glycosylated, even when the carbohydrate depleted form was used in order to have suitable free sites for glycosylation. A similar observation was also made in attempts to glycosylate carboxypeptidase by yeast microsomal preparations (HASILIK and TANNER 1978). Restrictions in secondary and/or tertiary structures could explain the failure to glycosylate exogenous acceptors. Biosynthesis of mannoprotein in *Candida albicans* using a purified plasma membrane fraction was reported by MARRIOTT (1977). About 12% of the incorporated mannose was sensitive to β-elimination.

4 Subcellular Sites of Glycosylation

There is now considerable evidence that the path of biogenesis of secretory glycoproteins starts at the rough endoplasmic reticulum and is a vectorial process. The peptide moiety is translated on ribosomes bound to the endoplasmic reticulum and in some way the nascent protein crosses the membrane and is discharged into the internal space. According to BLOBEL and DOBBERSTEIN (1975a, b) a "signal sequence" at the amino terminal end of the peptide causes binding of the ribosome to a receptor site on the endoplasmic reticulum, thereby creating a channel for passage of the peptide to the luminal side. Glycosylation has been shown to occur while the peptide is still attached to the ribosomes (LAWFORD and SCHACHTER 1966, KIELY et al. 1976, ROTHMAN and LODISH 1977), although conflicting evidence exists on this point (JAMIESON 1977). The glycoprotein migrates under progressive glycosylation via the Golgi apparatus and secretory vesicles to the plasma membrane, where its content is released by reverse pinocytosis. Studies addressing the intracellular location of mannoprotein biosynthesis in yeast have used different approaches, including cytological, biochemical, and genetic techniques.

Investigations on dividing yeast cells have revealed that bud formation is related to vesicles, probably deriving by proliferation from the endoplasmic reticulum (MOOR 1967) and fusing with the plasma membrane. MATILE et al. (1971) isolated vesicles by isopycnic centrifugation which contained mannan, protein, and glucanases. The involvement of vesicle intermediates for synthesis and secretion of mannoproteins has clearly been shown by NOVICK and SCHEKMAN (1979) and NOVICK et al. (1980) using a combined genetic and biochemical approach while studying conditional secretory and cell surface growth mutants. Two secretory enzymes, invertase and acid phosphatase, were shown to accumulate in intracellular vesicles, when the cells are brought to the restrictive temperature. The vesicles are removed and the accumulated enzymes are secreted when the cells are returned to a permissive temperature. At least 23 gene products are required for the transport of these secretory proteins through a series of distinct membrane-bounded organelles from the site of synthesis to the cell surface. Whether besides this exocytosis path there exists also a direct extrusion of secretory proteins after assembly at the plasmalemma or an eccrine mode of synthesis, as discussed by LINNEMANS et al. (1977), remains to be established. Evidence was presented in the latter study that in a cell cycle-independent mode, cell wall material is synthesized in invaginations of the plasmalemma.

KOŠINOVÁ et al. (1974) and FARKAŠ et al. (1974) carried out an autoradiographic study of mannan synthesis. Application of pulse-chase labeling technique using tritiated mannose revealed that the initial site of mannan biosynthesis is located in the cytoplasm, from where the new material is transported to the cell wall. The wall of the mother cell remained unlabeled. The mode of regional extension in the budding process is still controversial (tip versus neck growth). However, some of the discrepancies might be explained by the observation that the mode of wall growth depends largely on the position of the cell in the cell cycle (FARKAŠ et al. 1974). The results in terms of mannan incorporation showed that in the first stages of budding a nonpolarized or spherical extension occurs. Later, tip-growth is visible, followed again during the maturation phase, by incorporation of labeled material over the whole bud surface. Using fluorescein-labeled concanavaline A TKACZ and LAMPEN (1972) come to the conclusion that the major site of mannan insertion is the distal tip of the growing bud. The studies by JOHNSON and GIBSON (1966) dealing with the deposition of glucan, another yeast cell wall component, indicate a mode similar as for mannan. Moreover, FIELD and SCHEKMAN (1980) have emphasized that secretion is not a simple case of bud-limited secretion. Thus, although phosphatase activity was shown to be largely restricted to the bud portion, 95% of the cells are stainable for the enzyme within two generations after derepression suggesting a limited growth and secretion by the parent. In contradiction CHUNG et al. (1965), using cell wall-directed fluorescent antibodies, and in a more recent investigation SKUTELSKY and BAYER (1979), using a novel technique by labeling the cell surface with a biotin-avidin complex, favor neck growth. Independent of this, all the data agree with a conservative mode of surface growth in *S. cerevisiae;* i.e., that the cell wall glycoconjugates of the mother cell are stationary and are not transferred to the newly synthesized wall of the daughter cell.

To study intracellular location of enzymes involved in mannan formation, it is necessary to fractionate yeast cells. Although there are methods described to isolate particulate organelles, e.g., plasma membrane, vacuoles or mitochondria, the methodological progress in fractionating total homogenates in order to determine the recovery of enzyme activities is still very limited. Investigations on the intracellular localization of mannan synthetase activity was initiated by CORTAT et al. (1973). Upon subcellular fractionation using an urografin gradient mannosyl-transferase activity toward endogenous acceptor was found mainly to reside in a light membrane fraction (equivalent most likely to the endoplasmic reticulum) and a Golgi-like fraction, whereas only low activity was detected in the plasma membrane. Experiments on the cellular site of initial glycosylation were reported by RUIZ-HERRERA and SENTANDREU (1975). After short pulses of yeast cells with radioactive mannose label was associated with polysomes. Most of the radioactivity accounted for mannose and could be released by treatment of the polysomes with puromycin, suggesting that initial sugar binding occurs while the nascent polypeptide chains are still growing on the ribosomes. LARRIBA et al. (1976) presented evidence that Dol-P-Man participates in the glycosylation reaction at the polysomal level. Intracellular localization of the reactions involved in the formation of *O*-glycosidic bonds in yeast mannoproteins was studied by LEHLE et al. (1977) and MARRIOTT and TANNER (1979). The results strongly suggest that introduction of the carbohydrate chains starts at the endoplasmic reticulum with formation of Dol-P-Man and subsequent glycosyl transfer to the protein, but with increasing chain length mannosylation proceeds to an increasing extent also in the more "peripheral" membranes, i.e., Golgi-like vesicles and the plasma membrane. *N*-Acetylglucosaminyl transfer to Dol-P, the first reaction in the assembly of the lipid-oligosaccharide, is highest in terms of specific and total activity in the endoplasmic reticulum (MARRIOTT and TANNER 1979); the total inner core most likely is formed by the endoplasmatic reticulum.

The synthesis of the outer chain has not been studied in detail. However, it became evident that the percentage of radioactivity incorporated into alkali-stable positions of yeast mannan (*N*-glycosidically linked chains) is largest in the Golgi apparatus and the plasma membrane (LEHLE et al. 1977). It may indicate that these fractions contain increasing amounts of the outer chain. It will be important in the future to compare the composition of the mannoprotein fraction stable in weak alkali synthesized by the various membrane fractions. Additional evidence that also peripheral membrane regions have a function in glycosylation comes from studies by MARRIOTT (1977) and SANTOS et al. (1978). These authors have isolated purified plasma membrane fraction and found that it contains mannosyltransferase activity; in the latter case it was about 20% of the total activity of the cell and part of it is assumed to be exposed to the plasma surface. There is still considerable controversy as to the presence and function of surface glycosyltransferases, however. They may reflect residual transferase activities, originally facing the luminal side of the endomembrane system but getting to the outside during the fusion process of the vesicles with the plasma membrane. On the other hand, in cases where the secreted glycoprotein, as in yeast, becomes part of the cell wall, they may serve to complete the carbohydrate and integrate and anchor the molecule in question in the cell envelope.

From what has been said above, it is obvious that glycosylation of a secretory protein is a vectorial event requiring the coordination and cooperation of a well-defined biosynthetic machinery with highly organized subcellular structures. Whether the carbohydrate part plays a general role in this translocation process is still a matter of speculation. The proposal of EYLAR (1965) that attachment of sugars represents a recognition signal for transport, at least in its original sense, is not fully acceptable, since it is clear that many secretory proteins are nonglycosylated. One should, however, keep in mind the possibility that nonglycosylated and yet secreted proteins were synthesized as glycoproteins and have lost their carbohydrate part at some stage during their formation.

The sequence of events leading to synthesis of glycoproteins has another aspect not understood at all. How are sugar nucleotide precursors, synthesized in the cytoplasm, used by the glycosyl transferase, which are topographically separated by the membranes of the endoplasmic reticulum or Golgi elements? The idea put forward that the participating lipids mediate transfer of the hydrophilic sugar beyond the hydrophobic membranes has not been documented by any good evidence so far. Moreover, this would not solve the problem. As shown above, there are a number of sugar transfer reactions using directly sugar nucleotides as glycosyl donor.

5 Possible Control of Mannoprotein Biosynthesis

Seeing the complexity and diversity of the structure of mannoproteins or glycoproteins in general, one can only wonder how the cell regulates their biosynthesis. Biosynthesis of mannoproteins is dependent on continuous synthesis of protein. Studies on the incorporation of threonine and glucose into the wall of growing cells have revealed that addition of cycloheximide halted incorporation of peptide into the cell wall and this inhibition was parallel to mannan synthesis. Glucan formation was little influenced (ELORZA and SENTANDREU 1969). Also in cases using protoplasts, cycloheximide was shown to prevent formation and secretion of cell wall glycoproteins in general (FARKAŠ et al. 1970) or in a more specific way invertase (LIRAS and GASCÓN 1971) and phosphatase (VAN RIJN et al. 1972). The effect of cycloheximide on the synthesis of mannoproteins seems to reflect depletion of available mannosyl acceptors. The inhibition is not related to the turnover of the enzymes involved (ELORZA and SENTANDREU 1973) or directed to the activity of the mannosyltransferases (SENTANDREU and LAMPEN 1970, FARKAŠ et al. 1970). Also the possibility of an interference of cycloheximide with uptake and interconversion of hexoses, leading to the synthesis of GDP-Man, was ruled out; as a matter of fact, an accumulation of the amounts of GDP-Man and UDP-GlcNAc takes place in the presence of the drug (SENTANDREU and LAMPEN 1970).

The process, which attaches one polymer, the polypeptide, with another, the oligosaccharide, is probably a critical step in glycoprotein assembly. It is now believed that the oligosaccharide is assembled during and after completion of the polypeptide. The initiation of the Asn-GlcNAc-type oligosaccharide chain seems to be controlled by a particular amino acid sequence of the form ...(Asn-(X)-Ser/Thr)... Such a specific amino acid sequence has been proposed by MARSHALL (1974) after examination the amino acid sequence of many glycoproteins. Recently, this hypothesis could be experimentally proven in microsomal preparations from thyroid, rat liver, and also from yeast, using synthetic peptides with altered amino acid sequences as glycosyl acceptors (RONIN et al. 1978, BAUSE 1979, BAUSE and LEHLE 1979). Since, however, not all regions of a polypeptide chain having the asparagine "sequon" are glycosylated, additional structural requirements have to be claimed. No marker sequence seems to be a necessary prerequisite for the formation of the *O*-glycosidic linkage (BAUSE and LEHLE 1979). Presumably accessibility of the acceptor site for the appropriate glycosyltransferase rather than recognition of a specific amino acid sequence might be the key for *O*-glycosylation.

After initiation has occurred, elongation of the carbohydrate part takes place and one may ask how a mannan of characteristic size and degree of branching is built up. It looks as if this is controlled by a highly coordinated multiglycosyltransferase system, by the sequential action of spatially separated transferases on the basis of different affinities for the glycosyl donor and acceptors. The product of one glycosyltransferase becomes the substrate of the next transferase. NAKAJIMA and BALLOU (1975) have reported such differences. However, it is too early to be able to visualize a mechanism by which a polysaccharide component of defined size is assembled.

Action of the transferases may also depend on the ionic environment within the cell and on the presence of inhibitors or activators. Distinct differences for Mg^{2+} and Mn^{2+} ions of mannosyltransferases involved in *O*-glycosylation of mannoproteins have been demonstrated in *S. cerevisiae* (SHARMA et al. 1974) and *N. crassa* (GOLD and HAHN 1979).

Two inhibitors, 2-deoxyglucose (2-dGlc) and tunicamycin (TM), have been used in the past in studies related to several aspects of mannoprotein biosynthesis. Whereas the effect of 2-dGlc turns out to be the result of multiple effects on cell metabolism (KRÁTKÝ et al. 1975), TM seems to interfere in a more specific way only with the glycosylation process. TAKATSUKI et al. (1975), TKACZ and LAMPEN (1975) and LEHLE and TANNER (1976) have shown that TM inhibits in vitro the transfer of N-acetylglucosamine from UDP-GlcNAc to Dol-P. KUO and LAMPEN (1974) have observed that synthesis and secretion of invertase and acid phosphatase are inhibited in the presence of TM. Inhibition of the apparent formation of the vacuolar enzyme carboxypeptidase Y by TM was demonstrated by HASILIK and TANNER (1978). The enzyme synthesized under these conditions was smaller in size and, as expected, was devoid of sugar. Using protein-specific antibodies, it became evident that TM reduces also the amount of carbohydrate free protein. Since the product was metabolically stable, an inhibited rate of its biosynthesis was concluded. In addition the synthesis of other proteins was not affected. The results indicate a regulatory link between the glycosylation of the protein residue and its biosynthesis. A differential effect of TM for *alkaline* phosphatase has been reported recently by ONISHI et al. (1979). It was shown to be of glycoprotein nature having a small amount of asparagine-attached carbohydrate. However, the rate of alkaline phosphatase formation was not depressed by TM, although the enzyme, as those mentioned above, seems to lack carbohydrate under inhibiting conditions as judged from its failure to interact with concanavalin A. The authors infer that there is no regulatory link between glycosylation of a protein and its synthesis and the inhibition of other yeast glycoprotein enzymes may mean that as unglycosylated proteins, they cannot be correctly folded or processed and therefore escape detection. Further work is needed to gain more insight into the cause of this differential effect of TM.

Impediment of mannosylation by 2-dGlc results in inhibition of synthesis and secretion of total mannoprotein fraction (FARKAŠ et al. 1970, KRÁTKÝ et al. 1975) or in a more specific way of extracellular mannoprotein enzymes as invertase, acid phosphatase, or carboxypeptidase (LIRAS and GASCÓN 1971, KUO and LAMPEN 1972, VAN RIJN et al. 1972, HASILIK and TANNER 1978). 2-dGlc can be considered as an analog to mannose and glucose. 2-dGlc inhibition was discussed as interference of accumulated 2-dGlc-phosphate with the interconversion of hexose phosphates or with sugar uptake, thus leading to a shortage of substrates essential for glycosylation (KUO and LAMPEN 1972). Another, more direct and specific effect on glycoprotein synthesis could be a competition of sugar nucleotide derivatives of 2-dGlc with the corresponding physiological nucleotide sugars for glycosylation. This is suggested by the observations that 2-dGlc is converted in vivo both to GDP-dGlc and UDP-dGlc along conventional pathways (BIELY and BAUER 1968) and finally the 2-dGlc appears in cell wall glycoproteins (BIELY et al. 1972, 1974). LEHLE and SCHWARZ (1976) reported that yeast membranes catalyze formation of Dol-P-dGlc from GDP-dGlc and the subsequent transfer of the 2-dGlc residue to serine/threonine positions of a membrane-bound glycoprotein fraction. Further elongation to larger oligosaccharides via GDP-Man is blocked since 2-dGlc prevents formation of the following $(1 \rightarrow 2)$-α-linkage. Interference of 2-dGlc with lipid-linked oligosaccharides as shown for animal systems (SCHWARZ and DATEMA 1980) is also possible in yeast (L. LEHLE unpublished results). Summarizing, one can say, if experimental conditions are properly chosen so that no other cellular reactions are influenced by 2-dGlc, it can be used as a potential inhibitor for protein glycosylation.

6 Attempts to Study Solubilized Glycosyl Transferases

Most of the work on mannan biosynthesis described above was carried out with intact membranes and with both endogenous protein and lipid glycosyl acceptor, very often even in the presence of detergents. The actual enzyme specificity in such multiphasic systems might often be hidden by kinetic limitations like enzyme accessibility and amount of available acceptor. To overcome these problems and to understand the various reactions better from a biochemical point of view, it is necessary to isolate and purify the enzymes in question. The main obstacles for such investigations are the solubilization and stabilization of the enzyme, as well as the need for a test system with defined substrates. The solution of the latter problem seems in sight, since both low and high molecular weight mono- and oligosaccharide acceptors (LEHLE and TANNER 1974, FARKAŠ et al. 1976, LEHLE and TANNER 1978a, SCHUTZBACH and ANKEL

1971) as well as small peptides (BAUSE and LEHLE 1979) are available. The latter acceptors are useful in order to study protein glycosylation. Moreover polyprenols of different chain length and saturation can be isolated (MAŃKOWSKI et al. 1976) and may contribute to a better understanding of the role of the lipid moiety.

As discussed already in detail, NAKAJIMA and BALLOU (1975) described procedures and selective assays of four mannosyltransferase using GDP-Man and small mannooligosaccharide acceptors. The involvement of the enzymes in mannan outer chain biosynthesis was indicated, since the solubilized extract works in concert and makes hereby "mannan". KARSON and BALLOU (1978) reported on the properties of a 250-fold purified membrane-bound mannosylphosphate transferase that introduces the phosphate containing side branches. Recently, attempts have succeeded also to solubilize and partially characterize several glycosyltransferases of the dolichol pathway. BABCZINSKI et al. (1980) purified two membrane enzymes involved in the *O*-glycosylation of yeast mannoproteins. The enzyme forming Dol-P-Man from GDP-Man and Dol-P was purified 140-fold, and the enzyme transferring subsequently the mannose to protein-bound serine/threonine residues was purified 380-fold, respectively. PALAMARCZYK et al. (1980) have investigated the specificity of four solubilized glycosyl transferases toward a variety of polyprenols differing in chain length and saturation. The formation of Dol-PP-GlcNAc$_{1-2}$ from UDP-GlcNAc and Dol-P, as well as the synthesis of Dol-P-Man from GDP-Man and Dol-P, showed a strong requirement for α-saturated polyprenols, indicating that polyprenols of the dolichol type are the preferred substrates for glycosyltransferases in eukaryotes. Moreover a minimum chain length around seven isoprene units was found; those being shorter are inactive as glycosyl acceptor. Similar results concerning saturation and chain length were obtained also with the solubilized enzymes transferring the chitobiosyl and mannosyl residue, respectively, to peptide acceptors, a reaction thought to mimic protein glycosylation. At the moment the observations are still more descriptive and a detailed kinetic study has to await complete purification of the enzymes in question. Interestingly, the mannosyl transfer from Dol-P-Man to endogenous protein with intact membranes did not show a pronounced chain length effect (PLESS and PALAMARCZYK 1978) demonstrating the limitations in characterizing membrane-bound enzymes, as pointed out in the beginning of this chapter. The enzymes forming Dol-P-Man, Dol-PP-GlcNAc$_1$ and Dol-PP-GlcNAc$_2$ suffer from being easily irreversibly inactivated by detergents while solubilized (PALAMARCZYK et al. 1979). This inhibitory effect can be prevented by dolichyl phosphate. Further efforts have to be made to overcome the various difficulties in this new field of glycoprotein enzymology.

7 A Summary of Mannoprotein Biosynthesis; How It Might Occur: a Model

In this chapter an attempt has been made to summarize our present knowledge on a biosynthetic process as complicated as mannan biosynthesis. Although great progress has been achieved, it became clear that several detailed and

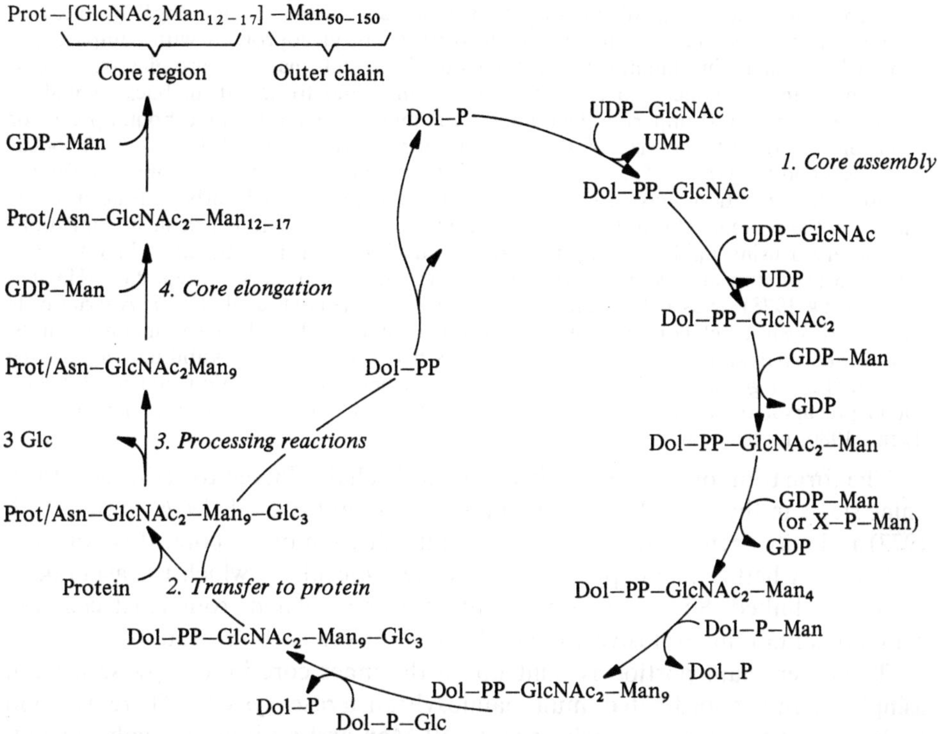

Fig. 2. Schematic pathway of biosynthesis of asparagine-linked carbohydrate component of yeast mannoproteins

important questions have to be answered in the future. However, it might not be too far-fetched trying to construct a model how this pathway might occur. Most of our understanding comes from studies with *S. cerevisiae*. Therefore the proposal outlined in Figs. 1 and 2 might be valid mainly for this species, but can also be considered as a working hypothesis for others.

It is assumed that the protein moiety is synthesized on membrane-bound ribosomes. The glycosylation process starts already at the nascent polypeptide chain. In case of the assembly of the *N*-linked carbohydrate chain the core region is built up via a lipid intermediate. An oligosaccharide with the composition $Glc_3Man_9GlcNAc_2$ is the immediate precursor (LEHLE 1980). The sequence starts with the transfer of a *N*-acetylglucosaminylphosphate residue onto Dol-P giving rise to the pyrophosphate derivative. Next, a second *N*-acetylglucosamine and one mannose residue are transferred directly from the level of the corresponding sugar nucleotides. Further mannose elongation uses DolPMan or another donor, at least in the final steps, from $Man_5GlcNAc_2$ to $Man_9GlcNAc_2$, they come from Dol-P-Man. All glucoses are transferred via Dol-P-Glc. Following assembly on the lipid an "en bloc" transfer of the whole oligosaccharide to protein occurs. The glucose residues may have the function of some kind of a recognition marker for the oligosaccharide: protein transferase, since subsequently all three glucose residues are excised by membrane-bound glucosidase(s).

There is no evidence available as yet to indicate a further "trimming" by cleaving some of the mannose units as shown for the synthesis of glycoproteins with complex-type carbohydrate chains in animal cells. In the case of the high mannose proteins from yeast such a mechanism, namely, to remove mannose units and to add them back, would be rather uneconomical. Comparing the number of mannose units in the lipid-bound precursor ($Man_9GlcNAc_2$) with the size of the protein-bound core region of $Man_{12-17}GlcNAc_2$ (NAKAJIMA and BALLOU 1975, LEHLE et al. 1979) a discrepancy is obvious. Therefore further mannosylation reactions have to occur to modify the lipid-linked carbohydrate after its transfer to the protein. This in fact was shown by in vitro studies, in which such elongations were observed using lipid free oligosaccharide and GDP-Man as substrates (LEHLE 1980). Extension may stop at this modified length, as in the case of carboxypeptidase (HASILIK and TANNER 1978) or as with some chains of invertase (LEHLE et al. 1979). Alternatively the cell proceeds to elaborate an additional outer chain with 100 to 150 mannose units. It would be interesting to find out what decides between the two possibilities of building up a small or large carbohydrate chain. In addition one should mention that no dolichol-bound precursor of the size $Glc_3Man_{12-17}GlcNAc$ was found, both in vivo and in vitro (LEHLE 1980, LEHLE et al. 1980).

The formation of the mannooligosaccharide chains linked to serine and threonine starts in the endoplasmic reticulum, too (LARRIBA et al. 1976, LEHLE et al. 1977) and may occur simultaneously with introduction of the core. It is not clear up to now, whether this happens on the same molecule to which the asparagine chains are linked. So far no component could be isolated from yeast cell wall mannan containing exclusively either *N*- or *O*-linked carbohydrate residues.

The outer chain portion is built up on the inner core in a stepwise fashion using a highly coordinated multimannosyltransferase system. There is good evidence that the donor is exclusively GDP-Man and no lipid is involved (BEHRENS and CABIB 1968, PARODI 1979, LEHLE 1980). Moreover, a separate set of mannosyltransferases seems to catalyze the reactions.

Side chain modification consists of addition of mannosylphosphate groups (KARSON and BALLOU 1978) and transfer of *N*-acetylglucosamine (SMITH et al. 1975), no lipid intermediate is involved in this stage.

In view of the recent findings other mechanisms of mannan biosynthesis discussed in the past cannot be considered as alternate possibilities. As a working hypothesis BALLOU and RASCHKE (1974) put forward the idea that the side-chain oligosaccharides are synthesized on a lipid carrier before being transferred to the growing chain by an $(1\rightarrow 6)$-α-polymerase. The finding that similar fragments are obtained by β-elimination and by acetolysis of mannoproteins led SENTANDREU and NORTHCOTE (1969) to postulate that *O*-glycosidically linked chains are transferred by a transglycosylation reaction to the growing polysaccharide part. This possibility has been ruled out by FARKAŠ et al. (1976). Following the kinetics of mannose incorporation into the *O*- and *N*-linked carbohydrate portion in combination with pulse-chase experiments it was demonstrated that the two parts are synthesized independently. This was supported further by experiments demonstrating a different thermal inactivation of the corresponding enzymes.

Mannoprotein synthesis is a vectorial process tightly coupled to membrane structure. After core synthesis in the endoplasmic reticulum the compound in question is channeled into secretory vesicles and transported either to the outside of the cell, as in the case of external invertase, acid phosphatase, asparaginase, and structural cell wall mannan, or into the vacuole as in the case of internal mannoproteins like carboxypeptidase. Transient on the way to their destination the mannoproteins are modified. After fusion of the vesicles with the plasma membrane the external mannoproteins become located in the cell wall. There is still some discussion in the literature about their structural arrangement and distribution. Based on the available biochemical, immunological, and cytochemical data (LAMPEN 1968, KIDBY and DAVIES 1970, LINNEMANS et al. 1977), one can assume that the mannan enzymes are located mainly in the periplasmic space and the innermost layer of the wall, whereas the structural mannan component is interspersed with the glucan layer covering the surface of the cell. The structural cell wall mannan seems to become metabolic inert once it is incorporated into the wall as judged from selective incorporation of 2-deoxy-glucose. The radioactivity in the wall persisted for at least three generations (KRÁTKÝ et al. 1975).

Yeast mannoproteins display many similarities in terms of structure and biosynthesis with glycoproteins from animal origin. Although yeasts have not been shown so far to contain complex-type oligosaccharide chains possessing in their peripheral regions *N*-acetylglucosamine, galactose or *N*-acetylneuraminic acid, they do modify instead, perhaps in analogy, the mannose-rich core by addition of the outer chain. The dolichol part of the biosynthetic pathway, however, leading to *N*-glycosylation seems to be very similar, if not identical, in yeast and animal cells implicating a very basic and important mechanism conserved during evolution. This may be underlined by recalling the many important and widespread roles glycoproteins are playing: such as in cell–cell recognition, cell differentiation, immune response, drug resistance, malignant transformation, to mention only some of them. Elucidation of the mechanism of glycoprotein biosynthesis may contribute to a better understanding of these phenomena.

References

Adya S, Elbein AD (1977) Glycoprotein enzymes secreted by *Aspergillus niger:* purification and properties of α-galactosidase. J Bacteriol 129:850–856

Algranati ID, Carminatti H, Cabib E (1963) The enzymic synthesis of yeast mannan. Biochem Biophys Res Commun 12:504–509

Ankel H, Ankel E, Schutzbach JS, Garancis JC (1970) Mannosyltransfer in *Cryptococcus laurentii.* J Biol Chem 245:3945–3955

Azuma I, Kimura H, Hirao F, Tsubura E, Yamamura Y, Misaki A (1971) Biochemical and immunological studies on *Aspergillus.* III. Chemical and immunological properties of glycopeptide obtained from A fumigatus. Jpn J Microbiol 15:237–246

Babczinski P (1980) Evidence against the participation of lipid intermediates in the in vitro biosynthesis of serine (threonine)-N-acetyl-D-galactosamine linkages in submaxillary mucin. FEBS Lett 117:207–211

Babczinski P, Tanner W (1973) Involvement of dolichol monophosphate in the formation of specific mannosyl linkages in yeast glycoproteins. Biochem Biophys Res Commun 54:1119–1124

Babczinski P, Haselbeck A, Tanner W (1980) Yeast mannosyl transferases requiring dolichyl phosphate and dolichyl phosphate mannose as substrate. Partial purification and characterization of the solubilized enzymes. Eur J Biochem 105:509–515

Ballou CE (1976) Structure and biosynthesis of the mannan component of the yeast cell envelope. Adv Microb Physiol 14:93–158

Ballou CE, Raschke WC (1974) Polymorphism of the somatic antigen of yeast. Science 184:127–134

Bardalaye PC, Nordin JH (1977) Chemical structure of the galactomannan from the cell wall of *Aspergillus niger*. J Biol Chem 252:2584–2591
Barr RM, Hemming FW (1972a) Polyprenols of *Aspergillus niger*. Their characterization, biosynthesis and subcellular distribution. Biochem J 126:1193–1202
Barr RM, Hemming FW (1972b) Polyprenol phosphate as an acceptor of mannose from guanosine diphosphate mannose in *Aspergillus niger*. Biochem J 126:1203–1208
Bause E (1979) Studies on the acceptor specificity of asparagine-N-glycosyl-transferase from rat liver. FEBS Lett 103:296–299
Bause E, Lehle L (1979) Enzymatic *N*-glycosylation and *O*-glycosylation of synthetic peptide acceptors by dolichol-linked sugar derivatives in yeast. Eur J Biochem 101:531–540
Behrens NH, Cabib E (1968) The biosynthesis of mannan in *Saccharomyces cerevisiae*. J Biol Chem 243:502–509
Behrens NH, Leloir LF (1970) Dolichol monophosphate glucose: an intermediate in glucose transfer in liver. Proc Natl Acad Sci USA 66:153–159
Behrens NH, Parodi AJ, Leloir LF (1971) Glucose transfer from dolichol monophosphate glucose: the product formed with endogenous microsomal acceptor. Proc Natl Acad Sci USA 68:2857–2860
Behrens NH, Carminatti H, Staneloni RJ, Leloir LF, Cantarella AI (1973) Formation of lipid-bound oligosaccharide containing mannose. Their role in glycoprotein synthesis. Proc Natl Acad Sci USA 70:3390–3394
Biely P, Bauer Š (1968) The formation of guanosine diphosphate 2-deoxy-D-glucose in yeast. Biochim Biophys Acta 156:432–434
Biely P, Kratky Z, Bauer Š (1972) Metabolism of 2-deoxy-D-glucose by baker's yeast. IV. Incorporation of 2-deoxy-D-glucose into cell wall mannan. Biochim Biophys Acta 255:631–639 (72)
Biely P, Kratky Z, Bauer Š (1974) Metabolism of 2-deoxy-D-glucose by baker's yeast. VI. A study on cell wall mannan. Biochim Biophys Acta 352:268–274
Blobel G, Dobberstein B (1975a) Transfer of proteins across membranes. I. Presence of proteolytically processed and unprocessed nascent immunoglobulin light chains on membrane-bound ribosomes of murine myeloma. J Cell Biol 67:835–851
Blobel G, Dobberstein B (1975b) Tranfer of proteins across membranes. II. Reconstitution of functional rough microsomes from heterologues components. J Cell Biol 67:852–862
Bretthauer RK, Wu S (1975) Synthesis of the mannosyl-*O*-serine (threonine) linkage of glycoproteins from polyisoprenyl phosphate mannose in yeast (*Hansenula holstii*). Arch Biochem Biophys 167:151–160
Bretthauer RK, Kozak LP, Irwin WE (1969) Phosphate and mannose transfer from guanosine diphosphate mannose to yeast mannan acceptors. Biochem Biophys Res Commun 37:820–827
Bretthauer RK, Wu S, Irwin WE (1973) Enzymatic transfer of mannose from guanosine diphosphate mannose to dolichol phosphate in yeast (*Hansenula holstii*). A possible step in mannan biosynthesis. Biochim Biophys Acta 304:736–747
Cabib E (1975) Molecular aspects of yeast morphogenesis. Annu Rev Microbiol 29:191–214
Cabib E, Leloir LF (1954) Guanosine diphosphate mannose J Biol Chem 206:779–790
Caccam JJ, Jackson JT, Eylar EH (1969) The biosynthesis of mannose-containing glycoproteins: a possible lipid intermediate. Biochem Biophys Res Commun 35:505–511
Chung K, Hawirko R, Isaak P (1965) Cell wall replication in *Saccharomyces cerevisiae*. Can J Microbiol 11:953–957
Cortat M, Matile P, Wiemken A (1973) Intracellular localization of mannan synthetase activity in budding baker's yeast. Biochem Biophys Res Commun 53:482–489
Elbein AD, Adya S, Lee YC (1977) Purification and properties of a β-mannosidase from *Aspergillus niger*. J Biol Chem 252:2026–2031
Elorza MV, Sentandreu R (1969) Effect of cycloheximide on yeast cell wall synthesis. Biochem Biophys Res Commun 36:741–747
Elorza MV, Sentandreu R (1973) The effect of cycloheximide on mannosyl transferase activity of a membrane preparation from *Saccharomyces cerevisiae*. In: Villanueva JR, Garcia-Acha I, Gascón S, Uruburu F (eds) Yeast, mould and plant protoplasts. Academic Press, London New York pp 205–209

Eylar EH (1965) On the biological role of glycoproteins. J Theor Biol 10:89–113
Farkaš V (1979) Biosynthesis of cell walls of fungi. Microb Rev 43:117–144
Farkaš V, Svoboda A, Bauer Š (1970) Secretion of cell wall glycoproteins by yeast protoplasts. The effect of 2-deoxy-D-glucose and cycloheximide. Biochem J 118:755–758
Farkaš V, Kovařik J, Kosinová A, Bauer Š (1974) Autoradiographic study of mannan incorporation into the growing cell walls of *Saccharomyces cerevisiae*. J Bacteriol 117:265–269
Farkaš V, Vagabov VM, Bauer Š (1976) Biosynthesis of yeast mannan. Diversity of mannosyltransferases in the mannan synthesizing enzyme system from yeast. Biochim Biophys Acta 428:573–582
Field C, Schekman R (1980) Localized secretion of acid phosphatase reflects the pattern of cell surface growth in *Saccharomyces cerevisiae*. J Cell Biol 86:123–128
Gander JE, Fang F (1976) The occurrence of ethanolamine and galactofuranosyl residues attached to *Penicillium charlesii* cell wall saccharides. Biochem Biophys Res Commun 71:719–725
Gander JE, Jentoft NH, Drewes LR, Rick PD (1974) The 5-O-β-D-galactofuranosyl-containing exocellular glycopeptide of *Penecillium charlesii*. Characterization of the phosphogalactomannan. J Biol Chem 249:2063–2072
Gander JE, Drewes LR, Fang F, Lui A (1977) 5-O-β-galactofuranosyl-containing exocellular glycopeptide of *Penicillium charlesii*. Incorporation of mannose from GDP-mannose into glycopeptide. J Biol Chem 252:2187–2193
Gold MH, Hahn HJ (1976) Role of a mannosyl lipid intermediate in the synthesis of *Neurospora crassa* glycoproteins. Biochemistry 15:1808–1813
Gold MH, Hahn HJ (1979) Effect of divalent metal ions on the synthesis of oligosaccharide chains of *Neurospora crassa*. Phytochemistry 18:1269–1272
Hanover JA, Lennarz WJ, Young JD (1980) Synthesis of *N*- and *O*-linked glycopeptides in oviduct membrane preparations. J Biol Chem 255:6713–6716
Hasilik A, Tanner W (1978) Carbohydrate moiety of carboxypeptidase Y and perturbation of its biosynthesis. Eur J Biochem 91:567–575
Hemming F (1974) Lipids in glycan biosynthesis. In: Goodwin TW (ed) Biochemistry of lipids, vol IV. Butterworths and Univ Park Press, London Baltimore, pp 39–97
Hunt LA, Etchinson JR, Summers DF (1978) Oligosaccharide chains are timmed during synthesis of the envelope glycoprotein of vesicular stomatitus virus. Proc Natl Acad Sci USA 75:754–758
Jamieson JC (1977) Studies on the site of addition of sialic acid and glucosamine to rat α_1-acid glycoprotein. Can J Biochem 55:408–414
Johnson BF, Gibson EJ (1966) Autoradiographic analysis of regional cell wall growth of yeasts. III. *Saccharomyces cerevisiae*. Exp Cell Res 41:580–591
Jung P, Tanner W (1973) Identification of the lipid intermediate in yeast mannan biosynthesis. Eur J Biochem 37:1–6
Karson EM, Ballou CE (1978) Biosynthesis of yeast mannan. Properties of a mannosylphosphate transferase in *S. cerevisiae*. J Biol Chem 253:6484–6492
Kauß H (1969) A plant mannosyl-lipid acting in reversible transfer of mannose. FEBS Lett 5:81–84
Kidby DK, Davies R (1970) Invertase and disulphide bridges in the yeast wall. J Gen Microbiol 61:327–333
Kiely ML, McKnight GS, Schimke RT (1976) Studies on the attachment of carbohydrate to ovalbumin nascent chains in the oviduct. J Biol Chem 252:4402–4408
Košinová A, Farkaš V, Machala S, Bauer Š (1974) Site of mannan synthesis in yeast. An autoradiographic study. Arch Microbiol 99:255–263
Kozak LP, Bretthauer RK (1970) Studies on the biosynthesis of *Hansenula holstii* mannans from guanosine diphosphate mannose. Biochemistry 9:1115–1122
Krátký Z, Biely P, Bauer Š (1975) Mechanism of 2-deoxy-D-glucose inhibition of cell wall polysaccharide and glycoprotein biosynthesis in *Saccharomyces cerevisiae*. Eur J Biochem 54:459–467
Kuo S-C, Lampen JO (1972) Inhibition by 2-deoxy-D-glucose of synthesis of glycoprotein

enzymes by protoplast of *Saccharomyces:* relation to inhibition of sugar uptake and metabolism. J Bacteriol 111:419–429
Kuo S-C, Lampen JO (1974) Tunicamycin – an inhibitor of yeast glycoprotein synthesis. Biochem Biophys Res Commun 58:287–295
Lampen JO (1968) External enzymes of yeast: their nature and formation. Antonie van Leeuwenhoek J Microbiol Serol 34:1–18
Larriba G, Elorza MV, Villanueva JR, Sentandreu R (1976) Participation of dolichol phosphomannose in the glycosylation of yeast wall mannoproteins at the polysomal level. FEBS Lett 71:316–320
Lawford GR, Schachter H (1966) Biosynthesis of glycoprotein by liver. The incorporation in vivo of ^{14}C-glucosamine into protein-bound hexosamine and sialic acid of rat liver subcellular fractions. J Biol Chem 241:5408–5418
Lehle L (1980) Biosynthesis of the core region of yeast mannoproteins. Formation of a glucosylated dolichol-bound oligosaccharide precursor, its transfer to protein and subsequent modification. Eur J Biochem 109:589–601
Lehle L, Schwarz RT (1976) Formation of dolichol monophosphate 2-deoxy-D-glucose and its interference with the glycosylation of mannoproteins in yeast. Eur J Biochem 67:239–245
Lehle L, Tanner W (1974) Membrane-bound mannosyl transferase in yeast glycoprotein biosynthesis. Biochim Biophys Acta 350:225–235
Lehle L, Tanner W (1975) Formation of lipid-bound oligosaccharide in yeast. Biochim Biophys Acta 399:364–374
Lehle L, Tanner W (1976) The specific site of tunicamycin inhibition in the formation of dolichol-bound N-acetylglucosamine derivatives. FEBS Lett 71:167–170
Lehle L, Tanner W (1978a) Glycosyl transfer from dolichyl-phosphate sugars to endogenous and exogenous glycoprotein acceptors in yeast. Eur J Biochem 83:563–570
Lehle L, Tanner W (1978b) Biosynthesis and characterization of large dolichylphosphate-linked oligosaccharides in *Saccharomyces cerevisiae*. Biochim Biophys Acta 539:218–229
Lehle L, Fartaczek F, Tanner W, Kauss H (1976) Formation of polyprenol mono- and oligosaccharides in *Phaseolus aureus*. Arch Biochem Biophys 175:419–426
Lehle L, Bauer F, Tanner W (1977) The formation of glycosidic bonds in yeast glycoproteins. Intracellular localisation of the reactions. Arch Microbiol 114:77–81
Lehle L, Cohen RE, Ballou CE (1979) Carbohydrate structure of yeast invertase. Demonstration of a form with only core oligosaccharides and a form with completed polysaccharide chains. J Biol Chem 254:12209–12218
Lehle L, Schulz I, Tanner W (1980) Dolichyl phosphate linked sugars as intermediates in the synthesis of yeast mannoproteins: an in vivo study. Arch Microbiol 127:231–237
Letoublon R, Got R (1974) Rôle d'un intermédiate dans le transfert du mannose à des accepteurs glycoprotéiques endogenes chez *Aspergillus niger*. FEBS Lett 46:214–217
Letoublon R, Got R (1977) Effet de proteins basiques de faible poids moleculaire sur le transfert de mannose dans les microsomes d'*Aspergillus niger*. FEBS Lett 80:343–347
Letoublon RM, Richard M, Louisot P, Got R (1971) Étude du transfert enzymatique de D-(^{14}C)mannose à partir de GDP-(^{14}C) mannose, sur un accepteur endogène dans les microsomes d'*Aspergillus niger*. Eur J Biochem 18:194–200
Letoublon RCP, Combe J, Got R (1973) Transfert du mannose à partir de GDP-mannose à des accepteurs lipidiques chez *Aspergillus niger*. Eur J Biochem 40:95–101
Li E, Tabas I, Kornfeld S (1978) The synthesis of complex-type oligosaccharides. I. Structure of the lipid-linked oligosaccharide precursor of the complex-type oligosaccharides of the Vesicular stomatitis virus G protein. J Biol Chem 253:7762–7770
Linnemans WAM, Boer P, Elbers PF (1977) Localization of acid phosphatase in *Saccharomyces cerevisiae:* a clue to cell wall formation. J Bacteriol 131:638–644
Liras P, Gascón S (1971) Biosynthesis and secretion of yeast invertase. Effect of cycloheximide and 2-deoxy-D-glucose. Eur J Biochem 23:160–165
Mańkowski T, Jankowski W, Chojnacki T, Franke P (1976) C_{55}-Dolichol: occurrence in pig liver and preparation by hydrogenation of plant undecaprenol. Biochemistry 15:2125–2130

Marriott MS (1977) Mannan-protein location and biosynthesis in plasma membranes from the yeast of *Candida albicans*. J Gen Microbiol 103:673–702

Marriott M, Tanner W (1979) Localization of dolichyl phosphate- and pyrophosphate dependent glycosyl transfer reactions in *S. cerevisiae*. J Bacteriol 139:565–572

Marshall RD (1974) The nature and metabolism of the carbohydrate-peptide linkages of glycoproteins. Biochem Soc Symp 40:17–26

Matile P, Cortat M, Wiemken A, Frey-Wyssling A (1971) Isolation of glucanase-containing particles from budding *Saccharomyces cerevisiae*. Proc Natl Acad Sci USA 68:636–640

Mayer RM (1971) The enzymatic synthesis of the phosphomannan of *Hansenula capsulata*. Biochim Biophys Acta 252:39–47

Moor H (1967) Endoplasmic reticulum as the initiator of bud formation in yeast. Arch Mikrobiol 57:135–146

Nakajima T, Ballou CE (1975) Microheterogeneity of the inner core region of yeast mannoprotein. Biochem Biophys Res Commun 66:870–879

Nakayama K, Araki Y, Ito E (1976) The formation of a mannose-containing trisaccharide on a lipid and its transfer to proteins in yeast. FEBS Lett 72:287–290

Novick P, Schekman R (1979) Secretion and cell surface growth are blocked in a temperature sensitive mutant of *Saccharomyces cerevisiae*. Proc Natl Acad Sci USA 76:1858–1862

Novick P, Field C, Schekman R (1980) The identification of 23 complementation groups required for posttranslational events in the yeast secretory pathway. Cell 21:205–215

Onishi HR, Tkacz JS, Lampen JO (1979) Glycoprotein nature of yeast alkaline phosphatase. Formation of active enzyme in the presence of tunicamycin. J Biol Chem 254:11943–11952

Palamarczyk G, Janczura E (1977) Lipid mediated glycosylation in yeast nuclear membranes. FEBS Lett 77:169–172

Palamarczyk G, Lehle L, Tanner W (1979) Polyprenyl phosphate prevents inactivation of yeast glycosyl transferase by detergents. FEBS Lett 108:111–115

Palamarczyk G, Lehle L, Mankowski T, Chojnacki T, Tanner W (1980) Specificity of solubilized yeast glycosyl transferases for polyprenyl derivatives. Eur J Biochem 105:517–523

Parodi AJ (1976) Protein glycosylation through dolichol derivatives in baker's yeast. FEBS Lett 71:283–286

Parodi AJ (1977) Synthesis of glycosyl-dolichol derivatives in baker's yeast and their role in protein glycosylation. Eur J Biochem 83:253–259

Parodi AJ (1978) Lipid intermediates in the synthesis of the inner core of yeast mannan. Eur J Biochem 83:253–259

Parodi AJ (1979) Synthesis of mannan outer chain and of dolichol derivatives. J Biol Chem 254:8343–8352

Parodi AJ, Leloir LF (1979) The role of lipid intermediates in the glycosylation of proteins in the eucaryotic cell. Biochim Biophys Acta 559:1–37

Pless DD, Palamarczyk G (1978) Comparison of polyprenyl derivatives in yeast glycosyl transfer reactions. Biochim Biophys Acta 529:21–28

Raizada MK, Kloepper HG, Schutzbach JS, Ankel H (1974) Biosynthesis of oligogalactosyl side chains of the cell envelope glycoprotein of *Cryptococcus laurentii*. J Biol Chem 249:6080–6086

Raizada MK, Schutzbach JS, Ankel H (1975) *Cryptococcus laurentii* cell envelope glycoprotein. Evidence for separate oligosaccharide side chains of different composition and structure. J Biol Chem 250:3310–3315

Reuvers F, Habets-Willems C, Reinking A, Boer P (1977) Glycolipid intermediates involved in the transfer of N-acetylglucosamine to endogenous proteins in a yeast membrane preparation. Biochim Biophys Acta 486:541–552

Reuvers F, Boer P, Hemming FW (1978) The presence of dolichol in a lipid diphosphate N-acetylglucosamine from *Saccharomyces cerevisiae* (baker's yeast). Biochem J 169:505–508

Richard M, Letoublon R, Louisot P, Got R (1971) Biosynthèse des glycoproteins. XVII. Characterisation d'une mannosyltransferase d'*Aspergillus oryzae*. Biochim Biophys Acta 230:603–609

Robbins PW, Hubbard SC, Turco SJ, Wirth DF (1977) Proposal for a common oligosaccharide intermediate in the synthesis of membrane glycoproteins. Cell 12:893–900
Ronin C, Bouchilloux S, Granier C, van Rietschoten J (1978) Enzymatic N-glycosylation of synthetic Asn-X-Thr containing peptides. FEBS Lett 96:179–182
Rothman JE, Lodish HF (1977) Synchronised transmembrane insertion and glycosylation of a nascent membrane protein. Nature (London) 269:775–780
Rudick MJ (1979) Mannosyl transfer by membranes of *Aspergillus niger:* mannosylation of endogenous acceptors and partial analysis of the products. J Bacteriol 137:301–308
Rudick MJ, Elbein AD (1973) Glycoprotein enzymes secreted by *Aspergillus fumigatus.* Purification and properties of β-glucosidase. J Biol Chem 248:6506–6513
Rudick MJ, Elbein AD (1974) Glycoprotein enzymes secreted by *Aspergillus fumigatus.* Purification and properties of α-glucosidase. Arch Biochem Biophys 161:281–290
Rudick MJ, Elbein AD (1975) Glycoprotein enzymes secreted by *Aspergillus fumigatus.* Purification and properties of a second β-glucosidase. J Bacteriol 124:534–541
Ruiz-Herrera J, Sentandreu R (1975) Site of glycosylation of mannoproteins from *Saccharomyces cerevisiae.* J Bacteriol 124:127–133
Sakaguchi O, Suzuki M, Yokota K (1968) Effect of partial acid hydrolysis on precipitin activity of Aspergillus fumigatus galactomannan. Jpn J Microbiol 12:123–124
Samuel O, Nordin JH (1971) An unique system for the study of the biosynthesis and secretion of a specific glycoprotein enzyme. Biochem Biophys Res Commun 45:1376–1383
Santos E, Villanueva JR, Sentandreu R (1978) The plasma membrane of *Saccharomyces cerevisiae.* Isolation and some properties. Biochim Biophys Acta 508:39–54
Schutzbach JS, Ankel H (1971) Multiple mannosyl transferases in *Cryptococcus laurentii.* J Biol Chem 246:2187–2194
Schutzbach JS, Ankel H (1972) Xylosyltransferases in *Cryptococcus laurentii.* J Biol Chem 247:6574–6580
Schutzbach JS, Raizada MK, Ankel H (1974) Heteroglycan synthesis in *Cryptococcus laurentii.* J Biol Chem 249:2953–2958
Schwarz RT, Datema R (1980) Inhibitors of protein glycosylation. TIBS 5:65–67
Sentandreu R, Lampen JO (1970) Biosynthesis of yeast mannan: inhibition of synthesis of mannose acceptor by cycloheximide. FEBS Lett 11:95–99
Sentandreu R, Lampen JO (1971) Participation of a lipid intermediate in the biosynthesis of *Saccharomyces cerevisiae* LK2G12 mannan. FEBS Lett 14:109–113
Sentandreu R, Lampen JO (1972) Biosynthesis of mannan in *Saccharomyces cerevisiae.* Isolation of a lipid intermediate and its identification as a mannosyl-1-phosphoryl-polyprenol. FEBS Lett 27:331–334
Sentandreu R, Northcote DH (1969) The characterization of oligosaccharide attached to threonine and serine in mannan glycopeptides obtained from the cell wall of yeast. Carbohydr Res 10:584–585
Sharma CB, Babczinski P, Lehle L, Tanner W (1974) The role of dolichol monophosphate in glycoprotein biosynthesis in *Saccharomyces cerevisiae.* Eur J Biochem 46:35–41
Skutelsky E, Bayer EA (1979) The ultrastructural delineation of cell growth and division processes using the avidin-biotin complex. Exp Cell Res 121:331–336
Slodki ME (1963) Structure of *Hansenula capsulata* NRRL Y-1842 phosphomannan. Biochim Biophys Acta 69:96–102
Smith WL, Nakajima T, Ballou CE (1975) Biosynthesis of yeast mannan. Isolation of *Kluyveromyces lactis* mannan mutants and a study of the incorporation of N-acetyl-D-glucosamine into the polysaccharide side chains. J Biol Chem 250:3426–3435
Soliday CL, Kolattukudy PE (1979) Introduction of O-glycosidically linked mannose into proteins via mannosyl phosphoryl dolichol by microsomes from *Fusarium solari* f. pisi. Arch Biochem Biophys 197:367–378
Spiro MJ, Spiro RG, Bhoyroo VD (1976) Lipid-saccharide intermediates in glycoprotein biosynthesis. I. Formation of an oligosaccharide-lipid by thyroid slices and evaluation of its role in protein glycosylation. J Biol Chem 251:6400–6408
Tabas I, Schlesinger S, Kornfeld S (1978) Processing of high mannose oligosaccharides to form complex type oligosaccharides on the newly synthesized polypeptides of the vesicular stomatitis virus G protein and the IgG heavy chain. J Biol Chem 253:716–722

Takatsuki A, Kohno K, Tamura G (1975) Inhibition of biosynthesis of polyisoprenol sugars in chick embryo microsomes by Tunicamycin. Agric Biol Chem 39:2089–2091

Tanner W (1969) A lipid intermediate in mannan biosynthesis in yeast. Biochem Biophys Res Commun 35:144–150

Tanner W, Jung P, Behrens NH (1971) Dolichol monophosphates: mannosyl acceptors in a particulate in vitro system of *S. cerevisiae*. FEBS Lett 16:245–248

Tkacz JS, Lampen JO (1972) Wall replication in *Saccharomyces* species: use of fluorescein-conjugates concanavalin A to reveal the site of mannan insertion. J Gen Microbiol 72:243–247

Tkacz JS, Lampen JO (1975) Tunicamycin inhibition of polyisoprenol N-acetyl-glucosaminyl pyrophosphate formation in calf-liver microsomes. Biochem Biophys Res Commun 65:248–257

Van Rijn HJM, Boer P, Steyn-Parvé E-P (1972) Biosynthesis of acid phosphatase of baker's yeast. Factors influencing its production by protoplasts and characterization of the secreted enzyme. Biochim Biophys Acta 268:431–441

Waechter CJ, Lennarz WJ (1976) The role of polyprenol-linked sugars in glycoprotein synthesis. Annu Rev Biochem 45:95–112

Yen PH, Ballou CE (1973) Composition of a specific intercellular agglutination factor. J Biol Chem 248:8316–8318

Zatz M, Barondes SH (1969) Incorporation of mannose into mouse brain lipid. Biochem Biophys Res Commun 36:511–517

III. Export of Carbohydrate Material

20 Secretory Processes – General Considerations and Secretion in Fungi

R. SENTANDREU, G. LARRIBA and M.V. ELORZA

1 Introduction

Since studies on the biosynthesis of complex polysaccharides as well as glycoproteins have been made possible, a great deal of effort has been devoted to elucidating the mechanisms by which most of these components finally appear in the extracellular fluid or in vesicles of different nature. It appears evident that the basic mechanism of the secretion process is the passage of molecules through a membrane, and as we shall see, this mechanism is very similar in prokaryotic and eukaryotic cells. In addition plant and animal cells have developed a complex machinery involving the participation in a highly ordered sequence of different organelles (rough endoplasmic reticulum, RER; smooth

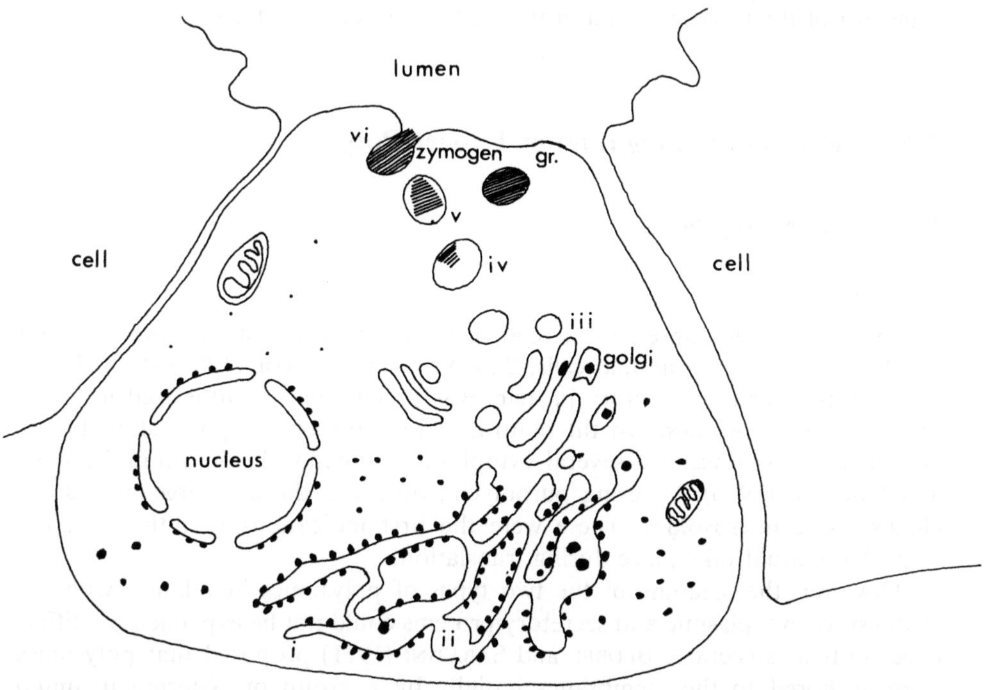

Fig. 1. Secretory pathway in the eukaryotic cells. Diagram summarizing the events in the elaboration of export proteins in a pancreatic exocrine cell. Secretory proteins are synthesized on RER and subsequently transferred into the cisternal space (Steps *i*, *ii*), then they are passed through the Golgi complex (Step *iii*) into condensing vacuoles (Step *iv*). Finally these vacuoles become converted into zymogen granules (Step *v*) which are released from the cell (Step *vi*). (Modified from PALADE et al. 1962)

endoplasmic reticulum, SER; Golgi complex, etc). Unfortunately, some of these organelles have not been unambiguously recognized or characterized in fungi, and consequently we lack a coherent model comparable to that described for higher organisms.

In view of the present difficulties in establishing a solid picture of the secretion process in fungi, we will first consider the mechanisms currently accepted for bacteria, plant, and mammalian cells. In connection with this, we will discuss the secretion of glycoproteins and polysaccharides in *Saccharomyces cerevisiae* and other fungi. We shall see that strong reasons exist for believing that the secretion pattern in fungi shares common steps with the basic model prevailing for bacterial and higher systems.

PALADE (1975) divided the secretory process of pancreatic exocrine cells, in six operational steps: (1) synthesis of macromolecules, (2) segregation, (3) intracellular transport, (4) concentration, (5) intracellular storage, and (6) discharge (Fig. 1).

Concentration and intracellular storage are found normally in cells whose discharge is discontinuous and controlled by hormones. In fungi both steps are almost unknown, so we shall reduce our discussion to the other steps. We will first review the synthesis and segregation of secretory proteins together as they normally take place concomitantly. Then we will concentrate on the transport of the molecules and in the mechanism of exocytosis.

2 Synthesis and Segregation of Export Polymers

2.1 The Signal Hypothesis

2.1.1 Mammalian Cells

Following PALADE's observation relative to the presence of membrane-bound and free polysomes in animal cells, SIEKEVITZ and PALADE (1960) showed that chymotrypsinogen, a secretory protein, was preferentially synthesized in membrane-bound ribosomes. On the contrary, free polysomes appeared to be involved in the synthesis of several cytoplasmic proteins. In addition, REDMAN and SABATINI (1966), by using puromycin, observed the discharge of nascent chains inside microsomes. These were the first indications that the secretion of proteins could take place during translation.

However, the meaning of the two types of polysomes in relation with the synthesis of cytoplasmic and secretory proteins could not be explained by differences in their structure. BLOBEL and SABATINI (1971) proposed that polysomes were anchored to the membranes initially by a group of *N*-terminal amino acids of the growing protein chain. This implies that proteins destined for secretion are synthesized with a sequence that guides them, and the polysomes, to specific sites of the endoplasmic reticulum.

Such a sequence was first demonstrated by MILSTEIN et al. (1972) in a study of in vitro synthesis of the light chain of immunoglobulin. The primary transla-

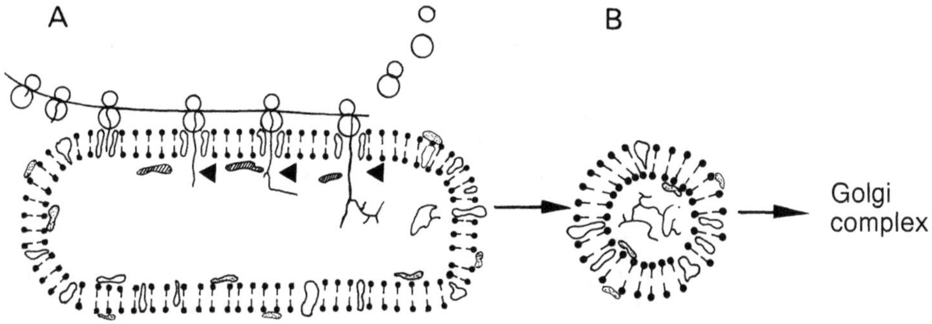

Fig. 2A, B. Synthesis and segregation of secretory proteins. The proteins to be secreted are translated from specific messenger RNA's. The growing polypeptide has a "signal sequence" which recognizes receptors in the ER and leads to ribosome attachment and co-translational translocation of the polypeptide into the cisternal space. Both initial glycosylation and signal sequence removal also take place during translocation (**A**). The protein is transported from the RER cisternal space to the Golgi complex via transitional elements (**B**). •— Phospholipids; integral membrane protein; glycosylating enzymes; ◄ endopeptidase; periphery membrane protein

tion product turned out to be a longer molecule than the mature protein in having 15 additional amino acids in their *N*-terminus. When translation of the mRNA was carried out in the presence of membranes, the precursor protein was no longer detectable, only mature molecules were found; but addition of membranes after precursor release from the polysomes failed to modify the precursor. These authors suggested that the globulin was synthesized as a pro-protein with an *N*-terminal signal which would allow it to cross the membrane. The *N*-terminal sequence would in turn be cleaved during the transference of the molecule to the lumen of the vesicles.

In addition, SCHECHTER et al. (1975) showed that the *N*-terminus signal of the immunoglobulin L chain was highly hydrophobic with leucine accounting for about 30% of the 20 amino acid residues.

These results were confirmed in several laboratories and led to the signal hypothesis. In particular, BLOBEL and DOBBERSTEIN (1975a, b) identified a large variety of secretion enzymes which are synthesized in vitro with a hydrophobic *N*-terminus composed of 15 to 30 amino acids.

The presence of a hydrophobic sequence fits in well with the model, since it would facilitate the initial anchoring of the polysomes to the membrane. It has also been proposed that a channel formed by specific membrane proteins expanding the bilayer would provide the path needed by the nascent peptide to cross the membrane (BLOBEL and DOBBERSTEIN 1975a). The protein is extruded through the channel and finally appears in the lumen of the RER. Although a cotranslational cleavage of the signal peptide was initially suggested, a post-translational processing has also been demonstrated (JACKSON and BLOBEL 1977). Recently, STRAUSS et al. (1979) have characterized an endopeptidase involved in the preprotein processing which cleaves the signal peptide at the –COOH side of an alanine residue. This enzyme acts during or shortly after translation of the secretory protein. (Fig. 2).

It is not always the case that the signal sequence is at the *N*-terminus or cleaved during maturation because ovalbumin, a secretory protein, failed to show a hydrophobic extra sequence in its *N*-terminus region (PALMITTER et al. 1978). However, an internal hydrophobic sequence, which can act as a signal, has recently been described (LINGAPPA et al. 1978). The isolated region competitively inhibited the transference of another secretory protein into vesicles. It was expressed only by nascent but not by the completed, presumably folded chain (LINGAPPA et al. 1979).

The signal hypothesis proposed, however, does not explain how some proteins are synthesized in the cytosol by free polysomes and then secreted or incorporated into different membrane-bound organelles. Such is the case of ribulose 1,5 diphosphate carboxylase, a soluble chloroplast enzyme composed of two subunits. One of the subunits is coded by the chloroplast's DNA, the other is synthesized in the cytosol from mRNA of nuclear origin. The complete enzyme finally appears inside the chloroplast. If chloroplasts are added after in vitro synthesis of the small subunit, the completed polypeptide is taken up and processed into the final product (HIGHFIELD and ELLIS 1978).

In conclusion it appears that in addition to the translocator(s) involved in the cotranslational secretion, there exist other signals involved in the post-translational recognition and export of proteins from the cytosol into organelles.

2.1.2 Bacteria

Recently, evidence for specific *N*-terminal sequences in proteins which have to be secreted or incorporated in membranes has been obtained in bacteria (INOUE and BECKWITH 1977; RANDALL et al. 1978, SARVAS et al. 1978, SUTCLIFFE 1978).

Experiments carried out by SMITH et al. (1977, 1979) have demonstrated that cotranslational secretion is a common phenomenon in prokaryotes. These authors took advantage of the possibility of labeling the nascent protein from the external side of the bacterial protoplasts, while still growing and protruding from the membrane. Part of the label was found to be associated with polysomes and was released by low Mg^{2+}, dilute alkali, puromycin, and by chain completion in vitro. One of the in vitro completed proteins was identified as alkaline phosphatase (SMITH et al. 1977). This finding has been followed up by a similar demonstration of cotranslational secretion for other enzymes including amylase from *Bacillus subtilis* (SMITH et al. 1977), the toxin of *Corynebacterium diphteriae,* and penicillinase of *Bacillus licheniformis* (DAVIS and TAI 1980).

In a manner similar to eukaryotic cells, secreted proteins (alkaline phosphatase, α-amylase, penicillinase and diphteria toxin), as well as membrane proteins, are synthesized on membrane-bound polysomes, while cytoplasmic proteins (elongation factor EFTu) appear to be synthesized on free polysomes (RANDALL and HARDY 1977, DAVIES and TAI 1980). Evidence for processing has also been obtained for penicillinase of *Bacillus licheniformis* and diphtheria toxin.

The role of the signal sequence in prokaryotic cells has been evaluated by using the gene fusion technique. SILHAVY et al. (1977) replaced the *N*-terminal sequence of β-galactosidase, a cytoplasmic protein, by the corresponding sequence of maltose binding protein, a periplasmic protein. The hybrid molecule was found in the cytoplasmic membrane. When hybridized with phage λ receptor, an outer membrane protein, the final location was dependent on the length of the latter in the hybrid (SILHAVY et al. 1977). When it was short, the β-galactosidase activity was found in the cytoplasm. A longer portion of the phage λ receptor drove the hybrid to the outer membrane. Thus the information necessary to direct the phage λ receptor to its outer membrane location is found on its *N*-terminus sequence. In addition, mutations presumably located in the signal sequences of either maltose binding protein or the phage λ receptor-β-galactosidase hybrid appearing in the outer membrane blocked the export of proteins (BASSFORD and BECKWITH 1979, BASSFORD et al. 1979). In the last case, hybrids were entrapped in the membrane and prevented the secretion of normal outer membrane proteins. These results suggest that secretory proteins compete for a limited number of recognition sites localized in the membrane.

The fact that β-galactosidase-maltose binding protein hybrids were found in the cytoplasm or cytoplasmic membrane and never in the periplasmic space indicates that signal sequences are not sufficient for secretion. Apparently some portions of the proteins cannot cross the membrane (SILHAVY et al. 1977, EMR 1978).

Post-translational secretion has also been described in bacteria. Such is the case of colicines E_1 and E_3 (JAKES and MODEL 1979), subunit A of colera toxin and several products of a sex factor (DAVIS and TAI 1980). The fact that mature proteins can be released from the cytoplasm after completion calls for a specific transport system involving special translocators as suggested in animal cells. It is possible that two energy-driven mechanisms, one depending on the elongation of the growing polypeptide chain, and the other involving an unknown transport machinery, exist in the cells. Competition experiments between co-translationally and post-translationally secreted proteins such as those described above might help to evaluate both types of secretory mechanisms.

Interestingly, bacteria recognize mammalian signal sequences as shown by the secretion of the ovoalbumin gen product (FRASER and BRUCE 1978, MERCEREAU-PUIJALON et al. 1978). This suggests that secretory mechanisms are universal and appeared early in the evolution of the living world.

2.1.3 Fungi

Definitive evidence for a signal sequence in secretory proteins in fungi is still lacking. However, processing of the vacuolar yeast glycoprotein carboxypeptidase *Y* has been shown by HASILIK and TANNER (1978b). Carboxypeptidase *Y* is synthesized via a larger precursor which is converted to the active form by proteolytic cleavage of a peptide with an apparent molecular weight of 6,000. Although longer than those found in bacterial and mammalian secretory proteins the extra peptide did contain a high amount of leucine, suggesting a hydrophobic nature. The subcellular location of the precursor and enzyme indicated that the former was preferentially bound to membranes, while the final product appeared in the supernatant of cell homogenates. This is consistent with synthesis of the precursor taking place at the RER and then being transported to the vacuole where it would undergo processing. However it is also consistent with the idea that it preferentially sticks to membranes due to the hydrophobic extra sequence. The fact that yeast proteinase B converted the precursor in a form electrophoretically indistinguishable from the active enzyme is consistent with an intravacuolar processing. The calculated half-life of the precursor was about 6 min and accordingly, the processing cannot be compared with the co-translational cleavage observed for some bacterial and mammalian preproteins; rather, it recalls the modifications undergone by some secretory prohormones (CAMPBELL and BLOBEL 1976, DEVILLERS-THIERY et al. 1975, HABENER et al. 1979). It also mimics the situation described for the small subunit of ribulose 1,5 diphosphate carboxylase (see above), and points to a post-translational secretion.

There exist another two examples of post-translational import and processing in yeast. F_1 -ATPase is a mitochondrial enzyme formed by five different subunits which are synthesized in cytoplasmic ribosomes. At least three subunits are synthesized as larger precursors, both in vitro and in vivo. The addition of yeast mitochondria to the products synthesized in vitro resulted in the uptake of the precursors together with their conversion to the mature form (MACCECCHINI et al. 1979).

Cytochrome oxidase is a component of the inner mitochondrial membrane. Four of its seven subunits are synthesized on cytoplasmic ribosomes as a single "polyprotein" precursor with a molecular weight of 8,000 in excess of that accounted for by the final products (POYTON and MCKEMMIE 1979 a). Pulse-chase experiments in vivo indicated that the polyprotein crosses the outer mitochondrial membrane without undergoing detectable modifications and is processed stepwise at the inner membrane. Polyprotein precursor was also transported into, and processed by, intact isolated mitochondria (POYTON and MCKEMMIE 1979 b).

In vitro translocation of ATPase and cytochrome oxidase precursors by mitochondria occurred in the presence of cycloheximide. This accordingly argues against a transport dependent on a co-translational event and again recalls the transport of the small unit of ribulose 1,5 diphosphate carboxylase.

The fact that the active form of yeast proteases as well as other hydrolases are localized in vacuoles calls for a molecular mechanism to prevent activation before reaching their final destination. Such a mechanism could be provided by an extra peptide, such as the one described for carboxypeptidase *Y*. In addition, this peptide might also act as signal in the recognition of receptor sites situated in the vacuole membrane. We do not know the role(s) played by the extra sequence of the mitochondrial enzyme precursors, but if we assume that special signal sequences, that may be cleaved or not, are involved in the recognition of different organelles, we should definitively individualize the concepts of signal sequence, vectorial translation, post-translational transport and processing in order to preserve a unified concept for a signal hypothesis.

2.2 Mechanisms of Glycosylation

2.2.1 Mammalian Cells

As the structure and biosynthesis of glycoproteins and polysaccharides of fungi have been dealt with in detail in previous chapters (Chaps. 15–19, this Vol.), we will only consider the aspect of glycosylation in connection with the secretion process.

It is clear that most secretory proteins are glycoproteins and that EYLAR's early suggestion (1965) that the carbohydrate moiety of glycoproteins may act as a signal for secretion has now been subjected to experimental analysis. Most of the work in this field concerns the role of carbohydrate *N*-glycosydically bound to the proteins. Glycosylation of these residues may be prevented by tunicamycin, an antibiotic that specifically blocks the synthesis of dolichyl pyrophosphate *N*-acetylglucosamine which is a key intermediate in the formation of core oligosaccharides in glycoproteins containing the *N*-acetylglucosaminyl-asparagine (Asn-GlcNAc) type of bond (TKACZ and LAMPEN 1975, TAKATSUKI et al. 1975, LEHLE and TANNER 1976).

Initial glycosylation of proteins takes place during translation (KIELY et al. 1976) and after chain completion and release from polysomes (BERGMAN and KUEHL 1978). Evidence indicates that at least some animal and fungal (see below) glycoproteins may be secreted when glycosylation is inhibited by tunicamycin (STRUCK and LENNARZ 1977, STRUCK et al. 1978, MIZRAHI et al. 1978, FUJISAWA et al. 1978, SCHREIBER et al. 1979). However, alterations in the secretion pattern for some glycoproteins have frequently been reported. HICKMAN et al. (1977) found that in plasma cells grown in the presence of tunicamycin, immunoglobulins did not become glycosylated, were not secreted, but accumulated in swollen vesicles of the rough endoplasmic reticulum. The extent of secretion inhibition correlated well with the carbohydrate content of various immunoglobulins (HICKMAN and KORNFELD 1978). In contrast, WILLIAMSON et al. (1980) did not find any effect of tunicamycin on the secretion of IgA or IgG, although secretion of the most heavily glycosylated IgM was inhibited relative to the nonglycosylated ones.

Like secretory proteins, membrane proteins of animal cells appear to be glycosylated at the level of nascent peptide but, again, glycosylation is not an absolute requirement for either insertion as a membrane protein or cleavage as shown for the membrane of vesicular stomatitis virus (ROTHMAN et al. 1978) or Semliki Forest virus (GAROFF and SCHWARTZ 1978). However, at least in one case, the absence of carbohydrate in the precursor prevented the processing (SCHULTZ and OROSZLAN 1979). In other glycoproteins migration of the unglycosylated product to the cell surface was impaired, probably due to intracellular aggregation (LEAVIT et al. 1977, GIBSON et al. 1979). Proteolysis may also account for some of the observed alterations (OLDEN et al. 1978).

2.2.2 Fungi

Several fungal enzymes are glycoproteins and some of them are exported across the plasma membrane and appear to have a role in cell economy (e.g., invertase, acid phosphatase, melibiase, etc.) or cell morphogenesis (β-glucanases), while others are retained in vacuoles (carboxypeptidase *Y*, alkaline phosphatase etc.). In addition to enzymes, fungal cells also export structural glycoproteins and polysaccharides (mannan, glucan, chitin, etc.), as well as survival products (killer, sexual factors, etc.).

Two types of saccharides bound to the protein moiety have so far been described in yeast glycoproteins. Mannose, mannobiose, mannotriose, and mannotetraose are linked to serine or threonine through an *O*-glycosydic bond. Another type of saccharide is bound to the protein moiety through a *N*-glycosydic linkage. It is composed of an inner core very similar to high mannose chains found in mammalian glycoproteins and may or may not be extended by the so-called outer chain. It should be noted that both types of bonds have never been described in the same molecule (BALLOU 1976).

As in animal systems, glycosylation of fungal glycoproteins occurs through intermediate lipids. Dolichol-phosphate-mannose (Dol-P-Man) is the donor of the first *O*-glycosydically bound residue (BABCZINSKI and TANNER 1973) while dolichol-diphosphate-*N*,*N′*-diacetyl chitobiose-(mannose)$_n$ (Dol-P-P-GlNAc$_2$-Man$_n$) is thought to be responsible for the transfer of the inner core as a single unit (LEHLE and TANNER 1978, PARODI 1978).

Several research groups have been working to find the location of the enzymes involved in the transfer of glycosyl residues. Studies carried out in our laboratory suggest that glycosylation of mannoproteins in *S. cerevisiae* is initiated at the nascent peptide level (RUIZ-HERRERA and SENTANDREU 1975). In the same line, LARRIBA et al. (1976) found that mannose from dolichol-P-^{14}C-man was transferred in vitro to a polysome-enriched fraction. Almost all the radioactivity could be released by mild alkaline treatment, indicating that mannose had been transferred to serine or threonine. In agreement with these data, LEHLE et al. (1977) reported that the first mannose of the *O*-glycosydically linked oligosaccharides was incorporated with the highest specific activity at the level of the endoplasmic reticulum.

Incorporation of the *N*-glycosydically linked core of yeast mannoproteins also appears to occur very early after or even during the synthesis of the protein portion, MARRIOT and TANNER (1979) have isolated a membrane preparation enriched in the transferases involved on the formation of both *N*- and *O*-glycosydic bonds and have suggested that the endoplasmic reticulum was the origin of this preparation. Following a different approach, HASILIK and TANNER (1978a) have shown that the precursor of carboxypeptidase *Y* is labeled immediately after a pulse with ^{3}H-mannose. However, no direct evidence has so far been obtained for a co-translational glycosylation in fungi.

The role of asparagine-linked oligosaccharides in the secretion of several individual fungal enzymes has been evaluated. KUO and LAMPEN (1974) reported that tunicamycin inhibited the synthesis, though not the secretion, of external yeast glycoproteins such as invertase, acid phosphatase, and mannan by yeast protoplasts. In contrast, it did not impair the synthesis of intracellular, nonglycosylated proteins (α-glucosidase) or the synthesis of wall polysaccharides (glucan and chitin). No accumulation of carbohydrate-free invertase or acid phosphatase was detected in treated cells.

The synthesis of carboxypeptidase *Y*, a yeast glycoprotein located in the vacuole, was only inhibited by 50% to 60% by tunicamycin under conditions in which the synthesis of external invertase was completely halted. The decrease in enzyme activity correlated well with the reduced amounts of carbohydrate-free protein detected by immunoprecipitation (HASILIK and TANNER 1976). Since maturation by cleavage of the nonglycosylated precursor to the active form was not impaired and the enzyme was stable, it was concluded that there is a regulatory link between the synthesis of the protein moiety of the molecule and its glycosylation (HASILIK and TANNER 1978b).

Yet another different behavior has been reported for alkaline phosphatase, which is also a lysosomal enzymatic glycoprotein (ONISHI et al. 1979). Yeast protoplasts synthesized the active enzyme in the presence of tunicamycin at the same rate as controls, even though the product was nonglycosylated. ONISHI et al. (1979) suggested that there is no regulatory link between the glycosylation of a protein and its synthesis.

We have studied the effect of tunicamycin on the synthesis and secretion of two external yeast enzymes: α-galactosidase and exo-β-glucanase. The synthesis of total active α-galactosidase (internal plus external) by yeast protoplasts was immediately stopped after the addition of tunicamycin, although the preexisting internal enzyme which contains reduced amount of carbohydrate was secreted. Interestingly enough, increasing concentrations of tunicamycin (10 to 30 μg ml^{-1}), which did not affect protein synthesis, did speed up the rate of secretion.

Yeast protoplasts also secrete at least one form of exo-(1→3)-β-glucanases. The addition of tunicamycin depressed the rate of secretion of the active enzyme by 50% and a new form, with molecular weight 14,000 d lower than the native enzyme, was found in the supernatant. Unlike the native enzyme, the glucanase formed in the presence of tunicamycin did not precipitate with concanavalin A (conA), suggesting that it was devoid of carbohydrate. Both molecular forms had a similar Km against laminarin or p-nitrophenyl-β-D-glucoside (PNPG), but the nonglycosylated form was less stable to pH or temperature changes (unpublished results).

Aspergillus niger secretes a number of glycosidases into the culture medium, and some of them, at least, are glycoproteins (ADYA and ELBEIN 1977). The presence of tunicamycin (40 μg ml^{-1}) decreased the activities of external β-*N*-acetyl-glucosaminidase, α-galactosidase, and β-glucosidase by 70%, and the secreted enzymes were devoid of carbohydrate. Similarly, intracellular activities were inhibited by 50%. At least for β-*N*-acetylglycosaminidase, the decrease in activity was due to a decrease in the immuno-detectable enzyme (SPEAKE et al. 1980).

It is evident that the effects in secretion produced by a blockage in glycosylation vary widely among the individual glycoproteins. We now know that secretion is a complex phenomenon involving biosynthesis, passage through a membrane and transport of the molecule to its final destination via a complex group of organelles. Any impairment in the steps involved will result in modifications of the secretion process. The above data suggest that transport of hydrolases is not prevented by the absence of carbohydrate. Nonglycosylated active exo-β-glucanase is transported and secreted, and no accumulation of the active molecule was observed in the cytoplasm. Neither does a defect in transport appear to account for the lack of secretion of nonglycosylated external invertase, acid phosphatase, or α-galactosidase. If these forms accumulate in internal membranes they should be inactive. However, in the case of invertase, no enzyme was immunodetectable (ONISHI et al. 1979). It may be concluded that either (1) the inhibition of glycosylation impairs synthesis of the protein moiety or (2) the nonglycosylated product is folded in such a way that it escapes detection by antibodies, or it is specially susceptible to proteases and is rapidly degraded. The first possibility may well also account for the behavior of exo-β-glucanase and carboxypeptidase *Y*. As mentioned before, a regulatory link between glycosylation and protein synthesis has been proposed for the latter. However, a partial proteolytic degradation at the nascent peptide level cannot be discarded for both enzymes. In any case, this regulatory link does not appear to be a general phenomenon, as shown by the behavior of alkaline phosphatase.

If our aim is to build a unified model for the different patterns observed, we would face the problem in terms of the effects of the carbohydrate moiety of a glycoprotein on its final configuration (PAZUR et al. 1970, WANG and HIRS 1977, MARGOLIS et al. 1978, SCHWARTZ et al. 1976, GIBSON et al. 1979). Glycosylation of nascent peptide may influence the tertiary and quaternary structure of the protein from this moment. Heavily glycosylated glycoproteins such as external invertase, acid phosphatase, and α-galactosidase may be affected more drastically than exo-β-glucanase and carboxypeptidase *Y* whose carbohydrate content is around 20%. Nonglycosylated alkaline phosphatase did not show detectable differences in either synthesis or properties with respect to the glycosylated enzyme which contains less than 10% carbohydrate. However the activity of α-galactosidase of *Aspergillus niger*, an enzyme containing less than 5% carbohydrate, showed a decrease of 70%. It appears that not only the amount but also the location of the saccharides on the polypeptide chain, as well as other structural or environmental factors, may determine the relative importance of the carbohydrate in the tertiary structure of a protein and, in turn, its susceptibility to proteolytic enzymes, stability, aggregation, etc.

3 Transport of Secretion Products

3.1 Mammalian Cells

Secretory proteins synthesized in the RER have to travel toward the cell periphery in order to be exported (Fig. 1). Initial evidence of such transport was provided by the work of CARO and PALADE (1964). Using ^{3}H-leucine in pulse-chase radioautographic experiments they found that the precursors appeared initially in the RER, then in the Golgi apparatus, and finally in zymogen granules. This scheme has since been extended to other systems by using the same technique and has become the currently accepted model (for a review see PALADE 1975).

Not all proteins that cross RER membranes become exported to the exterior of the cells. In particular, some enzymes are destined, instead, for lysosomes or other organelles. It appears that an additional signal is required in order to discern between the several possibilities. Evidence for this came from the hypothesis that lysosomal enzymes undergo a secretion-endocytosis process during their transfer into lysosomes (HICKMAN and NEUFELD 1972). These lysosomal enzymes bear a phosphomannosyl residue on the asparagine-bound oligosaccharides which may function either intracellularly or on the cell surface as a signal for their transfer into lysosomes (KAPLAN et al. 1977, VON FIGURA and KLEIN 1979, NATOWICZ et al. 1979). The addition of the marker appears to be necessary in order to segregate lysosomal enzymes. Genetic deficiency of the phosphorylated marker or interference with the glycosylation leads to increased secretion and the loss of lysosomal enzymes into the extracellular space (VON FIGURA et al. 1979).

Nothing is known, however, about the selection of proteins that should bear mannose-6-phosphate, the site where the addition of the marker occurs, or how it acts in order to segregate lysosomal enzymes from secretory proteins destined for export. Possibly related to these problems is the fact that asparagine-linked oligosaccharides of secretory as well as viral membrane glycoproteins undergo extensive modification. Both high-mannose or complex-type oligosaccharides of mammalian cells (see Chaps. 8 and 19, this Vol.) derive from a common precursor containing di-*N*-acetyl chitobiose and several hexoses, which are synthesized while linked to a polyisoprenoid lipid carrier (ROBBINS et al. 1977). At a later stage, membrane-bound glucosidases and mannosidases remove the glucosyl and part of the mannosyl residues to give a tri-to-pentamannose *N*-acetylchitobiose which, in turn, will be completed to the final form by the addition of other sugars (fucose, galactose, *N*-acetyl-glucosamine and sialic acid) to produce the complex-type oligosaccharides (CHEN and LENNARZ 1978, TABAS et al. 1978, HUNT et al. 1978, GRINNA and ROBBINS 1979, WHITE and WIRTH 1979). We do not know as yet whether "trimming" is necessary for further addition of mannose-6-phosphate, since the latter appears to be associated to high-mannose type oligosacccharides (VON FIGURA et al. 1979). TABAS et al. (1978) have suggested that oligosaccharide processing may be related to the evolution of high mannose to complex-type oligosaccharides which appear to have evolved later.

For the time being, we may assume that, apart from their specificity, location of glucosidases and/or transferases involved in carbohydrate modifications is very precise. Thus, it is not difficult to envisage an accumulation of specific molecules at specific places of the Golgi complex, which in turn, will be the precursors of vesicles destined for intracellular use. The same mechanism could account for secretory vesicles.

Further transport of secretory vesicles to the cell surface may be directed by cytoskeletal structures. The interaction of Golgi-derived vesicles with these structures would result in their transport to the cell surface. In support of this hypothesis is the fact that the process is accelerated by ATP, perhaps by phosphorylation of microtubule-associated proteins (NEWMARK 1979). It is also pertinent to mention that the lipid components of the endoplasmic reticulum undergo changes during transport of secretory vesicles to the plasma membrane. The content of cholesterol and sphingomyelin continuously increases with a concomitant decrease in nonsaturated fatty acids.

3.2 Fungi

Most work concerning the transport of secretory material in fungi is related to the mode of deposition of cell wall material during the apical growth in filamentous fungi or budding in yeast.

A large number of internal membranous structures that may be involved in the transport process have been described in yeast cells. They include the endoplasmic reticulum, Golgi or Golgi-like vesicles, and a variety of structures which appear to derive from the former, such as proliferating endoplasmic reticulum, vesicles derived form endoplasmic reticulum, flat vesicles, etc. In addition, vacuoles are to be seen in all the cells, although the term vacuole has more than one meaning. In some cases, it is used to describe the biggest compartment of the resting cell, whereas in others it represents an operational term standing for all kinds of vesicles of the manifold interconnected endomembrane system and not for a clearly delimited physiological unit (WIEMKEN et al. 1979). On the other hand not only the ontogeny but also the existence of some of the above-cited structure, such as Golgi apparatus, are controversial (ELORZA et al. 1977).

In spite of this, evidence exists that several membranous structures are involved in the transport of cell wall constituents and secretory proteins. In 1967 MOOR reported that the budding process in *S. cerevisiae* is initiated by a vesiculation of the endoplasmic reticulum. The resulting vesicles were transported to the place where the prospective bud was to be formed and the process continued until completion of the bud. It was suggested that the vesicles contained disulfide reductase, an enzyme thought to be involved in cell wall modifications. The same kind of vesicle was reported by SENTANDREU and NORTHCOTE (1968), though a different content was suggested; their vesicles contained cell wall material.

The isolation and biochemical characterization of the vesicles was accomplished by MATILE et al. (1971) who coined the term "glucanase vesicles" due to their content in exo-$(1 \rightarrow 3)$-β-glucanase. Microscopic observation and analytical data showed that they were abundant in early budding cells and almost absent from cells in the late budding phase, or stationary cells. In addition to exo-$(1 \rightarrow 3)$-β-glucanase, glucanase vesicles, contain mannan, mannan synthase activity, and acid phosphatase (CORTAT et al. 1973, LINNEMANS et al. 1977). However, these results should be interpreted cautiously (CABIB 1975).

A more precise picture was deduced by using acid phosphatase as a secretory marker in protoplasts or whole *S. cerevisiae* cells (VAN RIJN et al. 1975, LINNEMANS et al. 1977). In protoplasts the enzyme was found associated with the endoplasmic reticulum, flat vesicles (subsurface cisternae), a Golgi-like structure, the central vacuole and the nuclear membrane. The presence of acid phosphatase in the nuclear membrane is not surprising as it appears to derive from the endoplasmic reticulum (MATILE et al. 1969). More striking, however, is its location in the central vacuole. This opens new, intriguing questions on the role of this organelle. Regardless of these considerations, if only one place of synthesis is assumed, a temporal relationship for the acid phosphatase molecules present in these organelles might be proposed. This implies that a secretory transport, similar to that described for mammalian or plant cells, does occur in yeast.

Oligosaccharides bound to asparagine appear to undergo extensive processing in yeast glycoproteins (PARODI 1979, see also, Chap. 19, this Vol.). The oligosaccharides are transferred in bloc from dolichol-P-P derivatives containing 2 *N*-acetylglucosamines and 9 to 12 hexoses, including mannose and glucose. Following its transfer to the protein moiety, the oligosaccharide is "trimmed" by glucosidases that split off the glucose residues. Further processing involves the addition of mannose residues, but there is no information whether any of the mannose molecules originally present in the dolichyl-P-P derivative is previously removed.

We do not know if a signal equivalent to mannose-6-P on the oligosaccharide moiety of lysosomal yeast glycoproteins might be necessary to segregate them from export glycoproteins. As described above, lack of carbohydrate does not appear to affect the transport of exo-(1 → 3)-β-glucanase, a secretory glycoprotein. Neither does it prevent the conversion of pro-carboxypeptidase Y to the active enzyme which is thought to occur inside the lysosome. However, the precise location for the unglycosylated lysosomal enzymes, carboxypeptidase Y or alkaline phosphatase has not been determined yet.

Analytical data of several secretory glycoproteins indicate that during their transport they are associated to membranes. External invertase of *S. cerevisiae* is a glycoprotein containing 50% carbohydrate, formed by two subunits (TRIMBLE and MALEY 1977) and located outside the plasma membrane. *S. cerevisiae* cells which are actively synthesizing invertase contain a smaller form of the enzyme associated to internal membranes (BABCZINSKI and TANNER 1978, RODRÍGUEZ et al. 1978). This enzyme has been shown to be a precursor of the external one. The addition of tunicamycin inhibited further synthesis of both the external (KUO and LAMPEN 1974) and the membrane-bound enzyme, suggesting that the latter also bears the inner core (BABCZINSKI and TANNER 1978). The presence of carbohydrate in the molecule was also indicated by precipitation with concanavalin A (RODRÍGUEZ et al. 1978) and isopicnic density gradient centrifugation (RODRÍGUEZ et al. 1980).

Acid phosphatase also contains 50% carbohydrate and is actively secreted by derepressed protoplasts. Glycosylated forms of the enzyme, which appear to be precursors of the external enzyme, are also found either free or associated to membranes in a protoplast lysate. External acid phosphatase contains more carbohydrate than the membrane-bound enzyme as shown by isopicnic centrifugation (BOER et al. 1975, RODRÍGUEZ et al. 1980). The membrane-bound enzyme was heterogeneous in its content in carbohydrate, a small proportion of the molecules reaching the buoyant density of the external enzyme (BOER et al. 1975). Finally, the soluble internal form of the enzyme behaves similarly to the external acid phosphatase in isopicnic density gradients, suggesting a similar amount of carbohydrate (RODRÍGUEZ et al. 1980).

The synthesis and secretion of α-galactosidase closely resemble those of acid phosphatase. The enzyme is found outside the protoplasts and also inside, either soluble or associated to membranes. All three forms are glycosylated, the internal forms being precursors of the external enzyme (see above). The molecular weight increases from the membrane-bound enzyme to the external enzyme, the soluble internal form being of an intermediate size, almost the same as the external α-galactosidase (unpublished results).

Similarities in the secretory pathways of these secretory enzymes are evident. In all cases, a membrane-bound form with a reduced amount of carbohydrate is present. Since mannosyltransferases are also membrane-bound enzymes (ALGRANATTI et al. 1963, BEHRENS and CABIB 1968, see above) it may be assumed that the membrane-associated invertase, acid phosphatase, and α-galactosidase are being completed by the addition of mannosyl residues to the inner core. The glycosylated forms found in the supernatant of the protoplast lysate appear to be near completion or even finished, as shown for α-galactosidase and acid phosphatase, respectively. It seems reasonable to suppose that they are released from internal vesicles sensitive to the osmotic shock. The presence of acid phosphatase in the central vacuole may be relevant here. If we consider the central vacuole as a storage compartment, the intravacuolar acid phosphatase probably represents the finished product. Since no endocytosis has so far been described in yeast, a molecular mechanism selects and retains in the same compartment molecules destined for intra and extracellular use. Such a mechanism may, in turn, be related to the problem of how molecules destined for secretion are segregated from those whose final fate is internal structures, vacuoles, or lysosomes.

Studies on transport in filamentous fungi are also related to hyphal growth. Microscopic observations are consistent with a model in which the material used in the apical wall synthesis is firstly secreted into endoplasmic reticulum cisternae, which appear to be the origin of the vesicles commonly observed at the sites of hyphal growth (MCCLURE et al. 1968, GIRBARDT 1969, GROVE and BRACKER 1970). Stacks of cisternae or single cisternae resembling a Golgi body have also been frequently observed (GROVE and BRACKER 1970). The tip organization varies in the different types of fungi. Oomycetes present two kinds of vesicles randomly distributed. The larger ones contain an electron opaque material while the smaller vesicles are more compact. A similar situation is found in Zygomycetes, but in this case the large vesicles form a distinct ordered line near the plasma membrane. In septate fungi (but not in Zygomycetes or Oomycetes) a special structure known as Spitzenkörper is visible during the active extension of the hyphal growth. As soon as growth stops the Spitzenkörper fades, appearing again before growth is reinitiated (MCCLURE et al. 1968). High-resolution microscopy reveals that it is formed by microvesicles and, at least in Ascomycetes, also contains small membranous tubules associated with a few ribosomes (GROVE and BRACKER 1970). Although the precise role of the Spitzenkörper is not clear, its position may be related to the direction of extension of the hyphae, as noted by Girbard (MCCLURE et al. 1968). It would then act in directing the flow of vesicles.

The content of the apical vesicles is heterogenous. Some of them contain polysaccharides, as shown by cytochemical staining (MCCLURE et al. 1968,

HEATH et al. 1971, DARGENT 1975, see also GROVE 1978) and accordingly, it may be involved in the construction of the cell wall. Exoenzymes such as cellulase or alkaline phosphatase have also been visualized by similar methods in *Achlya ambisexualis* (NOLAN and BAL 1974) and *A. bisexualis* (DARGENT 1975) respectively. The different content of vesicles may thus be related, at least in part, to its different appearance under the electronmicroscope.

Vesicles appear to be formed in subapical areas. Which mechanisms direct them toward the apex? We lack a definitive answer though several hypotheses are currently being considered. SLAYMAN and SLAYMAN (1962) described in *Neuropora crassa* a decrease in the membrane potential from the old part of a hypha to the tip. Following this observation, BARTNICKI-GARCIA (1973) suggested that the potential gradient established might drive the vesicles to the apical area by electrophoresis. However, if the drop in potential is caused by a drop in the number of potassium pumps, as suggested by JENNINGS (1973), an osmotic gradient would also be produced. The flow of liquid to the tip would in turn carry the vesicles to the same place.

The involvement of microtubules and/or microfilaments in the transport or orientation of vesicles is a speculative subject. A ring of filamentous structures associated with the plasma membrane of budding cells of *S. cerevisiae* has been described by BYERS and GOETSCH (1976). It has been suggested that this ring is involved in a control mechanism which limits vesicle fusion at very determined places. Microfilaments have been described in filamentous fungi (GULL 1975), but evidence for their participation in the transport of secretory vesicles is still fragmentary.

In summary, the movement of vesicles to the tips of hyphae in filamentous fungi recalls the transport observed in *S. cerevisiae* budding cells. These observations, plus the fact that intermediate lipids (Dol-P-Man) have been found associated to fungal membranes (BARR and HEMMING 1972, LETOUBLON et al. 1973), suggested that the same basic mechanism underlines synthesis and transport of secretory molecules in both kinds of fungi.

4 Discharge of Secretion Products

Binding and fusion of secretory granules or vesicles with the plasma membrane is the last step in the secretory process. They lead to the externalization of the products contained into vesicles by exocytosis.

Intermixing of membrane phospholipids and an increase in the mobility of proteins appears to be a necessary event during the fusion process. It would permit a temporal separation of protein from contact areas.

Calcium ions appear to be involved in the mechanism that triggers both the access of secretory vesicles to the plasma membrane and their fusion. YIN and STOSSEL (1979) have purified a regulatory protein from macrophages, gelsolin, which dissolved actin networks in association with a limited breakdown of actin filaments. The expression of the regulatory function of gelsolin depends

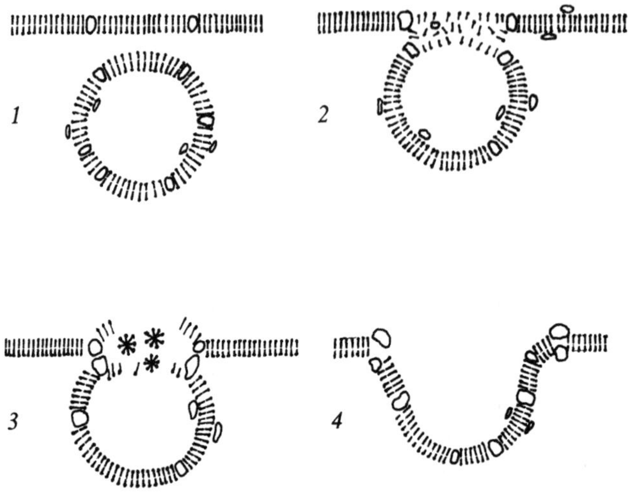

Fig. 3. Diagram of the process of membrane fusion. Freeze-fracturing has revealed complementary regions (*1*) in mucocyst and plasma membranes in *Tetrahymena*, which leads to the formation of the fusion pocket (*2*) and a reorganization of the membrane constitutents (*3*) so that a unique bilayer joins plasma and mucocyst membranes (*4*). (Modified from SATIR et al. 1973)

on variations in free calcium concentration which are likely to occur in living cells. These changes may be involved in locomotion, but also in secretion, by allowing vesicles to reach the plasma membrane. Another protein, synexin, has been implicated in the aggregation of secretory granules mediated by calcium. Finally, there is evidence that pancreatic acinar cells release calcium in response to adequate stimuli. This led to the formation of ion channels, depolarization, and secretion.

The existence of signals or recognition sites between membranes is of great significance in order to understand not only exocytosis but also other processes such as endocytosis, transport, etc. Some insight into this research area has been gained from the work of SATIR et al. (1973) with *Tetrahymena*. Freeze-fracture techniques have revealed the presence of complementary circles of particles, the fusion rosette, in both plasma membrane and packaged vesicles (mucocyst). The first event in the fusion of mucocyst and plasma membrane is the formation of a depression in the central area of the rosette to form the fusion pocket. It is followed by a fusion of rosettes which may provide not only recognition but also a reinforcing transient ring to avoid rupture of the membranes at the moment of contact. Subsequently, a reorganization of the membrane constituents takes place, such that a cohesive bilayer joins mucocyst and plasma membranes (Fig. 3).

The involvement of signals is also illustrated by the recent work of MEYER and BURGER (1979). These workers have isolated a protein from plasma membrane from the adrenal medulla which binds to secretory vesicles, and they suggest that such a molecule could function as a plasma membrane receptor for chromaffin granules during the secretory process.

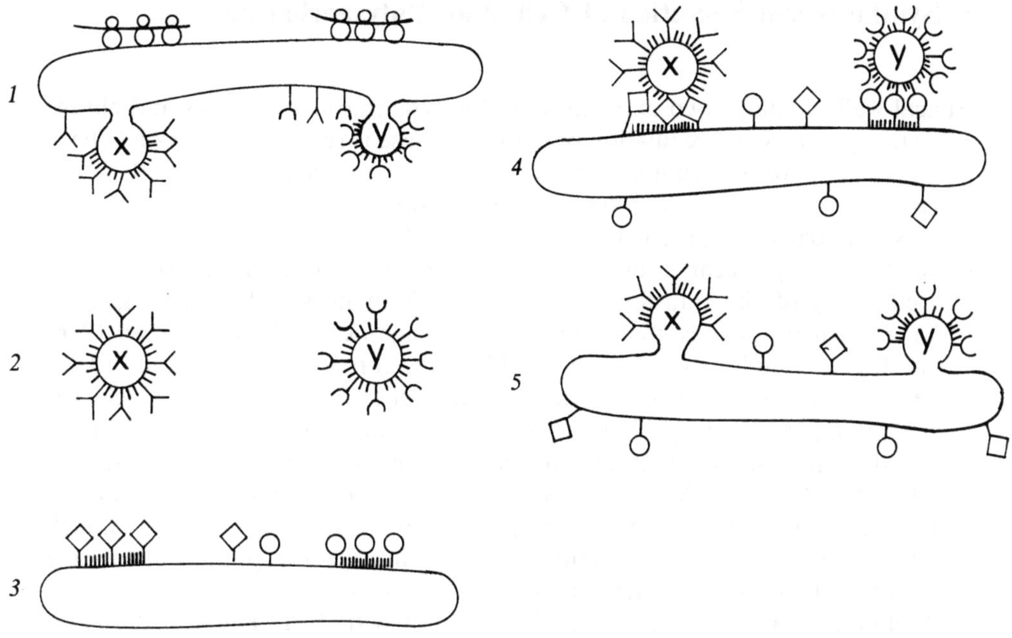

Fig. 4. Diagram of intracellular transport of secretory proteins. Movement of secretion products might be controlled by clathrin-coated vesicles. Coated regions in the RER (*1*) may allow concentration of both specific polymers (*X* or *Y*) and surface "signals" (Y Y) designed to pinch off vesicles (*2*). The vesicular signals would recognize receptors (⚲⚲) on the Golgi complex or other membranous organelles (*3, 4*) and finally, membrane fusion will take place as a result of clathrin release from their surface (*5*)

Receptors in the secretory vesicles appear necessary in order to be identified by the membranes with which they should fuse. Growing evidence in this area points to a picture in which vesicles of different nature are coated with specific proteins such as clathrin. The coat probably carries specific receptors that are not only recognized by complementary ones, but also avoid the mixing of different families of vesicles whose content should be kept separated (PEARSE 1980) (Fig. 4).

The fusion of vesicles located at the apex of hyphae, with the plasma membrane has been frequently visualized in fungi (GROVE and BRACKER 1970) and fusion profiles indicate continuity between the vesicle and the plasma membranes. However, to our knowledge no detailed studies such as those described above for mammalian cells have been carried out in fungi.

A genetic approach to the study of the secretory process in fungi has recently been developped by NOVICK and SCHEKMAN (1979). They have isolated a *S. cerevisiae* mutant that shows a conditional, reversible block in the secretory pathway. Vesicles containing invertase and acid phosphatase accumulate at the restrictive temperature. However the stage of the secretory process blocked in this mutant is still unknown. It is likely that vesicles are defective in the mechanism of transport or unable to fuse with the plasma membrane.

5 Synthesis and Secretion of Cell Wall Polysaccharides

Fungal walls contains two main structural polysaccharides: glucans and chitin. Since the synthesis and regulation of chitin synthase is reviewed by CABIB (Chap. 16, this Vol.), we will summarize recent findings relating to the synthesis of glucan and speculate on the mode of its secretion.

SENTANDREU et al. (1975) first detected synthesis in situ of small amounts of yeast $(1 \rightarrow 3)$-β-glucan using a toluenized preparation of *S. cerevisiae* cells. The efficiency of the transference of glucose from uridine-diphospho-glucose into the polymer has since then been improved significantly (BALINT et al. 1976, LOPEZ-ROMERO and RUIZ-HERRERA 1977, 1978, SHEMATEK and CABIB 1980). SHEMATEK and CABIB (1980) reported that the glucan synthetase was localized in the plasma membrane of yeast protoplasts and suggested that regulation occurred at that level. GTP and an endogenous compound that appears to be phosphorylated by ATP have been implicated in changes in the enzyme activity (SHEMATEK et al. 1980). It is possible to propose a secretory model, in which the nascent glucan chains protrude through the plasma membrane at the sites where the synthetase is located and activated.

Preliminary results obtained in collaboration with RUIZ-HERRERA differ in some aspects from those just described. In addition to the membrane-bound enzyme, an enzyme fraction not sedimentable at high speed can be obtained by mechanical disruption of whole cells in 1 M sucrose-containing buffer. The average chain length of microfibrils synthesized by this preparation estimated by electronmicroscopy was about 700 glucose residues. The membrane-bound enzyme prepared by disrupting the yeast in the presence of sucrose is not activated by ATP. In addition we have been unable to label the reducing end of glucan microfibrils synthesized by the nonsedimentable fraction with tritiated sodium borohydride even in the presence of 0.1 M sodium hydroxide. This suggests that the glucan may be synthesized while bound to a carrier molecule and opens new vistas in the field of polysaccharide secretion by yeast (LARRIBA, MORALES, and RUIZ-HERRERA in preparation).

The significance of these differences is not clear, but it is possible that the origin of the preparations, protoplasts or whole cells respectively, may account for them. LINNEMANS et al. (1977) in a study of the cytochemical localization of acid phosphatase, suggested that the cell wall is built up in two sequential modes. Mode I, in which "glucanase vesicles" are involved, occurs during the formation of the bud. Mode II does not involve "glucanase vesicles" and takes place in ageing yeast cells and probably also during the regeneration of protoplasts. Similar conclusions were reached by RODRÍGUEZ et al. (1979) using a completely different approach. Thus, it is possible that glucan synthesized by protoplasts is manufactured at the plasmalemma by mode II and glucan synthesized by exponentially growing cells is formed in vesicles or other cytoplasmic structures that are conveyed to the bud during the budding process. The fusion of glucan synthetase with the plasma membrane may result in alterations of its regulatory properties which in turn would explain its activation by effectors such as GTP or ATP.

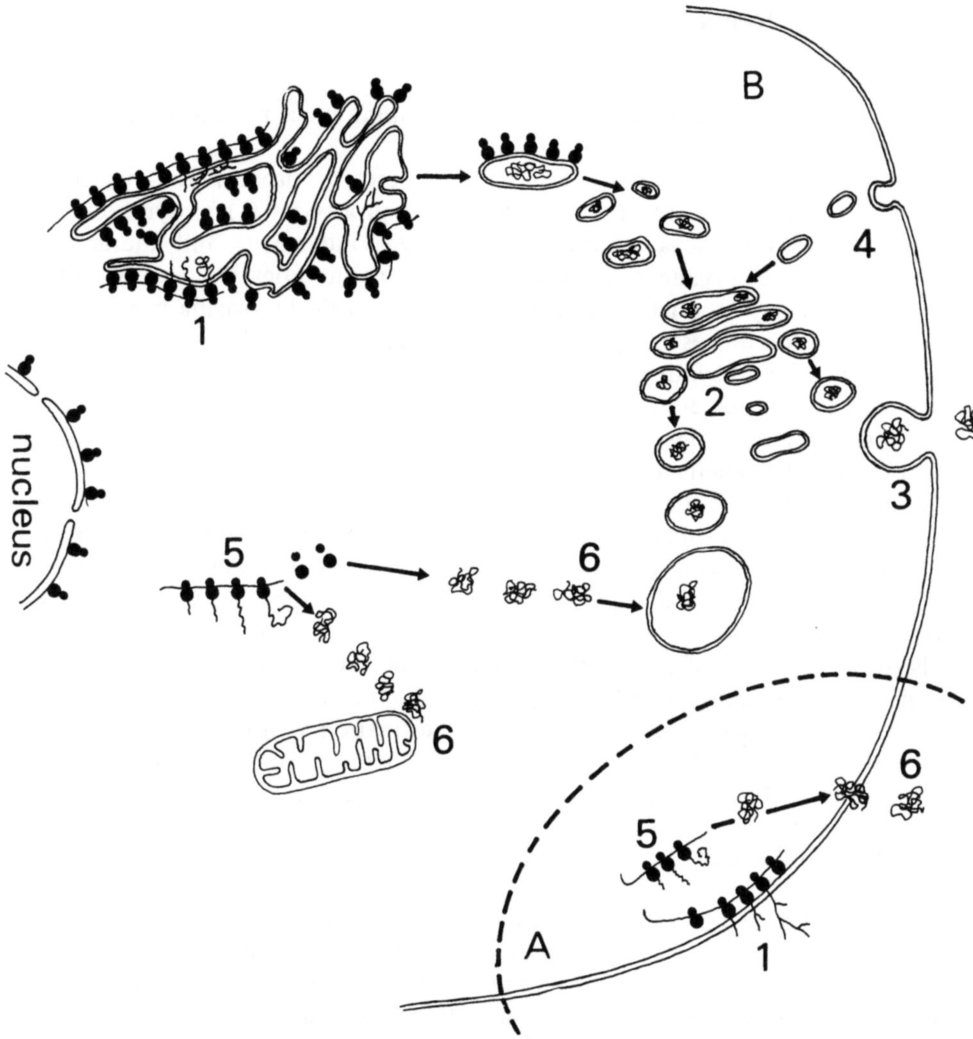

Fig. 5. Secretory processes taking place in prokaryotic (*A*) and eukaryotic (*B*) cells. Cotranslational translocation of nascent proteins from polysomes attached to membranes (*1*) to the cell exterior (*A*) or to the cisternal space of the RER (*B*) occurs in prokaryotic or eukaryotic cells. In the latter case the secretory proteins are processed in transit through the Golgi complex (*2*) to the plasma membrane (*3*) or different organelles [lysosomes, mitochondria, (*6*)]. Discharge takes place after fusion of the secretory vesicles and plasmalemma membranes (*3*). After exocytosis, part of the membrane is recovered and reutilized (*4*). Some proteins are synthesized on free polysomes (*5*) and secreted to the cell exterior (*6A*) or taken up by organelles (*6B*) possessing membrane receptors which recognized specific signals on the proteins. Maturation can occur either before or at the time of uptake of the protein

6 Concluding Remarks

A general view of different secretory processes taking place both in prokaryotic and eukaryotic cells is summarized in Fig. 5.

From an operational point of view, the secretory process in fungi may be considered as the result of three interrelated processes: (1) biosynthesis of macromolecules that cross membranes, (2) genesis, movement, and interaction of the membranous organelles involved in transport, modification, and processing of the secretory products, and finally, (3) fusion with the plasma membrane. Regulation of the process may therefore be exerted at many different levels.

The regulation of synthesis of inducible (melibiase, etc.) or repressible (invertase, acid phosphatase, etc.) secretory enzymes occurs at the genetic level. Fungal cells receive a continuous flow of information from the surrounding medium which reaches the genetic apparatus and triggers the synthesis of specific mRNA molecules. On the other hand, regulation of synthesis of cell wall polysaccharides (glucans and chitin) probably occurs at the level of polysaccharide synthetases.

Less is known about the second process. We assume that formation of the RER takes place as a result of the interaction of the signal sequence of a protein with specific receptors on the endoplasmic reticulum. The next step is the transformation of the RER to the SER as a result of polysome release. At this time the finished protein is set free in the lumen of the RER. The transport of secretory products to their final destination is initiated by breakage of the SER into vesicles that find their way to the Golgi body where they are probably packaged in different vesicles. The molecular bases of this process are unknown but the existence of a system of signals and receptors appears likely. Some results in this area have already been reported in mammalian systems.

The mechanisms that direct vesicles to their deposition sites is also a speculative subject. Electric forces, microtubules or microfilaments, or cytoplasmic streamings have been implicated in fungi. The timing of the cell division cycle and factors such as sex pheromones, and those influencing dimorphism may regulate both the orientated flow of vesicles and the nature of the secreted substances.

The fusion of the vesicles with the plasma membrane at specific sites (apex of the bud or hyphae) is presumably mediated by recognition of specific signals. The availability of conditional secretory mutants of *S. cerevisiae*, probably defective in such recognition, will help to elucidate the molecular events involved in fusion.

Finally some proteins are synthesized on free polysomes and secreted to the cell exterior or taken up by organelles possessing membrane receptors which recognized specific signals on the proteins.

The secretory process is an extremely complicated one. A great deal of information has recently been obtained relating to the synthesis of secretory products in fungi, specially the glycoproteins. However, the segregation of the molecules in different vesicles and the movement of these vesicles to specific places require further investigation. Fungal cells actively secrete a great number of substances, and this is mediated by membranous structures similar to those occurring in higher organisms. The study of the secretory process in fungi may contribute to our knowledge of the molecular mechanism of secretion not only in microorganisms but also in mammalian cells.

References

Adya S, Elbein AD (1977) Glycoprotein enzymes secreted by *Aspergillus niger*. Purification and properties of α-galactosidase. J Bacteriol 129:850–856

Algranatti ID, Carminatti H, Cabib E (1963) The enzymic synthesis of yeast mannan. Biochem Biophys Res Commun 12:504–509

Babczinski P, Tanner W (1973) Involvement of dolichol-monophosphate in the formation of specific mannosyl-linkages in yeast glycoproteins. Biochem Biophys Res Commun 54:1119–1124

Babczinski P, Tanner W (1978) A membrane associated isoenzyme of invertase in yeast precursor of the external glycoprotein. Biochim Biophys Acta 548:426–436

Balint Š, Farkaš V, Bauer Š (1976) Biosynthesis of β-glucans catalyzed by a particulate enzyme preparation from yeast. FEBS Lett 64:44–47

Ballou C (1976) Structure and biosynthesis of the mannan component of the yeast cell envelope. Adv Microb Physiol 14:93–158

Bartnicki-García S (1973) Fundamental aspects of hyphal morphogenesis. Symp Soc Gen Microbiol 23:245–267

Barr RH, Hemming FW (1972) Polyprenol phosphate as an acceptor of mannose from guanosine diphosphate mannose in *Aspergillus niger*. Biochem J 126:1203–1208

Bassford PJ, Beckwith J (1979) *Escherichia coli* mutants accumulating the precursor of a secreted protein in the cytoplasm. Nature (London) 277:538–541

Bassford PJ, Silhavy TJ, Beckwith J (1979) Use of gene fusion to study secretion of maltose-binding protein into *Escherichia coli* periplasm. J Bacteriol 139:19–31

Behrens NH, Cabib E (1968) The biosynthesis of mannan in *Saccharomyces cerevisiae*. J Biol Chem 243:502–509

Bergman LW, Kuehl WM (1978) Temporal relationship of translation and glycosylation of immunoglobulin heavy and light chain. Biochemistry 17:5174–5180

Blobel G, Dobberstein B (1975a) Transfer of proteins across membranes I. Presence of proteolytically processed and unprocessed nascent immunoglobulin light chains on membrane-bound ribosomes of murine myeloma. J Cell Biol 67:835–851

Blobel G, Dobberstein B (1975b) Transfer of proteins across membranes. II. Reconstitution of functional rough microsomes for heterologous components. J Cell Biol 67:852–862

Blobel G, Sabatini DD (1971) Biomembranes: Vol 2, Ribosome membrane interaction in eucaryotic cells. In: Manson LA (ed) Plenum Publishing Corporation, New York, London, pp 193–195

Boer P, Van Rijn JM, Reinking A, Steyn-Parvé EP (1975) Biosynthesis of acid phosphatase of baker's yeast characterization of a protoplast-bound fraction containing precursors of the exoenzyme. Biochim Biophys Acta 377:331–342

Byers B, Goetsch L (1976) A highly ordered ring of membrane associated filaments in budding yeast. J Cell Biol 69:717–721

Cabib E (1975) Molecular aspects of yeast morphogenesis. Annu Rev Microbiol 29:191–214

Campbell PN, Blobel G (1976) The role of organelles in the chemical modification of the primary translation products of secretory proteins. FEBS Lett:215–226

Caro LG, Palade GE (1964) Protein synthesis, storage and discharge in the pancreatic exocrine cells. J Cell Biol 20:473–495

Cortat M, Matile PH, Kopp F (1973) Intracellular localization of mannan synthetase activity in budding baker's yeast. Biochem Biophys Res Commun 53:482–489

Chen WW, Lennarz WJ (1978) Enzymatic excision of glucosyl units linked to the oligosaccharide chains of glycoproteins. J Biol Chem 253:5780–5785

Dargent R (1975) Sur l'ultrastructure des hyphes en croissance de l'*Achlya bisexualis* Coker. Mise en evidence d'une sécrétion polysaccharidique et d'une activité phosphatasique alkaline dans l'appareil de Golgi et au niveau des vésicules cytoplasmiques apicales. Comptes Residues 280:1445–1448

Davis BD, Tai RC (1980) The mechanism of protein secretion across membranes. Nature (London) 283:433–438

Devillers-Thiery A, Kindt T, Scheele G, Blobel G (1975) Homology in amino-terminal sequence of precursors to pancreatic sectretory proteins. Proc Natl Acad Sci USA 72:5016–5020

Elorza MV, Larriba G, Villanueva JR, Sentandreu R (1977) Budding in *Saccharomyces cerevisiae*: formation of the cross wall. Trans Br Mycol Soc 69:451–457

Emr SD (1978) Mutations altering the cellular localization of the phage receptor an *Escherichia coli* outer membrane protein. Proc Natl Acad Sci USA 75:5802–5806

Eylar EH (1965) On the biological role of glycoproteins. J Theor Biol 10:80–113

Fraser T, Bruce BJ (1978) Chicken ovoalbumin is synthesized and secreted by *Escherichia coli.* Proc Natl Acad Sci USA 75:5836–5940

Fujisawa F, Iwakura Y, Kawade Y (1978) Nonglycosylated mouse L cell interferon produced by the action of tunicamycin. J Biol Chem 253:8677–8679

Garoff H, Schwartz RT (1978) Glycosylation is not necessary for membrane insertion and cleavage of semliki forest virus membrane proteins. Nature (London) 274:487–490

Gibson R, Schlessinger S, Kornfeld S (1979) The nonglycosylated glycoproteins of vesicular stomatitis virus is temperature sensitive and undergoes intracellular agregation at elevated temperatures. J Biol Chem 254:3600–3607

Girbardt M (1969) Die Ultrastruktur der Anikalregion von Pilzhyphen. Protoplasma 67:413–441

Grinna LS, Robbins PW (1979) Glycoprotein biosynthesis. Rat liver microsomal glucosydases which process oligosaccharides. J Biol Chem 254:8814–8818

Grove SN (1978) The cytology of hyphal tip growth. In: Smith JE, Berry DR (eds) The filamentous fungi, vol. III. Edward Arnold, London

Grove SN, Bracker CE (1970) Protoplasmic organization of hyphal tips among fungi: vesicles and Spitzenkörper. J Bacteriol 104:989–1009

Gull K (1975) Cytoplasmic microfilament organization in two basidiomycete fungi. J Ultrastruct Res 50:226–232

Habener JF, Rosenblatt M, Dee PC, Potts JT (1979) Cellular processing of preproparathyroid hormone involves rapid hydrolysis of the leader sequence. J Biol Chem 254:10596–10599

Hasilik A, Tanner W (1976) Inhibition of the apparent rate of synthesis of the vacuolar glycoprotein. Carboxypeptidase and its protein antigen by tunicamycin in *Saccharomyces cerevisiae.* Antimicrob Agents Chemother 10:402–410

Hasilik A, Tanner W (1978a) Carbohydrate moiety of carboxypeptidase *Y* and perturbation of its biosynthesis. Eur J Biochem 91:567–575

Hasilik A, Tanner W (1978b) Biosynthesis of the vacuolar yeast glycoprotein carboxypeptidase *Y*. Conversion of precursor into the enzyme. Eur J Biochem 85:599–608

Heath IB, Gay JL, Greenwood AD (1971) Cell formation in the Saprolegniales: cytoplasmic vesicles underlying developing walls. J Gen Microbiol 65:225–232

Hickman S, Kornfeld S (1978) Effect of tunicamycin on IgM IgA and IgG secretion by mouse plasmacytoma cells. J Immunol 121:990–996

Hickman S, Neufeld EF (1972) A hypothesis for cell disease detective by hydrolases that not enter lysosomes. Biochem Biophys Res Commun 49:992–999

Hickman S, Kulcycky A, Lynch RG, Kornfeld S (1977) Studies of the mechanism of tunicamycin. Inhibition of IgA and IgE secretion by plasma cells. J Biol Chem 252:4402–4408

Highfield PE, Ellis RJ (1978) Synthesis and transport of the small subunit of chloroplast ribulose biphosphate carboxylase. Nature (London) 271:420–424

Hunt LA, Etchinson JR, Summers DF (1978) Oligosaccharide chains are trimmed during synthesis of the envelope glycoprotein of vesicular stomatitis virus. Proc Natl Acad Sci USA 75:754–758

Inoue H, Beckwith J (1977) Synthesis and processing of an *Escherichia coli* alkaline phosphatase precursor in vitro. Proc Natl Acad Sci USA 74:1440–1444

Jackson K, Blobel G (1977) Posttranslational cleavage of presecretory proteins with an extract of rough microsomes from dog pancreas containing signal peptidase activity. Proc Natl Acad Sci USA 74:5598–5602

Jakes KS, Model PJ (1979) Mechanism of export of colicin E_1 and colicin E_2. J Bacteriol 138:770–778

Jennings DH (1973) Cations and filamentous fungi: invasion of the sea and hyphal functioning. In: Anderson WP (ed) Ion transport in plants. Academic Press, London, New York, pp 323–335

Kaplan A, Achord DT, Sly WS (1977) Phosphohexosyl components of a lysosomal enzyme are recognized by pinocytosis receptors on human fibroblasts. Proc Natl Acad Sci USA 74:2026–2030

Kiely ML, McKnight S, Schimke RT (1976) Studies on the attachment of carbohydrate to ovalbumin nascent chains in hen oviduct. J Biol Chem 251:5490–5495

Kuo SC, Lampen JO (1974) Tunicamycin an inhibitor of yeast glycoprotein synthesis. Biochem Biophys Res Commun 58:287–295

Larriba G, Elorza MV, Villanueva JR, Sentandreu R (1976) Participation of dolichol phospho-mannose in the glycosylation of yeast wall mannoprotein at the polysomal level. FEBS Lett 71:316–320

Leavit R, Schlessinger S, Kornfeld S (1977) Impaired intracellular migration and altered solubility of nonglycosylated glycoprotein of vesicular stomatitis virus and sindbis virus. J Biol Chem 252:9018–9023

Lehle L, Tanner W (1976) The specific site of tunicamycin inhibition in the formation of dolichol-bound N-acetylglucosamine derivatives. FEBS Lett 71:167–170

Lehle L, Tanner W (1978) Glycosyl transfer from dolichyl phosphate sugars to endogenous and exogenous glycoprotein acceptor in yeast. Eur J Biochem 83:563–570

Lehle L, Bauer F, Tanner W (1977) The formation of glycosydic bonds in yeast glycoproteins. Intracellular localization of the reactions. Arch Microbiol 114:77–81

Letoublon RCP, Compte J, Got R (1973) Transfert du mannose à partir de GDP-mannose, à des accepteurs lipidiques chez *Aspergillus niger*. Eur J Biochem 40:95–101

Lingappa VR, Shields D, Woo SL, Blobel BJ (1978) Nascent chicken ovalbumin contains the functional equivalent of a signal sequence. J Cell Biol 79:567–572

Lingappa VR, Lingappa JR, Blobel G (1979) Chicken ovalbumin contains an internal signal sequence. Nature (London) 281:117–121

Linnemans WAM, Boer P, Elbers PF (1977) Localization of acid phosphatase in *Saccharomyces cerevisiae*: a clue to cell wall formation. J Bacteriol 131:638–644

López-Romero E, Ruiz-Herrera J (1977) Biosynthesis of β-glucans by cell-free extracts from *Saccharomyces cerevisiae*. Biochim Biophys Acta 500:372–384

López-Romero E, Ruiz-Herrera J (1978) Properties of β-glucan synthetase from *Saccharomyces cerevisiae*. Antonie van Leeuwenhoek J Microbiol Serol 44:329–339

Maccecchini ML, Rudin Y, Blobel G, Schatz G (1979) Import of proteins into mitochondria: Precursors forms of the extramitochondrially made F_1-ATP are subunits in yeast. Proc Natl Acad Sci USA 76:343–347

Margolis HC, Nakagawa Y, Douglas KT, Kaiser ET (1978) Multiple forms of carboxypeptidase Y from *Saccharomyces cerevisiae*. Kinetic demostration of effects of carbohydrate residues on the catalytic action of a glycoenzyme. J Biol Chem 253:7891–7897

Marriot M, Tanner W (1979) Localization of dolichyl phosphate and pyrophosphate-dependent glycosyl transfer reaction in *Saccharomyces cerevisiae*. J Bacteriol 139:565–572

Matile P, Moor H, Robinow C (1969) Yeast cytology. In: Rose AH, Harrison JS (eds) The yeast, vol I. Academic Press, London, New York, pp 123–238

Matile P, Cortat M, Wiemken A, Frey-Wyssling A (1971) Isolation of glucanase-containing particles from budding *Saccharomyces cerevisiae*. Proc Natl Acad Sci USA 68:636–640

McClure WK, Park D, Robinson PM (1968) Apical organization of the somatic hyphae of fungi. J Gen Microbiol 50:177–182

Mercereau-Puijalon O, Royal A, Cami B, Garapin A, Krust A, Gannon F, Kourilsky P

(1978) Synthesis of an ovoalbumin-like protein by *Escherichia coli* K12 harbouring a recombinant plasmid. Nature (London) 275:505–510
Meyer DI, Burger MM (1979) Isolation of a protein from plasma membrane of adrenal medulla which binds to secretory vesicles. J Biol Chem 254:9854–9859
Milstein C, Brownlee GG, Harrison TM, Mathews MB (1972) A possible precursor of immunoglobulin light chains. Nature (London) New Biol 239:117–120
Mizrahi A, O'Malley JA, Carter WA, Takatsuki A, Tamura G, Sulkowski E (1978) Glycosylation of interferons. Effects of tunicamycine on human immune interferon. J Biol Chem 253:7612–7615
Moor H (1967) Endoplasmic reticulum as the initiator of bud formation in yeast (*S. cerevisiae*). Arch Microbiol 57:135–146
Natowicz MR, Chi M M-Y, Lowry OH, Sly WS (1979) Enzymatic identification of mannose-6-phosphate on the recognition marker for receptor mediated pynocytosis of β-glucuronidase by human fibroblasts. Proc Natl Acad Sci USA 76:608–615
Newmark P (1979) Pathway to secretion. Nature (London) 281:629–630
Nolan RA, Bal AK (1974) Cellulase localization in hyphae of *Achlya ambisexualis*. J Bacteriol 117:840–843
Novick P, Schekman R (1979) Secretion and cell-surface growth are blocked in a temperature-sensitive mutant of *Saccharomyces cerevisiae*. Proc Natl Acad Sci USA 76:1858–1862
Olden L, Pratt RM, Yamada KM (1978) Role of carbohydrates in protein secretion and turnover. Effects of tunicamycin on the major cell surface glycoprotein of chick embryo fibroblast. Cell 13:461–473
Onishi HS, Tkacz JS, Lampen JO (1979) Glycoprotein nature of yeast alkaline phosphatase. Formation of active enzyme in the presence of tunicamycin. J Biol Chem 254:11943–11952
Palade GE (1975) Intracellular aspects of the process of protein synthesis. Science 189:347–358
Palade GE, Sieckevitz P, Caro LG (1962) Ciba Foundation Symposium on exocrine pancreas normal abnormal functions. In: Reuck AVS, Cameron MP (eds) Churchill, London, pp 23–49
Palmitter RD, Gagnon J, Walsh KA (1978) Ovoalbumin: A secreted protein without a transient hydrophobic leader sequence. Proc Natl Acad Sci USA 75:94–98
Parodi AJ (1978) Lipid intermediates in the synthesis of the inner core of yeast mannan. Eur J Biochem 83:253–259
Parodi AJ (1979) Biosynthesis of yeast glycoproteins. Processing of the oligosaccharides transferred from dolichol derivatives. J Biol Chem 254:10051–10060
Pazur JH, Knull HR, Limpsou DL (1970) Glycoenzymes: a note on the role for the carbohydrate moieties. Biochem Biophys Res Commun 40:110–116
Pearse B (1980) Coated vesicles. Trends Biochem Sci May 1980:131–134
Poyton RO, McKemmie (1979a) A polyprotein precursor to all four cytoplasmically translated subunits of cytochrome C oxidase from *Saccharomyces cerevisiae*. J Biol Chem 254:6763–6771
Poyton RO, McKemmie (1979b) Post-translational processing and transport of the polyprotein precursor to subunits IV to VII of yeast cytochrome C oxidase. J Biol Chem 254:6772–6780
Randall LL, Hardy STS (1977) Synthesis of exported proteins by membrane-bound polysomes from *Escherichia coli*. Eur J Biochem 75:43–53
Randall LL, Hardy STS, Josefsson LG (1978) Precursors of three exported protein in *Escherichia coli*. Proc Natl Acad Sci USA 75:1209–1212
Redman CM, Sabatini DD (1966) Vectorial discharge of peptides released by puromycin from attached ribosomes. Proc Natl Acad Sci USA 56:608–615
Robbins PW, Hubbard SC, Turco SJ, Wirth DF (1977) Proposal for a common oligosaccharide intermediate in the synthesis of membrane glycoproteins. Cell 12:893–900
Rodríguez L, Ruiz T, Villanueva JR, Sentandreu R (1978) Yeast invertase: subcellular distribution and possible relationship between the isoenzymes. Curr Microbiol 1:41–44
Rodríguez L, Laborda F, Sentandreu R (1979) Patterns of wall synthesis in *Saccharomyces cerevisiae*. Curr Microbiol 2:293–297

Rodríguez L, Ruiz T, Elorza MV, Villanueva JR, Sentandreu R (1980) Metabolic relationship between invertase and acid phosphatase isoenzymes in *Saccharomyces cerevisiae.* Biochim Biophys Acta 629:445–454

Rothman JE, Katz FN, Lodish HF (1978) Glycosylation of a membrane protein is restricted to the growing polypeptide chain but is not necessary for insertion a transmembrane protein. Cell 15:1447–1454

Ruiz-Herrera J, Sentandreu R (1975) Site of initial glycosylation of mannoproteins from *Saccharomyces cerevisiae.* J Bacteriol 124:127–133

Sarvas M, Hirth KP, Fuchs E, Simons K (1978) A precursor form of the penicillinase from *Bacillus licheniformis.* FEBS Lett 95:76–80

Satir B, Schooley C, Satir P (1973) Membrane fusion in a model system. Mucocyst secretion in *Tetrahymena.* J Cell Biol 53:153–159

Schechter I, McKean DJ, Guyer R, Terry W (1975) Partial aminoacid sequence of the precursor of immunoglobulin light chain programmed by messenger RNA in vitro. Science 188:160–162

Schreiber G, Dryburgh H, Milleship A, Matsuda Y, Inglis A, Phillips J, Edwards K, Maggs J (1979) The synthesis and secretion of rat transferrin. J Biol Chem 254:12013–12019

Schultz AH, Oroszlan S (1979) Tunicamycin inhibits glycosylation of precursor polyprotein encoded by env gene of Ranscher murine leukemia virus. Biochem Biophys Res Commun 86:1206–1213

Schwartz RT, Rohrscheider JM, Schmidt MFG (1976) Suppression of glycoprotein formation of Semliki Forest, influenza and avian sarcoma virus by tunicamycin. J Virol 19:782–791

Sentandreu R, Northcote DH (1968) The structure of a glycopeptide isolated from the yeast cell wall. Biochem J 109:419–432

Sentandreu R, Elorza MV, Villanueva JR (1975) Synthesis of yeast wall glucan. J Gen Microbiol 90:13–20

Shematek EM, Cabib E (1980) Biosynthesis of the yeast cell wall. II. Regulation of *β*-(1,3) glucan synthetase by ATP and GTP. J Biol Chem 255:895–902

Shematek EM, Braatz JA, Cabib E (1980) Biosynthesis of the yeast cell wall I. Preparation and properties of *β* (1,3) glucan synthetase. J Biol Chem 255:888–894

Siekevitz P, Palade GE (1960) A cytochemical study on the pancreas of the guinea pig. V. In vivo incorporation of leucine-1-C^{14} into the chymotrypsinogen of various cell fractions. J Biophys Biochem Cytol 7:619

Silhavy TJ, Shuman HA, Beckwith J, Schwartz M (1977) Use of gene fusions to study outer membrane protein localization in *Escherichia coli.* Proc Natl Acad Sci USA 74:5411–5415

Slayman CL, Slayman CW (1962) Measurements of membrane potentials in *Neurospora.* Science 136:876–877

Smith WP, Tai PC, Thompson RC, Davis BD (1977) Extracellular labeling of nascent polypeptides traversing the membrane of *Escherichia coli.* Proc Natl Acad Sci USA 74:2830–2834

Smith WP, Tai PC, Davis BD (1979) Extracellular labeling of growing secreted polypeptide chain in *Bacillus subtilis* with diazoiodo-sulfaniclic acid. Biochemistry 18:198–202

Speake BK, Hemming FW, White DA (1980) The effects of tunicamycin on protein glycosylation in mammalian and fungal systems. Biochem Soc Trans 8:166–168

Strauss AW, Zimmerman M, Boime I, Ashe B, Mumford RA, Alberts AW (1979) Characterization of an endopeptidase involved in pre-protein processing. Proc Natl Acad Sci USA 76:4225–4229

Struck DK, Lennarz WJ (1977) Evidence for the participation of saccharide lipids in the synthesis of the oligosaccharide chain of ovoalbumin. J Biol Chem 252:1007–1013

Struck DK, Sinta PB, Lane MD, Lennarz WJ (1978) Effect of tunicamycin on the secretion of serum proteins by primary cultures of rat and chick hepatocytes. Studies on transferrin very low density lipoprotein, and serum albumin. J Biol Chem 253:5332–5337

Sutcliffe JG (1978) Nucleotide sequence of the ampicillin resistance gene of *Escherichia coli* plasmid pBR322. Proc Natl Acad Sci USA 75:3737–3741

Tabas I, Schlesinger S, Kornfeld S (1978) Processing of high mannose oligosaccharides to form complex type oligosaccharides on the newly synthesized polypeptides of the vesicular stomatitis virus G protein and the IgG heavy chain. J Biol Chem 253:716–722

Takatsuki A, Kohno K, Tamura (1975) Inhibition of biosynthesis of polyisoprenol sugars in chick embryo microsomes by tunicamycin. Agric Biol Chem 39:2089–2091

Tkacz JS, Lampen JO (1975) Tunicamycin inhibition of poliisoprenol N-acetylglucosaminyl pyrophosphate formation in calf-liver microsomes. Biochem Biophys Res Commun 65:248–257

Trimble RB, Maley F (1977) Subunit structure of external invertase from *Saccharomyces cerevisiae*. J Biol Chem 252:4409–4412

Van Rijn HJM, Linnemans WAM, Boers P (1975) Localization of acid phosphatase in protoplasts from *Saccharomyces cerevisiae*. J Bacteriol 123:1144–1149

Von Figura K, Klein (1979) Isolation and characterization of phosphorylated oligosaccharides from β-N-acetylglucosaminidase that are recognized by cell-surface receptor. Eur J Biochem 94:347–354

Von Figura K, Rey M, Prinz R, Ullrich K (1979) Effect of tunicamycin on transport of lysosomal enzymes in cultured skin fibroblasts. Eur J Biochem 101:103–109

Wang FFC, Hirs CHW (1977) Influence of the heterosaccharides in porcine pancreatic ribonuclase on the conformation and stability of the protein. J Biol Chem 252:8358–8364

Wiemken A, Schellenberg M, Urech K (1979) Vacuoles: The sole compartment of digestive enzymes in yeast *Saccharomyces cerevisiae*. Arch Microbiol 123:23–35

Williamson AR, Singer HH, Singer PA, Mosmann (1980) The value of tunicamycin for studies on immunoglobulin biosynthesis. Biochem Soc Trans 8:168–170

White ON, Wirth DF (1979) Structure of the murine leukemia virus envelope glycoprotein precursor. J Virol 29:735–743

Yin HL, Stossel TP (1979) Control of cytoplasmic actin gelsol transformation by gelsolin, a calcium dependent regulatory protein. Nature (London) 281:583–586

21 Secretion of Cell Wall Material in Higher Plants

J.H.M. WILLISON

1 Introduction

Secretion is a process in which a product is released from an organ or cell. While most secretions flow away from their cellular source, it is generally true that the cell wall surrounding any cell has arisen as a result of the secretory activity of that cell alone (and most wall abutments arise during the separation of neighboring cells). Although sometimes quoted as an exception, this is probably as true of lignin as it is of other cell wall components (GROSS 1977). In order to cover its surface and control its morphogenesis, the cell must direct its secretions to the various parts of its surface. As examples: the protoplast of the unicellular alga *Pleurochrysis* rotates within its wall so that the scales released from its single dictyosome cover the entire cell with an even thickness of scaly wall (BROWN 1975); vesicles released from dictyosomes of tip-growing cells migrate toward the tip, following a gradient of increasing calcium concentration, to be released in the growth zone (REISS and HERTH 1978, 1979); and microfibril deposition during guard cell development is patterned after the arrangements of cortical cytoplasmic microtubules (PALEVITZ and HEPLER 1976).

Naively, one might expect that cell wall precursors (pectins, hemicelluloses, cell wall glycoproteins, etc.) are biosynthesized and then accumulated, packaged, and secreted from the cell, to be assembled into the completed wall at the surface of the protoplast. However, biosynthesis and secretion are not neatly separated like this, but are integrated and interdependent processes.

Necessarily, as a cell develops, not only are the secretions directed to defined regions of the cell surface, but the chemical and physical natures of these secretions vary. For example, in the developmental program of the cotton fiber there is a major change in the composition of the cell wall after the onset of secondary wall deposition, and during primary wall deposition the composition of the wall changes constantly (MEINERT and DELMER 1977, HUWYLER et al. 1979). While some of this change can be attributed to turnover (HUWYLER et al. 1979), most is due to changes in the pattern of secretion. Similarly, environmental factors may influence secretion in cell suspension cultures (as in the arabinosylation of the hydroxyproline-rich glycoprotein of bean cell cultures; KLIS and EELTINK 1979) and in whole plants (as in indole acetic acid-induced changes in cell wall composition; NISHITANI et al. 1979).

Since biochemical aspects of cell wall polysaccharide biosynthesis have been reviewed in other chapters in this volume (see particularly Chaps. 5 and 6, this Vol.), I shall concentrate on cytological aspects of secretion. However, the pertinent literature is so extensive (see reviews of: ROLAND 1973a, PRESTON 1974, CHRISPEELS 1976, MORRÉ and MOLLENHAUER 1976, ROBINSON 1977,

SCHNEPF and HERTH 1978, NORTHCOTE 1979, BROWN 1979) that only selected topics can be reviewed in the limited space available here. This chapter has therefore been planned as a supplement to a thorough review of the roles of endomembranes in secretion to be found in an earlier volume in this series (MORRÉ and MOLLENHAUER 1976). Aspects covered in detail here are: cell plate formation, microfibril secretion, and the secretion of lipidic adcrustations.

2 Cell Plate Formation

Most cell walls are initiated as cell plates within dividing cells in embryonic regions of plants. The events of cytokinesis are not all readily separable from those of karyokinesis, and the significance of several of the cytological features of cytokinesis remain controversial (BAJER and MOLÈ-BAJER 1972, O'BRIEN 1972, JONES and PAYNE 1977), controversies compounded by a residue of confusion in terminology. I shall use the term "phragmoplast" to refer to the microtubular apparatus (Fig. 1) which arises within the anaphase spindle and which is involved with cell plate formation (ESAU and GILL 1965). "Phragmosome" is used to refer to the layer of cytoplasm which develops in vacuolate cells in preparation for cell plate formation (SINNOTT and BLOCH 1940), not to ill-defined membrane-bound bodies associated with the forming cell plate (PORTER and MACHADO 1960).

2.1 Role of the Phragmoplast

Dynamic cytokinetic events in living *Haemanthus* endosperm, observed light microscopically, have been described by Bajer and colleagues (see BAJER and MOLÈ-BAJER 1972). In late anaphase, as the spindle contracts during phragmoplast formation, vesicles stream away from spindle poles between visible phragmoplast fibrils and appear to fuse with a plate of nodular thickenings of the phragmoplast fibrils located in the mid-interzone. The cell plate is usually (but not without exception: MORRISON and O'BRIEN 1976, KARAS and CASS 1976) initiated in the center of the cell, and grows centrifugally toward the periphery. In large or vacuolate cells, a ring-shaped zone of specialized cytoplasm containing the phragmoplast encircles the extending edge of the cell plate as it passes through the phragmosome (SINNOTT and BLOCH 1941, EVERT and DESHPANDE 1970, ROBERTS and NORTHCOTE 1970, GOOSEN-DE ROO and SPRONSEN 1978).

Clearly, just as the spindle plays a critical role in the segregation of sister chromatids, the phragmoplast is instrumental in determining the site at which the cell plate is deposited. Electronmicroscopy reveals that the phragmoplast microtubules probably arise as bundles from packets of amorphous material located in the mid-interzone during anaphase (BAJER 1968, HEPLER and JACKSON 1968) and later from similar structures as the phragmoplast progresses peripherally. The bundles of microtubules correspond with the phragmoplast fibers

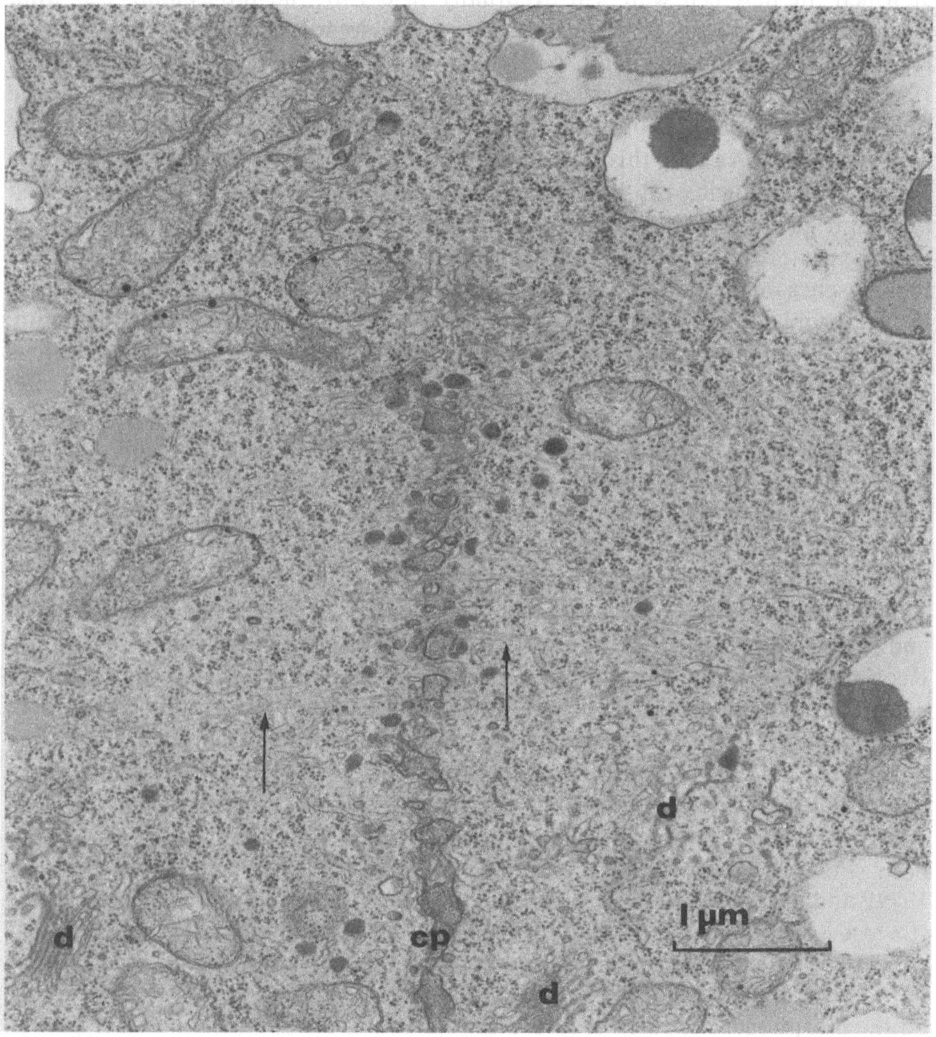

Fig. 1. The phragmoplast associated with a developing cell plate (*cp*) from a *Zea mays* root tip, glutaraldehyde-osmium fixed. Organelles, particularly dictyosomes (*d*), are excluded from the phragmoplast, which is visible as a series of microtubules (*arrows*). (Micrograph courtesy of Dr. H.H. MOLLENHAUER, previously unpublished)

of light microscopy, and the packets of amorphous material with the midzone swellings (BAJER 1968). These packets of amorphous material are now considered to be microtubule-organizing centers (MTOC's) (PICKETT-HEAPS 1969, but see MORRISON and O'BRIEN 1976 for an opposing view), which relocate or arise at the site of the future cell plate in preparation for the elaboration of the phragmoplast. The phragmoplast MTOC's do not themselves, however, determine the plane of division. In many cytokineses examined in detail, a preprophase band of microtubules lying next to the plasmalemma marks the site at which the new cell wall will eventually fuse with the pre-existing cell wall, despite the disappearance of this band once the spindle is elaborated (Pickett-Heaps and Northcote 1966a, b, Hepler and Palevitz 1974, Gunning et al. 1978a). This is so even in cases where the spindle axis lies obliquely to the future plane of the division wall, as in developing *Allium* guard cells (Palevitz and Hepler 1974a). In this case, the entire spindle/phragmoplast reorientates within the cell, during cell plate formation, to coincide with the plane predicted by the pre-prophase band of microtubules. Palevitz and Hepler (1974a) suggest that the pre-prophase band is not itself instrumental, but rather that it indicates the presence of a band of plasmalemma-associated MTOC's which organize the pre-prophase band of microtubules in response to increased levels of tubulin, synthesized in preparation for development of the spindle. This suggestion is supported by the detailed observations of Gunning et al. (1978a, b) on *Azolla* root development. Lintilhac (1974; see also Miller 1980) suggests that the cell plate arises in a plane of minimum stress within the dividing cell; perhaps this factor determines the positioning of the definitive MTOC's.

In addition to naturally occurring aberrations (Mahlberg et al. 1975, Lin 1977), cell plate formation may be arrested or modified by a number of agents applied experimentally. Colchicine (Whaley et al. 1966), Thuringiensin-A (Sharma et al. 1978), caffeine (López-Sáez et al. 1966, Paul and Goff 1973) and low calcium levels (Paul and Goff 1973) all prevent cell plate formation. Reorientation of the phragmoplast in developing guard cells is prevented by colchicine (which disrupts microtubules), but it continues in the presence of caffeine, which inhibits only the vesicle coalescence phase of cell plate formation (Palevitz and Hepler 1974b). Cytochalasin B (which intereferes with actin microfilaments) has unpredictable effects on both phragmoplast reorientation and cell plate formation (Palevitz and Hepler 1974b). Microfilaments have been shown to be a spindle component in plants (Forer and Jackson 1976, 1979) and cytochalasin B has been shown to inhibit cytoplasmic streaming (Lazar-Keul et al. 1978) and, more particularly, the movement of secretory vesicles in wheat coleoptiles and maize root tips (Mollenhauer and Morré 1976, Pope et al. 1979). These results, in combination with the many findings on the morphogenetic roles of microtubules and microfilaments in other groups of organisms (Hepler and Palevitz 1974, Palevitz 1980), indicate that the microtubules play a skeletal role, creating channels for vesicle movement among other things, while microfilaments are responsible for the movement itself.

2.2 Role of Organelles

A generalized description of cell plate secretion based on published findings is difficult, since the extent to which discrepancies are due to real differences between cells or organisms rather than to differences in electronmicroscopic technique or interpretation of micrographs, is difficult to determine. Inevitably, the following description is at variance with the views of some authors. The cell plate arises in the middle of the phragmoplast by fusion of vesicles which accumulate as a thin disk lying at right angles to the microtubules. Vesicles

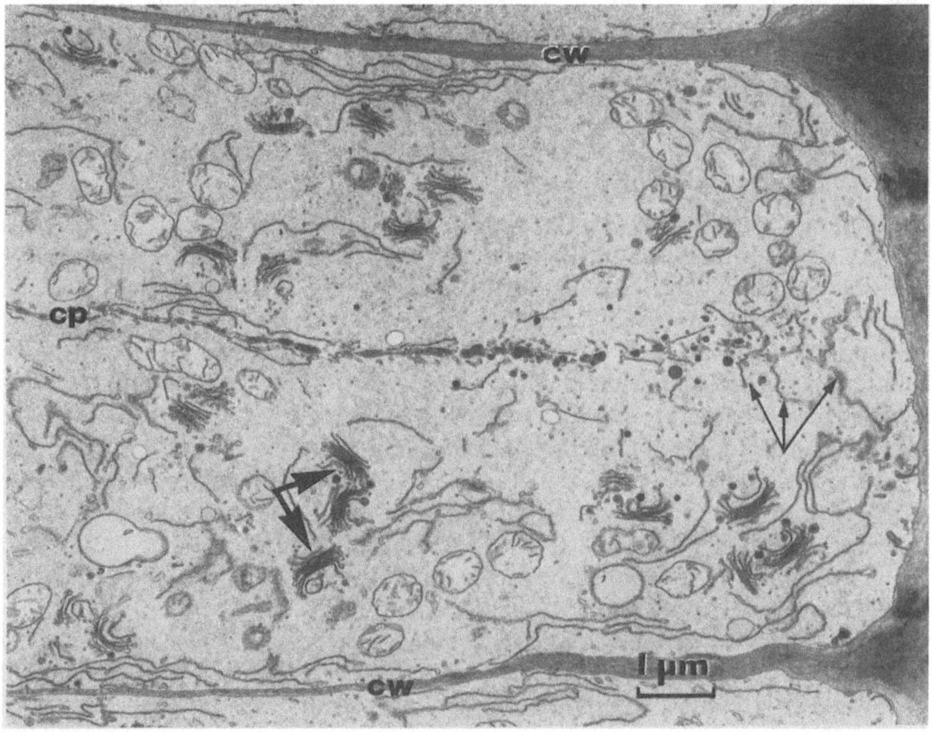

Fig. 2. A developing cell plate (*cp*) from an epidermal cell of a *Zea mays* root tip, permanganate-fixed. Note that there are no microtubules (as a result of permanganate fixation), but that the position of the phragmoplast is marked by a relatively clear area. Dictyosomes (*large arrows*) encircle the phragmoplast, and endoplasmic reticulum crosses the future path of the cell plate at its forming edge (*small arrows*). Note that the vesicles which appear to be fusing to form the cell plate are very similar in appearance to those which are associated with dictyosomes. Cell walls (*cw*). (Micrograph courtesy of Dr. H.H. MOLLENHAUER and reprinted from MOLLENHAUER and MOLLENHAUER 1978, with permission)

of two or three classes appear to be involved. In potassium permanganate-fixed preparations (Fig. 2), the plate appears to arise initially from smaller vesicles and later with the addition of larger vesicles (WHALEY et al. 1966, MOLLENHAUER and MOLLENHAUER 1978). In glutaraldehyde-osmium-fixed preparations, smaller vesicles later appear to be supplemented with coated vesicles (CRONSHAW and ESAU 1968, EVERT and DESHPANDE 1970, FRANKE and HERTH 1974) although strings of vesicles having the appearance of periodically constricted tubes have also been described (HEPLER and JACKSON 1968). Vesicle fusion occurs principally after the phragmoplast has been disassembled and is facilitated by the elaboration of stellate bodies having narrow branched radiating arms (HEPLER and NEWCOMB 1967, JONES and PAYNE 1977). Caffeine appears to inhibit cytokinesis by interfering with membrane fusion at the arms (JONES and PAYNE 1977). With the exception of the vesicles which contribute to the cell plate, organelles tend to be excluded from the phragmoplast (Fig. 1), although "drape-like extensions of endoplasmic reticulum" (PORTER and MACHADO 1960) usually penetrate the forming edge of the phragmoplast (Fig. 2) and lie across the future cell plate at an angle normal to its main plane. Such observations, together with inconclusive autoradiographic findings (PICKETT-HEAPS 1967a, b), have led to speculation that endoplasmic reticulum is the direct source of the vesicles which give rise to the cell plate (see O'BRIEN 1972, MORRISON and O'BRIEN 1976). It is more likely, however, that this endoplasmic reticulum is involved with the initiation of plasmodesmata (JONES and PAYNE 1977). More evidence supports the concept (WHALEY and MOLLENHAUER 1963, FREY-WYSSLING et al. 1964) that dictyosomes, which are found at the periphery of the phragmoplast, are the major source of the cell plate vesicles: autoradiography of ^{3}H-galactose incorporation indicates that the galactose is transferred from dictyosomes to the forming cell plate, presumably as a pectin (DAUWALDER and WHALEY 1974); the cell plate of osmium-impregnated tissue is not stained, like the distal face of Golgi apparatus but unlike the endoplasmic reticulum (GAZEAU 1975); and the pattern of activity of the Golgi apparatus varies with the cell cycle, distinctive small vesicles which arise during cell division appearing to be those involved with the initiation of the cell plate (WHALEY et al. 1966, MOLLENHAUER and MOLLENHAUER 1978). Coated vesicles probably arise from the distal faces of dictyosomes (i.e., from GERL: RYSER 1979).

It is generally assumed that the contents of the vesicles which associate to form the cell plate give rise to the middle lamella of the cell wall, a proposal supported by the affinity of the cell plate for basic dyes (BECKER 1938) and by the ^{3}H-galactose incorporation studies of DAUWALDER and WHALEY (1974). Less straightforward is the finding that forming cell plates elicit UV-fluorescence, similar to that of sieve plates, when stained with aniline blue (WATERKEYN 1967, FULCHER et al. 1976, LONGLY and WATERKEYN 1977). This fluorescence is transient, being typical of only the earliest stages of cell plate formation. SMITH and MCCULLY (1978) emphasize that although this fluorescence may be attributed to the presence of "callose", it cannot be assumed that the material is a pure $(1\rightarrow3)$-β-glucan. Since several vesicle types aggregate to form the cell plate, it is not improbable that each carries at least one different component toward the elaboration of the cell plate material. A typical microfibrillar primary wall begins to arise within the cell plate shortly after vesicle coalescence (Fig. 1), and may result in the masking of "callose" fluorescence.

3 Secretion of Primary Wall Matrix Substances

In the biphasic plant cell wall, the microfibrillar continuum is interspersed with a hydrated labyrinthine matrix (see Chaps. 2 and 4, this Vol.). It is now well established that most, if not all, of the matrix materials are secreted via the Golgi apparatus (it is becoming usual to use the term "Golgi apparatus" for the cellular complement of dictyosomes). The role of the Golgi apparatus in cell wall secretion was initially established for scale-bearing unicellular algae

(MANTON 1967, and see Chap. 3, this Vol.), and, although atypical of most plants, this probably remains the most fully characterized secretory system. In higher plants, the secretion of root cap slime (a fucosylated product closely related to cell wall pectin) is a model for the secretion of cell wall matrix materials and is treated separately by ROUGIER (Chap. 22, this Vol.). For information on the relationships between the endoplasmic reticulum, Golgi apparatus, and plasma membrane, the reader should refer to MORRÉ and MOLLENHAUER (1976). Secretory vesicles carrying products to the cell wall fuse selectively with the plasma membrane and, although there is evidence that the secretory vesicle membrane is modified to become similar to the plasma membrane (ROLAND 1969, VIAN 1974, MORRÉ and MOLLENHAUER 1974), little is known about the mechanism of this selectivity.

Evidence for the involvement of endomembranes in cell-wall secretion can be summarized as follows: (1) Electronmicroscope autoradiography of pulse-chase-treated tissues demonstrates that insoluble (i.e., polymer-incorporated) labels pass from an intracellular location to the cell wall. Unfortunately, the rapid metabolic turnover of most labels (galactose and fucose being exceptions) limits the applicability of the method (for discussion, see PICKETT-HEAPS 1972). (2) Dictyosomes and their presumed secretory vesicles are stained with polysaccharide-specific stains (Fig. 3) (ROLAND and SANDOZ 1969, RYSER 1979). The implicitly static nature of samples for electronmicroscopy makes dynamic interpretation difficult in such cases (see O'BRIEN 1972, for discussion), but with isolated protoplasts (which start with no wall) it is difficult to conclude other than that the stained vesicles have been released from the dictyosome and are in transit to the plasma membrane (ROLAND and PRAT 1973). (3) If labeled sugars (or proline for cell wall glycoprotein, GARDINER and CHRISPEELS 1975, KAWASAKI and SATO 1979) are supplied to intact tissues, the label may be recovered from subcellular fractions enriched for particular endomembranes (ROBINSON et al. 1976). Some of this label is found in products which resemble cell wall matrix substances (VAN DER WOUDE et al. 1971, PAULL and JONES 1975, BOWLES and NORTHCOTE 1976). (4) In some cases, appropriate enzymatic activities have been demonstrated in isolated endomembrane fractions (RAY et al. 1969, GARDINER and CHRISPEELS 1975, NAGAHASHI and BEEVERS 1978, GREEN and NORTHCOTE 1979a).

The most fundamentally contentious issue in matrix material secretion perhaps concerns the relationship between endoplasmic reticulum and the Golgi apparatus. In animal cells, it is well established that the Golgi apparatus arises at its forming face from smooth transitional vesicles which bud from rough endoplasmic reticulum. Though less clear in plants, this same model is generally assumed to apply (MORRÉ and MOLLENHAUER 1976) and electronmicrographs which are claimed to illustrate transitional elements have been published (BENBADIS and DEYSSON 1975, MORRÉ 1977). ROBINSON (1977, 1980), however, has built a case for the independence of these two endomembrane organelles, suggesting that the interrelationship exists only via a cytosolic pool of membrane precursors. Although no clear evidence exists for the initiation of matrix polysaccharide synthesis other than in the Golgi apparatus (BOWLES and NORTHCOTE 1976, RAY et al. 1976, ROBINSON 1980), it is probable that synthesis of lipid

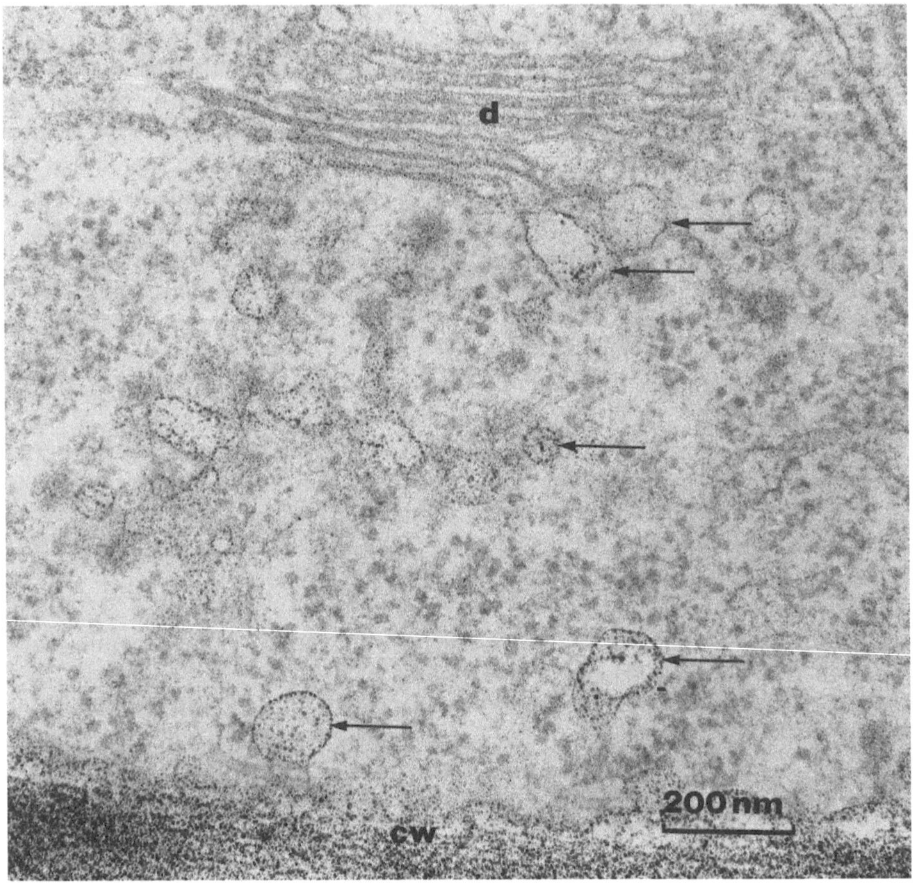

Fig. 3. Polysaccharides stained after periodate oxidation, showing localization in the cell wall (*cw*) and in dictyosome-derived vesicles (*arrows*). Dictyosome (*d*). Cotton fiber at time of primary wall deposition. (Micrograph courtesy of Dr. U. RYSER and reprinted from RYSER 1979, with permission)

intermediates (possible carrier glycolipids), supposed to be involved in root cap slime biosynthesis, occurs in endoplasmic reticulum (GREEN and NORTHCOTE 1979b). Similarly, the protein moiety of the precursor of cell wall glycoprotein may be deposited within rough endoplasmic reticulum (CHRISPEELS 1976). Transport of these precursors to the Golgi apparatus via transitional vesicles seems the most reasonable proposition.

The problem of the direction in which transitional vesicles between Golgi apparatus and plasma membrane are moving is made more complicated by the fact that, in primary-walled cells at least, there are at least two classes of vesicle: smooth and coated. In a careful study of developing cotton fibers, RYSER (1979) concluded that there were more dictyosome-associated coated vesicles during the stage of primary wall formation, but he was unable to conclude that the vesicles were involved in wall formation or even that those associated with the dictyosome and the plasma membrane belonged to the same population.

There has been recent progress toward understanding the vectorial movement of dictyosome-derived secretory vesicles. Cytochalasin B blocks vesicle transport, perhaps at the stage of loading onto the microfilamentous system (MOLLENHAUER and MORRÉ 1976, POPE et al. 1979). Work with the calcium ionophore A23187 indicates that in pollen tubes (REISS and HERTH 1979), pea stems (GRIFFING and RAY 1979), and moss rhizoids (CHEN and JAFFE 1979) the direction of movement is determined by a calcium gradient. That rhizoids are formed at the positive pole when moss spores are placed in an electrical field, suggests that this process is electrophysical (CHEN and JAFFE 1979).

4 Secretion of Microfibrils

4.1 Site of Synthesis

Although the sites of synthesis of the majority of cell wall polysaccharides appear to be in the endomembrane system, cellulose is an exception (see reviews of ROLAND 1973a, NORTHCOTE 1974, CHRISPEELS 1976, ROBINSON 1977, BROWN and WILLISON 1977). When isolated endomembrane fractions, notably Golgi apparatus or its secretory vesicles, have been shown to have $(1 \rightarrow 4)$-β-glucan synthetase activity, this has proven to have been (or has been reinterpreted as having been) associated with the synthesis of either hemicellulosic heteropolysaccharides (xyloglucan or glucomannan), mixed linked $(1 \rightarrow 3)$- and $(1 \rightarrow 4)$-β-glucans, cell wall glycoprotein, or root cap slime (RAY et al. 1969, VILLEMEZ 1974, WRIGHT and NORTHCOTE 1976, BROWN and KIMMINS 1977, DELMER 1977, HELSPER et al. 1977, RAYMOND et al. 1978). Indeed, DELMER (1977) implies that in vitro cellulose synthesis by cell-free systems of higher plant origin has never been achieved. Briefly stated, the following lines of reasoning point to the plasma membrane as the site of cellulose synthesis. That there appear to be no covalent links between cellulose and other polysaccharides (despite the covalent linking of most of the wall components) indicates that matrix and microfibrils are synthesized in different locations (VILLEMEZ 1974). The consistent failure to obtain in vitro cellulose synthesis suggests that the organelle involved in its synthesis is always disrupted by fractionation (SHORE and MACLACHLAN 1975). In the autoradiographic experiments of WOODING (1968), label from pulse-fed ^{3}H-glucose did not pass from a cytoplasmic location to the wall of secondarily thickened cells, but appeared initially in the wall. Similarly, in tissue fractionation experiments (BOWLES and NORTHCOTE 1972, 1974), most of the cell wall-associated label from ^{14}C-glucose fed to root tips appeared in glucans, but most of the label associated with endomembrane organelles had been turned over and was found in nonglucan polysaccharides. Furthermore, the kinetics of incorporation of label into β-glucans from UDP-glucose indicate that the plasma membrane is the site of incorporation (RAYMOND et al. 1978). In fine structural studies, no clear evidence of microfibrils within the endomembrane system of higher plants has ever been found, but specific elaborations of the plasma membrane are associated with the tips of microfibrils of secondary walls (see below).

4.2 Mechanism of Synthesis

The finding that cellulose exists in the form of microfibrils having constant cross-sectional dimensions but indefinite lengths (FRANZ et al. 1943, for review see PRESTON 1974) has long led to speculation about the mechanism of its synthesis. On purely biophysical grounds, it was proposed that synthesis of the glucan chains and their assembly into microfibrils occurred simultaneously through the agency of enzymes (or enzyme complexes) associated with microfibril tips (ROELOFSEN 1958, PRESTON 1959). To account for the regular arrangement of microfibrils in the walls of algae such as *Valonia*, PRESTON (1964, 1965) envisaged an array of regularly arranged globules, three layers deep, covering the surface of the cell. In this model, the problem of mobility of the site of synthesis is overcome by responsibility for the synthesis of a microfibril passing from particle to particle, and microfibril orientation is determined by the fixed positioning of the particle array. No such particle arrays have ever been found, although many authors (MÜHLETHALER 1965, PRESTON 1965, MARK 1971, NORTHCOTE 1974, PRESTON 1974, KREGER and KOPECKÁ 1976, ROBINSON 1977) have optimistically noted fibrils extending from patches of hexagonally packed particles on the plasmalemma of freeze-fractured *Saccharomyces cerevisiae*. However, the hexagonal packing of these particles probably has no relevance to wall formation (the arrangement is found only in nongrowing cells of certain strains TAKEO et al. 1976, WILLISON, unpublished results), and the fibrillar extensions from the particles are technical artifacts (SLEYTR and UMRATH 1974). The door to a quite different solution to the twin problems of mobility of the site of synthesis and control of microfibril orientation was opened by the evolution of the fluid-mosaic model of membrane structure (SINGER and NICOLSON 1972). HEATH (1974) proposed that complexes responsible for microfibril synthesis were embedded in the fluid lipid-bilayer, leaving them free to move, and that orientation was determined by linkage to underlying cytoplasmic microtubules (Fig. 4). Certain walls having randomly arranged microfibrils were considered to arise in the absence of microtubules.

Evidence supporting the existence of a membrane-linked particulate cellulose-synthesizing complex has come from structural studies, but remains otherwise unconfirmed. ROLAND (1968) described particle complexes, about 70 nm in diameter, attached to the ends of microfibrils in material dissociated with hydrogen peroxide. Although an attachment to microfibrils was not clearly shown, similar particles have been described in thin sections of the surfaces of cells depositing secondary walls (ROBARDS 1969, OLESEN 1980). Less equivocal are the freeze-fracture findings of membrane-impressed particle complexes associated with microfibril tips (MUELLER et al. 1976, WILLISON and BROWN 1977, WILLISON and GROUT 1978, MUELLER and BROWN 1980). Fracturing splits this complex into a large granule associated with the E fracture face (Fig. 5) and a hexagonal

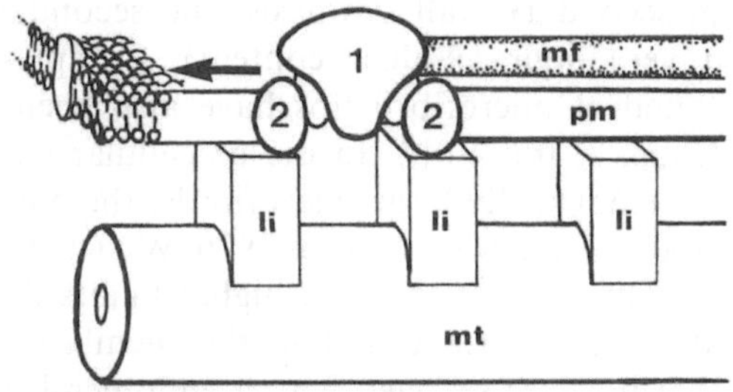

Fig. 4. A schematic and highly simplified model showing the relationships among: a cytoplasmic microtubule (*mt*); microtubule-membrane links (*li*); a plasma membrane (*pm*)-associated microfibril-synthesizing complex (*1* and *2*), and a microfibril (*mf*). For convenience, the links are shown as separate blocks associated with the rosette particles (*2*). The *arrow* indicates the direction of movement of the particle complex, driven by synthesis itself, along the microtubule-determined track

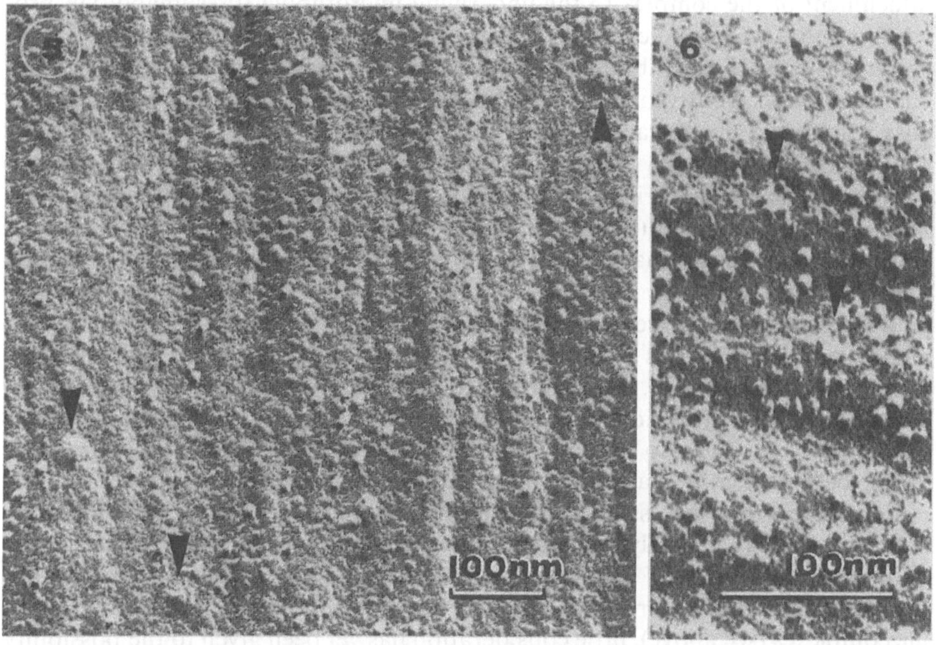

Figs. 5 and 6. E (Fig. 5) and P (Fig. 6) fracture faces of cotton fibers at the time of secondary wall deposition. Large granules impressed into the E fracture face (*arrows* Fig. 5) are presumed to complement rosettes of particles (*arrows* Fig. 6) on the P fracture face. (Fig. 5 is reprinted from WILLISON and BROWN 1977, with permission)

"rosette" associated with the P fracture face (Fig. 6). A tentative model for the relationship of these parts, in which the "rosette" surrounds the granule, has been proposed by MUELLER and BROWN (1980) from observations on corn (*Zea mays*) root tips. It should be noted that similar complexes of membrane particles (Figs. 5 and 6) in developing cotton fibers, described by WILLISON and BROWN (1977), were found on the plasma membranes of fibers involved in secondary wall synthesis (the secondary wall, unlike the primary wall, has a very high cellulose content). Complexes of membrane-associated particles found at microfibril tips have also been described in several green algae (see Chap. 3, this Vol.). In the unicellular genera *Oocystis* and *Glaucocystis*, which have unusually large microfibrils, the complexes are linear and quite distinctive (BROWN and MONTEZINOS 1976, WILLISON and BROWN 1978), but in *Micrasterias*, "rosettes", like those in higher plants described above, have been found both at the cell surface and in the membranes of cytoplasmic vesicles, suggesting that they are secreted via an endomembrane pathway (KIERMAYER and SLEYTR 1979, GIDDINGS et al. 1980). Roles for the components of the various complexes have sometimes been ascribed, but confirmation will be difficult to obtain. If the particle complexes are responsible for microfibril synthesis (either from nucleotide sugars, or from glucosyl- or cellobiosyl-phosphoryldolichol, or even from higher glucan intermediates, see Chap. 5, this Vol.), then there are several functions which might have to be fulfilled by the various parts, including: attachment of the complex to the underlying microtubule (presumably via proteins on the cytoplasmic surface of the plasma membrane); transfer of the substrate across the membrane (in the case of nucleotide sugar substrate); glucan chain initiation (controlling degree of polymerization); and transfer of the glucose residues to the glucan chains of the microfibril.

From the discussion above, it might appear that a single, unchallenged hypothesis for microfibril secretion exists. This is not so and the subject has long been controversial (see SHAFIZADEH and MCGINNIS 1971). Most notably, separation of glucan chain synthesis and microfibril crystallization has been proposed (MARX-FIGINI 1969, PALMA et al. 1976) on the basis of observations of weight-averaged degree of polymerization (dp_w) of cellulose, which varies among organisms and is constant with time during development of secondary walls in any one species. This has led to advocacy of a model for microfibril assembly in which the individual glucan chains are synthesized upon a cellular "template", of 7 μm to 22 μm in length according to species (PALMA et al. 1976), before their crystallization into natural cellulose. Some support for this notion comes from successful crystallization of cellulose I from solution (ATALLA and NAGEL 1974), but only under conditions which do not mimic those occurring naturally. Although for secondary wall cellulose this separate crystallization hypothesis is difficult to reconcile with descriptions of microfibril-terminal granule complexes (see above), little consideration has yet been given to the possibility that membrane-associated particles might rearrange during freezing and that the complexes of particles, as seen, may not represent the in vivo condition.

Not all crystalline $(1 \rightarrow 4)$-β-glucans need be assembled by the same mechanism, even within the same cell wall. The microfibrils of primary and secondary walls are outwardly similar, but there are substantial differences in their crystallinity, spectral characteristics, and appearance in the electronmicroscope (FREUNDLICH and ROBARDS 1974, NOWAK-OSSORIO et al. 1976, HERTH and MEYER 1977, CHANZY et al. 1978, 1979, and see MARCHESSAULT and SUNDARARAJAN 1977), which indicate that the crystalline core of primary wall microfibrils is considerably smaller and less orderly than that of secondary wall microfibrils and that there may be differences in the association between the cellulose and its encrusting hemicelluloses. Furthermore, the dp_w of primary wall cotton fiber cellulose is variable and about half that of the secondary wall (MARX-FIGINI and SCHULZ 1966). When isolated tobacco mesophyll protoplasts develop new cell walls, no evidence of microfibril-terminal particle-complexes has been found (WILLISON and COCKING 1975, GROUT 1975). Although such complexes could have been present without having been seen (WILLISON and GROUT 1978), it is also possible that the mechanism of assembly of these microfibrils, which are typical of primary walls (HERTH and MEYER 1977), is different from that of secondary wall microfibrils. Strings of membrane particles orientated parallel with microfibrils are found only on plasma membranes associated with primary walls (WILLISON 1976), and even microtubules associated with primary walls differ in their responses to chemical and physical treatments from those associated with secondary walls (JUNIPER and LAWTON 1979).

Even though cellulose synthetases may be active only upon, or within, the plasma membrane, we can expect that they arrive there by following a secretory pathway initiated in the endoplasmic reticulum (see discussion in NORTHCOTE 1977, 1979). Although the $(1 \rightarrow 4)$-β-glucan synthetases, present in the Golgi apparatus and its derivative secretory vesicles, may not be active cellulose synthetases (see earlier), it has been suggested that some of these may be either cellulose synthetases in transit to the plasma membrane or enzymes involved in the synthesis of precursors of cellulose (NORTHCOTE 1974, SHORE and MACLACHLAN 1975, FRANZ 1976, SATOH et al. 1976, HELSPER et al. 1977). Such evidence must be treated cautiously, however, in view of the presence of $(1 \rightarrow 4)$-β-oligoglucans in so many wall components. Nevertheless, the crystallographic evidence of ENGELS (1974) suggests that a component of primary wall microfibrils may be synthesized within secretory vesicles (but see ROBINSON 1977, for a criticism).

4.3 Control of Microfibril Orientation

Microtubules, as part of the cytoskeletal system (HEPLER and PALEVITZ 1974), are generally considered to be principal agents determining microfibril orientation. The following findings support this contention: (1) in developing systems, the orientations of microtubules and microfibrils are usually parallel (NEWCOMB 1969, HEPLER and PALEVITZ 1974, PALEVITZ and HEPLER 1976); (2) colchicine, but not cytochalasin B, appears to result in a loss of cellular control of microfibril orientation, which sometimes (but not always, see SCHNEPF and DEICHGRÄBER 1979) leads to aberrant development (PICKETT-HEAPS 1967c, PALEVITZ and HEPLER 1976, SRIVASTAVA et al. 1977, ROBINSON and HERZOG 1977, PALEVITZ 1980); (3) colchicine-induced loss of morphogenetic control is re-established once colchicine is removed and microtubules are reformed (QUADER et al. 1978, HARDHAM and GUNNING 1980); (4) cross-bridges sometimes link microtubules to the plasma membrane (ROBARDS 1969, HEPLER and PALEVITZ 1974, BROWER and HEPLER 1976, WILLISON and BROWN 1977); (5) patterns of cytoplasmic microtubules are responsive to factors such as growth hormones, which are determinants of morphogenesis (SAWHNEY and SRIVASTAVA 1975, VOLFOVÁ et al.

1977). Although the model (Fig. 4) suggests that microtubules are influential in aligning microfibrils by acting directly as tracks for microfibril-synthesizing complexes, there are other possibilities. Microtubules are not essential for maintaining the parallelism of microfibrils in walls, only for creating lamellation by changing orientations (SCHNEPF and DEICHGRÄBER 1979), a finding which is consistent with several observations that microtubule reorientation appears to precede a change in microfibril orientation, rather than accompanying it (see HEPLER and PALEVITZ 1974). Microfibril orientation is perhaps maintained simply by the stiffness of the crystalline rods. Changes in orientation, however, might be induced either by a reorientation of the track upon which the hypothetical synthesizing centers move, or, if synthesis and crystallization are separate events, microtubules might induce temporary alterations in membrane organization which could, conceivably, influence epitaxial crystallization of cellulose.

4.4 Cellulose Secretion by *Acetobacter xylinum*

Although prokaryotic and apparently out of place in this chapter, *Acetobacter* synthesizes a very pure $(1 \rightarrow 4)$-β-D-glucan (COLVIN 1972). No consideration of cellulose synthesis and secretion can be complete without reference to *Acetobacter,* however, for it has long been considered a model for the synthesis of cellulose from all sources. Earlier work had indicated that microfibril assembly proceeded at some distance from the cell, under the control of enzyme(s) located at microfibril tips, and that a lipidic intermediate molecule transferred an activated glucose oligomer from the cells to the growing microfibrils in the surrounding medium (COLVIN 1972). It is now clear that the site of cellulose synthesis lies within (and on) the bacterial cell wall (see below).

Each Gram-negative rod-shaped bacterium synthesizes a single strand of cellulose in the form of a flattened ribbon (BROWN et al. 1976). Because growth of the ribbon causes the bacterium to rotate slowly, the ribbon is twisted at intervals (BROWN et al. 1976, ZAAR 1977, BROWN and COLPITTS 1978) and the twisting tends to induce lateral rolling, making the ribbon appear more like a solid rod in places. The ribbon elongates at 2 to 7 µm min^{-1}, depending on temperature (COOPER and BROWN 1979), and the long axis of the ribbon always lies parallel with the long axis of the bacterium involved in its synthesis (BROWN et al. 1976, ZAAR 1977). The ribbon is an aggregate of parallel "microfibrils" (each 2 to 2.5 nm in diameter and perhaps equivalent to elementary fibrils) and each microfibril can be seen to terminate at a different site along the axis of the bacterium. Freeze-fracturing and etching reveal a row of particles in the bacterial envelope, lying parallel to (and immediately beneath) the ribbon of cellulose (Fig. 7) (BROWN et al. 1976, BROWN and WILLISON 1977, ZAAR 1979). Each particle is associated with, and perhaps inserted in, a small ring set in the outer membrane of the envelope (ZAAR 1979) rather similar in appearance to rings through which fimbriae have been claimed to arise in another bacterium (HAMILTON et al. 1975). Fractionation of the cell envelope of *Acetobacter* indicates that enzymes involved in cellulose biosynthesis are distributed throughout the layers of the envelope (COOPER and MANLEY 1975). This evidence,

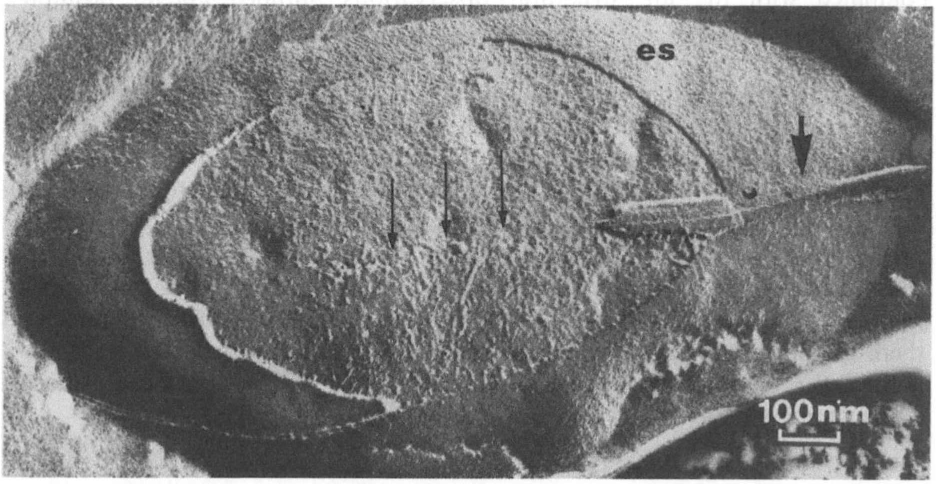

Fig. 7. Freeze-fractured and etched *Acetobacter xylinum* showing parallelism between the microfibrillar ribbon (*large arrow*) and a row of particles embedded in the outer membrane of the bacterium (*small arrows*). Etching has revealed the outer surface of the bacterium (*es*). (Micrograph reprinted from BROWN and WILLISON 1977, with the permission of Rockefeller University Press)

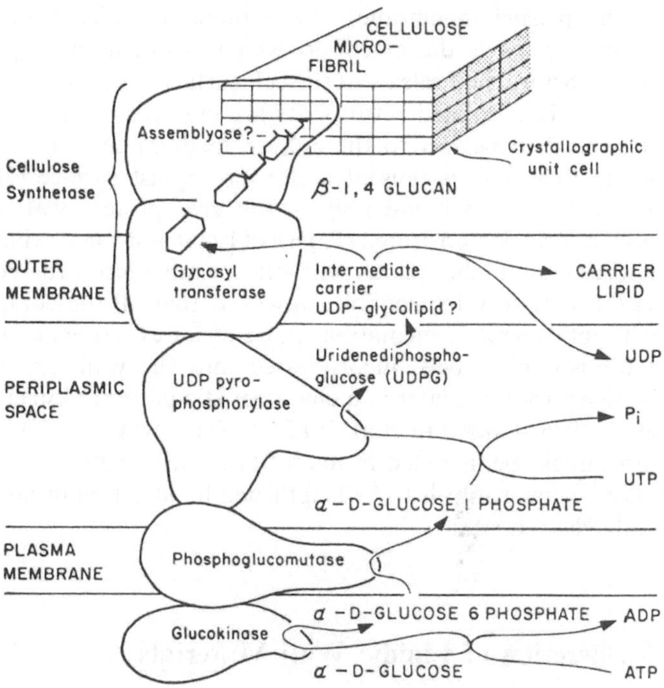

Fig. 8. A proposal for the location of enzymes involved in cellulose biosynthesis in *Acetobacter*. This multi-enzyme complex is presumed to correspond with one (or perhaps a small group) of the granules shown in Fig. 7. (Reprinted from BROWN and WILLISON 1977, with the permission of Rockefeller University Press)

combined with the ultrastructural observations outlined above, prompted BROWN and WILLISON (1977) to propose that an enzyme complex spans the cell envelope (Fig. 8). This model is consistent with evidence that EDTA will release lipid-linked diphosphate sugars from *Acetobacter* (GARCIA et al. 1974) and is readily modified to accept an intermediate polymer (KING and COLVIN 1976). ZAAR (1979) suggests that the surface pores (the "rings" described above) correspond with the "assemblyase" in this model. Some support for the proposal that hydrogen-bonding linking glucan chains in individual microfibrils arises very close to the site of chain elongation comes from the observation that cells grown in Calcofluor (which binds to cellulose) will produce microfibrils, but these will not assemble into ribbons (HAIGLER et al. 1980). Claims that there is evidence to support a hypothesis that a swollen, hydrated microfibril is formed prior to its consolidation (LEPPARD et al. 1975) have been criticized (WILLISON et al. 1980).

5 Cell Wall Assembly

Matrix substances of the cell wall appear to be secreted via the typical Golgi apparatus pathway, while cellulose is assembled as microfibrils at the surface of the plasma membrane. The manner in which these two groups associate is unclear, as is the extent to which covalent linking of matrix components occurs before their release at the cell surface, or even whether synthesis continues there. When isolated protoplasts are cultured, matrix substances (notably pectins) tend to be lost to the culture medium until microfibrillar material with which they bind is present at the protoplast surface (HANKE and NORTHCOTE 1974). In vitro self-assembly of the glycoprotein wall of *Chlamydomonas* has been clearly demonstrated (ROBERTS 1974) and some evidence exists that hemicelluloses from higher plants will self-assemble into ordered arrays (ROLAND et al. 1977). It is very improbable, however, that all the components could assemble without enzymatic mediation. Pulse-chase experiments demonstrate that certain radio-isotopic labels, incorporated into the wall, are turned over rapidly to be reincorporated later in a characteristic pattern (AMELUNXEN et al. 1976, SPIESS et al. 1976, TAKEUCHI et al. 1980). MACLACHLAN (1976) has suggested that cellulases might be included in microfibril-synthesizing complexes of primary walls, allowing microfibrils to be lengthened by insertion of new segments at enzymatically cleaved sites.

6 Secretion of Lipidic Wall Materials

In addition to the matrix and microfibrillar components which are typical of nearly all higher plant cell walls, other components are sometimes present and are secreted as part of the developmental program of particular cells. The lipidic materials, sometimes described as "adcrustations" (FREY-WYSSLING 1976), are examples. Relatively little is known about the secretion of the cuticle, epicuticular waxes, and suberinized layers associated with plant cell walls, despite recent progress toward unravelling their biosynthesis (KOLATTUKUDY 1970, 1977, and Chap. 10, this Vol.). Models of the molecular architecture of the cuticle have been little tested (DEAS and HOLLOWAY 1977) and the results of electronmicroscopic studies have led to conflicting opinions (see below).

6.1 The Cuticle

The outer layer of the bilayered cuticle (known as the "cuticle proper") is composed principally of wax and cutin and is visible in the electronmicroscope as a series of anastomosing electron-transparent lamellae which are interspersed with electron-dense material (JUNIPER and COX 1973, JARVIS and WARDROP 1974, HEIDE-JØRGENSEN 1978). As with the unit membranes, freeze-fracturing reveals face views of the lamellae (WILLISON 1980). The chemical constitution of the lamellae is unclear, although (as in the case of suberinized layers) some authors interpret the electron-lucent lamellae to be predominantly of wax and the electron-dense filler to be predominantly of cutin (JEFFREE et al. 1976, SARGENT 1976b). The "cuticular layer" underlies the cuticle proper. It sometimes appears to consist of a cutin-and-wax impregnated region of the cell wall (MARTIN and JUNIPER 1970, HALLAM and JUNIPER 1971), but in other cases it may contain no microfibrils and be separated from the cell wall by a thin pectic layer (FREY-WYSSLING and MÜHLETHALER 1965, SARGENT 1976a). The cuticle proper precedes the cuticular layer ontogenetically (JARVIS and WARDROP 1974, SARGENT 1976a, HEIDE-JØRGENSEN 1978).

The cutin polyester is formed largely by the esterification of C_{16} and C_{18} hydroxy fatty acids, a process which may be catalyzed in cell-free systems of epidermal origin by a particulate fraction originating from the cuticle itself (CROTEAU and KOLATTUKUDY 1975), suggesting that polymerization of the cuticle occurs in situ. Hydroxylation of the precursor acids occurs in the cytoplasm of the epidermal cells, partly in association with an unidentified endomembrane component (SOLIDAY and KOLATTUKUDY 1977). Dehydrogenases involved in the formation of constituent dicarboxylic acids are apparently soluble and presumably cytosolic (KOLATTUKUDY et al. 1975). No clear evidence is available as to the means by which these nonpolar precursor molecules are transported to the site of polymer formation. It has been suggested that the reticulate pattern in the cuticular layer results from the presence of precursors of wax and cutin flowing through fine channels (CHAFE and WARDROP 1973), although this conflicts with the view that such channels, like ectodesmata, are pectinaceous and implicated in the foliar absorption of polar molecules (HEIDE-JØRGENSEN 1978, HOCH 1979). During the development of the cuticle proper, lamellae may be seen to arise within dark-staining particles located at the inner face of the cuticle proper (HEIDE-JØRGENSEN 1978). FREY-WYSSLING and MÜHLETHALER (1965) suggested that such particles might migrate through the wall from the plasmalemma, but, as O'BRIEN (1967) pointed out, the particles are too large to pass through the mesh of a normal wall. It has also been suggested that spherosomes (lipid droplets), which have been described in the cytoplasm of epidermal cells, are intracellular accumulations of cutin and wax precursors (O'BRIEN 1967, HEIDE-JØRGENSEN 1978).

6.2 Epicuticular Waxes

Mature cuticles are commonly encrusted with epicuticular waxes, which are complex mixtures including long-chain alkanes, alkanoic acids, primary alcohols and their wax esters (KOLATTUKUDY 1970, see also Chap. 10, this Vol.). The waxes arise after development of the cuticle and take many species-characteristic forms (such as plates, rods, or tubes) which change during development and according to environmental factors (JEFFREE et al. 1976, FREEMAN et al. 1979). Secretion of epicuticular waxes presents problems similar to those encountered in the secretion of cutin precursors. Cell-free systems obtained from developing jojoba seeds will catalyze both chain elongation of stearoyl-acyl carrier proteins and esterification of the product (POLLARD et al. 1979). Most significantly, this activity appears to be associated with the "floating wax pad" (although it might have been contaminated with cellular membranes), suggesting that all the requirements for wax biosynthesis from precursor C_{16} or C_{18} fatty acids are secreted to an extracellular location. SARGENT (1976a, b) drew a similar conclusion on the basis of cytochemical evidence that (a) the cuticle proper stained with ruthenium red (interpreted as localizing fatty acids), and (b) the base of the cuticular layer showed peroxidase activity.

There are two major concerns which have led to debate over wax secretion: firstly, the means by which the distinctive forms and arrangements of surface waxes arise, and secondly, the means by which lipids that are solids at ambient temperatures are secreted to the surface of a relatively thick and impervious cuticle (JEFFREE et al. 1976). Some authors have proposed, on the basis of electronmicroscopical evidence, that there are direct wax channels (of 2.5 to 10 nm diameter) through the wall and cuticle (HALL 1967a, FISHER and BAYER 1972) which connect with pores at the surface of the cuticle (HALL 1967b, WELLS and FRANICH 1977). Others, however, have criticized the evidence and its interpretation and propose that waxes move through the anastomosing pathway between the cutin plates of the cuticle proper (HALLAM 1970a, HALLAM and JUNIPER 1971, JUNIPER and COX 1973), or more simply that the outer surface of the cuticle is progressively transformed into epicuticular wax (SARGENT 1976a, b). Clearly, the former proposal does not eliminate the possibility of pores at the surface of the cuticle. The proponents of wax pores suggest that the form of the epicuticular wax results partly from the pattern of arrangement of the pores from which the wax exudes (HALL and DONALDSON 1963, VON WETTSTEIN-KNOWLES 1974, WELLS and FRANICH 1977). Opponents have demonstrated, however, that leaf waxes will recrystallize from solution in forms characteristic of the plants from which they were isolated (HALLAM 1970b, JEFFREE et al. 1976) and therefore, that chemical composition is the principal determinant of wax form. Furthermore, JEFFREE and co-workers (1976) demonstrated, using an ingenious recrystallization apparatus in which the wax solution "transpired" through a porous disc, that wax crystal form is independent of the size or arrangement of pores. It is clear therefore that the only role for wax pores (if such exist) might be in determining the sites at which wax crystals arise (WELLS and FRANICH 1977). The mode of transport of wax is even less clear than the pathway taken. The major hypotheses invoke either the existence of a volatile solvent, or transition of the wax at the organ surface from a liquid to a solid form (MARTIN and JUNIPER 1970, JEFFREE et al. 1976). However, the problem does not arise if, as proposed by SARGENT (1976b), wax is synthesized within the lamellate cuticle, which is thereby transformed into epicuticular wax.

6.3 Suberinized Layers

Suberinized wax layers are found in internal cell walls at specific sites where the permeability of the apoplast is restricted, as in: cork (WATTENDORFF 1974a, SITTE 1975); various idioblasts (WATTENDORFF 1974b, 1976); salt glands (OLESEN 1979); bundle sheaths of leaves of C_4 plants (EVERT et al. 1977, ESPELIE and KOLATTUKUDY 1979); the periderm of wounded storage tubers (BARCKHAUSEN and ROSENSTOCK 1973, BARCKHAUSEN 1978); and in the endodermis and exodermis of roots (BONNETT 1968, ROBARDS et al. 1973, TIPPETT and O'BRIEN 1976, PETERSON et al. 1978, OLESEN 1978). Although these layers have been examined histochemically and ultrastructurally (references above), most have not been analyzed chemically. Where chemical analyses have been performed, the layers are distinct from, but related to, epidermal cuticles (Chap. 10, this Vol.). Ultrastructurally, suberinized layers are usually similar to the cuticle proper, having thin unstained lamellae separated by dark-staining filler, composed respectively of wax and suberin (SITTE 1975, SOLIDAY et al. 1979). The Casparian strip of the endodermis is distinct in having only a single lamella, tightly bound to the plasma membrane, which is associated with a region of the cell wall (possibly suberin-encrusted) that stains unusually homogeneously (BONNETT 1968, ROBARDS et al. 1973). In wound periderm, at least, it is the wax lamellae which function in controlling permeability (SOLIDAY et al. 1979).

Suberinized layers appear to arise rapidly (ROBARDS et al. 1973, WATTENDORFF 1974a) and, if the cells remain alive, are isolated from the plasma membrane by a tertiary layer of cell wall (WATTENDORFF 1974a). In wound periderm, the transcription of mRNA for at least one of the enzymes of suberin biosynthesis is wound-induced (AGRAWAL and KOLATTUKUDY 1977) and polysomes and rough endoplasmic reticulum are elaborated, presumably in association with the synthesis of these and other proteins (BARCKHAUSEN 1978). Although changes in the quantities of several organelles have been reported in association with the formation of suberinized layers, it is difficult to assign a role in suberin (or wax) precursor biosynthesis specifically to any one of them because many secretory and synthetic activities are going on simultaneously (see WATTENDORFF 1974a, BARCKHAUSEN 1978), but an unusual development of smooth tubular endoplasmic reticulum is noteworthy (BARCKHAUSEN and ROSENSTOCK 1973,

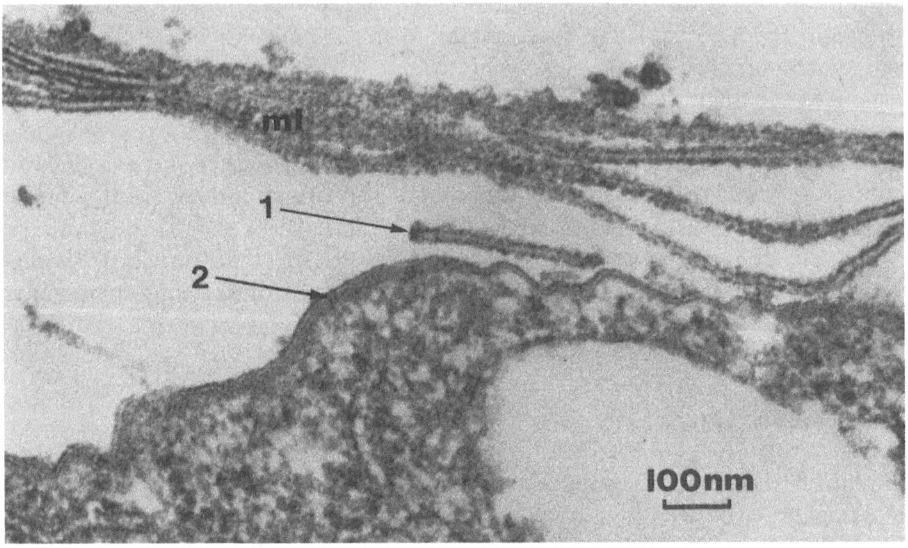

Fig. 9. A protoplast isolated from tomato fruit locule tissue and cultured for 4 days in a medium suitable for cell wall regeneration. The regenerated wall (*ml*) consists of a series of lamellae which appear to arise from the plasma membrane as discrete plates (*1*), which are assembled in close association with the plasma membrane (*2*)

WILLISON 1973, BARCKHAUSEN 1978). What little evidence is available suggests that wax-and-suberin lamellae may be secreted in a manner similar to the cuticle proper, except that the plates are assembled in close association with the plasma membrane as if this membrane were a template (OLESEN 1978, 1979). Sporopollenin has been claimed to be deposited in a similar manner (ROWLEY and SOUTHWORTH 1967). Acid-insoluble (but otherwise unidentified) lamellae, similar in appearance to suberinized lamellae, are formed by certain isolated tomato fruit protoplasts in culture (WILLISON and COCKING 1972, WILLISON 1973). The lamellae are assembled at the plasma membrane as discs (Fig. 9), which arise from the membrane to fuse into an anastomosing continuum (WILLISON 1973).

Acknowledgements. I am very grateful to Dr. D.G. Robinson for sending a pre-publication copy of an article (ROBINSON 1980), and to Drs. H.H. Mollenhauer and U. Ryser for their generosity in supplying electronmicrographs.

References

Agrawal VP, Kolattukudy P (1977) Biochemistry of suberization, ω-hydroxyacid oxidation in enzyme preparations from suberizing potato tuber discs. Plant Physiol 59:667–672

Amelunxen F, Spiess E, Thio Tiang Nio E (1976) Untersuchungen zur Zellwandbildung III. Autoradiographie von Dünnschnitten und biochemische Analyse isolierter Dictyosomen der Ranken von *Cucurbita maxima* L. Cytobiologie 13:260–278

Atalla RH, Nagel SC (1974) Cellulose: its regeneration in the native lattice. Science 185:522–523

Bajer A (1968) Fine structure studies on phragmoplast and cell plate formation. Chromosoma 24:383–417
Bajer AS, Molè-Bajer J (1972) Spindle dynamics and chromosome movement. Int Rev Cytol Suppl 3. Academic Press, New York, London
Barckhausen R (1978) Ultrastructural changes in wounded plant storage tissue cells. In: Kahl G (ed) Biochemistry of wounded plant tissues. Walter de Gruyter, Berlin, pp 391–417
Barckhausen R, Rosenstock G (1973) Feinstrukturelle Untersuchungen zur traumatogen Suberinisierung beim Knollenparenchym von *Solanum tuberosum* L. Z Pflanzenphysiol 69:193–203
Becker WA (1938) Recent investigations in vivo on the division of plant cells. Bot Rev 4:446–472
Benbadis M-C, Deysson G (1975) Morphologie ultrastructurale des dictyosomes dans les cellules des méristèmes radiculaires d'*Allium sativum* L. Planta 123:283–290
Bonnett HT Jr (1968) The root endodermis: fine structure and function. J Cell Biol 37:199–205
Bowles DJ, Northcote DH (1972) The sites of synthesis and transport of extracellular polysaccharides in the root tissues of maize. Biochem J 130:1133–1145
Bowles DJ, Northcote DH (1974) The amounts and rates of export of polysaccharides found within the membrane system of maize root cells. Biochem J 142:139–144
Bowles DJ, Northcote DH (1976) The size and distribution of polysaccharides during their synthesis within the membrane system of maize root cells. Planta 128:101–106
Brower DL, Hepler PK (1976) Microtubules and secondary wall deposition in xylem: the effects of isopropyl-N-phenylcarbamate. Protoplasma 87:91–111
Brown RG, Kimmins WC (1977) Glycoproteins. Int Rev Biochem 13:183–209
Brown RM Jr (1975) *Pleurochrysis scherffelii* (Chrysophyceae), vegetative development. Film E 1682. Inst Wiss Film, Göttingen
Brown RM Jr (1979) Biogenesis of natural polymer systems with special references to cellulose assembly and deposition. In: Walk EM (ed) Proc 3rd Philip Morris Science Symp. Philip Morris Inc. New York, pp 52–123
Brown RM Jr, Colpitts TJ (1978) Direct visualization of cellulose synthesis by high resolution darkfield microscopy and time-lapse cinematography. J Cell Biol 79:157a
Brown RM Jr, Montezinos D (1976) Cellulose microfibrils: visualization of biosynthetic and orienting complexes in association with the plasma membrane. Proc Natl Acad Sci USA 73:143–147
Brown RM Jr, Willison JHM (1977) Golgi apparatus and plasma membrane involvement in secretion and cell surface deposition, with special emphasis on cellulose biogenesis. In: Brinkley BR, Porter KR (eds) International cell biology 1976–1977. Rockefeller Univ Press, New York, pp 267–283
Brown RM Jr, Willison JHM, Richardson CL (1976) Cellulose biosynthesis in *Acetobacter xylinum*: visualization of the site of synthesis and direct measurement of the in vivo process. Proc Natl Acad Sci USA 73:4565–4569
Chafe SC, Wardrop AB (1973) Fine structural observations on the epidermis II. The cuticle. Planta 109:39–48
Chanzy H, Imada K, Vuong R (1978) Electron diffraction from the primary wall of cotton fibers. Protoplasma 94:299–306
Chanzy H, Imada K, Mollard A, Vuong R, Barnoud F (1979) Crystallographic aspects of sub-elementary cellulose fibrils occurring in the wall of rose cells cultured in vitro. Protoplasma 100:303–316
Chen T-S, Jaffe LF (1979) Forced calcium entry and polarized growth of *Funaria* spores. Planta 144:401–406
Chrispeels MJ (1976) Biosynthesis, intracellular transport, and secretion of extracellular macromolecules. Annu Rev Plant Physiol 27:19–38
Colvin JR (1972) The structure and biosynthesis of cellulose. CRC Crit Rev Macromol Sci 1:47–81
Cooper D, Manley RStJ (1975) Cellulose synthesis by *Acetobacter xylinum* II. investigation into the relation between cellulose synthesis and cell envelope components. Biochim Biophys Acta 381:97–108

Cooper KM, Brown RM Jr (1979) Cellulose assembly in *Acetobacter xylinum*: site of formation and variation of the rate of synthesis. J Cell Biol 83:69a
Cronshaw J, Esau K (1968) Cell division in leaves of *Nicotiana*. Protoplasma 65:1–24
Croteau R, Kolattukudy PE (1975) Biosynthesis of hydroxy-fatty acid polymers, enzymatic conversion of 18-hydroxyoleic acid to 18-hydroxy-cis-9,10-epoxystearic acid by a particulate preparation from spinach (*Spinacea oleracea*). Arch Biochem Biophys 170:61–72
Dauwalder M, Whaley WG (1974) Patterns of incorporation of [3H]-galactose by cells of *Zea mays* root tips. J Cell Sci 14:11–27
Deas AHB, Holloway PJ (1977) The intermolecular structure of some plant cutins. In: Tevini M, Lichtenthaler HK (eds) Lipids and lipid polymers in higher plants. Springer Berlin Heidelberg New York, pp 293–299
Delmer DP (1977) The biosynthesis of cellulose and other plant cell wall polysaccharides. Recent Adv Phytochem 11:45–77
Engels FM (1974) Function of Golgi vesicles in relating to cell wall synthesis in germinating *Petunia* pollen. IV. identification of cellulose in pollen tube walls and Golgi vesicles by x-ray diffraction. Acta Bot Neerl 23:209–215
Esau K, Gill RH (1965) Observations on cytokinesis. Planta 67:168–181
Espelie KI, Kolattukudy PE (1979) Composition of the aliphatic components of 'suberin' from the bundle sheaths of *Zea mays* leaves. Plant Sci Lett 15:225–230
Evert RF, Desphande BP (1970) An ultrastructural study of cell division in the cambium. Am J Bot 57:942–961
Evert RF, Eschrich W, Heyser W (1977) Distribution and structure of the plasmodesmata in mesophyll and bundle-sheath cells of *Zea mays* L. Planta 136:77–89
Fisher DA, Bayer DE (1972) Thin sections of plant cuticles demonstrating channels and wax platelets. Can J Bot 50:1509–1511
Forer A, Jackson WT (1976) Actin filaments in the endosperm mitotic spindles in a higher plant, *Heamanthus katherinae* Baker. Cytobiologie 12:119–214
Forer A, Jackson WT (1979) Actin in spindles of *Haemanthus katherinae* endosperm. J Cell Sci 37:323–347
Franke WW, Herth W (1974) Morphological evidence for de novo formation of plasma membrane from coated vesicles in exponentially growing cultured plant cells. Exp Cell Res 89:447–451
Franz E, Schiebold E, Weygand C (1943) Über den morphologischen Aufbau der Bakteriencellulose. Naturwissenschaften 31:350
Franz G (1976) The dependence of membrane-bound glucan synthetases on glycoprotein which can act as acceptor molecules. Appl Polymer Symp 28:611–621
Freeman B, Albrigo LB, Biggs RH (1979) Ultrastructure and chemistry of cuticular waxes of developing *Citrus* leaves and fruits. J Am Soc Hortic Sci 104:801–808
Freundlich A, Robards AW (1974) Cytochemistry of differentiating plant vascular cell walls with special reference to cellulose. Cytobiologie 8:355–370
Frey-Wyssling A (1976) The plant cell wall. Encyclopedia of plant anatomy vol III4, Borntraeger, Berlin, Stuttgart
Frey-Wyssling A, Mühlethaler K (1965) Ultrastructural plant cytology. Elsevier, Amsterdam, London, New York
Frey-Wyssling A, López-Sáez JF, Mühlethaler K (1964) Formation and development of the cell plate. J Ultrastruct Res 10:422–432
Fulcher RG, McCully ME, Setterfield G, Sutherland J (1976) β,1-3 glucans may be associated with cell plate formation during cytokinesis. Can J Bot 54:539–542
Garcia RC, Recondo E, Dankert M (1974) Polysaccharide biosynthesis in *Acetobacter xylinum*, enzymatic synthesis of lipid diphosphate and monophosphate sugars. Eur J Biochem 43:93–105
Gardiner M, Chrispeels MJ (1975) Involvement of the Golgi apparatus in the synthesis and secretion of hydroxyproline-rich cell wall glycoproteins. Plant Physiol 55:536–541
Gazeau C-M (1975) Caractères des structures imprégnées par l'osmium au cours de la formation de la plaque télophasique dans le méristème radiculaire de *Triticum vulgare* (graminées). J Microscopie 23:321–326
Giddings TH, Brower DL, Staehelin LA (1980) Visualization of particle complexes in

the plasma membrane of *Micrasterias denticulata* associated with the formation of cellulose fibrils in primary and secondary cell walls. J Cell Biol 84:327–339
Goosen-de Roo L, van Spronsen PC (1978) Electron microscopy of the active cambial zone of *Fraxinus excelsior* L. IAWA Bull 4:59–64
Green JR, Northcote DH (1979a) Location of fucosyl transferases in the membrane system of maize root cells. J Cell Sci 40:235–244
Green JR, Northcote DH (1979b) Polyprenyl phosphate sugars synthesized during slime-polysaccharide production by membranes of the root-cap cells of maize (*Zea mays*). Biochem J 178:661–671
Griffing LR, Ray PM (1979) Dependence of cell wall secretion on calcium. Plant Physiol Suppl 63:51
Gross GC (1977) Biosynthesis of lignin and related monomers. Recent Adv Phytochem 11:141–184
Grout BWW (1975) Cellulose microfibril deposition at the plasmalemma surface of regenerating tobacco mesophyll protoplasts: a deep-etch study. Planta 123:275–282
Gunning BES, Hardham AR, Hughes JE (1978a) Pre-prophase bands of microtubules in all categories of formative and proliferative cell division in *Azolla* roots. Planta 143:145–160
Gunning BES, Hardham AR, Hughes JE (1978b) Evidence for initiation of microtubules in discrete regions of the cell cortex in *Azolla* root-tip cells, and an hypothesis on the development of cortical arrays of microtubules. Planta 143:161–179
Haigler C, Brown RM Jr, Benzimen M (1980) Calcofluor white ST alters in vivo assembly of cellulose microfibrils. Science 210:903–906
Hall DM (1967a) Wax microchannels in the epidermis of white clover. Science 158:505–506
Hall DM (1967b) The ultrastructure of wax deposits on plant leaf surfaces. II Cuticular pores and wax formation. J Ultrastruct Res 17:34–44
Hall DM, Donaldson LA (1963) The ultrastructure of wax deposits on plant leaf surfaces. I Growth of wax on leaves of *Trifolium repens*. J Ultrastruct Res 9:259–267
Hallam ND (1970a) Leaf wax fine structure and ontogeny in *Eucalyptus* demonstrated by means of a specialized fixation technique. J Microscopy 92:137–144
Hallam ND (1970b) Growth and regeneration of waxes on the leaves of *Eucalyptus*. Planta 93:257–268
Hallam ND, Juniper BE (1971) The anatomy of the leaf surface In: Preece TF, Dickinson CH (eds) Ecology of leaf surface micro-organisms. Academic Press, London New York, pp 3–38
Hamilton RC, Bover FG, Mason TJ (1975) An association between fimbriae and pores in the wall of *Fusiformis nodosus*. J Gen Microbiol 91:421–424
Hanke DE, Northcote DH (1974) Cell wall formation by soybean callus protoplasts. J Cell Sci 14:29–50
Hardham AR, Gunning BES (1980) Some effects of colchicine on microtubules and cell division in roots of *Azolla pinnata*. Protoplasma 102:31–51
Heath IB (1974) A unified hypothesis for the role of membrane bound enzyme complexes and microtubules in plant cell wall synthesis. J Theor Biol 48:445–449
Heide-Jørgenson HS (1978) The xeromorphic leaves of *Hakea suaveolens* R. Br. II. Structure of epidermal cells, cuticle develpment and ectodesmata. Bot Tidsskr 72:227–244
Helsper JPFG, Veerkamp JH, Sassen MMA (1977) β-glucan synthetase activity in Golgi vesicles of *Petunia hybrida*. Planta 133:303–308
Hepler PK, Jackson WT (1968) Microtubules and early stages of cell-plate formation in the endosperm of *Haemanthus katherinae* Baker. J Cell Biol 38:437–446
Hepler PK, Palevitz BA (1974) Microtubules and microfilaments. Annu Rev Plant Physiol 25:309–362
Hepler PK, Newcomb EH (1967) Fine structure of cell plate formation in the apical meristem of *Phaseolus* roots. J Ultrastruct Res 19:498–513
Herth W, Meyer Y (1977) Ultrastructural and chemical analysis of the wall fibrils synthesized by tobacco mesophyll protoplasts. Biol Cell 30:33–40
Hoch HC (1979) Penetration of chemicals into the *Malus* leaf cuticle, an ultrastructural analysis. Planta 147:186–195

Huwyler HR, Franz G, Meier H (1979) Changes in the composition of cotton fibre cell walls during development. Planta 146:635–642
Jarvis LR, Wardrop AB (1974) The development of the cuticle in Phormium tenax. Planta 119:101–112
Jeffree CE, Baker EA, Holloway PJ (1976) Origins of the fine structure of plant epicuticular waxes. In: Dickinson CH, Preece TF (eds) Microbiology of aerial plant surfaces. Academic Press, London New York, pp 119–158
Jones MGK, Payne HL (1977) Cytokinesis in *Impatiens balsamina* and the effect of caffeine. Cytobios 20:79–92
Juniper BE, Cox GC (1973) The anatomy of the leaf surface: the first line of defence. Pestic Sci 4:543–547
Juniper BE, Lawton JR (1979) The effect of caffeine, different fixation regimes and low temperature on microtubules in the cells of higher plants. Planta 145:411–416
Karas I, Cass DD (1976) Ultrastructural aspects of sperm cell formation in rye: evidence for cell plate involvement in generative cell division. Phytomorphology 26:36–45
Kawasaki S, Sato S (1979) Isolation of the Golgi apparatus from suspension cultured tobacco cells and preliminary observations on the intracellular transport of extensin-precursor. Bot Mag 92:305–314
Kiermayer O, Sleytr UB (1979) Hexagonally ordered "rosettes" of particles in the plasma membrane of *Micrasterias denticulata* Bréb, and their significance for microfibril formation and orientation. Protoplasma 101:133–138
King GGS, Colvin JR (1976) Intermediate polymer(s) of cellulose biosynthesis. Appl Polymer Symp 28:623–636
Klis FM, Eeltink H (1979) Changing arabinosylation patterns of wall-bound hydroxyproline in bean cell cultures. Planta 144:479–484
Kolattukudy PE (1970) Biosynthesis of cuticular lipids. Annu Rev Plant Physiol 21:163–192
Kolattukudy PE (1977) Biosynthesis and degradation of lipid polymers. In: Tevini M, Lichtenthaler HK (eds) Lipids and lipid polymers in higher plants. Springer, Berlin Heidelberg New York, pp 271–292
Kolattukudy PE, Croteau R, Walton TJ (1975) Biosynthesis of cutin, enzymatic conversion of ω-hydroxy fatty acids to dicarboxylic acids by cell-free extracts of *Vicia faba* epidermis. Plant Physiol 55:875–880
Kreger PR, Kopecká M (1976) Assembly of wall polymers during the regeneration of yeast protoplasts. In: Peberdy JF, Rose AH, Rogers HJ, Cocking EC (eds) Microbial and plant protoplasts. Academic Press, London New York, pp 237–252
Lazar-Keul G, Keul M, Wagner G (1978) Reversible Hemmung der Protoplasmaströmung in Wurzelhaaren der Gerste (*Hordeum vulgare* L.) und Tomate (*Lycopersicum esculentum* Mill.) durch Cytochalasin B. Z Pflanzenphysiol 90:461–466
Leppard GG, Sowden LC, Colvin JR (1975) Nascent stages of cellulose biosynthesis. Science 189:1094–1095
Lin B-Y (1977) Ploidy variation in maize endosperm. J Hered 68:143–149
Lintilhac PM (1974) Differentiation, organogenesis, and the tectonics of cell wall orientation. III theoretical considerations of cell wall mechanics. Am J Bot 61:230–237
Longly B, Waterkeyn L (1977) Étude de la cytokinése. I. les stades callosiques de la plaque cellulaire somatique. Cellule 72:195–224
López-Sáez JF, Risueño MC, Gimenez-Martin (1966) Inhibition of cytokinesis in plant cells. J Ultrastruct Res 14:85–94
Maclachlan GA (1976) A potential role for endo-cellulase in cellulose biosynthesis. Appl Polymer Symp 28:645–658
Mahlberg PG, Turner FR, Walkinshaw C, Venketeswaran S, Mehrotra B (1975) Observations on incomplete cytokinins in callus cells. Bot Gaz 136:189–195
Manton I (1967) Further observations on scale formation in *Chrysochromulina chiton.* J Cell Sci 2:411–418
Marchessault RH, Sundararajan PR (1977) Bibliography of crystal structure of polysaccharides, 1967–1974. Adv Carbohydr Chem Biochem 33:387–404
Mark RE (1971) Mechanical behaviour of cellulose in relation to cell wall theories. J Polymer Sci Part Polymer Symp 36:393–406

Martin JT, Juniper BE (1970) The cuticles of plants. Edward Arnold, London
Marx-Figini M (1969) On the biosynthesis of cellulose in higher plants. J Polymer Sci Part Polymer Symp 28:57–67
Marx-Figini M, Schulz GV (1966) Über die Kinetik und den Mechanismus der Biosynthese der Cellulose in den höheren Pflanzen (nach Versuchen an den Samenhaaren der Baumwolle). Biochim Biophys Acta 112:81–101
Meinert MC, Delmer DP (1977) Changes in biochemical composition of the cell wall of the cotton fiber during development. Plant Physiol 59:1088–1097
Miller JH (1980) Orientation of the plane of cell division in fern gametophytes: the roles of cell shape and stress. Am J Bot 67:534–542
Mollenhauer HH, Mollenhauer BA (1978) Changes in the secretory activity of the Golgi apparatus during the cell cycle in root tips of maize (*Zea mays* L.) Planta 138:113–118
Mollenhauer HH, Morré DJ (1976) Cytochalasin B, but not colchicine, inhibits migration of secretory vesicles in root tips of maize. Protoplasma 87:39–48
Morré DJ (1977) Membrane differentiation and the control of secretion: a comparison of plant and animal Golgi apparatus. In: Brinkley BR, Porter KR (eds) International cell biology 1976–1977. Rockefeller Univ Press, New York, pp 293–303
Morré DJ, Mollenhauer HH (1974) The endomembrane concept: a functional integration of endoplasmic reticulum and Golgi apparatus. In: Robards AW (ed) Dynamic aspects of plant ultrastructure. McGraw-Hill, New York London, pp 84–137
Morré DJ, Mollenhauer HH (1976) Interactions among cytoplasm, endomembranes, and the cell surface. Encyclopedia of plant physiology, vol III. Springer, Berlin Heidelberg New York, pp 288–344
Morrison IN, O'Brien TP (1976) Cytokinesis in the developing wheat grain; division with and without a phragmoplast. Planta 130:57–67
Mühlethaler K (1965) Growth theories and the development of the cell wall. In: Côte WA Jr (ed) Cellular ultrastructure of woody plants. Syracuse Univ Press, Syracuse, pp 51–60
Mueller SC, Brown RM Jr (1980) Evidence for an intramembrane component associated with a cellulose microfibril-synthesizing complex in higher plants. J Cell Biol 84:315–326
Mueller SC, Brown RM Jr, Scott TK (1976) Cellulosic microfibrils: nascent stages of synthesis in a higher plant cell. Science 194:949–951
Nagahashi J, Beevers L (1978) Subcellular localization of glycosyl transferases involved in glycoprotein biosynthesis in the cotyledons of *Pisum sativum* L. Plant Physiol 61:451–459
Newcomb EH (1969) Plant microtubules. Annu Rev Plant Physiol 20:253–288
Nishitani K, Shibaoka H, Masuda Y (1979) Growth and cell wall changes in azuki bean epicotyls. II Changes in wall polysaccharides during auxin-induced growth of excised segments. Plant Cell Physiol 20:463–472
Northcote DH (1974) Sites of synthesis of the polysaccharides of the cell wall. In: Pridham JB (ed) Plant carbohydrate biochemistry. Academic Press, London New York, pp 165–181
Northcote DH (1977) The synthesis and assembly of plant cell walls: possible control mechanisms. Cell Surface Rev 4:717–739
Northcote DH (1979) The involvement of the Golgi apparatus in the biosynthesis and secretion of glycoproteins and polysaccharides. Biomembranes 10:51–76
Nowak-Ossorio M, Gruber E, Schurz J (1976) Untersuchungen zur Cellulose-Bildung in Baumwollsamen. Protoplasma 88:255–263
O'Brien TP (1967) Observations on the fine structure of the oat coleoptile. I. The epidermal cells of the extreme apex. Protoplasma 63:385–416
O'Brien TP (1972) The cytology of cell wall formation in some eukaryotic cells. Bot Rev 38:87–118
Olesen P (1978) Studies on the physiological sheaths in roots. I. Ultrastructure of the exodermis in *Hoya carnosa* L. Protoplasma 94:325–340
Olesen P (1979) Ultrastructural observations on the cuticular envelope in salt glands of *Frankenia pauciflora*. Protoplasma 99:1–9
Olesen P (1980) The visualization of wall-associated granules in thin sections of higher

plant cells: Occurrence, distribution, morphology and possible role in cell wall biogenesis. Z Pflanzenphysiol 96:35–48
Palevitz BA (1980) Comparative effects of phalloidin and cytochalasin B on motility and morphogenesis in *Allium*. Can J Bot 58:773–785
Palevitz BA, Hepler PK (1974a) The control of the plane of division during stomatal differentiation in *Allium*. I. Spindle reorientation. Chromosoma 46:297–326
Palevitz BA, Hepler PK (1974b) The control of the plane of division during stomatal differentiation in *Allium*. II. Drug studies. Chromosoma 46:327–341
Palevitz BA, Hepler PK (1976) Cellulose microfibril orientation and cell shaping in developing guard cells of *Allium*: the role of microtubules and ion accumulation. Planta 132:71–93
Palma A, Büldt G, Jovanović SM (1976) Absolutes Molekulargewicht der nativen Cellulose der Alge *Valonia*. Makromol Chem 177:1063–1072
Paul DC, Goff CW (1973) Comparative effects of caffeine, its analogues and calcium deficiency on cytokinesis. Exp Cell Res 78:399–413
Paull RI, Jones RL (1975) Studies on the secretion of maize root cap slime. Plant Physiol 56:307–312
Peterson CA, Peterson RL, Robards AW (1978) A correlated histochemical and ultrastructural study of the epidermis and hypodermis of onion roots. Protoplasma 96:1–21
Pickett-Heaps JD (1967a) The use of radioautography for investigating wall secretions in plant cells. Protoplasma 64:49–66
Pickett-Heaps JD (1967b) Further observations on the Golgi apparatus and its functions in cells of the wheat seedling. J Ultrastruct Res 18:287–303
Pickett-Heaps JD (1967c) The effects of colchicine on the ultrastructure of dividing plant cells, xylem wall differentiation, and distribution of cytoplasmic microtubules. Dev Biol 15:206–236
Pickett-Heaps JD (1969) The evolution of the mitotic apparatus: an attempt at comparative ultrastructural cytology in dividing plant cells. Cytobios 3:257–280
Pickett-Heaps JD (1972) Autoradiography with the electron microscope: experimental techniques and considerations using plant tissues. In: Lüttge U (ed) Microautoradiography and electron probe analysis, their application to plant physiology. Springer, Berlin Heidelberg New York, pp 168–190
Pickett-Heaps JD, Northcote DH (1966a) Organization of microtubules and endoplasmic reticulum during mitosis and cytokinesis in wheat meristems. J Cell Sci 1:109–120
Pickett-Heaps JD, Northcote DH (1966b) Cell division in the formation of the stomatal complex of the young leaves of wheat. J Cell Sci 1:121–128
Pollard MR, McKeon T, Gupta LM, Stumpf PK (1979) Biosynthesis of waxes by developing jojoba (*Simmondsia chinensis*) seeds. 2. The demonstration of wax biosynthesis by cell-free homogenates. Lipids 14:651–662
Pope DG, Thorpe JR, Al-Azzawi MJ, Hall JL (1979) The effect of cytochalasin B on the rate of growth and ultrastructure of wheat coleoptiles and maize roots. Planta 144:373–383
Porter KR, Machado RD (1960) Studies on the endoplasmic reticulum. IV. Its form and distribution during mitosis in cells of onion root tip. J Biophys Biochem Cytol 7:167–180
Preston RD (1959) Wall organization in plant cells. Int Rev Cytol 8:33–60
Preston RD (1964) Structural and mechanical aspects of plant cell walls with particular reference to synthesis and growth. In: Zimmerman MH (ed) Formation of wood in forest trees. Academic Press, London New York, pp 169–188
Preston RD (1965) The biosynthesis of cellulose. In: Pridham JB, Swain T (eds) Biosynthetic pathways in higher plants. Academic Press, London New York, pp 123–132
Preston RD (1974) The physical biology of plant cell walls. Chapman and Hall, London
Quader H, Wagenbreth I, Robinson DG (1978) Structure, synthesis and orientation of microfibrils. V. On the recovery of *Oocystis solitaria* from microtubule inhibitor treatment. Cytobiologie 18:39–51
Ray PM, Shininger TL, Ray MM (1969) Isolation of β-glucan synthetase particles from plant cells and identification with Golgi membranes. Proc Natl Acad Sci USA 64:605–612

Ray PM, Eisinger WR, Robinson DG (1976) Organelles involved in cell wall polysaccharide formation and transport in pea cells. Ber Dtsch Bot Ges 89:121–146

Raymond Y, Fincher GB, Maclachlan GA (1978) Tissue slice and particulate β-glucan synthetase activities from *Pisum* epicotyls. Plant Physiol 61:938–942

Reiss H-D, Herth W (1978) Visualization of Ca^{2+}-gradient in growing pollen tubes of *Lilium longiflorum* with chlorotetracycline fluorescence. Protoplasma 97:373–377

Reiss H-D, Herth W (1979) Calcium gradients in tip growing plant cells visualized by chlorotetracycline fluorescence. Planta 146:615–621

Robards AW (1969) Particles associated with developing plant cell walls. Planta 88:376–379

Robards AW, Jackson SM, Clarkson DT, Sanderson J (1973) The structure of barley roots in relation to the transport of ions into the stele. Protoplasma 77:291–311

Roberts K (1974) Crystalline glycoprotein cell walls of algae: their structure, composition and assembly. Philos Trans R Soc London Ser B 268:129–146

Roberts K, Northcote DH (1970) The structure of sycamore callus cells during division in a partially synchronized suspension culture. J Cell Sci 6:299–321

Robinson DG (1977) Plant cell wall synthesis. Adv Bot Res 5:89–151

Robinson DG (1980) The role of endomembrane organelles in plant cell wall synthesis. In: Nover L, Lynen F, Mothes K (eds) Cell compartmentation and metabolic channeling. VEB Gustav Fischer Verlag, Jena, in press

Robinson DG, Herzog W (1977) Structure, synthesis and orientation of microfibrils. III A survey of the action of microtubule inhibitors on microtubules and microfibril orientation in *Oocystis solitaria*. Cytobiologie 15:463–474

Robinson DG, Eisinger WR, Ray PM (1976) Dynamics of the Golgi system in wall matrix polysaccharide synthesis and secretion by pea cells. Ber Dtsch Bot Ges 89:147–161

Roelofsen PA (1958) Cell-wall structure as related to surface growth. Acta Bot Neerl 7:77–89

Roland J-C (1968) Recherches sur l'infrastructure de l'éspace membranaire des cellules végétales. CR Acad Sci Ser D 267:712–715

Roland J-C (1969) Mise en évidence sur coupes ultrafines de formation polysaccharidiques directement associés au plasmalemme. CR Acad Sci Ser D 269:939–942

Roland J-C (1973) The relationship between the plasmalemma and the plant cell wall. Int Rev Cytol 36:45–92

Roland J-C, Prat R (1973) Les protoplastes et quelques problèmes concernant le rôle et l'élaboration des parois. Colloq Int CNRS 212:243–271

Roland J-C, Sandoz D (1969) Détection cytochimique des sites de formation des polysaccharides pré-membranaires dans les cellules végétales. J Microscopie 8:263–268

Roland J-C, Vian B, Reis D (1977) Further observations on cell wall morphogenesis and polysaccharide arrangement during plant growth. Protoplasma 91:125–141

Rowley JR, Southworth D (1967) Deposition of sporopollenin on lamellae of unit membrane dimensions. Nature (London) 213:703–704

Ryser U (1979) Cotton fibre differentiation: occurrence and distribution of coated and smooth vesicles during primary and secondary wall formation. Protoplasma 98:223–239

Sargent C (1967a) The occurrence of a secondary cuticle in *Libertia elegans* (Iridaceae). Ann Bot 40:355–359

Sargent C (1976b) In situ assembly of cuticular wax. Planta 129:123–126

Satoh S, Matsuda K, Tamari K (1976) β-1,4-glucan occurring in homogenate of *Phaseolus aureus* seedlings. Possible nascent stage of cellulose biosynthesis in vivo. Plant Cell Physiol 17:1243–1254

Sawhney VK, Srivastava LM (1975) Wall fibrils and microtubules in normal and gibberellic-acid-induced growth in lettuce hypocotyl cells. Can J Bot 53:824–835

Schnepf E, Deichgräber G (1979) Elongation of setae of *Pellia* (Bryophyta): fine structural analysis. Z Pflanzenphysiol 94:283–297

Schnepf E, Herth W (1978) General and molecular cytology. Fortschr Bot 40:1–11

Shafizadeh F, McGinnis GD (1971) Morphology and biogenesis of cellulose and plant cell-walls. Adv Carbohydr Chem Biochem 26:297–349

Sharma CBSR, Sahu RK, Panigrahi S (1978) Inhibition of cytokinesis by a bacterial toxin: Thuringiensin-A. Caryologia 31:89–94

Shore G, Maclachlan GA (1975) The site of cellulose synthesis. Hormone treatment alters the intracellular location of intracellular β-1,4-glucan (cellulose) synthetase activities. J Cell Biol 64:557–571

Singer SJ, Nicolson L (1972) The fluid mosaic model of the structure of cell membranes. Science 175:720–731

Sinnot EW, Bloch R (1940) Cytoplasmic behaviour during division of vacuolate plant cells. Proc Natl Acad Sci USA 26:223–227

Sinnott EW, Bloch R (1941) Division in vacuolate plant cells. Am J Bot 28:225–232

Sitte P (1975) Die Bedeutung der molekularen Lamellenbauweise von Korkzellwänden. Biochem Physiol Pflanz 168:287–297

Sleytr UB, Umrath W (1974) A simple fracturing device for obtaining complementary replicas of freeze-fractured and freeze-etched specimens and tissue fragments. J Microscopy 101:177–186

Smith MM, McCully ME (1978) A critical evaluation of the specificity of aniline blue induced fluorescence. Protoplasma 95:229–254

Soliday CL, Kolattukudy PE (1977) Biosynthesis of cutin, ω-hydroxylation of fatty acids by a microsomal preparation from germinating *Vicia faba*. Plant Physiol 59:1116–1121

Soliday CL, Kolattukudy PE, Davis RW (1979) Chemical and ultrastructural evidence that waxes associated with the suberin polymer constitute the major diffusion barrier to water vapor in potato tuber (*Solanum tuberosum* L.). Planta 146:607–614

Spiess E, Thio Tiang Nio E, Amelunxen F (1976) Untersuchungen zur Zellwandbildung. II. Nachweis der Zellwandsynthese und der Synthese von Cellulose, Hemicellulose und Pektin in den Ranken von *Cucurbita maxima* L. Cytobiologie 13:251–259

Srivastava LM, Sawhney VK, Bonettemaker M (1977) Cell growth, wall deposition, and correlated fine structure of colchicine-treated lettuce hypocotyl cells. Can J Bot 55:902–917

Takeo K, Shigeta M, Takagi Y (1976) Plasma membrane ultrastructural differences between the exponential and stationary phases of *Saccharomyces cerevisiae* as revealed by freeze-etching. J Gen Microbiol 97:323–329

Takeuchi Y, Komamine A, Saito T, Watanabe K, Morikawa N (1980) Turnover of cell wall polysaccharides of a *Vinca rosea* suspension culture. Physiol Plant 48:536–541

Tippett JT, O'Brien TP (1976) The structure of *eucalypt* roots. Aust J Bot 24:619–632

Van Der Woude WJ, Morré DJ, Bracker CE (1971) Isolation and characterization of secretory vesicles in germinated pollen of *Lilium longiflorum*. J Cell Sci 8:331–351

Vian B (1974) Précisions fournies par le cryodécapage sur la restructuration et l'assimilation au plasmalemme des membranes des dérivés golgiens. CR Acad Sci Ser D 278:1483–1486

Villemez CL (1974) The relation of plant enzyme-catalyzed β-(1,4)-glucan synthesis to cellulose biosynthesis in vivo. In: Pridham JB (ed) Plant carbohydrate biochemistry. Academic Press, London New York, pp 183–189

Volfová A, Chvojka L, Haňkovská J (1977) The orientation of cell wall microtubules in wheat coleoptile segments subjected to phytohormone treatment. Biol Plant 19:421–425

Von Wettstein-Knowles P (1974) Ultrastructure and origin of epicuticular wax tubes. J Ultrastruct Res 46:483–498

Waterkeyn L (1967) Sur l'existence d'un 'stade callosique' présenté par la paroi cellulaire, au cours de la cytokinèse. CR Acad Sci Ser D 265:1792–1794

Wattendorff J (1974a) The formation of cork cells in the periderm of *Acacia senegal* Willd. and their ultrastructure during suberin deposition. Z Pflanzenphysiol 72:119–134

Wattendorff J (1974b) Ultrahistochemical reactions of the suberized cell walls in Acorus, Acacia, and Larix. Z Pflanzenphysiol 73:214–225

Wattendorff J (1976) Ultrastructure of the suberized styloid crystal cells in *Agave* leaves. Planta 128:163–165

Wells LG, Franich RA (1977) Morphology of epicuticular wax on primary needles of *Pinus radiata* seedlings. NZJ Bot 15:525–529

Whaley WG, Mollenhauer HH (1963) The Golgi apparatus and cell plate formation: a postulate. J Cell Biol 17:216–221

Whaley WG, Dauwalder M, Kephart J (1966) The Golgi apparatus and an early stage in cell plate formation. J Ultrastruct Res 15:169–180
Willison JHM (1973) Fine structural changes occurring during the culture of isolated tomato fruit protoplasts. Colloq Int CNRS 212:215–241
Willison JHM (1976) An examination of the relationship between freeze-fractured plasmalemma and cell-wall microfibrils. Protoplasma 88:187–200
Willison JHM (1980) Freeze-fractured cuticles of plants. Plant Sci Lett 18:121–126
Willison JHM, Brown RM Jr (1977) An examination of the developing cotton fiber: wall and plasmalemma. Protoplasma 92:21–41
Willison JHM, Brown RM Jr (1978) Cell wall structure and deposition in *Glaucocystis*. J Cell Biol 77:103–119
Willison JHM, Cocking EC (1972) The production of microfibrils at the surface of isolated tomato-fruit protoplasts. Protoplasma 75:397–403
Willison JHM, Cocking EC (1975) Microfibril synthesis at the surface of tobacco mesophyll protoplasts, a freeze-etch study. Protoplasma 84:147–159
Willison JHM, Grout BWW (1978) Further observations on cell-wall formation around isolated protoplasts of tobacco and tomato. Planta 140:53–58
Willison JHM, Brown RM Jr, Mueller SC (1980) A critical examination of the evidence for swollen hydrated cellulose fibrils in green plants. J Microscopy 118:177–186
Wooding FBP (1968) Radioautographic and chemical studies of incorporation into sycamore vascular tissue walls. J Cell Sci 3:71–80
Wright K, Northcote DH (1976) Identification of β-1-4 glucan chains as part of a fraction of slime synthesized within the dictyosomes of maize root caps. Protoplasma 88:225–239
Zaar K (1977) The biogenesis of cellulose by *Acetobacter xylinum*. Cytobiologie 16:1–15
Zaar K (1979) Visualization of pores (export sites) correlated with cellulose production in the envelope of the Gram-negative bacterium *Acetobacter xylinum*. J Cell Biol 80:773–777

22 Secretory Activity of the Root Cap

M. ROUGIER

1 Introduction

The root tips of all plants are covered by a specialized group of cells termed the root cap. During the last two decades, numerous studies attempted to elucidate the development, structure, and functioning of the root cap cells (see reviews of JUNIPER 1972, BARLOW 1975). According to HABERLANDT (1914), this tissue performs three functions: protection of the root meristem, facilitation of root penetration through the soil by secretion of a lubricating slime or "mucigel", and perception of gravitational stimuli. This chapter is concerned with the second function. In particular, the function of the root cap in slime production will be discussed in relation to available evidence from morphological, physiological, and biochemical studies.

2 Root Cap Architecture and Root Cap Secretory Activity

2.1 Occurrence and Localization of Secretory Cells in the Root Cap

Environmental conditions are known to affect both the structure of the root cap and its secretory function (GILL and TOMLINSON 1975). MOLLENHAUER (1967) compared the root cap architecture of epiphytic, terrestrial, and aquatic plants and clearly demonstrated the presence of some secretory activity in all terrestrial and epiphytic roots studied. In contrast, secretory activity was found to be absent from cap cells of roots growing in water. Thus, at least two general types of root caps may be defined: those with secretory activity and those without apparent secretory activity. The development of a root cap containing secretory cells may be described as follows: cell division, cell enlargement, dictyosome modification and initiation of secretory activity, secretion, degeneration of the secretory cells. In contrast, the development of a root cap without secretory activity may be summarized by the three following stages: cell division, cell vacuolization, cell degeneration. Recent studies on the root cap structure of roots grown in aquatic, aerial, or terrestrial media (SAMBIN 1978) are in agreement with this developmental scheme.

The secretory cells are always located at the periphery of the root cap (see JUNIPER 1972, JUNIPER and PASK 1973, JUNIPER et al. 1977). As seen in Fig. 1, the root cap is divided into three main regions: a basal region adjacent to the root apex or root cap meristem, a central region composed of core

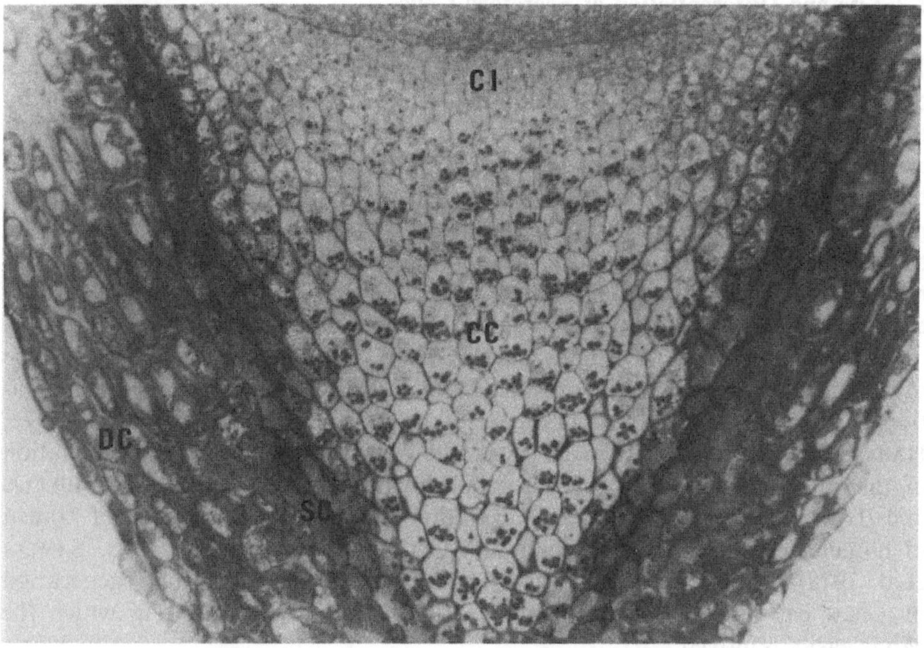

Fig. 1. Median longitudinal semi-thin section of the root cap of *Zea mays*. Section stained by PAS reaction. Abbreviations: *CI* cap initials; *CC* core cells; *SC* secretory cells; *DC* degenerating cells. Magnification ×300. (M. ROUGIER unpublished)

cells rich in starch grains, and a peripheral or outer region characterized by layers of active secretory cells with thickened outer tangential walls and surrounded by a zone of sloughing degenerating cells.

Several detailed light and electronmicroscopic studies have focused on the root cap system. Most of them are related to Angiosperm root caps (MOLLENHAUER et al. 1961, JUNIPER and ROBERTS 1966, NORTHCOTE and PICKETT-HEAPS 1966, MOLLENHAUER 1967, STREET et al. 1967, JUNIPER and PASK 1973, PHILLIPS and TORREY 1974, MARTIN and HARRIS 1976, SAMBIN 1978) and more recently to fern root caps (PETERSON and BRISSON 1977).

Among the root caps showing secretory cells, some systems such as those of maize (*Zea mays*) (MORRE et al. 1967) or wheat (*Triticum*) (NORTHCOTE and PICKETT-HEAPS 1966) exude a large amount of secretory material that passes through the cell wall and accumulates externally as slime droplets. Others, such as pea roots (SPINK and WILSON, personal communication reported by MOLLENHAUER 1967), cultured tomato roots (STREET et al. 1967), and cultured *Convolvulus* roots (PHILLIPS and TORREY 1974) secrete a smaller amount of material that accumulates between the cell wall and the protoplast and remains in this position throughout the subsequent stages of development. Thus, the occurrence of secretory cells within a root cap cannot always be correlated with the production of detectable slime. The exudation of the secretory product may also be related to the stage of development of the root cap and to various physiological and environmental parameters (see Sect. 5).

2.2 Life and Differentiation of Secretory Cells

The life of the peripheral secretory cells is directly related to the dynamics of the root cap. It is well known that cap cells are constantly generated by the meristematic activity of their initials which, when they cease dividing, differentiate and are displaced from the basal initial zone toward the periphery of the cap where they are constantly sloughed off. Thus the activity of the cap meristem serves to replace the lost cells. The time to renew all the cells of the cap represents the cap renewal time. According to Barlow (1975), the renewal time for the caps of primary roots depends on the number of initials, the duration of their mitotic cycle and the total number of cells.

Different methods have been developed to calculate the cap renewal time of various primary roots. The most frequently used are based on: (1) the estimation of the rate of cell production or of cell sloughing (Clowes 1971, 1976, Harkes 1976, Mac Leod 1976, Clowes and Woolston 1978), (2) the calculation of the rate of transit of cap cells labeled with tritiated thymidine (Phillips and Torrey 1971, Harkes 1973), (3) the calculation of the rate of transit of binucleate cells induced by caffeine (Clowes 1976, Barlow 1978, Sambin et al. 1978). The method of decapping and subsequent study of regeneration of a new cap as reported by Barlow (1974) is limited to roots in which the cap is easily removed.

Very different cap renewal times (1 to 9 days) are reported in the literature. High rates of cell production or of cell sloughing are recorded for the cap of maize primary roots by the studies of Clowes and coworkers (Clowes 1971, 1976, Clowes and Woolston 1978). In contrast, for the same system, lower rates of cell production or of cell transit are estimated by Barlow (1974, 1978). Low cap renewal times are also found for other species (Phillips and Torrey 1971, Harkes 1973, 1976, Mac Leod 1976, Sambin et al. 1978). From data on cap renewal time it is possible to calculate the rate of displacement of cells per day: for example, the cap cells of *Convolvulus* are displaced at a constant rate of 72 microns per day over a period of 6 to 9 days.

The consequences of such calculations on the cap renewal time and rate of cell displacement per day on the life duration of secretory cells must be considered. Thus, if the transit time of a cell along the length of the cap is one week with a constant rate of displacement, one can assume that the life of individual secretory cells equals 1 to 2 days. However, the life-time of individual secretory cells must be only a matter of hours in systems in which the cap renewal time is as short as one day. In addition, it appears that for maize root caps, axial outer secretory cells and lateral ones may not have the same duration of life in that cells take about 7 days to reach the tip of the cap columella and only 2 or 3 days to reach the flanks of the cap (Barlow 1978). Thus, rate of displacement of secretory cells and consequently their life duration seems to vary both between species and within the same species in relation to variations of growth or developmental conditions and also within different regions of a single root cap.

It has been shown that coordinated series of changes in fine structure and function occur during the displacement of cells from the proximal meristematic zone of the cap to the outer zones in which secretory cells are localized. *Zea mays* has been reported to be a particularly favorable plant to study the structure and function of cap cells (BARLOW 1975). Numerous detailed investigations are found in the literature on maize cap cytology (MOLLENHAUER et al. 1961, JUNIPER and ROBERTS 1966, BERJAK 1968, JUNIPER and FRENCH 1970, MAITRA and DE 1972 ...) and differentiation (CLOWES and JUNIPER 1964, JUNIPER and CLOWES 1965, MOLLENHAUER 1965a, BARLOW 1975, 1976, JUNIPER et al. 1977). In a recent review, BARLOW (1975) has compared the sequences of qualitative developmental changes that occur during cap cell displacement in *Zea* to the quantitative data of JUNIPER and CLOWES (1965). The more obvious changes in the cytoplasmic organelles in cap cells of *Zea* may be summarized as follows:

a) The plastids are maintained as a constant number (20 plastids per cell) but their appearance alters with the conversion of proplastids in the meristematic cells to amyloplasts in the central core geoperceptive zone. The amyloplasts lose their starch again and their capacity to sediment in the peripheral secretory cells of the cap.
b) The number of mitochondria per cell increases markedly from the cap initials to the most distal cap cells and concomitantly their cristae develop.
c) Both an increase in the total amount of endoplasmic reticulum and changes in the distribution of the membrane profiles are observed.
d) The dictyosomes increase in number (30 dictyosomes are present in a root cap initial compared with 200 in an outer root cap cell) and show marked differences in structure and secretory activity. Small vesicles only are produced in meristematic and core zones, but in active secretory cells the dictyosomes hypertrophe and produce much larger vesicles. In peripheral degenerescent cells, the organelles once more recover their former appearance.

Synchronous with these cytoplasmic events, changes in the cell wall thickness and stainability are also observed in *Zea* root cap (CLOWES and JUNIPER 1964, JUNIPER and FRENCH 1970, JUNIPER and PASK 1973, PAULL and JONES 1976b). The degradation of the primary cell wall, as well as the breakdown of the plasmodesmatal connections between cells (JUNIPER and BARLOW 1969), are concomitant with cell vacuolation and degeneration. Highly vacuolated moribund cells embedded in slime are sloughed out at the final stage (JUNIPER and ROBERTS 1966).

Similar basic patterns of changes in the fine structure of cap cells have been reported for seminal or adventitious roots of various terrestrial (NORTHCOTE and PICKETT-HEAPS 1966, STREET et al. 1967, PHILLIPS and TORREY 1974, MARTIN and HARRIS 1976, PETERSON and BRISSON 1977) or epiphytic plants (MOLLENHAUER 1967, SAMBIN 1978).

2.3 Ultrastructure of Secretory Cells

In maize all outer root cap cells are secretory but a well-defined developmental sequence can be observed between the initiation of secretory activity and final degeneration and sloughing of cells from the root (MOLLENHAUER et al. 1961, WHALEY et al. 1964).

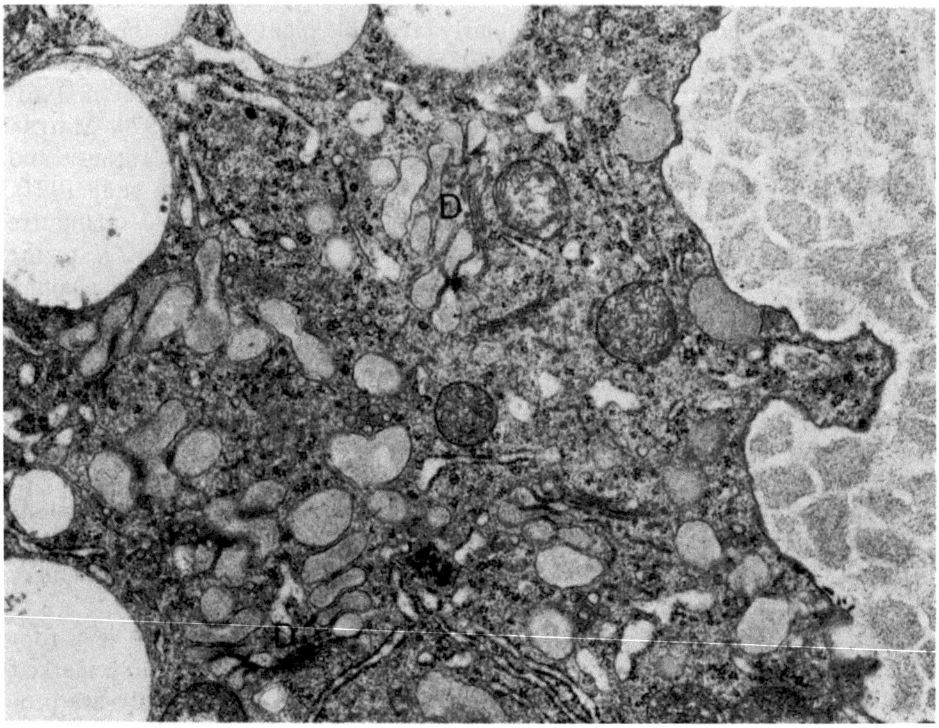

Fig. 2. Part of a secretory outer root cap cell of maize showing hypertrophied dictyosomes (*D*), invaginated plasmalemma and secretory product extruded into the periplasmic area. Magnification ×18,000. (M. ROUGIER unpublished)

The active secretory cells are mainly characterized by the peculiar hypersecretory form of their Golgi apparatus, first detected by WHALEY et al. (1959). Since this preliminary study, the morphology and secretory activity of this basic organelle has been worked out in detail by numerous ultrastructural investigations using various fixation and staining procedures (MOLLENHAUER et al. 1961, MOLLENHAUER and WHALEY 1963, MOLLENHAUER and MORRÉ 1966, DAUWALDER et al. 1969, ROUGIER 1971, DAUWALDER and WHALEY 1973, MOLLENHAUER and MORRE 1974, 1975, 1976). The partial hypertrophy of the Golgi cisternae represents a striking feature of the dictyosomes in secretory outer root cap cells (Fig. 2). All dictyosomes within a given cell are at the same developmental or functional state (MOLLENHAUER et al. 1975). According to MOLLENHAUER et al. (1961), the Golgi cisternae form two types of vesicles: small vesicles produced by the unhypertrophied cisternae and large elongated spherical vesicles, about 0.30 μ, formed from the hypertrophied distal regions. In all instances, the secretory vesicles have been observed to "pinch off" from tubular regions of the organelle (MOLLENHAUER and MORRÉ 1966). Changes in form and content of the large vesicles are observed to occur concomitantly with their movement to the cell surface (MOLLENHAUER and WHALEY 1963). The membranes of the vesicles are assumed to be incorporated into the plasma

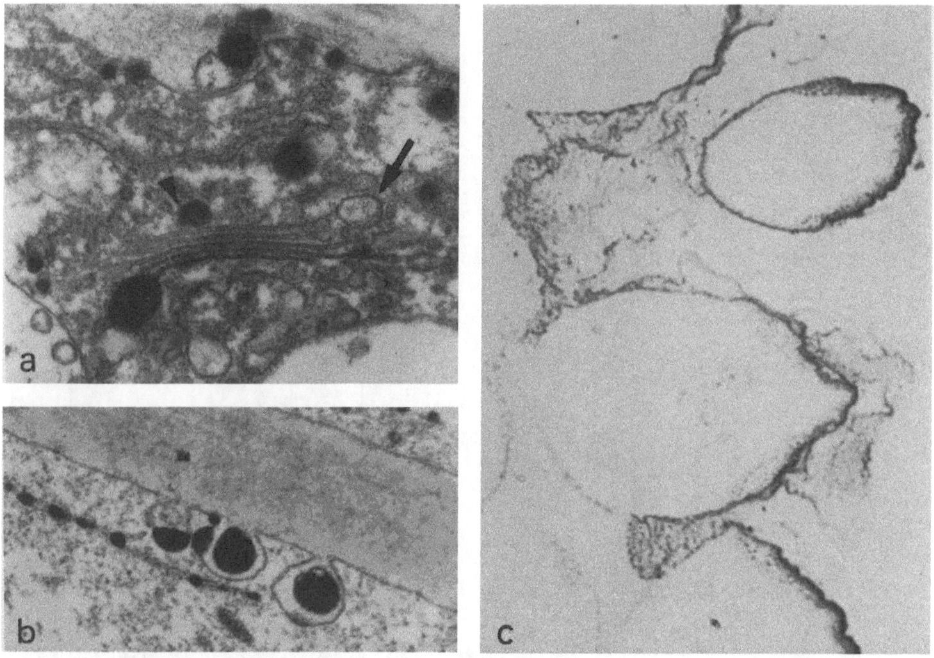

Fig. 3a–c. Peripheral cells of the root cap of *Ophioglossum petiolatum* at stage of secretory product synthesis and wall thickening. **a** represents a dictyosome pinching off two types of vesicles: electron-dense vesicles (*short arrows*) and vesicles with fibrillar contents (*long arrows*). Magnification ×29,000. **b** shows osmiophilic materials enclosed within invaginations of the plasmalemma. Osmiophilic deposits are also seen within a segment of endoplasmic reticulum. Magnification ×33,000. (Reproduced with kind permission from PETERSON and BRISSON 1977). **c** Peripheral root cap cells of *Ophioglossum petiolatum* with tanniferous vesicles (*dots*) within and between cell walls. Some vesicles are associated with carbohydrate secreted material. Prussian blue staining (light microscopy). Magnification ×1,300. (Reproduced with kind permission from BRISSON et al. 1977)

membrane and their contents released outside the protoplast. Intercisternal fibers closely associated with the secretory vesicles have been described (MOLLENHAUER 1965b, MOLLENHAUER and MORRE 1975). These fibers are found within the space between adjacent elongated vesicles even at early stages of vesicle formation. They increase and decrease in number as the vesicles mature and are no longer visible when vesicles separate from dictyosomes. These intercisternal elements are common to several cell types (MOLLENHAUER 1965b) and are assumed to help the organization and shaping of the secretory vesicles. In addition to studies on maize, the existence of two types of vesicles (with fibrillar or osmiophilic contents) fusing with the plasmalemma and releasing their products between the plasmalemma and the primary wall has also been noticed in peripheral active secretory cells of *Ophioglossum* (PETERSON and BRISSON 1977, Fig. 3a, b).

JUNIPER and ROBERTS (1966) suggested that the hypertrophied nature of the dictyosomes may be attributed to a sudden increase in carbohydrate supply

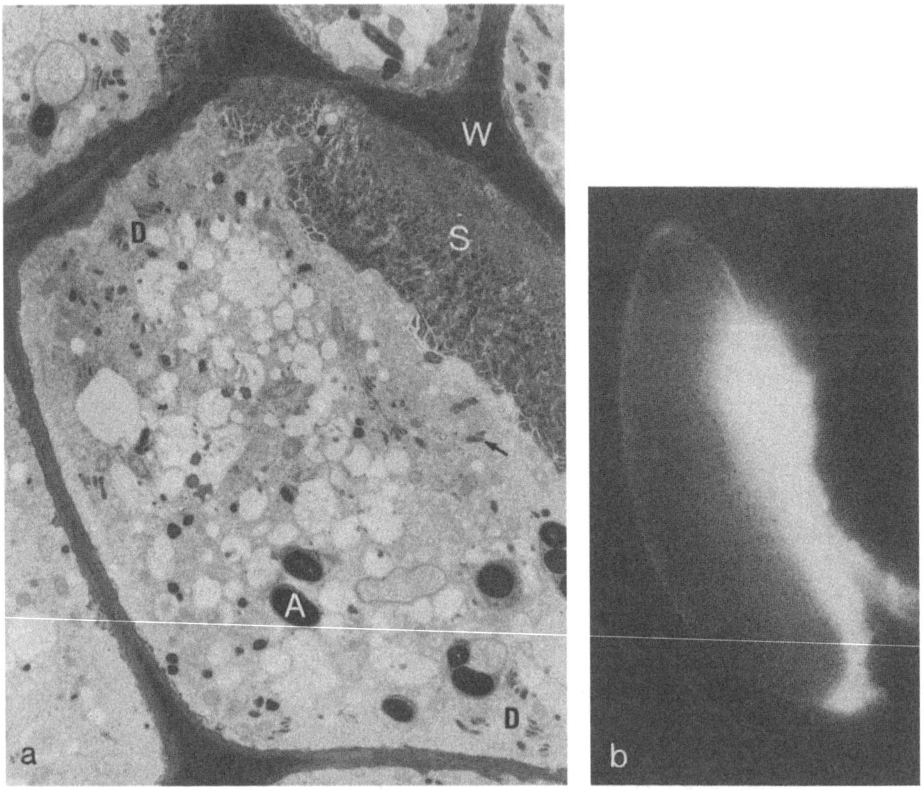

Fig. 4. a Section of a secretory outer root cap cell of *Zea mays* stained by the PATAG procedure. Polysaccharides are detected within dictyosomes (*D*), Golgi-derived vesicles (*arrows*), amyloplasts (*A*), cell wall (*W*) and secretory slime products (*S*) preferentially accumulated between the plasmalemma and the outer tangential wall. Magnification ×4,800. (M. ROUGIER unpublished). **b** Binding of Flu-UeA to the surface of an isolated outer root cap cell of maize. The lectin binding reveals a polarized pattern of distribution which corresponds to the pattern of discharge of the slime observed in **a**. Magnification ×1,800. (M. ROUGIER et al. 1979a)

within the cells, originating from the breakdown products of the stored starch grains in the amyloplasts of the peripheral cells. Except for *Cattleya* (MOLLENHAUER 1967), a similar loss of starch has been reported in the peripheral cells of some other roots (for example, PETERSON and BRISSON 1977, PHILLIPS and TORREY 1974, MAITRA and DE 1972). But the disappearance of starch grains has also been correlated to the production of lipid in *Medicago* root cap (MAITRA and DE 1972).

The endoplasmic reticulum of secretory root cap cells of maize is characterized by membrane-bound ribosomes, and although very abundant is not organized in any particular orientation in the cell (JUNIPER and FRENCH 1973). Very few dictyosomes appear to be directly associated with endoplasmic reticulum elements. However, after cold treatment, frequent endoplasmic reticulum-Golgi apparatus continuities have been observed (MOLLENHAUER et al. 1975) and were suggested to represent a general phenomenon although less frequently observed in roots at room temperature.

An additional striking feature of the peripheral cap cells of maize and other species is the presence of thickened cell walls (MOLLENHAUER 1967, STREET et al. 1967, JUNIPER 1972, JUNIPER and PASK 1973, PHILLIPS and TORREY 1974, MARTIN and HARRIS 1976, PETERSON and BRISSON 1977, SAMBIN 1978). This thickening appears to be due to the preferential accumulation, betweeen the plasmalemma and the outer tangential wall, of material of the same electron density and configuration as the material present in the vesicles (ROUGIER 1971 and Fig. 4a). The thickening of the wall increases in the most distal secretory root cap cells. The explanation of this phenomenon is detailed in Section 4.

When cells reach the edge of the root cap, they are still alive but cease their secretory activity: their dictyosomes revert to a flattened appearance, vacuolation increases markedly and cell wall degradation proceeds so that these cells may be considered virtually as protoplasts (JUNIPER et al. 1977). The secreted slime contains such sloughed cells.

3 Characteristics of the Secretory Products

As previously discussed, almost all the intensively growing primary and lateral roots of terrestrial and epiphytic plants produce slime. This section is devoted to the chemical characterization of the root cap slime and to the in situ localization and identification of some of its components by means of cytochemical methods.

3.1 Collection and Chemical Analysis of Root Cap Slimes

Some root caps exude profuse amounts of slime that accumulate exterior to the secreting region. In particular, the slime of maize is exuded as drops that are easily collected at the tip of the roots. Most of the detailed data available are related to the composition of maize slime secreted either by nodal roots (FLOYD and OHLROGGE 1970) or seminal roots (JUNIPER and ROBERTS 1966, JONES and MORRE 1967, HARRIS and NORTHCOTE 1970, JONES and MORRÉ 1973, PAULL et al. 1975). Composition of slime in other systems is not investigated. The collection of slime has been shown to be easier when roots are placed in water (FLOYD and OHLROGGE 1970) or in appropriate incubating medium (PAULL et al. 1975) since it is highly hydrophilic and swells in contact with water (SAMTSEVICH 1965).

There is only limited agreement from the studies reported by several investigators on the chemical composition and properties of the secreted polymer produced by maize roots. PAULL et al. (1975) have suggested that this could result from differences between the varieties of maize studied as well as from variations in culture method or purification procedures (see also Table 1). However, accord-

Table 1. Collection and chemical analysis of root cap slime of maize

Reference	JUNIPER and ROBERTS (1966)	JONES and MORRÉ (1967)	FLOYD and OHLROGGE (1970)	HARRIS and NORTHCOTE (1970)	JONES and MORRÉ (1973)	PAULL et al. (1975)	
Variety of maize used	"Orla 266"	WF-9 × M-14	WF-9 × 38-11	"Orla 266"	WF-9 × M-14	SX-17	
Conditions of root growth	Nonsterile conditions	Nonsterile conditions	Nonsterile conditions	Sterile conditions	Nonsterile conditions	Nonsterile conditions	
Slime purification	Purified material	Purified material	"Crude" material	Purified material	Purified material	(1) "Crude" material (2) Purified material	
Method of sugar determination	Paper chromatography	Paper chromatography	Paper chromatography	Gas liquid chromatography	Gas liquid chromatography	Gas liquid chromatography	
Major Findings:						(1) %	(2) %
Neutral sugars	% Undetermined	%	Molecular ratio	%	%		
Glucose	+	37 ± 4	(in young tissue)	22	22	19	5.7
Galactose	+	35 ± 4	7	21	39	35.5	30
Arabinose	+	–	8	15	7	9.5	14.5
Xylose	+	5 ± 2	5	4	–	10.5	7.5
Ribose	–	11 ± 2	–	–	5	–	–
Fucose	–	–	11	32	8	24	39
Mannose	–	–	–	6	6	1.8	3
Uronic acids	Small contents	Galacturonic acid 12 ± 3	3	Galacturonic + glucuronic acids	Galacturonic acid 12	—	+
Other components		Protein 6 to 8%	Protein 0.5 to 5% APase ATPase				

ing to WRIGHT (1975), the use of different maize varieties does not influence the overall composition of the slime. Estimation of the overall labeling patterns of slimes from five varieties demonstrates that the variations found are no greater than those noticed between different preparations of material from the same variety. This author emphasizes that sterilization of the starting material is very important and is the only way to avoid erroneous data, since many bacteria of different polysaccharide composition may be able to use slime as a food source. Unfortunately, very few studies mention the use of sterile conditions (Table 1) and the results must therefore be regarded with due caution. In general, two classes of secretory material have been investigated: either "crude material" or "purified secretory product". The "crude material" is obtained by wiping the root tips onto the surface of glass fiber disks (HARRIS and NORTHCOTE 1970, PAULL et al. 1975). In the preparation of the "purified secretory product", various additional steps are incorporated. These include: suction or washing of slime from root tips, dilution in water, centrifugation for removing sloughed cells and other debris, precipitation of the supernatant by ethanol, washings and lyophilization of the precipitate (JONES and MORRÉ 1967, 1973, PAULL et al. 1975). The "crude material" consists of a mixture of secretory product, sloughed cells, cell walls, and various debris, while the second consists only of pure root cap secretory product. PAULL et al. (1975) have compared results obtained from the analysis of "crude" and "purified" secreted material: the composition of the two materials is surprisingly very similar.

3.1.1 Monosaccharide Components

Investigations into the composition of root cap secretory product show that it is mainly composed of polysaccharides. This has been demonstrated either by quantitative data collected from the hydrolysis of slime material and the subsequent determination of sugar components by paper or gas liquid chromatography (see Table 1) or by qualitative data obtained from electrophoresis of radioactive slime (WRIGHT and NORTHCOTE 1974, 1975, WRIGHT 1975). The first chemical analysis performed by JUNIPER and ROBERTS (1966) has allowed the identification of the sugar content of maize slime as glucose, galactose, xylose, arabinose, and small amounts of uronic acids. In contrast, the results of JONES and MORRÉ (1967) indicated that the slime contained a high proportion of ribose (11%). However, except for this report and one later study performed by the same authors (1973) all subsequent analyses do not confirm the presence of ribose but reveal the presence of high proportions of fucose (FLOYD and OHLROGGE 1970, HARRIS and NORTHCOTE 1970, PAULL et al. 1975). The synthesis of large amounts of fucose by the cap tissues of maize roots has been confirmed by various studies using radioactive precursors (PAULL and JONES 1975a, WRIGHT 1975, WRIGHT and NORTHCOTE 1975, see also Sect. 4). The early discrepancies in composition may be explained by the similarity in R_{GLC} values of xylose and fucose using chromatograms run with the solvent system used by JONES and MORRÉ (1967). In later analyses (HARRIS and NORTHCOTE 1970), an additional solvent system was employed to specifically separate xylose from fucose. In addition to fucose, glucose and galactose represent the other major neutral sugar components of the slime produced by maize roots (Table 1). There are, however, variations in both glucose and galactose contents of the slime both with respect to the varieties of maize investigated and also in different preparations of slime from a single variety.

The presence of uronic acids in the maize slime has also been detected by several investigators (Juniper and Roberts 1966, Jones and Morré 1967, 1973, Harris and Northcote 1970, Paull et al. 1975). Their characterization as galacturonic acid has been reported by Jones and Morré (1967, 1973) and Wright and Northcote (1975).

3.1.2 Other Components

Maize root slime may also contain protein (Jones and Morré 1967, Floyd and Ohlrogge 1970 for example). The highest protein content has been reported by Wright and Northcote (1975). However, it is not yet clear whether the protein constitutes an endogenous component of the slime polymer or whether it is only released from the dying cells of the outer root cap zone.

Preliminary results reported by Hall et al. (1966) indicate the presence of degradative enzymes as components of maize root cap secretion. These enzymes were thought to help in separation and eventual breakdown of the cap cells. However, a clear distinction between secreted material proper and contents of dying cells was not attempted. The presence of ATPase and acid phosphatase has been reported in the stored as well as the fresh exudate of nodal roots (Floyd and Ohlrogge 1970). These authors suggest that if acid phosphatase is present in the mucilage layer at the root–soil interface, then probably organic phosphate esters in the soil would be hydrolyzed to inorganic phosphate which could then be used by the plant.

3.2 Structure of Root Cap Slimes

Estimations of the molecular weight of maize slime are given by some authors (Floyd and Ohlrogge 1971, Paull et al. 1975). Results presented by Floyd and Ohlrogge (1971) are based on the estimation of the molecular weight of the exudate of nodal roots by the use of a modification of the standard light scattering procedure. However, although the modified method is much simpler, it is less accurate and results should be regarded with caution. The calculated values obtained are 9×10^7 for the dry exudate and 18×10^7 for freshly collected exudate. The estimation related to the dry exudate is most probably the most accurate of the two since this value is the average of several determinations.

Slime secreted by roots of cultivar SX-17 of *Zea mays* has been reported by Paull et al. (1975) to have a molecular weight greater than 2×10^6 as obtained using techniques of exclusion chromatography, ultracentrifugal analysis, and relative viscosity of ethanol-precipitated material.

Information about the polymeric structure of maize slime has also been gained by the use of glass-fiber paper electrophoresis (Wright and Northcote

1974, WRIGHT 1975, WRIGHT et al. 1976). WRIGHT and NORTHCOTE (1974) first demonstrated a similarity between root cap slime and pectins, and considered that slime may be a form of pectin modified in such a way as to provide a hydrated protective coating around the tip. Thus the slime was found to consist of sequences of polygalacturonic acid interspersed with regions of glucose residues (WRIGHT 1975). However, further data obtained from the analyses of degraded material (WRIGHT and NORTHCOTE 1976) suggested the structure to consist of a central core region of $(1 \rightarrow 4)$-β-glucose linked chains. This confirmed earlier studies on $(1 \rightarrow 4)$-β-glucan which was found to be a component of mustard seed slime (GRANT et al. 1969). Thus, it is possible that the presence of a central core region of $(1 \rightarrow 4)$-β-glucan makes the slime fibrillar, while an outer matrix of hydrophilic polysaccharides renders it soluble. In order to determine the fucose position within the slime polysaccharide, WRIGHT et al. (1976) purified rat epididymal α-L-fucosidase and used it for analytical purposes. Since the enzyme removed only 17% of the fucose residues of the slime, a large proportion of the fucose residues is thought either to occupy an internal position or to be β-linked to other sugars within the polymer.

3.3 In Situ Identification of Slime Components

The identification and ultrastructural localization of carbohydrate components of the slime has been greatly improved by the use of various cytochemical methods. These methods are based upon either the detection of various chemical groups or the selective enzymatic or chemical extraction of various polysaccharides (see PICKETT-HEAPS 1967b and the reviews of ROUGIER et al. 1973, ROLAND 1974, ROUGIER 1976b). Using the PATAG procedure (test using periodic acid, thiocarbohydrazide, silver proteinate according to THIERY 1967), or analogous method (PICKETT-HEAPS 1967b), investigators have reported that there is a strongly staining material present in cell walls, exuded slime products, hypertrophied dictyosomes and associated vesicles (Fig. 4a). These observations have been recorded for secretory root cap cells of *Triticum vulgare* (PICKETT-HEAPS 1967b, 1968), *Cucurbita pepo* (COULOMB and COULON 1971), *Zea mays* (ROUGIER 1971, BREISCH 1974, GUCKERT et al. 1975), *Pisum sativum* (VIAN and ROLAND 1972, VIAN 1974a, LECHENE DE LA PORTE 1976), *Cattleya* sp. and *Cissus sicyoïdes* (SAMBIN 1978). Similarity between maize slime and pectin has also been suggested by ruthenium red staining and pectinase extraction of the secretory products (ROUGIER 1971). These results are useful in that they confirm both the polysaccharide nature of the slime (Figs. 4a and 5a) and also reveal the presence of polysaccharide precursors within the dictyosomes (Fig. 5a). The consequences of such data on an understanding of slime biosynthesis will be emphasized in Section 4.

Recently, lectins have been used to provide a more precise identification and cytochemical localization of carbohydrate components of the slime (ROUGIER et al. 1979a). Lectins are known to bind selectively to specific sugars and represent a new class of reagents for carbohydrate cytochemistry with a high level of specificity (see reviews of NICOLSON 1978 and ROTH 1978).

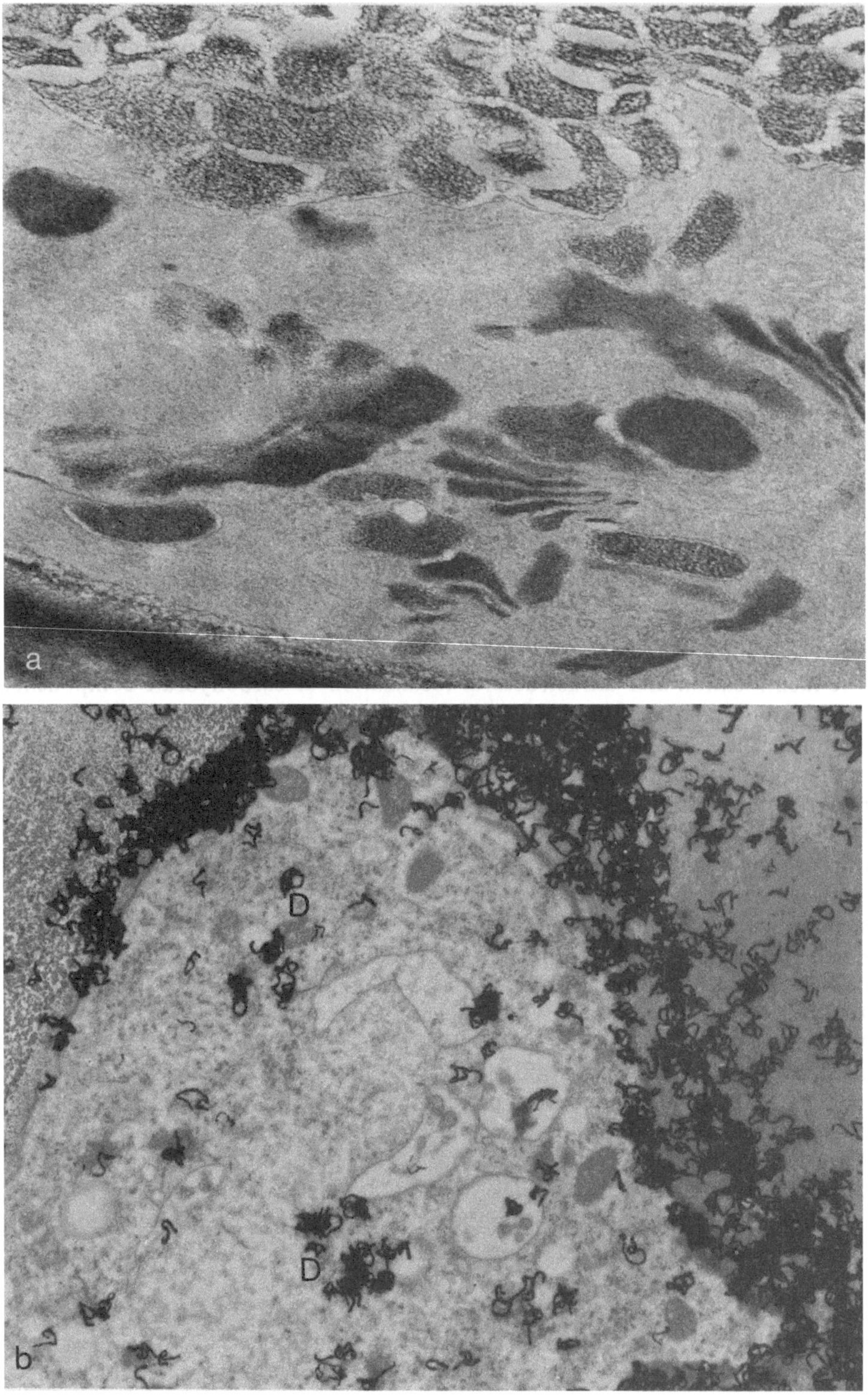
a
D
D
b

Experiments including agglutination inhibition with sonicated slime and labeling root cap cells with fluorescent lectins have resulted in the following findings: (1) the fluorescent lectins used are able to visualize in situ the main sugars of the slime as detected by biochemical analysis: fucose, galactose, and glucose; (2) the use of lectins provide some new data on the conformational organization of the slime polymers: e.g., fucose appears to be present at least partly in an external configuration since it is accessible to lectins in the native (nonsonicated) form of slime; (3) the lectins bind to the surface of secretory root cap cells and the lectin binding reveals a polarized pattern of distribution which corresponds to the pattern of discharge of the slime (Fig. 4b). These results confirm the ability of lectins to visualize the carbohydrates at the surface of secretory root cap cells and indicate their potential use for future cytochemical studies (ROUGIER et al. 1979b).

Apart from polysaccharides, polyphenols have also been detected in situ as components of the slime. Using ultraviolet fluorescence microscopy, HARRIS and HARTLEY (1976) have reported the presence of ferulic acid esterified to the polysaccharides of the root cap slime of *Zea mays*. They have suggested that Golgi cisternae may be the site of bonding between ferulic acid and the polysaccharides of the slime and the Golgi-derived vesicles the site of transport of the resulting esters. This phenolic acid is usually bound to the cell walls of the Gramineae where it may influence the degration of plant organic matter in the soil and inhibit the growth of plant pathogens (KOSUGE 1969). The in situ identification of phenolic substances as constituents of the secretory product of peripheral root cap cells of the fern *Ophioglossum* has been achieved by the use of phenolic histochemistry (BRISSON et al. 1977). Ferric chloride dissolved in a glutaraldehyde solution acts as a phenolic indicator that can be used for both light microscopy and transmission or scanning electronmicroscopy. A view of peripheral root cap cells with tanniferous vesicles associated with the cell walls and with carbohydrate-secreted material is seen in Fig. 3c.

The histochemical study of SUEIRO and FELIPE ANTON (1974) on the localization of acid phosphatase in the mucilaginous layer of seminal and nodal roots of *Zea mays* confirms FLOYD and OHLROGGE (1970) findings mentioned in Section 3.1.2.

Fig. 5. a Part of a secretory outer root cap cell of *Zea mays* after PATAG staining. The Golgi cisternae are seen to be uniformly stained (dense granular deposits). Some Golgi-derived secretory vesicles reveal a fibrillar content similar to the fibrillar aspect of the secretory material accumulated outside the plasmalemma. Magnification ×60,000. (M. ROUGIER in DUMAS et al. 1974). **b** Part of a secretory outer root cap cell of *Zea mays* labeled with [^{3}H]-fucose for 30 min. The dictyosomes (*D*) are labeled and a massive incorporation of labeled material into the secreted material may be observed. Magnification ×9,600. (M. ROUGIER 1976a)

4 Secretory Pathways

4.1 Biosynthesis of Slime Polysaccharides

BOWLES and NORTHCOTE (1972) stated that, since slime production is restricted to the root cap (HARRIS and NORTHCOTE 1970) and fucose can be used as a specific marker for the presence of slime (KIRBY and ROBERTS 1971), the maize root represents an ideal system for studying the involvement of different membranes in the production of slime. In the initial studies, roots were labeled in vivo with D-[U-^{14}C]-glucose, and after homogenization in glutaraldehyde and fractionation, isolated membrane fractions were washed with ethanol and relative incorporation of radioactivity into monosaccharides determined in total hydrolysates. They found that fucose-containing polymers were present only in the membrane system (dictyosomes, smooth membrane, and rough membrane fractions) of the tip region of the root. The results provided the first direct evidence that the Golgi apparatus and the endoplasmic reticulum could be involved in the synthesis of slime polysaccharide.

In a kinetic study (BOWLES and NORTHCOTE 1974), in vivo pulse-labeling indicated that radioactivity into polymeric material within the membrane fractions reached a steady-state level within 30 min, while incorporation into secretory material did not saturate. Using these results and the assumption that the increase in secretory material was made up exclusively from the radioactive precursor, the rate of increase (pmol min^{-1}) of the secreted slime sugar components was determined. Assuming that the only transport route of slime polysaccharides to the cell surface is via dictyosome-derived vesicles, the rate of turnover of the dictyosomes was calculated to be 20 s in order to maintain the observed rates of increase in secreted slime.

In later studies (BOWLES and NORTHCOTE 1976), after in vivo labeling and isolation of membranes, the fractions were extracted sequentially to give H_2O-soluble, lipid, and insoluble extracts. After degradation with specific lipases and proteases and separation of the extracted components according to size, the relative incorporation of radioactivity into monosaccharides in each fraction was again determined. The results provided some evidence that initiation of slime polymer synthesis may begin in the endoplasmic reticulum since there was a variation in the amounts of H_2O-soluble material and that material was released by protease treatment between the endoplasmic reticulum and dictyosome fractions. It was suggested that the low molecular weight substances released by proteases could be precursors formed during biosynthesis of slime polymers.

Further fractionation experiments, in which fucose was used instead of glucose for in vivo labeling of maize roots provided additional evidence that the membrane system is involved in the synthesis and secretion of slime (PAULL and JONES 1976a). Cesium chloride density gradient centrifugation was first used in this study of slime. Later, this method was adopted to compare the components of H_2O-soluble extracts of membranes and soluble cytoplasmic components with extracellular slime (GREEN and NORTHCOTE 1978). Using the gradient procedure, polysaccharides are found to sediment at 1.6 to 2.0 g cm^{-3}

whereas proteins sediment at 1.3 g cm^{-3}. Comparative in vivo labeling with L-[1-^{3}H]-fucose, D-[U-^{14}C]-glucose, L-[4.5-^{3}H]-leucine allowed the resolution of major components at 1.3, 1.37, 1.55, and 1.63 g cm^{-3}. The extracellular slime fraction was found to consist of only two components at 1.3 and 1.6, whereas the two intracellular fractions, those of the soluble components and the H_2O-soluble membrane extract, could be resolved into all of the four possible components. Importantly, when the three fractions were considered and analyzed as a whole, pulse-labeling of intact roots with tritiated fucose led to a saturation of radioactivity in the 1.37 and 1.55 g cm^{-3} whereas radioactivity in the 1.63 g cm^{-3} did not saturate. Pulse-chase in unlabeled fucose led to both a peak and fall of radioactive incorporation into 1.37 and 1.55 g cm^{-3} components and a steady increase in incorporation into 1.63. The authors stated that this indicated a precursor product relationship between the components, and suggested that glycoproteins may function as intermediates in slime biosynthesis.

Recent investigations have focused on an alternative procedure for determining the biosynthesis of slime and have involved in vitro incorporation of GDP-fucose into isolated membrane fractions prepared from maize roots (GREEN and NORTHCOTE 1979a, b, JAMES and JONES 1979a, b). In particular, the studies of GREEN and NORTHCOTE (1979a) have shown that fucose (and glucose) can be transferred to lipid and polymeric material by isolated fractions of the Golgi apparatus and the endoplasmic reticulum. The membrane fractions were characterized by the use of marker enzymes including IDPase and antimycin A-insensitive NADH cytochrome C reductase. The total fucosylated lipid has been found to contain both dolichol and sterol derivatives but the nature of the polymeric material synthesized has not yet been investigated. Glycosylated dolichol derivatives have been shown to be involved in the synthesis of asparagine-N-acetylglucosamine linkage type glycoproteins (HEMMING 1974, NORTHCOTE 1979). GREEN and NORTHCOTE (1979a) suggested that fucosylated lipids may be intermediates in the synthesis of the fucose-containing glycoproteins previously characterized by cesium chloride gradient centrifugation (GREEN and NORTHCOTE 1978). Optimum conditions for fucosyl transferase activities with respect to pH, metal ion dependance and effects of Triton X 100 were investigated for fucose incorporation from GDP-fucose (JAMES and JONES 1979a). It is interesting that the transferases responsible for incorporation of fucose into lipid and polymeric material seem to have two intracellular locations: the endoplasmic reticulum and the Golgi apparatus (GREEN and NORTHCOTE 1979b, JAMES and JONES 1979b). In contrast to the approach of GREEN and NORTHCOTE (1979b), JAMES and JONES (1979a) chose to focus on the synthesis and characterization of the polymeric acceptor material fucosylated in vitro. The authors (JAMES and JONES 1979b) demonstrated that there were two sites of fucosylation of polymeric material in maize roots: the endoplasmic reticulum and the Golgi apparatus, and thereby confirmed the earlier work of BOWLES and NORTHCOTE (1972) and GREEN and NORTHCOTE (1979b). In addition, using a membrane fraction which sedimented after centrifugation at 20,000 *g* for 1 h, the authors (JAMES and JONES 1979a) investigated the nature of the fucosylated polymer synthesized in vitro. Using several criteria, the fucosylated polymer was found to be very similar to slime. In particular, partially hydrolyzed slime (under hydrolytic conditions in which terminal fucose only is removed) was found to act as an exogenous acceptor for endogenous fucosyl transferase activity. The authors suggest, therefore, that the system is valid for the study of slime biosynthesis.

4.2 Transport via Granulocrine Process

Kinetic autoradiographic studies of the slime polysaccharide secretion have contributed to illustrate the granulocrine and membrane flow processes first detected by cytological studies (Mollenhauer et al. 1961) and involved in the collection, packaging, and export of the secretory material from the cell. Polysaccharide secretion was first monitored by radioautography after incorporation of tritiated glucose in roots of wheat (Northcote and Pickett-Heaps 1966, Pickett-Heaps 1967a), maize (Dauwalder et al. 1969) and onion (Goff 1973). In maize, radioactive galactose (Dauwalder and Whaley 1974) and fucose (Rougier 1975, 1976a, Paull and Jones 1975b, Fig. 5b) were also used for high-resolution autoradiographic studies of slime secretion. Fucose provides a specific labeling of the secreted slime (Kirby and Roberts 1971) compared to glucose or galactose which are readily metabolized by plant cells. Since an accumulation of silver grains over the dictyosomes and derived secretory vesicles has always been described, it seems likely that the radioactive sugars supplied to the roots incorporate subsequently into forming polysaccharides within the Golgi apparatus. The involvement of the endoplasmic reticulum could not be demonstrated by radioautography, since a diffuse scattering of grains was only observed on this part of the membrane system. To explain the discrepancy between the above data and those obtained by the use of biochemical techniques, Bowles and Northcote (1974) have suggested that the slime components in the endoplasmic reticulum are in a low molecular weight form that would not allow their detection by radioautography due to their loss during the fixation, washing, and dehydration procedures used.

This explanation is also in accordance with various cytochemical data (see Sect. 3) which indicate that concentration of slime polysaccharides occurs only at the Golgi apparatus. The characteristics of the polysaccharide material carried by the Golgi-derived secretory vesicles depends on the stage of vesicle differentiation, the cytochemical procedure used for the investigation and the choice of the experimental material. Dense granular contents characterize the Golgi saccules and newly formed vesicles of maize root cap cells (Rougier 1971, and Fig. 5a). As reported earlier from cytological studies (Mollenhauer and Whaley 1963), the staining characteristics of the vesicles undergo sequential modifications after they detach from dictyosomes. Since changes from a granular appearance to a fibrillar one similar to that of the slime accumulated outside the cell may be observed with the PATAG procedure (Fig. 5a), it is suggested that the maturation and polymerization of the secretory products take place inside the secretory vesicles. In contrast, only traces of polysaccharide content are evidenced in pea root cap within the dictyosomes and the secretory vesicles at their stage of formation. The polysaccharides are in contact with the membranes of the secretory vesicles and emerge further from them to constitute

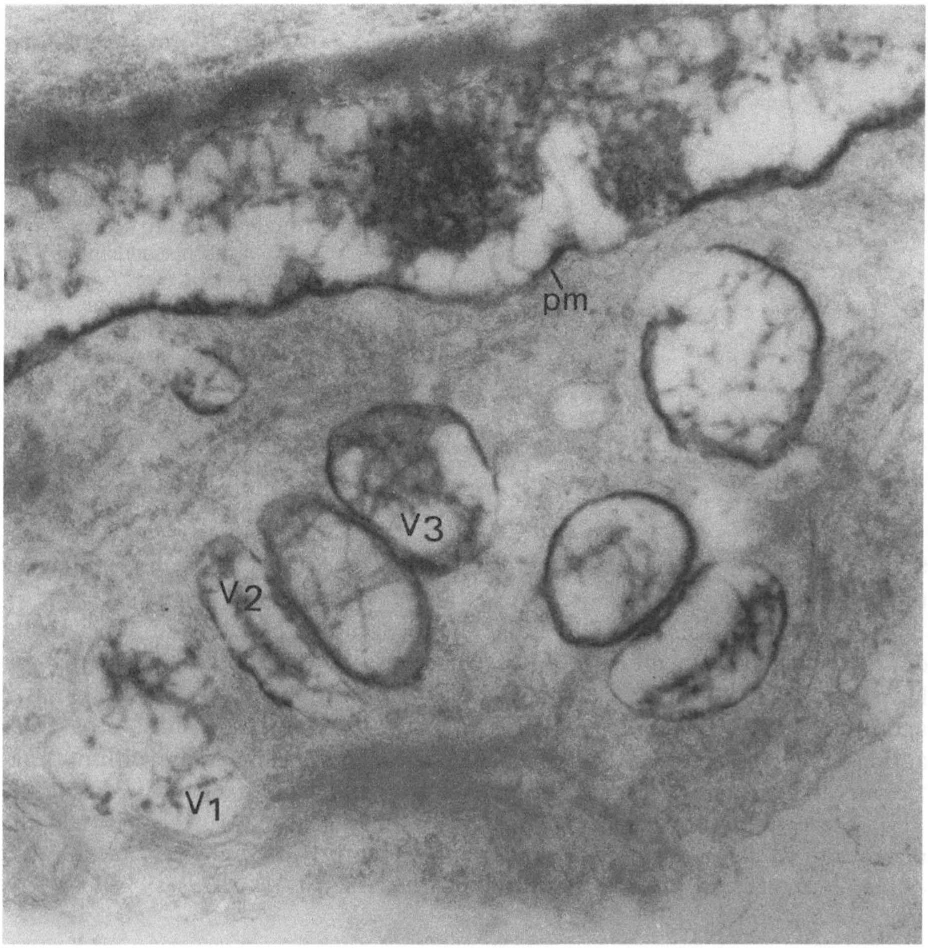

Fig. 6. Part of a secretory outer root cap cell of *Pisum sativum* after PTA-staining. The membranes of young Golgi vesicles (v_1) are poorly reactive. During the differentiation of the secretory vesicles, their staining characteristics appear to strongly modify (compare v_1 to v_2 and v_3). The membranes of mature vesicles (v_3) become cytochemically similar to the plasmalemma (*pm*) and their contents condense. Magnification ×50,000. (Reproduced with kind permission from VIAN and ROLAND 1972)

at the final stage a dense central nodule linked to the vesicle membrane by numerous polysaccharide tractus (VIAN and ROLAND 1972, 1974, VIAN 1974a, Fig. 6). Similar observations have been made of secretory vesicles in *Cattleya* and *Epidendrum* root caps (SAMBIN 1978). In correlation with development of secretory products, the "concentration vesicles" exhibit a gradual cytochemical transformation of their membranes (VIAN and ROLAND 1972, VIAN 1974a, Fig. 6). This membrane restructuring is initiated by the detachment of secretory vesicles from the Golgi cisternae and completed at the final stage of membrane

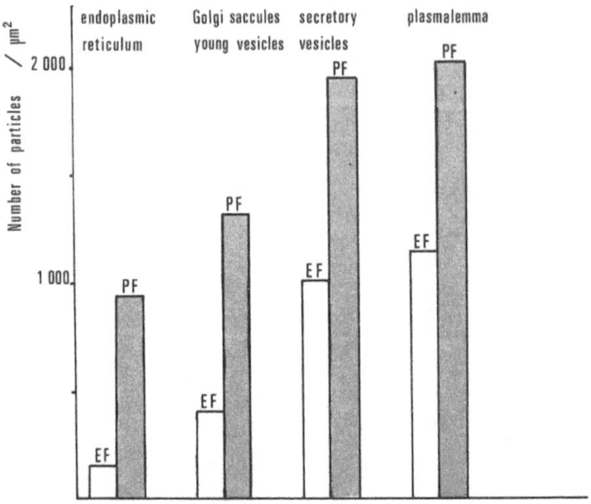

Fig. 7. Diagram showing the increase in particle number of membranes engaged in the membrane flow process in a secretory outer root cap cell of *Pisum sativum*. The number of particles increases strongly from the endoplasmic reticulum and Golgi cisternae or young vesicles to secretory vesicles and plasmalemma. *PF* protoplasmic face; *EF* external face of the various membranes. (With kind permission of B. VIAN; modification of Fig. 11 from VIAN 1974b)

flow when the membranes of the vesicles become cytochemically similar to the plasmalemma.

Additional ultrastructural details of the events preceeding and occurring during the fusion of membrane of the Golgi-derived vesicles to the plasmalemma have been reported by freeze-etching studies of root cap secretory cells (BRANTON and MOOR 1964, VIAN 1972, 1974b). These studies reveal the presence of particles at the surface of dictyosomes, secretory vesicles, and plasmalemma with correlative increasing number (Fig. 7) and demonstrate the process of vesicle discharge. The observations of VIAN (1972) strongly suggest a direct fusion between the membrane of the vesicles with the plasmalemma, but the exact mechanism of membrane fusion process is still unknown (NORTHCOTE 1979).

4.3 Slime Discharge

The fusion of vesicles with the cell surface leads to a release of contents between the plasmalemma and the cell wall. Two possibilities have been suggested for the release of the slime (GREEN and NORTHCOTE 1978): the polysaccharides may be separated from glycoprotein either intracellularly within the vesicles of the dictyosomes or extracellularly after the vesicles have fused with the cell surface. Slime accumulation takes place preferentially between the plasmalemma and what is or will be the outer tangential wall (JUNIPER 1972, JUNIPER and PASK 1973). It is not a consequence of an oriented migration of the secretory vesicles since an apparently random migration has been ascribed from cytological or autoradiographic studies (NORTHCOTE and PICKETT-HEAPS 1966, ROUGIER 1976a). Thus, it seems that peripheral cells possess a perception mechanism able to predict which wall will finally constitute the outer tangential wall (JUNIPER 1972). This perception phenomenon appears to be independent of the orientation of the root and consequently is different from the type of gravity perception observed in core cells.

During the final stages of discharge, the polysaccharide must pass through the outer cell wall. PAULL and JONES (1976b) have suggested that the cell wall of the root cap secretory cells may act as a final barrier to slime release. Progressive degradation of the cell wall leading to increased passage of the accumulated slime through the barrier zone has been suggested. Thus they have correlated changes in structure and staining of the cell walls from the center of the cap to the exterior with increased functioning of hydrolytic enzymes, in particular (1 → 3)-β-glucanase.

4.4 Model of Secretion

Recent biochemical findings detailed above suggest that the biosynthesis and intracellular transport of root cap slime may involve lipid and protein associated

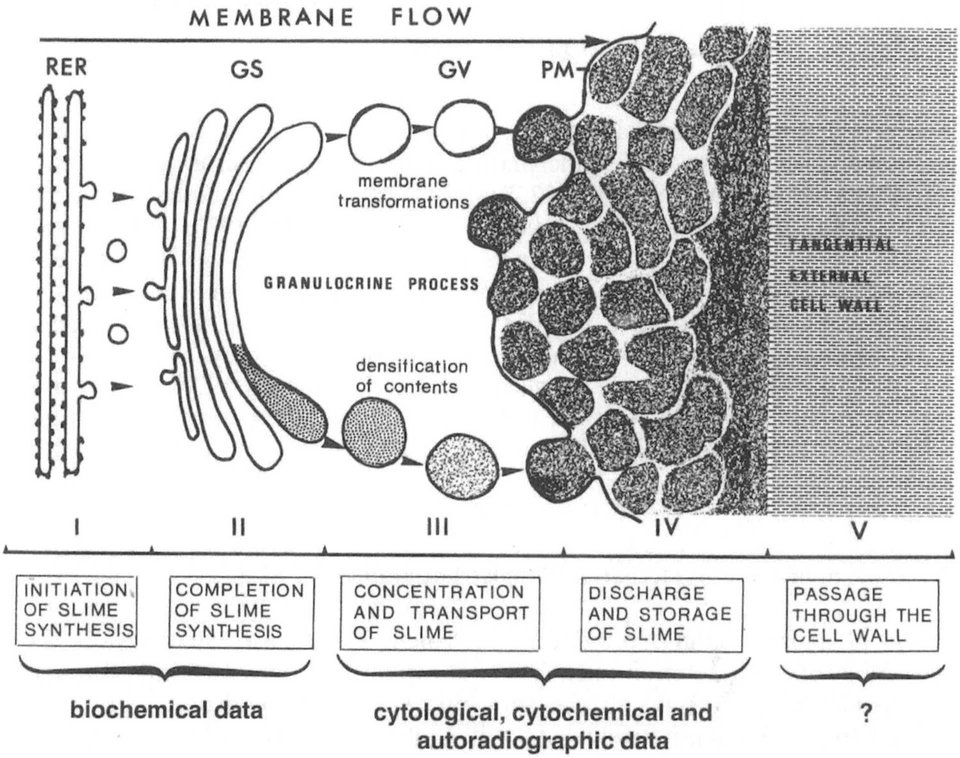

Fig. 8. Model of slime secretion. This general model indicates the intervention of the rough endoplasmic reticulum (*RER*) and Golgi saccules (*GS*) in slime synthesis and illustrates the transport of slime precursors in Golgi vesicles (*GV*) to the plasmalemma (*PM*) via a granulocrine process involving membrane modifications (in pea system) and densification of contents (in maize system). The exported slime is shown to be preferentially stored against the tangential external cell wall. The passage of the exported slime through the cell wall is still an unanswered question

with both the membranes of the endoplasmic reticulum and the Golgi apparatus. Many questions, however, concerning the function of the different membranes still remain to be answered. A granulocrine process via Golgi secretory vesicles is involved in the concentration and transport of the slime precursors from their site of synthesis to the cell surface. During their intracellular migration the secretory vesicles show striking transformations both of their membranes and contents. The secretory products are released from the cell by a reverse pinocytosis but the exact mechanisms of the vesicle membrane fusion to the plasmalemma and release of secretory products are still unknown. The exported material is preferentially stored between the plasmalemma and outer tangential cell wall. The final stage of the secretion process involves the passage of the secretory products through the cell wall facilitated by the progressive degradation of the cell wall. Finally, the secretory product is periodically released from the root cap with the sloughed cells embedded within it. These sequential events are summarized by the scheme in Fig. 8.

5 Physiology of the Secretion Processes

The production of a mucilaginous secretion by plant root caps is a general phenomenon observed either under natural conditions (FLOYD and OHLROGGE 1970, GREAVES and DARBYSHIRE 1972, SAMBIN 1978) or various culture conditions (SAMTSEVICH 1965, MORRÉ et al. 1967, JONES and MORRÉ 1973, PAULL et al. 1975). Formation of secretory products is reported to begin on the first day of seed germination whether the conditions are sterile or nonsterile and at temperatures ranging from 6 to 30° C (SAMTSEVICH 1965). For lateral roots, a massive secretion of polysaccharide on the outside of the root cap does not occur until the roots have emerged (BELL and MACCULLY 1970). The formation of an exudate requires moist conditions (MORRÉ et al. 1967, FLOYD and OHLROGGE 1970) and may be influenced by various factors, as will be considered below.

5.1 Control of the Polysaccharide Droplet Formation and Size

Roots of maize seedlings provide a convenient system for the study of the physiology of slime secretion since their root cap forms a polysaccharide containing droplet which adheres to the root tip. Therefore the droplet formation can be followed quantitatively before it is shed from the root. This system has been successfully utilized both for estimations of the amount of secretory product produced by the root cap (MORRÉ et al. 1967) and for investigations into the effects of various treatments on polysaccharide secretion (MORRÉ et al. 1967, JONES and MORRÉ 1973, MOLLENHAUER et al. 1975, PAULL et al. 1975, MOLLENHAUER and MORRÉ 1976, POPE et al. 1979).

However, both standard and controlled conditions of droplet formation are required for such physiological studies (MORRÉ et al. 1967). In experiments performed by these authors, roots of seedlings grown on miracloth (method II) did not form conspicuous polysaccharide droplets and roots of seedlings grown in petri plates (method I) ceased formation of the polysaccharide droplets if the root tips were not maintained in direct contact with water. Under conditions in which external droplet formation did not occur, the secretory product accumulated between the protoplasts and the cell walls.

Three parameters must be considered when investigations into the size of the polysaccharide droplet are performed: (1) the rate of polysaccharide extrusion, (2) the duration of polysaccharide extrusion, (3) the efficiency of droplet shedding. The rate and duration of polysaccharide extrusion reflect the amount of products secreted, whereas the duration directly influences the efficiency of droplet shedding (JONES and MORRÉ 1973).

Precise calculations of the size and weight of the polysaccharide droplet have been obtained (MORRÉ et al. 1967). In addition to these results, more approximate ones based on the average dimensions of "gel-like caps" and root caps of several crop plants grown in various media have also been reported (SAMTSEVICH 1965).

Both increases in droplet size and stimulation of the rate of slime production may be observed under conditions of increasing temperature or in the presence of various exogenous monosaccharides: e.g., glucose, fucose, and sucrose (JONES and MORRÉ 1973, PAULL et al. 1975). The greatest effect reported (JONES and MORRÉ 1973) is caused by incubation in mannose, since this sugar increases both the rate and duration of polysaccharide extrusion and also affects the droplet shedding. The concentration of mannose that is most effective in stimulating changes in droplet production is surprisingly that which is inhibitory to root growth. In contrast, a decrease in size of the polysaccharide droplet and the droplet formation may be caused by lowering the temperature, incubation in various gaseous mixtures or inhibitors (MORRÉ et al. 1967), and also incubation in certain sugars: galactose, arabinose, and galacturonic acid (JONES and MORRÉ 1973).

5.2 Control of Vesicle Production and Discharge

The rate of polysaccharide droplet formation can be directly correlated to the intensity of the secretory activity of the Golgi apparatus and to the vectorial migration of the secretory vesicles from their sites of formation at the Golgi apparatus to the cell surface. This is clearly demonstrated by ultrastructural observations on modifications of the formation, transport, and discharge of secretion vesicles induced by various treatments (MOLLENHAUER et al. 1975; MOLLENHAUER and MORRÉ 1976, POPE et al. 1979)

MOLLENHAUER et al. (1975) observed that after cold treatment (which reduces the rate of formation of polysaccharide droplets and decreases secretion by at least 75%), secretory vesicles although still attached to the cisternae of the Golgi apparatus, are reduced in size and altered in form. Both the smaller size of vesicles and the increase in their time of retention within the cytoplasm could explain the decrease in size of the polysaccharide droplet associated with cold treatment. Oxygen tension may also affect vesicle production and/or discharge of slime from the protoplast. Studies of root tips treated with cyanide, nitrogen, or CO_2 have indicated that vesicles are rapidly shed from the dictyosome cisternae and no further production of vesicles is evidenced until return to aerobic conditions (WHALEY et al. 1964). Inhibitors of ATP production reduce also the synthesis of polysaccharide and inhibit the formation of secretion vesicles in root cap cells (WHALEY 1966, MORRÉ et al. 1967).

Evidence that the movement of secretory vesicles to the surface is a directed phenomenon is provided by observations that the distribution of Golgi apparatus-derived vesicles become modified after the treatment of maize root caps with cytochalasin B (MOLLENHAUER and MORRÉ 1976, POPE et al. 1979). The vesicles no longer move to the surface but accumulate at or near their sites of formation while the general dictyosome structure and the intercisternal elements of dictyosomes remain unaffected. These findings imply that intracellular transport may be mediated via a cytochalasin B-sensitive mechanism. It seems unlikely, however, that microtubules are also involved, since colchicine does not affect the distribution of the secretory vesicles (MOLLENHAUER and MORRÉ 1976).

5.3 Characteristics of the Secretory Cycle

Secretion in the maize root tip follows a 3-h cycle.

The periodic nature of droplet formation was first observed in kinetic experiments performed by MORRÉ et al. (1967). The periodicity has been ascribed to the release of waves of Golgi-derived vesicles (JONES and MORRÉ 1973) or, more plausibly, to waves of cytokinesis in the cap initials which produce a periodicity in cell production (PAULL and JONES 1976b). The length of the period and operation of the cycle was found to be independent of temperature but associated with a cyanide-sensitive active phase of secretion. The existence of the cycle and its independence from temperature was later confirmed by MOLLENHAUER et al. (1975). The data reported by PAULL and JONES (1976b) also confirm the periodicity. They measured the release of carbohydrate material containing hexose and fucose residues from the root of single seedlings and the incorporation of L-fucose-[^{3}H] into alcohol-precipitable material of individual roots. In addition to cycles of slime secretion, this study also indicates that cell sloughing is cyclic having a 2 to 3 h periodicity. Hence a secretory cell must undergo several secretory cycles before it is sloughed off from the root, and the number of these cycles should be considered in a calculation of the total life time of the cells (see Sect. 2).

It has been possible to separate secretion into two distinct phases (MORRÉ et al. 1967). The active phase corresponds to vesicle production which is temperature-dependent and sensitive to metabolic inhibitors. The passive phase corresponds to the extrusion of polysaccharide following discharge from the protoplast. The passive process, which seems to be unaffected by temperature changes, is dependent upon the degree of hydration of the polysaccharide and the turgor pressure of the cell.

5.4 Participation of Enzymes in the Secretion Processes

Enzymes involved in polysaccharidic secretion are difficult to study in the root cap system since it is impossible to specifically isolate cell organelles of the secretory cells and to identify their characteristic enzymes. Nevertheless, some enzymes can be identified in situ cytochemically. Apart from those enzymes which exhibit cell wall degrading activities (see Sect. 3) or the enzymes acting at the root–soil interface (see Sect. 6), several others have been suggested to participate more or less directly in the secretion processes. DAUWALDER et al. (1969) and GOFF (1973) have localized inosine- and other nucleoside-diphosphatases and thiamine pyrophosphatase in the Golgi apparatus. It has been suggested that the nucleoside diphosphatase may represent inactivated glycosylsynthetases or transferases and therefore may reflect participation of that organelle in the synthesis of the polysaccharide (GOFF 1973).

6 Function of the Root Cap Slime

6.1 Slime as a Constituent of Mucigel

In contrast to slime which may be regarded as a secretory product of root cap cells, the epidermal cells of the whole tip region of the root have also been shown to synthesize a mucilaginous substance which extends backward from the root tip for several centimeters (JENNY and GROSSENBACHER 1963, DART and MERCER 1964, LEISER 1968, BRAMS 1969, GREAVES and DARBYSHIRE 1972, LEPPARD 1974, OLD and NICOLSON 1975, FOSTER and ROVIRA 1976, FELIPE ANTON and LOPEZ FANDO 1977, WERKER and KISLEV 1978). This mucilaginous layer which ensheaths the root appears to be a true gel (GREENLAND 1979) allowing free diffusion of water and ions through it; it is mainly composed of polysaccharides with a pectin-like composition (LEPPARD 1974). Both the slime and epidermal secretion have been discussed under the general term mucigel by most of the authors studying the root-soil interface (see the review of BALANDREAU and KNOWLES 1978). The mucigel is thought to be involved in the various processes taking place at the root–soil interface (GREENLAND 1979), in particular playing a prominent role in the rhizosphere (BALANDREAU and KNOWLES 1978).

6.2 Function of the Mucigel at the Root Level

HABERLANDT (1914) originally suggested that the root cap slime functions both as a lubricant and a protectant. The ability of mucigel to protect younger parts of the root from dessication during short periods of drought has been mentioned more recently (LEISER 1968, BOWEN and ROVIRA 1976, GREENLAND 1979).

The presence of mucigel around the roots is considered to have important implications on the transport of nutrients from the soil to the root (JENNY and GROSSENBACHER 1963, FLOYD and OHLROGGE 1970). For example, the mucigel may increase the availability of ions to the root since the COO^- groups of the pectin-like structures may play an important role in affecting cation exchange and diffusion and thereby indirectly affecting plant nutrition (JENNY and GROSSENBACHER 1963, LEPPARD and RAMAMOORTHY 1975). The finding of acid phosphatase in the mucigel (FLOYD and OHLROGGE 1971) may also favor plant nutrition since organic phosphate esters in the soil may be hydrolyzed by this enzyme into inorganic phosphate which could then be assimilated. CLARKSON and SANDERSON (1969) have also shown that the mucigel is able to immobilize certain toxic ions.

6.3 Function of Mucigel at the Root-Soil Interface

6.3.1 Slime and Sloughed Cells as a Source of Organic Carbon and Nitrogen in the Rhizosphere

The use of radioactive labeling of plants in vivo with ^{14}C (ROVIRA 1972, 1973) has allowed identification of carbon-containing materials in the root exudates. There are three major forms of exudation from seedlings: soluble compounds, water-insoluble material made up of polysaccharides and sloughed cells and volatile compounds exuded in the relative amounts of 1:4:8 (ROVIRA 1972). The polysaccharide material secreted by root cap cells constitutes a major component of the organic material released from roots.

Various estimations of sloughed organic material including calculation of the root cap sloughed cells and determination of their C, N, and H have been made in peanut plants growing in axenic cultures (GRIFFIN et al. 1976). The results indicate that 0.26 to 0.73 mg of sloughed organic matter are produced/plant/week, and approximately 0.15% of the total root carbon, nitrogen, and hydrogen is sloughed per week. Root caps have been identified as major components of the sloughed matter. The mass of root cap slime and associated sloughed cells produced by one plant growing in soil during its entire vegetation period per hectare has been calculated by SAMTSEVICH (1965, 1971). Using the assumption that slime is produced by all the root tips during their entire growth period, he calculated that the total mass per hectare is 300 m^3 for barley, 700 m^3 for winter wheat, and 1,250 m^3 for maize. Perennial plants whose roots continue growth for a longer period of the year than annuals are assumed to excrete more slime material into the soil.

More precise estimates of the amount of sloughed cells exclusively shed by a root cap are reported in several studies on cap renewal time (see Sect. 2). In the maize system studied by CLOWES (1971, 1976) and CLOWES and WOOLSTON (1978) the number of cells within the cap of a primary root 10 to 30 mm long and released by this tissue each day varies from 3,900 to 20,900 (CLOWES 1976). CLOWES calculated that an average of 12,080 cells are produced by a maize primary root cap during one day. MACLEOD (1976) has also investigated the number of cells sloughed from each secondary root cap of *Vicia faba* into the soil every day and estimated the number to be 420 to 636. From that result, they extrapolated further and suggested that one 11-day old *Vicia* plant would release 56,000 to 85,000 cap cells to the rhizosphere during its growth period from germination.

From these very varied conclusions drawn from studies on single, primary, secondary roots, and from entire root systems, it may be ascertained that sloughed cells embedded in slime enrich the rhizosphere soil.

6.3.2 Action of the Mucigel on Soil Aggregation and Stability

Intimate contact between the root mucigel and minerals was demonstrated in an early report (JENNY and GROSSENBACHER 1963) and more recently with the use of the electronmicroscope (CAMPBELL and ROVIRA 1973, BREISCH 1974, ROVIRA and CAMPBELL 1974, BREISCH et al. 1975, GUCKERT et al. 1975). The scanning electronmicroscope appears to be particularly useful for studying the root–soil interface (DART 1971, CAMPBELL and ROVIRA 1973, ROVIRA and CAMPBELL 1974). Its use reveals that soil particles (clay and silt) are embedded in the cell debris and polysaccharide substances produced by the wheat root cap (ROVIRA and CAMPBELL 1974). Detailed in situ relationships between the slime of maize and clays may also be observed by conventional transmission electronmicroscopic techniques (BREISCH 1974, BREISCH et al. 1975, GUCKERT et al. 1975). Under experimental conditions using a montmorillonite clay substrate (BREISCH 1974, GUCKERT et al. 1975) or under natural environmental conditions (BREISCH et al. 1975), clay layers are seen at the surface of the slime or embedded into it. These ultrastructural observations differ from the results of X-ray diffraction studies performed on films of clay and clay-maize nodal root exudate complexes (FLOYD and OHLROGGE 1971).

Such associations between slime products and soil particles are generally thought to provide better root aggregation and stability (ROVIRA 1972, GUCKERT et al. 1975 for example). TURCHENEK and OADES (1976) consider that the adsorption of fine clay particles to mucigel constituents is the first step in the formation of stable microaggregates around plant roots. According to JENNY and GROSSENBACHER (1963), the closeness of the mucigel and mineral particles suggest chemical interaction with the participation of Al-COO^- complexes and H^+ ions.

6.3.3 Role of Root Cap Slime on Microbial Colonization

Thorough reviews (BOWEN and ROVIRA 1976, ROVIRA 1976) have recently been published on the microbial colonization of plant roots. Consequently, this discussion will be limited to an examination of data related to the colonization of root cap slime by microorganisms and to possible significance of this colonization for the rhizosphere.

Unfortunately, the data collected during the last two decades is totally inconclusive. Some investigators have claimed that the root tip is almost devoid of microorganisms (ROVIRA 1973, ROVIRA and CAMPBELL 1974) whereas, in complete contrast, others have demonstrated that microorganisms may colonize

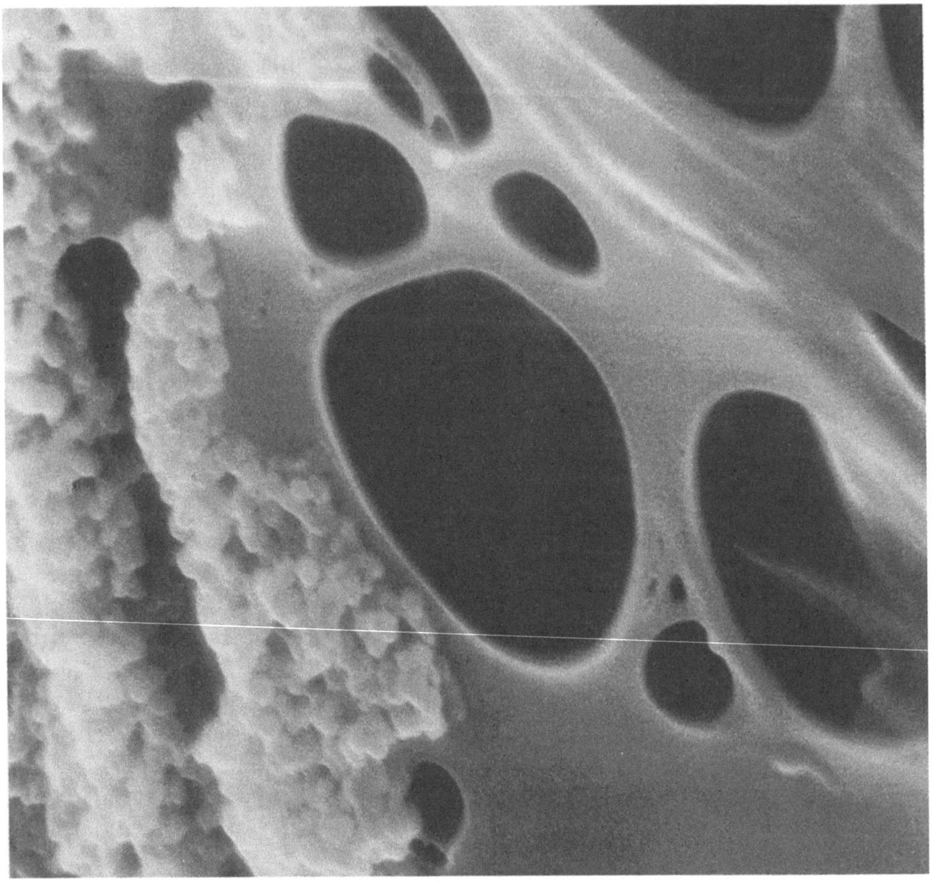

Fig. 9. A scanning electronmicroscope picture of the root cap slime of rice colonized by *Beijerinckia* sp. Magnification ×120,000. (Reproduced with kind permission from BALANDREAU 1975)

the slime at the root cap level (BREISCH 1974, BREISCH et al. 1975, GUCKERT et al. 1975, MOTA et al. 1975, BALANDREAU and KNOWLES 1978). The presence of microorganisms has been illustrated both at the surface of the root cap slime or within it. These observations involve roots collected directly from plants growing in soil (ROVIRA 1972, BREISCH 1974, BREISCH et al. 1975, GUCKERT et al. 1975) or from roots germinated under various experimental conditions and inoculated with one or several well known selected bacteria (DART and MERCER 1964, GREAVES and DARBYSHIRE 1972, HAMAD-FARES 1976, UMALI-GARCIA et al. 1979, and Fig. 9). From transmission electronmicroscopic studies, bacteria are seen to penetrate the mucigel and to be sometimes surrounded by an electron-transparent zone which has been suggested to result from the slime degradation (BREISCH 1974, BREISCH et al. 1975). One possible interpretation of this data is that slime can represent a food source for bacteria living in the rhizosphere.

7 Concluding Remarks

Although root cap secretory activity has been extensively studied during the last 20 years by means of complementary methods of investigation and with the use of *Zea mays* as a standard system, the subject is still open for further investigations. The knowledge resulting from various results obtained on maize needs to be extended to other plant systems in which information is still lacking on such topics as slime composition, slime biosynthesis and physiology of slime secretion.

Even using maize, which may indeed represent a favorable system for studying: (1) the cytology and functioning of the root cap, (2) the slime composition, (3) the biosynthetic events associated with the membrane flow process involved in slime secretion, future research needs to be done. Among potential future investigations, certain aspects are important to consider. These include those related to increasing our knowledge of the biosynthetic mechanism of slime and of its modifications and transport through the cell wall after release at the plasmalemma. In addition, the function of the root cap slime as a component of mucigel present at the root–soil interface needs to be reinvestigated and understood within the context of rhizosphere colonization.

Acknowledgment. Thanks are due to Dianna Bowles for correcting the English of the manuscript.

References

Balandreau J (1975) Activité nitrogénasique dans la rhizosphère de quelques graminées. Thèse Doct Etat, Univ Nancy I

Balandreau J, Knowles R (1978) The rhizosphere. In: Dommergues YR, Krupa SV (eds) Interactions between non-pathogenic soil microorganisms and plants. Elsevier Scientific Publishing Company, Amsterdam Oxford New York, pp 243–268

Barlow PW (1974) Regeneration of the cap of primary roots of *Zea mays*. New Phytol 73:937–954

Barlow PW (1975) The root cap. In: Torrey JG, Clarkson DT (eds) The development and function of roots. Academic Press, London New York, pp 21–54

Barlow PW (1976) The integrity and organization of nuclear DNA in cells of the root cap of *Zea mays* probed by terminal deoxynucleotidyl transferase and microdensitometry. Z Pflanzenphysiol 80:271–278

Barlow PW (1978) Cell displacement through the columella of the root cap of *Zea mays* L. Ann Bot 42:783–790

Bell JK, MacCully ME (1970) A histological study of lateral root initiation and development in *Zea mays*. Protoplasma 70:179–205

Berjak P (1968) A lysosome-like organelle in the root cap of *Zea mays*. J Ultrastruct Res 23:233–242

Bowen GD, Rovira AD (1976) Microbial colonization of plant roots. Annu Rev Plant Pathol 14:121–144

Bowles DJ, Northcote DH (1972) The sites of synthesis and transport of extracellular polysaccharides in the root tissues of maize. Biochem J 130:1133–1145

Bowles DJ, Northcote DH (1974) The amounts and rates of export of polysaccharides found within the membrane system of maize root cells. Biochem J 142:139–144

Bowles DJ, Northcote DH (1976) The size and distribution of polysaccharides during their synthesis within the membrane system of maize root cells. Planta 128:101–106

Brams E (1969) The mucilaginous layer of *Citrus* roots. Its delineation in the rhizosphere and removal from roots. Plant Soil 30:105–108

Branton D, Moor H (1964) Fine structure in freeze-etched *Allium cepa* L. root tips. J Ultrastruct Res 11:401–411

Breisch H (1974) Contribution à l'étude du rôle des exsudats racinaires dans les processus d'agrégation des sols. Doct Spécialité Agronomie, Univ Nancy I, pp 1–74

Breisch H, Guckert A, Reisinger O (1975) Etude au microscope électronique de la zone apicale des racines de maïs. Soc Bot Coll Rhizosphere 122:55–60

Brisson JD, Peterson RL, Robb J, Rauser WE, Ellis BE (1977) Correlated phenolic histochemistry using light, transmission, and scanning electron microscopy, with examples taken from phytopathological problems. Scanning Electron Microsc 2:667–676

Campbell R, Rovira AD (1973) The study of the rhizosphere by scanning electron microscopy. Soil Biol Biochem 5:747–752

Clarkson DT, Sanderson J (1969) The uptake of a polyvalent cation and its distribution in the root apices of *Allium cepa*: Tracer and autoradiographic studies. Planta 89:136–154

Clowes FAL (1971) The proportion of cells that divide in root meristem of *Zea mays*. Ann Bot 35:249–251

Clowes FAL (1976) Cell production by root caps. New Phythol 77:399–407

Clowes FAL, Juniper BE (1964) The fine structure of the quiescent centre and neighbouring tissues in root meristems. J Exp Bot 15:622–630

Clowes FAL, Woolston RE (1978) Sloughing of root cap cells. Ann Bot 42:83–89

Coulomb P, Coulon J (1971) Fonctions de l'appareil de Golgi dans les méristèmes radiculaires de la courge (*Cucurbita pepo* L. Cucurbitacée). J Microsc 10:203–214

Dart PJ (1971) Scanning electron microscopy of plant roots. J Exp Bot 22:163–168

Dart PJ, Mercer FV (1964) The legume rhizosphere. Arch Mikrobiol 47:344–378

Dauwalder M, Whaley WG (1973) Staining of cells of *Zea mays* root apices with the osmium-zinc iodide and osmium impregnation techniques. J Ultrastruct Res 45:279–296

Dauwalder M, Whaley WG (1974) Patterns of incorporation of (^{3}H) galactose by cells of *Zea mays* root tips. J Cell Sci 14:11–27

Dauwalder M, Whaley WG, Kephart JE (1969) Phosphatases and differentiation of the Golgi apparatus. J Cell Sci 4:455–497

Dumas C, Perrin A, Rougier M, Zandonella P (1974) Some ultrastructural aspects of different vegetal glandular tissues. Port Acta Biol 14:501–520

Felipe Anton MR, Lopez-Fando C (1977) Gel formation on seminal root surface as seen at electron microscope. Agrochimica 21:305–310

Floyd RA, Ohlrogge AJ (1970) Gel formation on nodal root surfaces of *Zea mays*. I. Investigation of the gel's composition. Plant Soil 33:331–343

Floyd RA, Ohlrogge AJ (1971) Gel formation on nodal root surface of *Zea mays*. Some observations relevant to understanding its action at the root-soil interface. Plant Soil 34:595–606

Foster RC, Rovira AD (1976) Ultrastructure of wheat rhizosphere. New Phytol 76:343–352

Gill AM, Tomlinson PB (1975) Aerial roots: an array of forms and functions. In: Torrey JG, Clarkson DT (eds) The development and function of roots. Academic Press, London New York, pp 238–260

Goff CW (1973) Localization of nucleoside diphosphatase in the onion root tip. Protoplasma 78:397–416

Grant GT, McNab C, Rees DA, Skerrett RJ (1969) Seed mucilages as examples of polysaccharide denaturation. Chem Commun 805–806

Greaves MP, Darbyshire JF (1972) The ultrastructure of the mucilaginous layer on plant roots. Soil Biol Biochem 4:443–449

Green JR, Northcote DH (1978) The structure and function of glycoproteins synthesized during slime-polysaccharide production by membranes of the root-cap cells of maize (*Zea mays*). Biochem J 170:599–608

Green JR, Northcote DH (1979a) Polyprenyl phosphate sugars synthesized during slime-polysaccharide production by membranes of the root-cap cells of maize (*Zea mays*). Biochem J 178:661–671

Green JR, Northcote DH (1979b) Location of fucosyl transferases in the membrane system of maize root cells. J Cell Sci 40:235–244
Greenland DJ (1979) The physics and chemistry of the soil-root interface: some comments. In: Harley JL, Russell RS (eds) The soil-root interface. Academic Press, London New York, pp 83–98
Griffin GJ, Hale MG, Shay FJ (1976) Nature and quantity of sloughed organic matter produced by roots of axenic peanut plants. Soil Biol Biochem 8:29–32
Guckert A, Breisch H, Reisinger O (1975) Interface sol-racine-I. Etude au microscope électronique des relations mucigel-argile-microorganismes. Soil Biol Biochem 7:241–250
Haberlandt G (1974) Physiological plant anatomy, 4th German edn (translation by M Dummond), Mac Millan, London
Hall D, Mollenhauer HH, Morre DJ (1966) Evidence for secretion of cell dispersing enzymes from maize root cap and epidermis. Am J Bot 75:p65
Hamad-Fares I (1976) La fixation de l'azote dans la rhizosphère du riz. Réalisation d'un modèle gnotobiotique. Thèse Doct Etat, Univ Nancy I
Harkes PAA (1973) Structure and dynamics of the root cap of *Avena sativa* L. Acta Bot Neerl 22:321–328
Harkes PAA (1976) Organization and activity of the root cap meristem of *Avena sativa* L. New Phytol 76:367–375
Harris PJ, Hartley RD (1976) Detection of bound ferulic acid in cell walls of the gramineae by ultraviolet fluorescence microscopy. Nature (London) 259:508–510
Harris PJ, Northcote DH (1970) Patterns of polysaccharide biosynthesis in differentiating cells of maize root-tips. Biochem J 120:479–491
Hemming FW (1974) Lipids in glycan biosynthesis. In: Goodwin TW (ed) Biochemistry of lipids, 4 series one: Univ Park Press Baltimore, pp 39–97
James DW Jr, Jones RL (1979a) Characterization of GDP-fucose polysaccharide fucosyl transferase in corn roots (*Zea mays* L.). Plant Physiol 64:909–913
James DW Jr, Jones RL (1979b) Intracellular localization of GDP-fucose polysaccharide fucosyl transferase in corn roots (*Zea mays* L.). Plant Physiol 64:914–918
Jenny H, Grossenbacher K (1963) Root-soil boundary zones as seen in the electron microscope. Proc Soil Sci Soc Am 27:273–277
Jones DD, Morré DJ (1967) Golgi apparatus mediated polysaccharide secretion by outer root cap cells of *Zea mays*. II. Isolation and characterization of the secretory product. Z Pflanzenphysiol 56:166–169
Jones DD, Morré DJ (1973) Golgi apparatus mediated polysaccharide secretion by outer root cap cells of *Zea mays*. III. Control by exogenous sugars. Physiol Plant 29: 68–75
Juniper BE (1972) Mechanisms of perception and pattern of organisation in root caps. In: Miller MW, Kuehnert CC (eds) The dynamics of meristem cell populations. Plenum Publishing Corporation, New York London, pp 119–131
Juniper BE, Barlow PW (1969) The distribution of plasmodesmata in the root tip of maize. Planta 89:352–360
Juniper BE, Clowes FAL (1965) Cytoplasmic organelles and cell growth in root caps. Nature (London) 208:864–865
Juniper BE, French A (1970) The fine structure of the cells that perceive gravity in the root tip of maize. Planta 95:314–329
Juniper BE, French A (1973) The distribution and redistribution of endoplasmic reticulum (ER) in geoperceptive cells. Planta 109:211–224
Juniper BE, Pask G (1973) Directional secretion by the Golgi bodies in maize root cells. Planta 109:225–231
Juniper BE, Roberts RM (1966) Polysaccharide synthesis and the fine structure of root cells. JR Microsc Soc 85:63–72
Juniper BE, Gilchrist AJ, Robins RJ (1977) Some features of secretory systems in plants. Histochem J 9:659–680
Kirby EG, Roberts RM (1971) The localized incorporation of ^{3}H-L-fucose into cell wall polysaccharides of the cap and epidermis of corn roots. Autoradiographic and biosynthetic studies. Planta 99:211–221

Kosuge T (1969) The role of phenolics in host response to infection. Annu Rev Phytopathol 7:195–222
Lechene De La Porte P (1976) Différenciation des cellules de la coiffe de pois et d'orge. Etude morphologique et cytochimique des modifications des dictyosomes. Ann Sci Nat 12 Ser 17:345–356
Leiser AT (1968) A mucilaginous root shealth in Ericaceae. Am J Bot 55:391–398
Leppard GG (1974) Rhizoplane fibrils in wheat: demonstration and derivation. Science 185:1066–1067
Leppard GG, Ramamoorthy S (1975) The aggregation of wheat rhizoplane fibrils and the accumulation of soil-bound cations. Can J Bot 53:1729–1735
MacLeod RD (1976) Cap formation during the elongation of lateral roots of *Vicia faba* L. Ann Bot 40:877–885
Maitra SC, De DN (1972) Ultrastructure of root cap cells: formation and utilization of lipid. Cytobios 5:111–118
Martin EM, Harris WM (1976) Adventitious root development from the coleoptilar zone in *Zea mays* L. Am J Bot 63:890–897
Mollenhauer HH (1965a) Transition forms of Golgi apparatus secretion vesicles. J Ultrastruct Res 12:439–446
Mollenhauer HH (1965b) An intercisternal structure in the Golgi apparatus. J Cell Biol 24:504–511
Mollenhauer HH (1967) A comparison of root cap cells of epiphytic, terrestrial and aquatic plants. Am J Bot 54:1249–1259
Mollenhauer HH, Morré DJ (1966) Tubular connections between dictyosomes and forming secretory vesicles in plant Golgi apparatus. J Cell Biol 29:373–376
Mollenhauer HH, Morré DJ (1974) Polyribosomes associated with the Golgi apparatus. Protoplasma 79:333–336
Mollenhauer HH, Morré DJ (1975) A possible role for intercisternal elements in the formation of secretory vesicles in plant Golgi apparatus. J Cell Sci 19:231–237
Mollenhauer HH, Morré DJ (1976) Cytochalasin B, but no colchicine, inhibits migration of secretory vesicles in root tips of maize. Protoplasma 87:39–48
Mollenhauer HH, Whaley WG (1963) An observation of the functioning of the Golgi apparatus. J Cell Biol 17:222–225
Mollenhauer HH, Whaley WG, Leech JH (1961) A function of the Golgi apparatus in outer root cap cells. J Ultrastruct Res 5:193–200
Mollenhauer HH, Morré DJ, Vanderwoude WJ (1975) Endoplasmic reticulum-Golgi apparatus associations in maize root tips. Mikroskopie 31:257–272
Morré DJ, Jones DD, Mollenhauer HH (1967) Golgi apparatus mediated polysaccharide secretion by outer root cap cells of *Zea mays*. I. Kinetics and secretory pathway. Planta 74:286–301
Mota M, Silva MT, Salema R (1975) Electron microscopic study of bacteria in the intercellular spaces of the root cap of *Luzula purpurea* Link. J Submicr Cytol 7:373–378
Nicolson GL (1978) Ultrastructural localization of lectin receptors. In: Koehler JK (ed) Advanced techniques in biological electron microscopy. Springer, Berlin Heidelberg New York, pp 1–30
Northcote DH (1979) The involvement of the Golgi apparatus in the biosynthesis and secretion of glycoproteins and polysaccharides. Biomembranes 10:51–76
Northcote DH, Pickett-Heaps JD (1966) A function of the Golgi apparatus in polysaccharide synthesis and transport in the root-cap cells of wheat. Biochem J 98:159–167
Old KM, Nicolson TH (1975) Electron microscopical studies of the microflora of roots of sand dune grasses. New Phytol 74:51–58
Paull RE, Jones RL (1975a) Studies on the secretion of maize root cap slime. II. Localization of slime production. Plant Physiol 56:307–312
Paull RE, Jones RL (1975b) Studies on the secretion of maize root cap slime. III. Histochemical and autoradiographic localization of incorporated fucose. Planta 127:97–110
Paull RE, Jones RL (1976a) Studies on the secretion of maize root cap slime. IV. Evidence for the involvement of dictyosomes. Plant Physiol 57:249–256
Paull RE, Jones RL (1976b) Studies on the secretion of maize root cap slime. V. The cell wall as a barrier to secretion. Z Pflanzenphysiol 79:154–164

Paull RE, Johnson CM, Jones RL (1975) Studies on the secretion of maize root cap slime. I. Some properties of the secreted polymer. Plant Physiol 56:300–306

Peterson RL, Brisson JD (1977) Root cap structure in the fern *Ophioglossum petiolatum*: light and electron microscopy. Can J Bot 55:1861–1878

Phillips HL, Torrey JG (1971) Deoxyribonucleic acid synthesis in root cap cells of cultured roots of *Convolvulus*. Plant Physiol 48:213–218

Phillips HL, Torrey JG (1974) The ultrastructure of the root cap in cultured roots of *Convolvulus arvensis* L. Am J Bot 61:879–887

Pickett-Heaps JD (1967a) The use of autoradiography for investigating wall secretion in plant cells. Protoplasma 64:49–66

Pickett-Heaps JD (1967b) Preliminary attempts at ultrastructural polysaccharide localization in root tip cells. J Histochem Cytochem 15:442–455

Pickett-Heaps JD (1968) Further ultrastructural observations on polysaccharide localization in plant cells. J Cell Sci 3:55–64

Pope DG, Thorpe JR, Al-Azzawi MJ, Hall JL (1979) The effect of cytochalasin B on the rate of growth and ultrastructure of wheat coleoptiles and maize roots. Planta 144:373–383

Roland JC (1974) Cytochimie des polysaccharides végétaux: détection et extraction sélectives. J Microsc 21:233–244

Roth J (1978) The lectins. Molecular probes in cell biology and membrane research. Exp Pathol Suppl 3:1–186

Rougier M (1971) Etude cytochimique de la sécrétion des polysaccharides végétaux à l'aide d'un matériel de choix: les cellules de la coiffe de *Zea mays*. J Microsc 10:67–82

Rougier M (1975) Incorporation de fucose tritié dans l'apex radiculaire de maïs. J Microsc 23:73a–74a

Rougier M (1976a) Sécrétion de polysaccharides dans l'apex radiculaire de maïs: étude radioautographique par incorporation de fucose tritié. J Microsc 26:161–166

Rougier M (1976b) Méthodes et perspectives d'étude des sécrétions végétales. Soc Bot Coll Sécrét Veg 123:7–18

Rougier M, Vian B, Gallant D, Roland JC (1973) Aspects cytochimiques de l'étude ultrastructurale des polysaccharides végétaux. Ann Biol 12:43–75

Rougier M, Kieda C, Monsigny M (1979a) Use of lectin to detect the sugar components of maize root cap slime. J Histochem Cytochem 27:878–881

Rougier M, Chaboud A, Kieda C, Monsigny M (1979b) Distribution and ultrastructural localization of fucose-containing glycoconjugates at the surface of slime secretory root cap cells. Biol Cell 36:9A

Rovira AD (1972) Studies on the interactions between plant roots and micro-organisms. J Austr Inst Agr Sc 6:91–94

Rovira AD (1973) Zones of exudation along plant roots and spatial distribution of micro-organisms in the rhizosphere. Pestic Sci 4:361–366

Rovira AD (1976) Biology of the soil-root interface. In: Harley JL, Scott Russel R (eds) The soil-root interface. Academic Press, London New York, pp 145–160

Rovira AD, Campbell R (1974) Scanning electron microscopy of micro-organisms on the roots of wheat. Microbiol Ecol 1:15–23

Sambin B (1978) Etude de l'influence du milieu sur la structure et le fonctionnement de la coiffe racinaire. Thèse Doct Spécialité, Univ Lyon I

Sambin B, Rougier M, Zandonella P (1978) Renewal of the root cap of *Cissus sicyoïdes*. In: Riedacker A, Gagnaire-Michard J (eds) Symposium Root Physiol Symbiosis, Nancy, 11–15, Septembre 1978, pp 160–170

Samtsevich SA (1965) Active excretions of plant roots and their significance. Fiziol Rast 12:837–846

Samtsevich SA (1971) Root excretions of plants. An important source of humus formation in the soil. Humus Planta 5:147–154

Street HE, Öpik H, James FEL (1967) Fine structure of the main axis meristems of cultured tomato roots. Phytomorphology 17:391–401

Sueiro SF, Felipe Anton MR (1974) Localizacion histoquimica de fosfatasa acida en el mucilago de raices de plantas de maiz crecidas en soluciones nutritivas. An Edaf Agrobiol 33:199–214

Thiery JP (1967) Mise en évidence des polysaccharides sur coupes fines en microscopie électronique. J Microsc 6:987–1017
Turchenek LW, Oades JM (1976) Organo-clay particles in soils. In: Emerson WW, Dexter AR (eds) Modification of soil structure. Wiley, Chichester, pp 137–144
Umali-Garcia M, Rubbell DH, Gaskins MH, Dazzo FB (1979) Adsorption and mode of entry of *Azospirillum brasilense* to grass roots. Int Workshop Assoc N_2-fixation 2–6 July 1979, Cena, Piracicaba, Brasil, p 6
Vian B (1972) Aspects, en cryodécapage, de la fusion des membranes des vésicules cytoplasmiques et du plasmalemme lors des phénomènes de sécrétion végétale. CR Acad Sc 275:2471–2474
Vian B (1974a) Recherches sur les relations ultrastructurales et ontogéniques entre plasmalemme et paroi dans les cellules végétales en croissance. Thèse Doct Etat, Université Paris VI
Vian B (1974b) Précisions fournies par le cryodécapage sur la restructuration et l'assimilation au plasmalemme des membranes des dérivés golgiens. CR Acad Sci 278:1483–1486
Vian B, Roland JC (1972) Différenciation des cytomembranes et renouvellement du plasmalemme dans les phénomènes de sécrétion végétales. J Microsc 13:119–136
Vian B, Roland JC (1974) Cytochemical and ultrastructural observations on polysaccharides during secretion and exocytosis. Port Acta Biol 14:1–6
Werker E, Kislev M (1978) Mucilage on the root surface and root hairs of *Sorghum*: heterogeneity in structure, manner of production and site of accumulation. Ann Bot 42:809–816
Whaley WG (1966) Proposals concerning replication of the Golgi apparatus. In: Sitte P (ed) Funktionelle und morphologische Organisation der Zelle. Probleme der biologischen Reduplikation, 3 Wiss Konf Ges Dtsch Naturforsch Ärzte. Springer, Berlin Heidelberg New York, pp 340–370
Whaley WG, Kephart JE, Mollenhauer HH (1959) Developmental changes in the Golgi apparatus of maize root cells. Am J Bot 46:743–751
Whaley WG, Kephart JE, Mollenhauer HH (1964) The dynamics of cytoplasmic membranes during development. In: Locke M (ed) Cellular membranes in development. 22nd Symp Soc Study Dev Growth. Academic Press, London New York pp 135–174
Wright K (1975) Polysaccharides of root-cap slime from five maize varieties. Phytochemistry 14:759–763
Wright K, Northcote DH (1974) The relationship of root-cap slimes to pectins. Biochem J 139:525–534
Wright K, Northcote DH (1975) An acidic oligosaccharide from maize slime. Phytochemistry 14:1793–1798
Wright K, Northcote DH (1976) Identification of β-1-4-glucan chains as part of a fraction of slime synthesized within the dictyosomes of maize root caps. Protoplasma 88:225–239
Wright K, Northcote DH, Davey R (1976) Preparation of rat epididymal α-L-fucosidase free from other glycosidases; its action on root-cap slime from *Zea mays* L. Carbohydr Res 47:141–150

IV. Cell Surface Phenomena

23 Defined Components Involved in Pollination[1]

A.E. CLARKE

1 Introduction

Our understanding of fertilization has progressed from the early microscopic observations which led to definition of the biology of pollen tube growth and fertilization, through cytochemical examination of pollen and stigmas which provided evidence for the presence of certain classes of components, and finally to the isolation and structural characterization of a few of these components.

At present the information is fragmentary; there is good data on the structure of a few components of a few pollination systems, but no complete analysis of the interacting surfaces of any one system. Ultimately, definition of the molecular basis of fertilization depends on this kind of information. It is a formidable task as it involves analysis of stigma surface and stylar secretions, which are comparable in complexity with gums and other plant exudates, but there is the additional problem of collecting sufficient material for examination. Similarly with the pollen grains, the wall components which make contact with the female sexual tissues are likely to be more complex than those of a primary cell wall. Even with the availability of large amounts of material for analysis, our view of the primary cell wall at the molecular level is far from complete, so it is not surprising that our knowledge of the composition of the pollen cell wall layers is less comprehensive. In this section, structural features of some components which have been isolated and subjected to detailed chemical analysis, are reviewed. In no case can any specific function be unequivocally ascribed to these components. Many other components have been identified cytochemically and partly characterized.

2 Arabinogalactans as Pistil Components of *Gladiolus* and *Lilium*

Arabinogalactans are major components of both the stigma surface secretion and the style canal mucilage of mature *Gladiolus* pistils. These components were initially detected cytochemically by staining with the β-glucosyl artificial carbohydrate antigen (CLARKE et al. 1978) which is prepared by coupling diazotized 4-amino phenyl glucoside to phloroglucinol (YARIV et al. 1962). This material precipitates arabinogalactans from a wide range of plant extracts (YARIV et al. 1967, JERMYN and YEOW 1975, CLARKE et al. 1978, 1979). The brilliant red color of this and other β-glycosyl artificial antigens makes them useful

1 A general treatment of the pollen-pistil interactions is reserved for a forthcoming volume of this Series

cytochemical reagents and gel stains for detecting arabinogalactans. The nature of the interaction between the β-glucosyl artificial antigen and arabinogalactans is not precisely defined; however the artificial antigen must bear a glucopyranose residue with a β-D-configuration at C(*O*)1 and the D-gluco-configuration at C(*O*)2 and the 1:4 orientation of the azo and glycosyl oxy groups to the phenyl ring is required (JERMYN 1978).

The requirements for binding of the arabinogalactan protein are less defined, but probably depend on the overall physical and chemical properties of the arabinogalactan protein rather than a specific binding site. All the molecules precipitated from plant extracts are high molecular weight polymers containing branched (3→6)-β-galactans with arabinosyl residues in terminal positions (CLARKE et al. 1979). The protein component (5 to 15%) is very resistant to proteolysis, indicating that the protein is buried within the molecule (GLEESON and JERMYN 1979). Removal of terminal arabinosyl residues does not affect the interaction, but alteration of the galactan component either by mild acid or enzymic degradation destroys the capacity for interaction.

The staining of stigma surfaces of *Gladiolus, Lilium, Primula, Petunia, Prunus,* and *Secale* by the β-glucosyl artificial carbohydrate antigen indicates the presence of arabinogalactans (CLARKE and GLEESON 1981). The arabinogalactans of the style canal and stigma surface of mature *Gladiolus* pistils have been isolated by affinity chromatography using an insolubilized galactosyl-binding lectin from *Tridacna maxima* (GLEESON et al. 1979, GLEESON and CLARKE 1980a). They represented 30% to 50% of the total dry weight of material washed from the stigma surface (less than 1 μg per pistil) and 40% of the dry weight of the soluble style material (more than 100 μg per pistil). The style material was polydisperse in the molecular weight range 150,000 to 400,000, and the arabinogalactans within this range were homogeneous with respect to monosaccharide composition and linkage type (GLEESON and CLARKE 1979).

Methylation analysis of the isolated stigma and style arabinogalactans is shown in Table 1. The analyses are similar, both preparations contain galactose and arabinose as major monosaccharides in similar proportions. Arabinose and glucose are present solely in terminal positions. Galactose is mainly (1,3,6)-

Table 1. Methylation analysis of stigma and style arabinogalactans

Linkage type	Linkage composition (mol %)		
	Gladiolus Stigma	*Gladiolus* Style	*Lilium* Stigma exudate[a]
Terminal rhamnosyl	0	0	7
Terminal arabinosyl	17	13	32
Terminal glucosyl	7	trace	0
Terminal galactosyl	16	29	11
1,3 linked galactosyl	13	14	10
1,6 linked galactosyl	6	6	5
1,3,6 linked galactosyl	41	39	30

[a] Data of ASPINALL and ROSELL, recalculated, excluding glucuronic acid

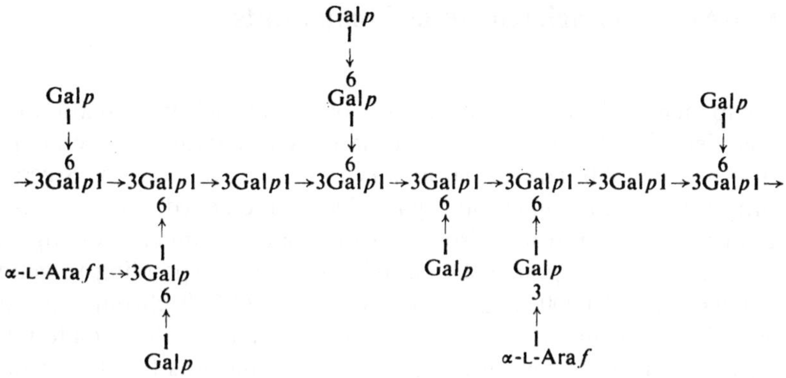

Fig. 1. A proposed model of the style arabinogalactan-protein

linked with smaller amounts of (1→3)-linked (1→6)-linked and terminal residues. Thus the molecules are highly branched. There are differences in terminal residues, the stigma arabinogalactan having a higher proportion of terminal arabinose and glucose than the style arabinogalactan. This and other data (GLEESON and CLARKE 1979) are consistent with a structure having a (1→3)-linked β-galactan backbone, branched through C(*O*)6 to β(1→6) galactan side chains some of which carry terminal L-arabinofuranosyl residues (Fig. 1). The style arabinogalactan is associated, probably covalently, with 3% protein (GLEESON and CLARKE 1979), but the small amounts of material available precluded a similar protein analysis of the stigma surface arabinogalactan.

A fraction from the stigma surface of *Lilium longiflorum* also contains a related arabino 3,6 galactan (ASPINALL and ROSELL 1978) which accounts for 35% of the dry weight of the stigma exudate. Methylation analysis (Table 1) shows this differs from the *Gladiolus* arabinogalactans in having a higher proportion of arabinose (32%) and a significant content of terminal rhamnose residues (7%). It also contains glucuronic acid (11%) both as terminal units and as (1→4)-linked residues.

Thus the arabinogalactans are major components of the stigma surface secretions of *Gladiolus* and *Lilium* and of the style mucilage of *Gladiolus;* the core structures (3,6-galactan) are similar in each case with differences being expressed in the side branch monosaccharides. They are also detected cytochemically as components of stigma surfaces of other plants, but their structure and quantitative importance in these cases is not established. The availability of both the artificial β-glucosyl antigen and antisera raised to the *Gladiolus* style arabinogalactan protein and directed to the side branch arabinosyl and galactosyl residues (GLEESON and CLARKE 1980b) makes both rapid screening and quantitative estimation of arabinogalactans in stigma and style extracts possible.

A high molecular weight fraction of the stigmatic exudate of *Aptenia cordifolia* differs in composition from that of *Lilium* in having a high proportion of both galacturonic and glucuronic acids as well as a range of neutral monosaccharides (Glc, Fru, Gal, Ara, Man, Xyl, and Rha) (KRISTEN et al. 1979).

3 S-Allele Associated Style Components

Components which correlate with a particular self-incompatibility group have been detected immunologically in the pollen or styles of several species. (KNOX and CLARKE 1980, HESLOP-HARRISON 1978). In two cases the S-genotype specific antigen has been isolated and partially characterized. In both cases only small amounts of material were isolated from "large numbers" of stigmas.

In *Brassica campestris* 230 μg of antigen corresponding to genotype S_7 was isolated from 10,000 stigmas, by Sephadex G-200 chromatography, followed by affinity chromatography on Con A-Sepharose. The material eluted with methyl-α-glucoside contained the S_7 antigen and was finally purified by isoelectric focusing. The purified material had pI 5.7, contained protein and carbohydrate in the ratio 2:1, and an apparent molecular weight of 57,000 determined by SDS-polyacrylamide gel electrophoresis (NISHIO and HINATA 1979).

A glycoprotein corresponding to *Brassica oleracea* genotype S_2 has been isolated from stigmas by gel-filtration and ion-exchange chromatography. The isolated material had a molecular weight 54,000, determined by velocity sedimentation analysis and contained 25% carbohydrate. (% monosaccharide composition Ara, 56; Gal, 18; Glc, 18; Man 8). Pretreatment of S_2S_2 pollen with this glycoprotein prevented pollen germinating on (normally compatible) stigmas, indicating that the preparation was biologically active (FERRARI et al. 1981).

In *Prunus avium,* two antigenic components were detected in mature styles, one was apparently specific to the stylar tissues of the genus *Prunus,* (P-antigen) and the other was specific to a particular S-allele group of *Prunus avium* (S-antigen) (RAFF et al. 1981, RAFF and CLARKE 1981). Both these antigens have been isolated (1 mg from 5,000 pistils) and are minor components (<5%) of the soluble material of the style. They are both glycoproteins (P, 17.2%; S, 16.3% carbohydrate as glucose) with different amino acid compositions and charges (P negatively charged and S positively charged at pH 8.9). Antisera raised to the purified glycoproteins show no cross reactivity. Because of the small amounts of material available, indirect analyses, relying on the behavior of the antigens after labeling with ^{125}I have been performed.

Immunoprecipitation (SDS-PAGE 10% gels) gives a molecular weight of 32,000 for P-antigen, and shows, as expected, that there are two components in the S-antigen preparation molecular weight 37,000 and 39,000, presumably corresponding to the S_3 and S_4 gene products in the diploid stylar material. Airfuge analysis gives molecular weight 20,000 (P-antigen) and 17,000 (S-antigen) reflecting the nonideal behavior of glycoproteins in SDS-PAGE gels. Both antigens bound the lectins Con A, PNA, RCA II, and *Lotus* lectin specifically and to similar extents; the S-antigen bound SBA more effectively than did the P-antigen. These data indicate that both antigens contain (1→2)-α-mannosyl and/or glucosyl residues as well as galactosyl and fucosyl residues and that there may be differences in the arrangement of the galactosyl residues in the two antigens. Unexpectedly, the major component of *Prunus avium* styles was not an arabinogalactan, but rather a glycoprotein, apparent molecular weight

68,000, containing 5.4% carbohydrate (as glucose). The arabinogalactan, detected by its binding to the β-glucosyl artificial antigen, was only a minor ($<5\%$) component of the style mucilage (MAU et al. 1981).

4 Callose as a Response to Self-Incompatible Matings

Callose is usually identified on the basis of its fluorescent staining with decolorized aniline blue, and there are few instances in which the staining material has been isolated and analyzed. During self-(incompatible) pollination of *Secale cereale* pollen tubes grow through the stigma surface, for a distance equivalent to four or five cells in the female tissues. Aniline blue staining droplets are detected in the tubes 20 min after pollination, and after 1 h coalesce into distinct plugs and tube growth is arrested. Deposition continues until 3 h after pollination when callose occludes about one third of the pollen grain surface and the basal part of the pollen tube. Sufficient of this incompatibility callose was collected from 2,000 self-pollinated stigmas for analysis (VITHANAGE et al. 1980). The material remaining after sequential treatment with $CHCl_3$:MeOH, hot buffer and acid hypochlorite retained the staining properties of the deposits in the whole tissue. This insoluble material was further fractionated into a DMSO-soluble fraction (15%) and a DMSO-insoluble fraction; both fractions retain the capacity to stain with aniline blue. The DMSO-insoluble fraction was difficult to hydrolyze and no total analysis was obtained. The DMSO-soluble fraction contained mainly glucose (90%) with arabinose as a minor component (10%). Most of the glucose was $(1 \rightarrow 4)$-linked but there was a low proportion (9%) of $(1 \rightarrow 3)$-linked glucosyl residues. All the arabinose was accounted for as terminal residues. The question of whether the DMSO-soluble material represents a mixture of a $(1 \rightarrow 4)$-β-glucan and a $(1 \rightarrow 3)$-β-glucan or a macromolecule containing both linkages, or a mixture of all three types of polymer, could not be resolved, from either the chemical or X-ray diffraction data. However, hydrolysis patterns with linkage-specific enzymes, including the *Bacillus subtilis* endo-$(1 \rightarrow 3)$-β-glucan hydrolase, indicated that at least part of the preparation contained $(1 \rightarrow 3)$-β-and $(1 \rightarrow 4)$-β-linkages in the same linear glucan, but the possibility that substantial amounts of $(1 \rightarrow 4)$-β-glucans and $(1 \rightarrow 3)$-β-glucans were also present, remained.

5 Callose as a Pollen Tube Wall Component

When a style is squashed in aniline blue stain after a compatible pollination, the pollen tube walls show faint staining with plug-like deposits of brilliantly staining material at intervals along the length of the tube, giving it a ladder-like appearance. These callose cross walls are laid down just behind the growing tip and are believed to isolate the living protoplast from the spent, empty tube and grain. There is limited growth of pollen germinated in vitro compared with that obtained in vivo; the tube walls fluoresce with aniline blue, and callose cross walls may also be observed. No attempts to isolate and define the cross wall callose have been recorded but the composition of whole in vitro-grown *Lilium longiflorum* pollen tubes has been examined. The alkali-insoluble residue (5 M NaOH at 100 °C) from a pollen tube wall preparation, retained the capacity to stain with aniline blue. The major monosaccharide was glucose and methylation showed (1 → 3)- and (1 → 4)-linked glucosyl residues in the ratio 2:1. The question of whether these residues originated from a mixture of (1 → 3) and (1 → 4) glucans or a mixed linkage (1 → 3); (1 → 4)-β-glucan was unresolved (HERTH et al. 1974). However, treatment with a (1 → 3)-β-glucan exo-hydrolase weakened the aniline blue fluorescence of the wall fragments but did not completely abolish the (1 → 3)-β-glucan X-ray pattern.

Staining of pollen tube walls with resorcin blue, another "callose-specific" stain, is abolished by treatment with a (1 → 3)-β-glucan exo-hydrolase, also indicating the presence of (1 → 3)-glucosidic linkages (REYNOLDS and DASHEK 1976). It seems likely that the tubes contain at least two separate crystalline β-glucans, one (1 → 3) linked and the other (1 → 4) linked.

References

Aspinall GO, Rosell KG (1978) Polysaccharide component in the stigmatic exudate from *Lilium longiflorum*. Phytochemistry 17:919–921

Clarke AE, Gleeson PA, Jermyn MA, Knox RB (1978) Characterization and localization of β-lectins in lower and higher plants. Aust J Plant Physiol 5:707–722

Clarke AE, Anderson RL, Stone BA (1979) Form and function of arabinogalactans and arabinogalactan-proteins. Phytochemistry 18:521–540

Clarke AE, Gleeson PA (1981). Molecular aspects of recognition and response in the pollen-stigma interaction. Adv Phytochem 15:161–211

Ferrari TE, Bruns D, Wallace DH (1981) Isolation of a plant glycoprotein involved with control of intercellular recognition. Plant Physiol 67:270–277

Gleeson PA, Clarke AE (1979) Structural studies on the major component of the *Gladiolus* style mucilage, an arabinogalactan-protein. Biochem J 181:607–621

Gleeson PA, Clarke AE (1980a) Comparison of the structures of the major components of the stigma and style secretions of *Gladiolus*: the arabino 3,6 galactans. Carbohydr Res 83:187–192

Gleeson PA, Clarke AE (1980b) Antigenic determinants of a plant proteoglycan, the *Gladiolus* style arabinogalactan-protein. Biochem J 189:437–447

Gleeson PA, Jermyn MA (1979) Alteration in the composition of β-lectins caused by chemical and enzymic attack. Aust J Plant Physiol 6:25–38

Gleeson PA, Jermyn MA, Clarke AE (1979) Isolation of an arabinogalactan-protein by lectin affinity chromatography on tridacnin-Sepharose 4B. Anal Biochem 92:41–45

Herth W, Frank WW, Bittiger H, Kuppel R, Keilich G (1974) Alkali-resistant fibrils of β-1,3- and β-1,4-glucans: structural polysaccharides in the pollen tube wall of *Lilium longiflorum*. Cytobiology 9:344–367

Heslop-Harrison J (1978) Genetics and physiology of angiosperm incompatibility systems. Proc. R. Soc. Lond. B 202:73–92

Jermyn MA (1978) Comparative specificity of Concanavalin A and the β-lectins. Aust J Plant Physiol 5:687–696

Jermyn MA, Yeow YM (1975) A class of lectins present in the tissues of seed plants. Aust J Plant Physiol 2:501–531
Knox RB and Clarke AE (1980) Discrimination of self and non-self in plants. Contemp Top Immunobiol 9:1–30
Kristen U, Biedermann M, Liebezeit G, Dawson R (1979) The composition of stigmatic exudate and the ultrastructure of the stigma papillae in *Aptenia cordifolia*. Eur J Cell Biol 19:281–287
Mau S-L, Raff J, Clarke AE (1981) Isolation and partial characterization of style-specific antigenic glycoproteins from *Prunus avium* style mucilage. Planta (submitted)
Nishio T, Hinata K (1979) Purification of an S-specific glycoprotein in self-incompatible *Brassica campestris* L. Jpn J Genet 54(4):307–311
Raff JW, Clarke AE (1981) Tissue specific antigens are secreted by suspension cultured callus cells of the sweet cherry *Prunus avium*. Planta (in press)
Raff JW, Knox RB, Clarke AE (1981) Style and S-genotype specific antigens of *Prunus avium*. Planta (in press)
Reynolds JD, Dashek WV (1976) Cytochemical analysis of callose localization in *Lilium longiflorum* pollen tubes. Ann Bot 40:409–416
Vithanage HIMV, Gleeson PA, Clarke AE (1980) Callose: its nature and involvement in self-incompatibility response in *Secale cereale*. Planta 148:49–509
Yariv J, Lis H, Katchalski E (1967) Precipitation of arabic acid and some seed polysaccharides by glycosylphenylazo dyes. Biochem J 105:1C–2C
Yariv J, Rapport MM, Graf L (1962) The interaction of glycosides and saccharides with antibody to the corresponding phenylazo glycosides. Biochem J 85:383–388

24 Carbohydrates in Plant-Pathogen Interactions[1]

T. Kosuge

1 Introduction

Since carbohydrates are major plant constituents, it is to be expected that they would be involved in plant–pathogen interactions in a number of ways. Disruption of carbohydrate metabolism is one of the major factors contributing to crop loss. Carbohydrate-containing compounds produced by plant pathogens function as toxins and evoke symptoms of disease in plants. Plants contain toxic carbohydrate derivatives that help make them resistant to ingress by potential pathogens. Pathogen enzymes that degrade plant structural polysaccharides function as principal agents of disease. Carbohydrates also appear to be involved in recognition between pathogens and their hosts.

Researchers in carbohydrates of plant–pathogen interactions are given many choices of systems to investigate. Some bacteria cause soft rot of storage organs; others cause foliage diseases or wilt; still others cause galls. Fungi constitute a very broad group of plant pathogens ranging from those that are obligate parasites, such as rusts and powdery mildews, to perthotrophs which kill plant tissue and then feed on constituents of dead cells. There are both RNA and DNA viruses which place heavy demands for pentose sugars during active stages of replication. Some pathogens have special attributes which not only permit them to survive saprophytically but also allow them to spend a portion of their life cycle in plants often under restricted or sophisticated conditions.

It is important to emphasize that not all phenomena described in the following sections will be found in one particular plant–pathogen interaction. Rather, certain phenomena are unique to a given plant–pathogen interaction.

2 Role of Polysaccharides in the Early Interactions Between Plant and Pathogen

2.1 Role of Recognition in Plant–Pathogen Interactions

Recognition mediated through macromolecules provides the basis for many fundamental reactions in biology. For example, communications in cell-to-cell interactions are important during the development of all multi-cellular organisms. Therefore it follows that mechanisms for recognition in many plant–pathogen interactions could have evolved from systems concerned with cell-to-

1 This Chapter (as well as Chaps. 25 and 26) provides a valuable extension to several contributions in Volume 4 of this Series (Physiological Plant Pathology, eds.: Heitefuss, R and Williams, PH)

cell communication through macromulecular interactions. Binding sites on macromolecules in the cell wall or membrane of the host, designed for cell-to-cell recognition during normal plant development, could become targets for establishing recognition between plant and pathogens. Although not all pathogens show species specificity for their host, in those interactions where specificity occurs, carbohydrates and glycoproteins may help provide a basis for recognition.

One of the most challenging problems concerned with studying plant–pathogen interactions is the identification of products of virulence genes in the pathogen and products of resistance genes in the host and to understand how they function in plant–pathogen interactions. In this connection the systems that exemplify Flor's gene-for-gene hypothesis may offer the best opportunities for study. The hypothesis states that "for each gene conditioning rust reaction in the host there is a specific gene conditioning pathogenicity in the parasite" (FLOR 1956). With two alleles at a locus controlling plant resistance and two at a complementary locus controlling pathogen virulence, only one of the four possible interactions, the combination of resistance and avirulence, leads to the expression of resistance and avirulence or a low infection type (ELLINGBOE 1972, DAY 1974). Further, virulence is recessive and resistance is dominant. ALBERSHEIM and ANDERSON-PROUTY (1975) and DAY (1974) pointed out that components found on cell walls of pathogens which interact with plant cell membranes could account for specificity in gene-for-gene systems. Products of a plant's resistance genes could be receptors for molecules whose synthesis is controlled by avirulence genes of pathogens. Receptors might be membrane-bound proteins which recognize carbohydrates or glycoproteins of the pathogen. Products of pathogen's avirulence genes could be glycosyltransferases. Since the loss of the function of one or more avirulence genes permits an avirulent organism to become virulent, it follows that loss of the product of avirulence genes should not be deleterious to the pathogen. Products of avirulence genes should nevertheless have some useful function for the pathogen otherwise they would be lost by selection.

Not all of the evidence presented in the following discussion will support Flor's hypothesis. Nevertheless, emphasis will be placed on recognition as it involves establishment of a particular plant–pathogen interaction and its culmination in resistance or susceptibility.

2.2 Recognition in the Infection Process

The *Rhizobium*-legume model (see SCHMIDT and BOHLOOL, Chap. 26, and KAUSS, Chap. 25, this volume) offers an attractive explanation for common antigen relationships that have been observed in plant–pathogen interactions (DEVAY and ADLER 1976). One important difference is that bacterial plant pathogens, unlike *Rhizobium* species, cannot penetrate plant tissues directly but must gain entry through natural openings such as stomata, hydrathodes, lenticels, or nectaries; some plant pathogenic bacteria are introduced directly into plant tissues by insect vectors. Consequently, any recognition between plant and pathogen must occur after the bacteria gain entry into plant tissue. The binding of *Agrobacterium tumefaciens* to host cells is necessary for the transfer of DNA (the T-region of the Ti plasmid) from the bacterial cell to the host cell in the conversion of a plant cell into a tumorous one (Watson et al. 1975). Recognition through binding was inferred by the work of DeVay et al. (1972), who noted that host plants had antigens that cross-reacted with antisera prepared from antigens extracted from virulent *A. tumefaciens* cells. However, in agar diffusion plates, antisera from avirulent *A. tumefaciens* cells formed no or very weak precipitin bands against host antigen preparations. Further, antigens from bacteria-free crown gall tissue reacted weakly against antisera of virulent *A. tumefaciens* antigens. It is now known that *A. tumefaciens* cells bind to host cells during early stages of infection (Lippincott et al. 1977, Ohyama et al. 1979) and also bind to cell walls prepared from their hosts. However, transformed plant cells did not bind *A. tumefaciens* cells (Lippincott et al. 1977, Lippincott and Lippincott 1978) and bacteria-free crown gall cells lost their antigenicity for antibacteria antisera. These results suggest that the binding sites concerned with bacterial recognition of host cells are lost as a result of conversion to the tumorous condition. Concanavalin A and soybean lectins agglutinated *A. tumefaciens* cells and enhanced tumor induction on potato tuber disks (Anand et al. 1977), but had no effect on attachment of the bacteria to the plant cell (Ohyama et al. 1979). Binding sites on the bacterium appear to occur in the lipopolysaccharide (Whatley et al. 1976) but the involvement of lectins in recognition between *A. tumefaciens* and host cells remains to be determined.

2.3 Recognition in Resistance Reactions

Water-soaked areas develop when leaves are inoculated with a bacterium pathogenic to that host; this provides an environment conducive to the multiplication of the pathogen. The same condition is produced if extracellular polysaccharides from the bacterium are infiltrated in the leaves of the host plant (El-Banoby and Rudolph 1979, 1980). The reaction is host-specific since plants which are not hosts for the bacterium will not react to infiltration with its extracellular polysaccharides (Rudolph 1978, El-Banoby and Rudolph 1979). There are also reports that water soaking is elicited by infiltrating leaves with lipopolysaccharides from plant-pathogen bacteria (Keen and Williams 1971). Since the reaction is specific, it has been suggested that recognition is involved between sites on the extracellular polysaccharide or lipopolysaccharide fractions and a receptor on the cell wall or plasma membrane of the host cell. Extracellular polysaccharides may fulfill dual functions of protecting virulent cells from binding by host cell walls (see below) and altering host cells or plasma membranes so as to cause water-soaked conditions to develop and assist with multiplication of the bacterium.

In contrast to their postulated roles in recognition for successful infection in *Rhizobium*–plant interactions, lectins may function as resistance factors. Thus, avirulent cells of *Pseudomonas solanacearum* appear to bind to and become immobilized by binding to lectins on the tobacco leaf cell walls; virulent cells remain unattached and multiply rapidly in intercellular spaces of host leaves. The attachment of avirulent bacteria initiates processes leading to a hypersensitive response (Graham et al. 1977).

Potato tuber lectin, purified to homogeneity, agglutinated strongly with avirulent isolates of *P. solanacearum;* virulent isolates either failed to agglutinate or agglutinated only weakly with purified lectin preparations. Virulent isolates produce an extracellular polysaccharide (EPS) which protects cells from binding with the lectin both in vitro and in host tissue; avirulent cells do not produce EPS and therefore are not protected from agglutination by the lectin. Purified lipopolysaccharide preparations from avirulent cells precipitated when mixed with potato lectin. Further, another type of avirulent *P. solanacearum,* that does not evoke the hypersensitive reaction, has the "smooth" (also called complete) lipopolysaccharide which characteristically is high in rhamnose and xylose but is low in glucose. Avirulent isolates which evoke the hypersensitive reaction have the rough (incomplete) form of lipopolysaccharide enriched with glucose but low in rhamnose and xylose content. The change from smooth to the rough type of lipopolysaccharide is associated with the loss of sugars associated with the *O*-antigen site (WHATLEY et al. 1980). Apparently lectin-binding sites are present in the *O*-antigen portion of the *Agrobacterium* lipopolysaccharide (SEQUEIRA and GRAHAM 1977).

Injection of tobacco leaves with purified protein lipopolysaccharide fractions from plant pathogenic bacteria prevented the hypersensitive reactions normally evoked by injection with incompatible pseudomonads (MAZZUCHI et al. 1979). The purified protein–lipopolysaccharide complex itself does not evoke the hypersensitive reaction. Live cells of the incompatible strain are needed to evoke the response. The isolated protein–lipopolysaccharide fraction could block the hypersensitive reaction by occupying host cell sites normally available for binding cells of the incompatible strain. Protein–lipopolysaccharide fractions also protect tobacco tissue against infection with compatible strains, but the mechanism involved remains to be determined.

Among pathogenic fungi, germ tubes of both compatible and incompatible isolates of *Phytophthora infestans* become attached to the plasmalemma of the host cells. Only the incompatible isolates, however, evoke the hypersensitive reaction in resistant host cells (NOZUE et al. 1979). Potato lectin mediated binding of germinated cytospores of compatible and incompatible races of *P. infestans* to membrane preparations from potato tubers but *N,N′*-diacetylchitobiose, the specific hapten of potato lectin, inhibited the binding. Concanavalin A also agglutinated germinated spores and the agglutination was inhibited by alpha methyl-D-mannoside and alpha methyl-D-glucoside, both specific haptens of Concanavalin A (FURUICHI et al. 1980). Thus lectins may be involved in the attachment of *P. infestans* hyphae to the plasmalemma of host plant cells.

Lectin-like substances from sweet potato roots also agglutinated germinated spores of *Ceratocystis fimbriata,* but agglutination of different isolates was modulated by low molecular weight materials present in sweet potato root extracts (KOJIMA and URITANI 1978). These may be functionally similar to the suppressors of elicitors described below. Differences in carbohydrate-containing spore surface structures were detected in lectin-binding studies on macroconidia of *Fusarium* sp. but the significance of these observations remains unclear (KLEINSCHUSTER and BAKER 1974).

In summary, it appears that polysaccharides are involved as recognition factors in *Rhizobium*–legume interactions and plant–pathogen interactions. However, a role for lectins in microorganism plant interactions remains to be conclusively demonstrated (see also, Chap. 25 and 26, this Vol.).

2.4 Specificity Through Suppressors and Protectors

The determinant of pathogen specificity in *P. infestans* may involve factors such as water-soluble glucans which suppress the hypersensitive reaction triggered on potato tissue by hyphal-components isolated from the fungus. Suppressor activity was greater if the preparation came from a *P. infestans* race capable of establishing a compatible interaction with the potato plant which yielded the tissue preparation (GARAS et al. 1979). The suppressors may function by modulating the interaction between carbohydrate portions of the fungal wall and receptors on host cell walls or membranes. As with potato tuber tissues, water-soluble glucans from the fungus also suppress the hypersensitive reaction evoked on protoplasts by the hyphal preparations (DOKE and TOMIYAMA 1980a, b).

Components responsible for suppressor activity appear to contain $(1 \rightarrow 3)$-β- and $(1 \rightarrow 6)$-β-linkages and are polymers consisting of 17 to 23 glucose units (DOKE et al. 1979). Membrane fractions mixed with elicitor preparations reduced terpenoid-accumulating activity and glucans from the compatible and incompatible races modulate that reaction. The hypersensitivity elicitor may bind to a receptor on the cell membrane to evoke the hypersensitive reaction; the suppressor from the compatible race may act by blocking the binding of the hypersensitivity elicitor with the receptor site in the plant cell. In this way the suppressor glucans function as determinants of host specificity.

Soluble glycoproteins (protectors) from cultures of incompatible races of *Phytophthora megasperma* f. sp. *glycinea* protected soybean seedlings from inoculation with compatible races; similar fractions from compatible races failed to give protection. Such observations imply that the soluble glycoproteins confer a specificity in the interactions of *P. megasperma* f. sp. *glycinea* with its host soybean (WADE and ALBERSHEIM 1979). This is in contrast to glucans from *P. infestans* which confer host specificity by suppressing the hypersensitive reaction. Protectors may also confer specificity in the bean–*Colletotrichum lindemuthianum* interaction (BERARD et al. 1972).

3 Carbohydrate Metabolism of Pathogens in Host Tissue

3.1 Production of Polysaccharide-Degrading Enzymes

3.1.1 Nature and Action in Host Tissue

The degradation of plant structural polysaccharides by extracellular enzymes of pathogens commonly occurs in disease. It is safe to say that for each type of polysaccharide known to occur in plants a corresponding pathogen enzyme that catalyzes its breakdown has been described (Table 1). In many cases, the

Table 1. Plant pathogen enzymes that act on plant structural polysaccharides

Plant component and linkage	Pathogen enzymes/regulation	References
Pectate (D-galacturonan) (1→4)-α-D-GalpA	*Polygalacturonase* Inducers: D-galacturonic acid Repressors: D-glucose, D-galacturonic acid	BASHAM and BATEMAN (1975a, b), CERVONE et al. (1978), COOPER (1977), COOPER and WOOD (1975, 1980), ENGLISH et al. (1972), PULHALLA and HOWELL (1975)
	Pectate lyase, pectate trans-eliminase Inducers: D-galacturonic acid Repressors: D-glucose, D-galacturonic acid; c-AMP reversed catabolite repression in plant pathogenic bacteria	BASHAM and BATEMAN (1975a, b), CHATTERJEE and STARR (1977), COOPER and WOOD (1975, 1980), HISLOP et al. (1979), HUBBARD et al. (1978), MOUNT et al. (1979), STACK et al. (1980)
Cellulose (1→4)-β-D-Glcp	*Cellulase* Inducers: Cellobiose Repressors: D-glucose	BATEMAN et al. (1969), COOPER and WOOD (1975), COOPER et al. (1978), ENGLISH et al. (1971), GUPTA and HEALE (1971)
Galactan (1→3)-β-D-Galp (1→4)-β-D-Galp (1→4)-α-D-Galp (1→6)-α-D-Galp	*Galactanase* Inducers: L-arabinose, D-galactose Repressors: D-glucose	BATEMAN et al. (1969), COOPER and WOOD (1975), COOPER et al. (1978), MULLEN and BATEMAN (1975)
Xylan (1→4)-β-D-Xylp	*Xylanase* Inducers: D-xylose Repressors: D-glucose	BATEMAN et al. (1969), COOPER and WOOD (1975), COOPER et al. (1978)
Arabinan (1→3)-α-L-Araf (1→5)-α-L-Araf	*Arabinanase* Inducers: L-arabinose Repressors: L-arabinose, D-glucose	BATEMAN et al. (1969), COOPER and WOOD (1975), COOPER et al. (1978)

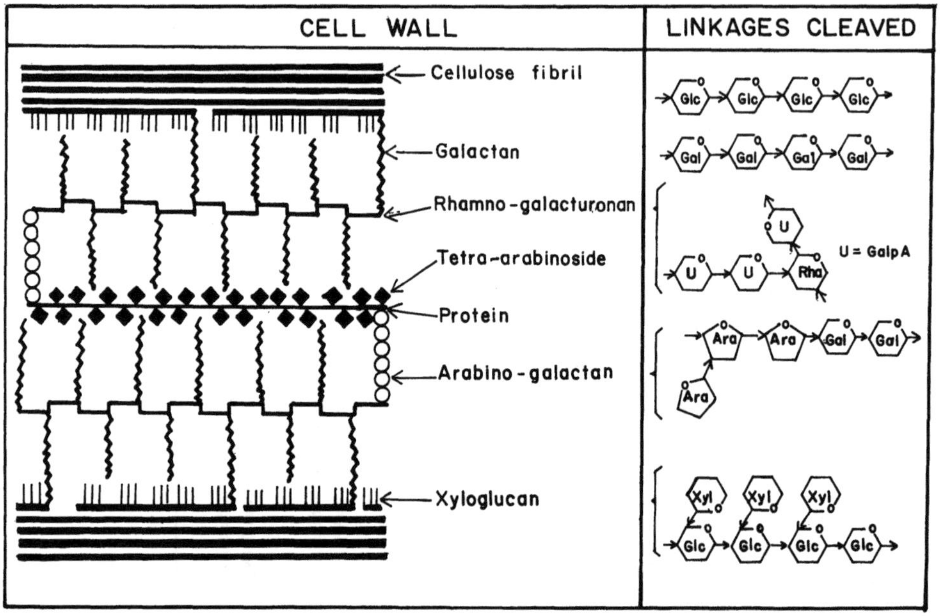

Fig. 1. A model of a plant cell wall (Keegstra et al. 1973). Polymers commonly degraded by enzymes secreted by plant pathogens are diagrammatically represented under *Linkages Cleaved*

role of the enzyme is often secondary with respect to pathogenesis. Nevertheless, the damage caused by such enzymes either interferes with normal development of the plant or reduces its economic value.

The model of a plant cell wall proposed by Keegstra et al. (1973) serves as a point of departure for this discussion on plant cell wall degradation by pathogens. The various carbohydrate polymers depicted in Fig. 1 are targets for depolymerization by enzymes produced by plant pathogens (Bateman 1976, Bateman and Basham 1976, Cooper 1977). The most widely studied are the enzymes which catalyze the breakdown of pectic substances by hydrolytic and eliminative reactions (Albersheim and Anderson-Prouty 1975). Enzymes of this group generally are the first polysaccharidases detected in infected plant tissue (Goodenough and Kempton 1976). Soft rots of storage organs are due to extensive degradation of pectic substances and storage carbohydrates. Degradation of plant structural components, catalyzed by pathogen enzymes, is a major factor contributing to symptoms of diseases such as "damping off" in which the stem collapses as the result of extensive degradation of structural components of the hypocotyl. Wilt of plants may be evoked by gels formed in vascular elements by action of pectin-degrading enzymes.

It is now well documented that treatment of cell walls with pectin-degrading enzymes causes release of sugars other than galacturonides. For example, homogeneous preparations of endopolygalacturonate lyase from *E. chrysanthemi* released arabinose, rhamnose, and galactose in addition to galacturonate (Basham and Bateman 1975a). Endopolygalacturonase preparations of *Colletotrichum lindemuthianum* reduced the amount of arabinose, galactose, rhamnose, and xylose in addition to galacturonate from bean cell walls (ENGLISH et al. 1972).

Cell wall preparations from some plants were not readily degraded by polysaccharidases unless polygalacturonide components were first degraded or modified (BAUER et al. 1973, KARR and ALBERSHEIM 1970). Such results implied that accessibility to certain cell wall constitutents was blocked by overlying polymers of polygalacturonides and other carbohydrates. Sequential production of various enzymes associated with cell wall degradation has been reported (COOPER and WOOD 1975, ENGLISH et al. 1971, GOODENOUGH and KEMPTON 1976, MULLEN and BATEMAN 1975). Degradation of cell walls was increased if the preparations were incubated with a mixture of several polysaccharidases rather than individual enzyme preparations. However, some plant cell walls were readily degraded by single polysaccharidases (COOPER et al. 1978).

Pectin-degrading enzymes of the "endo" type are concerned with maceration of tissues and cell death (BATEMAN 1976, BASHMAN and BATEMAN 1975a, b). Plant cell death may be caused by damage to the plasmalemma by destruction of pectin material in the intercellular space and cell walls; the weakened cell structure may no longer balance the pressure exerted by the protoplast. On the other hand, there could be a direct effect on the plasma membrane, causing an alteration in function and ion leakage (BATEMAN 1976, HISLOP et al. 1979). Decompartmentation by loss of membrane function can readily explain ultimate death of host cells.

Tissue maceration is but one of the several symptoms that can be evoked by pectin-degrading enzymes. Disease symptoms including vascular gel formation were reproduced by supplying susceptible tomato cuttings with partially purified preparations of endopolygalacturonase and endopectate lyase (COOPER and WOOD 1980). Peroxidase and other proteins were released from cotton plant cell walls by polygalacturonase from *Verticillium albo-atrum* (STRAND and MUSSELL 1975, STRAND et al. 1976, BARNETT 1974). Acid phosphatase was released from cultured apple cells and membrane permeability was altered by polygalacturonate lyase from *Monilinia fructigena* (HISLOP et al. 1979).

Polygalacturonases show differential affinity for substrates from tissues normally colonized by the pathogen that produced the enzymes. Thus, *Rhizoctonia fragariae,* a pathogen of strawberry, produced polygalacturonase which had a strong affinity for strawberry fruit tissue but not for cells in slices of potato tuber, carrot, and beet root slices (CERVONE et al. 1978).

3.1.2 Conditions Affecting Production

As expected, most, if not all, structural polysaccharide-degrading enzymes have been shown to be inducible and subject to catabolite repression by glucose or other monosaccharides. In *V. albo-atrum* and *Fusarium oxysporum* f. sp. *lycopersici,* polysaccharidase was specifically induced by the monosaccharide unit predominant in the polymeric substrate for the enzyme; the exception was cellulase which was induced by cellobiose but not by glucose (GUPTA and HEALE 1971). Conditions for enzyme synthesis and repression appear to be specific since induction occurred with low concentrations of the monomeric carbohydrate, but was repressed when the concentration of monomeric carbohydrate reached concentrations slightly in excess of requirements for growth. Almost complete repression occurred if the inducer carbohydrate was supplied at three times the optimal rate. The enzymes are also repressed by glucose (Cooper and Wood 1975). In most cases, there is a low level of constitutive synthesis that is catabolite-repressible.

Such control over enzyme synthesis in *V. albo-atrum* reflects adaptation of the pathogen to its nutritional environment in host tissue. Since it exists in the xylem, the fungus would be exposed to an environment deficient in carbon and may depend upon breakdown of vessel walls for some of its nutrients. However, since extensive breakdown of xylem vessels apparently does not occur by *Verticillium,* it appears that the organism is adapted to being confined to the vessels. Further, organisms adapted to living in the xylem encounter substrates such as vessel walls that are insoluble and therefore not available as inducers for enzyme synthesis. As with other organisms, it is probable that the basal level of enzyme activity provides sufficient breakdown of wall components and release of products which can in turn serve as inducers (Moran and Starr 1969, Cooper 1977, Cooper and Wood 1975). It is noteworthy that both endo and exo forms of pectin-degrading enzymes are produced by *V. albo-atrum* and by other pathogens; perhaps the exo-polygalacturonase provides low amounts of galacturonate which can serve as an inducer for the pectate lyase and endo polygalacturonase.

Certain enzymes such as cellulase may be produced only when the hyphae are in contact with the substrate (Gupta and Heale 1971); it is possible that such enzymes are wall-bound or mural enzymes (Gander 1974, Barash and Klein 1969) and offer protection from dilution by secretion into the surrounding medium. Mural enzymes also provide the advantage of releasing products at the site of transport across the plasmalemma.

The bacterial soft rotter, *Erwinia amylovora,* produced polygalacturonate lyase constitutively and inducibly (Moran and Starr 1969, Zucker et al. 1972). Certain plant extracts of unknown chemical composition repressed lyase production in one isolate of *E. carotovora* but glucose catabolite repression of the enzyme produced by this bacterium could not be demonstrated (Zucker et al. 1972). However, other *E. carotovora* isolates demonstrate typical catabolite repression of pectate lyase that is relieved by cyclic adenosine monophosphate (c-AMP) (Mount et al. 1979).

3.1.3 Role of Polysaccharide-Degrading Enzymes in Vivo

From the standpoint of pathology, the principal reason for studying the enzymes is to establish their importance in disease. In most cases, questions remain unanswered regarding their primary role in plant disease, whether they contribute to the virulence of the pathogen, and how they contribute to symptom expression in infected plants.

In most early studies, results obtained in artificial culture were assumed to represent activity properties of enzymes in diseased tissue. Results of such studies were relatively meaningless in explaining the role of the enzymes in disease, since conditions in culture had little resemblance to nutrition in plant tissue. With the knowledge now available on mechanisms which regulate synthesis and activity of the enzymes, researchers have turned to the use of mutants deficient in production of pectin-degrading enzymes. Mutants of *V. dahliae* deficient in production of endopolygalacturonase, pectin methylesterase, pectin and pectate lyases still produced normal verticillium wilt disease symptoms in cotton (HOWELL 1976). This supported earlier results that enzyme production in culture and virulence were unrelated among three isolates of *V. dahliae* (WIESE et al. 1970). On the other hand, MUSSELL and GREEN (1970) observed a correlation between polygalacturonase production and virulence among Verticillium isolates. Furthermore, symptoms produced in cotton cuttings by endopolygalacturonase preparations resembled those produced by infection with the fungus (MUSSELL 1973). Likewise, COOPER and WOOD (1980) reported that endopolygalacturonase or pectate lyase supplied individually to tomato cuttings reproduced symptoms of *Verticillium* infection. Interpretation of the conflicting results of the different authors is complicated by the observations that (a) production of polysaccharide-degrading enzymes in culture under the usual laboratory conditions cannot be equated to production in plants ; the various regulatory mechanisms controlling production and activity of the enzymes must be considered (COOPER and WOOD 1980). (b) The amount of macromolecules needed to initiate wilting in plants is very small (VAN ALFEN and ALLARD-TURNER 1979); thus, it is difficult to directly relate quantity of wilt-inducing polysaccharidases produced to virulence of a pathogen. (c) The "plugging" phenomenon is dependent more on size and less on chemical composition of the macromolecule (VAN ALFEN and ALLARD-TURNER 1979).

Among diseases caused by bacteria, genetic evidence by CHATTERJEE and STARR (1977) provides convincing arguments for the essential role of polygalacturonate lyase in soft rots caused by *Erwinia chrysanthemi*. Mutants which produced polygalacturonase but, deficient in synthesis of the lyase, failed to macerate plant tissue. Introduction by crossing of the pat (polygalacturonate lyase) gene restored in the recipient cells lyase production and capacity to macerate plant tissue.

Apart from their role in destruction of plant tissue, it is evident that polysaccharide-degrading enzymes provide no basis for specificity in plant–pathogen interactions. With the exception of cellobiose induction of cellulase, induction of polysaccharidases occurs by the monosaccharide component of the polymeric substrates to be degraded and not by the polysaccharides themselves (COOPER and WOOD 1980). Since the monosaccharide components occur ubiquitously in polysaccharides of plants, they would provide little specificity regarding the type of polysaccharide-degrading enzymes that could be produced by a pathogen.

3.2 Carbohydrates as a Source of Energy of Pathogens in Host Tissue

3.2.1 Catabolite Repression in Pathogens in Host Tissue

Hypotheses have been advanced that carbohydrate content in plants affects the virulence of the pathogen. Thus, low sugar diseases are those in which low sugar content of host tissue is associated with susceptibility; high sugar diseases are those in which high sugar content is related to susceptibility (HORSFALL and DIMOND 1957). High sugar diseases generally concern obligate parasites and low sugar diseases are characteristic of those caused by facultative parasites. The basis for the former is uncertain but may involve maintenance of an optimal osmotic environment for the pathogen. The latter cases are believed to involve glucose catabolite repression of enzymes associated with the degradation of structural polysaccharides, which would occur in the presence of high sugar content in plant tissue. Indeed, application of glucose to tomato plants reduced symptom development by *Fusarium oxysporum* f. sp. *lycopersici*. Since glucose in culture repressed polygalacturonase synthesis, it was suggested that the effect of glucose on symptom development was the in planta repression of enzyme synthesis by the pathogen (PATIL and DIMOND 1968). Similar experiments by HORTON and KEEN (1966) on onion seedlings inoculated with *Pyrenochaeta terrestris* also suggested in planta glucose repression of cellulase and polygalacturonase synthesis. Further, treatments that reduced glucose content in onion seedlings increased the amount of cellulase and polygalacturonase in inoculated seedlings and also increased rapidity of symptom development (HORTON and KEEN 1966). While it is quite likely that regulation of enzyme synthesis in pathogens would occur in plants by conditions detected in culture, convincing evidence is needed from experiments with appropriate regulatory mutants.

3.2.2 Acquisition of Energy Sources by Pathogens in Host Tissue

Plant pathogens in host tissue would be expected to have special attributes for acquiring carbon sources from their surroundings. Fungal plant pathogens such as *Verticillium dahliae* and pathogenic bacteria such as *Xanthomonas campestris* exist mainly in xylem vessels throughout most stages of disease development; *Pseudomonas phaseolicola* grows in extracellular spaces in bean leaves in a "water-soaked" environment perhaps evoked by its extracellular polysaccharides (EL-BANOBY and RUDOLPH 1979); soft rot bacteria such as *Erwinia amylovora,* would exist in a carbohydrate-rich environment provided by the breakdown of tissue by pectin-degrading enzymes. However, despite the obvious importance of systems for transport of carbohydrates from the surrounding media into the pathogen cells, little information is available on the properties of such systems in plant pathogens.

Growth rates calculated for plant pathogenic bacteria in plant leaves are much lower than rates calculated for cells in culture. However, there is little information about carbohydrates utilized in plant tissues by the bacteria. Further, in plant tissues, it is not known whether growth by the pathogen is transport-limited or nutrient-limited, as can be amply demonstrated in culture (BULL and TRINCI 1977, JANNASCH and MATELES 1974). Xylem-inhabiting pathogens might be expected to have high-affinity carbohydrate transport systems which would allow them to survive in carbohydrate-deficient environment of xylem fluid. The same organisms, however, survive saprophytically in plant debris or grow equally well on ordinary nutrient media. Adaptability to carbohydrate-poor and -rich media in part could be provided by transport systems of the types found in *Neurospora crassa* which has two glucose transport systems.

One is a low affinity system (K_m of 8 mM) which is active when the fungus is grown in high glucose; the other is a high affinity system, $K_m = 10\ \mu M$, which is repressed in a high glucose medium but is active when the organism is placed in a low glucose medium (SCARBOROUGH 1970, SCHULTE and SCARBOROUGH 1975).

3.2.3 Pathways of Carbon Catabolism in Host Tissue

Though no systematic studies have been performed on plant pathogens in general, it is generally accepted that plant pathogenic fungi utilize the Embden-Meyerhof and hexose monophosphate pathways for carbohydrate utilization. Plant pathogenic bacteria such as *A. tumefaciens* utilize the Entner-Doudoroff and pentose phosphate pathways for carbohydrate utilization but it is not known whether this is representative of plant pathogenic bacteria in general (ARTHUR et al. 1973). *A. tumefaciens* also utilizes sucrose by a unique pathway involving conversion to 3-ketosucrose (α-D-ribo-hexopyranosyl-3-ulose-β-D-fructofuranoside). The enzyme that catalyzes the reaction is not specific for sucrose and is active against glucose, glucose-1-phosphate, both α- and β-glucosides (FUKUI 1969, VAN BEEUMEN and DE LEY 1975) as well as several other monosaccharides and disaccharides (FUKUI and HOCHSTER 1963, FUKUI 1969, HAYANO and FUKUI 1967). The bacterium also produces a hydrolase (α-ketoglucosidase) which converts 3-ketosucrose to 3-ketoglucose and fructose. However, 3-ketosucrose accumulated in the cultures of the bacterium despite the presence of high levels of α-keto-glucosidase activity. Separation by compartmentation may explain this apparent anomaly (HAYANO and FUKUI 1970). It is interesting that 3-ketoglucose can be reduced to glucose by a NADPH-linked enzyme produced by this bacterium (HAYANO et al. 1973). The functional role of the reactions:

$$\begin{array}{l} \text{sucrose} \rightarrow \text{ketosucrose} \rightarrow \text{ketoglucose} + \text{fructose} \\ \qquad\qquad\qquad\qquad\qquad\quad \downarrow \\ \qquad\qquad\qquad\qquad\qquad\ \text{glucose} \end{array}$$

remains unclear.

Claviceps purpurea, which parasitizes grasses and cereals, possesses a β-fructofuranosidase with transfructosyl activity which converts sucrose to glucose and an oligofructoside. The enzyme catalyzes the formation of fructosyl sucrose and several other oligosaccharides of fructose and glucose. While the organism utilizes glucose or fructose as a sole cource of carbon, in a mixture of the two, glucose is preferentially utilized. Since fructose in high concentrations is inhibitory to the fungus, formation of oligosaccharides may be a mechanism for preventing accumulation of inhibitory concentrations of fructose. After glucose becomes limiting, fructose is utilized (DICKERSON 1972, DICKERSON et al. 1978). These mechanisms may help the fungus to grow in the presence of very high sucrose concentrations in the honeydew of parasitized tissue.

3.3 Production of Toxic Carbohydrates by Pathogens in Host Tissue

3.3.1 Structure of Toxins

Certain carbohydrates produced by plant pathogens are toxic to plants and are associated with disease symptom expression. Apart from their role in plant

Table 2. Some carbohydrate-containing phytotoxins produced by plant pathogens

Toxin	Composition	Source	Plants affected by toxin	References
Toxins associated with wilt				
Glycopeptide	21,400 MW glycopeptide with glucose, mannose, 2-keto-3-deoxy-gluconate	*Corynebacterium sepidonicum*	Tomato (*Lycopersicon esculentum*); potato (*Solanum tuberosum*)	STROBEL (1970), STROBEL et al. (1972)
Glycopeptide	Glycopeptide of MW 5×10^6; with L-fucose, mannose, glucose, galactose, keto-deoxyacid	*C. insidiosum*	Alfalfa (*Medicago sativa*)	REIS and STROBEL (1972a, b), VAN ALFEN and TURNER (1975b)
Glycopeptides (3 fractions)	Glycopeptides containing galactose, glucose, mannose, or fructose, glucose, galactose, mannose	*C. michiganense*	Tomato	RAI and STROBEL (1969)
Amylovorin	Polysaccharide	*Erwinia amylovora*	Apple (*Malus pumila*); Pear (*Pyrus communis*)	GOODMAN et al. (1974)
Polysaccharide	Mannose, glucose, glucuronolactone, (peptide?)[a]	*Xanthomonas campestris*	Cabbage (*Brassica oleracea*)	SUTTON and WILLIAMS (1970)
Mycolaminarin	$(1 \rightarrow 3)$-β-D-glucan with 6-linked glucan side chains	*Phytophthora cinnamomi*; *P. palmivora*; *P. megasperma*	*Persea indica*; tomato (*Lycoperiscon esculentum*); soybean (*Glycine max*); cacao (*Theobroma cacao*)	KEEN et al. (1975), WANG and BARTNICKI-GARCIA (1973, 1974)

Polysaccharides	(1 → 3)-β-D-glucan with (1 → 3) (1 → 6); (1 → 3, 1 → 6)-linked residues	*P. cinnamomi*, *P. cryptogea*	*Eucalyptus sieberi*; *E. cypellocarpa*	Woodward et al. (1980)
Protein-lipopoly-saccharide	Protein (15%); lipid (15%); carbohydrate (70%) mostly glucose, galactose, mannose, galacturonic acid	*Verticillium albo-atrum*	Cotton (*Gossypium hirsutum*)	Keen and Long (1972), Keen et al. (1972)
Malseccin	Glycopeptide with glucose, galactose, mannose	*Phoma tracheiphila*	*Citrus limon* (lemon)	Nachmias et al. (1979)
Glycopeptide	Rhamnomannan peptides	*Ceratocystis ulmi*	Elm (*Ulmus americana*) also nonhosts	Strobel et al. (1978), Van Alfen and Turner (1975a)
Toxins with membrane effects				
Glycopeptide	Glycopeptide possibly mannan with galactosyl and glucosyl residues	*Cladosporium fulvum*	Tomato (*Lycopersicon esculentum*); other plants	Lazarovits et al. (1979), Lazorovits and Higgins (1979)
Fusicoccin	Terpene glucoside (Fig. 5)	*Fusicocum amygdali*	Many plants	Ballio et al. (1968, 1971), Marre (1979), Turner and Graniti (1976)
Helmintho-sporoside	α-Galactoside (?)	*Helminthosporium sacchari*	*Saccharum officinalis*	Strobel (1974a), Strobel (1973), Strobel and Hess (1974), Steiner and Strobel (1971)

[a] Toxin preparations contained 0.04 to 0.07% nitrogen

disease, several of them have interesting physiologic effects and therefore can be used as probes for studying physiologic activities of plants.

Chemically, the sugar-containing toxins can be placed into three broad groups: glycosides, polysaccharides, and glycoproteins (Table 2). It is well to remember that the carbohydrate-containing toxins discussed below are but a few of the many toxins that have been described elsewhere (STROBEL 1977, SCHEFFER 1976, RUDOLPH 1976, PATIL 1974).

3.3.1.1 Glycosides

Fusicoccin (Fig. 2, Table 2) is a diterpene glucoside produced by *Fusicoccum amygdali* (BALLIO et al. 1968). The toxin, when administered to cuttings of peach and almonds, reproduces the symptoms of the disease caused on trees infected with the fungus. Several derivatives of fusicoccin have been prepared including dihydrofusicoccin, as well as fully acetylated and partially deacetylated forms of both fusicoccin and dihydrofusicoccin (BALLIO et al. 1971).

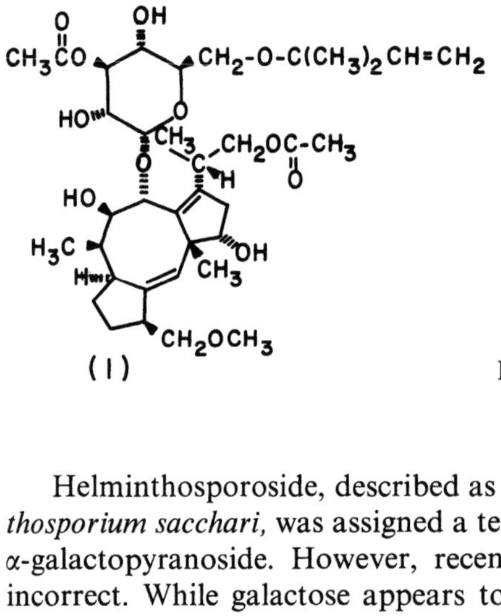

(I)

Fig. 2. The structure of fusicoccin

Helminthosporoside, described as a host-specific toxin produced by *Helminthosporium sacchari,* was assigned a tentative structure of 2-hydroxycyclopropyl-α-galactopyranoside. However, recent evidence suggests that the structure is incorrect. While galactose appears to be a component of helminthosporoside, there is considerable divergence in evidence concerning the composition of the remainder of the molecule (G.A. STROBEL personal communication). Additional work is needed for the final resolution of the structure.

A series of toxic glycosides appear to be produced by *Rhynchosporium secalis,* a fungus causing scald of barley. The name rhynchosporoside was given to one form of the toxin which was tentatively identified as a cellobioside of 1,2-propanediol (AURIOL et al. 1978). More recent evidence suggests that the preparation used in the earlier work consisted of a mixture of several toxin forms. It appears that the fungus produces a series of (1-*O*)α-D-glucosides of 1,2-propanediol. The presence of small amounts of 2- or 3- and 6-linked glucosyl residues were detected in some toxin preparations but it remains to be determined if any of them is associated with the toxin structure (BELTRAN et al. 1980).

3.3.1.2 Glycopeptides

Corynebacterium insidiosum toxin consists mainly of L-fucosyl residues with lesser amounts of glucosyl and galactosyl residues, an unknown ketodeoxy sugar, and trace quantities of mannosyl and rhamnosyl residues. The peptide comprises about 2.5% of the toxin. The toxin is estimated to be about 5×10^6 molecular weight. Copper is bound to the peptide portion but is not required for biological activity. The carbohydrate portion is linked to the hydroxyl group of threonine residues by *O*-glycoside linkages (REIS and STROBEL 1972a, b). A fucosyl transferase was detected associated with the membrane fraction of the bacterium and may be involved in the synthesis of the toxin (SADOWSKI and STROBEL 1973).

A glycopeptide from *C. sepedonicum*, also reported to be associated with wilt symptoms, is estimated to have a molecular weight of 20,000. Glucose and mannose are its major constituents and lesser quantities of rhamnose, galactose, ribose, and arabinose have been detected (STROBEL 1970, 1977, STROBEL et al. 1972). The toxin also contains about 10% 2-keto-3-deoxy-gluconate which accounts for its acidity. The toxin appears to have a highly branched structure with 1,2,6-linked glucosyl and mannosyl residues at branching points. Mannosyl and rhamnosyl residues occur at the terminals of the branched chains. The peptide comprises only 5% by weight of the toxin and is linked by seryl and threonyl residues to terminal mannosyl residues (STROBEL et al. 1972).

A glycoprotein, malseccin, has been isolated from cultures of *Phoma tracheiphila* which reproduces symptoms of the disease caused by the fungus on citrus trees. The toxin has a molecular weight of 9.3×10^5; it contains 29.5% carbohydrate and 36% peptide (NACHMIAS et al. 1979). The remaining portion of the toxin remains unidentified (Table 2). The carbohydrate portion appears to be linked to the peptide through seryl and threonyl residues. Removal of the peptide by proteinases abolishes toxin activity. *V. albo-atrum* produces a toxic extracellular lipoprotein polysaccharide complex with a carbohydrate composition reminiscent of the alkali-soluble, cell wall fraction of the fungus (KEEN and LONG 1972).

3.3.1.3 Polysaccharides

Amylovorin, a polysaccharide produced by *Erwinia amylovora*, consists of galactose (68%), glucuronate (19%) and glucose (8%). The galactosyl residues are primarily $(1 \rightarrow 3)$-linked with lesser amounts of $(1 \rightarrow 6)$ and $(1 \rightarrow 3,\ 1 \rightarrow 4)$-linkages; glucuronate residues are $(1 \rightarrow 4)$-linked and glucose units occur on the terminals of chains. The molecular weight is about 165,000 (unpublished results of M. MCNEIL, K. JOHNSON, and P. ALBERSHEIM, quoted in STROBEL 1977).

Xanthomonas campestris produces an extracellular polysaccharide which contains a backbone of $(1 \rightarrow 4)$-β-linked glucopyranosyl units with alternating glucose residues with side chains of glycopyranosyl units attached to the main chain through $(1 \rightarrow 3)$-linkages. The polymer size ranges from 1.4 to 3.6×10^6. The polysaccharide may provide a protective slime for the bacterium (SUTTON and WILLIAMS 1970). Wilt-inducing polysaccharides have been isolated from several *Phytophthora* species and shown to be branched $(1 \rightarrow 3)$-β-glucans high in (1 → 3, 1 → 6)-linked glucosyl and (1 → 3)-linked glucosyl units (Woodward et al. 1980, Keen et al. 1975).

3.3.2 Mode of Action of Toxins

Classification based on symptom production is at best tenuous since some toxins have more than one visible effect on plants. However, such classification has been useful since it directs the attention of the researcher to a possible mode of action of the toxin. Carbohydrate-containing toxins can be placed into broad groups on the basis of symptom production as noted below.

3.3.2.1 Toxins Interfering with Water Relations

Several carbohydrate-containing toxins reported to cause wilt in plants are listed in Table 2. Studies on structural requirements for biological activity reveal that the peptide portion of the toxin produced by *Corynebacterium sepedonicum* can be removed without loss of toxicity and the free carboxyl of 2-keto-2-deoxygluconate is necessary for toxicity (Johnson and Strobel 1970). Although the peptide portion of the glycopeptide isolated from *Corynebacterium insidiosum* appears to bind copper, chelation of copper is not essential for biological activity (Reis and Strobel 1972b). The toxin appears to interfere with stem conductance by plugging pit membranes (Van Alfen and Turner 1975a). Claims that amylovorin is host-specific (Goodman et al. 1974) has been challenged by Sjulin and Beer (1976), who reported that the material affects plants such as *Spirea* which is not a host for *Erwinia amylovora*. ^{14}C-labeled mycolaminarin administered to cuttings of host plants was extensively degraded and radioactive products widely distributed in the plant (Keen et al. 1972). Degradation could have occurred by action of β-glucanases which are known to increase in plants as a result of infection (Pegg and Vessey 1973) or ethylene treatment (Abeles et al. 1970). Radioactivity was widely distributed in cotton plant cuttings following administration of ^{14}C-labeled protein-lipopolysaccharide but degradation of the toxin was not studied (Keen et al. 1972). Such extensive degradation of administered toxin preparations complicates interpretation of their modes of action. Further, many wilt-evoking toxins produce other symptoms on plants. For example, malseccin causes necrosis, vein clearing and abscission of leaves in addition to wilt (Nachmias et al. 1979). Interference with water relations can trigger a variety of physiologic disturbances which could account for many of the symptoms besides wilting.

The studies of Van Alfen and Allard-Turner (1979) place size limits on polymers capable of causing wilt. Those which have dimensions in excess of pore sizes in pit membranes will function effectively as wilt toxins by plugging the pores and obstructing water movement in the xylem. Thus dextrans of molecular weights less than 250,000 were ineffective in obstructing water conductance in vascular elements. Further, only a small amount of the polysaccharide would be needed to interfere with water movement in the xylem; the quantity of dextran of 2×10^6 molecular weight required to obstruct vascular flow was 8 picomoles at petiole junctions and 0.4 picomole in leaflet veins. The action of a polysaccharide in wilting has no specificity but is related to its size and the size of the pores in pit membranes in the vascular elements of the host plant.

3.3.2.2 Toxins Affecting Membranes and Other Host Components

The most noticeable symptom caused by fusicoccin is wilt and desiccation of shoots. Stomata remain open and control of transpiration and gas exchange is lost. A major site of action appears to involve plasma membrane action since the toxin causes increases in net efflux of H^+, K^+ influx, and electrogenic cell membrane potential (MARRÈ 1979). As summarized by MARRÈ (1979) other membrane effects are stimulation of uptake of chloride and other anions, sodium ion extrusion, and uptake of carbohydrates and amino acids. Effects on metabolism include stimulation of respiration, dark CO_2 fixation, increase in pyruvate and glucose-6-phosphate levels. Effect on physiology and development include stimulation of cell enlargement, increase in stomatal opening against antagonism by abscisic acid, and promotion of seed germination. KURKDJIAN et al. (1979) reported that fusicoccin indirectly stimulated cell division by affecting distribution of growth substances through intra- and extracellular proton distribution. A drop in pH of nearly 2 units was observed in the extracellular fluid of sycamore cell cultures and concentration of intracellular 2,4-dichlorophenoxy acetic acid, which was used to study auxin distribution, increased eight fold. Cell division was stimulated when the intracellular auxin concentration reached a critical level.

GRONEWALD et al. (1979) noted certain similarities between effects of fusicoccin and the "washing response" of corn root tissue; increase in net efflux of H^+, K^+ influx, and electrogenic cell membrane potential were observed in both situations. The toxin stimulated the activity of isolated plasmalemma ATPase (BEFFAGNA et al. 1977).

Helminthosporoside appears to bind to a membrane-bound protein having a molecular weight of 48,000. The protein consists of four identical subunits and possesses at least two toxin-binding sites (STROBEL 1973). Binding of the toxin is reduced in the presence of α-galactosides when attached to the membrane-bound protein; the toxin may cause a conformational change which ultimately affects membrane permeability (STROBEL 1974a, STROBEL and HESS 1974).

Rhynchosporoside causes both chlorosis and necrosis to develop on leaves of susceptible barley plants. The toxin is not fully host-specific since it affects rye (*Secale cereale*) and certain wheat cultivars in addition to barley plants (AURIOL et al. 1978). The mechanism of action of this toxin and its related forms remains to be determined.

3.3.3 Role of Toxins in Disease

There are questions concerning the primary role of polysaccharides that cause wilting of plants. In some cases, wilting does not occur until advanced stages of disease are reached and only after other disease do symptoms become prominent. The work of VAN ALFEN and ALLARD-TURNER (1979) suggests that onset of plant wilting is probably a predictable outcome of most host–pathogen interactions involving xylem-inhabiting microorganisms that produce extracellular polysaccharides. In any case the interference with water flow has a major damaging influence on metabolic processes in the plant and effectively reduces yield of crop plants.

Toxins such as helminthosporoside, shown to be host-specific, have primary roles in disease. Similarly, toxins such as malseccin and fusicoccin, which are not host-specific, nevertheless reproduce symptoms of natural infection and are considered to be primary determinants of disease. Toxins such as fusicoccin and helminthosporoside, by their effects on host membrane function, will have extensive disruptive effects on cell compartmentation and ultimately host metabolism.

3.3.4 Conditions Affecting Production of Toxins

For the most part, toxic extracellular polysaccharides were produced in culture under rather unspecific conditions. An exception is amylovorin which is not produced in culture on laboratory media but must be isolated from infected plants or from cultures of tissues of the host. The basis for this phenomenon is unknown, but it is possible that host metabolites regulate production of the toxin in a manner reported for the *Helminthosporium sacchari*–sugar cane interaction. Continual production of helminthosporoside by subcultures of the fungus depends upon addition of certain constituents of the host. Otherwise, toxin production gradually diminishes (PINKERTON and STROBEL 1976). Sugar cane contains a compound, serinol, responsible for activating *H. sacchari* for toxin production. Sugar cane clones of the type resistant to *H. sacchari* do not produce activators of helminthosporoside production; those susceptible to the fungus produce the activator. Other examples of host regulation of pathogen toxin production undoubtedly will be uncovered in the future.

4 Carbohydrates in Host Response to Infection

4.1 Polysaccharide Elicitors of the Phytoalexin Response

4.1.1 Structure and Mode of Action of Elicitors

The term elicitor was first used to describe fungal preparations which stimulated accumulation of phytoalexins in plant tissues. Although a number of substances were found to exhibit elicitor activity (see Sect. 4.1.2), the most interest has been directed to carbohydrate-containing elicitors which could be extracted from cells of fungi, bacteria, and plants (Table 3).

Elicitor preparations evoke hypersensitive death of host cells in addition to causing accumulations of phytoalexins. Both phenomena constitute disease resistance reactions of the plant. Cell wall components in extracts of sonically disrupted mycelia of an incompatible isolate of *P. infestans* initiated the hypersensitive death of cells as well as rishitin accumulation in resistant potato tuber tissue (DOKE et al. 1979). Similar preparations of hyphal wall components of *P. infestans* also caused aggregation of protoplasm in potato tuber protoplasts (DOKE and TOMIYAMA 1980a).

Cell wall components isolated from *Colletotrichum lindemuthianum* elicited both browning and phytoalexin accumulation when applied to cotyledons of several different French bean varieties (ANDERSON 1978, ANDERSON-PROUTY and ALBERSHEIM 1975, THEODOROU and SMITH 1979). As with the previously described examples, the extent of their responses did not match the specific differential patterns of resistance and susceptibility of the varieties to the races. However, different bean varieties responded differently to a given race (THEODOROU and SMITH 1979).

Carbohydrates with elicitor activity are branched (1→3)-β-glucans; unbranched (1→3)-β-D-glucans, (1→4)-β-D-glucans, and (1→4)-α-D-glucans are inactive (AYERS et al. 1976a, b). Substituted (1→4)-β-D-glucans such as chitosans possess elicitor activity (HADWIGER et al. 1980). Mannan-enriched glycopeptides also show elicitor activity and terminal α-D-mannopyranosides may be essential for *P. megasperma* elicitor activity (AYERS et al. 1976a, b). Since compounds such as laminaribiose and methyl-β-D-glucopyranoside interfere with elicitor activity, it is suggested that the (1→3)-β-D-glucopyranoside configuration is concerned with binding of the *P. infestans* elicitor with the plant cell wall (MARCAN et al. 1979). Elicitor receptor sites may reside on plasma membranes of host cells since plant membrane preparations mixed with elicitor preparations reduced elicitor activity (DOKE et al. 1979).

The interaction between elicitor and plant cells initiates reactions culminating in the hypersensitive reaction and phytoalexin accumulation. Plant cells undergoing the hypersensitive reaction may release an "endogenous elicitor" which diffuses to and "switches on" phytoalexin synthesis in surrounding cells (HARGREAVES and BAILEY 1978). The "endogenous elicitor" concept helps explain how a variety of substances which injure cells can stimulate phytoalexin accumulation and offers an explanation for elicitor activity displayed by pectin-degrading enzymes. A glycoprotein possessing both endopolygalacturonase and elicitor activities has been purified to homogeneity from *Rhizopus stolonifer* (STEKOLL and WEST 1978, WEST and LEE 1980). Further, LYON and ALBERSHEIM (1980) reported that polygalacturonate lyase purified from *Erwinia carotovora* stimulates glyceollin synthesis in soybean hypocotyls, and suggested that the enzyme could release an endogenous elicitor from host cells either directly or indirectly by disrupting lysosomes and releasing degradative enzymes. Albersheim's group (HAHN et al. 1980) also reported that hydrolysates of wall polysaccharides from healthy soybean tissues elicited phytoalexin accumulation in soybean cells; those results lend further support to the idea that endogenous elicitors exist in plant cells and are released by reactions associated with cell injury.

4.1.2 Specificity of Elicitors

There has been considerable controversy regarding the specificity of elicitors since a variety of materials such as dyes, various DNA intercalating agents,

Table 3. Carbohydrate-containing elicitors of phytoalexins

Elicitor source	Composition	Phytoalexin elicited	Plants involved	References
Bacteria				
Erwinia carotovora	Polygalacturonase	Glyceollin	Soybean	LYON and ALBERSHEIM (1980)
Pseudomonas glycinea	Cell envelope, possibly carbohydrates involved	Glyceollin	Soybean	BRUEGGER and KEEN (1979)
Fungi				
Colletotrichum lindemuthianum	Glucan (1 → 3) and (1 → 4)-linked glucosyl residues; galactose, mannose	Phaseollin, Phaseollidin, Phaseollinisoflavan Kievitone	Bean (*Phaseolus vulgaris*)	ANDERSON-PROUTY and ALBERSHEIM (1975), THEODOROU and SMITH (1979)
C. trifoli	Glucan with mannose, galactose, rhamnose	Phaseollin, Phaseollidin, Phaseollinisoflavan, Kievitone	Bean	ANDERSON (1978)
C. destructivum	Glucan with mannose, galactose, rhamnose	Phaseollin, Phaseollidin, Phaseollinisoflavan, Kievitone	Bean	ANDERSON (1978)

Fulvia fulva	Glycopeptides with galactose, glucose, mannose, glucuronate, glucosamine, galactosamine	Rishitin	Tomato (*Lycopersicon esculentum*)	Dow and Callow (2979), De Wit and Roseboom (1980)
P. infestans	Glucan	Rishitin	Potato (*Solanum tuberosum*)	Doke et al. (1979), Marcan et al. (1979)
P. megasperma I	(1 → 3)-β-D-glucan; and (1 → 3, 1 → 6)-linked branched glucosyl residues	Glyceollins Rishitin	Soybean (*Glycine max*); potato	Ayers et al. (1976a, b), Cline et al. (1978)
P. megasperma II	Glycopeptide, branched mannan with glucose, glucosamine residues	Glyceollins	Soybean	Ayers et al. (1976a, b)
P. megasperma III	Glycopeptide, glucomannan	Glyceollins	Soybean	Ayers et al. (1976a, b)
P. megasperma IV	Glycopeptide; branched mannan with glucose, glucosamine	Glyceollins	Soybean	Ayers et al. (1976a, b)
R. stolonifer	Glycopeptide, MW. 32,000; 92% mannose, 8% glucosamine; has endopolygalacturonase activity	Casbene	Castor bean (*Ricinus communis*)	Stekoll and West (1978), West and Lee (1980)
Plants				
Soybean	Cell wall fractions	Glyceollin	Soybean	Hahn et al. (1980)

heavy metals, basic peptides, various enzymes, and fungal and bacterial polysaccharide preparations all have been reported to cause synthesis of phytoalexins. In most cases, only phytoalexin accumulation was measured. By measuring both rates of turnover and synthesis of glyceollin in soybean cotyledons treated with various substances, Yoshikawa (1978) observed that heavy metals caused decreased turnover but did not increase synthesis of glyceollin; the net effect was an increased accumulation of glyceollin. Fungal preparations, however, caused increased synthesis of glyceollin but had no effect on glyceollin turnover. If these results apply to elicitor-phytoalexin relationships in general, it would appear that the plant's reactions to microbial preparations are mechanistically distinct from those caused by heavy metals. Further, the endogenous elicitor concept will also help explain the apparent elicitor activity attributed to many noncarbohydrate compounds. The observations that elicitor preparations do not show the race specificity of live mycelia should not detract from the notion that carbohydrate polymers on pathogen cells represent host recognition sites. Undoubtedly, the procedures for extracting elicitors recover only a part of the structure concerned with interaction with the host cell. The suppressors found in *P. infestans* (Doke et al. 1979, Doke and Tomiyama 1980b), in *Fusarium solani* cultures (Daniels and Hadwiger 1976) and the protectors isolated from *P. megasperma* f. sp. *glycinea* (Wade and Albersheim 1979) appear to provide part of the specificity absent in elicitor preparations. Future studies should clarify the roles of these substances in providing specificity in the plant–pathogen interaction.

Little is known about the "receptor" sites on the plant cell. Current evidence supports the proposal that they occur on the plasma membrane. Since protoplasts from different cultivars showed differential reactivity with each elicitor-suppressor combination, it appears that receptor sites vary among the different cultivars and also may provide specificity in the recognition process (Doke and Tomiyama 1980b).

4.2 Altered Carbohydrate Metabolism in Host in Response to Infection

4.2.1 Altered Carbon Metabolism in Host

There is abundant evidence that disease interferes with carbohydrate metabolism in plants (Daly 1976). In virus-infected plants, for example, starch granules persist in chloroplasts and fail to undergo the usual light–darkness turnover (Tomlinson and Webb 1978, Carroll and Kosuge 1969). Photophosphorylation is altered (Magyarosy et al. 1973) and photoassimilation drops in many plant–pathogen interactions (Hall and Loomis 1972); transport of photoassimilates to sinks is reduced and in some cases reversed in leaves infected with biotrophic fungi. Respiration is increased in plant tissues as a result of infection and is accompanied by increased carbohydrate consumption by the Embden-Meyerhof and pentose phosphate pathways (Daly 1976). In some cases transient accumulation and turnover of starch occur (MacDonald and Strobel 1970). Unfortunately most of the studies on carbohydrate metabolism in infected plants were completed a decade or more ago, without benefit of the significant amount of information that has been developed during the past decade on regulation of carbohydrate metabolism in plants. In this connection studies on the biochemistry of plant–pathogen interactions should benefit from the following information.

Light regulation of enzymes of the chloroplast, including those in the reductive pentose phosphate pathway, is now well documented and insures that the cycle operates only if the plant has received adequate light (KELLEY et al. 1976, BASHAM 1979, SCHEIBE and BECK 1979). Light also indirectly inactivates certain enzymes of glycolysis and the oxidative pentose phosphate pathway (ANDERSON et al. 1974, HUBER 1979, KACHRU and ANDERSON 1975). Light inactivation of enzymes of glycolysis and the oxidative pentose phosphate pathway provides an effective switching system that turns on and off the systems for carbohydrate metabolism. For the most part, such mechanisms were ignored in studies on plant–pathogen interactions, despite the frequent observations that light affects symptom expression and pathogen development. Furthermore, the earlier suggestion by Preiss (PREISS and KOSUGE 1976) that orthophosphate provides effective control over carbohydrate metabolism now is well documented and should help explain many of the aberrations in carbohydrate metabolism in infected plants. For example, 3-phosphoglycerate (3 PGA) activates, but orthophosphate inhibits adenosine diphosphate glucose (ADPG) synthase, the key regulatory enzyme in starch biosynthesis. Thus a high 3 PGA to orthophosphate ratio promotes starch synthesis, but a low ratio favors starch turnover by inhibiting ADPG synthase and stimulating phosphorylase. Further, orthophosphate regulates photoassimilate export from the chloroplast since its uptake is coupled to efflux of triose phosphates via the phosphate translocator system (FLÜGGE et al. 1980, FLIEGE et al. 1978). During active photosynthesis, orthophosphate must be taken up by chloroplasts to replenish the supply depleted by the efflux of phosphate in the form of triose phosphates. Orthophosphate is needed to maintain the production of ATP via photophosphorylation which ultimately drives the reductive pentose phosphate cycle. However, in darkness when the reductive pentose phosphate pathway is inoperative, orthophosphate is consumed by phosphorolysis of starch and is necessary for starch turnover to occur (HELDT et al. 1977).

In virus-infected plants demands for viral nucleic acid synthesis may limit the availability of orthophosphate for carbohydrate metabolism. Shortage of phosphate would account for many of the aberrations in carbohydrate metabolism in virus-infected plants. Thus, starch turnover would be hampered and ATP production would be reduced in the chloroplast. This would in turn reduce photosynthesis, a common symptom of virus diseases of plants.

The sequestering of orthophosphate will also help explain impairment of source-sink movement of photoassimilates in virus-infected plants (PANOPOULUS et al. 1972). Without orthophosphate replenishment in the chloroplast, the movement of triose phosphate from the chloroplast to the cytoplasm is reduced and the initial processes of photoassimilate transport would be inhibited. Sinks such as storage organs are not filled.

4.2.2 Induction of Polysaccharide-Degrading Enzymes by Host Tissue in Response to Infection

The in vivo lysis of fungal hyphae in plant tissues has been observed in a number of plant–pathogen interactions and attributed to enzymes produced by host tissues. However, enzymes associated with the phenomenon were only recently identified. With the identification of (1 → 3)-β-D-glucanase and chitinase activities in plant tissues (Abeles et al. 1970), several researchers have shown correlations between hyphal lysis and the presence of hyphal wall-degrading enzymes in host tissue. Pegg and Vessey (1973) found chitinase activity to be constitutive in tomato plants, but found no positive correlation between possession of chitinase activity and resistance of tomato to *Verticillium albo-atrum*. Moreover, chitinase activity increased following infection. Further, they found the pH of xylem exudates of diseased plants to be high (6.8 to 7.1) and well beyond the pH optimum of 5.4 to 5.8 for chitinase. Lysis of hyphae by tomato chitinase preparations was not reported.

(1 → 3)-β-D-glucanase activity increased in melon plants infected with *Colletotrichum lagenarum* (Rabenantoando et al. 1976) as well as with *Fusarium oxysporum* f. sp. *melonis* (Netzer et al. 1979) but the increased activity at least in the former case was shown to be due to production in host tissue of glucanase by the pathogen. Interestingly, glucanase was either induced or activated in melon seedlings by dipping them in a solution of laminarin, a (1 → 3)-β-D-glucan. Seedlings normally susceptible to *Fusarium* became more resistant if pretreated with laminarin (Netzer et al. 1979).

4.2.3 Production of Polysaccharides by Host in Response to Infection

4.2.3.1 Production of Wall Appositions and Related Structures

It is commonly observed that plant cells respond to stress by forming appositions on the inner surface of walls opposite the point of stress applied on the outer wall (Aist 1976, Politis and Goodman 1978, Sherwood and Vance 1976, 1980). Wall appositions, called papillae, formed on barley epidermal cell walls in response to attempted penetration by the powdery mildew fungus may be a resistance mechanism (Israel et al. 1980). The papillae appear to be newly synthesized depositions on the inner wall of the plant cell and give positive reactions with callose-staining reagents. Hence, they are thought to be composed mainly of (1 → 3)-β-glucans; histochemical stains for other constituents gave positive reactions for lignin, cellulose, suberin, and protein (Aist 1976). Hydroxyproline-rich glycoproteins accumulate in cucumber (*Cucumis melo*) seedlings infected by *C. lagenarum* (Esquerrè-Tugaya and Lamport 1979) and correlated with increased seedling resistance to infection by the fungus (Esquerrè-Tugayè et al. 1979).

In plants in general, it is well documented that callose deposition is particularly stimulated by injury (Currier 1957); the enzyme, UDP-glucose: (1 → 3)-β-D-glucosyl transferase, which catalyzes the synthesis of (1 → 3)-β-glucans is widespread in plants (Brett 1978, Anderson and Ray 1978). However, more information is needed on the chemical composition of the appositions and the processes by which they are synthesized. In this connection, microsurgical procedures have been devised for collection of the appositions for microchemical analysis (KUNOH et al. 1979).

4.2.3.2 *Vascular Gelation Phenomenon*

Formation of gels in vascular elements is commonly seen in plants suffering from vascular diseases. Gel formation is stimulated by both pathogenic and nonpathogenic microorganisms. However, the relative persistence of the gels appears to determine whether the infection by pathogens is localized or becomes systemic (BECKMAN 1969). The gel appears along the walls of the xylem vessels and later fills the entire vessel lumen. Since administration of $^{14}CO_2$ to plants did not result in incorporation of radioactivity into gels during the most active part of their formation, it appears that distension of constituents of the primary wall and the middle lamella and not de novo synthesis contributes most to gel formation (VANDERMOLEN et al. 1977). The phenomenon can be triggered by introduction of polygalacturonase or polygalacturonate lyase into cuttings of tomato and other plants (COOPER and WOOD 1980). Gels collected from banana stems infected with *Fusarium oxysporum* contain high concentrations of galacturonate and other constituents of host cell walls (VANDERMOLEN and LABAVITCH unpublished results). The complete characterization of gel constituents and analysis of cell wall constituents of the healthy banana stem should help identify the origin of the material.

4.3 Toxic Host Glycosides in Plant–Pathogen Interactions

4.3.1 Structure and Metabolism of Toxic Glycosides

Plants contain a variety of glycosides which are either toxic to microorganisms or are converted by enzyme action to products that are antimicrobial. Potentially the compounds discussed below represent barriers to ingress by plant pathogens. For detailed additional discussions on the chemistry and biochemistry of this group of compounds, the reader is referred to articles in *Secondary Plant Products,* Volume 8 in this series.

4.3.1.1 *Saponins*

A number of saponins have been described which are either themselves toxic to microorganisms or are converted by enzymes to products with antibiotic activity. Two saponins, avenacosides A and B, occur in leaves of oat (*Avena sativa*) plants. Although not toxic themselves, they can be converted by oat leaf enzymes to toxic products, the 26-desglucoavenacosides (Fig. 3). The reaction appears to occur when plant tissue is injured and loss of compartmentation occurs (LÜNING and SCHLÖSSER 1976). The desglucoavenacosides have hemolytic activity and are implicated as disease-resistance factors.

Oat plants contain another saponin, avenacin (Fig. 4), which occurs in high concentrations in roots. The *N*-methylanthranilate moiety associated with

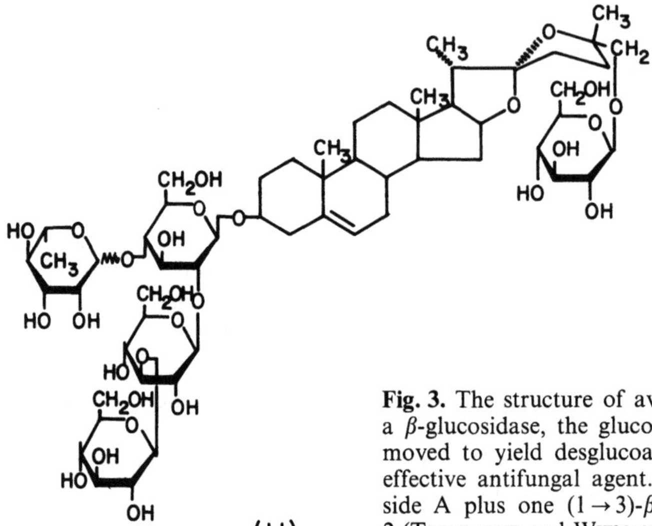

(II)

Fig. 3. The structure of avenacoside A. By action of a β-glucosidase, the glucose at the 26-position is removed to yield desglucoavenacoside A which is an effective antifungal agent. Avenacoside B = avenacoside A plus one (1→3)-β-Glcp attached to glucose 2 (Tschesche and Wiemann 1977)

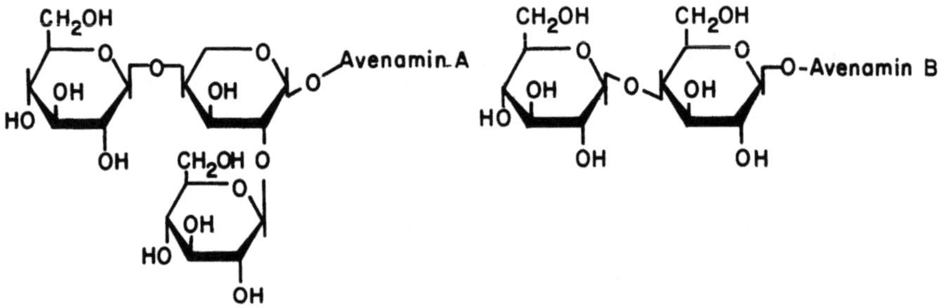

Avenacin A (IV) Avenacin B (V)

Fig. 4. The carbohydrate structure of avenacins A and B. The aglycones, avenamins A and B, are steroidal compounds containing *N*-methylanthranilate; but the structure of the aglycone for avenacin A proposed by Burkhardt et al. (1964) has been questioned by Tschesche et al. (1973)

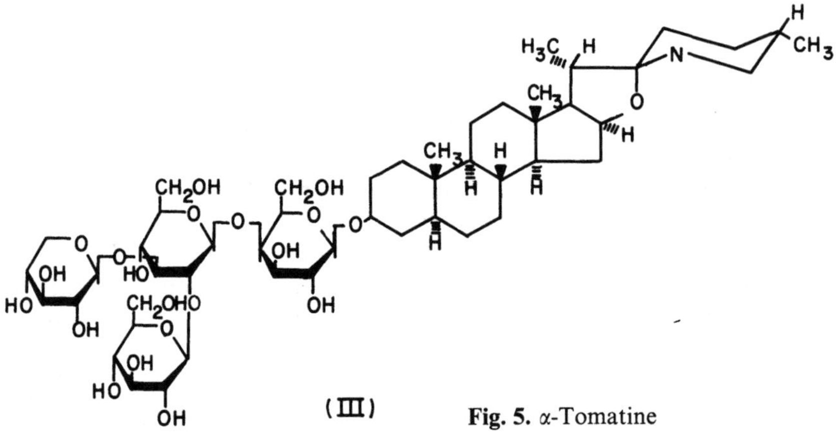

(III)

Fig. 5. α-Tomatine

avenacin imparts an intense fluorescence to the compound which is strongly visible when oat roots are examined under ultraviolet light. Removal of the anthranilate or a carbohydrate moiety reduces or abolishes the antibiotic properties of the compound.

Tomato leaves and green fruit contain a steroidal glycoalkaloid, α-tomatine (Fig. 5), which occurs in high concentration in leaves and green fruit but decreases in concentration in fruit during ripening (SCHLÖSSER 1975). In common with avenacin, removal of any one of the carbohydrate moieties reduces or abolishes the toxicity of α-tomatine.

4.3.1.2 Cyanogenic Glucosides

Because they release HCN and a toxic aldehyde upon decomposition, cyanogenic glucosides have been studied as possible resistance factors in plant–pathogen interactions. Cyanogenic compounds that occur in higher plants are described by CONN (1980). In sorghum plants, the cyanogenic compounds are deposited in vacuoles and are physically separated from the β-glucosidase which occurs in the cytoplasm (SAUNDERS and CONN 1978). Upon injury to the cell, compartmentation is lost, substrate and enzyme are mixed, and decomposition of the cyanogenic glucoside occurs. Decomposition of the compounds after cell injury is rapid and almost complete release of HCN may occur within a few hours (AKAZAWA et al. 1960).

4.3.1.3 Glucosinolates

These thioglucosides are secondary metabolites characteristic of Cruciferae and ten other plant families (BENN 1977, UNDERHILL 1980). As with other compounds in the foregoing sections these compounds remain essentially intact in uninjured tissue but undergo rapid, extensive degradation after plant tissue is injured. The glucosinolates themselves are nontoxic but the breakdown products isothiocyanates and nitriles have antibiotic or other biological activity. The glucosinolates described by UNDERHILL (1980) are representative of those that occur in food crops and studied in connection with plant–pathogen interactions.

4.3.1.4 Phenolic Glucosides

β-Glucosides of phenolic compounds occur commonly in plants and a number of them have been studied as possible resistance factors in plant–pathogen interactions (COUTURE et al. 1971, HILDEBRAND et al. 1969, HARBORNE 1980, RAA 1968). Phenolic aglycones, produced by β-glucosidase activity, are readily oxidized by the well-known ubiquitous polyphenol oxidase system to yield reactive, toxic quinones. The quinones self-polymerize or form colored adducts with proteins and other cellular constituents. For structures of several phenolic glucosides studied in plant–pathogen interactions, see HARBORNE (1980).

4.3.1.5 Other Glycosides

Pigments called betalains commonly occur as glycosides in the Centrospermae. These compounds are reported to possess antibiotic activity but their role, if any, in disease resistance needs further investigation.

Betamin, a common form of those pigments, occurs as the red pigment of table beets. Gomphrenin, another glucoside of betanidin, occurs in *Gomphrena globosa,* a plant commonly used for assays of plant virus titer (MABRY 1980).

The tuliposides are another group of interesting glycosides which yield toxic products by degradation (see Sect. 4.3.2).

4.3.2 Role of Toxic Glycosides in Host–Pathogen Interactions

A saponin possesses a polar carbohydrate portion and a steroidal, nonpolar end that allows it to function as a membrane lytic agent. Further, to function as an antifungal agent, the polar end must contain a certain structural arrangement of several carbohydrates. The removal of one carbohydrate from a "cluster" of four in α-tomatine, for example, abolishes antimicrobial activity (Arneson and Durbin 1967). It appears that such structural requirements for activity pertain to other steroidal glycosides with antimicrobial activity. Thus, reactions concerned with the removal of the carbohydrates are effective detoxifying mechanisms.

The avenacosides appear to be potentially important in resistance of oat plants to pathogens. The hydrolytic reaction removing glucose could be initiated by cell injury when leaf tissue is penetrated by a potential pathogen and the desglucoavenacosides would be released at the point of injury. The chemicals have antimicrobial activity as well as hemolytic activity (Lüning and Schlösser 1976) and might also disrupt membrane function in microorganisms. The conversion of an inactive form to an active antibiotic agent by a host glycoside is an attractive mechanism for studying disease resistance.

Avenacin appears to protect oat roots against wheat isolates of *Ophiobolus graminis*, the fungus-causing take-all disease of cereals. Growth of wheat isolates is fully inhibited by the saponin at 33 μg/ml on a minimal nutrient medium (Maizel et al. 1964). Oat isolates (var. *avenae*) produce an enzyme, avenacinase, which catalyzes the removal of the pentose moiety and thereby detoxifies avenacin (Turner 1961, Burkhardt et al. 1964). The oat isolate therefore can colonize oat roots by inactivating the toxin. Little is known about the mode of action of avenacin on fungi; since it is known to possess hemolytic activity, it is presumed to affect membranes of fungi.

Arneson and Durbin (1968) found a number of fungi to be inhibited by α-tomatine. Conversion of α-tomatine to β-tomatine by *Septoria lycopersici* effectively detoxifies the compound and permits the pathogen to breach the chemical barrier of resistance imposed by α-tomatine (Arneson and Durbin 1967, 1968). A fungal enzyme which catalyzes the conversion of α-tomatine to β_2-tomatine has been partially characterized (Durbin and Uchytil 1969). The protonated form of α-tomatine occurs at pH's above 6.0 (Arneson and Durbin 1968). It is claimed that the acidity of green fruits (pH 4.0 to 4.5) keeps α-tomatine in a protonated form which is nontoxic. Thus the potential fungicide barrier functions only if the pH of the inoculation site rises to 6.0 or higher (Schlösser 1975).

The release of isothiocyanates by degradation of glucosinolates during tissue disruption potentially is an important defense reaction. Isothiocyanates show toxicity toward a number of microorganisms, but studies on correlation between glucosinolate content and resistance have not yielded data that support this role in resistance to club root (Walker and Stahmann 1955). On the other hand, in both cultivated and wild cabbage (*Brassica oleracea*), high isothiocyanate concentration was correlated with resistance to downy mildew caused by *Peronospora parasitica* (Greenlagh and Mitchell 1976).

An interesting role of glucosinolates is implicated in club root of cabbage (*Brassica oleraceae*), a disease characterized by enlargement and proliferation cells in roots, ultimately giving rise to swollen, club-like roots. Glucobrassicin content increased nearly 3-fold and indoleacetonitrile increased 30-fold in infected tissue (BUTCHER et al. 1974). During infection indoleacetonitrile synthesis in this plant possibly occurs through degradation of glucobrassicin by the action of myrosinase. In the process of development of this organism in host tissue, it forms an intracellular, wall-less plasmodium which remains in intimate contact with host cytoplasm; thus there would be ample opportunity to cause sufficient changes in host cell compartmentation and metabolism to initiate the degradation of glucobrassicin. In this case, the presence of glucobrassicin might be necessary for the development of disease symptoms.

Injury or loss of compartmentation of plant tissues initiates the process of decomposition of cyanogenic glucosides culminating in the release of HCN. Both host and pathogen produce β-glucosidases which catalyze the decomposition of lotaustralin and linamarin in birdsfoot trefoil infected with *Stemphylium loti* (MILLAR and HIGGINS 1970). Pathogens, moreover, produce an inducible HCN-metabolizing system which converts HCN to formamide and ultimately allows utilization of HCN for a source of carbon and nitrogen. Further, pathogens of birdsfoot trefoil form an inducible cyanide-resistant pathway which allows the organism to survive and grow in the presence of HCN. Consequently pathogens are well adapted for colonizing cyanogenic plants. Nonpathogens lack the two adaptive systems (FRY and EVANS 1977).

Perhaps the best-documented case for a compound involved in disease resistance concerns the onion-*Colletotrichum circinans* interaction (WALKER and STAHMANN 1955). Catechol and protocatechuic acid were identified as the major toxic chemicals which protected the outer scales of colored onion bulbs against penetration by the fungus. White bulbs did not produce the chemicals and were susceptible to penetration and decay by the fungus. Protection was lost if the outer dried scales were removed from the colored bulbs, exposing the fleshy tissue which was readily penetrated and colonized by the fungus. Fleshy scales were low in catechol and protocatechuic acid. The source for catechol and protocatechuic acid could be degradation of the aglycones of the glycosides of the flavonoid compounds, quercitin and cyanidin, which are the pigments in the colored varieties.

The two acyl glucosides of hydroxymethylene butyric acid which occur in tulip plants undergo enzymatic decomposition to yield glucose and the corresponding aglycones (BEIJERSBERGEN and LEMMERS 1972, SLOB et al. 1975). The aglycones can undergo lactonization under acidic conditions or, above pH 7, remain in anionic form. The glucosides do not possess antimicrobial activity. However, lactones inhibit common plant pathogens. During developmental periods when tuliposide content is low, tulip bulbs are susceptible to infection by *Fusarium oxysporum* f. sp. *tulipae;* they are resistant to the fungus at other times when tuliposide content is high. The resistance of tulip bulbs to certain potential pathogens may be due to the formation of the toxic lactones during penetration. Another tulip pathogen, *Botrytis tulipae* is relatively insensitive to lactone B and may also escape inhibition by providing conditions, during infection, that maintain the aglycones as free acids (SCHÖNBECK and SCHROEDER 1972).

4.4 Concluding Remarks

In plant–pathogen interactions, involvement of carbohydrates begins with the acquisition of nutrients prior to initiation of infection, and continues during ingress and establishment of pathogens in host tissue. For some pathogens utilization of plant structural components is necessary for pathogenesis. Ingress may be halted if the potential pathogen encounters a toxic glycoside produced by the plant. The compounds are stored in vacuoles but are released and degraded when plant cells are injured by a potential pathogen or a herbivore. The toxic aglycones that are released can be deterrents to potential pathogens and herbivores. Pathogens overcome such barriers by either detoxifying or being resistant to the chemical.

Some pathogens produce carbohydrate derivatives that kill or weaken their host plants. Where host specificity is involved, the possession of certain carbohydrate structures may determine the outcome of a pathogen's initial ingress into plant tissues. Avirulent bacteria are immobilized by being bound to sites on plant cells; but extracellular polysaccharides protect virulent bacteria from being immobilized. On the other hand, capacity to initiate infection by *Agrobacterium* requires binding perhaps mediated by polysaccharides on the bacterial cells and sites on the host cell.

Recognition mediated by polysaccharides and host cell components triggers hypersensitive death of host cells and production of toxic compounds which deters further ingress of plant tissue by incompatible strains of fungi. Compatible strains of fungi produce glycopeptides which suppress the hypersensitive reaction. Once established, pathogens may interfere with metabolism in plants; disruption carbohydrate metabolism is a major factor reducing quality and yield of economically important plants.

A number of interesting problems await solution by researchers. Recently, it has been reported that binding of *Phytophthora* zoospores to plant roots may be mediated by root cap slime and L-fucose receptors on the zoospore surface (HINCH and CLARKE 1980). Since binding initiates steps that ultimately can result in successful infection of the root by the fungus, it will be important to understand the basis for this interaction. Soil-borne plant pathogens such as *Phytophthora* cause diseases of many economically important plants.

It will be interesting to further pursue the significance of the agglutination of plant protoplasts by fungal wall components (PETERS et al. 1978), and the agglutination of bacterial cells by plant products (ANDERSON and JASALAVICH 1979). Promise for future fruitful investigations is evident in the observations that elicitors affect metabolism in plant cells (KUĆ 1976, KUĆ et al. 1979), that binding of bacterial cells promotes induced resistance in plants (SEQUEIRA et al. 1977), and that growth substances such as ethylene stimulate secretion of polysaccharidases in plant tissue (ABELES and LEATHER 1971), particularly since ethylene evolution occurs commonly in infected plants. Since metabolism in diseased plants shows many characteristics of metabolism in aging or senescing tissue, it may be useful to apply the approaches of ADAMS and ROWAN (1970) to study carbohydrate turnover in host–pathogen interactions.

References

Abeles FB, Leather GR (1971) Abscission: Control of cellulase secretion by ethylene. Planta 97:87–91

Abeles FB, Bosshart RP, Forrence LE, Habig WH (1970) Preparation and purification of glucanase and chitinase from bean leaves. Plant Physiol 47:129–134

Adams PB, Rowan KS (1970) Glycolytic control of respiration during aging of carrot root tissue. Plant Physiol 45:490–494

Aist JR (1976) Papillae and related wound plugs of plant cells. Annu Rev Phytopathol 14:145–163

Akazawa T, Miljanich P, Conn EE (1960) Studies on cyanogenic glycoside of *Sorghum vulgare*. Plant Physiol 35:535–538

Albersheim P, Anderson-Prouty AJ (1975) Carbohydrates, proteins, cell surfaces, and the biochemistry of pathogenesis. Annu Rev Plant Physiol 26:31–52

Anand VK, Pueppke SG, Heberlein GT (1977) The effect of lectins on *Agrobacterium tumefaciens* caused crown gall tumor induction. Plant Physiol Suppl 59:109

Anderson AJ (1978) Isolation from three species of *Colletotrichum* of glucan-containing polysaccharides that elicit phytoalexin production in bean. Phytopathology 68:189–194

Anderson AJ, Jasalavich C (1979) Agglutination of pseudomanad cells by plant products. Physiol Plant Pathol 15:149–159

Anderson LE, Toh-Chin Lim Ng, Kyung-Eun Yoon Park (1974) Inactivation of pea leaf chloroplastic and cytoplasmic glucose 6-phosphate dehydrogenases by light and dithiothreitol. Plant Physiol 53:835–839

Anderson RL, Ray PM (1978) Labeling of the plasma membrane of pea cells by a surface-localized glucan synthetase. Plant Physiol 61:723–730

Anderson-Prouty AJ, Albersheim P (1975) Host-pathogen interactions VIII. Isolation of a pathogen-synthesized fraction rich in glucan that elicits a defense response in the pathogen's host. Plant Physiol 56:286–291

Arneson PA, Durbin RD (1967) Hydrolysis of tomatine by *Septoria lycopersici*: a detoxification mechanism. Phytopathology 57:1358–1360

Arneson PA, Durbin RD (1968) Studies on the mode of tomatine as a fungitoxic agent. Plant Physiol 43:683–686

Arthur LO, Bulla LA Jr, St Julian G, Nakamura L (1973) Carbohydrate metabolism in *Agrobacterium tumefaciens*. J Bacteriol 116:304–313

Auriol P, Strobel G, Pio Beltram J, Gray G (1978) Rhynchosporoside, a host-selective toxin produced by *Rhynchosporium secalis*, the causal agent of scald disease of barley. Proc Natl Acad Sci USA 73:4339–4343

Ayers AR, Ebel J, Valent B, Albersheim P (1976a) Host-pathogen interactions X. Fractionation and biological activity of an elicitor isolated from the mycelial walls of *Phytophthora megasperma* var. sojae. Plant Physiol 7:760–765

Ayers AR, Valent B, Ebel J, Albersheim P (1976b) Host-pathogen interactions XI. Composition and structure of wall-released elicitor fractions. Plant Physiol 57:766–774

Ballio A, Brufani M, Casinovi CG, Cerrini S, Fideli W, Pellicciari R, Vaciaqo A (1968) The structure of fusicoccin. Experentia 24:631–635

Ballio A, Pocchiari F, Russi S, Silano V (1971) Effects of fusicoccin and some related compounds on etiolated pea tissues. Physiol Plant Pathol 1:95–104

Barash I, Klein L (1969) The surface localization of polygalacturonase in spores of *Geotrichum candidum*. Phytopathology 59:319–324

Barnett NM (1974) Release of peroxidase from soybean hypocotyl cell walls by *Sclerotium rolfsii* culture filtrates. Can J Bot 52:265–271

Basham HG, Bateman DF (1975a) Killing of plant cells by pectic enzymes: the lack of direct injurious interaction between pectic enzymes or soluble reaction products and plant cells. Phytopathology 65:141–153
Basham HG, Bateman DF (1975b) Relationship of cell death in plant tissue treated with a homogeneous endopectate lyase to cell wall degradation. Physiol Plant Pathol 5:249–262
Bassham JA (1979) The reductive pentose phosphate cycle and its regulation. In: Gibbs M, Latzko E (eds) Encyclopedia of plant physiology, vol VI. Springer, Berlin Heidelberg New York pp 9–30
Bateman DF (1976) Plant cell wall hydrolysis by pathogens. In: Friend J, Threlfall DR (eds) Biochemical aspects of plant-parasite relationships. Academic Press, London New York, pp 79–103
Bateman DF, Basham HG (1976) Degradation of plant cell walls and membranes. In: Pirson A, Zimmerman MH (eds) Encyclopedia of plant physiology, new series vol.IV. Springer, Berlin Heidelberg New York, pp 316–355
Bateman DF, Van Etten HD, English PD, Nevins DJ, Albersheim P (1969) Susceptibility to enzymatic degradation of cell walls from bean plants resistant and susceptible to *Rhizoctonia solani* Kuhn. Plant Physiol 44:641–648
Bauer WD, Talmadge KW, Keegstra K, Albersheim P (1973) The structure of plant cell walls II. The hemicellulose of the walls of suspension-cultured sycamore cells. Plant Physiol 51:174–187
Beckman CH (1969) The mechanics of gel formation by swelling of simulated plant cell wall membranes and perforation plates of banana root vessels. Phytopathology 59:837–843
Beffagna N, Cocucci S, Marré E (1977) Stimulating effect of fusicoccin on K-activated ATPase in plasmalemma preparations from higher plant tissues. Plant Sci Lett 8:91–98
Beijersbergen JCM, Lemmers CBG (1972) Enzymic and nonenzymic liberation of tulipalin A (α-methylene butyrolactone) in extracts of tulip. Physiol Plant Pathol 2:265–270
Beltran JP, Strobel GA, Beier R, Mundy BP (1980) Some synthetic phytotoxins structurally related to Rhynchosporoside. Plant Physiol 65:554–556
Benn M (1977) Glucosinolates. Pure Appl Chem 49:197–210
Berard DF, Kuć J, Williams EB (1972) A cultivar-specific protection factor from incompatible interactions of green bean with *Colletotrichum lindemuthianum*. Physiol Plant Pathol 2:123–127
Brett CT (1978) Synthesis of β-(1→3)-glucan from extracellular uridine diphosphate glucose as a wound response in suspension-cultured soybean cells. Plant Physiol 62:377–382
Bruegger BB, Keen NT (1979) Specific elicitors of glyceollin accumulation in *Pseudomonas glycinea*-soybean host-parasite system. Physiol Plant Pathol 15:43–51
Bull AT, Trinci (1977) The physiology and metabolic control of fungal growth. Adv Microbiol Physiol 15:1–84
Burkhardt HJ, Maizel JV, Mitchell HK (1964) Avenacin, an antimicrobial substance isolated from *Avena sativa*. II. Structure. Biochemistry 3:426–431
Butcher DN, El-Tigani S, Ingram DS (1974) The role of indole glucosinolates in the club root disease of the Cruciferae. Physiol Plant Pathol 4:127–140
Carroll TW, Kosuge T (1969) Changes in structure of chloroplasts accompanying necrosis of tobacco leaves systemically infected with tobacco mosaic virus. Phytopathology 59:953–962
Cervone F, Scala A, Scala F (1978) Polygalacturonase from *Rhizoctonia fragariae*: further characterization of two isoenzymes and their action towards strawberry tissue. Physiol Plant Pathol 12:19–26
Chatterjee AK, Starr MP (1977) Donor strains of the soft-rot bacterium *Erwinia chrysanthemi* and conjugational transfer of the pectolytic capacity. J Bacteriol 132:862–869
Cline K, Wade M, Albersheim P (1978) Host-Pathogen interactions. XV. Fungal glucans which elicit phytoalexin accumulation in soybean also elicit the accumulation of phytoalexins in other plants. Plant Physiol 62:918–921
Conn EE (1980) Cyanogenic glucosides. In: Pirson A, Zimmerman MH (eds) Encyclopedia

of plant physiology, new series, vol VIII. Springer, Berlin Heidelberg New York, pp 461–492
Cooper RM (1977) Regulation of synthesis of cell wall-degrading enzymes of plant pathogens. In: Solheim B, Raa J (eds) Cell wall biochemistry related to specificity in host-pathogen interactions. Universitetsforlaget, Tromso, Oslo, Bergen
Cooper RM, Wood RKS (1975) Regulation of synthesis of cell wall degrading enzymes by *Verticillium albo-atrum* and *Fusarium oxysporum* f. sp. *lycopersici*. Physiol Plant Pathol 5:135–156
Cooper RM, Wood RKS (1980) Cell wall degrading enzymes of vascular wilt fungi. III. Possible involvement of endopectin lyase in *Verticillium* wilt of tomato. Physiol Plant Pathol 16:285–300
Cooper RM, Rankin B, Wood RKS (1978) Cell wall-degrading enzymes of vascular wilt fungi. II. Properties and modes of action of polysaccharidases of *Verticillium albo-atrum* and *Fusarium oxysporum* f. sp. *lycopersici*. Physiol Plant Pathol 13:101–134
Couture RM, Routley DG, Dunn GM (1971) Role of cyclic hydroxamic acids in monogenic resistance of maize to *Helminthosporium turcicum*. Physiol Plant Pathol 1:515–521
Currier HB (1957) Callose substance in plant cells. Am J Bot 44:478–488
Daly JM (1976) The carbon balance of diseased plants: changes in respiration, photosynthesis and translocation. In: Pirson A, Zimmermann MH (eds) Encyclopedia of Plant Physiology, new series, Vol. IV. Springer, Berlin Heidelberg New York, pp 450–479
Daniels DL, Hadwiger LA (1976) Pisatin-inducing components in filtrates of virulent and avirulent *Fusarium solani* cultures. Physiol Plant Pathol 8:9–19
Day PR (1974) Genetics of host-parasite interaction. WH Freeman and Company, San Francisco, 238
DeVay JE, Adler HE (1976) Antigens common to hosts and parasites. Annu Rev Microbiol 30:147–168
DeVay JE, Charudattan R, Wimalajeewa DLS (1972) Common antigenic determinants as a possible regulator of host-pathogen compatibility. Am Nat 106:185–194
De Wit PJGM, Roseboom PHM (1980) Isolation, partial characterization and specificity of glycoprotein elicitors from culture filtrates, mycelium and cell walls of *Cladosporium fulvum* (syn *Fulvia fulva*). Physiol Plant Pathol 16:391–408
Dickerson AG (1972 A β-D-fructofuranosidase from *Claviceps purpurea*. Biochem J 129:263–272
Dickerson AG, Mantle PG, Nisbet LJ, Shaw BI (1978) A role for β-glucanases in the parasitism of cereals by *Claviceps purpurea*. Physiol Plant Pathol 12:55–62
Doke N, Tomiyama K (1980a) Effect of hyphal wall components from *Phytophthora infestans* on protoplasts of potato tuber tissues. Physiol Plant Pathol 16:169–176
Doke N, Tomiyama K (1980b) Suppression of the hypersensitive response of potato tuber protoplasts to hyphal wall components by water soluble glucans isolated from *Phytophthora infestans*. Physiol Plant Pathol 16:177–186
Doke N, Garas NA, Kuć J (1979) Partial characterization and aspects of the mode of action of a hypersensitivity-inhibiting factor (HIF) isolated from *Phytophthora infestans*. Physiol Plant Pathol 15:127–140
Dow JM, Callow JA (1979) Partial characterization of glycopeptides from culture filtrates of *Fulvia fulva* (Cooke) Ciferri (syn *Cladosporium fulvum*), the tomato leaf mould pathogen. J Gen Microbiol 113:57–66
Durbin RD, Uchytil TF (1969) Purification and properties of a fungal glucosidase acting on α-tomatine. Biochim Biophys Acta 191:176–178
El-Banoby FE, Rudolph K (1979) Induction of water-soaking in plant leaves by extracellular polysaccharides from phytopathogenic pseudomonads and xanthomonads. Physiol Plant Pathol 15:341–349
El-Banoby FE, Rudolph K (1980) Purification of extracellular polysaccharides from *Pseudomonas phaseolicola* which induce water-soaking in bean leaves. Physiol Plant Pathol 16:425–437
Ellingboe AH (1972) Genetics and physiology of primary infection by *Erysiphe graminis*. Phytopathology 62:401–406
English PD, Jurale JB, Albersheim P (1971) Host-pathogen interactions. Parameters affect-

ing polysaccharide-degrading enzyme secretion by *Colletotrichum lindemuthianum* grown in culture. Plant Physiol 47:1–6
English PD, Maglothin A, Keegstra K, Albersheim P (1972) Cell wall-degrading endopolygalacturonase secreted by *Colletotrichum lindemuthianum*. Plant Physiol 49:293–297
Esquerré-Tugayé MT, Lamport DTA (1979) Cell surface in plant-microorganism interactions. I. A structural investigation of cell wall hydroxyproline-rich glycoproteins which accumulate in fungus-infected plants. Plant Physiol 64:314–319
Esquerré-Tugayé MT, Lafitte C, Mazau D, Toppan A, Touzé A (1979) Cell surfaces in plant-microorganism interactions. II. Evidence for the accumulation of hydroxyproline-rich glycoproteins in the cell wall of diseased plants as a defense mechanism. Plant Physiol 64:320–326
Fliege R, Flügge UI, Werdan K, Heldt HW (1978) Specific transport of inorganic phosphate, 3-phosphoglycerate and triosephosphates across the inner membrane of the envelope in spinach chloroplasts. Biochim Biophys Acta 502:232–247
Flor HH (1956) The complementary genetic systems in flax and flax rust. Adv Genet 8:29–54
Flügge UI, Freisl M, Heldt HW (1980) Balance between metabolite accumulation and transport relation to photosynthesis by isolated spinach chloroplasts. Plant Physiol 65:574–577
Fry WE, Evans PH (1977) Association of formamide hydro-lyase with fungal pathogenicity to cyanogenic plants. Phytopathology 67:1001–1006
Fukui S (1969) Conversion of glucose-1-phosphate to 3-ketoglucose-1-phosphate by cells of *Agrobacterium tumefaciens*. J Bacteriol 97:793–798
Fukui S, Hochster RM (1963) Conversion of disaccharides to their corresponding glycoside-3-ulose by intact cells of *Agrobacterium tumefaciens*. Can J Biochem Physiol 41:2361–2371
Furuichi N, Tomiyama K, Doke N (1980) The role of potato lectin in the binding of germ tubes of *Phytophthora infestans* to potato cell membrane. Physiol Plant Pathol 16:249–256
Gander JE (1974) Fungal cell wall glycoproteins and peptido-polysaccharides. Annu Rev Microbiol 28:103–119
Garas NA, Doke N, Kuć J (1979) Suppression of the hypersensitive reaction in potato tubers by mycelial components from *Phytophthora infestans*. Physiol Plant Pathol 15:117–126
Goodenough PW, Kempton RJ (1976) The activity of cell-wall degrading enzymes in tomato roots infected with *Pyrenochaeta lycopersici* and the effect of sugar concentrations in these roots on disease development. Physiol Plant Pathol 9:313–320
Goodman RN, Huang JS, Huang PY (1974) Host-specific polysaccharide from apple tissue infected by *Erwinia amylovora*. Science 183:1081–1082
Graham TL, Sequeira L, Huang TSR (1977) Bacterial lipopolysaccharide as inducers of disease resistance in tobacco. Appl Environ Microbiol 34:424–432
Greenlagh JR, Mitchell ND (1976) The involvement of flavour volatiles in the resistance to downy mildew of wild and cultivated forms of *Brassica oleracea*. New Phytol 77:391–398
Gronewald JW, Cheeseman JM, Hanson JB (1979) Comparison of the responses of corn root tissue to fusicoccin and washing. Plant Physiol 63:255–259
Gupta DP, Heale JB (1971) Induction of cellulase (C) in *Verticillium albo-atrum*. J Gen Microbiol 63:163–173
Hadwiger LA, Adams MJ, Beckman J (1980) The nuclear localization of hexoseamine molecules in relation to chotosan's role in disease resistance. Plant Physiol Suppl 65:136
Hahn MG, Darvill AG, Albersheim P (1980) Polysaccharide fragments from the walls of soybean cells elicit phytoalexin (antibiotic) accumulation in soybean cells. Plant Physiol Suppl 65:136
Hall AE, Loomis RS (1972) An explanation for the difference in photosynthetic capabilities of healthy and beet yellows virus-infected sugar beets (*Beta vulgaris* L.). Plant Physiol 50:576–580
Harborne JB (1980) Plant phenolics. In: Pirson A, Zimmerman MH (eds) Encyclopedia

of plant physiology, new series, vol VIII. Springer, Berlin Heidelberg New York, pp 329–402

Hargreaves JA, Bailey JA (1978) Phytoalexin production by hypocotyls of *Phaseolus vulgaris* in response to constitutive metabolites released by damaged bean cells. Physiol Plant Pathol 13:89–100

Hayano K, Fukui L (1967) Purification and properties of 3-ketosucrose-forming enzyme from the cells of *Agrobacterium tumefaciens*. J Biol Chem 242:3665–3672

Hayano K, Fukui S (1970) α-3-ketoglucosidase of *Agrobacterium tumefaciens*. J Bacteriol 101:692–697

Hayano K, Tsubouchi Y, Fukui S (1973) 3-ketoglucose reductase of *Agrobacterium tumefaciens*. J Bacteriol 113:652–657

Heldt HW, Chong Ja Chong, Maronde D, Herold A, Stankovic ZS, Walker DA, Kraminer A, Kirk MR, Heber U (1977) Role of orthophosphate and other factors in the regulation of starch formation in leaves and isolated chloroplasts. Plant Physiol 59:1146–1155

Hildebrand DC, Powell CC Jr, Schroth MN (1969) Fire blight resistance in Pyrus: Localization of arbutin and β-glucosidase. Phytopahtology 59:1534–1539

Hinch JM, Clarke AE (1980) Adhesion of fungal zoospores to root surface is mediated by carbohydrate determinants of the root slime. Physiol Plant Pathol 16:303–307

Hislop EC, Keon JPR, Fielding AH (1979) Effects of pectin lyase from *Monilinia fructigena* on viability, ultrastructure and localization of acid phosphatase of cultured apple cells. Physiol Plant Pathol 14:371–381

Horsfall JG, Dimond AE (1957) Interactions of tissue sugar, growth substances and disease susceptibility. Z Pflanzenkr Pflanzenschutz 64:415–421

Horton JC, Keen NT (1966) Sugar repression of endopolygalacturonase and cellulase synthesis during pathogenesis by *Pyrenochaeta terrestris* as a resistance mechanism in onion pink root. Phytopathology 56:908–916

Howell CR (1976) Use of enzyme-deficient mutants of *Verticillium dahliae* to assess the importance of pectolytic enzymes in symptom expression of *Verticillum* wilt of cotton. Physiol Plant Pathol 9:279–283

Hubbard JP, Williams JD, Niles RM, Mount MS (1978) The relation between glucose repression of endo-polygalacturonate trans-eliminase and adenosine 3′5′-cyclic monophosphate levels in *Erwinia carotovora*. Phytopathology 68:95–99

Huber SC (1979) Orthophosphate control of glucose-6-phosphate dehydrogenase light modulation in relation to the induction phase of chloroplast photosynthesis. Plant Physiol 64:846–851

Israel HW, Wilson RG, Aist JR, Kunoh H (1980) Cell wall appositions and plant disease resistance: acoustic microscopy of papillae that block fungal ingress. Proc Natl Acad Sci USA 77:2046–2049

Jannasch HW, Mateles RI (1974) Experimental bacterial ecology studied in continuous culture. In: Rose AH, Tempest DW (eds) Advances in Microbial Physiology vol 11, Academic Press, New York pp 165–210

Johnson TB, Strobel GA (1970) The active site on the phytotoxin of *Corynebacterium sepedonicum*. Plant Physiol 45:761–764

Kachru RB, Anderson LE (1975) Inactivation of pea leaf phosphofructokinase by light and dithiothreitol. Plant Physiol 55:199–202

Karr AL Jr, Albersheim P (1970) Polysaccharide-degrading enzymes are unable to attack plant cell walls without prior action by a "wall-modifying enzyme". Plant Physiol 46:69–80

Keegstra K, Talmadge KW, Bauer WD, Albersheim P (1973) The structure of plant cell walls. III. A model of the walls of suspension-cultured sycamore cells based on the interconnections of the macromolecular components. Plant Physiol 51:188–196

Keen NT, Long M (1972) Isolation of a protein-lipopolysaccharide complex from *Verticillium albo-atrum*. Physiol Plant Pathol 2:307–315

Keen NT, Williams PH (1971) Chemical and biological properties of a lipomucopolysaccharide from *Pseudomonas lachrymans*. Physiol Plant Pathol 1:247–264

Keen NT, Long M, Erwin DC (1972) Possible involvement of a pathogen-produced protein-lipopolysaccharide complex in *Verticillium* wilt of cotton. Physiol Plant Pathol 2:317–331

Keen NT, Wang MC, Bartnicki-Garcia S, Zentmyer GA (1975) Phytotoxicity of mycolaminarans-β-1,3-glucans from *Phytophthora* spp. Physiol Plant Pathol 7:91–97
Kelly GJ, Latzko E, Gibbs M (1976) Regulatory aspects of photosynthetic carbon metabolism. Ann Rev Plant Physiol 27:181–205
Kleinschuster SJ, Baker R (1974) Lectin-detectable differences in carbohydrate-containing surface moieties of macroconidia of *Fusarium roseum avenaceum* and *Fusarium solani*. Phytopathology 64:394–399
Kojima M, Uritani I (1978) Isolation and characterization of factors in sweet potato root which agglutinate germinated spores of *Ceratocystis fimbriata*, black rot fungus. Plant Physiol 62:751–753
Kuć J (1976) Phytoalexins and the specificity of plant-parasite interaction. In: Wood RKS, Graniti A (eds) Specificity in plant diseases. Plenum Publ Corp, New York London, pp 253–271
Kuć J, Henfling J, Garas N, Doke N (1979) Control of terpenoid metabolism in the potato-*Phytophthora infestans* interaction. J Food Protect 42:508–511
Kunoh H, Aist JR, Israel HW (1979) Microsurgical isolation of intact plant cell wall appositions for microanalyses. Can J Bot 57:1349–1353
Kurkdjian A, Leguay J, Guern J (1979) Influence of fusicoccin on the control of cell division by auxins. Plant Physiol 64:1053–1057
Lazarovits G, Higgins VJ (1979) Biological activity and specificity of a toxin produced by *Cladosporium fulvum*. Phytopathology 69:1056–1061
Lazarovits G, Bhullar BS, Sugiyama HJ, Higgins VJ (1979) Purification and partial characterization of a glycoprotein toxin produced by *Cladosporium fulvum*. Phytopathology 69:1062–1068
Lippincott BB, Whatley MH, Lippincott JA (1977) Tumor induction by *Agrobacterium* involves attachment of the bacterium to a site on the host plant cell wall. Plant Physiol 59:388–390
Lippincott JA, Lippincott BB (1978) Cell walls of crown-gall tumors and embryonic plant tissues lack *Agrobacterium* adherence sites. Science 199:1075–1077
Lüning HU, Schlösser E (1976) Role of saponins in antifungal resistance VI. Interactions *Avena sativa-Drechslera avenacea*. Z Pflanzenkr Pflanzenschutz 83(6):317–327
Lyon G, Albersheim P (1980) The nature of the phytoalexin elicitor of *Erwinia carotovora*. Plant Physiol 65:137
Mabry TJ (1980) Betalains. In: Pirson A, Zimmermann MH (eds) Encyclopedia of plant physiology new series, vol VIII. Springer, Berlin Heidelberg New York, pp 513–533
MacDonald PW, Strobel GA (1970) Adenosine diphosphate-glucose pyrophosphorylase control of starch accumulation in rust-infected wheat leaves. Plant Physiol 46:126–135
Magyarosy AC, Buchanan BB, Schurmann P (1973) Effect of a systemic virus infection on chloroplast function and structure. Virology 55:426–438
Maier RJ, Brill WJ (1978) Involvement of *Rhizobium japonicum* O antigen in soybean nodulation. J Bacteriol 133:1295–1299
Maizel JV, Burkhardt HJ, Mitchell HK (1964) Avenacin, an antimicrobial substance isolated from *Avena sativa*. I. Isolation and antimicrobial activity. Biochemistry 3:424–431
Marcan H, Jarvis MC, Friend J (1979) Effect of methyl glycosides and oligosaccharides on cell death and browning of potato tuber discs induced by mycelial components of *Phytophthora infestans*. Physiol Plant Pathol 14:1–9
Marrè E (1979) Fusicoccin: a tool in plant physiology. Annu Rev Plant Physiol 30:273–288
Mazzucchi U, Pupillo P (1976) Prevention of confluent hypersensitive necrosis in tobacco leaves by a bacterial protein-lipopolysaccharide complex. Physiol Plant Pathol 9:101–112
Mazzucchi U, Bazzi C, Pupillo P (1979) The inhibition of susceptible and hypersensitive reactions by protein-lipopolysaccharide complexes from phytopathogenic pseudomonads: relationship to polysaccharide antigenic determinants. Physiol Plant Pathol 14:19–30
Millar RL, Higgins VJ (1970) Association of cyanide with infection of birdsfoot trefoil by *Stemphylium loti*. Phytopathology 60:104–110
Moran F, Starr MP (1969) Metabolic regulation of polygalacturonic acid trans-eliminase in *Erwinia*. Eur J Biochem 11:291–295

Mount MS, Berman PM, Mortlock RP, Hubbard JP (1979) Regulation of endopolygalacturonate transeliminase in an adenosine 3′, 5′-cyclic monophosphate-deficient mutant of *Erwinia carotovora*. Phytopathology 69:117–120
Mullen JM, Bateman DF (1975) Polysaccharide degrading enzymes produced by *Fusarium roseum* "Avenaceum" in culture and during pathogenesis. Physiol Plant Pathol 6:233–246
Mussell HW (1973) Endopolygalacturonase: evidence for involvement in *Verticillium* wilt of cotton. Phytopathology 63:62–70
Mussell HW, Green RJ Jr (1970) Host colonization and polygalacturonase production by two tracheomycotic fungi. Phytopathology 60:192–195
Nachmias A, Barash I, Buchner V, Solel Z, Strobel GA (1979) A phytotoxic glycopeptide from lemon leaves infected with *Phoma tracheiphila*. Physiol Plant Pathol 14:135–140
Netzer D, Kritzman G, Chet I (1979) β-(1,3) Glucanase activity and quantity of fungus in relation to *Fusarium* wilt in resistant and susceptible near-isogenic lines of muskmelon. Physiol Plant Pathol 14:47–55
Nozue M, Tomiyama K, Doke N (1979) Evidence for adherence of host plasmalemma to infecting hyphae of both compatible and incompatible races of *Phytophthora infestans*. Physiol Plant Pathol 15:111–115
Ohyama K, Pelcher LE, Schaefer A, Fowke LC (1979) In vitro binding of *Agrobacterium tumefaciens* to plant cells from suspension culture. Plant Physiol 63:382–387
Panopoulos NJ, Faccioli G, Gold AH (1972) Translocation of photosynthate in curly top virus-infected tomatoes. Plant Physiol 50:266–270
Patil SS (1974) Toxins produced by phytopathogenic bacteria. Annu Rev Phytopathol 12:259–279
Patil SS, Dimond AE (1968) Repression of polygalacturonase synthesis in *Fusarium oxysporum* f. sp. by sugars and its effect on symptom reduction in infected tomato plants. Phytopathology 58:676–682
Pegg GF, Vessey JC (1973) Chitinase activity in *Lycopersicon esculentum* and its relationship to the in vivo lysis of *Verticillium albo-atrum* mycelium. Physiol Plant Pathol 3:207–222
Peters BM, Cribbs DH, Stelzig DA (1978) Agglutination of plant protoplasts by fungal cell wall glucans. Science 201:366–365
Pinkerton F, Strobel G (1976) Serinol as an activator of toxin production in attenuated cultures of *Helminthosporium sacchari*. Proc Natl Acad Sci USA 73:4007–4011
Politis DJ, Goodman RN (1978) Localized cell wall appositions: incompatibility response of tobacco leaf cells to *Pseudomonas pisi*. Phytopathology 68:309–316
Preiss J, Kosuge T (1976) Regulation of enzyme activity in metabolic pathways. From Plant Biochem 3rd edn. Academic Press, London New York, pp 277–336
Puhalla JE, Howell CR (1975) Significance of endopolygalacturonase activity to symptom expression of verticillium wilt in cotton, assessed by the use of mutants of *Verticillium dahliae* Kleb. Physiol Plant Pathol 7:147–152
Raa J (1968) Polyphenols and natural resistance of apple leaves against *Venturia inaequalis*. Neth J Plant Pathol Suppl 74:37–45
Rabenantoandro Y, Auriol P, Touzé A (1976) Implication of β-(1→3) glucanase in melon anthracnose. Physiol Plant Pathol 8:313–324
Rai PV, Strobel GA (1969) Phytotoxic glycopeptides produced by *Corynebacterium michiganense*. I. Methods of preparation, physical and chemical characterization. Phytopathology 59:47–52
Reis SM, Strobel GA (1972a) A phytotoxic glycopeptide from cultures of *Corynebacterium insidiosum*. Plant Physiol 49:676–684
Reis SM, Strobel GA (1972b) Biological properties and pathological role of a phytotoxic glycopeptide from *Corynebacterium insidiosum*. Physiol Plant Pathol 2:133–142
Rudolph K (1976) Non-specific toxins. In: Pirson A, Zimmermann MH (eds) Encyclopedia of plant physiology, new series, vol IV. Springer, Berlin Heidelberg New York, pp 270–315
Rudolph K (1978) A host specific principle from *Pseudomonas phaseolicola* (Burkh.) Dowson, inducing water-soaking in bean leaves. Phytopathol Z 93:218–226

Sadowski PL, Strobel GA (1973) Guanosine diphosphate-L-fucose glycopeptide fucosyltransferase activity in *Corynebacterium insidiosum*. J Bacteriol 115:668–672
Saunders JA, Conn E (1978) Presence of the cyanogenic glucoside dhurrin in isolated vacuoles from *Sorghum*. Plant Physiol 61:154–157
Scarborough GA (1970) Sugar transport in *Neurospora crassa*. J Biol Chem 245:1694–1698
Scheffer RP (1976) Host-specific toxins in relation to pathogenesis and disease resistance. In: Pirson A, Zimmerman MH (eds) Encyclopedia of plant physiology, new series, vol IV. Springer, Berlin Heidelberg New York, pp 247–269
Scheibe R, Beck E (1979) On the mechanism of activation by light of the NADP-dependent malate dehydrogenase in spinach chloroplasts. Plant Physiol 64:744–748
Schlösser E (1975) Role of saponins in antifungal resistance. III. Tomatin dependent development of fruit rot organisms on tomato fruits. Z Pflanzenkr Pflanzenschutz 84:476–484
Schönbeck F, Schroeder C (1972) Role of antimicrobial substances (tuliposides) in tulips attacked by *Botrytis* spp. Physiol Plant Pathol 2:91–99
Schulte TH, Scarborough GA (1975) Characterization of the glucose transport systems in *Neurospora crassa* sl. J Bacteriol 122:1076–1080
Sequeira L, Graham TL (1977) Agglutination of avirulent strains of *Pseudomonas solanacearum* by potato lectin. Physiol Plant Pathol 11:43–54
Sequeira L, Gaard G, De Zoeten GA (1977) Interaction of bacteria and host cell walls: its relation to mechanisms of induced resistance. Physiol Plant Pathol 10:43–50
Sherwood RT, Vance CP (1976) Histochemistry of papillae formed in reed canarygrass leaves in response to noninfecting pathogenic fungi. Phytopathology 66:503–510
Sherwood RT, Vance CP (1980) Resistance to fungal penetration in Gramineae. Phytopathology 70:273–279
Sjulin TM, Beer SV (1976) Evidence that amylovorin and *Erwinia amylovora* infection induce shoot wilt by different mechanisms. Proc Am Phytopathol Soc 3:360
Slob A, Jekel B, De Jong B, Schlatmann E (1975) On the occurrence of tuliposides in the Liliiflorae. Phytochemistry 14:1997–2005
Stack JP, Mount MS, Berman PM, Hubbard JP (1980) Pectic enzyme complex from *Erwinia carotovora*: a model for degradation and assimilation of host pectic fractions. Phytopathology 70:267–272
Steiner GW, Strobel GA (1971) Helminthosporside, a host-specific toxin from *Helminthosporium sacchari*. J Biol Chem 246:4350–4357
Stekoll M, West CA (1978) Purification and properties of an elicitor of castor bean phytoalexin from culture filtrates of the fungus *Rhizopus stolonifer*. Plant Physiol 61:38–45
Strand LL, Mussell H (1975) Solubilization of peroxidase activity from cotton cell walls by endopolygalacturonases. Phytopathology 65:830–831
Strand LL, Rechtoris C, Mussell H (1976) Polygalacturonases release cell-wall-bound proteins. Plant Physiol 58:722–725
Strobel GA (1970) A phytotoxin glycopeptide from potato plants infected with *Corynebacterium sepedonicum*. J Biol Chem 245:32–38
Strobel GA (1973) The helminthosporoside-binding protein of sugarcane. Its properties and relationship to susceptibility to the eye spot disease. J Biol Chem 248:1321–1328
Strobel GA (1974a) The toxin-binding protein of sugarcane, its role in the plant and disease development. Proc Natl Acad Sci USA 71:4232–4236
Strobel GA (1974b) Phytotoxins produced by plant parasites. Annu Rev Plant Physiol 25:541–566
Strobel GA (1977) Bacterial phytotoxins. Annu Rev Microbiol 31:205–204
Strobel GA, Hess WM (1974) Evidence for the presence of the toxin-binding protein on the plasma membrane of sugarcane cells. Proc Natl Acad Sci USA 71:1413–1417
Strobel GA, Talmadge KW, Albersheim P (1972) Observations on the structure of the phytotoxic glycopeptide of *Corynebacterium sepedonicum*. Biochim Biophys Acta 261:365–374
Strobel GA, Van Alfen N, Hapner KD, McNeil M, Albersheim P (1978) Some phytotoxic glycopeptides from *Ceratocystis ulmi*, the dutch elm disease pathogen. Biochim Biophys Acta 538:60–75
Sutton JC, Williams PH (1970) Comparison of extracellular polysaccharide of *Xanthomonas campestris* from culture and from infected cabbage leaves. Can J Bot 48:645–651

Theodorou MK, Smith IM (1979) The response of French bean varieties to components isolated from races of *Colletotrichum lindemuthianum*. Physiol Plant Pathol 15:297–309
Tomlinson JA, Webb MJW (1978) Ultrastructural changes in chloroplasts of lettuce infected with beet western yellows virus. Physiol Plant Pathol 12:13–18
Tschesche R, Wiemann W (1977) Desgluco-avenacosid-A und -B, biologisch aktive Nuatigeninglycoside. Chem Ber 110:2416–2423
Tschesche R, Jha HC, Wulff G (1973) Triterpenes XXIX. Structure of avanacines. Tetrahedron 29:623–633
Turner EMC (1961) An enzymatic basis for pathogenic specificity in *Ophiobolus graminis*. J Exp Bot 12:169–175
Turner NC, Graniti A (1976) Stomatal response of two almond cultivars to fusicoccin. Physiol Plant Pathol 9:175–182
Underhill EW (1980) Glucosinolates. In: Pirson A, Zimmermann MH (eds) Encyclopedia of plant physiology, new series, vol VIII. Springer, Berlin Heidelberg New York, pp 493–511
Van Alfen NK, Allard-Turner V (1979) Susceptibility of plants to vascular disruption by macromolecules. Plant Physiol 63:1072–1075
Van Alfen NK, Turner NC (1975a) Influence of *Ceratocystis ulmi* toxin on water relations of elm (*Ulmus americana*). Plant Physiol 55:312–316
Van Alfen NK, Turner NC (1975b) Changes in alfalfa stem conductance induced by *Corynebacterium insidiosum* toxin. Plant Physiol 55:559–561
Van Beeumen J, De Ley J (1975) Hexopyranoside: cytochrome c oxidoreductase from *Agrobacterium*. In: Colowick SP, Kaplan NO (eds) Methods in enzymology, vol XLI (B). Academic Press, London New York, pp 153–158
VanderMolen GE, Beckman CH, Rodehorst D (1977) Vascular gelation: a general response phenomenon following infection. Physiol Plant Pathol 11:95–100
Von Meyenburg K (1971) Transport-limited growth rates in a mutant of *Escherichia coli*. J Bacteriol 107:878–888
Wade M, Albersheim P (1979) Race-specific molecules that protect soybeans from *Phytophthora megasperma* var. sojae. Proc Natl Acad Sci USA 76:4433–4437
Walker JC, Stahmann MA (1955) Chemical nature of disease resistance in plants. Annu Rev Plant Physiol 6:351–366
Wang MC, Bartnicki-Garcia S (1973) Novel phosphoglucans from the cytoplasm of *Phytophthora palmivora* and their selective occurrence in certain life cycle stages. J Biol Chem 248:4112–4118
Wang MC, Bartnicki-Garcia S (1974) Mycolaminarians: storage β-1,3-D-glucans from the cytoplasm of the fungus *Phytophthora palmivora*. Carbohydr Res 37:331–338
Watson B, Currier TC, Gordon MP, Chilton M, Nester EW (1975) Plasmid required for virulence of *Agrobacterium tumefaciens*. J Bacteriol 123:255–264
West C, Lee SC (1980) Identity of casbene elicitor and endopolygalacturonase from culture filtrates of *Rhizopus stolonifer*. Plant Physiol 65:137
Whatley MH, Bodwin JS, Lippincott BB, Lippincott JA (1976) Role for *Agrobacterium* cell envelope lipopolysaccharide in infection site attachment. Infect Immun 13:1080–1083
Whatley MH, Hunter N, Cantrell MA, Hendrick C, Keegstra K, Sequeira L (1980) Lipopolysaccharide composition of the wilt pathogen, *Pseudomonas solanacearum*. Plant Physiol 65:557–559
Wiese MV, DeVay JE, Ravenscroft AV (1970) Relationship between polygalacturonase activity and cultural characteristics of *Verticillium* isolates pathogenic in cotton. Phytopathology 60:641–646
Woodward JR, Keane PJ, Stone BA (1980) Structures and properties of wilt-inducing polysaccharides from *Phytophthora* species. Physiol Pathol 16:439–454
Yoshikawa M (1978) Diverse modes of action of biotic and abiotic phytoalexin elicitors. Nature (London) 275:546–547
Zucker M, Hankin L, Sands D (1972) Factors governing pectate lyase synthesis in soft rot and non-soft rot bacteria. Physiol Plant Pathol 2:59–67

V. Lectin-Carbohydrate Interaction

25 Lectins and Their Physiological Role in Slime Molds and in Higher Plants

H. KAUSS

1 What Are Lectins?

It was recognized as early as the end of the last century that proteins ("Fermente") extracted from the seeds of certain plants exhibit the unusual property of agglutinating animal erythrocytes. Extracts from the seeds of various plant species showed a high degree of specificity when the appropriate source of red blood cells was chosen. For instance, pea extract was very effective in agglutinating erythrocytes from rabbit but far less so with those from sheep or pigeon, whereas human red blood cells were strongly agglutinated by bean and more weakly by pea or lentil extracts (reviewed by LIS and SHARON 1981). The phenomenon was not understood until SUMNER and HOWELL (1936) showed that a protein crystallized from the seeds of *Canavalia ensiformis* [concanavalin A (ConA) Table 1] was able to precipitate glycogen and starch out of solution and that agglutination of erythrocytes by the same protein was prevented by the presence of sucrose. Both findings taken together suggested that the agglutinins of plant origin ("phytohemagglutinins") may exert their action by binding to sugar residues located at the erythrocyte surface. It became evident later that certain phytohemagglutinins exhibit ABO blood group specificity. For this reason BOYD and SHAPLEIGH (1954) proposed the term "lectin" (lat. "legere" – to select, to pick out). As proteins or glycoproteins that exhibit hemagglutination activity can also be found in some animals and bacteria, the primary implication of "lectin" was broadened and this term is now in general use. Literature on lectins has increased so much in recent years that throughout this article citation must often be restricted to recent reviews or to the apparently most significant paper out of a series.

A protein/glycoprotein must fulfill several clear criteria to be classified as a lectin. It will be pointed out below (Sect. 2.1.2) that lectins are typically characterized by definite binding sites exhibiting a high degree of specificity for certain carbohydrates, a property best demonstrated when agglutination is inhibited on addition of saccharides or of soluble glycoproteins with the appropriate carbohydrate side chains. Although erythrocytes are frequently used, other cells (e.g., lymphocytes, yeasts, bacteria) may also be used, thus widening the range of naturally available "receptor" groups. Agglutination requires a minimum of two binding sites per lectin molecule to allow simultaneous complexing of at least two particles leading to a three-dimensional net. If monovalent lectin subunits (see Sect. 5.4) bind to the appropriate saccharides on cells, no agglutination can be observed, nevertheless the nature of the interaction is so similar that such monovalent subunits are generally classified as lectins.

Binding of proteins to polysaccharides, glycoproteins, or artificial matrices substituted with the appropriate carbohydrate residues (see Sect. 2.1) would be another criterion by which a protein may be designated a lectin, as long as binding is not followed by an enzymatic alteration of the bound ligand. With any assay method used, it is generally desirable to prove the carbohydrate-specific nature of the interaction by reversion or inhibition with low molecular weight hapten-like saccharides. It is often difficult to demonstrate this, however, as the binding to complex ligands may be so strong that inhibition is not possible.

The above cited criteria are all met by the well-studied lectins listed in Tables 1 and 2, a collection of what one could call "classical lectins". Definition of the term lectin becomes more difficult if biological properties are included. Classical lectins are considered to be devoid of any enzymatic activity. As will be discussed below (see Sect. 5.5) however, lectins might be structurally or evolutionarily related to enzymes, or may even show some enzymatic activity. In addition, some enzymes are able to stimulate mitosis in lymphocytes (O'BRIEN et al. 1978, PAUS and STEEN 1978), a property regarded as typical of lectins (see Sect. 3). A biochemist mainly interested in a clear-cut system for classifying enzymes and lectins will tend not to include any protein with catalytic properties in the lectin section. However, there is an increasing tendency to believe (LIENER 1976, KAUSS 1976, CALLOW 1977, RÜDIGER 1978, LIS and SHARON 1981) that the lectins represent an artificial group of proteins/glycoproteins which are grouped together solely on the basis of the assay method and which may have quite diverse functions in plants. This idea, although unorthodox, allows some overlap between enzymes and lectins. This does not mean that we should regard any enzyme of carbohydrate metabolism as a lectin because it binds carbohydrates at its active site prior to catalysis. However, from a physiological point of view it appears wise not to base the definition of a lectin on a strictly artificial system. One should keep an open mind to be able to evaluate physiological effects not to be explained solely on the basis of the catalytic property of an enzyme. This may in the future help to elucidate what functions lectins have in vivo.

As mentioned above, one of the major criteria for defining lectins is the demonstration of a binding site specific for distinct carbohydrate residues. The so-called "all-β-lectins" are macromolecular glycoproteins precipitated by "Yariv's artificial antigens" (copolymers prepared by coupling 4-aminophenyl glycosides to diazotized phloroglucinol). They do not appear to fulfill the above-cited criterion as they interact with the β-anomers of various sugars exhibiting quite different structures (CLARKE et al. 1978). Properties of these substances (arabinogalactan proteins of the wall) will be discussed in this Volume (see Chap. 7). Another hemagglutinin which will not be discussed further in this contribution is one isolated from wheat flour which is almost pure carbohydrate with D-xylose and L-arabinose being the major sugars (MINETTI et al. 1976). No haptenic sugar could be found to inhibit hemagglutination but D- and L-tryptophane did so at the rather high concentration of 75 mM. This property seems to be a general feature of several neutral polysaccharides which appear to be bound to cell surfaces (MINETTI et al. 1978). It is not clear whether or not the hemagglutinin or its binding characteristics are somehow related to the highly heat-stable carbohydrate-rich substance isolated from sweet potato which was shown to agglutinate germinated spores of a fungal plant pathogen (KOJIMA and URITANI 1978).

2 Biochemical Properties of Lectins

2.1 Assay and Isolation

Three-dimensional stable aggregates are the only products of the action of lectins on cells. To quantitate the reaction, most assay systems employ 1:2 serial dilutions of the lectin solution in tubes or on plates and the reciprocal value of that lectin dilution just sufficient to agglutinate the added cells is given as the titer. Due to the logarithmic nature of dilution series, such determinations only partly allow quantitative considerations. The agglutination endpoint is normally evaluated visually, thus subjective factors of reading have to be taken into account. More recently lectin titration has been performed in many laboratories with the microtiter system which employs 8 × 12 cm plastic trays, each exhibiting 8 rows of 12 wells. It has the advantage of rapid performance of the dilution steps and 50 μl quantities of a lectin solution are needed. On the V-shaped bottom of the wells nonagglutinated erythrocytes are collected in sharp central dots, whereas agglutinated ones settle in a diffuse way on the marginal planes. This method is of great value for routine work, but in two respects has serious drawbacks. The first is that it does not show enzymatic release of once agglutinated erythrocytes which occurs using a lectin from mung beans which exhibits additional enzymatic activity (see Sect. 5.5). The second is that it might give rise to low or negative results when heated plant or fungal extracts are present. Under that condition some component evidently sticks to the plastic surfaces of the wells which results in central dots of erythrocytes despite the fact that aggregates are formed and microscopically visible (YOUNG and KAUSS unpublished). The accuracy and reproducibility of the microtiter method can be greatly enhanced if the visual endpoint determination is followed by an electronic particle-counting step which allows quantitative determination of nonagglutinated cells (KÖHLE and KAUSS 1980). The choice of cells used in the assay depends to some extent on the viewpoint of the investigator; if special human blood groups are not required, then rabbit erythrocytes are frequently used as they appear to exhibit a rather broad range of diverse surface carbohydrate groups. For some lectins treatment of the cells with neuraminidase to remove terminal sialic acid is essential (e.g., for peanut lectin, LOTAN et al. 1975a) and treatment of erythrocytes with trypsin often improves the sensitivity of the assay. The latter treatment is thought to remove certain relatively large glycoprotein residues, thus exposing a greater number and variety of carbohydrate residues, mainly attached to membrane lipids (TUNIS et al. 1979).

An alternative procedure for lectin assay involves complex formation with specific polysaccharides or substituted proteins bearing several specific sugar residues in the same molecule. This enables quantitative measurements to be made, employing methods similar to those used for antibody determination, such as turbidimetry, quantitative precipitation, or double-diffusion on Ouchterlony plates. One example of this principle is the determination of Concanavalin A using glycogen (GOLDSTEIN and HAYES 1978).

The majority of lectins so far studied in detail are easily soluble in aqueous solutions. In such cases phosphate-buffered saline (PBS) or dilute buffers are sufficient for extraction. As will be discussed in Section 5.3 however, an increasing number of lectins have recently been found in association with membranes; such lectins need sonication, complexing solutions, or detergents for solubilization. These treatments are dangerous, however, in that they may partly or fully destroy the binding property. Lectins bound to endogenous carbohydrates may be solubilized with solutions of simple haptenic sugars, if the affinity of these sugars is sufficient to overcome the usually high affinity of lectins for complex oligosaccharides (see Sects. 2.2 and 4). Again with the risk of damage, such lectins bound with high affinity may be eluted with detergents.

Although over the years lectins have been purified by all the methods normally used for protein fractionation, more recently the method of choice is often affinity chromatography. In fact, as the affinity of lectins toward their specific sugars is often higher than that of enzymes toward their substrates, the lectins provide classical examples of this method (LIS and SHARON 1981, GOLDSTEIN and HAYES 1978). In most cases purification starts with a conventional step such as ammonium sulfate precipitation. Then the dialyzed protein mixture is applied to a column containing material with suitable carbohydrate groups. After washing out unbound proteins, elution is ideally performed with a haptenic soluble saccharide. Binding to the column may be so strong, however, that elution with acidic buffers is necessary; this procedure is less specific and may damage lectins. In some fortuitous cases natural absorbents with the appropriate specific sugar groups are available, such as Sephadex for Concanavalin A or Sepharose for *Ricinus communis* agglutinin. Preparation of specifically designed affinity material is more time-consuming and needs experience as well as luck; commercially available support material in an activated form (e.g., Sepharose CL, Sepharose AH, Epoxy-Sepharose) is extremely useful. It is generally advisable not to use the binding ligand with the highest affinity for the lectin in question, as subsequent elution of lectin with saccharides may be impossible. A selection of methods used for affinity chromatography of lectins may be found in recent reviews (SHARON and LIS 1972, LIS and SHARON 1973, GOLDSTEIN and HAYES 1978, LIS and SHARON 1981).

2.2 Carbohydrate Binding Specificity

Most information about the chemical structures required for optimal binding to a given lectin is derived from competition studies in which soluble sugars or sugar derivatives (including glycoproteins) are added to the lectin assays described in Section 2.1. Because of the close similarity to immunological methods, such inhibiting saccharides are often called "haptens" or "haptenic saccharides". In principle any of the above-mentioned assay systems can be used. With the microtiter system the lectin is diluted against a 10^{-2} M solution of the hapten (final concentration 5×10^{-3} M). Alternatively the hapten is serially diluted, and a constant small amount of lectin solution, just sufficient to agglutinate the cells when no hapten is present, is pipeted into each well. The results obtained are expressed either as the number of dilution steps or the concentration of hapten required to produce 50% inhibition of the agglutination activity. Although hapten inhibition gives relative values only, these give an accurate indication as to the structural features of a saccharide which determine specificity. Moreover it has been shown, e.g., for Con A (LOONTIENS et al. 1975), that the relative inhibition values for various glycosides correlate exactly with the association constants estimated either by equilibrium dialysis, by ^{13}C-NMR spectroscopy or by difference spectrum with p-nitrophenyl-α-mannopyranoside.

Table 1 contains a selection of well-studied seed lectins, arranged according to the dominant inhibiting sugar. Although a very convenient and often used criterion for classification, this is far from a reflection of the complexity of the various determinants of specificity. The group of D-Man-specific lectins appears to be relative homogenous, although definite differences exist in the affinity toward the various ligands. For instance, Con A is hardly inhibited by 3-*O*-methyl-D-Glc, whereas this sugar is one of the best inhibitors for pea lectin. On the other hand, the action of Con A is inhibited by 50% at 9 mM D-Man, whereas a concentration of 50 mM is needed to achieve the same inhibition for pea lectin (ALLEN et al. 1976). Some lectins listed as specific for D-Gal can also bind D-GalNAc to some extent, and vice versa. However some members of group b appear to be rather specific and bind only to D-Gal (e.g., PNA, Table 1). There appears to be no general rule regarding anomeric specificity. Some lectins are quite specific for only one anomeric form (e.g., Con A for the α-form) whereas others show practically no anomeric specificity (e.g., RCA_{120}). Another important complication is shown by the lectins with D-GlcNAc-specifity. In most cases, these lectins only bind to oligomers of β-D-GlcNAc and not to the monomer, with some preference of oligomers of more than two units. The lectin II from *Bandeiraea simplicifolia,* however, can also be inhibited by monomeric D-GlcNAc. For technical reasons the specificity of the lectins of group c, Table 1, is determined on the basis of their inhibition by (1→4)-β-oligomers of D-GlcNAc derived from chitin. Glycoproteins characterized by the presence of an internal $\text{-(D-GlcNAc)}_2\text{-}$ sequence are also found to be potent inhibitors of such lectins, as discussed, e.g., by YOKOYAMA et al. (1978) for the mitogenic pokeweed lectin. This fact, together with the examples discussed below and the finding that under certain conditions even Con A can bind to internal α-D-Man (GOLDSTEIN and HAYES 1978) raises doubts about generalization of the idea, often expressed, that lectins may prefer external glycoside residues when binding to natural carbohydrates.

Many lectins appear to have rather extended binding sites as they exhibit specificity for complex oligosaccharides and cannot be classified in the simple manner used in Table 1. The widely used lectin PHA from red kidney bean (*Phaseolus vulgaris*) provides a good example. This lectin has hemagglutinating as well as mitogenic activity (see Sect. 3). KORNFELD and KORNFELD (1970)

Table 1. Properties of some seed lectins. For references see GOLDSTEIN and HAYES (1978), LIS and SHARON (1981)

	Abbrev.	MW ($\times 10^{-3}$)	Subunits		Sugar binding sites	Metal content [atoms/mol]		Metal reversibly removed	Mito-gen	AB0-blood group specificity
			MW ($\times 10^{-3}$)	No		Mn^{2+}	Ca^{2+}			
a) Specific for D-mannose (D-glucose)										
Canavalia ensiformis (jack bean)	Con A	102.5	25.5	4	4	4	4	Mn^{2+}, Ca^{2+}	+	Nonspec.
Pisum sativum (garden pea)		49–53	5.8, 12–18	2+2	2	1	2.5	Ca^{2+}	+	Nonspec.
Vicia faba (broad bean)		47–53	9–14.3, 17–18	2+2					+	Nonspec.
Lens culinaris (lentil)	LA	42–63	6–18	2+2	2	0.6	3.8	Mn^{2+}	+	Nonspec.
b) Specific for D-galactose										
Ricinus communis (castor bean)	RCA_{120}	120	29.5–31, 34–37	2+2	2	<0.1	<0.1			Nonspec.
Abrus precatorius (jequirity bean)		134	33, 35–36	2+2	2				+	B, O > A
Arachis hypogaea (peanut)	PNA	110	27–28	4	2				(+)[a]	ABO[e]
Bandeiraea simplicifolia I		114	28.5	4	4		2[b]			$B \gg A_1$
c) Specific for D-GlcNAc										
Triticum vulgaris (wheat germ)[f]	WGA	36	18	2	4	None	Trace			Nonspec.
Bandeiraea simplicifolia II		113	30	4	4					ABO not aggl.[g]
d) Specific for D-GalNAc										
Glycine max (soybean)	SBA	120	30	4	2	1–1.7[c]	3.5–4.1	Mn^{2+}	(+)[a]	A > O > B
Phaseolus lunatis I (lima bean)		247–269	31	8	4	1	4	Mn^{2+}	+	$A_1 > A_2 \gg B$
Dolichos biflorus (horse gram)		110	26	4		1.6[d]	5.4			$A_1 \gg A_2$
e) Specific for L-fucose										
Lotus tetragonolobus I (asparagus pea)		120	27.8	4	4					$O \gg A_2$

[a] polymeric form after neuraminidase treatment; [b] $+1.2\ Mg^{2+}$; [c] $+0.3\ Zn^{2+}$; [d] $+2.0\ Zn^{2+}$; [e] after neuraminidase treatment; [f] inhibited by oligomeric β-D-GlcNAc only (see Sect. 2.2); [g] agglutinates acquired-B, T-activated and T_k polyagglutinable cells

Table 2. Source and properties of purified non-seed plant lectins

Plant	Localization	Specificity	Remarks	Reference
Agaricus bisporus (commercial mushroom)	Fruiting body	Complex	Inhibited by oligosacch. but barely by the respective monosacch. (see Sect. 2.2)	PRESANT and KORNFELD (1972)
Aloe arborescens (candelabra aloe)	Leaves	?	2 lectins, lectin 2 is mitogenic and also binds to α_2-macroglobulin and C3.	SUZUKI et al. (1979)
Cucurbita sativa (cucumber)	Phloem and fruit exudate	Oligomers of β-D-GlcNAc	See Sect. 5.2, similar or identical lectins are found in other Cucurbitaceae	GIETL et al. (1979) ALLEN (1979)
Dictyostelium discoideum (slime mold)	Cell surface and cytoplasm	D-GalNAc, D-Gal	See Sect. 4, two isolectins	FRAZIER et al. (1975)
Escherichia coli (bacteria)	Surface (pili?)	D-Man	High molecular weight; agglutinates yeast and human epithelial cells	ESHDAT et al. (1978)
Fomes fomentarius (touchwood)	Fruiting body	α-D-Gal	Ref. includes data on 2 other fungi	HOŘEJŠÍ and KOCOUREK (1978)
Ononis hircina (restharrow)	Root	D-GalNAc, D-Gal	Relatively spec. for human 0 blood groups, *O. spinosa* contains a similar lectin.	HOŘEJŠÍ et al. (1978a)
Phytolacca americana (pokeweed)	Root (+leaves and berry)	Oligomers of β-D-GlcNAc	Mixture of 5 isolectins, highly mitogenic	YOKOYAMA et al. (1978)
Robinia pseudoacacia (black locust)	Bark, phloem exudate	D-GalNAc	For distribution see Sect. 5.2	HOŘEJŠÍ et al. (1978b) GIETL et al. (1979)
Solanum tuberosum (potato)	Tuber	Oligomers of β-D-GlcNAc	Several features similar to cell wall glycoprotein (see Sects. 2.3, 5.3)	ALLEN et al. (1978)
Trifolium repens (white clover)	Root surface	2-deoxy-D-Glc	Agglutinates *Rhizobium trifolii* but not erythrocytes, also present in seeds.	DAZZO et al. (1978)
Ulmus glabra (scotch elm)	Phloem exudate	Oligomers of β-D-GlcNAc	Ref. includes data on lectins from phloem exudates of 3 other trees.	GIETL and ZIEGLER (1980b)

were the first to isolate its receptors present on human erythrocytes. A glycopeptide (molecular weight 2,000) with asparagine-linked oligosaccharide residues was split from the cell surface with trypsin and partially degraded. Comparison with known saccharides showed that its internal sugar sequence →D-Gal→D-GlcNAc→D-Man→ is the major inhibiting determinant. Although none of the constituent monomers shows any inhibition singly, the presence of both D-Gal and the penultimate D-Man was of importance; the additional presence of external sialic acid (N-acetyl-neuraminic acid, Neu5Ac) had little influence. Moreover, the peptide backbone of the glycopeptide also significantly affected the binding specifity. As discussed in more detail in Section 2.3, the situation with PHA is even more complicated. In addition to the erythrocyte-binding subunits discussed above, the same lectin molecules may contain lymphocyte-binding subunits in various proportions which apparently are specific for the following internal sequence (TOYOSHIMA et al. 1972):

$$\rightarrow (\text{D-Man})_3 \rightarrow \text{D-GlcNAc} \rightarrow \text{D-GlcNAc} \rightarrow.$$

A situation in many respects similar to that for PHA was demonstrated for the *Agaricus bisporus* lectin (PRESANT and KORNFELD 1972, Table 2). Agglutination of human erythrocytes is strongly inhibited by a carbohydrate residue *O*-glycosidically linked to serine or threonine. This residue was isolated; its most likely structure is:

$$\begin{array}{l}\text{Neu5Ac} \rightarrow \text{D-Gal} \rightarrow \text{D-GalNAc} \rightarrow \text{Ser-peptide.}\\ \qquad\qquad\qquad\qquad\qquad\quad \downarrow \\ \qquad\qquad\qquad\qquad\quad \text{Neu5Ac}\end{array}$$

In this case, removal of the terminal Neu5Ac-residues increased the haptene inhibitory activity eightfold, whereas additional removal of D-Gal destroyed all the activity. The intact glycopeptide was 15,000 times more potent than the simple sugars D-Gal and D-GalNAc, and 65 times more potent than the disaccharide D-Gal → DGalNAc. The isolated human erythrocyte receptor for *Phaseolus vulgaris* agglutinin (see above) was devoid of any haptene inhibitory activity toward the *A. bisporus* lectin. The various examples discussed above help to explain why lectins show such great selectivity toward complex carbohydrate residues.

2.3 Structure

To agglutinate cells or precipitate specific polysaccharides a lectin molecule requires at least two sugar-binding sites. The examples in Table 1 demonstrate that this is achieved in most cases by the composition of the molecule from four identical or nonidentical subunits, each bearing a binding site. In some cases (PNA, SBA, and *Phaseolus lunatus* agglutinin) half of the subunits appear to have no binding site. Wheat germ agglutinin, in contrast, has two binding sites per subunit. Many lectins show a tendency to aggregate in solution. In the case of Con A this aggregation is reversible and known to depend on pH and temperature (HUET et al. 1974), whereas soybean agglutinin undergoes irreversible self-association to higher molecular weight aggregates with increased hemagglutinating activity (LOTAN et al. 1975b). Postsynthetic fragmentation appears to be another cause of inhomogeneity. Thus certain subunits in an otherwise homogenous Con A population are apparently composed of two polypeptides. This splitting appears not to cause a significant change in hemagglutinating activity (GOLDSTEIN and HAYES 1978).

The various subunits of a lectin molecule must not exhibit the same binding site specificity; the *Phaseolus vulgaris* agglutinin is a classical example. It is made up of subunits E exhibiting an erythrocyte binding site responsible for hemagglutination and subunits L which are very potent lymphocyte mitogens. Receptor oligosaccharides to which the respective binding sites fit best have very different structures (Sect. 2.2). In the seeds the two types of subunits occur in the arrangements E_4, L_1E_3, L_2E_2, L_3E_1, L_4 representing a series of isolectins which have varying proportions of the two activities (LEAVITT et al. 1977). Although the two subunit types are very similar in their overall amino acid composition and tryptic peptide fragments they have different *N*-terminal amino acids and isoelectric points, and consequently can be separated. Nevertheless, the differences in primary structure appear to be relatively small, in spite of the rather drastic differences in binding specificity (LIS and SHARON 1981). Similar observations have been made with other lectins. Thus the *Bandeiraea simplicifolia I* isolectins consist of two types of subunits which bind preferentially to α-D-GalNAc and D-Gal, respectively, and make up five closely related proteins (GOLDSTEIN and HAYES 1978). In most cases it is not clear to what extent the numerous isolectins reported in the literature obey similar principles.

Many lectins have been shown to contain divalent ions, mostly Ca^{2+} and Mn^{2+}, which are required for activity (Table 1). The lectins' affinity for these metals is often so high that EDTA cannot remove them at neutral pH values; acidic conditions are then needed to show a metal ion requirement. Consequently there are less frequent reports of reactivation of lectins by readdition of metal ions following their removal (Table 1). Detailed information on the role of metal ions and their spatial orientation is available for Con A. The appropriate binding sites are situated near the sugar-binding site. The metal-binding site S1 has first to be occupied by Mn^{2+} (or another transition metal) before the Ca^{2+}-binding site S2 can be filled, and occupation of both metal sites is required for sugar binding (for details see LIS and SHARON 1981, GOLDSTEIN and HAYES 1978).

Many plant lectins are glycoproteins, the characteristics of which are discussed in Vol. 13A, Chap. 13 and 14. The carbohydrate content is usually in the range of 0% to 20% and can be as high as 50% for potato lectin (ALLEN et al. 1978). The fact that some lectins (Con A, WGA, PNA) contain no detectable carbohydrate indicates that this part of the molecule may not be very important for sugar binding. A similar conclusion can be drawn from the fact that *Dolichos biflorus* lectin showed only a slight decrease in hemagglutinating activity when up to 40% of its D-Man was removed by treatment with α-mannosidase (BIROC and ETZLER 1978).

The overall amino acid composition of lectins reflects that of typical plant proteins. Many of them are relatively rich in aspartic acid, serine, and threonine and low in sulfur-containing amino acids. Some lectins (wheat germ, potato, and pokeweed lectins), however, contain much cysteine (LIS and SHARON 1981), offering many possibilities for disulfide bridge formation. This explains the rather high stability of wheat germ agglutinin. It is worth noting some close similarity in composition between potato lectin and the so-called "extensin", glycoproteins assumed to be typical of plant cell walls (see Chap. 7, this Vol.). Both types of substances appear to be very rich in hydroxyproline to which short arabinoside side chains are bound; both of them have single D-Gal residues connected to serine and contain ornithine, an unusual amino acid for proteins (ALLEN et al. 1978).

Data are accumulating on the primary sequence of several lectins. The most interesting feature appears to be that several lectins isolated from leguminous plants possess extensive homology in amino acid sequence near the *N*-terminus. This might explain why several lectins exhibiting different binding specifity show immunological cross-reactivity which indicates that they have some antigenic determinants in common (HANKINS et al. 1979, HOWARD et al. 1979). The properties mentioned above, along with many other details about the primary, secondary, and tertiary structure of lectins, can be found in recent reviews (GOLDSTEIN and HAYES 1978, LIS and SHARON 1981).

3 Applications and Biological Properties of Plant Lectins in Animal Systems

In recent decades lectins have proved to be of unique value as tools in the fields of membrane biochemistry and cell biology and this, in turn, has greatly enhanced research on lectins. The very selective interaction of lectins with cell surfaces, glycoproteins, and glycolipids is based on the binding specificity described in Section 2.2. A few examples for the application of lectins will be given without going into detail; for more information and references the review of LIS and SHARON (1981) should be consulted.

One of the most obvious examples is the use of lectins in blood banks to type blood of the ABO-system and to detect "secretors", people who secrete soluble blood group substances A, B, or H in their saliva or other body fluids. A less well-known application is the determination of T-antigen on human erythrocytes with the help of peanut agglutinin. This procedure greatly simplifies the differential diagnosis of "polyagglutination". The appearance of the T-antigen is due to the alteration of surface carbohydrate groups by exogenous neuraminidases in the course of certain viral or bacterial infections.

Insolubilized lectins are used as a selective affinity absorbent for the isolation of membrane constituents, enzymes which are glycoproteins and tRNA species containing glycosylated bases. They are also used to separate cells bearing different surface carbohydrate groups such as lymphocyte subpopulations or cells liberated from different tissues. For the latter purposes the lectins can either be immobilized by covalent bonds, by group-specific binding to a glycoprotein-bearing column, or may be used in a soluble form (REISNER and SHARON 1980).

A further very fruitful field of lectin application is the characterization of cell or membrane vesicle surfaces, for which the lectins are mostly used in a conjugated form to allow fluorescence- or electronmicroscopical examination of receptor distribution. Indeed, the "fluid mosaic model" of membrane structure is to a significant extent founded on such experiments (NICOLSON 1976). A more recent and very promising application of lectins is the selection of cultured eukaryotic cell lines lacking certain sugars in their surface carbohydrate residues. Such cell lines are of great value for studies on the multistep mechanisms involved in the assembly of surface carbohydrate groups (KORNFELD and KORNFELD 1978). Another problem which has greatly stimulated interest in lectins is carcinogenesis. Agglutination of normal and malignant cells by lectins has been reported to be different in some instances which might be due to differences in the mobility of surface carbohydrate components (NICOLSON 1976, LIS and SHARON 1981).

Most of the applications of lectins in animal cell systems indicated above – although very useful – provide no apparent indication of a biological role of lectins. In this respect more promising results have been obtained in the field of immunobiology. When small lymphocytes from the peripheral blood, which represent cells arrested in G_0 or G_1 of the cell cycle, are treated with certain lectins they undergo profound internal changes, subsequently divide, and eventually differentiate to plasma cells or T-lymphocyte effector cells (O'BRIEN et al. 1978). Lectins which can cause this dramatic effect are called mitogenic. One of the mitogens used most frequently is *Phaseolus vulgaris* agglutinin (PHA, see Sects. 2.2 and 2.3), and some others can be found in Table 1. With mitogenic lectins known to be inhibited by simple haptenic saccharides, e.g., Con A, it has been demonstrated that these sugars can also inhibit the mitogenic action. The same is true for lectins which require more complex haptenic inhibitors, e.g., PHA, where oligosaccharides derived from appropriate glycoproteins are potent inhibitors of the mitogenic action (TOYOSHIMA et al. 1972). Thus it seems clear that the first and crucial event is a group-specific binding of the lectin to surface carbohydrate receptors.

Many experiments in this field have produced striking results in support of the idea that the cell surface plays a role as a primary control and recognition organelle whose stimulation, via secondary signals, can trigger internal metabolic and differentiation processes. It is not clear what mechanism is responsible for transmitting the signal produced by lectin binding, but the early effects observed after stimulation of lymphocytes include changes in Ca^{2+}, Na^+/K^+ and amino acid transport as well as in energy metabolism. The very dramatic resultant increases in protein or DNA-synthesis are often used to monitor mitogenic action. The fact that other agents can also be mitogenic may be important for elucidation of the primary action of lectins. Examples of such nonspecific mitogens are bacterial lipopolysaccharides, Ca^{2+}-ionophores in a narrow concentration range, heavy metal ions, enzymes which may alter the cell surface and antibodies against certain surface antigens. This renders the possibility likely that a "second messenger" is generated after cell surface stimulation which is then affecting multiple regulated steps.

The function of mitogenic lectins on lymphocytes in vitro is far more complex than indicated above. There is increasing evidence that the cell material used represents populations of functionally different cells and that its composition varies according to the source of the lymphocytes (peripheral blood, spleen, lymph nodes, thymus) and the animals used. The established major subpopulations, namely T- and B-lymphocytes, are mitogenically stimulated by various lectins to different extents (O'Brien et al. 1978, Sharon 1976). It is also clear that even in vitro cooperation between B- and T-lymphocytes as well as macrophages has to be taken into consideration. This cooperation occurs in part at the level of surface contact between these cells and is also due to soluble regulating factors produced selectively by the different cell subpopulations. Under in vivo conditions such cooperation is additionally influenced by the various host tissues to which lymphocytes have access. Thus the overall process of mitogenesis represents a complex physiological network and it appears that the action of lectins in vitro might be to mimic some binding step which in vivo results from contact between cell surface elements.

4 Role of Lectins in Slime Mold Aggregation

In their nonsocial phase the cellular slime molds can feed as unicellular amoebae on soil bacteria for many generations. Under unfavorable conditions the amoebae are drawn together and form a motile multicellular pseudoplasmodium which finally differentiates to form species-specific fruiting bodies. Cellular slime molds such as *Dictyostelium discoideum, D. purpureum,* or *Polysphondylium pallidum* have attracted much interest as one of the supposed simple model systems for studies on the molecular basis of development (Sussman 1976, Sussman and Brackenbury 1976). The critical step in our context is the initiation of cell aggregation which experimentally can be initiated by food deprivation. It is already known that this aggregation requires coordination of rather complex physiological and biochemical events, including long-range conditioning by cAMP-pulses and chemotactic attraction of the individual amoeba. Artificially mixed cells are capable of sorting themselves out, each group being able to form its specific fruiting body. The overall aggregation process seems to be only in part due to species specificity of chemotaxis. The crucial recognition step between individual amoeba of a certain species obviously resides to a great part in a cell surface property (Gerisch et al. 1975, Newell 1977, Bozarro and Gerisch 1978, McDonough et al. 1980). It is generally believed that recognition is intimately associated with cell-to-cell adhesion, both phenomena being mediated by the same complementary binding molecules on the cell surfaces. Although this is a very convenient assumption, one could also imagine that both functions reside on separate molecular entities, thus making the aggregation process even more complex. Such a mechanism, however, might explain some inconsistencies in interpretation of current experimental data.

Some cell aggregation factors of animal origin have hemagglutinating properties. This might have been one of the reasons that hemagglutination has also been used to monitor carbohydrate binding proteins (lectins) in cellular slime molds. BARONDES and coworkers (BARONDES and ROSEN 1976, FRAZIER et al. 1975, ROSEN et al. 1979, BARONDES and HAYWOOD 1979) have purified lectins mainly by affinity chromatography from several species of *Dictyostelium* and *Polysphondylium* and have characterized them in some detail. These lectins are generally named according to the species name of the slime mold (e.g., discoidin for *D. discoideum*); if several lectins are present in one species these are differentiated by roman numbers (e.g., discoidin I, discoidin II). Although most of the lectins from *Dictyostelium* species have similar subunit weights (around 25,000) and molecular weights around 100,000, they show slight but definite differences in physical as well as immunological properties (BARONDES and HAYWOOD 1979). The relative potency of various saccharides in inhibiting the hemagglutinating activity of the lectins isolated from different species of *Dictyostelium* and *Polysphondylium* is clearly different, although an overall tendency to bind to galactose-like molecules is evident (BARONDES and ROSEN 1976, BARONDES and HAYWOOD 1979). Thus the lectins isolated from slime molds appear to have the potential to detect slight differences between surface oligosaccharide chains which have a similar basic structure. The lectins represent a substantial part (1 to 5%) of the soluble intracellular protein. It has been reported, however, that they are also accessible to some extent at the surface of aggregation competent cells, as demonstrated with fluorescent antibodies (CHANG et al. 1975).

When the time course of total soluble hemagglutinating activity in extracts from *D. discoideum* (strain NC-4) after removal of bacterial food was followed, a striking increase was observed after about 6 to 9 h, at which time the cells become cohesive (REITHERMAN et al. 1975). Using glutaraldehyde-fixed cells of *P. pallidum* and *D. discoideum* and purified soluble pallidin and discoidin I and II, the presence of surface carbohydrate groups appropriate for agglutination by these soluble lectins was shown. In these experiments, agglutinability strongly increased 6 to 9 h after the induction of differentiation, indicating that surface carbohydrate groups to which discoidin can bind ("receptors") become more prominent in parallel with the increase in total cellular lectin and cell cohesiveness. On the basis of some reasonable assumptions REITHERMAN et al. (1975) showed that this enhanced agglutinability of glutaraldehyde fixed competent cells was apparently due to an increase in both the number of binding sites and in their affinity. For *D. discoideum* and discoidin I, for instance, the binding sites increased from 3×10^5 to 5×10^5 per cell with a concomitant increase of the apparent association constant (K_a) from 5×10^7 to $1.3 \times 10^9\ M^{-1}$, the K_a being distinctly higher than for the binding of discoidin I to rabbit erythrocytes (1.3×10^6). From these data the authors concluded that both the number of binding sites and their affinity for the soluble lectin from the respective slime mold species appears to be species-specific to some extent.

Thus the above-mentioned observations indicate the appearance of lectins at the cell surface, as well as their appropriate receptors in parallel to cell cohesiveness. It has been suggested, therefore, that the developmentally regulated slime mold lectins may play a role in cell-to-cell adhesion and recognition (BARONDES and ROSEN 1976). Although the data known are consistent with such a mechanism, the evidence remains only correlative. At present there is no experimental proof that cell-to-cell binding and sorting out in vivo is causally mediated by lectin-receptor interactions. One critical experiment would be to inhibit this process by addition of the haptenic saccharides. Although it has been reported that aggregation of heat-treated aggregation-competent cells of *P. pallidum* is inhibited by lactose and galactose at reasonable concentrations (BARONDES and ROSEN 1976), the binding between live slime mold cells can only be affected at such high sugar concentrations that nonspecific effects appear likely. These difficulties might be due to the rather high apparent affinity between slime mold lectins and their endogenous receptors as compared with the usually low affinity of simple haptenic saccharides. Based on the molarity of terminal D-Gal residues, the glycoprotein asialofetuin is about 1,000-fold more potent as an inhibitor of pallidin than lactose. However, also this high molecular weight inhibitor can interfere with cohesiveness only when cells are rendered "permissive" by treatment with 2.5 mM 2,4-DNP or under hypertonic conditions, procedures causing pronounced shrinkage and possibly also some damage to the cells (ROSEN et al. 1977). One could imagine that only under such conditions the binding sites of the surface-located pallidin become exposed enough to be accessible for the sparring asialofetuin molecule. But why should they then, under natural conditions, be accessible for endogenous receptors located at the surface of the other cell? Are there additional regulatory devices which control the properties of the binding site of surface-located pallidin? One possibility to solve these problems would be to isolate the carbohydrate residues of the endogenous receptors in sufficient amounts to enable their use as inhibitors.

An elegant alternative approach to the question of a molecular basis for cell-to-cell contact in cellular slime molds was made by GERISCH and coworkers with the help of univalent antibody fragments (Fab) directed against membrane antigens of aggregation-competent cells (MÜLLER and GERISCH 1978). Aggregation appears to involve at least two different binding mechanisms. During aggregation cells of *D. discoideum* elongate and adhere at their ends; this type of cell assembly is not affected by the presence of 10 mM EDTA. In contrast, side-by-side contact of the cells is fully inhibited by 10 mM EDTA. Both types of cell adhesion can also be clearly distinguished using Fab. Appearance of the antigenic determinants representing the end-to-end contact sites (cs-A) is induced by starvation, whereas those representing the side-by-side contact sites (cs-B) are also present in noninduced growth phase cells (MÜLLER and GERISCH 1978). The glycoprotein with the antigenic properties of cs-A is a minor constituent of the cell surface and has recently been isolated (MÜLLER et al. 1979). It is not identical with discoidin, and its carbohydrate residues do not appear to represent the receptors for discoidin. Thus the experimental results of the lectin studies with *D. discoideum* of BARONDES and coworkers and the studies of GERISCH and coworkers using Fab from anti-cs-A cannot be related at the moment.

More data are available for *P. pallidum* (ROSEN et al. 1977, BOZZARO and GERISCH 1978) which allow some attempt to correlate the present knowledge of developmentally regulated lectins with the results obtained using antibodies against contact sites. Anti-pallidin Fab exhibits a significant influence on cell adhesion only if the cells are treated with hypertonic solutions of glucose or

2.5 mM 2,4-dinitrophenol, treatments which appear to damage the cells to some extent and thus weaken or change the adhesive "apparatus". With aggregation-competent healthy cells, anti-pallidin Fab has a rather weak effect or none at all. In contrast, anti-contact site Fab inhibits cell adhesion strongly. These and some other results (GERISCH et al. 1980) provide strong arguments against a direct participation of pallidin as the binding molecule in cell aggregation. Interpretation of the overall results is, however, still controversial. Discussion has to take into consideration a set of yet not exactly known affinities of antibodies toward surface-located as well as soluble pallidin, and of pallidin toward possible receptors at red blood and slime mold cells. Another problem may result from the fact that the antibodies used are raised against soluble intracellular slime mold lectins. Such proteins possibly expose antigenic determinants different from the determinants which are accessible at membrane-located lectin molecules. More recently a model has been proposed in which a three-component system, composed of high-affinity complementary ligands and a third regulative component function together in cell adhesion (MÜLLER and GERISCH 1978, GERISCH 1980). Such a multi-component system possibly allows some space for a future synoptic view of the experiments on slime mold lectins and those on contact sites analyzed with the use of Fab.

5 Lectins in Higher Plants

5.1 Recognition Phenomena Possibly Mediated by Lectins

The cells of the immune system of warm-blooded animals appear to be designed for recognition of foreign molecules and for regulation of their development by means of surface contact (see Sect. 3). For this reason they presented a particularly good system for studying lectin-mediated effects such as mitosis. Plant physiologists are not in such a fortunate situation as immunobiologists were. In the vegetative phase of higher plants there are practically no situations in which cells freely meet and thus have to recognize each other. During differentiation plant cells of one developmental line stay spatially oriented. Although one can imagine that such neighboring cells influence each other by means of surface contact signals, no experimental plant system has so far been found which appears suitable for proving the existence of contact regulation and for investigating the possible involvement of lectins in this process. Early reports that lectins in high concentrations can increase the mitotic index in root tips of *Allium cepa* (NAGL 1972a) and growth in *Allium cepa* and *Phaseolus coccineus* (NAGL 1972b) were not confirmed; it appears now that the observed effects may have been due to bacterial infections. A short note that soybean agglutinin acts as a mitogen in soybean callus culture (HOWARD et al. 1976) has not been confirmed in detail.

In contrast to vegetative tissues, the reproductive phase of plants may offer better examples in which cells of the same species but at a different stage of development have to be recognized and either accepted or rejected. The overall process of fertilization in higher plants is rather complex, however, and will be discussed in Chapter 23, this Volume. As far as I can see there is not yet experimental evidence to conclude that certain lectins or lectin-like complementary proteins are clearly responsible for the recognition processes which without doubt are involved in the pollen–stigma interactions. In the mating reactions of algae such a clear-cut statement is also not yet possible, although with algae it is easier to control, synchronize, and localize the molecular structures involved in recognition. The best-understood experimental system appears to be that of *Chlamydomonas* (Goodenough 1977). Here the first steps of the mating interaction take place at the tips of the gametic flagella and indirect evidence suggests that proteins and surface carbohydrates may play some role. Thus this system may in the near future provide answers to some questions concerning the molecular basis for recognition in mating of the gametes of green plants. Considerable progress has also been made in elucidation of the biochemistry of the mating reaction in yeasts. In this process, described in Chapter 18, this Volume, recognition is also based on complex carbohydrates. Again, however, little is known about the receptors involved in the mechanism.

Root nodule formation in legumes is known to require specific species or strains of the genus *Rhizobium* and – vice versa – the host specificity forms the basis for species differentiation of this bacteria. The symbiosis is newly generated for every seedling using inoculum from the soil, thus recognition of the appropriate bacteria is a prerequisite. There is growing evidence that lectins are involved in the initial stage of this selection procedure; a detailed discussion can be found in Chapter 26, this Volume. Great progress has been made in this field, particularly with the combination *Rhizobium trifolii*/*Trifolium repens* (white clover; Dazzo 1979). This latter system provides the best available evidence that initial binding of infective strains of *Rhizobium* to the root hair surface occurs with the help of the lectin trifoliin (Dazzo et al. 1978, see Table 2). Although the good correlation of all available data is very convincing, it remains to be proven beyond doubt that surface located lectins are indeed a prerequisite for infection. At first glance it would appear that conclusive proof could be obtained by blocking the surface binding sites by the haptenic sugar (2-deoxy-D-glucose), by anti-clover root Fab or anti-lectin Fab, and observing whether formation of infection threads or root nodules is affected. It has, indeed, been demonstrated that both can be inhibited by 2-deoxy-D-glucose; significant effects are evident at 2.5 mM, but full inhibition requires concentrations up to 30 mM (Dazzo 1977). Evaluation of these results, however, is difficult as this sugar also inhibits glycan synthesis (F.B. Dazzo personal communication). It is also difficult to try blocking root hair infection and nodulation by Fab because these events occur days to weeks after inoculation. During this time the Fab added to the system would be greatly diluted out by bacterial multiplication or newly exuded lectin, respectively (see also Sect. 5.2). Nevertheless, in the complex legume–*Rhizobium* interaction there is additional indirect evidence that lectins might be of physiological significance. In white clover seedlings synthesis of surface located lectin appears to be repressed in the presence of low concentrations of NO_3^- or NH_4^+, conditions which are also known to inhibit infection and nodulation (Dazzo and Brill 1978). Heteropolysaccharides from strain

5053 of *Rhizobium japonicum* (which is infective for soybean) inhibit growth of soybean cells in suspension culture (OZAWA and YAMAGUCHI 1979). This observation might provide the means for future studies to explain how the morphological alterations during nodule formation are induced by homologous rhizobia bound to the surface of legume roots. Of these events the most spectacular, which occurs at the beginning of the infection process, is a very tight curling ("shepherd's crook") of root hair tips at the site of bacterial attachment (RAA et al. 1977). This effect appears to result from a local inhibition of cell extension growth and is reminiscent of a contact-regulated differentiation. Although hormones may also be involved, there is an obvious analogy between such contact regulation and the mitogenic effects of lectins on animal cells (see Sect. 3).

Another example in which cells of higher plants interact with other cells – in this case also of an entirely foreign type – is pathogenesis in higher plants. This field is covered in detail in Chapter 24 (this Volume). An attempt will be made here to summarize as far as possible to a participation of lectins in this process. When the successful results obtained using plant lectins for research into animal lymphocytes came to the attention of plant physiologists, several reviewers were fascinated by the idea that lectins may play a role as recognition molecules in pathogenesis (ALBERSHEIM et al. 1969, ALBERSHEIM and ANDERSON-PROUTY 1975, CALLOW 1975, CALLOW 1977, SEQUEIRA 1978). It has long been known that plant–pathogen interactions are highly specific. In addition, several phenomena suggest that carbohydrate residues at the surface of pathogens or at soluble products derived from them may be crucial in pathogenesis. Three examples of such phenomena will be mentioned. Branched glucans ("elicitors") derived from the walls of various fungi can induce plant tissues to produce the potent antibiotics called phytoalexins (KUĆ 1972, ALBERSHEIM and VALENT 1978). This effect, however, is not specific for material from fungi which are pathogenic for the plant in question and a given plant produces its own characteristic spectrum of phytoalexins, regardless of the origin of the elicitors. More recently, in soybean a race-specific resistance of unknown nature against *Phytophthora megasperma* var. *sojae* has been shown to be induced by glycoproteins isolated from culture fluid of this pathogen (WADE and ALBERSHEIM 1979). The great potential of plant lectins to detect slight differences in carbohydrate structure suggests their participation in these obvious recognition processes, although this has not yet been shown experimentally.

The only more direct evidence for lectin involvement is in the case of induced resistance of tobacco against *Pseudomonas solanearum*, and this evidence remains only correlative (SEQUEIRA 1978). It has been found that avirulent strains of this pathogen, when infiltrated into leaves, become attached to the host cell surface. This is followed by various ultrastructural changes such as vesicle formation and deposition of fibrillar material which appears to surround the bacteria. These changes are paralleled by induction of resistance by some unknown mechanisms. A lectin with properties similar to potato lectin (Table 2) was found in tobacco and this lectin is able to agglutinate the avirulent strains of *P. solanearum* in vitro. It has therefore been suggested that this lectin might reside at the plant cell surface and represent the determinant to which the bacteria by means of their surface lipopolysaccharides bind in the initial attachment process. Virulent strains, however, produce a soluble extracellular slime which inhibits bacterial agglutination in vitro and thus also appears to bind to the lectin. According to the hypothesis of SEQUEIRA (1978) blocking the lectin's binding site in vivo by slime prevents recognition of bacteria. In consequence, the virulent strains of *P. solanearum* are not bound and the defence mechanism is not triggered. The bacteria stay free, multiply, and cause the disease. This hypothesis is quite attractive as it suggests that the virulent pathogen can avoid the alarm system of the plant by producing substances which bind to the lectin-like recognition molecules without triggering the defence mechanism. Many details have to be clarified, however, before an involvement of bound lectins in the defence mechanism can be confirmed.

5.2 Distribution of Lectins in Various Tissues

Many of the classical lectins were originally purified from seeds, particularly those of legumes (Fabaceae, Table 1). In more recent years this picture has gradually changed as more and more lectins from other tissues and other plants have been described and purified (Table 2). Retrospectively, the fact that seeds were chosen as a source of lectins appears mainly to reflect the fact that for biochemists they represent an easily available commercial source of plant material and contain relatively high concentrations of plant proteins soluble in dilute buffers. The apparent preference for legumes may result from the assay cells originally used. Red blood cells of humans and other warm-blooded animals happen to be especially rich in surface carbohydrate groups containing the particular sugar residues for which many legume lectins by chance have a high affinity. A systematical search of similar intensity, but using other assay methods or cells, may have revealed many more lectins from nonlegume plants.

Several investigators have studied the appearance of lectins during seed formation or their disappearance during seed germination, in the hope of finding evidence of a physiological role. In the osage orange (*Maclura pomifera*) the level of buffer-soluble agglutinin activity is maximum in mature seeds, lower in young seedlings and lower still, sometimes barely detectable, in mature plants (JONES et al. 1967). Similar observations have been made for lentil, red kidney bean, soybean, and horse gram (HOWARD et al. 1972, MIALONIER et al. 1973, PUEPPKE et al. 1978, TALBOT and ETZLER 1978). In these plants there is a decline in lectin concentration more or less parallel with the disappearance of protein, possibly indicating that the bulk of seed lectins represents storage material. However, the reserve protein from the protein bodies in the endosperm of *Ricinus communis* disappears much faster than the agglutinin, and this led to the suggestion that in *Ricinus* the agglutinin is apparently not a storage protein (YOULE and HUANG 1976). Similar conclusions can be drawn from the finding that lectins could be extracted with phosphate buffered 0.14 M sodium chloride solution from all tissues of 6-year-old trees of *Robinia pseudoacacia* (GIETL and ZIEGLER 1980a). Activity was found in roots, root nodules, bark, phloem, wood, and leaves, and was exceptionally high in the phloem region near the cambium (or in adhering cambial cells). The activity in the extracts was inhibited by oligomers of β-D-GlcNAc and also by D-GalNAc.

Using affinity chromatography it was shown exemplarily for the wood extract, inhibition by the two sugars is due to the presence of two different lectins. In the context of lectin distribution in various tissues it is interesting that lectins are also excreted from sterile roots (GIETL and ZIEGLER 1979). This applies not only to *Phaseolus vulgaris*, for which a role in symbiosis appears possible, but also to *Helianthus annuus* and *Zea mays*. The exuded lectin activity from the former plant is inhibited by oligomers of β-D-GlcNAc and also by α-D-methylmannoside, whereas the lectin from the two other plants appears to recognize only oligomers of β-D-GlcNAc (see Sect. 5.3.3 for discussion of the possible significance of specificity).

Another argument against the generalization that seed lectins are merely storage proteins is the discovery that synthesis of pea lectin carbohydrate residues also appears to occur in membrane preparations from pea epicotyl (PONT-LEZICA et al. 1978). Lipid-linked sugars are involved in this process. As it is generally assumed that such glycosylations are co-translational events (HOPP et al. 1979) this would indicate de novo synthesis. Cells of developing plants obviously require the agglutinin for some unknown function. One explanation for the apparently controversial findings may be the possibility that the genes for a certain lectin may be read for several purposes. Thus in seeds of some plants the gene product may be deposited in a soluble form for storage purposes, whereas in other cells of the plant the same gene product – in smaller amounts – may be integrated into a membranous organelle serving a special metabolic function. For example, trifoliin can be isolated in rather large amounts from seeds of white clover, but is also located on the surface of root hairs (DAZZO 1979, Table 2). Similarly, in the cellular slime molds (see Sect. 4), the lectins represent a substantial part of the soluble intracellular proteins but are also located in the cell membrane in such a way as to expose at least parts of the molecule (or some antigenic determinants) to the outside of the cell. During future more refined studies on changes in lectin levels in the course of plant development lectin fractions which are not readily soluble in aqueous buffers should also, therefore, be examined. Also to be noted are the recent findings that in seed extracts the lectins can be partly associated to glycoproteins or proteins (GANSERA et al. 1979), a fact which may lead to false determinations of total activities. Although serious problems due to possible rebinding are encountered (see Sect. 5.3) further studies on distribution and changes of nonsoluble lectins might help to demonstrate a multifunctional role. Recent studies by BOWLES et al. (1979) indicate that during development from seedlings to mature plants the activity of the soluble lectin fraction in peanut and soybean plants is low, but that of membrane-associated lectins is always high; the latter lectin fractions appear to differ in carbohydrate binding specificity from the buffer-soluble lectins in the respective seeds.

5.3 Subcellular Localization of Lectins

5.3.1 Lectins Associated with Membranes

Another approach to the question of possible physiological functions of lectins in plants has been to study the subcellular localization of lectins. As mentioned above, the buffer-soluble lectin activity in vegetative tissues from mature plants tends to be low. However, when crude membrane fractions from a variety of plants and tissues were subsequently extracted by sonication in phosphate buffered 0.14 M NaCl, sonication in 0.5 M potassium phosphate buffer, pH 7.1, and sonication in 0.1 M EDTA/0.06% Triton-X-100 rather high lectin activities were found in the extracts (BOWLES and KAUSS 1975). These studies were extended by subfractionation of the membranes from hypocotyl segments of *Vigna radiata* (mung bean) to show that these extractants can solubilize proteins with lectin activity from subfractions enriched in mitochondria, plasma membrane, Golgi apparatus, and endoplasmic reticulum (BOWLES and KAUSS 1976). There was a tendency toward higher levels of activity, as well as specific activity based on protein content, in meristematic or growing tissues; the detergent extracts generally exhibited the highest activity. Inhibition studies with some available saccharide inhibitors showed considerable differences in the carbohydrate specifity of the lectins in the different extracts, indicating the presence of different lectins in the various membranes or tissues. In addition, it appeared that some of the extracts possibly contained more than one lectin. These results were taken as an indication that the lectins in metabolically active cells are mainly either integrated in the lipid bilayer or bound via their carbohydrate binding site to membrane constituents containing carbohydrate residues. As a working hypothesis, the idea was put forward that these bound lectins might either function in recognition and fusion of internal membranes, or in the stabilization of membrane protein complexes (for further discussion see contribution of D. BOWLES, Vol. 13A, Chapter 14).

To evaluate the observations on membrane-bound lectins, further studies were performed on lectins contained in mitochondria from the endosperm of *Ricinus communis* (castor bean). Mitochondria were chosen as the function of their membranes is relatively well known and they do not take part in intracellular membrane flow. It was found that the level of lectin activity is highest in the mitochondrial inner membrane (BOWLES et al. 1976). Part of the lectin could also be eluted from this membrane with lactose and was identified as the well-known *Ricinus communis* agglutinin of molecular weight 120,000 (RCA_{120}); further lectin activity could be extracted by subsequent treatment with detergent. The hemagglutinating activity of this latter fraction was only partly inhibited by lactose, indicating that in addition to RCA_{120} some other lectin (or lectins) of unknown identity was present. At first glance it appeared reasonable to assume that the RCA_{120} in the mitochondrial inner membrane was an integral membrane component, as this membrane was envisaged as having no contact with the soluble RCA_{120} which is also present in large amounts in the crude extract after homogenization of the endosperm. Further studies, however, established that this view was erroneous. The endosperm was homogenized in the presence of ^{3}H-labeled RCA_{120} with or without added lactose (KÖHLE and KAUSS 1979). In preparations without the binding-site specific sugar the subfraction representing the mitochondrial inner membrane contained sufficient labeled agglutinin to account for the RCA_{120} which was previously reported to be associated with this membrane. Obviously this lectin fraction found in the membrane results from binding of soluble RCA_{120} to appropriate carbohydrate residues. With the methods employed one cannot say whether the labeled lectin really binds to mitochondrial inner membranes, possibly as the result of damage to the outer membrane during preparation, or whether the subfraction was contaminated by small amounts of other membrane material with the appropriate saccharide groups. The large amounts of soluble RCA_{120} in endosperm homogenates most probably result from the rupture of protein bodies in aqueous buffers (TULLY and BEEVERS 1976, YOULE and HUANG 1976). The results suggest that further searches for a role of RCA_{120} in the organization of the mitochondrial membrane of *Ricinus* endosperm are not promising. Nevertheless the unidentified lectins that were solubilized with detergent along with RCA_{120} might still be of functional importance. However, work on this problem should be performed preferentially with plant tissues devoid of soluble lectins to prevent artificial binding as far as possible.

5.3.2 Lectins Associated with Walls

Wall subfractions of certain plants also appear to be rich in lectins which are not readily soluble. Hypocotyls of mung beans (*Vigna radiata*) were homogenized and the walls subsequently extracted firstly with 0.5 M phosphate buffer, pH 7.1, and then with 0.1 M EDTA/0.06% Triton-X-100. After dialysis, both extracts showed agglutinating activity toward trypsinized rabbit erythrocytes (KAUSS and GLASER 1974, KAUSS and BOWLES 1976). Based on fresh tissue weight, the upper growing part of the hypocotyl showed lower activity than the lower nonextending part. Both types of extract appear to contain lectins of similar binding specifity, as the agglutinating activities are strongly inhibited by D-galactose and para-nitrophenyl-α-D-galactoside and to a lesser extent by γ-D-galactonolactone and D-xylose. The titer of the lectins is strongly dependent on the pH value; particularly for the buffer-extractable lectin a sharp decrease in titer between pH 6.5 and 5.5 is evident. The above-mentioned properties suggested as a working hypothesis that the wall-associated lectin may play a role in extension growth, the molecular basis of which, unfortunately, is poorly understood (see Chap. 11, this Vol.). Speculation about the possible participation of lectins in this process suggested either a direct contribution toward binding forces between different polysaccharide chains, the formation of membrane-wall contact or a mechanism for guiding vesicle-delivered wall material to the proper place in the extending wall. Alternatively, the lectin or at least the detergent-soluble part of it could conceivably be remnants of special parts of the plasma membrane (e.g., at plasmodesmata) which are not readily removed from the wall even under plasmolyzing conditions. The experimental evidence presently available is so limited that a detailed discussion of the various possibilities appears premature. More recent attempts to purify the buffer-soluble part of the wall-associated lectin from mung bean hypocotyl (HAASS et al. 1981) indicate that the lectins do not contain hydroxyproline, a marker amino acid for "extensin" (see Chap. 7, this Vol.). A great surprise was the discovery that during extensive purification most hydrolases are either destroyed or removed from the wall extract, but α-galactosidase remains together with the lectin through all the purification steps, indicating that both activities might be present in the same molecule. In this, and in binding specificity, the wall-associated lectin from mung beans appears to be similar to a lectin which was highly purified from seeds of the same plant (HANKINS and SHANNON 1978, see also Sect. 5.5). It was also demonstrated (HAASS et al. 1981) that part of the wall-associated lectin/α-galactosidase can be eluted by salt solutions other than the complexing phosphate buffer. This lectin fraction can bind back to the walls even when D-galactose is present at concentrations sufficient to inhibit hemagglutination. These results may indicate that at least part of the lectin is not bound to the wall via its carbohydrate binding site. The full significance of these observations, however, can only be properly evaluated once more is known about the complex processes involved in wall formation and extension.

5.3.3 Phloem Lectins

Another way to extract proteins selectively from a specialized cell type is to cut phloem cells and collect the sap which flows out due to the slightly positive pressure. Sieve tube sap from *Robinia pseudoacacia* was shown to contain a lectin specific for D-GalNAc (KAUSS and ZIEGLER 1974). This lectin has been purified by affinity chromatography; it is not inhibited by oligomers of β-D-GlcNAc (GIETL et al. 1979). The same paper reports lectin activity in the sieve tube sap from a great number of other tree species. Sugar specificity of the lectins in 15 species was examined and they were all found to be inhibited by oligomers of β-D-GlcNAc. The same specificity was demonstrated for the lectin present in the phloem exudate of *Cucurbita pepo,* for which hemagglutination activity had previously been reported but no inhibiting simple sugar found (SABNIS and HART 1978). Some of the phloem lectins specific for oligomers of β-D-GlcNAc have since been purified by affinity chromatography. ALLEN (1979) could demonstrate that the lectin from the phloem exudate of the fruit of the vegetable marrow (*Cucurbito pepo*), which is similar to the lectins from other Cucurbitaceae, is not a glycoprotein and consists of a single polypeptide chain of about 20,000 molecular weight. No hydroxyproline is present, thus the lectin is unlike the potato lectin which exhibits a similar specificity (see Table 2). In contrast to the cucumber phloem lectin, the lectins from phloem sap of *Quercus robur, Salix alba, Tilia cordata,* and *Ulmus glabra* trees belonging to four quite different families, appear to have a molecular weight of about 30,000 and to be composed of rather small subunits with a molecular weight of about 5,000 (GIETL and ZIEGLER 1980b).

One can only speculate about the possible functions of phloem lectins. The carbohydrate specificity of the lectins characterized so far does not appear to be related to the main sugars transported in the phloem, therefore a direct role in sugar transport is unlikely. On the other hand, it is rather interesting that many phloem lectins appear to have similar oligomer-β-D-GlcNAc binding specificity in spite of the fact that the respective plants are completely unrelated systematically. In addition, in *Robinia pseudoacacia* where the phloem sap lectin exhibits D-GalNAc specificity, high activity of oligomer-β-D-GlcNAc specific lectins has been demonstrated in cells adjacent to phloem cells (GIETL and

ZIEGLER 1980a). This might indicate preservation of a property important for function of phloem cells.

It has been suggested by MIRELMAN et al. (1975) and BARKAI-GOLAN et al. (1978) that wheat germ agglutinin, which also exhibits oligomer-β-D-GlcNAc specificity, plays a role in inhibiting the growth of some fungi containing exposed chitin in their hyphal walls. However, the lectin concentrations applied to show such an effect appear to be rather high. In addition, affinity-purified preparations of wheat germ agglutinin may have contained some chitinase which is especially active on nascent chitin (MOLANO et al. 1979) and thus caused the growth inhibiting effect. It is also uncertain how high lectin concentration in individual cells, e.g., of the seed or root, might be. Thus, the suggestion that lectins have a general function as antibiotics against phytopathogens in the above-mentioned cells cannot be proven or disproven in the moment. However, crude phloem sap or isolated phloem lectins at concentrations normally found in vivo do not inhibit growth of *Trichoderma viride, Tr. homingii,* or *Fusarium solani,* the fungi used in the above-mentioned experiments with wheat germ lectin (GIETL and ZIEGLER 1980b). Although similar experiments have not apparently been done with the rare microorganisms known to infect the phloem specifically, these observations render the assumption doubtful that phloem lectins merely function as antibiotics. A more promising line of research may originate from the fact that all lectins with specificity for oligomers of β-D-GlcNAc also bind very strongly to glycoproteins such as fetuin or thyroglobulin. These glycoproteins are characterized by an internal → D-GlcNAc → D-GlcNAc-→ sequence, typical of the core structure of carbohydrate residues linked *N*-glycosidically to asparagine. An increasing number of lectins is being described which can bind to that structure. It appears possible that the natural receptors for oligomer-β-D-GlcNAc specific lectins (see Table 2 and Sect. 5.2) are unknown glycoproteins, the binding of which may be of significance in recognition or regulation phenomena.

5.4 Plant Toxins Containing Lectin Subunits

Several seeds used as a lectin source have long been known to be toxic. The toxic principles are apparently of two different types, one much more toxic than the other. Extracts of raw seeds of *Glycine max* (soybean) or *Vicia faba* (broad bean) are of low toxicity. The resultant symptoms develop only after prolonged feeding periods and are recognized by a retardation of animal growth or a hemolytic anemia ("favism") found among certain human population groups in eastern mediterranean regions. It is not entirely clear to what extent the lectins from the respective seeds contribute to the low toxicity of a diet based on raw seeds. It appears, however, that most lectins are only slightly toxic to animals (GOLDSTEIN and HAYES 1978, LIENER 1976).

In contrast, some lectin-containing seeds are extremely toxic and appear to contain the most potent toxins known. Uncooked extracts of *Ricinus communis* (castor bean) and *Abrus precatorius* (jequirity bean) can be lethal within a few hours. The respective toxins are proteins and have been called ricin and abrin. These toxins are present at about 1 mg/g in seeds and are lethal at about 1 µg/kg body weight in mice and dogs (OLSNES and PIHL 1978). Both ricin and abrin served as models for toxic proteins during investigation of the fundamental principles of immunology in the late 19th and early 20th century. More recently a quite similar toxic protein was isolated from the modecca flower (*Adenia digitata,* Passifloraceae, OLSNES et al. 1978). The root of this plant, which grows wild in Southern Africa, resembles edible roots of *Coccinia* species (Cucurbitaceae) and has caused several reported instances of lethal intoxication.

Abrin and ricin are retained on affinity columns during purification of lectins from the respective seeds, and are eluted from the column together with the lectin on application of the haptenic saccharide solution. Both toxins thus exhibit carbohydrate binding specificity. The toxins have molecular weights of 65,000 and can be separated by gel chromatography from the lectins which have higher molecular weights (Table 1). Abrin and ricin are glycoproteins and consist of two polypeptide chains which are connected by a single -S-S-bond. The smaller polypeptide (A-chain, molecular weight 30,000 in abrin and 32,000 in ricin) has been called "effectomer" as it is responsible for the toxic effects (OLSNES and PIHL 1978). Although isolated A-chains alone are not toxic to intact cells they are very potent inhibitors of protein synthesis in cell-free systems due to an irreversible inactivation of ribosomes. After treatment the potency for peptide chain elongation and the ability to bind the factor EF_2 is lost. There are several arguments which suggest that the action of the "effectomer" is of a catalytic nature and that the target site at the 60S ribosomal subunit is enzymatically altered in some, as yet unknown, manner (OLSNES and PIHL 1978).

The toxic "effectomer" can enter live cells only with the help of the greater toxin subunit (B-chain, "haptomer", molecular weight 35,000 in abrin and 34,000 in ricin). This subunit exhibits a single carbohydrate binding site for D-galactose residues and can bind with high affinity to cell surface receptors, e.g., on HeLa cells. Isolated monovalent B-chains appear to aggregate in solution to some extent as they exhibit some agglutinating activity and this property clearly allows classification of the "haptomer" as a lectin subunit. This lectin obviously serves as a carrier which binds to the cell surface and thus facilitates the entrance of the "effectomer" into the cells. Only preliminary data are available on the actual mechanism by which the toxic A-chains are transferred into the cells (SANDVIG et al. 1979). A similar trick entering cells is observed with diphtheria or cholera toxins, in which a covalently bound part of the molecule first binds at the cell surface allowing the toxic part of the molecule to enter the cells. In this context it appears of interest that highly toxic hybrid molecules have been prepared in which the binding fragment of diphtheria toxin has been replaced by a subunit of Concanavalin A (GILLILAND et al. 1978) or of the lectin from *Wistaria floribunda* (UCHIDA et al. 1978). Thus the naturally occurring toxins ricin and abrin may represent a model system demonstrating a possible general function of lectins, namely the facilitation of membrane transfer for macromolecules. Although this might appear at the moment to be a rather vague hypothesis one should bear in mind that receptor-mediated endocytosis is of general importance, as demonstrated for the cholesterol-transporting low density lipoprotein in animals (BROWN and GOLDSTEIN 1979).

Generally, the presence of toxic or bitter constituents in plant organs is regarded as providing some protection against animal attack. In the case of classical lectins and that of toxins containing lectin subunits there is little experimental evidence to show that a particular animal avoids one type of seeds because of its lectin or toxin content and prefers others. The only relevant report shows that addition of a hemagglutinin from black bean (*Phaseolus vulgaris*) to the normal diet of bruchid beetle will kill the larvae (JANZEN et al. 1976). These larvae can feed on hemagglutinin-free cowpeas (*Vigna unguiculata*) but die if they feed on black beans. It was concluded that the major adaptive significance of black bean hemagglutinin may be to protect the seeds from insect predators.

5.5 Possible Relations Between Lectins and Enzymes

It has been discussed in Section 2.3 that the carbohydrate binding site of lectins resembles in many respects the substrate binding site of enzymes. Due to the lack of conclusive evidence of a functional role of lectins in plants, several authors mention the possibility that some lectins may be partly denatured enzymes or represent enzymes of complex carbohydrate metabolism which can bind to carbohydrate but cannot perform their catalytic action as another substrate is lacking under agglutination assay conditions (e.g., CALLOW 1977, KAUSS 1976, LIS and SHARON 1981). Such a possibility should be borne in mind and may apply to few lectins, however there is little experimental evidence to sustain this view. It was reported in a short note that pea lectin and PHA (red kidney bean hemagglutinin) can degrade the lipopolysaccharides of their symbiont rhizobia but not that of nonsymbiont bacteria (ALBERSHEIM and WOLPERT 1976). This statement was not documented later by a detailed report; it appears possible that the affinity-purified lectins used (WOLPERT and ALBERSHEIM 1976) may have contained contaminating hydrolases. Chemical cross-linking with glutaraldehyde of the univalent enzyme lysozyme has been reported to result in the formation of a hemagglutinin (HOŘEJŠÍ and KOCOUREK 1974). Agglutination can be inhibited by a low molecular weight glycopeptide isolated from the human erythrocyte surface; however, the concentrations needed ($1.25\ mg \cdot ml^{-1}$) appear rather high. Although the cross-composition of the haptenic substance indicates the presence of D-GlcNAc, no detailed structure is known. Studies performed after the above cited paper was published showed that D-GlcNAc-oligomers derived from chitin were only slightly inhibitory and agglutination was dependent on ionic strength (J. KOCOUREK personal communication). It appears possible, therefore, that the very basic protein lysozyme might cause unspecific hemagglutination not involving the carbohydrate binding site.

The only example in which a highly purified lectin also exhibits enzymatic properties concerns a lectin from seeds of *Vigna radiata* (mung bean, HANKINS and SHANNON 1978). Hemagglutination caused by this lectin is inhibited by D-galactose and para-nitrophenyl-α-D-galactose. The enzymatic specificity can only be tested using the latter substance which is a chromogenic substrate. The α-galactosidase activity appears to be highly specific as, e.g., raffinose or stachyose are not split. Inhibition studies with several saccharides indicate that the carbohydrate-binding site of the lectin and the α-galactosidase have a rather similar or identical carbohydrate specificity. A similar or identical lectin is associated with the walls of hypocotyl segments from mung beans (see Sect. 5.3.2). Further lectin/α-galactosidases with similar properties were found in seeds of four other legume species (HANKINS et al. 1980a). These proteins appear to belong to a family of immunologically related α-galactosidases found in a large number of legumes; in only a few of these plants does the enzyme also show hemagglutinating activity (HANKINS et al. 1980b). In this context it appears of interest that a protein has been isolated from stems and leaves of *Dolichos biflorus* which itself does not exhibit lectin activity but is immunologically cross-reactive with the seed lectin from the same plant (TALBOT and ETZLER 1977). The authors feel this material may possibly represent a precursor of the lectin. It also appears possible, however, that it is a degradation product or – in the light of the above suggestions – an enzyme related to a lectin.

Another type of lectin–enzyme relationship appears to be represented by the Cl^--ATPase from *Limonium* leaf microsomes (HILL and HANKE 1979). Activity of this enzyme is increased in the presence of D-galactose and D-GalNAc and the enzyme can be enriched on a column of D-GalNAc-Sepharose. These findings are considered to be evidence for a carbohydrate binding site on the Cl^--ATPase, which might help to regulate or assemble this enzyme complex.

Acknowledgements. I would like to thank D. YOUNG and H. KÖHLE for very valuable help with the manuscript and G. GERISCH for critically reading Section 4. Many colleagues have kindly provided me with reprints or preprints which is very much appreciated.

Note Added in Proof: Readers may also consult the recent review: BARONDES SH (1981) Lectins: Their multiple endogenous cellular functions. Ann Rev Biochem 50:201–231.

References

Albersheim P, Anderson-Prouty AJ (1975) Carbohydrates, proteins, cell surfaces, and the biochemistry of pathogenesis. Annu Rev Plant Physiol 26:31–52

Albersheim P, Valent BS (1978) Host-pathogen interactions in plants. J Cell Biol 78:627–643

Albersheim P, Wolpert JS (1976) The lectins of legumes are enzymes which degrade the lipopolysaccharides of their symbiont rhizobia. Plant Physiol Suppl 57:79

Albersheim P, Jones TM, English PD (1969) Biochemistry of the cell wall in relation to infective processes. Annu Rev Phytopathol 7:171–194

Allen AK (1979) A lectin from the exudate of the fruit of the vegetable marrow (*Cucurbita pepo*) that has a specificity for β-1,4-linked N-acetylglucosamine oligosaccharides. Biochem J 183:133–137

Allen AK, Desai NN, Neuberger A (1976) The purification of the glycoprotein lectin from the broad bean (*Vicia faba*) and a comparison of its properties with lectins of similar specificity. Biochem J 155:127–135

Allen AK, Desai NN, Neuberger A, Creeth JM (1978) Properties of potato lectin and the nature of its glycoprotein linkages. Biochem J 171:665–674

Barkai-Golan R, Mirelman D, Sharon N (1978) Studies on growth inhibition by lectins of penicillia and aspergilli. Arch Microbiol 116:119–124

Barondes SH, Haywood PL (1979) Comparison of developmentally regulated lectins from three species of cellular slime mold. Biochim Biophys Acta 550:297–308

Barondes SH, Rosen SD (1976) Cell surface carbohydrate-binding proteins: role in cell recognition. In: Barondes SH (ed) Neuronal Recognition. Chapman and Hall, London, pp 331–356
Biroc SL, Etzler ME (1978) The effect of periodate oxidation and α-mannosidase treatment on *Dolichos biflorus* lectin. Biochim Biophys Acta 544:85–92
Bowles DJ, Kauss H (1975) Carbohydrate-binding proteins from cellular membranes of plant tissue. Plant Sci Lett 4:411–418
Bowles DJ, Kauss H (1976) Characterization, enzymatic and lectin properties of isolated membranes from *Phaseolus aureus*. Biochim Biophys Acta 443:360–374
Bowles DJ, Schnarrenberger C, Kauss H (1976) Lectins as membrane components of mitochondria from *Ricinus communis*. Biochem J 160:375–382
Bowles DJ, Lis H, Sharon N (1979) Distribution of lectins in membranes of soybean and peanut plants; I. General distribution in root, shoot and leaf tissue at different stages of growth. Planta 145:193–198
Boyd WC, Shapleigh E (1954) Specific precipitating activity of plant agglutinins (lectins). Science 119:419
Bozzaro S, Gerisch G (1978) Contact sites in aggregating cells of *Polysphondylium pallidum*. J Mol Biol 120:265–279
Brown MS, Goldstein JL (1979) Receptor-mediated endocytosis: insights from the lipoprotein receptor system. Proc Natl Acad Sci USA 76:3330–3337
Callow JA (1975) Plant lectins. Curr Adv Plant Sci 7:181–193
Callow JA (1977) Recognition, resistance and the role of plant lectins in host-parasite interactions. Adv Bot Res 4:1–49
Chang CM, Reitherman RW, Rosen SD, Barondes SH (1975) Cell surface localization of discoidin, a developmentally regulated carbohydrate-binding protein from *Dictyostelium discoideum*. Exp Cell Res 95:136–142
Clarke AE, Gleeson PA, Jermyn MA, Knox RB (1978) Characterization and localization of β-lectins in lower and higher plants. Aust J Plant Physiol 5:707–722
Dazzo FB (1977) Cross reactive antigens and lectin as determinants of host specifity in the *Rhizobium*-clover symbiosis. Diss Abstr Int 37:7
Dazzo FB (1979) Adsorption of microorganisms to roots and other plant surfaces. In: Bitton G, Marshall K (eds) Adsorption of microorganisms to surfaces. Wiley and Sons, New York, pp 253–316
Dazzo FB, Brill WJ (1978) Regulation by fixed nitrogen of host-symbiont recognition in the *Rhizobium*-clover symbiosis. Plant Physiol 62:18–21
Dazzo FB, Yanke WE, Brill WJ (1978) Trifoliin: a *Rhizobium* recognition protein from white clover. Biochem Biophys Acta 539:272–286
Eshdat Y, Ofek I, Yashouv-Gan Y, Sharon N, Mirelmann D (1978) Isolation of a mannose-specific lectin from *Escherichia coli* and its role in the adherence of the bacteria to epithelial cells. Biochem Biophys Res Commun 85:1551–1559
Frazier WA, Rosen SD, Reitherman W, Barondes SH (1975) Purification and comparison of two developmentally regulated lectins from *Dictyostelium discoideum*. Discoidin I and II. J Biol Chem 250:7714–7721
Gansera R, Schurz H, Rüdiger H (1979) Lectin associated proteins from the seeds of Leguminosae. Z Physiol Chemie 360:1579–1585
Gerisch G (1980) Univalent antibody fragments as tools for the analysis of cell interactions in *Dictyostelium*. In: Monroy A, Moscona AA, Friedlander M (eds) Current Topics in Developmental Biology, "Developmental Immunology" 14, Part II. Academic Press, New York, pp 243–270
Gerisch G, Fromm H, Hoesgen A, Wick U (1975) Control of cell-contact sites by cyclic AMP pulses in differentiating *Dictyostelium* cells. Nature 255:547–549
Gerisch G, Krelle H, Bozzaro S (1980) Analysis of cell adhesion in *Dictyostelium* and *Polysphondylium* by the use of Fab. In: Cell Adhesion and Motility. Curtis ASG, Pitts JD (eds) Cambridge Univ. Press, pp 293–307
Gietl C, Ziegler H (1979) Lectins in the excretion of intact roots. Naturwissenschaften 66:161–162
Gietl C, Ziegler H (1980a) Distribution of carbohydrate-binding proteins in different tissues of *Robinia pseudoacacia* L. Biochem Physiol Pflanzen 175:58–66

Gietl C, Ziegler H (1980b) Affinity chromatography of carbohydrate binding proteins in the phloem exudate from several tree species. Biochem Physiol Pflanzen 175:50–66
Gietl C, Kauss H, Ziegler H (1979) Affinity chromatography of a lectin from *Robinia pseudoacacia* L. and demonstration of lectins in sieve-tube sap from other tree species. Planta 144:367–371
Gilliland DG, Collier RJ, Moehring JM, Moehring TJ (1978) Chimeric toxins: toxic, disulfide-linked conjugate of concanavalin A with fragment A from diphtheria toxin. Proc Natl Acad Sci USA 75:5319–5323
Goldstein IJ, Hayes CE (1978) The lectins: Carbohydrate-binding proteins of plants and animals. Adv Carboh Chem Biochem 35:128–340
Goodenough U (1977) Mating interactions in *Chlamydomonas*. In: Reissig JL (ed) Microbiol Interactions. Receptors and Interaction, Series B, Vol 3. Chapman and Hall, London, pp 324–350
Haass D, Frey R, Thiesen M, Kauss H (1981) Partial purification of a hemagglutinin associated with the cell walls from hypocotyls of *Vigna radiata*. Planta 151:490–496
Hankins CN, Shannon LM (1978) The physical and enzymatic properties of a phytohemagglutinin from mung beans. J Biol Chem 253:7791–7797
Hankins CN, Kindinger JI, Shannon LM (1979) Legume lectins: I. Immunological cross reactions between the enzymic lectin from mung beans and other well characterized legume lectins. Plant Physiol 64:104–107
Hankins CN, Kindinger, JI, Shannon LM (1980a) Legume α-galactosidases which have hemagglutinin properties. Plant Physiol 65:618–622
Hankins CN, Kindinger JI, Shannon LM (1980b) Legume α-galactosidase forms devoid of hemagglutinin activity. Plant Physiol 66:375–378
Hill BS, Hanke DE (1979) Properties of the chloride-ATPase from *Limonium* salt glands: activation by, and binding to, specific sugars. J Membrane Biol 51:185–194
Hopp HE, Romero P, Pont-Lezica R (1979) Subcellular localization of glucosyl transferases involved in lectin glucosylation in *Pisum sativum* seedlings. Plant Cell Physiol 20:1063–1069
Hořejší V, Kocourek J (1974) Studies on phytohemagglutinins XXI. The covalent oligomers of lysozyme – first case of semisynthetic hemagglutinins. Experientia 30:1348–1349
Hořejší V, Kocourek J (1978) Studies on Lectins. XXXVI. Properties of some lectins prepared by affinity chromatography on O-glycosyl polyacrylamide gels. Biochim Biophys Acta 538:299–315
Hořejší V, Chaloupecká O, Kocourek J (1978a) Studies on lectins. XLIII. Isolation and characterization of the lectin from restharrow roots (*Ononis hircina* jacq.). Biochim Biophys Acta 539:287–293
Hořejší V, Haskovec C, Kocourek J (1978b) Studies on lectins. XXXVIII. Isolation and characterization of the lectin from black locust bark (*Robinia pseudoacacia* L.). Biochim Biophys Acta 532:98–104
Howard IK, Sage HJ, Horton CB (1972) Studies on the appearance and location of hemagglutinins from a common lentil during the life cycle of the plant. Arch Biochem Biophys 149:323–326
Howard J, Shannon L, Oki L, Murashige T (1976) Soybean agglutinin as a mitogen for soybean callus cells. Fed Proc 35:1402
Howard J, Kindinger J, Shannon LM (1979) Conservation of antigenic determinants among different seed lectins. Arch Biochem Biophys 192:457–465
Huet C, Lonchampt M, Huet M, Bernadac A (1974) Temperature effects on the concanavalin A molecule and on concanavalin A binding. Biochim Biophys Acta 365:28–39
Janzen DH, Juster HB, Liener IE (1976) Insecticidal action of phytohemagglutinin in black beans on a bruchid beetle. Science 192:795–796
Jones JM, Cawley LP, Teresa GW (1967) The lectins of *Maclura pomifera*: Zymographic studies, distribution in the developing plant and production in tissue cultures of epicotyl. J Immunol 98:364–367
Kauss H (1976) Plant lectins (phytohemagglutinins). In: Ellenberg H, Esser K, Merxmüller H, Schnepf E, Ziegler H (eds) Progress in Botany 38. Springer, Berlin Heidelberg New York, pp 58–70

Kauss H, Bowles D (1976) Some properties of carbohydrate-binding proteins (lectins) solubilized from cell walls of *Phaseolus aureus*. Planta 130:169–174
Kauss H, Glaser Ch (1974) Carbohydrate-binding proteins from plant cell walls and their possible involvement in extension growth. FEBS Lett 45:304–307
Kauss H, Ziegler H (1974) Carbohydrate-binding proteins from the sieve-tube sap of *Robinia pseudoacacia* L. Planta 121:197–200
Köhle H, Kauss H (1979) Binding of *Ricinus communis* agglutinin to the inner mitochondrial membrane as an artifact during preparation. Biochem J 184:721–723
Köhle H, Kauss H (1980) Improved analysis of hemagglutination assays for quantitation of lectin activity. Analyt Biochem 103:227–229
Kojima M, Uritani I (1978) Isolation and characterization of factors in sweet potato root which agglutinate germinating spores of *Ceratocystis fimbriata*, black rot fungus. Plant Physiol 62:751–753
Kornfeld R, Kornfeld S (1970) The structure of a phytohemagglutination receptor site from human erythrocytes. J Biol Chem 245:2536–2545
Kornfeld S, Kornfeld R (1978) Use of lectins in the study of mammalian glycoproteins. In: Horowitz MI, Pigmen W (eds) The glycoconjugates. Vol II. Academic Press, New York, pp 437–449
Kuć J (1972) Phytoalexins. Annu Rev Phytopath 10:207–232
Leavitt RD, Felsted RL, Bachur NR (1977) Biological and biochemical properties of *Phaseolus vulgaris* isolectins. J Biol Chem 252:2961–2966
Liener IE (1976) Phytohemagglutinins (phytolectins). Annu Rev Plant Physiol 27:291–319
Lis H, Sharon N (1973) The biochemistry of plant lectins (phytohemagglutinins). Annu Rev Biochem 42:541–574
Lis H, Sharon N (1981) Lectins in higher plants. In: Stumpf PK, Conn EE (eds) The biochemistry of plants. Vol VI. Academic Press, New York
Loontiens FG, Van Wauwe JP, De Bruyne CK (1975) Concanavalin A: Relation between hapten inhibition indexes and association constants for different glycosides. Carbohydr Res 44:150–153
Lotan R, Skutelsky E, Danon D, Sharon N (1975a) The purification, composition, and specifity of the anti-T lectin from peanut (*Arachis hypogaea*). J Biol Chem 250:8518–8523
Lotan R, Lis H, Sharon N (1975b) Aggregation and fragmentation of soybean agglutinin. Biochem Biophys Res Commun 62:144–150
McDonough JP, Springer WR, Barondes SH (1980) Species-specific cell cohesion in cellular slime molds. Demonstration by several quantitative assays and with multiple species. Exp Cell Res 125:1–4
Mialonier G, Privat JP, Monsigny M, Kahlem G, Durand R (1973) Isolement, propriétés physico-chimiques et localisation *in vivo* d'une phytohémagglutine (lectine) de *Phaseolus vulgaris* L. (Var. rouge). Physiol Veg 11:519–537
Minetti M, Aducci P, Teichner A (1976) A new agglutinating activity from wheat flour inhibited by tryptophan. Biochim Biophys Acta 437:505–517
Minetti M, Teichner A, Aducci P (1978) Interaction of neutral polysaccharides with human erythrocyte membrane: involvement of phospholipid bilayer. Biochem Biophys Res Commun 80:46–55
Mirelman D, Galun E, Sharon N, Lotan R (1975) Inhibition of fungal growth by wheat germ agglutinin. Nature 256:414–416
Molano J, Polacheck J, Duran A, Cabid E (1979) An endochitinase from wheat germ. J Biol Chem 254:4901–4907
Müller K, Gerisch G (1978) A specific glycoprotein as the target site of adhesion blocking Fab in aggregating *Dictyostelium* cells. Nature 274:445–449
Müller K, Gerisch G, Fromme I, Mayer H, Tsugita A (1979) A membrane glycoprotein of aggregating *Dictyostelium* cells with the properties of contact sites. Eur J Biochem 99:419–426
Nagl W (1972a) Phytohemagglutinin: Temporäre Erhöhung der mitotischen Aktivität bei *Allium*, und partieller Antagonismus gegenüber Colchicin. Exp Cell Res 74:599–602
Nagl W (1972b) Phytohemagglutinin: Transitory enhancement of growth in *Phaseolus* and *Allium*. Planta 106:269–272

Newell PC (1977) How cells communicate: the system used by slime molds. Endeavour, New Series 1:63–68

Nicolson GL (1976) Transmembrane control of receptors on normal and tumor cells. I. Cytoplasmic influence over cell surface components. Biochim Biophys Acta 457:57–108

O'Brien RL, Parker JW, Dixon JFP (1978) Mechanisms of lymphocyte transformation. Progr Mol Subcell Biol 6:201–270

Olsnes S, Pihl A (1978) Abrin and ricin – two toxic lectins. Trends Bioc 3:7–10

Olsnes S, Haylett T, Refsnes K (1978) Purification and characterization of the highly toxic lectin modeccin. J Biol Chem 253:5069–5073

Ozawa T, Yamaguchi M (1979) Inhibition of soybean cell growth by the adsorption of *Rhizobium japonicum*. Plant Physiol 64:65–68

Paus E, Steen HB (1978) Mitogenic effect of α-mannosidase on lymphocytes. Nature 272:452–454

Pont-Lezica R, Romero P, Hopp HE (1978) Glycosylation of membrane-bound proteins by lipid-linked glucose. Planta 140:177–183

Presant CA, Kornfeld S (1972) Characterization of the cell surface receptor for the *Agaricus bisporus* hemagglutinin. J Biol Chem 247:6937–6945

Pueppke SG, Bauer WD, Keegstra K, Ferguson AL (1978) Role of lectins in plant-microorganism interactions. II. Distribution of soybean lectin in tissues of glycine max. Plant Physiol 61:779–784

Raa J, Robertsen B, Solheim B, Tronsmo A (1977) Cell surface biochemistry related to specificity of pathogenesis and virulence of microorganisms. In: Solheim B, Raa J (eds) Cell Wall Biochemistry. Universitetsforlaget Oslo – Bergen – Tromsø, Columbia Univ Press, New York, pp 11–28

Reisner Y, Sharon N (1980) Cell fractionation by lectins. Trends Biochem Sci 5:29–31

Reitherman RW, Rosen SD, Frazier WA, Barondes SH (1975) Cell surface species-specific high affinity receptors for discoidin: Developmental regulation in *Dictyostelium discoideum*. Proc Natl Acad Sci USA 72:3541–3545

Rosen SD, Chang CM, Barondes SH (1977) Intercellular adhesion in the cellular slime mold *Polysphondylium pallidum* inhibited by interaction of asialofetuin or specific univalent antibody with endogenous cell surface lectin. Dev Biol 61:202–213

Rosen SD, Kaur J, Clark DL, Pardos BT, Frazier WA (1979) Purification and characterization of multiple species (isolectins) of a slime mold lectin implicated in intercellular adhesion. J Biol Chem 254:9408–9415

Rüdiger H (1978) Lectine, pflanzliche zuckerbindende Proteine. Naturwissenschaften 65:239–244

Sabnis DD, Hart JW (1978) The isolation and some properties of a lectin (hemagglutinin) from *Cucurbita* phloem exudate. Planta 142:97–101

Sandvig K, Olsnes S, Pihl A (1979) Inhibitory effect of ammonium chloride and chloroquine on the entry of the toxic lectin modeccin into HeLa cells. Biochem Biophys Res Commun 90:648–655

Sequeira L (1978) Lectins and their role in host-pathogen specificity. Annu Rev Phytopathol 16:453–81

Sharon N (1976) Lectins as mitogens. In: Oppenheim JJ, Rosenstreich DL (eds) Mitogens in immunobiology. Academic Press, New York, pp 31–41

Sharon N, Lis H (1972) Lectins: cell-agglutinating and sugar-specific proteins. Science 177:307–314

Sumner JB, Howell SF (1936) The identification of the hemagglutinin of the jack bean with concanavalin A. J Bacteriol 32:227–237

Sussmann M (1976) The genesis of multicellular organization and the control of gene expression in *Dictyostelium discoideum*. Progr Mol Subcell Biol 4:103–131

Sussmann M, Brackenbury R (1976) Biochemical and molecular-genetic aspects of cellular slime mold development. Annu Rev Plant Physiol 27:229–65

Suzuki I, Saito H, Inoue S, Migita S, Takahashi T (1979) Purification and characterization of two lectins from *Aloe arborescens* Mill. J Biochem (Tokyo) 85:163–171

Talbot CF, Etzler ME (1977) Isolation and characterization of an inactive form of lectin from stem and leaves of *Dolichos biflorus*. Fed Proc 36:795

Talbot CG, Etzler ME (1978) Development and distribution of *Dolichos biflorus* lectin as measured by radioimmunoassay. Plant Physiol 61:847–850

Toyoshima S, Funkuda M, Osawa T (1972) Chemical nature of the receptor site for various phytomitogens. Biochemistry 11:4000–4005

Tully RE, Beevers H (1976) Protein bodies of castor bean endosperm. Plant Physiol 58:710–716

Tunis M, Lis H, Sharon N (1979) Participation of the carbohydrate moieties of glycolipids and glycoproteins in agglutination of erythrocytes by lectins. In: Peeters H (ed) Proceedings of the 27th Colloquium on Protides of the Biological Fluids, Pergamon Press, Oxford, New York, pp 521–524

Uchida T, Yamaizumi M, Mekada E, Okada Y, Tsuda M, Kurokawa T, Sugino Y (1978) Reconstitution of hybrid toxin from fragment A of diphtheria toxin and a subunit of *Wistaria floribunda* lectin. J Biol Chem 253:6307–6310

Wade M, Albersheim P (1979) Race-specific molecules that protect soybeans from *Phytophthora megasperma* var. sojae. Proc Natl Acad Sci USA 76:4433–4437

Wolpert J, Albersheim P (1976) Host-symbiont interactions. I. The lectins of legumes interact with the O-antigen-containing lipopolysaccharides of their symbiont *Rhizobia*. Biochem Biophys Res Commun 70:729–737

Yokoyama K, Terao T, Osawa T (1978) Carbohydrate-binding specificity of pokeweed mitogens. Biochim Biophys Acta 538:384–396

Youle RJ, Huang AHC (1976) Protein bodies from the endosperm of castor bean. Subfractionation, protein components, lectins, and changes during germination. Plant Physiol 58:703–709

26 The Role of Lectins in Symbiotic Plant-Microbe Interactions

E.L. SCHMIDT and B.B. BOHLOOL

1 Introduction

Symbiosis is usually understood to be a constant and intimate association between two organisms which results in mutual benefit to both. Inherent to the concept of symbiosis is the view that the association is a specific one, involving only certain partners matched by evolution. Symbiosis between higher plants and microorganisms occurs in a wide range of expressions, reflecting both the diversity of plants and protists, and the opportunity for partner selection provided by continuous and close contact between plants and the multitude of microorganisms sharing their niche. Those interactions that have achieved symbiosis include mycorrhizae, lichens, various eukaryotes in association with blue-green bacteria (blue-green algae), and the legume and nonlegume nitrogen-fixing associations.

Mechanisms whereby two such biologically diverse partners as a higher plant and a microorganism interact to initiate a symbiotic association are not known. Typically the plant component initiates growth in soil, to the accompaniment of a burgeoning, plant-selective, satellite population of heterogeneous microorganisms about its root system. The microbial partner-to-be must be a part of this newly forming rhizosphere population, and must be capable of competing successfully within that population. It seems most likely that it must also respond to some sort of highly specific mutual recognition mechanism (or mechanisms) which insures selection of the appropriate rhizosphere microorganism as the first step in the consolidation of the partnership. None of these preliminary events in the establishment of a symbiotic interaction are known for even a single association, but such events must occur, and they must be critical to the success of the interaction. The most advanced model for the symbiotic mode of plant–microbe interactions, and the only one for which the early formative stages have been considered, is the legume–*Rhizobium* symbiosis.

Legume–*Rhizobium* symbioses result when soil bacteria of the genus *Rhizobium* invade the roots of developing legume plants and induce the formation of root nodules capable of nitrogen fixation. The interaction is generally specific, involving only legume plants and rhizobia bacteria. Often the specificity is such as to limit the partnership to a single species of legume and a single species of *Rhizobium*. Whether nodulated by a single *Rhizobium* or by one of several rhizobia, the legume plant must participate in the selection of the appropriate bacterial partner from a large and diverse population of nonrhizobial bacteria in its rhizosphere. Whether or not the roots of a single species of legume are nodulated by a given *Rhizobium*, the bacterium must also participate at the root surface in the partner selection process. Some mechanism for mutual recognition between plant root and *Rhizobium* must be presumed, for there is no evidence that bacteria other than rhizobia invade intact legume roots, and only the single example (TRINICK 1973) of a nonlegume plant root system invaded by a *Rhizobium*.

There is evidence that mutual recognition between potential partners in the legume–*Rhizobium* symbioses may be mediated by lectins synthesized by the host plant (BROUGHTON 1978, SCHMIDT 1979). Lectins are proteins or glycoproteins that specifically bind saccharides (LIS and SHARON 1973). They occur in many plants, especially in the seeds of legumes, and have been referred to also as phytohemagglutinins because of their ability to agglutinate red blood cells. Lectins have also been found in bacteria, fungi, viruses, invertebrates, and vertebrates (LIS and SHARON 1973, KAWASAKI and ASHWELL 1976). Many possible roles have been suggested for lectins in plants (LIENER 1976) but their natural function has never been resolved. Some evidence has implicated lectins in plant–pathogen interactions involving bacteria (SEQUEIRA 1978) and fungi (WHEELER 1976, BARKAI-GOLAN et al. 1978), and lectins of fungal origin may be involved in the initial stages of establishing nitrogen-fixation symbioses in lichens (LOCKHART et al. 1978). Most of the interest in the role of plant lectins in nature, however, has focused on the possibility of a recognition function in the legume–*Rhizobium* symbiosis. The postulate is that lectins present a recognition site on the legume roots that interacts specifically with a distinctive saccharide associated with the surface of the appropriate *Rhizobium* cell as a prelude to nodulation (HAMBLIN and KENT 1973, BOHLOOL and SCHMIDT 1974).

2 Pre-Recognition Events of Concern

2.1 Rhizosphere Competence

If plant lectin mediates the recognition between *Rhizobium* and root, some preparation on the part of both partners may be expected in the rhizosphere. Although the pre-recognition events are of obvious importance to the establishment of the symbiosis, and provide the best opportunity for practical management of symbiotic nitrogen fixation, very few details are known. Rhizosphere research is very difficult and the methodology has not been adequate to examine the complex interactions at microscopic habitats where soil, root, and bacteria interface.

Rhizobia are considered to be soil bacteria that first establish in a given soil as a result of the introduction of a host legume (VEST et al. 1973). The diversity of rhizobia in a soil, the density and stability of the rhizobial soil community, the permanency of their residence in the absence of an appropriate legume partner, and substrates for their growth in the soil at large, are all important but virtually unknown aspects of the pre-symbiosis. Methodology for the study of rhizobia in soil has been indirect, with reliance on the plant nodule as the natural enrichment culture from which to isolate a *Rhizobium*. This was necessary because rhizobia look like most other bacteria in the soil and selective artificial media have never been successful. Immunofluorescence, the only approach capable of specific detection and enumeration of individual strains of rhizobia directly in the soil, is of fairly recent application (BOHLOOL and SCHMIDT 1970, SCHMIDT 1973, SCHMIDT 1974).

Response of soil rhizobia to the presence of a legume plant may commence with germination of a seed. Inhibitory diffusates from the germinating seed were sometimes observed to repress rhizobia in the vicinity (VINCENT 1965, HALA and MATHERS 1977), but such inhibition has not been documented for other than artificial systems. Initial root development and extension in contact with the soil is accompanied by selective stimulation of some members of the soil population in the immediate vicinity. This is the initiation of the rhizosphere, an interface of soil and developing root system where the dominant influence of the root is expressed in the heightened activity of a microbial population qualitatively different from and quantitatively greater than the soil population outside the rhizosphere zone. Transition from soil to rhizosphere by bacteria appears to be controlled by root exudates (ROVIRA 1969), soil properties (DOMMERGUES 1978) and whatever still unknown attributes such as growth rates, growth efficiencies, and substrate concentration responses contribute to the success of certain bacteria in competition for the rhizosphere zone.

All known rhizobia were originally isolated from a root nodule and hence were obviously competent rhizosphere bacteria under some circumstances. On the other hand, nodulation often fails even though the legume seed is inoculated with appropriate rhizobia. This suggests the possibility that some rhizobia may not adapt to the rhizosphere and hence not even reach a recognition stage. Plant variety (VEST et al. 1973), soil properties (MARSHALL 1965, DIXON 1969), and microbial antagonisms (CASS-SMITH and HOLLAND 1958, HOLLAND 1970) are among the factors reported to limit nodulation, perhaps because the *Rhizobium* failed to survive and grow in the rhizosphere. It seems likely, however, that nodulation may fail in many instances despite rhizosphere colonization by appropriate rhizobia, because there is considerable evidence that rhizobia generally are good rhizosphere bacteria. Comparisons between rhizosphere soil and nearby nonrhizosphere soil demonstrate the rhizosphere competence of rhizobia (ROVIRA 1961, TUZIMURA and WATANABE 1962a, b, REYES and SCHMIDT 1979). Numbers of rhizobia were 10 to 200 times greater in rhizosphere than nonrhizosphere soil. Good rhizosphere competence on the part of rhizobia holds for nonlegume plants as well. DIATLOFF (1969) found that inoculation of seeds of nonhost legumes and even cereals with *R. japonicum* showed promise as a means to introduce soybean rhizobia into some Australian soils.

Rhizosphere competence differs with different strains of the same *Rhizobium* species. Certain rhizobia become established in soil and it is these indigenous strains that are usually successful in establishing a partnership with a suitable host legume, to the exclusion of potentially better nitrogen-fixing strains that may be added as inoculum (HOLLAND 1970, WEAVER and FREDERICK 1974).

A much-cited and practically important example of this phenomenon is found in *R. japonicum* serotype 123. This strain was introduced along with soybeans in soils of midwestern USA, about 50 years ago. It is now the dominant serotype in these soils (HAM et al. 1971) where it consistently out-competes more desirable strains used as seed inoculum (KAPUSTA and ROUWENHORST 1973). Nothing is known as to the attributes for success of strain 123 or other indigenous rhizobia, but presumably their excellent rhizosphere competence could stem from superior growth rates, growth efficiencies, antagonistic capabilities, or recognition mechanisms.

2.2 Host Rhizosphere Stimulation

Essentially nothing is known about the way in which the legume plant presumably directs the pre-recognition events. The marked increases of rhizobia that have often been noted in the rhizosphere of host legumes compared to nonrhizosphere soil (ROVIRA 1961, TUZIMURA and WATANABE 1962a, ROBINSON 1967) have led to suggestions that a specific *Rhizobium* may be selectively favored by a substance or substances excreted into the rhizosphere by the host plant. The concept of specific stimulation, the strategy whereby the appropriate *Rhizobium* symbiont is selectively favored by its host root, has been stated in considerable detail by NUTMAN (1965). While the idea that a rather precise rhizosphere relationship of this type is appealing, as having evolved to help insure success in attaining the symbiotic state, the evidence is not convincing. Careful experimental work is needed to compare populations of rhizobia of different partner preferences in the same legume rhizosphere and to compare such responses also to those observed in nonlegume rhizospheres. Such experiments must be conducted in normal soil: root systems, since a host of nonrhizobia bacteria also in the rhizosphere may be competitors for any "specific" stimulator substance.

EGERAAT (1972) studied the root zone of sterile pea plants in solution culture in attempts to identify root exudates that influence the pea symbiont *R. leguminosarum*. Homoserine was quantitatively the most important amino acid found in analyses of ninhydrin-positive compounds in the root exudates. Interestingly *R. leguminosarum* grew very well in pure culture with homoserine as the only carbon and nitrogen source, whereas homoserine was toxic to *R. trifolii* and *R. phaseoli*, neither of which can nodulate the pea. Conditions of this experiment were necessarily highly artificial in order to identify components of the exudate, and the lead should be examined further under nonsterile conditions. For any such signal compound or specific substrate it must be shown that the substance is produced during normal root growth, and that it somehow is not available to the nonrhizobia that vastly outnumber any specific *Rhizobium* in the rhizosphere.

In the first, and as yet the only, direct examination of the population dynamics of rhizobia in developing rhizospheres in soil, REYES and SCHMIDT (1979) found no evidence for specific stimulation of *R. japonicum* by soybeans. Rhizosphere stimulation of strain 123 was only slightly, if at all, greater for host than for nonhost plants, and, although strain 123 was responsive to rhizosphere conditions in general, it was still numerically negligible in the total rhizosphere population. The ability, alluded to previously, of strain 123 to out-compete other *R. japonicum* strains in soybean rhizospheres, was found not to reside in massive increases in population in the developing rhizosphere. As good rhizosphere bacteria, the rhizobia may merely be capable of competing with other rhizosphere bacteria for common, nonselective exudates of most plants, host or nonhost. But perhaps the recognition mechanism in the legume host is so sensitive and specific that even a modest and nonspecific rhizosphere response provides enough rhizobia at the root surfaces to accomplish partner selection.

2.3 Arrival at the Recognition Site

Rhizobia usually are considered to invade legumes by way of root hair cells, although this seems not to be the case for many tropical legumes (DART 1974). The recognition site therefore, commonly should be associated with root hair surfaces, or perhaps the surfaces of only certain root hairs. The bacterial partner must not only be in the rhizosphere, but must be in this particular part of the rhizosphere. It is not known if there are some additional mechanisms beyond rhizosphere growth response to assure the presence of rhizobia in the vicinity of the recognition site, but chemotaxis has been considered.

In their review of the literature on ecological aspects of chemotaxis, CHET and MITCHELL (1976) concluded that the chemotactic response may play an important role as a mechanism for attracting fungi and bacteria to the root zone. The first report on chemotaxis as a mechanism for attraction of rhizobia was that of CURRIER and STROBEL (1976). Six strains of rhizobia were labeled with ^{14}C, and tested for response to plant extracts. Responses were positive for all strains, but the pattern of the responses was nonspecific in that both homologous and heterologous plant extracts were chemotactic for a given species of *Rhizobium*. Subsequent work by the same authors (CURRIER and STROBEL 1977) concentrated on the trefoil rhizobia. Again there was no pattern of response related to the kind of legume extract studied. Confirmation of such chemotaxis on the part of rhizobia, even though nonspecific and only in pure culture, would be extremely interesting. It could, for example, be suggestive of a distinct step among pre-recognition events, to be examined further at the level of the whole root.

Even though a chemotactic capability is eventually demonstrated for rhizobia in culture, there are a number of difficulties in relating this to the rhizosphere situation. The highly particulate maze of discontinuous soil pores found in the rhizosphere will certainly modify, if not block completely, any chemotactic response. Chemotactic agents are usually diffusible substances such as amino acids, sugars, and alcohols. It is difficult to picture such readily available materials as escaping microbial degradation and persisting long enough to diffuse further than the very surface of a root cell.

3 Lectin-Mediated Recognition

3.1 Lectin Recognition Hypothesis

FRED and STEVENS (1923) performed agglutination tests on 41 cultures of rhizobia and were able to separate these into a far greater number of strains than had been possible by plant host specificity grouping. The use of serological methods for examining the legume symbiosis was later extended by BALDWIN et al. (1927) with attempts to group legume plants on the basis of precipitin tests and anaphylactic reactions of their seed proteins. Interaction between legume seed proteins and rhizobia was not demonstrated until 1973. HAMBLIN and KENT (1973) examined the reaction of a single strain of *R. phaseoli* to a crude lectin (phytohemagglutinin) fraction obtained from the seeds of the bean *Phaseolus vulgaris*. Rhizobia treated with lectin were shown to be capable of agglutinating red blood cells, whereas untreated bacteria were incapable of hemagglutination. When the root systems of very young (6-day) plants were dipped in suspensions of erythrocytes, red blood cells bound to root hairs in some instances but not others, to suggest that lectin may have been present at some root surfaces. These results, along with the finding that hemagglutinating material occurred only in the nodules and roots below the nodules, suggested to the authors that bean lectin was capable of binding rhizobia at a site suitable for infection.

Whereas the experiments of HAMBLIN and KENT did not address the specificity of lectin–rhizobia interactions, those of BOHLOOL and SCHMIDT (1974) sought to examine this point. They studied the soybean–*R. japonicum* system with the use of fluorescence microscopy to visualize the binding of fluorescein isothiocyanate labeled soybean lectin to rhizobia. Lectin was observed to bind to all but 3 of 25 strains of *R. japonicum,* and not to bind to any of 23 other strains representative of rhizobia that nodule legumes other than soybeans. The microscopic appearance of rhizobia after reaction with the fluorescent labeled soybean lectin is shown in Fig. 1.

Confirmation of specific binding of soybean seed lectin to cells of *R. japonicum* was reported by WOLPERT and ALBERSHEIM (1976), BHUVANESWARI et al. (1977), BHUVANESWARI and BAUER (1978), BROUGHTON (1978), KATO et al. (1978). The specificity element of the lectin-binding hypothesis was also observed to pertain in the white clover-*R. trifolii* system (DAZZO and HUBBELL 1975, DAZZO and BRILL 1977). Lectin present in clover seed extracts agglutinated infective but not noninfective strains of *R. trifolii,* and did not agglutinate any representatives of several different groups of heterologous rhizobia.

Specificity of interaction between legume lectin and *Rhizobium* symbiont is obviously important to the lectin recognition hypothesis. Much the greatest attention to specificity has been devoted to the soybean–*R. japonicum* symbiosis, wherein, because the partnership is highly specific, the lectin–rhizobia reaction must necessarily be highly specific as well, to support the hypothesis. Evidence is in favor of adequate specificity. BOHLOOL and SCHMIDT (1974) found that 23 of 25 *R. japonicum* strains bound soybean lectin; a fraction from one of the three negative binding strains of BOHLOOL and SCHMIDT was subsequently found to bind specifically to a soybean lectin affinity column (WOLPERT and ALBERSHEIM 1976). BHUVANESWARI et al. (1977) examined 22 *R. japonicum* strains with fluorescent and tritium tagged soybean lectin, to find that 15 of the 22 were positive for lectin binding. In none of the above studies were any heterologous rhizobia observed to react with soybean lectin, with the exception of two strains from the cowpea miscellany group of rhizobia (BHUVANESWARI and BAUER 1978). The two cowpea strains were found not to nodulate soybean in this study, although certain cowpea strains apparently have sufficient affinity to the *R. japonicum* group to nodulate soybeans (LEONARD 1923). The specificity data reported initially by BOHLOOL and SCHMIDT (1974) has since been extended with respect to heterologous rhizobia and non-rhizobia bacteria obtained from culture collections and isolated from soil (KOWALSKY and SCHMIDT unpublished data). Among approximately 150 such cultures the cowpea strain 3G464, previously reported to bind by BHUVANESWARI and BAUER (1978), was the only one to bind soybean lectin-FITC.

Not all investigators have found the same significant correlations between lectin binding and host–symbiont specificity noted above for the soybean system. LAW and STRIJDOM (1977) surveyed numerous legume lectins for reactivity with a wide range of homologous and heterologous rhizobia. Their data, based on microscopic examination of rhizobia-FITC labeled lectin preparations, indicated that most of the strains were unable to bind lectin from their normal host plants. Many of the strains, however, were observed to bind heterologous lectin. Only two of ten *R. japonicum* strains reacted with soybean lectin, and one of those reacted with two other lectins. Moreover, it was found that strains of *R. leguminosarum, R. phaseoli,* and *R. trifolii* bound soybean lectin. It would appear desirable to resolve those instances of direct contradiction of data, as in the case of *R. japonicum,* following an exchange of strains. This has not been attempted as yet. CHEN and PHILLIPS (1976) found that fluorescent soybean lectin bound to five of seven strains of rhizobia heterologous for soybean. Several of the strains were the same as those reported not to bind by BOHLOOL and SCHMIDT (1974). It is possible that the low fluorescence intensities reported by CHEN and PHILLIPS were indicative of nonspecific trapping of the lectin in the viscous polysaccharides formed by the strains in question. Resolution of discrepancies in binding specificity is unlikely in the near future because various investigators use different strains and different culture conditions, and because no lectin-binding analysis is standardized, and none is free of numerous procedural and technical pitfalls.

3.2 Lectin-Binding Site of Symbiont

There is no agreement as to the nature of the lectin-binding site at or near the cell surface of rhizobia. Evidence has been presented for either lipopolysaccharide (LPS) or exopolysaccharide (EPS) as carrying the recognition sugar formed by the rhizobia cell.

The first reports of lectin binding by *R. japonicum* (BOHLOOL and SCHMIDT 1974, 1976) were accompanied by photomicrographs which showed the FITC-

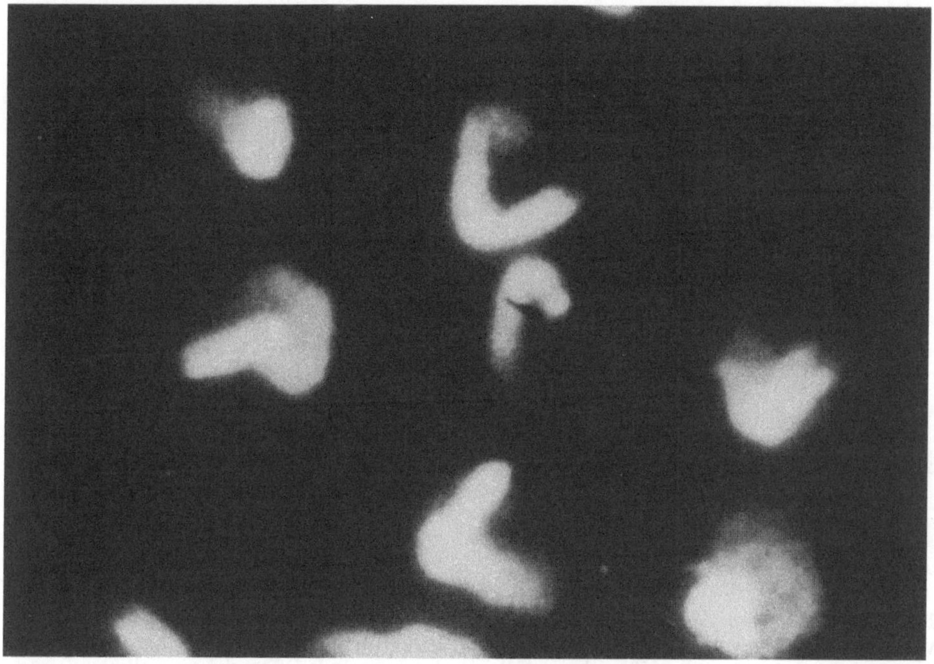

Fig. 1. Cells of *R. japonicum*, strain 138, reacted with FITC-labeled soybean lectin, as viewed in black and white prints of fluorescence micrograph. Note diffuse binding of lectin about cell, commonly with polar accumulation of the lectin

labeled lectin as a diffuse, often fan-shaped, zone commonly distributed around one end of the cell. As seen by fluorescence microscopy the lectin is clearly associated with extracellular material and not with the cell wall; unfortunately, black and white prints (Fig. 1) show that both the cell outline and the bound lectin in white and the color distinctions between the two, easily seen by fluorescent examination, are lost. Nonetheless the diffuse nature of the binding is obvious even in black and white, and it is not reasonable to conceive of LPS or *O*-antigens components of LPS diffusing away from the cell to this extent. The polysaccharide complex that binds lectin occurs not only associated with *R. japonicum* cells as capsular EPS seen by fluorescence microscopy, but also as diffusible EPS which can be found in cell-free culture supernatant (TSIEN and SCHMIDT 1980).

Additional morphological evidence of EPS as including the lectin-binding material of *R. japonicum* was presented in the electronmicrographs of CALVERT et al. (1978). Ferritin-labeled lectin was found clearly associated with the EPS capsular material around the cell walls rather than with the cell wall itself. Similar distribution of ferritin-soybean lectin around a whole cell is shown in an electronmicrograph of a negatively stained preparation (Fig. 2). Ouchterlony double gel diffusion, while not particularly sensitive, has been used to demonstrate that *R. japonicum* EPS forms precipitin bands with soybean lectin (TSIEN and SCHMIDT 1977, 1980, KAMBERGER 1979). KAMBERGER used both EPS

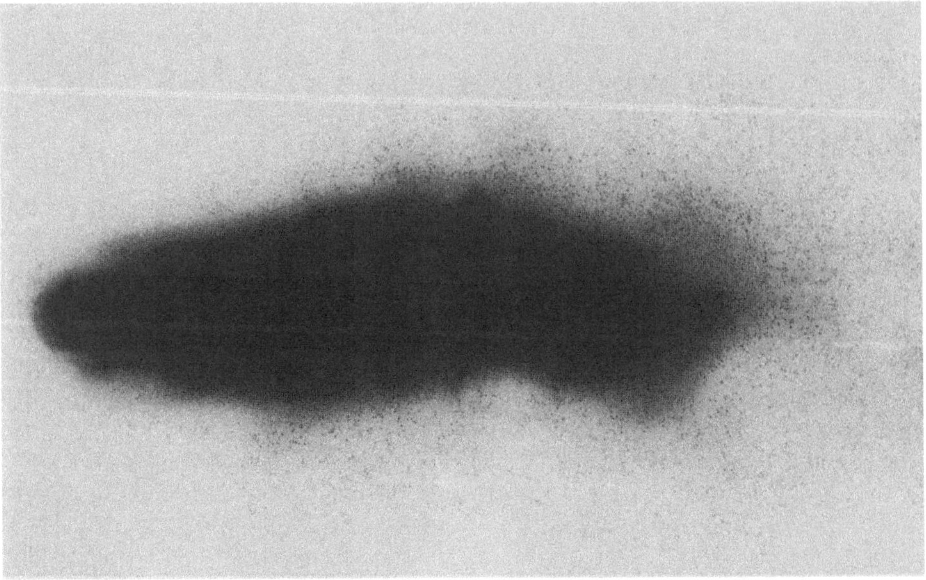

Fig. 2. *R. japonicum,* strain 138, cell labeled with soybean lectin-ferritin conjugate and visualized directly with electronmicroscopy with no further staining or treatment. Note association of electron-dense ferritin particles with extracellular material partially surrounding cell

and LPS preparations, and found that the LPS fraction of *R. japonicum* was negative with respect to lectin binding.

Lipopolysaccharide was reported as the lectin-binding site in two studies of *R. japonicum*. WOLPERT and ALBERSHEIM (1976) reported that LPS extracts from *R. japonicum* and three other rhizobia bound specifically to homologous lectins on affinity columns, but EPS fractions did not. KATO et al. (1979) also isolated LPS and EPS fractions from *R. japonicum* and *R. leguminosarum,* and found that the LPS of each was most active in inhibiting the hemagglutinating activity of the corresponding lectins. The same fractions inhibited the binding reaction between the lectins and homologous rhizobia. TSIEN and SCHMIDT (unpublished data) use chemical approaches generally similar to those of KATO et al. but find all the lectin-binding activity of *R. japonicum* in purified EPS fractions which are devoid of 2-keto-3-deoxyoctonate (KDO), a specific indicator of LPS. Still additional evidence against the involvement of LPS in lectin binding by *R. japonicum* is the observation that bacteroids bind antibodies raised against the LPS of vegetative cells with characteristic intensity and strain specificity, but bacteroids do not bind lectin (E.L. SCHMIDT and B.B. BOHLOOL unpublished data).

Evidence is also inconclusive with respect to EPS or LPS as the lectin receptor site for rhizobia other than *R. japonicum*. A lectin extracted from clover seed agglutinated infective *R. trifolii,* and this agglutination was inhibited by 2-deoxyglucose, a component of the capsular polysaccharides of *R. trifolii* (DAZZO and HUBBELL 1975). DAZZO and BRILL (1977) labeled *R. trifolii* capsular polysaccharide with FITC and used fluorescence microscopy to demonstrate selected localization of the EPS on clover root hairs, where presumably lectin is localized. The capsular polysaccharide involved in lectin binding is an acidic heteropolysaccharide lacking KDO (DAZZO and BRILL 1979). According to DAZZO et al. (1979) the appropriate capsular material is of transient occurrence during culture of *R. trifolii*, and is found only in late log and early stationary phases of growth in liquid culture. In the case of *R. leguminosarum*, PLANQUÉ and KIJNE (1977) reported evidence for a glycan polysaccharide of cell wall but not LPS origin, as responsible for binding pea lectin. Other workers (WOLPERT and ALBERSHEIM 1976, KAMBERGER 1979, KATO et al. 1979) favor LPS as the carrier of the recognition sugar in *R. leguminosarum*. SANDERS et al. (1978) found that an EPS-minus mutant of *R. leguminosarum* was unable to nodulate the pea, but the lectin-binding capacity of this mutant was not reported. Fractions of LPS from *R. meliloti* were observed to bind alfalfa lectin, but EPS preparations reacted nonspecifically with lectins from pea and lentil seeds.

Subsequent work in lectin recognition will undoubtedly resolve some of the inconsistencies now evident in present characterizations of the binding site presented by rhizobia. It will be informative to have more information based on highly purified and strictly homogeneous cell fractions, but such fractions are difficult to attain. It will be even more informative to have well-characterized root lectin preparations to react with the rhizobial fractions, but indications are that characterization of lectins as they occur in the roots of most legume plants is also extremely difficult.

3.3 Lectin-Binding Site of Host

The great majority of the study of lectins in the *Rhizobium*–legume interaction has made use of lectins obtained from the seed, with the realization that those that occur in the root, at the site of first interaction, may be different from seed lectins. Lectins have been found in the seeds of several hundreds of species and varieties of legumes (LIENER 1976), their frequently high concentrations and accessibility in the seed has made them the material of choice for most investigations.

Occurrence of lectins in legume roots is well known (LIS and SHARON 1973, LIENER 1976), but they are poorly characterized as to their relationship with the corresponding seed lectins. Root extracts frequently have much lower hemagglutinating activity than seed preparations from the same legume (BOHLOOL and SCHMIDT 1974, ROUGE 1974, KATO et al. 1979). Such is not the case for all legumes since KRUPE and ENSGRABER (1958) found very high levels of hemagglutinating lectin in *Vicia cracca*, and lectin extracted from the roots of *Lotononis bainesii*, indigenous to South Africa, had a much higher hemagglutinating titer than that from the seed (LAW and STRIJDOM 1977).

Distribution of lectins in root tissues of soybeans was studied by PUEPPKE et al. (1978). They used radioimmunoassay to examine lectin occurrence in plant tissues during the life cycle of the Beeson soybean cultivar. Lectin was found in all tissues of very young plants at concentrations 100-fold less than that of the cotyledons, and fell to nondetectible levels (0.1 µg/g fresh weight) by

21 days, just prior to cotyledon loss. Thereafter no lectin activity was demonstrable in any plant tissue until newly formed seeds were observed to have the sole localization of lectin. TALBOT and ETZLER (1978) also using radioimmunoassay, studied lectin distribution in the legume *Dolichos biflorus;* they found a low but measurable amount that reacted with antibodies to the seed lectin in leaves, stem, and pod, but no lectin was detectable in the roots of the plant at any stage of development.

In general the paucity of information on the nature and distribution of root lectins seems to stem from difficulties associated with their isolation in any but trace amounts. Lectins were shown to be membrane components of plant tissue, requiring rather extensive fractionation and extraction procedures (BOWLES and KAUSS 1975, KAUSS and BOWLES 1976). BOWLES et al. (1979) found lectin activity in roots, shoots, and leaves of soybean and peanut in all stages of plant development. In soybeans the root lectins were generally membrane-bound, whereas in peanuts lectin activity was present in membrane-associated and soluble cytoplasmic fractions. Preliminary work on soybean root lectin by JACK et al. (1979) gave low yields, but indicated similarity to the seed lectin with respect to hemagglutination, affinity chromatography, and reaction to anti-seed lectin antibodies.

Apparently some legume root lectins are more readily extracted and isolated than those mentioned above. GIETL and ZIEGLER (1979), for example, noted in a preliminary report that intact roots of seedlings of *Phaseolus vulgaris* excreted lectin of good hemagglutination titer. DAZZO et al. (1978) were able to elute the clover lectin from seedling roots with the aid of 2-deoxyglucose. Both root and seed lectin agglutinated *R. trifolii* cells, inhibited by 2-deoxyglucose, were similar in electrophoretic mobilities, and reacted similarly to anti-seed lectin serum.

The only success reported relative to identification of the cellular site of lectin localization in the root was for the *R. trifolii*-clover system by DAZZO and BRILL (1977). Capsular polysaccharide preparations of *R. trifolii* labeled with FITC were found to localize selectively on clover root hairs, and more abundantly on the tips of the root hairs than at basal portions. Since the polysaccharide bound to most root hairs, and infection in a given region must be restricted to one or a few root hairs, further specificity barriers must be encountered. There are either post-recognition steps that determine which of the infection sites develop, or else recognition sites must greatly outnumber the specific rhizobia to be encountered under normal rhizosphere conditions.

3.4 Lectin-Binding Competence in Rhizobia

The weight of the evidence, in the *R. japonicum*–soybean symbiosis at least, is in favor of specific binding of soybean lectin by *R. japonicum.* Nevertheless a major unresolved problem in the specific recognition hypothesis in general and in its application to the soybean legume in particular remains with regard to the *R. japonicum* strains that nodulate well but have not been demonstrated to bind lectin. BOHLOOL and SCHMIDT (1974) in the first report on specific recognition found, for example, that 3 of 25 strains of *R. japonicum* bound no soybean lectin, and 7 others were positive only for a small percentage of cells. A possible explanation is that lectin-binding competence is a function of cultural conditions, and that certain strains synthesize lectin-binding material in response to the nutritional circumstances of the natural rhizosphere but not in usual culture media. However, attempts to induce lectin binding in the three negative strains or to enhance the percentage of positive cells in several low-incidence binders were uniformly negative (B.B. BOHLOOL and E.L. SCHMIDT unpublished data). The tests included growth with various carbohydrate sources at several pH levels, varying concentrations of ammonia and nitrate, and the use of soil and root extracts and sterile soybean roots.

Variability in soybean lectin binding by *R. japonicum* strains as a function of culture conditions was explored by BHUVANESWARI et al. (1977). The ability of strains to bind lectin was reported to depend greatly on the growth phase of the bacteria in artificial culture media. In subsequent work (BHUVANESWARI and BAUER 1978) the question of nonbinding strains was addressed. They found that five of six strains completely negative for lectin binding as grown in the usual synthetic medium developed lectin-binding competence in 1% to 5% of the population when grown in a medium containing soybean root extract. All six strains developed lectin binding competence to a similar degree after growth in the presence of soybean seedlings in solution culture.

The lectin receptor of *R. japonicum* seems clearly to be extracellular, and like extracellular products of bacteria in general, may be expected to vary with the nutritional and environmental circumstances of growth. Much more investigation is necessary to define the growth conditions that affect lectin-binding competence, and to relate those conditions to the normal rhizosphere. That special conditions at the very root surface in nature might elicit lectin-binding competence in negative lectin binders however, is not the only plausible hypothesis. The root lectin may be sufficiently different from the seed lectin used in lectin-binding studies, so that certain strains recognize only the root lectin. No soybean root lectin is yet available to test this possibility.

The presence of fixed nitrogen in the soil is an environmental factor long linked to both the initiation and the expression of the nitrogen-fixing symbiosis. Fixed nitrogen in soil beyond poorly defined threshold values restricts nodulation and/or limits the nitrogen-fixing process. DAZZO and BRILL (1978) examined effects of fixed nitrogen in a clover rooting medium on the occurrence of the clover lectin on seedling roots, using the immunocytofluorometric assay of DAZZO et al. (1978). Either nitrate at 16 mM or ammonia at 1 mM inhibited infection and nodulation of white clover seedlings inoculated with *R. trifolii*. Increases in either ammonia or nitrate led to decreases in detectable lectin on roots and in numbers to *R. trifolii* adsorbed to roots. These data suggest for the first time a direct role for the plant root in lectin-binding competence.

4 Post-Recognition Events

Lectin binding, if actually the mechanism of recognition, may or may not also accomplish an attachment event to consolidate the recognition. Lectin binding alone may account for firm attachment of rhizobia at the infection site or lectin binding may conceivably result in a loose attachment of one or more appropriate *Rhizobium* cells, to be followed by a separate firm attachment event.

4.1 Attachment

If rhizobia are first recognized by lectin and then attached by a separate mechanism, they could make use of one or more of the strategies used by other bacteria for firm attachment. Adherence to a substrate or host is accomplished among bacteria by means of fimbriae, LPS, proteins and lipoproteins, capsular and slime secretions, deposition of inorganic cements, and holdfast microcapsular areas often localized on special structures or prostheca. Many pathogenic bacteria exhibit an affinity for certain tissues of an animal host, and attachment at those tissue sites is a specific prelude to infection (SMITH 1977). The specificity in animal infections can involve attachment receptors on both bacteria and host. The great potential for specific receptors on bacteria is shown by the functional sensitivity of some outer membrane polypeptides. These have been purified and identified as bacteriophage and colicin receptors (BRAUN and HANTKE 1977).

Localization of specific rhizobia at their hypothetical binding sites on roots in soil environments has not been seen. Electronmicrographs have been taken in studies of artificial systems with pure culture of rhizobia and axenic legume roots. SAHLMAN and FAHRAEUS (1963) provided the first such micrographs when they showed cells of *R. trifolii* attached perpendicularly to root hairs of clover. Artificial systems may provide a suitable model for the mode of attachment – apparently polar for many rhizobia – but do not necessarily provide a good model for localization since Gram-negative bacteria (like rhizobia) attach nonspecifically to virtually any surface. MARSHALL et al. (1975) for example, found that several species of rhizobia attached with polar orientation to aseptic clover root hairs irrespective of their plant specificity.

DAZZO et al. (1976) examined specific vs. nonspecific attachment by means of microscopic observation of the density of attachment to clover root hairs of infective as compared to noninfective strains of *R. trifolii*. Selective adsorption involving about a tenfold increase in infective over noninfective strains was reported. Treatment of clover roots with 2-deoxyglucose as a specific inhibitor of lectin binding decreased attachment of the infective strains. Considerable additional microscopic data are needed for attachment data of this type where pure cultures of plant and *Rhizobium* are used to avoid the severe complications of less artificial experimental systems. Simple competition experiments which include a specific *Rhizobium* and some morphologically distinguishable Gram-negative non-*Rhizobium* would be highly desirable to clarify distinctions between specific and nonspecific attachment to loci on legume roots.

A model involving lectin as responsible for both the recognition and attachment events was proposed by DAZZO and HUBBELL (1975). According to this hypothesis, lectin acts as a bridge to bind the *R. trifolii* cell at a capsular polysaccharide receptor site while anchored to a host root cell wall polysaccharide which is antigenically similar to the capsular polysaccharide of the bacterium. The antigenic carbohydrate found both on the host cell and on the homologous rhizobial cell has been termed the "cross-reactive antigen." In later work DAZZO and BRILL (1977) reported that clover root hairs preferentially adsorbed the cross-reactive antigen as prepared from *R. trifolii* exopolysaccharide. Antigen sharing between microbial invaders and plant hosts has been observed (DEVAY and ADLER 1976), but confirmation of the cross-reactive antigen model in legume symbioses would be very desirable because of its possible significance from both mechanistic and evolutionary viewpoints. Meanwhile questions relative to the existence of an association-specific, plant-microbe common antigen must be raised in the light of the marked serological specificity evident among different strains of the same *Rhizobium* species (GRAHAM 1963, DUDMAN and BROCKWELL 1968, DUDMAN 1977). An antigen common to a host plant and to all rhizobia capable of nodulating that plant should result in serological cross-reactivity among those strains. This is apparently not the case since antisera prepared against one *Rhizobium* strain commonly fail to react with related strains of the same species (GRAHAM 1963). In experiments by DUDMAN and BROCKWELL (1968) three clover nodule isolates showed either major or minor bands of identity by immunodiffusion with inoculate strain TAI of *R. trifolii,* but a fourth isolate showed no precipitation lines whatsoever with anti-TAI.

Special cellular adaptations have been observed in *R. japonicum,* and it has been suggested that one such feature may be concerned with post-recognition attachment. BOHLOOL and SCHMIDT (1976) described a marked polarity of the bacterial cell as viewed by immunofluorescence and fluorescent lectin binding. Polarity was evident in the localization of FITC-lectin bound loosely to the cell and in a concentration of strain-homologous or strain-heterologous fluorescent antibody on a site at one end of the cell. The latter polar tip region is a site for attachment, since cells attached to each other in a tip-to-tip orientation. The attachment tips were subsequently termed "extracellular polar bodies" (EPB) and characterized as extracellular polar accumulations of granular and fibrillar material having properties consistent with polysaccharide and LPS (TSIEN and SCHMIDT 1977). As a special attachment structure, the EPB may serve to consolidate the highly specific lectin-binding reaction by means of a firm polar binding at the site for nodule initiation. Electronmicrographs of KUMARASINGHE and NUTMAN (1977) show rhizobia attached endwise to clover root hairs by means of an attachment structure on the bacterium that closely resembled the EPB of *R. japonicum.*

4.2 Post-Recognition Function of Lectins

Consideration of lectins in symbiotic plant–microbe interactions has concentrated on the recognition properties of plant lectins, but other properties of these biologically active substances might also be expressed. The mitogenic action of lectins on lymphocytes and other animal cells has been studied extensively since 1960 (LIS and SHARON 1973, LIENER 1976). There is some evidence that lectins are mitogenic for plant cells (HOWARD et al. 1977). Nodule development in legume roots is characterized by rapid cell division after the rhizobia reach the specialized (polyploid) root cells. The signals that trigger host cells to proliferate into the nodule mass are unknown, but it is tempting to consider that mitogenic lectins along with plant hormones may play a role. Lectin–plant receptor cell interactions could be involved in stimulation of cell division in the root cortex in advance of the infection thread.

Infection of nodule tissue is thought to proceed by an endocytotic process involving fusion and invagination of host cell membrane and infection thread membrane leading to encapsulation of the rhizobia in the peribacteroid envelope. Lectins are known to mediate analogous processes within certain lymphoid cells (EDELSON and COHN 1974). Uptake of *R. leguminosarum* by pea leaf protoplasts and subsequent stabilization in the form of vesicles with host cell cytoplasm was observed by DAVEY and COCKING (1972). Again the endocytotic action of lectins may have been implicated.

Lectin has been found associated with cell surfaces of the cellular slime mold, *Dictyostelium discoideum,* and implicated in its morphogenic conversion from unicellular amebae to a multicellular organism (BARONDES 1978). The Gram-negative bacterium *Myxococcus xanthus* has been an attractive microbial system for the study of cell–cell interactions during a complex life cycle involving aggregation and cellular morphogenesis. CUMSKY and ZUSMAN (1979) reported synthesis of a major new development-specific with lectin-like activity during fruiting body formation by *M. xanthus.* Relatively large amounts of the lectin were formed by aggregating cells, possibly to be exported to the membrane or secreted outside the cell for possible binding to a developmentally regulated receptor. These instances of cell-to-cell communication mediated by lectin suggest the possibility of a communication role for lectins in legume symbioses. Following highly specific recognition between *Rhizobium* and root at the nodule initiation site, other events are required to communicate the successful recognition and to signal the next step toward entry into the root tissue.

5 Concluding Comments

The function of lectins in plants has been highly conjectural, and remains so. The strongest evidence thus far for a natural role of lectins relates to those of the legumes, in support of the hypothesis that such lectins recognize appropriate symbiont rhizobia. But this evidence is still circumstantial and much of it somewhat contradictory. There are many elements of the problem to suggest that both its circumstantial and controversial aspects will prevail for many years.

Much of the appeal of the lectin recognition hypothesis derives from its sweeping simplicity. But information bearing on the lectin model is still very recent and restricted to a small number of legumes and a relatively few strains of rhizobia. Eventually lectins may be shown to account for the very precise and sensitive recognition intrinsic to some legume–rhizobia associations, but even if this proves to be the case the model may not extrapolate to all legume-nitrogen fixing symbioses. Legumes constitute a large and diverse group of plants and few have been studied as to the mode of infection. Rhizobia are diverse as well, and their present groupings based on common host nodulation may lump together some very different bacteria. An evident challenge to the simple lectin recognition model occurs in the cases of some legume hosts nodulated by apparently quite different rhizobia. *Lotus, Lupinus, Vigna, Acacia,* and *Macroptilium,* for example, establish symbioses with both fast- and slow-growing rhizobia. Are there separate lectin recognition sites on the root systems of such plants, or common receptor sites on all endosymbionts, or is there no involvement of lectin whatsoever?

Unequivocal, or even preponderant, evidence that lectin functions in specific recognition for selected legume–*Rhizobium* associations awaits the resolution of numerous aspects of the problem. It is critical that the nature, distribution, and sugar-binding properties of lectins as they occur in the roots, rather than in seeds, be clarified. Study of the interactions of root lectins with rhizobia even in in vitro may resolve some current controversies ranging from specificity of interaction to the nature of the lectin receptor sites of the bacteria. Much more effort should go to the repetition and confirmation of key points, with careful attention to the mimicking of experimental details and the use of extensive controls. Experiments directly concerned with confirmation have been infrequent and often indirect, using strains, plants, and methodologies different from those of the original investigators. Much more effort also should be directed to in situ experiments, but these require not only special techniques which are seriously limited in their capacity to deal with the technical difficulties involved, but also full awareness on the part of investigators of the complexities of the system. Attachment of rhizobia to roots as viewed in pure culture systems can be thoroughly misleading, but viewing this event in situ, while clearly desirable, requires detection of the specific *Rhizobium* in the rhizosphere and its distinction from biological and nonbiological artifacts.

If lectins serve a recognition function in legume symbioses, the partner must have evolved attributes to insure a meeting in the rhizosphere and means to consolidate the relationship. Little information is available to suggest special devices on the part of the plant to present a lectin at a microsite accessible to a *Rhizobium,* and still protect it from degradation by rhizosphere bacteria. Similarly, little is known of the evolutionary devices that combine in the rhizobia to aid survival in the soil and insure occurrence, recognition, and attachment in the rhizosphere. The suggestion of a "cross-reactive antigen" common to both root and *Rhizobium* (DAZZO and HUBBELL 1975) demands an elaborate evolutionary device achieved independently by both partners. Such elaborate co-evolutionary development would appear to be remarkable where (for lectin recognition) only a sugar moeity in the proper position is needed, and not the evolution of a complete antigen. Nevertheless study of the special adaptations of each partner as both converge to initiate the symbiotic interaction could contribute greatly to clarification of the recognition mechanism, whether it be by lectins, or by another devious means.

Acknowledgment. Much of the published and unpublished work of the authors cited here was supported by National Science Foundation grant DEB 77-10172.

References

Baldwin IB, Fred EB, Hastings EG (1927) Grouping of legumes according to biological reactions of their seed proteins. Possible explanation of the phenomenon of cross inoculation. Bot Gaz 83:217–243

Barkai-Golan R, Mirelamn D, Sharon N (1978) Studies on growth inhibition by lectins of Penicillia and Aspergilli. Arch Microbiol 116:119–124

Barondes SH (1978) Developmentally regulated slime mold lectins and specific cell cohesion. In: Lerner RA, Bergsma D (eds) Molecular basis of cell-cell interaction. Alan R Liss Inc, New York, pp 491–496

Bhuvaneswari TV, Bauer WD (1978) Role of lectins in plant-microorganism interactions. III. Influence of rhizosphere/rhizoplane culture conditions on the soybean lectin-binding properties of rhizobia. Plant Physiol 62:71–74

Bhuvaneswari TV, Pueppke SG, Bauer WD (1977) Role of lectins in plant-microorganism interactions. I. Binding of soybean lectin to rhizobia. Plant Physiol 60:486–491

Bohlool BB, Schmidt EL (1970) Immunofluorescent detection of *Rhizobium japonicum* in soils. Soil Sci 110:229–236

Bohlool BB, Schmidt EL (1974) Lectins: a possible basis for specificity in the *Rhizobium*-legume root nodule symbiosis. Science 185:269–271

Bohlool BB, Schmidt EL (1976) Immunofluorescent polar tips of *Rhizobium japonicum*: possible site of attachment or lectin binding. J Bacteriol 125:1188–1194

Bowles DJ, Kauss H (1975) Carbohydrate-binding proteins from cellular membranes of plant tissue. Plant Sci Lett 4:411–418

Bowles DJ, Lis H, Sharon N (1979) Distribution of lectins in membranes of soybean and peanut plants. I. General distribution in root, shoot and leaf tissue at different stages of growth. Planta 145:193–198

Braun V, Hantke K (1977) Bacterial receptors for phages and colicins as constituents of specific transport systems. In: Reissig JL (ed) Microbial interactions: receptors and recognition. Ser B, Vol 3. Chapman and Hall, London, pp 101–137

Broughton WJ (1978) Control of specificity in legume-*Rhizobium* associations. J Appl Bacteriol 45:165–194

Calvert HE, Lalonde M, Bhuvaneswari TV, Bauer WD (1978) Role of lectins in plant-microorganism interactions. IV. Ultrastructural localization of soybean lectin binding sites on *Rhizobium japonicum*. Can J Microbiol 24:785–793

Cass-Smith WP, Holland AA (1958) The effect soil fungicides and fumigants on the growth of subterranean clover on new light land. J Dep Agric West Aust 7:225–231

Chen APT, Phillips DA (1976) Attachment of *Rhizobium* to legume roots as the basis for specific interactions. Physiol Plant 38:83–88

Chet I, Mitchell R (1976) Ecological aspects of microbial chemotactic behaviour. Ann Rev Microbiol 30:221–239

Cumsky M, Zusman DR (1979) Myxobacterial hemagglutinin: a development-specific lectin of *Myxococcus xanthus*. Proc Natl Acad Sci USA 76:5505–5509

Currier WW, Strobel GA (1976) Chemotaxis of *Rhizobium* spp. to plant root exudates. Plant Physiol 57:820–823

Currier WW, Strobel GA (1977) The chemotactic behaviour of trefoil *Rhizobium*. FEMS Microbiol Lett 1:243–246

Dart PJ (1974) Development of root-nodule symbioses. I. The infection process. In: Quispel A (ed) The biology of nitrogen fixation. North Holland Publishing, Amsterdam, pp 381–429

Davey MR, Cocking EC (1972) Uptake of bacteria by isolated higher plant protoplasts. Nature 239:455–456

Dazzo FB, Brill WJ (1977) Receptor site on clover and alfalfa roots for *Rhizobium*. Appl Environ Microbiol 33:132–136

Dazzo FB, Brill WJ (1978) Regulation by fixed nitrogen of host-symbiont recognition in the *Rhizobium*-clover symbiosis. Plant Physiol 62:18–21

Dazzo FB, Brill WJ (1979) Bacteroid polysaccharide which binds *Rhizobium trifolii* to clover root hairs. J Bacteriol 137:1362–1373

Dazzo FB, Hubbell DH (1975) Cross-reactive antigens and lectins as determinants of symbiotic specificity in the *Rhizobium*-clover association. Appl Microbiol 30:1017–1033

Dazzo FB, Napoli CA, Hubbell DH (1976) Adsorption of bacteria to roots as related to host specificity in the *Rhizobium*-clover symbiosis. Appl Environ Microbiol 32:166–171

Dazzo FB, Yanke WE, Brill WJ (1978) Trifoliin: a *Rhizobium* recognition protein from white clover. Biochim Biophy Acta 539:276–286

Dazzo FB, Urbano MR, Brill WJ (1979) Transient appearance of lectin receptors on *Rhizobium trifolii*. Current Microbiol 2:15–20

DeVay JE, Adler HE (1976) Antigens common to hosts and parasites. Annu Rev Microbiol 30:147–168

Diatloff A (1969) The introduction of *Rhizobium japonicum* to soil by seed inoculation of non-host legumes and cereals. Aust J Exp Agric Anim Husb 9:357–360

Dixon ROD (1969) Rhizobia (with particular reference to host plants). Annu Rev Microbiol 23:137–158

Dommergues YD (1978) The plant-microorganism system. In: Dommergues YR, Krupa SV (eds) Interactions between non-pathogenic soil microorganisms and plants. Elsevier, Amsterdam, pp 1–37

Dudman WF (1977) Serological methods and their application to dinitrogen-fixing organisms. In: Hardy RWF, Gibson AH (eds). A treatise on dinitrogen fixation, Vol 4. Wiley, New York, pp 487–508

Dudman WF, Brockwell J (1968) Ecological studies of root nodule bacteria introduced into field environments. I. A survey of field performance of clover inoculants by gel immune diffusion serology. Aust J Agric Res 19:739–747

Edelson P, Cohn Z (1974) Effects of concavalin A on mouse peritoneal macrophages. I. Stimulation of endocytic activity and inhibition of phago-lysosome formation. J Exp Med 140:1364–1386

Egerraat AWSM van (1972) Pea root exudates and their effect upon root nodule bacteria. Meded. Landbouwhogesch. Wageningen 72–27, H Veenman en Zonen N.V., Wageningen, 90 pp

Fred EB, Stevens JW (1923) Grouping legume nodule bacteria. Wis Agric Exp Stn Spec Bull 352:76–77

Gietl C, Ziegler H (1979) Lectins in the excretion of intact roots. Naturwissenschaften 66:161–162

Graham PH (1963) Antigenic affinities of the root nodule bacteria of legumes. Antonie van Leeuwenhoek; J Microbiol Serol 30:68–72

Hala CN, Mathers DG (1977) Toxicity of white clover seed diffusate and its effect on survival of *Rhizobium trifolii*. N Z J Agric Res 20:69–73

Ham GE, Frederick LR, Anderson LC (1971) Serogroups of *Rhizobium japonicum* in soybean nodules samples in Iowa. Agron J 63:69–72

Hamblin J, Kent SP (1973) Possible role of phytohaemagglutinin in *Phaseolus vulgaris* L. Nature (London) New Biol 245:28–30

Holland AA (1970) Competition between soil- and seed-borne *Rhizobium trifolii* in nodulation of introduced *Trifolium subterraneum*. Plant Soil 32:293–302

Howard J, Shannon L, Oki L, Murashige T (1977) Soybean agglutinin. A mitogen for soybean callus cells. Exp Cell Res 107:448–450

Jack MA, Schmidt EL, Wold F (1979) Studies of the *in vivo* function of soybean lectin (Abstract) Fed Proc 38:411

Kamberger W (1979) An Ouchterlony double diffusion study on the interaction between legume lectins and rhizobial cell surface antigens. Arch Microbiol 121:83–90

Kapusta G, Rouwenhorst DL (1973) Interaction of selected pesticides on *Rhizobium japonicum* in pure culture and under field conditions. Agron J 65:112–115

Kato G, Maruyama Y, Nakumara M (1979) Role of lectins and lipopolysaccharides in

the recognition process of specific legume-*Rhizobium* symbiosis. Agric Biol Chem 43:1085–1092
Kauss H, Bowles DJ (1976) Some properties of carbohydrate-binding proteins (lectins) solubilized from cell walls of *Phaseolus aureus*. Planta 130:169–174
Kawasaki T, Ashwell G (1976) Chemical and physical properties of an hepatic membrane protein that specifically binds asialoglycoproteins. J Biol Chem 251:296–1302
Krupe M, Ensgraber A (1958) Untersuchungen über die Natur des "Phytagglutinins" in chemischer, immunochemischer und pflanzenphysiologischer Sicht. I. Anatomische Studien über den Ort ihres Vorkommens in höheren Samenpflanzen. Planta 50:371–378
Kumarasinghe RMK, Nutman PS (1977) *Rhizobium* stimulated callose formation in clover root hairs and its relation to infection. J Exp Bot 28:961–976
Law IJ, Strijdom BW (1977) Some observations on plant lectins and *Rhizobium* specificity. Soil Biol Biochem 9:79–84
Leonard LT (1923) Nodule-production kinship between the soy bean and the cowpea. Soil Sci 5:277–283
Liener IE (1976) Phytohemagglutinins (phytolectins) Annu Rev Plant Physiol 27:291–319
Lis H, Sharon N (1973) The biochemistry of plant lectins (phytohemagglutinins). Annu Rev Biochem 42:541–574
Lockhart CM, Rowell P, Stewart WDP (1978) Phytohaemagglutinins from the nitrogen-fixing lichens *Peltigera canina* and *P. polydactyla*. FEMS Microbiol Lett 3:127–130
Marshall KC (1964) Survival of root nodule bacteria in dry soils exposed to high temperatures. Aust J Agric Res 15:273–281
Marshall KC, Cruickshank RH, Bushby HVA (197) The orientation of certain root-nodule bacteria at interfaces, including legume root-hair surface. J Gen Microbiol 91:198–200
Nutman PS (1965) The relation between nodule bacterial and the legume host in the rhizosphere and the process of infection. In: Baker KF, Snyder WC (eds) Ecology of soil-borne plant pathogens. Univ California Press, Berkeley, pp 231–247
Planqué K, Kijne JW (1977) Binding of pea lectins to a glycan type polysaccharide in the cell walls of *Rhizobium leguminosarum*. FEBS Lett 73:64–66
Pueppke SG, Bauer WD, Keegstra K, Ferguson AL (1978) The role of lectins in plant-microorganism interactions. II. Distribution of soybean lectin in tissues of *Glycine max* (L) Merr. Plant Physiol 61:779–784
Reyes VG, Schmidt EL (1979) Population densities of *Rhizobium japonicum* strain 123 estimated directly in soil and rhizospheres. Appl Environ Microbiol 37:854–858
Robinson AC (1967) The influence of host on soil and rhizosphere populations of clover and lucerne nodule bacteria in the field. J Aust Inst Agric Sci 33:207–209
Rouge MP (1974) Étude de la phytohemagglutine de Lentille au cours de la germination et des premiers stades du developpement de la plante. Évolution dans les racines, les tiges, et les feuilles. CR Acad Sci 278:3083–3086
Rovira AD (1961) Rhizobium numbers in the rhizosphere of red clover and paspalum in relation to soil treatment and the numbers of bacteria and fungi. Aust J Agric Res 12:77–83
Rovira AD (1969) Plant root exudates. Bot Rev 35:35–57
Sahlman K, Fahraeus G (1963) An electron microscope study of root hair infection by *Rhizobium*. J Gen Microbiol 33:425–427
Sanders RE, Carlson RW, Albersheim P (1978) A *Rhizobium* mutant incapable of nodulation and normal polysaccharide secretion. Nature 271:240–242
Schmidt EL (1973) Fluorescent antibody techniques for the study of microbial ecology. Bull Ecol Res Comm (Stockholm) 17:67–76
Schmidt EL (1974) Quantitative autecological study of microorganisms in soil by immunofluorescence. Soil Sci 118:141–149
Schmidt EL (1979) Initiation of plant root-microbe interactions. Annu Rev Microbiol 33:355–376
Sequeira L (1978) Lectins and their role in host-pathogen specificity. Annu Rev Phytopathol 16:453–481
Smith H (1977) Microbial surface in relation to pathogenicity. Bact Rev 41:475–500
Talbot CF, Etzler ME (1978) Development and distribution of *Dolichos biflorus* lectin as measured by radioimmunoassay. Plant Physiol 61:847–859

Trinick MJ (1973) Symbiosis between *Rhizobium* and the non-legume *Trema aspera*. Nature 244:459–460

Tsien HC, Schmidt EL (1977) Polarity in the exponential-phase *Rhizobium japonicum* cell. Can J Microbiol 123:1274–2384

Tsien HC, Schmidt EL (1980) Accumulation of soybean lectin binding polysaccharide during the growth of *Rhizobium japonicum* as determined by hemagglutionation inhibition assay. Appl Environ Microbiol 39:1100–1104

Tuzimura K. Watanabe I (1962a) The growth of *Rhizobium* in the rhizosphere of the host plant. Ecological studies of root nodule bacteria. Soil Sci Plant Nutr 8:19–24

Tuzimura K, Watanabe I (1926b) The effect of rhizosphere of various plants on the growth of *Rhizobium*. Ecological studies of root nodule bacteria (Part III) Soil Sci Plant Nutr 8:13–17

Vest G, Weber DF, Sloger C (1973) Nodulation and nitrogen fixation. In: Caldwell BE (ed) Soybeans: improvement, production and uses. Am Soc Agron Madison, Wisconsin, pp 353–390

Vincent JM (1965) Environmental factors in the fixation of nitrogen by the legume. In: Bartholomew WV, Clark FE (eds) Soil nitrogen. Am Soc Agron Madison, Wisconsin, pp 384–435

Weaver RW, Frederick LR (1974) Effect inoculation rate on competitive nodulation of *Glycin max* L. Merrill. II. Field studies. Agron J 66:233–236

Wheeler H (1976) Role of phytotoxin in specificity, In: Wood RKS, Graniti A (eds) Specificity in plant diseases. Plenum Press, New York, pp 217–236

Wolpert JS, Albersheim (1976) Host-symbiont interactions. I. The lectins of legumes interact with the O-antigen-containing lipopolysaccharides of their symbiont *Rhizobia*. Biochem Biophys Res Commun 70:729–737

Author Index

Page numbers in *italics* refer to the references

Aach HG, see Braun E 347, *349*
Aach HG, see Touet G 345, *351*
Aaronson S 144, *156*
Abd-El-Al ATH, Phaff HJ 420, 421, 422, *432*
Abdel-Fattah AF, Edrees M 300, *306*
Abdel-Fattah AF, Hussein MM-D, Fouad ST 296, *306*
Abdel-Fattah AF, Hussein MM-D, Salem HM 296, 297, *306*
Abdel-Fattah AF, see Hussein MM-D 296, *310*
Abdul-Baki AA, Ray PM 59, *61*
Abel WO, see Meyer Y 60, *65*
Abeles FB 102, *114*
Abeles FB, Bosshart RP, Forrence LE, Habig WH 600, 608, *615*
Abeles FB, Leather GR 102, *114*, 614, *615*
Abercrombie M, Jones J, Lock M, Perry M, Stoodley R 420, *432*
Acerbo SN, Schubert WJ, Nord FF 201, *221*
Achord DT, see Kaplan A 497, *509*
Ackman RG, see Craigie JS 302, *308*
Adair WL Jr, see Grange DK 167, 168, *190*
Adair WL Jr, see Keller RL 168, *191*
Adamany AM, see Mills JJ 168, 187, *191*
Adamich M, Hemmingsen BB 347, *348*
Adams BG, see Buckholz RG 449, *455*
Adams DO, see Yang SF 143, *165*
Adams E, see Rao NV 143, *162*
Adams MJ, see Hadwiger LA 603, *618*
Adams PA, Montague MJ, Tepfer M, Rayle DL, Ikuma H, Kaufman PB 262, 263, *269*
Adams PB, Rowan KS 615, *615*
Addicott FT, Wiatr SM 102, *114*
Adler E 198, 201, *221*
Adler E, Björkvist KJ, Haggroth S 196, *221*
Adler E, Larsson S, Lundquist K, Miksche GE 210, *222*
Adler HE, see DeVay JE 585, *617*, 671, *675*
Aducci P, see Minetti M 628, *655*
Adya S, Elbein AD 467, *477*, 495, *507*
Adya S, see Elbein AD 467, *478*
Agrawal VP, Kolattukudy PE 235, 236, *248*, 531, *532*
Aist JR, 111, 112, *114*, 247, *248*, 608, *615*
Aist JR, see Israel HW 111, *121*, 608, *619*
Aist JR, see Kunoh H 111, *122*, 609, *620*
Akazawa T, Miljanich P, Conn EE 611, *615*
Akazawa T, see Okamoto K 106, *126*
Akiyama Y, Katō K 35, *42*, 135, *156*
Akiyama Y, Mori M, Katō K 135, *156*
Akiyama Y, see Mani UV 135, 153, *161*
Alam SS, Hemming FW 57, *61*
Al-Azzawi MJ, see Pope DG 516, 521, *538*, 562, 563, 564, *573*
Albersheim P 29, 32, 33, 35, 36, 37, 38, 40, *42*, 68, *114*, 264, 265, *269*
Albersheim P, Anderson-Prouty AJ 364, *384*, 585, 590, *615*, 643, *652*
Albersheim P, Jones TM, English PD 643, *652*
Albersheim P, Killias U 40, *42*
Albersheim P, Mühlethaler K, Frey-Wyssling A 83, *114*
Albersheim P, Valent BS 113, *114*, 643, *652*
Albersheim P, Wolpert JS 651, *652*
Albersheim P, see Anderson-Prouty AJ 603, 604, *615*
Albersheim P, see Ayers AR 603, 605, *615*
Albersheim P, see Bateman DF 589, *616*
Albersheim P, see Bauer WD 591, *616*
Albersheim P, see Burke D 264, *269*
Albersheim P, see Cline K 605, *616*
Albersheim P, see Darvill AG 6, *8*, 34, *42*
Albersheim P, see Delmer DP 50, *62*
Albersheim P, see English PD 589, 591, *617*, *618*
Albersheim P, see Grimes WJ 50, *63*
Albersheim P, see Hahn MG 603, 605, *618*
Albersheim P, see Karr AL Jr 591, *619*
Albersheim P, see Keegstra K 6, *8*, 133, *160*, 151, 264, *271*, 357, *389*, 590, *619*
Albersheim P, see Lyon G 603, 604, *620*
Albersheim P, see McNab JM 71, *124*

Albersheim P, see McNeil M 8, *8*, 29, 31, 32, 33, 34, 35, 37, 38, 41, *44*, 68, 86, *124*, 134, *161*, 264, *271*
Albersheim P, see Sanders RE 667, *676*
Albersheim P, see Strobel GA 596, 598, 599, *622*, 597
Albersheim P, see Talmadge KW 138, *164*
Albersheim P, see Valent BS 92, *131*, 265, *273*
Albersheim P, see Wade M 588, 606, *623*, 643, *657*
Albersheim P, see Wolpert JS 651, *657*, 663, 664, 666, 667, *677*
Alberts AW, see Strauss AW 489, *511*
Albrigo LB, see Freeman B 530, *534*
Al-Delaimy KA, Borgstrom G, Bedford CL 104, *114*
Alexander M, see Ballesta J-PG 361, *384*
Alexander M, see Bloomfield BJ 373, 374, *385*
Alexander M, see Gunnison D 145, *159*
Alexander M, see Kuo MJ 374, *389*
Alexander M, see Skujins JJ 374, *393*
Algranatti ID, Carminatti H, Cabib E 459, *477*, 500, *507*
Ali ZM, see Brady CJ 104, *115*
Alibert G, Ranjeva R, Boudet A 204, *222*
Allan GG, Lewin J, Johnson PG 303, 305, *306*
Allard-Turner V, see Van Alfen NK 593, 600, 601, *623*
Allen AK 633, 648, *652*
Allen AK, Desai NN, Neuberger A 631, *652*
Allen AK, Desai NN, Neuberger A, Creeth JM 135, 136, *156*, 633, 635, 636, *652*
Allen AK, Neuberger A 133, *156*
Allen CM, Kalin JR, Sack J, Veruzzo D 168, *188*
Allen DM, Northcote DH 318, *329*
Allin D, see Selitrennikoff CP 397, 398, 410, *415*
Allison AV, Shalla TA 112, *114*
Allsobrook AJR, Nunn JR, Parolis H 291, *306*
Alviano CS, Gorin PAJ, Travassos LR 364, *384*
Amelunxen F, Spiess E, Thio Tiang Nio E 528, *532*
Amelunxen F, see Speiss E 528, *540*
Anand VK, Pueppke SG, Heberlein GT 586, *615*
Andersen G, see Thiele H 287, *316*
Anderson AJ 603, 604, *615*
Anderson AJ, Jasalavich C 614, *615*
Anderson DB, Kerr T 76, *114*
Anderson FB, Millbank JW 424, *432*
Anderson KL, Ray PM 48, 55, *61*
Anderson LC, see Ham GE 661, *675*
Anderson LE, Toh-Chin Lim Ng, Kyung-Eun Yoon Park 607, *615*
Anderson LE, see Kachru RB 607, *619*
Anderson MA, see Anderson RL 106, *114*
Anderson NS, Dolan TCS, Lawson CJ, Penman A, Rees DA 293, *307*
Anderson RL, Anderson MA, Stone BA 106, *114*
Anderson RL, Clarke AE, Jermyn MA, Knox RB, Stone BA 92, *114*, 154, *156*
Anderson RL, Ray PM 73, 86, 111, *114*, 608, *615*
Anderson RL, Stone BA 106, *114*
Anderson RL, see Clarke AE 6, *8*, 34, 35, *42*, 90, 109, *116*, 152, 153, 155, *157*, 577, 578, *582*
Anderson-Prouty AJ, Albersheim P 603, 604, *615*
Anderson-Prouty AJ, see Albersheim P 364, *384*, 585, 590, *615*, 643, *652*
Angyal SJ, Bender VJ, Ralph BJ 362, *384*
Ankel E, see Ankel H 79, *114*, 466, *477*
Ankel H, Ankel E, Feingold DS 79, *114*
Ankel H, Ankel E, Schutzbach JS, Garancis JC 466, *477*
Ankel H, see Raizada MK 466, *481*
Ankel H, see Schutzbach JS 466, *482*, 473
Anno K, Seno N, Ota M 297, *307*
Anno K, Terahata H, Hayashi Y, Seno N 296, *307*
Antalis C, see Raschke WC 444, 446, 447, *457*
Appiano A, see Pennazio S 112, *126*
Arai K, see Araki C 291, *307*
Arai K, see Nakamura T 262, 263, *271*
Arai M, Lee TH, Murao S 419, *432*
Arai M, Murao S 419, *432*
Arai M, see Nakamura S 185, *191*
Araki C 291, *307*
Araki C, Arai K, Hirase S 291, *307*
Araki Y, see Nakayama K 462, 463, *481*
Archer DB 380, *384*, 402, 409, *411*
Arima K, see Takatsuki A 183, *193*
Arneson PA, Durbin RD 612, *615*
Arnold WN 421, *432*
Arnott HJ, see Horner HT Jr 107, *121*
Arnott S, Fulmer A, Scott WE, Moorhouse R, Rees DA 294, *307*
Arnott S, Scott WE, Rees DA, McNab CGA 293, *307*
Aronson JM 353, 355, 363, *384*, 426, *432*
Aronson JM, Fuller MS 358, *384*
Aronson JM, Klapprott JA, Lin CC 144, *156*
Aronson JM, Machlis L 354, 355, *384*
Aronson JM, Preston RD 355, 366, *384*
Aronson JM, see Cooper BA 358, 365, *386*

Aronson JM, see Lin CC 355, *389*
Arrigoni O, Arrigoni-Liso R, Calabrese G 143, *156*
Arrigoni O, DeSantis A, Arrigoni-Liso R, Calabrese G 142, *156*
Arrigoni O, see Zacheo G 141, *165*
Arrigoni-Liso R, see Arrigoni O 142, *156*, 143
Arrigoni-Liso R, see Zacheo G 141, *165*
Arroyo-Begovich A, Herrera JR 402, *411*
Arroyo-Begovich A, Ruiz-Herrera J 380, *384*
Arthur LO, Bulla LA Jr, St Julian G, Nakamura L 595, *615*
Asahi K, see Isono K 400, *413*
Asano N, see Katō Y 32, *44*
Asensio C, see Gancedo JM 361, *387*
Ashe B, see Strauss AW 489, *511*
Ashford AE, Jacobsen JV 106, *114*
Ashford D, Neuberger A 134, *156*
Ashida J, see Yoda S 261, 262, 263, *273*
Ashwell G, see Kawasaki T 659, *676*
Aspinall GO 3, 6, 7, 8, *8*, 31, 33, 34, 35, *42*
Aspinall GO, Jiang K-S 35, *42*
Aspinall GO, Molloy JA, Craig JWT 32, *42*
Aspinall GO, Rosell K-G 578, 579, *582*
Atalla RH, Nagel SC 524, *532*
Atkins EDT, Mackie W, Parker KD, Smolko EE 286, *307*
Atkins EDT, Mackie W, Smolko EE 286, *307*
Atkins EDT, Parker KD 282, *307*
Atkins EDT, Parker KD, Preston RD 282, *307*
Atkins EDT, see Nieduszynski IA 280, *313*
Atkinson AW, Gunning BES, John PCL 322, *329*, 345, 346, *349*
Atkinson AW, Gunning BES, John PCL, McCullough W 322, *329*
Auriol P, Strobel G, Pio Beltram J, Gray G 598, 601, *615*
Auriol P, see Rabenantoandro Y 608, *621*
Avants JK, see Pressey R 104, *127*, 150, *162*
Awad M, Young RE 104, 105, *114*
Axelos M, Péaud-Lenoël C 50, 58, *61*, 81, *114*
Axelos M, see Péaud-Lenoël C 51, 52, *65*, 72, *126*
Axelos M, see Smith MM 71, *129*
Ayers AR, Ebel J, Valent B, Albersheim P 603, 605, *615*
Ayers AR, Valent B, Ebel J, Albersheim P 603, 605, *615*
Azuma I, Kimura H, Hirao F, Tsubura E, Yamamura Y, Misaki A 466, *477*
Azuma I, see Kanetsuna F 372, *388*, 428, 429, *435*

Baardseth E, see Haug A 287, *310*
Babczinski P 461, *477*
Babczinski P, Haselbeck A, Tanner W 474, *477*
Babczinski P, Tanner W 461, *477*, 494, 499, *507*
Babczinski P, see Sharma CB 461, 472, *482*
Bachur NR, see Leavitt RD 635, *655*
Bacic A 107, *114*
Bacic A, Stone BA 106, *114*
Bacon JSD 424, *432*
Bacon JSD, Davidson ED, Jones D, Taylor IF 356, 359, *384*
Bacon JSD, Farmer VC, Jones D, Taylor IF 358, 359, *384*, 417, 418, *433*
Bacon JSD, Gordon AH, Jones D, Taylor IF, Webley DM 424, 425, *433*
Bacon JSD, Gordon AH, Morris EJ, Farmer VC 107, *114*
Bacon JSD, Jones D, Farmer VC, Webley DM 360, 367, *384*, 419, *433*
Bailey DS, Northcote DH 17, *22*
Bailey DS, see Dürr M 70, 71, 81, 83, *117*
Bailey JA, see Hargreaves JA 603, *619*
Bailey RW, Haq S, Hassid WZ 55, 58, *62*, 83, *115*
Bailey RW, Hassid WZ 71, *114*
Bailey RW, Kauss H 137, 140, *156*
Bailey RW, see Monro JA 38, *44*, 137, 140, *161*, 264, *271*
Bailey RW, see Parsons MJ 294, *313*
Bailey RW, see Pickmere SE 294, *314*
Bajer A 514, 515, *533*
Bajer AS, Molè-Bajer J 514, *533*
Baker DA 151, *156*
Baker DB, Ray PM 264, *269*
Baker EA, Procopiou J 230, 231, *248*
Baker EA, see Deas AHB 228, 231, *249*
Baker EA, see Holloway PJ 226, 231, *250*
Baker EA, see Jeffree CE 529, 530, 531, *536*
Baker R, see Kleinschuster SJ 587, *620*
Bal AK, Payne JF 96, *115*
Bal AK, see Nolan RA 501, *510*
Balandreau J 568, *569*
Balandreau J, Knowles R 565, 568, *569*
Baldrian J, see Beran K 356, 359, 368, *385*, 397, *411*
Baldwin IB, Fred EB, Hastings EG 663, *674*
Bálint Š, Farkaš V, Bauer Š 432, *433*, 504, *507*
Ballance GM, Meredith WOS, Laberge DE 106, *115*
Ballesta J-PG, Alexander M 361, *384*
Ballio A, Brufani M, Casinovi CG, Cerrini S, Fideli W, Pellicciari R, Vaciaqo A 597, 598, *615*

Ballio A, Pocchiari F, Russi S, Silano V 597, 598, *615*
Ballou C 352, 362, 381, *384*, 494, *507*
Ballou CE 441, 444, 447, *455*, 459, *477*
Ballou CE, Lipke PN, Raschke WC 449, *455*
Ballou CE, Maitra SK, Walker JW, Whelan WL 406, *411*
Ballou CE, Raschke WC 444, *455*, 476, *477*
Ballou CE, see Ballou DL 444, 447, *455*
Ballou CE, see Burke D 441, 453, 454, *455*
Ballou CE, see Cohen RE 443, 447, *455*, 445
Ballou CE, see Douglas RH 444, *456*
Ballou CE, see Hashimoto C 442, 452, *456*
Ballou CE, see Jones GH 443, *456*
Ballou CE, see Karson EM 444, 447, *456*, 464, 474, 476, *479*
Ballou CE, see Lee YC 445, *456*
Ballou CE, see Lehle L 441, 443, 445, 449, 450, 451, 452, *456*, 462, 476, *480*
Ballou CE, see Lipke PN 453, *457*
Ballou CE, see Nakajima T 441, 442, 446, 452, *457*, 443, 462, 464, 472, 474, 476, *481*
Ballou CE, see Raschke WC 444, 445, *457*, 446, 447
Ballou CE, see Rosenfeld L 442, 444, *457*
Ballou CE, see Sing V 443, 453, *457*
Ballou CE, see Smith WL 445, 449, *458*, 465, 476, *482*
Ballou CE, see Thieme TR 442, 443, 445, *458*
Ballou CE, see Tung K-K 76, *130*
Ballou CE, see Yen PH 441, 443, 447, 453, 454, *458*, 465, *483*
Ballou DL, Cohen RE, Ballou CE 444, 447, *455*
Ballou DL, see Cohen RE 443, 447, *455*
Ban D, see Dunlop PC 449, *456*
Banás E, see Burczyk J 346, 347, *349*
Banás J, see Burczyk J 346, 347, *349*
Bandurski RS, see Chang CW 150, 151, *157*
Bandurski RS, see Kivilaan A 92, *122*
Bandurski RS, see Lee S 92, *123*
Bandurski RS, see Lee SH 150, *161*
Bandurski RS, see Morohashi Y 58, *65*
Banno Y, see Kitazima Y 355, *389*
Bannon MW 255, *269*
Bara M, Gordon SA 109, *115*
Barash I, Klein L 592, *615*
Barash I, see Nachmias A 597, 599, 600, *621*
Barber GA, Elbein AD, Hassid WZ 49, 51, 52, *62*
Barber GA, Hassid WZ 51, 52, *62*
Barber GA, see Elbein AD 51, 52, *63*
Barber GA, see Jiang K 145, *159*
Barckhausen R 110, *115*, 236, *248*, 531, 532, *533*
Barckhausen R, Rosenstock G 531, *533*
Bardalaye PC, Nordin JH 466, *478*
Barkai-Golan R, Mirelman D, Sharon N 649, *652*, 659, *674*
Barlow AJE, see Chattaway FW 419, *434*
Barlow PW 542, 544, 545, *569*
Barlow PW, see Juniper BE 545, *571*
Barnes HA, see Dea ICM 37, *43*
Barnett JR, Preston RD 321, *329*
Barnett NM 143, *156*, 591, *615*
Barnett RW, see Medcalf DG 296, 297, *312*
Barnoud F, see Chanzy H 524, *533*
Barnoud F, see Higuchi T 203, *223*
Barnoud F, see Joseleau JP 35, *43*, 94, *122*
Barnoud F, see Ruel K 198, *224*
Barondes SH 672, *674*
Barondes SH, Haywood PL 639, *652*
Barondes SH, Rosen SD 639, 640, *653*
Barondes SH, see Chang CM 639, *653*
Barondes SH, see Frazier WA 633, 639, *653*
Barondes SH, see McDonough JP 638, *655*
Barondes SH, see Reitherman RW 639, *656*
Barondes SH, see Rosen SD 640, *656*
Barondes SH, see Zatz M 170, *193*, 460, *483*
Barr RH, Hemming FW 501, *507*
Barr RM, Hemming FW 467, 468, *478*
Barras DR 378, *385*, 422, 423, *433*
Barras DR, Moore AE, Stone BA 426, 430, *433*
Barrett AJ, Northcote DH 34, 37, *42*
Barrett AJ, see Stoddart RW 37, *46*
Barshad I, see Fuller MS 355, *387*
Barthlott W, see Herth W 328, *330*
Bartley IM 104, *115*
Bartnicki-García S 91, *115*, 353, 354, 356, 358, 363, 366, 369, 372, 379, 380, 383, *385*, 395, 397, 401, 404, *411*, 426, 427, 430, *433*, 501 *507*
Bartnicki-García S, Bracker CE, Reyes E, Ruiz-Herrera J 48, *62*, 380, *385*, 399, 402, 404, 405, *411*
Bartnicki-García S, Lindberg B 361, *385*
Bartnicki-García S, Lippman E 379, 382, *385*, 400, 408, *411*
Bartnicki-García S, McMurrough I 367, *385*, 421, *433*
Bartnicki-García S, Nickerson WJ 357, *385*
Bartnicki-García S, Reyes E 355, 357, 361, 366, *385*

Bartnicki-García S, see Bracker CE 21, *22*, 380, *385*, 404, *411*
Bartnicki-García S, see Keen NT 596, 600, *620*
Bartnicki-García S, see López-Romero E 384, *389*, 403, *413*
Bartnicki-García S, see McMurrough I 379, *390*, 398, 399, 400, 402, *414*
Bartnicki-García S, see Reid ID 356, 360, 362, *392*
Bartnicki-García S, see Ruiz-Herrera J 21, *24*, 380, 383, *392*, 399, 402, 403, 404, 409 *414*
Bartnicki-García S, see Tokunaga J 358, 365, 369, *393*, 426, *439*
Bartnicki-García S, see Wang MC 355, 357, 381, *394*, 427, 431, *439*, 596, *623*
Bartnicki-García S, see Zevenhuizen L 426, 427, *440*
Bartnicki-García S, see Zevenhuizen LPTM 355, 358, *394*
Basha SM, Beevers L 169, *188*
Basham HG, Bateman DF 589, 590, 591, *616*
Basham HG, see Bateman DF 590, *616*
Basham JA 607, *616*
Basile DV 144, *156*
Bassford PJ, Beckwith J 491, *507*
Bassford PJ, Silhavy TJ, Beckwith J 491, *507*
Bastide JM, see Bastide M 374, *385*
Bastide M, Trave P, Bastide JM 374, *385*
Bateman DF 139, *156*, 590, 591, *616*
Bateman DF, Basham HG 590, *616*
Bateman DF, Van Etten HD, English PD, Nevins DJ, Albersheim P 589, *616*
Bateman DF, see Basham HG 589, 590, 591, *616*
Bateman DF, see Mullen JM 589, 591, *621*
Bates GW, Ray PM 268, *269*
Bathgate GN, see Palmer GH 106, *126*
Batra KK, Hassid WZ 52, *62*, 72, *115*
Batra KK, Nordin JH, Kirkwood S 75, *115*
Batra KK, see Flowers HM 51, 52, 56, *63*, 72, 73, *119*
Bauer F, see Lehle L 470, 476, *480*, 494, *509*
Bauer H, Bush DA, Horisberger M 425, *433*
Bauer H, Horisberger M, Bush DA, Sigarlakie E 368, *385*, 406, *411*
Bauer H, Sigarlakie E 451, *455*
Bauer Š, see Bálint Š 432, *433* 504, *507*
Bauer Š, see Biely P 449, *455*, 473, *478*
Bauer Š, see Farkaš V 378, *386*, 421, 423, *434*, 464, 473, 476, *479*, 469, 471
Bauer Š, see Košinová A 469, *479*
Bauer Š, see Krátký Z 472, 473, 477, *479*
Bauer WD, Talmadge KW, Keegstra K, Albersheim P 591, *616*
Bauer WD, see Bhuvaneswari TV 663, 664, 669, *674*, 663
Bauer WD, see Calvert HE 665, *674*
Bauer WD, see Keegstra K 6, *8*, 133, *160*, 264, *271*, 357, *389*, 590, *619*
Bauer WD, see Pueppke SG 644, *656*, 667, *676*
Bauer WD, see Talmadge KW 138, *164*
Bause E 472, *478*
Bause E, Lehle L 461, 472, 474, *478*
Bayer DE, see Fisher DA 530, *534*
Bayer EA, see Skutelsky E 469, *482*
Bayley ST, see Setterfield G 19, *24*
Baynes JW, Hsu AF, Heath EC 170, *188*
Baynes JW, see Hsu A-F 175, *190*
Bazzi C, see Mazzucchi U 587, *620*
Beasley CA, see Delmer DP 54, *63*
Beck E 34, *42*
Beck E, see Scheibe R 607, *622*
Becker WA 518, *533*
Beckman CH 609, *616*
Beckman CH, see VanderMolen GE 113, *131*, 609, *623*
Beckman J, see Hadwiger LA 603, *618*
Beckwith J, see Bassford PJ 491, *507*
Beckwith J, see Inoue H 490, *509*
Beckwith J, see Silhavy TJ 491, *511*
Bedford CL, see Al-Delaimy KA 104, *114*
Beer SV, see Sjulin TM 600, *622*
Beevers H, see Tully RE 647, *657*
Beevers L, Mense RM 57, *62*
Beevers L, see Basha SM 169, *188*
Beevers L, see Nagahashi J 70, *125*, 519, *537*
Beffagna N, Cocucci S, Marré E 601, *616*
Behrens NF, Parodi AJ, Leloir LF 174, *188*
Behrens NH, Cabib E 459, 476, *478*, 500, *507*
Behrens NH, Carminatti H, Staneloni RJ, Leloir LF, Cantarella AI 175, *188*, 461, *478*
Behrens NH, Leloir LF 171, *188*, 460, *478*
Behrens NH, Parodi AJ, Leloir LF 461, *478*
Behrens NH, see Dallner G 171, 187, *189*
Behrens NH, see Levy JA 174, *191*
Behrens NH, see Tanner W 460, *483*
Beier R, see Beltran JP 598, *616*
Beijersbergen JCM, Lemmers CBG 613, *616*
Bell DJ, Northcote DH 417, *433*
Bell JK, MacCully ME 562, *569*

Bell JK, McCully ME 97, *115*
Belogortseve NI, see Pavlenko AF 297, 298, *313*
Beltrán JP, Carbonell J 48, 53, *62*, 72, *115*
Beltran JP, Strobel GA, Beier R, Mundy BP 598, *616*
Beltran JP, see Carbonell J 81, *116*
Ben-Arie R, Kislev N, Frenkel C 103, *115*
Ben-Arie R, Lavee S 110, *115*
Benbadis M-C, Deysson G 519, *533*
Bender J, see Ralph BJ 429, *438*
Bender VJ, see Angyal SJ 362, *384*
Benitez T, Villa TG, García Acha I 400, 408, *411*
Benitez T, see Notario V 420, 423, *437*
Benitez T, see Villa TG 420, *439*
Benn M 611, *616*
Benziman M, Haigler C, Brown RM, White A, Cooper K 25, *28*
Benziman M, see Haigler C 25, 26, *28*, 528, *535*
Beran F 422, *433*
Beran K, Holan Z, Baldrian J 356, 359, 368, *385*, 397, *411*
Beran K, see Seichertová O 406, *415*
Berard DF, Kuć J, Williams EB 588, *616*
Berger RK, Reid PD 102, *115*
Bergman LW, Kuehl WM 493, *507*
Berjak P 545, *569*
Berliner MD 347, *349*
Berliner MD, Wenc KA 347, *349*
Berman PM, see Mount MS 589, 592, *621*
Berman PM, see Stack JP 589, *622*
Bernadac A, see Huet C 634, *654*
Bernardi G, Springer GF 296, *307*
Besford RT, Hobson GE 104, *115*
Betz H, Hinze H, Holzer H 401, *411*
Beveridge RJ, see Dea ICM 39, *42*
Bewley JD, see Halmer P 107, *120*
Bewley JD, see Reid JSG 107, *127*
Bhatnagar RS, see Rapaka RS 143, *163*
Bhattacharjee SS, Yaphe W, Hamer GK 292, 293, *307*
Bhattacharjee SS, see Gorin PAJ 445, *456*
Bhattacharjee SS, see Hamer GK 291, *309*
Bhavanandan VP, Bouveng HO, Lindberg B 428, *433*
Bhoyroo VD, see Spiro MJ 175, 178, 179, 181, *192*, 463, *482*
Bhullar BS, see Lazarovits G 597, *620*
Bhuvaneswari TV, Bauer WD 663, 664, 669, *674*
Bhuvaneswari TV, Pueppke SG, Bauer WD 663, 664, 669, *674*
Bhuvaneswari TV, see Calvert HE 665, *674*
Bidwell RGS 302, *307*
Bidwell RGS, Ghosh NR 298, *307*
Bidwell RGS, Percival Elizabeth, Smestad B 288, 289, *307*
Biedermann M, see Kristen U 579, *583*
Bieleski RL 151, *157*
Biely P, Bauer Š 473, *478*
Biely P, Krátky Z, Bauer Š 449, *455*, 473, *478*
Biely P, see Farkaš V 378, *386*
Biely P, see Farkaš V 421, 423, *434*
Biely P, see Krátký Z 472, 473, 477, *479*
Biely P, see Vrsanska M 406, *415*
Bierbach M 346, *349*
Biggins J 347, *349*
Biggs RH, see Freeman B 530, *534*
Biguet J, see Poulain DG 419, *438*
Binneman PH, see Brieskorn CH 230, *248*
Biquet J, see Poulain DG 367, *391*
Birk Y, see Huberman M 102, *121*
Biroc SL, Etzler ME 635, *653*
Bisalputra T, Weier TE 346, *349*
Bishop CT, Blank F, Gardner PE 419, *433*
Bishop CT, see Yu RJ 419, *440*
Bishop CT, see Zitko V 34, *46*
Bittiger H, see Herth W 86, *120*, 144, *159*, 318, *330*, 582, *582*
Björkman A 198, *222*
Björkvist KJ, see Adler E 196, *221*
Bjorndal H, Eriksson KE, Garegg PJ, Lindberg B, Swan B 282, *307*
Bjorndal H, Lindberg B 428, *433*
Bjorndal H, see Manners DJ 417, 418, 427, *436*
Black WAP 279, 280, *307*
Black WAP, Dewar ET, Woodward FN 296, *307*
Blackwell J, see Gardner KH 13, *23*, 47, *63*
Blackwell J, see Minke R 395, *414*
Blake JD, Richards GN 38, 39, *42*
Blakely LM, see Wu JH 112, *132*
Blakeman JP, see Parbery DG 248, *252*
Bland DE, see Wardrop AB 196, 197, *224*
Blank F, see Bishop CT 419, *433*
Blank F, see Yu RJ 419, *440*
Bleiss W, see Göring H 268, *270*
Blobel BJ, see Lingappa VR 490, *509*
Blobel G, Dobberstein B 468, *478*, 489, *507*
Blobel G, Sabatini DD 488, *507*
Blobel G, see Campbell PN 492, *507*

Blobel G, see Devillers-Thiery A 492, *508*
Blobel G, see Jackson K 489, *509*
Blobel G, see Katz FN 180, *191*
Blobel G, see Lingappa VR 181, *191*, 490, *509*
Blobel G, see Maccecchini ML 492, *509*
Bloch R, see Sinnot EW 514, *540*
Blomquist GJ, see Jackson LL 248, *250*
Bloodgood RA, see Coss RA 60, *62*
Bloom HL, see Wallner SJ 104, *131*
Bloomfield BJ, Alexander M 373, 374, *385*
Blumenthal HJ, see Davidson EA 398, *412*
Blumenthal HJ, see Ghosh S 398, *412*
Blumenthal HJ, see Hashimoto T 365, *387*
Bobitt TF, Nordin JH, Roux M, Revol JF, Marchessault RH 360, 363, 365, *385*
Bodwin JS, see Whatley MH 586, *623*
Boer P, Steyn-Parvé EP 449, *455*
Boer P, Van Rijn JM, Reinking A, Steyn-Parvé EP 499, *507*
Boer P, see Linnemans WAM 367, *389*, 469, 477, *480*, 498, 504, *509*
Boer P, see Reuvers F 460, 462, 463, *481*
Boer P, see Van Rijn HJM 362, *392*, 471, 473, *483*, 498, *512*
Boffey SA, Northcote DH 39, *42*
Bohlool BB, Schmidt EL 659, 660, 663, 664, 667, 668, 671, *674*
Boime I, see Strauss AW 489, *511*
Boldrini L, see Cassone A 17, *23*
Bonettemaker M, see Srivastava LM 525, *540*
Bonner BA, see Goeschl JD 143, *158*
Bonner J 258, *269*
Bonner J, see Cooil B 265, *270*
Bonner J, see Olson AC 259, 262, *272*
Bonner J, see Tagawa T 258, 265, *273*
Bonnett HT Jr 531, *533*
Bonnett HT Jr, Newcomb EH 84, *115*
Bopp M, see Mader M 150, *161*
Borgstrom G, see Al-Delaimy KA 104, *114*
Bornman CH 102, *115*
Bornstein P, see Duskin D 184, *189*
Bosshart RP, see Abeles FB 600, 608, *615*
Botkin DB, see Woodwell GM 11, *24*
Bouchilloux S, see Ronin C 472, *482*
Boudet A, see Alibert G 204, *222*
Bourke ME, see Price CA 347, *351*
Bourne EJ, Johnson PG, Percival Elizabeth 299, *307*
Bourne EJ, Megarry ML, Percival Elizabeth 280, 283, 301, *307*
Bourne EJ, Percival Elizabeth, Semstad B 283, *307*
Bouveng HO, Kiessling H, Lindberg B, McKay J 427, *433*
Bouveng HO, Meier H 109, *115*
Bouveng HO, see Bhavanandan VP 428, *433*
Bover FG, see Hamilton RC 526, *535*
Bowden JK, Hodgson B 417, 418, *433*
Bowen DV, see Kieras JH 304, *311*
Bowen GD, Rovira AD 565, 567, *569*
Bowers B, Levin G, Cabib E 400, 406, *411*
Bowers B, see Cabib E 89, *116*, 368, *386*, 395, 397, 401, 406, *411*, 397, 400, 423, *434*
Bowers B, see Durán A 380, *386*, 403, 404, *412*
Bowers B, see Molano J 397, *414*
Bowker DM, Turvey JR 291, *307*
Bowler E, see Scott FM 103, *129*
Bowles DJ, Kauss H 646, *653*, 668, *674*
Bowles DJ, Lis H, Sharon N 645, *653*, 668, *674*
Bowles DJ, Northcote DH 81, 83, 84, *115*, 519, 521, *533*, 556, 557, 558, *569*
Bowles DJ, Schnarrenberger C, Kauss H 646, *653*
Bowles DJ, see Kauss H 36, 38, *44*, 87, *122*, 647, *655*, 668, *676*
Bowles DJ, see Lehle L 187, *191*
Boyd J, Turvey JR 284, 285, *307*
Boyd J, see Morris ER 286, *313*
Boyd WC, Shapleigh E 627, *653*
Bozzaro S, Gerisch G 638, 640, *653*
Bozzaro S, see Gerisch G 641, *653*
Braatz JA, see Shematek EM 504, *511*
Brackenbury R, see Sussmann M 638, *656*
Bracker CE 353, *385*
Bracker CE, Ruiz-Herrera J, Bartnicki-García S 21, *22*, 380, *385*, 404, *411*
Bracker CE, see Bartnicki-García S 48, *62*, 380, *385*, 399, 402, 404, 405, *411*
Bracker CE, see Grove SN 382, *387*, 500, 503, *508*
Bracker CE, see Van der Woude WJ 82, 83, 85, *131*, 519, *540*
Bradford KJ, see Yang SF 143, *165*
Brading JWE, Georg-Plant MMT, Hardy DM 300, *308*
Bradler G, see Eckhardt K 185, *189*
Brady CJ 104, *115*
Brady CJ, Ali ZM, MacAlpine GA, McGlasson WB 104, *115*

Brams E 565, *570*
Brannon JH, see Medcalf DG 304, *312*
Branton D, Bullivant S, Gilula NB, Karnovsky MJ, Moor H, Mühlethaler K, Northcote DH, Packer L, Satir B, Satir P, Speth V, Staehelin LA 321, *329*
Branton D, Moor H 560, *570*
Braun A, see Galun M 408, *412*
Braun E, Aach HG 347, *349*
Braun PC, Calderone RA 380, *385*, 399, 402, 404, 409, *411*
Braun V, Hantke K 670, *674*
Brawley SH, see Quatrano RS 296, 298, 306, *314*
Brawley V, see Schekman R 382, *392*, 404, 407, *415*
Bray D, see Robbins PW 74, *128*
Breddam K, see Johansen JT 442, 452, *456*
Breisch H 553, 567, 568, *570*
Breisch H, Guckert A, Reisinger O 567, 568, *570*
Breisch H, see Guckert A 553, 567, 568, *571*
Brett CT 49, 55, *62*, 72, 111, *115*, 151, *157*, 608, *616*
Brett CT, Leloir LF 173, *189*
Brett CT, Northcote DH 53, 58, *62*, 70, 74, 75, *115*
Brett CT, see Pont Lezica R 58, *65*
Bretthauer RK, Kozak LP, Irwin WE 465, *478*
Bretthauer RK, Wu S 465, *478*
Bretthauer RK, Wu S, Irwin WE 465, *478*
Bretthauer RK, see Kozak LP 465, *479*
Brieskorn CH, Binneman PH 230, *248*
Brieskorn CH, Kabelitz L 230, *248*
Briggs DE, see Morrall P 106, *124*
Bright SWJ, Northcote DH 30, *42*
Brill WJ, see Dazzo FB 633, 642, *653*, 663, 666, 667, 668, 669, 671, *674*, *675*
Brill WJ, see Maier RJ *620*
Brisson JD, Peterson RL, Robb J, Rauser WE, Ellis BE 547, 555, *570*
Brisson JD, Robb J, Peterson RL 232, *248*
Brisson JD, see Peterson RL 543, 545, 547, 548, 549, *573*
Brock TD 377, *385*, 420, 421, *433*
Brock TD, see Crandall MA 441, 453, 454, *455*
Brockwell J, see Dudman WF 671, *675*
Brody S 372, *385*
Brody S, see Hanson B 372, 373, *387*
Brodzki P 109, *115*
Brooks J, see Kokkinakis DM 149, *160*
Broughton WJ 659, 663, *674*
Brower DL, Hepler PK 525, *533*
Brower DL, see Coss RA 60, *62*
Brower DL, see Giddings TH 325, 326, *330*, 524, *534*
Brower DL, see Giddings TH Jr 94, *119*
Brown AJ, Kolattukudy PE 231, 237, *248*
Brown CL, see Wodzicki TJ 94, *131*
Brown DH, see Glaser L 379, *387*, 398, 399, *412*
Brown EG, Short KC 49, *62*
Brown HT, Morris GH 106, *115*
Brown JP 424, *433*
Brown L, see Kolattukudy PE 230, 245, *251*, 243, 244
Brown MS, Goldstein JL 651, *653*
Brown MS, see Goldstein JL 168, *190*, 187
Brown O, see Jilka R 83, 84, *122*
Brown RG, Hanic LA, Hsiao M 428, *433*
Brown RG, Kimmins WC 521, *533*
Brown RG, see Kimmins WC 111, *122*, 141, *160*
Brown RM 320, *329*
Brown RM, Franke WW, Kleinig H, Falk H, Sitte P 318, *329*, 346, *349*
Brown RM, Herth W, Franke WW, Romanovicz D 320, *329*
Brown RM Jr 513, 514, *533*
Brown RM Jr, Colpitts TJ 526, *533*
Brown RM Jr, Franke WW, Kleinig M, Sitte P 144, *157*
Brown RM Jr, Montezinos D 55, *62*, 524, *533*
Brown RM Jr, Willison JHM 521, 526, 527, 528, *533*
Brown RM Jr, Willison JHM, Richardson CL 526, *533*
Brown RM Jr, see Cooper KM 526, *534*
Brown RM Jr, see Haigler C 528, *535*
Brown RM Jr, see Herth W 144, *159*
Brown RM Jr, see Mueller SC 94, *125*, 522, 523, *537*
Brown RM Jr, see Westafer JM 94, *131*
Brown RM Jr, see Willison JHM 522, 523, 524, 525, *541*, 528, *541*
Brown RM, Montezinos D 325, *329*
Brown RM, Romanovicz DK 318, *329*
Brown RM, Willison JHM, Richardson CL 20, *22*
Brown RM, see Benziman M 25, *28*
Brown RM, see Franke WW 318, *330*
Brown RM, see Haigler C 25, 26, *28*
Brown RM, see Herth W 318, *330*
Brown RM, see Montezinos DK 325, *331*
Brown RM, see Romanovicz DK 318, *332*
Brown RM, see Romanovicz PK 59, *66*
Brown RM, see Willison JHM 17, *24*
Brown SA 203, *222*
Brown SA, Neish AC 201, *222*
Brown SA, see Higuchi T 205, *223*
Brownlee GG, see Milstein C 488, *510*
Bruce BJ, see Fraser T 491, *508*

Bruegger BB, Keen NT 604, *616*
Brufani M, see Ballio A 597, 598, *615*
Bruinsma J, see Hänish Ten Cate ChH 102, 103, *120*
Brummond DO, Gibbons AP 51, 52, *62*
Brummond DO, see Larsen GL 51, 53, 54, *64*, 72, *123*
Bruni A, see Fasulo MP 60, *63*
Bruns D, see Ferrari TE 580, *582*
Bryant JA 377, *385*
Brysk MM, Chrispeels MJ 137, 140, *157*
Buchala AJ 31, *42*
Buchala AJ, Franz G *42*, 48, *62*
Buchala AJ, Fraser CG, Wilkie KCB 31, *42*
Buchala AJ, Wilkie KCB 31, *42*, 48, *62*, 93, 94, *115*, *116*
Buchanan BB, see Magyarosy AC 606, *620*
Buchbinder BO, see Sun SM 181, *193*
Buchner V, see Nachmias A 597, 599, 600, *621*
Buck KW, Chen AW, Dickerson AG, Chain EB 358, *385*, 427, *433*
Buck KW, see Obaidah MA 360, *391*
Buckholz RG, Adams BG 449, *455*
Buckner JS, see Kolattukudy PE 243, *251*, 244, 245, 246
Buescher RW, Tigchelaar EC 104, 105, *116*
Buffinga H, see Winter H 139, *164*
Büldt G, see Palma A 524, *538*
Bulger J, see Hultin HO 104, *121*
Bull AT 373, 374, *386*, 428, 429, *433*
Bull AT, Chesters CGC 426, *433*
Bull AT, Trinci AJP 594, *616*
Bull AT, see Chesters CGC 420, *434*
Bulla LA Jr, see Arthur LO 595, *615*
Bullivant S, see Branton D 321, *329*
Burczyk J, Grzybek H, Banás J, Banás E 346, 347, *349*
Burg EA, see Burg SP 150, *157*
Burg SP, Burg EA 150, *157*
Burger MM, see Meyer DI 502, *510*
Burgess J 13, 21, *23*
Burgess J, Fleming EN *23*, 30, *42*, 85, *116*
Burgess J, Linstead PJ 60, *62*
Burke D, Kaufman P, McNeil M, Albersheim P 264, *269*
Burke D, Previato LM, Ballou CE 441, 453, 454, *455*
Burkhardt HJ, Maizel JV, Mitchell HK 610, 612, *616*
Burkhardt HJ, see Maizel JV 612, *620*
Burnett JH 353, 355, 365, 369, 370, 379, *386*
Burnett JH, see Hunsley D 353, 355, 358, 363, 364, 365, 369, 371, *388*, 396, 397, 408, *413*
Burström H 256, 258, 264, *269*
Burström H, Uhrström I, Wurscher R 260, *269*
Bush DA, Horisberger M 426, 428, *433*
Bush DA, Horisberger M, Horman I, Wursch P 419, *433*
Bush DA, see Bauer H 368, *385*, 406, *411*, 425, *433*
Bushby HVA, see Marshall KC 670, *676*
Butcher DN, El-Tigani S, Ingram DS 613, *616*
Butcher WI, see Schwertner HA 346, *351*
Butt VS, see Vaughan PFT 204, *224*
Butterfield BG, Meylan BA 95, *116*
Butterfield BG, see Meylan BA 95, *124*
Buvat R 94, 96, *116*
Byers B, Goetsch L 501, *507*
Byhan O, see Schuster G 112, *129*
Byrne H, see Verma DPS 267, *273*
Bystrom BG, see Scott FM 103, *129*
Cabib E 381, *386*, 459, *478*, 498, 504, *507*
Cabib E, Bowers B 368, *386*, 397, 400, 406, *411*
Cabib E, Durán A, Bowers B 397, 401, *411*
Cabib E, Farkas V 380, *386*, 401, 402, 406, 408, *411*
Cabib E, Keller FA 401, *411*
Cabib E, Leloir LF 459, *478*
Cabib E, Ulane RE 401, *411*
Cabib E, Ulane RE, Bowers B 89, *116*, 395, 397, 401, 406, *411*, 423, *434*
Cabib E, see Algranatti ID 459, *477*, 500, *507*
Cabib E, see Behrens NH 459, 476, *478*, 500, *507*
Cabib E, see Bowers B 400, 406, *411*
Cabib E, see Durán A 380, *386*, 399, 400, 401, 402, 403, *412*, 404
Cabib E, see Huang KP 403, *413*
Cabib E, see Keller FA 399, 400, *413*
Cabib E, see Molano J 397, *414*, 405, 649, *655*
Cabib E, see Shematek EM 504, *511*
Cabib E, see Ulane RE 401, 402, 406, *415*
Caccam JF, Jackson JJ, Eyler EH 170, *189*
Caccam JJ, Jackson JT, Eylar EH 460, *478*
Cahfe SC, Wardrop AB 529, *533*
Calabrese G, see Arrigoni O 142, *156*, 143
Calderone RA, see Braun PC 380, *385*, 399, 402, 404, 409, *411*
Caldicott AB, Eglinton G 226, *248*
Caldicott AB, Simoneit BRT, Eglinton G 228, 230, *248*
Callow JA 364, *386*, 628, 643, 651, *653*
Callow JA, see Dow JM 605, *617*
Callow ME, Coughlan SJ, Evans LV 317, *329*
Callow ME, Evans LV 278, 298, *308*, 317, *329*

Callow ME, see Evans LV 298, *308*, *309*, 303, 317, *330*
Calvert HE, Lalonde M, Bhuvaneswari TV, Bauer WD 665, *674*
Cameron AC, see Yang SF 143, *165*
Cameron DS, see Taylor IEP 353, *393*, 416, *439*
Cami B, see Mercereau-Puijalon O 491, *509*
Campbell JMcA, Peberdy JF 402, *412*
Campbell PN, Blobel G 492, *507*
Campbell R, Rovira AD 567, *570*
Campbell R, see Rovira AD 567, *573*
Canney M, see Lang A 112, *123*
Cantarella AI, see Behrens NH 175, *188*, 461, *478*
Cantarella AI, see Levy JA 174, *191*
Cantino EC, Myers RB 410, *412*
Cantino EC, see Mills GL 400, *414*
Cantino EC, see Norman J 398, *414*
Cantrell MA, see Whatley MH 587, *623*
Carbonel LM, see San Blass G 361, *392*
Carbonell LM, see Kanetsuna F 429, *435*
Carbonell J, Beltran JP, Conejero V 81, *116*
Carbonell J, see Beltrán JP 48, 53, *62*, 72, *115*
Carbonell LM, Kanetsuna F, Gil F 361, 365, *386*
Carbonell LM, see Kanetsuna F 372, *388*, 428, *435*
Carbonell LM, see San-Blas G 428, 429, *438*
Carbonell M, see Kanetsuna F 428, *435*
Carceller M, see Lavinthman N 55, *64*
Cardemil L, Pincheira G 362, *386*
Cardini CE, see Lavinthman N 55, *64*
Cardini CE, see Leloir LF 398, *413*
Carefoot JR 348, *349*
Carlberg GE, Percival Elizabeth, 283, 301, 302, *308*
Carlberg GE, Percival Elizabeth, Rahman MA 297, *308*
Carlier AR, see Laere A van 380, *389*
Carlier AR, see Van Laere AJ 402, *415*
Carlo PL, Villamez CL 170, *189*
Carlson RW, see Sanders RE 667, *676*
Carlström D 395, *412*
Carminatti H, see Algranatti ID 459, *477*, 500, *507*
Carminatti H, see Behrens NH 175, *188*, 461, *478*
Carminatti H, see Levy JA 174, *191*
Caro LG, Palade GE 496, *507*
Caro LG, see Palade GE 487, *510*
Carpenter WD, see Jaworski EG 398, *413*
Carpita N, Sabularse D, Montezinos D, Delmer DP 68, 101, *116*, 154, *157*
Carpita NC, see Maltby D 93, *123*
Carr DJ, Carr SGM 237, *249*
Carr DJ, see Carr SGM 15, *23*
Carr DJ, see O'Brien TP 225, *252*
Carr SGM, Carr DJ 15, *23*
Carr SGM, see Carr DJ 237, *249*
Carroll TW, Kosuge T 606, *616*
Carson DD, Lennarz WJ 187, *189*
Carter GA, see Lampard JF 248, *251*
Carter WA, see Mizrahi A 493, *510*
Casinovi CG, see Ballio A 597, 598, *615*
Cass DD, see Karas I 514, *536*
Cassagne C 244, *249*
Cassagne C, Lessire R 243, 244, 245, *249*
Cassone A, Kerridge D, Gale EF 418, *434*
Cassone A, Mattia E, Boldrini L 17, *23*
Cass-Smith WP, Holland AA 660, *674*
Catley BJ 429, *434*
Catley BJ, Kelly DJ 429, *434*
Catt JW, Hills GJ, Roberts K 86, *116*, 145, 146, *157*, 328, *329*, *330*
Cawley LP, see Jones JM 644, *654*
Cerezo AS 282, 292, *308*
Cerezo AS, Lezerovich A, Labriola R, Rees DA 282, *308*
Cerrini S, see Ballio A 597, 598, *615*
Cervone F, Scala A, Scala F 589, 591, *616*
Chaboud A, see Rougier M 555, *573*
Chadwick CM, Northcote DH 81, *116*
Chafe SC 15, *23*
Chain EB, see Buck KW 358, *385*, 427, *433*
Chaloupecká O, see Hořejši V 633, *654*
Chambat G, see Joseleau JP 35, *43*
Chambers J, Elbein AD 52, *62*, 73, *116*
Chambers J, Forsee WT, Elbein AD 177, *189*
Chang CM, Reitherman RW, Rosen SD, Barondes SH 639, *653*
Chang CM, see Rosen SD 640, *656*
Chang CW, Bandurski RS 150, 151, *157*
Chang PLY, Trevithick JR 377, 378, *386*
Chanzy H, Imada K, Mollard A, Vuong R, Barnoud F 524, *533*
Chanzy H, Imada K, Vuong R 524, *533*
Chao H, see Molner J 172, *191*
Chapman A, Fujimoto K, Kornfeld S 177, 186, *189*
Chapman A, Li E, Kornfeld S 175, 176, *189*
Chapman DJ, see Heaney-Kieras J 144, *159*, 304, *310*
Charudattan R, see DeVay JE 586, *617*

Chattaway FW, Chenolikar S, O'Reilly J, Barlow AJE 419, *434*
Chatterjee AK, Starr MP 589, 593, *616*
Cheadle VI, see Esau K 83, 94, *118*
Cheeseman JM, see Gronewald JW 601, *618*
Chen APT, Phillips DA 664, *674*
Chen AW, see Buck KW 358, *385*, 427, *433*
Chen MS, see Loewus F 40, *44*
Chen T-S, Jaffe LF 521, *533*
Chen WW, Lennarz WJ 174, *189*, 497, *507*
Cheng CKC, Marsh HV 220, *222*
Chenolikar S, see Chattaway FW 419, *434*
Chenouda MS 352, *386*
Chernel J, see Harada H 60, *64*
Chessen JH, see Medcalf DG 304, *312*
Chesters CGC, Bull AT 420, *434*
Chesters CGC, see Bull AT 426, *433*
Chet I, Mitchell R 662, *674*
Chet I, see Kritzman G 378, *389*, 430, *436*
Chet I, see Netzer D 608, *621*
Chi M M-Y, see Natowicz MR 497, *510*
Child JJ, see Sietsma JH 355, 358, 363, 365, 368, *392*, 426, 428, *439*
Chilton M, see Watson B 586, *623*
Cho YP, Chrispeels MJ 135, 138, *157*
Chojnacki T, see Mańkowski T 474, *480*
Chojnacki T, see Palamarczyk G 474, *481*
Chong Ja Chong, see Heldt HW 607, *619*
Chrispeels MJ 513, 520, 521, *533*
Chrispeels MJ, see Brysk MM 137, 140, *157*
Chrispeels MJ, see Cho YP 135, 138, *157*
Chrispeels MJ, see Ericson MC 169, *189*
Chrispeels MJ, see Gardiner M 137, *158*, 519, *534*
Chrispeels MJ, see Sadava D 142, 147, *163*, 143
Christ B 237, *249*
Christensen RA, see Reichert TA 148, *163*
Chung K, Hawirko R, Isaak P 469, *478*
Churms SC, Merrifield EH, Stephen AM 6, *8*, 75, *116*
Chvojka L, see Volfová A 525, *540*
Ciarlante D, Curtis RW 142, *157*
Claes H 335, 338, 340, 341, *349*
Claire JWL, Dawes CJ 295, *308*
Clark AF, Villemez CL 52, *62*, 73, *116*
Clark CF, see Villamez CL 170, 173, *193*
Clark DL, see Rosen SD 639, *656*
Clark JE, see Heath OVS 143, *159*
Clarke AE, Gleeson P, Harrison S, Knox RB 155, *157*
Clarke AE, see Knox RB 237, *250*
Clarke AE, Anderson RL, Stone BA 6, *8*, 34, 35, *42*, 90, 109, *116*, 152, 153, 155, *157*, 577, 578, *582*
Clarke AE, Gleeson PA 578, *582*
Clarke AE, Gleeson PA, Jermyn MA, Knox RB 577, *582*, 628, *653*
Clarke AE, Knox RB 81, 111, 113, *116*, 151, *157*
Clarke AE, Stone BA 74, 99, 111, *116*, 426, *434*
Clarke AE, see Anderson RL 92, *114*, 154, *156*
Clarke AE, see Gleeson PA 155, *158*, 578, 579, *582*,
Clarke AE, see Hinch JM 614, *619*
Clarke AE, see Knox RB 580, *583*
Clarke AE, see Mau S-L 581, *583*
Clarke AE, see Raff JW 580, *583*
Clarke AE, see Vithanage HIMV 110, *131*, 581, *583*
Clarke JA, Lisker N, Ellingboe AH, Lamport DTA 142, *157*
Clarkson DT, Sanderson J 566, *570*
Clarkson DT, see Ferguson IB 225, *250*
Clarkson DT, see Robards AW 225, *253*, 531, *539*
Clarkson DT, see Torrey JG 247, *253*
Cleare M, Percival Elizabeth 304, *308*
Cleland RE 91, 109, *116*
Cleland R, Karlsnes AM 142, 143, *157*
Cleland RE, see Gotelli IB 144, 147, *159*
Cleland RE 256, 257, 258, 259, 260, 261, 262, 265, 267, *269*
Cleland RE, Haughton PM 259, 260, 262, *269*
Cleland RE, Rayle DL 256, 261, 265, 266, 268, *269*, *270*
Cleland RE, Thompson M, Rayle DL, Purves WK 262, 263, *270*
Cleland RE, see Ray PM 255, *272*
Cleland RE, see Rayle DL 258, 261, 264, 265, 266, 268, *272*
Cleland RE, see Tepfer M 265, *273*
Cleland RE, see Waaland SD 255, *273*
Clermont S, Percheron F 81, *116*
Cline K, Wade M, Albersheim P 605, *616*
Clowes FAL 544, 566, *570*
Clowes FAL, Juniper BE 85, 97, 98, *116*, 545, *570*
Clowes FAL, Woolston RE 544, 566, *570*
Clowes FAL, see Juniper BE 545, *571*
Cocking EC, see Davey MR 672, *674*
Cocking EC, see Willison JHM 18, *24*, 30, *46*, 525, 532, *541*
Cocucci S, see Beffagna N 601, *616*
Code J, see Santos T 423, *438*

Coggins CW Jr, Lewis LN, Knapp JCF 103, *116*
Cohen RE *455*
Cohen RE, Ballou CE 445, *455*
Cohen RE, Ballou DL, Ballou CE 443, 447, *455*
Cohen RE, see Ballou DL 444, 447, *455*
Cohen RE, see Hashimoto C 442, 452, *456*
Cohen RE, see Lehle L 441, 443, 445, 449, 450, 451, 452, *456*, 462, 476, *480*
Cohn Z, see Edelson P 672, *675*
Cole GT, Sekiya T, Kasai R, Yokoyama T, Nozawa Y 365, *386*
Coleman A 347, *349*
Collier RJ, see Gilliland DG 650, *654*
Collinge AJ, Trinci APJ 382, *386*
Collinge AJ, see Trinci APJ 370, 382, *393*
Colpitts TJ, see Brown RM Jr 526, *533*
Colvin JR 13, *23*, 25, *28*, 526, *533*
Colvin JR, Leppard GG 15, 20, *23*
Colvin JR, see Kahn AW 58, *64*
Colvin JR, see King GGS 528, *536*
Colvin JR, see Kiobakken J 178, *191*
Colvin JR, see Kjosbakken J 20, *23*
Colvin JR, see Leppard GG 15, 20, *23*, 528, *536*
Colvin JR, see Manocha MS 355, 363, 365, 366, *390*
Compte J, see Letoublon RCP 467, *480*, 501, *509*
Conchie J, Percival EGV 297, *308*
Conde J, see Santos T 377, *392*
Conejero V, see Carbonell J 81, *116*
Conn E, see Saunders JA 611, *622*
Conn EE 611, *616*
Conn EE, see Akazawa T 611, *615*
Conn EE, see Koukol J 203, *223*
Conte MV, Pore RS 280, *308*
Cooil B, Bonner J 265, *270*
Cook GA, see Stirling JL 373, *393*
Cook GMW, Stoddart RW 90, *116*
Cook JA, Fincher GB, Keller F, Stone BA 72, 88, *116*
Cook WH, see Smith DB 292, *315*
Coombe BG 151, *157*
Coombs J, Volcani BE 144, *157*
Cooper BA, Aronson JM 358, 365, *386*
Cooper D, Manley RStJ 526, *533*
Cooper FP, see Yu RJ 419, *440*
Cooper K, see Benziman M 25, *28*
Cooper KM, Brown RM Jr 526, *534*
Cooper R, see Levy EC 248, *251*
Cooper RM 589, 590, 592, *617*
Cooper RM, Rankin B, Wood RKS 589, 591, *617*
Cooper RM, Wood RKS 589, 591, 592, 593, 609, *617*
Cortat M, Matile P, Kopp F 498, *507*
Cortat M, Matile P, Wiemken A 377, 378, *386*, 422, 423, *434*, 470, *478*
Cortat M, see Matile P 422, *436*, 469, *481*, 498, *509*
Coss RA, Bloodgood RA, Brower DL, Pickett-Heaps JD, McIntosh JR 60, *62*
Cote RH 297, *308*
Coughlan SJ, Evans LV 317, *330*
Coughlan SJ, see Callow ME 317, *329*
Coughlan SJ, see Evans LV 298, *309*
Coulomb P, Coulon J 83, *117*, 553, *570*
Coulon J, see Coulomb P 83, *117*, 553, *570*
Courtney JS, Morré DJ 261, 262, 264, 265, 268, *270*
Courtney JS, Morré DJ, Key JL 262, *270*
Courtois JE, see Petek F 139, *162*
Cousin MT 112, *117*
Couture RM, Routley DG, Dunn GM 611, *617*
Cox GC, see Juniper BE 529, 530, *536*
Crafts AS, Currier HB 110, *117*
Craig GD, Newsam RJ, Gull K, Wood DA 409, *412*
Craig JWT, see Aspinall GO 32, *42*
Craigie JS, McLachlan J, Majak W, Ackman RG, Tocher CS 302, *308*
Craigie JS, Wong KF 295, *308*
Craigie JS, see Hanic LA 278, *309*
Craigie JS, see McCandless EL 294, *312*
Craigie JS, see Wong F 295, 305, *316*
Crandall M 364, *386*
Crandall M, Egel R, Mackay VL 421, *434*
Crandall MA, Brock TD 441, 453, 454, *455*
Crasnier M, Noat G, Ricard J 151, *157*
Crawford JK, see Hasegawa S 104, 105, *120*
Crayton MA, Wilson E, Quatrano RS 298, *308*
Crayton MA, see Quatrano RS 298, *314*
Creeth JM, see Allen AK 135, 136, *156*, 633, 635, 636, *652*
Cribbs DH, see Peters BM 614, *621*
Crisp CE 230, *249*
Cronshaw J, Esau K 517, *534*
Cronshaw J, Myers A, Preston RD 279, 280, *308*
Cronshaw J, see Esau K 112, *118*
Croteau R, Fagerson IS 230, *249*
Croteau R, Kolattukudy PE 234, *249*, 529, *534*
Croteau R, see Kolattukudy PE 230, 245, *251*, 244, *536*
Cruickshank RH, see Marshall KC 670, *676*
Cumming DF 49, *62*
Cumsky M, Zusman DR 672, *674*

Cunningham RL, see Medcalf DG 304, *312*
Currier HB 112, *117*, 608, *617*
Currier HB, Strugger S 112, *117*
Currier HB, Webster DH 111, 112, *117*
Currier HB, see Crafts AS 110, *117*
Currier HB, see Eschrich W 112, *118*
Currier HB, see McNairn RB 112, *124*
Currier HB, see Shih CY 112, *129*
Currier HB, see Webster DH 112, *131*
Currier TC, see Watson B 586, *623*
Currier WW, Strobel GA 662, *674*
Curtis RW, see Ciarlante D 142, *157*
Czaninski Y 95, 97, *117*
Czurda V 334, *349*

D'Agostino G, see Pennazio S 112, *126*
Dalbec JM, see Lenney JF 401, *413*
Daleo CR, Pont Lezica R 168, *189*
Daleo GR, Hopp HE, Romero PA, Pont Lezica R 168, *189*
Daleo GR, Pont Lezica R 58, *62*
Daleo GR, see Hopp HE 48, 51, 53, 56, 57, 58, 59, 64, 74, *121*
Dalessandro G, Northcote DH 40, *42*, 76, 77, 79, 81, *117*
Dallner G, Behrens NH, Parodi A, Leloir LF 171, 187, *189*
Dal'Olio G, see Fasulo MP 60, *63*
Daly JM 606, *617*
Daniels D, see Johnson KD 139, 150, *159*, 266, *271*
Daniels DL, Hadwiger LA 606, *617*
Dankert M, see Garcia RC 528, *534*
Dankert M, see Robbins PW 74, *128*
Dankert MA, see Pont Lezica R 58, *65*, 58, *66*, 171, *192*
Danon D, see Lotan R 629, *655*
Darbyshire JF, see Greaves MP 562, 565, 568, *570*
Dargent R 501, *508*
Darken MA 408, *412*
Darley WM, Porter D, Fuller MS 317, *330*
Dart PJ 567, *570*, 662, *674*
Dart PJ, Mercer FV 565, 568, *570*
Darvill AG, McNeil M, Albersheim P 6, *8*, 34, *42*
Darvill AG, Smith CJ, Hall MA 92, *117*, 140, *157*, 264, 267, 268, *270*
Darvill AG, see Hahn MG 603, 605, *618*
Darvill AG, see McNeil M 8, *8*, 29, 31, 32, 33, 34, 35, 37, 38, 41, *44*, 68, 86, *124*, 134, *161*
Darvill AG, see Valent BS 92, *131*
Dashek WV 137, *157*
Dashek WV, Erickson SS, Hayward DM, Lindbeck G, Mills RR 150, *157*
Dashek WV, Rosen WG 83, 84, *117*
Dashek WV, see Reynolds JD 582, *583*
Datema R, Ende H van den, Wessels JGH 353, 355, 357, 361, 363, 366, *386*
Datema R, Wessels JGH, Ende H van den 353, 355, 357, 366, *386*
Datema R, see Schwarz RT 186, *192*, 473, *482*
Datko AH, Maclachlan GA 267, *270*
Dauwalder M, Whaley WG 518, *534*, 546, 558, *570*
Dauwalder M, Whaley WG, Kephart JE 546, 558, 565, *570*
Dauwalder M, see Whaley WG 516, 517, 518, *541*
Davey MR, Cocking EC 672, *674*
Davey R, see Wright K 553, *574*
Davidson E, see Ghosh S 398, *412*
Davidson EA 398, *412*
Davidson EA, Blumenthal HJ, Roseman S 398, *412*
Davidson ED, see Bacon JSD 356, 359, *384*
Davies C, see Reid JSG 107, *128*
Davies DR 328, *330*
Davies DR, Lyall V 22, *23*, 146, *157*
Davies DR, Roberts K 146, *157*
Davies MD, Dickinson DB 79, *117*
Davies R, see Kidby DK 367, *389*, 477, *479*
Davignon ML 104, *117*
Davis BD 201, *222*
Davis BD, Tai RC 490, 491, *508*
Davis BD, see Smith WP 490, *511*
Davis R, see Shaykh M 230, 239, *253*
Davis RW, see Dean BB 225, *249*
Davis RW, see Espelie KE 225, 226, 230, 231, *250*
Davis RW, see Soliday CL 242, 247, *253*, 529, 531, *540*
Davis TE, Domer JE 363, *386*
Davis TE, Domer JE, Li YT 431, *434*
Dawes CJ 293, *308*
Dawes CJ, see Claire JWL 295, *308*
Dawson R, see Kristen U 579, *583*
Day PR 585, *617*
Dayhoff MO, Orcutt BC 141, *157*
Dazzo FB 152, *158*, 642, 645, *653*
Dazzo FB, Brill WJ 642, *653*, 663, 666, 667, 668, 669, 671, *674*, *675*
Dazzo FB, Hubbell DH 663, 666, 670, 673, *675*
Dazzo FB, Napoli CA, Hubbell DH 670, *675*
Dazzo FB, Urbano MR, Brill WJ 667, *675*
Dazzo FB, Yanke WE, Brill WJ 633, 642, *653*, 668, 669, *675*
Dazzo FB, see Umali-Garcia M 568, *574*
De Bokx JA 112, *117*

De Bruyne CK, see Loontiens FG 631, *655*
De DN, see Maitra SC 545, 548, *572*
De Jong B, see Slob A 613, *622*
De Ley J, see Van Beeumen J 595, *623*
De Renobales M, see Jones THD 430, *435*
De Vries H, see Heinen W 237, *250*
De Vries OMH, Wessels JGH *434*
De Vries OMH, see Wessels JGH 427, 429, *440*
De Wit PJGM, Roseboom PHM 605, *617*
De Zoeten GA, see Sequeira L 614, *622*
Dea ICM 68, *117*
Dea ICM, Morris ER, Rees DA, Welsh EJ, Barnes HA, Price J 37, *43*
Dea ICM, Rees DA, Beveridge RJ, Richards GN 39, *42*
Dean BB, Kolattukudy PE 225, 236, *249*
Dean BB, Kolattukudy PE, Davis RW 225, *249*
Dean BB, see Espelie KE 229, 230, *249*
Dean BB, see Kolattukudy PE 235, 236, *251*
Dean BB, see Soliday CL 236, *253*
Deas AHB, Baker EA, Holloway PJ 228, 231, *249*
Deas AHB, Holloway PJ 231, *249*, 529, *534*
Deas AHB, see Holloway PJ 229, *250*, 230, *250*
Deason TR, see O'Kelley JC 348, *350*
Dee PC, see Habener JF 492, *508*
Deichgräber G, see Schnepf E 318, *332*, 326, 525, 526, *539*
Dekazos ED 112, *117*
Dekazos ED, Worley JF 112, *117*
Del Rey FD, Santos T, Garcia-Acha I, Nombela C 378, *386*
Del Rey FD, see Santos T 377, *392*, 423, *438*
Del Rey FD, Garcia-Acha I, Nombela C 423, 430, 431, *434*
Delmer DP 47, 50, *62*, 281, *308*, 521, *534*
Delmer DP, Albersheim P 50, *62*
Delmer DP, Beasley CA, Ordin L 54, *63*
Delmer DP, Heininger U, Kulow C 54, 56, *63*
Delmer DP, Kulow C, Ericson MC 170, *189*
Delmer DP, Lamport DTA 147, 148, *158*
Delmer DP, see Carpita N 68, 101, *116*, 154, *157*
Delmer DP, see Ericson MC 57, *63*, 70, *118*, 175, *189*
Delmer DP, see Heiniger U 53, 54, *64*, 72, *120*
Delmer DP, see Maltby D 93, *123*
Delmer DP, see Meinert MC 32, 40, *44*, 48, *65*, 513, *537*
Delwiche CC, see Woodwell GM 11, *24*
Den Hartog-Adams I, see Parke M 347, *351*
Dennis DT, Preston RD 279, *308*
Derrenbacker EC, see Punnett T 144, 147, *162*
Des S, see Thomas D 72, *130*
Desai NN, see Allen AK 135, 136, *156*, 631, *652*, 633, 635, 636
DeSantis A, see Arrigoni O 142, *156*
Desphande BP, see Evert RF 514, 517, *534*
DeVay JE, Adler HE 585, *617*, 671, *675*
DeVay JE, Charudattan R, Wimalajeewa DLS 586, *617*
DeVay JE, see Wiese MV 593, *623*
Deven JM, see Letourneau DR 356, 363, 366, *389*
Devillers-Thiery A, Kindt T, Scheele G, Blobel G 492, *508*
Dewar ET, see Black WAP 296, *307*
Dey PM 139, *158*
Dey PM, Pridham JB 139, *158*
Deysson G, see Benbadis M-C 519, *533*
Dias HF, see Stobbs LW 112, *130*
Diatloff A 660, *675*
Diboll AG, see Parker BC 278, *313*
Dickerson AG 595, *617*
Dickerson AG, Mantle PG, Nisbet LJ, Shaw BI 595, *617*
Dickerson AG, see Buck KW 358, *385*, 427, *433*
Dickinson DB, see Davies MD 79, *117*
Dickinson DB, see Hopper JE 79, *121*
Dickinson DB, see Pharr DM 105, *126*
Dickinson DB, see Southworth D 48, 53, *66*, 72, *129*
Dickson JG, Shands HL 106, *117*
Dietrich CP, see Plessmann Camargo E 398, 399, 410, *414*
Digby J, Firn RD 267, *270*
Dijkstra J, see Shimomura T 112, *129*
Dimitman JE, see Wu JH 112, *132*
Dimond AE, see Horsfall JG 594, *619*
Dimond AE, see Patil SS 594, *621*
DiNinno VL, McCandless EL 294, *308*
DiNinno VL, see McCandless EL 294, *312*
Diwald K 334, *349*
Dixon JFP, see O'Brien RL 628, 637, 638, *656*
Dixon PS, see Parke M 279, *313*
Dixon ROD 660, *675*
Dobberstein B, Kiermayer O 325, *330*
Dobberstein B, see Blobel G 468, *478*, 489, *507*
Dobberstein B, see Kiermayer O 81, *122*, 325, *331*
Dodel A 334, 347, *349*
Dodge JD 144, *158*
Doi A, see Doi K 424, *434*, 425
Doi E, see Hata T 401, *412*, 451, *456*

Doi K, Doi A, Fukui T 425, *434*
Doi K, Doi A, Ozaki T, Fukui T 424, *434*
Doke N, Garas NA, Kuć J 588, 603, 605, 606, *617*
Doke N, Tomiyama K 588, 603, 606, *617*
Doke N, see Furuichi N 587, *618*
Doke N, see Garas NA 588, *618*
Doke N, see Kuć J 614, *620*
Doke N, see Nozue M 587, *621*
Dolan TCS, see Anderson NS 293, *307*
Domer JE, see Davis TE 363, *386*, 431, *434*
Dominguez JB, Goni FM, Uruburu F 428, *434*
Dommergues YD 660, *675*
Donaldson LA, see Hall DM 530, *535*
Doner LW, Whistler RL 298, *308*
Dooijewaard-Kloosterziel AMP, Sietsma JH, Wouters JTM 374, *386*
Dörfel H, see Fischer FG 287, *309*
Dortland JF, see Hänish Ten Cate ChH 102, 103, *120*
Doty MS, see Santos GA 293, *315*
Dougal DK, Shimbayashi K 134, *158*
Douglas KT, see Margolis HC 451, 452, *457*, 496, *509*
Douglas LJ, see Ramsay AM 417, *438*
Douglas RH 444, *455*
Douglas RH, Ballou CE 444, *456*
Dow JM, Callow JA 605, *617*
Dowler MJ, see Johnson KD 139, 150, *159*, 266, *271*
Dreher TW, Grant BR, Wetherbee R 281, 306, *308*
Drewes LR, see Gander JE 467, *479*, 467, *479*
Dryburgh H, see Schreiber G 493, *511*
Dubremetz JF, see Poulain DG 367, *391*, 419, *438*
Duckworth M, Yaphe W 291, *308*
Duckworth M, see Young K 291, *316*
Duda CT, see Maclachlan GA 92, *123*
Dudman WF 671, *675*
Dudman WF, Brockwell J 671, *675*
Duff RB 34, *43*, 360, *386*
Dumas C, Perrin A, Rougier M, Zandonella P 555, *570*
Dumas C, see Fèvre M 432, *434*
Dunlop PC, Meyer GM, Ban D, Roon RJ 449, *456*
Dunn GM, see Couture RM 611, *617*
Durán A, Bowers B, Cabib E 380, *386*, 403, 404, *412*
Durán A, Cabib E 399, 400, 401, 402, 403, *412*
Durán A, Cabib E, Bowers B 404, *412*
Durán A, see Cabib E 397, 401, *411*
Durán A, see Molano J 405, *414*, 649, *655*
Durand R, see Mialonier G 644, *655*
Durbin ML, see Sexton R 103, *129*
Durbin RD, Uchytil TF 612, *617*
Durbin RD, see Arneson PA 612, *615*
Durell LW 373, 374, *386*
Durham DR, see Merkel GJ 148, *161*
Dürr M, Bailey DS, Maclachlan GA 70, 71, 81, 83, *117*
Duskin D, Bornstein P 184, *189*
Duskin D, see Mahoney WC 183, *191*
Dustin P 324, *330*
Dutton CGS, Kabir MS 31, *43*

Eagon PC, Heath EC 179, 186, *189*
Ebel J, Grisebach H 207, 208, *222*
Ebel J, see Ayers AR 603, 605, *615*
Eberhardt G, Schubert WJ 201, *222*
Ebert E, Zenk MH 427, *434*
Ebner KE 80, *117*
Ebner KE, see Lingappa VR 181, *191*
Echlin P, Godwin H 99, *117*
Eckhardt K, Thrum H, Bradler G, Tonew E, Tonew M 185, *189*
Eda S, Katō K 32, 34, *43*
Eda S, Ohnishi A, Katō K 31, *43*
Eda S, see Kato K 7, *8*
Eda S, see Mori M 32, *44*
Eddy AA, Woodhead JS 418, *434*
Edelman J, Ginsburg V, Hassid WZ 54, *63*
Edelson P, Cohn Z 672, *675*
Edrees M, see Abdel-Fattah AF 300, *306*
Edwards K, see Schreiber G 493, *511*
Edwards TE, see Peat S 417, *437*, 441, *457*
Eeltink H, see Klis FM 135, *160*, 513, *536*
Egel R, see Crandall M 421, *434*
Egel R, see Kröning A 377, *389*, 422, *436*
Egerraat AWSM van 661, *675*
Eglinton G, Hunneman DH 227, *249*
Eglinton G, see Caldicott AB 226, *248*, 228, 230
Eglinton G, see Hunneman DH 226, 228, 230, *250*
Eguchi J, see Hori M 398, 400, *413*
Ehmann C, see Wolf DH 403, *415*
Eisinger WR, see Morré DJ 260, 263, *271*
Eisinger WR, see Ray PM 71, 81, 83, *127*, 519, 521, *539*
Eisinger WR, see Robinson DG 83, 84, 85, *128*, 519, *539*
Elahi M, see Tahir MA 104, *130*
El-Banoby FE, Rudolph K 586, 594, *617*
Elbein AD 70, 71, 73, *118*, 168, *189*
Elbein AD, Adya S, Lee YC 467, *478*
Elbein AD, Barber GA, Hassid WZ 51, 52, *63*

Elbein AD, Forsee WT 47, 48, 53, *63*, 72, *118*
Elbein AD, Gafford J, Kang MS 185, *189*
Elbein AD, see Adya S 467, *477*, 495, *507*
Elbein AD, see Barber GA 49, 51, 52, *62*
Elbein AD, see Chambers J 52, *62*, 73, *116*, 177, *189*
Elbein AD, see Ericson MC 73, *118*, 173, 183, 186, *189*
Elbein AD, see Forsee WT 58, *63*, 170, 171, 173, 175, *190*
Elbein AD, see Heifetz A 170, 173, 175, *190*
Elbein AD, see James DW Jr 185, *190*
Elbein AD, see Kang MS 177, 186, *190*
Elbein AD, see Laine R 172, *191*
Elbein AD, see Lin T-Y 71, *123*
Elbein AD, see Pan YT 184, *192*
Elbein AD, see Rudick MJ 467, *482*
Elbein AD, see Spencer JP 177, *192*, 186, *192*
Elbers PF, see Linnemans WAM 367, *389*, 469, 477, *480*, 498, 504, *509*
Ellias EA, see Mullins JT 377, *390*
Ellingboe AH 585, *617*
Ellingboe AH, see Clarke JA 142, *157*
Ellis BE, see Brisson JD 547, 555, *570*
Ellis RJ, see Highfield PE 490, *508*
Elnaghy MA, Nordin P 49, *63*
Elorza MV, Larriba G, Villanueva JR, Sentandreu R 498, *508*
Elorza MV, Ruiz EM, Villanueva JR 424, *434*
Elorza MV, Sentandreu R 471, *478*
Elorza MV, see Larriba G 470, 476, *480*, 494, 504, *509*
Elorza MV, see Rodríguez L 499, *511*
Elorza MV, see Sentandreu R 432, *438*, 504, *511*
Elstner EF, Heupel A 150, *158*, 213, *222*
Elstner EF, see Gross GG 41, *43*, 150, *159*, 214, *222*
El-Tigani S, see Butcher DN 613, *616*
Emori M, see Yoshioka Y 362, *394*
Emr SD 491, *508*
Ende H van den, see Datema R 353, 355, 357, 361, 363, 366, *386*, 353, 355, 357, 366, *386*
Endo A, Kakiki K, Misato T 400, 408, *412*
Endo A, Kuroda M, Tanzawa K 187, *189*
Endress AG, see Thomson WW 225, *253*
Engels FM 83, *118*, 525, *534*
Englar JR, see Whyte JNC 281, *316*
Engleman EM 112, *118*
English PD, Jurale JB, Albersheim P 589, 591, *617*
English PD, Maglothin A, Keegstra K, Albersheim P 589, 591, *618*
English PD, see Albersheim P 643, *652*
English PD, see Bateman DF 589, *616*
Ensgraber A, see Krupe M 667, *676*
Ephritikhine G, see Kemp J 56, *64*, 74, *122*
Erickson SS, see Dashek WV 150, *157*
Ericson MC, Chrispeels MJ 169, *189*
Ericson MC, Delmer DP 57, *63*, 70, *118*, 175, *189*
Ericson MC, Gafford J, Elbein AD 73, *118*, 173, 183, 186, *189*
Ericson MC, see Delmer DP 170, *189*
Erikson KE, Hamp SG 373, *386*
Eriksson KE, see Bjorndal H 282, *307*
Erwin DC, see Keen NT 597, 600, *619*
Esau K 9, *23*, 97, 99, *118*
Esau K, Cheadle VI, Gill RH 83, 94, *118*
Esau K, Cronshaw J 112, *118*
Esau K, Gill RH 514, *534*
Esau K, see Cronshaw J 517, *534*
Eschrich W 48, *63*, 98, 99, 100, 110, 111, 112, *118*
Eschrich W, Currier HB, Yamaguchi S, McNairn RB 112, *118*
Eschrich W, Hüttermann A, Heyser W, Tammes PML, Van Die J 98, *118*
Eschrich W, see Evert RF 531, *534*
Eshdat Y, Ofek I, Yashouv-Gan Y, Sharon N, Mirelmann D 633, *653*
Espelie KE, Davis RW, Kolattukudy PE 225, 226, 230, 231, *250*
Espelie KE, Dean BB, Kolattukudy PE 229, 230, *249*
Espelie KE, Kolattukudy PE 228, 229, 230, 231, 236, *249*, 531, *534*
Espelie KE, Sadek NZ, Kolattukudy PE 239, 242, 247, *249*
Esquerré-Tugayé MT, Lafitte C, Mazau D, Toppan A, Touzé A 111, *118*, 141, *158*, 608, *618*
Esquerré-Tugayé MT, Lamport DTA 135, 141, *158*, 608, *618*
Esquerré-Tugayé MT, Mazau D 141, *158*
Etchinson JR, see Hunt LA 463, *479*, 497, *508*
Ettl H 347, *349*
Ettl H v 145, *158*
Etzler ME, see Biroc SL 635, *653*
Etzler ME, see Talbot CF 644, 652, *656*, *657*, 668, *676*
Evans JM, see Peat S 417, *437*
Evans LV 144, *158*
Evans LV, Callow ME 298, *308*, 317, *330*
Evans LV, Callow ME, Coughlan SJ 298, *309*
Evans LV, Callow ME, Percival E, Fareed V 317, *330*
Evans LV, Callow ME, Percival Elizabeth, Sheik-Fareed V 303, *308*

Evans LV, Holligan MS 278, 287, *308*, 317, *330*
Evans LV, Simpson M, Callow ME 317, *330*
Evans LV, see Callow ME 278, 298, *308*, 317, *329*, 317
Evans LV, see Coughlan SJ 317, *330*
Evans ML 139, *158*, 266, *270*
Evans PH, see Fry WE 613, *618*
Evans PJ, Hemming FW 170, *190*
Evans RB, Manners DJ 419, *434*
Evelegh MJ, see McCandless EL 294, *312*
Eveleigh DE, see Sietsma JH 354, 355, 358, *392*
Evert RF 98, 99, *118*
Evert RF, Desphande BP 514, 517, *534*
Evert RF, Eschrich W, Heyser W 531, *534*
Ewald CM, Perlin AS 74, *118*
Ewings D, see Verma DPS 267, *273*
Eyden-Emons van A, see Sassen A 324, *332*
Eylar EH 471, *479*, 493, *508*
Eylar EH, see Caccam JJ 460, *478*
Eyler EH, see Caccam JF 170, *189*

Faccioli G, see Panopoulos NJ 607, *621*
Facey V 102, *118*
Fagerson IS, see Croteau R 230, *249*
Fahn A 101, *118*
Fahraeus G, see Sahlman K 670, *676*
Faigle H, see Kratzl K 201, *223*
Falk H, see Brown RM 318, *329*, 346, *349*
Falk M, see MacLachlan J 326, *331*
Falk SO, Hertz CH, Virgin HI 261, *270*
Fan DF, Feingold DS 81, *118*
Fan DF, Maclachlan GA 150, *158*, 267, *270*
Fang F, see Gander JE 467, *479*
Fanshier D, see Neufeld EF 81, *125*
Fareed V, see Evans LV 317, *330*
Farkaš V 353, 379, 381, *386*, 409, *412*, 431, *434*, 441, *456*, 459, *479*
Farkaš V, Biely P, Bauer Š 378, *386*, 421, 423, *434*
Farkaš V, Kovařik J, Kosinová A, Bauer Š 469, *479*
Farkaš V, Svoboda A, Bauer Š 471, 473, *479*
Farkaš V, Vagabov VM, Bauer Š 464, 473, 476, *479*
Farkaš V, see Balint Š 504, *507*
Farkaš V, see Bálint Š 432, *433*
Farkas V, see Cabib E 380, *386*, 401, 402, 406, 408, *411*
Farkaš V, see Košinová A 469, *479*
Farmer VC, see Bacon JSD 107, *114*, 358, 359, *384*, 360, 367, *384*, 417, 418, *433*, 419, *433*
Fartaczek F, see Lehle L 173, *191*, 462, *480*
Fasulo MP, Vannini GL, Bruni A, Dal'Olio G 60, *63*
Fehrenbacher G, Perry K, Thorner J 453, *456*
Feingold DS, Neufeld EF, Hassid WZ 48, 51, 52, *63*, 70, 73, 81, *118*
Feingold DS, see Ankel H 79, *114*
Feingold DS, see Fan DF 81, *118*
Feingold DS, see Neufeld EF 81, *125*
Felipe Anton MR, Lopez-Fando C 565, *570*
Felipe Anton MR, see Sueiro SF 555, *573*
Felsted RL, see Leavitt RD 635, *655*
Fencl Z 375, *386*
Fene RE, see Rothman JE 182, *192*
Fergus BJ, see Scott JAN 197, *224*
Ferguson AL, see Pueppke SG 644, *656*, 667, *676*
Ferguson IB, Clarkson DT 225, *250*
Ferrari TE, Bruns D, Wallace DH 580, *582*
Ferry JD 257, 259, *270*
Fèvre M 377, 378, *387*, 430, *434*
Fèvre M, Dumas C 432, *434*
Feyen V 347, *349*
Fideli W, see Ballio A 597, 598, *615*
Fiedler F, Glazer L 255, 264, *270*
Field C, Schekman R 469, *479*
Field C, see Novick P 469, *481*
Fielding AH, see Hislop EC 589, 591, *619*
Fincher GB 106, *119*
Fincher GB, Sawyer WH, Stone BA 135, 152, 155, *158*
Fincher GB, Stone BA 106, *119*
Fincher GB, see Cook JA 72, 88, *116*
Fincher GB, see Raymond Y 48, 54, *66*, 73, 111, *127*, 521, *539*
Fink GR 447, *456*
Fink GR, see Wolf DH 452, *458*
Finkle BJ, Masri MS 205, *222*
Finkle BJ, Nelson RF 205, *222*
Firn RD, see Digby J 267, *270*
Fischer FG, Dörfel H 287, *309*
Fisher DA, Bayer DE 530, *534*
Fisher IS, Percival Elizabeth 299, *309*
Fleet GH, Manners DJ 359, *387*, 417, 418, 425, *434*
Fleet GH, Phaff HJ 379, *387*, 416, 420, 421, 422, 423, 424, 425, *435*
Fleet GH, see Kopecká M 16, *23*, 418, 425, *436*
Fleet GH, see Rombouts FM 424, 425, *438*
Fleming EN, see Burgess J *23*, 30, *42*, 85, *116*
Flemming M, see Schuster G 112, *129*
Fliege R, Flügge UI, Werdan K, Heldt HW 607, *618*

Fling M, see Horowitz NH 374, *388*
Flor HH 585, *618*
Flores-Carreon A, Gomez-Villaneuva A, San-Blas G 431, *435*
Flores-Carreon A, see McMurrough I 398, *414*
Flowers HM, Batra KK, Kemp J, Hassid WZ 51, 52, 56, *63*, 72, 73, *119*
Floyd RA, Ohlrogge AJ 549, 550, 551, 552, 555, 562, 566, 567, *570*
Flügge UI, Freisl M, Heldt HW 607, *618*
Flügge UI, see Fliege R 607, *618*
Foda MS, Phaff HJ 419, *435*
Fogarty WM, Ward OP 104, *119*
Fogg GE, see Parker BC 358, *391*
Ford CW, Percival Elizabeth 283, *309*
Forer A, Jackson WT 516, *534*
Forrence LE, see Abeles FB 600, 608, *615*
Forrest IS, Wainwright T 92, *119*
Forsee WT, Elbein AD 58, *63*, 170, 171, 173, 175, *190*
Forsee WT, Valkovich G, Elbein AD 173, *190*
Forsee WT, see Chambers J 177, *189*
Forsee WT, see Elbein AD 47, 48, 53, *63*, 72, *118*
Fosket DE, see Rubery PH 220, *224*
Foster RC, Rovira AD 565, *570*
Fouad ST, see Abdel-Fattah AF 296, *306*
Fouad ST, see Hussein MM-D 296, *310*
Fowell R 453, *456*
Fowke HC, Pickett-Heaps JD 83, *119*
Fowke LC 30, *43*
Fowke LC, see Marchant HJ 347, *350*
Fowke LC, see Ohyama K 586, *621*
Foyle RAJ, see Percival Elizabeth 304, *314*
Franich RA, see Wells LG 530, 531, *540*
Franke H, see Mansell RL 209, *223*
Franke P, see Mańkowski T 474, *480*
Franke WW, Brown RM 318, *330*
Franke WW, Herth W 517, *534*
Franke WW, see Brown RM 318, *329*, 320, *329*, 346, *349*
Franke WW, see Brown RM Jr 144, *157*
Franke WW, see Herth W 86, *120*, 144, *159*, 318, *330*, 582, *582*
Franke WW, see Morré DJ 82, *125*
Frankel R, Izhar S, Nitsan J 100, *119*
Frankel R, see Izhar S 100, 101, *121*
Franz E, Schiebold E, Weygand C 522, *534*
Franz G 32, *43*, 48, 49, 53, 56, *63*, 71, 74, 93, *119*, 525, *534*
Franz G, Meier H 52, 54, 55, *63*
Franz G, see Buchala AJ *42*, 48, *62*
Franz G, see Heiniger U 50, *64*, 79, *120*
Franz G, see Huwyler HR 48, *64*, 93, *121*, 513, *536*
Franz G, see Villemez CL Jr 48, 51, 52, *67*
Fraser CG, Wilkie KCB 31, *43*
Fraser CG, see Buchala AJ 31, *42*
Fraser T, Bruce BJ 491, *508*
Frazier WA, Rosen SD, Reitherman W, Barondes SH 633, 639, *653*
Frazier WA, see Reitherman RW 639, *656*
Frazier WA, see Rosen SD 639, *656*
Fred EB, Stevens JW 663, *675*
Fred EB, see Baldwin IB 663, *674*
Frederick LR, see Ham GE 661, *675*
Frederick LR, see Weaver RW 660, *677*
Freeman B, Albrigo LB, Biggs RH 530, *534*
Frei E, Preston RD 277, 278, 280, 281, 282, 283, *309*
Freisl M, see Flügge UI 607, *618*
French A, see Juniper BE 545, 548, *571*
Frenkel C, see Ben-Arie R 103, *115*
Frensdorff A, see Galun M 408, *412*
Freudenberg K 209, 210, 212, *222*
Freudenberg K, Neish AC 36, *43*, 210, 215, 217, *222*
Freundlich A, Robards AW 524, *534*
Frey R, see Haass D 647, 648, *654*
Frey-Wyssling A 9, 10, 12, 15, 19, *23*, 29, *43*, 529, *534*
Frey-Wyssling A, López-Sáez JF, Mühlethaler K 518, *534*
Frey-Wyssling A, Mühlethaler K 529, *534*
Frey-Wyssling A, see Albersheim P 83, *114*
Frey-Wyssling A, see Matile P 422, *436*, 469, *481*, 498, *509*
Friedmann I, see Ocampo-Paus R 346, *350*
Friend J 111, *119*
Friend J, see Mankarios AT 150, *161*
Friend J, see Marcan H 603, 605, *620*
Fromm H, see Gerisch G 638, *653*
Fromme I, see Müller K 640, *655*
Fry DJ, see Wishart GJ 82, *131*
Fry SC 149, 150, *158*
Fry WE, Evans PH 613, *618*
Fu Y-L, Gutmann PJ, Timell TE 109, *119*
Fuchs E, see Sarvas M 490, *511*
Fuchs Y, see Pesis E 103, *126*
Fuensanta R, see Santamaria F 427, *438*
Fujihara S, Yamamoto R, Masuda Y 259, 260, *270*
Fujii T, Shimokoriyama M 142, *158*

Fujikawa T, Nakashima K 296, *309*
Fujikawa T, Wada M 296, *309*
Fujimoto K, see Chapman A 177, 186, *189*
Fujimura T, see Komamine A 41, *44*
Fujioka K, see Obata T 424, 425, *437*
Fujisawa F, Iwakura Y, Kawade Y 493, *508*
Fujita M, Harada H 197, *222*
Fukazawa K, see Imagawa H 195, 196, *223*
Fukuda M, Yamaguchi T 151, *158*
Fukui L, see Hayano K 595, *619*
Fukui S 595, *618*
Fukui S, Hochster RM 595, *618*
Fukui S, see Hayano K 595, *619*
Fukui T, see Doi K 424, *434*, 425
Fukumoto J, see Siegel S 109, *129*
Fukuoka F, see Nishikawa Y 427, *437*
Fukuoka F, see Yoshioka Y 362, *394*
Fukushima T, Yamazaki M 110, *119*
Fulcher RG, McCully ME, Setterfield G, Sutherland J 90, 91, *119*, 518, *534*
Fulcher RG, O'Brien TP, Lee JW 107, *119*
Fuller MS 355, *387*
Fuller MS, Barshad I 355, *387*
Fuller MS, see Aronson JM 358, *384*
Fuller MS, see Darley WM 317, *330*
Fulmer A, see Arnott S 294, *307*
Fulton AB 145, *158*
Fultz SA, Sussman AS 370, *387*
Fultz SA, Woolf RA 369, *387*
Funkuda M, see Toyoshima S 634, 637, *657*
Furtado JS, Olive LS 317, *330*
Furuichi N, Tomiyama K, Doke N 587, *618*
Fushiki H, see Nakamura Y 218, *223*

Gaard G, see Sequeira L 614, *622*
Gabriel M 347, *349*
Gafford J, see Elbein AD 185, *189*
Gafford J, see Ericson MC 73, *118*, 173, 183, 186, *189*
Gagnon J, see Palmitter RD *510*
Gale EF, see Cassone A 418, *434*
Gallant D, see Rougier M 553, *573*
Galling G 348, *349*
Galston AW, see Kinnersley AM 18, *23*
Galun E 408, *412*
Galun E, see Galun M 408, *412*
Galun E, see Mirelman D 649, *655*
Galun M, Braun A, Frensdorff A, Galun E 408, *412*
Gamborg OL 201, *222*
Gamborg OL, Neish AC 202, *222*
Gancedo C, see Gancedo JM 361, *387*
Gancedo JM, Gancedo C, Asensio C 361, *387*
Gander JE 352, 362, 364, 381, *387*, 592, *618*
Gander JE, Drewes LR, Fang F, Lui A 467, *479*
Gander JE, Fang F 467, *479*
Gander JE, Jentoft NH, Drewes LR, Rick PD 467, *479*
Gannon F, see Mercereau-Puijalon O 491, *509*
Gansera R, Schurz H, Rüdiger H 645, *653*
Garancis JC, see Ankel H 466, *477*
Garapin A, see Mercereau-Puijalon O 491, *509*
Garas NA, Doke N, Kuć J 588, *618*
Garas NA, see Doke N 588, 603, 605, 606, *617*
Garas NA, see Kuć J 614, *620*
Garcia RC, Recondo E, Dankert M 528, *534*
Garcia Acha I, see Villanueva JR 368, *393*
Garciá Acha I, see Ramos S 369, *391*
García Acha I, see Benitez T 400, 408, *411*
Garcia-Acha I, see Del Rey FD 378, *386*, 423, 430, 431, *434*
Garcia-Arenal F, Sagasta EM 111, 112, *119*
Garcia Mendoza C, Leal JA, Novaes-Ledieu M 374, *387*
Gardiner M, Chrispeels MJ 137, *158*, 519, *534*
Gardner KH, Blackwell J 13, *23*, 47, *63*
Gardner PE, see Bishop CT 419, *433*
Garegg PJ, see Bjorndal H 282, *307*
Garoff H, Schwarz RT 184, *190*, 493, *508*
Garzuly-Janke R 371, *387*
Gascoigne JA, Gascoigne MM 13, *23*
Gascoigne MM, see Gascoigne JA 13, *23*
Gascón S, see Liras P 471, 473, *480*
Gaskins MH, see Umali-Garcia M 568, *574*
Gassmann G, see Müller DG 347, *350*
Gawlik SR, Millington WF 346, 347, *349*
Gay JL, see Heath IB 382, *388*, 501, *508*
Gay MR, see Hills GJ 22, *23*, 86, *120*, 145, *159*, 328, *330*
Gay MR, see Roberts K 146, *163*
Gazeau C-M 518, *534*
Geissmann T, Neukom H 76, *119*
Geitler L 347, *349*
Georg-Plant MMT, see Brading JWE 300, *308*
Gerisch G 641, *653*
Gerisch G, Fromm H, Hoesgen A, Wick U 638, *653*
Gerisch G, Krelle H, Bozzaro S 641, *653*
Gerisch G, see Bozzaro S 638, 640, *653*

Gerisch G, see Müller K 640, 641, *655*
Gertel ET, Green PB 13, *23*
Ghosh AP, see Irving RA 180, *190*
Ghosh HP, see Toneguzzo F 180, *193*
Ghosh NR, see Bidwell RGS 298, *307*
Ghosh S, Blumenthal HJ, Davidson E, Roseman S 398, *412*
Ghuysen JM, see Strominger JL 340, *351*
Gibbons AP, see Brummond DO 51, 52, *62*
Gibbons GC 106, *119*
Gibbs M, see Kelly GJ 607, *620*
Gibor A, see Ray DA 145, *163*, 341, *351*
Gibor A, see Solter KM 341, *351*
Gibson EJ, see Johnson BF 469, *479*
Gibson R, Schlessinger S, Kornfeld S 493, 496, *508*
Giddings TH, Brower DL, Staehelin LA 325, 326, *330*, 524, *534*
Giddings TH Jr, Brower DL, Staehelin LA 94, *119*
Giddings TH, see Norman J 398, *414*
Giebel J, Stobiecka M 141, *158*
Gietl C, Kauss H, Ziegler H 633, 648, *654*
Gietl C, Ziegler H 633, 644, 645, 648, 649, *653*, *654*, 668, *675*
Gil F, see Carbonell LM 361, 365, *386*
Gil F, see Kanetsuna F 372, *388*
Gilbert WA, O'Kelley JC 348, *349*
Gilchrist AJ, see Juniper BE 542, 545, 549, *571*
Gilkes NR, Hall MA 33, 35, 40, *43*, 59, *63*, 92, *119*, 151, *158*
Gill AM, Tomlinson PB 542, *570*
Gill F, see Kanetsuna F 429, *435*
Gill RH, see Esau K 83, 94, *118*, 514, *534*
Gilliland DG, Collier RJ, Moehring JM, Moehring TJ 650, *654*
Gilula NB, see Branton D 321, *329*
Gimenez-Martin, see López-Sáez JF 516, *536*
Ginsburg V, see Edelman J 54, *63*
Ginsburg V, see Neufeld EF 81, *125*
Ginzburg BZ 90, *119*
Girbardt M 382, *387*, 500, *508*
Glaser C, see Kauss H 36, *44*, 86, *122*, 141, *160*, 647, *655*
Glaser L 50, 52, *63*
Glaser L, Brown DH 379, *387*, 398, 399, *412*
Glazer L, see Fiedler F 255, 264, *270*
Gleeson P, see Clarke AE 155, *157*
Gleeson PA, Clarke AE 155, *158*, 578, 579, *582*
Gleeson PA, Jermyn MA 153, *158*, 578, *582*
Gleeson PA, Jermyn MA, Clarke AE 578, *582*
Gleeson PA, see Clarke AE 577, *582*, 578, *582*, 628, *653*
Gleeson PA, see Vithanage HIMV 110, *131*, 581, *583*
Glover R, see Smidsrod O 286, *315*
Godwin H, see Echlin P 99, *117*
Goeschl JD, Pratt HK, Bonner BA 143, *158*
Goetsch L, see Byers B 501, *507*
Goff CW 558, 565, *570*
Goff CW, see Paul DC 516, *538*
Gold AH, see Panopoulos NJ 607, *621*
Gold MH, Hahn HJ 468, 472, *479*
Gold MH, Mitzel D, Segel I 370, *387*
Goldberg M, see Hackman RH 355, *387*, 395, *412*
Goldberg R 266, 268, *270*
Goldschmid HR, Perlin AS 74, *119*
Goldstein D, see Katz D *389*
Goldstein IJ, Hayes CE 630, 631, 632, 635, 636, 649, *654*
Goldstein JL, Brown MS 168, *190*
Goldstein JL, Hegelson JAS, Brown MS 187, *190*
Goldstein JL, see Brown MS 651, *653*
Gomez-Villaneuva A, see Flores-Carreon A 431, *435*
Goni FM, see Dominguez JB 428, *434*
Gooday GW 317, *330*, 356, 366, 369, 379, 380, 382, 383, *387*, 400, 402, 408, 409, *412*
Gooday GW, Rousset-Hall A de 398, 399, 409, *412*
Gooday GW, Rousset-Hall A, de Hunsley D 400, 408, *412*
Gooday GW, Trinci APJ 255, 264, *270*
Gooday GW, see Hunsley D 363, 367, 370, *388*
Gooday GW, see Rousset-Hall A de 398, 399, *414*
Gooday GW, see Rousset-Hall A de 379, *392*
Goodenough DA, see Weiss RL 340, *351*
Goodenough PW, Kempton RJ 590, 591, *618*
Goodenough U 642, *654*
Goodenough UW, Weiss RL 340, *349*
Goodenough UW, see Minami SA 329, *331*
Goodenough UW, see Weiss RL 340, *351*
Goodman RN, Huang JS, Huang PY 596, 600, *618*
Goodman RN, see Politis DJ 608, *621*
Goodwin RH, Taves C 60, *63*
Goosen-de Roo L, Van Spronsen PC 514, *535*
Gordon AH, see Bacon JSD 107, *114*, 424, 425, *433*
Gordon EM, McCandless EL 278, *309*
Gordon MP, see Watson B 586, *623*
Gordon SA, see Bara M 109, *115*

Gordon-Mills EM, McCandless EL 278, 294, 295, *309*
Gordon-Mills EM, see McCandless EL 294, 295, *312*
Goren R, see Huberman M 102, 103, *121*
Gorin PAJ 445, *456*
Gorin PAJ, Mazurek M, Spencer JFT 445, *456*
Gorin PAJ, Perlin AS 442, *456*
Gorin PAJ, Spencer JFT 284, *309*, 352, 358, 362, 367, *387*, 426, 427, 428, 429, *435*
Gorin PAJ, Spencer JFT, Bhattacharjee SS 445, *456*
Gorin PAJ, Spencer JFT, MacKenzie S 420, *435*
Gorin PAJ, see Alviano CS 364, *384*
Gorin PAJ, see Previato JO 372, *391*
Gorin PAJ, see Spencer JFT 441, *458*
Goring DAI, see Musha Y 197, *223*
Goring DAI, see Ruel K 198, *224*
Goring DAI, see Scott JAN 197, *224*
Göring H, Bleiss W, Schenk D, Kretschmer H 268, *270*
Got R, see Letoublon R 467, *480*
Got R, see Letoublon RCP 467, *480*, 501, *509*
Got R, see Letoublon RM 467, *480*
Got R, see Richard M 467, *481*
Gotelli IB, Cleland RE 144, 147, *159*
Gould SEB, Rees DA 108, *119*
Gould SEB, Rees DE, Richardson NG, Steele IW 37, *43*
Goulding KJ, see Northcote DH 144, *162*, 280, *313*
Graf L, see Yariv J 152, *165*, 577, *583*
Graham PH 671, *675*
Graham TL, Sequeira L, Huang TSR 586, *618*
Graham TL, see Sequeira L 587, *622*
Grambow GE, see Grambow HJ 248, *250*
Grambow HJ, Grambow GE 248, *250*
Grange DK, Adair WL Jr 167, 168, *190*
Granier C, see Ronin C 472, *482*
Graniti A, see Turner NC 597, *623*
Grant BR, see Dreher TW 281, 306, *308*
Grant BR, see Hawthorne DB 281, *310*
Grant BR, see Howard RJ 281, *310*
Grant GT, McNab C, Rees DA, Skerrett RJ 86, 101, *119*, 553, *570*
Grant GT, Morris ER, Rees DA, Smith JC, Thom D 287, *309*
Grant GT, Morris ER, Rees DA, Smith P, Thom D 37, *43*
Grasdalen H, Larsen B, Smidsrod O 285, 287, 288, *309*
Gray G, see Auriol P 598, 601, *615*
Greaves MP, Darbyshire JF 562, 565, 568, *570*
Green JR, Northcote DH 73, 101, *119*, *120*, 519, 520, *535*, 556, 557, 560, *570*
Green PB 255, *270*, 382, 383, *387*
Green PB, see Gertel ET 13, *23*
Green PB, see Ray PM 255, *272*
Green RJ Jr, see Mussell HW 593, *621*
Greenlagh JR, Mitchell ND 612, *618*
Greenland DJ 565, *571*
Greenwood AD, see Heath IB 382, *388*, 501, *508*
Greiling H, Vogele P, Kisters R, Ohlenbusch HD 449, *456*
Griesbach H 149, *159*
Griffin GJ, Hale MG, Shay FJ 566, *571*
Griffing LR, Ray PM 521, *535*
Griffiths DA, Griffiths DJ 346, *349*
Griffiths DJ, see Griffiths DA 346, *349*
Grimes WJ, Jones BL, Albersheim P 50, *63*
Grimm I, Sachs H, Robinson DG 324, *330*
Grimm I, see Robinson DG 324, *332*
Grimm I, see Sachs H 322, 324, *332*
Grinna LS, Robbins PW 497, *508*
Grisebach H, Hahlbrock K 221, *222*
Grisebach H, see Ebel J 207, 208, *222*
Grisebach H, see Hahlbrock K 208, 220, *222*
Groleau D, Yaphe W 291, *309*
Gronewald JW, Cheeseman JM, Hanson JB 601, *618*
Gross GG 149, 150, *159*, 207, 208, 209, 218, 220, *222*, 513, *535*
Gross GG, Janse C 150, 151, *159*
Gross GG, Janse C, Elstner EF 41, *43*, 150, *159*, 214, *222*
Gross GG, Mansell RL, Zenk MH 220, *222*
Gross GG, see Mansell RL 209, *223*
Gross GG, see Stöckigt J 208, *224*
Gross KC, Wallner SJ 104, *120*
Grossenbacher K, see Jenny H 565, 566, 567, *571*
Grout BWW 18, 20, *23*, 525, *535*
Grout BWW, see Willison JHM 522, 525, *541*
Grove SN 501, *508*
Grove SN, Bracker CE 382, *387*, 500, 503, *508*
Groves ST, see Ramus J 304, *314*
Gruber E, see Nowak-Ossorio M 48, 56, *65*, 524, *537*
Gruen HE, see Kitamoto Y 376, *389*
Grünwald J, Seemüller E 247, *250*

Grzybek H, see Burczyk J 346, 347, *349*
Guckert A, Breisch H, Reisinger O 553, 567, 568, *571*
Guckert A, see Breisch H 567, 568, *570*
Guern J, see Kurkdjian A 601, *620*
Guiseley KB 291, *309*
Gull K 366, *387*, 501, *508*
Gull K, Trinci PJ 370, *387*
Gull K, see Craig GD 409, *412*
Gunawardena P, Williamson FB 295, *309*
Gunja-Smith Z, Patil NB, Smith EE 419, *435*
Gunja-Smith Z, Smith EE 419, *435*
Gunning BES, Hardham AR, Hughes JE 515, 516, *535*
Gunning BES, Steer MW 19, 20, *23*, 82, *120*
Gunning BES, see Atkinson AW 322, *329*, 345, 346, *349*
Gunning BES, see Hardham AR 525, *535*
Gunning BES, see Hughes JE 112, *121*
Gunnison D, Alexander M 145, *159*
Gupta DP, Heale JB 589, 591, 592, *618*
Gupta LM, see Pollard MR 530, *538*
Gurevite E, see Levy EC 248, *251*
Gurney-Smith M, see Hills GJ 22, *23*
Gurney-Smith M, see Roberts K 22, *24*, 145, *163*, 328, *332*
Guseva TF, see Ovodov YS 296, *313*
Gutmann PJ, see Fu Y-L 109, *119*
Guyer R, see Schechter I 489, *511*

Haass D, Frey R, Thiesen M, Kauss H 647, 648, *654*
Habener JF, Rosenblatt M, Dee PC, Potts JT 492, *508*
Haberlandt G 542, 565, *571*
Habets-Willems C, see Reuvers F 462, 463, *481*
Habig WH, see Abeles FB 600, 608, *615*
Hackman RH, Goldberg M 355, *387*, 395, *412*
Haddon LE, Northcote DH 80, *120*
Hadwiger LA, Adams MJ, Beckman J 603, *618*
Hadwiger LA, see Daniels DL 606, *617*
Hager A, Menzel H, Krauss A 142, *159*, 256, 258, *270*
Hägglund E 109, *120*
Haggroth S, see Adler E 196, *221*
Hahlbrock K, Grisebach H 208, 220, *222*
Hahlbrock K, see Grisebach H 221, *222*
Hahn HJ, see Gold MH 468, 472, *479*
Hahn MG, Darvill AG, Albersheim P 603, 605, *618*
Haigh FC, see Johnson DC 41, *43*
Haigler C, Brown RM, Benziman M 25, 26, *28*
Haigler C, Brown RM Jr, Benziman M 528, *535*
Haigler C, see Benziman M 25, *28*
Hala CN, Mathers DG 660, *675*
Hale MG, see Griffin GJ 566, *571*
Hall AE, Loomis RS 606, *618*
Hall CW, see Neufeld EF 79, *125*
Hall D, Mollenhauer HH, Morré DJ 552, *571*
Hall DM 530, *535*
Hall DM, Donaldson LA 530, *535*
Hall JL, see Pope DG 516, 521, *538*, 562, 563, 564, *573*
Hall MA, Ordin L 59, *63*
Hall MA, see Darvill AG 92, *117*, 140, *157*, 264, 267, 268, *270*
Hall MA, see Gilkes NR 33, 35, 40, *43*, 59, *63*, 92, *119*, 151, *158*
Hall MA, see Mankarios AT 150, *161*
Hall MA, see Ordin L 51, 52, 59, *65*, 71, 72, *126*
Hall MA, see Thomas B 111, 112, *130*
Hall TC, see Sun SM 181, *193*
Hallam ND 530, *535*
Hallam ND, Juniper BE 529, 530, *535*
Halmer P, Bewley JD, Thorpe TA 107, *120*
Halvorson HO, see Tingle MA 420, *439*
Ham GE, Frederick LR, Anderson LC 661, *675*
Hamad-Fares I 568, *571*
Hamblin J, Kent SP 659, 663, *675*
Hamer GK, Bhattacharjee SS, Yaphe W 291, *309*
Hamer GK, see Bhattacharjee SS 292, 293, *307*
Hamilton RC, Bover FG, Mason TJ 526, *535*
Hammerschmidt R, Kuc J 152, *159*
Hammond JBW, see Wood DA 400, *415*
Hamp SG, see Erikson KE 373, *386*
Handa N 303, *309*
Hanic LA, Craigie JS 278, *309*
Hanic LA, see Brown RG 428, *433*
Hänish Ten Cate ChH, Van Netten J, Dortland JF, Bruinsma J 102, 103, *120*
Hanke DE, Northcote DH 18, *23*, 30, *43*, 528, *535*
Hanke DE, see Hill BS 652, *654*
Hankin L, Kolattukudy PE 237, *250*
Hankin L, see Zucker M 592, *623*
Hankins CN, Kindinger JI, Shannon LM 636, 652, *654*
Hankins CN, Shannon LM 648, 651, *654*
Haňkovská J, see Volfová A 525, *540*
Hanover JA, Lennarz WJ, Young JD 461, *479*
Hansen JE, see McCandless EL 294, *312*

Hanson B, Brody S 372, 373, *387*
Hanson JB, see Gronewald JW 601, *618*
Hantke K, see Braun V 670, *674*
Hapner KD, see Strobel GA 597, 598, *622*
Haq QN, Percival Elizabeth 301, *309*, *310*
Haq S, see Bailey RW 55, 58, *62*, 83, *115*
Hara C, see Ukai S 362, *393*
Hara M, Umetsu N, Miyamoto C, Tamari K 60, *63*
Hara S, see Obata T 424, 425, *437*
Harada H, Ohyama K, Chernel J 60, *64*
Harada H, see Fujita M 197, *222*
Harada T, see Saito H 427, *438*
Harborne JB 611, *618*
Hardham AR, Gunning BES 525, *535*
Hardham AR, see Gunning BES 515, 516, *535*
Hardy DM, see Brading JWE 300, *308*
Hardy STS, see Randall LL 490, *510*
Harel E, see Mayer AM 111, *124*
Hargreaves JA, Bailey JA 603, *619*
Harkes PAA 544, *571*
Harkin JM, Obst TR 210, 213, *222*
Harold FM 352, *387*
Harris PJ, Hartley RD 150, *159*, 555, *571*
Harris PJ, Northcote DH 83, *120*, 549, 550, 551, 552, 556, *571*
Harris PJ, see Hartley RD 111, *120*
Harris PJ, see Juniper BE 94, *122*
Harris WM, see Martin EM 543, 545, 549, *572*
Harrison S, see Clarke AE 155, *157*
Harrison S, see Knox RB 237, *250*
Harrison TM, see Milstein C 488, *510*
Hart DA, Kindel PK 34, *43*
Hart JW, see Sabnis DD 648, *656*
Hartley RD 36, *43*, 76, *120*
Hartley RD, Harris PJ, Russell GE 111, *120*
Hartley RD, see Harris PJ 150, *159*, 555, *571*
Hase E, see Mihara S 335, 341, *350*
Hasegawa M, Higuchi T, Ishikawa H 201, *222*
Hasegawa S, Maier VP, Kaszychi HP, Crawford JK 104, 105, *120*
Haselbeck A, see Babczinski P 474, *477*
Hasenclever HF, see Yu RJ 419, *440*
Hashimoto C, Cohen RE, Zhang W-J, Ballou CE 442, 452, *456*
Hashimoto T, Wu-Yuan CF, Blumenthal HJ 365, *387*
Hasilik A 403, 408, *412*
Hasilik A, Muller H, Holzer H 401, 402, *412*
Hasilik A, Tanner W 185, *190*, 442, 445, 451, 452, *456*, 468, 471, 473, *479*, 491, 494, 495, *508*
Haskins RH, see Sietsma JH 354, 355, 358, 363, 365, 368, *392*, 426, 428, *439*
Haskovec C, see Hořejší V 633, *654*
Hassid WZ 47, *64*
Hassid WZ, see Bailey RW 55, 58, *62*, 71, *114*, 83, *115*
Hassid WZ, see Barber GA 49, 51, 52, *62*
Hassid WZ, see Batra KK 52, *62*, 72, *115*
Hassid WZ, see Edelman J 54, *63*
Hassid WZ, see Elbein AD 51, 52, *63*
Hassid WZ, see Feingold DS 48, 51, 52, *63*, 70, 73, 81, *118*
Hassid WZ, see Flowers HM 51, 52, 56, *63*, 72, 73, *119*
Hassid WZ, see Kauss H 94, *122*
Hassid WZ, see Lin TY 288, 289, *312*
Hassid WZ, see Neufeld EF 81, *125*
Hassid WZ, see Nikaido H 47, *65*, 281, *313*
Hassid WZ, see Solms J 49, *66*
Hassid WZ, see Tsai CM 51, 52, *67*, 72, 75, *130*
Hassid WZ, see Villemez CL 71, 75, *131*
Hassid WZ, see Villemez CL Jr 48, 51, 52, *67*
Hastings EG, see Baldwin IB 663, *674*
Hata K, see Yamasaki T 217, 218, *224*
Hata T, Hayashi R, Doi E 401, *412*, 451, *456*
Haug A 277, 285, 287, 302, *310*
Haug A, Larsen B 289, *310*
Haug A, Larsen B, Baardseth E 287, *310*
Haug A, Larsen B, Smidsrod O 284, 287, 288, *310*
Haug A, Myklestad S 303, *310*
Haug A, Smidsrod O 286, 287, *310*
Haug A, see Hellebust JA 288, *310*
Haug A, see Larsen B 289, 296, 297, *311*
Haug A, see Madgwick J 289, 305, *312*
Haug A, see Smestad B 305, *315*
Haug A, see Smidsrod O 286, 287, *315*
Haughton PM, Sellen DB 259, *270*
Haughton PM, Sellen DB, Preston RD 259, *270*
Haughton PM, see Cleland RE 259, 260, 262, *269*
Haughton PM, see Rayle DL 264, 266, *272*
Hawirko R, see Chung K 469, *478*
Hawker LW 353, *388*
Hawthorne DB, Sawyer WH, Grant BR 281, *310*
Hayano K, Fukui L 595, *619*
Hayano K, Fukui S 595, *619*
Hayano K, Tsubouchi Y, Fukui S 595, *619*
Hayashi R 441, *456*

Hayashi R, Moore S, Stein WH 452, *456*
Hayashi R, see Hata T 401, *412*, 451, *456*
Hayashi R, see Morikawa H 21, *24*, 68, *124*
Hayashi Y, see Anno K 296, *307*
Hayashibe M, Katohda S 406, *412*
Hayes CE, see Goldstein IJ 630, 631, 632, 635, 636, 649, *654*
Haylett T,.see Olsnes S 650, *656*
Hayward DM, see Dashek WV 150, *157*
Haywood PL, see Barondes SH 639, *652*
Heale JB, see Gupta DP 589, 591, 592, *618*
Healey PL, see Unzelman JM 85, *131*
Heaney-Kieras J, Chapman DJ 304, *310*
Heaney-Kieras J, Roden L, Chapman DJ 144, *159*, 304, *310*
Heath EC, see Baynes JW 170, *188*
Heath EC, see Eagon PC 179, 186, *189*
Heath EC, see Hsu A-F 175, *190*
Heath EC, see Turco SJ 174, *193*
Heath IB 321, *330*, 522, *535*
Heath IB, Gay JL, Greenwood AD 382, *388*, 501, *508*
Heath MF, Northcote DH 136, *159*
Heath OVS, Clark JE 143, *159*
Heber U, see Heldt HW 607, *619*
Heberlein GT, see Anand VK 586, *615*
Hegelson JAS, see Goldstein JL 187, *190*
Hegewald E, Schnepf E 346, *350*
Hegnauer H, Hohl HR 365, 369, *388*
Heide-Jørgenson HS 529, 530, *535*
Heifetz A, Elbein AD 170, 173, 175, *190*
Heimke JW, Starr RC 338, 340, *350*
Heinen W, De Vries H 237, *250*
Heinen W, Linskens HF 238, *250*
Heinen W, see Linskens HF 238, *252*
Heiniger U, Delmer DP 53, 54, *64*, 72, *120*
Heiniger U, Franz G 50, *64*, 79, *120*
Heininger U, see Delmer DP 54, 56, *63*
Heinz CE, see Mann JW 424, *436*
Heldt HW, Chong Ja Chong, Maronde D, Herold A, Stankovic ZS, Walker DA, Kraminer A, Kirk MR, Heber U 607, *619*
Heldt HW, see Fliege R 607, *618*
Heldt HW, see Flügge UI 607, *618*
Hellebust JA, Haug A 288, *310*
Heller JS, Villemez CL 71, *120*
Heller JS, see Villemez CL Jr 52, *67*
Hellerqvist CG, Lindberg B, Samuelsson K 427, *435*
Helsper JPFG 58, *64*, 73, *120*
Helsper JPFG, Veerkamp HH, Sassen MMA 53, *64*, 72, *120*, 521, 525, *535*
Hemerick GA, see Medcalf DG 304, *312*
Hemmes DE, Kojima-Buddenhagen ES, Hohl HH 426, *435*
Hemming FW 166, 167, *190*, 459, *479*, 557, *571*
Hemming FW, see Alam SS 57, *61*
Hemming FW, see Barr RH 501, *507*
Hemming FW, see Barr RM 467, 468, *478*
Hemming FW, see Evans PJ 170, *190*
Hemming FW, see Oliver GJA 175, *192*
Hemming FW, see Reuvers F 460, *481*
Hemming FW, see Richards JB 170, *192*
Hemming FW, see Speake BK 495, *511*
Hemming FW, see Wellburn AR 57, *67*
Hemmingsen BB, see Adamich M 347, *348*
Hendrick C, see Whatley MH 587, *623*
Hendriks T 72, *120*
Henfling J, see Kuć J 614, *620*
Henis Y, see Kritzman G 378, *389*, 430, *436*
Hepler PK, Jackson WT 514, 517, *535*
Hepler PK, Newcomb EH 85, 98, *120*, 517, *535*
Hepler PK, Palevitz BA 60, *64*, 516, 525, 526, *535*
Hepler PK, see Brower DL 525, *533*
Hepler PK, see Palevitz BA 513, 515, 516, 525, *538*
Hergent HL 231, *250*
Herndon WR, see O'Kelley JC 348, *350*
Herold A, see Heldt HW 607, *619*
Herrera JR, see Arroyo-Begovich A 402, *411*
Herskowitz I, see McCullough J 455, *457*
Hertel C, Quader H, Robinson DG, Marmé D 324, *330*
Herth W 25, 26, *28*, 319, 326, 328, *330*
Herth W, Barthlott W 328, *330*
Herth W, Franke WW, Bittiger H, Kuppel A, Keilich G 86, *120*
Herth W, Franke WW, Bittiger H, Kuppel R, Keilich G 582, *582*
Herth W, Franke WW, Stadler J, Bittiger H, Keilich G, Brown RM 318, *330*
Herth W, Franke WW, Stadler J, Bittiger H, Keilick G, Brown RM Jr 144, *159*
Herth W, Kuppel A, Schnepf E 326, *330*
Herth W, Meyer Y 524, 525, *535*
Herth W, Zugenmaier P 328, *330*

Herth W, see Brown RM 320, *329*
Herth W, see Franke WW 517, *534*
Herth W, see Meyer Y 60, 61, *65*
Herth W, see Reiss H-D 513, 521, *539*
Herth W, see Schnepf E 326, *332*, 514, *539*
Hertz CH, see Falk SO 261, *270*
Herzog W, see Robinson DG 324, *332*, 525, *539*
Heslop-Harrison J 580, *582*
Hess WM, see Strobel GA 597, 601, *622*
Heupel A, see Elstner EF 150, *158*, 213, *222*
Heussler P 346, *350*
Heyes JK, see Penny P 262, *272*
Heyn ANJ 138, 150, *159*, 256, 258, 262, 266, *270*
Heyser W, see Eschrich W 98, *118*
Heyser W, see Evert RF 531, *534*
Hibberd DJ 317, *330*
Hickman S, Kornfeld S 493, *508*
Hickman S, Kulcycky A, Lynch RG, Kornfeld S 493, *508*
Hickman S, Kulczycki A Jr, Lynch RG, Kornfeld S 184, *190*
Hickman S, Neufeld EF 497, *508*
Higgins LW, see Tanaka H 416, 418, 424, *439*
Higgins VJ, see Lazarovits G 597, *620*
Higgins VJ, see Millar RL 613, *620*
Highfield PE, Ellis RJ 490, *508*
Higuchi T 203, 204, 209, 210, 216, 221, *222*
Higuchi T, Barnoud F 203, *223*
Higuchi T, Brown SA 205, *223*
Higuchi T, Ito Y 212, *223*
Higuchi T, Ito Y, Kawamura I 204, *223*
Higuchi T, Shimada M, Nakatsubo F, Tanahashi M 203, 206, 218, 219, *223*
Higuchi T, see Hasegawa M 201, *222*
Higuchi T, see Kawamura I 196, *223*
Higuchi T, see Kuroda H 207, *223*
Higuchi T, see Nakamura Y 218, *223*
Higuchi T, see Shimada M 205, 206, *224*
Higuchi T, see Tanahashi M 216, *224*
Higuchi T, see Yamasaki T 217, 218, *224*
Hildebrand DC, Powell CC Jr, Schroth MN 611, *619*
Hill AC, Rhodes MJC 204, *223*
Hill BS, Hanke DE 652, *654*
Hill RC, see Rhodes MJC 220, *224*
Hills GJ 22, *23*, 86, *120*, 145, *159*
Hills GJ, Gurney-Smith M, Roberts K 22, *23*
Hills GJ, Phillips JM, Gay MR, Roberts K 22, *23*, 86, *120*, 145, *159*, 328, *330*
Hills GJ, see Catt JW 86, *116*, 145, 146, *157*, 328, *329*, *330*
Hills GJ, see Roberts K 22, *24*, 145, 146, *163*
Hills GS, see Roberts K 328, *332*
Hinata K, see Nishio T 580, *583*
Hinch JM, Clarke AE 614, *619*
Hines ER, see Marchant HJ 60, *65*
Hinman MB, Villemez CL 71, *120*, *131*
Hinton DM, Pressey R 105, *120*
Hinton DM, see Pressey R 104, *127*
Hinze H, see Betz H 401, *411*
Hirao F, see Azuma I 466, *477*
Hirase S 291, *310*
Hirase S, Watanabe K 292, *310*
Hirase S, see Araki C 291, *307*
Hirn KE 334, *350*
Hirokawa T, see Takeda H 280, *315*
Hirose K, see Ukai S 362, *393*
Hirs CHW, see Wang FFC 496, *512*
Hirst EL, Jones JKN 34, *43*
Hirst Sir E, Mackie W, Percival Elizabeth 299, 300, *310*
Hirth KP, see Sarvas M 490, *511*
Hiruki C, Shukla P 112, *121*
Hiruki C, Tu JC 111, 112, *121*
Hislop EC, Keon JPR, Fielding AH 589, 591, *619*
Hobson GE 104, 105, *121*
Hobson GE, see Besford RT 104, *115*
Hoch HC 529, *535*
Hochster RM, see Fukui S 595, *618*
Hodgson B, see Bowden JK 417, 418, *433*
Hoesgen A, see Gerisch G 638, *653*
Hoffman NE, see Yang SF 143, *165*
Hoffmann GC, Simson BW, Timell TE 427, *435*
Hoffmann GC, Timell TE 109, *121*
Hofmann T, see Irving RA 180, *190*
Hogetsu T, Shibaoka H, Shimokoriyama M 60, 61, *64*
Hogsett WE, see Quatrano RS 296, 298, 306, *314*
Hogsett WS, see Quatrano RS 296, 298, 306, *314*
Hohl HH, see Hemmes DE 426, *435*
Hohl HR, see Hegnauer H 365, 369, *388*
Hohl HR, see Michalenko GO 363, 365, *390*
Holan Z, see Beran K 356, 359, 368, *385*, 397, *411*
Holan Z, see Seichertová O 406, *415*
Holland AA 660, *675*
Holland AA, see Cass-Smith WP 660, *674*
Holleman J 143, *159*

Holligan MS, see Evans LV 278, 287, *308*, 317, *330*
Holloway PJ 227, 230, 231, 236, *250*
Holloway PJ, Baker EA 226, *250*
Holloway PJ, Baker EA, Martin JT 226, 231, *250*
Holloway PJ, Deas AHB 229, *250*
Holloway PJ, Deas AHB, Kabarra AM 230, *250*
Holloway PJ, see Deas AHB 228, 231, *249*, 529, *534*
Holloway PJ, see Jeffree CE 529, 530, 531, *536*
Holzer H, see Betz H 401, *411*
Holzer H, see Hasilik A 401, 402, *412*
Holzer H, see Maier K 401, *414*
Homer RB, Roberts K 136, 146, 147, *159*
Hong KC, see Young K 291, *316*
Honjo G, Watanabe M 280, *310*
Hopp HE, Romero P, Pont Lezica R 645, *654*
Hopp HE, Romero PA, Daleo GR, Pont Lezica R 48, 51, 53, 56, 57, 58, 59, *64*, 74, *121*
Hopp HE, Romero PA, Pont Lezica R 61, *64*, 74, *121*
Hopp HE, Romero PA, Pont Lezica R 170, 171, 178, *190*
Hopp HE, see Daleo GR 168, *189*
Hopp HE, see Pont Lezica R 56, 58, *66*, 70, 74, 87, *127*
Hopp HE, see Pont Lezica R 178, 180, *192*, 645, *656*
Hopper JE, Dickinson DB 79, *121*
Hořejší V, Chaloupecká O, Kocourek J 633, *654*
Hořejší V, Haskovec C, Kocourek J 633, *654*
Hořejší V, Kocourek J 633, 651, *654*
Hori H 153, *159*
Hori H, Sato S 35, *43*
Hori M, Eguchi J, Kakiki K, Misato T 398, 400, *413*
Hori M, Kakiki K, Misato T 398, 400, *413*
Hori M, Kakiki K, Suzuki S, Misato T 400, *413*
Horikoshi K 424, *435*
Horine RK, Ruesink AW 18, *23*
Horisberger M, Lewis BA, Smith F 429, *435*
Horisberger M, Vonlanthen M 367, *388*, 397, *413*
Horisberger M, see Bauer H 368, *385*, 406, *411*, 425, *433*
Horisberger M, see Bush DA 419, 426, 428, *433*
Horman I, see Bush DA 419, *433*
Horne RW, see Northcote DH 144, *162*, 280, *313*, 359, *391*, 417, *437*
Horner HT Jr, Arnott HJ 107, *121*
Horner HT, Rogers MA 101, *121*
Horowitz NH, Fling M, Maclead H, Watanabe Y 374, *388*
Horsfall JG, Dimond AE 594, *619*
Horsford SPC, McCandless EL 294, *310*
Horton CB, see Howard IK 644, *654*
Horton JC, Keen NT 594, *619*
Horton RF, Osborne DJ 102, *121*
Hough JS, see Maddox IS 421, 422, *436*
Houwink AL, Kreger DR 355, 356, 358, 359, 368, *388*
Howard IK, Sage HJ, Horton CB 644, *654*
Howard J, Kindinger J, Shannon LM 636, *654*
Howard J, Shannon L, Oki L, Murashige T 641, *654*, 672, *675*
Howard RJ, Wright SW, Grant BR 281, *310*
Howell CR 593, *619*
Howell CR, see Puhalla JE 589, *621*
Howell SF, see Sumner JB 627, *656*
Hsaio TC 109, *121*
Hsiao M, see Brown RG 428, *433*
Hsu A-F, Baynes JW, Heath EC 175, *190*
Hsu AF, see Baynes JW 170, *188*
Huang AHC, see Youle RJ 644, 647, *657*
Huang JS, see Goodman RN 596, 600, *618*
Huang KP, Cabib E 403, *413*
Huang PY, see Goodman RN 596, 600, *618*
Huang TSR, see Graham TL 586, *618*
Hubbard JP, Williams JD, Niles RM, Mount MS 589, *619*
Hubbard JP, see Mount MS 589, 592, *621*
Hubbard JP, see Stack JP 589, *622*
Hubbard SC, Robbins PW 181, *190*
Hubbard SC, see Robbins PW 463, *482*, 497, *510*
Hubbell DH, see Dazzo FB 663, 666, 670, 673, *675*
Huber DJ, Nevins DJ 150, *159*, 267, *270*
Huber DJ, see Nevins DJ 92, *125*, 264, *271*
Huber SC 607, *619*
Huberman M, Goren R 102, 103, *121*
Huberman M, Goren R, Birk Y 102, *121*
Huet C, Lonchampt M, Huet M, Bernadac A 634, *654*
Huet M, see Huet C 634, *654*
Hughes JE, Gunning BES 112, *121*
Hughes JE, see Gunning BES 515, 516, *535*
Hughes RC 80, *121*
Huizing HJ, Rietema H 280, *310*
Huizing HJ, Rietema H, Sietsma JH 283, *310*
Hultin HO, Sun B, Bulger J 104, *121*
Hunneman DH, Eglinton G 226, 228, 230, *250*
Hunneman DH, see Eglinton G 227, *249*
Hunsley D, Burnett JH 353, 355, 358, 363, 364, 365, 369, 371, *388*, 396, 397, 408, *413*
Hunsley D, Gooday GW 363, 367, 370, *388*

Hunsley D, see Gooday GW 400, 408, *412*
Hunt LA, Etchinson JR, Summers DF 463, *479*, 497, *508*
Hunter N, see Whatley MH 587, *623*
Huotari F, Nelson T, Smith F, Kirkwood S 425, *435*
Hurkman WJ, Pedersen K, Smith LD, Larkins BA 181, *190*
Hussein MM-D, Fouad ST, Abdel-Fattah AF 296, *310*
Hussein MM-D, see Abdel-Fattah AF 296, 297, *306*
Hüttermann A, see Eschrich W 98, *118*
Huwyler HR, Franz G, Meier H 48, *64*, 93, *121*, 513, *536*

Igari M, see Tanimoto E 139, *164*, 266, *273*
Ikehara Y, see Molner J 172, *191*
Ikekawa T, see Yoshioka Y 362, *394*
Ikuma H, see Adams PA 262, 263, *269*
Ikuma H, see Montague MJ 59, *65*
Imada K, see Chanzy H 524, *533*
Imagawa H, Fukazawa K, Ishida S 195, 196, *223*
Imai S, see Suzuki H 105, *130*
Ingledew WM, see Patel GB 419, *437*
Inglis A, see Schreiber G 493, *511*
Ingram DS, see Butcher DN 613, *616*
Inoue H, Beckwith J 490, *509*
Inoue S, see Suzuki I 633, *656*
Inoue Y, see Mitani M 400, *414*
Iriki Y, Miwa T 283, *310*
Iriki Y, Suzuki T, Nisizawa K, Miwa T 282, *311*
Iriki Y, see Miwa T 282, 283, *313*
Irino T, see Miyazaki T 361, *390*
Irving RA, Toneguzzo F, Rhee SH, Hofmann T, Ghosh AP 180, *190*
Irwin WE, see Bretthauer RK 465, *478*
Isaac S, Ryder NS, Peberdy JF 402, 409, *413*
Isaak P, see Chung K 469, *478*
Ishaaya I, see Levy EC 248, *251*
Isherwood FA, Selvendran RR 49, *64*
Isherwood FA, see Selvendran RR 49, *66*
Isherwood FA, see Jermyn MA 104, *122*
Ishida S, see Imagawa H 195, 196, *223*
Ishikawa H, see Hasegawa M 201, *222*
Isono K, Asahi K, Suzuki S 400, *413*
Israel HW, Wilson RG, Aist JR, Kunoh H 111, *121*, 608, *619*
Israel HW, see Kunoh H 111, *122*, 609, *620*
Israel HW, see Steward FC 134, *164*
Iswasaki T, see Shibuya N 106, *129*
Itakura K, see Tanaka H 424, 425, *439*
Iten W, Matile P 375, *388*
Ito E, see Nakayama K 462, 463, *481*
Itô K 112, *121*
Ito T, Takatsuki A, Kawamura K, Sato K, Tamura G 183, *190*
Ito T, see Sakurada I 13, *24*
Ito T, see Takatsuki A 183, *193*
Ito Y, see Higuchi T 204, 212, *223*
Ito Y, see Kitazima Y 355, *389*
Itoh T 60, *64*
Iwakura Y, see Fujisawa F 493, *508*
Iwami S, Masuda Y 261, *270*
Izhar S, Frankel R 100, 101, *121*
Izhar S, see Frankel R 100, *119*
Izumi K 291, *311*

Jabbar Mian A, Percival Elizabeth 284, 296, 297, *311*
Jaccard M, Pilet PE 263, *270*
Jack MA, Schmidt EL, Wold F 668, *675*
Jackson JJ, see Caccam JF 170, *189*
Jackson JT, see Caccam JJ 460, *478*
Jackson K, Blobel G 489, *509*
Jackson LL, Blomquist GJ 248, *250*
Jackson SM, see Robards AW 225, *253*, 531, *539*
Jackson WT, see Forer A 516, *534*
Jackson WT, see Hepler PK 514, 517, *535*
Jacobs M, Ray PM 92, *121*, 268, *271*
Jacobsen JV, Pressman E 107, *121*
Jacobsen JV, Pressman E, Pyliotis NA 107, *122*
Jacobsen JV, see Ashford AE 106, *114*
Jaenicke L, Waffenschmidt S 340, 342, 344, *350*
Jaffe LF, see Chen T-S 521, *533*
Jaffe LF, see Peng HB 321, *331*
Jakes KS, Model PJ 491, *509*
James DW Jr, Elbein AD 185, *190*
James DW Jr, Jones RL 73, *122*, 557, *571*
James FEL, see Street HE 543, 545, 549, *573*
James MJ, Kandutsch AA 187, *190*
Jamieson JC 468, *479*
Jan YN 380, *388*, 398, 399, 400, *413*
Janczura E, see Palamarczyk G 462, *481*
Janeücek F, see Karácsonyi Š 35, *43*
Jankowski W, see Mańkowski T 474, *480*
Jannasch HW, Mateles RI 594, *619*
Janse C, see Gross GG 41, *43*, 150, 151, *159*, 214, *222*
Jantzen E, see Nelson TE 425, *437*
Janzen DH, Juster HB, Liener IE 651, *654*

Jarvis LR, Wardrop AB 529, *536*
Jarvis MC, see Mankarios AT 150, *161*
Jarvis MC, see Marcan H 603, 605, *620*
Jasalavich C, see Anderson AJ 614, *615*
Jaworski EG, Wang LC, Carpenter WD 398, *413*
Jaworski EG, see Porter CA 398, 399, *414*
Jeanes A 428, *435*
Jeanloz RW, see Warren CD 170, *193*
Jeffree CE, Baker EA, Holloway PJ 529, 530, 531, *536*
Jeffs RA, Northcote DH 39, *43*
Jekel B, see Slob A 613, *622*
Jelsma J 360, *388*
Jelsma J, Kreger DR 359, 360, *388*
Jennings DH 383, *388*, 501, *509*
Jenny H, Grossenbacher K 565, 566, 567, *571*
Jensen JW, see Schutzbach JS 177, *192*
Jentoft NH, see Gander JE 467, *479*
Jermyn MA 152, 156, *159*, 578, *582*
Jermyn MA, Isherwood FA 104, *122*
Jermyn MA, Yeow YM 152, 153, 154, *159*, 577, *583*
Jermyn MA, see Anderson RL 92, *114*, 154, *156*
Jermyn MA, see Clarke AE 577, *582*, 628, *653*
Jermyn MA, see Gleeson PA 153, *158*, 578, *582*
Jermyn MA, see Woods EF 152, *164*
Jha HC, see Tschesche R 610, *623*
Jiang K, Barber GA 145, *159*
Jiang K-S, see Aspinall GO 35, *42*
Jilka R, Brown O, Nordin P 83, 84, *122*
Jimenez-Martinez A, see Novaes-Ledieu M 358, *391*
Johansen JT, Breddam K, Ottesen M 442, 452, *456*
John PCL, see Atkinson AW 322, *329*, 345, 346, *349*
Johnson B 379, *388*
Johnson BF, Gibson EJ 469, *479*
Johnson CM, see Paull RE 549, 550, 551, 552, 562, 563, *573*
Johnson DC, Nicholson MD, Haigh FC 41, *43*
Johnson J, Kirkwood S, Misaki A, Nelson TE, Scaletti JV, Smith F 427, *435*
Johnson J, see Misaki A 417, *437*
Johnson J Jr, see Misaki A 359, *390*
Johnson J, see Nelson TE 425, *437*
Johnson KD, Daniels D, Dowler MJ, Rayle DL 139, 150, *159*, 266, *271*
Johnson PG, Percival Elizabeth 299, *311*
Johnson PG, see Allan GG 303, 305, *306*
Johnson PG, see Bourne EJ 299, *307*
Johnson TB, Strobel GA 600, *619*
Johnston IR 360, *388*, 429, *435*
Johnston KH, McCandless EL 294, *311*
Jones BL, see Grimes WJ 50, *63*
Jones D, see Bacon JSD 356, 358, 359, 360, 367, *384*, 417, 418, 419, 424, 425, *433*
Jones DD, Morré DJ 549, 550, 551, 552, 562, 563, 564, *571*
Jones DD, see Morré DJ 101, *125*, 543, 562, 563, 564, *572*
Jones EW 452, *456*
Jones EW, see Zubenko GS 403, *415*
Jones GH, Ballou CE 443, *456*
Jones J, see Abercrombie M 420, *432*
Jones JKN, see Hirst EL 34, *43*
Jones JM, Cawley LP, Teresa GW 644, *654*
Jones MGK 98, *122*
Jones MGK, Payne HL 85, 91, *122*, 514, 517, 518, *536*
Jones RF 144, *159*, 304, *311*
Jones RF, Wiese L 348, *350*
Jones RF, see Wiese L 348, *351*
Jones RL 107, *122*
Jones RL, see James DW Jr 73, *122*, 557, *571*
Jones RL, see McNeil M 264, *271*
Jones RL, see Paull RE 83, 101, *126*, 154, *162*, 519, *538*, 545, 551, 556, 558, 561, 564, *572*, 549, 550, 552, 562, 563, *573*
Jones RL, see Taiz L 106, *130*
Jones RL, see Terry M 268, *273*
Jones RS, see Linker A 284, *312*
Jones THD, De Renobales M, Pon N 430, *435*
Jones TM, see Albersheim P 643, *652*
Josefsson LG, see Randall LL 490, *510*
Joseleau JP, Barnoud F 94, *122*
Joseleau JP, Chambat G, Vignon M, Barnoud F 35, *43*
Jotterand-Dolivo MC, Pilet PE 143, *160*
Jovanović SM, see Palma A 524, *538*
Jung P, Tanner W 167, 170, *190*, 460, *479*
Jung P, see Tanner W 460, *483*
Juniper BE 542, 549, 560, *571*
Juniper BE, Barlow PW 545, *571*
Juniper BE, Clowes FAL 85, 97, 98, *116*, 545, *571*
Juniper BE, Cox GC 529, 530, *536*
Juniper BE, French A 545, 548, *571*
Juniper BE, Gilchrist AJ, Robins RJ 542, 545, 549, *571*
Juniper BE, Lawton JR 525, *536*
Juniper BE, Lawton JR, Harris PJ 94, *122*
Juniper BE, Pask G 542, 543, 545, 549, 560, *571*

Juniper BE, Roberts RM 543, 545, 547, 549, 550, 551, 552, *571*
Juniper BE, see Clowes FAL 545, *570*
Juniper BE, see Hallam ND 529, 530, *535*
Juniper BE, see Martin JT 225, 226, 236, 239, *252*, 529, 531, *537*
Jurale JB, see English PD 589, 591, *617*
Juster HB, see Janzen DH 651, *654*

Kabarra AM, see Holloway PJ 230, *250*
Kabelitz L, see Brieskorn CH 230, *248*
Kabir MS, see Dutton CGS 31, *43*
Kachru RB, Anderson LE 607, *619*
Kahlem G, see Mialonier G 644, *655*
Kahn AW, Colvin JR 58, *64*
Kaiser ET, see Margolis HC 451, 452, *457*, 496, *509*
Kakiki K, see Endo A 400, 408, *412*
Kakiki K, see Hori M 398, 400, *413*
Kakiki K, see Ohta N 398, 400, *414*
Kakiuchi N, see Yamaki S 104, 105, *132*
Kalb AJ, see Sterling C 104, *130*
Kalin JR, see Allen CM 168, *188*
Kalinovskiai, see Pavlenko AF 297, 298, *313*
Kamberger W 665, 667, *675*
Kamisaka S, Sano H, Katsumi M, Masuda Y 262, 263, *271*
Kamisaka S, see Kawamura H 262, 263, *271*
Kandler O, see Tanner W 56, *67*
Kandutsch AA, see James MJ 187, *190*
Kanegasaki S, see Wright A 399, *415*
Kaneko T, see Kitamura K 424, *436*
Kanetsuna F, Carbonell LM 428, *435*
Kanetsuna F, Carbonell LM, Azuma I, Yamamura Y 372, *388*
Kanetsuna F, Carbonell LM, Gil F, Azuma I 372, *388*, 429, *435*
Kanetsuna F, see Carbonell LM 361, 365, *386*
Kanetsuna F, Carbonell M, Azuma I, Yamamura Y 428, *435*
Kang MS, Spencer JP, Elbein AD 177, 186, *190*
Kang MS, see Elbein AD 185, *189*
Kang MS, see Spencer JP 186, *192*
Kapil RN, Tiwari SC 101, *122*
Kaplan A, Achord DT, Sly WS 497, *509*
Kapusta G, Rouwenhorst DL 661, *675*
Karácsonyi Š, Toman R, Janeček F, Kubačková M 35, *43*
Káračsonyi Š, see Toman R 35, *46*
Karas I, Cass DD 514, *536*
Karasawa K, see Nakamura S 185, *191*
Karlsnes AM, see Cleland R 142, 143, *157*
Karnovsky MJ, see Branton D 321, *329*
Karr AL Jr, Albersheim P 591, *619*
Karson EM, Ballou CE 444, 447, *456*, 464, 474, 476, *479*
Kartenbeck J, see Morré DJ 82, *125*
Kasai R, see Cole GT 365, *386*
Kashiwabara Y, see Min KH 284, *312*
Kaszychi HP, see Hasegawa S 104, 105, *120*
Katchalski E, see Yariv J 577, *583*
Kato G, Maruyama Y, Nakumara M 663, 666, 667, *675*
Kato K, Watanabe F, Eda S 7, *8*
Katō K, see Akiyama Y 35, *42*, 135, *156*
Katō K, see Eda S 31, 32, 34, *43*
Katō K, see Mori M 32, *44*
Kato M, see Lin K 248, *252*
Katō Y, Asano N, Matsuda K 32, *44*
Katohda S, see Hayashibe M 406, *412*
Katona L, see Lamport DTA 35, *44*, 135, 138, 139, *161*
Katsumi M, Kazama H 262, 263, *271*
Katsumi M, see Kamisaka S 262, 263, *271*
Katz D, Goldstein D, Rosenberger RF *389*
Katz D, Rosenberger RF 371, 382, 384, *388*, 408, 409, *413*
Katz D, see Sternlicht E 384, *393*
Katz FN, Rothman JE, Lingappa VR, Blobel G, Lodish HF 180, *191*
Katz FN, see Rothman JE 493, *511*
Katz M, Ordin L 92, *122*, 267, *271*
Kaufman P, see Burke D 264, *269*
Kaufman PB, see Adams PA 262, 263, *269*
Kaur J, see Rosen SD 639, *656*
Kauss H 170, *191*, 628, 651, *654*
Kauß H 460, *479*
Kauss H, Bowles DJ 36, 38, *44*, 87, *122*, 647, *655*, 668, *676*
Kauss H, Glaser C 36, *44*, 86, *122*, 141, *160*, 647, *655*
Kauss H, Hassid WZ 94, *122*
Kauss H, Swanson AL 75, *122*
Kauss H, Ziegler H 648, *655*
Kauss H, see Bailey RW 137, 140, *156*
Kauss H, see Bowles DJ 646, *653*, 668, *674*
Kauss H, see Gietl C 633, 648, *654*
Kauss H, see Haass D 647, 648, *654*
Kauss H, see Köhle H 629, 646, *655*
Kauss H, see Lehle L 173, *191*, 462, *480*

Kauss H, see Odzuck W 71, 75, *126*
Kawade Y, see Fujisawa F 493, *508*
Kawai H, see Moriguchi M 398, *414*
Kawai H, see Yamamoto K 398, *415*
Kawamura H, Kamisaka S, Masuda Y 262, 263, *271*
Kawamura H, see Masuda Y 261, 262, *271*
Kawamura I, Higuchi T 196, *223*
Kawamura I, see Higuchi T 204, *223*
Kawamura K, see Ito T 183, *190*
Kawamura K, see Takatsuki A 183, *193*
Kawasaki S, Sato S 137, *160*, 519, *536*
Kawasaki T, Ashwell G 659, *676*
Kawase M 97, *122*
Kazama H, see Katsumi M 262, 263, *271*
Keane PJ, see Woodward JR 111, 113, *132*, 597, 600, *623*
Keegstra K, Albersheim P 151, *160*
Keegstra K, Talmadge KW, Bauer WD, Albersheim P 6, *8*, 133, *160*, 264, *271*, 357, *389*, 590, *619*
Keegstra K, see Bauer WD 591, *616*
Keegstra K, see English PD 589, 591, *618*
Keegstra K, see Pueppke SG 644, *656*, 667, *676*
Keegstra K, see Talmadge KW 138, *164*
Keegstra K, see Whatley MH 587, *623*
Keen NT, Long M 597, 599, *619*
Keen NT, Long M, Erwin DC 597, 600, *619*
Keen NT, Wang MC, Bartnicki-Garcia S, Zentmyer GA 596, 600, *620*
Keen NT, Williams PH 586, *619*
Keen NT, see Bruegger BB 604, *616*
Keen NT, see Horton JC 594, *619*
Keilich G, see Herth W 86, *120*, 318, *330*, 582, *582*
Keilick G, see Herth W 144, *159*
Keller F, see Cook JA 72, 88, *116*
Keller FA, Cabib E 399, 400, *413*
Keller FA, see Cabib E 401, *411*
Keller RL, Adair WL Jr, Ness G 168, *191*
Kelly DJ, see Catley BJ 429, *434*
Kelly GJ, Latzko E, Gibbs M 607, *620*
Kemp J, Loughman BC 53, 56, *64*
Kemp J, Loughman BC, Ephritikhine G 56, *64*, 74, *122*
Kemp J, see Flowers HM 51, 52, 56, *63*, 72, 73, *119*
Kempton RJ, see Goodenough PW 590, 591, *618*
Kenchinton W, see Rudall KM 355, *392*
Kent SP, see Hamblin J 659, 663, *675*
Keon JPR, see Hislop EC 589, 591, *619*
Kephart JE, see Dauwalder M 546, 558, 565, *570*
Kephart JE, see Whaley WG 516, 517, 518, *541*, 545, 546, 564, *574*
Kern KA, see Raschke WC 444, 446, 447, *457*
Kerr T, see Anderson DB 76, *114*
Kerridge D, see Cassone A 418, *434*
Kessler G, Nickerson WJ 418, *436*
Keul M, see Lazar-Keul G 516, *536*
Keusch L 107, *122*
Key JL, see Courtney JS 262, *270*
Khan AA, Kolattukudy PE 244, 245, *250*
Khomenko VA, see Ovodov YS 296, *313*
Kidby DK, Davies R 367, *389*, 477, *479*
Kieda C, see Rougier M 548, 553, 555, *573*
Kiely ML, McKnight GS, Schimke RT 178, *191*, 468, *479*, 493, *509*
Kieras FJ, see Kieras JH 304, *311*
Kieras JH, Kieras FJ, Bowen DV 304, *311*
Kiermayer O 325, 326, *330*, *331*
Kiermayer O, Dobberstein B 81, *122*, 325, *331*
Kiermayer O, Sleytr UB 325, *331*, 524, *536*
Kiermayer O, see Dobberstein B 325, *330*
Kiessling H, see Bouveng HO 427, *433*
Kiho T, see Ukai S 362, *393*
Kijne JW, see Planqué K 667, *676*
Killias U, see Albersheim P 40, *42*
Kimmins WC, Brown RG 111, *122*, 141, *160*
Kimmins WC, see Brown RG 521, *533*
Kimura H, see Azuma I 466, *477*
Kindel PK, see Hart DA 34, *43*
Kindinger J, see Howard J 636, *654*
Kindinger J, see Musolan C 73, *125*
Kindinger JE, see Musolan C 73, 110, *125*
Kindinger JI, see Hankins CN 636, 652, *654*
Kindinger JI, see Van Der Woude WJ 51, 53, *67*, 72, *131*
Kindt T, see Devillers-Thiery A 492, *508*
King GGS, Colvin JR 528, *536*
Kinnersley AM, Racusen RH, Galston AW 18, *23*
Kiobakken J, Colvin JR 178, *191*
Kirby EG, Roberts RM 556, 558, *571*
Kirk JTO 347, *350*
Kirk MR, see Heldt HW 607, *619*
Kirk TK, see Vance CP 140, 142, 152, *164*

Kirkwood S 354, 383, *389*
Kirkwood S, see Batra KK 75, *115*
Kirkwood S, see Huotari F 425, *435*
Kirkwood S, see Johnson J 427, *435*
Kirkwood S, see Misaki A 359, *390*, 417, *437*
Kirkwood S, see Nelson TE 425, *437*
Kislev M, see Werker E 565, *574*
Kislev N, see Ben-Arie R 103, *115*
Kisters R, see Greiling H 449, *456*
Kitamoto Y, Gruen HE 376, *389*
Kitamura K, Kaneko T, Yamamoto Y 424, *436*
Kitamura K, Yamamoto Y 424, 425, *436*
Kitano H, see Okamoto K 106, *126*
Kitazima Y, Banno Y, Noguchi T, Nozawa Y, Ito Y 355, *389*
Kivilaan A, Bandurski RS, Schulze A 92, *122*
Kivilaan A, see Lee SH 92, *123*, 150, *161*
Kivirikko KI, see Myllyla R 143, *162*
Kjosbakken J, Colvin JR 20, *23*
Klapprott JA, see Aronson JM 144, *156*
Klebs G 347, *350*
Klein, see Von Figura K 497, *512*
Klein L, see Barash I 592, *615*
Kleinig H, see Brown RM 318, *329*, 346, *349*
Kleinig M, see Brown RM Jr 144, *157*
Kleinschuster SJ, Baker R 587, *620*
Klis FM 139, 143, *160*
Klis FM, Eeltink H 135, *160*, 513, *536*
Kloepper HG, see Raizada MK 466, *481*
Knapp JCF, see Coggins CW Jr 103, *116*
Knee M 104, *122*
Knee M, Sargent JA, Osborne DJ 103, *122*
Knorr LC, see Rouse AH 104, *128*
Knowles R, see Balandreau J 565, 568, *569*
Knox RB, Clarke AE 580, *583*
Knox RB, Clarke AE, Harrison S, Smith P, Marchalonis JJ 237, *250*
Knox RB, see Anderson RL 92, *114*, 154, *156*
Knox RB, see Clarke AE 155, *157*
Knox RB, see Clarke AE 81, 111, 113, *116*, 151, *157*, 577, *582*, 628, *653*
Knox RB, see Raff JW 580, *583*
Knull HR, see Pazur JH 496, *510*
Kochert G 145, *160*, 344, 347, *350*
Kocourek J, see Hořejší V 633, 651, *654*
Kodama Y, see Takatsuki A 183, *193*
Koffler H, see Troy FA 353, 355, 357, 362, 363, *393*, 428, *439*
Köhle H, Kauss H 629, 646, *655*
Kohn R, Larsen B 286, *311*
Kohno K, see Takatsuki A 472, *483*, 493, *512*
Kojima M, Uritani I 587, *620*, 629, *655*
Kojima-Buddenhagen ES, see Hemmes DE 426, *435*
Kokkinakis DM, Brooks J 149, *160*
Kolattukudy PE 225, 226, 228, 230, 231, 232, 235, 238, 242, 244, 245, 246, *250*, *251*, 529, 530, *536*
Kolattukudy PE, Brown L 243, 244, *251*
Kolattukudy PE, Buckner JS 243, *251*
Kolattukudy PE, Buckner JS, Brown L 245, *251*
Kolattukudy PE, Buckner JS, Liu TYJ 246, *251*
Kolattukudy PE, Croteau R, Brown L 230, 245, *251*
Kolattukudy PE, Croteau R, Buckner JS 244, 245, *251*
Kolattukudy PE, Croteau R, Walton TJ *536*
Kolattukudy PE, Dean BB 235, 236, *251*
Kolattukudy PE, Kronman K, Poulose AJ 227, 230, 231, 236, *251*
Kolattukudy PE, Liu TJ 246, *251*
Kolattukudy PE, Walton TJ 230, 233, 239, 242, 243, 245, *251*
Kolattukudy PE, Walton TJ, Kushwaha RPS 234, *251*
Kolattukudy PE, see Agrawal VP 235, 236, *248*, 531, *532*
Kolattukudy PE, see Brown AJ 231, 237, *248*
Kolattukudy PE, see Croteau R 234, *249*, 529, *534*
Kolattukudy PE, see Dean BB 225, 236, *249*
Kolattukudy PE, see Espelie KE 225, 226, *250*, 228, 229, 230, 231, 236, 239, 242, 247, *249*, 531, *534*
Kolattukudy PE, see Hankin L 237, *250*
Kolattukudy PE, see Khan AA 244, 245, *250*
Kolattukudy PE, see Lin T-S 237, 238, 239, *252*
Kolattukudy PE, see Maiti IB 239, 247, *252*
Kolattukudy PE, see Purdy RE 237, 239, *252*
Kolattukudy PE, see Riley RG 231, *253*
Kolattukudy PE, see Shaykh M 230, 239, 247, *253*
Kolattukudy PE, see Soliday CL 232, 234, 237, 238, 236, *253*, 242, 247, 468, *482*, 529, 531, *540*
Kolattukudy PE, see Walton TJ 226, 227, 228, 229, 230, 231, 233, *254*
Koltin Y, see Wessels JGH 375, 376, *394*, 430, *439*
Komamine A 221, *223*
Komamine A, Morigaki T, Fujimura T 41, *44*
Komamine A, see Takeuchi Y 32, *46*, 92, *130*, 528, *540*
Kominato Y, see Morita Y 149, *161*

Kopecká M, Phaff HJ, Fleet GH 16, *23*, 418, 425, *436*
Kopecká M, see Kreger DR 356, 360, 369, 380, 381, *389*, 522, *536*
Kopp F, see Cortat M 498, *507*
Korn ED, Northcote DH 418, *436*
Kornfeld R 398, *413*
Kornfeld R, Kornfeld S 168, *191*, 631, *655*
Kornfeld R, see Kornfeld S 398, *413*, 637, *655*
Kornfeld RD, Kornfeld S 381, *389*
Kornfeld S, Kornfeld R 637, *655*
Kornfeld S, Kornfeld R, Neufeld EF, O'Brien PJ 398, *413*
Kornfeld S, see Chapman A 175, 176, 177, 186, *189*
Kornfeld S, see Gibson R 493, 496, *508*
Kornfeld S, see Hickman S 184, *190*, 493, *508*
Kornfeld S, see Kornfeld R 168, *191*, 631, *655*
Kornfeld S, see Kornfeld RD 381, *389*
Kornfeld S, see Leavit R 493, *509*
Kornfeld S, see Li E 169, 175, 177, 178, *191*, 463, *480*
Kornfeld S, see Presant CA 633, 634, *656*
Kornfeld S, see Tabas I 182, *193*, 463, *482*, 497, *512*
Kornmann P 334, *350*
Koshijima T, see Yaku F 36, *46*
Košinová A, Farkaš V, Machala S, Bauer Š 469, *479*
Kosinová A, see Farkaš V 469, *479*
Kosuge T 555, *572*
Kosuge T, see Carroll TW 606, *616*
Kosuge T, see Preiss J 607, *621*
Koukol J, Conn EE 203, *223*
Kourilsky P, see Mercereau-Puijalon O 491, *509*
Kovařik J, see Farkaš V 469, *479*
Kozak LP, Bretthauer RK 465, *479*
Kozak LP, see Bretthauer RK 465, *478*
Krag SS, see Robbins PW 175, 178, *192*
Krähmer H 225, *251*
Kraminer A, see Heldt HW 607, *619*
Krátký Z, Biely P, Bauer Š 472, 473, 477, *479*
Krátky Z, see Biely P 449, *455*, 473, *478*
Krátky Z, see Vrsanska M 406, *415*
Kratzl K, Faigle H 201, *223*
Krauss A, see Hager A 142, *159*, 256, 258, *270*
Kreger DR 354, 355, 356, 357, 358, 360, 361, 367, *389*, 419, *436*
Kreger DR, Kopecká M 356, 360, 369, 380, 381, *389*, 522, *536*
Kreger DR, see Houwink AL 355, 356, 358, 359, 368, *388*
Kreger DR, see Jelsma J 359, 360, *388*
Kreger DR, see Wessels JGH 358, 360, 362, 363, 364, 365, *394*, 427, 429, *440*
Kreger-Van Rij NJW, Veenhuis M 368, *389*
Krelle H, see Gerisch G 641, *653*
Kretschmer H, see Göring H 268, *270*
Kristen U, Biedermann M, Liebezeit G, Dawson R 579, *583*
Kritzman G, Chet I, Henis Y 378, *389*, 430, *436*
Kritzman G, see Netzer D 608, *621*
Kröning A, Egel R 377, *389*, 422, *436*
Kronman K, see Kolattukudy PE 227, 230, 231, 236, *251*
Krupe M, Ensgraber A 667, *676*
Krust A, see Mercereau-Puijalon O 491, *509*
Kubačková M, see Karácsonyi Š 35, *43*
Kubačková M, see Toman R 35, *46*
Kuboi T, see Yamada Y 220, *224*
Kuć J 614, *620*, 643, *655*
Kuć J, Henfling J, Garas NA, Doke N 614, *620*
Kuć J, see Berard DF 588, *616*
Kuć J, see Doke N 588, 603, 605, 606, *617*
Kuć J, see Garas NA 588, *618*
Kuc J, see Hammerschmidt R 152, *159*
Kuehl WM, see Bergman LW 493, *507*
Kuhl A, Lorenzen H 335, 346, *350*
Kuhn RW, Walsh KA, Neurath H 452, *456*
Kulcycky A, see Hickman S 493, *508*
Kulczycki A Jr, see Hickman S 184, *190*
Kulow C, see Delmer DP 54, 56, *63*, 170, *189*
Kulow C, see Maltby D 93, *123*
Kumarasinghe RMK, Nutman PS 671, *676*
Kunoh H, Aist JR, Israel HW 111, *122*, 609, *620*
Kunoh H, see Israel HW 111, *121*, 608, *619*
Kuo J, see O'Brien TP 225, *252*
Kuo MJ, Alexander M 374, *389*
Kuo S-C, Lampen JO 186, *191*, 472, 473, *479*, *480*, 494, 499, *509*
Kuppel A, see Herth W 86, *120*, 326, *330*
Kuppel R, see Herth W 582, *582*
Kurkdjian A, Leguay J, Guern J 601, *620*
Kuroda H, Shimada M, Higuchi T 207, *223*
Kuroda H, see Namba H 432, *437*
Kuroda H, see Shimada M 205, 206, *224*
Kuroda M, see Endo A 187, *189*
Kurokawa T, see Uchida T 650, *657*

Kusano S 112, *123*
Kushwaha RPS, see Kolattukudy PE 234, *251*
Kuutti-Savolainen ER, see Myllyla R 143, *162*
Kyung-Eun Yoon Park, see Anderson LE 607, *615*

Labarca C, see Loewus F 155, *161*
Labavitch JM, Ray PM 32, 33, 38, 40, *44*, 92, *123*, 151, *160*, 268, *271*
Laberge DE, see Ballance GM 106, *115*
Laborda F, see Rodríguez L 504, *510*
Labriola R, see Cerezo AS 282, *308*
Lachance MA, Villa TG, Phaff HJ 420, *436*
Lachance MA, see Villa TG 420, 421, 422, 423, 424, *439*
Laere A van, Carlier AR 380, *389*
Lafitte C, see Esquerré-Tugayé MT 111, *118*, 141, *158*, 608, *618*
Lahoz R, see Santamaria F 427, *438*
Laine R, Elbein AD 172, *191*
Lalonde M, see Calvert HE 665, *674*
Lamberti F, see Zacheo G 141, *165*
Lamers A, see Sassen A 324, *332*
Lampard JF, Carter GA 248, *251*
Lampen JO 359, 362, 367, *389*, 441, 449, *456*, 477, *480*
Lampen JO, see Kuo S-C 186, *191*, 494, 499, *509*, 4723, 473, *479*, *480*
Lampen JO, see Neumann NP 442, 449, 450, *457*
Lampen JO, see Onishi HR 451, *457*, 473, *481*
Lampen JO, see Onishi HS 495, 496, *510*
Lampen JO, see Sentandreu R 460, 471, *482*
Lampen JO, see Tkacz JS 183, *193*, 469, 472, *483*, 493, *512*
Lamport DTA 29, 35, 38, 40, 41, *44*, 87, 92, *123*, 133, 134, 135, 136, 137, 138, 139, 140, 141, 144, 145, 147, 148, 149, 150, 151, 152, 154, *160*, 344, *350*
Lamport DTA, Katona L, Roerig S 35, *44*, 135, 138, 139, *161*
Lamport DTA, Miller DH 135, 136, *160*
Lamport DTA, Northcote DH 134, 149, 151, *160*, *161*
Lamport DTA, see Clarke JA 142, *157*
Lamport DTA, see Delmer DP 147, 148, *158*
Lamport DTA, see Esquerré-Tugayé MT 135, 141, *158*, 608, *618*
Lamport DTA, see Mani UV 135, 153, *161*
Lamport DTA, see Miller DH 145, *161*
Lamport DTA, see Mort AJ 38, *45*, 140, 153, *162*
Lamport DTA, see Pope DG 35, *45*, 152, *162*
Lane GR, see Mepham RH 99, *124*
Lane MD, see Struck DK 184, *193*, 493, *511*
Lang A, Canney M, Whitty R 112, *123*
Lang W 141, 143, *161*
Larkin PJ 154, *161*
Larkins BA, see Hurkman WJ 181, *190*
Larm O, Lindberg B 74, *123*
Larriba G, Elorza MV, Villanueva JR, Sentandreu R 470, 476, *480*, 494, 504, *509*
Larriba G, see Elorza MV 498, *508*
Larsen B 296, *311*
Larsen B, Haug A 289, 296, 297, *311*
Larsen B, Haug A, Painter TL 296, *311*
Larsen B, see Grasdalen H 285, 287, 288, *309*
Larsen B, see Haug A 284, 288, 287, 289, *310*
Larsen B, see Kohn R 286, *311*
Larsen B, see Madgwick J 289, 305, *312*
Larsen B, see Medcalf DG 296, 297, *312*
Larsen GL, Brummond DO 51, 53, 54, *64*, 72, *123*
Larsson S, see Adler E 210, *222*
Latzko E, see Kelly GJ 607, *620*
Lavee S, see Ben-Arie R 110, *115*
Lavie D, see Levy EC 248, *251*
Lavinthman N, Tandecarz J, Carceller M, Mendiara S, Cardini CE 55, *64*
Law IJ, Strijdom BW 664, 667, *676*
Lawford GR, Schachter H 468, *480*
Lawson CJ, Rees DA 295, 305, *311*
Lawson CJ, Rees DA, Stancioff DJ, Stanley NF 293, *311*
Lawson CJ, see Anderson NS 293, *307*
Lawton JR, see Juniper BE 94, *122*, 525, *536*
Lazar-Keul G, Keul M, Wagner G 516, *536*
Lazarovits G, Bhullar BS, Sugiyama HJ, Higgins VJ 597, *620*
Lazarovits G, Higgins VJ 597, *620*
Leadbeater BSC 318, *331*
Leal JA, see Garcia Mendoza C 374, *387*
Leather GR, see Abeles FB 102, *114*, 614, *615*
Leavit R, Schlessinger S, Kornfeld S 493, *509*
Leavitt RD, Felsted RL, Bachur NR 635, *655*
Lechene De La Porte P 553, *572*
Lee JW, see Fulcher RG 107, *119*
Lee SC, see West C 603, 605, *623*
Lee SD, see Rapaka RS 143, *163*
Lee SH, Kivilaan A, Bandurski RS 92, *123*, 150, *161*
Lee TH, see Arai M 419, *433*
Lee YC, Ballou CE 445, *456*
Lee YC, see Elbein AD 467, *478*

Leech JH, see Mollenhauer HH 543, 545, 546, 558, *572*
Leguay J, see Kurkdjian A 601, *620*
Lehle L 462, 463, 475, 476, *480*
Lehle L, Bauer F, Tanner W 470, 476, *480*, 494, *509*
Lehle L, Bowles DJ, Tanner W 187, *191*
Lehle L, Cohen RE, Ballou CE 441, 443, 445, 449, 450, 451, 452, *456*, 462, 476, *480*
Lehle L, Fartaczek F, Tanner W, Kauss H 173, *191*, 462, *480*
Lehle L, Schulz I, Tanner W 463, 476, *480*
Lehle L, Schwarz RT 473, *480*
Lehle L, Tanner W 56, *64*, 183, *191*, 400, *413*, 461, 462, 463, 464, 472, 473, *480*, 493, 494, *509*
Lehle L, see Bause E 461, 472, 474, *478*
Lehle L, see Palamarczyk G 474, *481*
Lehle L, see Schwarz RT 184, 185, *192*
Lehle L, see Sharma CB 461, 472, *482*
Lehmann H, Schulz D 83, *123*
Leiser AT 565, *572*
Leive L 10, *23*
Leloir LF 49, *65*
Leloir LF, Cardini CE 398, *413*
Leloir LF, see Behrens NH 171, 174, 175, *188*, 460, 461, *478*
Leloir LF, see Brett CT 173, *189*
Leloir LF, see Cabib E 459, *478*
Leloir LF, see Dallner G 171, 187, *189*
Leloir LF, see Levy JA 174, *191*
Leloir LF, see Parodi AJ 399, *414*, 459, *481*
Lembi CA, see Van Der Woude WJ 51, 53, 59, *67*, 72, *131*
Lemmers CBG, see Beijersbergen JCM 613, *616*
Leney L, see Saleh TM 197, *224*
Lennarz WJ, see Carson DD 187, *189*
Lennarz WJ, see Chen WW 174, *189*, 497, *507*
Lennarz WJ, see Hanover JA 461, *479*
Lennarz WJ, see Lucas JJ 175, *191*
Lennarz WJ, see Pless DD 179, *192*
Lennarz WJ, see Schneider EG 185, *192*
Lennarz WJ, see Struck DK 183, 187, 184, *193*, 493, *511*
Lennarz WJ, see Waechter CJ 48, *67*, 166, 173, *193*, 381, *394*, 459, *483*
Lenney JF, Dalbec JM 401, *413*
Lenney JF, Matile P, Wiemken A, Schellenberg M, Meyer J 401, 472, *413*
Leonard LT 664, *676*
Leonard SJ, see Postlmayr HL 104, *127*
Leonard TJ 374, *389*
Leopold AC, see Majumder SK 112, *123*
Leopold AC, see Webster BD 102, *131*
Leppard GG 565, *572*
Leppard GG, Colvin JR 15, 20, *23*
Leppard GG, Ramamoorthy S 566, *572*
Leppard GG, Sowden LC, Colvin JR 20, *23*, 528, *536*
Leppard GG, see Colvin JR 15, 20, *23*
Lessire R 243, *251*
Lessire R, see Cassagne C 243, 244, 245, *249*
Lestang G De, Quillet M 298, 299, 306, *311*
Letoublon R, Got R 467, *480*
Letoublon R, see Richard M 467, *481*
Letoublon RCP, Combe J, Got R 467, *480*
Letoublon RCP, Compte J, Got R 501, *509*
Letoublon RM, Richard M, Louisot P, Got R 467, *480*
Letourneau DR, Deven JM, Manocha MS 356, 363, 366, *389*
Levin G, see Bowers B 400, 406, *411*
Levy EC, Ishaaya I, Gurevite E, Cooper R, Lavie D 248, *251*
Levy JA, Carminatti H, Cantarella AI, Behrens NH, Leloir LF, Tabora E 174, *191*
Lew F, see Reid PD 102, *128*
Lew FT, Lewis LN 102, *123*
Lewin J, see Allan GG 303, 305, *306*
Lewin RA 348, *350*
Lewis BA, see Horisberger M 429, *435*
Lewis BG, see Speakman JB 111, 113, *129*
Lewis DR, see Northcote DH 81, *125*
Lewis LN, Linkins AE, O'Sullivan S, Reid PD 105, *123*
Lewis LN, Varner JE 103, *123*
Lewis LN, see Coggins CW Jr 103, *116*
Lewis LN, see Lew FT 102, *123*
Lewis LN, see Linkins AE 103, *123*
Lewis LN, see Reid PD 102, *128*
Lewis LN, see Sexton R 103, *129*
Lezerovich A, see Cerezo AS 282, *308*
Li E, Kornfeld S 169, 175, 177, 178, *191*
Li E, Tabas I, Kornfeld S 463, *480*
Li E, see Chapman A 175, 176, *189*
Li Y-T 443, *457*
Li YT, see Davis TE 431, *434*
Lian B, see Smidsrod O 287, *315*
Liang JN, Stevens ES, Morris ER, Rees DA 294, *311*
Lieberman M 111, *123*, 143, *161*
Liebezeit G, see Kristen U 579, *583*
Liener IE 628, 649, *655*, 659, 667, 671, *676*
Liener IE, see Janzen DH 651, *654*

Liese W 95, *123*
Liga A, see Simonescu CI 284, *315*
Likens GE, see Woodwell GM 11, *24*
Lilley GG, see Woods EF 152, *164*
Limpsou DL, see Pazur JH 496, *510*
Lin B-Y 516, *536*
Lin CC, Aronson JM 355, *389*
Lin CC, Sicher Jr RC, Aronson JM 355, *389*
Lin CC, see Aronson JM 144, *156*
Lin K, Yamada H, Kato M 248, *252*
Lin T-S, Kolattukudy PE 237, 238, 239, *252*
Lin T-Y, Elbein AD, Su J-C 71, *123*
Lin T-Y, Hassid WZ 288, 289, *312*
Lin T-Y, see Villemez CL 71, 75, *131*
Lindbeck G, see Dashek WV 150, *157*
Lindberg B, Rosell K-G, Svensson S 33, *44*
Lindberg B, Svensson S 353, *389*
Lindberg B, see Bartnicki-Garcia S 361, *385*
Lindberg B, see Bhavanandan VP 428, *433*
Lindberg B, see Bjorndal H 282, *307*, 428, *433*
Lindberg B, see Bouveng HO 427, *433*
Lindberg B, see Hellerqvist CG 427, *435*
Lindberg B, see Larm O 74, *123*
Lindberg B, see Manners DJ 417, 418, 427, *436*
Lingappa JR, see Lingappa VR 181, *191*, 490, *509*
Lingappa VR, Lingappa JR, Blobel G 490, *509*
Lingappa VR, Lingappa JR, Prasad R, Ebner KE, Blobel G 181, *191*
Lingappa VR, Shields D, Woo SL, Blobel BJ 490, *509*
Lingappa VR, see Katz FN 180, *191*
Linker A, Jones RS 284, *312*
Linkins AE, Lewis LN, Palmer RL 103, *123*
Linkins AE, see Lewis LN 105, *123*
Linnemans WAM, Boer P, Elbers PF 367, *389*, 469, 477, *480*, 498, 504, *509*
Linnemans WAM, see Van Rijn HJM 498, *512*
Linskens HF, Heinen W 238, *252*
Linskens HF, see Heinen W 238, *250*
Linskens HF, see Van den Ende G 247, *253*
Linstead PJ, see Burgess J 60, *62*
Lintilhac PM 516, *536*
Lipke PN, Taylor A, Ballou CE 453, *457*
Lipke PN, see Ballou CE 449, *455*
Lippincott BB, Whatley MH, Lippincott JA 586, *620*
Lippincott BB, see Lippincott JA 586, *620*
Lippincott BB, see Whatley MH 586, *623*
Lippincott JA, Lippincott BB 586, *620*
Lippincott JA, see Lippincott BB 586, *620*
Lippincott JA, see Whatley MH 586, *623*
Lippman E, see Bartnicki-Garcia S 379, 382, *385*
Lippman E, see Bartnicki-García S 400, 408, *411*
Liras P, Gascón S 471, 473, *480*
Lis H, Sharon N 169, *191*, 627, 628, 630, 632, 635, 636, 637, 651, *655*, 659, 667, 671, *676*
Lis H, see Bowles DJ 645, *653*, 668, *674*
Lis H, see Lotan R 635, *655*
Lis H, see Sharon N 168, *192*, 630, *656*
Lis H, see Tunis M 629, *657*
Lis H, see Yariv J 577, *583*
Lisker N, see Clarke JA 142, *157*
Liu T, see Robbins PW 175, 178, *192*
Liu TJ, see Kolattukudy PE 246, *251*
Liu TYJ, see Kolattukudy PE 246, *251*
Liu TYJ (1972) M.S. Thesis, Washington State University, Pullman, Washington, USA 243, *252*
Lizada C, see Yang SF 143, *165*
Lloyd KO 442, *457*
Lock M, see Abercrombie M 420, *432*
Lockhart CM, Rowell P, Stewart WDP 659, *676*
Lockhart JA 255, 262, 263, *271*
Lodder J (ed) 420, *436*
Lodish HF, see Katz FN 180, *191*
Lodish HF, see Rothman JE 180, *192*, 468, *482*, 493, *511*
Loerch JD, see Vidra JD 50, *67*
Loescher WH, Nevins DJ 92, *123*
Loescher WH, Nevins DJ 32, 40, *44*, 264, 268, *271*
Loescher WH, see Nevins DJ 92, *125*, 150, *162*, 264, *271*
Loewus F, Chen MS, Loewus MW 40, *44*
Loewus F, Labarca C 155, *161*
Loewus F, Wagner G, Schiff JA, Weistrop J 295, *312*
Loewus FA, see Rosenfield CL 155, *163*
Loewus MW, see Loewus F 40, *44*
Lonchampt M, see Huet C 634, *654*
Long M, see Keen NT 597, 599, 600, *619*
Longly B, Waterkeyn L 518, *536*
Loomis RS, see Hall AE 606, *618*
Loomis WD 142, *161*
Loontiens FG, Van Wauwe JP, De Bruyne CK 631, *655*
Lopez-Fando C, see Felipe Anton MR 565, *570*
Lopez-Romero E, Ruiz-Herrera J 380, *389*

López-Romero E, Ruiz-Herrera J 398, 399, 400, 402, *413*, 432, *436*, 504, *509*
López-Romero E, Ruiz-Herrera J, Bartnicki-García S 384, *389*, 403, *413*
Lopez-Romero E, see Ruiz-Herrera J 399, 402, 404, *414*
López-Sáez JF, Risueño MC, Gimenez-Martin 516, *536*
López-Sáez JF, see Frey-Wyssling A 518, *534*
Lord JM, see Mellor RB 180, *191*
Lorenzen H, see Kuhl A 335, 346, *350*
Lotan R, Lis H, Sharon N 635, *655*
Lotan R, Skutelsky E, Danon D, Sharon N 629, *655*
Lotan R, see Mirelman D 649, *655*
Lott JNA, see McCandless EL 294, 295, *312*
Loughman BC, see Kemp J 53, 56, *64*, 74, *122*
Louisot P, see Letoublon RM 467, *480*
Louisot P, see Richard M 467, *481*
Love J, Percival Elizabeth 283, 300, *312*
Lovett JS 410, *413*
Lowry OH, see Natowicz MR 497, *510*
Lucas JJ, Waechter CW, Lennarz WJ 175, *191*
Lucas JJ, see Waechter CJ 173, *193*
Ludwig CH, see Sarkanen KV 215, 216, *224*
Luh BS, see Postlmayr HL 104, *127*
Lui A, see Gander JE 467, *479*
Lundquist K, see Adler E 210, *222*
Lüning HU, Schlösser E 609, 612, *620*
Lüning K, Müller DG 347, *350*
Lüning K, see Müller DG 347, *350*
Lyall V, see Davies DR 22, *23*, 146, *157*
Lynch RG, see Hickman S 184, *190*, 493, *508*
Lyon G, Albersheim P 603, 604, *620*
Lyons JM 110, *123*
Lystash JC, see Pike CS 143, *162*

Ma Y, see Sun SM 181, *193*
Mabry TJ 611, *620*
MacAlpine GA, see Brady CJ 104, *115*
Maccecchini ML, Rudin Y, Blobel G, Schatz G 492, *509*
MacCully ME, see Bell JK 562, *569*
MacDonald K, Pickett-Heaps JD 322, *331*
MacDonald PW, Strobel GA 606, *620*
Macey MJK, Stumpf PK 243, *252*
Machado RD, see Porter KR 514, 518, *538*
Machala S, see Košinová A 469, *479*
Machala S, see Vrsanska M 406, *415*
Machida Y, see Yamaki S 104, 105, *132*
Machlis L, see Aronson JM 354, 355, *384*
Mackay VL, see Crandall M 421, *434*
MacKay VL, see Tkacz JS 453, *458*
MacKenzie S, see Gorin PAJ 420, *435*
Mackie IM, Percival Elizabeth 282, 300, *312*
Mackie W 286, *312*
Mackie W, Preston RD 10, 11, 12, *24*, 277, 283, *312*
Mackie W, Sellen DB 282, 283, *312*
Mackie W, see Atkins EDT 286, *307*
Mackie W, see Hirst Sir E 299, 300, *310*
Maclachlan GA 56, *65*, 267, *271*, 528, *536*
Maclachlan GA, Duda CT 92, *123*
Maclachlan GA, see Dürr M 70, 71, 81, 83, *117*
Maclachlan GA, see Datko AH 267, *270*
Maclachlan GA, see Fan DF 150, *158*, 267, *270*
Maclachlan GA, see Raymond Y 48, 54, *66*, 73, 111, *127*, 521, *539*
Maclachlan GA, see Shore G 53, 54, 59, *66*, 521, 525, *540*
Maclachlan GA, see Spencer FS 52, 56, 59, *66*
Maclachlan GA, see Verma DPS 267, *273*
MacLachlan J, McInnes AG, Falk M 326, *331*
Maclead H, see Horowitz NH 374, *388*
MacLeod RD 544, 566, *572*
Macmillan JD, see Mann JW 424, *436*
MacWilliam IC 359, *390*
Maddox IS, Hough JS 421, 422, *436*
Mader M, Meyer Y, Bopp M 150, *161*
Madgwick J, Haug A, Larsen B 289, 305, *312*
Madgwick JC, Ralph BJ 296, *312*
Maggs J, see Schreiber G 493, *511*
Maglothin A, see English PD 589, 591, *618*
Magyarosy AC, Buchanan BB, Schurmann P 606, *620*
Mahadevan PR, Mahadkar UR 377, 378, *390*, 430, *436*
Mahadevan PR, Tatum EL 353, 363, 365, 372, *390*, 428, *436*
Mahadkar UR, see Mahadevan PR 377, 378, *390*, 430, *436*
Mahlberg PG, Turner FR, Walkinshaw C, Venketeswaren S, Mehrotra B 516, *536*
Mahoney WC, Duskin D 183, *191*
Maier K, Müller H, Tesch R, Trolp R, Witt I, Holzer H 401, *414*
Maier RJ, Brill WJ *620*
Maier VP, see Hasegawa S 104, 105, *120*
Maiti IB, Kolattukudy PE 247, *252*
Maiti IB, Kolattukudy PE, Shaykh M 239, *252*
Maitra SC, De DN 545, 548, *572*

Maitra SK, see Ballou CE 406, *411*
Maitra SK, see Nakajima T 443, *457*
Maizel JV, Burkhardt HJ, Mitchell HK 612, *620*
Maizel JV, see Burkhardt HJ 610, 612, *616*
Majak W, see Craigie JS 302, *308*
Majumder SK, Leopold AC 112, *123*
Makai K, see Yamamoto R 261, *273*
Maley F, see Tarentino AL 442, 443, 449, 450, 451, *458*
Maley F, see Trimble RB 442, 443, 449, 450, 452, *458*, 499, *512*
Malhotra OP, Singh H 139, *161*
Maltby D, Carpita NC, Montezinos D, Kulow C, Delmer DP 93, *123*
Mandels M 430, *436*
Mangin ML 112, *123*
Mani UV, Akiyama Y, Mohrlok S, Lamport DTA 135, 153, *161*
Mankarios AT, Hall MA, Jarvis MC, Threlfall DR, Friend J 150, *161*
Mańkowski T, Jankowski W, Chojnacki T, Franke P 474, *480*
Mankowski T, see Palamarczyk G 474, *481*
Manley RStJ, see Cooper D 526, *533*
Mann JW, Heinz CE, Macmillan JD 424, *436*
Manners DJ, Masson AJ, Patterson JC 359, 367, *390*, 417, 418, *436*
Manners DJ, Masson AJ, Patterson JC, Bjorndal H, Lindberg B 417, 418, 427, *436*
Manners DJ, Meyer MT 419, *436*
Manners DJ, see Evans RB 419, *434*
Manners DJ, see Fleet GH 359, *387*, 417, 418, 425, *434*
Manners DJ, see Rombouts FM 424, 425, *438*
Manney TR, Meade JH 453, *457*
Manocha MS, Colvin JR 355, 363, 365, 366, *390*
Manocha MS, see Letourneau DR 356, 363, 366, *389*
Manocha MS, see Stobbs LW 112, *130*
Mansell RL, Gross GG, Stöckigt J, Franke H, Zenk MH 209, *223*
Mansell RL, Stöckigt JT, Zenk MH 207, *223*
Mansell RL, see Gross GG 220, *222*
Mansell RL, see Stöckigt J 208, *224*
Manshard E, see Mix M 326, *331*
Mantle PG, see Dickerson AG 595, *617*
Manton I 318, 320, *331*, 519, *536*
Marcan H, Jarvis MC, Friend J 603, 605, *620*
Marchalonis JJ, see Knox RB 237, *250*
Marchant HJ 60, *65*
Marchant HJ, Fowke LC 347, *350*
Marchant HJ, Hines ER 60, *65*
Marchant HJ, Pickett-Heaps JD 322, *331*
Marchant R 366, 369, *390*
Marchant R, Smith DG 370, *390*, 406, *414*, 422, *436*
Marchant R, Wessels JGH 376, *390*
Marchant R, see Patton AM 382, *391*
Marchant R, see Valk P van der 353, 355, 356, 363, 365, 366, 367, 370, 374, *393*, 397, *415*
Marchant R, see Wessels JGH 358, 360, 362, 363, 364, 365, 366, 371, 374, 375, 376, 377, *394*, 427, 429, *440*
Marchessault RH, Sundararajan PR 524, *536*
Marchessault RH, see Bobitt TF 360, 363, 365, *385*
Marchessault RH, see Pérez S 361, *391*
Mares DJ, Stone BA 32, 39, *44*, 106, *123*
Maret R 356, 357, 362, 263, 365, 367, *390*
Margolis HC, Nakagawa Y, Douglas KT, Kaiser ET 451, 452, *457*, 496, *509*
Margulis L 134, *161*
Mark RE 10, 13, *24*, 522, *536*
Markwalder HU, Neukom H 39, 41, *44*, 86, *123*, 150, *161*
Marmé D, see Hertel C 324, *330*
Maronde D, see Heldt HW 607, *619*
Marré E 263, *271*, 597, 601, *620*
Marré E, see Beffagna N 601, *616*
Marriot KM, Tanner W 494, *509*
Marriott KM, Tanner W 188, *191*, 470, *481*
Marriott MS 468, 470, *481*
Marsh HV, see Cheng CKC 220, *222*
Marshall DC, see Penny P 262, *272*
Marshall DC, see Penny D 262, *272*
Marshall JJ 31, *44*, 424, *436*
Marshall KC 660, *676*
Marshall KC, Cruickshank RH, Bushby HVA 670, *676*
Marshall RD 74, *124*, 472, *481*
Martin EM, Harris WM 543, 545, 549, *572*
Martin HG, Thorne KJ 167, 187, *191*
Martin JT 248, *252*
Martin JT, Juniper BE 225, 226, 236, 239, *252*, 529, 531, *537*
Martin JT, see Holloway PJ 226, 231, *250*
Martinez PR, see Pont Lezica R 58, *65*
Maruyama Y, see Kato G 663, 666, 667, *675*
Marx-Figini M 13, *24*, 30, *44*, 47, *65*, 524, *537*
Marx-Figini M, Schulz GV 30, *44*, 47, *65*, 525, *537*
Mascarenhas JP 152, *161*
Mason TJ, see Hamilton RC 526, *535*

Masri MS, see Finkle BJ 205, *222*
Masson AJ, see Manners DJ 359, 367, *390*, 417, 418, 427, *436*
Masuda Y 256, 259, 260, 262, 267, 268, *271*
Masuda Y, Yamamoto R 266, 267, *271*
Masuda Y, Yamamoto R, Kawamura H, Yamagata Y 261, 262, *271*
Masuda Y, see Fujihara S 259, 260, *270*
Masuda Y, see Iwami S 261, *270*
Masuda Y, see Kamisaka S 262, 263, *271*
Masuda Y, see Kawamura H 262, 263, *271*
Masuda Y, see Nishitani K 262, 268, *272*, 513, *537*
Masuda Y, see Sakurai N 59, *66*, 92, *128*, 262, 267, 268, *272*, *273*
Masuda Y, see Tanimoto E 262, 267, *273*
Masuda Y, see Wada S 267, *273*
Masuda Y, see Yamagata Y 261, 262, 263, *273*
Masuda Y, see Yamamoto R 259, 262, 261, 267, *273*
Masuda Y, see Zarra I 92, *132*, 262, *273*
Mateles RI, see Jannasch HW 594, *619*
Mathers DG, see Hala CN 660, *675*
Matheson NK, see McCleary BV 107, *124*
Mathews MB, see Milstein C 488, *510*
Matic M 227, *252*
Matile P 105, *124*, 375, *390*
Matile P, Cortat M, Wiemken A, Frey-Wyssling A 422, *436*, 469, *481*, 498, *509*
Matile P, Moor H, Robinow C 498, *509*
Matile P, see Cortat M 377, 378, *386*, 422, 423, *434*, 470, *478*, 498, *507*
Matile P, see Iten W 375, *388*
Matile P, see Lenney JF 401, 402, *413*
Matile P, see Wiemken-Gehrig V 105, *131*
Matsuda K, see Katō Y 32, *44*
Matsuda K, see Nakajima T 428, *437*
Matsuda K, see Satoh H 56, 60, *66*
Matsuda K, see Satoh S 74, *128*, 525, *539*
Matsuda K, see Shida M 360, *392*
Matsuda K, see Yamagishi T 152, *164*
Matsuda K, see Yamaki S 104, 105, *132*
Matsuda Y, Tamaki S, Tsubo Y 340, *350*
Matsuda Y, see Schreiber G 493, *511*
Mattia E, see Cassone A 17, *23*
Mau S-L, Raff J, Clarke AE 581, *583*
Mayer AM, Harel E 111, *124*
Mayer H, see Müller K 640, *655*
Mayer RM 465, *481*
Mazau D, see Esquerré-Tugayé MT 111, *118*, 141, *158*, 608, *618*
Mazelis M, see Oputa CO 109, *126*
Mazliak P 244, *252*
Mazurek M, see Gorin PAJ 445, *456*
Mazzucchi U, Bazzi C, Pupillo P 587, *620*
Mazzucchi U, Pupillo P *620*
McCalla DR, Neish AC 202, *223*
McCandless EL, Craigie JS, Hansen JE 294, *312*
McCandless EL, Craigie JS, Walter JA 294, *312*
McCandless EL, Evelegh MJ, Vollmer CM, DiNinno VL 294, *312*
McCandless EL, Okada WT, Lott JNA, Volmer CM, Gordon-Mills EM 294, 295, *312*
McCandless EL, see DiNinno VL 294, *308*
McCandless EL, see Gordon EM 278, *309*
McCandless EL, see Gordon-Mills EM 278, 294, 295, *309*
McCandless EL, see Horsford SPC 294, *310*
McCandless EL, see Johnston KH 294, *311*
McCleary BV, Matheson NK 107, *124*
McClendon JH, Nolan WG, Wenzler HF 107, *124*
McClure K, Park D, Robinson PM 382, *390*
McClure WK, Park D, Robinson PM 500, *509*
McCullough J, Herskowitz I 455, *457*
McCullough W, see Atkinson AW 322, *329*
McCully ME 278, 299, *312*
McCully ME, see Bell JK 97, *115*
McCully ME, see Fulcher RG 90, 91, *119*, 518, *534*
McCully ME, see Smith MM 48, *66*, 90, 112, *129*, 518, *540*
McCully ME, see Sutherland J 90, 97, 98, *130*
McDonough JP, Springer WR, Barondes SH 638, *655*
McDowell RH, see Percival Elizabeth 277, 278, 284, 291, 299, 303, 306, *314*
McGinnis GD, see Shafizadeh F 29, 30, *45*, 47, *66*, 524, *539*
McGlasson WB, see Brady CJ 104, *115*
McInnes AG, see MacLachlan J 326, *331*
McIntosh JR, see Coss RA 60, *62*
McKay J, see Bouveng HO 427, *433*
McKean DJ, see Schechter I 489, *511*
McKemmie , see Poyton RO 492, *510*
McKeon T, see Pollard MR 530, *538*
McKinnell JP, Percival Elizabeth 280, 300, 301, *312*
McKnight GS, see Kiely ML 178, *191*, 468, *479*, 493, *509*
McLachlan J, see Craigie JS 302, *308*

McLean RJ 346, *350*
McMurrough I, Bartnicki-Garcia S 379, *390*, 398, 399, 400, 402, *414*
McMurrough I, Flores-Carreon A, Bartnicki-Garcia S 398, *414*
McMurrough I, Rose AH 417, *436*
McMurrough I, see Bartnicki-Garcia S 367, *385*, 421, *433*
McNab C, see Grant GT 86, 101, *119*, 553, *570*
McNab CGA, see Arnott S 293, *307*
McNab JM, Villemez CL, Albersheim P 71, *124*
McNairn RB 112, *124*
McNairn RB, Currier HB 112, *124*
McNairn RB, see Eschrich W 112, *118*
McNeil M, Albersheim P, Taiz L, Jones RL 264, *271*
McNeil M, Darvill AG, Albersheim P 8, *8*, 29, 31, 32, 33, 34, 35, 37, 38, 41, *44*, 68, 86, *124*, 134, *161*
McNeil M, see Burke D 264, *269*
McNeil M, see Darvill AG 6, *8*, 34, *42*
McNeil M, see Strobel GA 597, 598, *622*
McNeil M, see Valent BS 92, *131*
Meade JH, see Manney TR 453, *457*
Medcalf DG, Larsen B 296, 297, *312*
Medcalf DG, Schneider TL, Barnett RW 296, 297, *312*
Medcalf DG, Scott JR, Brannon JH, Hemerick GA, Cunningham RL, Chessen JH, Shah J 304, *312*
Megarry ML, see Bourne EJ 280, 283, 301, *307*
Mehrotra B, see Mahlberg PG 516, *536*
Meier H 76, 93, 107, 109, *124*
Meier H, Reid JSG 85, 107, *124*
Meier H, see Bouveng HO 109, *115*
Meier H, see Franz G 52, 54, 55, *63*
Meier H, see Huwyler HR 48, *64*, 93, *121*, 513, *536*
Meier H, see Reid JSG 107, *127*, *128*, 107, *128*
Meinert MC, Delmer DP, 32, 40, *44*, 48, *65*, 513, *537*
Mekada E, see Uchida T 650, *657*
Melchers F 178, *191*
Mellman IS, see Miller DH 145, *161*
Mellor RB, Roberts LM, Lord JM 180, *191*
Mendgen K 113, *124*
Mendiara S, see Lavinthman N 55, *64*
Mense RM, see Beevers L 57, *62*
Menzel H, see Hager A 142, *159*, 256, 258, *270*
Mepham RH, Lane GR 99, *124*
Mercer FV, see Dart PJ 565, 568, *570*
Mercereau-Puijalon O, Royal A, Cami B, Garapin A, Krust A, Gannon F, Kourilsky P 491, *509*
Meredith WOS, see Ballance GM 106, *115*
Merkel GJ, Durham DR, Perry JJ 148, *161*
Merrifield EH, see Churms SC 6, *8*, 75, *116*
Mertz J, Nordin P 83, *124*
Métraux JP, Taiz L 265, *271*
Meyer DI, Burger MM 502, *510*
Meyer GM, see Dunlop PC 449, *456*
Meyer J, see Lenney JF 401, 402, *413*
Meyer MT, Phaff HJ 421, *436*
Meyer MT, see Manners DJ 419, *436*
Meyer Y, Abel WO 60, *65*
Meyer Y, Herth W 60, 61, *65*
Meyer Y, see Herth W 524, 525, *535*
Meyer Y, see Mader M 150, *161*
Meylan BA, Butterfield BG 95, *124*
Meylan BA, see Butterfield BG 95, *116*
Mialonier G, Privat JP, Monsigny M, Kahlem G, Durand R 644, *655*
Michalenko GO, Hohl HR, Rast D 363, 365, *390*
Middlebrook MJ, Preston RD 366, *390*
Migita N, Nakano J, Okada T, Ohashi G, Noguchi M 196, *223*
Migita S, see Suzuki I 633, *656*
Mihara S, Hase E 335, 341, *350*
Mikkelsen JD 246, *252*
Miksche GE, see Adler E 210, *222*
Miljanich P, see Akazawa T 611, *615*
Mill PJ 417, *436*
Millar RL, Higgins VJ 613, *620*
Millbank JW, see Anderson FB 424, *432*
Miller DH 145, *161*
Miller DH, Mellman IS, Lamport DTA, Miller M 145, *161*
Miller DH, see Lamport DTA 135, 136, *160*
Miller JH 516, *537*
Miller M, see Miller DH 145, *161*
Miller MW, Yoneyama M, Soneda M 420, *437*
Milleship A, see Schreiber G 493, *511*
Millington WF, see Gawlik SR 346, 347, *349*
Mills GL, Cantino EC 400, *414*
Mills JJ, Adamany AM 168, 187, *191*
Mills RR, see Dashek WV 150, *157*
Milstein C, Brownlee GG, Harrison TM, Mathews MB 488, *510*
Min KH, Sasaki SF, Kashiwabara Y, Umekawa M, Nisizawa K 284, *312*
Minami SA, Goodenough UW 329, *331*
Minetti M, Aducci P, Teichner A 628, *655*
Minetti M, Teichner A, Aducci P 628, *655*

Minke R, Blackwell J 395, *414*
Mirelman D, Galun E, Sharon N, Lotan R 649, *655*
Mirelman D, see Barkai-Golan R 649, *652*, 659, *674*
Mirelmann D, see Eshdat Y 633, *653*
Misaki A, Johnson J, Kirkwood S, Scaletti JV, Smith F 417, *437*
Misaki A, Johnson J Jr, Kirkwood S, Scaletti JV, Smith F 359, *390*
Misaki A, Tsumuraya Y, Takaya S 429, *437*
Misaki A, see Azuma I 466, *477*
Misaki A, see Johnson J 427, *435*
Misaki A, see Saito H 427, *438*
Misaki A, see Shibuya N 32, 33, 39, *45*, 106, *129*
Misaki A, see Sone Y 428, *439*
Misato T, see Endo A 400, 408, *412*
Misato T, see Hori M 398, 400, *413*
Misato T, see Ohta N 398, 400, *414*
Mishra NC 372, *390*
Mishra NC, Tatum EL 372, *390*, 432, *437*
Mitani M, Inoue Y 400, *414*
Mitchell A, Taylor IF 353, *390*
Mitchell AP, see Zubenko GS 403, *415*
Mitchell B 145, 148, *161*
Mitchell HK, see Burkhardt HJ 610, 612, *616*
Mitchell HK, see Maizel JV 612, *620*
Mitchell ND, see Greenlagh JR 612, *618*
Mitchell R, see Chet I 662, *674*
Mitzel D, see Gold MH 370, *387*
Miwa T, Iriki Y, Suzuki T 282, 283, *313*
Miwa T, see Iriki Y 282, *311*, 283, *310*
Miwa T, see Takeda H 283, *316*
Mix M 325, *331*
Mix M, Manshard E 326, *331*
Miyamoto C, Tamari K 52, *65*, 73, *124*
Miyamoto C, see Hara M 60, *63*
Miyazaki T, Irino T 361, *390*
Miyazaki, see Yamada M 365, *394*
Mizrahi A, O'Malley JA, Carter WA, Takatsuki A, Tamura G, Sulkowski E 493, *510*
Mizuno M, Shimojima Y, Sugawara T, Takeda J 185, *191*
Model PJ, see Jakes KS 491, *509*
Moehring JM, see Gilliland DG 650, *654*
Moehring TJ, see Gilliland DG 650, *654*
Moericke V 112, *124*
Moestrup Ø, Walne PL 320, *331*
Moestrup O, Thomsen HA 320, *331*
Mohrlok S, see Maŋi UV 135, 153, *161*
Moir GFJ, see Sheldrake AR 97, *129*
Molano J, Bowers B, Cabib E 397, *414*
Molano J, Polacheck J, Durán A, Cabib E 405, *414*, 649, *655*
Molè-Bajer J, see Bajer AS 514, *533*
Mollard A, see Chanzy H 524, *533*
Mollenhauer BA, see Mollenhauer HH 517, 518, *537*
Mollenhauer HH 542, 543, 545, 547, 548, 549, *572*
Mollenhauer HH, Mollenhauer BA 517, 518, *537*
Mollenhauer HH, Morré DJ 60, *65*, 516, 521, *537*, 546, 547, 562, 563, 564, *572*
Mollenhauer HH, Morré DJ, Vanderwoude WJ 546, 548, 562, 563, 564, *572*
Mollenhauer HH, Whaley WG 546, 558, *572*
Mollenhauer HH, Whaley WG, Leech JH 543, 545, 546, 558, *572*
Mollenhauer HH, see Hall D 552, *571*
Mollenhauer HH, see Morré DJ 80, 81, 82, 83, 85, *124*, 101, *125*, 513, 514, 519, *537*, 543, 562, 563, 564, *572*
Mollenhauer HH, see Whaley WG 83, *131*, 518, *540*, 545, 546, 564, *574*
Molloy JA, see Aspinall GO 32, *42*
Molner J, Chao H, Ikehara Y 172, *191*
Monro JA, Bailey RW, Penny D 38, *44*, 137, 140, *161*
Monro JA, Penny D, Bailey RW 38, *44*, 264, *271*
Monsigny M, see Mialonier G 644, *655*
Monsigny M, see Rougier M 548, 553, *573*, 555, *573*
Montague MJ, Ikuma H 59, *65*
Montague MJ, see Adams PA 262, 263, *269*
Montezinos D, Brown RM 325, *331*
Montezinos D, see Brown RM 325, *329*
Montezinos D, see Brown RM Jr 55, *62*, 524, *533*
Montezinos D, see Carpita N 68, 101, *116*, 154, *157*
Montezinos D, see Maltby D 93, *123*
Moor H 422, *437*, 469, *481*, 498, *510*
Moor H, see Branton D 321, *329*, 560, *570*
Moor H, see Matile P 498, *509*
Moore AE, Stone BA 105, *124*
Moore AE, see Barras DR 426, 430, *433*
Moore AT, see Tripp VW 134, *164*
Moore PM, Peberdy JF 380, *390*, 398, 399, *414*
Moore PM, see Peberdy JF 398, 399, *414*
Moore S, see Hayashi R 452, *456*
Moorhouse R, see Arnott S 294, *307*
Moran F, Starr MP 592, *620*
Mori M, Eda S, Katō K 32, *44*

Mori M, see Akiyama Y 135, *156*
Morigaki T, see Komamine A 41, *44*
Moriguchi M, Yamamoto K, Kawai H, Tochikura T 398, *414*
Moriguchi M, see Yamamoto K 398, *415*
Morikawa H, Hayashi R, Senda M 21, *24*, 68, *124*
Morikawa H, Senda M 68, *124*
Morikawa N, see Takeuchi Y 92, *130*, 528, *540*
Morita Y, Kominato Y, Shimizu K 149, *161*
Morohashi Y, Bandurski RS 58, *65*
Morozova AB, see Stepanenko BN 52, *67*
Morrall P, Briggs DE 106, *124*
Morré DJ 102, *124*, 519, *537*
Morré DJ, Eisinger WR 260, 263, *271*
Morré DJ, Jones DD, Mollenhauer HH 101, *125*, 543, 562, 563, 564, *572*
Morré DJ, Kartenbeck J, Franke WW 82, *125*
Morré DJ, Mollenhauer HH 80, 81, 82, 83, 85, *124*, 513, 514, 519, *537*
Morré DJ, Van der Woude WJ 83, 84, 86, *124*
Morré DJ, see Courtney JS 261, 262, 264, 265, 268, *270*
Morré DJ, see Hall D 552, *571*
Morré DJ, see Jones DD 549, 550, 551, 552, 562, 563, 564, *571*
Morré DJ, see Mollenhauer HH 60, *65*, 516, 521, *537*, 546, 547, 562, 563, 564, *572*, 548
Morré DJ, see Olson AC 259, 262, *272*
Morré DJ, see Van der Woude WJ 51, 53, *67*, 59, 72, *131*, 82, 83, 85, *131*, 519, *540*
Morris EJ, see Bacon JSD 107, *114*
Morris EO 420, *437*
Morris ER, Rees DA, Thom D 286, *313*
Morris ER, Rees DA, Thom D, Boyd J 286, *313*
Morris ER, see Dea ICM 37, *43*
Morris ER, see Grant GT 37, *43*, 287, *309*
Morris ER, see Liang JN 294, *311*
Morris GH, see Brown HT 106, *115*
Morris MR, Northcote DH 84, *125*
Morrison IM 31, 36, 41, *45*
Morrison IN, O'Brien TP 90, *125*, 514, 515, 518, *537*
Mort AJ 140, *162*
Mort AJ, Lamport DTA 38, *45*, 140, 153, *162*
Mortlock RP, see Mount MS 589, 592, *621*
Mosmann, see Williamson AR 493, *512*
Mota M, Silva MT, Salema R 568, *572*
Mount MS, Berman PM, Mortlock RP, Hubbard JP 589, 592, *621*
Mount MS, see Hubbard JP 589, *619*
Mount MS, see Stack JP 589, *622*
Mozer TV, see Stuart DA 135, *164*
Mueller SC, Brown RM Jr 94, *125*, 522, 523, *537*
Mueller SC, Brown RM Jr, Scott TK 522, *537*
Mueller SC, see Willison JHM 528, *541*
Muggli R, see Sarko A 280, *315*
Mühlethaler K 522, *537*
Mühlethaler K, see Albersheim P 83, *114*
Mühlethaler K, see Branton D 321, *329*
Mühlethaler K, see Frey-Wyssling A 518, *534*, 529
Mullen JM, Bateman DF 589, 591, *621*
Müller DG, Gassmann G, Lüning K 347, *350*
Müller DG, see Lüning K 347, *350*
Muller H, see Hasilik A 401, 402, *412*
Müller H, see Maier K 401, *414*
Müller K, Gerisch G 640, 641, *655*
Müller K, Gerisch G, Fromme I, Mayer H, Tsugita A 640, *655*
Mullick DB 225, *252*
Mullins JT, Ellias EA 377, *390*
Mullins JT, see Thomas D des S 377, *393*
Mullis RH, Thompson NS, Parham RA 21, *24*
Mumford RA, see Strauss AW 489, *511*
Mundkur B 419, *437*
Mundy BP, see Beltran JP 598, *616*
Murachi T, see Suzuki H 105, *130*
Murao S, see Arai M 419, *432*
Murashige T, see Howard J 641, *654*, 672, *675*
Muray RHA, Northcote DH 141, *162*
Musha Y, Goring DAI 197, *223*
Musolan C, Kindinger J 73, *125*
Musolan C, Ordin L, Kindinger JE 73, 110, *125*
Mussell H, Strand LL 152, *162*
Mussell H, see Strand LL 591, *622*
Mussell HW 593, *621*
Mussell HW, Green RJ Jr 593, *621*
Muzzarelli RA 395, *414*
Muzzarelli RAA 356, *390*
Myers A, see Cronshaw J 279, 280, *308*
Myers RB, see Cantino EC 410, *412*
Myklestad M, see Smestad B 305, *315*
Myklestad S, see Haug A 303, *310*
Myllyla R, Kuutti-Savolainen ER, Kivirikko KI 143, *162*
Myllyla R, Schubotz LM, Weser U, Kivirikko KI 143, *162*
Myllyla R, Tuderman L, Kivirikko KI 143, *162*

Nachmias A, Barash I, Buchner V, Solel Z, Strobel GA 597, 599, 600, *621*
Nagahashi J, Beevers L 70, *125*, 519, *537*
Nagasaki S, see Yamamoto S 424, *440*, 425
Nagata T, Takebe I 30, *45*
Nagel CW, Patterson ME 104, *125*
Nagel SC, see Atalla RH 524, *532*
Nagl W 641, *655*
Nakagawa Y, see Margolis HC 451, 452, *457*, 496, *509*
Nakajima T, Ballou CE 441, 442, 446, 452, *457*, 462, 464, 472, 474, 476, *481*
Nakajima T, Maitra SK, Ballou CE 443, *457*
Nakajima T, Tamari K, Matsuda K, Tanaka H, Ogasawara N 428, *437*
Nakajima T, see Smith WL 465, 476, *482*
Nakamura L, see Arthur LO 595, *615*
Nakamura N, Tanabe O 427, *437*
Nakamura S, Arai M, Karasawa K, Yonehara H 185, *191*
Nakamura T, Sekine S, Arai K, Takahashi N 262, 263, *271*
Nakamura W, Nozu Y 213, *223*
Nakamura Y, Fushiki H, Higuchi T 218, *223*
Nakano J, see Migita N 196, *223*
Nakashima K, see Fujikawa T 296, *309*
Nakatsubo F, see Higuchi T 203, 206, 218, 219, *223*
Nakayama K, Araki Y, Ito E 462, 463, *481*
Nakumara M, see Kato G 663, 666, 667, *675*
Nakumura W 212, *223*
Namba H, Kuroda H 432, *437*
Namba Y, see Obata T 424, 425, *437*
Napoli CA, see Dazzo FB 670, *675*
Nasada Y, see Shimoda C 422, *439*
Natowicz MR, Chi M M-Y, Lowry OH, Sly WS 497, *510*
Necas O 368, 369, *391*
Neish AC 54, *65*, 69, *125*, 203, *223*
Neish AC, see Brown SA 201, *222*
Neish AC, see Freudenberg K 36, *43*, 210, 215, 217, *222*
Neish AC, see Gamborg OL 202, *222*
Neish AC, see McCalla DR 202, *223*
Neish AC, see Young MR 203, *224*
Nelmes BJ, Preston RD 104, *125*
Nelson RF, see Finkle BJ 205, *222*
Nelson T, see Huotari F 425, *435*
Nelson TE 425, *437*
Nelson TE, Johnson J, Jantzen E, Kirkwood S 425, *437*
Nelson TE, see Johnson J 427, *435*
Nesbitt LR, see Sietsma JH 355, 358, 363, 365, 368, *392*, 426, 428, *439*
Ness G, see Keller RL 168, *191*
Nester EW, see Watson B 586, *623*
Netzer D, Kritzman G, Chet I 608, *621*
Neuberger A, see Allen AK 133, *156*, 135, 136, 631, *652*, 633, 635, 636
Neuberger A, see Ashford D 134, *156*
Neufeld EF, Feingold DS, Hassid WZ 81, *125*
Neufeld EF, Ginsburg V, Putman EW, Fanshier D, Hassid WZ 81, *125*
Neufeld EF, Hall CW 79, *125*
Neufeld EF, see Feingold DS 48, 51, 52, *63*, 70, 73, 81, *118*
Neufeld EF, see Hickman S 497, *508*
Neufeld EF, see Kornfeld S 398, *413*
Neukom H, see Geissmann T 76, *119*
Neukom H, see Markwalder HU 39, 41, *44*, 86, *123*, 150, *161*
Neumann NP, Lampen JO 442, 449, 450, *457*
Neurath H, see Kuhn RW 452, *456*
Nevins DJ, Loescher W 150, *162*
Nevins DJ 40, *45*, 139, 150, *162*, 267, *271*
Nevins DJ, Huber DJ, Yamamoto R, Loescher WH 92, *125*, 264, 271
Nevins DJ, see Bateman DF 589, *616*
Nevins DJ, see Huber DJ 150, *159*, 267, *270*
Nevins DJ, see Loescher WH 32, 40, *44*, 92, *123*, 264, 267, 268, *271*
Nevins DJ, see Sakurai N 92, *128*, 267, 268, *273*
Nevins DJ, see Yamamoto R 266, 267, *273*
Nevo Z, Sharon N 144, *162*
Newcomb EH 525, *537*
Newcomb EH, see Bonnett HT Jr 84, *115*
Newcomb EH, see Hepler PK 85, 98, *120*, 517, *535*
Newell PC 638, *656*
Newmark P 497, *510*
Newsam RJ, see Craig GD 409, *412*
Nguyen HT, see Schneider EG 185, *192*
Nicholson MD, see Johnson DC 41, *43*
Nickerson WJ 374, *391*
Nickerson WJ, see Bartnicki-Garcia S 357, *385*
Nickerson WJ, see Kessler G 418, *436*
Nicolai E, Preston RD 321, *331*
Nicolson GL 553, *572*, 637, *656*
Nicolson L, see Singer SJ 522, *540*
Nicolson TH, see Old KM 565, *572*
Niederpruem DJ 376, *391*
Niederpruem DJ, Wessels JGH 375, *391*

Niederpruem DJ, see Wessels JGH 375, 376, *394*, 430, *440*
Nieduszynski IA, Atkins EDT 280, *313*
Nikaido H, Hassid WZ 47, *65*, 281, *313*
Niles RM, see Hubbard JP 589, *619*
Nisbet LJ, see Dickerson AG 595, *617*
Nishikawa Y, Tanaka M, Shibata S, Fukuoka F 427, *437*
Nishio T, Hinata K 580, *583*
Nishitani K, Masuda Y 268, *272*
Nishitani K, Shibaoka H, Masuda Y 262, *272*, 513, *537*
Nishitani K, see Sakurai N 92, *128*, 268, *273*
Nisizawa K, see Iriki Y 282, *311*
Nisizawa K, see Min KH 284, *312*
Nisizawa K, see Suzuki H 105, *130*
Nisizawa K, see Takeda H 283, *316*
Nitsan J, see Frankel R 100, *119*
Noat G, see Crasnier M 151, *157*
Noguchi M, see Migita N 196, *223*
Noguchi T, see Kitazima Y 355, *389*
Nolan RA, Bal AK 501, *510*
Nolan WG, see McClendon JH 107, *124*
Nombela C, see Del Rey FD 378, *386*, 423, 430, 431, *434*
Nombela C, see Santos T 377, *392*, 423, *438*, 430, 431
Noodén LD, Thimann KV 258, *272*
Nord FF, see Acerbo SN 201, *221*
Nordin JH, see Bardalaye PC 466, *478*
Nordin JH, see Batra KK 75, *115*
Nordin JH, see Bobitt TF 360, 363, 365, *385*
Nordin JH, see Samuel O 467, *482*
Nordin P, see Elnaghy MA 49, *63*
Nordin P, see Jilka R 83, 84, *122*
Nordin P, see Mertz J 83, *124*
Norman J, Giddings TH, Cantino EC 398, *414*
Northcote DH 29, 30, 31, 33, 35, 36, 39, 40, 41, *45*, 70, 82, 83, 90, 98, 99, *125*, 419, *437*, 514, 521, 522, 525, *537*, 557, 560, *572*
Northcote DH, Goulding KJ, Horne RW 144, *162*, 280, *313*
Northcote DH, Horne RW 359, *391*, 417, *437*
Northcote DH, Lewis DR, 81, *125*
Northcote DH, Pickett-Heaps JD 83, *125*, 543, 545, 558, 560, *572*
Northcote DH, Wooding FBP 40, *45*, 80, 85, 98, 110, *125*
Northcote DH, see Allen DM 318, *329*
Northcote DH, see Bailey DS 17, *22*
Northcote DH, see Barrett AJ 34, 37, *42*
Northcote DH, see Bell DJ 417, *433*
Northcote DH, see Boffey SA 39, *42*
Northcote DH, see Bowles DJ 81, 83, 84, *115*, 519, 521, *533*, 556, 557, 558, *569*
Northcote DH, see Branton D 321, *329*
Northcote DH, see Brett CT 53, 58, *62*, 70, 74, 75, *115*
Northcote DH, see Bright SWJ 30, *42*
Northcote DH, see Chadwick CM 81, *116*
Northcote DH, see Dalessandro G 40, *42*, 76, 77, 79, 81, *117*
Northcote DH, see Green JR 73, 101, *119*, *120*, 519, 520, *535*, 556, 557, 560, *570*
Northcote DH, see Haddon LE 80, *120*
Northcote DH, see Hanke DE 18, *23*, 30, *43*, 528, *535*
Northcote DH, see Harris PJ 83, *120*, 549, 550, 551, 552, 556, *571*
Northcote DH, see Heath MF 136, *159*
Northcote DH, see Jeffs RA 39, *43*
Northcote DH, see Korn ED 418, *436*
Northcote DH, see Lamport DTA 134, 149, 151, *160*, *161*
Northcote DH, see Morris MR 84, *125*
Northcote DH, see Muray RHA 141, *162*
Northcote DH, see Pickett-Heaps JD 83, 85, *126*, 516, *538*
Northcote DH, see Roberts K 83, 85, 91, 98, *128*, 137, *163*, 514, *539*
Northcote DH, see Rubery PH 39, *45*, 203, *224*
Northcote DH, see Sentandreu R 407, *415*, 422, *438*, 476, *482*, 498, *511*
Northcote DH, see Stoddart RW 37, *46*, 39, *45*
Northcote DH, see Thornber JP 39, *46*, 76, 93, *130*
Northcote DH, see Wooding FBP 40, *46*, 83, *132*
Northcote DH, see Wright K 101, *132*, 521, *541*, 551, 552, 553, *574*
Notario V, Villa TG, Benitez T, Villanueva JR 420, 423, *437*
Notario V, Villa TG, Villanueva JR 420, *437*
Notario V, see Villa TG 420, 421, 423, *439*
Notario V, see Villanueva JR 373, *393*
Novaes-Ledieu M, Jimenez-Martinez A 358, *391*
Novaes-Ledieu M, see Garcia Mendoza C 374, *387*
Novick P, Field C, Schekman R 469, *481*
Novick P, Schekman R 469, *481*, 503, *510*
Nowak-Ossorio M, Gruber E, Schurz J 48, 56, *65*, 524, *537*
Nozawa Y, see Cole GT 365, *386*

Nozawa Y, see Kitazima Y 355, *389*
Nozu Y, see Nakamura W 213, *223*
Nozue M, Tomiyama K, Doke N 587, *621*
Nukaya A, see Poovaiah BW 104, 105, *127*
Nukushina Y, see Sakurada I 13, *24*
Nunn JR, Parolis H 291, *313*
Nunn JR, Parolis H, Russell I 291, *313*
Nunn JR, see Allsobrook AJR 291, *306*
Nutman PS 661, *676*
Nutman PS, see Kumarasinghe RMK 671, *676*

Oades JM, see Turchenek LW 567, *574*
Obaidah MA, Buck KW 360, *391*
Obata T, Fujioka K, Hara S, Namba Y 424, 425, *437*
O'Brien PJ, see Kornfeld S 398, *413*
O'Brien RL, Parker JW, Dixon JFP 628, 637, 638, *656*
O'Brien TP 82, 94, 95, 97, *125*, *126*, 514, 518, 519, 529, 530, *537*
O'Brien TP, Carr DJ 225, *252*
O'Brien TP, Kuo J 225, *252*
O'Brien TP, Thimann KV 94, 95, *126*
O'Brien TP, see Fulcher RG 107, *119*
O'Brien TP, see Morrison IN 90, *125*, 514, 515, 518, *537*
O'Brien TP, see Tippett JT 531, *540*
O'Brien TP, see Zee S-Y 225, *254*
Obst TR, see Harkin JM 210, 213, *222*
Ocampo-Paus R, Friedmann I 346, *350*
Odoardi M, see Palmieri S 40, *45*, 149, *162*
O'Donnell JJ, Percival Elizabeth 300, *313*
Odzuck W, Kauss H 71, 75, *126*
Ofek I, see Eshdat Y 633, *653*
Ogasawara N, see Nakajima T 428, *437*
Ohashi G, see Migita N 196, *223*
Ohashi H, see Shimada M 205, *224*
Ohiwa T 347, *350*
Ohlenbusch HD, see Greiling H 449, *456*
Ohlrogge AJ, see Floyd RA 549, 550, 551, 552, 555, 562, 566, 567, *570*
Ohnishi A, see Eda S 31, *43*
Ohta N, Kakiki K, Misato T 398, 400, *414*
Ohyama K, Pelcher LE, Schaefer A, Fowke LC 586, *621*
Ohyama K, see Harada H 60, *64*
Okada T, see Migita N 196, *223*
Okada WT, see McCandless EL 294, 295, *312*
Okada Y, see Uchida T 650, *657*
Okamoto K, Kitano H, Akazawa T 106, *126*
O'Kelley JC, Deason TR 348, *350*
O'Kelley JC, Herndon WR 348, *350*
O'Kelley JC, see Gilbert WA 348, *349*
Oki L, see Howard J 641, *654*, 672, *675*
Okina M, see Takatsuki A 183, *193*
Old KM, Nicolson TH 565, *572*
Olden L, Pratt RM, Yamada KM 493, *510*
Olesen P 225, *252*, 522, 531, 532, *537*
Olive LS, see Furtado JS 317, *330*
Oliver GJA, Hemming FW 175, *192*
Olsnes S, Haylett T, Refsnes K 650, *656*
Olsnes S, Pihl A 650, *656*
Olsnes S, see Sandvig K 650, *656*
Olson AC, Bonner J, Morré DJ 259, 262, *272*
O'Malley JA, see Mizrahi A 493, *510*
O'Neill AN 296, 297, *313*
O'Neill MA, Selvendran RR 135, 140, *162*
Onishi HR, Tkacz JS, Lampen JO 451, *457*, 473, *481*
Onishi HS, Tkacz JS, Lampen JO 495, 496, *510*
Opheim DJ 451, *457*
Öpik H, see Street HE 543, 545, 549, *573*
Oputa CO, Mazelis M 109, *126*
Orcutt BC, see Dayhoff MO 141, *157*
Ordin L 109, *126*
Ordin L, Hall MA 51, 52, 59, *65*, 71, 72, *126*
Ordin L, see Delmer DP 54, *63*
Ordin L, see Hall MA 59, *63*
Ordin L, see Katz M 92, *122*, 267, *271*
Ordin L, see Musolan C 73, 110, *125*
Ordin L, see Pinsky A 58, *65*
Ordin L, see Plaut Z 109, *126*
Ordin L, see Van Der Woude WJ 51, 53, *67*, 72, *131*
O'Reilly J, see Chattaway FW 419, *434*
Orgell WH 227, *252*
Oroszlan S, see Schultz AH 493, *511*
Orton WL, see Taylor NW 453, 454, *458*
Osawa T, see Toyoshima S 634, 637, *657*
Osawa T, see Yokoyama K 631, 633, *657*
Osborn MJ 166, 170, *192*
Osborne DJ 262, 263, *272*
Osborne DJ, see Horton RF 102, *121*
Osborne DJ, see Knee M 103, *122*
Osborne DJ, see Ridge I 141, 143, 150, *163*
Oschman JL 346, *350*
O'Sullivan S, see Lewis LN 105, *123*
Ota M, see Anno K 297, *307*
Ottesen M, see Johansen JT 442, 452, *456*
Ottolenghi P 449, *457*
Ouellette GB 16, 18, *24*
Outka DE, Williams DC 320, *331*

Overman MA, Warmke HE 101, *126*
Ovodov YS, Khomenko VA, Guseva TF 296, *313*
O'Vodov YS, see Pavlenko AF 296, *313*, 297, 298
Ozaki T, see Doi K 424, *434*
Ozawa T, Yamaguchi M 643, *656*

Packer L, see Branton D 321, *329*
Painter TJ 293, *313*
Painter TL, see Larsen B 296, *311*
Palade GE 488, 496, *510*
Palade GE, Sieckevitz P, Caro LG 487, *510*
Palade GE, see Caro LG 496, *507*
Palade GE, see Siekevitz P 488, *511*
Palamarczyk G, Janczura E 462, *481*
Palamarczyk G, Lehle L, Mankowski T, Chojnacki T, Tanner W 474, *481*
Palamarczyk G, Lehle L, Tanner W 474, *481*
Palamarczyk G, see Pless DD 474, *481*
Palevitz BA 516, 525, *538*
Palevitz BA, Hepler PK 513, 515, 516, 525, *538*
Palevitz BA, see Hepler PK 60, *64*, 516, 525, 526, *535*
Palma A, Büldt G, Jovanović SM 524, *538*
Palmer GH, Bathgate GN 106, *126*
Palmer RL, see Linkins AE 103, *123*
Palmieri S, Odoardi M, Soressi GP, Salamini F 40, *45*, 149, *162*
Palmitter RD, Gagnon J, Walsh KA *510*
Pan YT, Schmitt JW, Sanford BA, Elbein AD 184, *192*
Panayotatos N, Villemez CL 71, 81, *126*
Panigrahi S, see Sharma CBSR 516, *540*
Panopoulos NJ, Faccioli G, Gold AH 607, *621*
Parameswaren N, Wilhelm GE 247, *252*
Parbery DG, Blakeman JP 248, *252*
Pardos BT, see Rosen SD 639, *656*
Parham RA, see Mullis RH 21, *24*
Park D, Robinson PM 384, *391*
Park D, see McClure K 382, *390*
Park D, see McClure WK 500, *509*
Parke M, den Hartog-Adams I 347, *351*
Parke M, Dixon PS 279, *313*
Parker BC, Diboll AG 278, *313*
Parker BC, Preston RD, Fogg GE 358, *391*
Parker JW, see O'Brien RL 628, 637, 638, *656*
Parker KD, see Atkins EDT 282, *307*, 286
Parodi A, see Dallner G 171, 187, *189*
Parodi AJ 462, 463, 464, 476, *481*, 494, 499, *510*
Parodi AJ, Leloir LF 399, *414*, 459, *481*
Parodi AJ, see Behrens NF 174, *188*
Parodi AJ, see Behrens NH 461, *478*
Parolis H, see Allsobrook AJR 291, *306*
Parolis H, see Nunn JR 291, *313*
Parsons MJ, Pickmere SE, Bailey RW 294, *313*
Parsons MJ, see Pickmere SE 294, *314*
Parsons RT, Stephens K, Strickland JDH 303, *313*
Parthasarathy MV, see Setia RC 102, *129*
Pascher A 334, *351*
Pask G, see Juniper BE 542, 543, 545, 549, 560, *571*
Patel GB, Ingledew WM 419, *437*
Patil NB, see Gunja-Smith Z 419, *435*
Patil SS 598, *621*
Patil SS, Dimond AE 594, *621*
Patterson JC, see Manners DJ 359, 367, *390*, 417, 418, 427, *436*
Patterson ME, see Nagel CW 104, *125*
Patton AM, Marchant R 382, *391*
Paul DC, Goff CW 516, *538*
Paull RE, Johnson CM, Jones RL 549, 550, 551, 552, 562, 563, *573*
Paull RE, Jones RL 83, 101, *126*, 154, *162*, 519, *538*, 545, 551, 556, 558, 561, 564, *572*
Paus E, Steen HB 628, *656*
Pavlenko AF, Belogortseve NI, Kalinovskiai, Ovodov YS 297, 298, *313*
Pavlenko AF, O'Vodov YS 296, *313*
Payne HL, see Jones MGK 85, 91, *122*, 514, 517, 518, *536*
Payne JF, see Bal AK 96, *115*
Pazur JH, Knull HR, Limpsou DL 496, *510*
Pearse B 503, *510*
Peat S, Turvey JR, Evans JM 417, *437*
Peat S, Whelan WJ, Edwards TE 417, *437*, 441, *457*
Péaud-Lenoël C, Axelos M 51, 52, *65*, 72, *126*
Péaud-Lenoël C, see Axelos M 50, 58, *61*, 81, *114*
Péaud-Lenoël C, see Smith MM 71, *129*
Peberdy JF 368, 379, *391*
Peberdy JF, Moore PM 398, 399, *414*
Peberdy JF, see Campbell JMcA 402, *412*
Peberdy JF, see Isaac S 402, 409, *413*
Peberdy JF, see Moore PM 380, *390*, 398, 399, *414*
Peberdy JF, see Ryder NS 380, *392*, 399, 402, *415*
Pedersen K, see Hurkman WJ 181, *190*
Pegg GF 143, *162*
Pegg GF, Vessey JC 600, 608, *621*
Pelcher LE, see Ohyama K 586, *621*
Pellicciari R, see Ballio A 597, 598, *615*
Peng HB, Jaffe LF 321, *331*
Penman A, Rees DA 293, *313*

Penman A, Sanderson GR 285, 286, *313*
Penman A, see Anderson NS 293, *307*
Pennazio S, D'Agostino G, Appiano A, Redolfi P 112, *126*
Penny D, Penny P, Marshall DC 262, *272*
Penny D, see Monro JA 38, *44*, 137, 140, *161*, 264, *271*
Penny D, see Penny P 256, *272*, 262
Penny D, see Perley JE 139, *162*
Penny P, Penny D 256, *272*
Penny P, Penny D, Marshall DC, Heyes JK 262, *272*
Penny P, see Penny D 262, *272*
Percheron F, see Clermont S 81, *116*
Percival E, see Evans LV 317, *330*
Percival EGV, Ross AG 281, 296, *313*, *314*
Percival EGV, see Conchie J 297, *308*
Percival Elizabeth 289, 290, 295, 297, 298, 299, *314*
Percival Elizabeth, Foyle RAJ 304, *314*
Percival Elizabeth, McDowell RH 277, 278, 284, 291, 299, 303, 306, *314*
Percival Elizabeth, Rahman MA, Weigel H 304, *314*
Percival Elizabeth, Smestad B 301, 302, *314*
Percival Elizabeth, Turvey JR 279, *314*
Percival Elizabeth, Wold JK 302, *314*
Percival Elizabeth, Young M 297, 299, *314*
Percival Elizabeth, see Bidwell RGS 288, 289, *307*
Percival Elizabeth, see Bourne EJ 280, 283, 301, *307*, 283, 299
Percival Elizabeth, see Carlberg GE 283, 297, 301, 302, *308*
Percival Elizabeth, see Cleare M 304, *308*
Percival Elizabeth, see Evans LV 303, *308*
Percival Elizabeth, see Fisher IS 299, *309*
Percival Elizabeth, see Ford CW 283, *309*
Percival Elizabeth, see Haq QN 301, *309*, *310*
Percival Elizabeth, see Hirst Sir E 299, 300, *310*
Percival Elizabeth, see Jabbar Mian A 284, 296, 297, *311*
Percival Elizabeth, see Johnson PG 299, *311*
Percival Elizabeth, see Love J 283, 300, *312*
Percival Elizabeth, see Mackie IM 282, 300, *312*
Percival Elizabeth, see McKinnell JP 280, 300, 301, *312*
Percival Elizabeth, see O'Donnell JJ 300, *313*
Pérez S, Roux M, Revol JF, Marchessault RH 361, *391*
Perley JE, Penny D 139, *162*
Perlin AS, Taber WA 427, *437*
Perlin AS, see Ewald CM 74, *118*
Perlin AS, see Goldschmid HR 74, *119*
Perlin AS, see Gorin PAJ 442, *456*
Perrin A, see Dumas C 555, *570*
Perry JJ, see Merkel GJ 148, *161*
Perry K, see Fehrenbacher G 453, *456*
Perry M, see Abercrombie M 420, *432*
Pesis E, Fuchs Y, Zauberman G 103, *126*
Peska S, see Tucker P 180, *193*
Petek F, Villarroya E, Courtois JE 139, *162*
Peters BM, Cribbs DH, Stelzig DA 614, *621*
Peterson CA, Peterson RL, Robards AW 225, *252*, 531, *538*
Peterson CA, Rauser WE 112, *126*
Peterson RL, Brisson JD 543, 545, 547, 548, 549, *573*
Peterson RL, see Brisson JD 232, *248*, 547, 555, *570*
Peterson RL, see Peterson CA 225, *252*, 531, *538*
Peterson RL, see Scott MG 231, *253*
Peveling E, see Robenek H 85, *128*
Phaff HJ 352, 353, 354, 358, 360, 367, *391*, 417, 420, 421, 424, 425, 430, *438*, 441, *457*
Phaff HJ, see Abd-El-Al ATH 420, 421, 422, *432*
Phaff HJ, see Fleet GH 379, *387*, 416, 420, 421, 422, 423, 424, 425, *435*
Phaff HJ, see Foda MS 419, *435*
Phaff HJ, see Kopecká M 16, *23*, 418, 425, *436*
Phaff HJ, see Lachance MA 420, *436*
Phaff HJ, see Meyer MT 421, *436*
Phaff HJ, see Rombouts FM 424, 425, *438*
Phaff HJ, see Tanaka H 416, 418, 424, *439*
Phaff HJ, see Villa TG 420, 421, 422, 423, 424, *439*
Pharr DM, Dickinson DB 105, *126*
Phelps CF, see Winterburn PJ 140, *164*
Philippi ML 53, *65*
Phillips DA, see Chen APT 664, *674*
Phillips HL, Torrey JG 543, 544, 545, 548, 549, *573*
Phillips J, see Schreiber G 493, *511*
Phillips JM, see Hills GJ 22, *23*, 86, *120*, 145, *159*, 328, *330*
Pickett-Heaps JD 60, *65*, 83, 84, *126*, 221, *223*, 515, 518, 519, 525, *538*, 553, 558, *573*
Pickett-Heaps JD, Northcote DH 83, 85, *126*, 516, *538*
Pickett-Heaps JD, Staehelin LA 324, *331*, 346, *351*
Pickett-Heaps JD, see Coss RA 60, *62*
Pickett-Heaps JD, see Fowke HC 83, *119*
Pickett-Heaps JD, see MacDonald K 322, *331*
Pickett-Heaps JD, see Marchant HJ 322, *331*
Pickett-Heaps JD, see Northcote DH 83, *125*, 543, 545, 558, 560, *572*

Pickett-Heaps JD, see Staehelin LA 324, *332*
Pickmere SE, Parsons MJ, Bailey RW 294, *314*
Pickmere SE, see Parsons MJ 294, *313*
Pihl A, see Olsnes S 650, *656*
Pihl A, see Sandvig K 650, *656*
Pike CS, Pike CS, Un H, Lystash JC, Showalter AM 143, *162*
Pike CS, see Pike CS 143, *162*
Pilet PE 256, *272*
Pilet PE, Roland J-C 262, *272*
Pilet PE, see Jaccard M 263, *270*
Pilet PE, see Jotterand-Dolivo MC 143, *160*
Pilet PE, see Tanimoto E 150, *164*
Pilnik W, Rombouts FM 104, *126*
Pilnik W, Voragen AGJ 104, *126*
Pilnik W, see Versteeg C 104, *131*
Pincheira G, see Cardemil L 362, *386*
Pinkerton F, Strobel G 602, *621*
Pinsky A, Ordin L 58, *65*
Pio Beltram J, see Auriol P 598, 601, *615*
Pirson A 347, *351*
Pirt SJ, see Rowley BI 374, *392*
Planqué K, Kijne JW 667, *676*
Platt-Aloia KA, see Thomson WW 225, *253*
Plattner RD, see Seymour FR 445, *457*
Plaut Z, Ordin L 109, *126*
Pless DD, Lennarz WJ 179, *192*
Pless DD, Palamarczyk G 474, *481*
Plessmann Camargo E, Dietrich CP, Sonneborn D, Strominger JL 398, 399, 410, *414*
Plummer TH, see Tarentino AL 442, 443, 449, 450, 451, *458*
Pocchiari F, see Ballio A 597, 598, *615*
Pokorny V, see Seichertová O 406, *415*
Polacheck I, Rosenberger RF 405, *414*, 430, *438*
Polacheck J, see Molano J 405, *414*, 649, *655*
Polacheck Y, Rosenberger RF 363, 365, 371, 374, 376, 378, *391*
Politis DJ, Goodman RN 608, *621*
Pollard MR, McKeon T, Gupta LM, Stumpf PK 530, *538*
Pomeranz Y 106, *126*
Pon N, see Jones THD 430, *435*
Pont Lezica R 57, *65*
Pont Lezica R, Brett CT, Martinez PR, Dankert MA 58, *65*
Pont Lezica R, Romero PA, Dankert MA 58, *66*, 171, *192*
Pont Lezica R, Romero PA, Hopp HE 56, 58, *66*, 70, 74, 87, *127*, 178, 180, *192*, 645, *656*
Pont Lezica R, see Daleo GR 58, 62, 168, *189*
Pont Lezica R, see Hopp HE 58, 61, *64*, 74, *121*, 170, 171, 178, *190*, 645, *654*
Pooley HM, see Shockman GD 378, *392*
Poovaiah BW, Nukaya A 104, 105, *127*
Popa VI, see Simonescu CI 284, *315*
Pope AMS, see Stirling JL 373, *393*
Pope DG 92, *127*, 135, 137, 140, 152, 154, *162*
Pope DG, Lamport DTA 35, *45*, 152, *162*
Pope DG, Thorpe JR, Al-Azzawi MJ, Hall JL 516, 521, *538*, 562, 563, 564, *573*
Pore RS, see Conte MV 280, *308*
Porter CA, Jaworski EG 398, 399, *414*
Porter D, see Darley WM 317, *330*
Porter KR, Machado RD 514, 518, *538*
Postlmayr HL, Luh BS, Leonard SJ 104, *127*
Potgieter HJ, see Skujins JJ 374, *393*
Potts JT, see Habener JF 492, *508*
Poulain DG, Tronchin G, Dubremetz JF, Biquet J 367, *391*, 419, *438*
Poulose AJ, see Kolattukudy PE 227, 230, 231, 236, *251*
Powell CC Jr, see Hildebrand DC 611, *619*
Poyton RO, McKemmie 492, *510*
Prasad R, see Lingappa VR 181, *191*
Prat R 18, *24*
Prat R, Roland J-C 112, *127*
Prat R, see Roland J-C 519, *539*
Pratt HK, see Goeschl JD 143, *158*
Pratt RM, see Olden L 493, *510*
Preece IA 106, *127*
Preiss J, Kosuge T 607, *621*
Presant CA, Kornfeld S 633, 634, *656*
Pressey R, Avants JK 104, *127*, 150, *162*
Pressey R, Hinton DM, Avants JK 104, *127*
Pressey R, see Hinton DM 105, *120*
Pressman E, see Jacobsen JV 107, *121*, *122*
Preston EW, see Thompson RD 144, *164*
Preston RD 9, 15, 17, 19, *24*, 29, 30, 31, 33, 37, *45*, 94, *127*, 256, *272*, 320, 321, *331*, 513, 522, *538*
Preston RD, see Aronson JM 355, 366, *384*
Preston RD, see Atkins EDT 282, *307*
Preston RD, see Barnett JR 321, *329*
Preston RD, see Cronshaw J 279, 280, *308*
Preston RD, see Dennis DT 279, *308*
Preston RD, see Frei E 277, 278, 280, 281, 282, 283, *309*
Preston RD, see Haughton PM 259, *270*
Preston RD, see Mackie W 10, 11, 12, *24*, 277, 283, *312*

Preston RD, see Middlebrook MJ 366, *390*
Preston RD, see Nelmes BJ 104, *125*
Preston RD, see Nicolai E 321, *331*
Preston RD, see Parker BC 358, *391*
Preston RD, see Probine MC 259, *272*
Preston RD, see Robinson DG 52, *66*, 280, *315*, 321, 324, 325, *332*
Preston RD, see Thompson EW 279, *316*
Previato JO, Gorin PAJ, Travassos LR 372, *391*
Previato LM, see Burke D 441, 453, 454, *455*
Price CA, Bourke ME 347, *351*
Price J, see Dea ICM 37, *43*
Pridham JB, see Dey PM 139, *158*
Pringle JR, see Sloat BF 382, *393*
Pringle JT, see Sloat BF 407, *415*
Pringsheim EG 346, *351*
Prinz R, see Von Figura K 497, *512*
Privat JP, see Mialonier G 644, *655*
Probine MC, Preston RD 259, *272*
Procopiou J, see Baker EA 230, 231, *248*
Procter AR, see Scott JAN 197, *224*
Prosser JI, Trinci APJ 383, *391*
Pueppke SG, Bauer WD, Keegstra K, Ferguson AL 644, *656*, 667, *676*
Pueppke SG, see Anand VK 586, *615*
Pueppke SG, see Bhuvaneswari TV 663, 664, 669, *674*
Puhalla JE, Howell CR 589, *621*
Punnett T, Derrenbacker EC 144, 147, *162*
Pupillo P, see Mazzucchi U 587, *620*
Purdy RE, Kolattukudy PE 237, 239, *252*
Purves WK, see Cleland RE 262, 263, *270*
Putman EW, see Neufeld EF 81, *125*
Pyliotis NA, see Jacobsen JV 107, *122*

Quader H 25, *28*, *331*
Quader H, Robinson DG 324, *332*
Quader H, Wagenbreth I, Robinson DG 322, 324, *332*, 525, *538*
Quader H, see Hertel C 324, *330*
Quader H, see Robinson DG 25, *28*, 324, *332*
Quail PH 82, *127*, 137, *162*
Quatrano RS, Brawley SH, Hogsett WE 296, 298, 306, *314*
Quatrano RS, Crayton MA 298, *314*
Quatrano RS, Hogsett WS, Roberts M 296, 298, 306, *314*
Quatrano RS, Stevens P 278, 281, 289, 298, *314*
Quatrano RS, see Crayton MA 298, *308*
Quatrano RS, see Stevens PT 281, *315*
Quillet M 296, *314*
Quillet M, see Lestang G De 298, 299, 306, *311*

Raa J 611, *621*
Raa J, Robertsen B, Solheim B, Tronsmo A 152, *162*, 643, *656*
Rabenantoandro Y, Auriol P, Touzé A 608, *621*
Racusen RH, see Kinnersley AM 18, *23*
Raeven MBJM, see Vermeulen CA 380, 383, *393*, 403, 404, 409, *415*
Raff J, see Mau S-L 581, *583*
Raff JW, Clarke AE 580, *583*
Raff JW, Knox RB, Clarke AE 580, *583*
Rahmann MA, see Carlberg GE 297, *308*
Rahman MA, see Percival Elizabeth 304, *314*
Rai PV, Strobel GA 596, *621*
Raizada MK, Kloepper HG, Schutzbach JS, Ankel H 466, *481*
Raizada MK, Schutzbach JS, Ankel H 466, *481*
Raizada MK, see Schutzbach JS 466, *482*
Ralph BJ, Bender J 429, *438*
Ralph BJ, see Angyal SJ 362, *384*
Ralph BJ, see Madgwick JC 296, *312*
Ramamoorthy S, see Leppard GG 566, *572*
Ramos S, Garciá Acha I 369, *391*
Ramsay AM, Douglas LJ 417, *438*
Ramus J 304, *314*, 317, *332*, 347, *351*
Ramus J, Groves ST 304, *314*
Ramus J, Robins DM 317, *332*
Randall LL, Hardy STS 490, *510*
Randall LL, Hardy STS, Josefsson LG 490, *510*
Ranjeva R, see Alibert G 204, *222*
Rankin B, see Cooper RM 589, 591, *617*
Rao NV, Adams E 143, *162*
Rao VSR, see Sathyanarayana BK 360, *392*
Rapaka RS, Sorensen KR, Lee SD, Bhatnagar RS 143, *163*
Raper CA, see Raper JR 376, *391*
Raper JR, Raper CA 376, *391*
Rapport MM, see Yariv J 152, *165*, 577, *583*
Raschke WC, Ballou CE 444, 445, *457*
Raschke WC, Kern KA, Antalis C, Ballou CE 444, 446, 447, *457*
Raschke WC, see Ballou CE 444, *455*, 449, 476, *477*
Rasmussen GK 102, *127*
Rasmussen HP 102, *127*

Rast D, see Michalenko GO 363, 365, *390*
Rast D, see Sietsma JH 352, 370, 375, 381, *392*
Raudaskoski M 376, *391*
Rauser WE, see Brisson JD 547, 555, *570*
Rauser WE, see Peterson CA 112, *126*
Ravenscroft AV, see Wiese MV 593, *623*
Ray DA, Gibor A 341, *351*
Ray DA, Solter KM, Gibor 145, *163*
Ray MM, see Ray PM 52, *66*, 86, *127*, 519, *538*
Ray PM 53, 59, *66*, 71, 92, *127*, 256, 264, *272*
Ray PM, Eisinger WR, Robinson DG 71, 81, 83, *127*, 519, 521, *539*
Ray PM, Green PB, Cleland RE 255, *272*
Ray PM, Ruesink AW 256, 258, *272*
Ray PM, Shininger MM, Ray MM 52, *66*
Ray PM, Shininger TL, Ray MM 86, *127*, 519, *538*
Ray PM, see Abdul-Baki AA 59, *61*
Ray PM, see Anderson KL 48, 55, *61*
Ray PM, see Anderson RL 73, 86, 111, *114*, 608, *615*
Ray PM, see Baker DB 264, *269*
Ray PM, see Bates GW 268, *269*
Ray PM, see Griffing LR 521, *535*
Ray PM, see Jacobs M 92, *121*, 268, *271*
Ray PM, see Labavitch JM 32, 33, 38, 40, *44*, 92, *123*, 151, *160*, 268, *271*
Ray PM, see Robinson DG 83, 84, 85, *128*, 519, *539*
Ray PM, see Wada S 32, 33, 38, 39, *46*, 264, *273*
Rayburn WR 348, *351*
Rayle DL 265, *272*
Rayle DL, Cleland RE 258, 261, 264, 265, 266, 268, *272*
Rayle DL, Haughton PM, Cleland RE 264, 266, *272*
Rayle DL, see Adams PA 262, 263, *269*
Rayle DL, see Cleland RE 256, 261, 265, 266, 268, *269*, *270*, 262, 263
Rayle DL, see Johnson KD 139, 150, *159*, 266, *271*
Raymond Y, Fincher GB, Maclachlan GA 48, 54, *66*, 73, 111, *127*, 521, *539*
Raymond Y, see Shore G 53, 54, *66*
Rechtoris C, see Strand LL 591, *622*
Recondo E, see Garcia RC 528, *534*
Redman CM, Sabatini DD 488, *510*
Redolfi P, see Pennazio S 112, *126*
Rees DA 33, 37, *45*, 136, *163*, 277, 289, 292, 293, 295, 305, *315*, 354, 359, *391*
Rees DA, Samuel JWB 284, *315*
Rees DA, Scott WE 360, *391*
Rees DA, Welsh EJ 68, 86, *127*, 277, 293, *315*
Rees DA, see Anderson NS 293, *307*
Rees DA, see Arnott S 293, *307*, 294
Rees DA, see Cerezo AS 282, *308*
Rees DA, see Dea ICM 37, *43*, 39, *42*
Rees DA, see Gould SEB 108, *119*
Rees DA, see Grant GT 37, *43*, 86, 101, *119*, 287, *309*, 553, *570*
Rees DA, see Lawson CJ 293, *311*, 295, 305
Rees DA, see Liang JN 294, *311*
Rees DA, see Morris ER 286, *313*
Rees DA, see Penman A 293, *313*
Rees DE, see Gould SEB 37, *43*
Reese ET 373, 374, *391*, 430, *438*
Reeve RM 103, 104, *127*
Refsnes K, see Olsnes S 650, *656*
Regensburg BA, see Wessels JGH 358, 360, 362, 363, 364, 365, *394*, 427, 429, *440*
Reichelt B 422, 423, *438*
Reichert TA, Yu JMC, Christensen RA 148, *163*
Reid ID, Bartnicki-Garcia S 356, 360, 362, *392*
Reid JSG 107, *127*
Reid JSG, Bewley JD 107, *127*
Reid JSG, Davies C, Meier H 107, *128*
Reid JSG, Meier H 107, *127*, *128*
Reid JSG, Wilkie KCB 31, *45*
Reid JSG, see Meier H 85, 107, *124*
Reid PD, Strong HG, Lew F, Lewis LN 102, *128*
Reid PD, see Berger RK 102, *115*
Reid PD, see Lewis LN 105, *123*
Reiners WA, see Woodwell GM 11, *24*
Reinking A, see Boer P 499, *507*
Reinking A, see Reuvers F 462, 463, *481*
Reis D, Roland J-C 21, *24*
Reis D, Vian B, Roland J-C 68, 86, *128*
Reis D, see Roland J-C 528, *539*
Reis SM, Strobel GA 596, 599, 600, *621*
Reisinger O, see Breisch H 567, 568, *570*
Reisinger O, see Guckert A 553, 567, 568, *571*
Reisner Y, Sharon N 636, *656*
Reiss H-D, Herth W 513, 521, *539*
Reissig JL 398, *414*
Reitherman RW, Rosen SD, Frazier WA, Barondes SH 639, *656*
Reitherman RW, see Chang CM 639, *653*
Reitherman W, see Frazier WA 633, 639, *653*
Reuvers F, Boer P, Hemming FW 460, *481*
Reuvers F, Habets-Willems C,

Reinking A, Boer P 462, 463, *481*
Revol JF, see Bobitt TF 360, 363, 365, *385*
Revol JF, see Pérez S 361, *391*
Rey M, see Von Figura K 497, *512*
Reyes E, see Bartnicki-Garcia S 48, *62*, 355, 357, 361, 366, *385*, 380, 399, 402, 404, 405, *411*
Reyes VG, Schmidt EL 660, 661, *676*
Reymond O 346, *351*
Reynolds JD, Dashek WV 582, *583*
Rhee SH, see Irving RA 180, *190*
Rhodes MJC, Hill RC, Wooltorton LSC 220, *224*
Rhodes MJC, Wooltorton LSC 207, 208, 220, *223*
Rhodes MJC, see Hill AC 204, *223*
Ricard J, see Crasnier M 151, *157*
Richard M, Letoublon R, Louisot P, Got R 467, *481*
Richard M, see Letoublon RM 467, *480*
Richards EL, see Whistler RL 31, *46*
Richards GN, see Blake JD 38, 39, *42*
Richards GN, see Dea ICM 39, *42*
Richards JB, Hemming FW 170, *192*
Richardson B, see Schwertner HA 346, *351*
Richardson CL, see Brown RM 20, *22*
Richardson CL, see Brown RM Jr 526, *533*
Richardson NG, see Gould SEB 37, *43*
Rick PD, see Gander JE 467, *479*
Ridge I, Osborne DJ 141, 143, 150, *163*
Rietema H, see Huizing HJ 280, *310*, 283
Riley RG, Kolattukudy PE 231, *253*
Ring SG, Selvendran RP 35, 41, *45*
Risueño MC, see López-Sáez JF 516, *536*
Robards AW 98, *128*, 522, 525, *539*
Robards AW, Jackson SM, Clarkson DT, Sanderson J 225, *253*, 531, *539*
Robards AW, see Freundlich A 524, *534*
Robards AW, see Peterson CA 225, *252*, 531, *538*
Robb J, see Brisson JD 232, *248*, 547, 555, *570*
Robbins PW, Bray D, Dankert M, Wright A 74, *128*
Robbins PW, Hubbard SC, Turco SJ, Wirth DF 463, *482*, 497, *510*
Robbins PW, Krag SS, Liu T 175, 178, *192*
Robbins PW, see Grinna LS 497, *508*
Robbins PW, see Hubbard SC 181, *190*
Robbins PW, see Turco SJ 179, *193*, 181
Robenek H, Peveling E 85, *128*
Robert JC 376, *392*
Roberts C, see Winge O 449, *458*
Roberts K 145, 146, *163*, 328, *332*, 528, *539*
Roberts K, Gay MR, Hills GJ 146, *163*
Roberts K, Gurney-Smith M, Hills GJ 22, *24*, 145, *163*
Roberts K, Gurney-Smith M, Hills GS 328, *332*
Roberts K, Northcote DH 83, 85, 91, 98, *128*, 137, *163*, 514, *539*
Roberts K, see Catt JW 86, *116*, 145, 146, *157*, 328, *329*, *330*
Roberts K, see Davies DR 146, *157*
Roberts K, see Hills GJ 22, *23*, 86, *120*, 145, *159*, 328, *330*
Roberts K, see Homer RB 136, 146, 147, *159*
Roberts LM, see Mellor RB 180, *191*
Roberts M, see Quatrano RS 296, 298, 306, *314*
Roberts RM 173, *192*
Roberts RM, see Juniper BE 543, 545, 547, 549, 550, 551, 552, *571*
Roberts RM, see Kirby EG 556, 558, *571*
Robertsen B, see Raa J 152, *162*, 643, *656*
Robertsen BK, see Valent BS 92, *131*
Robinow C, see Matile P 498, *509*
Robins DM, see Ramus J 317, *332*
Robins RJ, see Juniper BE 542, 545, 549, *571*
Robinson AC 661, *676*
Robinson DG 9, 14, 20, *24*, 47, *66*, 82, 83, 84, *128*, 324, *332*, 513, 519, 521, 522, 525, *539*
Robinson DG, Eisinger WR, Ray PM 83, 84, 85, *128*, 519, *539*
Robinson DG, Grimm I, Sachs H 324, *332*
Robinson DG, Herzog W 324, *332*, 525, *539*
Robinson DG, Preston RD 52, *66*, 280, *315*, 321, 324, 325, *332*
Robinson DG, Quader H 25, *28*, 324, *332*
Robinson DG, Ray PM 84, *128*
Robinson DG, Schlösser UG 328, *332*, 348, *351*
Robinson DG, White RK, Preston RD 321, *332*
Robinson DG, see Grimm I 324, *330*
Robinson DG, see Hertel C 324, *330*
Robinson DG, see Quader H 322, 324, *332*, 525, *538*
Robinson DG, see Ray PM 71, 81, 83, *127*, 519, 521, *539*
Robinson DG, see Sachs H 322, 324, *332*
Robinson DG, see Schlösser UG 348, *351*
Robinson GB 178, *192*
Robinson PM, see McClure K 382, *390*
Robinson PM, see McClure WK 500, *509*
Robinson PM, see Park D 384, *391*
Robyt JF 47, *66*, 74, *128*

Rodehorst D, see VanderMolen GE 609, *623*
Rodehorst E, see VanderMolen GE 113, *131*
Rodén L, Schwartz NB 74, 86, *128*
Roden L, see Heaney-Kieras J 144, *159*, 304, *310*
Röderer G, see Schnepf E 326, *332*
Rodríguez L, Laborda F, Sentandreu R 504, *510*
Rodríguez L, Ruiz T, Elorza MV, Villanueva JR, Sentandreu R 499, *511*
Rodríguez L, Ruiz T, Villanueva JR, Sentandreu R 499, *510*
Roelofsen PA 13, *24*, 29, *45*, 91, *128*, 255, *272*, 417, 418, *438*, 522, *539*
Roelofson PA 356, 366, 369, *392*
Roerig S, see Lamport DTA 35, *44*, 135, 138, 139, *161*
Rogers MA, see Horner HT 101, *121*
Rohrschneider JM, see Schwarz RT 184, *192*, 496, *511*
Roland J-C 20, *24*, 513, 519, 521, 522, *539*, 553, *573*
Roland J-C, Prat R 519, *539*
Roland J-C, Sandoz D 519, *539*
Roland J-C, Vian B 144, *163*, 255, *272*
Roland J-C, Vian B, Reis D 528, *539*
Roland J-C, see Pilet PE 262, *272*
Roland J-C, see Prat R 112, *127*
Roland J-C, see Reis D 21, *24*, 68, 86, *128*
Roland J-C, see Rougier M 553, *573*
Roland J-C, see Vian B 553, 559, *574*
Rollins ML, see Tripp VW 134, *164*
Romanovicz DK, Brown RM 318, *332*
Romanovicz DK, see Brown RM 318, 320, *329*
Romanovicz PK, Brown RM 59, *66*
Rombouts FM, Fleet GH, Manners DJ, Phaff HJ 424, 425, *438*
Rombouts FM, Phaff HJ 424, 425, *438*
Rombouts FM, see Pilnik W 104, *126*
Rombouts FM, see Versteeg C 104, *131*
Romero PA, see Pont Lezica R 645, *656*
Romero PA, see Daleo GR 168, *189*
Romero PA, see Hopp HE 48, 51, 53, 56, 57, 58, 59, 61, *64*, 74, *121*, 170, 171, 178, *190*, 645, *654*
Romero PA, see Pont Lezica R 56, 58, *66*, 70, 74, 87, *127*
Romero PA, see Pont-Lezica R 171, *192*, 178, 180
Ronin C, Bouchilloux S, Granier C, Van Rietschoten J 472, *482*
Roon RJ, see Dunlop PC 449, *456*
Rose AH, see McMurrough I 417, *436*
Roseboom PHM, see De Wit PJGM 605, *617*
Rosell K-G, see Aspinall GO 578, 579, *582*
Rosell K-G, see Lindberg B 33, *44*
Roseman S 80, *128*
Roseman S, see Davidson EA 398, *412*
Roseman S, see Ghosh S 398, *412*
Rosen SD, Chang CM, Barondes SH 640, *656*
Rosen SD, Kaur J, Clark DL, Pardos BT, Frazier WA 639, *656*
Rosen SD, see Barondes SH 639, 640, *653*
Rosen SD, see Chang CM 639, *653*
Rosen SD, see Frazier WA 633, 639, *653*
Rosen SD, see Reitherman RW 639, *656*
Rosen WG, see Dashek WV 83, 84, *117*
Rosenberger RF 353, 358, 360, *392*, 397, *414*
Rosenberger RF, see Katz D 371, 382, 384, *388*, *389*, 408, 409, *413*
Rosenberger RF, see Polacheck I 405, *414*, 430, *438*
Rosenberger RF, see Polacheck Y 363, 365, 371, 374, 376, 378, *391*
Rosenberger RF, see Sternlicht E 384, *393*
Rosenblatt M, see Habener JF 492, *508*
Rosenfeld L, Ballou CE 442, 444, *457*
Rosenfield CL, Loewus FA 155, *163*
Rosenstock G, see Barckhausen R 531, *533*
Rosness PA 430, *438*
Rosness PA, Wright BE 426, 428, *438*
Ross AG 279, *315*
Ross AG, see Percival EGV 281, 296, *313*, *314*
Roth J 553, *573*
Rothman JE, Fene RE 182, *192*
Rothman JE, Katz FN, Lodish HF 493, *511*
Rothman JE, Lodish HF 180, *192*, 468, *482*
Rothman JE, see Katz FN 180, *191*
Rouge MP 667, *676*
Rougier M 83, *128*, 546, 549, 553, 555, 558, 560, *573*
Rougier M, Chaboud A, Kieda C, Monsigny M 555, *573*
Rougier M, Kieda C, Monsigny M 548, 553, *573*
Rougier M, Vian B, Gallant D, Roland J-C 553, *573*
Rougier M, see Dumas C 555, *570*
Rougier M, see Sambin B 544, *573*
Rouse AH, Knorr LC 104, *128*
Rousset-Hall A de, Gooday GW 379, *392*, 398, 399, *414*
Rousset-Hall A de, see Gooday GW 398, 399, 409, *412*, 400, 408
Routley DG, see Couture RM 611, *617*
Rouwenhorst DL, see Kapusta G 661, *675*

Roux M, see Bobitt TF 360, 363, 365, *385*
Roux M, see Pérez S 361, *391*
Rovira AD 566, 567, 568, *573*, 660, 661, *676*
Rovira AD, Campbell R 567, *573*
Rovira AD, see Bowen GD 565, 567, *569*
Rovira AD, see Campbell R 567, *570*
Rovira AD, see Foster RC 565, *570*
Rowan KS, see Adams PB 615, *615*
Rowell P, see Lockhart CM 659, *676*
Rowley BI, Pirt SJ 374, *392*
Rowley JR, Southworth D 532, *539*
Royal A, see Mercereau-Puijalon O 491, *509*
Royle DJ 247, *253*
Rubbell DH, see Umali-Garcia M 568, *574*
Rubery PH, Fosket DE 220, *224*
Rubery PH, Northcote DH 39, *45*, 203, *224*
Rubinstein B, see Terry M 268, *273*
Rudall KM, Kenchinton W 355, *392*
Rudick MJ 467, *482*
Rudick MJ, Elbein AD 467, *482*
Rüdiger H 628, *656*
Rüdiger H, see Gansera R 645, *653*
Rudin Y, see Maccecchini ML 492, *509*
Rudolph K 586, 598, *621*
Rudolph K, see El-Banoby FE 586, 594, *617*
Ruel K, Barnoud F, Goring DAI 198, *224*
Ruesink AW 267, *272*
Ruesink AW, see Horine RK 18, *23*
Ruesink AW, see Ray PM 256, 258, *272*
Ruiz EM, see Elorza MV 424, *434*
Ruiz Herrera J, see Bartnicki-Garcia S 380, *385*
Ruiz T, see Rodríguez L 499, *510*, *511*
Ruiz-Herrera J, Bartnicki-Garcia S 380, 383, *392*
Ruiz-Herrera J, Bartnicki-García S 21, *24*, 402, 403, 409, *414*
Ruiz-Herrera J, López-Romero E, Bartnicki-García S 399, 402, 404, *414*
Ruiz-Herrera J, Sentandreu R 470, *482*, 494, 504, *511*
Ruiz-Herrera J, Sing VO, Van der Woude WJ, Bartnicki-García S 21, *24*, 380, *392*, 399, 404, *414*
Ruiz-Herrera J, see Arroyo-Begovich A 380, *384*
Ruiz-Herrera J, see Bartnicki-Garcia S 48, *62*
Ruiz-Herrera J, see Bartnicki-García S 399, 402, 404, 405, *411*
Ruiz-Herrera J, see Bracker CE 21, *22*, 380, *385*, 404, *411*
Ruiz-Herrera J, see López-Romero E 380, *389*, 384, 398, 399, 400, 402, *413*, 403, 432, *436*, 504, *509*
Rusan V, see Simonescu CI 284, *315*
Russel DW 204, *224*
Russell GE, see Hartley RD 111, *120*
Russell I, see Nunn JR 291, *313*
Russi S, see Ballio A 597, 598, *615*
Ryder NS, Peberdy JF 380, *392*, 399, 402, *415*
Ryder NS, see Isaac S 402, 409, *413*
Ryser U 94, *128*, 518, 519, 520, *539*

Sabatini DD, see Blobel G 488, *507*
Sabatini DD, see Redman CM 488, *510*
Sabnis DD, Hart JW 648, *656*
Sabularse D, see Carpita N 68, 101, *116*, 154, *157*
Sachs H, Grimm I, Robinson DG 322, 324, *332*
Sachs H, see Grimm I 324, *330*
Sachs H, see Robinson DG 324, *332*
Sachs H, see Schlösser UG 348, *351*
Sack J, see Allen CM 168, *188*
Sadava D, Chrispeels MJ 142, 147, *163*
Sadava D, Volcani BE 144, *163*
Sadava D, Walker F, Chrispeels MJ 143, *163*
Sadek NZ, see Espelie KE 239, 242, 247, *249*
Sadowski PL, Strobel GA 599, *622*
Sagasta EM, see Garcia-Arenal F 111, 112, *119*
Sage HJ, see Howard IK 644, *654*
Sahlman K, Fahraeus G 670, *676*
Sahu RK, see Sharma CBSR 516, *540*
Saito H, Misaki A, Harada T 427, *438*
Saito H, see Suzuki I 633, *656*
Saito T, see Suzuki S 444, *458*
Saito T, see Takeuchi Y 92, *130*, 528, *540*
Sakaguchi O, Suzuki M, Yokota K 466, *482*
Sakakibara A 36, *45*
Sakurada I, Nukushina Y, Ito T 13, *24*
Sakurai N, Masuda Y 59, *66*, 92, *128*, 262, 267, 268, *272*
Sakurai N, Nevins DJ, Masuda Y 92, *128*, 267, 268, *273*
Sakurai N, Nishitani K, Masuda Y 92, *128*, 268, *273*
Salamini F, see Palmieri S 40, *45*, 149, *162*
Saleh TM, Leney L, Sarkanen KV 197, *224*
Salem HM, see Abdel-Fattah AF 296, 297, *306*
Salema R, see Mota M 568, *572*
Salpeter MM, see Steward FC 134, *164*
Salton MRJ 10, *24*
Sambin B 542, 543, 545, 549, 553, 559, 562, *573*
Sambin B, Rougier M, Zandonella P 544, *573*

Samtsevich SA 549, 562, 563, 566, *573*
Samuel JWB, see Rees DA 284, *315*
Samuel O, Nordin JH 467, *482*
Samuelsson K, see Hellerqvist CG 427, *435*
San Blass G, Carbonel LM 361, *392*
San-Blas G, Carbonell LM 428, 429, *438*
San-Blas G, see Flores-Carreon A 431, *435*
Sanchez M, see Santos T 430, 431, *438*
Sanders RE, Carlson RW, Albersheim P 667, *676*
Sanderson GR, see Penman A 285, 286, *313*
Sanderson J, see Clarkson DT 566, *570*
Sanderson J, see Robards AW 225, *253*, 531, *539*
Sandoz D, see Roland J-C 519, *539*
Sands D, see Zucker M 592, *623*
Sandvig K, Olsnes S, Pihl A 650, *656*
Sanford BA, see Pan YT 184, *192*
Sano H, see Kamisaka S 262, 263, *271*
Santamaria F, Fuensanta R, Lahoz R 427, *438*
Santos E, Villanueva JR, Sentandreu R 470, *482*
Santos GA, Doty MS 293, *315*
Santos T, Del Rey FD, Conde J, Villanueva JR, Nombela C 377, *392*, 423, *438*
Santos T, Sanchez M, Villanueva JR, Nombela C 430, 431, *438*
Santos T, Villanueva JR, Nombela C 430, 431, *438*
Santos T, see Del Rey F 378, *386*
Santos T, see Villanueva JR 373, *393*
Sargent C 529, 530, 531, *539*
Sargent JA, see Knee M 103, *122*
Sarkanen KV, Ludwig CH 215, 216, *224*
Sarkanen KV, see Saleh TM 197, *224*
Sarko A, Muggli R 280, *315*
Sarvas M, Hirth KP, Fuchs E, Simons K 490, *511*
Sasaki SF, see Min KH 284, *312*
Sassen A, Eyden-Emons van A, Lamers A, Wanka F 324, *332*
Sassen MMA 95, 96, 97, *128*
Sassen MMA, see Helsper JPFG 53, *64*, 72, *120*, 521, 525, *535*
Sathyanarayana BK, Rao VSR 360, *392*
Satir B, Schooley C, Satir P 502, *511*
Satir B, see Branton D 321, *329*
Satir P, see Branton D 321, *329*
Satir P, see Satir B 502, *511*
Sato K, see Ito T 183, *190*
Sato S, see Hori H 35, *43*
Sato S, see Kawasaki S 137, *160*, 519, *536*
Satoh H, Matsuda K, Tamani K 56, *66*
Satoh H, Takahara M, Matsuda K 60, *66*
Satoh S, Matsuda K, Tamari K 74, *128*, 525, *539*
Saunders JA, Conn E 611, *622*
Saunders PT, see Trinci APJ 382, *393*
Sawhney VK, Srivastava LM 525, *539*
Sawhney VK, see Srivastava LM 525, *540*
Sawyer WH, see Fincher GB 135, 152, 155, *158*
Sawyer WH, see Hawthorne DB 281, *310*
Scala A, see Cervone F 589, 591, *616*
Scala F, see Cervone F 589, 591, *616*
Scaletti JV, see Johnson J 427, *435*
Scaletti JV, see Misaki A 359, *390*, 417, *437*
Scarborough GA 403, *415*, 595, *622*
Scarborough GA, see Schulte TH 595, *622*
Schachter H, see Lawford GR 468, *480*
Schaefer A, see Ohyama K 586, *621*
Schatz G, see Maccecchini ML 492, *509*
Schechter I, McKean DJ, Guyer R, Terry W 489, *511*
Scheele G, see Devillers-Thiery A 492, *508*
Scheffer RP 598, *622*
Scheibe R, Beck E 607, *622*
Scheirer DC 95, *128*
Schekman R, Brawley V 382, *392*, 404, 407, *415*
Schekman R, see Field C 469, *479*
Schekman R, see Novick P 469, *481*, 503, *510*
Schellenberg M, see Lenney JF 401, 402, *413*
Schellenberg M, see Wiemken A 498, *512*
Schenk D, see Göring H 268, *270*
Schiebold E, see Franz E 522, *534*
Schiff JA, see Loewus F 295, *312*
Schilde-Rentschler L 18, *24*
Schiller J 346, *351*
Schimke RT, see Kiely ML 178, *191*, 468, *479*, 493, *509*
Schlatmann E, see Slob A 613, *622*
Schlesinger S, see Tabas I 463, *482*, 497, *512*
Schlessinger S, see Gibson R 493, 496, *508*
Schlessinger S, see Leavit R 493, *509*
Schlösser E 611, 612, *622*
Schlösser E, see Lüning HU 609, 612, *620*
Schlösser UG 145, *163*, 334, 335, 336, 337, 338, 340, 342, 343, 346, *351*
Schlösser UG, Sachs H, Robinson DG 348, *351*
Schlösser UG, see Robinson DG 328, *332*, 348, *351*
Schmidt EL 659, 660, *676*
Schmidt EL, see Bohlool BB 659, 660, 663, 664, 667, 668, 671, *674*
Schmidt EL, see Jack MA 668, *675*

Schmidt EL, see Reyes VG 660, 661, *676*
Schmidt EL, see Tsien HC 665, 666, 671, *677*
Schmidt MFG, see Schwarz RT 184, 185, *192*, 496, *511*
Schmitt JW, see Pan YT 184, *192*
Schnarrenberger C, see Bowles DJ 646, *653*
Schneider EF, Wardrop AB 16, *24*, 363, 366, 369, *392*
Schneider EG, Nguyen HT, Lennarz WJ 185, *192*
Schneider TL, see Medcalf DG 296, 297, *312*
Schnepf E 280, *315*
Schnepf E, Deichgräber G 318, *332*, 525, 526, *539*
Schnepf E, Deichgräber G, Röderer G, Herth W 326, *332*
Schnepf E, Herth W 514, *539*
Schnepf E, Röderer G, Herth W 326, *332*
Schnepf E, see Hegewald E 346, *350*
Schnepf E, see Herth W 326, *330*
Scholtyseck E 10, *24*
Schönbeck F, Schroeder C 614, *622*
Schönherr J, Ziegler H 247, *253*
Schooley C, see Satir B 502, *511*
Schreiber G, Dryburgh H, Milleship A, Matsuda Y, Inglis A, Phillips J, Edwards K, Maggs J 493, *511*
Schroeder C, see Schönbeck F 614, *622*
Schroth MN, see Hildebrand DC 611, *619*
Schubert WJ, see Acerbo SN 201, *221*
Schubert WJ, see Eberhardt G 201, *222*
Schubotz LM, see Myllyla R 143, *162*
Schulte TH, Scarborough GA 595, *622*
Schultz AH, Oroszlan S 493, *511*
Schulz D, see Lehmann H 83, *123*
Schulz GV, see Marx-Figini M 30, *44*, 47, *65*, 525, *537*
Schulz I, see Lehle L 463, 476, *480*
Schulze A, see Kivilaan A 92, *122*
Schurmann P, see Magyarosy AC 606, *620*
Schurz H, see Gansera R 645, *653*
Schurz J, see Nowak-Ossorio M 48, 56, *65*, 524, *537*
Schuster G 112, *129*
Schuster G, Byhan O 112, *129*
Schuster G, Flemming M 112, *129*
Schutzbach JS, Ankel H 473, *482*
Schutzbach JS, Raizada MK, Ankel H 466, *482*
Schutzbach JS, Springfield JD, Jensen JW 177, *192*
Schutzbach JS, see Ankel H 466, *477*
Schutzbach JS, see Raizada MK 466, *481*
Schwaiger H, Tanner W 184, *192*
Schwartz M, see Silhavy TJ 491, *511*
Schwartz NB, see Rodén L 74, 86, *128*
Schwarz RT, Datema R 186, *192*, 473, *482*
Schwarz RT, Rohrschneider JM, Schmidt MFG 184, *192*, 496, *511*
Schwarz RT, Schmidt MFG, Lehle L 184, 185, *192*
Schwarz RT, see Garoff H 184, *190*, 493, *508*
Schwarz RT, see Lehle L 473, *480*
Schweiger RG 303, *315*
Schwertner HA, Butcher WI, Richardson B 346, *351*
Scott FM, Bystrom BG, Bowler E 103, *129*
Scott JAN, Procter AR, Fergus BJ, Goring DAI 197, *224*
Scott JR, see Medcalf DG 304, *312*
Scott MG, Peterson RL 231, *253*
Scott TK, see Mueller SC 522, *537*
Scott WE, see Arnott S 293, *307*, 294
Scott WE, see Rees DA 360, *391*
Scurfield G 108, *129*, 356, 366, 367, *392*
Seemüller E, see Grünwald J 247, *250*
Segel I, see Gold MH 370, *387*
Seichertová O, Beran K, Holan Z, Pokorny V 406, *415*
Seiler A 107, *129*
Seipp D 104, *129*
Sekine S, see Nakamura T 262, 263, *271*
Sekiya T, see Cole GT 365, *386*
Selitrennikoff CP, Allin D, Sonneborn DR 397, 398, 410, *415*
Selitrennikoff CP, Sonneborn DR 398, 410, *415*
Sellen DB, see Haughton PM 259, *270*
Sellen DB, see Mackie W 282, 283, *312*
Selvendran RP 38, *45*
Selvendran RP, see Ring SG 35, 41, *45*
Selvendran RR, Isherwood FA 49, *66*
Selvendran RR, see Isherwood FA 49, *64*
Selvendran RR, see O'Neill MA 135, 140, *162*
Semstad B, see Bourne EJ 283, *307*
Senda M, see Morikawa H 21, *24*, 68, *124*
Seno N, see Anno K 296, *307*, 297
Sentandreu R, Elorza MV, Villaneuva JR 432, *438*
Sentandreu R, Elorza MV, Villanueva JR 504, *511*
Sentandreu R, Lampen JO 460, 471, *482*
Sentandreu R, Northcote DH 407, *415*, 422, *438*, 476, *482*, 498, *511*
Sentandreu R, see Elorza MV 471, *478*, 498, *508*
Sentandreu R, see Larriba G 470, 476, *480*, 494, 504, *509*

Sentandreu R, see Rodríguez L 499, 504, *510*, *511*
Sentandreu R, see Ruiz-Herrera J 470, *482*, 494, 504, *511*
Sentandreu R, see Santos E 470, *482*
Sequeira L 141, *163*, 643, 644, *656*, 659, *676*
Sequeira L, Gaard G, De Zoeten GA 614, *622*
Sequeira L, Graham TL 587, *622*
Sequeira L, see Graham TL 586, *618*
Sequeira L, see Whatley MH 587, *623*
Setia RC, Parthasarathy MV, Shah JJ 102, *129*
Setterfield G, Bayley ST 19, *24*
Setterfield G, see Fulcher RG 90, 91, *119*, 518, *534*
Sexton R 102, *129*
Sexton R, Durbin ML, Lewis LN, Thomson WW 103, *129*
Seymour FR, Slodki ME, Plattner RD, Stodola RM 445, *457*
Shafizadeh F, McGinnis GD 29, 30, *45*, 47, *66*, 524, *539*
Shah J, see Medcalf DG 304, *312*
Shah JJ, see Setia RC 102, *129*
Shalla TA, see Allison AV 112, *114*
Shands HL, see Dickson JG 106, *117*
Shannon L, see Howard J 641, *654*, 672, *675*
Shannon LM, see Hankins CN 636, 648, 651, 652, *654*
Shannon LM, see Howard J 636, *654*
Shapleigh E, see Boyd WC 627, *653*
Sharma CB, Babczinski P, Lehle L, Tanner W 461, 472, *482*
Sharma CB, Sharma TN 139, *163*
Sharma CBSR, Sahu RK, Panigrahi S 516, *540*
Sharma TN, see Sharma CB 139, *163*
Sharon N 638, *656*
Sharon N, Lis H 168, *192*, 630, *656*
Sharon N, see Barkai-Golan R 649, *652*, 659, *674*
Sharon N, see Bowles DJ 645, *653*, 668, *674*
Sharon N, see Eshdat Y 633, *653*
Sharon N, see Lis H 169, *191*, 627, 628, 630, 632, 635, 636, 637, 651, *655*, 659, 667, 671, *676*
Sharon N, see Lotan R 629, *655*, 635
Sharon N, see Mirelman D 649, *655*
Sharon N, see Nevo Z 144, *162*
Sharon N, see Reisner Y 636, *656*
Sharon N, see Tunis M 629, *657*
Shashkov AS, see Yarotskii SV 293, *316*
Shaw BI, see Dickerson AG 595, *617*
Shay FJ, see Griffin GJ 566, *571*
Shaykh M, Kolattukudy PE, Davis R 230, 239, *253*
Shaykh M, Soliday C, Kolattukudy PE 247, *253*
Shaykh M, see Maiti IB 239, *252*
Sheik-Fareed V 303, *315*
Sheik-Fareed V, see Evans LV 303, *308*
Sheldrake AR 96, 97, 102, *129*
Sheldrake AR, Moir GFJ 97, *129*
Shematek EM, Braatz JA, Cabib E 504, *511*
Shematek EM, Cabib E 504, *511*
Shepard JF, Totten RE 30, *45*
Sherwood RT, Vance CP 608, *622*
Sherwood RT, see Vance CP 140, 142, 152, *164*
Shibaoka H, see Hogetsu T 60, 61, *64*
Shibaoka H, see Nishitani K 262, *272*, 513, *537*
Shibata H, see Tanaka M 156, *164*
Shibata S, see Nishikawa Y 427, *437*
Shibuya N, Iswasaki T 106, *129*
Shibuya N, Misaki A 32, 33, 39, *45*, 106, *129*
Shida M, Uchida T, Matsuda K 360, *392*
Shields D, see Lingappa VR 490, *509*
Shigeta M, see Takeo K 522, *540*
Shih CY, Currier HB 112, *129*
Shimada M, Kuroda H, Higuchi T 205, 206, *224*
Shimada M, Ohashi H, Higuchi T 205, *224*
Shimada M, see Higuchi T 203, 206, 218, 219, *223*
Shimada M, see Kuroda H 207, *223*
Shimahara H, Sugiyama N 293, *315*
Shimbayashi K, see Dougal DK 134, *158*
Shimizu K, see Morita Y 149, *161*
Shimoda C, Nasada Y, Yanagishima N 422, *439*
Shimoda C, Yanagishima N 422, *439*
Shimojima Y, see Mizuno M 185, *191*
Shimokoriyama M, see Fujii T 142, *158*
Shimokoriyama M, see Hogetsu T 60, 61, *64*
Shimomura T, Dijkstra J 112, *129*
Shininger MM, see Ray PM 52, *66*
Shininger TL, see Ray PM 86, *127*, 519, *538*
Shinozyki K, see Yamamoto R 259, 262, *273*
Shiraishi T, see Yamamoto S 425, *440*
Shiraki D, see Siegel S 109, *129*
Shockman GD, Pooley HM, Thompson JS 378, *392*
Shore G, Maclachlan GA 53, 59, *66*, 521, 525, *540*
Shore G, Raymond Y, Maclachlan GA 53, 54, *66*
Shore G, see Spencer FS 59, *66*

Short KC, see Brown EG 49, *62*
Showalter AM, see Pike CS 143, *162*
Shukla P, see Hiruki C 112, *121*
Shuman HA, see Silhavy TJ 491, *511*
Sicher Jr RC, see Lin CC 355, *389*
Siddiqui IR, Wood P 34, *45*
Sieckevitz P, see Palade GE 487, *510*
Siegel S, Speitel T, Shiraki D, Fukumoto J 109, *129*
Siegel SM 220, *224*
Siekevitz P, Palade GE 488, *511*
Sietsma JH, Child JJ, Nesbitt LR, Haskins RH 355, 358, 363, 365, 368, *392*, 426, 428, *439*
Sietsma JH, Eveleigh DE, Haskins RH 354, 355, 358, *392*
Sietsma JH, Rast D, Wessels JGH 352, 370, 375, 381, *392*
Sietsma JH, Wessels JGH 352, 356, 357, 359, 360, 362, 363, 375, 380, 381, 383, *392*, 427, 428, 429, *439*
Sietsma JH, see Dooijewaard-Kloosterziel AMP 374, *386*
Sietsma JH, see Huizing HJ 283, *310*
Sietsma JH, see Wessels JGH 363, 375, *394*
Sievers A 83, *129*
Sigarlakie E, see Bauer H 368, *385*, 406, *411*, 451, *455*
Silano V, see Ballio A 597, 598, *615*
Silhavy TJ, Shuman HA, Beckwith J, Schwartz M 491, *511*
Silhavy TJ, see Bassford PJ 491, *507*
Silva MT, see Mota M 568, *572*
Silva PC 279, *315*
Simoneit BRT, see Caldicott AB 228, 230, *248*
Simonescu CI, Popa VI, Rusan V, Liga A 284, *315*
Simons K, see Sarvas M 490, *511*
Simpson M, see Evans LV 317, *330*
Simson BW, Timell TE 154, *163*
Simson BW, see Hoffmann GC 427, *435*
Sing V, Yeh YF, Ballou CE 443, 453, *457*
Sing VO, see Ruiz-Herrera J 21, *24*, 380, *392*, 399, 404, *414*
Singer HH, see Williamson AR 493, *512*
Singer PA, see Williamson AR 493, *512*
Singer SJ, Nicolson L 522, *540*
Singh H, see Malhotra OP 139, *161*
Sinnot EW, Bloch R 514, *540*
Sinta PB, see Struck DK 184, *193*, 493, *511*
Sitte P 227, *253*, 531, *540*
Sitte P, see Brown RM 318, *329*, 346, *349*
Sitte P, see Brown RM Jr 144, *157*
Sjulin TM, Beer SV 600, *622*
Skerrett RJ, see Grant GT 86, 101, *119*, 553, *570*
Skuja H 334, *351*
Skujins JJ, Potgieter HJ, Alexander M 374, *393*
Skutelsky E, Bayer EA 469, *482*
Skutelsky E, see Lotan R 629, *655*
Slayman CL, Slayman CW 501, *511*
Slayman CW, see Slayman CL 501, *511*
Sleytr UB, Umrath W 522, *540*
Sleytr UB, see Kiermayer O 325, *331*, 524, *536*
Sloat BF, Pringle JR 382, *393*
Sloat BF, Pringle JT 407, *415*
Slob A, Jekel B, De Jong B, Schlatmann E 613, *622*
Slodki ME 465, *482*
Slodki ME, see Seymour FR 445, *457*
Sloger C, see Vest G 659, 660, *677*
Sly WS, see Kaplan A 497, *509*
Sly WS, see Natowicz MR 497, *510*
Smestad B, Haug A, Myklestad M 305, *315*
Smestad B, see Bidwell RGS 288, 289, *307*
Smestad B, see Percival Elizabeth 301, 302, *314*
Smidsrod O, Glover R, Whittington SG 286, *315*
Smidsrod O, Haug A 286, *315*
Smidsrod O, Haug A, Lian B 287, *315*
Smidsrod O, see Grasdalen H 285, 287, 288, *309*
Smidsrod O, see Haug A 284, 287, 288, *310*, 286
Smith CJ, see Darvill AG 92, *117*, 140, *157*, 264, 267, 268, *270*
Smith DB, Cook WH 292, *315*
Smith DG, see Marchant R 370, *390*, 406, *414*, 422, *436*
Smith EE, see Gunja-Smith Z 419, *435*
Smith F, see Horisberger M 429, *435*
Smith F, see Huotari F 425, *435*
Smith F, see Johnson J 427, *435*
Smith F, see Misaki A 359, *390*, 417, *437*
Smith H 670, *676*
Smith IM, see Theodorou MK 603, 604, *623*
Smith JC, see Grant GT 287, *309*
Smith JE, Stanley RG 52, *66*
Smith JE, see Thomas D 72, *130*
Smith LD, see Hurkman WJ 181, *190*
Smith M 134, 137, 140, *163*
Smith MM, Axelos M, Péaud-Lenoël C 71, *129*
Smith MM, McCully ME 48, *66*, 90, 112, *129*, 518, *540*
Smith MM, Stone BA 32, *45*, 51, 52, *66*, 72, 106, *129*
Smith P, see Grant GT 37, *43*
Smith P, see Knox RB 237, *250*
Smith WL, Ballou CE 445, 449, *458*
Smith WL, Nakajima T, Ballou CE 465, 476, *482*

Smith WP, Tai PC, Davis BD 490, *511*
Smith WP, Tai PC, Thompson RC, Davis BD 490, *511*
Smolko EE, see Atkins EDT 286, *307*
Sobotka FE, Stelzig DA 105, *129*
Soeder CJ 346, *351*
Solel Z, see Nachmias A 597, 599, 600, *621*
Solheim B, see Raa J 152, *162*, 643, *656*
Soliday C, see Shaykh M 247, *253*
Soliday CL, Dean BB, Kolattukudy PE 236, *253*
Soliday CL, Kolattukudy PE 232, 234, 237, 238, *253*, 468, *482*, 529, *540*
Soliday CL, Kolattukudy PE, Davis RW 242, 247, *253*, 529, 531, *540*
Solms J, Hassid WZ 49, *66*
Solter KM, Gibor A 341, *351*
Solter KM, see Ray DA 145, *163*
Sone Y, Misaki A 428, *439*
Soneda M, see Miller MW 420, *437*
Sonneborn D, see Plessmann Camargo E 398, 399, 410, *414*
Sonneborn DR, see Selitrennikoff CP 397, 398, 410, *415*
Sorensen KR, see Rapaka RS 143, *163*
Soressi GP, see Palmieri S 40, *45*, 149, *162*
South GR 284, *315*
Southworth D, Dickinson DB 48, 53, *66*, 72, *129*
Southworth D, see Rowley JR 532, *539*
Sowden LC, see Leppard GG 20, *23*, 528, *536*
Sowokinos JR 50, *66*
Speake BK, Hemming FW, White DA 495, *511*
Speakman JB, Lewis BG 111, 113, *129*
Speiss E, Thio Tiang Nio E, Amelunxen F 528, *540*
Speitel T, see Siegel S 109, *129*
Spencer AK, see Stahmann MA 150, *163*
Spencer FS, Maclachlan GA 52, *66*
Spencer FS, Shore G, Ziola B, Maclachlan GA 59, *66*
Spencer FS, Ziola B, Maclachlan GA 52, 56, 59, *66*
Spencer JFT, Gorin PAJ 441, *458*
Spencer JFT, see Gorin PAJ 284, *309*, 352, 358, 362, 367, *387*, 420, *435*, 426, 427, 428, 429, 445, *456*
Spencer JP, Elbein AD 177, *192*
Spencer JP, Kang MS, Elbein AD 186, *192*
Spencer JP, see Kang MS 177, 186, *190*
Speth V, see Branton D 321, *329*
Spiess E, see Amelunxen F 528, *532*
Spiro MJ, Spiro RG, Bhoyroo VD 463, *482*
Spiro MJ, Spiro RJ, Bhoyroo VD 175, 178, 179, 181, *192*
Spiro RG, see Spiro MJ 463, *482*
Spiro RJ, see Spiro MJ 175, 178, 179, 181, *192*
Sprau F 112, *129*
Springer GF, see Bernardi G 296, *307*
Springer WR, see McDonough JP 638, *655*
Springfield JD, see Schutzbach JS 177, *192*
Sprinson DB 201, *224*
Srivastava LM, Sawhney VK, Bonettemaker M 525, *540*
Srivastava LM, see Sawhney VK 525, *539*
St Julian G, see Arthur LO 595, *615*
Stack JP, Mount MS, Berman PM, Hubbard JP 589, *622*
Stadler J, see Herth W 144, *159*, 318, *330*
Staehelin LA 324, *332*
Staehelin LA, Pickett-Heaps JD 324, *332*
Staehelin LA, see Branton D 321, *329*
Staehelin LA, see Giddings TH 325, 326, *330*, 524, *534*
Staehelin LA, see Giddings TH Jr 94, *119*
Staehelin LA, see Pickett-Heaps JD 324, *331*, 346, *351*
Stafford HA 221, *224*
Stahmann MA, Spencer AK 150, *163*
Stahmann MA, see Walker JC 612, 613, *623*
Stancioff DJ 292, *315*
Stancioff DJ, see Lawson CJ 293, *311*
Staneloni RJ, see Behrens NH 175, *188*, 461, *478*
Stankovic ZS, see Heldt HW 607, *619*
Stanley NF, see Lawson CJ 293, *311*
Stanley RG, see Smith JE 52, *66*
Stanley RG, see Thomas D 72, *130*
Starr MP, see Chatterjee AK 589, 593, *616*
Starr MP, see Moran F 592, *620*
Starr RC 344, 347, *351*
Starr RC, see Heimke JW 338, 340, *350*
Steele IW, see Gould SEB 37, *43*
Steen HB, see Paus E 628, *656*
Steer MW, see Gunning BES 19, 20, *23*, 82, *120*
Stein WH, see Hayashi R 452, *456*
Steiner GW, Strobel GA 597, *622*
Stekoll M, West CA 603, 605, *622*
Stelzig DA, see Peters BM 614, *621*
Stelzig DA, see Sobotka FE 105, *129*
Stepanenko BN, Morozova AB 52, *67*
Stephan AM, see Churms SC 6, *8*
Stephen AM 6, *8*
Stephen AM, see Churms SC 75, *116*
Stephens GJ, Woods RKS 150, *163*
Stephens K, see Parsons RT 303, *313*

Sterling C 103, *130*
Sterling C, Kalb AJ 104, *130*
Sterling C, see Wilson LA 231, *254*
Stern H, see Stieglitz H 99, *130*
Sternlicht E, Katz D, Rosenberger RF 384, *393*
Stetson B, see Turco SJ 179, *193*
Stevens ES, see Liang JN 294, *311*
Stevens JW, see Fred EB 663, *675*
Stevens P, see Quatrano RS 278, 281, 289, 298, *314*
Stevens PT, Quatrano RS 281, *315*
Steward FC, Israel HW, Salpeter MM 134, *164*
Stewart WDP, see Lockhart CM 659, *676*
Steyn-Parvé EP, see Boer P 449, *455*, 499, *507*
Steyn-Parvé EP, see Van Rijn HJ 362, *392*, 471, 473, *483*
Stieglitz H 99, *130*
Stieglitz H, Stern H 99, *130*
Stieglitz HB 99, *130*
Stirling JL, Cook GA, Pope AMS 373, *393*
Stobbs LW, Manocha MS, Dias HF 112, *130*
Stobiecka M, see Giebel J 141, *158*
Stöckigt J, Mansell RL, Gross GG, Zenk MH 208, *224*
Stöckigt J, see Mansell RL 207, 209, *223*
Stöckmann VE 358, *393*
Stoddart RW, Barrett AJ, Northcote DH 37, *46*
Stoddart RW, Northcote DH 39, *45*
Stoddart RW, see Cook GMW 90, *116*
Stodola RM, see Seymour FR 445, *457*
Stone BA, see Anderson RL 92, *114*, 106, 154, *156*
Stone BA, see Bacic A 106, *114*
Stone BA, see Barras DR 426, 430, *433*
Stone BA, see Clarke AE 6, *8*, 34, 35, *42*, 74, 99, 111, *116*, 90, 109, 152, 153, 155, *157*, 426, *434*, 577, 578, *582*
Stone BA, see Cook JA 72, 88, *116*
Stone BA, see Fincher GB 106, *119*, 135, 152, 155, *158*
Stone BA, see Mares DJ 32, 39, *44*, 106, *123*
Stone BA, see Moore AE 105, *124*
Stone BA, see Smith MM 32, *45*, 51, 52, *66*, 72, 106, *129*
Stone BA, see Woodward JR 111, 113, *132*, 597, 600, *623*
Stone KJ, Strominger JL 186, *192*
Stoodley R, see Abercrombie M 420, *432*
Stosch HA v 347, *351*
Stossel TP, see Yin HL 501, *512*
Strand LL, Mussell H 591, *622*
Strand LL, Rechtoris C, Mussell H 591, *622*
Strand LL, see Mussell H 152, *162*
Strasburger's Textbook of Botany 76, *130*
Strauss AW, Zimmerman M, Boime I, Ashe B, Mumford RA, Alberts AW 489, *511*
Street HE, Öpik H, James FEL 543, 545, 549, *573*
Strickland JDH, see Parsons RT 303, *313*
Strijdom BW, see Law IJ 664, 667, *676*
Strobel G, see Auriol P 598, 601, *615*
Strobel G, see Pinkerton F 602, *621*
Strobel GA 596, 597, 598, 599, 601 *622*
Strobel GA, Hess WM 597, 601, *622*
Strobel GA, Talmadge KW, Albersheim P 596, 598, 599, *622*
Strobel GA, Van Alfen N, Hapner KD, McNeil M, Albersheim P 597, 598, *622*
Strobel GA, see Beltran JP 598, *616*
Strobel GA, see Currier WW 662, *674*
Strobel GA, see Johnson TB 600, *619*
Strobel GA, see MacDonald PW 606, *620*
Strobel GA, see Nachmias A 597, 599, 600, *621*
Strobel GA, see Rai PV 596, *621*
Strobel GA, see Reis SM 596, 599, 600, *621*
Strobel GA, see Sadowski PL 599, *622*
Strobel GA, see Steiner GW 597, *622*
Strominger JL, Ghuysen JM 340, *351*
Strominger JL, see Plessmann Camargo E 398, 399, 410, *414*
Strominger JL, see Stone KJ 186, *192*
Strominger JL, see Wedgewood JF 168, *193*
Strong HG, see Reid PD 102, *128*
Struck DK, Lennarz WJ 183, 187, *193*, 493, *511*
Struck DK, Sinta PB, Lane MD, Lennarz WJ 184, *193*, 493, *511*
Strugger S, see Currier HB 112, *117*
Strunk C 369, *393*
Stuart DA, Mozer TV, Varner JE 135, *164*
Stuart DA, Varner JE 135, 137, 140, 141, *164*
Stumpf PK, see Macey MJK 243, *252*
Stumpf PK, see Pollard MR 530, *538*
Su J-C, see Lin T-Y 71, *123*
Su JS *67*
Subramanian D, see Thirupathaiah V 104, 105, *130*
Sueiro SF, Felipe Anton MR 555, *573*
Sugawara T, see Mizuno M 185, *191*
Sugino Y, see Uchida T 650, *657*
Sugiyama HJ, see Lazarovits G 597, *620*
Sugiyama N, see Shimahara H 293, *315*

Suhadolnik RJ 400, *415*
Sulkowski E, see Mizrahi A 493, *510*
Sumiya K, Yamada T 262, *273*
Summers DF, see Hunt LA 463, *479*, 497, *508*
Sumner JB, Howell SF 627, *656*
Sun B, see Hultin HO 104, *121*
Sun SM, Ma Y, Buchbinder BO, Hall TC 181, *193*
Sunayama H, see Suzuki S 444, *458*
Sundaralingam M 360, *393*
Sundararajan PR, see Marchessault RH 524, *536*
Sussman AS, see Fultz SA 370, *387*
Sussmann M 638, *656*
Sussmann M, Brackenbury R 638, *656*
Sutcliffe JG 490, *511*
Sutherland J, McCully ME 90, 97, 98, *130*
Sutherland J, see Fulcher RG 90, 91, *119*, 518, *534*
Sutton JC, Williams PH 596, 599, *622*
Suzuki H, Imai S, Nisizawa K Murachi T 105, *130*
Suzuki I, Saito H, Inoue S, Migita S, Takahashi T 633, *656*
Suzuki M, see Sakaguchi O 466, *482*
Suzuki S, Sunayama H, Saito T 444, *458*
Suzuki S, see Hori M 400, *413*
Suzuki S, see Isono K 400, *413*
Suzuki T, see Iriki Y 282, *311*
Suzuki T, see Miwa T 282, 283, *313*
Svensson S, see Lindberg B 33, *44*, 353, *389*
Svensson S-B 60, *67*
Svoboda A, see Farkaš V 471, 473, *479*
Swain T 111, *130*
Swan B, see Bjorndal H 282, *307*
Swan EP 231, *253*
Swanson AL, see Kauss H 75, *122*
Swanson AL, see Villemez CL 71, 75, *131*

Tabas I, Kornfeld S 182, *193*
Tabas I, Schlesinger S, Kornfeld S 463, *482*, 497, *512*
Tabas I, see Li E 463, *480*
Taber WA, see Perlin AS 427, *437*
Tabora E, see Levy JA 174, *191*
Tagawa T, Bonner J 258, 265, *273*
Tahir MA, Elahi M 104, *130*
Tai PC, see Smith WP 490, *511*
Tai RC, see Davis BD 490, 491, *508*
Taiz L, Jones RL 106, *130*
Taiz L, see McNeil M 264, *271*
Taiz L, see Métraux JP 265, *271*
Takagi Y, see Takeo K 522, *540*
Takahara M, see Satoh H 60, *66*
Takahashi N, see Nakamura T 262, 263, *271*
Takahashi T, see Suzuki I 633, *656*
Takatsuki A, Arima K, Tamura G 183, *193*
Takatsuki A, Kawamura K, Okina M, Kodama Y, Ito T, Tamura G 183, *193*
Takatsuki A, Kohno K, Tamura G 472, *483*, 493, *512*
Takatsuki A, see Ito T 183, *190*
Takatsuki A, see Mizrahi A 493, *510*
Takaya S, see Misaki A 429, *437*
Takebe I, see Nagata T 30, *45*
Takeda H, Hirokawa T 280, *315*
Takeda H, Nisizawa K, Miwa T 283, *316*
Takeda J, see Mizuno M 185, *191*
Takeo K, Shigeta M, Takagi Y 522, *540*
Takeuchi Y, Komamine A 32, *46*, 92, *130*
Takeuchi Y, Komamine A, Saito T, Watanabe K, Morikawa N 92, *130*, 528, *540*

Talbot CF, Etzler ME 644, 652, *656*, *657*, 668, *676*
Talmadge KW, Keegstra K, Bauer WD, Albersheim P 138, *164*
Talmadge KW, see Bauer WD 591, *616*
Talmadge KW, see Keegstra K 6, *8*, 133, *160*, 264, *271*, 357, *389*, 590, *619*
Talmadge KW, see Strobel GA 596, 598, 599, *622*
Tamaki S, see Matsuda Y 340, *350*
Tamani K, see Satoh H 56, *66*
Tamari K, see Hara M 60, *63*
Tamari K, see Miyamoto C 52, *65*, 73, *124*
Tamari K, see Nakajima T 428, *437*
Tamari K, see Satoh S 74, *128*, 525, *539*
Tammes PML, see Eschrich W 98, *118*
Tamura G, see Ito T 183, *190*
Tamura G, see Mizrahi A 493, *510*
Tamura G, see Takatsuki A 183, *193*, 472, *483*, 493, *512*
Tanabe O, see Nakamura N 427, *437*
Tanahashi M, Higuchi T 216, *224*
Tanahashi M, see Higuchi T 203, 206, 218, 219, *223*
Tanaka H, Itakura K, Toda K 424, 425, *439*
Tanaka H, Phaff HJ 424, *439*
Tanaka H, Phaff HJ, Higgins LW 416, 418, 424, *439*
Tanaka H, see Nakajima T 428, *437*
Tanaka M, Shibata H, Uchida T 156, *164*
Tanaka M, Uchida T 139, *164*
Tanaka M, see Nishikawa Y 427, *437*
Tandecarz J, see Lavinthman N 55, *64*
Tanimoto E, Igari M 139, *164*, 266, *273*
Tanimoto E, Masuda Y 262, 267, *273*
Tanimoto E, Pilet PE 150, *164*

Tanimoto E, see Wada S 267, *273*
Tanner W 170, *193*, 460, *483*
Tanner W, Jung P, Behrens NH 460, *483*
Tanner W, Kandler O 56, *67*
Tanner W, see Babczinski P 461, 474, *477*, 494, 499, *507*
Tanner W, see Hasilik A 185, *190*, 442, 445, 451, 452, *456*, 468, 471, 473, *479*, 491, 494, 495, *508*
Tanner W, see Jung P 167, 170, *190*, 460, *479*
Tanner W, see Lehle L 56, *64*, 173, 183, 187, *191*, 400, *413*, 461, 462, 463, 464, 470, 472, 473, 476, *480*, 493, 494, *509*
Tanner W, see Marriott KM 188, *191*, 470, *481*, 494, *509*
Tanner W, see Palamarczyk G 474, *481*
Tanner W, see Schwaiger H 184, *192*
Tanner W, see Sharma CB 461, 472, *482*
Tanzawa K, see Endo A 187, *189*
Tarentino AL, Plummer TH, Maley F 442, 443, 449, 450, 451, *458*
Tatum EL, see Mahadevan PR 353, 363, 365, 372, *390*, 428, *436*
Tatum EL, see Mishra NC 372, *390*, 432, *437*
Tatum EL, see Wrathall CR 372, *394*
Tautvydas KJ 145, *164*
Taves C, see Goodwin RH 60, *63*
Taylor A, see Lipke PN 453, *457*
Taylor IEP 144, *164*
Taylor IEP, Cameron DS 353, *393*, 416, *439*
Taylor IF, see Bacon JSD 356, 358, 359, *384*, 417, 418, *433*, 424, 425
Taylor IF, see Mitchell A 353, *390*
Taylor NS 454, *458*
Taylor NW, Orton WL 453, 454, *458*
Teichner A, see Minetti M 628, *655*
Tepfer M, Cleland RE 265, *273*
Tepfer M, see Adams PA 262, 263, *269*
Terahata H, see Anno K 296, *307*
Terao T, see Yokoyama K 631, 633, *657*
Teresa GW, see Jones JM 644, *654*
Terry M, Rubinstein B, Jones RL 268, *273*
Terry W, see Schechter I 489, *511*
Tesch R, see Maier K 401, *414*
Theodorou MK, Smith IM 603, 604, *623*
Thiele H, Andersen G 287, *316*
Thieme TR, Ballou CE 442, 443, 445, *458*
Thiéry J-P 83, *130*, 553, *574*
Thiesen M, see Haass D 647, 648, *654*
Thimann KV, see Noodén LD 258, *272*
Thimann KV, see O'Brien TP 94, 95, *126*
Thio Tiang Nio E, see Amelunxen F 528, *532*
Thio Tiang Nio E, see Speiss E 528, *540*
Thirupathaiah V, Subramanian D 104, 105, *130*
Thom D, see Grant GT 37, *43*, 287, *309*
Thom D, see Morris ER 286, *313*
Thomas B, Hall MA 111, 112, *130*
Thomas B, Webb JA 139, 151, *164*
Thomas D des S, Mullins JT 377, *393*
Thomas D, Des S, Smith JE, Stanley RG 72, *130*
Thomas D, Des S, Stanley RG 72, *130*
Thompson EW, Preston RD 279, *316*
Thompson JS, see Shockman GD 378, *392*
Thompson M, see Cleland RE 262, 263, *270*
Thompson NS, see Mullis RH 21, *24*
Thompson RC, see Smith WP 490, *511*
Thompson RD, Preston EW 144, *164*
Thomsen HA, see Moestrup O 320, *331*
Thomson WW, Platt-Aloia KA, Endress AG 225, *253*
Thomson WW, see Sexton R 103, *129*
Thornber JP, Northcote DH 39, *46*, 76, 93, *130*
Thorne KJ, see Martin HG 167, 187, *191*
Thorner J, see Fehrenbacher G 453, *456*
Thorpe JR, see Pope DG 516, 521, *538*, 562, 563, 564, *573*
Thorpe TA, see Halmer P 107, *120*
Threlfall DR, see Mankarios AT 150, *161*
Thrum H, see Eckhardt K 185, *189*
Tigchelaar EC, see Buescher RW 104, 105, *116*
Timell TE 6, 7, *8*, 31, 33, 34, *46*, 93, *130*
Timell TE, see Fu Y-L 109, *119*
Timell TE, see Hoffmann GC 109, *121*, 427, *435*
Timell TE, see Simson BW 154, *163*
Tingle MA, Halvorson HO 420, *439*
Tippett JT, O'Brien TP 531, *540*
Tiwari SC, see Kapil RN 101, *122*
Tkacz JS, Lampen JO 183, *193*, 469, 472, *483*, 493, *512*
Tkacz JS, MacKay VL 453, *458*
Tkacz JS, see Onishi HR 451, *457*, 473, *481*
Tkacz JS, see Onishi HS 495, 496, *510*
Tocher CS, see Craigie JS 302, *308*
Tochikura T, see Moriguchi M 398, *414*
Tochikura T, see Yamamoto K 398, *415*
Toda K, see Tanaka H 424, 425, *439*

Toh-Chin Lim Ng, see Anderson LE 607, *615*
Tokunaga J, Bartnicki-Garcia S 358, 365, 369, *393*, 426, *439*
Toman R, Káračsonyi Š, Kubačková M 35, *46*
Toman R, see Karácsonyi Š 35, *43*
Tomiyama K, see Doke N 588, 603, 606, *617*
Tomiyama K, see Furuichi N 587, *618*
Tomiyama K, see Nozue M 587, *621*
Tomlinson JA, Webb MJW 606, *623*
Tomlinson PB, see Gill AM 542, *570*
Toneguzzo F, Ghosh HP 180, *193*
Toneguzzo F, see Irving RA 180, *190*
Tonew E, see Eckhardt K 185, *189*
Tonew M, see Eckhardt K 185, *189*
Toppan A, see Esquerré-Tugayé MT 111, *118*, 141, *158*, 608, *618*
Torrey JG, Clarkson DT 247, *253*
Torrey JG, see Phillips HL 543, 544, 545, 548, 549, *573*
Totten RE, see Shepard JF 30, *45*
Touet G, Aach HG 345, *351*
Touzé A, see Esquerré-Tugayé MT 111, *118*, 141, *158*, 608, *618*
Touzé A, see Rabenantoandro Y 608, *621*
Towers GHN, see Young MR 203, *224*
Toyoshima S, Funkuda M, Osawa T 634, 637, *657*
Tracey MV 12, *24*
Travassos LR, see Alviano CS 364, *384*
Travassos LR, see Previato JO 372, *391*
Trave P, see Bastide M 374, *385*
Trevithick JR, see Chang PLY 377, 378, *386*
Trimble RB, Maley F 442, 443, 449, 450, 452, *458*, 499, *512*
Trinci APJ 408, *415*
Trinci APJ, Collinge AJ 370, 382, *393*
Trinci APJ, Saunders PT 382, *393*
Trinci APJ, see Bull AT 594, *616*
Trinci APJ, see Collinge AJ 382, *386*
Trinci APJ, see Gooday GW 255, 264, *270*
Trinci APJ, see Gull K 370, *387*
Trinci APJ, see Prosser JI 383, *391*
Trinick MJ 659, *677*
Tripp VW, Moore AT, Rollins ML 134, *164*
Trolp R, see Maier K 401, *414*
Tronchin G, see Poulain DG 367, *391*, 419, *438*
Tronsmo A, see Raa J 152, *162*, 643, *656*
Troy FA, Koffler H 353, 355, 357, 362, 363, *393*, 428, *439*
Tsai CM 53, *67*
Tsai CM, Hassid WZ 51, 52, *67*, 72, 75, *130*
Tschesche R, Jha HC, Wulff G 610, *623*
Tschesche R, Wiemann W 610, *623*
Tsien HC, Schmidt EL 665, 666, 671, *677*
Tsubo Y, see Matsuda Y 340, *350*
Tsubouchi Y, see Hayano K 595, *619*
Tsubura E, see Azuma I 466, *477*
Tsuda M, see Uchida T 650, *657*
Tsugita A, see Müller K 640, *655*
Tsuji S, see Yaku F 36, *46*
Tsumuraya Y, see Misaki A 429, *437*
Tu JC, see Hiruki C 111, 112, *121*
Tucker P, Peska S 180, *193*
Tuderman L, see Myllyla R 143, *162*
Tulloch AP 239, 242, *253*
Tully RE, Beevers H 647, *657*
Tung K-K, Ballou CE 76, *130*
Tunis M, Lis H, Sharon N 629, *657*
Turchenek LW, Oades JM 567, *574*
Turco SJ, Heath EC 174, *193*
Turco SJ, Robbins PW 181, *193*
Turco SJ, Stetson B, Robbins PW 179, *193*
Turco SJ, see Robbins PW 463, *482*, 497, *510*
Turner EMC 612, *623*
Turner FR, see Mahlberg PG 516, *536*
Turner NC, Graniti A 597, *623*
Turner NC, see Van Alfen NK 596, 597, 600, *623*
Turvey JR 277, 279, *316*
Turvey JR, Williams EL 282, 291, *316*
Turvey JR, see Bowker DM 291, *307*
Turvey JR, see Boyd J 284, 285, *307*
Turvey JR, see Peat S 417, *437*
Turvey JR, see Percival Elizabeth 279, *314*
Tuzimura K, Watanabe I 660, 661, *677*

Uchida T, Yamaizumi M, Mekada E, Okada Y, Tsuda M, Kurokawa T, Sugino Y 650, *657*
Uchida T, see Shida M 360, *392*
Uchida T, see Tanaka M 139, 156, *164*
Uchytil TF, see Durbin RD 612, *617*
Uhrström I 261, *273*
Uhrström I, see Burström H 260, *269*
Ukai S, Hirose K, Kiho T, Hara C 362, *393*
Ulane RE, Cabib E 401, 402, 406, *415*
Ulane RE, see Cabib E 89, *116*, 395, 397, 401, 406, *411*, 423, *434*
Ullrich K, see Von Figura K 497, *512*
Ullrich W 112, *131*

Umali-Garcia M, Rubbell DH, Gaskins MH, Dazzo FB 568, *574*
Umekawa M, see Min KH 284, *312*
Umetsu N, see Hara M 60, *63*
Umrath W, see Sleytr UB 522, *540*
Un H, see Pike CS 143, *162*
Underhill EW 611, *623*
Unzelman JM, Healey PL 85, *131*
Urbano MR, see Dazzo FB 667, *675*
Urech K, see Wiemken A 498, *512*
Uritani I, see Kojima M 587, *620*, 629, *655*
Uruburu F, see Dominguez JB 428, *434*
Usov AS, see Yarotskii SV 293, *316*

Vaciaqo A, see Ballio A 597, 598, *615*
Vagabov VM, see Farkaš V 464, 473, 476, *479*
Valent B, see Ayers AR 603, 605, *615*
Valent BS, Albersheim P 265, *273*
Valent BS, Darvill AG, McNeil M, Robertsen BK, Albersheim P 92, *131*
Valent BS, see Albersheim P 113, *114*, 643, *652*
Valk P van der, Marchant R 366, *393*
Valk P van der, Marchant R, Wessels JGH 353, 355, 356, 363, 365, 366, 367, 370, 374, *393*, 397, *415*
Valk P van der, Wessels JGH 356, 360, 361, 365, 368, 369, 380, 382, *393*
Valk P van der, see Wessels JGH 368, 379, *394*
Valkovich G, see Forsee WT 173, *190*
Van Alfen N, see Strobel GA 597, 598, *622*
Van Alfen NK, Allard-Turner V 593, 600, 601, *623*
Van Alfen NK, Turner NC 596, 597, 600, *623*
Van Beeumen J, De Ley J 595, *623*
Van den Ende G, Linskens HF 247, *253*
Van der Woude WJ, Lembi CA, Morré DJ 59, *67*
Van der Woude WJ, Lembi CA, Morré DJ, Kindinger JI, Ordin L 51, 53, *67*, 72, *131*
Van der Woude WJ, Morré DJ, Bracker CE 82, 83, 85, *131*, 519, *540*
Van der Woude WJ, see Morré DJ 83, 84, 86, *124*
Van der Woude WJ, see Ruiz-Herrera J 21, *24*, 380, *392*, 399, 404, *414*
Van Die J, see Eschrich W 98, *118*
Van Etten HD, see Bateman DF 589, *616*
Van Laere AJ, Carlier AR 402, *415*
Van Netten J, see Hänish Ten Cate ChH 102, 103, *120*
Van Rietschoten J, see Ronin C 472, *482*
Van Rijn HJ, Boer P, Steyn-Parvé EP 362, *392*
Van Rijn HJM, Boer P, Steyn-Parvé E-P 471, 473, *483*
Van Rijn HJM, Linnemans WAM, Boer P 498, *512*
Van Rijn JM, see Boer P 499, *507*
Van Spronsen PC, see Goosen-de Roo L 514, *535*
Van Wauwe JP, see Loontiens FG 631, *655*
Vance CP, Kirk TK, Sherwood RT 140, 142, 152, *164*
Vance CP, see Sherwood RT 608, *622*
VanderMolen GE, Beckman CH, Rodehorst D 609, *623*
VanderMolen GE, Beckman CH, Rodehorst E 113, *131*
Vanderwoude WJ, see Mollenhauer HH 546, 548, 562, 563, 564, *572*
Vannini GL, see Fasulo MP 60, *63*
Varner JE, see Lewis LN 103, *123*
Varner JE, see Stuart DA 135, 137, 140, 141, *164*
Vaughan PFT, Butt VS 204, *224*
Veenhuis M, see Kreger-Van Rij NJW 368, *389*
Veerkamp HH, see Helsper JPFG 53, *64*
Veerkamp JH, see Helsper JPFG 72, *120*, 521, 525, *535*
Venketeswaren S, see Mahlberg PG 516, *536*
Verma DPS, Maclachlan GA, Byrne H, Ewings D 267, *273*
Vermeulen CA, Raeven MBJM, Wessels JGH 380, 383, *393*, 403, 404, 409, *415*
Versteeg C, Rombouts FM, Pilnik W 104, *131*
Veruzzo D, see Allen CM 168, *188*
Vessey JC, see Pegg GF 600, 608, *621*
Vest G, Weber DF, Sloger C 659, 660, *677*
Vian B 519, *540*, 553, 559, 560, *574*
Vian B, Roland J-C 553, 559, *574*
Vian B, see Reis D 68, 86, *128*
Vian B, see Roland J-C 144, *163*, 255, *272*, 528, *539*
Vian B, see Rougier M 553, *573*
Vidra JD, Loerch JD 50, *67*
Vignon M, see Joseleau JP 35, *43*
Villa TG, Lachance MA, Phaff HJ 420, 421, 422, 423, 424, *439*
Villa TG, Notario V, Benitez T, Villanueva JR 420, *439*
Villa TG, Notario V, Villanueva JR 420, 421, 423, *439*
Villa TG, Phaff HJ 421, *439*
Villa TG, see Benitez T 400, 408, *411*
Villa TG, see Lachance MA 420, *436*
Villa TG, see Notario V 420, 423, *437*
Villa TG, see Villanueva JR 373, *393*
Villamez CL, Clark CF 170, 173, *193*
Villamez CL, see Carlo PL 170, *189*

Villanueva JR, Garcia Acha I 368, *393*
Villanueva JR, Notario V, Santos T, Villa TG 373, *393*
Villanueva JR, see Elorza MV 424, *434*, 498, *508*
Villanueva JR, see Larriba G 470, 476, *480*, 494, 504, *509*
Villanueva JR, see Notario V 420, 423, *437*
Villanueva JR, see Rodríguez L 499, *510*
Villanueva JR, see Santos E 470, *482*
Villanueva JR, see Santos T 377, *392*, 423, *438*, 430, 431
Villanueva JR, see Sentandreu R 432, *438*, 504, *511*
Villanueva JR, see Villa TG 420, 421, 423, *439*
Villarroel LH, Zanlungo AB 296, *316*
Villarroya E, see Petek F 139, *162*
Villemez CL 71, 75, *131*, 521, *540*
Villemez CL, Hinman MB 71, *131*
Villemez CL, Lin T-Y, Hassid WZ 71, 75, *131*
Villemez CL, Swanson AL, Hassid WZ 71, 75, *131*
Villemez CL, see Clark AF 52, *62*, 73, *116*
Villemez CL, see Heller JS 71, *120*
Villemez CL, see Hinman MB 71, *120*
Villemez CL, see McNab JM 71, *124*
Villemez CL, see Panayotatos N 71, 81, *126*
Villemez CL Jr 51, 53, 56, *67*
Villemez CL Jr, Franz G, Hassid WZ 48, 51, 52, *67*
Villemez CL Jr, Heller JS 52, *67*
Vincent JM 660, *677*
Virgin HI, see Falk SO 261, *270*
Vithanage HIMV, Gleeson PA, Clarke AE 110, *131*, 581, *583*
Vogele P, see Greiling H 449, *456*
Volcani BE, see Coombs J 144, *157*
Volcani BE, see Sadava D 144, *163*
Volfová A, Chvojka L, Haňkovská J 525, *540*
Vollmer CM, see McCandless EL 294, *312*
Volmer CM, see McCandless EL 294, 295, *312*
Von Figura K, Klein 497, *512*
Von Figura K, Rey M, Prinz R, Ullrich K 497, *512*
Von Meyenburg K *623*
Von Wettstein-Knowles P 244, 245, 246, *253*, 530, *540*
Vonlanthen M, see Horisberger M 367, *388*, 397, *413*
Voragen AGJ, see Pilnik W 104, *126*
Vreeland V 278, 287, 299, *316*
Vries OMH de 360, *393*
Vries OMH de, Wessels JGH 356, 368, 371, 373, 374, 384, *393*
Vries OMH de, see Wessels JGH 358, 360, 362, 363, 364, 365, 368, 379, *394*
Vrsanska M, Krátky Z, Biely P, Machala S 406, *415*
Vuong R, see Chanzy H 524, *533*

Waaland JR 294, *316*
Waaland JR, see Waaland SD 255, *273*
Waaland SD, Waaland JR, Cleland RE 255, *273*
Wada M, see Fujikawa T 296, *309*
Wada S, Ray PM 32, 33, 38, 39, *46*, 264, *273*
Wada S, Tanimoto E, Masuda Y 267, *273*
Wade M, Albersheim P 588, 606, *623*, 643, *657*
Wade M, see Cline K 605, *616*
Waechter CJ, Lennarz WJ 48, *67*, 166, *193*, 381, *394*, 459, *483*
Waechter CJ, Lucas JJ, Lennarz WJ 173, *193*
Waechter CW, see Lucas JJ 175, *191*
Waffenschmidt S, see Jaenicke L 340, 342, 344, *350*
Wagenbreth I, see Quader H 322, 324, *332*, 525, *538*
Wagner G, see Lazar-Keul G 516, *536*
Wagner G, see Loewus F 295, *312*
Wainwright T, see Forrest IS 92, *119*
Walbrecht DT, see Winter H 139, *164*
Walker DA, see Heldt HW 607, *619*
Walker F, see Sadava D 143, *163*
Walker JC, Stahmann MA 612, 613, *623*
Walker JE, see Wallner SJ 139, 150, *164*
Walker JW, see Ballou CE 406, *411*
Walkinshaw C, see Mahlberg PG 516, *536*
Wallace DH, see Ferrari TE 580, *582*
Wallner SJ, Bloom HL 104, *131*
Wallner SJ, Walker JE 139, 150, *164*
Wallner SJ, see Gross KC 104, *120*
Walne PL, see Moestrup Ø 320, *331*
Walsh KA, see Kuhn RW 452, *456*
Walsh KA, see Palmitter RD *510*
Walter JA, see McCandless EL 294, *312*
Walton TJ, Kolattukudy PE 226, 227, 228, 229, 230, 231, 233, *254*
Walton TJ, see Kolattukudy PE 230, 233, 239, 242, 243, 245, *251*, 234, *536*
Wang FFC, Hirs CHW 496, *512*
Wang LC, see Jaworski EG 398, *413*
Wang MC, Bartnicki-Garcia S 355, 357, 381, *394*, 427, 431, *439*, 596, *623*
Wang MC, see Keen NT 596, 600, *620*
Wanka F 346, *351*
Wanka F, see Sassen A 324, *332*

Ward OP, see Fogarty WM 104, *119*
Wardrop AB 78, 94, 108, *131*
Wardrop AB, Bland DE 196, 197, *224*
Wardrop AB, see Cahfe SC 529, *533*
Wardrop AB, see Jarvis LR 529, *536*
Wardrop AB, see Schneider EF 16, *24*, 363, 366, 369, *392*
Warmke HE, see Overman MA 101, *126*
Warren CD, Jeanloz RW 170, *193*
Watanabe F, see Kato K 7, *8*
Watanabe I, see Tuzimura K 660, 661, *677*
Watanabe K, see Hirase S 292, *310*
Watanabe K, see Takeuchi Y 92, *130*, 528, *540*
Watanabe M, see Honjo G 280, *310*
Watanabe T, see Yamagishi T 152, *164*
Watanabe Y, see Horowitz NH 374, *388*
Waterkeyn L 90, 99, 100, *131*, 518, *540*
Waterkeyn L, see Longly B 518, *536*
Watkins WM 449, *458*
Watson B, Currier TC, Gordon MP, Chilton M, Nester EW 586, *623*
Wattendorf J 225, *254*
Wattendorff J 531, *540*
Weaver RW, Frederick LR 660, *677*
Webb JA, see Thomas B 139, 151, *164*
Webb MJW, see Tomlinson JA 606, *623*
Weber DF, see Vest G 659, 660, *677*
Webley DM, see Bacon JSD 360, 367, *384*, 419, 424, 425, *433*
Webster BD 102, *131*
Webster BD, Leopold AC 102, *131*
Webster DH, Currier HB 112, *131*
Webster DH, see Currier HB 111, 112, *117*
Wedgewood JF, Strominger JL 168, *193*
Weier TE, see Bisalputra T 346, *349*
Weigel H, see Percival Elizabeth 304, *314*
Weigl J, Yaphe W 293, *316*
Weiss RL, Goodenough DA, Goodenough UW 340, *351*
Weiss RL, see Goodenough UW 340, *349*
Weistrop J, see Loewus F 295, *312*
Wellburn AR, Hemming FW 57, *67*
Wells LG, Franich RA 530, 531, *540*
Welsh EJ, see Dea ICM 37, *43*
Welsh EJ, see Rees DA 68, 86, *127*, 277, 293, *315*
Wenc KA, see Berliner MD 347, *349*
Wenzler HF, see McClendon JH 107, *124*
Werdan K, see Fliege R 607, *618*
Werker E, Kislev M 565, *574*
Weser U, see Myllyla R 143, *162*
Wessels JGH 359, 360, 374, 375, 376, 377, *394*, 430, *439*
Wessels JGH, De Vries OMH 365, *394*
Wessels JGH, Koltin Y 375, 376, *394*, 430, *439*
Wessels JGH, Kreger DR, Marchant R, Regensburg BA, De Vries OMH 358, 360, 362, 363, 364, 365, *394*, 427, 429, *440*
Wessels JGH, Marchant R 366, 371, 374, 375, 376, 377, *394*
Wessels JGH, Niederpruem DJ 375, 376, *394*, 430, *440*
Wessels JGH, Sietsma JH 363, 375, *394*
Wessels JGH, Valk P van der, De Vries OMH 368, 379, *394*
Wessels JGH, see Datema R 353, 355, 357, 361, 363, 366, *386*
Wessels JGH, see De Vries OMH 356, 368, 371, 373, 374, 384, *393*, *434*
Wessels JGH, see Marchant R 376, *390*
Wessels JGH, see Niederpruem DJ 375, *391*
Wessels JGH, see Sietsma JH 352, 356, 357, 359, 360, 362, 363, 370, 375, 380, 381, 383, *392*, 427, 428, 429, *439*
Wessels JGH, see Valk P van der 353, 355, 356, 360, 361, 363, 365, 366, 367, 368, 369, 370, 374, 380, 382, *393*
Wessels JGH, see Van der Valk P 397, *415*
Wessels JGH, see Vermeulen CA 380, 383, *393*, 403, 404, 409, *415*
West C, Lee SC 603, 605, *623*
West CA, see Stekoll M 603, 605, *622*
Westafer JM, Brown RM Jr 94, *131*
Wetherbee R, see Dreher TW 281, 306, *308*
Weygand C, see Franz E 522, *534*
Whaley WG 564, *574*
Whaley WG, Dauwalder M, Kephart JE 516, 517, 518, *541*
Whaley WG, Kephart JE, Mollenhauer HH 545, 546, 564, *574*
Whaley WG, Mollenhauer HH 83, *131*, 518, *540*
Whaley WG, see Dauwalder M 518, *534*, 546, 558, 565, *570*
Whaley WG, see Mollenhauer HH 543, 545, 546, 558, *572*
Whatley MH, Bodwin JS, Lippincott BB, Lippincott JA 586, *623*
Whatley MH, Hunter N, Cantrell MA, Hendrick C, Keegstra K, Sequeira L 587, *623*
Whatley MH, see Lippincott BB 586, *620*
Wheeler H 659, *677*
Whelan WJ 73, *131*
Whelan WJ, see Peat S 417, *437*, 441, *457*

Whelan WL, see Ballou CE 406, *411*
Whistler RL, Richards EL 31, *46*
Whistler RL, see Doner LW 298, *308*
White A, see Benziman M 25, *28*
White DA, see Speake BK 495, *511*
White ON, Wirth DF 497, *512*
White RK, see Robinson DG 321, *332*
Whitmore FW 39, 41, *46*, 140, 141, 150, *164*
Whittaker RH, see Woodwell GM 11, *24*
Whittington SG, see Smidsrod O 286, *315*
Whitty R, see Lang A 112, *123*
Whyte JNC, Englar JR 281, *316*
Wiatr SM, see Addicott FT 102, *114*
Wick U, see Gerisch G 638, *653*
Wiemann W, see Tschesche R 610, *623*
Wiemken A, Schellenberg M, Urech K 498, *512*
Wiemken A, see Cortat M 377, 378, *386*, 422, 423, *434*, 470, *478*
Wiemken A, see Lenney JF 401, 402, *413*
Wiemken A, see Matile P 422, *436*, 469, *481*, 498, *509*
Wiemken A, see Wiemken-Gehrig V 105, *131*
Wiemken-Gehrig V, Wiemken A, Matile P 105, *131*
Wiersema ICM, see Winter H 139, *164*
Wiese L, Jones RF 348, *351*
Wiese L, see Jones RF 348, *350*
Wiese MV, DeVay JE, Ravenscroft AV 593, *623*
Wilhelm GE, see Parameswaren N 247, *252*
Wilkie KCB 7, *8*, 31, 32, *46*
Wilkie KCB, Woo S-L 31, *46*
Wilkie KCB, see Buchala AJ 31, *42*, 48, *62*, 93, 94, *115*, *116*
Wilkie KCB, see Fraser CG 31, *43*
Wilkie KCB, see Reid JSG 31, *45*
Wilkins MB 143, *164*
Williams DC, scc Outka DE 320, *331*
Williams EB, see Berard DF 588, *616*
Williams EL, see Turvey JR 282, 291, *316*
Williams JD, see Hubbard JP 589, *619*
Williams PH, see Keen NT 586, *619*
Williams PH, see Sutton JC 596, 599, *622*
Williamson AR, Singer HH, Singer PA, Mosmann 493, *512*
Williamson FB, see Gunawardena P 295, *309*
Willison JHM 20, *24*, 525, 529, 532, *541*
Willison JHM, Brown RM 17, *24*
Willison JHM, Brown RM Jr 522, 523, 524, 525, *541*
Willison JHM, Brown RM Jr, Mueller SC 528, *541*
Willison JHM, Cocking EC 18, *24*, 30, *46*, 525, 532, *541*
Willison JHM, Grout BWW 522, 525, *541*
Willison JHM, see Brown RM 20, *22*
Willison JHM, see Brown RM Jr 521, 526, 527, 528, *533*
Wilson E, see Crayton MA 298, *308*
Wilson LA, Sterling C 231, *254*
Wilson RG, see Israel HW 111, *121*, 608, *619*
Wimalajeewa DLS, see DeVay JE 586, *617*
Winge O, Roberts C 449, *458*
Winter H, Wiersema ICM, Walbrecht DT, Buffinga H 139, *164*
Winterburn PJ, Phelps CF 140, *164*
Wirth DF, see Robbins PW 463, *482*, 497, *510*
Wirth DF, see White ON 497, *512*
Wishart GJ, Fry DJ 82, *131*
Witt I, see Maier K 401, *414*
Wloch W 109, *131*
Wodzicki TJ, Brown CL 94, *131*
Wold F, see Jack MA 668, *675*
Wold JK, see Percival Elizabeth 302, *314*
Wolf DH, Ehmann C 403, *415*
Wolf DH, Fink GR 452, *458*
Wolpert JS, Albersheim P 651, *657*
Wolpert JS, Albersheim 663, 664, 666, 667, *677*
Wolpert JS, see Albersheim P 651, *652*
Wong F, Craigie JS 295, 305, *316*
Wong KF, see Craigie JS 295, *308*
Woo SL, see Lingappa VR 490, *509*
Woo S-L, see Wilkie KCB 31, *46*
Wood DA, Hammond JBW 400, *415*
Wood DA, see Craig GD 409, *412*
Wood P, see Siddiqui IR 34, *45*
Wood RKS, see Cooper RM 589, 591, 592, 593, 609, *617*
Woodhead JS, see Eddy AA 418, *434*
Wooding FBP 521, *541*
Wooding FBP, Northcote DH 40, *46*, 83, *132*
Wooding FBP, see Northcote DH 40, *45*, 80, 85, 98, 110, *125*
Woods EF, Lilley GG, Jermyn MA 152, *164*
Woods RKS, see Stephens GJ 150, *163*
Woodward FN, see Black WAP 296, *307*
Woodward JR, Keane PJ, Stone BA 111, 113, *132*, 597, 600, *623*
Woodwell GM, Whittaker RH, Reiners WA, Likens GE, Delwiche CC, Botkin DB 11, *24*
Woolf RA, see Fultz SA 369, *387*

Woolston RE, see Clowes FAL 544, 566, *570*
Wooltorton LSC, see Rhodes MJC 207, 208, 220, *223*, *224*
Worley JF, see Dekazos ED 112, *117*
Worth HGJ 31, *46*
Wouters JTM, see Dooijewaard-Kloosterziel AMP 374, *386*
Wrathall CR, Tatum EL 372, *394*
Wright A, Kanegasaki S 399, *415*
Wright A, see Robbins PW 74, *128*
Wright BE, see Rosness PA 426, 428, *438*
Wright K 551, 553, *574*
Wright K, Northcote DH 101, *132*, 521, *541*, 551, 552, 553, *574*
Wright K, Northcote DH, Davey R 553, *574*
Wright SW, see Howard RJ 281, *310*
Wu JH 112, *132*
Wu JH, Blakely LM, Dimitman JE 112, *132*
Wu JH, Dimitman JE 112, *132*
Wu S, see Bretthauer RK 465, *478*
Wulff G, see Tschesche R 610, *623*
Wursch P, see Bush DA 419, *433*
Wurscher R, see Burström H 260, *269*
Wurtz M, Zetsche K 280, *316*
Wutz M, Zetsche K 144, *164*
Wu-Yuan CF, see Hashimoto T 365, *387*

Yaku F, Tsuji S, Koshijima T 36, *46*
Yamada H, see Lin K 248, *252*
Yamada KM, see Olden L 493, *510*
Yamada M, Miyazaki 365, *394*
Yamada T, see Sumiya K 262, *273*
Yamada Y, Kuboi T 220, *224*
Yamagata Y, Masuda Y 261, 262, 263, *273*
Yamagata Y, Yamamoto R, Masuda Y 261, *273*
Yamagata Y, see Masuda Y 261, 262, *271*
Yamagishi T, Matsuda K, Watanabe T 152, *164*
Yamaguchi M, see Ozawa T 643, *656*
Yamaguchi S, see Eschrich W 112, *118*
Yamaguchi T, see Fukuda M 151, *158*
Yamaizumi M, see Uchida T 650, *657*
Yamaki S, Kakiuchi N 104, 105, *132*
Yamaki S, Machida Y, Kakiuchi N 104, 105, *132*
Yamaki S, Matsuda K 104, 105, *132*
Yamamoto K, Kawai H, Moriguchi M, Tochikura T 398, *415*
Yamamoto K, see Moriguchi M 398, *414*
Yamamoto R, Makai K, Masuda Y 261, *273*
Yamamoto R, Masuda Y 267, *273*
Yamamoto R, Nevins DJ 266, 267, *273*
Yamamoto R, Shinozyki K, Masuda Y 259, 262, *273*
Yamamoto R, see Fujihara S 259, 260, *270*
Yamamoto R, see Masuda Y 261, 262, 266, 267, *271*
Yamamoto R, see Nevins DJ 92, *125*, 264, *271*
Yamamoto R, see Yamagata Y 261, *273*
Yamamoto S, Nagasaki S 424, *440*
Yamamoto S, Shiraishi T, Nagasaki S 425, *440*
Yamamoto Y, see Kitamura K 424, 425, *436*
Yamamura Y, see Azuma I 466, *477*
Yamamura Y, see Kanetsuna F 372, *388*, 428, *435*
Yamasaki T, Hata K, Higuchi T 217, 218, *224*
Yamazaki M, see Fukushima T 110, *119*
Yanagishima N 422, *440*
Yanagishima N, see Shimoda C 422, *439*
Yang SF, Adams DO, Lizada C, Yu Y, Bradford KJ, Cameron AC, Hoffman NE 143, *165*
Yanke WE, see Dazzo FB 633, 642, *653*, 668, 669, *675*
Yaphe W 291, 293, *316*
Yaphe W, see Bhattacharjee SS 292, 293, *307*
Yaphe W, see Duckworth M 291, *308*
Yaphe W, see Groleau D 291, *309*
Yaphe W, see Hamer GK 291, *309*
Yaphe W, see Weigl J 293, *316*
Yaphe W, see Young K 291, *316*
Yariv J, Lis H, Katchalski E 577, *583*
Yariv J, Rapport MM, Graf L 152, *165*, 577, *583*
Yarotskii SV, Shashkov AS, Usov AS 293, *316*
Yashouv-Gan Y, see Eshdat Y 633, *653*
Ycas M 147, *165*
Yeh YF, see Sing V 443, 453, *457*
Yen PH, Ballou CE 441, 443, 447, 453, 454, *458*, 465, *483*
Yeow YM, see Jermyn MA 152, 153, 154, *159*, 577, *583*
Yin HL, Stossel TP 501, *512*
Yipp C 150, *165*
Yoda S, Ashida J 261, 262, 263, *273*
Yokota K, see Sakaguchi O 466, *482*
Yokoyama K, Terao T, Osawa T 631, 633, *657*
Yokoyama T, see Cole GT 365, *386*
Yonehara H, see Nakamura S 185, *191*
Yoneyama M, see Miller MW 420, *437*
Yoshikawa M 606, *623*
Yoshioka Y, Emori M, Ikekawa T, Fukuoka F 362, *394*
Youle RJ, Huang AHC 644, 647, *657*

Young JD, see Hanover JA 461, *479*
Young K, Duckworth M, Yaphe W 291, *316*
Young K, Hong KC, Duckworth M, Yaphc W 291, *316*
Young M, see Percival Elizabeth 297, 299, *314*
Young MR, Towers GHN, Neish AC 203, *224*
Young RE 104, *132*
Young RE, see Awad M 104, 105, *114*
Yu JMC, see Reichert TA 148, *163*
Yu RJ, Bishop CT, Cooper FP, Blank F, Hasenclever HF 419, *440*
Yu Y, see Yang SF 143, *165*
Yuen S 428, *440*

Zaar K 20, *24*, 150, *165*, 526, 528, *541*
Zacheo G, Lamberti F, Arrigoni-Liso R, Arrigoni O 141, *165*
Zalokar M 428, *440*
Zandonella P, see Dumas C 555, *570*
Zandonella P, see Sambin B 544, *573*
Zanlungo AB, see Villarroel LH 296, *316*
Zarra I, Masuda Y 92, *132*, 262, *273*
Zatz M, Barondes SH 170, *193*, 460, *483*
Zauberman G, see Pesis E 103, *126*
Zee S-Y, O'Brien TP 225, *254*
Zenk MH, see Ebert E 427, *434*
Zenk MH, see Gross GG 220, *222*
Zenk MH, see Mansell RL 207, 209, *223*
Zenk MH, see Stöckigt J 208, *224*
Zentmyer GA, see Keen NT 596, 600, *620*
Zetsche K, see Wurtz M 280, *316*
Zetsche K, see Wutz M 144, *164*
Zevenhuizen L, Bartnicki-Garcia S 426, 427, *440*
Zevenhuizen LPTM, Bartnicki-Garcia S 355, 358, *394*
Zhang W-J, see Hashimoto C 442, 452, *456*
Ziegler H, see Gietl C 633, 644, 645, 648, 649, *653*, *654*, 668, *675*
Ziegler H, see Kauss H 648, *655*
Ziegler H, see Schönherr J 247, *253*
Zimmerman M, see Strauss AW 489, *511*
Ziola B, see Spencer FS 52, 56, 59, *66*
Zitko V, Bishop CT 34, *46*
Zonneveld BJM 362, 371, 376, *394*, 428, 429, 431, *440*
Zubenko GS, Mitchell AP, Jones EW 403, *415*
Zucker M, Hankin L, Sands D 592, *623*
Zugenmaier P, see Herth W 328, *330*
Zusman DR, see Cumsky M 672, *674*

Subject Index

Page numbers in **bold face** refer to figures and tables

Abies alba 198
Abrin 650
Abrus precatorius **632**, 649
Abscisic acid 102, 111, 236, 601
Abscission 102, 103
Abscission zone 90, 102, 103
Acacia 673
Acanthamoeba 170
Acetabularia calyculus 283
Acetabularia crenulata 283, 301
Acetabularia mediterranea 283
Acetobacter xylinum **26**, 50, 60, 178, 526, **527**
Acetolysis 442, 446, 449, 464, 466, 476
Acetylated polysaccharides 107
Acetylation 88
di-*N*-acetylchitobiose 441, 452
Acetyl-CoA **168**
N-acetylgalactosamine 461
N-acetylglucosamine 357, 359, 371, 379, 395, 399, 442, 446, 452, 460, 463, 465, 472, 476, 497
N-Acetylglucosamine-6-P **398**
β-*N*-acetyl-glucosaminidase 495
N-acetylneuraminic acid 443, 634
Achlya ambisexualis 501
Achlya bisexualis 501
Acid phosphatase 376, 441, 449, 469, 473, 476, 494, 496, 498, 499, 500, 503, 504, 506, 552, 555, 566, 591
Acid-induced growth 142
Acoustic microscopy 111
Acrosiphonia **334**
Acrosiphonia centralis 300
Acrosiphoniales 300, 306
Actin 501
Actin microfilament 516
Acyl carrier protein 243
Acyl CoA 76
Acyl-CoA reductase 244
Adenia digitata 650
S-adenosyl L-methionine: catechol *O*-methyltransferase 205
S-Adenosylmethionine **203**, 206
ADPG synthase 607
ADP-glucose 50
Aeodes orbitosa 291
Aerenchyma 94, 97
Affinity chromatography 630, 648
Agar 291 ff., 292, 305
Agarase 291
Agardheilla tenera 293
Agaricus bisporus 365, 409, **633**
Agaricus bisporus lectin 634
Agaropectin 291
Agarose **291**, 294
Agave americana **229**
Agglutination 340, 341, 347, 348, 422, 453, 454, 455, 627
5-agglutinin 443, 453, 454
Agrobacterium tumefaciens 585, 586, 595
Alaria esculenta 347
Alcian blue 278
Alcian yellow 278
Aldobiuronic acid 361
Aleurone 90, 106, 107, 184
Alfalfa lectin 667
Algae, alginic acid 284
  autolysine 333
  cell wall 17, 277 ff.
  cellulose and glucans 279 ff., 320 ff.
  cytology of wall formation 317
  fucan 295
  functions of polysaccharide 305
  galactan 289
  glycoprotein wall 328
  mannan 283
  microfibril deposition 320 ff.
  polysaccharide secretion 303
  polysaccharides, taxonomy 306
  slime production 317
  sporopollenin 322, 345
  sulfated polysaccharide of chlorophyceae 299 ff.
  wall-degrading enzymes 333
  xylan 282
Algal rhizoid 255
Alginate 277, 278, 298, 299, 305
  biosynthesis 288 ff.
  block distribution **288**
  conformation 286
  constitution and structure 284 ff.
  degrading lyases 284
  egg box model 287
  ion exchange properties 285

Alginate
  occurrence 284
  use of proton magnetic resonance 285, 286
  variations in structure 287
Alkali-labile oligosaccharides **446**
Alkaline phosphatase 151, 451, 473, 490, 494, 495, 499, 501
Alkanes 242, **245**
iso Alkanes **240**
n-Alkanes **240**
Alkanoic acids **242**
Alkenes **240**
Allium 516
Allium cepa 641
Allomyces 356
Allomyces macrogynus 354, 369
Aloe arborescens **633**
Amidotransferase 410
Amino acid transport 637
5-amino 5-deoxyglucose 150
Amiprophosmethyl (APM) 324
Ammonia-lyase 220
Amoebae 638
Amphomycin 177, 186
Amygdalus persica **229**
α-amylase 184, 490
Amyloplast 545, **548**
Amylose 74
Amylose-like glucan 419
Amylovorin **596**, 599, 600, 602
Anatheca dentata 291
Anemia phyllitidis **229**
Angiosperm 76, 77, 93, 99, 108, 196, 200, 201, 205, **207**, 212, 217, 218, 226, 543
  megasporogenesis 101
3,6-anhydro sugar 290
3,6-Anhydro-D-galactose 2-sulfate **292**
3,6-anhydrogalactose 291, **292**, 295, 305
Aniline blue 90, 97, 99, 109, 110, 111, **112**, **113**, 518, 581
Anteiso-Alkanes **240**
Anther 99, 100
Anthracene **240**
Antibiotic 185, 649
Anti-freeze 154
O-antigen 587, 665
T-antigen 636
Antitranspirant 154
Apiogalacturonan 34
Apiose 6, 34
Aplanospores **334**
Apodachlya brachynema 355
Apoplast, movement of ions 68, 247
  restriction of permeability 531
Apple 103, 104
Aptenia cordifolia 579
Arabinan **4**, 6, 34ff., 37, 39, **71**, 75, **589**
Arabinanase **589**
Arabinase 102
Arabinofuranosidase 106
Arabinofuranosyl hydroxyproline 134
Arabinogalactan 3, 37, 88, 137, 169, 264, 577, **578**, **579**, 580, 581, 628
  in cell plate formation 90
  in coniferous woods 6, 77
  4-linked arabinogalactan 6
  3,6-linked arabinogalactan 6, 34
  metabolism of 92
  proteins 133, 135, 136, 147, 152ff.
  type I **4**, 6
  type II 3, **4**, 6
Arabino-3;6-galactan 75, 92
Arabinoglucuronoxylan 94
Arabinose 93, 107, 268, 299, 551, 563, 578, 581, 590, 591, 599, 628
L-arabinose 3-sulfate 299
β-arabinosidases 139
β-L-arabinosides 139
Arabinosyl hydroxyproline 145
Arabinoxylan 5, 7, 88, 264
Arachis hypogaea **632**
Archaebacteria 133, 147, 148
Arginine 344
Armillaria mellea 359
Aromatic amino acid 201
Aromatic hydrocarbons **240**
Arthrobacter 406, 424, 425
Ascomycetes **354**, 355, 366, 380, 382, 500
Ascophyllan 295
Ascophyllum nodosum 288, 296, 297, 298
Ascopore 421
Ascorbate 133, **204**, 206, 221, 245
Ascorbic acid oxidase 156
Ascosporulation 423
Ascus 422
Asexual reproduction in algae 333ff.
Asparaginase II 449, 476
Asparagus pea **632**
Aspergillus 360, 362
Aspergillus aculeatus 370
Aspergillus awamori 365
Aspergillus flavus 380
Aspergillus nidulans 365, 371, 374, 376, 378, 380, 405, 408, 409, 428, 430, 431
Aspergillus niger 141, 360, 365, 429, 466, 495, 496
Aspergillus oryzae 467
Asterionella socialis 303
ATPase 383, 492, 552
Atriplex 226
Aureobasidium pullulans 428
Autolysines 150, 151, 333ff., 378
Autolysis 150, 152, 375, 416, 421, 422, 430
Autoradiography 82, 83, 317

Autospore **322**
Auxin 32, 40, 50, 59, 91, 102, 109, 139, 143, 150, 220, 221, 261, 263, 264, 265, 266, 267, 422, 601
Auxin oxidase 149
Avena coleoptile 261, **262**, 264, 265, 266, 267, 268
Avena sativa 609
Avena stem internode **263**
Avenacin 609, **610**, 611, 612
Avenacosides 609
Avocado 103, 104, 105
Azolla 516
Azotobacter vinelandii 289

Bacillariophyceae 283, 303, 305
Bacillus circulans 424
Bacillus licheniformis 490
Bacillus subtilis 490, 581
Bacitracin 186
Bacteria 166, 170, 255, 264, 284, 340, 398, 488, 491, 526, 594, 627, **633**, 643, 658, 659, 670
Banana 104, 110
Band growth 255
Bandeiraea simplicifolia 631, **632**, 635
Bangia fuscopurpurea 282, 283
Bangiales 281
Barley coleoptile **262**, 267
Basidiomycetes **354**, 355, 359, 362, 366, 376, 380, 382, 404, 409, 426
Batophora oerstedi 283
Beijerinckia **568**
Bending technique 258, **262**
Benzamidine 344
Benzyladenine 59
Beta vulgaris **230**
Betalains 611
Betamin 611
Betula papyrifera 197
Betula pendula **230**
Betulin **242**
Betulinus 360
Bifurcaria bifurcata 296, 297
Black locust **633**
Blastocladiella 398, 399, 400, 406
Blastocladiella emersonii 397, 410
Blastomyces dermatitides 428, 429
Blastomyces dermatitidis 431
Blastospores 17
Blood group 627
Bordered pits 14
Botrytis tulipae 613
Brassica 208
Brassica campestris 580
Brassica napobrassica **230**
Brassica oleracea 244, **596**, 612
Brassica oleraceae 613
Brassica rapa **230**
Broad bean **632**, 649
Bryophyte 148
Bryopsidella neglecta 283
Bryopsis maxima 283
Bryopsis plumosa 283
Bud formation 377, 406ff., 422, 469
Bud scar 356, 359, **364**, 368, 397, 406, **407**
Bullera alba 420
Bundle sheath 225, 228, 247, 531

Caffeic acid **203**, **204**, **206**, 207
Caffeine 516, 517, 544
Calcification 320
Calcite 317
Calcium 37, 84
  binding to alginate 286
  chelating agents 3
  dependent protease 344
  deregulation of intracellular levels 324
  effect on chitin synthase 410
  in cell plate formation 516
  in sulfated polysaccharides 302
  inhibition of growth 265
  inhibition of secretory vesicles 86
  mitogenic 637
  role in secretion 341, 501, 521
Calcofluor 370, 528
Calcofluor white 25, **26**, 404, 406
Callose **4**, 49, 55, 85, 86, 89, 98, 99, **100**, 110, 111, **112**, **113**, 377
  see also (1→3)-β-glucan
  dissolution 99, 101
  in megasporogenesis 101
  in pollination 581
  synthesis 48
Callosities 111
Callus 39, 111
Cambium 15, 39, 76, 154, 197, 644
Canavalia ensiformis 627, **632**
Candelabra aloe **633**
Candida albicans 367, 380, 399, 404, 409, 418, 419, 468
Candida curvata 420
Candida humicola 420
Candida utilis 420, 421
Capsicum 104, 105
Capsular polysaccharides 74, 166
Capsular wall material **364**, 365
Carboxymethylation 443, 450
Carboxypeptidase Y 185, 441, 442, 451, 452, 472, 473, 476, 491, 492, 494, 496
Carcinogenesis 637
Carposporophyl 295
Carrageen 305
λ-carrageenan 278, 292ff., 293, 294, 295

Carrageenan-type galactan 290
Casbene **605**
Casparian band 225, 228, 247, 531
Castor bean **632**
Catabolite repression 591, 593
Catechol 613
Cattleya 548, 553, 559
Caulerpa anceps 282
Caulerpa brachypus 282
Caulerpa filiformis 282, 300
Caulerpa racemosa 282, 300
Caulerpa sertularoides 300
Caulerpa simpliciuscula 281
CDP-glucose 50
Celery petiole **262**, 263
Cell adhesion 639, 641
Cell aggregation 638
Cell cycle 341, 378, 381, 406, 423, 469, 518, 637
Cell extension 138, 139, 149, 150, 255
Cell metabolism, during environmental stress 109
Cell plate 40, 69, 83, 85, 90, 514, **515**, 516, **517**, 518
Cell recognition 133, 154, 155
Cell wall, and cellular morphology 370ff.
  and taxonomy of fungi 354
  assembly 528
  attachment to suberin 231
  autolysis 150, 333ff.
  chitin, of the 405ff.
  definition 9
  degrading enzymes 333ff., 373ff., 588ff., 608
  enzymes, of the 133ff., 149ff.
  extensibility 255ff.
  extraction procedure 3, 140, 278
  galactosidases 138
  glycoproteins, of the 133ff., 328
  hydrophobic layers of 225ff.
  inter-polymeric linkages 7, 383
  interrupted lamella model of wood **198**
  lectin association 647
  loosening 40, 109, 256ff., 374, 377, 421
  mechanical properties 257
  model of plant cell wall **590**
  of algae 17, 277ff., 317ff.
  of bacillariophyceae 303
  of *Chlamydomonas* 145, 328
  of diatoms 303, 326
  of fungi 352ff., 364
  of *Fusarium sulphureum* 16
  of protoplasts 18, 368
  of yeasts 16, 354ff.
  phosphatase 151
  polysaccharide composition, quant. 37
  protease 151, 152
  structure of a wood fiber **196**
  sugar composition, quant. 107
  wall-loosening enzymes in fungi 377ff.
Cell wall metabolism, autolysines of algae 334ff.
  autolysis in the life cycle of plants 333
  change from primary to secondary wall 93
  chilling response 110
  during cell division 87, 90
  during differentiation 39, 87, 93, 94ff.
  during environmental changes 108ff.
  during germination 105ff.
  during growth 39, 87
  during pathogenesis 111ff., 588ff.
  during senescence 102ff.
  in fungal development 421ff.
  in morphogenesis 89
  methodological consideration 87ff.
  wounding response 110
Cell-cell interaction 151, 154
Cellobiase 150
Cellobiose 47, 56, 321, 432, 591, 593
Cellodextrins 55, 56
Cellulase 56, 96, 102, 105, 109, 139, 150, 267, 347, 373, 426, 430, 501, 528, **589**, 591, 592, 593, 594
  anti-cellulase 103
Cellulose 3, **4**, 6, 12, 14, 18, 19, 20, 29, 37, 40, 68, 88, 93, 94, **100**, 106, 108, 148, 169, 264, 267, 277, 278, 279, 281, 298, 318, 320, 353, **354**, **364**, 366, **589**, 608
  biosynthesis 47ff., **57**, 58, 59, 61, 70, 522
  biosynthesis in algae 281
  biosynthesis in bacteria 26, 58, 178
  carboxymethyl-cellulose 56, 96, 97
  α-cellulose 56
  degree of polymerization 30, 524
  from cotton 56
  hormonal control of biosynthesis 59
  hydrogen bonds 32, 528
  in *Dictyostelium discoideum* 426, 430
  in fungi 354ff., 357ff., 374, 426
  in wood **198**
  inhibitors of formation 60
  lipid intermediate 57, 170, 178
  membrane protein as primer 56
  metabolism 47, 105
  of algae 279ff.
  polymerization 28
  primer in biosynthesis 55ff.
  Prototheca zopfii 56
  X-ray diagram 280, 426
  scales 59, 513
  site of synthesis 521
  synthesis in *Acetobacter* 25, 178, 526, **527**
  synthesis in *Oocystis* 25, 324ff.

synthesis in *Prototheca zopfii* 178
synthetase 28, 525
Cellulose microfibril 12ff., 17, 19, 26, 85, 102, 197, 198
assembly 25ff.
colchicine 60, 324
digestion 105
enzyme complex 19, 20, 55, 56, **523**
formation 81, 324, 522
in algae 279ff.
in fungi 358, 369
orientation 525ff.
Cellulose I 25
Cellulose II 25
Centrospermae 611
Ceramium rubrum 291
Ceratocystis fimbriata 587
Ceratocystis ulmi **597**
Cereal β-D-glucan **4**
Cetyltrimethylammonium borate 442
Cetyltrimethylammoniumbromide 378
Chaetamorpha melagonium 302
Chaetangium fastigiatum 282
Chaetoceros affinis 305
Chaetoceros curvisetus 305
Chaetoceros desipiens 305
Chaetomium globosum 365
Chaetomorpha 280, 299, 322
Chaetomorpha capillaris 300
Chaetomorpha linum 300
Chaetomorpha melagonium 321
Chalaza **227**
Charophyta 279
Chemotaxis 638, 662
Cherry 104
Chilli 105
Chilling 110, **112**
Chitin 12, 19, 148, 353, **354**, 355, **364**, 367, 370, 371, 372, 374, 381, **396**, 427, 494, 504, 631, 649
biosynthesis 379, 397ff.
chain length 395
chitosomes **401**, 405
degradation 405
feedback inhibition **398**
formation of precursor 397, **398**
in *Blastocladiella emersonii* 397, 410
in hyphal growth 408ff.
in vivo, regulation of chitin synthesis 405ff.
lipid intermediate 399
localization 397
microfibril 16, 21, 26, 326ff., 355ff., 367, 368, 380, 397, **401**
occurrence in 355ff.
properties of chitin synthetase 398ff.
regulation of chitin synthetase 401ff.
solubility 395
structure 395ff., **396**
subcellular distribution of chitin synthetase 403ff.
synthesis in *Poterioochromonas* 25
synthesis in *Thalassiosira fluviatilis* **326**
X-ray diffraction 395
Chitin ring 406, 408
Chitin synthase 89, 379, 383
Chitinase 355, 359, 373, 374, 375, 376, 399, 405, 608, 649
Chitobiose 463
Chitodextrin 399
Chitosan **354**, 355, 357, 603
Chitosomes 21, 380, **401**, 405, 409
Chlamydomonas 133, 136, 147, 334, **339**, 528, 642
Chlamydomonas angulosa **339**
Chlamydomonas capensis 338
Chlamydomonas debaryana **339**
Chlamydomonas eugametos 338
Chlamydomonas gigantea 338
Chlamydomonas globosa **339**
Chlamydomonas gymnogama 145, 338
Chlamydomonas incerta **339**
Chlamydomonas komma **339**
Chlamydomonas pseudogigantea 338
Chlamydomonas reinhardtii 145, 146, 148, 328, 334, **335**, **336**, **337**, 338, **339**, **342**, 344
Chlamydomonas smithii **339**
Chlamydomonas zimbabwiensis 338
Chlorella 279, 322, 324, 334, 335, 346, 347
Chlorella ellipsoidea 280
Chlorella fusca 344, 348
Chlorella pyrenoidosa 280
Chlorella vulgaris 148
Chlorococcales 322, 344, 346
Chlorococcum 346
Chlorococcum echinozygotum 348
Chlorococcum oleofaciens 145
Chlorodesmis formosana 282
Chlorogenic acid 205, **206**
Chlorogonium 346
Chlorophyceae 145, 148, 279, 280, 282, 283, 306, 334, 343, 344, 346
Chloroplast 490
Chlorosis 601
Chlorothecium inaequale **334**
Choanephora cucurbitarum 356
Cholera toxin 491, 650
Cholesterol **168**, 186, 498
Cholesterol biosynthesis 167
Choline 372, 373
Chondrus crispus 292, 294, 295
Chorismic acid **202**
Chromatids 514
Chrysochromulina chiton 318, 320
Chrysochromulina megacylindrica 320

Chrysophyceae 317, 318
Chrysophyta 279
Chymotrypsinogen 488
Chytridiomycetes **354**, 366
Cinnamate-4-hydroxylase **204**, 220
Cinnamic acid **203**, **204**, 221
Cinnamoyl-CoA reductase 208, **209**, 220
Cinnamyl alcohol 36, 41, 149
Cinnamyl alcohol dehydrogenase **209**, 220
Circular dichroism spectra 147
Cissus sicyoïdes 553
Citrus deliciosa **229**
Citrus limon **229**, **597**
Citrus paradisi **227**, **229**, **230**
Citrus reticulata **229**
Citrus sinensis **229**
Cladophora 280
Cladophora rupestris 299, 300, 302, 321
Cladophorales 299, 300, 306, 320, 322
Cladosporium fulvum **597**
Clathrin **503**
Claviceps fusiformis 427
Claviceps purpurea 427, 595
Cleistothecia 376
Climacteric fruit 103
Coated vesicles **503**, 517, 518, 520
Coccolith 317
Coccolithosomes 320
Cochliobolus, miyabeanus 432
Codiales 300
Codiolum pusillum 280, 283, 300, 301, 302
Codium adherens 283
Codium fragile 283, 300
Codium intricata 283
Codium latum 283
Coelastrum 346, 348
Coffea arabica **229**
Colcemid 324
Colchicine 60, 320, 324, 325, 326, 516, 525, 564
Colicine 491
Collagen 136, 147, 155, 184
Collenchyma 69, 93, 103
Colletotrichum 141
Colletotrichum circinans 613
Colletotrichum destructivum **604**
Colletotrichum lagenarium 152, 608
Colletotrichum lindemuthianum 588, 590, 603, **604**
Colletotrichum trifoli **604**
Compactin 187
Companion cell 98
Compression wood 108, 109, 197
Concanavalin A 341, 344, 380, 403, 404, 442, 469, 495, 499, 586, 587, 627, 630
Congo red 25, **27**
Coniferyl alcohol 205, **210**, 212, 213, 215, 216
Coniferyl aldehyde 196, **209**, 209
Conjugales 325
Conjugation 377, 422
Connective tissue 86
Contractile vacuole 146
Convolvulus 543, 544
Copper-complex 362
Coprinus 356, 399, 402, 409
Coprinus cinereus 369, 380
Coprinus congregatus 376
Coprinus lagopus 375
Corn coleoptile 267
Corolla 105
Corynebacterium diphteriae 490
Corynebacterium insidiosum **596**, 599, 600
Corynebacterium michiganense **596**
Corynebacterium sepedonicum **596**, 599, 600
Coscinodiscus nobilis 305
Cotranslational cleavage 489, 492
Cotton fiber 30, 40, 49, 54, **78**, 175, 513, **520**, **523**, 525
Cotton hair 89, 91, 93, 94
Cottonwood 197
Cotyledon 105
p-coumaraldehyde 209
p-coumarate 3-hydroxylase **204**
p-coumaric acid 200, 203, **204**, **206**, 208, 219
Coumarin 60
p-coumaroyl-CoA 219
p-coumaryl alcohol 212, 219
Creep 258, **259**, 260, 261, **262**, 263
Cross and Bevan reaction 196
Crown gall 586
Cryptate 211, 221 324
Cryptococcus 419
Cryptococcus albidus 420, 421
Cryptococcus informo-miniatus 420
Cryptococcus laurentii 420, 466
Cryptococcus macerans 420
Cucumber 110, **633**
Cucumber hypocotyl **262**, **263**
Cucumis lagenarum 608
Cucumis melo 608
Cucumis sativa 152
Cucurbita pepo 151, 553, 648
Cucurbita sativa **633**
Cunninghamella elegans 380
Cupric ion complexation 441
Cuticle 9, 69, 225, 226, **227**, 237, 247, 265, 278, 283, 529, 530, 531
Cutin, aliphatic components of **229**
  amount per surface 226
  biodegradation of 237ff.
  biosynthesis of 232ff.
  comparison of cutinases **239**
  composition and structure 227ff.
  fruit of *Lycopersicon esculentum* **226**

function of 247
isolation of 227
location and ultrastructure of 225
methods of depolymerization **228**, 231
monomers of 228, **232**
secretion of 529ff.
site of synthesis 236
Cutin acids 232
Cutinase 234, 237, 247, 468
Cyanide-insensitive respiration 142
Cyanidin 613
Cyanobacteria 147
Cyanogenic glucosides 611
Cyanophyceae 347
Cyclic AMP 221, 592, 638
Cycloheximide 59, 107, 267, 324, 325, 346, 384, 409, 471, 492
Cyclotella meneghiniana **326**
Cylindrotheca fusiformis 303
Cymopolia barbata 283
Cytochalasin B 516, 521, 525, 564
Cytochrome P_{450} 233, 234
Cytokinesis 408, 514, 515, 517, 564
Cytokinin 40, 102, 263
Cytophaga johnsonii 424
Cytoplasmic streaming 506
Cytoskelet 497, 525

2,4-D 59
Dasycladales 299
Dasycladus clavaeformis 283
Dasycladus vermicularis 283
Date 104
Daucus carota **230**
Deacetylation 359
De-arabinosylation 134
De-esterification 104
Dehydroascorbate 143
2,3-dehydrodolichyl-pyrophosphate 167
Delignifcation 97
2-deoxyglucose 84, 179, 185, 186, 371, 379, 472, 477, 642, 666, 668, 670
2-deoxyglucose-phosphate 473
Derbesia 283
Derbesia lamourouxii 283
Derbesia marina-Halicystis ovalis 280
Derbesia tenuissima 283
Derbesia-Halicystis 144
Desglucoavenacoside 609, **610**, 612
Desmarestia aculeata 297
Desmarestia firma 297
Desmarestia ligulata 297
Development during microsporogenesis **100**
Dextran 74
Diacetylchitobiose 379
N,N'-diacetylchitobiose 587
N,N'-diacetylchitobiosyl-pyrophosphoryl-dolichol 173, 174
Dianemycin 324
Diatoms 144, 283, 303, 305, 326
Dicarboxylic acids **242**
Dichlorobenzonitrile 61
Dicotyledons 7, 36, 105, 136, 148, 150, **207**, 264, 268
Dictyosome 59, 60, **73**, 81, 83, 85, 318, 513, **515**, **517**, 518, **520**, 545, **546**, **547**, **548**, 553
see also Golgi
Dictyostelium 426, 639
Dictyostelium discoideum 426, 428, 430, **633**, 638, 640, 672
Dictyostelium purpureum 638
Dictyota dichotoma 296
Diethylpyrocarbonate 378
Dihydrofusicoccin 598
Dihydroxypalmitic acid 228
Dihydroxyproline 144
Dikaryon 375, 376, 377
β-Diketones **240**, **246**
Dilignol 215, 216
Dinophyceae 346
Diols **242**
Diphtheria toxin 490, 650
α,α'-dipyridyl 143
Discoidin 639, 640
Disease resistance 133, 142, 151
Dolichol 167, 170, 186, 460, 468, 557
Dolichol pathway **166**, **177**
in yeast 461ff., **475**
subcellular location 187, 468ff.
sulibilization of enzymes 473ff.
Dolichos biflorus **632**, 652, 668
Dolichos biflorus lectin 635
Dolichyl, identification of 460
Dolichyl diphosphate oligosaccharide 461
Dolichyl pyrophosphoryl *N*-acetylglucosamine **172**, 173, **174**, 183, 185, 186, 400, 462, 493
Dolichyl-phosphate 58, **167**, 171, 173, 400
inhibitors of synthesis 186
protection of enzymes 493
synthesis of **168**
Dolichyl-phosphate synthetase 168
Dolichylphosphoryl 2-deoxyglucose 185, 473
Dolichyl-phosphoryl-glucose 171, 178, 463
Dolichyl-phosphoryl-mannose **167**, 170, 176, 177, 178, 186, 187, 494
Dolichyl-pyrophosphoryl-chitobiose 462
Dormancy 99
Durvillea antarctica 284

Earlywood 197
Ectodesmata 237, 529
Effectomer 650
Egg-box model 37

Elastic extensibility 259
Elastic extension 257
Elementary fibril 29
Elicitor 113, 155, 587, 588, 602, 603, **604**, 606, 643
β-elimination 37, 56, 135, 139, 362, 442, 443, 446, 447, 452, 453, 461, 465, 468, 476
Elodea densa 220
Elsinan 429
Elsinoe leucospila 429
Embryo 105, 106, 107
Embryo sac 101
En bloc transfer 475
Enchylemma 85
Endocytosis 502, 672
Endodermis 225, 247, 531
Endo-(1→3)-β-glucanase 421, 422, 423, 424
Endo-(1→6)-β-glucanase 424
Endolysis 56
Endo-(1→6)-α-mannanase 443, 450, **451**
Endomembrane system 80, 81, 86, 89, 317, 329
Endomyces 419
Endo-*N*-acetyl-β-D-glucosaminidase 442, 443, 445, 450, 451
Endopeptidase 489
Endoplasmic reticulum 59, 73, 81, 82, 84, 85, 89, 91, 94, 96, 98, 101, 137, 178, 182, 187, 188, 221, 232, 233, 237, 245, 295, 317, 318, 468, 470, 476, 487, 493, 494, 498, **517**, 518, 519, 525, 531, 545, **547**, 548, 556, 557, 561, 646
Endopolygalacturonase 590, 591, 603, **605**
Endosperm 85, 93, 105, 106, 107, 647
Enteromorpha 280, 317
Enteromorpha compressa 300, 301
Enteromorpha intestinalis 302
Entner-Doudoroff pathway 595
Ephedra 206
Epicuticular Waxes 530
Epidendrum 559
Epidermis 225, 226
Epimerase 79
4-epimerase 77
Epiphytism 151
Epoxidation 234
Eremosphaera viridis 324
Erwinia amylovora 592, 594, **596**, 599, 600
Erwinia carotovora 592, 603, **604**
Erwinia chrysanthemi 590, 593
Erysiphe graminis 142
Erythrocytes 627
D-Erythrose-4-phosphate **202**
Escherichia coli **633**
Esterase 151
Esterification 69
Etherification 69
Ethylene 102, 111, 113, 141, 142, 143, 220, 263, 600
Eucalyptus cypellocarpa **597**
Eucalyptus papuana **226**
Eucalyptus sieberi **597**
Eucheuma nudum 295
Eucheuma spinosum 293
Eudorina 144, 145
Euglena 60, 347
Euglena gracilis 244
Euglenophyta 279, 360
Evolution 146, 148, 194, 206, 354
Co-evolution 673
H^+-excretion 268
Exine **100**
Exocytosis 410, 501, 502, **505**
Exodermis 531
Exoglucanase 420, 421, 495, 496
Exo-(1→3)-β-glucanase 422, 423, 425, 498
Exo-α-mannanase 443, 446
Exo-polygalacturonase 592
Extensin 133ff., 136, 137, 150, 154, 155, 636, 647
Extensin hypothesis 138
Extensin precursor 137
Extension growth 643
Extracellular starch 420
HF extraction 139
Exudate gum 3
Exudation 154

Fabaceae 644
α-factor 382, 455
Farnasene **240**
Farnesyl-pyrophosphate 167
Fatty acid synthetase 243, **245**
Fatty alcohols 242, 244
Favism 649
Fehling's solution 442
Fern root cap 543
Fertilization 151, 155, 289, 298, 577, 642
Ferulate-5-hydroxylase 218
Ferulic acid 40, 76, 107, 150, 197, **203**, **205**, 206, **207**, **208**, 218, 231, 555
Ferulic acid dimerization 86
Feruloyl adenylate 208
Feruloyl-CoA reductase 220
Feryloyl-CoA 208
Fiber cells 83
Ficaprenol 171
Ficaprenyl-phosphate 171
Fimbriae 670
Flagella 335, 340, 341, 347, 410, 642
Flammulina velutipes 376
Flavonoid biosynthesis 208, 220, 221
Flavonol glycosides 152
Flocculation 419

Fluid-mosaic model 522, 637
Fluorosugar 185
Fomes fomentarius **633**
Forsythia 207, 208, 215
Frankenia grandifolia **227**
Frost hardiness 154
Fructans 70
β-fructofuranosidase 595
Fructose-6-phosphate 397
Fruit, climacteric 103
  nonclimacteric 103
  senescence of 102, 103
  softening of 103, 104
Fruiting bodies 356, 375, 633, 638, 672
Fucan 278, 279, 283, 306
  amount of in algae 296
  biosynthesis 298
  definition 296
  fractionation 296
  location and function 298
  of brown algae 295
  protein in 296
  secretion 303
  site of sulfate 297
  structure 297
  synonyms 295
Fucoidan 295, 296, 317
Fucoidin 295
Fucose **73**, 93, 169, 295, 297, 303, 304, 305, 317, 361, 443, 497, 551, 553, 555, 556, 558, 563, 564, **596**
Fucose 4-sulfate 297
α-L-fucosidase 553
Fucosyl transferase 557, 558, 599
Fucus 278, 281
Fucus distichus 281, 298
Fucus inflatus 298
Fucus gardneri 288, 289
Fucus serratus 298
Fucus spiralis 296
Fucus vesiculosus 288, 289, 296, 297, 298
Fucus zygote 317
Fulvia fulva **605**
Fungal Cutinase 237, 238
Fungal infection **112**
Fungi, cell walls 352ff.
  cellular morphology 370ff.
  synthesis of the wall 379
  wall loosening enzymes 377ff.
  walls of taxonomic groups **354**
Furcellaran 293
Furcellaria fastigiata 293
Fusarium 587
Fusarium oxysporum 591, 594, 608, 609, 613
Fusarium solani 468, 606, 649
Fusicoccin 263, 597, **598**, 601, 602
Fusicoccum amygdali 360, **597**, 598
Fusion rosette 502

Galactan 6, 34ff., 37, 39, 40, **71**, 80, 91, 139, 277, **589**
Galactanase **589**
Galactans of Algae 289ff.
(1→4)-galactan 92, 108
(1→3);(1→6)-β-galactans 109
(3→6)-β-galactan 578
Galactoarabinoxylan 31
Galactoglucomannan **5**, 7, 77
Galactomannan 85, 107, 169, **354**, 362, 367, 466, 467
Galactonolactone 139, 266, 647
Galactosamine 372, **605**
Galactose 107, 355, 362, 461, 467, 497, 551, 552, 555, 563, 578, 590, 591, **596**, **597**, 599, **604**, 647
D-Galactose 2,6-disulfate **290, 292**
D-Galactose 2-sulfate **292**
D-Galactose 4-sulfate **292**
D-Galactose 6-sulfate **290**, 291, **292**, 295, 299, 305
L-galactose 290
L-galactose 6-sulfate 295
α-galactosidase 495
β-galactosidase 107, 139, 266, 449, 491, 496, 500, 647, 652
  inhibitor of 139
Galactosyl hydroxyproline 145
Galactosyl serine 135, 138, 139
Galactosyl transferase 466
Galactosylsucrose oligosaccharides 151
Galacturonan 6, 39, **71**
Galacturonic acid 552, 563, 579, 590, 591, 592, **597**
Gallic acid **206**
Gametangia **334**
Gamete autolysine 335, 336, 338, 340, 341, 344, 348
Gametogenesis **100**
Gas transport 97
GDP-deoxyglucose 185
GDP-fucose 557
GDP-glucose 49, 51, **52**, 57, 70, 71, 173
GDP-guluronic acid 288, 289
GDP-mannose 51, **71**, 170, 176, 459
GDP-mannuronic acid 288
Gelidium amansii 291
Gelsolin 501
Geminella 334, **345**, 346
Gene fusion 491
Geotropism 108, 143
Gerl 518
Germ tube 356
Germination 105, 107

Gibberellin 50, 59, 102, 106, 107, 111, 220, **262**, **263**
Gigartina atropurpurea 293, 294
Gigartina stellata 295
Ginkgo biloba **229**
Gladiolus 155, 577, 578, 579
Glaucocystis 280, 524
Glaucocystis nostochinearum 324
Glenodinium lubiniensiforme **334**
Glucan 3, 4, 6, 12, 19, 87, 155, 169, 268, **354**, 355, 356, 357, 359, 364, 366, 406, 416ff., 471, 494, 504, **604**
R-glucan 375, 376, 377, 381, 427, 430
(1→3);(1→4)-β-glucan endohydrolase 90
(1→3)-β-glucan exo-hydrolase 582
(1→3)-β-glucan hydrolase 99, 101, 106, 111
  development during microsporogenesis 100
(1→4)-β-glucan hydrolase 106
(1→3);(1→4)-β-glucan hydrolase 106
Glucan synthetase 50, 59, 75, 86, 94, 110, 111
β-glucanase 374, 377, 406, 416, 420, 454, 494, 600
Glucanase vesicles 498, 504
R-glucanase 375, 377
(1→3)-β-glucanase 150, 266, 267, 269, 356, 373, 374, 375, 376, 377, 378, 430, 449, 561
(1→3)-α-glucanase 421, 608
(1→6)-β-glucanase 373, 421, 430
Glucans in fungi, biosynthesis 431ff.
  degradation 420ff.
  degree of polymerization 417
  glucose repression of (1→3)-β-glucanase 431
  structure 417ff.
  X-ray diffraction 419
(1→3)-β-glucan 17, 48, 58, 74, 86, 90, 93, 109, 111, 281, 353, **354**, 359, 360, 360ff., 362, 365, 367, 368, 370, 371, 372, 374, 376, 377, 379, 381, 417ff., 504, 518, 582, **596**, **597**, 599, 603, 608
  see also Callose
(1→3)-β/(1→6)-β-D-Glucan **354**, 355, 358, **364**, 366, 367, 368, 371, 372, 374, 375, 381
(1→4)-β-glucan 13, 19, 20, 56, 279, 553
(1→6)-β-D-glucan 417ff.
(1→3);(1→4)-β-glucan 68, **71**, 74, 75, 88, 91, 92, 106, 582
(1-3)-α-D-glucan **364**
Glucobrassicin 613
Glucolipid 54, 58, 171, 172
Glucomannan **5**, 7, 33, 51, 68, **71**, 75, 80, 88, 93, 94, 106, 169, 419, 521
Gluconolactone 266, 422
Glucoprotein 56, **57**, 180
Glucosamine 185, 186, 280, 355, 357, 361, 366, 371, 372, **605**
Glucosamine auxotroph 406
Glucosamine-6-phosphate 397, **398**
Glucose repression 594
Glucose transport system 594
Glucose-6-phosphate dehydrogenase 372
β-glucosidase 106, 150, 181, 266, 463, 467, 475, 494, 495, 613
(1→6)-α-glucosidase 266
β-glucosidase inhibitor 150
Glucosinolates 611, 612
β-glucosyl artificial antigen 577, 578, 581
Glucosyl inositol 56, 74
Glucosyl-pyrophosphoryl-polyprenol 171
Glucoxylan 280
Glucuronan 361ff.
Glucuronic acid 94, 238, 291, 295, 297, 298, 299, 300, 301, 303, 305, 355, 361, 466, 579, **605**
Glucuronoarabinoxylan 31, 37
Glucuronolactone **596**
Glucuronomannan **5**, 7
Glucuronomanno-protein **354**
Glucuronosyl-GlcNAc-pyrophosphoryl-dolichol 174
Glucuronoxylan 39
Glucuronoxylofucans 295
Glucurono-xylo-mannan 428
Glucuronoxylorhamnan 301, 302, 305
Glutathione 221
Glyceollin 603, **604**, **605**, 606
Glycerol 291
Glycine max **596**, **605**, **632**, 649
  suspension culture 207
Glycogen 360, 417, 419, 428
  glycoprotein as primer 55
Glycogen synthetase 403
Glycoprotein 8, 22, 70, 74, 80, 81, 86, 91, 317, 318, 352
  amino acid sequon for glycosylation 472
  and hemagglutination 627
  biosynthesis of mannoproteins 459ff.
  carboxypeptidase 451
  Chlamydomonas 22, 133, 145ff., 328, 340
  complex type 168, **169**, 182, 443, 497
  core glycosylation **166**, 461ff., **475**
  core region of 168
  deglycosylation 153
  formation 57, 178ff., 459ff.
  fungal cutinase 238, 468
  N-glycosylation **166**, 461ff., **475**, 494
  O-glycosidic bond 238, 443, **446**, 453, 460, 461ff., 466, 467, 468, 470
  high mannose type 168, **169**, 182, 476, 494, 497
  hydroxyproline-rich 35, 37, 106, 111, 133ff., 150, 328, 344, 608, 636
  in fungi 355, 362, 365, 441ff.
  in plant-pathogen/host interaction 588

in pollination 580
in root cap slime 557
inhibitors of glycosylation 183 ff.
invertase 449 ff.
lectins 635
mannoprotein 441 ff.
of the cell wall 133, 328
phylogenetic considerations 146 ff.
processing 138, 181 ff.
regulatory function of glycosylation 185
role of carbohydrate 184 ff.
sexual agglutination factors 453 ff.
structures 168
trimming 181 ff., 497
viral coats 176, 180, 184
(1 → 4)-β-D-Glycosaminoglycan 355
Glycosidases 149, 150, 151, 495
Glycosidases, inhibitors of 139, 150, 266, 267, 422
Glycosulfatase 293
Glycosylamine bond 168
N-glycosylation 461 ff.
Glycosyldiglyceride 400
Glycuronans 357, 366
Glyoxysome 188
Gnetum 206
Gnetum gnemon **229**
Golgi 40, **72**, 82, 83, 84, 85, 94, 101, 137, 171, 182, 187, 221, 295, 298, 299, 303, 304, 317, 318, **319**, 320, 325, 468, 470, 471, **487**, 488, 496, 497, 498, 500, **505**, 518, 519, 521, 525, 528, 555, 556, 557, 558, **559**, **561**, 564, 646
see also Dictyosome
Gomphrena globosa 611
Gossypium hirsutum **597**
Gracilaria 291
Graft hybrid 98
Gramineae 7, 148, 155, 555
ammonialyase 203
cell walls 31
germination of seeds 106
hydroxyproline-rich glycoprotein 136
lignin 200, 212, 219
Granule bands 325
Granulocrine secretion 558, **561**
see also secretion
Grapes 103
Grateloupiaceae 291
Griffithsia pacifica 255, 347
Griseofulvin 324
Growth rings 76
Guaiacyl lignin 197, **205**, 207, 218
Guaiacyl type 198
Guard cell 516
Guggul 102
Guluronic acid 284, **285**, 286, 305
Gum 113
Gum duct 102
Gum-secreting cells 101
Guttation 154
Gymnodinium 346
Gymnosperms 108, 196, 205, **207**, 212, 218, 226
cambium 93
glucomannan of 33
megasporogenesis 101
precursors for polysaccharide synthesis 77
secondary wall of 33
seeds 105
xylans of 7

Haemanthus 514
Halicystis 280
Halimeda cuneata 282
Halosphaera 347
Halycoryne wrightii 283
Hanseniaspora uvarum 421
Hanseniaspora valbyensis 421, 422
Hansenula anomala 418, 420, 421, 422
Hansenula capsulata 465
Hansenula holstii 465
Hansenula wingei 377, 420, 421, 441, 443, 453, 465
Hapten 628, 630, 631
Haptomer 650
Haptophyceae 317, 318, 320
Hardwood 7, 197, 200, 218
Haustorium 107
Headward growth 74
Helianthus annuus 645
Helix pomatia 424
Helminthosporium sacchari **597**, 598, 602
Helminthosporoside **597**, 598, 601, 602
Hemagglutinin 87, 628
Hemicellulase 102
Hemicellulose 12, 21, 30, 31 ff., 37, 40, 59, 134, 148, 150, **198**, 357, 521, 528
biosynthesis 61
composition 31
extraction procedures 3
turnover 40
Hemolytic anemia 649
Herbicide amiprophosmethyl 324
Heteroglucan 68, 74, 80, 91
Heteroglucuronan 361
Heteroglycuronans 362, 363
Heteromannan 68, 74
Heteroxylan 68, 74, 80, 91, 93, 106
Hexosamine 397, 408, 410
Hexosamine auxotroph 408
HF-deglycosylation 153
Himanthalia lorea 296, 297
Histoplasma 428
Histoplasma capsulatum 372, 429, 431

Histoplasma farciminosum 429
HMG-CoA reductase 168, 186, 187
Homo-arabinans 35
Homogalacturonan 34, 37
Homokaryon 376
Homoserine 661
Hordeum vulgare **229**
Horse gram **632**, 644
Host-pathogen interaction 365, 584ff.
Hyalogonium 346
Hydrathode 585
Hydraulic conductivity 255
Hydrazine-labile serine 139
Hydrazinolysis 139, 442, 452
Hydrodictyon 322, 346, 347
Hydrodictyon reticulatum **334**
Hydrogen bond 32, 37, 142, 150, 264
Hydrogen fluoride 140, 153
Hydrogen ions 261, 264
Hydrogen peroxide 213, **214**
Hydroglucan 358, 359, 360, 381
ω-hydroxy acids 235, **242**
ω-hydroxy fatty acid dehydrogenase 235, 236
p-hydroxybenzaldehyde 200, 231
25-hydroxycholesterol 186
Hydroxycinnamate: CoA ligase 207, **208**, 220
Hydroxycinnamoyl-CoA reductase 207
Hydroxycinnamyl alcohol oxidoreductase 207, 208
5-hydroxyferulic acid 205
Hydroxylamineferric chloride 83, 102, 107
Hydroxylation 232ff.
Hydroxymethylglutaryl-CoA 167, 186
Hydroxyoleic acid 228
Hydroxypalmitic acid 228
β-hydroxyphenylalanine 238
Hydroxyproline 35, 133, 134, 136, 141, 143, 144, 147, 151, 152, 328, 647, 648
Hydroxyproline arabinoside 134, 135, 138, 141
Hydroxyproline arabinosylation 148
Hydroxyproline-*O*-galactose 135
Hydroxyproline-*O*-glucose 135
β-hydroxytyrosine 238
Hymenomonas carterae 318, 320
Hypersensitivity 586, 587, 602
Hyphochytridiomycetes **354**, 355
Hypnea charoides 293
Hypodermis 225
Hypogravity 109

Idioblast 225
Immunoglobulin 88, 488, 489, 493
Immunoglobulin light chain 179
Incompatibility 581
Indoleacetonitrile 613
Infection 111, **112**
Influenza virus 184
Infra red analysis 68
Inhibitor of autolysine 343
Inosine-diphosphatase 565
Inositol 40, 56, 302, 372, 373
myo-inositol pathway 79
Instron technique 258, **259**, 260, 261, **262**, **263**
Integument 107
Intine **100**
Intussusception 264
Inulin 299
Invertase 151, 441, 442, 443, 445, 449, **451**, 469, 471, 472, 473, 476, 493, 494, 496, 499, 500, 503, 506
Ionophore X-537 A 324
Ionophore A 23187 324
Ca^{2+}-ionophore 637
Ipomoea batatas **230**
Iridaea cordata 294
Islandic acid 427
Isolectins 635
Isopentyl-pyrophosphate 167
Isopropyl *N*-phenylcarbamate 60
Isothiocyanate 612

Jack bean **632**
Japanese black pine hypocotyl **262**
Jequirity bean **632**, 649

Kelps 317
Keto-deoxyacid **596**
2-keto-2-deoxygluconate 600
2-keto-3-deoxy-gluconate **596**, 599
2-keto-3-deoxyoctonate 666
3-ketoglucose 595
α-ketoglucosidase 595
α-Ketols **240**
Ketones **240**, **245**, 246
3-ketosucrose 595
Kievitone **604**
Killer factor 494
Kinetin 111, 220
see also Cytokinin
Kinks 294
Klebsiella aerogenes 284
Klebsormidium 346, 347
Kloeckera apiculata 418
Kluyveromyces aestuarii 420
Kluyveromyces fragilis 418, 420, 421, 422
Kluyveromyces lactis 420, 444, 465
Kluyveromyces phaseolosporus 420, 421, 422, 423

Laccase 209, 210, 212
α-lactalbumin 179, 180
γ-Lactones **242**
Lactose synthetase 80

Lactuca sativa **229**
Laetiporus sulphureus 360
Laminaran 278, 281, 289
Laminaria 278
Laminaria digitata 287, 288, 298, 347
Laminaria hyperborea 287, 288, 347
Laminaria saccharina 347
Laminaribiose 56, 423
Laminarin 420, 421, 422, 424, 495, 608
Larix leptolepis **195**, 196
Latewood 197
Latex 97
Laticifer 89, 91, 94, 96
Laurencia pinnatifida 282
Leader sequence 180
Lectin 36, 88, 133, 141, 442, 444, **548**, 553, 555, 578, 580, 586
  all-β-lectins 628
  applications 636
  as part of toxins 649ff.
  assay 629
  binding competence in rhizobia 668
  binding specificity 630
  distribution in plant tissues 644ff.
  functional aspects in higher plants 641ff.
  in legume-*Rhizobium* symbioses 658ff.
  in slime mold aggregation 638ff.
  name and definition 627
  properties of non-seed plant lectins **633**
  properties of seed lectins **632**
  protection of seeds 651
  recognition hypothesis in nodulation 663ff.
  structure 634
  subcellular localization in plants 645ff.
β-lectins 147, 152, 344
Legume symbiosis 658
Legumin 169
Leiocarpan A 7
Lemon 103, 104
Lens culinaris **632**
Lenticel 585
Lentil **632**, 644
Lessonia flavicans 296
Lettuce hypocotyl **263**
Lichen 658
Lignification 93, 149, 152
Lignin 11, 12, 15, 36, 41, 69, 76, 86, 111, 140, 142, 155, 513, 608
  biogenesis of precursors 201ff.
  biosynthesis of 194, **203ff.**
  carbohydrate complexes 210
  chemistry 198, **199**, **200**, 201
  content of wood, % 196, 197
  Cross and Bevan color reaction **194**
  differences in biosynthesis between gymnosperms and angiosperms 218ff.
  formation of dilignols **211**
  growth of the polymer **212**, **213**
  linkage types in 198, **199**, **200**, 201
  Mäule reaction 196, 206, **207**
  morphological distribution 197
  nitrobenzene oxidation 206
  occurrence in plants 194
  organelles in lignification 221
  phloroglucinol reaction **194**
  polymerization of hydroxycinnamyl alcohols 209ff.
  precursors 201ff.
  regulation of biosynthesis 220ff.
  role of peroxidase 212
  $S_1-S_2-S_3$ layers **196**
  tests for 194, 196
Lignituber 111
Lilium 577, 578, 579
Lilium longiflorum 579, 582
Lima bean **632**
Limonium 652
Linamarin 613
Linoleic acid 228, 234, 373
Linolenic acid 234
Lipid intermediate 48, 57ff., 70, 81, 166ff., 460ff., 475, 520
Lipid-linked glucose 178
Lipid-linked monosaccharides 170ff.
Lipid-linked oligosaccharides 174ff., **175**, 176ff., 183
Lipomyces 420
Lipopolysaccharide 76, 166, 170, 586, 587, 637, 644, 651, 664, 666
Liverworts 226
Long-chain alcohols 230
Lotaustralin 613
Lotononis bainesii 667
Lotus 580, 673
Lotus tetragonolobus **632**
Lupin hypocotyl **262**
Lupinus 673
Luteic acid 427
Lycopersicon esculentum **596**, **597**, **605**
Lycorine 143
Lymphocytes 628, 636, 637, 638
Lysine 344
Lysogeny 97
Lysosome 182, 375, 497, 499, 500, **505**, 603
Lysozyme 344, 347, 651
Lysyl aldehyde 150

Maclura pomifera 644
Macrocystis pyrifera 288, 298, 303, 347
Macrophage 501
Macroptilium 673
Malabar papaiarnasium **229**
Malate dehydrogenase **214**, 214, 215
Malformin 141

Malic dehydrogenase 150
Malonic acid 427
Malonyl-CoA 243, **245**
Malseccin **597**, 599, 602
Maltose binding protein 491
Malus pumila **229**, **596**
Mannan 5, 7, 9, 12, 17, 19, **71**, 107, 169, 278, 280, 282, 283, 301, 362, 367, 371, 372, 406, 418, 494
(1→4)-β-mannan hydrolase 107
D-mannanase 443
(1→3)-α-mannan 283
(1→4)-β-mannan 283
Mannitol 268, 288
Mannobiose 443, 446, 461, 463, 467, 494
(1→3)-α-mannobiose 445
Mannobiosylphosphate 444
Mannolipid 170, 172, 173, 460ff.
Mannoprotein 367, 381
  alkali labile oligosaccharide 449, 460ff.
  biosynthesis 447, 459ff.
  carboxypeptidase Y 451ff.
  cell wall mannoprotein of *S. cerevisiae* 445ff.
  control of biosynthesis 471
  core **446**, 447, 462, 494
  definition 441
  homogeneity 442ff.
  immunogenicity 444
  invertase 449ff., 451
  methods for structural analysis 442ff.
  mutants 443, 447, **448**
  occurrence 441
  of *S. cerevisiae* 441ff.
  outer chain **446**, 447, 463, 470, 474, 494
  sexual agglutination factors 453ff.
  structures 445ff.
  subcellular sites of glycosylation 468
β-mannosidase 107, 175, 182, 266, 443, 451
Mannosyl transferase 177
Mannosylphosphate 444, 447, 464, 476
Mannosylphosphate transferase 464, 474
mannosyl-1-phosphate 465
mannosyl-6-phosphate 465, 497
Mannosyl-phosphoryl-dolichol **167**, 460
Mannotetraose 446, 464, 465, 494
Mannotetraose phosphate 446
Mannotriose 446, 461, 463, 494
D-mannuronic acid 284, **285**, 286, 287, 305
Mating 340, 341, 347, 365, 376, 378, 421, 453, 642
α mating factor 407
Matrix proteins 134
Mäule reaction 196, 206
Medicago 548
Medicago sativa **596**
Megasporocyte 101
Megasporogenesis 101
Meiocyte 90, 99, 101
Meiospores **335**, 347
Melanin 373, 374
Melibiase 494, 506
Membrane flow 82, 85, **561**
Membrane fusion **502**
Methoxy pectic polysaccharides 110
Methyl esterification 84, 88
4-*O*-methyl glucuronic acid 94
Methylbenzimidazole-2-yl-carbamate 324
2-*O*-methylfucose 6, 34
3-*O*-methylgalactose 291
4-*O*-methylgalactose 301, 304
4-*O*-methyl-L-galactose 291
6-*O*-Methyl-D-galactose **290**
β-methyl-glucose 56
2-*O*-methylglucuronic acid 304
4-*O*-methyl-glucuronoxylans **5**, 7
O-Methyltransferase **203**, 218
2-*O*-methylxylose 6, 34
3-*O*-methylxylose 303, 304
Mevalonic acid 167, 187
Micrasterias 325, 347, 524
Micrasterias denticulata **326**
Microconidia 365
Microfibril 12ff., 14, 16, 18, 19, 20, 25, **26**, 28, 29, 47, 68, 295, 317, 318, 320, 321, 324, 325, 345, 361, 363, 365, 366, 369, 379, 504
Microfilament 501
Microsomes 170, 171, 174, 181, 182, 187, 221, 245
Microspora quadrata **334**
Microsporogenesis **100**
Microtiter 629, 631
Microtubule **27**, 28, 40, 221, 320, 321, **322**, 324, 325, 326, 497, 501, 513, 514, **515**, 516, **517**, 522, **523**, 524, 525, 526, 564
  colchicine 60
  isopropyl *N*-phenylcarbamate 60
Microtubule-organizing center 515ff.
Middle lamella 11, 15, 40, 69, 76, 90, 93, 95, 102, 103, 154, 197, 295, 299, 518, 609
Milled wood lignin 198
Mitochondria 188, **505**, 545
Mitochondrial membrane 171, 187
Mitogenic 631, 635, 637, 641, 671
Mitosis 628
Mixed function oxidase 233, 234
Mixed β-Glucan 31ff., 38
Monilinia fructigena 427, 591
Monocotyledons 38, 93, 105, 150, **207**, 264
Monokaryon 375, 376, 377
Monolignol 212, 213, 216
Monooxygenase 204
Monostroma fuscum 302
Morphoregulation 144

Moss 226
Moss rhizoid 521
Mougeotia 60, 347
Mougeotia oedogonioides **334**
Mucilage 3, 83, 85, 86, 101, 352
Mucilage-secreting cells 101
Mucocyst **502**
Mucor 399, 403, 405
Mucor mucedo 357, 361
Mucor rouxii 357, 361, 372, 379, 380, 409
Mucorales 352, 379
Mucoran 361
Mucoric acid 361
multi-net growth theory 369
Mung bean hypocotyl **262**
Murein 340
Mycodextran 429
Mycodextranase 467, 468
Mycolaminaran 427
Mycolaminarin **596**, 600
Mycoplasma **112**
Mycorrhiza 658
Mycospocidin 185
Myo-inositol pathway 79
Myristyl lecithin 344
Myxococcus xanthus 672

N-acetyl galactosamine 153
N-acetylglucosamine **166**, 168, 172, 355
Nasturtium 238
Necrosis 601
Nectaries 585
Nemacystus decipiens 296
Neomeris annulata 283
Neuraminidase 629, 636
Neurospora 362, **396**, 398, 403
Neurospora crassa 352, 360, 365, 367, 369, 372, 382, 428, 430, 468, 501, 594
Nigeran 360, 361, 365, 370, 429
Nitella 265
Nitroblue tetrazolium 143
Nitrogen fixation 658ff.
p-nitrophenyl-β-D-glucoside 420, 421, 495
Nitrous acid 357, 359, 361
Nitzschia angularis 303
Nitzschia frustulum 303, 305
Nodulation 642
Nojirimycin 267
Noncellulosic glucan 48
Noncellulosic polysaccharide, activation of monosaccharides 76
  assembly 70, 82ff., 85
  biosynthesis of glycosyl ester 75
  biosynthesis of glycosyl ether 75
  chain elongation 74
  chain initiation 73
  chain termination 75
  composition 30ff.
  control of synthesis 76
  definition 30
  degradation 95, 105ff.
  direction of growth 74
  insertion of side branches 74
  interconversion of monosaccharides 76
  location of synthetases 81ff.
  metabolic changes of the primary wall 91
  metabolism 68ff.
  polymerizing system 70
  pulse labeling studies 84
  sequences of linkages 74
  subcellular routes of assembly 82ff.
  sugar composition 93
  supply of precursors 76
  synthetases, specificity 71
Nucellus 105, 107
Nuclear membrane 171, 498
Nuclear migration 376
Nucleoside diphosphate sugars 47, 69
Nucleotide sugar 69, 70, 76, 80

Oedogonium concatenatum **334**
Oerskovia 424
O-galactosyl hydroxyproline 153
Oleanolic acid **242**
Oleic acid 228, 234
Oligogalactosyl chain 466
Oligolignol 216
Oligosaccharide-lipid 175, 176, 178
Ononis hircina **633**
Oocystis 59, 346, 524
Oocystis lacustris 324
Oocystis solitaria 322, **322**, 324
  cell wall **27**
Oomycete 500
Oomycetes **354**, 355, 358, 365, 397
Opaque-2 maize 181
Ophiobolus graminus 612
Ophioglossum 555
Ophioglossum petiolatum **547**
Orange 103, 104
Ordered granule hypothesis 321
Oryzalin 324
Osmotic potential 255
Ouchterlony 630, 665
Ovalbumin 178, 179, 467, 490
α-oxidation 245
ω oxoacid dehydrogenase 235
Oxygen-sensitive growth 142

Pachyman 426
Padina pavonia 284, 296, 297
Palladium ions 143
Pallidin 639, 640, 641
Palmitic acid 228, 232

Pandorina 145
Pandorina morum 347
Pandorina unicocca 348
Papillae 111, 113
Paracoccidioides brasiliensis 372, 428, 429, 431
Paramylon A 360
Parenchyma 69, 93, 103
Pastinaca sativa **230**
Pathogenesis 111, 151, 584, 643
Pathogenic fungi 139, 237, 247
Pea epicotyl 261, **262**, **263**, 266, 268
Pea lectin 180, 631, 645, 651
Peach 103, 104, 105
Peanut **632**, 668
Peanut agglutinin 636
Pear 103, 104, 105
Pectate **589**
Pectate lyase **589**, 592
Pectate trans-eliminase **589**
Pectic galactan 34
Pectic galactans 109
Pectic polysaccharides 15, 18, 30, 37, 40, 75, 90, 93, 102, 104, 106, 107, 134, 148
  esterases 102, 104, 110, 150, 593
  hydrolases 90, 102
  low methoxy 110
  polysaccharide composition of 33
  sugar composition 34
  transeliminase 102
Pectic substances, see pectic polysaccharides
Pectin 39, 77, 83, **100**, 357, 553
  see also pectic polysaccharides
Pectinase 553
Pediastrum 346, 347, 348
Pelvetia 321
Pelvetia canaliculata 280, 289, 296, 298, 299
Pelvetia wrightii 296, 297
Pelvetian 295
Penicillinase 490
Penicillium 360, 362
Penicillium charlesii 467
Penicillium chrysogenum 428
Penicillum islandicum 427
Penicillium italicum 430
Penicillium luteum 427
Penicillium melinii 467
Penicillus dumetosus 282
Penium spirostriolatum **326**
Pentose phosphate pathway 595, 607
Peptidogalactomannan 466
Peptidoglycan 147, 166, 170
Perforation plate **95**
Pericycle cells 97
Periderm 225, 235, 242, 247, 531
Periodate oxidation 83
Periodate-Schiff stain 107, 278
Periodic acid-silver proteinate 97
Periplasma 154, 441, 477, **546**
Periplasmic bodies 97
Periplasmic space 101, 149, 151, 382, 423
Perisperm 105, 107
Peronospora parasitica 612
Peroxidase 86, 149, 150, 152, **203**, **210**, 212, 213, **214**, 215, 216, 220, 221, 591
Peroxidatic deamination 150
Persea indica **596**
Petunia 578
Phaeodactylum tricornutum 283, 303
Phaeophyceae 277, 279, 280, 281, 282, 283, 284, 295, 303, 306
Phaffia rhodozyma 420
Phaseollidin **604**
Phaseollin **604**
Phaseollinsioflavan **604**
Phaseolus 32
Phaseolus aureus 49, 56, **71**, 173, 187, 268
  $(1\rightarrow3)$-β-glucan formation 51
  polyprenol 58
  sterol glucoside 58
Phaseolus coccineus 641
Phaseolus lunatis **632**
Phaseolus lunatus agglutinin 634
Phaseolus vulgaris 181, **604**, 631, 645, 651, 663, 668
Phaseolus vulgaris agglutinin 634, 635, 637
Phenanthrene **240**
Phenol oxidase 209, 212
Phenolase **204**
Phenolic acid 76, 86, 92, 111
Phenolic esters **240**
Phenolic glucosides 611
Phenolics 140, 152
Phenoxy radicals **210**, 214, **216**, 218
Phenylalanine 201, **203**, 220, 221
  biosynthesis of **202**
Phenylalanine ammonia-lyase 203
Phenyl-propane 12
Phenylpropanoids 201
Phloem 40, 48, 93, 96, 99, 194, 220, **633**, 644
Phloem lectins 648
Phloroglucinol reaction 196
Phoma tracheiphila **597**, 599
Phormium tenax **227**
Phosphate deficiency 151
Phosphate translocator 607
Phosphoenol pyruvic acid **202**
Phosphoglucomutase 81, 372
6-phosphogluconate dehydrogenase 372
3-phosphoglycerate 607
Phosphomannan 465
Phosphomannose 497
Phosphorylase 607
Photophosphorylation 606
Phototropism 108

Phragmoplast 514, **515**, 516, **517**, 518
Phycomyces 356, 366, 369
Phycomyces blakesleeanus 357, 361, 380
Phyllocladene **240**
Phythium acanthicum 365
Phythium debaryanum 365
Phytoalexin 111, 113, 602, 603, **604**, 606, 643
Phytochrome 142, 143
Phytohemagglutinin 627
Phytohormones 59, 91, 220, 255, 261
Phytolacca americana **633**
Phytophthora cinnamomi **113**, 381, 426, 427, 431, **596**, **597**
Phytophthora cryptogea **597**
Phytophthora infestans 587, 588, 602, 603, **605**, 606
Phytophthora megasperma 588, **596**, 603, **605**, 606, 643
Phytophthora palmivora 356, 365, 369, 426, 427, **596**
Phytotoxins **596**
Picea abies **226**
Picea mariana 197
Pichia amethionina 453, 454
Pichia pastoris 421
Pichia polymorpha 420
Pilei 375, 376
Pineapple 103, 105
Pinocytosis 468
Pinus sylvestris **229**
Piptoporus betulinus 360
Piricularia oryzae 428
Pistil 577, 578
Pisum sativum 55, **71**, 180, **229**, 245, 247, 553, **559**, **560**, **632**
Pit chamber 14
Pit field 110
Pit membranes 95
Pit-fields 13
Plant hormones 59, 91, 220, 255, 261
Plant-pathogen, and lectin 643
  elicitors and phytoalexin 602ff., 643
  gene-for-gene hypothesis 585
  polysaccharide-degrading enzymes 588ff.
  recognition of the host 584ff.
  toxic host glycosides 609
  toxins 595ff.
Plasma membrane 9, 18, 20, 82, 94, 97, 149, 154, 171, 236, 245, 303, 318, 320, 324, 326, 352, 367, 376, 380, 397, 403, **404**, 405, 409, 468, 470, 501, **505**, 519, **546**, **561**, 601, 647
Plasmalemma see Plasma membrane
Plasmid 586
Plasmodesmata 13, 89, 97, 99, 518, 545, 647
Plasmolysis **112**, 321
Plasmometric method 258
Plastic extensibility 259
Plastic extension 257
Plastids 545
Platymonas tetrahele 317
Pleurochrysis 144, 320, 346, 513
Pleurochrysis scherffelii 318, **319**
Plum 103
Podocarpus 206
Podophyllotoxin 324
Pokeweed **633**
Pokeweed lectin 631
Pollen 99, **100**, 577, 580, 582, 642
Pollen cutinase 238
Pollen mother cell 90
Pollen tube 48, **73**, 83, 84, 85, 86, 89, 91, 150, 155, 237, 247, 255, 521, 577, 581, 582
Pollen wall enzymes 152
Pollination 151, 155, 577, 581
Polyene antibiotics 418
Polyesters **240**
Polygalacturonase 104, 150, **589**, 592, **604**, 609
Polygalacturonate ester 104
Polygalacturonate lyase 592, 593, 603, 609
Polygalacturonate synthetase 75
Polygalacturonic acid 38, 553
Polyglucuronic acid **354**
Polyhydroxyproline 136
Polyisoprenol 170
Polyisoprenyl-phosphate **167**
C_{55} polyisoprenyl-pyrophosphate 186
Polyoxin D 371, 379, **400**, **404**, 406, 408
Polyphenol 555
Polyphenolic substances 111, 140
Polyphosphate 352, 357
Polyporus 360
Polyporus fomentarius 428
Polyporus giganteus 428
Polyporus igniarius 428
Polyporus mylittae 356
Polyporus tumulosus 429
Polyprenol 48, 57, 61, 70, 167, 171, 460
Polyprenyl-phosphate 81, 171, 176, 467
Polyproline 143, 156
Polyproline II 136, 146, 147
Polyribosomes, see polysomes
Polysaccharidase 373
Polysomes 178, 180, 181, 182, 470, 488, 489, 490, 493, **505**, 531
Polysphondylium pallidum 638, 640
Polyuronan 12, 40
Poria cocos 426
Porphyra umbilicalis 278, 282, 283, 295
Porphyridium aerugineum 304
Porphyridium cruentum 144, 303, 304
Post-translational modification 133
Post-translational segregation 489, 492
Potato lectin 133, 136, 141, 587, 635, 636, 643, 648

Potato tuber 235, 236, 247
Potato-type lectins 147
Poterioochromonas 326
Poterioochromonas malhamensis **26**
Poterioochromonas stipitata **326**
Powdery mildew 608
P-protein 98
Prasinophyceae 317, 318, 320
Prasiola japonica 283
Preprophase band 515, 516
Preproteins 489, 492
Primary wall 12ff., 14, 16, 17, 29, 40, 47, 69, 76, 83, 93, 102, 103, 108, 133, 149, **196**, 197, **520**, 525, 528, 609
  assembly of macromolecules 137
  degradation of 545
  deposition and expansion 91ff.
  dicotyledons 36
  monocotyledons 38
  protein content 134ff.
  secretion of matrix substances 518ff.
Primula 578
Primulin 406
Pristane **240**
Pro-carboxypeptidase Y 499
Procollagen 148
Procutin 236
Pro-cutinase 238
Proline hydroxylation, inhibition of 143
Prolyl hydroxylase 133, 142, 147, 154, 156
Pronase 179, 340, 344, 357
Proplastids 545
Protease of autolysines 340, 344
Protein bodies 647
Protein glycosylation 178ff., 459ff.
Proteinase B 401, 403, 406, 407, 408, 492
Proteoglycan 74, 296, 299
Proteolytic activation 401
Prothotheca zopfii 48
Protocatechuic acid **206**, 613
Protocatechuic aldehyde 205, **206**
Protoplasts 18, 20, 335, 336, 338, 348, 367, 380, 382, 384, 403, 405, 425, 495
  cell wall 18, 30
  cell wall regeneration 18, 60, 85, 328, 329, 356, 360, 368ff., 371, 379, 525, 528
  in root cap degeneration 549
  of tomato fruit **532**
  of *Trichoderma* 373ff.
  of yeast 356, 360, 424, 462
  peroxidase 150
Protosiphon botryoides 348
Prototheca 280
Prototheca zopfii 58, 178
Prunus 578
Prunus avium 580
Prymnesiophyceae 317
Pseudodichotomosiphon constricta 282
Pseudomonas 284
Pseudomonas atlantica 291
Pseudomonas carrageenovora 293
Pseudomonas glycinea **604**
Pseudomonas phaseolicola 594
Pseudomonas solanacearum 141, 586, 587
Pseudomonas solanearum 643
Pseudo-nigeran 360, 419, 429
Pseudoplasmodium 638
Psilophytales 194
Psilotum nudum **229**
Pteridine **204**
Pteridophyta **207**
Pteromonas 347
Pterygophora californica 347
Pullulan 428, 429
Pullularia pullulans 427, 428
Puromycin 470, 488, 490
Pustulan 420, 421, 424, 427
Pyramimonas amylifera 318
Pyramimonas tetrarhynchus 320
Pyrenochaeta terristris 594
Pyrophosphorylase 49, 50, 81
Pyrrophyta 279
Pyrus communis **229**, **596**
Pyruvic acid **202**, 289, 292
Pythium acanthicum 368, 426, 428

Quercitin 613
Quercus robur 648
Quercus suber **230**
Quinone 150, 152
Quinone methide 210, **211**

Radioimmunoassay 667, 668
Radiosphaera 346
Radish cotyledon 263
Raffinose 56, 652
X-ray method 277, 278, 282, 295
Reaction wood 108
Receptor-mediated endocytosis 650
Recognition phenomena 151
Red kidney bean 644
Red kidney bean hemagglutinin 651
Reproductive cycle 306
Resonance frequency technique 260, 261
Resorcin blue 582
Restharrow **633**
Rhamnan 301
Rhamnogalacturonan 3, **4**, 6, 33ff., 37, 68, 80, 264
Rhamnomannan 597
Rhamnose 93, 299, 300, 301, 303, 304, 305, 579, 587, 590, 591, 599, **604**
Rheological analysis 259
Rhizidiomyces 355

Rhizobia, fast- and slow-growing 673
lectin-binding site 664
Rhizobium 152, 585, 642, 651, 658ff.
Rhizobium japonicum 643, 660, 661, 663, 664, **665**, **666**, 668, 669, 671
Rhizobium leguminosarum 661, 664, 666, 667, 672
Rhizobium meliloti 667
Rhizobium phaseoli 661, 663, 664
Rhizobium trifolii 642, 661, 663, 664, 666, 667, 668, 669, 671
Rhizoclonium 299
Rhizoctonia fragariae 591
Rhizoid 298, 321, 356
Rhizopus 425
Rhizopus chinensis 424
Rhizopus stolonifer 603
Rhizosphere 565, 566, 567, 568, 658, 659ff.
Rhodella maculata 144, 303
Rhodochorton floridulum 282
Rhodophyceae 277, 280, 281, 282, 283, 289, **290**, 303, 306
Rhodophycean galactans, agar 291
agarose **291**
antibodies 294
biosynthesis 295
carrageenan **292**, **293**
κ-carrageenan **292**
cell wall localization 294
conformation 293ff.
galactose residues of **290**
possible substitutions 289
Rhodotorula glutinis 419
Rhodotorula minuta 421
Rhodotorula peneaus 420
Rhodymenia palmata 282
Rhodymenia pertusa 281
Rhynchosporium secalis 598
Rhynchosporoside 598, 601
Ribes grossularia **229**, **230**
Ribes nigrum **230**
Ribonuclease A 179
Ribose 551, 599
Ribulose 1,5 diphosphate carboxylase 490
Rice coleoptile **262**
Ricin 649, 650
Ricinus communis **605**, **632**, 644, 646, 649
Ricinus communis agglutinin 630
Rishitin 603, **605**
Robinia pseudoacacia **633**, 644, 648
Root **113**
hairs 101, 150, 255, 642, 643, 645, 662
lateral root initiation 97
lectins **633**, 667ff.
lignification 220
model of slime secretion **561**
nodule formation 642, 658ff.
pericycle cells 97
secretion of the cap 542ff.
suberin and ion movement 247
Root cap 83, 85, 101, 519
cap columella 544
cap renewal time 544
chemical analysis of slime 549ff.
control of slime formation 562
function of the slime 565ff.
localization of secretory cells 542
secretory activity 542ff.
secretory pathways 556
sections **543**, **546**, **547**, **548**, **555**, **559**
slime discharge 560
ultrastructure of secretory cells 545ff.
Root nodule 644
Root tip **515**, **517**
Roots, microbial colonization 567
Root-soil interface 565, 566, 567
Rosette particles **523**
Rosettes 325, 326
Rosmarinus officinalis **229**
Ruthenium red 102

Saccharomyces 354, 403, 406
Saccharomyces carlsbergensis 399, 401, 419, 459
Saccharomyces cerevisiae 356, 358, 360, 362, 367, 369, 377, 380, 381, **396**, 399, 400, 401, 405, **407**, 417ff., 427, 432, 441ff., 459ff., 465, 488, 498, 499, 501, 503, 504, 522
Saccharomyces commune 356
Saccharomyces ellipsoideus 422
Saccharomyces fermentati 418
Saccharomyces kluyveri 453, 454
Saccharomyces rosei 421
Saccharum officinalis **597**
S-Adenosyl L-methionine 75
Salix alba 648
S-Allele 580
Sapindus saponaria **229**
Saponin 609, 612
Saprolegnia 426
Saprolegnia monoica 378, 430, 432
Sapromyces 426
Sargassan 295
Sargassum linifolium 296, 297
Sargassum pallidum 296, 297
Scales 317, 318, **319**, 320
Scenedesmus 324, 346, 348
Scenedesmus obliquus **322**
Schizogeny 97
Schizophyllum *403*, 404, 409
Schizophyllum commune 356, 357, 359, 360, 362, 365, 366, 368, 370, 371, 374, 375, 380, 381, 409, 427, 429, 430
Schizosaccharomyces 354, 367, 379, 419
Schizosaccharomyces japonicus 421

Schizosaccharomyces malidevorans 421
Schizosaccharomyces octosporus 421
Schizosaccharomyces pombe 377, 421, 422
Schizosaccharomyces versatilis 420, 421, 422
Schweitzer's reagent 358, 426
Sclerotia 426
Sclerotinia libertiana 266, 267
Sclerotium rolfsii 378, 427, 430
Scotch elm **633**
Scotiella 346
Scutellum 106
Sea urchin 185, 187
Secale 578
Secale cereale 581, 601
Secondary alcohols **240**
Secondary wall 15, 19, 29, 36, 47, 69, 76, 94, 134, **196**, 525
  deposition and growth 93ff.
  formation 40, **326**
  lignin concentration 197
  particle complexes 522, **523**
  S_1-S_2-S_3 14, 16, 197
Secretion 154
  and glycosylation 493ff.
  and intracellular transport 496ff., **503**
  discharge of product 501ff.
  genetic approach 503
  glycosidases 467
  in bacteria 490ff., **505**
  model of slime secretion **561**
  mycodextranase 467
  of autolysines 340ff.
  of cell wall polysaccharides 504, 513ff.
  of cellulose by *Acetobacter* 526ff.
  of cellulose microfibrils 521
  of lipidic wall material 528ff.
  of mannoproteins 469, 472
  of primary wall substances 518ff.
  of proteins 184, 186
  of the root cap 542ff.
  peptidophosphogalactomannan 467
  receptors in 502, 503
  secretory pathway **487**, **505**, 556
  signal sequence 468, 489ff.
  ultrastructure of secretory cells 545ff.
Secretory vesicles 83, 84, 89
Seed coat 225, **227**
Seed globulin 181
Seed lectins 631
Segregation 181, 488, **489**
Selfincompatible mating 110
Semliki Forest virus 184
Senescence 102
  flower 105
  fruit 103
  leaf 105
Septoria lycopersici 612
Septum **364**, 366, 367, 370, 374, 376, 380, 381, **396**, 397, 406, **407**, 408
Serine galactosylation 140
Serine protease 340
Serinol 602
Sexual reproduction in algae 333ff.
Shepherd's crook 643
Shikimate dehydrogenase 220
Shikimic acid 201, **202**, **203**
Shmoo formation 407
Sialic acid 169, 497, 629, 634
Sieve cell 377
Sieve plate 86, 89, 97, 110, 111
  differentiation 98
Sieve tube 83, 98, 648
Signal peptide 180, 181
Signal sequence 180, **489**
Silica 317, 318
Silica deposits 136
Silver hexamine 83
Silver proteinate 83
Sinapate: CoA ligase 218
Sinapic acid **203**, **205**, 206, **207**, 208, 218
Sinapyl alcohol 205, 210, 212, 217, 218
Sinapyl aldehyde 209, 218
Siphonocladales 320
β-sitosterol **242**
Skeletonema costatum 303
Slime 69, 83, 101, 352
Slime mold **633**, 638ff., 645
Solanum tuberosum **229**, **230**, **596**, **605**, **633**
Solieriaceae 291
Sorastrum 347
Sorbose 372
Sorghum bicolor **230**
Sorghum vulgare 221
Soybean 30, 33, 55, **632**, 644, 649, 661, 663, 664, 668, 669
  cell culture 49, 643
Soybean agglutinin 634, 641
Soybean hypocotyl **262**
Soybean lectin 586, **665**, **666**
Spermatozoid 344, 347
Sphingomyelin 498
Spinacia oleracea **230**
Spirea 600
Spirodela oligorrhiza 151
Spirogyra 347
Spitzenkörper 500
Spondylomorum 346
Spongiochloris 346
Spongomorpha arcta 300
Sporangia **334**, **336**, 341, **342**
Sporangiophore 356, 364, 366, 369
Sporangium **337**, **343**, 344, 345
Sporangium autolysine 335, **337**, 338, **339**, 340, 341, 348

Spores **342**, 346, 364, 365, 375
Sporobolomyces 367
Sporopollenin **100**, 322, 345, 532
Sporothrix schenckii 372
Sporulation 377, 378
Spruce lignin 210, **214**
Squalene 167, **168**
Squalene synthetase 168
Stachyose 652
Starch 49, 70, 76, 278, 299, 428
Stearic acid 228
Stemphylium loti 613
Steroids **242**
Steryl glucoside 54, 172
Steryl glycoside 81
Stigeoclonium 347
Stigma 151, 237, 247, 577, 578, 579, 581, 642
Stigma mucilage 155
Stigma papillae 110
Stomata 247, 585
Stone cells 76
Strawberry 103
Streptococci 378
Streptomyces 355, 424
Streptomyces cacoi 400
Streptomyces lysosuperificus 183
Streptovirudin 185
Stress 109, **112**
Stress relaxation 258, **259**, 260, 261, **262**, **263**
Style 155, 577, 578, 579, 580, 581
Suberin 76, 110, 531, 608
  aliphatic components of **229**, **235**
  biodegradation of 237ff.
  biosynthesis of 232ff.
  composition and structure 227ff.
  formation caused by injury 225
  function of 247
  isolation of 227
  lamellar **227**
  location and ultrastructure of 225
  methods for depolymerization **228**
  monomers of 228, **232**
  phenolic materials in 231
  regulation of formation 236
  secondary structure 231
  site of synthesis 236
Sucrose 49, 76
Sucrose synthetase 49, 50
Sugar phosphate 76, 80, 84
Sugar transport 648
Sulfate esters 277, 283, 299, 300, 303, 305
Sulfated polysaccharides 296ff., 317, 277ff., 299
  biosynthesis 302
  degradation 301
  occurrence and composition 299ff.
  of chlorophyceae 299ff.
  protein in 299ff.
Sulfated sugars 146
Sulfohydrolase 295
Sulfotransferase 295
Surface contact 638
Surface growth 255
Sweet potato 110
Sycamore 135, 139
Sycamore cells 49, 81, 601
  cell wall composition 37
  primary wall 31
  rhamnogalacturonan 34
  structural model of the wall 36
  xyloglucan 32
Symbiosis 151, 658ff.
Symplastic connection 98
Synexin 502
Syringaldehyde 200, **207**
Syringaresinol 217, 218
Syringyl lignin **194**, 196, 197, **205**, 207, 217, 218

Tailward growth 74
Tapetal cell 90
Tapetum 99
TDP-GalU **71**
TDP-glucose 50, 51
Teichoic acid 166
Tension wood 108
Terminal complexes 28, 325
Terpene glucoside **597**
Terpenes **242**
Terpenoid hydrocarbons **240**
Terpenoids 588
Tertiary wall 14
Tetra-arabinoside 135
Tetrahymena **502**
Thecae 317
Theobroma cacao **596**
Thermomicrobium roseum 148
Thiamine pyrophosphatase 565
Thiocarbamates **245**
Thuringiensin-A 516
Thyroglobulin 182
Tilia cordata 648
Tissue polysaccharides 74
Tmesipteris viellardii **230**
Tobacco cells 7
Toluidine blue 107, 278
α-Tomatine **610**, 611, 612
Tomato 104, 105
Touchwood **633**
Tracheid 14, 69, 89, 90, 94, 96, 97, 109, 194, **195**, 196, 197, 198
Transglucosylase 267
Transglycosylation 56, 265, 266, 269, 476

Transpiration 247
Trehalose 49
Tremella fuciformis 428
Tremella mesenterica 360
Tribonema **334**
Tribonema aquale 304
Tribonemataceae 304
Trichoderma harzianum 373
Trichoderma homingii 649
Trichoderma viride 430, 431, 649
Trichome 85, 101, 225
Trichomycetes 397
Trichophyton mentagrophytes 365
Trichosporon cutaneum 420
Tridacna maxima 578
Trifluralin 324
Trifoliin 642, 645
Trifolium repens **633**, 642
Trihydroxystearic acid 228
Trilaminar cell wall layer **322**
Trimming 497
Tripeptide sequence, asn-X-ser (thr) 179, 472
Triticum 543
Triticum vulgare 553, **632**
Tropaeolum majus **230**, 238
Tropism 108
Trypsin 401, **402**, 431, 454
L-tryptophan 201
Tsushimycin 186
Tubulin 324, 516
Tuliposides 611, 613
Tunicamine 183
Tunicamycin 173, 181, **183**, 184, 185, 400, 452, 472, 493, 495, 499
Turgor 144, 255, 258, 261, 268, 321, 343, 377, 382, 383, 564
Turnover 48, 89, 151, 375
  metabolism of cell walls 89ff.
  of cell wall polysaccharides 40
  of chitin 405
  of glyceollin 606
  of mannolipid 170
  of wall galactose 268
Tyloses 97
Tyrosine ammonia-lyase 203
L-tyrosine 201, **203**, 221
  biosynthesis of **202**

Ubiquinone **168**
Udotea orientalis 282
UDP-arabinose **71**, 77, 173
UDP-Gal 4-epimerases 81
UDP-galactose 71, 77, 173
UDP-GalU **71**
UDP-Glc dehydrogenase 77, 79
UDP-Glc glucan synthetase 79
UDP-Glc pyrophosphorylase 79
UDP-GlcA decarboxylase 77
UDP-GlcNAc 172, 173
UDP-glucose 49, 51, **52**, 55, 56, 57, 70, **71**, 77, 81, 98, 171, 179, 372
  externally supplied 54, 55, 111
  in extracellular space 151
UDP-glucuronic acid 173
UDP-methyl-galacturonate 75
UDP-*N*-acetylglucosamine 21, 379, **398**, 399, **400**, 410
UDP-xylose **71**, 77, 173
Ulmus americana **597**
Ulmus glabra **633**, 648
Ulothrix 346, 347
Ulothrix zonata **334**
Ulotrichales **345**, 346
Ulva 280
Ulva lactuca 300, 302, 305
Ulvales 300, 306
Umbilicaria pustulata 427
Uronic acids 93, 107, 109, 264, 361, 551, 552
Urospora filiformis 280
Urospora penicilliformis 283, 300, 301
Urospora wormskioldii 280, 283, 300, 301, 302
Ursolic acid **242**

Vaccinium macrocarpon **230**
Vacuolation 545, 549
Vacuole 96, 375
Vacuum-ultraviolet circular dichroism 294
Valonia 265, 280, 320, 322, 522
Valonia ventricosa 265
Vanillin 200, 201, **207**, 231
Verticillium 143
Verticillium albo-atrum 591, 592, **597**, 599, 608
Verticillium dahliae 593, 594
F-vesicles 325, **326**
Vesicular stomatitis virus 180
Vicia cracca 667
Vicia faba **230**, 567, **632**, 649
Vicillin 169
Vigna 673
Vigna angularis 268
Vigna angularis epicotyl **262**
Vigna radiata 646, 647, 651
Vigna unguiculata 651
Virus 176, 180, 184, 606
Virus infection **112**
Viscoelastic extension 257, 258
Vitis vinifera **230**
Volvocales 144, 148, 302, 328, 340, 343
Volvox 144, 148, 334, 343, 344, 347
Volvox africanus 344
Volvox aureus 344
Volvox carteri 145, 344
Volvox dissipatrix 344
Volvox gigas 344

Volvox pocockiae 344
Volvox spermatioides 344
Volvulina steinii 348

Wall apposition 608
Wall degrading enzymes, developmental stage
  specificity 338
  in algae 333ff.
  in fungi 373ff.
  inhibitor of 343
  in the life cycle of plants 333
  protease 344
  secretion of 340
  serine protease 340
  species specificity 338
Wall extensibility 255, **259**, **262**, **263**
Wall extension 255ff.
Wall loosening 109, 256
  chemical creep 265
  enzymatic 265
  enzymes in fungi 377ff., 421
  inhibitors of 267
  mechanism of auxin-induced 263ff.
  nonenzymatic 265
  polysaccharide hydrolases 266
  transglycosylase 266
Wall loosening factor 256, 261
Wall network 140
Wall-less mutants 146
Water potential 109, 255
Wax crystals 225
Wax esters 244
Waxes, amount per surface 226
  biosynthesis of 243ff.
  composition of 239ff.
  function of 247
  general structures **240ff.**
  isolation and analysis 239
  location and ultrastructure of 225
  of the leaf surface **226**
  secretion of 529ff.
Wheat germ **632**
  cell-free system 180, 181
Wheat germ agglutinin 408, 634, 636, 649
Wickerhamia fluorescence 425
Wiesner test 196
Wilting 600, 601, 602
Wisteria floribunda 650
Woloszynskia 347
Wound-healing 235, 236, 247
Wounding response 86, 110, **112**

Xanthomonas campestris 594, **596**, 599
Xanthophyceae 304
Xanthophyta 279
Xenopus laevis 181
X-ray diagram 426
Xylan **5**, 7, 12, 31, **71**, 75, 94, 280, 282, **589**
β-xylan hydrolase 106
Xylanase **589**
(1→4)-β-xylanase 346
Xylem 39, 40, 76, 83, 93, 94, **95**, 113, 154,
  194, 197, 220, 600, 609
Xylogenesis 221
Xyloglucan **4**, 6, 32ff., 37, **71**, 88, 92, 103,
  106, 264, 268, 269, 521
Xylomannan 283, 362
Xylo-manno-protein **354**
(1→4)-β-xylosan 346
Xylose 93, 268, 299, 300, 301, 303, 355, 466,
  551, 587, 591, 628, 647
Xylosidase 150
Xylosyl threonine 144
Xylosyl-phosphoryl-polyprenol 173
Xylosyl serine 144

Yariv's artificial antigen 90, 152, 154, 628
Yeast 255, 354, 355, 358, 362, 367, 371, 377,
  397, 398, 401, 402, 406, 642
  see also *S. cerevisiae* and other species
  biosynthesis of mannoproteins 459ff.
  bud formation 406ff.
  cell wall 16, 396ff., 417ff., 441ff.
  chitin of 396ff.
  dolichols 460ff.
  glucans of 417ff.
  isolation 441ff.
  mannoproteins 441ff.
  mating 453ff.
  septum initiation 89, 406
Yeast vacuole 401, 406, 441, 451, 492, 495,
  499, 500
Yield stress 255

Zea mays **230**, **262**, 264, **515**, **517**, 523, **543**,
  545, **548**, 552, 553, **555**, 555, 569, 645
Zea mays coleoptile 268
Zeatin 221
Zein proteins 181
Zoospore release, modes of **343**
Zoospores 318, 321, 322, **334**, **335**, **336**, 338,
  340, 341, 346, **364**, 368, 397, 410
Zygnema 347
Zygomycetes **354**, 355, 357, 358, 361, 366,
  500
Zygospore 146
Zygote 289, 298, 306, 329, **335**, 338, 343
Zymogen 89, 381, 401, **402**, 405, 409, 423,
  431, 496
Zymogen granules **487**

Volume 9

Hormonal Regulation of Development I

Molecular Aspects of Plant Hormones
Editor: J. MacMillan
With contributions by numerous experts

1980. 126 figures. XVII, 681 pages
ISBN 3-540-10161-6

In Preparation

Volume 10

Hormonal Regulation of Development II

From the Cell to the Whole Plant
Editor: T. K. Scott
With contributions by numerous experts

ISBN 3-540-10196-9

Volume 11

Hormonal Regulation of Development III

Role of Environmental Factors
Editors: R. P. Pharis, D. M. Reid
With contributions by numerous experts

ISBN 3-540-10197-7

Physiological Plant Ecology

Editors of Volume 12 A–D: O. L. Lange, P. S. Nobel, C. B. Osmond, H. Ziegler

Volume 12 A

Responses to the Physical Environment

1981. 109 figures, 40 tables.
Approx. 670 pages
ISBN 3-540-10763-0

Volume 12 B

Water Relations and Carbon Assimilation

ISBN 3-540-10906-4

Volume 12 C

Responses to the Chemical and Biological Environment

ISBN 3-540-10907-2

Volume 12 D

Ecosystem Processes: Mineral Cycling, Productivity and Pollution

ISBN 3-540-10908-0

Volume 13 A

Plant Carbohydrates I

Intracellular Carbohydrates
Editors: F. A. Loewus, W. Tanner

1982. Approx. 103 figures,
approx. 98 tables. Approx. 900 pages
ISBN 3-540-11060-7

Springer-Verlag
Berlin
Heidelberg
New York

MIX
Papier aus verantwortungsvollen Quellen
Paper from responsible sources
FSC® C105338

If you have any concerns about our products,
you can contact us on
ProductSafety@springernature.com

In case Publisher is established outside the EU,
the EU authorized representative is:
Springer Nature Customer Service Center GmbH
Europaplatz 3, 69115 Heidelberg, Germany

Printed by Libri Plureos GmbH
in Hamburg, Germany